ENCYCLOPEDIA OF PHYSICS

CHIEF EDITOR

S. FLÜGGE

VOLUME XLVI/2

COSMIC RAYS II

EDITOR

K. SITTE

WITH 286 FIGURES

SPRINGER-VERLAG
BERLIN · HEIDELBERG · NEW YORK
1967

HANDBUCH DER PHYSIK

HERAUSGEGEBEN VON

S. FLÜGGE

BAND XLVI/2

KOSMISCHE STRAHLUNG II

BANDHERAUSGEBER
K. SITTE

MIT 286 FIGUREN

SPRINGER-VERLAG
BERLIN · HEIDELBERG · NEW YORK
1967

Alle Rechte, insbesondere das der Übersetzung in fremde Sprachen, vorbehalten

Ohne ausdrückliche Genehmigung des Verlages ist es auch nicht gestattet, dieses Buch oder Teile daraus auf photomechanischem Wege (Photokopie, Mikrokopie) oder auf andere Art zu vervielfältigen

© by Springer-Verlag / Berlin · Heidelberg 1967

Library of Congress-Catalog-Card Number A 56-2942

Printed in Germany

Die Wiedergabe von Gebrauchsnamen, Handelsnamen, Warenbezeichnungen usw. in diesem Werk berechtigt auch ohne besondere Kennzeichnung nicht zu der Annahme, daß solche Namen im Sinne der Warenzeichen- und Markenschutz-Gesetzgebung als frei zu betrachten wären und daher von jedermann benutzt werden dürften

Titel-Nr. 5789

Contents.

Theory of Cascade Showers. By Dr. Jun Nishimura, Associate Professor, Doctor of Physics, Institute for Nuclear Physics, University of Tokyo (Japan). (With 35 Figures) 1

  A. Introduction 1

  B. The elementary processes 3
    I. Corrections to the Bethe-Heitler formula 4
    II. Radiation length and critical energy 10

  C. The diffusion equations for electron showers, and the limitations of the approximations 11
    I. The diffusion equations 11
    II. Limitations of the approximations 15

  D. Behaviour of the shower functions and elementary solutions 18

  E. One-dimensional electron shower theory 22
    I. Shower functions in Approximation A 23
    II. Shower functions in Approximation B 28
    III. Solution in other approximations 33

  F. Three-dimensional electron shower theory 38
    I. Three-dimensional electron shower theory in the Landau approximation . . 38
    II. Three-dimensional electron shower theory without the Landau approximation 44
    III. Behaviour of the structure function and its numerical evaluation 49

  G. Application of the theory to the shower phenomena 69

Appendix 85

References 113

Cosmic Rays and High-Energy Physics. By Dr. Yoichi Fujimoto, Professor, Science and Engineering Research Laboratory, Waseda University, Tokyo (Japan), and Dr. Satio Hayakawa, Professor, Department of Physics, Nagoya (Japan). (With 36 Figures) . . 115

    I. Discovery of multiple production 115
    II. Direct observations of high-energy interactions 118
    III. Behaviour of high-energy cosmic rays 138
    IV. Models of multiple production 153
    V. Summary of cosmic-ray information 169

References 177

The Spectrum and Charge Composition of the Primary Cosmic Radiation. By William R. Webber, Director, World Data Center A. for Cosmic Rays; Associate Professor, Department of Physics, University of Minnesota, Minneapolis, Minnesota (United States of America). (With 49 Figures) 181

    I. Introduction 181

    II. Solar influences on the spectrum 184
      a) Solar produced changes in the total radiation 186
      b) Solar produced changes in the proton component 189
      c) Solar produced changes in the helium component 191
      d) Solar produced changes in the S nuclei ($Z \geq 6$) 192
      e) From a spectrum of the solar modulation 192

III. Spectrum of the total radiation 195
  a) Primary spectrum derived from ion chamber and counter measurements in the latitude sensitive region 195
  b) Primary spectrum at energies of 100—1000 GeV derived from the spectrum of γ-rays at high altitudes 198
  c) Primary spectrum at energies from 100—10 000 GeV derived from the spectra of muons and protons at sea level 198
IV. The spectrum of the singly charged component 198
  a) Integral spectrum in the latitude sensitive region 199
  b) Differential spectrum at low energies 200
  c) The proton spectrum in 1963 and its extrapolation to sunspot minimum 202
  d) Presence of deuterium in the primary radiation 204
V. The spectrum of the helium component 204
  a) Integral measurements in the latitude sensitive region 205
  b) Differential spectrum of helium nuclei at low energies 205
  c) Differential spectrum of helium nuclei at higher energies 207
  d) Isotopic composition of helium nuclei 208
VI. Nuclei heavier than helium 209
  a) Determination of fragmentation parameters 210
  b) Interaction mean free paths 211
VII. The spectrum of the L nuclei 213
  a) Integral intensities 213
  b) Differential spectrum of L nuclei 214
VIII. The spectrum of the $S(M+H)$ nuclei 215
  a) Integral measurements in the latitude sensitive region 216
  b) The differential spectrum of the S nuclei at low energies 216
  c) Differential spectrum of S nuclei at higher energies 218
  d) The H/M ratio as a function of rigidity 218
IX. Intensity and spectrum of the electronic component 219
X. Charge composition and spectrum of the cosmic radiation between 30 and 10 000 GeV 223
XI. The energy spectrum of the total radiation at energies $> 10^{14}$ eV/Nuc 224
XII. Detailed charge features of the radiation 227
  a) Isotopic composition of the primary radiation 231
XIII. A comparison of the spectra of the different charge groups 232
  a) Charge composition at energies $> 10^{14}$ eV 233
  b) Spectra at energies $< 3 \times 10^9$ eV 234
  c) Effects of solar modulation on the low energy end of the spectrum 237
XIV. Astrophysical consequences of the above results in relation to the acceleration and propagation of cosmic rays 238
  a) Propagation of the primary radiation in the galaxy 239
  b) The process of acceleration and escape from the source region 254
  c) Source regions for primary cosmic rays 257
References 260

High Energy Photons and Neutrinos from Cosmic Sources. By ROBERT J. GOULD, Assistent Professor, Department of Physics, University of California, La Jolla, and Geoffrey R. Burbidge, Professor, Department of Physics, University of California, La Jolla, California (United States of America). (With 11 Figures) 265

I. Introduction 265
II. Production in the interstellar gas, the galactic halo, and the intergalactic medium 267
  a) Meson production in cosmic ray nuclear collisions 268
  b) The electron production spectrum 270
  c) Electron energy losses in the galaxy 270
  d) Electron production and energy losses in the intergalactic medium 273
  e) The electron energy spectrum in the halo and intergalactic medium 274

f) High energy photon flux from various processes 276
g) Comparison with observations 279
h) Tests of cosmological theories 284

III. Discrete sources of high energy photons 285
  a) General summary of the observations 285
  b) Possible galactic sources 286
  c) Mechanisms for x-ray production in discrete sources 290
  d) The Crab nebula 294
  e) The galactic center 299
  f) Solar system sources 299
  g) Hard radiation from stellar coronae 300
  h) Extragalactic discrete sources 301
IV. Neutrino sources 302
V. Conclusion 305
References 307

The Time Variations of the Cosmic Ray Intensity. By Dr. JOHN JAMES QUENBY, Lecturer in Physics. Physics Department, Imperial College, London (Great Britain). (With 37 Figures) 310

I. Introduction 310
II. Geophysical effects of cosmic radiation 311
  a) Geomagnetic threshold rigidities 311
  b) Effect of magnetic fields due to external current systems 317
  c) Asymptotic directions and the analysis of anisotropies in the incident particle flux 325
  d) Specific yield functions 328
III. Experimental evidence on the modulation mechanism 330
  a) Cosmic ray time variation 330
  b) Solar energetic particles 344
IV. Modulation mechanisms 349
  a) Introductory survey 349
  b) Modulation by regular magnetic fields 351
  c) Modulation by disordered magnetic fields 357
  d) Summary 368
References 368

Nukleonen in der Atmosphäre. Von Professor Dr. ERWIN SCHOPPER, Direktor des Instituts für Kernphysik der Universität Frankfurt, Frankfurt a.M. (Deutschland), Dr. ERICH LOHRMANN, Deutsches Elektronen-Synchrotron, Hamburg-Grossflottbek (Deutschland) und Dr. GÜNTER MAUCK, Institut für Kernphysik der Universität Frankfurt, Frankfurt a.M. (Deutschland). (Mit 88 Figuren) 372

A. Einleitung 372
B. Die Primärkomponente der kosmischen Strahlung 377
C. Kernwechselwirkungen der Nukleonen-Komponente in der Atmosphäre . . . 383
D. Protonen in der Atmosphäre 410
E. Neutronen in der Atmosphäre 424
  a) Direktmessungen 441
  b) Sterne in Kernemulsion 453
  c) Neutronen-Monitor 458
  d) Neutronenmessungen (direkt und indirekt) mit speziellen kosmischen Strahlungsspektrographen 461
  e) Die Winkelverteilung schneller Neutronen 463
  f) Neutronenproduktion und C^{14}-Produktion durch solare Protonen . . . 510
  g) Verlustfluß im Sonnenfleckenzyklus. (Theoretische Ergebnisse.) 518
  h) Experimentelle Verlustfluß-Intensitäten 527
  i) Aperiodisch emittierter Verlustfluß. (Solare Teilcheneinbrüche.) . . . 532
Anhang 1: Kosmische Strahlung — Registrierstationen 535
Anhang 2: Das Erdmagnetfeld; geomagnetische Koordinaten; spezielle Wahl . . 537
Literatur 538

Cosmic Ray Produced Radioactivity on the Earth. By Devendra Lal, Professor, Head of the Geophysics Research Group, Tata Institute of Fundamental Research, Ph. D. Bombay (India), and Bernard Peters, Professor of Physics, Niels Bohr Institute, University of Copenhagen, Copenhagen (Denmark). (With 24 Figures) 551
    A. Introduction 551
    B. Corpuscular radiation 555
        I. Primary particles 555
        II. Propagation of cosmic radiation in the atmosphere 557
    C. Production of isotopes in the atmosphere 561
        I. Methods of evaluating the source functions 562
        II. Altitude and latitude distribution of nuclear disintegrations in the atmosphere 563
        III. Rates of production of isotopes in the atmosphere 566
        IV. Time variations in isotope production 573
        V. Global averages of nuclear disintegrations and of isotope production in the atmosphere 581
        VI. Possible additional sources of isotopes in the earth's atmosphere 583
    D. Terrestrial isotopes of non-atmospheric origin 585
        I. Production of isotopes in the lithosphere 585
        II. Accretion of isotopes contained in extraterrestrial matter 586
    E. Circulation of isotopes in the geosphere 587
        I. Dispersion of isotopes in the atmosphere and fall-out 589
        II. Introduction of isotopes into the deep ocean 595
        III. Conservation laws for a system of well mixed reservoirs 597
        IV. The distribution of isotopes in the principal terrestrial reservoirs 599
        V. The applicability of various isotopes to the study of particular geophysical processes 601
    F. Some important observations on cosmic ray produced radio nuclides in the geosphere 602
        I. Global inventory studies 602
        II. Studies in the atmosphere 603
        III. Oceanographic studies 606
        IV. Studies in the biosphere 607
    G. Appendix 607
References 608

Effects of Cosmic Rays on Meteorites. By Professor Dr. Masatake Honda, The University of Tokyo, Tokyo (Japan) and Dr. James R. Arnold, Professor of Chemistry, University of California, San Diego, La Jolla, California, (United States of America). (With 6 Figures) 613
    Course of radiation in a meteorite 614
    History of meteorites 615
    Secular equilibrium 616
    Time variations 617
    Meteorite samples 618
    Measurement 619
    Cosmic-ray age 622
    Production rates 625
    Recovered artificial satellites 630
References 631

Sachverzeichnis (Deutch-Englisch) 635

Subject Index (English-German) 645

Theory of Cascade Showers.

By

J. Nishimura.

With 35 Figures.

A. Introduction.

When a fast electron passes through matter, it emits photons due to its acceleration in the Coulomb field of the atomic nuclei of the material. Each of the photons again reproduces electrons through the pair creation process, so that as a result of these interactions, the numbers of both electrons and photons increase. This phenomenon is called an electron shower.

The probability of the emission of photons or of pair creation can be calculated on the basis of Dirac's theory of the electron. This calculation was carried out by Bethe and Heitler [1].

The electron shower theory has been developed since 1937, when Bhabha and Heitler (1), and Carlson and Oppenheimer (2) first published their papers on this subject. The purpose of their work was to interpret cosmic ray phenomena like showers and bursts which are characterised by the production of many particles in the material by a single incident particle, and to test the applicability of quantum electrodynamics at the high energies involved.

Qualitatively at least they succeeded in explaining those effects, and especially the intensity-altitude curve of cosmic rays with its peak intensity around 100 mb as observed by balloon experiments (3) was well interpreted by the electron showers if one assumed that the primary cosmic rays were all electrons.

However, the later development of the cosmic ray studies indicated that the primary cosmic rays are predominantly of nucleonic nature, and that for this transition effect the multiple production of mesons plays an essential role. The cosmic-ray electrons and photons in the atmosphere are now interpreted as due to electron showers mainly started by the two γ-rays from the decay of π^0-mesons originated in the multiple production of mesons by the nucleonic component. Therefore the cosmic ray phenomena present, in general, a complicated mixture of nucleonic and electronic processes. Thus, the electron shower theory has become an indispensable tool for the analysis of these phenomena, and with the improvement of the experimental techniques and the accumulation of the data, more accurate theories have become necessary.

In the theories of Bhabha and Heitler, and Carlson and Oppenheimer, approximate expressions for the shower functions were derived using the Bethe-Heitler formulae [1] for the radiation and pair creation cross sections. As early as in 1938, Landau and Rumer [2] have shown that the exact solution for the shower functions can be derived by applying the method of functional transform if one ignores the effect of the ionization loss of the shower electrons. Almost at the same time, Snyder (4) and Serber (5) derived approximate expressions for the shower functions including the effect of the ionization loss, by using the complete screening cross sections in the Bethe-Heitler formulae with a mathematical procedure similar to that introduced by LANDAU and RUMER. TAMM and BELENKY (6) also gave solutions under similar assumptions with a slightly different mathematical approach. Finally, the exact solutions for the shower function under these assumptions were derived by SNYDER [3], and by SCOTT [4].

All these papers are confined to the development of a one-dimensional theory of electron showers. Further progress along these lines is summarized in a number of excellent review articles [5], [6], [7].

The shower electrons are scattered by the Coulomb field of the atoms of the material, and the particles spread sidewise as observations in an actual electron cascade demonstrate. Since the spread of the shower particles plays a rather essential role in cosmic-ray phenomena such as extensive air showers and the bursts observed in certain detectors, attempts have also been made since 1940 to develop three-dimensional theories. The problem was first treated by EULER and WERGELAND (7), and afterwards MOLIÈRE (8) derived the structure function at the shower maximum by numerical methods. Around 1950 his function was widely used for the analysis of the data on extensive air showers. MOLIÈRE's treatment, however, is not as complete as that of the one-dimensional shower theory, so that further efforts were made to obtain more accurate solutions. Among these ROBERG and NORDHEIM (9), EYGES and FERNBACH (10), and GREEN and MESSEL (11) have derived expressions giving the exact solutions in terms of the moments of the spread of the shower particles under certain assumptions.

A direct derivation of the structure function was also attempted by several authors (10), (12) to (15), [8]. Among them KAMATA and NISHIMURA derived formulae for the structure functions with an accuracy corresponding to that given by SNYDER and by SCOTT in the theory of the one-dimensional electron shower. This is seen from the fact that integration of the of KAMATA-NISHIMURA expressions with respect to the lateral spread yields exactly the function obtained by SNYDER and SCOTT.

The three-dimensional theory was originally developed to provide the basis for an analysis of the extensive air showers. Now an extensive air shower is known to be a mixture of nucleonic and electronic cascades. The nucleonic cascade develops as a result of the multiple production of mesons [9], and one of the purposes of the study of extensive air showers is the investigation of these multiplication processes in the region of extremely high energies. The complexities of the problem made it desirable to check the essential features of meson production from the data on extensive air showers without entering into the mathematical details of the theory of nucleonic cascades. Some results along this line results are reported elsewhere in this volume [9], where of course the theories of electron showers are applied as the basis for the analysis of those phenomena.

Recent developments in the emulsion technique make it possible to observe the full development of an electron shower in a nuclear emulsion. These events most clearly reveal the lateral spread, and therefore permit the application of the theory for the energy determination of the primary particles.

Fluctuations in the number of shower particles also play an essential role in the analysis of the shower phenomena. Their importance has already been pointed out in the original work of BHABHA and HEITLER (1), but in view of its mathematical difficulties the problem was attacked mostly by the Monte Carlo method (16), and a more sophisticated mathematical treatment [10] was also attempted.

In this article, the solutions of the several shower theories are reviewed under a unified mathematical treatment, i.e., functional transformations and integral representations of the infinite series, developed by KAMATA and NISHIMURA [8] and CHAKRABARTY and GUPTA (17), and special attention is paid to clear exposition of the physical basis and the limitation of each theory. The elementary processes, the basic phenomena for the electron shower, are shortly summarized, and the general behaviour of the shower functions expected from these processes are discussed. Starting from the Landau-Rumer solution [2], the results of SNYDER [3] and SCOTT [4] are derived, and three-dimensional shower theories with and without consideration of the effect of large angle single scattering are developed along the line of KAMATA and NISHIMURA, as a general extension of the theories of SNYDER and SCOTT. Limitation of the applicability of those shower functions are critically examined, and the applications of the shower theory to the analysis of the cosmic ray phenomena are described.

B. The elementary processes.

When a charged particle passes through matter, interactions with atoms may arise, both through nuclear and electromagnetic processes. Among them, those caused by electromagnetic interactions are the most familiar, and as a result of the development of quantum electrodynamics, the cross sections of the processes are known with a considerable accuracy.

A fast-moving charged particle suffers collisions with atomic electrons in the material through the Coulomb force, and by knocking out electrons or by exciting the atoms it loses energy. It is convenient to summarize these phenomena as "ionization loss process". The calculation shows [5], [6], [7] that the energy loss of fast charged particles in the relativistic region is almost independent of the energy of the moving particle, and is likewise almost independent of the material traversed if its thickness is measured in g/cm^2.

In addition to this process, during the passage of a fast charged particle near the atomic nuclei the path of the particle is deflected by the strong Coulomb field of these nuclei. Due to the resulting acceleration the proper electromagnetic field of the particle is shaken off, and is emitted in the form of a photon. Thus the incident particle undergoes a radiation process as well. Its magnitude was calculated by BETHE and HEITLER [1], who showed that the amount of radiation loss is proportional to the particle energy. The two competing processes, ionization and radiation losses, are therefore seen to differ crucially in their energy dependence. Consequently, if one considers the total energy loss of a fast charged particle, radiation loss always gives the main contribution at sufficiently high energies, and ionization is important in the region of lower energies.

During their passage through matter high-energy photons are absorbed by pair creation, and lose their energy by Compton scattering. The probability of the occurrence of Compton scattering is given by the Klein-Nishina formula, and is roughly inversely proportional to the photon energy. At higher energy, therefore, Compton scattering becomes less important, and the energy losses through pair creation process are dominant.

The theory of pair creation process is closely connected to that of the radiation process. The former can be described as the transition of an electron from a state of negative energy to a state of positive energy through the absorption of a high-energy photon. On the other hand, the radiation process can be considered as a transition of an electron between two states of positive energy with the emission of a high-energy photon. In fact, the cross sections of the two processes have quite similar form, as will be shown in the following section. Since the theories of the ionization process, the Compton process, and the radiation and pair creation processes are fully described in the review papers [1], [5], [6] we shall not present them in detail here, but only summarize the corrections to the Bethe-Heitler formulae for the radiation and pair creation processes which should be taken into account for the actual evaluation of the shower functions.

I. Corrections to the Bethe-Heitler formula.

1. Screening effect. In calculating the probability of the radiation process one must take into account the screening effect of the Coulomb field by the outer atomic electrons if the average impact parameter effective for the radiation loss is of the same order of magnitude as the atomic radius. This really happens for a particle of high energy. The matrix element of this process is given by

$$V_{Fi} = \int \psi_F V \psi_{\text{in}} \, d\tau, \tag{1.1}$$

where V is the Coulomb field, and ψ_F and ψ_{in} are the wave functions of initial and final states, respectively. Then (1.1) becomes

$$V_{Fi} \cong \int \frac{e^{i(\mathbf{q} \mathbf{r})/\hbar c}}{r} r^2 \, dr, \tag{1.2}$$

where

$$\left. \begin{array}{l} q_\| \cong \sqrt{E^2 - m^2 c^4} - \sqrt{(E-E')^2 - m^2 c^4} - E' \\ \sim \dfrac{m^2 c^4 E'}{2E(E-E')}, \quad \text{for} \quad E, E-E' \gg mc^2, \end{array} \right\} \tag{1.3}$$

E is the energy of the electron in the initial state, and E' is the energy of the emitted photon. Therefore the main contribution to the integral (1.2) comes from collisions at the distance

$$r_{\text{eff}} \sim \frac{1}{q_\|} \sim \frac{2E(E-E')}{mc^2 E'} \cdot \frac{\hbar}{mc}. \tag{1.4}$$

For the atomic radius, we may assume the value $137 \dfrac{\hbar}{mc} \dfrac{1}{Z^{\frac{1}{3}}}$ given by the Thomas-Fermi gas model, where Z is the atomic number of the material.

Thus distinguishing the three cases of an atomic radius larger, nearly equal, or smaller than the value of r given in (1.4), i.e.,

$$\frac{137}{Z^{\frac{1}{3}}} \gtreqless \frac{E(E-E')}{mc^2 E'},$$

or

$$1 \gtreqless \frac{137 \cdot mc^2 E'}{E(E-E')} \cdot \frac{1}{Z^{\frac{1}{3}}},$$

or approximately

$$1 \gtreqless \frac{50 \text{ MeV} \cdot E'}{E(E-E') Z^{\frac{1}{3}}}, \tag{1.5}$$

screening is called complete, intermediate and not effective. At high energy the complete screening cross section must be applied, and in the low-energy limit the cross section for no screening. The complete screening cross section for radiation and pair creation processes were given by BETHE [1], and later revised by KIRPICHOV and POMERANCHUK (18). The probability for emission of a photon in the energy interval $(E', E'+dE')$ by an electron of energy E after traversing a medium of thickness dx g/cm^2, $\Phi_0 \dfrac{dE'}{E} dx$, is given by

$$\left. \begin{array}{l} \Phi_{0,\text{rad}} \dfrac{dE'}{E} = \dfrac{4}{137} \cdot \dfrac{N}{A} \cdot z^2 r_0^2 \dfrac{dE'}{E'} \left[\left\{ 1 + \left(1 - \dfrac{E'}{E}\right)^2 - \dfrac{2}{3}\left(1 - \dfrac{E'}{E}\right) \right\} \times \right. \\ \left. \times \log(191 \, Z^{-\frac{1}{3}}) + \dfrac{1}{9}\left(1 - \dfrac{E'}{E}\right) \right], \end{array} \right\} \tag{1.6}$$

and the probability of pair creation by a photon of energy E, giving rise to an electron in $(E', E'+dE')$

$$\Psi_{0,\text{pair}} \frac{dE'}{E} = \frac{4}{137} \cdot \frac{N}{A} \cdot z^2 r_0^2 \frac{dE'}{E} \left[\left\{ \left(\frac{E'}{E}\right)^2 + \left(1 - \frac{E'}{E}\right)^2 + \frac{2}{3} \frac{E'}{E}\left(1 - \frac{E'}{E}\right) \right\} \times \right. \\ \left. \times \log(191 Z^{-\frac{1}{3}}) - \frac{1}{9} \frac{E'}{E}\left(1 - \frac{E'}{E}\right) \right], \tag{1.7}$$

where Z and A are the atomic number and atomic weight of the traversed matter, respectively, N is AVOGADRO's number, and r_0 the classical electron radius $\frac{e^2}{mc^2}$. The corresponding radiation loss is obtained as

$$\int_0^E \Phi_0(E,E') E' \frac{dE'}{E} = X^{-1} E,$$

so that the energy loss per g/cm² due to radiation processes is given by

$$-\frac{dE}{dx} = X^{-1} E,$$

where we put

$$X^{-1} = \frac{4}{137} \frac{N}{A} z^2 r_0^2 \log 191 Z^{-\frac{1}{3}}.$$

Simple approximate formulae for the probabilities of the radiation and pair creation processes are now derived from (1.6) and (1.7):

$$\Phi_0 \frac{dE'}{E} \simeq X^{-1} \frac{dE'}{E'}, \tag{1.6'}$$

$$\Psi_0 \frac{dE'}{E} \simeq \frac{7}{9} X^{-1} \frac{dE'}{E}. \tag{1.7'}$$

Thus we can see that in the radiation process mostly low-energy photons are emitted, while pair electrons are distributed uniformly in the energy region from zero to the full energy of parent γ-ray.

The exact expressions for the incomplete screening cross-sections are summarized in the references [1], [5], [6]. Here only the approximate formulae for these cross sections will be presented.

BERNSTEIN (19) has found expressions which fit the probabilities over the whole range of the screening effect with a considerable accuracy. They can be written in the form

$$\Phi = \Phi_0 \frac{1}{1 + U\{E'/E(E-E')\}}, \tag{1.8}$$

$$\Psi = \Psi_0 \frac{1}{1 + U\{E/E'(E-E')\}}, \tag{1.9}$$

where

$$U = (mc^2/Z^{\frac{1}{3}}) 255/(15.6 - \frac{1}{3} \cdot 4 \log Z). \tag{1.10}$$

The numerical values of U are listed in Table 1. A comparison of the approximated expressions with the results of an accurate calculation is shown in Fig. 1.

Table 1. *Energy units U as a function of Z.*

z	1	5	10	20	30	40	50	60	70	80	90
U (Mev)	8.30	5.63	4.81	4.11	3.74	3.53	3.37	3.27	3.15	3.08	3.00

2. Landau effect of the radiation and pair creation processes.

As described in the proceeding section, the effective region of the radiation and pair processes is represented by

$$r_{\text{eff}} \sim \frac{\hbar c}{q_\parallel} \sim \frac{2E(E-E')}{mc^2 E'} \cdot \frac{\hbar}{mc} \qquad (1.4)$$

If E is so large, or E' is so small, that r_{eff} becomes larger than the average distance between the neighbouring atoms, an interference effect may arise for these processes. If the atoms are aligned, as in a crystal, the radiation is expected to be

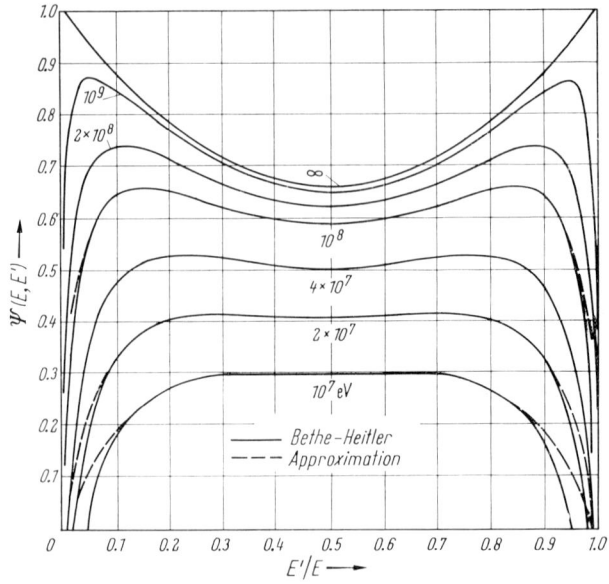

Fig. 1. The probability $\psi(E, E') dE'/E$ for the occurrence of pair creation in one radiation length of Pb. (From I. B. BERNSTEIN [19].)

sharply collimated in a certain direction which is determined by (1.2). In the case of an amorphous medium, no such interference phenomena occur, but the effect of multiple scattering of the electrons plays an important role for these processes.

The mean square angle of the multiple scattering of an electron of energy E after traversing a distance r in the material is

$$\langle \theta_s^2 \rangle \sim \left(\frac{K}{E} \right)^2 \frac{r}{X}, \qquad (2.1)$$

where the "scattering energy" K has a value of about 20 MeV. The derivation of this formula is given in Appendix 3. Since the average emission angle or opening angle for the radiation or pair processes is mc^2/E, the $q_\perp$ relating to $(\boldsymbol{q}\boldsymbol{r})$ in (1.2) is mc^2 for an isolated atom. If the angle due to the multiple scattering becomes larger than during the passage through the region r_{eff}, $q_\perp$ increases during the interactions with the atoms, and then due to the fast variation of $e^{i(\boldsymbol{q}\boldsymbol{r})}$ in the integral (1.2), the cross section should be reduced. The condition under which the effect becomes predominant,

$$\langle \theta_s^2 \rangle^{\frac{1}{2}} > \frac{mc^2}{E}, \qquad (2.2)$$

becomes, with the help of (2.1),

$$K \left(\frac{\hbar c}{q_\parallel X} \right)^{\frac{1}{2}} > mc^2. \qquad (2.3)$$

Then we have from (1.4)

$$\frac{E(E-E')}{E} > \left(\frac{mc^2}{K}\right)^2 mc^2 \cdot \frac{mc}{\hbar} X , \qquad (2.4)$$

or in a numerical form

$$\frac{E(E-E')}{E'} \gtrsim 4 \cdot 10^{12} \left(\frac{X}{\text{cm}}\right) \text{eV} . \qquad (2.5)$$

Thus the effect is most significant at high energy, and in heavy elements. For Pb, $X = 0.6$ cm, (2.5) gives

$$E \gtrsim 2 \cdot 10^{12} \text{ eV} \quad \text{at} \quad E' \sim \frac{E}{2} .$$

The cross sections for the two processes under the above conditions were calculated by LANDAU and POMERANCHUK (20) on a classical basis, and in quantum mechanical treatment by MIGDAL (21). In his calculation, Migdal introduces the parameter s defined by

$$s = \frac{1}{2} \frac{mc^2}{E} \frac{1}{\langle \theta_s^2 \rangle^{\frac{1}{2}}} = 10^6 \left(\frac{E' X}{E(E-E')}\right)^{\frac{1}{2}} \left(\frac{\text{eV}}{\text{cm}}\right)^{\frac{1}{2}}, \qquad (2.6)$$

representing the effect of multiple scattering as mentioned above. When s is larger than unity, the contribution of this effect is expected to be negligible. On the other hand, if s is smaller than unity, the effect becomes predominant.

The ratio of the probability for the radiation and the pair creation processes to the one given by BETHE and HEITLER are now written as a functions of s:

$$\frac{\Phi_{\text{rad}} dv}{\Phi_{\text{B.H}} dv} = \frac{\xi(s) \{G(s) v^2 + 2[1 + (1-v)^2] \phi(s)\} dv}{3 \{1 + (1-v)^2 - \frac{2}{3}(1-v)\} dv} \qquad (2.7)$$

and

$$\frac{\Psi_{\text{pair}} du}{\Psi_{\text{B.H}} du} = \frac{\xi(s) \{G(s) + 2[u^2 + (1-u)^2] \phi(s)\} du}{3 \{u^2 + (1-u)^2 + \frac{2}{3} u(1-u)\} du} , \qquad (2.8)$$

where the notation is the same as in (1.6) and (1.7), and u, v, the functions ξ, G and ϕ are defined by

$$u = \frac{E'}{E}, \quad v = \frac{E'}{E},$$

$$\left. \begin{array}{ll} \xi(s) = 1 + (\log s/\log s_1) & 1 \geq s \geq s_1 \\ = 1 & s > 1 \\ = 2 & s < s_1 \end{array} \right\} \qquad (2.9)$$

with

$$s_1^{\frac{1}{2}} = \frac{Z^{\frac{1}{3}}}{191}, \qquad (2.10)$$

$$G(s) = 12 \pi s^2 - 48 s^3 \sum_{k=0}^{\infty} \frac{1}{(k+s+\frac{1}{2})^2 + s^2} , \qquad (2.11)$$

and

$$\phi(s) = 6s - 6\pi s^2 + 24 s^3 \sum_{k=1}^{\infty} \frac{1}{(k+s)^2 + s^2} . \qquad (2.12)$$

The quantity s_1 has the following physical meaning: for $s = s_1$ the multiple scattering angle of the electron is equal to the angle of diffraction by the atomic nuclei of the medium traversed. In lead, for $E' = \frac{1}{2} E_0$, $s = s_1$ corresponds to an energy $E_0 \cong 2 \times 10^{18}$ eV. The numerical values of G and ϕ given by MIGDAL are shown in Table 2, from which it is seen that G and ϕ approach unity with in-

Table 2. *Values of the functions* $\phi(s)$ *and* $G(s)$.

s	$\phi(s)$	$G(s)$	s	$\phi(s)$	$G(s)$	s	$\phi(s)$	$G(s)$
0	0	0	0.4	0.880	0.800	0.9	0.985	0.975
0.05	0.258	0.094	0.5	0.931	0.875	1.0	0.990	0.985
0.1	0.446	0.206	0.6	0.954	0.917	1.5	0.998	0.994
0.2	0.686	0.475	0.7	0.965	0.945	2.0	0.999	0.998
0.3	0.805	0.695	0.8	0.975	0.963			

creasing values of s. Because of this asymptotic behaviour of ξ, G, and ϕ for $s \to \infty$, the ratio of the probabilities (2.7) and (2.8) also becomes unity as expected. For $s \to 0$, we obtain for the same ratio the values $\sqrt{v}$ and $\sqrt{u}$, a feature which is

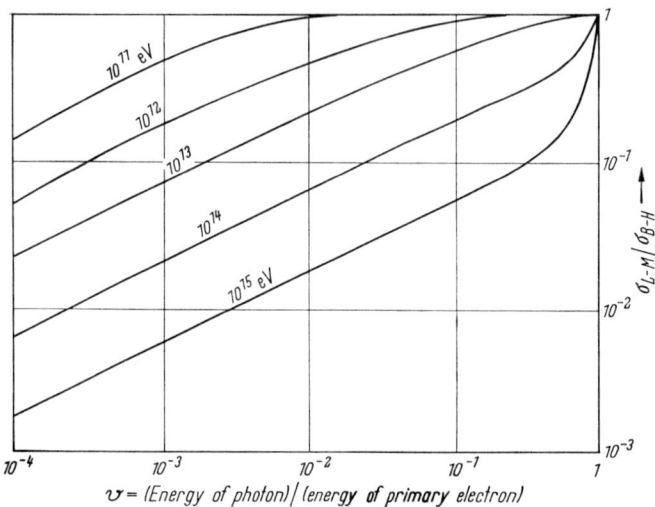

Fig. 2. Deviations from the Bethe-Heitler radiation cross section in Pb due to the Landau effect. $\sigma_{\text{L-M}}$ and $\sigma_{\text{B-H}}$ are the cross sections calculated by MIGDAL [*21*] and BETHE-HEITLER [*1*].

also expected from the classical treatment. The numerical value of the ratio (2.7) is shown for Pb in Fig. 2, which demostrates that the effect is predominant only at extremely high energies.

3. Radiation and pair creation probabilities by the atomic electrons. Radiation and pair creation processes are also expected to occur through collisions with the outer electrons of the atoms. WHEELER and LAMB (*22*) have shown that in the limit of complete screening, this additional collision probability becomes

$$\Phi_{\text{el}} \cong \frac{\zeta}{Z} \Phi_0 \tag{3.1}$$

and

$$\Psi_{\text{el}} \cong \frac{\zeta}{Z} \Psi_0, \tag{3.2}$$

where

$$\zeta = \frac{\log 1440 \, Z^{-\frac{2}{3}}}{\log 191 \, Z^{-\frac{1}{3}}}. \tag{3.3}$$

Thus the total probability is given by replacing Z^2 by $Z(Z+\zeta)$ in (1.6) and (1.7). Numerical values of ζ are listed in Table 3.

Table 3. *Numerical values of ζ as a function of Z.*

Z	1	5	10	20	30	40	50	60	70	80	90
ζ	1.38	1.31	1.27	1.24	1.21	1.19	1.18	1.17	1.16	1.15	1.14

4. Corrections for deviations from the Born approximation.

Since the Bethe-Heitler cross section was derived under the first Born approximation, it gives correct results only if

$$\frac{Ze^2}{\hbar c} \ll 1.$$

For heavy elements such as Pb, this value is about 0.6. Thus in such cases corrections should be applied for the deviation from this approximation.

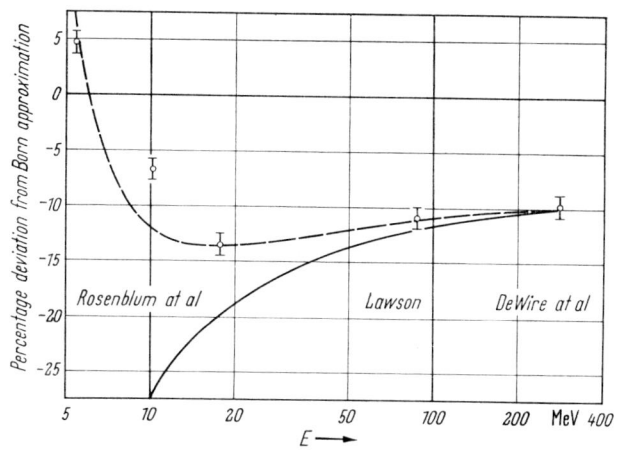

Fig. 3. Relative deviation (in percent) of the pair creation cross section from the values calculated with the Born approximation. The solid curve represents the uncorrected results of DAVIES, BETHE and MAXIMON (23), and the dashed curve the results after the correction proposed by these authors. Experimental data and the authors are also indicated.

DAVIS, BETHE and MAXIMON (23), and OLSEN (24) have carried out calculations for the radiation and pair creation processes without using the Born approximation. The results show that the probabilities decrease by an amount $\Delta \Phi$ and $\Delta \Psi$ for the two processes, respectively, which can be written in the form

$$\Delta \Phi_{\text{rad}} = \Phi_0 \frac{f(Z)}{\log 191 Z^{-\frac{1}{3}}}, \tag{4.1}$$

$$\Delta \Psi_{\text{pair}} = \Psi_0 \frac{f(Z)}{\log 191 Z^{-\frac{1}{3}}}, \tag{4.2}$$

and

$$\frac{\Delta \Phi}{\Phi_0} = \frac{\Delta \Psi}{\Psi_0} \cong \frac{f(Z)}{\log 191 Z^{-\frac{1}{3}}}, \tag{4.3}$$

no matter whether screening is absent, partial or complete, with

$$f(Z) = \alpha^2 \sum_{n=1}^{\infty} \frac{1}{n(n^2 + \alpha^2)} \tag{4.4}$$

and

$$\alpha = \frac{Ze^2}{\hbar c}, \tag{4.5}$$

so that for small values of α

$$f \cong \left(\frac{Ze^2}{\hbar c}\right)^2. \tag{4.6}$$

Thus the correction for Pb is about 10%.

Since the accuracy of their calculation is satisfactory only in the high-energy region, DAVIS et al. constructed a correction curve applicable to the low-energy part, by using experimental data and the results of exact calculations of JAEGER and HULME (25), (26). This correction curve is shown in Fig. 3.

II. Radiation length and critical energy.

5. Definition and numerical values. Since the energy loss of an electron by radiation is expressed by

$$\frac{dE}{dX} = -X_0^{-1} E,$$

Table 4. *Radiation length and critical energy for various materials.*
(From O. I. DOBDGENKO and A. A. POMANSKY, private communication.)

Material	Z	A	Radiation length (g/cm²)	Radiation length (cm)	Critical energy*, ** MeV
H	1	1.008	62.8	7500 (m)	350
He	2	4.003	93.1	5600 (m)	250
Li	3	6.940	83.3	15.6	138
C	6	12.010	43.3	16.9 (Graphite)	79
N	7	14.008	38.6	331 (m)	85
O	8	16.000	34.6	258 (m)	75
Al	13	26.980	24.3	9.10	40
Si	14	28.090	22.2	9.52	37.5
Fe	26	55.85	13.9	1.77	20.7
Cu	29	63.54	13.0	1.46	18.8
Br	35	79.916	11.5	3.71	15.7
Ag	47	10.988	9.0	0.86	11.9
I	53	126.910	8.5	1.74	10.7
W	74	183.92	6.8	0.35	8.08
Pb	82	207.21	6.4	0.57	7.40
Compound materials					
Air N 75.52% O 23.14% Ar 1.3%			37.1	308 (m)	81
SiO$_2$			27.4	10.3	47.3
H$_2$O			36.4	36.4	73.0
LiH			80.0	113	157
Nuclear Emulsion G$_5$ (58% RH) recalculated			11.4	2.98	16.4

* The density effect is included except for gaseous material. The data for gases are valid for a temperature of 20° C at atmospheric pressure.
** The critical energy is approximately given by 500 MeV/Z.

it is convenient to measure the thickness of matter in units of X_0. In this unit of length, the formula becomes

$$\frac{dE}{dt} = -E. \tag{5.1}$$

The unit X_0 is called the radiation length, and is defined by the relation

$$X_0^{-1} = \frac{4}{137} \frac{N}{A} r_0^2 Z(Z+\zeta) [\log 191 Z^{-\frac{1}{3}} - f(Z)]. \tag{5.2}$$

The radiation length thus introduced differs from the one used in earlier work by the correction for the contribution of atomic electrons, and for the deviation from the Born approximation. For small Z the correction due to ζ is effective, and for large Z that due to $f(Z)$ becomes important.

The probabilities for the radiation and pair creation processes per radiation length will be written in the small letters φ and ψ respectively. From (1.7), the total probability of pair creation is given by

$$\int_0^1 \psi_0(u)\,du = \frac{7}{9}. \tag{5.3}$$

In addition to radiation losses, an electron loses energy by ionization. The ionization loss during the passage through one radiation length is called the critical energy of the material traversed. Denoting this constant value by ε, the total energy loss of an electron can now be written in the form

$$\frac{dE}{dt} = E + \varepsilon, \tag{5.4}$$

where t is measured in radiation lengths. The values of radiation length and critical energy for various materials are shown in Table 4.

C. The diffusion equations for electron showers, and the limitations of the approximations.

I. The diffusion equations.

6. The approximations used in the shower theory. If we limit ourselves to electrons and photons of energy large compared with the critical energy in the material, the effect of the ionization loss of the electrons need not be included in the considerations on shower development. Then a shower theory can be constructed by taking into account radiation and pair creation processes only. If, in addition, the calculations of the shower function throughout use the complete screening cross sections for the radiation and pair creation processes, the procedure is known as "Approximation A". The treatment in which the effects of ionization loss are included, using complete screening cross sections for radiation and pair creation processes but neglecting the effect of Compton scattering, is called "Approximation B" [5].

The evaluation of the shower function including the effect of the deviation from the complete screening cross sections involves a great amount mathematical difficulties, and the calculation can be carried through only by numerical methods, or with the help of rather crude mathematical approximations.

The methods mentioned above describe only the one-dimensional development of the shower. However, the evidence of extensive air showers as well as emulsion observations demonstrate that in an electron shower, the particles are also displaced laterally, and in some instances, the lateral spread of a shower plays a rather essential role in its development.

There are several reasons for the lateral spread of the shower particles: the opening angle of an electron pair, the emission angle of a photon in the radiation process, and the deflection due to the multiple Coulomb scattering of an electron.

The ratio of the opening angle or the emission angle to the deflection angle due to multiple scattering was derived in the section on the Landau effect. After a traversal of t radiation lengths, this ratio is

$$\frac{\theta}{\langle\theta_s\rangle} \sim \frac{mc^2}{Kt^{\frac{1}{2}}} \sim 2.5 \times 10^{-2} t^{-\frac{1}{2}}, \qquad (6.1)$$

since

$$\theta \sim \frac{mc^2}{E}, \qquad (6.2)$$

and

$$\langle\theta_s\rangle \sim \frac{Kt^{\frac{1}{2}}}{E}. \qquad (6.3)$$

Thus the contribution of the opening angle or the emission angle may be neglected compared with that of multiple scattering, except at the very beginning of the development of an electron shower. A satisfactory three-dimensional shower theory can therefore be constructed by including only the contribution of the effect of multiple scattering.

Table 5. *Approximations used in the shower theories.*

Shower theory	Approximation	Cross section of radiation and pair creation process	Ionization loss	Compton effect
One-dimensional theory	Approximation A	complete screening	neglected	neglected
	Approximation B	complete screening	constant loss	neglected

Shower theory	Approximation	Scattering theory	Radiation, pair creation, ionization loss and Compton effect
Three-dimensional theory	Landau approximation	Fokker-Planck approximation	situation is similar to that of the one-dimensional theory
	without Landau approximation	Molière's scattering theory	situation is similar to that of the one-dimensional theory

In the customary terminology, a three-dimensional shower theory using the Fokker-Planck approximation (Appendix III) for multiple scattering is called a "treatment in the Landau approximation", while the use of Molière's theory (Appendix III) leads to a "treatment without the Landau approximation". The situation is summarized in Table 5.

7. Basic equations. The basic equations in these approximations are constructed in the following way: Let $\pi(E, t)\, dE$ and $\gamma(E, t)\, dE$ be the average number of electrons and of photons with energies between E and $E+dE$ in a shower at a depth t in radiation lengths. To describe the change in the numbers of electrons and photons after traversal of a layer Δt, we can then sum up the various effects discussed above, and proceed to the formally simple diffusion equations for these spectra. This will be shown in the following paragraphs.

α) *Change of the number of electrons in* Δt. Three processes contribute to this:

(i) Effects of the radiation process, $A'\pi$.

Electrons with energy between E and $E+dE$ which suffer radiation losses are shifted to a lower energy region. Then the number of electrons removed from this energy interval is

$$\pi(E, t) \int_0^E \varphi(E, E') \frac{dE'}{E} \Delta t. \qquad (7.1)$$

On the other hand, by the same process electrons of higher energy may be moved into the energy region between E and $(E+dE)$. This increase in $\pi(E, t)$ amounts to

$$\int_0^\infty \pi(E+E', t)\varphi(E+E', E')\frac{dE'}{E+E'}\Delta t. \tag{7.2}$$

Therefore the net change of the number of electrons is given by

$$-\int_0^\infty \left\{\pi(E', t)\varphi(E, E') - \pi(E+E', t)\varphi(E+E', E')\cdot\frac{E}{E+E'}\right\}\frac{dE'}{E}\Delta t. \tag{7.3}$$

If one uses the complete screening cross section, the probability $\varphi\frac{dE'}{E}$ can be written in the fractional form (1.6),

$$\varphi(E, E')\frac{dE'}{E} \to \varphi_0(v)\,dv, \tag{7.4}$$

and (7.3) becomes

$$-\int_0^1 \left[\pi(E) - \frac{1}{1-v}\pi\left(\frac{E}{1-v}\right)\right]\varphi_0(v)\,dv, \tag{7.5}$$

Written in operator form, (7.5) can be expressed by

$$-A'\pi = -\int_0^1 \left[\pi(E) - \frac{1}{1-v}\pi\left(\frac{E}{1-v}\right)\right]\varphi_0(v)\,dv. \tag{7.6}$$

(ii) Effects of the pair creation process, $B'\gamma$.

A photon with energy higher than E may produce electrons in this energy interval $(E, E+dE)$. Their number is

$$2\int_0^\infty \gamma(E+E', t)\psi(E+E', E)\frac{dE'}{E+E'}\Delta t. \tag{7.7}$$

Again using the complete screening cross sections (1.7), the increase is

$$B'\gamma = 2\int_0^1 \gamma\left(\frac{E}{u}\right)\psi_0(u)\frac{du}{u}. \tag{7.8}$$

(iii) Effect of the ionization loss, $\varepsilon\frac{\partial\pi}{\partial E}$.

Electrons lose energy by an amount $-\varepsilon\Delta t$ through ionization in Δt. Since electrons arriving at the depth $t+\Delta t$ with energy $E, E+dE$ should have had energies in the range $(E+\varepsilon\Delta t, E+dE+\varepsilon\Delta t)$ before traversing Δt, the corresponding change in the number of electrons is

$$\left.\begin{array}{l}\pi(E, t+\Delta t)\,dE - \pi(E, t)\,dE \\ = \pi(E+\varepsilon\Delta t)\,dE - \pi(E, t)\,dE \\ = \varepsilon\dfrac{\partial\pi}{\partial E}\Delta t.\end{array}\right\} \tag{7.9}$$

Summing up all the effects (i) to (iii), the diffusion equation for the electrons in a one-dimensional shower theory is given by

$$\frac{\partial\pi}{\partial t} = -A'\pi + B'\gamma + \varepsilon\frac{\partial\pi}{\partial E}. \tag{7.10}$$

β) *Change in the number of the photons.* Here only two processes contribute:
(i) Absorption of photons by the pair creation process, $\sigma_0 \gamma$.
The total losses can be written in the form

$$-\gamma \int_0^\infty \psi(E, E') \frac{dE'}{E} \Delta t, \qquad (7.11)$$

and with the complete screening cross section (1.7)

$$-\int_0^1 \psi_0(u) du = -\sigma_0 = -0.7733 \ldots . \qquad (7.12)$$

(ii) Effect of the radiation process, $C'\pi$.
An electron of higher energy may emit a photon with energy between E and $E+dE$, and thus gives a contribution

$$\int_0^\infty \pi(E+E') \varphi(E+E', E) \frac{dE'}{E+E'}. \qquad (7.13)$$

Again expressed in fractional form, (7.13) becomes

$$C'\pi = \int_0^1 \pi\left(\frac{E}{v}, t\right) \varphi_0(v) \frac{dv}{v}, \qquad (7.14)$$

under the assumption of complete screening. Thus for the change in number of the photons, we have

$$\frac{\partial \gamma}{\partial t} = -\sigma_0 \gamma + C'\pi. \qquad (7.15)$$

(7.10) and (7.15) serve as the basic equations for derivation of the shower functions in the various approximations.

γ) *One-dimensional electron shower theory.*
(i) Approximation A.
By neglecting ionization losses, the diffusion equations are reduced to

$$\frac{\partial \pi}{\partial t} = -A'\pi + B'\gamma, \qquad (7.10')$$

$$\frac{\partial \gamma}{\partial t} = -\sigma_0 \gamma + C'\pi, \qquad (7.15)$$

(ii) Approximation B.
In this treatment the unsimplified eqs. (7.10) and (7.15) must be solved.

δ) *Three-dimensional electron shower theory.* Let $\pi(E, \boldsymbol{r}, \boldsymbol{\theta}) dE\, d\boldsymbol{r}\, d\boldsymbol{\theta}$ and $\gamma(E, \boldsymbol{r}, \boldsymbol{\theta}) dE\, d\boldsymbol{r}\, d\boldsymbol{\theta}$ be the average number of electrons and photons of energy $(E, E+dE)$, with lateral deviation between $\boldsymbol{r}$ and $\boldsymbol{r}+d\boldsymbol{r}$, and deflection angle between $\boldsymbol{\theta}$ and $\boldsymbol{\theta}+d\boldsymbol{\theta}$ from the shower axis. Including the contribution of multiple scattering (A.3.13) and (A.3.27) described in Appendix 3, we then obtain the following equations:
(i) Landau approximation

$$\frac{\partial \pi}{\partial t} + \boldsymbol{\theta} \frac{\partial \pi}{\partial \boldsymbol{r}} = -A'\pi + B'\gamma + \varepsilon \frac{\partial}{\partial E}\pi + \frac{E_s^2}{4E^2}\left(\frac{\partial^2}{\partial \theta_1^2} + \frac{\partial^2}{\partial \theta_2^2}\right)\pi, \qquad (7.16)$$

$$\frac{\partial \gamma}{\partial t} + \boldsymbol{\theta} \frac{\partial \gamma}{\partial \boldsymbol{r}} = -\sigma_0 \gamma + C'\pi, \qquad (7.17)$$

where θ_1 and θ_2 are the components of $\boldsymbol{\theta}$ in the directions x and y, respectively; i.e., $(\boldsymbol{r}\boldsymbol{\theta}) = x\theta_1 + y\theta_2$.

(ii) Without the Landau approximation

$$\frac{\partial \pi}{\partial t} + \boldsymbol{\theta}\frac{\partial \pi}{\partial \boldsymbol{r}} = -A'\pi + B'\gamma + \varepsilon\frac{\partial}{\partial E}\pi + \int[\pi(\boldsymbol{\theta}-\boldsymbol{\theta}') - \pi(\boldsymbol{\theta})']\sigma(\boldsymbol{\theta}')d\boldsymbol{\theta}', \qquad (7.18)$$

$$\frac{\partial \gamma}{\partial t} + \boldsymbol{\theta}\frac{\partial \gamma}{\partial \boldsymbol{r}} = -\sigma_0\gamma + C'\pi. \qquad (7.19)$$

The details about the last term of the right-hand side of (7.18) are described in Appendix 3.

II. Limitations of the approximations.

8. Limitations in the one-dimensional electron shower theory.

α) *Approximation A.* Since we are taking the complete screening cross sections for the radiation and pair creation processes, and are also assuming the deflection angle to be very small, the application of the theory to the analysis of shower phenomena may, in this case, be subject to serious limitations.

In Approximation A, the total number of shower particles increases infinitely with the depth of matter traversed, since no energy dissipation term is included in the considerations. The calculation will give satisfactory results only for particle energies higher than the critical energy, the screening energy $137\,Z^{-\frac{1}{3}}\,mc^2$, and the scattering energy $K \sim 20$ MeV. For light elements, the first of these restrictions is the most serious one, since the critical energy is approximately given by $\frac{1000}{Z}\,mc^2$, and therefore reaches the largest values for low Z. The variation of the three quantities with the atomic number Z is shown in Fig. 4.

β) *Approximation B.* By including the effect of the ionization loss, the infinite multiplication inherent in Approximation A is avoided in this case. From the energy conservation law, the total ionization loss by the shower particles should be equal to the energy of the incident particles E_0, so that

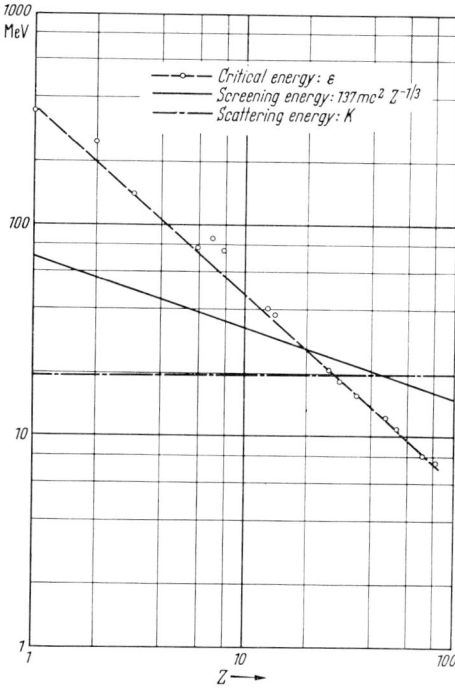

Fig. 4. Variation of the critical, the screening, and the scattering energy as a function of the atomic number Z.

$$\left.\begin{array}{c}\varepsilon\int\limits_0^\infty\int\limits_0^\infty \pi\, dE\, dt = E_0 \\[1em] \int\limits_0^\infty \Pi\, dt = \dfrac{E_0}{\varepsilon},\end{array}\right\} \qquad (8.1)$$

or

where Π is the total number of shower electrons at depth t. The integral of (8.1) is called the total track length of the shower electrons. The relation holds whatever the cross sections of radiation and pair creation processes are.

As the primary energy E_0 is shared among the shower particles, the average energy of shower particles decreases during the development of a shower, and at the shower maximum becomes a little smaller than the critical energy, as will be proved in Sect. 7. In order to study the average features of the shower particles, it is therefore necessary to use a partial screening cross section for the electrons and photons of effective energies around the critical energy. The decrease of the probability of pair creation due to complete screening is partly compensated by the contribution of Compton scattering near the critical energy, but the probability of the radiation processes also decreases towards low energies. The situation is illustrated in Fig. 5. The correction to the shower function resulting from these effects can be described, at least in a qualitative way, as equivalent to using a slightly longer radiation unit than that derived under the assumption of complete screening, and thus slightly increasing the critical energy. Considering relation (8.1), it can therefore be seen that the average number of shower particles at any point is in fact smaller, and shower particles penetrate to a larger depth than that evaluated in Approximation B.

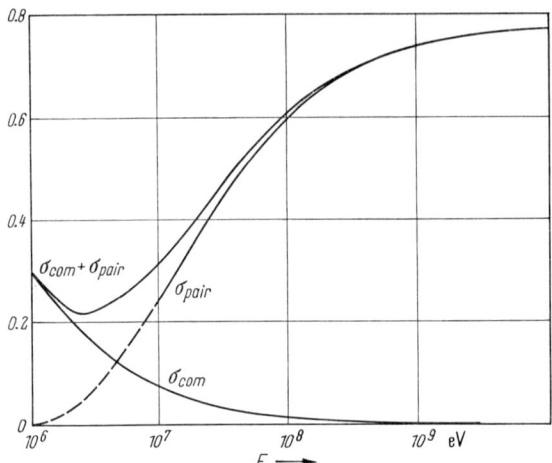

Fig. 5. Probabilities of the occurrence of pair creation and of Compton scattering in one radiation length of Pb. (From B. Rossi and K. Greisen [5].)

At the tail of the shower, the average energies of the shower electrons are so much reduced that the absorption of the photons determines the main features. As seen in Fig. 5, combination of Compton and pair creation processes gives the absorption coefficient a minimum value $\sigma_{\min}$ at a certain low energy. The shower absorption curve is represented by $e^{-\sigma_{\min} t}$ rather than expressed by the shower function of Approximation B. The theories of electron showers with accurate cross sections for radiation, pair creation and Compton scattering face great mathematical difficulties. Solutions can be obtained only by numerical methods, or derived analytically under rather crude mathematical approximations [6]. Some details will be discussed later.

Apart from these limitations owing to the screening cross sections, the effect of the lateral spread is quite important, especially when the behaviour of a shower developed in heavy elements is considered. As the average energy of shower particles is approximately given by the critical energy, the average deflection angle is

$$\frac{K}{\varepsilon} \simeq \frac{20 \text{ MeV}}{1000\, m c^2} Z. \tag{8.2}$$

If this value is larger than unity, some particles are strongly displaced, and some are even scattered backwards. In this case, the relation (8.1) should be replaced by

$$2\pi \int_0^1 d\cos\theta \int_0^\infty \pi(\theta)\, \frac{dt}{\cos\theta} \cong \frac{E_0}{\varepsilon}, \tag{8.3}$$

Sect. 9. Limitations of the approximation in the three-dimensional electron shower theory. 17

and the average number of the shower particles under t radiation units is obtained by multiplying the former solutions by a factor

$$\frac{\int\limits_0^1 d\cos\theta \int\limits_0^\infty \pi(\theta)dt}{\int\limits_0^1 d\cos\theta \int\limits_0^\infty \frac{\pi(\theta)}{\cos\theta} dt} < 1. \tag{8.4}$$

As seen from (8.2) the effect is most important for heavy elements, and was evaluated by several authors [6], [7]. At the end of this section we shall examine the contribution of the Landau effect to the shower function. Since the Landau effect is most significant of extremely high energies as shown in Fig. 2, it plays a major role only at the beginning of a shower of very large incident energy. As the shower develops, and the energy is divided among the particles, the Landau effect soon ceases to be important. Even if the energy of the primary particle is quite high, the development of the shower is influenced by it only in its very first stage.

9. Limitations of the approximation in the three-dimensional electron shower theory. Since the three-dimensional electron shower theories are constructed by a combination of one-dimensional shower theory and multiple scattering theory, restrictions to their validity arise from the approximations used in both its parts. In addition to the limitations described in the preceding sections, those due to the small angle approximations used in the multiple scattering theories must be considered.

Table 6. *Limitations of the approximations.*

Shower theory	Approximation	Applicable region
One-dimensional theory	Approximation A	$E \gg \varepsilon$, K and $137\ m\ c^2/Z^{\frac{1}{3}}$
	Approximation B	$E \gg K$ and $137\ m\ c^2/Z^{\frac{1}{3}}$
Three-dimensional theory	Landau approximation	Limitations as above; slight deviations from the theory without Landau approximation even in the region $\theta \lesssim 1$ $r \lesssim 1 \sim 2$ r. u.
	without Landau approximation	valid in the region $\theta \lesssim 1$ $r \lesssim 1 \sim 2$ r. u.

This would imply that the solutions hold only for the region $\theta \lesssim 1$, and as to the lateral spread, for $r \lesssim 1$, where r is measured in radiation units. But this does not seem to be the case. The argument would be correct only if it were true that the lateral spread of the particles is essentially determined by the contributions from the last radiation unit above the observation plane, since the mean free path of the energy degradation of the electrons equals one radiation length, and the magnitude of the deflection by multiple scattering is inversely proportional to the energy of the electrons. The precise calculation [8], shows, however, that in Approximation B the average height effective for the spread is $2/\sigma_0$ radiation lengths. Therefore the solution seems to be accurate even in the region beyond $r \sim 1$ radiation length.

In the theory of multiple scattering for a singly charged particle, the solution under the Fokker-Planck approximation deviates from the accurate one even in the small-angle region, as shown in Appendix 3. The situation is quite similar with respect to the solutions in the three-dimensional electron shower theory with and without the Landau approximations. To close this paragraph, the limitations here discussed are summarised in Table 6.

D. Behaviour of the shower functions and elementary solutions.

10. Track length, simplified model. Without entering into the mathematical details of the solutions of the basic equations, we shall derive some characteristic features of the shower functions from simple considerations, or by examining the shower function for a simplified model of the shower process.

If the energy dissipation of the particles occurs only through the constant ionization loss, the energy conservation principle (8.1), more explicitly written

$$\int_0^\infty \Pi(E_0, 0, t)\, dt = \frac{E_0}{\varepsilon},$$

holds irrespective of any assumptions on the probabilities of radiation, pair creation and Compton scattering processes. Thus, obviously, it must apply to — correctly given — solutions in Approximation B. In Sect. C we have shown that $\int \Pi\, dt$ represents the electron track length. The relation, therefore, also implies that the incident energy of the shower can be obtained without reference to any details of the shower function, if one determines the total track length. A practical application of this method is described in Sect. G.

The differential electron track length is defined by

$$\int_0^\infty \pi(E_0, E, t)\, dt, \tag{10.1}$$

and gives the average energy spectrum of the electrons in a shower. A photon track length can be introduced in complete analogy.

To reveal more details of the shower curve, we shall examine its behaviour under a very crude model in which every particle changes after one radiation length into two particles of equal energies. After t radiation lengths the number of particles is given by

$$\Pi = 2^t. \tag{10.2}$$

The number of particles increases up to a depth T defined by

$$\int_0^T 2^t\, dt = \frac{E_0}{\varepsilon}, \tag{10.3}$$

and should abruptly drop to zero for $t > T$. Now (10.3) gives for T,

$$T \cong \log \frac{E_0}{\varepsilon} \bigg/ \log 2, \tag{10.4}$$

and at this depth the particle number becomes

$$2^T \cong \frac{E_0}{\varepsilon}. \tag{10.5}$$

Although this model is quite primitive, it can be expected to demonstrate the qualitative features of the shower. Moreover, the logarithmic energy dependence of the depth of the shower maximum,

$$T \sim \log \frac{E_0}{\varepsilon}, \qquad (10.4')$$

and the almost linear relation between the particle number at shower maximum and the primary energy,

$$N \propto \frac{E_0}{\varepsilon}, \qquad (10.5')$$

are also found in the exact solutions, as will be shown in the later sections.

11. Elementary solutions. Returning to the basic equations we shall now examine their mathematical features. Since we use complete screening cross sections for the radiation and pair creation probabilities, the integral operators in the basic equations are, in Approximation A, represented by fractional forms of the energy of the shower particles. This means that for instance, shower particles of energy 10^{10} eV in a shower due to a primary electron or photon of energy 10^{12} eV are described by the same function as particles of energy 10^{11} eV in a shower with an incident energy 10^{13} eV. In other words, the shower is represented by a function which depends only on (E_0/E) and the depth t.

Now supposing that the solution can be expressed by

$$\left. \begin{aligned} \pi \, dE &= a \left(\frac{E_0}{E}\right)^s \frac{dE}{E} e^{-\mu t}, \\ \gamma \, dE &= b \left(\frac{E_0}{E}\right)^s \frac{dE}{E} e^{-\mu t}. \end{aligned} \right\} \qquad (11.1)$$

Substituting (11.1) into the Eqs. (7.10') and (7.15), we have

$$\left. \begin{aligned} +\mu &= A(s) - \frac{b}{a} B(s), \\ \frac{b}{a} \mu &= \frac{b}{a} \sigma_0 - C(s), \end{aligned} \right\} \qquad (11.2)$$

with

$$A(s) = \int_0^1 [1-(1-v)^s] \, \varphi_0(v) \, dv = 1.3603 \frac{d}{ds} \log \Gamma(s+2) - \frac{1}{(s+1)(s+2)} - 0.07513,$$

$$B(s) = 2 \int_0^1 u^s \psi_0(u) \, du = 2 \left[\frac{1}{s+1} - \frac{1.3603}{(s+2)(2+3)} \right],$$

$$C(s) = \int_0^1 u^s \varphi_0(u) \, du = \frac{1}{s+2} + \frac{1.3603}{s(s+1)}.$$

Combining these two Eqs. (11.2), we obtain

$$[\mu + \lambda_1(s)][\mu + \lambda_2(s)] = 0,$$

where

$$\lambda_1(s) = -\frac{A(s) + \sigma_0}{2} + \frac{1}{2} [\{A(s) - \sigma_0\}^2 + 4 B(s) C(s)]^{\frac{1}{2}}, \qquad (11.3)$$

$$\lambda_2(s) = -\frac{A(s) + \sigma_0}{2} - \frac{1}{2} [\{A(s) - \sigma_0\}^2 + 4 B(s) C(s)]^{\frac{1}{2}}, \qquad (11.4)$$

and corresponding to these two parameters;

$$\frac{b}{a} = \frac{C(s)}{\sigma_0 + \lambda_1}, \quad \text{for} \quad \mu = -\lambda_1 \qquad (11.5)$$

or

$$\frac{b}{a} = \frac{C(s)}{\sigma_0 + \lambda_2}, \quad \text{for} \quad \mu = -\lambda_2. \tag{11.6}$$

Thus the solutions are given by

$$\pi(E) \, dE = \frac{a E_0^s \, dE}{E^{s+1}} e^{\lambda_1(s) t}, \tag{11.7}$$

$$\gamma(E) \, dE = \frac{a C(s)}{\sigma_0 + \lambda_1} \frac{E_0^s \, dE}{E^{s+1}} e^{\lambda_1(s) t}, \tag{11.8}$$

or

$$\pi(E) \, dE = \frac{a E_0^s \, dE}{E^{s+1}} e^{\lambda_2(s) t}, \tag{11.9}$$

$$\gamma(E) \, dE = \frac{a C(s)}{\sigma_0 + \lambda_2} \frac{E_0^s \, dE}{E^{s+1}} e^{\lambda_2(s) t}, \tag{11.10}$$

or by a combination of these two solutions. The numerical value of A, B, C, λ_1 and λ_2 are shown in Table 7.

Table 7. *The notation is the same as that of reference [5]; the numerical values are recalculated**.

s	$A(s)$	$B(s)$	$C(s)$	$\lambda_2(s)$	$\lambda_1(s)$	$\lambda_1'(s)$	$\lambda_1''(s)$
0	0.0000	1.547	∞	$-\infty$	$+\infty$	$-\infty$	$+\infty$
0.1	0.1522	1.400	12.84	-4.715	3.789	-25.01	355.8
0.2	0.2865	1.280	6.123	-3.340	2.280	-9.457	65.43
0.3	0.4067	1.180	3.923	-2.749	1.569	-5.415	24.69
0.4	0.5154	1.095	2.846	-2.414	1.126	-3.655	12.48
0.5	0.6147	1.022	2.214	-2.201	0.8125	-2.693	7.418
0.6	0.7062	0.9593	1.802	-2.055	0.5754	-2.092	4.878
0.7	0.7909	0.9041	1.514	-1.952	0.3877	-1.684	3.428
0.8	0.8699	0.8554	1.302	-1.878	0.2348	-1.389	2.550
0.9	0.9439	0.8121	1.140	-1.825	0.1075	-1.166	1.965
1.0	1.014	0.7733	1.014	-1.787	0.0000	-0.9908	1.653
1.1	1.079	0.7383	0.9115	-1.761	-0.0917	-0.8497	1.276
1.2	1.142	0.7067	0.8278	-1.744	-0.1708	-0.7332	1.059
1.3	1.201	0.6778	0.7580	-1.735	-0.2391	-0.6358	0.8910
1.4	1.257	0.6515	0.6990	-1.732	-0.2984	-0.5532	0.7607
1.5	1.311	0.6273	0.6485	-1.734	-0.3501	-0.4824	0.6614
1.6	1.363	0.6049	0.6048	-1.741	-0.3952	-0.4216	0.5733
1.7	1.413	0.5843	0.5666	-1.751	-0.4347	-0.3691	0.4901
1.8	1.460	0.5651	0.5331	-1.764	-0.4693	-0.3236	0.4275
1.9	1.506	0.5473	0.5033	-1.780	-0.4996	-0.2840	0.3684
2.0	1.550	0.5360	0.4767	-1.797	-0.5263	-0.2498	0.3201
2.2	1.634	0.5004	0.4313	-1.837	-0.5704	-0.1943	0.2391
2.4	1.713	0.4737	0.394	-1.882	-0.6049	-0.1523	0.1823
2.6	1.787	0.4499	0.3627	-1.929	-0.6320	-0.1205	0.1375
2.8	1.857	0.4289	0.3362	-1.977	-0.6536	-0.0963	0.1079
3.0	1.924	0.4093	0.3134	-2.026	-0.6709	-0.0777	0.0974
4.0	2.213	0.3352	0.2349	-2.265	-0.7206	-0.0306	
5.0	2.449	0.2848	0.1882	-2.480	-0.7419	-0.0145	

* The author is indebted to Mr. MISAKI for communication of the numerical results listed above.

Since λ_2 is always negative, and as (11.4) shows, its absolute value is larger than σ_0, the coefficient $\frac{1}{\sigma_0 + \lambda_2}$ in (11.10) should always be negative, and in an actual case, the solution (11.10), can only be realized in combination with a solution (11.9).

On the other hand, λ_1 is positive for $s<1$, zero for $s=1$, negative for $s>1$ and approaches $-\sigma_0$ with increasing s, as shown in Table 7. Thus a spectrum dE/E^{s+1} with $s<1$ characterizes the behaviour of the developing stage of the shower, $s=1$ presents the situation at shower maximum, while beyond it, $s>1$ leads to the tail end of the shower. The spectral index s, therefore, indicates the shower age. In a "young" shower ($s<1$) the number of shower particles increases with age, while in an "old" one ($s>1$) it decreases. The situation is illustrated in Fig. 6.

It is noteworthy that the simplified model of the preceding section shows the same feature. If each shower particle changes into two particles of equal energy, the number of shower particles, $1/E^s$, is changed to $2/(2E)^s$. Thus the shower size does not vary for $s=1$, but increases or decreases for $s<1$, or $s>1$, respectively.

It may seem rather curious that with increase of the depth t, the number of shower particles with shower age $s<1$ increase infinitely. This derives from the fact that the total energy contained in the shower particles with energy higher than E, given by the integral

$$\int_E^\infty E \frac{dE}{E^{s+1}} \qquad (11.11)$$

becomes infinite in the case of $s<1$.

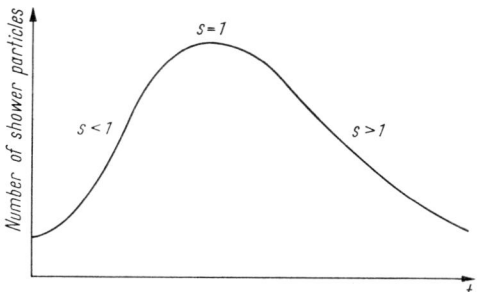

Fig. 6. Physical meaning of the "shower age" s.

12. Behaviour of the shower functions. In Approximation B, the situation becomes somewhat more complicated. The ionization term $\varepsilon \dfrac{\partial \pi}{\partial E}$ in Eq. (7.10) now destroys the fractional form with repect to the energy. At high energies ($E \gg \varepsilon$) the term $\varepsilon \dfrac{\partial \pi}{\partial E}$ may be neglected, and the situation is similar to that in Approximation A, but at low energies the number of shower particles is diminished by the effect of ionization loss.

One can, however, predict the shape of the spectrum of photons and electrons in the low-energy region in Approximation B in the following way. For electrons with energy higher than E', the radiation probability is approximately given by

$$\varphi_0 \frac{\partial E'}{E} \sim \frac{dE'}{E'}. \qquad (12.1)$$

Assuming the number of low-energy electrons to be depleted so much that the spectrum $\pi(E')\,dE'$ is less steep than $\dfrac{1}{E'}\,dE'$ at a certain region where $E' \lesssim E_c$, and further assuming that the absorption coefficient of the photons is independent of the energy, we then obtain the spectrum of low-energy photons by using (7.15),

$$\gamma\,dE' \cong \frac{dE'}{E'} e^{-\sigma_0 t} \int_0^t dt'\, e^{\sigma_0 t'} \int_0^\infty \pi(E, t')\, dE. \qquad (12.2)$$

These photons produce electrons through pair creation processes, with a spectrum which is similar to that of the photons. Since the low-energy electrons thus produced are stopped immediately by ionization loss, the electrons are in equilibrium with the source function determined by (12.2), i.e.;

$$-\varepsilon \frac{\partial \pi}{\partial E} = \text{source}, \qquad (12.3)$$

and
$$\varepsilon \frac{\partial \pi}{\partial E} \cong \frac{1}{E}. \tag{12.3'}$$

The solution of eq. (12.3') is given by
$$\pi \sim \log \frac{\varepsilon}{E}, \tag{12.4}$$

and the spectrum of photons and electrons in the low-energy region is represented by
$$\gamma \, dE \sim \frac{dE}{E}, \tag{12.5}$$
and
$$\pi \, dE \sim \log \frac{\varepsilon}{E} \, dE, \tag{12.6}$$

a conclusion which can also be proved by the exact solutions in Approximation B (Sect. 18). The result should, of course, be revised if one takes into account the effect of Compton scattering.

In the three-dimensional electron shower theory, shower electrons are scattered and spread out, and photons persist in the direction of the parent electrons. Electrons of energy E have lost most of their energy in traversing the last radiation unit, so that the average lateral spread of these particles can be expected to be (see Appendix 4)
$$\langle r \rangle \sim \frac{K}{E}, \tag{12.7}$$

where K is the "characteristic energy of scattering". Then the average behaviour of the lateral distribution function of the electrons with energy E in shower of age s will be given by
$$\frac{dE}{E^{s+1}} \sim \frac{r \, dr}{r^{2-s}}, \tag{12.8}$$
if $E \gtrsim E'$.

The angular spread of the electrons is also expected to follow a similar law. On the other hand, a slight modification is necessary for the lateral and angular distribution of the photons in the shower, because high-energy photons are concentrated nearer to the shower axis than the energetic shower electrons, as will be discussed later.

The lateral distribution of the energy flow in a shower, defined as the energy contained in an annular ring $(r, r+dr)$, is also obtained from a similar argument. One finds
$$\frac{E \, dE}{E^{s+1}} \sim \frac{r \, dr}{r^{3-s}}. \tag{12.9}$$

The resulting distribution is $1/r$ times steeper than the lateral spread of the number of shower particles, as will be expected from the fact that particles of higher energies are less deflected outward from the shower axis by Coulomb scattering.

E. One-dimensional electron shower theory.

We shall here derive the solution of the basic equations in Approximation A, and then treat the problems in Approximation B. Since in these approximations the complete screening cross sections are adopted for the radiation and pair creation processes, corrections for the deviation from these assumptions are necessary, and they will be discussed at the end of this section.

In Approximation A, we follow the theory of LANDAU and RUMER [2] in which the exact solutions are given in the form of complex integrals. At extremely high energies (say 10^{13} eV or more for Pb) one will have to take into account the Landau effect, as already discussed in Sect. 2.

In Approximation B, the solution is derived in terms of a power series in ε/E starting from the Landau-Rumer solutions of Approximation A. The expressions thus found are deformed into the double complex integrals which were obtained already by SNYDER [3] and SCOTT [4], and represent the exact solution for the basic equations in Approximation B. We shall also compare the track length calculated in this solution with that obtained by other authors under different approximations. Finally, the shower curves of BERNSTEIN (19), calculated with more accurate screening cross sections, will be briefly summarized in order to illustrate the deviation of the solution in Approximation B from his exact solution.

I. Shower functions in Approximation A.

13. Analytic solutions. As shown in Sect. 7, the elementary solutions of the basic Eqs. (7.10') and (7.15) can be written as power functions of the energy, since the integral operators A', B', C' are of fractional form with respect to the energy. The general form of the elementary solutions is, according to (11.7) to (11.10),

$$\pi \, dE = \frac{dE}{E^{s+1}} \left[a_1 \, e^{\lambda_1(s)t} + a_2 \, e^{\lambda_2(s)t} \right], \tag{13.1}$$

$$\gamma \, dE = \frac{dE}{E^{s+1}} \left[\frac{a_1 C}{\sigma_0 + \lambda_1} e^{\lambda_1(s)t} + \frac{a_2 C}{\sigma_0 + \lambda_2} e^{\lambda_2(s)t} \right]. \tag{13.2}$$

Solutions corresponding to a certain boundary condition are obtained by summing up these functions with respect to s, weighting them with an appropriate amplitude. This procedure, known as the Mellin transformation, is described in Appendix 1.

Let $\mathfrak{M}_e$ be the Mellin transform of the function π,

$$\mathfrak{M}_e = \int_0^\infty E^s \, \pi \, dE. \tag{13.3}$$

Then the inverse transformation is

$$\pi = \frac{1}{2\pi i} \int_c ds \, \frac{\mathfrak{M}_e}{E^{s+1}}, \tag{13.4}$$

where s is a complex parameter, and the path of the integral is running parallel to the imaginary axis. Extending (13.1) and (13.2) to complex values of the parameter s, one gets the solutions

$$\pi = \frac{1}{2\pi i} \int_c ds \, \frac{1}{E^{s+1}} \left[a_1 \, e^{\lambda_1(s)t} + a_2 \, e^{\lambda_2(s)t} \right], \tag{13.5}$$

$$\gamma = \frac{1}{2\pi i} \int_c ds \, \frac{1}{E^{s+1}} \left[\frac{a_1 C}{\sigma_0 + \lambda_1} e^{\lambda_1(s)t} + \frac{a_2 C}{\sigma_0 + \lambda_2} e^{\lambda_2(s)t} \right], \tag{13.6}$$

where a_1 and a_2 are functions of the complex parameter s, and are determined by the boundary condition imposed by the physical model.

Here we shall derive the solution for a shower initiated by a primary electron of energy E_0. The boundary conditions are

$$\pi = \delta(E_0 - E),$$
and
$$\gamma = 0$$
at $t = 0$. \tag{13.7}

Then we have

and

$$\left.\begin{array}{l}\mathfrak{M}_e = E_0^s \\ \mathfrak{M}_\gamma = 0\end{array}\right\} \text{ at } t=0, \qquad (13.7')$$

which means

and

$$\left.\begin{array}{l}\pi = \dfrac{1}{2\pi i}\int_c ds \left(\dfrac{E_0}{E}\right)^s \dfrac{1}{E} \\ \gamma = 0\end{array}\right\} \text{ at } t=0. \qquad (13.7'')$$

Table 8. *The notation is the same as that of reference* [5]; *the numerical values are recalculated**.

s	$H_1(s)$	$H_2(s)$	$L(s)$	$M(s)$	s	$H_1(s)$	$H_2(s)$	$L(s)$	$M(s)$
0	0.5000	0.5000	0.4689	0.5333	1.4	0.3312	0.6688	0.5768	0.3840
0.1	0.5365	0.4635	0.4776	0.5207	1.5	0.3057	0.6943	0.5737	0.3700
0.2	0.5433	0.4567	0.4872	0.5093	1.6	0.2809	0.7191	0.5684	0.3554
0.3	0.5424	0.4576	0.4975	0.4989	1.7	0.2572	0.7428	0.5612	0.3404
0.4	0.5364	0.4636	0.5084	0.4891	1.8	0.2348	0.7653	0.5523	0.3253
0.5	0.5263	0.4737	0.5195	0.4799	1.9	0.2138	0.7862	0.5419	0.3102
0.6	0.5128	0.4872	0.5306	0.4709	2.0	0.1944	0.8057	0.5304	0.2952
0.7	0.4962	0.5038	0.5412	0.4619	2.2	0.1602	0.8398	0.5050	0.2663
0.8	0.4771	0.5229	0.5511	0.4527	2.4	0.1319	0.8681	0.4781	0.2395
0.9	0.4559	0.5442	0.5599	0.4428	2.6	0.1090	0.8910	0.4511	0.2152
1.0	0.4327	0.5673	0.5672	0.4328	2.8	0.0905	0.9096	0.4251	0.1935
1.1	0.4083	0.5917	0.5728	0.4218	3.0	0.0755	0.9244	0.4005	0.1744
1.2	0.3830	0.6171	0.5763	0.4100	4.0	0.0341	0.9659	0.3039	0.1085
1.3	0.3571	0.6429	0.5777	0.3974	5.0	0.0181	0.9819	0.2421	0.0733

* The author is indebted to Mr. MISAKI for communication of the numerical results listed in this Table.

The Eqs. (13.7) to (13.7''), together with (13.5) and (13.6), completely determine a_1 and a_2: Evidently

$$a_1 + a_2 = E_0^s \qquad (13.8)$$

and

$$\dfrac{a_1 C}{\sigma_0 + \lambda_1} + \dfrac{a_2 C}{\sigma_0 + \lambda_2} = 0. \qquad (13.9)$$

Introducing

$$a_1 = H_1(s) E_0^s, \qquad (13.10)$$

and

$$a_2 = H_2(s) E_0^s, \qquad (13.11)$$

we can also write

$$H_1 = \dfrac{\sigma_0 + \lambda_1}{\lambda_1 - \lambda_2}, \qquad (13.12)$$

and

$$H_2 = -\dfrac{\sigma_0 + \lambda_2}{\lambda_1 - \lambda_2}. \qquad (13.13)$$

The solutions for the differential spectrum of electrons and photons are now given by

$$\pi = \dfrac{1}{2\pi i}\int ds \dfrac{E_0^s}{E^{s+1}} \{H_1(s) e^{\lambda_1(s)t} + H_2 e^{\lambda_2(s)t}\}, \qquad (13.14)$$

$$\gamma = \dfrac{1}{2\pi i}\int ds \dfrac{E_0^s}{E^{s+1}} \dfrac{L(s)}{s^{\frac{1}{2}}} [e^{\lambda_1(s)t} - e^{\lambda_2(s)t}], \qquad (13.15)$$

where

$$L(s) = \dfrac{s^{\frac{1}{2}} C(s)}{\lambda_1 - \lambda_2}. \qquad (13.16)$$

The numerical values of H_1, H_2 and L are shown in Table 8.

Correspondingly, we have for the integral spectrum of electrons or photons

$$\Pi = \frac{1}{2\pi i} \int \frac{ds}{s} \left(\frac{E_0}{E}\right)^s [H_1(s) e^{\lambda_1(s)t} + H_2(s) e^{\lambda_2(s)t}], \qquad (13.17)$$

$$\Gamma = \frac{1}{2\pi i} \int \frac{ds}{s^{\frac{3}{2}}} \left(\frac{E_0}{E}\right)^s L(s) [e^{\lambda_1(s)t} - e^{\lambda_2(s)t}]. \qquad (13.18)$$

The track length Z_Π is given by

$$Z_\Pi = \int_0^\infty \Pi \, dt = \frac{1}{2\pi i} \int \frac{ds}{s} \frac{1}{\lambda_1(s)} \left(\frac{E_0}{E}\right)^s \left[H_1(s) + \frac{\lambda_1(s)}{\lambda_2(s)} H_2(s)\right]. \qquad (13.19)$$

Since $\lambda_1(s=1)=0$, the integrand has a pole at $s=1$, and the integral (13.19) is the residue at this pole. We therefore have accurately

$$Z_\Pi = \frac{H_1(s)}{\lambda_1'(s)} \frac{E_0}{E} \bigg|_{s=1} = 0.437 \frac{E_0}{E}, \qquad (13.20)$$

— the track length is proportional to the primary energy.

Similar arguments can also be made for the differential track length of the electrons, and lead to the result

$$Z_\pi = 0.437 \frac{E_0}{E^2}. \qquad (13.21)$$

For the various boundary conditions, the solutions in Approximation A are summarized in Table 9 by giving the explicit expressions of a_1 and a_2 in the formulae (13.1) and (13.2) for an incident particle of energy E_0. Apart from the slight change of definition of a and b, we have generally

and
$$\pi = \frac{1}{2\pi i} \int ds \left(\frac{E_0}{E}\right)^s \frac{1}{E} (a_1 e^{\lambda_1(s)t} + a_2 e^{\lambda_2(s)t}), \qquad (13.1')$$

$$\gamma = \frac{1}{2\pi i} \int ds \left(\frac{E_0}{E}\right)^s \frac{1}{E} (b_1 e^{\lambda_1(s)t} + b_2 e^{\lambda_2(s)t}), \qquad (13.2')$$

or in the integral form

$$\Pi = \frac{1}{2\pi i} \int \frac{ds}{s} \left(\frac{E_0}{E}\right)^s (a_1 e^{\lambda_1(s)t} + a_2 e^{\lambda_2(s)t}), \qquad (13.22)$$

$$\Gamma = \frac{1}{2\pi i} \int \frac{ds}{s} \left(\frac{E_0}{E}\right)^s (b_1 e^{\lambda_1(s)t} + b_2 e^{\lambda_2(s)t}). \qquad (13.23)$$

The track length of each function is given by

$$Z_\pi = \frac{a_1}{\lambda_1'} \frac{E_0}{E^2} \bigg|_{s=1}, \qquad (13.24)$$

$$Z_\gamma = \frac{b_1}{\lambda_1'} \frac{E_0}{E^2} \bigg|_{s=1}, \qquad (13.25)$$

$$Z_\Pi = \frac{a_1}{\lambda_1'} \frac{E_0}{E} \bigg|_{s=1}, \qquad (13.26)$$

$$Z_\Gamma = \frac{b_1}{\lambda_1'} \frac{E_0}{E} \bigg|_{s=1}. \qquad (13.27)$$

14. Numerical evaluation. For a numerical evaluation of these functions the saddle point method can now be applied. Some details of this procedure are described in Appendix 1.

As an example, let us consider the average number of electrons of energy higher than E in a shower due to a primary electron of energy E_0. We start, therefore, from eq. (13.17) of the preceding section.

Now Table 7 shows that the relation $\lambda_1 > \lambda_2$ holds for all values of s, so that the second term in the integral is negligible compared with the first everywhere except for quite small values of t. Therefore (13.17) can be simplified to

$$\Pi = \frac{1}{2\pi i} \int ds\, H_1(s)\, e^{\lambda_1(s)t + ys - \log s} \tag{14.1}$$

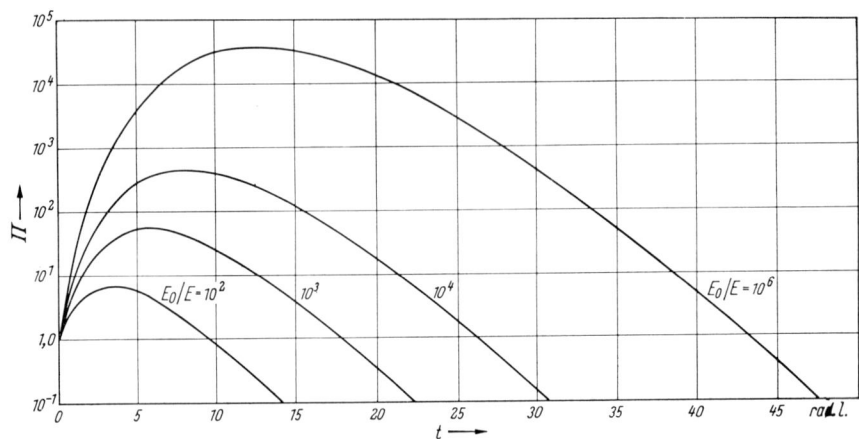

Fig. 7. Shower functions in Approximation A.

with $y = \log \frac{E_0}{E}$, and the saddle point method can be applied to the function in the exponent, as $H_1(s)$ is a slowly varying function of s. Thus we have

$$\Pi = \frac{H_1\left(\frac{E_0}{E}\right)^s e^{\lambda_1(s)t}}{s\sqrt{2\pi\left(\lambda_1'' t + \frac{1}{s^2}\right)}}\bigg|_{s=\bar{s}}, \tag{14.2}$$

where $\bar{s}$ is given by

$$\lambda_1'(\bar{s})t + \log \frac{E_0}{E} - \frac{1}{\bar{s}} = 0. \tag{14.3}$$

The shower functions calculated from (14.2) are reproduced in Fig. 7.

An estimate of the depth T at which the number of shower particles reaches its maximum can be obtained according to the considerations of Sect. 11. The elementary solutions lead us to expect, at the shower maximum, an energy spectrum proportional to $1/E$, that is, a parameter $\bar{s}=1$. This is confirmed by a more rigorous treatment outlined in the following.

Differentiation with respect to t of (14.2) yields

$$\frac{\partial \bar{s}}{\partial t} \cdot \frac{\partial \Pi}{\partial \bar{s}} + \frac{\partial \Pi}{\partial t} = 0, \tag{14.4}$$

where the first term must be retained since $\bar{s}$, by definition, is a function t.

As $\dfrac{H_1}{\sqrt{\lambda_1'' t + \frac{1}{s^2}}}$ is a slowly varying function of s, one may approximate (14.4) by

$$\frac{\partial \bar{s}}{\partial t}\left(\lambda_1'(\bar{s})t + \log \frac{E_0}{E} - \frac{1}{\bar{s}}\right) + \lambda_1(\bar{s}) = 0. \tag{14.5}$$

Here the first term vanishes according to (14.3), and we arrive at the condition
$$\lambda_1(\bar{s}) = 0, \tag{14.6}$$
which implies
$$\bar{s} = 1. \tag{14.7}$$
Using the relation $\lambda_1'(1) \approx -1$, we have, therefore, from (14.3) the solution for T
$$T = \log \frac{E_0}{E} - 1, \tag{14.8}$$
The total number of shower particles at the maximum is thus given by
$$\begin{aligned}\Pi &= H_1(1) \left(\frac{E_0}{E}\right) \Big/ \sqrt{2\pi(1.56\, T + 1)} \\ &\approx \frac{H_1(1)(E_0/E)}{\sqrt{2\pi\left(1.56 \log \frac{E_0}{E} - 0.56\right)}}.\end{aligned} \tag{14.9}$$

In a good approximation, the maximum shower size, reached at $\bar{s} = 1$, is therefore proportional to the primary energy E_0.

The corresponding depth T for a photon-primary shower differs only slightly from the value derived above. Formulae for the determination of T are also summarized in Table 9.

The slight differences in the values of T can be explained by simple physical reasons. Firstly, considering secondaries of various energies E, we note that the depths T at which they reach their maximum is the smaller the higher their

Table 9. *Explicit expression for $a_{1,2}$ and $b_{1,2}$, and the depth of the shower maximum.*
(From Rossi and Greisen [5].)

Primary particle		Electron		Photon	
Secondary particle		electron	photon	electron	photon
a_1, b_1		$H_1(s)$	$L(s)/s^{\frac{1}{2}}$	$M(s)\,s^{\frac{1}{2}}$	$H_2(s)$
a_2, b_2		$H_2(s)$	$-L(s)/s^{\frac{1}{2}}$	$-M(s)\,s^{\frac{1}{2}}$	$H_1(s)$
Depth of the shower maximum	$T_{\pi,\gamma}$	$\log \frac{E_0}{E}$	$\log \frac{E_0}{E} - \frac{1}{2}$	$\log \frac{E_0}{E} + \frac{1}{2}$	$\log \frac{E_0}{E}$
	$T_{\Pi,\Gamma}$	$\log \frac{E_0}{E} - 1$	$\log \frac{E_0}{E} - \frac{3}{2}$	$\log \frac{E_0}{E} - \frac{1}{2}$	$\log \frac{E_0}{E} - 1$

$$H_1(s) = \frac{\sigma_0 + \lambda_1(s)}{\lambda_1(s) - \lambda_2(s)}, \qquad H_2(s) = -\frac{\sigma_0 + \lambda_2(s)}{\lambda_1(s) - \lambda_2(s)}$$
$$L(s)/s^{\frac{1}{2}} = \frac{C(s)}{\lambda_1(s) - \lambda_2(s)}, \qquad M(s)\, s^{\frac{1}{2}} = \frac{B(s)}{\lambda_1(s) - \lambda_2(s)}.$$

energies are. Now $\Pi(E)$ comprises the particles of energies higher than the lower limit E, so that T_Π is slightly smaller than T_π. Secondly, for a photon-primary shower, the photon penetrates, on the average, about $1/\sigma_0$ radiation lengths before producing its first electron pair. Therefore the cascade maximum for a photon-initiated shower occurs at a depth larger than that of an electron-primary shower of the same primary energy by an amount

$$\Delta T = \frac{1}{\sigma_0} - \log 2 \cong 0.6. \tag{14.10}$$

The term $\log 2$ derives from the assumption that the energy of the primary photon is divided equally among the pair electrons.

II. Shower functions in Approximation B.

15. Analytic solutions. In deriving solutions for the basic eqs. (7.10) and (7.15), we keep in mind that the effect of ionization loss is of dominant importance for shower electrons of energies below the critical energy, and that the additional term $\varepsilon \frac{\partial \pi}{\partial E}$ in (7.10) is not of a fractional form in the energy.

Applying the operation $\frac{\partial}{\partial t} + \sigma_0$ on the both sides of Eq. (7.10), we have

$$\left\{ \frac{\partial^2}{\partial t^2} + (A' + \sigma_0) \frac{\partial}{\partial t} + (A'\sigma_0 - B'C') \right\} \pi = \left(\frac{\partial}{\partial t} + \sigma_0 \right) \varepsilon \frac{\partial \pi}{\partial E}. \tag{15.1}$$

The solution of this equation is obtained in the form of a power series in ε/E, in which the first term is the solution in Approximation A. Furthermore we introduce the abbreviation for the solution

$$\frac{1}{2\pi i} \int \left(\frac{E_0}{E} \right)^s \frac{dE}{E} \phi_0(s, t), \tag{15.2}$$

where $\phi_0(s, t)$ stands for the function derived in Approximation A, and is shown in Table 9 for our simple boundary conditions. As an example, if one considers the case of a primary electron the differential spectrum of the shower electrons (13.14) implies

$$\phi_{0e} = H_1(s) e^{\lambda_1(s)t} + H_2 e^{\lambda_2(s)t}. \tag{15.3}$$

The solution in the form of a series in ε/E,

$$\pi = \frac{1}{2\pi i} \int ds \left(\frac{E_0}{E} \right)^s \frac{dE}{E} \sum_{n=0}^{\infty} \left(-\frac{\varepsilon}{E} \right)^n \phi_n(s, t), \tag{15.4}$$

demands that the $\phi_n(s, t)$ satisfy the following difference equation, as will be seen by substitution into (15.1),

$$\left\{ \frac{\partial^2}{\partial t^2} + (A(s+n) + \sigma_0) \frac{\partial}{\partial t} + (A\sigma_0 - BC) \right\} \phi_n(s, t) \\ = (s+n) \left(\frac{\partial}{\partial t} + \sigma_0 \right) \phi_{n-1}(s, t). \tag{15.5}$$

Since this equation can be re-written as

$$\left(\frac{\partial}{\partial t} - \lambda_1(s+n) \right) \left(\frac{\partial}{\partial t} - \lambda_2(s+n) \right) \phi_n \\ = (s+n) \left(\frac{\partial}{\partial t} + \sigma_0 \right) \phi_{n-1}, \tag{15.6}$$

or in an operator form[1]

$$\phi_n = \left[\frac{H_1(s+n)}{\left(\frac{\partial}{\partial t} - \lambda_1(s+n) \right)} + \frac{H_2(s+n)}{\left(\frac{\partial}{\partial t} - \lambda_2(s+n) \right)} \right] (s+n) \phi_{n-1}(s, t), \tag{15.7}$$

---

[1]
$$\frac{\frac{\partial}{\partial t} + \sigma_0}{\left(\frac{\partial}{\partial t} - \lambda_1(s+n) \right) \left(\frac{\partial}{\partial t} - \lambda_2(s+n) \right)} \\ = \frac{1}{\lambda_1(s+n) - \lambda_2(s+n)} \left[\frac{\lambda_1(s+n) + \sigma_0}{\left(\frac{\partial}{\partial t} - \lambda_1(s+n) \right)} - \frac{\lambda_2(s+n) + \sigma_0}{\left(\frac{\partial}{\partial t} - \lambda_2(s+n) \right)} \right].$$

ϕ_n can be obtained from a recursion expression

$$\phi_n = (s+n) \int_0^t dt' \phi_{0e}(s+n, t-t') \phi_{n-1}(s, t'). \tag{15.8}$$

In order to prove the convergence of the series (15.4), we first expand the integrand for small values of t.

Since
$$\phi_{0e} = 1 - A(s)t + 0(t^2), \tag{15.9}$$
we have
$$|\phi_{0e}| \lesssim 1, \quad \text{at} \quad t \ll 1 \tag{15.10}$$

as the value of $A(s)$ is zero or positive for any positive values of s. Thus the relation (15.8) gives

$$|\phi_n| \lesssim \frac{\Gamma(s+n+1)}{\Gamma(s+1)} \frac{1}{\Gamma(n+1)} t^n \lesssim 2^{s+n} t^n, \tag{15.11}$$

and the series in the integrand in the formula (15.4),

$$\sum_{n=0}^{\infty} \left(-\frac{\varepsilon}{E}\right)^n \phi_n(s, t), \tag{15.12}$$

is uniformly convergent, if $\frac{2\varepsilon t}{E} < 1$.

Using the method of analytic continuation to deform the series (15.12), BHABHA and CHAKRABARTY (27) obtained another series with a different convergence domain. Instead, we shall apply here the method of integral representation (see Appendix 2) to the series (15.12), which leads to

$$\sum_{n=0}^{\infty} \left(-\frac{\varepsilon}{E}\right)^n \phi_n(s, t) = \frac{1}{2\pi i} \int dq \, \Gamma(-q) \left(\frac{\varepsilon}{E}\right)^q \mathfrak{M}(s, q, t), \tag{15.13}$$

where
$$\mathfrak{M}(s, n, t) = n! \, \phi_n(s, t). \tag{15.14}$$

$\mathfrak{M}(s, q, t)$ is thus defined, by substitution of (15.14) into (15.5), by the equation

$$\left[\frac{\partial}{\partial t} - \lambda_1(s+q)\right]\left[\frac{\partial}{\partial t} - \lambda_2(s+q)\right] \mathfrak{M}(s, q, t) \\ = (s+q) q \left(\frac{\partial}{\partial t} + \sigma_0\right) \mathfrak{M}(s, q-1, t) \tag{15.15}$$

with $\mathfrak{M}(s, 0, t) = \phi_0(s, t)$.

Then the solution of (7.10) and (7.15) is given by

$$\pi = -\frac{1}{4\pi^2} \iint ds \, dq \left(\frac{E_0}{E}\right)^s \frac{1}{E} \left(\frac{\varepsilon}{E}\right)^q \Gamma(-q) \mathfrak{M}(s, q, t), \tag{15.16}$$

and for the integral spectrum of shower electrons, the integration with respect to E yields

$$\Pi(E_0, E, t) = -\frac{1}{4\pi^2} \iint \frac{ds \, dq}{s+q} \left(\frac{E_0}{E}\right)^s \left(\frac{\varepsilon}{E}\right)^q \Gamma(-q) \mathfrak{M}(s, q, t). \tag{15.17}$$

The total number of shower electrons, $\Pi(E_0, 0, t)$, obtained as the limit of $E = 0$ in (15.17), is accurately

$$\Pi(E_0, 0, t) = \frac{1}{2\pi i} \int ds \left(\frac{E_0}{\varepsilon}\right)^s \Gamma(s) \mathfrak{M}(s, -s, t), \tag{15.18}$$

where we have used the fact that the integration with respect to q is equal to the residue of the integrand at the pole $q = -s$.

Similarly the differential and the integral spectra of the photons are obtained as

$$\gamma = -\frac{1}{4\pi^2} \iint ds\, dq \left(\frac{E_0}{E}\right)^s \frac{1}{E} \left(\frac{\varepsilon}{E}\right)^q \Gamma(-q)\, \mathfrak{N}(s, q, t), \tag{15.19}$$

and

$$\Gamma(E_0, E, t) = \frac{-1}{4\pi^2} \iint \frac{ds\, dq}{s+q} \left(\frac{E_0}{E}\right)^s \left(\frac{\varepsilon}{E}\right)^q \Gamma(-q)\, \mathfrak{N}(s, q, t), \tag{15.20}$$

where

$$\mathfrak{N}(s, q, t) = e^{-\sigma_0 t} C(s+q) \int_0^t e^{+\sigma_0 t'} \mathfrak{M}(s, q, t-t')\, dt'. \tag{15.21}$$

These expressions represent the solution of Eq. (7.15) for which we may write

$$\gamma = e^{-\sigma_0 t} \int_0^t e^{\sigma_0 t'} C' \pi(t')\, dt'. \tag{15.22}$$

16. Explicit expressions of the shower function. It remains now to determine the explicit form of $\mathfrak{M}(s, q, t)$ or $\mathfrak{N}(s, q, t)$ from (15.15). Using (15.8), $\mathfrak{M}(s, q, t)$ is expanded in the form (see also Appendix 4)

$$\mathfrak{M}(s, q, t) = \sum_{n=0}^{\infty} \{\mathfrak{M}_{1,n}(s, q) e^{\lambda_1(s+n)t} + \mathfrak{M}_{2,n}(s, q) e^{\lambda_2(s+n)t}\}, \tag{16.1}$$

where $\mathfrak{M}_{1,0}(s, q)$ and $\mathfrak{M}_{2,0}(s, q)$ are the functions determined by the boundary condition. In the case of a primary electron, we have from (15.3)

$$\mathfrak{M}_{1,0}(s, 0) = H_1(s) \tag{16.2}$$

and

$$\mathfrak{M}_{2,0}(s, 0) = H_2(s). \tag{16.3}$$

The differential spectrum of the shower electrons, (15.17), is now written in the form

$$\begin{aligned}\pi &= \frac{-1}{4\pi^2} \iint ds\, dq \left(\frac{E_0}{E}\right)^s \frac{1}{E} \left(\frac{\varepsilon}{E}\right)^q \Gamma(-q)\, \mathfrak{M}(s, q, t) \\ &= -\frac{1}{4\pi^2} \iint ds\, dq \left(\frac{E_0}{E}\right)^s \frac{1}{E} \left(\frac{\varepsilon}{E}\right)^q \Gamma(-q) \times \\ &\quad \times \sum_{n=0}^{\infty} \{\mathfrak{M}_{1,n} e^{\lambda_1(s+n)t} + \mathfrak{M}_{2,n} e^{\lambda_2(s+n)t}\}. \end{aligned} \tag{16.4}$$

The m-th term of the above series is

$$-\frac{1}{4\pi^2} \iint ds\, dq \left(\frac{E_0}{E}\right)^s \frac{1}{E} \left(\frac{\varepsilon}{E}\right)^q \Gamma(-q) \times \\ \times \{\mathfrak{M}_{1,m}(s, q) e^{\lambda_1(s+m)t} + \mathfrak{M}_{2,m}(s, q) e^{\lambda_2(s+m)t}\}, \tag{16.5}$$

and changing the integral variables to

$$s+m = s^*, \quad q-m = q^*,$$

the term becomes

$$-\frac{1}{4\pi^2} \left(\frac{\varepsilon}{E_0}\right)^m \iint ds^*\, dq^* \left(\frac{E_0}{E}\right)^{s^*} \frac{1}{E} \left(\frac{\varepsilon}{E}\right)^{q^*} \Gamma(-q^* - m) \times \\ \times \{\mathfrak{M}_{1,m}(s^* - m, q^* + m) e^{\lambda_1(s^*)t} + \mathfrak{M}_{2,m}(s^* - m, q^* + m) e^{\lambda_2(s^*)t}\}. \tag{16.6}$$

Then the differential spectrum of the electrons, written as a power series in ε/E_0, is

$$\pi = -\frac{1}{4\pi^2} \iint ds\, dq \left(\frac{E_0}{E}\right)^s \frac{1}{E} \left(\frac{\varepsilon}{E}\right)^q \sum_{n=0}^{\infty} \left(\frac{\varepsilon}{E_0}\right)^n \Gamma(-q-n) \times \\ \times \{\mathfrak{M}_{1,n} e^{\lambda_1(s)t} + \mathfrak{M}_{2,n} e^{\lambda_2(s)t}\}. \tag{16.7}$$

Accordingly, the integral spectrum (15.17) becomes

$$\Pi(E_0, E, t) = -\frac{1}{4\pi^2}\iint \frac{ds\,dq}{s+q}\left(\frac{E_0}{E}\right)^s\left(\frac{\varepsilon}{E}\right)^q \sum_{n=0}^{\infty}\left(\frac{\varepsilon}{E_0}\right)^n \Gamma(-q-n) \times \\ \times \{\mathfrak{M}_{1,n}(s-n, q+n)e^{\lambda_1(s)t} + \mathfrak{M}_{2,n}(s-n, q+n)e^{\lambda_2(s)t}\}, \quad (16.8)$$

and the total number of shower electrons is given by

$$\Pi(E_0, 0, t) = \frac{1}{2\pi i}\int ds\left(\frac{E_0}{\varepsilon}\right)^s \sum_{n=0}^{\infty}\left(\frac{\varepsilon}{E_0}\right)^n \Gamma(s-n) \times \\ \times \{\mathfrak{M}_{1,n}(s-n, -s+n)e^{\lambda_1(s)t} + \mathfrak{M}_{2,n}(s-n, -s+n)e^{\lambda_2(s)t}\}. \quad (16.9)$$

The spectrum of photons is obtained in a similar form.

These are the exact solutions in Approximation B, and were first obtained by SNYDER [3] and SCOTT [4]. Since they are given in the form of a power series in ε/E_0, the second and higher terms may practically be neglected if, as usual, the primary energy is much higher than the critical energy. In this case, the solutions can be approximated by

$$\Pi(E_0, E, t) = -\frac{1}{4\pi^2}\iint ds\,dq\left(\frac{E_0}{E}\right)^s\frac{1}{E}\left(\frac{\varepsilon}{E}\right)^q \Gamma(-q) \times \\ \times \{\mathfrak{M}_{1,0}(s, q)e^{\lambda_1(s)t} + \mathfrak{M}_{2,0}(s, q)e^{\lambda_2(s)t}\}, \quad (16.10)$$

$$\Pi(E_0, E, t) = -\frac{1}{4\pi^2}\iint \frac{ds\,dq}{s+q}\left(\frac{E_0}{E}\right)^s\left(\frac{\varepsilon}{E}\right)^q \Gamma(-q) \times \\ \times \{\mathfrak{M}_{1,0}(s, q)e^{\lambda_1(s)t} + \mathfrak{M}_{2,0}(s, q)e^{\lambda_2(s)t}\}, \quad (16.11)$$

$$\Pi(E_0, 0, t) = \frac{1}{2\pi i}\int ds\left(\frac{E_0}{\varepsilon}\right)^s \Gamma(s) \times \\ \times \{\mathfrak{M}_{1,0}(s, -s)e^{\lambda_1(s)t} + \mathfrak{M}_{2,0}(s, -s)e^{\lambda_2(s)t}\}. \quad (16.12)$$

These are the results already obtained by SNYDER (4), and by SERBER (5).

17. Numerical evaluation. The numerical values of these functions can be calculated by the saddle point method. More specifically, the evaluation for $\pi(E_0, E, t)$ and $\Pi(E_0, E, t)$ requires the use of the double saddle point method, while for the evaluation of $\Pi(E_0, 0, t)$ the calculation is similar to that of the shower function in Approximation A. Thus we have for the total number of shower electrons

$$\Pi = \frac{1}{\sqrt{2\pi}}\left(\frac{E_0}{\varepsilon}\right)^{\bar{s}}\Gamma(\bar{s})\frac{\mathfrak{M}_{1,0}(\bar{s}, -\bar{s})e^{\lambda_1(s)t}}{\sqrt{\lambda_1''(\bar{s})t + \left(\frac{\mathfrak{M}_{1,0}'}{\mathfrak{M}_{1,0}}\right)' + \frac{d^2}{ds^2}\log\Gamma(\bar{s})}}, \quad (17.1)$$

where $\bar{s}$ is given by

$$\lambda_1'(s)t + \log\frac{E_0}{\varepsilon} + \frac{\mathfrak{M}_{1,0}'}{\mathfrak{M}_{1,0}} + \frac{d}{ds}\log\Gamma(s)\Big|_{s=\bar{s}} = 0. \quad (17.2)$$

Putting now for the function $\mathfrak{M}_{1,0}$

$$\mathfrak{M}_{1,0}(s, -q) = H_1(s)\frac{\Gamma(s+q+1)}{\Gamma(s+1)}K_{1,0}(s_1-q), \quad (17.3)$$

and

$$\mathfrak{M}_{1,0}(s, -s) = \frac{H_1(s)}{\Gamma(s+1)}K_{1,0}(s, -s), \quad (17.4)$$

for a shower due to a primary electron, the equation for $K_{1,0}(s, -q)$ is according to (15.15),

$$[\lambda_1(s) - \lambda_1(s+q)][\lambda_1(s) - \lambda_2(s+q)]K_{1,0}(s,q) = q[\lambda_1(s) + \sigma_0]K_{1,0}(s, q-1)$$
$$\text{with} \quad K_{1,0}(s, 0) = 1. \qquad (17.5)$$

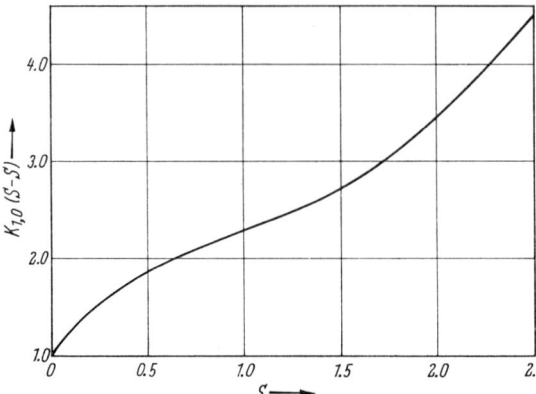

Fig. 8. Numerical values of $K_{1,0}(s, -s)$. (From B. Rossi and K. Greisen [5].)

The solution of this equation is discussed in Appendix 4, and the numerical values of $K_{1,0}(s, -s)$ are shown in Fig. 8.

The functions which should be substituted for $H_1(s)$ under different conditions are the same as in Approximation A, and are tabulated in Table 9.

Since $K_{1,0}(s, -s)$ is a slowly varying function of s, this term may be omitted in the calculation determining the saddle point. The depth of the shower maximum for the total number of shower particles is, thus, the same as that obtained by putting $E = \varepsilon$ in the solution of Approximation A. Numerical values of the function are listed in Table 10.

Table 10. *Number of electrons in an electron-initiated shower.* (From H. S. Snyder [3].)

t	log E_0/ε						
	2	3	4	5	6	7	8
0.5	2.38						
1	2.32	3.75	4.87	5.68	6.33	6.7	8.5
2	1.760	4.55	8.86	14.46	26.7	33.1	44.3
3	1.170	3.85	10.13	21.17	41.0	68.7	115.8
4	0.729	2.81	9.07	23.6	53.42	106.3	201.5
5		1.94	7.09	22.01	57.33	135.0	287.5
6		1.22	5.23	18.40	54.20	142.1	335.4
7		0.68	3.52	13.97	46.42	135.2	357.5
8		0.48	2.39	9.96	37.25	119.2	346.0
10			0.90	4.72	20.16	75.6	261.3
12			0.31	1.99	9.67	41.1	155.2
14				0.79	4.0	20.2	85.5
16				0.30	1.67	9.1	42.2
18				0.12	0.67	3.9	18.9
20					0.1	1.6	8.0

18. Accuracy of the shower function given by Snyder and Serber. To check the accuracy of the functions of (16.10), (16.11) and (16.12), and to check the results with regard to the properties of the shower particles, the track length and the spectrum of the shower particles are examined in the following.

α) *The total track length.* The integral of the exact formula (16.8) with respect to t is performed with the help of the pole at $s = 1$, and gives

$$Z_\pi = \int_0^\infty \Pi(E_0, 0, t)\, dt$$
$$\cong \left(\frac{E_0}{\varepsilon}\right) \sum_{n=0}^{\infty} \left(\frac{\varepsilon}{E_0}\right)^n \frac{\Gamma(1-n)}{\lambda_1'(1)} \{\mathfrak{M}_{1,n}(1-n, -1+n)\}. \qquad (18.1)$$

Since, as shown in Appendix 4,

$$\Gamma(1-n)\mathfrak{M}_{1,n} = 0 \quad \text{for } n \geq 1, \atop = H_1(1)K_{1,0}(1,-1) \quad \text{for } n = 0 \quad \} \quad (18.2)$$

only the first term contributes to the track length[1], and its value is

$$Z_\pi = \frac{H_1(1)K_{1,0}(1,-1)}{\lambda_1'(s)} \frac{E_0}{\varepsilon} \cong 0.437 \times 2.29 \frac{E_0}{\varepsilon} \cong \frac{E_0}{\varepsilon}. \quad (18.3)$$

The higher terms in the series do not contribute at all to the area under the shower curve, but will slightly modify the shape of the shower curve. Therefore this approximation can be expected to give results sufficiently accurate for practical applications. More details will be discussed in the following section.

β) *The spectrum at the low-energy end.* As it was shown at the beginning of this section, the shape of the differential spectrum of the shower particles at low energies is expected to be of the form $\log\frac{\varepsilon}{E}$ or $\frac{1}{E}$ for electrons or photons, respectively. Consider now the function π and γ,

$$\pi \cong -\frac{1}{4\pi^2}\iint ds\,dq \left(\frac{E_0}{E}\right)^s \frac{1}{E}\left(\frac{\varepsilon}{E}\right)^q H_1(s) \frac{\Gamma(-q)\Gamma(s+q+1)}{\Gamma(s+1)} K_{1,0}(s,-q)e^{\lambda_1(s)t} \quad (18.4)$$

and

$$\gamma \cong -\frac{1}{4\pi^2}\iint ds\,dq \left(\frac{E_0}{E}\right)^s \frac{1}{E}\left(\frac{\varepsilon}{E}\right)^q \frac{C(s+q)}{\sigma_0+\lambda_1(s)} \times \\ \times H_1(s)\frac{\Gamma(-q)\Gamma(s+q+1)}{\Gamma(s+1)} K_{1,0}(s,-q)e^{\lambda_1(s)t}. \quad \} \quad (18.5)$$

Expanding them to the order of $\frac{E}{\varepsilon}$, assuming $\frac{E}{\varepsilon} \ll 1$, deforming each integration path of the integral parameter to the left, and using the double poles at $q=-s-1$ or a simple pole at $q=-s$ in the function $\Gamma(s+q+1)$, $K_{1,0}(s,-q)$ or $C(s+q)$, the formulae become

$$\pi \propto \frac{1}{2\pi i}\int ds \left(\frac{E_0}{\varepsilon}\right)^s \left\{\log\frac{\varepsilon}{E} + \cdots\right\}, \quad (18.4')$$

$$\gamma \propto \frac{1}{2\pi i}\int ds \left(\frac{E_0}{\varepsilon}\right)^s \frac{\varepsilon}{E}, \quad (18.5')$$

in confirmation of the predictions from physical arguments.

III. Solution in other approximations.

In the preceding section we applied the complete screening cross sections for the radiation and pair creation processes, and neglected the effect of Compton scattering.

However, several attempts have also been made to derive solutions using more accurate cross sections. Among these, BERNSTEIN (*19*) obtained the shower functions with the help of the approximate formulae (1.8) and (1.9) for the radiation and pair creation processes. RICHARD and NORDHEIM (*28*) calculated the track length using accurate cross sections, and also including the effect of the Compton scattering and of high-energy knock-on electrons. In the following, the results derived from the solutions (16.10), (16.11), (16.12) will be compared with those obtained by their theories.

---

[1] This is not the case for the showers originated by primary photons with a spectrum dE_0/E_0 [*3*].

19. Solution of BERNSTEIN. BERNSTEIN starts from the approximate expressions (1.8) and (1.9) for the probabilities Φ and Ψ of the radiation and pair creation processes. Since these probabilities are not of fractional form with respect to the energy, the solutions of the basic equations cannot be obtained as simply as in the procedure described in the preceding sections. His mathematical treatment is essentially as follows.

First the probabilities Φ and Ψ are split into two terms,

$$\Phi = \Phi_0 + \Phi_1 \tag{19.1}$$

and

$$\Psi = \Psi_0 + \Psi_1, \tag{19.2}$$

where Φ_1 and Ψ_1 represent the non-fractional parts of those quantities. Then the operator A', B', C' and σ in the eqs. (7.10) and (7.15) are similarly divided:

$$A' = A'_0 + A'_1, \tag{19.3}$$

$$B' = B'_0 + B'_1, \tag{19.4}$$

$$C' = C'_0 + C'_1, \tag{19.5}$$

and

$$\sigma = \sigma_0 + \sigma_1, \tag{19.6}$$

where A'_0, B'_0, C'_0, σ_0 and A'_1, B'_1, C'_1, σ_1, are the sets of operators corresponding to the probabilities Φ_0, Ψ_0 and Φ_1, Ψ_1, respectively.

Writing the differential spectrum of the electrons and photons in the form

$$\pi = \pi_0 + \pi_1 + \cdots \tag{19.7}$$

$$\gamma = \gamma_0 + \gamma_1 + \cdots, \tag{19.8}$$

the basic Eqs. (7.10) and (7.15) can be re-written, after substitution of (19.3) into (19.7), and of (19.6) into (19.8), as the following sets of the equations

$$\frac{\partial \pi_0}{\partial t} = -A'_0 \pi_0 + B'_0 \gamma_0 + \varepsilon \frac{\partial \pi_0}{\partial E}, \tag{19.9}$$

$$\frac{\partial \gamma_0}{\partial t} = -\sigma_0 \gamma_0 + C'_0 \pi_0, \tag{19.10}$$

$$\frac{\partial \pi_1}{\partial t} = -A'_0 \pi_1 + B'_0 \gamma_1 + \varepsilon \frac{\partial \pi_1}{\partial E} - A'_1 \pi_0 + B'_1 \gamma_0, \tag{19.11}$$

$$\frac{\partial \gamma_1}{\partial t} = -\sigma_0 \gamma_1 + C'_0 \pi_1 - \sigma_1 \gamma_0 + C'_1 \pi_0, \tag{19.12}$$

$$\frac{\partial \pi_2}{\partial t} = \cdots, \tag{19.13}$$

$$\frac{\partial \gamma_2}{\partial t} = \cdots, \tag{19.14}$$

....

Since the Eqs. (19.9) and (19.10) are the same as the basic equations in Approximation B, their solutions should be given by the formulae (15.16) and (15.17).

Now multiplying by $\left(\frac{\partial}{\partial t}+\sigma_0\right)$ on both sides of (19.11), we have

$$\left\{\frac{\partial^2}{\partial t^2}+(A'_0-\sigma_0)\frac{\partial}{\partial t}+(A'_0\sigma_0-B'_0C'_0)\right\}\pi_1$$
$$-\left(\frac{\partial}{\partial t}+\sigma_0\right)\varepsilon\frac{\partial\pi_1}{\partial E}=\left(\frac{\partial}{\partial t}+\sigma_0\right)S,$$
(19.15)

where

$$S=(-A'_1+B'_1C'_1+B'_0C'_1)\pi_0-\sigma_1\gamma_0. \tag{19.16}$$

As shown in Appendix 4, π_0 is the Green function of the equation on the left hand side of (19.15), the solution of which is, therefore,

$$\pi_1=\int_0^t dt'\int_E^{E_0}\pi_0(E',E,t-t')S(E_0,E',t')\,dE'. \tag{19.17}$$

The higher-order solutions $\pi_2,\pi_3\ldots$ could also be obtained in a similar way.

Numerical evaluations up to the second term, i.e., to π_1, were carried out for showers in air and in Pb. The results are shown in Fig. 9.

Because of the omission of higher terms, some doubt must remain as to the accuracy of BERNSTEIN's results. But the differences between their characteristic features and those of Approximation B are worth noting. They can be understood as due to a decrease of the radiation and pair creation probabilities. Thus the features of the shower correspond to those calculated in Approximation B for a material of slightly larger radiation length, and therefore larger critical energy.

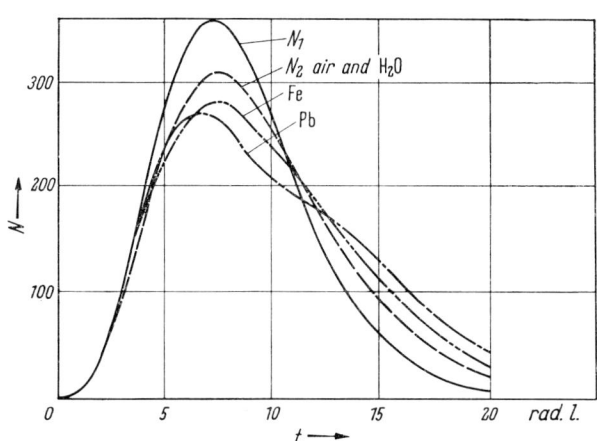

Fig. 9. Shower curves for a primary electron of energy $E_0 = e^s\cdot\varepsilon:N_1$ for calculation in Approximation B, N_2 according to BERNSTEIN (19).

20. Track length. The track length of the shower particles are obtained by the integration of the shower function with respect to t from 0 to infinity.

As it was shown in Sect. 13, the existence of a pole at $s=1$ in the function $1/\lambda_1(s)$ enables us to evaluate the complex integrals of the functions (16.10), (16.11) and (16.12) with respect to s by the residue at $s=1$. Since at the shower maximum the value of age parameter becomes unity, the track length represents the features of the shower at the shower maximum.

Integration of the basic eqs. (7.10) and (7.15) with respect to t from 0 to infinity leads to

$$-\pi(t=0)=-A'Z_\pi+B'Z_\gamma+\varepsilon\frac{\partial Z_\pi}{\partial E}, \tag{20.1}$$

$$-\gamma(t=0)=-\sigma Z_\gamma+C'Z_\pi. \tag{20.2}$$

Since these equations are simple compared with the equations for the shower functions, several attempts have been made to calculate the track length without the restrictions of simplifying assumptions. Thus, RICHARD and NORDHEIM (28) derived numerical values of the track length using the accurate cross sections for the radiation and pair creation processes, and also including the effects of Compton scattering and of the high-energy knock-on electrons. They started from the solutions of the track length in Approximation A, which is only accurate in the high-energy region, and then found solutions applicable at low energies by an iteration method. As already shown in Sect. 13, the differential track length of the electrons is given in Approximation A by

$$Z_n = 0.437 \frac{E_0}{E^2}.$$

The exact solution valid in the low-energy region may be written as

$$Z_n = 0.437 \frac{E_0}{E^2} f(E). \tag{20.3}$$

In their theory, RICHARD and NORDHEIM approximated the term $A'Z_n$ in (20.1) by $\frac{\partial}{\partial E} \varepsilon_r Z_n$, representing the change in the number of electrons by the radiation process. The validity of this substitution can be proved by the following argument:

As shown in Sect. 7, the term $A'Z_n$ is given by

$$-A'Z_n = \int_E^\infty Z_n(E+E') \varphi(E+E', E') \frac{dE'}{E+E'} \\ - \int_0^E Z_n(E) \varphi(E, E') \frac{dE'}{E}, \tag{20.4}$$

and if one assumes a fractional form for $\varphi(E, E')$,

$$\varphi(E, E') \frac{dE'}{E} = \varphi\left(\frac{E'}{E}\right) \frac{dE'}{E}, \tag{20.5}$$

(20.4) becomes

$$-A'Z_n = -\int_0^1 \left\{ Z_n(E) - \frac{1}{1-v} Z_n\left(\frac{E}{1-v}\right) \right\} \varphi(v) \, dv. \tag{20.6}$$

Substituting (20.3) into this relation, one has

$$-A'Z_n = -0.437 \frac{E_0}{E^2} \int_0^1 \left\{ f(E) - (1-v) f\left(\frac{E}{1-v}\right) \right\} \varphi(v) \, dv. \tag{20.7}$$

Since $\varphi(v)$ has a large value for small values of v, one may expand $(1-v)f\left(\frac{E}{1-v}\right)$ around $v=0$. Then (20.6) becomes

$$-A'Z_n = -0.437 E_0 \left\{ \frac{f(E)}{E^2} - \frac{1}{E} \frac{\partial}{\partial E} f(E) + \cdots \right\} \int_0^1 v \varphi(v) \, dv, \tag{20.8}$$

$$\cong \frac{\partial}{\partial E} \varepsilon_r Z_n, \tag{20.9}$$

where we put

$$\varepsilon_r = E \int_0^1 v \, \varphi(v) \, dv,$$

which is just the energy loss due to the radiation process. The solution of (20.1) is now given by

$$Z_n = \frac{1}{\varepsilon_r + \varepsilon} \int_E^\infty B' Z_\gamma \, dE'. \tag{20.10}$$

and (20.2) can be re-written:

$$Z_\gamma = \frac{1}{\sigma}[C'Z_\pi + \sigma_1 Z_\gamma], \qquad (20.11)$$

where σ is the total absorption coefficient of the photons including the effects of Compton scattering and pair creation, and σ_1 is the operator representing the contribution to the photons after the Compton scattering.

Starting from the solution (13.21), and using Eqs. (20.10) and (20.11), the iterations were made numerically. Actual calculations were carried out for air and Pb. The results are summarized in Table 11. In this table the track lengths calculated with the solutions (16.10), (16.11), (16.12) are also shown for comparison, together with the results obtained by TAMM and BELENKY (6)[1].

Table 11. *Differential and integral track lengths of electrons and photons.*

E/ε	Z_π [1]	Z_π T—B[2]	Z_π R—N[3] air	Z_π R—N[3] Pb	Z_Π [1]	Z_Π R—N[3] air
0					2.3	2.3
0.05	5.6	5.9	5.5		1.6	1.9
0.1	3.6	3.8	3.7		1.3	1.6
0.20	2.1	2.2	2.25		1.02	1.4
0.30	1.43	1.5	1.61		0.86	1.2
0.40	1.05	1.1	1.27		0.75	1.04
0.50	0.80	0.89	0.89	0.90	0.67	0.92
0.75	0.48	0.53	0.62		0.53	0.73
1.0	0.230	0.36	0.42	0.43	0.45	0.61
2.0	0.120	0.14	0.145	0.17	0.29	0.37
3.0	0.065	0.073	0.073	0.089	0.23	0.27

E/ε	Z_γ [1]	Z_γ R—N[3] air	Z_γ R—N[3] Pb	Z_Γ [1]	Z_Γ R—N[3] air
0.05	48	45		5.6	6.6
0.1	20	23		4.0	5.0
0.20	7.5	11		2.8	3.5
0.30	4.1	6.0		2.2	2.7
0.40	2.7	3.9		1.8	2.2
0.50	1.84	2.8	6.4	1.48	1.8
0.75	0.98	1.4		1.08	1.3
1.0	0.62	0.79	2.0	0.85	1.05
2.3	0.195	0.25	0.49	0.48	0.60
3.0	0.0965	0.12	0.21	0.34	0.44

[1] Evaluated from (18.4) and (18.5).
[2] I. TAMM and S. BELENKY: J. Phys. U.S.S.R. **1**, 177 (1939).
[3] J. A. RICHARD and NORDHEIM: Phys. Rev. **74**, 1106 (1948).

It seems rather surprising that these results for the electron track length agree quite well each other, in spite of the difference in the approximations used. The reason may be found in the following argument.

---
[1] The relation derived by these authors is

$$Z_\pi = (2.3)^2 \left\{ \left(1 - 2.3\frac{E}{\varepsilon}\right) e^{2.3\frac{E}{\varepsilon}} \left[E_i\left(-2.3\frac{E}{\varepsilon}\right) - 1 + \frac{\varepsilon + 2.3E}{2.3E_0} \right] - \left(1 + 2.3\frac{E}{\varepsilon}\right) e^{2.3\frac{E}{\varepsilon}} \left[-E_i\left(-\frac{E_0}{\varepsilon}\right)\right] \right\}$$

and is based on the use of simplified cross sections even for the low-energy region.

According to (20.1) and (20.2), the equation for the electron track length in a shower initiated by a primary electron can be written in the form

$$-\pi(t=0) = \left(-A' + B'\frac{1}{\sigma}C' + \varepsilon\frac{\partial}{\partial E}\right) Z_\pi. \tag{20.12}$$

The operators indicating the effects of pair creation and of Compton scattering are B' and σ, and changes in these operators are partly compensated in eq. (20.12), since they appear in this equation in the combination $B'\frac{1}{\sigma}C'$. Furthermore, the total track length of the electrons should satisfy the relation (8.1),

$$Z_\Pi(E=0) = E_0/\varepsilon,$$

in any solution, so that for all theories agreement must be reached at $E=0$.

The situation is slightly different for the photon track length.

We can apply (20.1) to express it in operator form, using the known electron track length Z_π,

$$Z_\gamma = \frac{1}{\sigma}[C' Z_\pi]. \tag{20.13}$$

Evidently no compensation effect such as in the case of the electron track length exists for this case. Consequently the photon track lengths calculated under different assumptions do not agree well with each other, in spite of the fair agreement of the electron track lengths obtained for all theories.

At the end of this section, it may be important to point out that at the shower maximum most of the electrons in a shower have energies lower than the critical energy. As stated above, the track length reflects the features of the shower at the shower maximum. Thus Table 11 can also be read as a representation of the shower characteristics at this stage. As to the number distributions of the shower electrons, it is seen that at the shower maximum only about 25% of the total have energies higher than the critical energy, while about one-half of all electrons lie in the energy range below $\frac{1}{5}$ of the critical energy even when the primary energy is quite high.

F. Three-dimensional electron shower theory.

In this Section we shall describe theories based on the use of the Landau approximation, as well as those developed without this method. Since in the first only the effects of multiple scattering are considered, while the second includes not only these but also the contributions of single and plural scattering, the theories without the Landau approximation give more accurate results.

As outlined in Sect. 9, the limitations of all approximations must be given carefull attention when shower functions for large angular or lateral spread are considered. The mathematical difficulties inherent in the theories applicable to these regions are formidable. A number of authors attempted to overcome them by Monte Carlo calculations, or else by the use of mathematically rather crude techniques.

To facilitate the presentation, we shall start with a discussion of the less complicated case of the three-dimensional theory in the Landau approximation.

I. Three-dimensional electron shower theory in the Landau approximation.

21. Derivation of the analytic solutions. LANDAU's basic equations for the three-dimensional shower theory were derived in Sect. 7, and written in the form

Sect. 21. Derivation of the analytic solutions.

(7.16) and (7.17),

$$\left(\frac{\partial}{\partial t} + \theta \frac{\partial}{\partial \boldsymbol{r}}\right)\pi = -A'\pi + B'\gamma + \frac{E_s}{4E^2}\left(\frac{\partial^2}{\partial \theta_1^2} + \frac{\partial^2}{\partial \theta_2^2}\right)\pi + \varepsilon \frac{\partial \pi}{\partial E},$$

$$\left(\frac{\partial}{\partial t} + \theta \frac{\partial}{\partial \boldsymbol{r}}\right)\gamma = -\sigma_0 \gamma + C'\pi,$$

where $\pi(E_0, E, \boldsymbol{r}, \boldsymbol{\theta}, t) \, d\boldsymbol{r} \, d\boldsymbol{\theta} \, dE$ or $\gamma(E_0, E, \boldsymbol{r}, \boldsymbol{\theta}, t) \, d\boldsymbol{r} \, d\boldsymbol{\theta} \, dE$ are the structure functions of the electrons or photons of energy $(E, E+dE)$, and the constant E_s has the value 21 MeV as shown in Appendix 3.

Applying a Fourier transformation, and multiplying on both sides of the above equations by $\exp(i(\boldsymbol{r}\boldsymbol{x}) + i(\boldsymbol{\zeta}\boldsymbol{\theta}))$, one gets

$$\left(\frac{\partial}{\partial t} - \boldsymbol{x} \frac{\partial}{\partial \boldsymbol{\zeta}}\right)f = -A'f + B'g - \frac{E_s^2 \zeta^2}{4E^2} f, \tag{21.1}$$

$$\left(\frac{\partial}{\partial t} - \boldsymbol{x} \frac{\partial}{\partial \boldsymbol{\zeta}}\right)g = -\sigma_0 g + C'f, \tag{21.2}$$

where f and g are the Fourier transform of π and γ, respectively, and are defined by

$$f(\boldsymbol{x}, \boldsymbol{\zeta}) = \frac{1}{4\pi^2} \int\!\!\!\int\!\!\!\int\!\!\!\int_{-\infty}^{+\infty} e^{i(\boldsymbol{r}\boldsymbol{x}) + i(\boldsymbol{\zeta}\boldsymbol{\theta})} \pi \, d\boldsymbol{r} \, d\boldsymbol{\theta}, \tag{21.3}$$

$$g(\boldsymbol{x}, \boldsymbol{\zeta}) = \frac{1}{4\pi^2} \int\!\!\!\int\!\!\!\int\!\!\!\int_{-\infty}^{+\infty} e^{i(\boldsymbol{r}\boldsymbol{x}) + i(\boldsymbol{\zeta}\boldsymbol{\theta})} \gamma \, d\boldsymbol{r} \, d\boldsymbol{\theta}. \tag{21.4}$$

Eliminating g from the above Eqs. (21.1) and (21.2), we get

$$\left\{\left(\frac{\partial}{\partial t} - \boldsymbol{x} \frac{\partial}{\partial \boldsymbol{\zeta}}\right)^2 + (A' + \sigma_0)\left(\frac{\partial}{\partial t} - \boldsymbol{x} \frac{\partial}{\partial \boldsymbol{\zeta}}\right) + (A'\sigma_0 - B'C')\right\} f \\ = \left(\frac{\partial}{\partial t} - \boldsymbol{x} \frac{\partial}{\partial \boldsymbol{\zeta}} + \sigma_0\right)\left(-\frac{E_s^2 \zeta^2}{4E^2} + \varepsilon \frac{\partial}{\partial E}\right) f. \tag{21.5}$$

Since the lateral structure function of the electrons $\pi_2(E_0, E, r \cdot t) 2\pi r \, dr$ is given by

$$\pi_2 = \int\!\!\!\int_{-\infty}^{+\infty} \pi(E_0, E, \boldsymbol{r}, \boldsymbol{\theta}, t) \, d\boldsymbol{\theta},$$

it follows from (21.3) and (21.4) that

$$\pi_2(E_0, E, r, t) = \int\!\!\!\int_{-\infty}^{+\infty} e^{-i(\boldsymbol{x}\boldsymbol{r})} f(\boldsymbol{x}, 0) \, d\boldsymbol{x} \\ = \int_0^\infty f(x, 0) J_0(r x) 2\pi x \, dx, \tag{21.6}$$

where we used the relation for the Bessel function

$$2\pi J_0(r x) = \int_0^{2\pi} e^{i r x \cos\varphi} d\varphi,$$

and φ is the angle between the vectors $\boldsymbol{r}$ and $\boldsymbol{x}$.

For the angular structure function, we obtain in a similar way

$$\pi_1(E_0, E, \theta, t)) = \int_0^\infty f(0, \zeta) J_0(\theta \zeta) 2\pi \zeta \, d\zeta. \tag{21.7}$$

The equation for the angular structure function is obtained by putting $x=0$ in eq. (21.5). However, the equation for the lateral structure function cannot similarly be derived from (21.5) by taking $\zeta=0$, because it is necessary to know the value $\dfrac{\partial}{\partial \zeta} f \Big|_{\zeta=0}$.

Let the parallel and perpendicular components of the vector ζ with respect to x be ζ_1 and ζ_2. For $\zeta_2=0$, we have from (21.5),

$$\left\{\left(\frac{\partial}{\partial t}-x\frac{\partial}{\partial \zeta_1}\right)^2+(A'+\sigma_0)\left(\frac{\partial}{\partial t}-x\frac{\partial}{\partial \zeta_1}\right)+(A'\sigma_0-B'C')\right\}f(\zeta_2=0) \\ =\left(\frac{\partial}{\partial t}-x\frac{\partial}{\partial \zeta_1}+\sigma_0\right)\left(-\frac{E_s^2\zeta_1^2}{4E^2}+\varepsilon\frac{\partial}{\partial E}\right)f(\zeta_2=0). \qquad (21.8)$$

If one puts $\zeta_1=0$ in the solution of this equation, the solution becomes the Hankel transform of the lateral structure function. Now introducing the variable t' and ξ just as we did in the usual multiple scattering theory:

$$t=t', \qquad \zeta_1=x(\xi-t'),$$

we have

$$\frac{\partial}{\partial t'}=\frac{\partial}{\partial t}-x\frac{\partial}{\partial \zeta_1}. \qquad (21.9)$$

Using these new variables t' and ξ, Eq. (21.8) becomes

$$\left\{\frac{\partial^2}{\partial t'^2}+(A'+\sigma_0)\frac{\partial}{\partial t'}+(A'\sigma_0-B'C')\right\}f \\ =\left(\frac{\partial}{\partial t'}+\sigma_0\right)\left\{\frac{-E_s^2 x^2}{4E^2}(\xi-t')^2+\varepsilon\frac{\partial}{\partial E}\right\}f. \qquad (21.8')$$

The solution of this equation is obtained in the form of a power series in $\dfrac{E_s^2 x^2}{4E^2}$ and ε/E, in similar way as described in Sect. 15. Thus, the solution[1] is

$$f_2=\frac{1}{4\pi^2 i}\int_{\delta-i\infty}^{\delta+i\infty}\left(\frac{E}{E_0}\right)^s\frac{ds}{E}\left\{\sum_{m=n}^{\infty}\sum_{n=0}^{\infty}\left(\frac{-E_s^2 x^2}{4E^2}\right)^m\left(\frac{-\varepsilon}{E}\right)^n\phi_{mn}(s,t)\right\}, \qquad (21.10)$$

where $\phi_{m,n}$ is given by the equation

$$\left[\frac{\partial^2}{\partial t^2}(A_{(s+2m+n)}+\sigma_0)\frac{\partial}{\partial t}+(A\sigma_0-BC)_{s+2m+n}\right]\Phi_{m,n} \\ =\left(\frac{\partial}{\partial t}+\sigma_0\right)\{(\xi-t)^2\Phi_{m-1,n}+(s+2m+n)\Phi_{m,n-1}\} \qquad (21.11) \\ \text{with } \phi_{m,n}=\lim_{(\xi-t)\to 0}\Phi_{m,n},$$

and $\Phi_{0,0}$ is the solution in Approximation A for a given boundary condition.

If one takes a shower due to a primary electron of energy E_0, $\Phi_{0,0}$ is defined by

$$\Phi_{0,0}=H_1(s)e^{\lambda_1(s)t}+H_2(s)e^{\lambda_2(s)t}. \qquad (21.11')$$

The solution of (21.11),

$$\Phi_{m,n}(s,\xi-t,t)=\int_0^t \Phi_{00}(s+2m+n,t-t')\times \\ \times\{(\xi-t')^2\Phi_{m-1,n}(s,\xi-t',t')+(s+2m+n)\Phi_{m,n-1}(s,\xi-t',t')\}dt', \qquad (21.12)$$

---

[1] In the following we write t' as t, and $f(\zeta=0)$ as f_2.

Sect. 21.   Derivation of the analytic solutions.   41

can be derived by arguments similar to those presented in the Sect. 15, or it can be proved simply by the substitution of (21.12) into (21.11). Now, the convergence of the series (21.10) is proved as follows.

From (21.11') and (21.12), we find[1]

$$\begin{aligned}\phi_{m,n} &\lesssim |\phi_{0,0}(s,t)| \frac{2^n \Gamma(s+m+n+1)\, t^{3m+n}}{3^m \Gamma(s+1)\Gamma(m+1)\Gamma(n+1)} \\ &\lesssim |\phi_{0,0}(s,t)|\, 2^s \left(\tfrac{4}{3}\right)^m 8^n t^{3m+n}\end{aligned} \quad (21.13)$$

for any values of s and t, in which $|\phi_{00}(s,t)|$ is the largest value of $\Phi_{00}(s,t')$ in the region where t' varies between 0 and t.

Then it follows that the double series in (21.10) is uniformly and absolutely convergent if $E > \dfrac{E_s \times t^{\frac{8}{3}}}{3}$ and $E > 8\varepsilon t$, and the above argument is merely a general extension of the one given in the section on Approximation B.

---

[1] The relation (21.13) is obtained as follows: Let $|\phi_{0,0}(s,t)|$ be the largest value of $\Phi_{0,0}(s,t')$ in the region where t' varies between 0 and t. Now defining the function $\Psi_{m,n}$ as

$$\begin{aligned}\Psi_{m,n} &= \int_0^t |\phi_{0,0}(s+2m+n,\, t-t')| \times \\ &\quad \times \{(s+m+n)(\xi-t')^2 \Psi_{m-1,m}(s,\xi-t',t') + 2(s+m+n)\Psi_{m,n-1}(s,\xi-t',t')\}\, dt', \\ \Psi_{00} &= |\phi_{0,0}|\end{aligned} \quad (21.12')$$

and comparing the above formula with (21.12), the relation

$$\Psi_{m,n} \geq \Phi_{m,n} \quad (21.12'')$$

is obtained. Putting

$$\psi_{m,n} = \frac{2^n \Gamma(s+m+n+1)}{\Gamma(s+1)} \Psi_{m,n}, \quad (21.12''')$$

an expression for $\psi_{m,n}$ is derived from (21.12'):

$$\psi_{m,n} = \int_0^t |\phi_{0,0}(s+2m+n,\, t-t')| \{(\xi-t')^2 \psi_{m-1,n}(s,\xi-t',t') + \psi_{m,n-1}(s,\xi-t',t')\}\, dt'.$$

Then the equation for the generating function of $\psi_{m,n}$ is

$$\frac{\partial G}{\partial t} = g_1 (\xi-t)^2 G + g_2 G$$

and

$$G = \sum_{m=0}^{\infty} \sum_{n=0}^{\infty} g_1^m g_2^n \psi_{m,n}.$$

G is obtained easily from the above equation as

$$G = \lim_{(\xi-t)\to 0} e^{\frac{-g_1(\xi-t)^3}{3} + \frac{g_1 \xi^3}{3} + g_2 t} = \sum_{m=0}^{\infty} \sum_{n=0}^{\infty} g_1^m g_2^n \frac{t^{3m} t^n}{3^m m!\, n!}.$$

Then we have

$$\psi_{mn} = \frac{t^{3m+n}}{3^m \Gamma(m+1)\Gamma(n+1)}.$$

From comparison with (21.12'''), the coefficients are seen to satisfy the condition

$$\frac{\Gamma(s+m+n+1)}{\Gamma(s+1)\Gamma(m+1)\Gamma(n+1)} = \frac{\Gamma(s+m+n+1)}{\Gamma(s+1)\Gamma(m+n+1)} \cdot \frac{\Gamma(m+n+1)}{\Gamma(m+1)\Gamma(n+1)}.$$

$$\lesssim 2^{s+2m+2n}$$

Here we used the relation

$$2^{s+m+n} = (1+1)^{s+m+n} \gtrsim \frac{\Gamma(s+m+n+1)}{\Gamma(s+1)\Gamma(m+n+1)}.$$

Now the double series (21.10) can be replaced by an integral representation (Appendix 2),

$$\left.\begin{array}{l}\sum_{m=0}^{\infty}\sum_{n=0}^{\infty}\left(-\frac{E_s^2 x^2}{4E^2}\right)^m\left(-\frac{\varepsilon}{E}\right)^n \Phi_{m,n}(s,t) \\ = -\frac{1}{4\pi^2}\iint dp\,dq\,\Gamma(-p)\,\Gamma(-q)\left(\frac{E_s^2 x^2}{4E^2}\right)^p\left(\frac{\varepsilon}{E}\right)^q \mathfrak{M}_2(p,q,s,t),\end{array}\right\} \quad (21.14)$$

where $\mathfrak{M}_2(p,q,s,t)$ is defined by

$$\frac{\mathfrak{M}_2(m,n,s,t)}{\Gamma(n+1)\,\Gamma(m+1)} = \Phi_{m,n}(s,t). \quad (21.15)$$

Thus we have, from (21.11), a recurrence equation for the function $\mathfrak{M}_2(p,q,s,t)$:

$$\left.\begin{array}{l}\left\{\dfrac{\partial}{\partial t}-\lambda_1(s+2p+q)\right\}\left\{\dfrac{\partial}{\partial t}-\lambda_2(s+2p+q)\right\}M(p,q,s,\xi-t,t) \\ =\left(\dfrac{\partial}{\partial t}+\sigma_0\right)\{(p(\xi-t)^2 M(p-1,q,s,\xi-t,t)+ \\ +q(s+2p+q)M(p,q-1,s,\xi-t',t)\},\end{array}\right\} \quad (21.11'')$$

with

$$\mathfrak{M}_2(p,q,s,t) = \lim_{(\xi-t)\to 0} M(p,q,s,\xi-t,t).$$

The solution for the lateral structure function of the shower electrons is now given by

$$\pi_2 = \int_0^\infty J_0(rx)\,f(x,0)\,x\,dx,$$

which can be written

$$\left.\begin{array}{l}\pi_2 = -\dfrac{1}{8\pi^4 i}\displaystyle\int\!\!\int\!\!\int_{-i\infty}^{+i\infty} ds\,dp\,dq\left(\dfrac{E_0}{E}\right)^s\dfrac{1}{E}\left(\dfrac{E}{E_s}\right)^2\times \\ \times\left(\dfrac{\varepsilon}{E}\right)^q\left(\dfrac{E^2 r^2}{E_s^2}\right)^{-p-1}\Gamma(p+1)\,\Gamma(-q)\,\mathfrak{M}_2(p,q,s,t).\end{array}\right\} \quad (21.16)$$

Here we make use of a property of the Bessel function

$$\int_0^\infty J_0(rx)\left(\frac{x^2}{4}\right)^p x\,dx = \frac{2\,\Gamma(p+1)}{\Gamma(-p)}\,x^{-2p-2}.$$

The restrictions $E > \dfrac{E_s x t^{\frac{3}{2}}}{3}$ and $E > 8\varepsilon t$ can now be dropped because (21.16) exists for any value of E and r. Thus it can be stated that (21.16) is the exact solution of the differential lateral structure function for any values of E and r derived by the principle of analytic continuation.

The expression for the integral lateral structure function, $\Pi(E_0, E, r, t)$, i.e. the lateral structure function for the electrons with energy larger than E, is obtained by integration with respect to the energy:

$$\left.\begin{array}{l}\Pi_2(E_0, E, r, t) = -\dfrac{1}{8\pi^4 i}\displaystyle\int\!\!\int\!\!\int_{-i\infty}^{i\infty}\dfrac{ds\,dp\,dq}{s+2p+q}\left(\dfrac{E_0}{E}\right)^s\left(\dfrac{E_s}{E}\right)^2\left(\dfrac{\varepsilon}{E}\right)^p\left(\dfrac{E^2 r^2}{E_s^2}\right)^{-p-1}\times \\ \times \Gamma(p+1)\,\Gamma(-q)\,\mathfrak{M}_2(p,q,s,t).\end{array}\right\} \quad (21.17)$$

Sect. 22.   Structure functions of electrons and photons.   43

At the limit of $E=0$, the integration with respect to q is equal to the residue of the integrand at $q=-s-2p$, and we have accurately

$$\Pi_2(E_0, 0, r, t) = -\frac{1}{4\pi^3}\int\int_{-i\infty}^{i\infty} ds\, dp \left(\frac{E_0}{\varepsilon}\right)^s \left(\frac{\varepsilon}{E_s}\right)^2 \left(\frac{\varepsilon^2 r^2}{E_s}\right)^{-p-1} \times \\ \times \Gamma(p+1)\Gamma(s+2p)\mathfrak{M}_2(p, -s-2p, s, t). \qquad (21.18)$$

The total number of the electrons is now obtained by the integration of Π_2 with respect to r, and we find

$$\int_0^\infty \Pi_2(E_0, 0, r, t)\, 2\pi r\, dr = \frac{1}{2\pi i}\int ds \left(\frac{E_0}{\varepsilon}\right)^s \Gamma(s)\, \mathfrak{M}_2(0, -s, s, t), \qquad (21.19)$$

in agreement with the expression derived in Sect. 5.

A similar treatment is possible for the analytical solution of the structure function for photons, which is sometimes important for the analysis of the spread of extensive air showers detected by an apparatus under a thick absorber. It can be shown easily from (21.2), using the transformation (21.9), that

$$g(x, \zeta=0) = \lim_{(\xi-t)\to 0}\int_0^t C' e^{-\sigma_0(t-t')} f(x, \zeta_1=x(\xi-t), \zeta_2=0)\, dt. \qquad (21.20)$$

Expressions similar to (21.16), (21.17) and (21.18) are obtained for the differential and integral structure functions, γ_2 and Γ_2, respectively, of photons.

22. Structure functions of electrons and photons.
It will be helpful here to summarize the results so far derived:

α) *Lateral structure function*

$$\pi_2(E_0, E, r, t) = -\frac{1}{8\pi^4 i}\int\int\int_{-i\infty}^{i\infty} ds\, dp\, dq \left(\frac{E_0}{E}\right)^s \frac{1}{E}\left(\frac{E}{E_s}\right)^2 \left(\frac{\varepsilon}{E_s}\right)^q \left(\frac{E^2 r^2}{E_s}\right)^{-p-1} \times \\ \times \Gamma(p+1)\Gamma(-q)\mathfrak{M}_2(p, q, s, t), \qquad (22.1)$$

$$\Pi_2(E_0, E, r, t) = -\frac{1}{8\pi^4 i}\int\int\int_{-i\infty}^{i\infty} ds\, dp\, dq\, \frac{1}{s+2p+q}\left(\frac{E_0}{E}\right)^s \left(\frac{E}{E_s}\right)^2 \left(\frac{\varepsilon}{E}\right)^q \times \\ \times \left(\frac{E^2 r^2}{E_s}\right)^{-p-1} \Gamma(p+1)\Gamma(-q)\mathfrak{M}_2(p, q, s, t), \qquad (22.2)$$

$$\Pi_2(E_0, 0, r, t) = -\frac{1}{4\pi^3}\int\int ds\, dp \left(\frac{E_0}{E}\right)^s \left(\frac{\varepsilon}{E_s}\right)^2 \left(\frac{\varepsilon^2 r^2}{E_s}\right)^{-p-1} \Gamma(p+1)\Gamma(s+2p) \times \\ \times \mathfrak{M}_2(p, -s-2p, s, t), \qquad (22.3)$$

where $\mathfrak{M}_2$ is given by Eq. (21.11''), and the boundary condition corresponding to a primary electron or photon is

$$M(0, 0, s, 0, t) = H_1(s) e^{\lambda_1(s)t} + H_2(s) e^{\lambda_2(s)t}, \qquad (22.4)$$

$$M(0, 0, s, 0, t) = \frac{B(s)}{\lambda_1(s) - \lambda_2(s)}\{e^{\lambda_1(s)t} - e^{\lambda_2(s)t}\}. \qquad (22.5)$$

β) *Angular structure function.* Similar expressions are also obtained for the angular structure functions, in which we substitute θ and $\mathfrak{M}_1$ for r and $\mathfrak{M}_2$. The function is given by the following equation with the same boundary condition as above:

$$\left[\frac{\partial}{\partial t}-\lambda_1(s+2p+q)\right]\left[\frac{\partial}{\partial t}-\lambda_2(s+2p+q)\right]\mathfrak{M}_1(p,q,s,t)$$
$$=\left(\frac{\partial}{\partial t}+\sigma_0\right)[p\,\mathfrak{M}_1(p-1,q,s,t)+q(s+2p+q)\,\mathfrak{M}_1(p,q-1,s,t)].\quad(22.6)$$

γ) *Structure function of photons.* To obtain the structure functions of photons, we substitute $\mathfrak{M}'_1$ or $\mathfrak{M}'_2$ for $\mathfrak{M}_1$ or $\mathfrak{M}_2$. These functions are defined with the help of (21.20):

$$\mathfrak{M}'_1=\int_0^t C(s+2p+q)\,e^{-\sigma_0(t-t')}\mathfrak{M}_1(p,q,s,t)\,dt, \quad (22.7)$$

$$\mathfrak{M}'_2=\lim_{(\xi-t)\to 0}\int_0^t C(s+2p+q)\,e^{-\sigma_0(t-t')}M(p,q,s,\xi-t,t)\,dt, \quad (22.8)$$

with the same boundary conditions, (22.4) and (22.5), as before.

The structure function in Approximation A can be derived from the solution under Approximation B, by going to the limit of $\varepsilon/E \to 0$. The integral with respect to q is essentially the same as the residues at the pole $q=0$ in $\Gamma(-q)$, so that the structure function in Approximation A is obtained by omitting the term $\Gamma(-q)$ and multiplying by $2\pi i$ the expression in Approximation B, applying it at $q=0$. The explicit expression of these structure functions, and the details of the behaviour of these functions will be given in the following section.

II. Three-dimensional electron shower theory without the Landau approximation.

23. Derivation of the analytic solutions. Before entering into details of the numerical results of the analytical solutions in the Landau approximation, we shall derive the solutions without this approximation. Since according to MOLIÈRE's scattering theory (see Appendix 3) the contributions of scattering are given in the form

$$-\int_0^\infty d\zeta\,J_0(\zeta\theta)\,\frac{K^2\zeta^2}{4E^2}\left[1-\frac{1}{\Omega}\left(\log\frac{K^2\zeta^2}{4E^2}\right)\right]f(x,\zeta), \quad (23.1)$$

the basic equation for the Fourier transform of the structure function of the electrons is

$$\left\{\left(\frac{\partial}{\partial t}-x\frac{\partial}{\partial \zeta}\right)^2+(A'+\sigma_0)\left(\frac{\partial}{\partial t}-x\frac{\partial}{\partial \zeta}\right)+(A'\sigma_0-B'C')\right\}f(x,\zeta)$$
$$=\left(\frac{\partial}{\partial t}-x\frac{\partial}{\partial \zeta}+\sigma_0\right)\left\{-\frac{K^2\zeta^2}{4E^2}\left(1-\frac{1}{\Omega}\log\frac{K^2\zeta^2}{4E^2}\right)+\varepsilon\frac{\partial}{\partial E}\right\}f(x,\zeta), \quad (23.2)$$

where the values K and Ω are defined in Appendix 3, and are shown for various materials in Table 12.

Table 12.

Material	C	Al	Fe	Pb	G₅ emulsion	Air
Z	6	13	26	82		
K (MeV) . .	19.2	19.5	19.6	19.7	19.7	19.3
Ω	15.4	14.9	14.3	12.9	14.0	15.2

Sect. 23. Derivation of the analytic solutions. 45

The solution of eq. (23.2) can now be obtained in a similar way as in the preceding section. As an example, we present here the derivation of the angular structure function in Approximation A. The extension of this treatment to the solutions in Approximation B, and to the lateral structure function, can easily be made.

Putting $\varepsilon = 0$ and $\boldsymbol{x} = 0$ in eq. (23.2), we get

$$\boldsymbol{L}' f = -\frac{K^2 \zeta^2}{4 E^2}\left(1 - \frac{1}{\Omega} \log \frac{K^2 \zeta^2}{4 E^2}\right) f, \tag{23.3}$$

where f is the Bessel transform of the angular structure function, and

$$\boldsymbol{L}' = \frac{1}{\frac{\partial}{\partial t} + \sigma_0} \left[\frac{\partial^2}{\partial t^2} + (A' + \sigma_0) \frac{\partial}{\partial t} + (A' \sigma_0 - B' C')\right]. \tag{23.4}$$

The solution in a power series of $K^2 \zeta^2 / 4 E^2$ and $K^2 \zeta^2 / 4 E^2 \log K^2 \zeta^2 / 4 E^2$, is

$$f = \frac{1}{4\pi^2 i} \int_{-i\infty}^{+i\infty} ds \left(\frac{E_0}{E}\right)^s \frac{1}{E} \sum_{n=0}^{\infty} \sum_{m=0}^{\infty} (-)^n \left(\frac{K^2 \zeta^2}{4 E^2}\right)^{n+m} \left(\frac{1}{\Omega} \log \frac{K^2 \zeta^2}{4 E^2}\right)^m \psi_{n,m}, \tag{23.5}$$

where the function $\psi_{n,m}$ can be determined by the following procedure: Substitution of (23.5) into (23.3) yields,

$$\left(\frac{K^2 \zeta^2}{4 E^2}\right)^{n+m} \left(\frac{1}{\Omega} \log \frac{K^2 \zeta^2}{4 E^2}\right)^m = \left(\frac{1}{\Omega} \frac{\partial}{\partial n'}\right)^m \left(\frac{K^2 \zeta^2}{4 E^2}\right)^{n'+m}\bigg|_{n'=n},$$

since

$$\left.\begin{aligned}
& \boldsymbol{L}' \frac{(-)^n}{E^{s+1}} \left(\frac{K^2 \zeta^2}{4 E^2}\right)^{n+m} \left(\frac{1}{\Omega} \log \frac{K^2 \zeta^2}{4 E^2}\right)^m \psi_{n',m} \\
&= \left(-\frac{1}{\Omega} \frac{\partial}{\partial n'}\right)^m \boldsymbol{L}' \frac{1}{E^{s+1}} \left(\frac{-K^2 \zeta^2}{4 E^2}\right)^{n'+m} \psi_{n,m}\bigg|_{n'=n} \\
&= \frac{1}{E^{s+1}} \left(\frac{-K^2 \zeta^2}{4 E^2}\right)^{n+m} \left(-\frac{1}{\Omega}\right)^m \left(\frac{\partial}{\partial n'} + \log \frac{K^2 \zeta^2}{4 E^2}\right)^m \boldsymbol{L}^{(0)}(s + 2n' + 2m)\bigg|_{n'=n} \psi_{n,m} \\
&= \frac{1}{E^{s+1}} \left(-\frac{K^2 \zeta^2}{4 E^2}\right)^{n+m} \left(-\frac{1}{\Omega}\right)^m \sum_{l=0}^{m} {}_m C_l \left(\log \frac{K^2 \zeta^2}{4 E^2}\right)^l \left(\frac{\partial}{\partial n'}\right)^{m-l} \boldsymbol{L}^{(0)}(s + 2n' + 2m)\bigg|_{n'=n} \psi_{n,m},
\end{aligned}\right\} \tag{23.6}$$

where

$$\boldsymbol{L}^{(0)}(s + 2n' + 2m) = \frac{1}{\frac{\partial}{\partial t} + \sigma_0} \left(\frac{\partial}{\partial t} + \lambda_1(s + 2n' + 2m)\right) \left(\frac{\partial}{\partial t} + \lambda_2(s + 2n' + 2m)\right).$$

Putting $n + m = N$ and $(\partial/\partial n')^{m-l} \boldsymbol{L}(s + 2n' + 2m) = \boldsymbol{L}^{(m-l)}$, the series part of $\boldsymbol{L}' f$ can be written as

$$\sum_{m=0}^{\infty} \sum_{N=m}^{\infty} \sum_{l=0}^{m} \left(\frac{-K^2 \zeta^2}{4 E^2}\right)^N \left(-\frac{1}{\Omega}\right)^m {}_m C_l \left(\log \frac{K^2 \zeta^2}{4 E^2}\right)^l \boldsymbol{L}^{(m-l)} \psi_{N-m,m}. \tag{23.7}$$

Now changing the order of the triple summations appearing in the above formula, as

$$\sum_{m=0}^{\infty} \sum_{N=m}^{\infty} \sum_{l=0}^{m} = \sum_{N=0}^{\infty} \sum_{m=0}^{N} \sum_{l=0}^{m} = \sum_{N=0}^{\infty} \sum_{l=0}^{N} \sum_{m=l}^{N},$$

the formula (23.7) becomes

$$\sum_{N=0}^{\infty} \sum_{l=0}^{N} \sum_{m=0}^{N-l} \left(-\frac{K^2 \zeta^2}{4 E^2}\right)^N \left(-\frac{1}{\Omega}\right)^{m'+l} {}_{m'+l} C_l \left(\log \frac{K^2 \zeta^2}{4 E^2}\right)^l \boldsymbol{L}^{(m')} \psi_{N-m'-l,m'+l}, \tag{23.8}$$

where we have replaced $m - l$ by m'. Substituting the solution (23.5) into Eq. (23.3), and referring to the results (23.8) and equating the coefficients of the term of the products of respective powers of $K^2 \zeta^2 / 4 E^2$ and $\log (K^2 \zeta^2 / 4 E^2)$, we have the recurrence formula, by putting m' to n,

$$\sum_{n=0}^{N-l} \left(-\frac{1}{\Omega}\right)^n {}_{n+l} C_l \boldsymbol{L}^{(n)} \psi_{N-n-l,n+l} = \psi_{N-l-1,l} + \psi_{N-l,l-1}, \tag{23.9}$$

with a boundary condition given by ψ_{00}. Now (23.5) is written in an integral form as

$$\left.\begin{array}{l}\displaystyle\sum_{n=0}^{\infty}\sum_{m=0}^{\infty}\left(\frac{-K^2\zeta^2}{4E^2}\right)^n\frac{1}{\Omega^m}\left(\frac{K^2\zeta^2}{4E^2}\right)^m\left(\log\frac{K^2\zeta^2}{4E^2}\right)^m\psi_{m,n}\\ \displaystyle =\frac{1}{2\pi i}\int_C dp\,\Gamma(-p)\left(\frac{K^2\zeta^2}{4E^2}\right)^p\sum_{m=0}^{\infty}\frac{1}{m!\Omega^m}\left(\frac{K^2\zeta^2}{4E^2}\right)^m\left(\log\frac{K^2\zeta^2}{4E^2}\right)^m\mathfrak{M}_1(p,m),\end{array}\right\} \quad (23.10)$$

where we put

$$\psi_{n,m}=\frac{\mathfrak{M}_1(n,m)}{n!\,m!}. \quad (23.11)$$

Then the difference equation defining $\mathfrak{M}_1(p,m)$ is obtained by the substitution of (23.11) into (23.9):

$$\left.\begin{array}{l}\displaystyle\sum_{n=0}^{p}{}_{n+m}C_n\left(-\frac{1}{\Omega}\right)^n L^{(n)}(s+2p+2m)\,\mathfrak{M}_1(p-n,m+n)\\ =p\,\mathfrak{M}_1(p-1,m)+m\,\mathfrak{M}_1(p,m-1),\end{array}\right\} \quad (23.12)$$

with appropriate initial conditions for $\mathfrak{M}_1(0,0)$ which are the same as those given in the preceding section.

By performing the inverse Hankel transform of (23.10), we get the angular structure function

$$\pi_1=\pi_1^{(0)}+\frac{1}{\Omega}\pi_1^{(1)}+\frac{1}{\Omega^2}\pi_1^{(2)}+\cdots, \quad (23.13)$$

where

$$\pi_1^{(0)}=-\frac{1}{4\pi^3}\int\!\!\!\int_{-i\infty}^{i\infty}ds\,dp\left(\frac{E_0}{E}\right)^s\frac{1}{E}\left(\frac{E}{K}\right)^2\left(\frac{E^2\theta^2}{K^2}\right)^{-p-1}\Gamma(p+1)\cdot\mathfrak{M}_1(p,0,s,t) \quad (23.14)$$

and

$$\left.\begin{array}{l}\displaystyle\pi_1^{(1)}=\frac{1}{4\pi^3}\int\!\!\!\int_{-i\infty}^{i\infty}ds\,dp\left(\frac{E_0}{E}\right)^s\frac{1}{E}\left(\frac{E}{K}\right)^2\left(\frac{E^2\theta^2}{K^2}\right)^{-p-1}\Gamma(p+1)p\times\\ \displaystyle\times\left\{\Psi(p+1)+\Psi(-p)-\log\frac{E^2\theta^2}{K^2}\right\}\mathfrak{M}_1(p-1,1,s,t).\end{array}\right\} \quad (23.15)$$

Here we make use of the relation

$$\int_0^{\infty}\left(\frac{K^2\zeta^2}{4E^2}\right)^P J_0(\zeta\theta)\zeta\,d\zeta=\frac{2\Gamma(p+1)}{\Gamma(-p)}\left(\frac{E^2\theta^2}{K^2}\right)^{-p-1}\cdot\frac{E^2}{K^2} \quad (23.16)$$

and

$$\left.\begin{array}{l}\displaystyle\int_0^{\infty}\left(\frac{K^2\zeta^2}{4E^2}\right)^P\left(\frac{K^2\zeta^2}{4E^2}\right)^m\log\left(\frac{K^2\zeta^2}{4E^2}\right)^m J_0(\zeta,\theta)\zeta\,d\zeta\\ \displaystyle=\frac{\partial^m}{\partial p^m}\int_0^{\infty}\left(\frac{K^2\zeta^2}{4E^2}\right)^{p+m}J_0(\zeta\theta)\zeta\,d\zeta\\ \displaystyle=\frac{2\partial^m}{\partial p^m}\frac{\Gamma(p+m+1)}{\Gamma(-p-m)}\left(\frac{E^2\theta^2}{K^2}\right)^{-p-m-1}\cdot\frac{E^2}{K^2}.\end{array}\right\} \quad (23.17)$$

For $m=1$ this expression becomes

$$\frac{E^2}{K^2}\frac{2\Gamma(p+2)}{\Gamma(-p-1)}\left(\frac{E^2\theta^2}{K^2}\right)^{-p-2}\left[\Psi(P+2)+\Psi(-p-1)-\log\frac{E^2\theta^2}{K^2}\right], \quad (23.18)$$

where Ψ is the logarithmic derivative of the Gamma function, and (23.15) is obtained by shifting the parameter p to the left by one unit. As shown in

Appendix 3, the series solution, (23.13), just corresponds to the one given in the scattering theory of MOLIÈRE for a single charged particle. As in MOLIÈRE's theory, the first term of the series (23.13) represents the spread due to multiple scattering of electrons when they are traversing the material, and is just the same function as the solution derived under the Landau approximation except for the slight differences in the definitions of K and $\mathfrak{M}$.

In the following sections the analytic solutions of the lateral and angular structure functions are summarized.

24. The structure functions of electrons and photons.

α) *Angular structure functions for electrons and photons in Approximation B.* The structure functions are expanded as follows;

$$\pi_1(E_0, E, \theta, t) = \pi_1^{(0)} + \frac{1}{\Omega} \pi_1^{(1)} + \cdots, \tag{24.1}$$

$$\gamma_1(E_0, E, \theta, t) = \gamma_1^{(0)} + \frac{1}{\Omega} \gamma_1^{(1)} + \cdots, \tag{24.2}$$

$$\Pi_1(E_0, 0, \theta, t) = \Pi_1^{(0)} + \frac{1}{\Omega} \Pi_1^{(1)} + \cdots, \tag{24.3}$$

where π_1 and γ_1 are the angular structure functions of electrons and photons with energy $(E, E+dE)$ as defined in the Sect. 21.

These structure functions are given by:

$$\left.\begin{aligned}\pi_1^{(0)} = -\frac{1}{8\pi^4 i} \int\!\!\!\int\!\!\!\int_{-i\infty}^{+i\infty} ds\, dp\, dq \left(\frac{E_0}{E}\right)^s \frac{1}{E}\left(\frac{E}{K}\right)^2 \left(\frac{\varepsilon}{E}\right)^q \left(\frac{E^2\theta^2}{K^2}\right)^{-p-1} \times \\ \times \Gamma(p+1)\Gamma(-q)\, \mathfrak{M}_1(p, 0, q, s, t),\end{aligned}\right\} \tag{24.4}$$

$$\left.\begin{aligned}\gamma_1^{(0)} = -\frac{1}{8\pi^4 i} \int\!\!\!\int\!\!\!\int_{-i\infty}^{+i\infty} ds\, dp\, dq \left(\frac{E_0}{E}\right)^s \frac{1}{E}\left(\frac{E}{K}\right)^2 \left(\frac{\varepsilon}{E}\right)^q \left(\frac{E^2\theta^2}{K^2}\right)^{-p-1} \times \\ \times \Gamma(p+1)\Gamma(-q)\, \mathfrak{M}_1'(p, 0, q, s, t),\end{aligned}\right\} \tag{24.5}$$

$$\left.\begin{aligned}\Pi_1^{(0)} = -\frac{1}{4\pi^3} \int\!\!\!\int_{-i\infty}^{+i\infty} ds\, dp \left(\frac{E_0}{\varepsilon}\right)\left(\frac{\varepsilon}{K}\right)^2 \left(\frac{\varepsilon^2\theta^2}{K^2}\right)^{-p-1} \times \\ \times \Gamma(p+1)\Gamma(2p+s)\, \mathfrak{M}_1(p, 0, s-2p, s, t),\end{aligned}\right\} \tag{24.6}$$

and

$$\left.\begin{aligned}\pi_1^{(1)} = \frac{1}{8\pi^4 i} \int\!\!\!\int\!\!\!\int_{-i\infty}^{+i\infty} ds\, dp\, dq \left(\frac{E_0}{E}\right)^s \frac{1}{E}\left(\frac{E}{K}\right)^2 \left(\frac{\varepsilon^2\theta^2}{K^2}\right)^{-p-1}\left(\frac{\varepsilon}{E}\right)^q \times \\ \times (p)\Gamma(p+1)\Gamma(-q) \times \\ \times \left\{\Psi(p+1)+\Psi(-p)-\log\left(\frac{E^2\theta^2}{K^2}\right)\right\} \mathfrak{M}_1(p-1, 1, q, s, t),\end{aligned}\right\} \tag{24.7}$$

$$\left.\begin{aligned}\gamma_1^{(1)} = \frac{1}{8\pi^4 i} \int\!\!\!\int\!\!\!\int_{-i\infty}^{+i\infty} ds\, dp\, dq \left(\frac{E_0}{E}\right)^s \frac{1}{E}\left(\frac{E}{K}\right)^2 \left(\frac{E^2\theta^2}{K^2}\right)^{-p-1}\left(\frac{\varepsilon}{E}\right)^q \times \\ \times (p)\Gamma(p+1)\Gamma(-q) \times \\ \times \left\{\Psi(p+1)+\Psi(-p)-\log\left(\frac{E^2\theta^2}{K^2}\right)+\frac{\partial}{\partial p}\log(C(s+2p+q))\right\} \times \\ \times \mathfrak{M}_1'(p-1, 1, q, s, t),\end{aligned}\right\} \tag{24.8}$$

$$\Pi_1^{(1)} = \frac{1}{4\pi^3} \int\!\!\int_{-i\infty}^{+i\infty} ds\, dp \left(\frac{E_0}{\varepsilon}\right)^s \left(\frac{\varepsilon}{K}\right)^2 \left(\frac{\varepsilon^2\theta^2}{K^2}\right)^{-p-1} \times$$
$$\times (p)\, \Gamma(p+1)\, \Gamma(2p+s) \times$$
$$\times \left[\left\{\Psi(p+1) + \Psi(-p) - \log\left(\frac{E^2\theta^2}{K^2}\right) + 2\Psi(2p+s)\right\}\right.$$
$$\left. \times \mathfrak{M}_1(p-1, 1, -2p-s, s, t) - 2 \frac{\partial}{\partial q}\mathfrak{M}_1(p-1, 1, q, s, t)\Big|_{q=-2p-s}\right],$$
\hfill (24.9)

where $\mathfrak{M}_1$ and $\mathfrak{M}_1'$ are defined by

$$\sum_{n=0}^{P} {}_{n+u,\,n}C \left(-\frac{1}{\Omega}\right)^n L^{(n)}(s+2p+2u+q)\, \mathfrak{M}_1(p-n, n+u, q, s, t)$$
$$= p\, \mathfrak{M}_1(p-1, u, q, s, t) + u\, \mathfrak{M}_1(p, u-1, q, s, t) + (s+2p+2u+q) \times$$
$$\times q\, \mathfrak{M}_1(p, u, q-1, s, t) - \frac{1}{\Omega} p q\, \mathfrak{M}_1(p-1, u+1, q-1, s, t),$$
\hfill (24.10)

and

$$\mathfrak{M}_1'(p, u, q, s, t) = \int_0^t C(s+2p+2u+q)\, e^{-\sigma_0(t-t')}\, \mathfrak{M}_1(p, u, q, s, t')\, dt' \quad (24.11)$$

with following boundary conditions:

for a shower from a primary electron

$$\mathfrak{M}_1(0, 0, 0, s, t) = H_1(s)\, e^{\lambda_1(s)t} + H_2(s)\, e^{\lambda_2(s)t}, \quad (24.12)$$

and for a shower from a primary photon

$$\mathfrak{M}_1(0, 0, 0, s, t) = \frac{B(s)}{\lambda_1(s) - \lambda_2(s)} \left\{e^{\lambda_1(s)t} - e^{\lambda_2(s)t}\right\}. \quad (24.13)$$

β) *Lateral structure functions for electrons and photons in Approximation B.*
The structure functions are also expanded as shown above, and are given by

$$\pi_2 = \pi_2^{(0)} + \frac{1}{\Omega} \pi_2^{(1)} + \cdots, \quad (24.14)$$

$$\gamma_2 = \gamma_2^{(0)} + \frac{1}{\Omega} \gamma_2^{(1)} + \cdots, \quad (24.15)$$

$$\Pi_2 = \Pi_2^{(0)} + \frac{1}{\Omega} \Pi_2^{(1)} + \cdots, \quad (24.16)$$

where π_2 and γ_2 are the structure functions of electrons and photons with energy $(E, E+dE)$, and Π_2 the integral lateral structure functions of the electrons.
These functions are given by

$$\pi_2^{(0)} = -\frac{1}{8\pi^4 i} \int\!\!\int\!\!\int_{-i\infty}^{+i\infty} ds\, dp\, dq \left(\frac{E_0}{E}\right)^s \frac{1}{E}\left(\frac{E}{K}\right)^2 \left(\frac{\varepsilon}{E}\right)^q \left(\frac{E^2 r^2}{K^2}\right)^{-p-1} \times$$
$$\times \Gamma(p+1)\, \Gamma(-q)\, \mathfrak{M}_2(p, 0, q, s, t),$$
\hfill (24.17)

$$\gamma_2^{(0)} = -\frac{1}{8\pi^4 i} \int\!\!\int\!\!\int_{-i\infty}^{+i\infty} ds\, dp\, dq \left(\frac{E_0}{E}\right)^s \frac{1}{E}\left(\frac{E}{K}\right)^2 \left(\frac{\varepsilon}{E}\right)^q \left(\frac{E^2 r^2}{K^2}\right)^{-p-1} \times$$
$$\times \Gamma(p+1)\, \Gamma(-q)\, \mathfrak{M}_2'(p, 0, q, s, t),$$
\hfill (24.18)

$$\Pi_2^{(0)} = -\frac{1}{4\pi^3} \int\!\!\int_{-i\infty}^{+i\infty} ds\, dp \left(\frac{E_0}{E}\right)^s \left(\frac{\varepsilon}{K}\right)^2 \left(\frac{\varepsilon^2 r^2}{K^2}\right)^{-p-1} \times$$
$$\times \Gamma(p+1)\, \Gamma(2p+s)\, \mathfrak{M}_2(p, 0, -s-2p, s, t),$$
\hfill (24.19)

Sect. 24. The structure functions of electrons and photons.

and

$$\pi_2^{(1)} = \frac{1}{8\pi^4 i} \int_{-i\infty}^{+i\infty}\int\int ds\, dp\, dq \left(\frac{E_0}{E}\right)^s \frac{1}{E}\left(\frac{E}{K}\right)^2 \left(\frac{E^2 r^2}{K^2}\right)^{-p-1} \left(\frac{\varepsilon}{E}\right)^q \times$$
$$\times (p)\Gamma(p+1)\Gamma(-q) \times$$
$$\times \left\{\Psi(p+1)+\Psi(-p)-\log\left(\frac{E^2 r^2}{K^2}\right)\right\} \mathfrak{M}_2(p-1, 1, q, s, t), \qquad (24.20)$$

$$\gamma_1^{(1)} = \frac{1}{8\pi^4 i} \int_{-i\infty}^{+i\infty}\int\int ds\, dp\, dq \left(\frac{E_0}{E}\right)^s \frac{1}{E}\left(\frac{E}{K}\right)^2 \left(\frac{E^2 r^2}{K^2}\right)^{-p-1} \left(\frac{\varepsilon}{E}\right)^q \times$$
$$\times (p)\Gamma(p+1)\Gamma(-q) \times$$
$$\times \left\{\Psi(p+1)+\Psi(-p)-\log\left(\frac{E^2 r^2}{K^2}\right) + \right.$$
$$\left. + \frac{\partial}{\partial p}\log C(s+2p+q)\right\} \mathfrak{M}_2'(p, 1, q, s, t), \qquad (24.21)$$

$$\Pi_2^{(1)} = \frac{1}{4\pi^3} \int_{-i\infty}^{+i\infty}\int ds\, dp \left(\frac{E_0}{\varepsilon}\right)^s \left(\frac{\varepsilon}{K}\right)^2 \left(\frac{\varepsilon^2 r^2}{K^2}\right)^{-p-1} \times$$
$$\times (p)\Gamma(p+1)\Gamma(2p+s) \times$$
$$\times \left[\left\{\Psi(p+1)+\Psi(-p)-\log\left(\frac{\varepsilon^2 r^2}{K^2}\right)+2\Psi(2p+s)\right\} \times \right.$$
$$\left. \times \mathfrak{M}_2(p-1, 1, -2p-s, s, t) - 2\frac{\partial}{\partial q}\mathfrak{M}_2(p-1, 1, q, s, t)\right] \qquad (24.22)$$
$$q = -2p-s,$$

where $\mathfrak{M}_2$ and $\mathfrak{M}_2'$ are defined by

$$\sum_{n=0}^{P} {}_{n+u}C_n\left(-\frac{1}{\Omega}\right)^n L^{(n)}(s+2p+2u+q) M(p-n, n+u, q, s, \xi-t, t)$$
$$= p(\xi-t)^2\left(1-\frac{1}{\Omega}\log(\xi-t)^2\right) M(p-1, u, q, s, \xi-t, t) +$$
$$+ u(\xi-t)^2 M(p, u-1, q, s, \xi-t, t) +$$
$$+ (s+2p+2u+q) q M(p, u, q-1, s, \xi-t, t) -$$
$$- \frac{2}{\Omega} p q M(p-1, u+1, q-1, s, \xi-t, t), \qquad (24.23)$$

with

$$\mathfrak{M}_2(p, u, q, s, t) = \lim_{(\xi-t)\to 0} M(p, u, q, s, \xi-t, t) \qquad (24.24)$$

and

$$\mathfrak{M}_2' = \int_0^t C(s+2p+2u+q) e^{-\sigma_0(t-t')} M(p, u, q, s, t-t', t')\, dt', \qquad (24.25)$$

with the boundary conditions (24.12) and (24.13), as in the previous case.

III. Behaviour of the structure function and its numerical evaluation.

Since the analytic solutions for the structure function were derived in the preceding sections, all the features of the lateral and the angular structure of the shower are now determined with in the limits of our approximation. However, for practical applications it is desirable that the analytic solutions be approximated

Handbuch der Physik, Bd. XLVI/2.

by simpler expressions. In the following section we shall describe and discuss a number of attempts in this direction, referring to

(i) simplified expressions for the structure functions,

(ii) the mean square deviations of the lateral and angular structure functions,

(iii) a numerical evaluation of the integral lateral structure function of electrons for $E_0 = \infty$,

(iv) the behaviour of the structure function near the shower axis, and

(v) the behaviour of the structure function at large distances from the shower axis.

25. Simplified expressions for the structure functions (contribution of terms other than $e^{\lambda_1(s)t}$).

Since the functions $\mathfrak{M}$ and $\mathfrak{M}'$ appearing in the complex integral of the lateral structure function are given by difference equations of the form (21.11), they can be expressed by a series, each term of which contains the exponentials $e^{\lambda_1(s)t}$, $e^{\lambda_1(s+2)t}, \ldots e^{\lambda_2(s)t}, \ldots$ This can be shown by an argument similar to that used in the section on one-dimensional electron shower theory under Approximation B.

In order to compare their contributions, we shall examine them by considering the first term of the lateral structure function, $\pi_2^{(0)}$. To avoid mathematical complications, we shall treat it in Approximation A. From (24.17) we obtain this term as[1]

$$\pi_2^{(0)} = -\frac{1}{4\pi^3} \iint ds\, dp \left(\frac{E_0}{E}\right)^s \frac{1}{E} \left(\frac{E}{K}\right)^2 \left(\frac{E^2 r^2}{K^2}\right)^{-p-1} \Gamma(p+1)\, \mathfrak{M}(p, 0, s, t), \quad (25.1)$$

and $\mathfrak{M}(p, 0, s, t)$ is given by a solution of the difference equation

$$\boldsymbol{L}'(s+2p)\, M(p, 0, s, t) = p(\xi - t)^2 M(p, s, t)$$

with

$$\lim_{(\xi - t) \to 0} M = \mathfrak{M},$$

where we have used the abbreviation $M(p, 0, s, \xi - t, t) = M(p, s, t)$. Now we write $M(p, s, t)$ in a series of $e^{\lambda_1(s)t}, e^{\lambda_1(s+2)t}, \ldots, e^{\lambda_2(s)t}, e^{\lambda_2(s+2)t}, \ldots$, as

$$\begin{aligned}\mathfrak{M}(p, 0, s, t) &\cong \mathfrak{M}_{1,0} e^{\lambda_1(s)t} + \mathfrak{M}_{1,1} e^{\lambda_1(s+2)t} + \cdots + \\ &+ \mathfrak{M}_{2,0} e^{\lambda_2(s)t} + \mathfrak{M}_{2,1} e^{\lambda_2(s+2)t} + \cdots\end{aligned} \quad (25.2)$$

As already mentioned in the Sect. 11, the following relation holds:

$$\lambda_1(s) > \lambda_1(s+2) > \lambda_1(s+4) > \cdots > \lambda_2(s) > \lambda_2(s+2) > \cdots.$$

Therefore the main contribution is expected to derive from the term with $e^{\lambda_1(s)t}$, the next in importance from the term with $e^{\lambda_1(s+2)t}$, and so on.

Thus we have approximately

$$\begin{aligned}\pi_2^{(0)} = -\frac{1}{4\pi^3} \iint ds\, dp \left(\frac{E_0}{E}\right)^s \frac{1}{E} \left(\frac{E}{K}\right)^2 \left(\frac{E^2 r^2}{K^2}\right)^{-p-1} \Gamma(p+1) \times \\ \times [\mathfrak{M}_{1,0}(p, 0, s)\, e^{\lambda_1(s)t} + \mathfrak{M}_{1,1}(p, 0, s,)\, e^{\lambda_1(s+2)t} + \cdots] \\ \text{for } t \geq 1.\end{aligned} \quad (25.3)$$

As shown in Appendix 4, $\mathfrak{M}_{1,1}$ can be written in the form

$$\mathfrak{M}_{1,1}(p, 0, s) = p F_{1,1}(s, t)\, \mathfrak{M}_{1,0}(p-1, 0, s+2), \quad \text{(A.4.50)}$$

---

[1] Here we replace the symbol $\mathfrak{M}_2$ in (24,17) by $\mathfrak{M}$ for simplicity.

where $F_{1,1}(s,t)$ is a function determined by the initial condition, and for an electron-initiated shower is given by

$$F_{1,1}(s,t) = \left(t + \frac{d}{d\mu}\right)^2 \left[\frac{\mu + \sigma_0}{(\mu - \lambda_1(s))(\mu - \lambda_2(s))}\right]_{\mu = \lambda_1(s+2)}. \quad (A.4.53)$$

The corresponding $F(s,t)$ for different boundary conditions are listed in Appendix 4.

Substituting (A.4.50) into the second term of the integral of (25.3), and using $(p+1)$ instead of the integration parameter p in the second term, we can rewrite (25.3) as

$$\pi_2^{(0)} = -\frac{1}{4\pi^3} \iint ds\,dp \left(\frac{E_0}{E}\right)^s \frac{1}{E} \left(\frac{E}{K}\right)^2 \left(\frac{E^2 r^2}{K^2}\right)^{-p-1} \Gamma(p+1) \times$$
$$\times \left[\mathfrak{M}_{1,0}(p,0,s) e^{\lambda_1(s)t} + \frac{K^2(p+1)^2}{E^2 r^2} \mathfrak{M}_{1,0}(p\ s+2) F_{1,1}(s,t) e^{\lambda_1(s+2)t} + \cdots\right]. \quad (25.3')$$

Similarly replacing the integral variable s in the second term by $(s-2)$, our expression becomes.

$$\pi_2^{(0)} = -\frac{1}{4\pi^3} \iint ds\,dp \left(\frac{E_0}{E}\right)^s \frac{1}{E} \left(\frac{E}{K}\right)^2 \left(\frac{E^2 r^2}{K^2}\right)^{-p-1} \Gamma(p+1) \times$$
$$\times \mathfrak{M}_{1,0}(p,0,s) e^{\lambda_1(s)t} \left[1 + \left(\frac{K}{E_0 r}\right)^2 (p+1)^2 F_{1,1}(s-2,t) + \cdots\right]. \quad (25.3'')$$

The numerical value of $F_{1,1}(s,t)$ in (25.3″) is shown in Fig. 10. The result leads to the conclusion that the values of the second term in the bracket are of the order of $10^{-2} \cdot t^2 (K/E_0 r)^2$.

The average energy of shower particles at a distance r from the shower axis is about K/r. Since at the shower maximum the number of shower particles, given by (14.9), is smaller than E_0/E, it follows that in order to find at least several electrons within a circle of radius r, we must have $E_0 r/K$ larger than a certain value, say at least 10. Now the depth of the shower maximum is approximately given by

$$t \sim \log E_0 r/K \quad (25.4)$$

so that the contribution of the second term becomes of the order of

$$10^{-2} (p+1)^2 \times$$
$$\times \left(\frac{K}{E_0 r} \log \frac{E_0 r}{K}\right)^2. \quad (25.5)$$

This is practically negligible compared with unity, and in sufficient accuracy, we can express (25.3) in the much simpler form

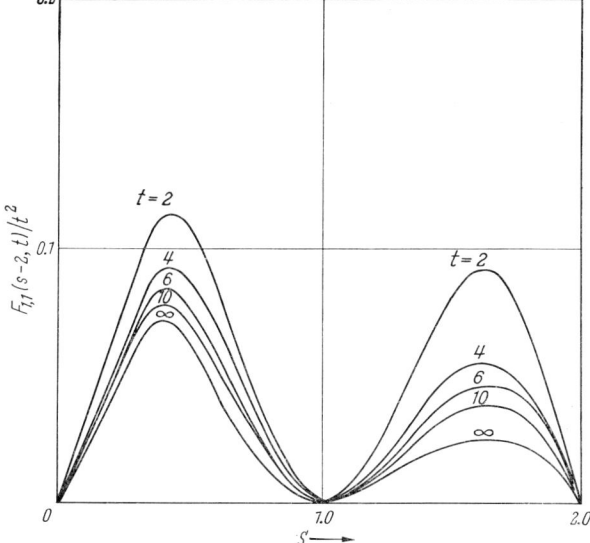

Fig. 10. Numerical values of $F_{1,1}(s-2,t)/t^2$.

$$\pi_2^{(0)} \cong -\frac{1}{4\pi^3} \iint ds\,dp \left(\frac{E_0}{E}\right)^s \frac{1}{E} \left(\frac{E_0 r}{K}\right)^{-p-1} \Gamma(p+1) \mathfrak{M}_{1,0}(p,0,s) e^{\lambda_1(s)t}. \quad (25.6)$$

However, this expression cannot be applied to the problem at the very beginning of the shower, where the contribution of the term with $e^{\lambda_2(s)t}$ must not be neglected.

A similar expression is also obtained for the solution in Approximation B, as shown in Appendix 4, in which we have in place of (25.3″)

$$\pi_2^{(0)} = -\frac{1}{8\pi^4 i}\iiint ds\, dp\, dq \left(\frac{E_0}{E}\right)^s \frac{1}{E} \frac{E^2}{K^2}\left(\frac{E^2 r^2}{K^2}\right)^{-p-1}\left(\frac{\varepsilon}{E}\right)^q \times \\
\times \Gamma(p-1)\Gamma(-q)\,\mathfrak{M}_{1,0}(p,q,s)\, e^{\lambda_1(s)t} \times \\
\times \left[1+O\left(\frac{\varepsilon}{E_0}\right)+\cdots\right]\left[1+\left(\frac{K}{E_0 r}\right)^2 (p+1)^2 F_{1,1}(s-2,t)+\cdots\right], \qquad (25.7)$$

and for the second term of the solution without the Landau approximation

$$\pi_2^{(1)} = \frac{1}{8\pi^4 i}\iiint ds\, dp\, dq \left(\frac{E_0}{E}\right)^s \frac{1}{E}\left(\frac{E^2}{K^2}\right)\left(\frac{E^2 r^2}{K^2}\right)^{-p-1}\left(\frac{\varepsilon}{E}\right)^q \Gamma(p+1)\,(p)\times \\
\times \Gamma(-q)\left[\Psi(p+1)+\Psi(-p)-\log\frac{E^2 r^2}{K^2}\right]\mathfrak{M}_{1,0}(p-1,1,q,s)\, e^{\lambda_1(s)t}\times \\
\times \left[1+O\left(\frac{\varepsilon}{E_0}\right)+\cdots\right]\left\{\left[1+\left(\frac{K}{E_0 r}\right)^2 (p+1)^2 F_{1,1}(s-2,t)+\right.\right. \\
\left.\left.+\left(\frac{K}{E_0 r}\right)^2 \frac{p+1}{p} F_{1,1}(s-2,t)+\cdots\right]\right\}. \qquad (25.8)$$

As in the previous case, it can be argued that the first term of these integrals predominates, and gives sufficiently accurate approximations to the structure functions. Finally, it may be important to point out that if one integrates the structure function with respect to the lateral deviations r, i.e., in the integrals

$$\int_0^\infty \pi_{(2)}^{(0)} 2\pi r\, dr, \quad \int_0^\infty \pi_{(2)}^{(1)} 2\pi r\, dr, \quad \int_0^\infty \pi_{(2)}^{(2)} 2\pi r\, dr, \ldots \qquad (25.9)$$

the second, third, ..., integrals of the series give essentially no contributions[1]. Furthermore, we have from (25.3″)

$$\lim_{\delta \to 0} \int_\delta^\infty 2\pi r\, dr\, \pi_2^{(0)} \\
= 1\text{st term} + \lim_{\delta \to 0}\left\{\frac{1}{8\pi^3 i}\iiint ds\, dp\, dq \left(\frac{E_0}{E}\right)^s \frac{1}{E}\left(\frac{E}{E_0}\right)^2\left(\frac{E^2\delta^2}{K^2}\right)^{-p-1}\times \right. \\
\left. \times \left(\frac{\varepsilon}{E}\right)^q \frac{(p+1)^2}{p+1}\Gamma(p+1)\,\mathfrak{M}_{1,0}(p,q,s)\, F_{1,1}e^{\lambda_1(s)t}\right\}+\cdots \\
= 1\text{st term}+O(\delta^2), \qquad (25.10)$$

where the integral with respect to p in the second term is evaluated with the help of the pole at $p=-2$ of $(p+1)^2\Gamma(p+1)=(p+1)\Gamma(p+2)$. Therefore the first term in the structure function, $\pi_2^{(0)}$, gives the exact result (except for the term in $e^{\lambda_2(s)t}$) to the function of one-dimensional shower theory, which can be expected from the fact that in the one-dimentional shower theory no effect characteristic of the three-dimentional theory should appear, at least within the frame of our approximations.

---

[1] It must be remembered that

$$\int_0^\infty \pi_2^{(1)} 2\pi r\, dr = \lim_{\zeta \to 0} f_2^{(1)} \Rightarrow 0.$$

26. Mean square deviations of the lateral and angular structure function.

The mean square deviations of the lateral and angular structure functions give us measures of the spread of the shower particles of a certain energy. As all structure functions can be represented in a similar form, we shall illustrate the treatment by taking as an example the lateral electron structure functions.

As repeatedly mentioned, the second term in the series solution without the Landau approximation, $\pi^{(1)}$, represents the contribution of single scattering. We are taking the scattering probability without large-angle cut-off, so that it is approximately given by $\dfrac{\theta}{\theta^4} d\theta$ at large angles. Therefore the function $\pi^{(1)}$ behave should just as $1/\theta^3$ or $1/r^3$ at the tail of the shower, and the mean square deviation of this term diverges in a logarithmic way.

The first term, on the other hand, gives the contribution of multiple scattering and it gives finite values for the mean square deviations. Except for the slight differences of the definitions among the quantities of K, E_s, $\mathfrak{M}(p, u, q, s, t)$ and $\mathfrak{M}(p, q, s, t)$, our formula is the same as that of the Landau approximation. Therefore we use the solutions of the Landau approximation, and write the lateral structure functions (22.1) and (22.2) as

$$\pi_2 = -\frac{1}{8\pi^4 i} \iiint ds\, dp\, dq \left(\frac{E_0}{E}\right)^s \frac{1}{E} \left(\frac{E}{E_s}\right)^2 \left(\frac{E^2 r^2}{E_s^2}\right)^{-p-1} \left(\frac{\varepsilon}{E}\right)^q \times$$
$$\times \Gamma(p+1)\Gamma(-q)\mathfrak{M}_2(p, q, s, t),$$

$$\Pi_2 = -\frac{1}{8\pi^4 i} \iint \frac{ds\, dp\, dq}{s+2p+q} \left(\frac{E_0}{E}\right)^s \left(\frac{E}{E_s}\right)^2 \left(\frac{E^2 r^2}{E_s}\right)^{-p-1} \left(\frac{\varepsilon}{E}\right)^q \times$$
$$\times \Gamma(p+1)\Gamma(2p+s)\mathfrak{M}_2(p, -s-2p, s, t).$$

Then the mean square deviation of each function is

$$\langle r^2 \rangle = \frac{\int_0^\infty r^2 \Pi_2 2\pi r\, dr}{\int_0^\infty \Pi_2 2\pi r\, dr} = \frac{E_s^2}{E^2} \frac{\iint ds\, dq \left(\frac{E_0}{E}\right)^s \frac{1}{E} \left(\frac{\varepsilon}{E}\right)^q \Gamma(-q) \mathfrak{M}_2(1, q, s, t)}{\iint ds\, dq \left(\frac{E_0}{E}\right)^s \frac{1}{E} \left(\frac{\varepsilon}{E}\right)^q \Gamma(-q) \mathfrak{M}_2(0, q, s, t)}, \quad (26.1)$$

$$\langle r_T^2 \rangle = \frac{\int_0^\infty r^2 \Pi_2 2\pi r\, dr}{\int_0^\infty \Pi_2 2\pi r\, dr} = \frac{E_s^2}{E^2} \frac{\iint \frac{ds\, dq}{2+s+q} \left(\frac{E_0}{E}\right)^s \left(\frac{\varepsilon}{E}\right)^q \Gamma(-q) \mathfrak{M}_2(1, q, s, t)}{\iint \frac{ds\, dq}{s+q} \left(\frac{E_0}{E}\right)^s \left(\frac{\varepsilon}{E}\right)^q \Gamma(-q) \mathfrak{M}_2(0, q, s, t)}. \quad (26.2)$$

Since π_2 is expressed by (25.6), we have approximately

$$\langle r^2 \rangle \cong \frac{E_s^2}{E^2} \frac{\iint ds\, dq \left(\frac{E_0}{E}\right)^s \frac{1}{E} \left(\frac{\varepsilon}{E}\right)^q \Gamma(-q) \mathfrak{M}_{1,0}(1, q, s) e^{\lambda_1(s)t}}{\iint ds\, dq \left(\frac{E_0}{E}\right)^s \frac{1}{E} \left(\frac{\varepsilon}{E}\right)^q \Gamma(-q) \mathfrak{M}_{1,0}(0, q, s) e^{\lambda_1(s)t}}, \quad (26.3)$$

$$\langle r_T^2 \rangle \cong \frac{E_s^2}{E^2} \frac{\iint \frac{ds\, dq}{2+s+q} \left(\frac{E_0}{E}\right)^s \left(\frac{\varepsilon}{E}\right)^q \Gamma(-q) \mathfrak{M}_{1,0}(1, q, s) e^{\lambda_1(s)t}}{\iint \frac{ds\, dq}{s+q} \left(\frac{E_0}{E}\right)^s \left(\frac{\varepsilon}{E}\right)^q \Gamma(-q) \mathfrak{M}_{1,0}(0, q, s) e^{\lambda_1(s)t}} \quad (26.4)$$

from the same argument as before.

At the high-energy limit $E \gg \varepsilon$, where Approximation A is applicable, the integration with respect to q can be performed with the help of a pole in the function $\Gamma(-q)$, and gives

$$\langle r^2 \rangle = \frac{E_s^2}{E^2} \frac{\int ds \left(\frac{E_0}{E}\right)^s \frac{1}{E} \mathfrak{M}_{1,0}(1,0,s) e^{\lambda_1(s)t}}{\int ds \left(\frac{E_0}{E}\right)^s \frac{1}{E} \mathfrak{M}_{1,0}(0,0,s) e^{\lambda_1(s)t}}, \tag{26.5}$$

$$\langle r_T^2 \rangle = \frac{E_s^2}{E^2} \frac{\int \frac{ds}{2+s} \left(\frac{E_0}{E}\right)^s \mathfrak{M}_{1,0}(1,0,s) e^{\lambda_1(s)t}}{\int \frac{ds}{s} \left(\frac{E_0}{E}\right)^s \mathfrak{M}_{1,0}(1,0,s) e^{\lambda_1(s)t}}. \tag{26.6}$$

If, furthermore, one assumes $E_0 \to \infty$, we have exactly

$$\langle r^2 \rangle = \frac{E_s^2}{E^2} \frac{\mathfrak{M}_{1,0}(1,0,s)}{\mathfrak{M}_{1,0}(0,0,s)}, \tag{26.7}$$

$$\langle r_T^2 \rangle = \frac{s}{2+s} \frac{E_s^2}{E^2} \frac{\mathfrak{M}_{1,0}(1,0,s)}{\mathfrak{M}_{1,0}(0,0,s)}. \tag{26.8}$$

The formulae for the moments $\langle r^{2n} \rangle$ can also be obtained in a similar way, with the results

$$\langle r^{2n} \rangle = n! \left(\frac{E_s}{E}\right)^{2n} \frac{\mathfrak{M}_{1,0}(n,0,s)}{\mathfrak{M}_{1,0}(0,0,s)}, \tag{26.9}$$

$$\langle r_T^{2n} \rangle = n! \frac{s}{s+2n} \left(\frac{E_s}{E}\right)^{2n} \frac{\mathfrak{M}_{1,0}(n,0,s)}{\mathfrak{M}_{1,0}(0,0,s)}, \tag{26.10}$$

for $E_0 \gg E$, and $E \gg \varepsilon$.

The expressions for $\langle r^2 \rangle$ and $\langle r_T^2 \rangle$ have also been derived and numerically evaluated by ROBERG and NORDHEIM (9), BORSELLINO (29), EYGES and FERNBACH (10), GREEN and MESSEL (11) and GUZAVIN and IVANENKO (15), each with a different approach[1]. Numerical evaluations were carried out for the cases

(i) $E_0 = \infty$ in Approximation A, (26.7) and (26.8), as well as in Approximation B, (26.3) and (26.4).

and

(ii) $E_0 =$ finite in Approximation A, (26.6), using the saddle point method for numerical evaluation.

The results are shown in Table 13 and 14, and in Fig. 11.

Table 13. $\langle \theta^{2n} \rangle$ and $\langle r^{2n} \rangle$ values for electrons in Approximation A, in units of $(E_s/E)^{2n}$. [From GUZAVIN and I. P. IVANENKO (15)].

s	$\langle \theta^2 \rangle$	$\langle \theta^4 \rangle$	$\langle \theta^6 \rangle$	$\langle r^2 \rangle$	$\langle r^4 \rangle$	$\langle r^6 \rangle$
0.2	$2.61 \cdot 10^{-1}$	$2.31 \cdot 10^{-1}$	$4.21 \cdot 10^{-1}$	$3.82 \cdot 10^{-2}$	$2.72 \cdot 10^{-2}$	$8.90 \cdot 10^{-2}$
0.4	$3.65 \cdot 10^{-1}$	$4.28 \cdot 10^{-1}$	1.01	$1.15 \cdot 10^{-1}$	$1.91 \cdot 10^{-1}$	1.48
0.6	$4.46 \cdot 10^{-1}$	$6.18 \cdot 10^{-1}$	1.68	$2.37 \cdot 10^{-1}$	$7.35 \cdot 10^{-1}$	$1.19 \cdot 10^1$
0.8	$5.13 \cdot 10^{-1}$	$7.95 \cdot 10^{-1}$	2.39	$4.27 \cdot 10^{-1}$	2.35	$7.98 \cdot 10^1$
1.0	$5.69 \cdot 10^{-1}$	$9.57 \cdot 10^{-1}$	3.10	$7.23 \cdot 10^{-1}$	7.23	$4.96 \cdot 10^2$
1.2	$6.16 \cdot 10^{-1}$	1.10	3.76	1.21	$2.22 \cdot 10^1$	$2.86 \cdot 10^3$
1.4	$6.54 \cdot 10^{-1}$	1.23	4.36	2.02	$6.70 \cdot 10^1$	$1.52 \cdot 10^4$
1.6	$6.85 \cdot 10^{-1}$	1.33	4.89	3.41	$2.01 \cdot 10^2$	$7.66 \cdot 10^4$
1.8	$7.11 \cdot 10^{-1}$	1.42	5.34	5.82	$5.89 \cdot 10^2$	$3.66 \cdot 10^5$
2.0	$7.32 \cdot 10^{-1}$	1.49	5.69	9.70	$1.59 \cdot 10^3$	$1.55 \cdot 10^6$

---

[1] $\mathfrak{M}_{1,0}(1,0,s,t)$ is given by (21.12) as

$$\mathfrak{M}_{1,0}(1,0,s,t) = \int_0^t \mathfrak{M}_{1,0}(0,0,s+2,t-t')(t-t')^2 \mathfrak{M}_{1,0}(0,0,s,t') dt'.$$

It can easily be proved that the formulae derived here are identical with those obtained by ROBERG and NORDHEIM.

Sect. 26. Mean square deviations of the lateral and angular structure function. 55

Table 14. *Mean square average of the structure function for $s=1.0$ in Approximation B.*
Comparison of the result of ROBERG and NORDHEIM [9] (R-N) with those of KAMATA and NISHIMURA [8], (K-N), for $E_0 = \infty$.

$$\langle r^2 \rangle_{\text{av}} = \frac{\int_0^\infty r^2 \, \pi_2 \, 2\pi r \, dr}{\int_0^\infty \pi_2 \, 2\pi r \, dr} \quad \text{and} \quad \langle r_T^2 \rangle_{\text{av}} = \frac{\int_E^\infty dE \int_0^\infty r^2 \, \pi_2 \, 2\pi r \, dr}{\int_E^\infty dE \int_0^\infty \pi_2 \, 2\pi r \, dr}.$$

E/ε	$\langle r^2 \rangle_{\text{av}}$		$\langle r_T^2 \rangle_{\text{av}}$	
	R-N	K-N	R-N	K-N
10	0.642 $(E_s/E)^2$ 0.49	0.723 $(E_s/E)^2$ 0.58	0.214 $(E_s/E)^2$ 0.18	0.241 $(E_s/E)^2$ 0.20
7	0.46	0.53	0.17	0.18
5	0.43	0.46	0.16	0.17
3	0.40	0.36	0.14	0.14
2	0.33	0.28	0.125	0.11
1.5	0.30	0.22	0.105	0.090
1.0	0.25 $(E_s/E)^2$	0.15 $(E_s/E)^2$	0.085 $(E_s/E)^2$	0.064 $(E_s/E)^2$
0.75	0.28	0.21	0.12	0.087
0.5	0.39	0.32	0.165	0.125
0.4	0.46	0.40	0.19	0.16
0.3	0.06	0.53	0.235	0.19
0.2	0.86	0.82	0.31	0.26
0.15	1.13	1.02	0.48	0.31
0.10	1.6	1.31	0.46	0.39
0.05	2.8	1.58	0.64	0.53

For case (ii), finite E_0, the evaluation is carried out in the following way. Here we have from (26.5) and (26.6),

$$\langle r^2 \rangle = \frac{E_s^2}{E^2} \left(\frac{E_0}{E}\right)^{s_1-s_2} e^{\{\lambda_1(s_1)-\lambda_2(s_2)\}t} \frac{\mathfrak{M}_{1,0}(1, 0, s_1)}{\mathfrak{M}_{1,0}(0, 0, s_2)} \times \sqrt{\frac{\lambda_1''(s_2) t}{\lambda_1''(s_1) t + \left(\frac{\mathfrak{M}'}{\mathfrak{M}}\right)'_{p=1, s=s_1}}}, \quad (26.11)$$

$$\langle r_T^2 \rangle = \frac{E_s^2}{E^2} \frac{s_2}{s_1+2} \left(\frac{E_0}{E}\right)^{s_1-s_2} e^{\{\lambda_1(s_1)-\lambda_2(s_2)\}t} \frac{\mathfrak{M}_{1,0}(1, 0, s_1)}{\mathfrak{M}_{1,0}(0, 0, s_2)} \times$$

$$\times \sqrt{\frac{\lambda_1''(s_2) t + \frac{1}{(s_2)^2}}{\lambda_1''(s_1) t + \frac{1}{(s_1+2)^2} + \left(\frac{\mathfrak{M}'}{\mathfrak{M}}\right)'_{p=1, s=s_1}}}, \quad (26.12)$$

where s_1 and s_2 are determined by the equations

$$\left.\begin{aligned}\log \frac{E_0}{E} + \lambda_1'(s_1) t &= 0, \\ \log \frac{E_0}{E} + \lambda_1'(s_2) t + \frac{\mathfrak{M}_{1,0}'(1, 0, s_2)}{\mathfrak{M}_{1,0}(1, 0, s_2)} &= 0\end{aligned}\right\} \quad \text{for } \langle r^2 \rangle, \quad (26.13)$$

$$\left.\begin{aligned}\log \frac{E_0}{E} + \lambda_1'(s_1) t - \frac{1}{s_1} &= 0, \\ \log \frac{E_0}{E} + \lambda_1'(s_2) t - \frac{1}{s_2+2} + \frac{\mathfrak{M}_{1,0}'(1, 0, s)}{\mathfrak{M}_{1,0}(1, 0, s)} &= 0\end{aligned}\right\} \quad \text{for } \langle r_T^2 \rangle. \quad (26.14)$$

Since between the quantities $\mathfrak{M}_{1,0}(1, 0, s)$, $\mathfrak{M}_{1,0}(0, 0, s)$ and $\langle r^2 \rangle_{E_0=\infty}$ the relations

$$\mathfrak{M}_{1,0}(1, 0, s) = \mathfrak{M}_{1,0}(0, 0, s) \frac{E^2}{E_s^2} \langle r^2 \rangle_{E_0=\infty} \quad (26.15)$$

and
$$\mathfrak{M}_{1,0}(0, 0, s) = H_1(s), \tag{26.16}$$

hold, the behaviour of $\mathfrak{M}_{1,1}(1, s)$ vs s is quite similar to $\langle r^2 \rangle$ which is shown in Table 13, because $H_1(s)$ is a slowly varying function of s. From this Table it can be seen that the numerical value of $\mathfrak{M}_{1,0}(1, 0, s)$ is approximately proportional to s^2. Thus, $\mathfrak{M}'/\mathfrak{M}$ in (26.13) and (26.14) can be replaced by $2/s_2$, and we have for a fixed value of t,

$$\left. \begin{aligned} \{\lambda_1'(s_2) - \lambda_1'(s_1)\}t + \\ + \frac{2}{s_2} = 0 \quad \text{for} \quad \langle r^2 \rangle, \end{aligned} \right\} \tag{26.13'}$$

$$\left. \begin{aligned} \{\lambda_1'(s_2) - \lambda_1'(s_1)\}t + \frac{1}{s_1} - \frac{1}{s_2+2} + \\ + \frac{2}{s_2} = 0 \quad \text{for} \quad \langle r^2 \rangle_T . \end{aligned} \right\} \tag{16.14'}$$

These expressions demonstrate the validity of the relation $s_2 < s_1$, which is a simple consequence of the behaviour of the function $\lambda_1'(s)$ shown in Table 7. Therefore, the shower age of the mean square deviation is slightly smaller than the usual shower age, and with decreasing values of E_0/E we find smaller mean square deviations. This may be interpreted as insufficient scattering of the shower electrons, since the thickness of the matter traversed is not large enough in the case of the shower with small values of E_0/E. This situation is clearly observed in the results of a Monte Carlo calculation for a three-dimensional shower. The numerical results from (26.8) and those obtained by Monte Carlo calculations of ADACHI et al. (16) are compared in Fig. 11 together with the results of more elaborate calculations (30) based on the accurate expression (26.2) in which all the terms proportional to $e^{\lambda_1(s)t}$, $e^{\lambda_1(s+2)t}$, $e^{\lambda_2(s)t}$, $e^{\lambda_2(s+2)t}$ are included.

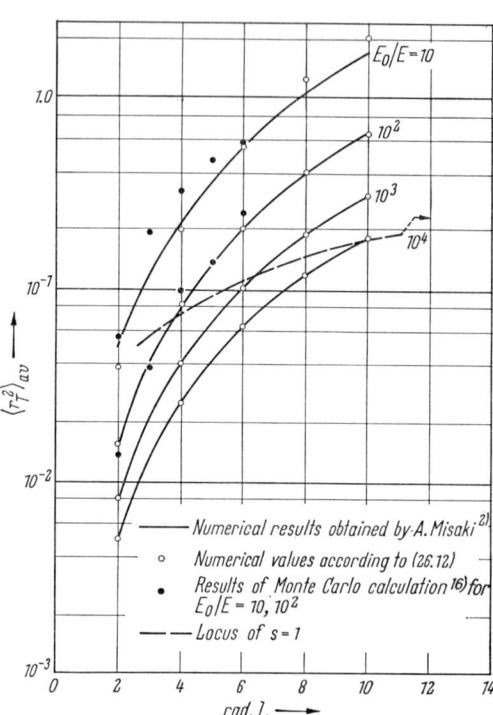

Fig. 11. $\langle rT \rangle_{\text{av}}$ of electrons in a photon-initiated shower (in Approximation A).

27. Numerical evaluation of integral structure functions of electrons for $E_0 = \infty$.

Since we are most interested in the integral structure functions of electrons, we shall show how they can be evaluated by the saddle point method. Other functions can also be determined in the same way. For the sake of simplicity, we shall here limit ourselves to the lateral structure function of electrons under the Landau approximation, and also to showers of infinitely high primary energy. Solutions without the Landau approximation are evaluated by the same procedure, and especially the numerical values of the first terms of the solutions are obtained approximately by replacing the E_s of the formulae in the Landau approximation, by corresponding terms in K.

It is sometimes convenient to introduce the normalized structure function, which is defined by

$$P_{\Pi_2}(E_0, 0, r, t) = \frac{\Pi_2}{\int_0^\infty \Pi_2 \, 2\pi r \, dr}, \tag{27.1}$$

so that

$$\int_0^\infty P_{\Pi_2}(E_0, 0, r, t) \, 2\pi r \, dr = 1. \tag{27.2}$$

Sect. 27.   Numerical evaluation of integral structure functions of electrons for $E_0 = \infty$.

It follows from (21.18) that

$$P_{\Pi_2}(E_0, 0, r, t) = \frac{\frac{1}{2\pi^2 i} \int\int_{-i\infty}^{i\infty} ds\, dp \left(\frac{E_0}{\varepsilon}\right)^s \left(\frac{\varepsilon}{E_s}\right)^2 \left(\frac{\varepsilon^2 r^2}{E_s^2}\right)^{-p-1} \Gamma(p+1)\Gamma(s+2p) \mathfrak{M}_2(p, -s-2p, s, t)}{\int_{-i\infty}^{i\infty} ds \left(\frac{E_0}{\varepsilon}\right)^s \Gamma(s) \mathfrak{M}_2(0, -s, s, t)}. \quad (27.3)$$

If one puts $\frac{E_0}{\varepsilon} = \infty$, the second term of (25.7) gives no contribution, and $\mathfrak{M}_2(p, q, s, t)$ is accurately given by $\mathfrak{M}_{1,0}(p, q, s, t)$. Furthermore, the integral with respect to s can now be carried out by the saddle point method independent of the integration with respect to p. Then we have instead of (27.3),

$$P_{\Pi_2}(\infty, 0, r, t) \cong \frac{\frac{1}{2\pi^2 i} \int_{-i\infty}^{+i\infty} dp \left(\frac{\varepsilon}{E_s}\right)^2 \left(\frac{\varepsilon^2 r^2}{E_s^2}\right)^{-p-1} \Gamma(\bar{s}+1)\Gamma(\bar{p}+2p) \mathfrak{M}_{1,0}(p, -\bar{s}-2p, \bar{s})}{\Gamma(\bar{s}) \mathfrak{M}_{1,0}(0, -\bar{s}, \bar{s})}, \quad (27.4)$$

where $\bar{s}$ is the shower age defined by

$$\lim_{\frac{E_0}{\varepsilon} \to \infty} \log\left(\frac{E_0}{\varepsilon}\right) + \lambda_1'(\bar{s})t = 0. \quad (27.5)$$

The integral with respect to p can also be evaluated by the saddle point method. Writing

$$e^{u(\bar{s}, p, t)} = \Gamma(p+1)\Gamma(\bar{s}+2p) \mathfrak{M}_{1,0}(p, -\bar{s}-2p, \bar{s}), \quad (27.6)$$

the saddle point $\bar{p}$ is determined by the equation

$$\therefore -2\log\left(\frac{\varepsilon r}{E_s}\right) + \frac{\partial u(\bar{s}, p, t)}{\partial p}\bigg|_{p=\bar{p}} = 0.$$

To carry out this calculation, it is necessary to have the numerical values of $\mathfrak{M}_{1,0}(p, -s-2p, s)$, which are derived in Appendix 4. Then we obtain

$$P_{\Pi_2}(\infty, 0, r, t) = \frac{1}{\pi \sqrt{2\pi u''(s, p)}} \cdot \frac{\left(\frac{\varepsilon}{E_s}\right)^2 \left(\frac{\varepsilon^2 r^2}{E_s^2}\right)^{-\bar{p}-1} e^{u(\bar{s}, \bar{p})}}{\Gamma(\bar{s}) \mathfrak{M}(0, -\bar{s}, \bar{s})}. \quad (27.7)$$

Since the function P_{Π_2} depends on t through the variable s, it appears advantageous to express it by $P_{\Pi_2}(\infty, 0, r, s)$. Furthermore, as at the limit $r \to 0$, P_{Π_2} is represented by $\frac{1}{r^{2-\bar{s}}}$ with the help of the pole at $p = -\frac{\bar{s}}{2}$ in (27.4), it is convenient to derive the numerical value of this function by introducing the function

$$P'_{\Pi_2} = \left(\frac{\varepsilon r}{E_s}\right)^{2-\bar{s}} P_{\Pi_2}.$$

The numerical results of the normalized distribution functions are presented in Table 15 and 16, and the behaviour of the integral structure functions for a few different shower ages are shown in Fig. 12—16, in comparison with the function given by MOLIÈRE.

In order to obtain the numerical values of the lateral structure functions, a simple approximate formula introduced by GREISEN (31) is useful:

$$P_{\Pi_2} = \left(\frac{E_s}{\varepsilon r}\right)^{\bar{s}-2} c(\bar{s}) \left(1 + \frac{\varepsilon r}{E_s}\right)^{\bar{s}-4.5}, \quad (27.8)$$

where $c(\bar{s})$ is a normalization factor. The accuracy of this formula is also shown in Figs. 13—16.

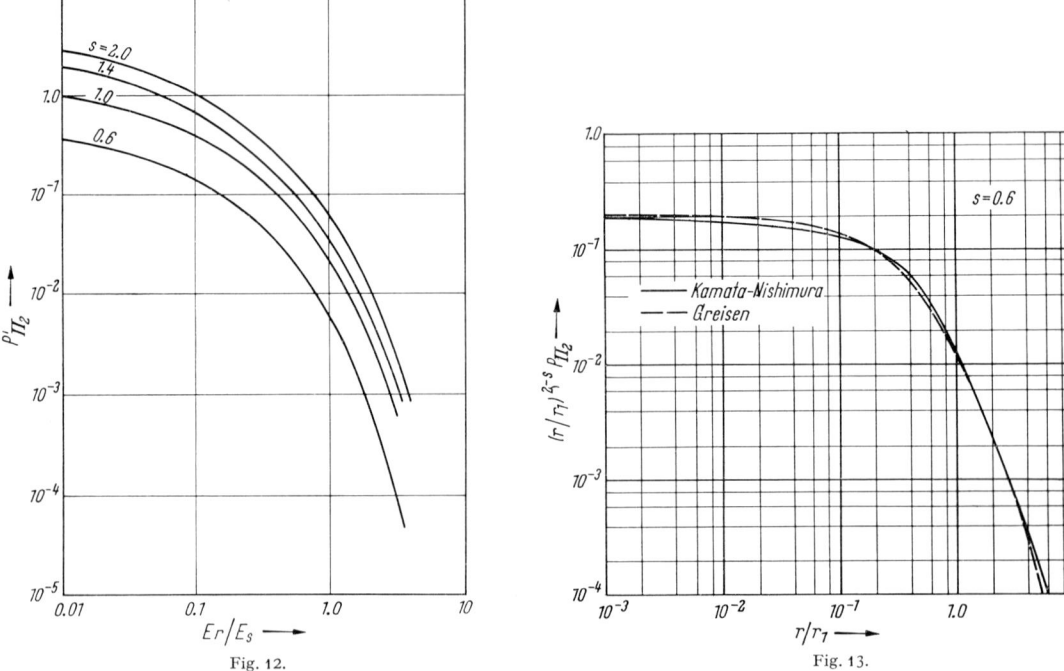

Fig. 12. Fig. 13.

Fig. 12. Normalized integral structure functions P'_{Π_2} at various shower ages, in Approximation A. (See also Table 15.) P'_{Π_2} is normalized by the condition

$$\int_0^\infty (Er/E_s)^{s-2} P'_{\Pi_2} 2\pi (Er/E_s) d(E_s r/E) = 1.$$

(From K. Kamata and J. Nishimura [8].)

Fig. 13. Normalized lateral structure function for $s=0.6$ in Approximation B. The Molière unit is defined as $r_1 = E_s/\varepsilon$ radiation lengths. (From K. Kamata and J. Nishimura [8].)

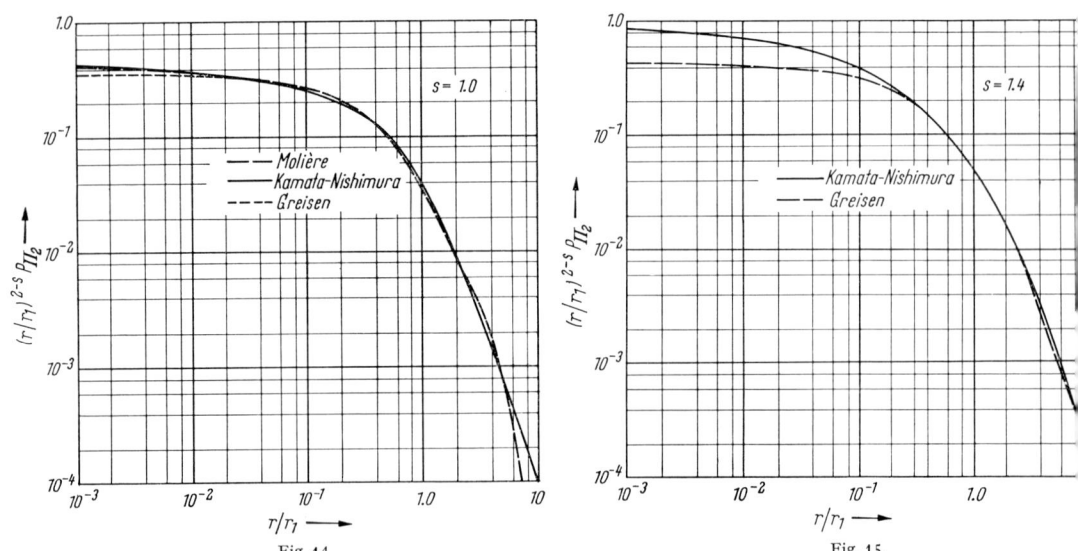

Fig. 14. Fig. 15.

Fig. 14. Normalized lateral structure function for $s=1.0$ in Approximation B. (From K. Kamata and J. Nishimura [8].)

Fig. 15. Normalized lateral structure function for $s=1.4$ in Approximation B. (From K. Kamata and J. Nishimura [8].)

Table 15. *Integral structure functions P_{Π_2} of electrons in Approximation A.*
P'_{Π_2} *is defined by* $P_{\Pi_2} = \left(\frac{Er}{E_s}\right)^{2-s} P'_{\Pi_2}$, *and is normalized as*

$$\int_0^\infty \left(\frac{Er}{E_s}\right)^{s-2} P'_{\Pi_2} 2\pi \left(\frac{Er}{E_s}\right) d\left(\frac{Er}{E_s}\right) = 1.$$

Er/E_s	0	0.01	0.03	0.1	0.3	1.0	3.0
$P'_{\Pi_2}(s=0.6)$	0.38	0.34	0.27	0.17	0.05	0.0044	0.00009
$P'_{\Pi_2}(s=1.0)$	1.01	0.91	0.70	0.37	0.15	0.019	0.00065
$P'_{\Pi_2}(s=1.4)$	2.5	1.9	1.35	0.68	0.27	0.041	0.0019
$P'_{\Pi_2}(s=2.0)$	$\lim_{r/r_1 \to 0}\left(-0.53\log\frac{Er}{E_s}\right)$	2.5	1.8	0.99	0.37	0.060	0.0037

From KAMATA and NISHIMURA [8].

Table 16. *Normalized lateral structure function* $(r/r_1)^{2-s} P_{\Pi_2}(r/r_1, s)$ *for* $s=0.6, 1.0, 1.4$ *and* 2.0 *in Approximation B, where* $r_1 = E_s/\varepsilon$ *radiation lengths.*

r/r_1	\multicolumn{4}{c}{s}
	0.6	1.0	1.4	2.0
0.001	$1.95 \cdot 10^{-1}$	$4.4 \cdot 10^{-1}$	$8.5 \cdot 10^{-1}$	2.4
0.003	$1.90 \cdot 10^{-1}$	$4.2 \cdot 10^{-1}$	$8.0 \cdot 10^{-1}$	1.92
0.01	$1.81 \cdot 10^{-1}$	$3.9 \cdot 10^{-1}$	$7.1 \cdot 10^{-1}$	1.40
0.03	$1.58 \cdot 10^{-1}$	$3.4 \cdot 10^{-1}$	$5.8 \cdot 10^{-1}$	$9.3 \cdot 10^{-1}$
0.1	$1.34 \cdot 10^{-1}$	$2.65 \cdot 10^{-1}$	$3.8 \cdot 10^{-1}$	$5.3 \cdot 10^{-1}$
0.2	$1.08 \cdot 10^{-1}$	$2.03 \cdot 10^{-1}$	$2.6 \cdot 10^{-1}$	$3.2 \cdot 10^{-1}$
0.5	$4.9 \cdot 10^{-2}$	$1.09 \cdot 10^{-1}$	$1.22 \cdot 10^{-1}$	$1.21 \cdot 10^{-1}$
1.0	$1.35 \cdot 10^{-2}$	$4.1 \cdot 10^{-2}$	$5.0 \cdot 10^{-2}$	$4.4 \cdot 10^{-2}$
2.0	$2.64 \cdot 10^{-3}$	$8.8 \cdot 10^{-3}$	$1.48 \cdot 10^{-2}$	$1.18 \cdot 10^{-2}$
5.0	$2.16 \cdot 10^{-4}$	$7.7 \cdot 10^{-4}$	$1.55 \cdot 10^{-3}$	$1.32 \cdot 10^{-3}$

From KAMATA and NISHIMURA [8].

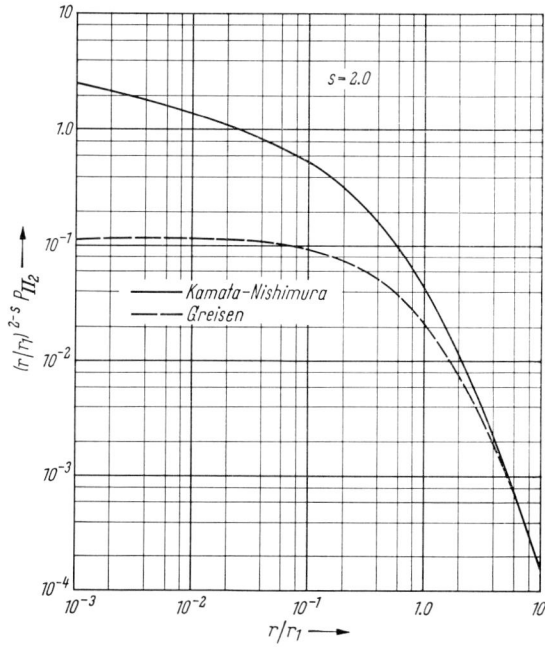

Fig. 16. Normalized lateral structure function for $s=2.0$ in Approximation B. (From K. KAMATA and J. NISHIMURA [8].)

28. Behaviour of the structure functions near the shower axis.

The behaviour of the structure functions near the shower axis can easily be derived by expanding the structure functions in a power series of r.

For the higher order terms, $\frac{1}{\Omega}\Pi^{(1)}$, $\frac{1}{\Omega^2}\Pi^{(2)}$, ..., without the Landau approximation, the singularities near the shower axis are of the same order as the one of the first term, as will be illustrated at the end of this section. Thus the contribution of the higher-order terms is of the order of $\frac{1}{\Omega}$ ($\Omega \sim 20$), compared with that of the first term. The second and the higher terms in $(K/E_0 r)^2$ in (25.3''), (25.7) and (25.8), however, may seem to give considerable contributions for small values of $\left(\frac{E_0 r}{K}\right)^1$. We shall, therefore, first examine the behaviour of the shower particles near the shower axis according to the first term only, and then discuss the contribution of the second and the higher-order terms in $\left(\frac{K}{E_0 r}\right)^1$.

The expansions with respect to r of the formulae for $\pi_2^{(0)}$, $\Pi_2^{(0)}$, $\gamma_2^{(0)}$, $\Gamma_2^{(0)}$ are carried out by using the poles of the integration parameter p in each function. These poles in the expressions $\Gamma(p+1)$, $\frac{1}{2p+s}$, $C(s+2p)$ and $\mathfrak{M}(s, p, 0, t)^2$ appear at $p=-1, -2, \ldots, p=-\frac{s}{2}$ and $p=-\frac{1}{2}\left(s+\frac{2}{3}\right)$ in each function, respectively. Concerning the position of the pole in $\mathfrak{M}$, details are discussed in Appendix 4. Thus the first terms of the expansion in r of the structure functions are the residues at these poles, and the integrations with respect to s by the saddle point method yields

$$\left.\begin{array}{l}\pi_2 \sim \dfrac{1}{r^{2-\bar{s}-\frac{2}{3}}} \quad \text{for} \quad 2-\bar{s}-\dfrac{2}{3}>0 \\[6pt] \pi_2 \sim \text{constant} \quad \text{for} \quad 2-\bar{s}-\dfrac{2}{3}<0\end{array}\right\} \text{in Approximation } A, \qquad (28.1)$$

where $\bar{s}$ is defined by the equation

$$\left.\frac{\partial \lambda_1(s) t}{\partial s}\right|_{\bar{s}} + \log\left(\frac{E_0}{E}\right) + \log\left(\frac{E r}{K}\right) \approx 0, \qquad (28.2)$$

with

$$\left.\begin{array}{l}\Pi_2 \sim \dfrac{1}{r^{2-\bar{s}}} \quad \text{for} \quad 2-\bar{s}>0 \\[6pt] \Pi_2 \sim \text{constant} \quad \text{for} \quad 2-\bar{s}<0\end{array}\right\} \text{in Approximation } B, \qquad (28.3)$$

$$\left.\frac{\partial \lambda_1(s) t}{\partial s}\right|_{\bar{s}} + \log\left(\frac{E_0}{\varepsilon}\right) + \log\left(\frac{\varepsilon r}{K}\right) \approx 0, \qquad (28.4)$$

$$\left.\begin{array}{l}\gamma_2 \sim \dfrac{1}{r^{2-\bar{s}}} \quad \text{for} \quad 2-\bar{s}>0 \\[6pt] \gamma_2 \sim \text{constant} \quad \text{for} \quad 2-\bar{s}<0\end{array}\right\} \text{in Approximation } A \qquad (28.5)$$

---

[1] For $\frac{E_0 r}{K} < 1$, one expects less than one shower particle within a circle of radius r, in a shower initiated by a single primary particle. But if one considers a shower initiated by many particles of the same energy, many particles will be found even in a region $\frac{E_0 r}{K} < 1$.

[2] In this section we use the abbreviation;

$$\mathfrak{M}(p, u=0, q, s, t) \Rightarrow \mathfrak{M}(p, q, s, t).$$

Behaviour of the structure functions near the shower axis.

with

$$\frac{\partial \lambda_1(s)t}{\partial s}\bigg|_{\bar{s}} + \log\left(\frac{E_0}{E}\right) + \log\left(\frac{E r}{K}\right) \approx 0, \tag{28.6}$$

$$\left.\begin{array}{l}\Gamma_2 \sim \dfrac{-\log r}{r^{2-\bar{s}}} \quad \text{for} \quad 2-\bar{s}>0 \\ \Gamma_2 \sim \text{constant} \quad \text{for} \quad 2-\bar{s}<0\end{array}\right\} \text{ in Approximation B} \tag{28.7}$$

with

$$\frac{\partial \lambda_1(s)t}{\partial s}\bigg|_{\bar{s}} + \log\left(\frac{E_0}{\varepsilon}\right) + \log\left(\frac{\varepsilon r}{K}\right) = 0. \tag{28.8}$$

The shape $\dfrac{1}{r^{2-\bar{s}}}$ for the integral structure function of the electrons is just what we expected from the discussion in Sect. 12.

At the limit $E_0 \to \infty$, all $\bar{s}$ here defined agree with the usual shower age in the one-dimensional shower theory. The properties of the structure function near the shower axis derived by MOLIÉRE (8), POMERANCHUK (32), and MIGDAL (33) are included in the above formulae as particular cases.

α) *Numerical evaluation of the structure functions near the shower axis using the first term in the expressions for* $\Pi_2^{(0)}$ *without the Landau approximation.*

Here we shall present numerical values for the integral lateral structure of a shower due to a primary electron of energy E_0 near the shower axis, in order to demonstrate the deviations from the results under the assumption $E_0 = \infty$ made in the preceding section.

As shown in (27.3), the normalized structure function P_{Π_2} is given in this case by

$$P_{\Pi_2} = \frac{\int ds \left(\dfrac{E_0}{\varepsilon}\right)^s \left(\dfrac{\varepsilon r}{K}\right)^{s-2} \Gamma\left(1-\dfrac{s}{2}\right) \mathfrak{M}_{1,0}\left(-\dfrac{s}{2}, 0, s\right) e^{\lambda_1(s)t}}{\int ds \left(\dfrac{E_0}{\varepsilon}\right)^s \Gamma(s) \mathfrak{M}_{1,0}(0, -s, s) e^{\lambda_1(s)t}}$$

$$\approx \sqrt{\frac{\lambda_2''(s_2)t + \dfrac{1}{s_2^2}}{\lambda_1''(s_1)t + \dfrac{1}{4}\Psi'\left(1-\dfrac{s_1}{2}\right)}} \cdot \frac{\Gamma\left(1-\dfrac{s_1}{2}\right)}{\Gamma(s_2)} \cdot \frac{\mathfrak{M}_{1,0}\left(-\dfrac{s_1}{2}, 0, s_1\right)}{\mathfrak{M}_{1,0}(0, -s_2, s_2)} \times \tag{28.9}$$

$$\times \left(\frac{\varepsilon r}{\varepsilon}\right)^{s_1-2} \left(\frac{E_0}{\varepsilon}\right)^{\Delta s} e^{\{\lambda_1(s_1) - \lambda_2(s_2)\}t},$$

where $\mathfrak{M}_{1,0}$ is defined in Sect. 25, and Δs, s_1 and s_2 are defined by

$$\Delta s = s_1 - s_2, \tag{28.10}$$

$$\lambda_2'(s_1)t + \log\frac{E_0}{\varepsilon} + \log\frac{\varepsilon r}{K} - \frac{1}{2}\Psi\left(1 - \frac{s_1}{2}\right) = 0, \tag{28.11}$$

$$\lambda_1'(s_2)t + \log\frac{E_0}{\varepsilon} - \frac{1}{s_2} = 0, \tag{28.12}$$

and s_2 is the usual shower age.

For a fixed value of t, we obtain from these equations the relation between s_1 and s_2:

$$-\frac{\lambda_1'(s_1)}{\lambda_1'(s_2)}\left[\log\frac{E_0}{\varepsilon} - \frac{1}{s_2}\right] + \log\frac{E_0}{\varepsilon} + \log\frac{\varepsilon r}{K} - \frac{1}{2}\Psi\left(1 - \frac{s_1}{2}\right) = 0. \tag{28.13}$$

The value of s_1 should be the same as that of s_2, if E_0/ε is infinite. This situation is also described in the preceding section.

For finite values of E_0/ε, however, we have different values for s_1 and s_2, and they coincide only if

$$\log \frac{\varepsilon r}{K} + \frac{1}{s_2} - \frac{1}{2} \Psi\left(1 - \frac{s_1}{2}\right) = 0, \tag{28.14}$$

which can easily be derived by putting $\dfrac{\lambda_1'(s_1)}{\lambda_2'(s_2)} = 1$ in (28.13).

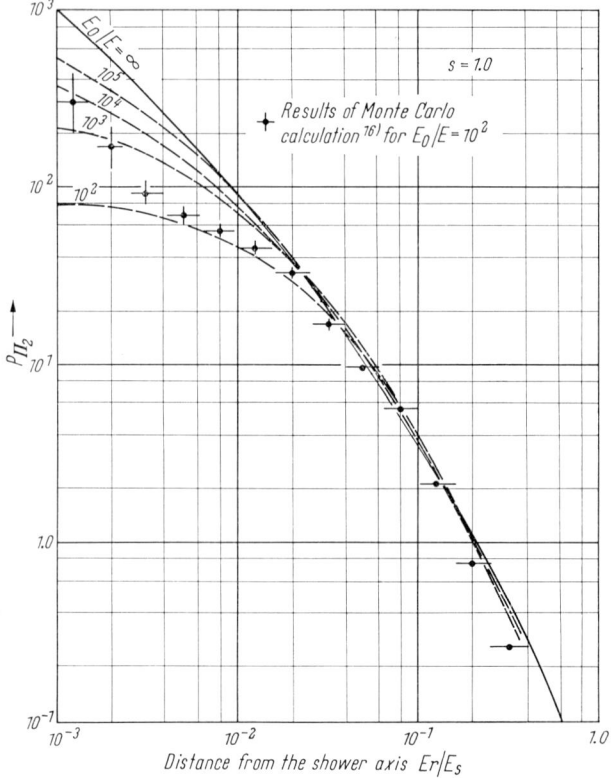

Fig. 17. Lateral distribution of shower electrons for finite primary energy in Approximation A. Normalized as

$$\int_0^\infty P_{\Pi_2} 2\pi (Er/E_s) \, d(Er/E_s) = 1.$$

At a certain distance from the shower axis satisfying the above formula, we expect the same values of $P_{\Pi_2}(E_0 = \text{finite})$ and $P_{\Pi_2}(E_0 = \infty)$ for the same shower age. Inside of this distance, the function $P_{\Pi_2}(E_0 = \text{finite})$ is less steep, and outside it becomes steeper than $P_{\Pi_2}(E_0 = \infty)$, since s_1 defined in (28.13) is a decreasing function of the distance r.

A numerical evaluation was carried out for $s_2 = 1$ both in Approximation A and in Approximation B, and the results are shown in the Fig. 17 and 18, together with the results obtained by the Monte Carlo calculations of ADACHI et al. (16).[1]

β) *Contribution of the second term proportional to $(K/E_0 r)^2$.* In order to discuss the contribution of this term to the structure functions $\pi_2^{(0)}$, $\Pi_2^{(0)}$, $\gamma_2^{(0)}$ and $\Gamma_2^{(0)}$,

---

[1] These authors pointed out that the numerical results in reference [8] seem to be inaccurate in comparison with the results obtained by themselves. Some mistakes in the numerical evaluation to Fig. 16 in reference [8] were thus found. The author is indebted to the group for their kind comment, and for informing him of their results before publication.

Sect. 28. Behaviour of the structure functions near the shower axis.

consider, for instance, the expression (25.3″)

$$\pi_2^{(0)} = -\frac{1}{4\pi^2} \iint ds\, dp \left(\frac{E_0}{E}\right)^s \frac{1}{E}\left(\frac{E}{K}\right)^2 \left(\frac{E^2 r^2}{K^2}\right)^{-p-1} \Gamma(p+1)\, \mathfrak{M}_{1,0}(p,0,s) \times \\ \times e^{\lambda_1(s)t}\left[1+\left(\frac{K}{E_0 r}\right)^2 (p+1)^2 F_{1,1}(s-2,t)+\cdots\right], \qquad (25.3'')$$

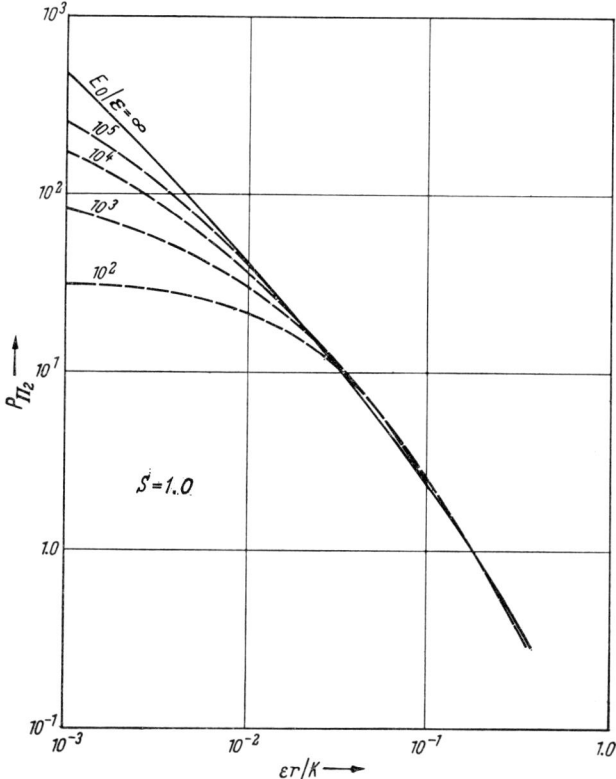

Fig. 18. Lateral structure function near the core for finite E_0, in Approximation B. Normalized as
$$\int_0^\infty P_{\Pi_2}\, 2\pi(\varepsilon r/K)\, d(\varepsilon r/K) = 1.$$

Developing in powers of r under the condition of $\frac{E_0 r}{K} < 1$, we have, with the help of the pole[1] at $p = -\frac{1}{3} - \frac{s}{2}$ in $\mathfrak{M}_{1,0}(p,0,s)$,

$$\pi_2^{(0)} \sim \frac{1}{2\pi i} \int ds \left(\frac{E_0}{E}\right)^s \frac{1}{E}\left(\frac{E}{K}\right)^2 \left(\frac{E^2 r^2}{K^2}\right)^{+1-\frac{1}{3}-\frac{s}{2}} \Gamma\left(1-\frac{1}{3}-\frac{s}{2}\right) e^{\lambda_1(s)t} \times \\ \times \left[1+\left(\frac{K}{E_0 r}\right)^2 \left(1-\frac{1}{3}-\frac{s}{2}\right)^2 F_{1,1}(s-2,t)+\cdots\right]. \qquad (28.15)$$

Here the integration path of s runs to the left of the position of the pole $s=\frac{4}{3}$ in the function $\Gamma\left(1-\frac{1}{3}-\frac{s}{2}\right).$

Assuming $\frac{E_0 r}{K} < 1$, one can develop the above integral into a power series of $\frac{E_0 r}{K}$, and with the help of the poles in $\Gamma\left(1-\frac{1}{3}-\frac{s}{2}\right)$ for the first term and in

---
[1] See Appendix 4.

$$\left(1-\frac{1}{3}-\frac{s}{2}\right)^2 \Gamma\left(1-\frac{1}{3}-\frac{s}{2}\right) = \left(1-\frac{1}{3}-\frac{s}{2}\right)\Gamma\left(2-\frac{1}{3}-\frac{s}{2}\right) \text{ for the second term,}$$

we have

$$\pi_2^{(0)} \sim \left[e^{\lambda_1(\frac{4}{3})t} - \left(\frac{E}{E_0}\right)^2 F_{1,1}\left(\frac{4}{3}, t\right) e^{\lambda_1(2+\frac{4}{3})t} + \cdots \right]. \tag{28.16}$$

Contrary to the expectation mentioned earlier in this section, the second term does not present the shape of a peaked function.

The ratio of the second term to the first is shown in Fig. 19 for different values of t. From this figure it may be concluded that the second term is practically negligible compared with the first one only if $E_0 \geq 10\,E$. Therefore the numerical values calculated in the preceding section give accurate results even in the region of quite small distances from the shower axis, provided only that $E_0 \gtrsim 10\,E$.

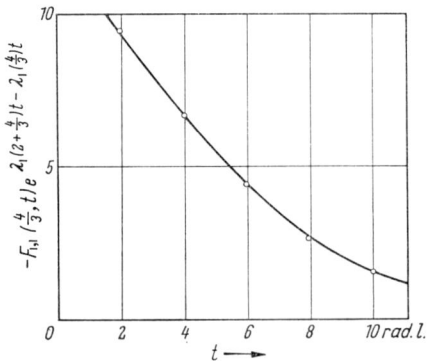

Fig. 19. Numerical values of $F_{1,1} e^{\lambda_1(2+4/3)t - \lambda_1(4/3)t}$.

γ) *Solutions without the Landau approximation.* It remains now to show the behaviour of the structure functions near the shower axis, with the second terms evaluated without the Landau approximation.

As an example we shall treat the problem of the integral lateral structure function of electrons, given in (24.22):

$$\left.\begin{aligned}\Pi_2^{(1)} = \frac{1}{4\pi^3} \int_{-i\infty}^{i\infty}\!\!\int ds\,dp \left(\frac{E_0}{\varepsilon}\right)^s \left(\frac{\varepsilon}{K}\right)^2 \left(\frac{\varepsilon^2 r^2}{K^2}\right)^{-p-1} p \cdot \Gamma(p+1)\Gamma(2p+s) \times \\ \times \left[\{\Psi(p+1)+\Psi(-p)-\log\frac{\varepsilon^2 r^2}{K^2}+2\Psi(2p+s)\} \times \right. \\ \left. \times \mathfrak{M}_2(p-1,1,-2p-s,s,t) - 2\frac{\partial}{\partial q}\mathfrak{M}_2(p-1,1,q,s,t)_{q=-2p-s}\right]. \end{aligned}\right\} \tag{24.22}$$

Developing the above formula in a power series of r, and assuming $r<1$, we have

$$\left.\begin{aligned}\Pi_2^{(1)} = \frac{1}{4\pi^2 i}\int ds \left(\frac{E_0}{\varepsilon}\right)^s \left(\frac{\varepsilon}{K}\right)^2 \left(\frac{\varepsilon^2 r^2}{K^2}\right)^{-1+\frac{s}{2}} \cdot \frac{s}{2}\cdot\Gamma\left(1-\frac{s}{2}\right)\times \\ \times \left[\Psi\left(1+\frac{s}{2}\right)\mathfrak{M}_2\left(-1-\frac{s}{2},1,0,s,t\right) - \right. \\ \left. -\frac{\partial}{\partial p}\mathfrak{M}_2(p-1,1,0,s,t)\Big|_{p=-\frac{s}{2}}\right]. \end{aligned}\right\} \tag{28.17}$$

Here the integration with respect to p is performed by using the simple and double poles at $p=-\frac{s}{2}$ in the functions $\Gamma(2p+s)$ and $\Gamma(2p+s)\Psi(2p+s)$ in the above integrand[1].

As described in Appendix 4, we may apply the relation

$$\mathfrak{M}(p, u=1, s, t) = \mathfrak{M}(p+1, u=0, s, t) + O\left(\frac{1}{\Omega}\right).$$

Then the structure function near the shower axis without Landau approximation is obtained in the simple form

$$\Pi_2 \cong \frac{1}{4\pi^2 i} \int ds\, \Gamma\left(1-\frac{s}{2}\right)\left(\frac{E_0 r}{K}\right)^s \frac{1}{r^2}\, \mathfrak{M}_2\left(-\frac{s}{2}, 0, s, t\right) \times$$
$$\times \left[1+\frac{1}{\Omega}\frac{s}{2}\left\{\Psi\left(1+\frac{s}{2}\right)-\frac{\mathfrak{M}_2'}{\mathfrak{M}_2}\Big|_{p=-\frac{s}{2},\, q=0}\right\}\right]. \quad (28.18)$$

Furthermore, if one wishes to include the contribution of the term of $e^{\lambda_1(s+2)t}$ in (25.7), the expression becomes

$$\Pi_2 = \frac{1}{4\pi^2 i}\int ds\,\Gamma\left(1-\frac{s}{2}\right)\left(\frac{E_0 r}{K}\right)^s \frac{1}{r^2}\,\mathfrak{M}_{1,0}\left(-\frac{s}{2}, 0, s\right)\left[1+\frac{1}{\Omega}\frac{s}{2}\times\right.$$
$$\left.\times\left\{\Psi\left(1+\frac{s}{2}\right)+\frac{\mathfrak{M}_{1,0}'}{\mathfrak{M}_{1,0}}\Big|_{p=-\frac{s}{2}}\right\}+\left(1-\frac{s}{2}\right)^2\left(\frac{K}{E_0 r}\right)^2 F_{1,1}(s-2,t)+\cdots\right]e^{\lambda_1(s)t}. \quad (28.19)$$

It appears that the last term in the bracket gives the main contribution at $\frac{E_0 r}{K} < 1$, but again this contribution is not significant compared with other terms, as already discussed in the preceding section.

δ) *Similarity relation of the structure function near the shower axis.* It is an important consequence that the structure function near the shower axis, (28.19), is a function of $E_0 r/K$ and t only, and not of the separate variables E_0 and r^2. This is due to the fact that the shower function in the one-dimensional theory is represented by the variables E_0/E and t only. Substituting $E \sim K/r$ leads, therefore, to the functional form mentioned above, the similarity relation. Thus, if we have obtained numerical values of the structure function near the shower axis for a shower of a certain primary energy E_0, we can predict the numerical values for a shower of a different primary energy. Furthermore, if one is interested in the number of shower particles within a circle of a certain radius r for a fixed value of t, the

---

[1] The calculation proceeds as follows

$$\left(\frac{\varepsilon^2 r^2}{K}\right)^{-p-1} p\,\Gamma(p+1)\left[\Psi(p+1)+\Psi(-p)-\log\left(\frac{\varepsilon^2 r^2}{K^2}\right)+2\Psi(2p+s+1)\right]\times$$
$$\times \mathfrak{M}_2(p-1, 1, -2p-s, s, t) - \frac{\partial}{\partial p}\log\left\{\left(\frac{\varepsilon^2 r^2}{K^2}\right)^{-p-1}\Gamma(p+1)\Gamma(2p+s+1)-\right.$$
$$\left.- p\,\mathfrak{M}_2(p-1, 1, -2p-s, s, t)\right\} - 2\frac{\partial}{\partial q}\mathfrak{M}_2(p-1, 1, q, s, t)_{q=p-\frac{s}{2},\, p=\frac{s}{2}}$$
$$= -\left(\frac{\varepsilon^2 r^2}{K^2}\right)^{\frac{s}{2}-1}\Gamma\left(1-\frac{s}{2}\right)\left[\Psi\left(1+\frac{s}{2}\right)\mathfrak{M}_2\left(-1-\frac{s}{2}, 1, 0, s, t\right)-\right.$$
$$\left.- \frac{\partial}{\partial p}\mathfrak{M}_2(p-1, 1, 0, s, t)\Big|_{p=-\frac{s}{2}}\right].$$

[2] One should remember that the complete definition of the function (28.19) is $\Pi_2\, 2\pi r\, dr$.

function is expressed by the variable $\frac{E_0 r}{K}$ only, i.e.,

$$\int_0^r \Pi_2^{(0)} 2\pi r \, dr = N\left(\frac{E_0 r}{K}, t\right). \tag{28.20}$$

This similarity relations make it quite simple to perform the actual analysis of the shower phenomena near the shower axis. The details of the application of this formula will be discussed in Sect. 30. Numerical evaluations were carried out for the two cases which are most important in the analysis of showers observed in nuclear emulsions: for a primary photon, and for an electron pair. The results are shown in Tables 16 and 17, and also in Fig. 20. The errors involved in our calculations are expected to be of the order of 20~30%, according to the arguments given in Appendix 6.

Table 17. *Number of electrons for a photon-initiated shower within a circle of radius r in emulsions*.* 1 rad. $l. = 2.98$ cm*.

t	$E_0 r$ GeV μ											
	$2 \cdot 10^3$	$5 \cdot 10^3$	10^4	$2 \cdot 10^4$	$5 \cdot 10^4$	10^5	$2 \cdot 10^5$	$5 \cdot 10^5$	10^6	$2 \cdot 10^6$	$5 \cdot 10^6$	10^7
0												
1	1.68	1.88	2.08	2.29	2.63	2.88	3.16	3.58	3.95	4.36	4.88	5.35
2	2.89	4.30	5.8	7.6	10.3	12.7	15.9	21.2	26	31.8	38.6	43.8
3	1.95	4.84	8.2	13.2	22.2	31.2	42.3	62	80	104	147	190
4	1.26	4.26	8.7	16.0	32.4	51.6	77.0	127	184	255	380	495
5	0.88	3.38	8.0	15.7	37.0	65.0	107	204	317	480	770	1060
6	0.58	2.50	6.45	14.2	37.0	73.5	130	273	445	680	1220	1780
8	0.168	1.08	3.3	8.10	25.6	59.0	123	300	570	1040	2140	3520
10		0.43	1.75	4.80	16.0	39.0	84.5	226	490	1020	2630	4750
12			0.58	1.90	8.2	20.0	48.0	160	375	800	2330	4700
14			0.245	0.78	3.7	10.8	29.0	100	235	560	1720	3800
16				0.288	1.45	4.45	13.0	50	131	340	1160	2800
18					0.66	2.08	6.0	25.0	66	183	670	1700
20					0.275	0.92	2.88	11.5	33	91	235	900

Table 18. *Number of electrons for a shower initiated by an electron pair within a circle of radius r in emulsions.* 1 rad. $l. = 2.98$ cm*.

t	$E_0 r$ GeV μ											
	$2 \cdot 10^3$	$5 \cdot 10^3$	10^4	$2 \cdot 10^4$	$5 \cdot 10^4$	10^5	$2 \cdot 10^5$	$5 \cdot 10^5$	10^6	$2 \cdot 10^6$	$5 \cdot 10^6$	10^7
0	2	2	2	2	2	2	2	2	2	2	2	2
1	2.93	5.7	6.75	7.7	9.4	10.8	12.3	14.3	16.2	18.0	20.6	23.0
2	1.75	5.27	10.6	16.0	25	32.4	42	56.5	72.5	88	113	137
3	1.06	4.40	9.7	19.0	38	57.5	86	137	188	250	360	466
4	0.66	3.1	7.7	17.0	42	75	124	230	340	510	790	1110
5	0.41	2.13	5.8	14.3	40	80	146	306	487	780	1330	2000
6	0.245	1.28	3.7	10.0	34	74	146	335	585	1020	2020	3220
8		0.46	1.58	4.4	18.8	48.6	112	315	680	1230	2840	5020
10		0.165	0.66	2.0	9.1	27.0	66	213	585	1120	2760	5500
12			0.226	0.78	4.0	12.0	34.5	126	326]	800	2300	4750
14				0.30	1.60	5.5	16.8	66	180	440	1480	3530
16				0.11	0.64	2.3	7.4	32.6	91	242	870	2150
18					0.218	0.82	2.8	13.0	39.5	118	450	1200
20						0.31	1.20	5.8	17.6	57	244	670

* In the Tables prepared for ICEF project [J. KIDD and J. NISHIMURA: Nuov. Cim., Suppl. (1964)] 1 radiation length is taken as 2.83 cm, and the numerical values are recalculated in Table 17 and 18.

29. Behaviour of the structure functions at large distances from the shower axis. As in the case of the scattering of a charged particle, the contribution of single scattering becomes important at large distances from the shower axis, or at a large angle from the direction of the shower axis. Its effects on the structure

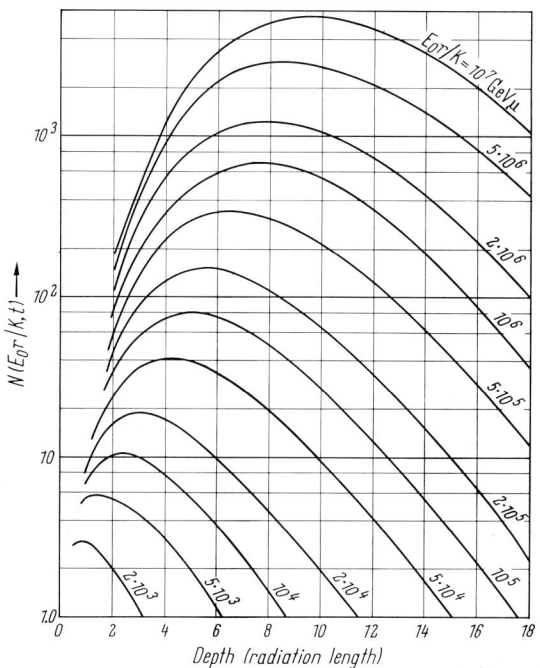

Fig. 20. Number of electrons within a circle of radius r for a shower initiated by an electron pair in emulsions.

function have been calculated by several authors, among whom EYGES (34) derived the following expressions for the contribution of single scattering:

$$\pi_2(r \to \infty) \sim \frac{\langle r^2 \rangle_{AV}}{4\pi \log 191 \, Z^{-\frac{1}{3}}} \frac{1}{r^4} \, \pi, \tag{29.1}$$

$$\Pi_2(r \to \infty) \sim \frac{\langle r_T^2 \rangle_{AV}}{4\pi \log 191 \, Z^{-\frac{1}{3}}} \frac{1}{r^4} \cdot \Pi, \tag{29.2}$$

$$\pi_1(\theta \to \infty) \sim \frac{\langle \theta^2 \rangle_{AV}}{4\pi \log 191 \, Z^{-\frac{1}{3}}} \frac{1}{\theta^4} \cdot \pi, \tag{29.3}$$

and

$$\Pi_1(\theta \to \infty) \sim \frac{\langle \theta_T^2 \rangle_{AV}}{4\pi \log 191 \, Z^{-\frac{1}{3}}} \frac{1}{\theta^4} \cdot \Pi, \tag{29.4}$$

where π and Π in the right-hand side are the one-dimensional shower cascade functions given in Sect. 15.

As an example, the derivation of these formulae is explained for the case of the angular structure function in the following. The contribution is given by

$$\pi_0(\theta \to \infty) = \int_0^t \int_0^{E_0} \Pi(E', E, t-t') \, \sigma(\theta) \, \pi(E_0, E', t') \, dt' \, dE'. \tag{29.5}$$

where $\sigma(\theta)\, dt'$ represents the probability of scattering after traversing the material dt', and is approximately given by

$$\sigma(\theta) = \frac{1}{4\pi \log 191\, Z^{-\frac{1}{3}}} \left(\frac{E_s}{E'}\right)^2 \frac{1}{\theta^4}, \quad (A.3.14)$$

as shown in Appendix 3.

Since $(E_s/E')^2\, dt$ is the mean square value of the scattered angle of a particle of energy E' after traversing the material dt', (29.5) can also be written in the form

$$\left. \begin{array}{l} \Pi_2(\theta \to \infty) \\ = \dfrac{\langle \theta_T^2 \rangle_{AV}}{4\pi \log 191\, Z^{-\frac{1}{3}}} \dfrac{1}{\theta^4}\, \Pi(E_0, E, t). \end{array} \right\} \quad (29.5')$$

The expressions (29.1), (29.2) and (29.3) are derived in a similar manner.

Since the second term of the solution without the Landau approximation represents the contribution of single and plural scattering, these solutions should become identical with the distributions obtained by EYGES if r or $\theta \to \infty$. Accordingly, the integrations with respect to p in the function $\pi_{(1)}^{(1)}, \pi_{(2)}^{(1)}, \gamma_2^{(1)}, \Pi_1^{(1)}\Pi_2^{(1)}, \gamma_{(1)}^{(1)}$, (24.7), (24.8), (24.9), (24.20), (24.21), (24.22) can be performed using the poles at $p=1$, 2, ... of the function $\Psi(-p)$ in the integrands. The first terms are then found to be proportional to $1/r^4$ or $1/\theta^4$. As an example, the differential angular structure function of the electrons may be considered. In this case the integration in (24.7) with respect

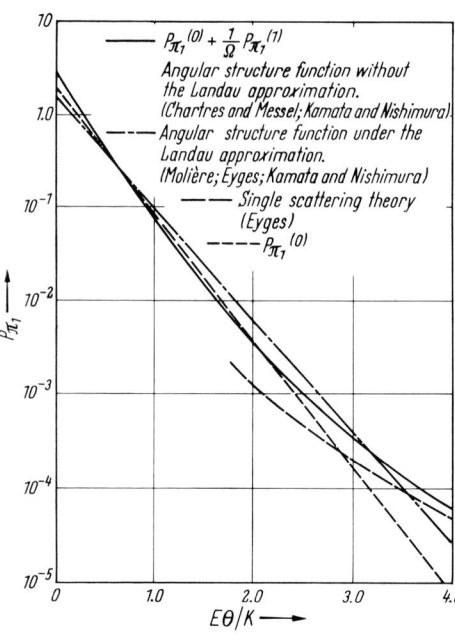

Fig. 21. Normalized differential angular structure function $P_{\pi_1}(s=1)$ in air, in Approximation A. P_{π_1} is normalized by the condition

$$\int_0^\infty P_{\pi_1} 2\pi (E\theta/K)\, d(E\theta/K) = 1$$

with $K=19.3$ MeV and $\Omega=15.3$. The difference between the results of CHARTRES and MESSEL, and of KAMATA and NISHIMURA, is so small that it cannot be represented in this graph. (From K. KAMATA and J. NISHIMURA [8].)

to p is carried out under the condition $\dfrac{E\theta}{K} \gg 1$, and yields, using the poles of $p\, \Psi(-p)$ at $p=1, 2, \ldots$,

$$\pi_1^{(1)} \cong \frac{1}{2\pi i} \int_{-i\infty}^{i\infty} ds \left(\frac{E_0}{E}\right)^s \frac{1}{E} \left\{ \Gamma(2) \left(\frac{E^2\theta^2}{K^2}\right)^{-2} \mathfrak{M}_1(0, 1, s, t) + \right. \\ \left. + 2\Gamma(3) \left(\frac{E^2\theta^2}{K^2}\right)^{-3} \mathfrak{M}_1(1, 1, s, t) + \cdots \right\}. \quad (29.6)$$

The first term of this series represents the effect of single scattering, with $\mathfrak{M}_1(0, 1, s, t)$ given by

$$\mathfrak{M}_1(0, 1, s, t) \sim \mathfrak{M}_1(1, 0, s, t)$$

as shown in Appendix 4. Since, as demonstrated in Sect. 2, $\mathfrak{M}_1(1, 0, s, t)$ is proportional to $\langle \theta^2 \rangle$, it is seen that (29.6) agrees with the function (29.3) derived by EYGES.

A similar argument also holds for the lateral structure function. In this case we have only to replace θ and $\mathfrak{M}_1$ by r and $\mathfrak{M}_2$, respectively.

Numerical calculations of the integrals in (29.6) prove that the series (29.6) converges very rapidly. For practical applications it is sufficient to confine the calculations to the first two terms, as in MOLIÈRE's theory for the scattering of a single charged particle. Numerical results are shown in Fig. 21, in comparison with those obtained in other theories.

Other structure functions, such as the lateral distribution in Approximation B, can be obtained in a way similar to that indicated above.

G. Application of the theory to the shower phenomena.

The theory of cascade showers has been widely applied in the analysis of cosmic ray phenomena, such as extensive air showers and the energetic interactions observed in nuclear emulsions and in other detectors. In this section we shall discuss its applications to energy measurements on electron showers, to the study of high-energy electrons and photons in the atmosphere, and to extensive air showers.

30. Energy measurement on electron showers.

α) *The calorimetric method.* Since the energy of a shower-initiating particle is eventually dissipated through ionization losses, it can be determined by a measurement of the total amount of the ionization produced by all the shower particles. The procedure is, of course, unaffected by the fluctuation phenomena in the electron shower, and independent of the detailed features of the shower development.

Ionization chambers, Čerenkov detectors, and scintillators sandwiched with metal plates were used for this purpose. Arrangements of this kind, often called "calorimeters", were widely used for the study of the cosmic ray phenomena (*35*). Large-size detectors are required in order that only a small fraction of the shower particles might escape through the sides or penetrate through the bottom. The practical limitations in the size of the detector make it important to evaluate the fraction of the energy missed by the arrangement. An analytical evaluation, however, is difficult since those escaping particles are mostly low-energy electrons and photons that have suffered large-angle scattering, thus involving a situation surely outside the scope of the approximations used in the theory.

An experiment to determine these losses for different materials has been performed by Kantz and Hofstadter (*36*). They inserted a small probe scintillator inside a well of shower-developing material to observe the distribution of the shower particles in this material. Recently, Murata et al. (*37*) carried out a similar experiment using X-ray films sandwiched with metal plates. They placed three types of film with different sensitivities under each layer of the Pb plates, and exposed this chamber to the 200 MeV electron beam of the I.N.S. electron synchrotron. The distribution of the shower particles in the detector was determined from the blackening of the lead-shielded films measured by a photometric device.

The total ionization loss thus observed agrees with the incident energy with a precision of $\sim 1\%$. The data obtained by Kantz and Hofstadter, however, led to a value higher than the energy of the incident electron. In both experiments the transition curve of the ionization of the shower particles was also determined. Here the work of Murata et al. shows the tail of the shower curve dropping off rather more rapidly than in the experiment of Kantz and Hofstadter.

The evaluation of the fraction of undetected energy given by Kantz and Hofstadter is, therefore, probably an overestimate. Their result might be explained as due to the effect of the background low energy electrons and photons produced by the surrounding material.

An example of the distribution of the shower particles observed in the X-ray films is shown in the photograph of Fig. 22. The fraction of energy absorbed in the detector is illustrated in Fig. 23. Of course these figures do not hold exactly for a shower of different energy, or for a shower developed in a different material, but they exhibit the general trend of the shower development in the material.

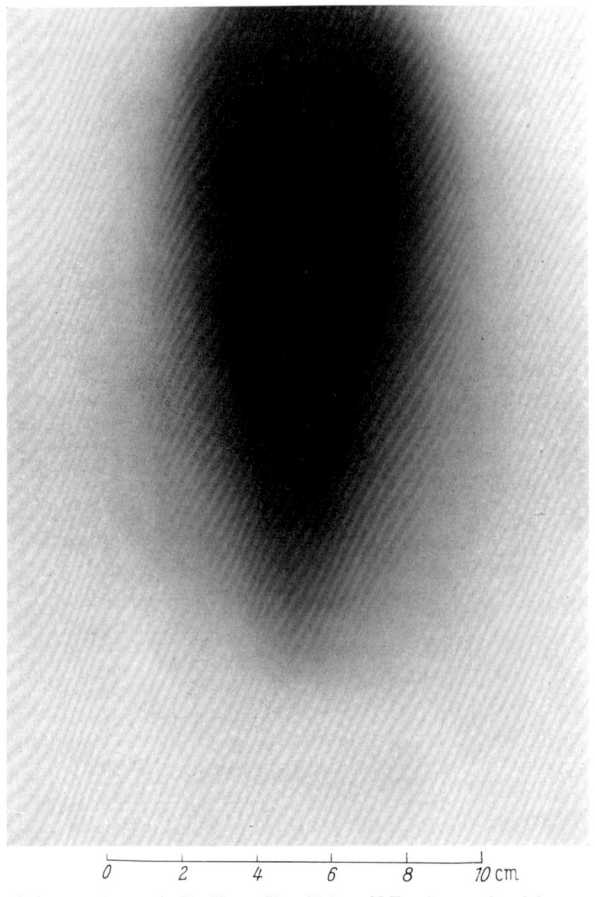

Fig. 22. Development of electron showers in Pb. X-ray films (Sakura N Type) were placed between thick Pb-plates, and exposed to the 200 MeV γ-ray beam of the I.N.S. Electron Synchrotron. The plane of the films was inclined by about 2° to the direction of the beam.

β) *Showers observed in nuclear emulsions.* Electron showers have been studied extensively in nuclear emulsion stacks and in emulsion chambers [9]. An emulsion stack is a block of many sheets of emulsions without backing glass plates. In recent experiments stacks as large as 96 l (40 × 60 × 40 cm) have been exposed to cosmic rays at balloon altitude. Since the thickness of this detector is about 20 radiation lengths, the full development of the electron showers can be followed in the nuclear emulsions. An emulsion chamber, on the other hand, is a sandwich of many sheets of metal plates and nuclear emulsion plates. Sizes as large as 10 m² have been used, and the thickness sometimes reaches several tens of radiation lengths. The electron showers developed in these detectors are observed under a microscope as parallel tracks running within several hundred microns from the shower axis.

The first systematic study of electron showers in nuclear emulsions was carried out by PINKAU (38). He measured the energy of the individual electron tracks by the scattering method, and thus obtained the energy spectrum of the electrons within a circle of a certain radius from the shower axis. Since most of the electrons of energy higher than E are confined to the interior of a circle of radius $r \sim K/E$ [1], the one-dimensional shower theory in Approximation A can be applied for the analysis of the electron shower. With this method PINKAU studied several showers observed in G stacks ($12 \times 16''$ G_5 emulsions), and determined the energy of the incident photons and electrons.

Another approach now widely adopted in the analysis of electron showers consists in determining the transition curve of the number of shower electrons

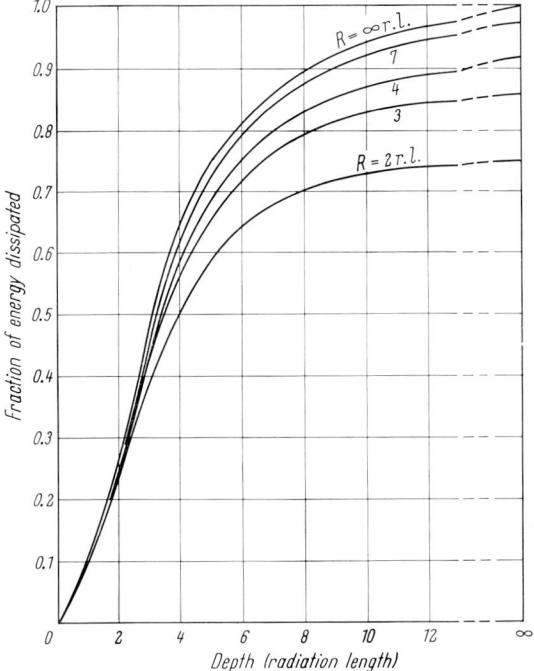

Fig. 23. Fraction of energy dissipated by a 200 MeV electron within a lead cylinder of radius R. (From Y. MURATA [37].)

within a circle of a certain radius r. As shown in Sect. 28, this number $N(r, t)$, expressed as a function of the depth in t cascade units, satisfies the relation

$$N(r, t) = N\left(\frac{E_0 r}{K}, t\right) \tag{28.20}$$

provided that $\frac{E_0 r}{K} \ll 1$, where E_0 is the energy of the incident electron or photon. Numerical results based on (28.19) and (28.20) were listed in Tables 17 and 18, in which we assumed the incident particle to be either an electron pair or a photon. Comparing the observed transition curves with the theoretical ones, the energy of the primary can be determined without recourse to the tedious procedure of scattering measurements.

In Figs. 24 and 25, examples of showers observed both in emulsion stacks[2] and emulsion chambers are presented. Furthermore, the lateral distribution of the

---

[1] If one takes $E=1$ GeV, the relation $r=K/E$ gives $r=600$ μ.
[2] The author is indebted to Prof. M. KOSHIBA for permitting use of the data observed in his emulsion stack.

showers observed at various depths in the emulsion chamber (*39*) is shown in Fig. 26.

γ) *Energy measurement on electron showers observed in multiple chambers.* Experiments on electron showers were also performed in multiple chambers in which shower electrons are strongly scattered so that the shower functions in the one-dimensional theory cannot be applied to the energy determination of the incident particles. In that case, the angular structure function provides the most important tool for the analysis.

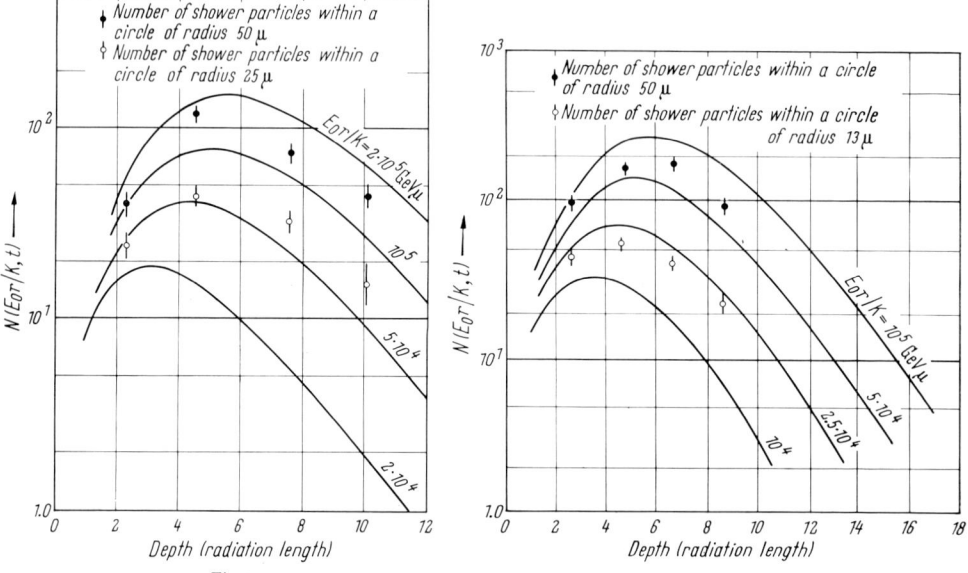

Fig. 24.   Fig. 25.

Fig. 24. Examples of showers initiated by an electron pair in nuclear emulsions. The energy of the electron pair is estimated to be 2.5 TeV. [From the data of F. ABRAHAM et al., Nuovo Cim. **29**, 315 (1963)].

Fig. 25. Example of showers observed in the emulsion chamber exposed at Mt. Norikura. The primary energy is estimated to be 1.5 TeV. [From the data of M. AKASHI, K. SHIMIZU, Z. WATANABE, N. OGITA, A. MISAKI, I. MITO, S. OYAMA, S. TOKUNAGA, M. TAMURA, Y. FUJIMOTO, S. HASEGAWA, J. NISHIMURA, K. NIU and K. YOKOI, J. Phys. Soc. Japan **17**, Suppl. III A, 427 (1962)].

The number of the shower particles under the plates within a certain angle θ from the direction of shower axis is counted. If the angle θ remains smaller than 1 radian, the angular structure functions derived in (24.6) and (24.9) are approximately valid, and we expect the observed number of shower electrons to be given by

$$\int_0^\theta \left[\Pi_1^{(0)} + \frac{1}{\Omega} \Pi_1^{(1)} + \frac{1}{\Omega^2} \Pi_1^{(2)} + \cdots \right] 2\pi \theta \, d\theta. \tag{30.1}$$

The track length of these particles is

$$\begin{aligned}
&\int_0^\infty dt \int_0^\theta \left[\Pi_1^{(0)} + \frac{1}{\Omega} \Pi_1^{(1)} + \frac{1}{\Omega^2} \Pi_1^{(2)} + \cdots \right] 2\pi \theta \, d\theta. \\
&= \int_0^\theta 2\pi \theta \, d\theta \int_0^\infty dt \left[\Pi_1^{(0)} + \frac{1}{\Omega} \Pi_1^{(1)} + \cdots \right] \\
&= \int_0^\theta 2\pi \theta \, d\theta \int_0^\infty dt \, \Pi_1^{(0)},
\end{aligned} \tag{30.2}$$

where we make use of the relation

$$\int_0^\infty \Pi_1^{(1)} dt = 0, \quad \int_0^\infty \Pi_1^{(2)} dt = 0 \dots . \tag{30.3}$$

The relation (30.3) can easily be proved by the residues at the pole $s=1$ in the complex integrals of the variable s.

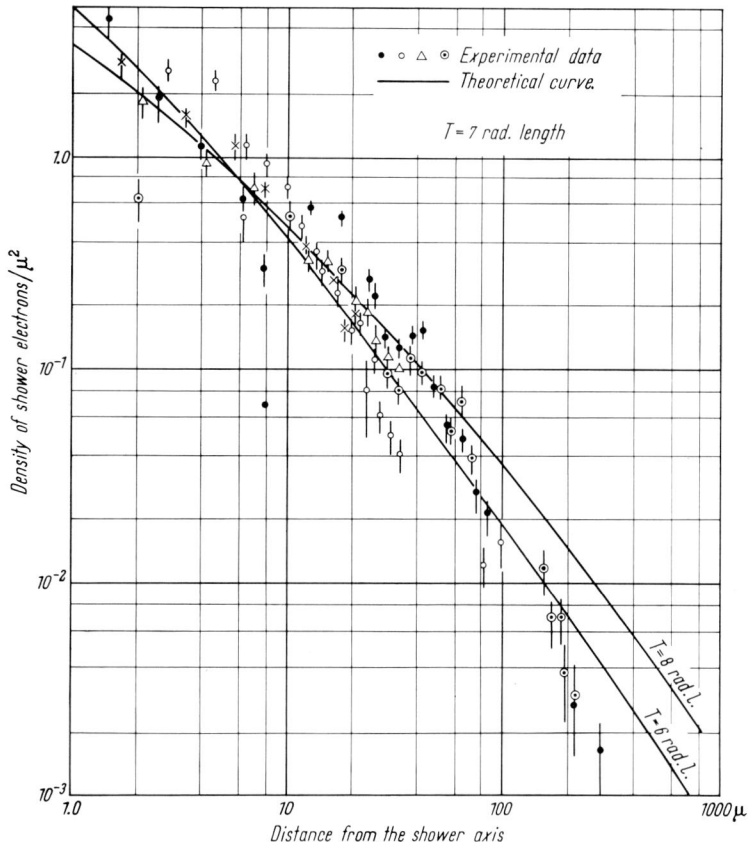

Fig. 26. Lateral distribution of shower electrons in an emulsion chamber. The experimental data are normalized to $E_0 = 10^{13}$ eV. [From E. Mikumo et al. (39)].

Now defining the function $\alpha(\theta)$ giving the fraction of the number of shower particles with in the angle θ,

$$\alpha(\theta) = \frac{\int_0^\theta 2\pi\,\theta\,d\theta \int_0^\infty \Pi_1^{(0)}(E_0, \theta, t)\,dt}{\int_0^\infty \Pi(E_0, t)\,dt} = \frac{\varepsilon}{E_0} \int_0^\theta 2\pi\,\theta\,d\theta \int_0^\infty \Pi_1^{(0)}(E_0, \theta, t)\,dt,$$

we have, from (24.6),

$$\alpha(\theta) = \frac{1}{2\pi i} \int \frac{dp}{p} \left(\frac{\varepsilon^2 \theta^2}{K^2}\right)^{-p} \Gamma(p+1)\Gamma(2p+1) \frac{\mathfrak{M}_{1,0}(p, 0, -1-2p, 1)}{\mathfrak{M}_{1,0}(0, 0, -1, 1)}, \tag{30.4}$$

where the integration path of p is running in the domain of $\text{Re}(p) > 0$. One can determine[1] the energy of the primary particle by referring to the numerical values of $\alpha(\theta)$. Some numerical results are shown in Table 19.

Table 19.

	For Pb			For Fe		
θ	15°	30°	60°	15°	30°	60°
$\alpha(\theta)$	11%	20%	27%	26%	42%	58%

31. High-energy photons and electrons in the atmosphere.

High-energy photons and electrons in the atmosphere can be studied by means of nuclear emulsions as described in the preceding section. Thus the energy spectra and absolute intensities of these particles were observed at various altitudes. The energy distribution proved to be well represented by a single power function [9],

$$\frac{dE}{E^{\gamma+1}}, \tag{31.1}$$

where $\gamma = 2.2 \pm 0.2$ for $E \gtrsim 10^{12}$ eV.

Since these particles are due to the electron showers mainly initiated by the jets in the atmosphere, a calculation of their intensity must be based on the assumption of sources distributed over the entire depth of the atmosphere.

Since the primary energy concerned is quite high compared with the critical energy in air ($\varepsilon = 81$ MeV), the problem will be treated in the frame of Approximation A. Thus, the energy spectrum of the source function is expected to be a power function with the same exponent as the observed spectrum.

An exponential depth distribution of the sources,

$$e^{-\frac{t}{\Lambda}} dt,$$

can be assumed [9], writing Λ for the absorption mean free path of the nuclear-active component.

Representing the absolute vertical intensity of the photons resulting from the decay of the π^0 mesons produced at a depth between t' and $t' + dt'$ by

$$J(E, t') \, dt' = J_0 \frac{dE}{E^{\gamma+1}} e^{-\frac{t'}{\Lambda}} dt', \quad (/\text{cm}^2 \cdot \text{sec} \cdot \text{sterad}), \tag{31.2}$$

the vertical intensity of the observed electrons and photons at depth t is given by

$$I = \int_0^t dt' \int_E^\infty dE_0 \, J(E_0, t') \, \{\Pi(E_0, E, t-t') + \Gamma(E_0, E, t-t')\}, \tag{31.3}$$

where Π and Γ are the shower functions for a photon-primary shower in Approximation A. The function Π and Γ can be taken from (13.22) and (13.23) together with Table 9:

$$\Pi = \frac{1}{2\pi i} \int ds \, \frac{M(s)}{s^{\frac{1}{2}}} \left(\frac{E_0}{E}\right)^s (e^{\lambda_1(s)t} - e^{\lambda_2(s)t}),$$

$$\Gamma = \frac{1}{2\pi i} \int ds \left[\frac{H_2(s)}{s} \left(\frac{E_0}{E}\right)^s e^{\lambda_1(s)t} + \frac{H_1(s)}{s} \left(\frac{E_0}{E}\right)^s e^{\lambda_2(s)t}\right],$$

so that

$$\Pi + \Gamma = \frac{1}{2\pi i} \int ds \, \frac{1}{s} \left(\frac{E_0}{E}\right)^s \{N_1(s) e^{\lambda_1(s)t} + N_2 e^{\lambda_2(s)t}\}, \tag{31.4}$$

---

[1] If one calculates the area under the transition curve of the observed shower particles with directions within an angle θ, the incident energy of the shower is obtained by multiplying that area with $\varepsilon/d(\theta)$.

where we put

$$N_1 = M(s)^{\frac{1}{2}} + H_2(s), \qquad (31.5)$$

$$N_2 = -M(s)^{\frac{1}{2}} + H_1(s). \qquad (31.6)$$

Numerical values of N_1 and N_2 are shown in Table 20.

Table 20.

s	1.6	1.8	2.0	2.1	2.4	2.6	2.8	3.0
N_1	1.169	1.201	1.224	1.234	1.240	1.238	1.234	1.226
N_2	−0.169	−0.201	−0.224	−0.234	−0.240	−0.238	−0.234	−0.226

Substituting the expressions (31.2) and (31.4) into the integrand of (31.3), we have

$$
\begin{aligned}
I &= \frac{J_0}{2\pi i} \int_0^t dt' \int_E^\infty dE_0 \frac{1}{E_0^{\gamma+1}} e^{-\frac{t'}{\Lambda}} \int \frac{ds}{s} \left(\frac{E_0}{E}\right)^s \left\{ N_1 e^{\lambda_1(s)(t-t')} + N_2(s) e^{\lambda_2(s)(t-t')} \right\} \\
&= \frac{J_0}{2\pi i} \int \frac{ds}{s} \frac{1}{\gamma - s} \frac{1}{E^\gamma} \left\{ N_1 \frac{e^{\lambda_1(s)t} - e^{-\frac{t}{\Lambda}}}{\frac{1}{\Lambda} + \lambda_1(s)} + N_2 \frac{e^{\lambda_2(s)t} - e^{-\frac{t}{\Lambda}}}{\frac{1}{\Lambda} + \lambda_2(s)} \right\}.
\end{aligned}
\qquad (31.7)
$$

where the integration path of s is crossing the real axis between the poles $s=0$ and $s=\gamma$. The integral (31.7) is now the residue at $s=\gamma$, and we obtain exactly

$$I = \frac{1}{\gamma} \frac{1}{E^\gamma} \left\{ N_1 \frac{e^{\lambda_1(\gamma)t} - e^{-\frac{t}{\Lambda}}}{\frac{1}{\Lambda} + \lambda_1(\gamma)} + N_2 \frac{e^{\lambda_2(\gamma)t} - e^{-\frac{t}{\Lambda}}}{\frac{1}{\Lambda} + \lambda_2(\gamma)} \right\}.$$

This is the intensity-depth relation of the high-energy electrons and photons which can be directly compared with the experimental data. A numerical evaluation, its comparison with the experimental data, and a discussion of the results will be found in reference [9].

32. Extensive air showers[1]. Extensive air showers are interpreted as a mixture of many electron cascades starting mainly from the decay photons of π^0-mesons originating in jets produced in the atmosphere by an extremely energetic primary cosmic ray particle. Therefore the three-dimensional electron shower theory cannot be applied in a straightforward way. Models of extensive air showers can be constructed on the basis of the angular spread and of the energy spectrum of the secondaries emitted in jets of various energies.

Since the transverse momenta of secondary particles are almost independent [9] of the primary as well as the secondary particle energy, nucleons of higher energy tend to concentrate near to the core of an extensive air shower. These collimated high-energy nucleons produce the jets, the secondaries of which will be the most effective source for the further development of the extensive air shower.

α) *Lateral structure function of the shower particles at extremely high energy.* Before considering the problem in some detail, we shall treat the simple phenomena which are expected to occur in the core of extensive air showers.

Occasionally several electron showers with parallel axes close to each other were observed in the emulsion chambers exposed at Mt. Norikura or at Mt.

---

[1] For a more complete description of this subject see, for instance, the article by G. Cocconi, Handbuch der Physik, Bd. 46/1, S. 215. Berlin-Göttingen-Heidelberg: Springer 1961.

Chacaltaya [9]. Such events are called "families". There are two types of families: those of nucleonic, and those of electronic origin. The former are due to a jet produced in the atmosphere above the chamber. In this case the pairs of γ-rays from π^0-meson decay will develop into electron showers which enter the chamber with an average lateral separation of the order of

$$r_N \sim \frac{\langle P_T \rangle c}{2E_0} \frac{L_N}{X_0}$$

radiation lengths, where $\langle P_T \rangle$ is the average transverse momentum of the π^0-mesons in the jet, known to be 0.4 GeV/c [9], L_N the production height of the original jet, and E_0 the energy of each electron shower. — The latter, the families of electronic origin, are electron shower already developed in the atmosphere, and the average lateral separation between the members of the group observed in the chamber is given by

$$r_e \sim \frac{K}{E_0},$$

(r_e in radiation lengths). Since the ratio of these two separations is

$$\frac{r_N}{r_e} \sim \frac{\langle P_T \rangle c}{2K} \cdot \frac{L_N}{X_0} \sim 10 \frac{L_N}{X_0},$$

one can usually assign the family to its proper category. With regard to the families of electronic origin, we can directly compare the observed lateral distribution with the electron shower theory. Electron showers of primary energy higher than 10^{12} eV can easily be detected by naked-eye scanning of the emulsions, and the average behaviour of the observed lateral distributions of those high-energy electrons and γ-rays in the atmosphere should be represented by the function

$$\int_0^t dt' \boldsymbol{J}(E_0, t') \{\boldsymbol{\Pi}_2(E_0, E, r, t-t') + \boldsymbol{\Gamma}_2(E_0, E, r, t-t')\}, \qquad (32.1)$$

where Π_2 and Γ_2 are the lateral distribution of electrons and photons in Approximation A, and $\boldsymbol{J}(E_0, t') dt'$ is the spectrum of γ rays produced in the jets at a depth $(i', t'+dt')$ radiation lengths. As shown in the preceding section,

$$\boldsymbol{J}(E_0, t) \cong e^{-\frac{t}{\Lambda}} \frac{dE_0}{E_0^{\gamma+1}},$$

where [9]

$$\Lambda \cong (110 \pm 10) \text{ g/cm}^2, \gamma = 2.2 \pm 0.2.$$

The integral with respect to t' in (32.1) is of the form

$$\int_0^t dt' \, e^{-\frac{t'}{\Lambda} + \lambda_1(s)(t-t')},$$

and remembering that the observation level is rather deep ($t=13$ radiation lengths at Mt. Chacaltaya), it can be approximated by

$$e^{\lambda_1(s)t} \int_0^\infty e^{-\frac{t}{\Lambda} - \lambda_1(s)t'} dt' = e^{\lambda_1(s)t} \frac{1}{\lambda_1(s) + \frac{1}{\Lambda}}.$$

The complex integral with respect to s appearing in (32.1) through Π_2 and Γ_2 is the residue of the pole at $\lambda_1(\bar{s}) = 1/\Lambda$, and the observed structure function has the shape of the function at a shower age $\bar{s}$.

The numerical value of s is obtained by substituting the numerical value of Λ into the expression above. From Table 6 we see that

$$\bar{s} = 1.4.$$

A comparison between the experimental data obtained by summing up the structure functions of the individual members of a family, and the theoretical predictions according to the argument presented above, is shown in Fig. 27. The agreement between the two seems to be fairly good (*40*).

β) *Lateral structure function observed in extensive air showers.*

(i) General remarks. Shower particles in an extensive air shower are observed by various detectors such as Geiger, scintillation, or Čerenkov counters, suitably placed with separations of several meters or several tens of meters. From the particle density determined in these detectors the lateral distribution function of the shower particles in an extensive air shower is constructed.

If the shower was initiated by a galactic γ-ray (*41*), the structure function obtained in the cascade theory can be applied directly to the analysis of the experimental data. Usually, however, the showers are started by a primary cosmic-ray nucleon giving rise to a complicated mixture of showers of nucleonic (*42*) and electronic origin.

Consider a simple model for the development of an air shower in which all π^0-mesons produced in the air travel in the direction of the shower axis, and also the decay γ-rays follow the primary direction exactly [*8*]. Let $F(E_0, t)\, dE_0\, dt$ be the number of such γ-rays with energy $(E_0, E_0 + dE_0)$ resulting from the decay of the π^0-mesons produced at a depth between t and $t + dt$ cascade units below the top of the atmosphere. Then the structure function G at a depth T is given by

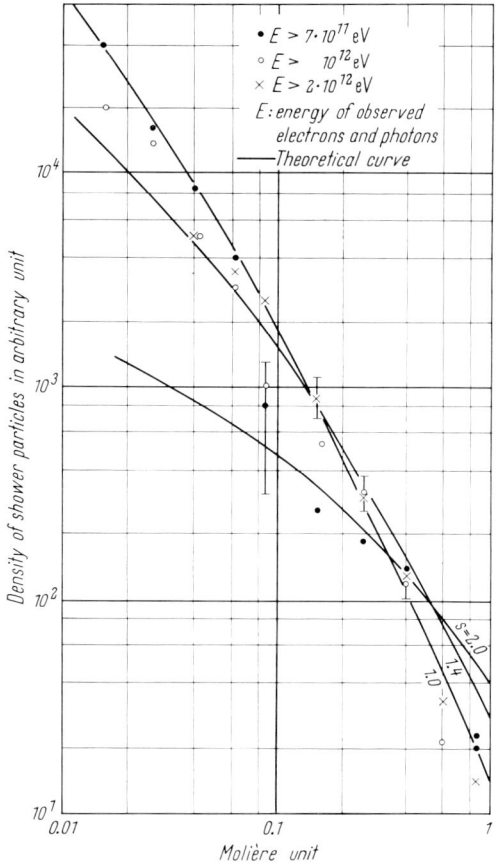

Fig. 27. Lateral distribution of high-energy shower electrons observed in emulsion chambers. [From M. OGAWA and T. MOROOKA (*40*)].

$$G = \int_0^\infty dE_0 \int_0^T dt\, F(E_0, t)\, \boldsymbol{\Pi}_2(E_0, 0, r, T-t). \tag{32.2}$$

Substituting (24.19) into (32.2) and integrating with respect to E_0 and t, we get

$$G \sim L_F(s, T) f(r, s)\, e^{\lambda_1(s) T}, \tag{32.3}$$

where

$$L_F(s, \tau) = \int_0^\infty dE_0 \left(\frac{E_0}{\varepsilon}\right)^s \int_0^T dt\, F(E_0, t)\, e^{\lambda_1(s) t}, \tag{32.4}$$

and $f(r, s)$ is the structure function at a shower age s defined by

$$\left(\frac{1}{L_F}\right)\frac{dL_F(s)}{ds} + \lambda_1(s) T = 0. \tag{32.5}$$

The absorption mean free path of these particles, L, at the depth T can be derived from the Eq. (32.3). Using (32.3) one finds

$$-\left(\frac{1}{L}\right) = \left(\frac{1}{L_F}\right)\frac{dL_F}{dT} + \lambda_1(s). \tag{32.6}$$

As L_F is a monotonously increasing function of T, we get from (32.6)

$$-\frac{1}{L} \gtrsim \lambda_1(s), \tag{32.7}$$

Taking 150 g/cm² of air as the value of L at the observation level we have

$$\lambda_1(s) \lesssim -0.25,$$

and therefore

$$s \gtrsim 1.3.$$

Thus it becomes evident that in this case the structure function can approximately be represented by that of a single cascade of a shower age s which is determined with the aid of the absorption mean free path L.

More detailed calculations for the lateral distribution of the compound cascade of the air showers are needed in order to show the slight dependence of the age parameter with changing distance from the shower axis. Such calculations, assuming a certain model for the multiple production of mesons in the atmosphere, were carried out, for example, by SITTE[1]. According to his calculation the age parameter increases slightly with increasing the distance from the shower axis, which means qualitatively that the lateral distribution becomes rather flatter than that expected from the single shower theory. Variations of about 0.1 in the age parameter between distances of 0.1 to 1.5 Molière unit were obtained for a shower of size about 10^5 particles. This result may be due to the fact that the shower particles far from the shower axis are preferentially contributed by the nucleonic jets produced at higher levels with energies higher than those of the jets responsible for the average behaviour of the shower particles in an extensive air shower. The mean free path of these high-energy nucleons may be slightly shorter than that cited above, thus making the age parameter of their contribution slightly bigger than that of the lateral distribution near the shower axis.

This, and the contribution of μ-mesons in the extensive air showers (39), account for the observation of similar shower age parameters for the lateral distribution of the extensive air shower with different shower sizes, because large showers are usually studied with detectors spaced over rather large distances, while less extensive arrays are used for small showers.

Approaching the problem from the opposite direction we can proceed as follows: Let the energy and the angular distributon of the source γ-rays be

$$\frac{\theta \, d\theta}{\theta^n} \delta\left(\theta - \frac{P_t}{E_0}\right) d\left(\frac{P_t}{E_0}\right), \tag{32.8}$$

where θ, P_t and E_0 are the angle at production, the transverse momentum, and the energy of the γ-rays. Suppose these γ-rays are produced t radiation lengths

---
[1] Private communication.

above the observation level. Then the γ-rays arrive at a distance $r(=t\theta)$ from the shower axis, and their laterial distribution is given by

$$F(r, E_0)\, r\, dr\, dE_0 = t^{n-2}\, \frac{r\, dr}{r^n}\, \delta\left(\frac{r}{t} - \frac{P_t}{E_0}\right) d\frac{P_t}{E_0}. \tag{32.9}$$

Therefore the lateral distribution of the shower electrons started by these γ-rays is given by

$$G(r) = \int d\mathbf{r}' \int dE_0 F(\mathbf{r}-\mathbf{r}', E_0)\, \Pi_2(E_0, 0, \mathbf{r}', t), \tag{32.10}$$

where Π_2 is the lateral structure function of a γ-ray iniciated shower of energy E_0. Since the integral (32.10) has just the same form as that treated in Appendix 1, the Hankel transformation is applicable, and one gets

$$J_G = 2\pi \int dE_0\, J_F\, J_{\Pi_2}, \tag{32.11}$$

where J_G, J_F and J_{Π_2} are the Hankel transforms of the respective functions. Π_2 is given by (24.19), and for small values of r can be written in the form

$$\Pi_2 \sim \int ds \left(\frac{E_0}{\varepsilon}\right)^s \frac{1}{r^{2-s}}\, e^{\lambda_1(s)t}. \tag{32.12}$$

The integral with respect to E_0 in (32.10) gives

$$\int dE_0\, F(\mathbf{r}-\mathbf{r}', E_0)\, \Pi_2(E_0, 0, \mathbf{r}, t) \sim \int ds \left(\frac{t P_t}{\mathbf{r}-\mathbf{r}'}\right)^s \frac{1}{(\mathbf{r}-\mathbf{r}')^n}\, \frac{1}{r^{2-s}}. \tag{32.13}$$

Thus we have

$$J_G \sim \int ds \left(\frac{t P_t}{\varepsilon}\right)^s \zeta^{n-2}\, e^{\lambda_1(s)t}, \tag{32.14}$$

and finally

$$G \sim \frac{1}{r^n} \int ds \left(\frac{t P_t}{\varepsilon}\right)^s e^{\lambda_1(s)t}. \tag{32.15}$$

Therefore the lateral distribution of the shower particles from sources with an angular spread $\dfrac{\theta\, d\theta}{\theta^n}$ is not properly described as a function of the shower age, but is directly affected by the angular spread of the source γ-rays.

(ii) *The experimental evidence.* The comparison of the experimental results averaged over a large amount of data (43) with the theoretical lateral structure function[1] of a single cascade is shown in Fig. 28. Beyond 20 m, the agreement is quite good between the curve calculated for $s=1.4$, and the observed data. For smaller distances the experimental data seem to be better fitted by a theoretical curve of rather younger shower.

If this is correct, the deviation from the structure function described by a unique shower age warrants serious discussion. There is, however, some evidence that especially within 10 m from the shower axis the record of a plastic scintillator shows particle numbers in excess of the true ones. This result was obtained by comparing the number of shower particles registered in a neon tube hodoscope with that counted at the same distance by scintillators. The discrepancy may be

---

[1] For the comparison with experimental data, it is necessary to use in the calculations in Approximation B the value of the Molière unit at about 2 radiation length above the observation level [8]. This stems from the fact that the density of the air decreases at higher altitude where the observed showers are developed.

The corresponding figures in Approximation A, depending on the shower age, are [8]

s	0.6	1.0	1.4	2.0	
ΔT	0.98	1.75	3.3	7.43	rad. lengths

explained by the fact that, as stated in Sect. 28, high-energy photons are concentrated near the shower axis, and some of them will contribute to the scintillator record by producing an electron pair within the plastic material. The deviations may thus be more apparent than real, and it may be concluded that the average features of the observed lateral distribution are not much affected by the spread of the source γ-rays, at least at distances beyond a few meters from the shower axis. As a result of such an analysis, it could also be shown that the transverse momenta of the π^0-mesons produced in air showers are of the order of several hundred MeV/c (*44*).

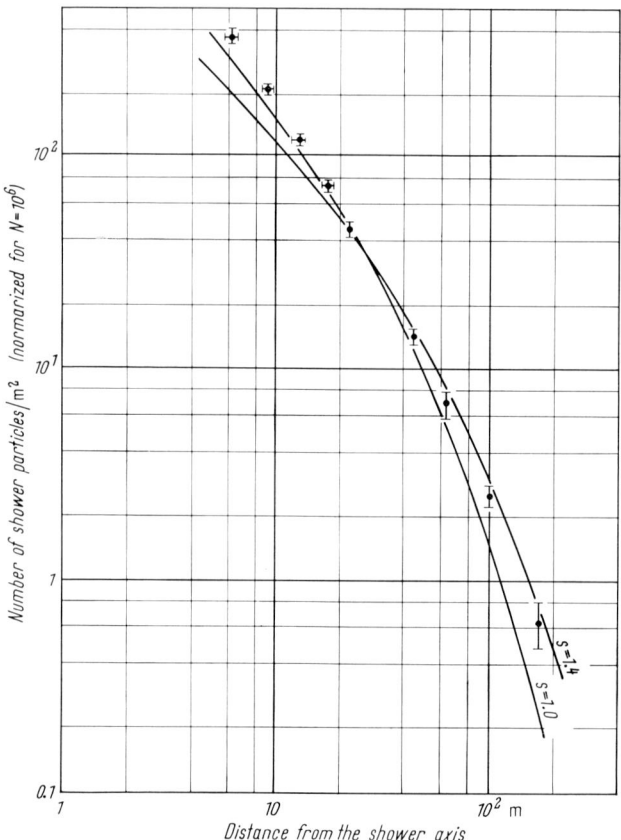

Fig. 28. Average lateral distribution of shower particles observed at sea level for showers with a total particle number $\geq 10^5$. [From ref. (*43*)].

By fitting to the experimentally observed lateral distribution of an individual air shower a theoretical structure function which a properly adjusted age parameter s, it is also possible to estimate the total number of shower particles arriving at the observation level. This is the procedure by which generally the energy of the shower primary is determined. Attempts were also made by several authors (*45*) to use the distribution of the values of the shower age s required for a fit with the observations on individual events, for the study of the principal features of the shower development in the atmosphere.

In the vicinity of the shower axis, say within a few meters, large fluctuations in the lateral distribution of the shower particles are often observed. There the spread of the source γ-rays can no longer be neglected. Furthermore, if an energetic nuclear-active particle near the shower axis produces a jet at a certain distance above the observation level, the electron showers started from this jet could give a dominant contribution to the lateral distribution near the center. The amount of this contribution depends on the energy of the nucleonic particle, and on the height at which the jet was produced. The situation can again be represented by our relation (32.15).

Examples of the lateral distribution near the core registered with a neon hodoscope [*46*] are shown in Fig. 29, in which various features of the distribution can be observed. Occasionally the records show a double-peaked structure from which information concerning nuclear interactions at extremely high energies may be derived [*9*].

Sect. 32. Extensive air showers.

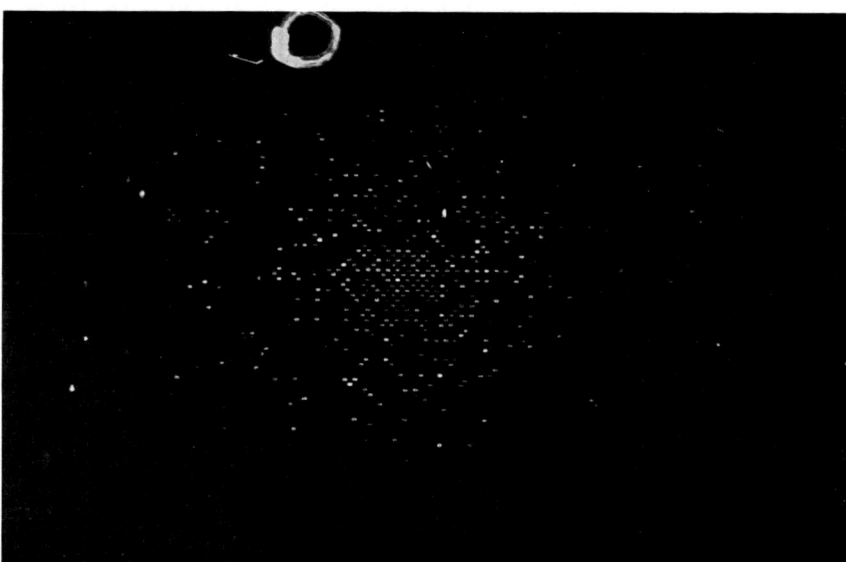

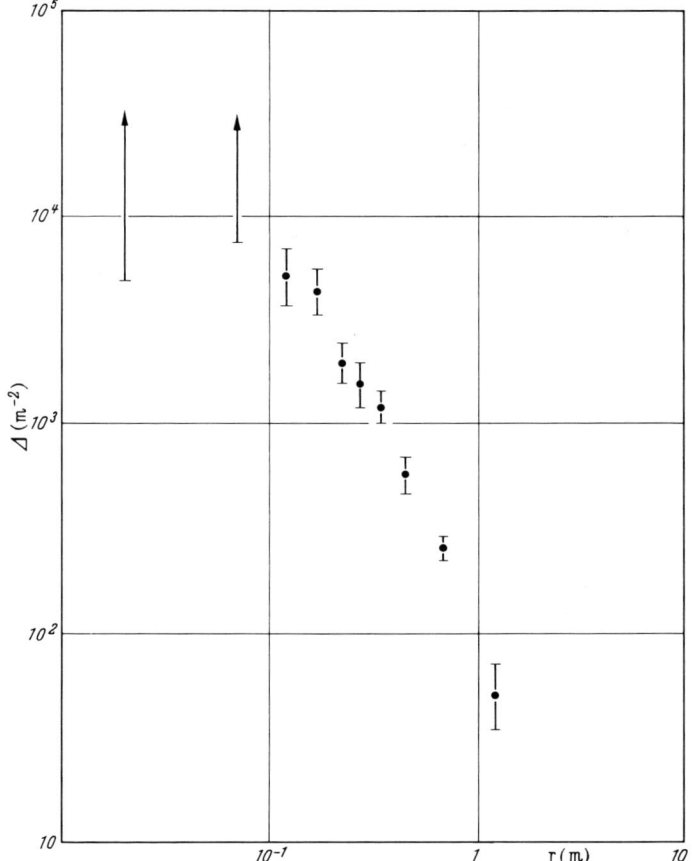

Fig. 29a Example of the lateral distribution of shower electrons near the shower axis observed with a neon hodoscope. [From M. ODA and Y. TANAKA (46)].

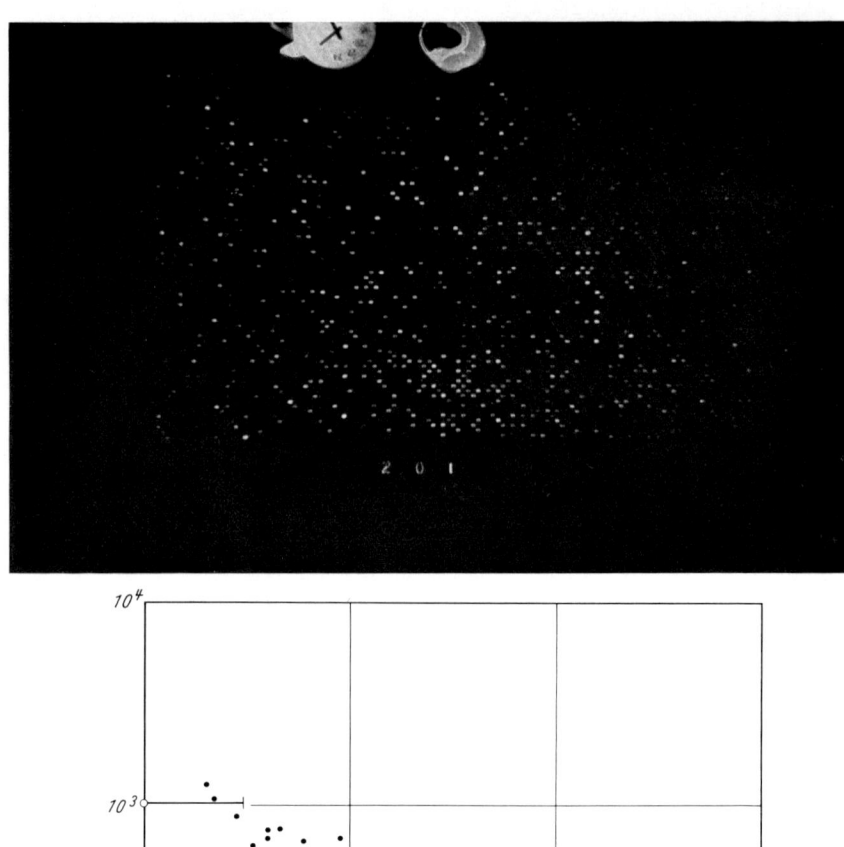

Fig. 29b. Example of the lateral distribution of shower electrons near the shower axis observed with a neon hodescope. [From M. Oda and Y. Tanaka (46)].

γ) *Lateral distribution of the energy flow of shower particles.* Apart from the lateral distribution in the number of shower particles, the lateral distribution of the energy flow carried by the shower particles also yields significant information on extensive air showers. As mentioned in Sect. 12, the lateral distribution of the energy flow is expected to be approximately $1/r$ times steeper than the number distribution. This is a simple physical consequence of the fact that the more energetic shower particles are less scattered, and tend to propagate nearer to the shower axis.

Quantitative calculations can be carried out as follows. Let $\Pi_E\, 2\pi r\, dr$ and $\Gamma_E\, 2\pi r\, dr$ be the lateral distributions of the energy flow of electrons and photons respectively. They are represented by the following expressions:

$$\Pi_E = \int_0^\infty dE\, E\, \pi_2(E_0, E, r, t), \tag{32.16}$$

$$\Gamma_E = \int_0^\infty dE\, E\, \gamma_2(E_0, W, r, t), \tag{32.17}$$

where π_2 and γ_2 are the lateral structure functions defined in Sect. 24.

In order to simplify the calculation, the solutions in the Landau approximation may be used. Then we have

$$\Pi_E = -\frac{\varepsilon}{4\pi^2} \int\!\!\!\int_{-i\infty}^{+i\infty} ds\, dp \left(\frac{E_0}{\varepsilon}\right)^s \left(\frac{\varepsilon}{E_s}\right)^2 \left(\frac{\varepsilon^2 r^2}{E_s^2}\right)^{-p-1}$$
$$\times \Gamma(p+1)\,\Gamma(s-1+2p)\,\mathfrak{M}_2(p, -2p-s+1, s, t), \tag{32.18}$$

$$\Gamma_E = -\frac{\varepsilon}{4\pi^2} \int\!\!\!\int_{-i\infty}^{+i\infty} ds\, dp \left(\frac{E_0}{\varepsilon}\right)^s \left(\frac{\varepsilon}{E_s}\right)^2 \left(\frac{\varepsilon^2 r^2}{E_s^2}\right)^{-p-1}$$
$$\times \Gamma(p+1)\,\Gamma(s-1+2p)\,\mathfrak{M}'_2(p, -2p-s+1, s, t). \tag{32.19}$$

It is sometimes convenient to represent Π_E and Γ_E in terms of the structure function Π_2 given in Sect. 22. The average energy of the electrons at a distance r from the shower axis is given by Π_E/Π_2, and we put

$$\frac{\Pi_E}{\Pi_2} = \varepsilon \frac{r_1}{r} g_e\!\left(\frac{r}{r_1}\right)$$

and

$$\frac{\Gamma_E}{\Pi_2} = \varepsilon \frac{r_1}{r} g_\gamma\!\left(\frac{r}{r_1}\right).$$

Numerical calculations of g_e and g_γ at the shower maximum were carried out, and the results are summarized in Table 21 taken from reference [8].

Table 21.

r/r_1	0	0.01	0.4	0.1	1.0	2.0	4.0
$g_e(r/r_1)$	0.15	0.15	0.12	0.13	0.14	0.20	0.30
$g_\gamma(r/r_1)$	0.21	0.26	0.45	0.82	0.83	1.8	3.7

They show that g is almost constant near the shower axis; thus Π_E and Γ_E are approximately given by $1/r^{3-s}$ as predicted at the beginning of this section.

At a large distance, however, g_e and g_γ increase nearly in proportion to the distance from the shower axis. Therefore the average energy of the shower particles remains almost constant in this region. The result may be interpreted in the following way. The mean free path of the photons for pair creation remains almost constant at all energies. Thus low-energy photons can travel up to a large distance from the shower axis, and the electrons found in this region are produced by these low-energy photons. But the resulting low-energy electrons can travel only a short distance because they lose their energy by the ionization process. As a result the average energy of the shower particles at varying distances from the axis remains constant, and the abundance of the photons compared with that of the electrons is much enhanced at a large distance from the core. For a comparison with the experimental information, arguments similar to those presented in the section on the lateral distribution should be applied.

The lateral distribution of the energy flow in an extensive shower was determined in an arrangement using large Čerenkov counters and scintillation detectors under lead plates. The results are shown in Fig. 30, for extensive air showers observed at sea level (43), and at Mt. Chacaltaya Laboratory (41).

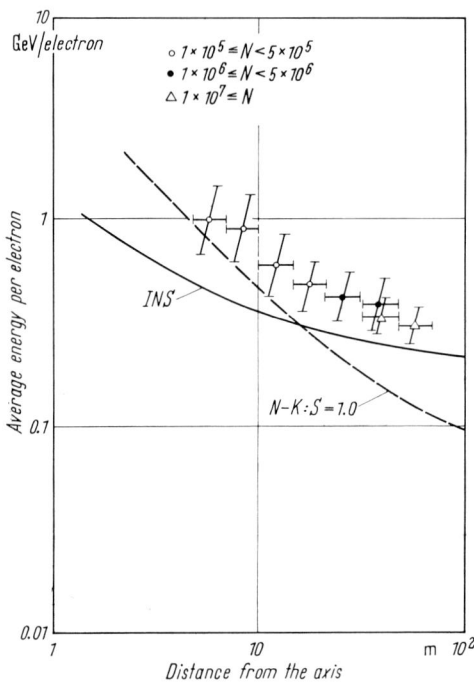

Fig. 30. Lateral distribution of the energy flow of electrons and photons in extensive air showers. The circles and triangles show the experimental data on the energy flow observed for a particle number N at Mt. Chacaltaya (5,200 m a.s.l.), the full curve "INS" the energy flow observed at sea level, and the dashed curve "N-K" the result of calculations for $s=1.0$. [From ref. (41) and (43)].

Summary. Since an electron shower is a phenomena which plays a considerable part in interactions at high and ultra-high energies, and furthermore the elementary processes involved are rather accurately known, the theory has been developed to a fair degree of exactness in the last three decades.

In this review article, the basic ideas of the electron shower theory are described, and one-dimensional electron shower theories with and without consideration to the contribution of the ionization losses are summarized, based on the mathematical technique of Mellin transformations and the method of integral representation. Three-dimensional shower theories with and without account of the contribution of the large-angle scattering of the shower electrons are reviewed on the basis of similar mathematical arguments, and it is shown that the terms proportional to $e^{\lambda_1(s)t}$ of the shower functions in those theories already give results accurate enough for practical purposes, in spite of the complex features of those functions.

The shower functions are applied to the analysis of electron showers observed in cosmic ray phenomena using various detectors, and it is shown that the theories are satisfactory for this purpose, within the frame set by the approximations used in each theory. A further improvement of the theories is, however, necessary in some instances, in spite of the mathematical difficulty inherent to this refinement. These points are described individually in this paper.

The mathematical procedures here summarized are a powerful tool for the solution of other problems of similar type. As an example, a treatment of the range fluctuation problem of high-energy μ-mesons is presented in Appendix 5.

Appendix.

1. Functional transformations. It has been well recognized that the functional transformation is one of the most powerful methods to derive solutions of differential or integral equations. The idea underlying this technique is to get a solution satisfying certain boundary conditions by summing up special solutions of a given equation. The transformation functions should form a complete, orthogonal set. Trigonometric, exponential and power functions with complex parameters, and the Bessel functions are widely used, and corresponding to these functions, the transforms are called the Fourier, Laplace, Mellin, and Hankel transforms.

Among these transformations, the Fourier transformation is historically the oldest, and others may be interpreted as its extensions. Changing the parameter of the Fourier transforms to a complex parameter, one gets the Laplace transforms. Replacing the variable in the function of the Laplace transforms by a logarithmic variable, the Mellin transforms are obtained. Double Fourier transformation with variables x and y can be reduced to the Hankel transformation, if the function is represented by the variable $r^2 = (x^2 + y^2)$ only.

The Fourier transformations are used for diffusion-type equations, the Laplace, Mellin and Hankel transformations for linear differential and integral equations with a kernel of certain types, i.e., integral equations with a kernel of a fractional form, and two-dimensional diffusion-type equations with axial symmetry, respectively.

In deriving the shower functions from the basic equations, these transformations are widely used, and some of the features of these transformations will be briefly summarized in this section.

A.1.1. Laplace transformation. The Laplace transform of a given function, $f(x)$, is defined by the integral

$$\int_0^\infty e^{-\mu x} f(x) = \mathfrak{L}_f(\mu), \qquad (A.1.1)$$

where μ is a complex variable. If this integral exists in a certain domain of μ, e.g., the domain with $R(\mu) < R(\bar{\mu})$, $\bar{\mu}$ being a certain fixed value, the inverse transform is given by

$$f = \frac{1}{2\pi i} \int_c d\mu \, \mathfrak{L}_f(\mu), \qquad (A.1.2)$$

where the path of the integration is running parallel to the imaginary axis in the convergence domain. The relation can be proved if one considers the Fourier transform of the function f_1 defined by

$$f_1(x) = e^{-\lambda x} f(x), \qquad (A.1.3)$$

where one assumes λ to be a real variable and equal to the real part of μ.

Example: The solution of the equation

$$\frac{df}{d\chi} = -\alpha f \quad \text{with the boundary condition } f(x=0) = 1, \qquad (A.1.4)$$

is given by $f = e^{-\alpha x}$.

The solution can also be obtained by the method of the Laplace transformation. Multiplying by $e^{-\mu x}$ on both sides of the above equation and integrating with respect to x, one gets

$$(\mu + \alpha) \mathfrak{L}_f = 1,$$

as

$$\int_0^\infty e^{-\mu x} \frac{df}{dx} dx = e^{-\mu x} f \Big|_0^\infty + \mu \int_0^\infty e^{-\mu x} f(x) = -1 + \mu L_f. \tag{A.1.5}$$

Then, from the inverse transformation, we have

$$f = \frac{1}{2\pi i} \int \frac{e^{\mu x}}{\mu + \alpha} d\mu, \tag{A.1.6}$$

and performing the complex integral with the help of a simple pole at $\mu = -\alpha$, we finally have the solution

$$f = e^{-\alpha x}.$$

A.1.2. Mellin transformation. If the kernel of the integral equation is of a fractional form, as in the case of the basic equation of the cascade theory, the Mellin transformation is known to be useful in deriving the solution.

Given a certain function $f(x)$, its Mellin transform is defined by

$$\mathfrak{M}_f = \int_0^\infty x^s f(x) dx, \tag{A.1.7}$$

where s is a complex variable.

The inverse transformation is

$$f(x) = \frac{1}{2\pi i} \int_c \frac{\mathfrak{M}_f(s)}{x^{s+1}} ds, \tag{A.1.8}$$

where the integration path c is running in the convergence domain parallel to the imaginary axis.

If one replaces the variables e^{-x} and μ in the Laplace transform by x and s, the Laplace transform is reduced to the Mellin transform.

Suppose an integro-differential equation with a kernel of a fractional form such as

$$\frac{df(x,t)}{dt} = \int_0^\infty K\left(\frac{x}{y}\right) f(y) \frac{dy}{y} \tag{A.1.9}$$

is given, with $f(x, t) = \delta(x - x_0)$ for $t = 0$. Multiplying by x^s on both sides of the equation, we have

$$\frac{d\mathfrak{M}_f(s,t)}{dt} = \mathfrak{M}_K(s) \mathfrak{M}_f(x,t) \quad \text{with } \mathfrak{M}_f = x_0^s, \quad \text{for } t = 0, \tag{A.1.10}$$

since the relation

$$\int_0^\infty dx\, x^s \int_0^\infty K\left(\frac{x}{y}\right) f(y) \frac{dy}{y} = \int_0^\infty y^s f(y) dy \int_0^\infty \left(\frac{x}{y}\right)^s K\left(\frac{x}{y}\right) \frac{dx}{y}.$$

holds. From (A.1.10) and the inverse transformation, we have finally the solution as

$$f(x) = \frac{1}{2\pi i} \int ds \left(\frac{x_0}{x}\right)^s \frac{1}{x} e^{\mathfrak{M}_K(s) t}. \tag{A.1.11}$$

Example: The Mellin transform of the function $f(x) = e^{-\lambda x}$, is given by

$$\mathfrak{M}_f = \int_0^\infty x^s e^{-\lambda x} dx = \frac{\Gamma(s+1)}{\lambda^{s+1}}.$$

Applying the inverse transformation, one gets

$$f = \frac{1}{2\pi i} \int_{c-i\infty}^{c+i\infty} ds \, \frac{\Gamma(s+1)}{\lambda^{s+1}},$$

where the integration path is running in the region where $|s| > -1$.

Since $\Gamma(s+1)$ has simple poles at $s = -1, -2, \ldots$, we have with the help of the residues

$$f = \sum \frac{(-\lambda x)^n}{n!},$$

evidently identical with the initial function

$$f(x) = e^{-\lambda x}.$$

A.1.3. Hankel transformation. Let $f(x, y)$ be a function of x and y. Then the Fourier transform with respect to these two variables is

$$F(\xi, \eta) = \frac{1}{2\pi} \int\!\!\int_{-\infty}^{+\infty} dx \, dy \, e^{ix\xi + iy\eta} f(x, y), \tag{A.1.12}$$

with an inverse transform

$$f = \frac{1}{2\pi} \int\!\!\int_{-\infty}^{+\infty} d\xi \, d\eta \, e^{-ix\xi - iy\eta} F(\xi, \eta). \tag{A.1.13}$$

The formula can also be written in the following form, with the substitutions

$$r \cos\theta = x, \quad r \sin\theta = y, \quad \zeta \cos\varphi = \xi, \quad \zeta \sin\varphi = \eta,$$

$$F(\zeta \cos\varphi, \zeta \sin\varphi) = \frac{1}{2\pi} \int_0^{2\pi} d\theta \int_0^{\infty} r \, dr \, e^{ir\zeta \cos(\theta - \varphi)} f(r \cos\theta, r \sin\theta), \tag{A.1.14}$$

$$f = \frac{1}{2\pi} \int_0^{2\pi} d\varphi \int_0^{\infty} \zeta \, d\zeta \, e^{-ir\zeta \cos(\theta - \varphi)} f(\zeta \cos\varphi, \zeta \sin\varphi). \tag{A.1.15}$$

If the function is azimuthally symmetric, or in other words a function of the variable $r = (x^2 + y^2)^{\frac{1}{2}}$ only, then (A.1.14) becomes

$$F = \frac{1}{2\pi} \int_0^{\infty} r \, dr \, f(r) \int_0^{2\pi} e^{ir\zeta \cos(\theta - \varphi)} d\theta. \tag{A.1.16}$$

Because of the periodic nature of $\cos(\theta - \varphi)$, (A.1.16) may be reduced to

$$F = \frac{1}{2\pi} \int_0^{\infty} r \, dr \, f(r) \int_0^{2\pi} e^{ir\zeta \cos\theta} d\theta. \tag{A.1.17}$$

Thus we have

$$F = \int_0^{\infty} r \, dr \, \boldsymbol{J}_0(\zeta, r) f(r), \tag{A.1.18}$$

instead of (A.1.12), where we have made use of the relation for the Bessel function

$$\int_0^{2\pi} e^{i\zeta r \cos\theta} d\theta = 2\pi \boldsymbol{J}_0(\zeta, r). \tag{A.1.19}$$

In the same way the inverse transform gives us

$$f = \int_0^\infty \zeta \, d\zeta \, F(\zeta) \, J_0(\zeta r). \qquad (A.1.20)$$

This is what we call the Hankel transformation, and is widely applied to two-dimensional problems with azimuthally symmetric variables.

Example: Let $f(x, y)$ and $g(x, y)$ be the axially symmetric function. Then the integral

$$h(r) = \int_0^\infty d\mathbf{r}' f(\mathbf{r} - \mathbf{r}') g(\mathbf{r}') \qquad (A.1.21)$$

can be evaluated with the help of the Hankel transformation. Multiplying by $\frac{1}{2\pi} e^{ix\xi + iy\eta}$ on both sides of (A.1.21), and integrating with respect to x and y from $-\infty$ to $+\infty$, the right hand side of this relation is given by

$$\begin{aligned}
& \frac{1}{2\pi} \int\!\!\!\int\!\!\!\int\!\!\!\int_{-\infty}^{+\infty} e^{ix\xi + ix\eta} f(x - x', y - y') \cdot g(x', y') \, dx \, dy \, dx' \, dy' \\
& = \frac{1}{2\pi} \int\!\!\!\int_{-\infty}^{+\infty} dx \, dy \, e^{ix\xi + iy\eta} f(x, y) \int\!\!\!\int_{-\infty}^{+\infty} dx' \, dy' \, e^{ix'\xi + iy'\eta} g(x', y') \\
& = 2\pi \int_0^\infty J_0(r\zeta) f(r) \, r \, dr \int_0^\infty J_0(r'\zeta) g(r') \, r' \, dr'.
\end{aligned} \qquad (A.1.22)$$

Then the Hankel transform of the function h is 2π times the product of the Hankel transform of each function f and g. Thus with the help of the inverse transformation, one gets

$$h(r) = 2\pi \int_0^\infty J_0(r\zeta) \, \zeta \, d\zeta \, F(\zeta) \, G(\zeta), \qquad (A.1.23)$$

where F and G are the Hankel transform of the function f and g.

Expressions such as (A.1.21) appear frequently in the problem of the spread from axially symmetric sources. An actual application is, for instance, made in the theory of multiple scattering.

A.1.4. Numerical evaluation of complex integral (saddle-point method). If the solution of an equation is obtained with a help of the Laplace or Mellin transforms, the solution is generally given in the form of an integral with respect to complex variables. The next problem is how to evaluate this complex integral. Sometimes the integration is performed with the help of poles as shown in the examples in the preceding sections. In most cases, however, the transform, $\mathfrak{L}_f$ or $\mathfrak{M}_f$, is so complicated that the evaluation of the inverse integral cannot be carried out in this simple manner. In such cases, the saddle-point method is applied to obtain the numerical values of the integral.

Let $\phi(z)$ be a regular function of complex variable, $z = x + iy$. Then from the Cauchy-Riemann equation,

$$\frac{\partial \phi}{\partial x} = -i \frac{\partial \phi}{\partial y}, \qquad (A.1.24)$$

one gets

$$\frac{\partial^2 \phi}{\partial x^2} = -\frac{\partial^2 \phi}{\partial y^2}. \qquad (A.1.25)$$

That is, if the first derivative along the real axis is zero at a certain point $z = \bar{x}$, the first derivative along the imaginary axis should also be zero at the same point, while the second derivatives along the real and the imaginary axis have the same absolute values with opposite signs.

If the integral

$$f = \frac{1}{2\pi i} \int dz\, \psi(z) \tag{A.1.26}$$

is written in the form

$$f = \frac{1}{2\pi i} \int dz\, e^{\phi(z)}, \tag{A.1.27}$$

and $\frac{\partial \phi}{\partial x} = 0$, $\frac{\partial^2 \phi}{\partial x^2} > 0$, at $z = \bar{x}$, then the integrand may be approximated by a Gaussian function on the path parallel to the imaginary axis crossing the real axis at $x = \bar{x}$. For expanding the function $\phi(z)$ around $z = \bar{x}$, where $\frac{\partial \phi}{\partial x} = 0$, one gets

$$\begin{aligned}\phi(z) &= \phi(x) + iy\, \frac{\partial \phi}{\partial i y} + \frac{1}{2!}(iy)^2 \frac{\partial \phi^2}{\partial i y^2} + \cdots \bigg|_{x=\bar{x}} \\ &= \phi(x) - \frac{1}{2} y^2 \frac{\partial^2 \phi}{\partial x^2} + \cdots \bigg|_{x=\bar{x}}. \end{aligned} \tag{A.1.28}$$

Then f is given by

$$f = \frac{1}{2\pi} \int_{-\infty}^{\infty} dy\, e^{\phi(\bar{x}) - \frac{1}{2} \frac{\partial^2 \phi}{\partial x^2} y^2 + \cdots} = \frac{1}{\sqrt{2\pi \frac{\partial^2 \phi}{\partial x^2}\big|_{x=\bar{x}}}}\, \psi(\bar{x}) + \cdots, \tag{A.1.29}$$

where $\bar{x}$ is given as a solution of the equation

$$\frac{\partial \phi}{\partial x} = 0. \tag{A.1.30}$$

The example shown in the section on the Mellin transform, (A.1.11)

$$f = \frac{1}{2\pi i} \int ds \left(\frac{x_0}{x}\right)^s \frac{1}{x} e^{\mathfrak{M}_K(s) t} \tag{A.1.11}$$

is now evaluated as

$$f = \frac{1}{\sqrt{2\pi \frac{\partial^2 \mathfrak{M}_K t}{ds^2}\big|_{s=\bar{s}}}} \left(\frac{x_0}{x}\right)^{\bar{s}} \frac{1}{x} e^{\mathfrak{M}_K(\bar{s}) t}, \tag{A.1.31}$$

where $\bar{s}$ is determined by putting the first derivation equal to zero, that is by the equation

$$\log \frac{x_0}{x} + \frac{\partial \mathfrak{M}_k}{\partial s} t = 0. \tag{A.1.32}$$

In order to get more accurate numerical values, one should include the contributions of the higher-order derivatives. In that case, the function is developed [7] as

$$\phi(z) = \phi(x) - \frac{1}{2} \frac{\partial^2 \phi}{\partial x^2} y^2 - \frac{1}{3!} \frac{\partial^3 \phi}{\partial x^3} (iy)^3 + \frac{1}{4!} \frac{\partial^4 \phi}{\partial x^4} y^4 \cdots \bigg|_{x=\bar{x}},$$

and

$$\begin{aligned} f &= \frac{1}{2\pi} \int_{-\infty}^{+\infty} dy\, e^{\phi(\bar{x}) - \frac{1}{2} \frac{\partial^2 \phi}{\partial x^2} y^2 - \frac{1}{3!} \frac{\partial^3 \phi}{\partial x^3}(iy)^3 + \frac{1}{4!} \frac{\partial^4 \phi}{\partial x^4} y^4 + \cdots}\bigg|_{x=\bar{x}} \\ &= \frac{1}{2\pi} \int_{-\infty}^{+\infty} dy\, e^{\phi(\bar{x}) - \frac{1}{2} \frac{\partial^2 \phi}{\partial x^2} y^2} \left[1 + \frac{1}{3!} \frac{\partial^3 \phi}{\partial x^3} (iy)^3 + \frac{1}{4!} \frac{\partial^4 \phi}{\partial x^4} y^4 + \cdots \right]\bigg|_{x=\bar{x}} \\ &= \frac{1}{\sqrt{2\pi \frac{\partial^2 \phi}{\partial x^2}}}\, \psi(\bar{x}) \left[1 + \frac{3}{4!} \frac{\partial^4 \phi}{\partial x^4} \bigg/ \left(\frac{\partial^2 \phi}{\partial x^2}\right)^2 + \cdots \right]\bigg|_{x=\bar{x}}. \end{aligned} \tag{A.1.33}$$

Though the approximation up to the second derivatives may seem to give only approximate numerical values, it is known that the inaccuracy due to this procedure is in general less than a few percent in the evaluation of the shower functions [3], (47).

If the function is given by a double complex integral, i.e.

$$f(x) = -\frac{1}{4\pi^2} \iint ds\, dp\, e^{\phi(s,p,x)}, \qquad (A.1.34)$$

where s and p are complex variables, the evaluation is obtained from

$$f = \frac{1}{2\pi \left(\frac{\partial^2 \phi}{\partial s^2} \frac{\partial^2 \phi}{\partial p^2} - \left(\frac{\partial^2 \phi}{\partial p\, \partial s}\right)^2\right)^{\frac{1}{2}}\Big|_{\substack{s=\bar{s}\\p=\bar{p}}}} e^{\phi(\bar{s},\bar{p},x)}, \qquad (A.1.35)$$

where $\bar{s}$ and $\bar{p}$ are defined by the equations

$$\frac{\partial \phi}{\partial s}\bigg|_{\substack{s=\bar{s}\\p=\bar{p}}} = 0 \qquad \frac{\partial \phi}{\partial p}\bigg|_{\substack{s=\bar{s}\\p=\bar{p}}} = 0. \qquad (A.1.36)$$

This expression is derived as a general extension of the saddle-point method for a single complex integral (27).

2. Integral representation of infinite series. If the solution $f(x)$ of a certain equation is given in a series form such as

$$f(x) = \sum_{n=0}^{\infty} a_n (-x)^n, \qquad (A.2.1)$$

with a convergence domain $0 \le x \le c$, the problem may arise to compute the numerical values of f outside of this domain. One must find another series having a different convergence domain by changing the form of the series (A.2.1), and the procedure is called the analytic continuation of the series (A.2.1.)

In the theory of electron showers, an exact solution in Approximation A is easily derived with the help of the Mellin transformation. This was shown in Sect. 13. In Approximation B, however, the ionization effect $\varepsilon \frac{\partial \pi}{\partial E}$ in the basic equation gives a term of non-fractional form with respect to the energy, and the solution is given by a power series in ε/E. Since generally the series does not converge for $\frac{\varepsilon}{E} > 1$, the method of analytic continuation was introduced in the electron shower theory by BHABHA and CHAKRABARTY (27), and afterwards widely used by KAMATA and NISHIMURA [8] and by CHAKRABARTY and GUPTA (17).

For the analytic continuation of a series, the integral representation of the infinite series seems to be most promising. As is usually recognized, the series (A.2.1) can be regarded as an expansion of complex integrals with simple poles at $p = 0, 1, \ldots$, and the term $a_n (-x)^n$ is just its residue at the pole $p = n$. An example of a function having such poles with residues $(-1)^p$ is the product $-\Gamma(-p)\Gamma(p+1)$, and thus the series can be represented by

$$f(x) = \frac{1}{2\pi i} \int_c dp\, \Gamma(-p) \Gamma(p+1) a_p\, x^p, \qquad (A.2.2)$$

where the integration path runs between the poles $p = 0$ and $p = -1$. Since the above function gives numerical values for any value of x, this integral representation seems to be quite useful to derive the numerical values of the shower function.

One may construct the series from (A.2.2), which may be applicable for $x > c$. In this case the integration path should be traced counter-clockwise, thus yielding

$$f = -\sum_{n=1}^{\infty} a_{-n}\left(\frac{-1}{x}\right)^n. \tag{A.2.3}$$

Example: Consider a function given by

$$f(x) = \frac{1}{1+x}, \tag{A.2.4}$$

The function is expanded in the following form for $x < 1$,

$$f(x) = 1 - x + x^2 - \cdots. \tag{A.2.5}$$

The series may be regarded as a form of (A.2.1), with the coefficients $a_n = 1$, for any value of n. Applying (A.2.3), one has for the above series

$$f(x) = \frac{1}{x} - \frac{1}{x^2} + \frac{1}{x^3} - \cdots, \quad \text{for} \quad x \gg 1.$$

This is just the series expansion of the function (A.2.4) for $x \gg 1$.

3. Theory of multiple scattering. Whenever a charged particle passes through matter, it suffers deflections of its path because of the Coulomb field of the atomic nuclei. It was shown in Sect. 6 that this is the main source of the lateral and angular spreads of electrons in a cascade shower.

The number of deflections of the path of the particle when it travels through a layer of thickness 1 g/cm² is roughly given by $(N/A) \cdot d^2$, where N, A and d are Avogadro's number, the atomic weight, and the diameter of the atoms of the material, respectively. Thus in dense material a deflection occurs $10^7 \sim 10^8$ times per g/cm². Unless the atoms are aligned regularly, the individual deflections are statistically independent from each other, so that in a first approximation they can be treated as multiple scattering. However, plural or single scattering may contribute, though with a rather small probability, occasional large deflections.

Let $f(\theta) \, 2\pi \theta \, d\theta$ be the probability of an electron to be deflected into an angle between θ and $\theta + d\theta$ after traversing a layer of matter of finite thickness x. Then the transport equation of the distribution function is given by

$$\frac{\partial f}{\partial x} = \int \{f(\boldsymbol{\theta} - \boldsymbol{\theta}') - f(\boldsymbol{\theta})\} \sigma(\boldsymbol{\theta}') \, d\boldsymbol{\theta}' \tag{A.3.1.}$$

with an initial condition $f = \dfrac{\delta(\theta)}{2\pi\theta}$ at $x = 0$, where $\sigma(\theta') \, 2\pi \theta' \, d\theta' \, dx$ is the probability of deflection of the particle by an angle between θ' and $\theta' + d\theta'$ in an infinitesimal layer of thickness dx.

As described in the preceding section, it is convenient to apply to this transport equation the Hankel transform with respect to the angle θ. Multiplying on both sides of the equation by the Bessel function, $J_0(\zeta \theta)$, one gets

$$\frac{\partial F}{\partial x} = 2\pi F \int_0^\infty \{J_0(\zeta \theta) - 1\} \sigma(\theta) \, \theta \, d\theta, \tag{A.3.2}$$

with the initial condition $F = \dfrac{1}{2\pi}$ at $t = 0$, where F is the Hankel transform of the distribution function. The distribution function is now given by

$$f = \frac{1}{2\pi} \int_0^\infty J_0(\zeta \theta) \, e^{2\pi x \int_0^\infty \{J_0(\xi \theta) - 1\} \sigma(\theta) \, d\theta} \zeta \, d\zeta. \tag{A.3.3}$$

and can be obtained by performing the integral

$$\int_0^\infty (J_0-1)\sigma(\theta)\theta\,d\theta. \tag{A.3.3'}$$

Since the scattering probability per g/cm², according to the MOTT formula [1]

$$\sigma(\theta)\,d\omega = N\frac{Z^2}{A}\frac{r_0^2}{4}\frac{m_e^2 c^2}{p^2\beta^2}\left(1-\beta^2\sin^2\frac{\theta}{2}\right)\frac{d\omega}{\sin^4\tfrac{1}{2}\theta}, \tag{A.3.4}$$

has a peaked shape for small θ, it may be approximated by

$$\sigma(\theta)\,2\pi\theta\,d\theta\,dx = 4N\frac{Z^2}{A}r_0^2\frac{m_e^2 c^2}{p^2\beta^2}\frac{2\pi\theta\,d\theta}{\theta^4}\,dx, \tag{A.3.5}$$

where z, p and β are the momentum and velocity divided by c of the charged particle, respectively, r_0 is the classical radius of the electron, and x the thickness of matter traversed measured in g/cm². Expressing x in radiation lengths, we have

$$\sigma(\theta)\,2\pi\theta\,d\theta \sim \frac{2\pi}{\log 191\,Z^{-\tfrac{1}{3}}}\cdot\frac{m_e^2 c^2}{\alpha p^2\beta^2}\cdot\frac{\theta\,d\theta}{\theta^4}. \tag{A.3.6}$$

where α is the fine structure constant.

Because of the steep peak for small θ, the integral (A.3.3') may be approximated by

$$\left.\begin{aligned}
&2\pi\int_0^\infty \{J_0(\zeta\theta)-1\}\sigma(\theta)\,\theta\,d\theta \\
&= 2\pi\int_0^\infty\left[-\frac{\zeta^2\theta^2}{4}+\frac{\zeta^4\theta^4}{(2!)^2(2)^2}+\cdots\right]\sigma(\theta)\,\theta\,d\theta \\
&\cong -\frac{1}{4}\int_{\theta_{\min}}^{\theta_{\max}}\theta^2\sigma(\theta)\,2\pi\theta\,d\theta = -\frac{\langle\theta^2\rangle}{4},
\end{aligned}\right\} \tag{A.3.7}$$

where

$$\langle\theta^2\rangle = \frac{2\pi\log\dfrac{\theta_{\max}}{\theta_{\min}}}{\log 191\,Z^{-\tfrac{1}{3}}}\cdot\frac{137\,m_e^2 c^4}{p^2\beta^2 c^2} = \frac{E_s^2}{p^2\beta^2 c^2}. \tag{A.3.8}$$

Here we have made use of the fact that in a good approximation, the relation $\log\theta_{\max}/\theta_{\min} = 2\log 191\,Z^{-\tfrac{1}{3}}$ holds, and have introduced the scattering energy E_s [5],

$$E_s = (4\pi\,137)^{\tfrac{1}{2}}\,m_e c^2 \approx 21\text{ MeV}. \tag{A.3.9}$$

The integral of (A.3.3) is now performed in this approximation, and we get

$$f = \frac{p^2\beta^2 c^2}{\pi E_s^2 t}\,e^{-\dfrac{p^2\beta^2 c^2\theta^2}{E_s^2 t}}. \tag{A.3.10}$$

In order to ensure the accuracy to this approximation, $\langle\theta^4\rangle$ must be negligibly small compared with $\langle\theta^2\rangle$. This is correct only if $2\pi\sigma(\theta)\,\theta\,d\theta$ has the form

$$2\pi\sigma(\theta)\,\theta\,d\theta = \lim_{\delta\to 0}\frac{E_s^2}{p^2\beta^2 c^2}\frac{\delta(\theta-\delta)}{\delta^2}\,d\theta. \tag{A.3.11}$$

In other words, the approximation gives only the effect of an infinite number of infinitely small deflections, that is, it gives only the effect of multiple scattering, but it does not include the effect of plural and single scattering.

If we limit ourselves to the multiple-scattering approximation, the transport equation itself can be re-written by applying a Taylor expansion of $f(\boldsymbol{\theta}-\boldsymbol{\theta}')$ around $\boldsymbol{\theta}$, assuming the values of $\boldsymbol{\theta}'$ to be small. Since $2\pi\,\sigma(\theta)\,\theta\,d\theta$ is azimuthally symmetric, we have for the right-hand side of Eq. (A.3.1),

$$\int_0^\infty \{f(\boldsymbol{\theta}-\boldsymbol{\theta}')-f(\boldsymbol{\theta})\}\sigma(\theta')\,d\theta' = \frac{1}{4}\langle\theta^2\rangle\left\{\frac{\partial^2}{\partial\theta^2}+\frac{1}{\theta}\frac{\partial}{\partial\theta}\right\}f+\cdots. \quad (A.3.12)$$

This leads to

$$\frac{\partial f}{\partial t} = \frac{1}{4}\frac{E_s^2}{p^2\beta^2 c^2}\left(\frac{1}{\theta}\frac{\partial}{\partial\theta}\,\theta\,\frac{\partial}{\partial\theta}\right)f. \quad (A.3.13)$$

The method here described is called the Fokker-Planck approximation, and a Gaussian-type solution equal to (A.3.10) is obtained from this equation.

The distribution functions including higher-order effects ($\langle\theta^4\rangle\ldots$) have been investigated by several authors, in particular by SNYDER and SCOTT (*48*), MOLIÈRE (*49*), and NIGAM et al. (*50*). All these theories are very similar in their mathematical basis, but differ in the single-scattering cross sections used in the calculations. The details are summarized in a review article by SCOTT (*51*).

SNYDER and SCOTT adopt an exponential screening potential for the modified Coulomb field of the atoms, and derive the scattering cross section under the first BORN approximation. MOLIÈRE uses an improved screening potential, and calculates the scattering cross section including the effect of higher-order terms. NIGAM, SUNDARESAN and WU start from the cross section for electron scattering calculated by DALITZ (*52*) in which a relativistic procedure, correct and complete within the limits of the Born approximation, is used.

Here we shall briefly summarize, as an example, the scattering theory of MOLIÈRE. The idea of MOLIÈRE's theory is to expand the distribution function in an inverse power series of the number of collisions suffered by the particle while passing through the material. Thus the first term represents the effect of multiple scattering, the second the effect of single scattering, the third and the higher the effects of plural scattering. Instead of using the scattering probability in (A.3.6)

$$2\pi\,\sigma(\theta)\,\theta\,d\theta\,dt = \frac{1}{2\log 191\,Z^{-\frac{1}{3}}}\frac{E_s^2}{E^2}\frac{1}{\theta^3}\,d\theta\,dt \quad \text{for}\quad \theta_{\min}\leq\theta\leq\theta_{\max}, \quad (A.3.6')$$

MOLIÈRE takes the more accurate expression

$$2\pi\sigma(\theta)\,\theta\,d\theta\,dt = \frac{1}{2\log 191\,Z^{-\frac{1}{3}}}\cdot\frac{E_s^2}{E^2}\cdot\frac{q(\theta)}{\theta^3}\,d\theta\,dt, \quad (A.3.14)$$

where $q(\theta)$ describes the effects of screening of the outer electrons of the atoms, and represents the deviation from the simple Rutherford cross section. $q(\theta)$ is expected to be zero for $\theta\cong 0$, and to increase to unity around $\theta=\theta_{\min}$.

Then the problem in obtaining the distribution function is to evaluate the integral, as shown by (A.3.3),

$$\int_0^\infty \frac{q(\theta)}{\theta^3}[1-J_0(\zeta\theta)]\,d\theta. \quad (A.3.15)$$

The integration with respect to θ is performed by dividing this integral into two parts

$$\int_0^\infty = \int_{\theta_0}^\infty + \int_0^{\theta_0}, \quad (A.3.16)$$

and choosing the value of θ_0 so that we may put $q(\theta)=1$ for $\theta \geq \theta_0$. Then the first term of the right-hand side of (A.3.15) can be calculated by integration by parts and gives

$$\begin{aligned}\int_{\theta_0}^{\infty} \frac{[1-J_0(\zeta\theta)]}{\theta^3} d\theta &= \int_{\zeta\theta_0}^{\infty} [1-J_0(x)] \frac{dx}{x^3} \\ &= \zeta^2 \left[\frac{1}{2(\zeta\theta_0)^2} \{1-J_0(\zeta\theta_0)\} + \frac{J_1(\zeta\theta_0)}{4\zeta\theta_0} + \frac{1}{4} \int_{\zeta\theta_0}^{\infty} \frac{J_0(x)}{x} dx \right] \\ &= \frac{\zeta^2}{4} [1+\log 2 - \log \zeta\theta_0 - c + O(\zeta\theta_0)^2 + \cdots],\end{aligned} \qquad (A.3.17)$$

where we make use of the relations

$$\frac{dJ_0(x)}{dx} = -J_1(x), \qquad \frac{dx J_1(x)}{dx} = x \cdot J_0, \qquad (A.3.18)$$

$$\int_y^{\infty} \frac{J_0(x)}{x} dx = \log 2 - c - \log y + O(y)^2 + \cdots, \qquad (A.3.19)$$

and $c=0.577\ldots$ is EULER's constant.

For the second term, we simply expand the Bessel function by assuming θ to be small, and obtain

$$\int_0^{\theta_0} \frac{[1-J_0(\zeta\theta)]}{\theta^3} q(\theta) d\theta = \frac{1}{4} \zeta^2 \int_0^{\theta_0} \frac{q(\theta)}{\theta} d\theta + O(\zeta^4) + \cdots, \qquad (A.3.20)$$

where $\int_0^{\theta_0} \frac{q(\theta)}{\theta} d\theta$ should have a value close to $\log \theta_0/\theta_{\min}$.

MOLIÈRE evaluates q more accurately, and puts

$$\int_0^{\theta_0} \frac{q(\theta)}{\theta} d\theta = \log \frac{\theta_0}{X_a} - \frac{1}{2}, \qquad (A.3.21)$$

where X_a, with the use of the scattering cross sections derived by the W.K.B. method, is given by

$$X_a^2 = \theta_{\min}^2 (1.13 + 3.76\,\alpha^2), \quad \text{with} \quad \alpha = Ze^2/\hbar c. \qquad (A.3.22)$$

The second term of this expression is a correction for the deviation from the Born approximation. Now the integral (A.3.3') becomes

$$2\pi \int_0^{\infty} \sigma(\theta)[1-J_0(\zeta\theta)] \theta\, d\theta \simeq \frac{E_s'^2 \zeta^2}{4E^2} \left[b - \log \frac{E_s'^2 \zeta^2}{4E^2} \right], \qquad (A.3.23)$$

where we introduce

$$E_s' = \frac{E_s}{2(\log 191 Z^{-\frac{1}{3}})^{\frac{1}{2}}}, \qquad (A.3.24)$$

and

$$b = 1 - 2c + \log \left(\frac{E_s'}{EX_a}\right)^2 = \log \left(\frac{E_s'}{EX_a}\right)^2 - 0.154. \qquad (A.3.25)$$

Since the total number of deflections in one radiation length is

$$\int_0^{\infty} \sigma(\theta) 2\pi\,\theta\, d\theta \simeq \frac{1}{4\log 191 Z^{-\frac{1}{3}}} \frac{E_s^2}{E^2 X_a^2} = \frac{E_s'^2}{E^2 X_a^2},$$

the value of b approximately represents the logarithm of the number of deflections during the passage of the particle through the material.

For further evaluation, MOLIÈRE introduces a new parameter defined by the transcendental equation

$$\Omega - \log \Omega = b. \tag{A.3.26}$$

Then (A.3.23) becomes

$$\frac{K^2 \zeta^2}{4 E^2} \left(1 - \frac{1}{\Omega} \log \frac{K^2 \zeta^2}{4 E^2}\right), \tag{A.3.27}$$

where we put

$$K = \Omega^{\frac{1}{2}} E_s'. \tag{A.3.28}$$

Now the Hankel transform of the distribution function is given by

$$F = e^{-\frac{K^2 \zeta^2}{4 E^2}\left(-\frac{1}{\Omega} \log \frac{K^2 \zeta^2}{4 E^2}\right) t}. \tag{A.3.29}$$

The values of Ω and K in several materials of interest are shown in Table 12. Since Ω has a value similar to that of b, or roughly $\log 10^7 \simeq 16$, one may expand (A.3.29) in a power series of $1/\Omega$. After expanding and applying the inverse Hankel transformation, one gets

$$f(\theta) = f^{(0)} + \frac{1}{\Omega} f^{(1)} + \frac{1}{\Omega^2} f^{(2)} + \cdots, \tag{A.3.30}$$

where

$$f^{(0)} = \frac{1}{\pi} \frac{E^2}{K^2 t} e^{-\frac{E^2 \theta^2}{K^2 t}} \tag{A.3.31}$$

and

$$f^{(n)} = \frac{1}{2\pi n!} \frac{E^2}{K^2} \int_0^\infty y \, dy \, J_0\left(\frac{E\theta}{K} y\right) e^{-\frac{y^2 t}{4}} \left(\frac{y^2}{4} \log \frac{y^2}{4}\right)^n t^n. \tag{A.3.32}$$

For the function $f^{(1)}$, MOLIÈRE gives the simple analytical formula

$$f^{(1)} = \frac{1}{\pi} \frac{E^2}{K^2 t} \left[e^{-\frac{E^2 \theta^2}{K^2 t}} \left(\frac{E^2 \theta^2}{K^2 t} - 1\right) \left\{ E_i\left(\frac{E^2 \theta^2}{K^2 t}\right) - \log \frac{E^2 \theta^2}{K^2 t}\right\} - \left(1 - 2 e^{-\frac{E^2 \theta^2}{K^2 t}}\right) \right], \tag{A.3.33}$$

where E_i is defined by

$$E_i(x) = \int_\infty^{-x} \frac{e^{-u}}{u} \, du.$$

The function $E_i(x)$ has the series representation

$$E_i(x) = c + \log x + x + \frac{1}{2} \frac{x^2}{2!} + \frac{1}{3} \frac{x^3}{3!} + \cdots \quad \text{for} \quad x \leq 17.$$

and

$$E_i(x) = \frac{e^x}{x}\left(1 + \frac{1!}{x} + \frac{2!}{x^2} + \cdots\right). \quad \text{for} \quad x > 17$$

Thus for large values of θ, $f^{(1)}$ is developed as

$$f^{(1)} = \frac{1}{\pi} \frac{K^2}{E^2} \frac{t}{\theta^4} + \cdots. \tag{A.3.34}$$

Substituting the expressions (A.3.28) and (A.3.24) into (A.3.34), we obtain the relation

$$\frac{1}{\Omega} f^{(1)} \cong \frac{1}{\pi \Omega} \frac{K^2}{E^2} \frac{t}{\theta^4} = \frac{1}{\pi} \left(\frac{E'_s}{E}\right)^2 \frac{t}{\theta^4}$$
$$= \frac{1}{4\pi \log 191 Z^{-\frac{1}{3}}} \left(\frac{E_s}{E}\right)^2 \frac{t}{\theta^4} \cong \sigma(\theta) t, \tag{A.3.35}$$

in which the right-hand side indicates the contribution of single scattering.

Except for a slight difference in the definitions of the scattering energies E_s and K, the first term (A.3.31) is the same as that obtained under the Fokker-Planck approximation, and represents the contribution of the effect of multiple scattering. The second term (A.3.33) is the contribution of single scattering, as (A.3.35) proves. Fig. 31 illustrates the contributions of the various terms, and it is seen that those of the third and the higher terms can be neglected in all practical applications.

As we have mentioned in Sect. 29, the considerations outlined above also hold in the three-dimensional electron shower theory.

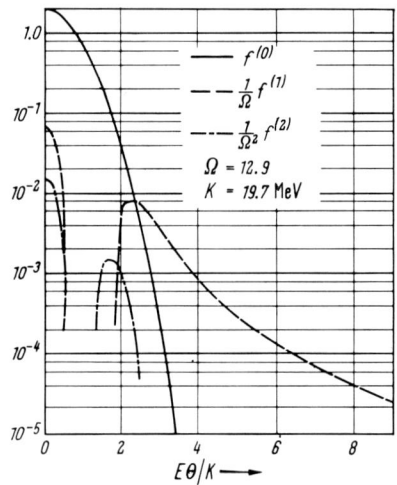

Fig. 31. Contribution of the successive terms to the scattering under the depth of one radiation length in Pb according to MOLIÈRE's theory.

4. Solution of difference equations. In the course of the calculations in the electron shower theory, it is sometimes necessary to derive solutions of difference equations. The simplest type of such difference equations will be

$$\mathfrak{A}(s+q)\mathfrak{M}(s,q) = q\mathfrak{M}(s,q-1), \tag{A.4.1}$$

where $\mathfrak{A}(s+q)$ is a function of $(s+q)$, and the initial condition is given for the function $\mathfrak{M}(s, q=0)$. Then the solution $\mathfrak{M}(s, n)$ is obtained as

$$\mathfrak{M}(s,n) = \Gamma(n+1) \prod_{i=1}^{n} \frac{1}{\mathfrak{A}(s+i)} \mathfrak{M}(s,0)$$

for integral values of n.

Assuming $\lim\limits_{n \to \infty} \frac{\mathfrak{A}(s+n+1)}{\mathfrak{A}(s+n)} = 1$, and further assuming that $\mathfrak{M}(s, q)$ is an analytic function of q, we have[1]

$$\mathfrak{M}(s,q) = \Gamma(q+1) \lim_{n \to \infty} \prod_{i=1}^{n} \frac{\mathfrak{A}(s+q+1)\mathfrak{A}(s+q+2)\ldots\mathfrak{A}(s+n+\delta)}{\mathfrak{A}(s+i)\{\mathfrak{A}(s+n)\}^\delta} \mathfrak{M}(s,0), \tag{A.4.2}$$

where δ is defined in such a way that $q-\delta$ is an integer. From this relation the numerical value of $\mathfrak{M}(s, q)$ can be found for any value of q.

The situation is more complicated in the theory of electron showers, and in the following we shall discuss the equations in Approximation B, in the Landau approximation, and without the Landau approximation.

---

[1] Another useful representation is

$$\mathfrak{M}(s,q) = \Gamma(q+1) \lim_{n \to \infty} \prod_{i=1}^{n} \frac{\mathfrak{A}(s+q+i)}{\mathfrak{A}(s+i)\{\mathfrak{A}(s+n)\}^q} \mathfrak{M}(s,0).$$

A.4.1. The difference equation in Approximation B.

In Approximation B, we have derived a difference equation of the type (15.15)

$$\left[\frac{\partial}{\partial t} - \lambda_1(s+q)\right]\left[\frac{\partial}{\partial t} - \lambda_2(s+q)\right]\mathfrak{M}(s, q) = (s+q)q\left[\frac{\partial}{\partial t} + \sigma_0\right]\mathfrak{M}(s, q-1), \qquad (A.4.3)$$

with an initial condition

$$\mathfrak{M}(s, 0) = H_1(s) e^{\lambda_1(s)t} + H_2(s) e^{\lambda_2(s)t}, \qquad (A.4.4)$$

taking as an example the case of an electron-initiated shower.

There are several ways to obtain the solution of this equation, such as operator calculus, the Green function method, and the method of the Laplace transforms. These will be described in the following.

α) *Operator calculus.*

Multiplying by the operator,

$$\frac{1}{\left[\frac{\partial}{\partial t} - \lambda_1(s+q)\right]\left[\frac{\partial}{\partial t} - \lambda_2(s+q)\right]},$$

on both sides of the Eq. (A.4.3), the right hand side of the equation becomes

$$\frac{(s+q)q\left[\frac{\partial}{\partial t} + \sigma_0\right]}{\left[\frac{\partial}{\partial t} - \lambda_1(s+q)\right]\left[\frac{\partial}{\partial t} - \lambda_2(s+q)\right]} = (s+q)q\left[\frac{H_1(s+q)}{\left[\frac{\partial}{\partial t} - \lambda_1(s+q)\right]} + \frac{H_2(s+q)}{\left[\frac{\partial}{\partial t} - \lambda_2(s+q)\right]}\right]. \qquad (A.4.5)$$

Thus we have

$$\mathfrak{M}(s, q) = (s+q)q \int_0^t [H_1(s+q) e^{\lambda_1(s+q)t'} + H_2(s+q) e^{\lambda_2(s+q)t'}] \\ \times \mathfrak{M}(s, q-1, t-t') \, dt'. \qquad (A.4.6)$$

Starting from $\mathfrak{M}(s, q=0)$, we can derive $\mathfrak{M}(s, q=n)$ if q is an integer. According to (A.4.6), the function $\mathfrak{M}(s, q=n)$ is given by a combination of terms proportional to $e^{\lambda_1(s)t}, e^{\lambda_1(s+1)t}, \ldots e^{\lambda_1(s+n)t}, e^{\lambda_2(s)t} \ldots e^{\lambda_2(s+n)t}$, of which the term $e^{\lambda_1(s)t}$ is the most important.

β) *Method of the Green function.*

The Green function, G, defined by the equation

$$\left[\frac{\partial}{\partial t} - \lambda_1(s)\right]\left[\frac{\partial}{\partial t} - \lambda_2(s)\right] G = \left[\frac{\partial}{\partial t} + \sigma_0\right] \delta(t), \qquad (A.4.7)$$

should be the same as the solution in Approximation A for a shower due to a primary electron, because (A.4.7) is of the same form as the equation for this problem. Therefore G is obtained as

$$G = H_1(s) e^{\lambda_1(s)t} + H_2(s) e^{\lambda_2(s)t} \quad \text{for} \quad t \geq 0, \qquad (A.4.8)$$

and the solution of (A.4.3) is given by

$$\mathfrak{M}(s, q) = (s+q)q \int_0^t G(s+q, t') \mathfrak{M}(s, q-1, t-t') \, dt'.$$

The solution is identical with the expression (A.4.6) derived in the preceding section.

γ) *Contribution of the terms proportional to $e^{\lambda_1(s)t}, e^{\lambda_1(s+1)t}, \ldots$*

Since, the term proportional to $e^{\lambda_1(s)t}$ is the predominant one, its details will be examined in the following.

Writing the function $\mathfrak{M}(s, q)$ in the form

$$\mathfrak{M}(s, q) = H_1(s) \frac{\Gamma(s+q+1)}{\Gamma(s+1)} K_{1,0}(s, q) e^{\lambda_1(s)t} + \cdots, \qquad (A.4.9)$$

and substituting this expression into (A.4.3), we have

$$\frac{[\lambda_1(s) - \lambda_1(s+q)][\lambda_1(s) - \lambda_2(s+q)]}{[\sigma_0 + \lambda_1(s)]} K_{1,0}(s, q) = q K_{1,0}(s, q-1), \\ \text{with } K_{1,0}(s, 0) = 1. \qquad (A.4.10)$$

Since this equation is similar to (A.4.1), and the condition

$$\lim_{q \to \infty} \frac{q[\lambda_1(s) - \lambda_1(s+q+1)][\lambda_1(s) - \lambda_2(s+q+1)]}{(q+1)[\lambda_1(s) - \lambda_2(s+q)][\lambda_1(s) - \lambda_2(s+q)]} = 1, \qquad (A.4.11)$$

holds, the solution for any values of q takes a form similar to (A.4.2):

$$K_{1,0}(s, q) = \Gamma(q+1)[\sigma_0 + \lambda_1(s)]^q \lim_{n \to \infty} \prod_{i=1}^{n} \frac{\{\lambda_1(s) - \lambda_1(s+q+i)\}\{\lambda_1(s) - \lambda_2(s+q+i)\}}{\{\lambda_1(s) - \lambda_1(s+i)\}\{\lambda_1(s) - \lambda_2(s+i)\}} \times \\ \times \frac{1}{[\{\lambda_1(s) - \lambda_1(s+n)\}\{\lambda_1(s) - \lambda_2(s+n)\}]^q}. \qquad (A.4.12)$$

As the value of $K_{1,0}(s, -s)$ is important for the calculation of the total number of the electrons and photons in Approximation B, the numerical values of $K_{1,0}(s, -s)$ are shown in Fig. 8. Next we shall examine the term proportional to $e^{\lambda_1(s+m)t}$, which we put in the form

$$H_1(s) \frac{\Gamma(s+q+1)}{\Gamma(s+1)} \frac{\Gamma(q+1)}{\Gamma(q-m+1)} K_{1,m}(s, q) F_{1,m}(s) e^{\lambda_1(s+m)t}, \qquad (A.4.13)$$

taking the initial condition as

$$K_{1,m}(s, q=m) = 1, \qquad (A.4.14)$$

where $F_{1,m}$ is a function which can be derived from (A.4.6). Substituting (A.4.13) into the difference equation (A.4.3), we have

$$[\lambda_1(s+m) - \lambda_1(s+q)][\lambda_1(s+m) - \lambda_2(s+q)] K_{1,m}(s, q) \\ = (q-m)[\lambda_1(s+m) + \sigma_0] K_{1,m}(s, q-1). \qquad (A.4.15)$$

Introducing $q - m = q^*$, the equation becomes

$$[\lambda_1(s+m) - \lambda_1(s_1+q^*+m)][\lambda_1(s+m) - \lambda_2(s+q^*+m)] K_{1,m}(s, q^*+m) \\ = q^*[\lambda_1(s+m) + \sigma_0] K_{1,m}(s, q^*+m-1), \\ \text{with } K_{1,m}(s, q^*=0) = 1. \qquad (A.4.16)$$

Since except for the difference in the variables, this equation has a form similar to that defining the function $K_{1,0}$, the solution for (A.4.16) will be

$$K_{1,m}(s, q+m) = K_{1,0}(s+m, q) \qquad (A.4.17)$$

or

$$K_{1,m}(s, q) = K_{1,0}(s+m, q-m). \qquad (A.4.18)$$

The terms proportional to $e^{\lambda_2(s+m)t}$ can be derived in the same way, giving solutions in which $K_{1,m}(s, q)$ and $F_{1,m}(s)$ should be replaced by $K_{2,m}(s, q)$ and

$F_{2,m}(s)$. Summarizing these arguments, we arrive at the final expression

$$\mathfrak{M}(s, q, t) = \sum_{i=0}^{\infty} \frac{\Gamma(s+q+1)}{\Gamma(s+1)} \frac{\Gamma(q+1)}{\Gamma(q-i+1)} \times \\ \times [H_1(s) K_{1,i}(s+i, q-i) F_{1,i}(s) e^{\lambda_1(s+i)t} + \\ + H_2(s) K_{2,i}(s+i, q-i) F_{2,i}(s) e^{\lambda_2(s+i)t}] \quad (A.4.19)$$

first derived by SNYDER [3], and by SCOTT [4]. The explicit forms of the functions $F_{1,i}$ and $F_{2,i}$ will be discussed in the following section. For different boundary conditions, the function $H_1(s)$ and $H_2(s)$ should be replaced by the functions given in Table 8.

δ) *Method of the Laplace transformation.*

The solution given in (A.4.19) can be evaluated if one derives the explicit expression of $F(s)$ according to (A.4.6). They can also be obtained in a different approach by application of the Laplace transformation described in Appendix 1. Multiplying by $e^{-\mu t}$ on both sides of Eq. (A.4.3), and intergrating with respect to t from zero to infinite, one gets

$$[\mu - \lambda_1(s+q)][\mu - \lambda_2(s+q)] \mathfrak{L}_\mathfrak{M}(s, q) = (s+q) q (\mu+\sigma_0) \mathfrak{L}_\mathfrak{M}(s, q-1) + \mathfrak{L}_{in}, \quad (A.4.20)$$

where $\mathfrak{L}_\mathfrak{M}(s, q)$ and $\mathfrak{L}_\mathfrak{M}(s, q-1)$ are the Laplace transforms of $\mathfrak{M}(s, q)$ and $\mathfrak{M}(s, q-1)$, defined by

$$\int_0^\infty e^{-\mu t} \mathfrak{M}(s, q) dt = \mathfrak{L}_\mathfrak{M}(s, q) \quad (A.4.21)$$

and

$$\int_0^\infty e^{-\mu t} \mathfrak{M}(s, q-1) dt = \mathfrak{L}_\mathfrak{M}(s, q-1). \quad (A.4.22)$$

$\mathfrak{L}_{in}$ is given by the initial condition:

$$\mathfrak{L}_{in} = [\mu + \sigma_0] \quad \text{for} \quad q=0, \quad (A.4.23)$$

and

$$\mathfrak{L}_{in} = 0 \quad \text{for} \quad q \neq 0, \quad (A.4.24)$$

again treating, as an example, showers of a primary electron.

Now the solution of (A.4.20) is

$$\mathfrak{L}_\mathfrak{M}(s, n) = \frac{\Gamma(s+n+1) \Gamma(n+1) (\mu+\sigma_0)^{n+1}}{\Gamma(s+1)} \times \prod_{i=0}^n \frac{1}{[\mu - \lambda_1(s+i)][\mu - \lambda_2(s+i)]}, \quad (A.4.25)$$

and applying the inverse transformation, we have

$$\mathfrak{M}(s, n) = \frac{1}{2\pi i} \frac{\Gamma(s+n+1) \Gamma(n+1)}{\Gamma(s+1)} \int_c d\mu [\mu+\sigma_0]^{n+1} e^{\mu t} \prod_{i=0}^n \frac{1}{[\mu - \lambda_1(s+i)][\mu - \lambda_2(s+i)]},$$

where the integration path of μ is running parallel to the imaginary axis, to the right of the pole $\mu = \lambda_1(s)$.

The term proportional to $e^{\lambda_1(s+m)t}$ in the function $\mathfrak{M}(s, n)$, obtained by using the pole at $\mu + \lambda_1(s+m) = 0$, is equal to

$$\frac{\Gamma(s+n+1) \Gamma(n+1)}{\Gamma(s+1)} \frac{[\lambda_1(s+m) + \sigma_0]^{n+1}}{[\lambda_1(s+m) - \lambda_2(s+m)]} \times \\ \times \prod_{i=0}^{m-1} \frac{1}{[\lambda_1(s+m) - \lambda_1(s+i)][\lambda_1(s+m) - \lambda_2(s+i)]} \times \\ \times \prod_{j=m+1}^{n} \frac{1}{[\lambda_1(s+m) - \lambda_1(s+j)][\lambda_1(s+m) - \lambda_2(s+j)]} \\ \text{for} \quad n \geq m, \quad (A.4.26)$$

and
$$= 0 \quad \text{for} \quad n < m. \tag{A.4.27}$$

Thus, putting now $m=0$ in (A.4.26), we get for the term proportional to $e^{\lambda_1(s)t}$

$$\frac{\Gamma(s+n+1)\Gamma(n+1)}{\Gamma(s+1)} \frac{1}{\lambda_1(s)-\lambda_2(s)} \prod_{i=1}^{n} \frac{[\lambda_1(s)+\sigma_0]^{n+1}}{[\lambda_1(s)-\lambda_1(s+i)][\lambda_1(s)-\lambda_2(s+i)]}. \tag{A.4.28}$$

This is just the function defined by

$$\frac{\Gamma(s+n+1)}{\Gamma(s+1)} H_1(s) K_{1,0}(s, n),$$

in the preceding section, with $K_{1,0}$ given by (A.4.12), or

$$K_{1,0}(s, n) = \Gamma(n+1)[\lambda_1(s)+\sigma_0]^n \prod_{i=1}^{n} \frac{1}{[\lambda_1(s)-\lambda_1(s+i)][\lambda_1(s)-\lambda_2(s+i)]}, \tag{A.4.29}$$

because $H_1(s)$ is defined by

$$H_1(s) = \frac{\lambda_1(s)+\sigma_0}{\lambda_1(s)-\lambda_2(s)}.$$

The function $F_{1,m}$ is obtained in the following way. For the term proportional to $e^{\lambda_1(s+m)t}$ in $\mathfrak{M}(s, n)$ we had found the relation (A.4.26). Substituting from (A.4.29), we have

$$\frac{\Gamma(s+n+1)\Gamma(n+1)}{\Gamma(s+1)\Gamma(n-m+1)} H_1(s+m) K_{1,0}(s+m, n-m) \times$$

$$\times [\lambda_1(s+m)+\sigma_0]^m e^{\lambda_1(s+m)t} \prod_{i=0}^{m-1} \frac{1}{[\lambda_1(s+m)-\lambda_1(s+i)][\lambda_1(s+m)-\lambda_2(s+i)]}.$$

Making use of the relation (A.4.18) between $K_{1,0}$ and $K_{1,m}$, and of the definition of $F_{1,m}$ in (A.4.13), we find

$$\left.\begin{array}{r} F_{1,m} = \dfrac{H_1(s+m)}{H_1(s)} [\lambda_1(s+m)+\sigma_0]^m \times \\ \times \displaystyle\prod_{i=0}^{m-1} \dfrac{1}{[\lambda_1(s+m)-\lambda_1(s+i)][\lambda_1(s+m)-\lambda_2(s+i)]} \end{array}\right\} \tag{A.4.30}$$

Again $H_1(s+m)$ and $H_1(s)$ should be replaced by the corresponding functions listed in the Table 8 if the shower is not initiated by a single electron.

A.4.2. *The difference equation for the angular structure function in the Landau approximation.* The difference equation directly from derived (22.6) is[1]

$$\left[\frac{\partial}{\partial t} - \lambda_1(s+2p)\right]\left[\frac{\partial}{\partial t} - \lambda_2(s+2p)\right]\mathfrak{M}(p, s) = p\left[\frac{\partial}{\partial t} + \sigma_0\right]\mathfrak{M}(p-1, s), \tag{A.4.31}$$

with a given initial condition for $\mathfrak{M}(0, s)$. Its solution can be obtained in a way similar to that described in the preceding section. Here we need only mention that $\mathfrak{M}(p, s)$ has a pole at $p = -1 - \dfrac{s}{2}$, because $\lambda_1(s+2p)$ and $\lambda_2(s+2p)$ have singularities at $p = -\dfrac{s}{2}$.

A.4.3. *The difference equation for the lateral structure function in the Landau approximation.* This case leads to a somewhat complicated difference equation,

---

[1] Here we use the abbreviation

$$\mathfrak{M}(p, q, s, t) = \mathfrak{M}(p, s, t).$$

given in Approximation A by[1]

$$\left[\frac{\partial}{\partial t}-\lambda_1(s+2p)\right]\left[\frac{\partial}{\partial t}-\lambda_2(s+2p)\right]M(p,s,\xi-t,t) \tag{A.4.32}$$
$$=\left(\frac{\partial}{\partial t}+\sigma_0\right)p(\xi-t)^2 M(p-1,s,\xi-t,t)$$

with

$$\mathfrak{M}(p,s,t)=\lim_{(\xi-t)\to 0}M(p,s,\xi-t,t), \tag{A.4.33}$$

and the initial condition determined by the function $M(0,s,0,t)$.

The solution is

$$M(p,s,\xi-t,t)=p\int_0^t \{H_1(s+2p)e^{\lambda_1(s+2p)(t-t')}+H_2(s+2p)e^{\lambda_2(s+2p)(t-t')}\}\times \\ \times(\xi-t')M(p-1,s,\xi-t',t')\,dt', \tag{A.4.34}$$

with a given initial condition for $M(0,s,\xi-t,t)$ which, in the example of a the shower due to a primary electron, becomes

$$M(0,s,0,t)=H_1(s)e^{\lambda_1(s)t}+H_2(s)e^{\lambda_2(s)t}.$$

Starting from this relation the numerical values of $\mathfrak{M}(p,s,t)$ can be derived from Eq. (A.4.32) with the help of (A.4.33), if p is an interger. As in the case of one-dimensional cascade theory, a different approach such as the use of Laplace transforms, the Green function and operator calculus is also possible.

Here we shall describe the method of the Laplace transformation.

α) *The Laplace transforms.*

Multiplying by $e^{-\mu t}$ on both sides of (A.4.32), as in the case of the difference equation in the one-dimensional shower theory, we get

$$[\mu-\lambda_1(s+2p)][\mu-\lambda_2(s+2p)]\mathfrak{L}(p)=p(\mu+\sigma_0)\left(\xi+\frac{\partial}{\partial\mu}\right)^2\mathfrak{L}(p-1)+\mathfrak{L}_{in}, \tag{A.4.35}$$

where

$$\mathfrak{L}(p)=\int_0^\infty e^{-\mu t}M(p,s,(\xi-t),t)\,dt,$$

and we have used the relation

$$\int_0^\infty e^{-\mu t}t^n M(p,s,(\xi-t),t)\,dt=\left(\frac{-\partial}{\partial\mu}\right)^n\mathfrak{L}(p).$$

The function $\mathfrak{L}_{in}$ is determined by the initial condition. Let us again consider the case of a shower originated by a primary electron. Then we have

$$\mathfrak{L}_{in}=\mu+\sigma_0 \quad \text{for} \quad p=0$$
$$\mathfrak{L}_{in}=0 \quad \text{for} \quad p\neq 0.$$

Substituting the term $\dfrac{\mu+\sigma_0}{[\mu-\lambda_1(s+2p)][\mu-\lambda_2(s+2p)]}$ for $f(p)$, Eq. (A.4.35) can be rewritten in the form

$$\mathfrak{L}(p)=pf(p)\left(\xi+\frac{\partial}{\partial\mu}\right)^2\mathfrak{L}(p-1)+f(0)\delta_{p,0} \tag{A.4.36}$$

---

[1] The equation is derived from (21.11″), using the abbreviations

$\mathfrak{M}(p,q,s,t)\to\mathfrak{M}(p,s,t)$ and $M(p,q=0,s,(\xi-t),t)\to M(p,s,(\xi-t),t)$.

and if p is an integer, n, we have the solution

$$\mathfrak{L}(n) = \Gamma(n+1) f(n) \left(\xi + \frac{\partial}{\partial \mu}\right)^2 f(n-1) \ldots \left(\xi + \frac{\partial}{\partial \mu}\right)^2 f(0). \tag{A.4.37}$$

From the inverse Laplace transformation we obtain

$$\mathfrak{M}(n, s, t) = \lim_{(\xi-t) \to 0} \frac{1}{2\pi i} \int_c d\mu \, e^{-\mu t} \mathfrak{L}(n), \tag{A.4.38}$$

where the integration path crosses the real axis at a point with Re $(p) >$ Re $(\lambda_1(s))$. The formula can be written in a simple form, using the symbols μ_e for the variable μ in the function $f(l)$, and Δ_l for $\partial/\partial \mu_l$. Then (A.4.37) becomes

$$\left.\begin{aligned}
& f(n) \left(\xi + \frac{\partial}{\partial \mu}\right)^2 f(n-1) \ldots \left(\xi + \frac{\partial}{\partial \mu}\right)^2 f(0) \\
&= f(n) (\xi + V_0 + V_1 \cdots + V_{n-1})^2 f(n-1) \ldots (\xi + V_0)^2 f(0) |_{\mu_0 = \mu_1 = \cdots = \mu_n = \mu} \\
&= \prod_{j=0}^{n-1} \left(\xi + \sum_{i=0}^{j} V_i\right)^2 \prod_{l=0}^{n} f(l) \bigg|_{\mu_0 = \cdots = \mu_n = \mu}.
\end{aligned}\right\} \tag{A.4.39}$$

The terms proportional to $e^{\lambda_1(s+2m)t}$ can now be obtained with the help of the poles of the type $\dfrac{1}{[\mu + \lambda_1(s+2m)]}$ in (A.4.39). Since $f(l)$ has the representation

$$f(l) = \frac{H_1(s+2l)}{\mu - \lambda_1(s+2l)} + \frac{H_2(s+2l)}{\mu - \lambda_2(s+2l)}, \tag{A.4.40}$$

these poles at $\mu = \lambda_1(s+2m)$ appear in (A.4.39) only through the function $f(m)$ and the derivatives of this function.

Consider a complex integral,

$$I = \frac{1}{2\pi i} \oint e^{\mu t} G(\mu) V_m^k f(m) \, d\mu, \tag{A.4.41}$$

where $G(\mu)$ is assumed to be a function having no singularities at $\mu = \lambda_1(s+2m)$ or $\lambda_2(S+2m)$. Since

$$V_m^k f = (-)^k k! \left[\frac{H_1(s+2m)}{[\mu - \lambda_1(s+2m)]^k} + \frac{H_2(s+2m)}{[\mu - \lambda_2(s+2m)]^k} \right], \tag{A.4.42}$$

we have, using Goursat's theorem,

$$I = H_1(s+2m) \left(-\frac{\partial}{\partial \mu}\right)^k e^{\mu t} G(\mu) \bigg|_{\mu = \lambda_1(s+2m)} + H_2(s+2m) \left(-\frac{\partial}{\partial \mu}\right)^k e^{\mu t} G(\mu) \bigg|_{\mu = \lambda_2(s+2m)},$$

$$\left.\begin{aligned}
&= H_1(s+2m) e^{\lambda_1(s+2m)t} \left[-t - \frac{\partial}{\partial \mu}\right]^k G(\mu) \bigg|_{\mu = \lambda_1(s+2m)} \\
&+ H_2(s+2m) e^{\lambda_2(s+2m)t} \left[-t - \frac{\partial}{\partial \mu}\right]^k G(\mu) \bigg|_{\mu = \lambda_2(s+2m)}.
\end{aligned}\right\} \tag{A.4.44}$$

In other words, if one regards the term proportional to $e^{\lambda_1(s+2m)t}$, the application of the operator (A.4.41) changes the operator $V_m^k f(m)$ into $H_1(s+2m) \left(-\dfrac{\partial}{\partial \mu}\right)^k$. Regarding the term proportional to $e^{\lambda_1(s+2m)t}$, the inverse Laplace transformation

Sect. A.4. Solution of difference equations.

changes the operator $\left(\sum_{i=0}^{j} V_i\right)^k f(m)$ in (A.4.39) into

$$\left(\sum_{i=0}^{j} V_i\right)^k f(m) = [V_0 + V_1 + \cdots + V_m + \cdots + V_j]^k f(m)$$

$$\Rightarrow H_1(s+2m) \left\{\sum_{i=0}^{j} V_i - V_m + \left(-\sum_{l=0}^{n} V_l + V_m\right)\right\}^k$$

$$= H_1(s+2m) \left\{-\sum_{i=j+1}^{n} V_i\right\}^k \quad \text{if} \quad m \leq j,$$
(A.4.45)

or

$$\left(\sum_{i=0}^{j} V_i\right)^k f(m) = H_1(s+2m) \left(\sum_{i=0}^{j} V_i\right)^k \quad \text{if} \quad m > j. \quad (\text{A.4.46})$$

Therefore we have the solution for the term proportional to $e^{\lambda_1(s+2m)t}$,

$$\frac{\mathfrak{M}_{1,m}(n,s,t)}{\Gamma(n+1)} = \lim_{(\xi-t)\to 0} H_1(s+2m) \prod_{j=m}^{n} \left(\xi - t - \sum_{i=j+1}^{n} V_i\right)^2 \prod_{j'=0}^{m-1} \left(\xi + \sum_{i'=0}^{j'} V_{j'}'\right)^2 \times$$

$$\times \prod_{l=0}^{n} \frac{f(l)}{f(m)}\bigg|_{\mu_0 = \cdots = \mu_n = \lambda_1(s+2m)} e^{\lambda_1(s+2m)t}$$
(A.4.47)

$$= H_1(s+2m) \prod_{j=m}^{n} \left(\sum_{i=j+1}^{n} V_i\right)^2 \prod_{j'=0}^{m-1} \left(t + \sum_{i'=0}^{j'} V_{j'}'\right)^2 \prod_{l=0}^{n} \frac{f(l)}{f(m)}\bigg|_{\mu_0 = \cdots = \mu_n = \lambda_1(s+2m)} e^{\lambda_1(s+2m)t}$$

or in a more explicit form,

$$\mathfrak{M}_{1,m} = \Gamma(n+1) H_1(s+2m) \left\{\frac{\partial^2}{\partial\mu^2} f(n) \frac{\partial^2}{\partial\mu^2} f(n-1) \frac{\partial}{\partial\mu^2} \cdots \frac{\partial}{\partial\mu^2} f(m+1)\right\} \times$$

$$\times \left\{\left(t + \frac{\partial}{\partial\mu}\right)^2 f(m-1) \cdots \left(t + \frac{\partial}{\partial\mu}\right)^2 f(0)\right\}\bigg|_{\mu=\lambda_1(s+2m)} e^{\lambda_1(s+2m)t}.$$
(A.4.48)

If one puts

$$\left\{\left(t+\frac{\partial}{\partial\mu}\right)^2 f(m+1) \cdots \left(t+\frac{\partial}{\partial\mu}\right)^2 f(0)\right\}_{\mu=\lambda_1(s+2m)} = F_{1,m}(s+2m,t), \quad (\text{A.4.49})$$

and substitutes it into (A.4.48), the function $\mathfrak{M}_{1,m}$ can be reduced to the form

$$\mathfrak{M}_{1,m} = \frac{\Gamma(n+1)}{\Gamma(n-m+1)} \mathfrak{M}_{1,0}(n-m, s+2m) F_{1,m}(s+2m, t), \quad (\text{A.4.50})$$

in view of the relation

$$\frac{\partial^2}{\partial\mu^2} f(n) \frac{\partial^2}{\partial\mu^2} f(n-1) \frac{\partial^2}{\partial\mu^2} \cdots \frac{\partial^2}{\partial\mu^2} f(m+1)\bigg|_{\mu=\lambda_1(s^*+2m)}$$

$$= \frac{\partial^2}{\partial\mu^2} f(n-m) \frac{\partial^2}{\partial\mu^2} f(n-m-1) \frac{\partial^2}{\partial\mu^2} \cdots \frac{\partial^2}{\partial\mu^2} f(1)\bigg|_{\mu=\lambda_1(s)}\bigg|_{s=s^*+2m}.$$
(A.4.51)

Here we make use of the relation

$$f(n-m)\bigg|_{\mu=\lambda_1(s)} = \frac{H_1(s+2n-2m)}{\lambda_1(s) - \lambda_1(s+2n-2m)} + \frac{H_2(s+2n-2m)}{\lambda_1(s) - \lambda_2(s+2m-2n)}$$

$$= \frac{H_1(s^*+2n)}{\lambda_1(s^*+2m) - \lambda_1(s^*+2n)} + \frac{H_2(s^*+2n)}{\lambda_1(s^*+2m) - \lambda_2(s^*+2n)}$$
(A.4.52)

$$= f(n)\bigg|_{s=s^*}.$$

With the help of (A.4.51) the lateral structure function can be developed into a power series in $(K/E_0 r)^2$ as described in the Sect. 25. The situation is quite similar to the case of one-dimensional cascade theory in Approximation B, in which the solution is developed in powers of ε/E_0. To estimate the contributions of the terms of the higher order in $(K/E_0 r)^2$, it is necessary to know the numerical value of $F_{1,m}$. For example, the explicit expression for $F_{1,1}$ is

$$F_{1,1} = \left(t + \frac{\partial}{\partial \mu}\right)^2 f(0)\bigg|_{\mu = \lambda_1(s+2)} =$$
$$= t^2 \left[\frac{H_1(s)}{A_{11}} + \frac{H_2(s)}{A_{12}}\right] - 2t \left[\frac{H_1(s)}{A_{11}^2} + \frac{H_2(s)}{A_{12}^2}\right] + 2 \left[\frac{H_1(s)}{A_{11}^3} + \frac{H_2(s)}{A_{12}^3}\right], \quad \text{(A.4.53)}$$

where we put

$$A_{11} = \lambda_1(s+2) - \lambda_1(s),$$
$$A_{12} = \lambda_1(s+2) - \lambda_2(s).$$

The expression (A.4.53) is valid for showers due to a primary electron. Two of the most important cases demand the following initial conditions:

$$\mathfrak{L}_{in} = (\mu + \sigma_0) \quad \text{for a primary electron,} \quad \text{(A.4.54)}$$

$$\mathfrak{L}_{in} = B(s) \quad \text{for a primary photon.} \quad \text{(A.4.55)}$$

Numerical values of $F_{1,1}$ are shown in the Fig. 10. — It remains, then, to evaluate the function $\mathfrak{M}_{1,0}(p, s)$ in (A.4.50).

A.4.4. Evaluation of the function $\mathfrak{M}_{1,0}(p, 0, s, t)$[1].

α) *The method of Kalos and Blatt* (13). Since the function $\mathfrak{M}_{1,0}(s, p=n, t)$ is defined by the more general expression (A.4.48), which for $m=0$ takes the form

$$\mathfrak{M}_{1,0}(p=n, s, t) = \Gamma(n+1) H_1(s) \left[\frac{\partial^2}{\partial \mu^2} f(n) \frac{\partial}{\partial \mu^2} f(n-1) \frac{\partial^2}{\partial \mu^2} \cdots \frac{\partial^2}{\partial \mu^2} f(1)\right]_{\mu = \lambda_1(s)} e^{\lambda_1(s)t} \quad \text{(A.4.56)}$$

with $f(n) = \dfrac{\sigma_0 + \mu}{[\mu + \lambda_1(s+2n)][\mu - \lambda_2(s+2n)]}$, its numerical values for all n can be determined according to the procedures already described.

Eq. (A.4.56) can also be written in the form:

$$\mathfrak{M}_{1,0}(n, s, t) = \Gamma(n+1) H_1(s) \left[\frac{\partial^2}{\partial \alpha^2} f'(n-1) \frac{\partial^2}{\partial \alpha^2} f'(n) \frac{\partial^2}{\partial \alpha^2} \cdots \frac{\partial^2}{\partial \alpha^2} f'(1)\right]_{\alpha=0} e^{\lambda_1(s)t}, \quad \text{(A.4.57)}$$

where

$$f'(n) = \frac{\sigma_0 + \alpha + \lambda_1(s)}{[\alpha - A_{11}(s+2n)][\alpha - A_{21}(s+2n)]}$$
$$= \frac{H_1(s+2n)}{\alpha - A_{11}(s+2n)} + \frac{H_2(s+2n)}{\alpha - A_{21}(s+2n)}, \quad \text{(A.4.58)}$$

by substituting

$$A_{11}(s+2n) = \lambda_1(s) - \lambda_1(s+2n),$$
$$A_{12}(s+2n) = \lambda_1(s) - \lambda_2(s+2n).$$

---

[1] In the following the abbreviation
$$\mathfrak{M}_{1,0}(p, q=0, s, t) = \mathfrak{M}(p, s, t)$$
will be used throughout.

In order to study the behaviour of the function $\mathfrak{M}_{1,0}(s, p, t)$ for arbitrary values of p, KALOS and BLATT (13) examined the expression (A.4.57) by assuming[1]

$$H_1(s+2m) = 0 \quad \text{except} \quad H_1(s),$$

and
$$H_2(s) = H_2(s+2) = \cdots = H_2(s+2m) = H,$$

$$\Lambda_{21}(s) = \Lambda_{21}(s+2) = \cdots = \Lambda_{21}(s+2n) = \Lambda.$$

Then (A.4.58) becomes

$$\mathfrak{M}(n, s, t) \cong \Gamma(n+1) H_1(s) H^n \underbrace{\frac{\partial^2}{\partial \alpha^2} \frac{1}{\alpha - \Lambda} \frac{\partial^2}{\partial \alpha^2} \cdots \frac{\partial^2}{\partial \alpha^2} \frac{1}{\alpha - \Lambda}}_{n \text{ terms}} \bigg|_{\alpha = 0}. \quad (A.4.59)$$

Since

$$\frac{1}{\alpha - \Lambda} = \int_0^\infty e^{-\beta(\alpha - \Lambda)} d\beta, \quad (A.4.60)$$

we can write for (A.4.59)

$$\mathfrak{M}(n, s, t) = \Gamma(n+1) H_1(s) H^n \int_0^\infty d\beta_1 \int_0^\infty d\beta_2 \cdots \int_0^\infty d\beta_n \, e^{-(\beta_1 + \cdots + \beta_n)\Lambda}$$
$$\times \beta_1^2 (\beta_1 + \beta_2)^2 \cdots (\beta_1 + \beta_2 + \cdots + \beta_n)^2. \quad (A.4.61)$$

Now replacing the variables $\beta_1, \beta_2, \ldots \beta_n$ by $Z_1, Z_2, \ldots Z_n$ defined by

$$\beta_1 = Z_1$$
$$\beta_1 + \beta_2 = Z_2$$
$$\beta_1 + \beta_2 + \cdots + \beta_n = Z_n,$$

the integral in (A.4.61) becomes

$$\mathfrak{M}(n, s, t) \cong \Gamma(n+1) H_1(s) H^n \int_0^\infty dZ_n \, e^{-Z_n \Lambda} \int_0^{Z_n} Z_{n-1}^2 \, dZ_{n-1} \cdots \int_0^{Z_2} Z_1^2 \, dZ_1$$
$$\cong \frac{\Gamma(n+1) H_1(s) H^n}{3^{n-1} \Gamma(n)} \int_0^\infty dZ_n \, e^{-Z_n \Lambda} Z_n^{3n-1} \quad (A.4.62)$$
$$\cong \frac{n \Gamma(3n) H_1(s)}{3^{n-1}} \frac{H^n}{\Lambda^{3n}}.$$

The restriction that the value of p is an integer is now dropped, and (A.4.62) indicates that under the assumptions made by KALOS and BLATT, a singularity of $\mathfrak{M}(s, p, t)$ exists at $p = -\frac{1}{3}$. Though their assumptions seem to be accurate only for large values of p, this result would suggest the existence of singularities of $\mathfrak{M}$ in the regions of negative values of p. A more direct way to test the existence of a singularity in the function $\mathfrak{M}_{1,0}(p, s, t)$ consists in the use of the method of generating functions.

β) *Method of the generating function* [8]. Let

$$G(x) = \sum_{n=0}^\infty \frac{(-x^2)^n}{n!} M(n, 0, s, \xi - t, t) = \frac{1}{2\pi i} \int dp \, \Gamma(-p) \, x^{2p} \, M(p, 0, s, \xi - t, t), \quad (A.4.63)$$

---

[1] Since $\lim_{p \to \infty} \lambda_1(s+2p) = -\sigma_0$ and $H_1(s+2n) = \dfrac{\sigma_0 + \lambda_1(s+2n)}{\lambda_1(s+2n) - \lambda_2(s+2n)}$, their assumption seems to be accurate in the limit for $p \to \infty$.

be the generating function of $\mathfrak{M}_2(p, 0, s, t)$ in the limit $\lim_{(\xi-t)\to 0} G$. Then, according to (A.4.32), the function $G(x)$ satisfies the following equation:

$$\left[\frac{\partial}{\partial t'} + A' - \frac{B'C'}{\frac{\partial}{\partial t}+\sigma_0}\right] G = -x^2(\xi-t)^2 G. \tag{A.4.64}$$

The solution of this equation is

$$G = e^{\frac{x^2(\xi-t)^3}{3}} \int_{\xi-t}^{\infty} e^{-\frac{x^2(\xi-t')^3}{3}} \left[A' - \frac{B'C'}{\frac{\partial}{\partial t}+\sigma_0}\right] G\, d(\xi-t'),$$

where the upper and the lower limits of the integral are determined using the property that G must be zero at the limit $\xi-t=\frac{\zeta}{x}=\infty$.

Now the first term in the above integrand implies a factor

$$A'G = \int_0^1 [G(x) - (1-v)\, G\{x(1-v)\}]\,\varphi_0(v)\, dv. \tag{A.4.65}$$

This is reduced to

$$A'G = \int_0^1 [G(x) - v^s G(xv)]\,\varphi_0(v)\, dv$$

$$= \int_0^x \left[G(x) - \left(\frac{y}{x}\right)^s G(y)\right]\varphi_0\left(1-\frac{y}{x}\right)\frac{dy}{x}.$$

At the limit $x=\infty$

$$A'G = \int_0^\infty \left[G(x) - \left(\frac{y}{x}\right)^s G(y)\right]\varphi_0\left(1-\frac{y}{x}\right)\frac{dy}{x},$$

and remembering

$$\varphi_0\left(1-\frac{y}{x}\right) \approx \frac{1}{1-(y/x)} = 1 + \frac{y}{x} + \cdots,$$

we get

$$\lim_{x\to\infty} A'G = \lim_{x\to\infty} G(x) = \lim_{x\to\infty} \frac{1}{x^{s+1}} \sum_n a_n(\xi-t)^n. \tag{A.4.66}$$

The second term of the integrand contains the expression

$$B'C'G(x(\xi-t), x) = 2\int_0^1 \psi_0(u)\,du \int_0^u G(xv(\xi-t), xv)\, v^s\, \varphi_0\left(\frac{v}{u}\right)\frac{dv}{u}$$

$$= 2\int_0^1 \psi_0(u)\,du \int_0^{xu} G(y(\xi-t), y)\, y^s\, x^s\, \varphi_0\left(\frac{y}{ux}\right)\frac{dy}{ux}.$$

This becomes at the limit $x=\infty$

$$B'C'G = 1.36\,\frac{B(0)}{x^s}\int_0^\infty G(y(\xi-t), y)\, y^{s-1}\, dy. \tag{A.4.66'}$$

Putting

$$\int_0^\infty G y^{s-1} dy = \sum b_n(\xi-t)^n,$$

Sect. A.4.   Solution of difference equations.   107

we get

$$\lim_{x\to\infty} \frac{B'C'G}{\frac{\partial}{\partial t}+\sigma_0} = \frac{1.36\, B(0)\, e^{\sigma_0(\xi-t)}}{x^s} \int_{(\xi-t)}^{\infty} e^{-\sigma_0(\xi-t')}\, d(\xi-t') \int_0^{\infty} G\, y^{s-1}\, dy. \tag{A.4.67}$$

To examine the behaviour of G at $x=\infty$, we consider the lowest order in the powers of $1/x$ of the expressions (A.4.66) and (A.4.67). Thus the solution of (A.4.64) becomes at the limit of $x=\infty$

$$\lim_{x\to\infty} G = e^{\frac{x^2(\xi-t)^3}{3}} \int_{(\xi-t)}^{\infty} e^{-\frac{x^2(\xi-t')^3}{3}} \sum_n \frac{c_n}{x^s} (\xi-t')^n\, d(\xi-t'). \tag{A.4.68}$$

Putting $(\xi-t')^3 = z$, we get

$$\left.\begin{array}{l} \displaystyle\lim_{(\xi-t)\to 0} G = \lim_{(\xi-t)\to 0} e^{\frac{x^2(\xi-t)^3}{3}} \int_{(\xi-t)}^{\infty} \frac{e^{-\frac{x^2 z}{3}}}{3\, x^s} z^{-\frac{2}{3}} \sum C_n z^{\frac{n}{3}}\, dz \\[2mm] \displaystyle = \sum_n \frac{1}{3} C_n \Gamma\!\left(\frac{1}{3}\right)\!\left(\frac{x^3}{3}\right)^{-\left(\frac{1}{3}\right)-\left(\frac{n}{3}\right)} x^{-s}. \end{array}\right\} \tag{A.4.69}$$

Remembering the definition (A.4.63), one can conclude that $\mathfrak{M}_{1,0}(p,0,s,t)$ has a simple pole at $p = -\left(\frac{s}{2}\right) - \left(\frac{1}{3}\right)$.

Since we can calculate the numerical values of $\mathfrak{M}_{1,0}(p,s,t)$ for integral values of p, and also know of the existence of the poles at $p = -\frac{s}{2} - \frac{1}{3}$, we can derive the numerical values of $\mathfrak{M}_{1,0}$ for any value of p larger than $-\frac{s}{2}-\frac{1}{3}$, by smooth interpolations between these exactly known values. Although this treatment seems to introduce some uncertainty into $\mathfrak{M}$ for non-integral values of p, it is rather difficult to draw two different curves giving values with deviations of more than 20 or 30% from each other. In the present situation when we have not yet succeeded in deriving a general solution of $\mathfrak{M}_{1,0}(s,p,t)$, for arbitrary values of p, there seems to be no alternative to this method, despite its unsatisfactory nature. Though the errors inherent in the procedure here adopted directly affects the numerical values of the shower functions, the resulting uncertainties are expected to be of the order of 10%, or a few times this value, as discussed above.

A.4.5. Difference equations without the Landau approximation. The essential features of the equation can be demonstrated in a treatment, for example, of the problem of the angular structure functions in Approximation A. The basic equation was derived in Sect. 24 where we found

$$\left.\begin{array}{l} \displaystyle\sum_{n=0}^{p}{}_{n+u}C_n\!\left(-\frac{1}{\Omega}\right)^n L^{(n)}(s+2p+2u)\, \mathfrak{M}_1(p,u,s,t) \\[2mm] = p\, \mathfrak{M}_1(p-1,u,s,t) + u\, \mathfrak{M}_1(p,u-1,s,t), \end{array}\right\} \tag{A.4.70}$$

to be used together with a given initial condition for $\mathfrak{M}_1(0,0,s,t)$.

If one wishes to determine the structure function accurately up to the order of $1/\Omega$, the evaluation must be carried through up to the second term of the solution (24.1) which was written in the form

$$\pi_1 = \pi^{(0)} + \frac{1}{\Omega}\pi^{(1)} + \cdots. \tag{24.1}$$

The solutions of the difference equation (A.4.70) to be used are $\mathfrak{M}_1(u=0)$ and $\mathfrak{M}_1(u=1)$ in the functions $\pi_1^{(0)}$ and $\pi_1^{(1)}$, respectively. Thus $\mathfrak{M}_1(u=0)$ should be evaluated up to the order of $(1/\Omega)$, and $\mathfrak{M}_1(u=1)$ up to the order of $(1/\Omega)^0$.

Developing the solution of the difference equation (A.4.70) as

$$\mathfrak{M}_1(p, u, s, t) = \mathfrak{M}_1^{(0)} + \frac{1}{\Omega}\mathfrak{M}_1^{(1)} + \cdots, \quad (A.4.71)$$

the equation for $\mathfrak{M}_1^{(0)}$ is obtained by substitution into (A.4.70):

$$\left.\begin{aligned}&\boldsymbol{L}^{(0)}(s+2p+2u)\,\mathfrak{M}_1^{(0)}(p, u, s, t) \\ &= p\,\mathfrak{M}_1^{(0)}(p-1, u, s, t) + u\,\mathfrak{M}_1^{(0)}(p, u-1, s, t),\end{aligned}\right\} \quad (A.4.72)$$

again with a given initial condition for $\mathfrak{M}_1(0, 0, s, t)$. Since this equation is symmetric with respect to p and u, it is easily proved that

$$\mathfrak{M}_1^{(0)}(p, u, s, t) = \mathfrak{M}_1^{(0)}(p+u, 0, s, t), \quad (A.4.73)$$

and that the equation for $\mathfrak{M}_1^{(0)}(p+u, 0, s, t)$ is given by

$$\boldsymbol{L}^{(0)}(s+2p+2u)\,\mathfrak{M}_1^{(0)}(p+u, 0, s, t) = (p+u)\,\mathfrak{M}_1^{(0)}(p+u-1, 0, s, t). \quad (A.4.74)$$

The equation is of exactly the same form as that derived in the Landau approximation, and therefore its solution is found along the lines discussed in Sect. A.4.2.

Now the second term $\frac{1}{\Omega}\mathfrak{M}_1^{(1)}$ is obtained by including the term of the order $\frac{1}{\Omega}$ (A.4.70),

$$\left.\begin{aligned}&\boldsymbol{L}^{(0)}(s+2p+2u)\,\mathfrak{M}_1(p, u) - \frac{(u+1)\,\boldsymbol{L}^{(1)}(s+2p+2u)}{\Omega}\,\mathfrak{M}_1(p-1, u+1) \\ &= p\,\mathfrak{M}_1(p-1, u) + u\,\mathfrak{M}_1(p, u-1).\end{aligned}\right\} \quad (A.4.75)$$

Substituting from (A.4.71), and using the relation (A.4.73), the equation becomes

$$\left.\begin{aligned}&\left\{\boldsymbol{L}^{(0)}(s+2p+2u) - \frac{u+1}{\Omega}\boldsymbol{L}^{(1)}(s+2p+2u)\right\}\mathfrak{M}_1^{(0)}(p+u, 0) + \\ &+ \frac{\boldsymbol{L}^{(0)}(s+2p+2u)}{\Omega}\mathfrak{M}_1^{(1)}(p, u) \\ &= (p+u)\,\mathfrak{M}_1^{(0)}(p+u-1, 0) + \frac{1}{\Omega}\{p\,\mathfrak{M}_1^{(1)}(p-1, u) + u\,\mathfrak{M}_1^{(1)}(p, u-1)\},\end{aligned}\right\} \quad (A.4.76)$$

and with the help of (A.4.74) we have, finally,

$$\left.\begin{aligned}&\boldsymbol{L}^{(0)}(s+2p+2u)\,\mathfrak{M}_1^{(1)}(p, u) = p\,\mathfrak{M}_1^{(1)}(p-1, u) + u\,\mathfrak{M}_1^{(1)}(p, u-1) + \\ &+ (u+1)\,\boldsymbol{L}^{(1)}(s+2p+2u)\,\mathfrak{M}_1^{(0)}(p+u, 0).\end{aligned}\right\} \quad (A.4.77)$$

Solutions of this equation are now obtained by recursion.

The diffusion equation for the lateral distribution is somewhat more complicated than that for the angular structure function. The procedures for deriving the solution, however, are quite similar to that described above, and need not be repeated here.

5. The range fluctuation problem of high-energy μ-mesons. A high-energy μ-meson penetrating a large amount of matter, like those observed in underground experiments, dissipates its energy by the processes of radiation, pair creation of electrons, photo-nuclear reaction, and ionization losses.

Since the radiation length of the μ meson is about 4×10^6 times larger than that of electrons, its critical energy is also higher by about this factor. Therefore radiation, pair creation, and photo-nuclear losses are important only at extremely high energies. The fractional energy losses due to these processes as a function of the meson energy are schematically illustrated in Fig. 32.

Observations (53) of the high-energy μ-meson flux far underground now enable us to compare the spectrum of these mesons with that of γ-rays of energy up to 10^{13} eV [9]. Since in this energy region the catastrophic losses, i.e. the radiation, pair creation, and photo-nuclear losses, are predominant, the range fluctuation problem of these μ-mesons becomes important in deriving the energy spectrum from the observed data of the range spectrum of μ-mesons.

Let $\boldsymbol{I}(E, t)\, dE$ be the differential spectrum of μ-mesons at a depth t (in units of the μ-meson radiation length). Then the diffusion equation for this function is obtained in a manner similar to that of the electron shower theory. Evidently we can write

$$\frac{\partial \boldsymbol{I}}{\partial t} = -A' \boldsymbol{I} + \varepsilon \frac{\partial \boldsymbol{I}}{\partial E}, \quad (A.5.1)$$

where A' is the operator representing the effect of radiation, pair creation, and photo-nuclear losses, and $\varepsilon \frac{\partial \boldsymbol{I}}{\partial E}$ is the term accounting for the effect of ionization losses.

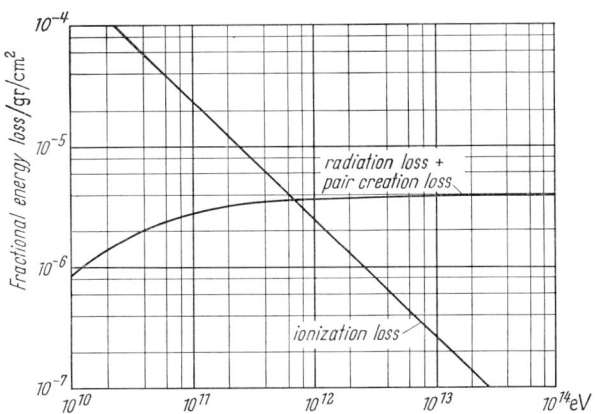

Fig. 32. Fractional energy loss of a μ-meson in a material with $Z=12.9$, and $A=26.3$ (photo-nuclear losses are neglected).

Since this equation is formally identical with the basic equation in Approximation B, its solution is given in exactly the same form,

$$\boldsymbol{I} = -\frac{1}{4\pi^2} \iint ds\, dq \left(\frac{E_0}{E}\right)^s \frac{1}{E} \left(\frac{\varepsilon}{E}\right)^q \Gamma(-q)\, \mathfrak{M}(s, q, t), \quad (A.5.2)$$

in which the initial condition is defined by the function $\mathfrak{M}(s, 0, 0)$. Considering a single μ-meson of energy E_0 incident at $t=0$, the initial condition is given by

$$\mathfrak{M}(s, 0, 0) = 1.$$

The total probability of finding μ-mesons at a depth t is now obtained by integration of (A.5.2) with respect to the energy from zero to infinity. With the help of the pole at $s=q$, one finds

$$\boldsymbol{J} = \frac{1}{2\pi i} \int ds \left(\frac{E_0}{\varepsilon}\right)^s \Gamma(s)\, \mathfrak{M}(s, -s, t). \quad (A.5.3)$$

Similarly, the total intensity of μ-mesons at a depth t, starting with a μ-meson spectrum

$$\pi(E_0)\, dE_0 = \frac{dE_0}{E_0^{\gamma+1}} \quad \text{at} \quad t=0, \quad (A.5.4)$$

is, from (A.5.2) and (A.5.3),

$$\boldsymbol{J}_\gamma = \int_0^\infty \frac{dE_0}{E_0^{\gamma+1}} \boldsymbol{J} = \left(\frac{1}{\varepsilon}\right)^\gamma \Gamma(\gamma)\, \mathfrak{M}(\gamma, -\gamma, t), \quad (A.5.5)$$

where the integration with respect to s is performed with the help of the pole at $s=\gamma$. According to this relation, the ratio of the observed intensities of μ-mesons under two different materials of the same thickness measured in radiation lengths is given by

$$\frac{J_{\gamma_1}}{J_{\gamma_2}} = \left(\frac{\varepsilon_2}{\varepsilon_1}\right)^\gamma, \tag{A.5.5'}$$

where we denoted the intensities under the materials of critical energies ε_1 and ε_2 by J_{γ_1} and J_{γ_2}, respectively. Therefore it may be concluded that by observing these two intensities, the exponent of the energy spectrum of μ-mesons can be determined without detailed analysis of the fluctuation phenomena, and without knowledge of the individual probabilities of the catastrophic energy losses.

The equation for the $\mathfrak{M}(s, p, t)$ is obtained by substitution of (A.5.2) into (A.5.1):

$$\left[\frac{\partial}{\partial t} + A(s+q)\right] \mathfrak{M}(s, q, t) = (s+q)\, q\, \mathfrak{M}(s, q-1, t) \tag{A.5.6}$$

with the initial condition for $\mathfrak{M}(s, 0, 0) = 1$, where

$$A(s+q) = \int_0^1 \{1 - (1-v)^{s+q}\}(\sigma_\gamma + \sigma_p + \sigma_{pN})\, dv$$

and σ_γ, σ_p and σ_{pN} are the probabilities of the radiation, pair creation and photo-nuclear processes, respectively. Solutions are again derived by a procedure similar to that used in the electron shower theory (Sect. 15). Multiplying by $e^{-\mu t}$ on both sides of (A.5.6), we have

$$\{\mu + A(s+q)\}\mathfrak{L}(s, q) = (s+q)\, q\, \mathfrak{L}(s, q-1) + \delta_{q,0}, \tag{A.5.7}$$

where $\mathfrak{L}$ is the Laplace transform of the function $\mathfrak{M}$.

In solving this equation we find, with reference to (A.4.1),

$$\mathfrak{L} = \lim_{m\to\infty} \frac{\Gamma(s+q+1)\Gamma(q+1)}{(\mu+A(s))\Gamma(s+1)} \prod_{i=1}^{m} \frac{\{\mu+A(s+q+i)\}}{\{\mu+A(s+i)\}\{\mu+A(s+m)\}^q}. \tag{A.5.8}$$

From the inverse transformation we have

$$\mathfrak{M}(s, q, t) = \frac{1}{2\pi i}\int d\mu\, e^{\mu t}\, \mathfrak{L}. \tag{A.5.9}$$

Following the procedure described in Appendix 4, and putting

$$\mathfrak{M}(s, q, t) = \sum_{n=0}^{\infty} \mathfrak{M}_n(s, q)\, e^{-A(s+n)t}, \tag{A.5.10}$$

$\mathfrak{M}_n(s, q)$ is determined by the residue at the pole $\mu + A(s+n)$ in (A.5.8), which gives

$$\mathfrak{M}_n(s, q) = \frac{\Gamma(s+q+1)\Gamma(q+1)}{\Gamma(s+1)} \times$$

$$\times \lim_{m\to\infty} \frac{\{A(s+q+n) - A(s+n)\}}{\{A(s) - A(s+n)\}\{A(s+m) - A(s+n)\}^q} \prod_{\substack{i=1 \\ i\neq n}}^{m} \frac{\{A(s+q+i) - A(s+n)\}}{\{A(s+i) - A(s+n)\}}. \tag{A.5.11}$$

For the evaluation of the total flux it is necessary to know the function $\mathfrak{M}(s, -s, t)$. Putting $q = -s$ in (A.5.11) we get

$$\mathfrak{M}_n(s, -s) = \frac{\Gamma(1-s)}{\Gamma(s+1)} \lim_{n\to\infty} \frac{\{A(n) - A(s+n)\}\{A(s+m) - A(s+n)\}^s}{\{A(s) - A(s+n)\}} \prod_{\substack{i=1 \\ i\neq n}}^{m} \frac{\{A(i) - A(s+n)\}}{\{A(s+i) - A(s+n)\}}.$$

Sect. A.5.   The range fluctuation problem of high-energy μ-mesons.

Assuming the exponent γ in (A.5.5) to be an integer, we find for $\mathfrak{M}_n(\gamma, -\gamma)$ according to (A.5.12)[1],

$$\begin{aligned}\mathfrak{M}_n(\gamma, -\gamma) &= \frac{1}{\Gamma(\gamma+1)} \lim_{\delta \to 0} \Gamma(1-\gamma+\delta)\{A(\gamma+n)-A(\gamma+n-\delta)\}\prod_{i=1}^{\gamma-1}\{A(i)-A(\gamma+n)\} \\ &= \frac{1}{\Gamma(\gamma+1)\Gamma(\gamma)} \frac{\partial A(\gamma+n)}{\partial \gamma}\prod_{i=1}^{\gamma-1}\{A(\gamma+n)-A(i)\}.\end{aligned} \quad (A.5.13)$$

Substituting into (A.5.5) we obtain, therefore,

$$J_\gamma = \left(\frac{1}{\varepsilon}\right)^\gamma \sum_{n=0}^\infty \frac{1}{\Gamma(\gamma+1)} \frac{\partial A(\gamma+n)}{\partial \gamma} e^{-A(\gamma+n)t} \prod_{i=1}^{\gamma-1}\{A(\gamma+n)-A(i)\}. \quad (A.5.14)$$

This is the depth-intensity relation of the μ-meson flux deep underground, and if one ignores the effect of fluctuations this should be of the form

$$J_0 = \frac{1}{\gamma}\left(\frac{b}{\varepsilon}\right)^\gamma \frac{1}{(e^{bt}-1)^\gamma}, \quad (A.5.15)$$

where bE is the energy loss of a μ-meson of energy E by radiation, pair creation and photo-nuclear processes. A numerical evaluation of (A.5.14) was carried out for values of the exponent $\gamma=2, 3$ and 4, using the complete screening cross sections for radiation [1] and pair creation [54].

Observations on the μ-meson flux far underground were made at Kolar Gold Field (53). There the values $\langle Z \rangle = 12.9$ and $\langle A \rangle = 26.3$ may be taken for the average atomic number and weight of the material. With them, the coefficients b are

$$b_r = 2.2 \cdot 10^{-6}/\text{g cm}^2$$
$$b_\text{pair} = 1.70 \cdot 10^{-6}/\text{g cm}^2.$$

Table 22. *Relative intensity $\frac{J_\gamma - J_0}{J_0}$ of underground μ-mesons as a function of the depth t.*

T / γ	0.6	0.8	1.0	1.5	2.0
2	0.04	0.075	0.12	0.27	0.46
3	0.23	0.38	0.55	1.26	2.21
4	0.52	0.89	1.36	3.63	7.89

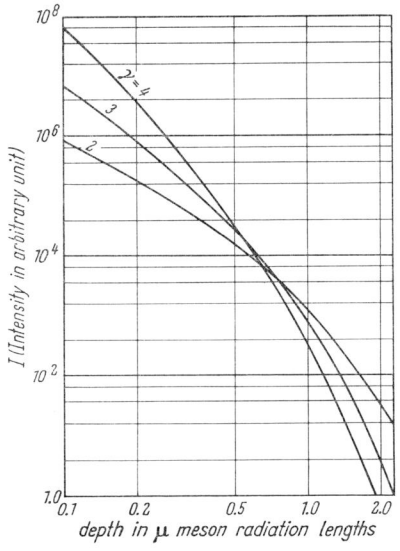

Fig. 33. Depth — intensity relation for μ-mesons.

The results of a numerical evaluation of (A.5.14) are shown in the Table 22, which lists the relative intensities $\frac{J_\gamma - J_0}{J_0}$ as a function of the depth. The depth-intensity relation (A.5.5) is also presented in Fig. 33. For a practical application of the analysis to the experimental data, however, the following effects must be

---

[1] Since $\Gamma(1-\gamma+\delta) = \frac{1}{1-\gamma+\delta}\Gamma(-\gamma+2+\delta) = \cdots = \frac{\Gamma(1)}{\delta(-1+\delta)\cdots(1-\gamma+\delta)}$

$= \frac{(-1)^{\gamma-1}}{\delta\Gamma(\gamma)},$

$\lim_{\delta \to 0} \Gamma(1-\gamma+\delta)[A(\gamma+n)-A(\gamma+n-\delta)] = \frac{(-)^{\gamma-1}}{\Gamma(\gamma)}\frac{\partial A(\gamma+n)}{\partial \gamma}.$

included: the effects of the photo-nuclear process, the deviations from complete screening for the radiation and pair creation processes, and also the dependence of the ionization loss[1] of the energy of μ-mesons. When all of these effects are taken into account, it can be concluded that the observed data on the μ-meson flux far underground seem to be compatible with the results on high-energy electrons and photons observed at high altitudes [*9*], (*55*).

6. Precision of the energy determination of electron showers in nuclear emulsions. The precision of the theoretical values in Tables 17 and 18 can be checked by comparing them with the observed number of electrons in a shower started by a

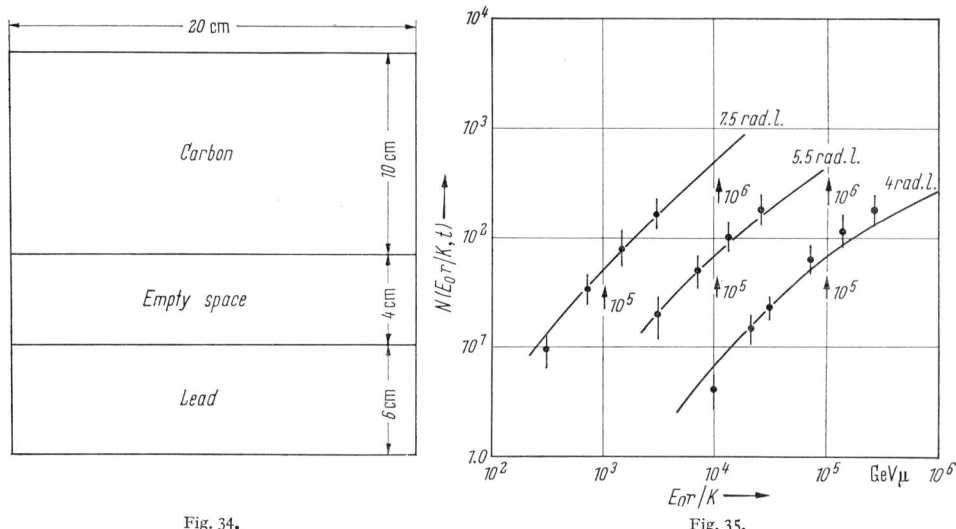

Fig. 34.   Fig. 35.

Fig. 34. Schematical view of an emulsion chamber.

Fig. 35. Comparison of the experimental data with the theoretical curves of the core approximation. The full circles show the energies determined by PINKAU's method. (K. PINKAU, private communication.)

photon or an electron pair of a given energy. For this purpose it is necessary to have at least one other independent way of determining the incident particle energy which does not rely upon the three-dimensional shower theory. One example of such an independent energy determination of the shower is provided by PINKAU's method (*38*). A more widely adopted method is the use of the kinematical relation between the opening angle and the energies of the two γ-rays

Table 23. *Comparison of the energy determined by the opening angle of γ-ray pairs, and by reference to the three-dimensional shower theory.*

Event	M.K. 1 (*56*)	T.O. 6 (*56*)	Ya.N. 2 (*56*)	P20 (*57*)	192 (*57*)
Energy of the π^0-meson determined by the opening angle (10^{11} eV)	2.6±0.6	1.8±0.3	2.2±0.7	19±5	38±7
Energy of the π^0 mesons determined from the three-dimensional shower theory (10^{11} eV)	3.0±0.7	3.5±0.5	2.7±0.4	19±5	33±7

---

[1] The ionization loss, in the calculation assumed to be constant, is more precisely expressed by the relation (*53*)
$$(dE/dx)_i = 1.88 + 0.0766 \log E/m_\mu c^2 \text{ MeV/gr cm}^2.$$

from a π^0-meson applicable in cases where the cascades of the two initiating γ-rays can be identified.

In 1956, emulsion chambers [9] were flown to study the transverse momentum of π^0-mesons. In this experiment a few examples of pairs of electron showers were observed which could be shown to have originated from a single π^0-meson in a jet occurring in the producer part of this chamber. The construction of the chamber is schematically reproduced in Fig. 34. Another example was found in the emulsions of the Bristol group. A comparison of the experimental results with the theory is given in Fig. 35, and in Table 23. The result seems to prove that the error involved in the numerical evaluation of the theories does not exceed 20~30%.

Because the of similarity relation between the particle number and $E_0 r/K$, it can be expected that the error in the relative energy determination for two different showers will be quite small. In this case, any discrepancy between the experimental and the theoretical results may be ascribed mainly to the fluctuation of the number of shower particle in actual showers.

References.

General references.

[1] HEITLER, W.: The Quantum Theorie of the Radiation. Oxford 1956.
[2] LANDAU, L., and G. RUMER: Proc. Roy. Soc. Lond. **166**, 213 (1938).
[3] SNYDER, H. S.: Phys. Rev. **76**, 1563 (1949).
[4] SCOTT, W. T.: Phys. Rev. **80**, 611 (1950).
[5] ROSSI, B., and GREISEN: Rev. Mod. Phys. **13**, 240 (1941).
[6] ROSSI, B.: High-Energy Particles. Englewood Cliffs, N. J.: Precentice-Hall Inc. 1956.
[7] BELENKY, S. Z.: Cascade Processes in Cosmic Rays. Gostekhizdat 1948. — BELENKY, S. Z., i. I. P. IVANENKO: Uspekhi Fiz. Nauk **69**, 591 (1958). — Soviet Phys. Usp. **2**, 912 (1960).
[8] KAMATA, K., and J. NISHIMURA: Suppl. Progr. Theoret. Phys. **6**, 93 (1958).
[9] FUJIMOTO, Y., u. S. HAYAKAWA: This volume, p. 115.
[10] SCOTT, W. T.: Phys. Rev. **82**, 893 (1951).

Literature cited.

(1) BHABHA, H. J., and W. HEITLER: Proc. Roy. Soc. Lond. (A) **159**, 432 (1937).
(2) CARLSON, J. F., and J. R. OPPENHEIMER: Phys. Rev. **51**, 220 (1937).
(3) For references, see for instance L. JANOSSY: Cosmic Rays (Oxford University Press 1950), and W. HEISENBERG: Kosmische Strahlung. Berlin-Göttingen-Heidelberg: Springer 1953.
(4) SNYDER, H. S.: Phys. Rev. **53**, 960 (1938).
(5) SERBER, R.: Phys. Rev. **54**, 317 (1938).
(6) TAMM, I. E., i S. BELENKY: J. Phys. U.S.S.R. **1**, 177 (1939).
(7) EULER, H., and H. WERGELAND: Astrophys. Norv. **3**, 163 (1940).
(8) MOLIÈRE, G.: In: W. HEISENBERG, Kosmische Strahlung. Berlin-Göttingen-Heidelberg: Springer 1953.
(9) ROBERG, J., and L. W. NORDHEIM: Phys. Rev. **75**, 444 (1949).
(10) EYGES, L., and S. FERNBACH: Phys. Rev. **82**, 23 (1951).
(11) GREEN, H. S., and H. MESSEL: Phys. Rev. **88**, 331 (1952).
(12) BELENKY, S.: J. Phys. U.S.S.R. **8**, 347 (1944).
(13) KALOS, M. H., and J. M. BLATT: Austral. J. Phys. **7**, 543 (1954).
(14) CHARTRES, B. A., and H. MESSEL: Phys. Rev. **104**, 517 (1956).
(15) GUZAVIN, V. V., e I. P. IVANENKO: Nuovo Cim. **8**, 749 (1958).
(16) In Approximation B, the problem was treated by H. MESSEL, D. F. CRAWFORD, A. D. SMISNOV and A. A. KARFOLOMEEV, J. Phys. Soc. Japan **17**, Suppl. A III, 444 (1962); a three-dimensional theory of shower fluctuations was given by A. ADACHI, Y. FUJIMOTO, N. OGITA, S. TAKAGI and A. UEDA, Suppl. Progr. Theoret. Phys. (to be published).
(17) CHAKRABARTY, S. K., and M. R. GUPTA: Phys. Rev. **101**, 813 (1956).
(18) KIRPICHEV, A., i. I. POMERANCHUK: Dokl. Akad. Nauk. U.S.S.R. **45**, 285 (1944).
(19) BERNSTEIN, I. B.: Phys. Rev. **80**, 995 (1956).
(20) LANDAU, L. D., i. I. POMERANCHUK: Dokl. Akad. Nauk U.S.S.R. **92**, 535, 735 (1953).

(21) MIGDAL, A. B.: Phys. Rev. **103**, 1811 (1956).
(22) WHEELER, J. A., and W. E. LAMB: Phys. Rev. **55**, 858 (1939).
(23) DAVIES, H., H. A. BETHE, and L. C. MAXIMON: Phys. Rev. **93**, 788 (1954).
(24) OLSEN, H.: Phys. Rev. **99**, 155 (1954).
(25) JAEGER, J. C., and H. R. HULME: Proc. Roy. Soc. Lond. A **153**, 443 (1936).
(26) JAEGER, J. C.: Nature, Lond. **148**, 86 (1941).
(27) BHABHA, H. J., and S. K. CHAKRABARTY: Phys. Rev. **74**, 1352 (1948).
(28) RICHARDS, J. A., and L. W. NORDHEIM: Phys. Rev. **74**, 1106 (1948).
(29) BORSELLINO, A.: Nuovo Cim. **7**, 323 (1950).
(30) MISAKI, A.: Suppl. Progr. Theoret. Phys. (to be published).
(31) GREISEN, K.: Progress of Cosmic Ray Physics (ed. J. G. WILSON), Vol. III,1. North Holland Publ. Co. 1956.
(32) POMERANCHUK, I.: J. Phys. U.S.S.R. **8**, 17 (1944).
(33) MIGDAL, A. B.: J. Phys. U.S.S.R. **9**, 183 (1945).
(34) EYGES, L.: Phys. Rev. **74**, 1810 (1948).
(35) GUSEVA, V. V., N. A. DOBROTIN, N. V. ZELEVINSKAYA, K. A. KOTELNIKOV, A. M. LEBEDEV, and S. A. SLAVATINSKY: J. Phys. Soc. Japan **17**, Suppl. A III, 175 (1962).
(36) KANTZ, A., and R. HOFSTADTER: Phys. Rev. **89**, 607 (1953). — Nucleonics **12**, 36 (1954).
(37) MURATA, Y.: J. Phys. Soc. Japan **20**, 209 (1965); also MURATA, Y., J. NISHIMURA, A. KUSUMEGI, K. NIU, A. MASAIKE, and R. KAJIKAWA: Nuovo Cim. (to be published).
(38) PINKAU, K.: Phil. Mag. **2**, 1389 (1957).
(39) MIKUMO, E., T. MOROOKA, and Y. OGAWA: Privata communication.
(40) OGAWA, M., and T. MOROOKA: Private communication.
(41) SUGA, K., I. ESCOBAR, K. MURAKAMI, V. DOMINGO, Y. TOYODA, C. CLARK, and M. LA POINTE: Proc. Intern. Conf. on Cosmic Rays, Jaipur, Vol. 4, 9 (1964).
(42) See for instance A. UEDA and N. OGITA: Progr. Theoret. Phys. **19**, 582 (1958).
(43) Private communication from the Air Shower Group of the Institute of Nuclear Study, University of Tokyo; also S. FUKUI, H. HASEGAWA, T. MATANO, I. MIURA, M. ODA, K. SUGA, G. TANAHASHI and Y. TANAKA: Progr. Theoret. Phys., Suppl. **16**, 1 (1960).
(44) NISHIMURA, J.: Soryusiron Kenkyu **12**, 24 (1956).
(45) See for instance S. MIYAKE, T. KANEKO and N. ITO: J. Phys. Soc. Japan **18**, 1094 (1963). — KAMATA, K., S. KAWASAKI, and K. MURAKAMI: Proc. Intern. Conf. on Cosmic Rays, Jaipur, Vol. 4, 214 (1964).
(46) ODA, M., and Y. TANAKA: J. Phys. Soc. Japan **17**, Suppl. III A, 282 (1962).
(47) EYGES, L.: Phys. Rev. **75**, 264 (1949).
(48) SNYDER, H. S., and W. T. SCOTT: Phys. Rev. **76**, 220 (1949).
(49) MOLIÈRE, G.: Z. Naturforsch. **3a**, 78 (1948).
(50) NIGAM, B. P., M. K. SUNDARESAN, and TA-YOU WU: Phys. Rev. **115**, 491 (1959).
(51) SCOTT, W. T.: Rev. Mod. Phys. **35**, 231 (1963).
(52) DALITZ, R. H.: Proc. Roy. Soc. Lond. (A) **260**, 509 (1951).
(53) MIYAKE, S., V. S. NARASIMHAN, and P. V. RAMANA MURTHY: Proc. Intern. Conf. on Cosmic Rays, Jaipur, Vol. 6, 249, 250 (1964).
(54) MUROTA, T., A. UEDA, and H. TANAKA: Progr. Theoret. Phys. **16**, 482 (1956).
(55) NISHIMURA, J.: Proc. Intern. Conf. on Cosmic Rays, Jaipur, Vol. 6, 224 (1964).
(56) MINAKAWA, O., Y. NISHIMURA, M. TSUZUKI, H. AIZU, H. HASEGAWA, Y. ISHII, S. TOKUNAGA, Y. FUJIMOTO, S. HASEGAWA, J. NISHIMURA, K. NIU, K. NISHIKAWA, K. IMAEDA e M. KUAZNO: Nuovo Cim., Suppl. **11**, 125 (1958).
(57) FOWLER, P. H., D. H. PERKINS, and K. PINKAU: Proc. Intern. Conf. on Cosmic Rays, Moscow, II, 302 (1960).

Cosmic Rays and High-Energy Physics.

By

Y. Fujimoto and S. Hayakawa.

With 36 Figures.

I. Discovery of multiple production.

1. Early theoretical attempts. Around 1930 a number of cosmic-ray phenomena were discovered which did not seem to fit within the framework of the existing theory of elementary particles, and were thought by some physicists to indicate the need for a new theory. Among them, the multiple production of secondary particles by a cosmic-ray particle passing through matter was the most typical one; the bursts observed with ionization chambers and the showers observed with counter coincidences and cloud chambers were regarded as evidence for multiple production. At about the same time many other cosmic-ray phenomena were successfully explained by quantum electrodynamics, which predicts the occurrence of a higher order process only with a probability as small as a power of the fine structure constant, $\alpha = 1/137$. It was, therefore, felt that multiple production was a phenomenon contradicting quantum field theory.

This point of view was emphasized by several authors, among them Heisenberg [1]. He was motivated by his own theory of nuclear forces in which the exchange of a pair of fermions between two nucleons was assumed to be the origin of the force. There he noticed that the coupling constant of this interaction had the dimension of a length, so that at very high energies the interaction energy could become greater than the energy of the free fields. Multiple production was believed to be associated with this particular feature, because the perturbation method which was considered as an inherent part of quantum electrodynamics would no longer be valid in a field theory of this kind. The model adopted by Heisenberg was found to be inadequate after the proposal of Yukawa's meson theory, and furthermore most of those phenomena which had been attributed to multiple production were successfully interpreted as due to the cascade showers and evaporation stars from nuclei. Nevertheless, the perturbation method was considered to be not applicable for the strong interaction of the meson with the nucleon, and the discovery of the penetrating showers offered stronger evidence than before for the existence of the multiple production process. Therefore an essential part of Heisenberg's idea could not be abandoned. On the basis of meson theory Heisenberg [2] pointed out that the strong interaction should lead to a large probability for large momentum transfer, for which the existing theory would break down, and a new theory with a universal length should be introduced. Although the original forms of Heisenberg's proposal have now become merely of historical interest, their essential points still remain valid.

The strong interaction will show a characteristic behaviour when the interaction energy becomes greater than the energy of a free meson. The distance within which this situation happens is defined as a characteristic length l, and in the region of dimension smaller than l a conventional quantum mechanical treatment would become inadequate. Within the distance $\leq l$ from the center of a nucleon, the energy associated with the meson field is so large that this causes a great inertia for the motion of the nucleon, thus suppressing the production of mesons of energies higher than $\hbar c/l$. In an analogy to hydrodynamics this situation may also be described in the following way: A high concentration of energy within such a small volume corresponds in the wave picture to a large amplitude of meson waves, and to a high quantum state in the quantum picture. Since a state with high quantum number can be described by the classical wave theory according to the correspondence principle, the dynamical behaviour of meson fields may be solved by the classical wave theory. A wave of large amplitude is characterized by non-linearity, and it dissipates into many incoherent waves of longer wave lengths within a short time. Then the energy spectrum will be represented by the Planck distribution with temperature of the order $\hbar c/l$.

When the spectrum is expressed as $f(\boldsymbol{k})\,d\boldsymbol{k}$, where $\boldsymbol{k}$ is the wave number, the average number of mesons emitted is

$$n = \int f(\boldsymbol{k})\,d\boldsymbol{k}/k, \tag{1.1}$$

and the probability of emitting n mesons is proportional to

$$w_n \propto \frac{1}{n!} \left(\int \frac{f(\boldsymbol{k})}{k}\,d\boldsymbol{k} \right)^n. \tag{1.2}$$

If $f(\boldsymbol{k})$ falls off rapidly as k increases, the multiplicity increases rapidly with energy, and most of the mesons produced are of low energies.

However, an objection against the above idea was expressed on the basis of the theory of radiation damping, according to which the strong interaction would result in the reduction of the effective interaction, so that meson production would be suppressed at high energies (3). The multiple production as observed in penetrating showers with cloud chambers and counter hodoscopes could be interpreted as due to the plural collision of an impinging nucleon interacting simultaneously with several nucleons in a nucleus (4). This model was successful in explaining the proportionality of the nucleon-nucleus cross section to the area of a nucleus, or to $A^{\frac{2}{3}}$. It was not easy to find evidence for genuine multiple production, because experiments with hydrogen targets were prohibitively difficult in those days.

Although the plural production was able to account for the general features of cosmic-ray phenomena such as the nucleon cascade (5), it was not possible to present definite evidence against genuine multiple production. Moreover, the development of the renormalization theory and the closer examination of meson theory showed up difficulties for the adoption of the damping theory as it stood. In quantum electrodynamics divergent parts of the inertia and damping terms cancel each other, and the result after renormalization is found identical with that expected from the perturbation theory if constants are appropriately renormalized.

On the other hand, quantum mechanical methods which deal with multiple production were developed in various directions. Mesons strongly interacting with the nucleon form a cloud surrounding a bare nucleon, and the cloud is excited by the collision with an impinging particle. The excited cloud emits mesons successively, as it comes down to the ground state. The multiple production was formulated in this way with use of the intermediate coupling theory (6) as well as of the perturbation theory (7). These theories demonstrated that it was possible to understand multiple production in the framework of quantum field theory, although their results were not suitable for quantitative purposes.

2. High-energy cosmic-ray phenomena.

When the theories of multiple production as above were proposed, the nature of the meson was not clear yet. The meson was recognized first as a charged particle of an intermediate mass, and was considered to be produced in cosmic rays by energetic γ-rays associated with cascade showers initiated by primary electrons. When the primary cosmic radiation was found to consist mainly of protons, the question arose whether or not the Pfotzer maximum could be accounted for only in terms of the decay electrons of mesons produced by primary cosmic rays. However, those attempts failed. Reference was then made to the information on nuclear forces which required the existence of a neutral meson, and also to the theoretical possibility that the neutral meson decayed into two γ-rays (8). It was, in fact, found that the introduction of this γ-ray source was responsible for the electrons forming the Pfotzer maximum (9).

On the other hand, the abundant production of such neutral mesons was doubted in view of the presumed rare association of cascade showers with penetrating showers. It was then suggested that the internal bremsstrahlung associated with meson production could contribute to γ-rays, in particular at high energies because of its increasing probability, so that extensive air showers could be initiated by this mechanism (10). However, the frequent association of cascade showers with penetrating showers was observed in later investigations. Soon afterwards accelerator experiments also confirmed the existence of the neutral pion, and its γ-decay.

Concerning charged mesons, inconsistencies were observed in various respects. The two-meson hypothesis (11) was then proposed, and the existence of π- and μ-mesons was indeed verified experimentally (12). This provided a basis of investigating the behaviour of high-energy muons and their relation to γ-rays, as will be discussed in III.

The production of γ-rays together with mesons initiates a large shower consisting of both electronic and nuclear cascades. Extensive air showers were considered to be examples of such electro-nuclear cascade processes. Thus, a considerable number of nuclear-active particles, both nucleons and pions, survive down to low altitudes and generate the electronic component. Therefore the structure of an extensive air shower is different from what we expect from a pure electronic cascade process, and depends essentially on the features of multiple production.

3. Evidence for multiple production.

Experimental evidence for multiple production was obtained around 1950 when proper techniques had been developed.

The quality of nuclear emulsions had been improved in such a way that relativistic charged particles produced by high-energy interactions are detectable (13). With such nuclear emulsions several spectacular events were observed, in each of which the number of relativistic secondary particles was greater than 50 (14), (15). Since such collisions could have taken place in heavy nuclei, such as Ag and Br, the large multiplicity could, at least in part, be attributed to plural collisions in a nucleus. However, the number of relativistic secondary particles, denoted by n_s, was too large to explain all of them in these terms. Since the collisions with nucleons in a nucleus result in the production of a certain number N_h of non-relativistic particles through knock-on and evaporation processes, an event with few non-relativistic particles would indicate that the number of nucleons having participated in the event under consideration should be small. An interesting event with two non-relativistic protons, $N_h = 2$, was considered as an example of such a processes (16). The fifteen relativistic particles, $n_s = 15$, associated with this event were, therefore, thought to be produced by a single collision. Moreover, these relativistic particles could be clearly divided into two groups, one forming a narrow cone and the other a diffuse cone of a wide opening angle. They were interpreted as due to mesons emitted, respectively, forward and backward in the center-of-mass system of the two colliding nucleons. Later observation proved this narrow cone to be so characteristic of a shower of very high energy that these events were called "jets" for their appearance.

Although emulsion events such as the one mentioned above provided strong support for the theory of genuine multiple production, it was pointed out that the ejection of few non-relativistic particles should not be regarded as conclusive evidence for a single collision. At high energies all secondary particles including recoil nucleons should be emitted within a narrow angular region, so that the whole process of a nucleon-nucleus collision could take place in a cylindrical hole bored by the traversing primary. The disturbance left in the tunneled nucleus is so weak that only a few particles come out as a result of evaporation (17). It was further argued that most of the multiple production processes could be accounted for in terms of the nucleon-meson cascade inside a nucleus, even when secondary particles were emitted at considerable angles (18).

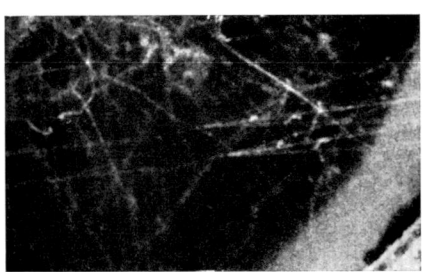

Fig. 1. The first photograph of multiple production events observed in a high-pressure cloud chamber filled with hydrogen gas (Osaka City University group).

In order to settle this question, studies of penetrating showers were made with hydrogen targets. Because a pure hydrogen target was hard to obtain and difficult to handle, most of them made use of the subtraction method, employing graphite (C) and paraffine (CH_2) targets. However, this method failed to give unambiguous results. Even where liquid hydrogen was used as a target, a contribution from the vessel was unavoidable.

In an experiment with a high-pressure hydrogen cloud chamber, however, a multiple event was found whose starting point was located in the gas, and this was regarded as an undisputable example of genuine multiple production, because the contribution of alcohol mixed with the hydrogen gas was negligible (19). The event is reproduced in Fig. 1.

Further evidence was obtained by an accelerator experiment in which double pion production was observed in a hydrogen diffusion chamber exposed to a proton beam (20).

Strong but indirect evidence for multiple production was also obtained from the analysis of the structure of extensive air showers. This suggested that the energy of a primary particle was shared among a considerable number of pions of relatively low energies, so that the development of the extensive air shower was controled by the electron-nuclear cascade (21), (22). It also indicated that the nuclear collision was not totally inelastic, and the multiplicity of neutral pions was rather high. A more detailed analysis of the altitude dependence and other properties of extensive air showers showed that the multiplicity of pions in such events is as high as twenty, and its energy dependence is rather weak (23).

II. Direct observations of high-energy interactions.

4. Difficulties in direct observations. It is needless to state that observations of extremely high energy interactions in cosmic rays provide an important means of exploring the behaviour of interactions of elementary particles in the high-energy region. Since the energy region we are mainly interested in will not be reached by artificial means within ten years in spite of the rapid progress in the construction of giant accelerators, the investigation of interactions above 1 TeV will, for the time being, have to depend exclusively on the information from cosmic rays.

However, there exist serious difficulties inherent in cosmic-ray work. First of all, the flux of cosmic rays decreases rapidly with increasing energy and with increasing atmospheric depth, so that the direct observation of high-energy nuclear interactions is possible only by means of a huge detector exposed for a long time at a high altitude, because otherwise the number of events suitable for our purpose would be too small to yield a statistically reliable result. This situation is illustrated in Table 1, in which statistics obtained with several typical experiments are shown. In order to study phenomena at still higher energies, we shall have to rely upon indirect means, such as investigations of extensive air showers.

Table 1. *Typical examples of high-energy experiments.*

Experiment	Detector	Exposure	Statistics	Remark
Brawley stack	solid emulsion stack, 80 litre	36 hrs. at 15 g/cm²	200 jets with $\Sigma E_\gamma \geq 400$ GeV	See II.9.
Japanese E.C.C.	emulsion chamber	24.8 m² year at 730 g/cm²	2000 γ-rays with $E_\gamma \geq 1$ TeV	See III.12.
Japan-Brasil collaboration	emulsion chamber	4.4 m² year at 550 g/cm²	15 γ-ray families with $\Sigma E_\gamma \geq 50$ TeV	See III.13.

Secondly, cosmic-ray particles are neither monenergetic nor collimated. The determination of the energy of an incident particle is a serious problem in the technology of cosmic-ray studies, whereas the incident direction of a charged particle can be measured with comparative ease. Difficulties in the energy determination as well as in the identification of particles hold also for secondary particles. Such difficulties are inherent to the detection technique now available for cosmic-ray studies.

Particle detectors, such as counters, nuclear emulsions, cloud chambers and so forth, are operated on the basis of the ionization caused by charged particles. The degree of ionization, as is well known, decreases with increasing energy in the non-relativistic region and reaches a minimum at a kinetic energy comparable to the rest energy of the particle concerned. As the energy increases further, the degree of ionization increases slowly in proportion to the logarithm of the total energy, but soon saturates to a constant value because of the density effect. The saturation value is called the plateau. The difference between the minimum and the plateau of the ionization is so small in solid and liquid detectors that the energy discrimination in this region is a prohibitively difficult task. In general, significant information on the energy of a relativistic particle can hardly be obtained from an ionization measurement.

As a typical example, we shall discuss the observation of charged particles with nuclear emulsions in a little more detail[1]. The energy of a relativistic particle is determined by means of the measurement of multiple scattering of a track. The magnitude of the multiple scattering is usually represented by $\bar{d}$, the average value of the second differences of track coordinates from a reference line measured

---

[1] For details of the emulsion technique, see the book of C. F. POWELL, P. H. FOWLER and D. H. PERKINS [3].

along the track in certain fixed distances l, l being called the cell size. The magnitude of $\bar{d}$ is given through the scattering constant K by

$$\bar{d} = K l^{\frac{3}{2}}/p\beta. \tag{4.1}$$

where p is the momentum and $c\beta$ the velocity of the particle concerned. In the relativistic case $p\beta$ is essentially equal to the energy E in units of $c=1$, where c is the light velocity. A difficult arises from the fact that the value of $\bar{d}$ decreases inversely proportional to E.

The measurement of $\bar{d}$ is associated with various errors, such as those due to reading, the stage noise of a microscope, the finite size of silver grains, and the distortion of the emulsions. Distortion is the most serious, because this error cannot be removed successfully after developing of the emulsions either by technical achievements in the measurement or by mathematical analysis. In fact, a considerable part of the distortion effect on the track is turbulent-like and thus in its character similar to multiple scattering. Hence it is called spurious scattering. Even in the most favourable case where a track is found nearly parallel to the plane of an emulsion plate and the developing is carried out with great caution, a reliable measurement of the energy beyond several GeV is hardly possible. For those jets of interest which are produced by a particle of energy as great as or greater than 1 TeV, therefore, the measurement of $\bar{d}$ is applicable neither to the primary particle nor to secondary particles, except to those emitted at large angles to the incident direction.

For tracks nearly parallel and close to each other, however, the method of relative scattering is applicable in which one measures the relative positions between these tracks. It is reported that the energy can be measured with this technique up to about 100 GeV, since the error due to large-scale distortion disappears in this case (24). Unfortunately, however, there are few such cases, because only a slight difference in dip angles prevents cancellation of the distortion effect. In practice, the method of relative scattering is applicable for a pair of electrons emitted with a small opening angle, or for an interaction where a reference track is available among surviving protons or nuclei produced by the break-up of a heavy primary nucleus.

As another example, we briefly discuss the cloud chamber technique. Since the spatial resolution is poorer in cloud chambers than in nuclear emulsions, the primary energies suitable for cloud chamber study are lower than those for emulsions, at most a fraction of TeV. On the other hand, there are several advantages; in particular the density effect is so small that at higher energies, ionization measurements can sometimes serve to identify particles according to their mass and energy (25), and the momenta and charges of particles can be measured from the curvatures in a magnetic field.

Here again several errors occur which become serious for the curvature measurement of high-energy particle tracks, because the curvature decreases inversely proportional to the momentum. Errors come from reading, the finite size of drops, the diffusion of drops, and the turbulence in the gas caused by the expansion. In a cloud chamber with a conventional magnetic field, the upper limit of measurable energy does not extend beyond several tens of GeV. If a magnetic field is applied in the space between two cloud chambers, arranged one above the other, the upper limit may be raised up to 100 GeV or so.

Turbulence in the gas does not take place in spark chambers. It is, therefore, hoped that eventually the limit of energy measurements will be pushed up by some factor (26).

5. Emulsion stacks and measurement of K/π-ratio.

In the following sections we shall describe a number of typical experimental methods, and the characteristic quantities of high-energy interactions observed by these.

5. Emulsion stacks and measurement of K/π-ratio. The emulsion-stack method has been widely used for the study of high-energy interactions because of its superiority in permitting observation of charged secondary particles in full detail. As described in the previous section, however, the identification of a secondary particle in a jet shower is in most cases impossible with the emulsion technique available at present through the measurement of its mass, but can be made through its characteristic decay mode and interactions. Ironically enough, more reliable identification can be made not for charged particles directly observed but for neutral particles, in particular for neutral pions.

The neutral pion has a mean lifetime as short as 2×10^{-16} sec and its mean flight length is only about 1 mm even at an energy of about 1 Tev, so that it can be regarded as decaying immediately. With the probability of 99% the decay of a neutral pion gives rise to two photons, and only a minor fraction goes into a photon and a pair of electrons. Since the conversion length of γ-rays is theoretically estimated as 3.5 cm in nuclear emulsion (27), one is able to observe electron pairs produced by these γ-rays and their cascade development with an emulsion stack of moderate size.

In practice, one usually detects high-energy interactions by scanning for large cascade showers and following them back to their origin. Jets associated with cascade showers of energies higher than 0.5 TeV may be found with a good probability by naked-eye scanning for cascade showers.

Investigations of neutral pions produced in jets are made on the basis of electron pair scanning. The scanning for pairs is usually done in the forward cone of the jet showers under a certain criterion. The electron pairs thus found by the scanning are composed of the following two kinds: the primary electron pairs which are originated directly from the decay γ-rays of neutral pions, and the secondary pairs which are generated in the course of the cascade process starting from the primary pairs. If the scanning is restricted to a short distance from the jet, one cascade unit or less, there is no ambiguity in selecting the primary pairs from the rest, since the angular divergence of a pair at the initial stage of the cascade multiplication is much smaller than the angular divergence between two primary pairs[1]. When the distance from the origin of the jet shower is increased, the angular divergence of the cascade multiplication becomes so large that a discrimination between the primary and the secondary pairs turns out to be impossible.

The neutral-charge ratio, R,

$$R = \text{no. of neutral pions/no. of charged particles} \\ \equiv \frac{n_{\pi^0}}{n_s} \quad (5.1)$$

is obtained from the observed number of primary electron pairs and the total length of charged secondary particle tracks in the scanned volume. In the calculation of R one assumes the theoretical value of the conversion length of γ-rays, and furthermore that the angular distribution of γ-rays from neutral pions and that of charged particles are the same. In Table 2, the experimental results of the neutral-charge ratio, R, are summarized.

---

[1] See the discussion in II.6. Emulsion chamber and transverse momentum.

As is seen from Table 2, the value of R is essentially energy independent, and its weighted average value is

$$R = 0.40 \pm 0.03. \tag{5.2}$$

It should be remarked that the whole procedure of obtaining R is based on the assumption that γ-rays originate mainly in the decay of neutral pions. There are several sources of γ-rays other than neutral pions, such as the internal bremsstrahlung associated with the production of charged particles, and the radiative decays of excited states of mesons and baryons including $\Sigma^0 \to \Lambda^0 + \gamma$. But the contributions from all the known sources of γ-rays other than neutral pions are

Table 2. *Neutral-charge ratio,* $R = n_{\pi^0}/n_s$

Estimated primary energy eV	R	Method	Reference	Estimated primary energy eV	R	Method	Reference
$1-2 \times 10^{10}$	0.50 ± 0.10	CC*	1	$1-5 \times 10^{12}$	0.46 ± 0.09	Em	4
2.5×10^{10}	0.48 ± 0.11	Em**	2	$1-5 \times 10^{12}$	0.40 ± 0.04	Em	5
$2-15 \times 10^{10}$	0.40 ± 0.04	CC	3	1×10^{13}	0.50 ± 0.11	Em	6
5×10^{11}	0.33 ± 0.08	Em	2				

* CC: cloud chamber. ** Em: emulsion.

[1] G. SALVINI, and Y. KIM: Phys. Rev. **85**, 921 (1952); **88**, 40 (1952).
[2] R. R. DANIEL, J. H. DAVIES, J. H. MULVEY, and D. H. PERKINS: Phil. Mag. **43**, 753 (1952).
[3] S. LAL, Y. PAL, and R. RAGHAVAN: J. Phys. Soc. Japan **17**, Suppl. A III, 393 (1962).
[4] M. F. KAPLON, W. D. WALKER, and M. KOSHIBA: Phys. Rev. **93**, 1424 (1954).
[5] F. A. BRISBOUT, C. DAHANAYAKE, A. ENGLER, Y. FUJIMOTO, and D. H. PERKINS: Phil. Mag. **1**, 605 (1956). — B. EDWARDS, J. LOSTY, K. PINKAU, D. H. PERKINS, and J. REYNOLD: Phil. Mag. **3**, 237 (1958).
[6] M. KOSHIBA, and M. F. KAPLON: Phys. Rev. **97**, 193 (1955).

estimated to be negligible compared to the statistical error of the present experimental value of R. However, it is an open question whether there exist some unknown sources of γ-rays with appreciable contribution in the extremely high energy region[1].

If all charged secondaries were charged pions, and if we assume charge independence of the strong interaction, the value of R should be $\frac{1}{2}$. The observed value, appreciably smaller than $\frac{1}{2}$, suggests that a not negligible part of the charged secondaries are not pions but may be called X-particles, which are supposed to consist of kaons, nucleons and anti-nucleons, and others. As a counterpart of charged X-particles, there will be neutral X-particle. Since X-particles are presumed to be strongly interacting like pions and to have a longer lifetime than neutral pions, they may be found through secondary interactions.

Scanning for secondary interactions can be made in the narrow cone of a jet over a long distance. One is able to see whether the secondary interaction is initiated by a neutral or a charged particle. But in some cases the identification of a track initiating the secondary interaction is not possible because of the high density of tracks or some other reasons. Let N_n, N_{ch}, and $N_?$ be the observed total numbers of neutral and charged secondary interactions and unidentified ones. Then the neutral-charge ratio of the secondary interactions, Q, which is defined as

$$Q = \frac{N_n + \frac{1}{2}N_? - \frac{1}{2}N_N[1 - \exp(-L/\lambda_p)]}{N_{ch} + \frac{1}{2}N_? - \frac{1}{2}N_N[1 - \exp(-L/\lambda_p)]} \tag{5.3}$$

---

[1] Attempts have been made to identify the two γ-rays coming from a single neutral pion. For this purpose one needs information on the energies of the individual γ-rays, which are not easily obtained in the emulsion stack work. See the discussion in II.6. Emulsion chambers and transverse momentum.

can be calculated. N_N is the expected number of surviving nucleons, λ_p is the collision mean free path of the nucleon in emulsions, and L is the distance over which the scanning is made. As we are interested for the moment only in particles created but not in surviving nucleons, we substract the expected number of their interactions in the denominator as well as in the numerator in (5.3). The expected number of surviving nucleon interactions and the number of unidentified secondary interactions are shared equally between neutral and charged interactions.

If one makes the assumption that the interaction mean free paths of all secondary particles are essentially equal, one finds that the neutral-charge ratio

Table 3. *Interaction mean free path in nuclear emulsions.*

Energy eV	Particle	Mean free path cm	Authors
6×10^9	p	38.2 ± 1.5	H. Winzeler et al.: Nuovo Cim. **17**, 8 (1960)
		34.7 ± 3.4	M. V. K. Appa Rao et al.: Proc. Ind. Acad. Sci. A **43**, 181 (1956)
9×10^9	p	34.0 ± 2.0	G. B. Zhdanov: J. Exp. Theoret. Phys. **34**, 856 (1958)
		37.1 ± 1.0	N. Bogachev et al.: DAN **121**, 615 (1958)
14×10^9	p	29.5 ± 1.3	N. P. Bricman et al.: Nuovo Cim. **20**, 1017 (1961)
23.5×10^9	p	36.6 ± 1.0	G. Cvijanovich et al.: Nuovo Cim. **20**, 1012 (1961)
27×10^9	p	37.9 ± 0.7	Y. Baudinet-Robinet: Nuclear Phys. **32**, 452 (1962)
28×10^9	p	37.9 ± 1.2	P. L. Jain et al.: Nuovo Cim. **21**, 859 (1961)
$\sim 10^{11}$	secondaries	41 ± 8	A. Barkow et al.: Phys. Rev. **122**, 617 (1961)
1.7×10^{11}	break-up p	41 ± 10	E. Lohrmann et al.: Phys. Rev. **122**, 672 (1961)
7×10^{12}	primaries	27^{+14}_{-7}	R. R. Daniel et al.: Internat. Conf. on Cosmic Rays, Jaipur, 1963. Vol. 5, 17
$\sim 10^{12}$	secondaries	26 ± 9	M. Koshiba: Internat. Conf. on Cosmic Rays, Jaipur, 1963. Vol. 5, 293
4.2×10^9	π	38.7 ± 2.3	J. O. Clark: Phil. Mag. **2**, 37 (1957)
4.3×10^9	π	33.7 ± 4.7	A. Margnes: Nuovo Cim. **5**, 291 (1957)
7×10^9	π	35 ± 3	V. A. Belyakov et al.: J. Exp. Theoret. Phys. **39**, 937 (1960)
7.5×10^9	π	38.6 ± 1.4	L. C. Grote et al.: Nuclear Phys. **24**, 300 (1961)
16×10^9	π	39.2 ± 1.0	A. Baldassare et al.: Aix-en-Provance Conf. 1961, p. 427

of the secondary interactions, Q, is equal to the ratio of number of neutral X-particles, n_{X^0}, to number of charged particles, n_s:

$$Q = n_{X^0}/n_s. \tag{5.4}$$

The value of Q thus obtained for primary energies around several TeV is (28), (29), (30):

$$Q = 0.24 \pm 0.04. \tag{5.5}$$

The assumption concerning the interaction mean free path may be checked for primary and secondary particles in a wide energy range, the latter being supposed to consist mainly of pions. The results thus far obtained on the mean free path are summarized in Table 3. This shows that the mean free path converges to a constant value at high energy, and is likely to be independent of the kind of particles, about 30 cm in G-5 emulsions. This does not necessarily contradict the evidence in the total $p-p$ and $\pi-p$ cross sections observed with accelerators, because the total cross section involves elastic and quasi-elastic collisions which are not recognized as nuclear interactions in emulsion experiments.

Once the experimental values of R and Q are obtained, the relative abundances of secondary particles can be evaluated with reference to the isovector nature of the pion:

$$n_{\pi^0}/n_{\pi^\pm} = \tfrac{1}{2}. \tag{5.6}$$

From (5.2.), (5.5) and (5.6) one has

$$\left.\begin{array}{l}\dfrac{n_{X^\pm}}{n_{\pi^\pm}+n_{X^\pm}} = 1-2R = 0.20 \pm 0.05, \\[6pt] \dfrac{n_{X^\pm}}{n_{\pi^\pm}} = \dfrac{1-2R}{2R} = 0.025 \pm 0.06, \\[6pt] \dfrac{n_{X^0}}{n_{\pi^\pm}} = \dfrac{Q}{2R} = 0.30 \pm 0.06, \\[6pt] \dfrac{n_{X^0}}{n_{X^\pm}} = \dfrac{Q}{1-2R} = 1.2 \pm 0.4, \\[6pt] \dfrac{n_{X^\pm}+n_{X^0}}{n_{\pi^\pm}+n_{\pi^0}+n_{X^\pm}+n_{X^0}} = 1 - \dfrac{3R}{1+R+Q} = 0.27 \pm 0.05.\end{array}\right\} \tag{5.7}$$

It should be noted that there are at least two factors which might affect the results (5.7) to a considerable degree. One is the possible effect of a selection bias which is naturally in favour of interactions with large cascade showers. The other is the assumption made throughout the procedure that the angular distribution is essentially the same for all kinds of particles produced. Although we are not at present in a position to discuss these questions in a quantitative way, they should be kept in mind when a comparison is made with other indirect information on the relative abundances of produced particles[1].

Now let us compare the cosmic-ray results with the data obtained from accelerator experiments.

It is interesting to see that the results (5.7) are consistent with those obtained with proton beams of energies above 10 GeV from accelerators. The energy independence of the relative abundances is indicated also by cosmic-ray results as shown in Table 2. They seem to suggest that the relative abundances of particles emitted are determined essentially by the final interactions, and that the emission of particles arises from a system of low temperatures (*31*).

Concerning the nature of the X-particles, accelerator experiments on the proton-proton collisions at 30 Gev give (*32*):

$$n(\pi^+) : n(K^+) : n(K^-) : n(\bar{p}) = 1 : 0.15 : 0.05 : 0.01,$$

and the abundance of hyperons is as small as that of K^-.

In the cosmic-ray energy region no extensive studies on the nature of X-particle with good accuracy have been reported yet. For interactions of energy about 100 Gev, cloud chamber observations show results on X-particles similar to the accelerator data above. At still higher energies the particles emitted backward in the center-of-mass system are identified by measuring the ionization-scattering as well as the ionization-range relations in emulsions (*33*). The X/π-ratio thus obtained is in good agreement with (5.7). These results indicate that most of the X-particles are kaons, and a few are baryons. However, the relative abundances in the extremely forward or backward directions appear to be different. For particles emitted within 5° in the backward direction in the center-of-mass system, kaons are found to be as frequent as pions, and one hyperon decay is observed in

---

[1] The K/π-ratio can be obtained in an indirect way by comparison between the muon flux and the γ-ray flux. See the discussion in III.15. Inter-relationship between γ-rays and muons.

20 events (34). It should also be noted that among these particles ten protons were identified, which suggests that the surviving nucleons have high energies and are equally divided into protons and neutrons.

6. Emulsion chambers and transverse momentum.

As has been described above, among the various secondary particles in a jet shower neutral pions are most easily and most reliably identified. If, in addition, the energies of individual γ-rays or neutral pions can be measured, the value of the information will be increased considerably. However, this measurement is in general subject to the difficulty that the cascade showers developing from the several γ-rays overlap and mix with each other as they proceed through the material.

One way of overcoming this difficulty is to observe these γ-rays as the primary electron pairs before they develop into cascade showers. Suppose a pair of electrons is found. The opening angle of the pair is about $m_e c^2/E$, where $m_e c^2$ is the electron rest energy. Comparing this opening angle with the multiple scattering angle, $(K/E)\sqrt{t}$, K being the scattering constant and t being the length measured in the radiation units, one sees that the opening angle dominates over the multiple scattering only up to a distance of $(m_e c^2/K)^2$ radiation units from the origin of the pair. This distance is of the order of 10^{-3} radiation length. Therefore an accurate opening angle measurement is not possible for γ-rays above several Gev due to the finite size of silver grains.

The relative scattering measurement on an electron pair gives us the energy value of the lower-energy electron in the pair, since the higher-energy track is used as a reference. This information will be used as an order-of-magnitude estimate of energy of the pair. But the high-energy measurement around 1 Tev or more is subject to a large error, because the scattering measurement has to be made over a considerable track length so that the result will be affected by energy loss of the electrons due to bremsstrahlung.

It was suggested that the energy of a high-energy electron pair can be determined by observing the decrease of the ionization density produced by the pair due to the destructive interference of their Coulomb fields (35). However, this method is applicable only for a pair with energy greater than 1 Tev.

Thus all of the methods with emulsion stacks are neither extensive nor accurate enough. Furthermore, if one is trying to pick up unambiguously only primary electron pairs, one has to restrict the place of observation for the pair to a certain distance from the jet, so that one is not able to detect all γ-rays from the jet.

A new approach was proposed by Japanese Emulsion Group, taking advantage of the fact that the cascade shower provides a means of the energy measurement which becomes more favourable as the energy increases, but getting rid of the mixing of cascade showers initiated from different γ-rays by separating them from each other before their development into showers.

Let us consider a cascade shower initiated by a γ-ray of energy E_γ. After traversing the matter of thickness t radiation units, the cascade contains many electrons whose average energy is given by

$$\langle E \rangle = E_\gamma \exp(-t), \tag{6.1}$$

before the shower reaches its maximum. On account of the mean lateral spread of electrons given in (4.1), these electrons will spread over a distance d, in radiation units,

$$d = \text{const}\,(K/E_\gamma)\exp(t). \tag{6.2}$$

The constant is of the order of unity. This is compared with the lateral spread of two γ-rays produced by the decay of a neutral pion. If the two γ-rays have equal energies, E_γ, their opening angle is given by

$$\vartheta_{\pi^0} = m_{\pi^0} c^2 / 2 E_\gamma, \qquad (6.3)$$

where $m_{\pi^0} c^2$ is the pion rest energy. After traversing t, the distance between these γ-rays is

$$l = \vartheta_{\pi^0} t = (m_{\pi^0} c^2 / 2 E_\gamma) t. \qquad (6.4)$$

Comparing (6.2) with (6.4), we see that the ratio d/l is independent of the energy and also of the properties of the medium. The two γ-rays can be distinguished from each other only for small t, $t \lesssim 1$. This qualitative consideration demonstrates that the measurement of two separated showers is practically impossible, as long as a homogeneous medium is used.

It is, therefore, required to use two different media, in one of which an interaction takes place and γ-rays thus produced are well separated without appreciable cascade development, and in the other individual cascade showers are developed to their full sizes. This can be achieved with an emulsion chamber which consists of a low-Z material as the producer of interactions, a high-Z material as the developer of cascade showers, and nuclear emulsions sandwiched in the latter for measuring the numbers of electrons in various stages of the cascade shower development.

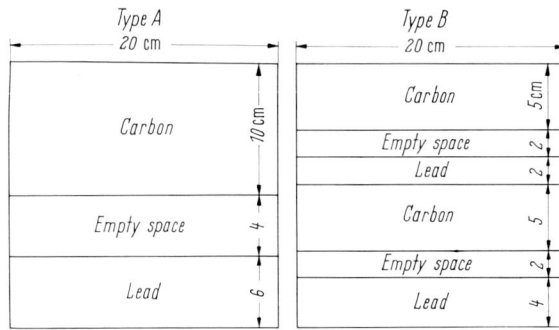

Fig. 2. Emulsion chambers flown by balloons in 1956 by the Japanese emulsion group.

A typical design of the emulsion chamber for a balloon-borne experiment in 1956 is shown in Fig. 2 (36). The chamber consists of a carbon layer in which jet showers are produced, an empty space in which γ-rays produced are separated, and a lead block in which cascade showers develop. Nuclear emulsions are sandwiched in the carbon and lead layers so that electrons in cascade showers as well as charged secondary tracks in jets can be observed. With this type of emulsion chambers an extensive measurement of the energies of individual γ-rays has become possible with reasonable reliability. Moreover, the starting point of a jet can be located in the carbon layer by observing the convergence of the axes of cascade showers.

Groups of cascade showers could also be studied in this manner. The energy of each cascade shower, E_γ, is estimated by measuring the shower size and comparing the result with the theoretical calculation by NISHIMURA and KAMATA[1]. In many cases a pair of cascade showers can be "married off" into a neutral pion, so that the method of energy estimation is checked by reference to the opening angle of the pair. When this is possible, the angle and the energy of the neutral pion are determined with better accuracy.

The main purpose of the Emulsion Chamber Project in 1956 was to measure the transverse momenta of γ-rays and neutral pions. The axis of a jet is located

---

[1] Discussions on the energy measurement of cascade showers will be found in III.12. Observation of high energy γ-rays. For further details, particularly the theoretical calculations, see J. NISHIMURA: Handbuch der Physik, this volume, Chapter 1.

as the center of gravity of all γ-rays in the event, and the emission angle of each γ-ray, ϑ_γ, is determined. Then the transverse momentum of a γ-ray is given as

$$P_T(\gamma) = E_\gamma \vartheta_\gamma. \tag{6.5}$$

Likewise, the transverse momentum of a neutral pion is obtained by use of a "married" γ-ray pair.

The distribution of $P_T(\gamma)$ and $P_T(\pi^0)$ thus obtained are shown in Fig. 3. It is interesting to see that no pion has a transverse momentum greater than 1 Gev/c. The distribution has a sharp peak near 240 Mev/c, and the value of P_T is independent of the energies of primary and secondary particles as well as of the emis-

Table 4. *Transverse momenta of particles produced in 10 GeV $\pi^- - p$ collisions.*

Particle	Average transverse momentum (GeV/c)
π^+	0.30 ± 0.01
K^0	0.39 ± 0.02
p	0.49 ± 0.05
Λ^0	0.46 ± 0.05
Σ^+	0.51 ± 0.04
Ξ^-	0.56 ± 0.08

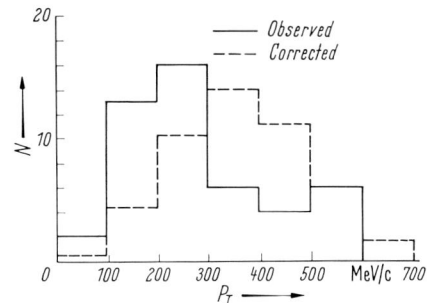

Fig. 3. Distribution of transverse momenta of neutral pions observed with emulsion chambers. The ordinate represents the number of neutral pions in every 100 MeV/c range of the transverse momentum. The full line represents the distribution of transverse momenta with respect to the center of gravity of neutral pions; the dotted line represents that with respect to the actual shower axis.

sion angles within the region where the measurements were made. The average values obtained after statistical correction with respect to the actual shower axis are

$$\langle P_T(\gamma) \rangle = 150 \pm 15 \text{ Mev}, \tag{6.6}$$

$$\langle P_T(\pi^0) \rangle = 390 \pm 20 \text{ Mev}. \tag{6.7}$$

A measurement of transverse momenta was carried out also with an emulsion stack (29). In the case of charged particles their $P\beta$ values were measured with the multiple scattering method for those emitted backward in the center-of-mass system, while for γ-rays from neutral pions their energies were measured by the relative scattering of electron pairs. The results for pions are in agreement with those obtained with the emulsion chamber. In comparison between them, attention should be paid to the difference in the selection criterion. In the emulsion chamber experiment P_T is measured for γ-rays with energies greater than a certain value, while in the emulsion stack experiment the measurement is made for γ-rays with emission angles smaller than a certain value. These criteria are derived from the method of detection.

In the accelerator energy region, accurate measurements of transverse momenta were also made, being stimulated by the cosmic ray work. Table 4 shows the average transverse momenta for various kinds of particles produced in $\pi^- - p$ collisions at 10 Gev (37). It is seen that the average P_T increases slightly with the mass of the particle.

If this continues to hold at very high energies, the difference between the transverse momenta of pions and X-particles would be insignificant. In fact, a cloud chamber observation of cosmic-ray showers produced at primary energies of around 10^{11} eV has given essentially equal transverse momenta of both kinds of particles, 310 MeV/c (25). Essentially the same result has been obtained with the ICEF stack, by measuring the grain densities and the scattering of charged particles emitted backward in the center-of-mass system (34).

On the other hand, transverse momenta as large as a few GeV/c are obtained for those charged particles emitted with the smallest and the largest angles, which are respectively identified as the surviving protons of the incident and the target nucleons (38), (39). The result is in agreement with that reported by the Bristol group from their measurement on the secondary jets in the forward narrow cone (29), although their value is subject to ambiguity in the energy measurement of secondary jets.

With emulsion chambers exposed at Mt. Norikura as well as at Mt. Chacaltaya[1] an appreciable fraction of events with $\Sigma E_\gamma \geq 10$ TeV are found to have large transverse momenta between γ-ray groups and surviving nucleons. This would also indicate a large transverse momentum of the surviving nucleon.

In the highest energy region of $>10^{15}$ eV, the general features of extensive air showers are found consistent with the assumption that the transverse momentum is the same as the one at low energies, as was made clear by the original analysis of P_T by NISHIMURA (40). Recently, however, several examples of extensive air showers have been reported to have multiple cores which indicate the existence of cases with large transverse momenta. In particular, air showers with double core structures observed with a neon hodoscope (41) and a cloud chamber (42) are interpreted as due to such secondary particles which have transverse momenta as high as 10 GeV/c[2].

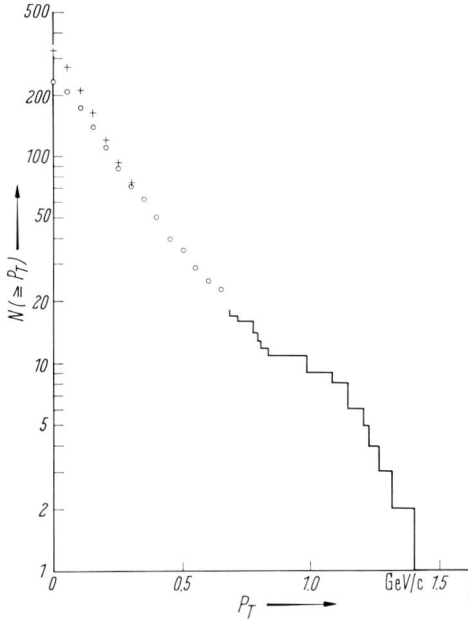

Fig. 4. Integral spectrum of transverse momenta of γ-rays obtained by the Bristol-Indian group. γ-rays of energies greater than 50 GeV are selected from the events of $\Sigma E_\gamma \geq 1.2$ TeV, as shown by open circles. Crossed points represent the number of events corrected so as to include all γ-rays with emission angles smaller than 10^{-2} rad.

It is, therefore, considered important to study the distribution of transverse momenta and its dependence on the primary energy. At accelerator energies the distribution of transverse momenta of produced pions as well as of scattered protons is given by an exponential law (43), (44)

$$f(P_T) 2\pi P_T d P_T = \exp(-P_T/P_{T0}) 2\pi P_T d P_T/P_{T0}^2 \qquad (6.8)$$

up to $P_T = 4$ GeV/c. The average transverse momentum is given by

$$\langle P_T \rangle = 2 P_{T0} = \begin{cases} 0.30 \text{ GeV for elastically scattered protons,} \\ 0.34 \text{ GeV/c for secondary pions.} \end{cases} \qquad (6.9)$$

---

[1] The emulsion chamber experiment carried out at Norikura and Chacaltaya will be described in detail in III.12, "Observation on high-energy γ-rays", and III.13, "γ-ray families'

[2] Of course, the possibility cannot be ruled out that these events were originated by heavy primaries, and that the explanation for the large separation must be sought in the development of the nucleonic cascade. This was, for instance, emphasized by McCUSKER (Intern. Conf. on Cosmic Rays, Jaipur (1963), Vol. 4, 35, ref. [12]).

A recent experiment of the Bristol-India collaboration with a balloon-borne emulsion chamber gives the distribution of the transverse momentum of γ-rays for $\Sigma E_\gamma \geq 1.2$ TeV, as shown in Fig. 4, which can be approximated by an exponential law, at least up to 0.7 GeV, with

$$\langle P_T(\gamma) \rangle = 0.24 \text{ GeV/c}. \tag{6.10}$$

A similar P_T-distribution was obtained in a higher energy region with emulsion chambers exposed at mountains, as shown in Fig. 5.

However, the distribution beyond 1 GeV appears to be less steep. It may not be implausible to infer that the probability for large transverse momenta increases with energy.

Observing that the low-energy data, (6.8) and (6.9), are essentially the same as the high-energy ones for the majority of pions, we may assume that the following standard value of P_T will hold for pions in a wide range of energy:

$$\langle P_T \rangle = 0.4 \text{ GeV/c}. \tag{6.11}$$

For the minority of pions and surviving nucleons, however, there are indications that cases of large transverse momentum, $P_T \approx$ several GeV/c, appear more frequently in the higher-energy region, though further experiments are necessary before a final conclusion can be reached.

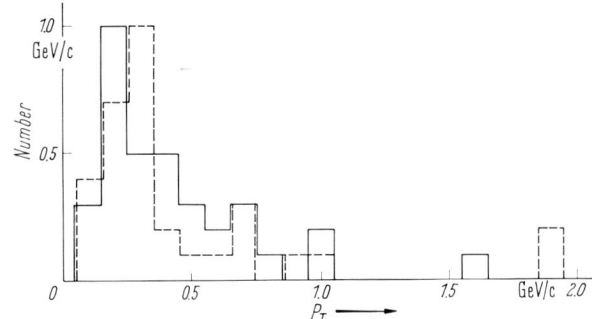

Fig. 5. Distributions of transverse momenta of neutral pions obtained at Mt. Chacaltaya by the Brazilian-Japanese group for events of $\Sigma E_\gamma \geq 20$ TeV. The full line represents the P_T-distribution for $\Sigma E_\gamma \geq 50$ TeV, the dotted line that for 50 TeV $> \Sigma E_\gamma \geq 20$ TeV.

7. Angular distribution. A "jet" shower is, as its name indicates, characterized by a sharply collimated angular distribution of the secondary particles. This can be explained as a direct result of a Lorentz transformation from the center-of-mass system with a large Lorentz factor, even if the angular distribution in this system were isotropic. Accordingly, the angular distribution is generally used for estimating the Lorentz factor, and consequently the primary energy.

In applying this procedure, it is essential to assume that the angular distribution in the center-of-mass system is symmetric in the forward and backward directions. This implies that the center-of-mass system can be defined unambiguously, assuming that the collisions take place between two single nucleons. However, in most of cosmic-ray experiments the target material is not hydrogen, and furthermore primary particles are not solely protons.

If the simplification above is taken for granted, the Lorentz factor γ_c for the transformation between the center-of-mass system (C.M.S.) and the laboratory system (L.S.) can be approximately obtained from the median angle, $\vartheta_\frac{1}{2}$, within which half of the secondary particles are contained, as

$$\gamma_c = \frac{1}{\tan \vartheta_\frac{1}{2}}. \tag{7.1}$$

The value of γ_c thus estimated is known to be inaccurate, because of large statistical fluctuations in the angular distribution.

The accuracy is increased if information from all the emission angles is used. The emission angle of the i-th particle in L.S., ϑ_i, is related to that in C.M.S.,

ϑ_i^*, through

$$\log \tan \vartheta_i = -\log \gamma_c + \log \tan\left(\frac{\vartheta_i^*}{2}\right). \tag{7.2}$$

The last term in the right-hand side vanishes owing to the symmetry, if summed over all secondary particles. This yields

$$\log \gamma_c = -\frac{1}{n_s} \sum_i \log \tan \vartheta_i, \tag{7.3}$$

and the primary energy is given by

$$E_C = (2\gamma_c^2 - 1) M c^2, \tag{7.4}$$

where $M c^2$ is the nucleon rest energy. This procedure of obtaining the primary energy is called the CASTAGNOLI method (45), and has been widely used in practice.

Both of these methods are based on the forward-backward symmetry of secondary particles. But the validity of this assumption is becoming increasingly doubtful. As will be described in later sections[1], the symmetry argument is found inapplicable to individual jet showers even in the case of a nucleon-nucleon collision. A serious disadvantage of these methods is that they give rise to systematic errors in the energy dependence of some quantities concerned, as for example the inelasticity coefficient.

Although the absolute value of γ_c is subject to an error, the relative angular distribution expressed by (7.2) is invariant under the Lorentz transformation. The plot of $\log \tan \vartheta_i$ on a line is shifted by an error in $\log \gamma_c$, but the distances between $\log \tan \vartheta_i$ remain unchanged. The usefulness of the log-tan ϑ plot as a guide for inventing various models of multiple production will be seen in the discussions of IV.

A new method not relying upon the Lorentz transformation could be devised when the transverse momentum was observed to be constant and independent of other parameters[2]. On this basis the energy of a relativistic particle can be obtained as

$$E_i = P_T \csc \vartheta_i, \tag{7.5}$$

from the emission angle ϑ_i in L.S. The energy transferred to all charged secondaries is thus estimated as

$$E_{ch} = P_T \sum_i \csc \vartheta_i, \tag{7.6}$$

and knowing the fraction of energy transferred to charged secondaries, K_{ch}, one can obtain the primary energy

$$E_0 = \frac{E_{ch}}{K_{ch}} = \left(\frac{P_T}{K_{ch}}\right) \sum_i \csc \vartheta_i. \tag{7.7}$$

This provides an independent method of estimating the primary energy, together with the measurement of ΣE_γ, the total energy transferred into γ-rays.

8. Calorimeter method and asymmetry.

The main experimental apparatus employed at present for the observation of interactions at around 100 GeV is the cloud chamber. Nuclear emulsions are found not suitable for investigations in this lower energy region, because the convenient method of scanning for secondary cascade showers cannot be applied in this energy range.

---

[1] See discussion in II.8. and II.9. "Calorimeter method and asymmetry", and "Large emulsion stack and inelasticity".

[2] This idea was originally proposed by NISHIMURA (40).

Most of the cloud chamber studies use large magnets for the momentum measurement. Thus the experiments have to be carried out either at sea level or at mountain altitudes. At these comparatively low altitudes, and with the restricted sensitive volume of the apparatus, cloud chamber studies cannot be extended into the energy region higher than several hundreds of GeV. Only future progress in the technique will make possible the investigation of interactions at higher energy with cloud chambers or track chambers of other kind.

As a typical example of cloud chamber work, let us describe the calorimeter experiment of a Soviet group (46). The apparatus used in this experiment is shown in Fig. 6.

A nuclear interaction takes place in the top layer of the apparatus. A cloud chamber above this layer distinguishes whether the primary particle is charged or neutral. The energies and the emission angles of the individual secondary particles are measured in a cloud chamber with a magnetic field.

The calorimeter is placed below the magnetic cloud chamber. It is a large block of material (Pb and/or Fe) with several layers of ionization chambers inserted in it. The calorimeter is designed in such a way that a large part of secondary particles originated in the LiH producer will enter and dissipate most of their energy in the calorimeter through nuclear and/or electromagnetic processes. The total amount of ionization produced in the calorimeter is estimated from the signals of the trays of ionization chambers. One now obtains the total energy dissipated in the calorimeter from the amount of ionization produced with help of a calibration of the ionization-energy relation. The total energy dissipated can be equated to the primary energy, since the energy dissipation in the calorimeter will be almost complete in most of the caes.

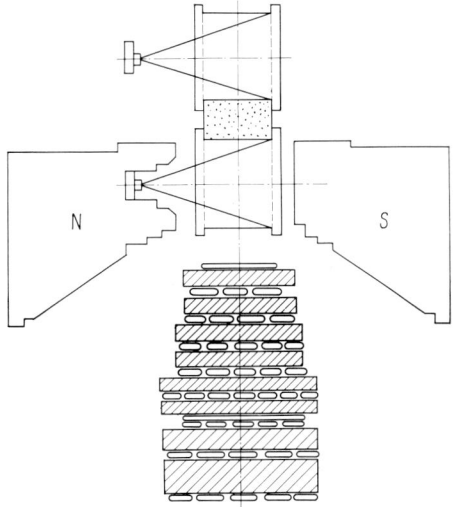

Fig. 6. Calorimeter used by the Soviet group. The apparatus consists of two cloud chambers with a LiH shower producer inserted between them. Below the magnetic cloud chamber the "calorimeter", a large material block with several layers of ionization chambers, is placed.

The examination of the photographs of the upper cloud chamber shows that about half of primary particles are neutral, and consequently most of the primary particles are nucleons. Since the shower producer is LiH, most of collisions may be assumed to take place between two nucleons. Under this assumption, the center-of-mass system of each collision is obtained from the primary energy measured with the calorimeter. It is also possible to perform a Lorentz transformation into the mirror system, in which the incident nucleon is at rest. The energies and the emission angles of the individual secondary particles can also be obtained in both the center-of-mass and the mirror systems.

If one collects all events observed, one finds good evidence for symmetry between the laboratory system and the mirror system in both the energy and angular distributions. This fact shows that there is no systematic error in the estimated Lorentz factor, that is, in the primary energy measurement with the calorimeter.

It is remarkable that the individual events do not always show the symmetry. The distributions of the momenta of the secondary particles in the center-of-mass system exhibit evidence of asymmetry; more than one-third of the events are found to be asymmetric. The asymmetry is expressed quantitatively by means of the inelasticity coefficients in the laboratory and the mirror systems, respectively denoted as K_L and K_M[1]. An event of $K_L > K_M$ represents predominance of forward emission, whereas $K_M > K_L$ means predominance of backward emission. The values of K_L and K_M of respective events are plotted in Fig. 7. A similar asymmetric property is observed also in the data of the angular distributions.

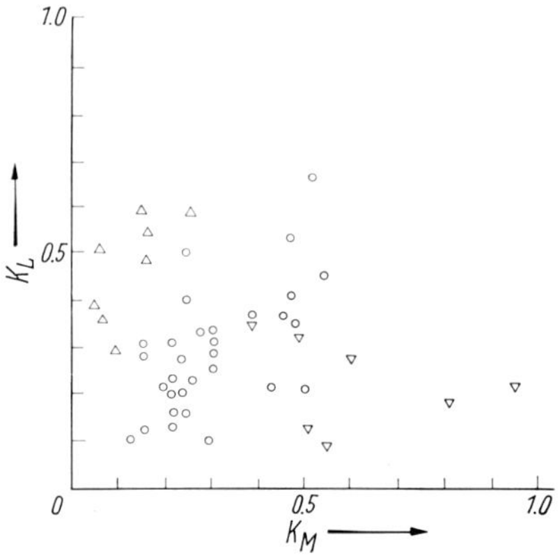

9. Large emulsion stacks.
When the size of the emulsion stack is larger than the nuclear interaction mean free path, secondary particles in a jet have a large probability to produce nuclear interactions again, and cascade showers from neutral pions in the jet will dissipate most of their energy in the stack. In this way all relevant features of a jet can now be observed by means of a single large emulsion stack. The large size has the further advantage of increasing the statistics, and of collecting events of greater energies.

Fig. 7. Correlation between the inelasticities in the laboratory system, K_L, and the mirror system, K_M, based on 46 events observed with the calorimeter apparatus shown in Fig. 6. ▲: forward asymmetry, ▽: backward asymmetry, and ○: symmetry in the angular distribution.

On the other hand, the cost for a great amount of emulsions and the flight of a heavy pay load is enormous, and the labour for scanning and analyzing a great number of events is beyond the manpower of a single group. However, this has been overcome, for example, under the International Cooperative Emulsion Flight (ICEF)[2]. The stack exposed consists of 80-liter Ilford G-5 emulsions, with 500 sheets of 600 μ × 60 cm × 45 cm, and the number of events accumulated is about 200 with dissipated energies greater than 400 GeV. It should be remembered that the scanning procedure for cascade showers by naked eyes gives a detection bias which must be corrected for. As an example, ICEF event No. 1 is reproduced in Fig. 8.

In this experiment, the energy of the primary particle for every jet shower is in general estimated with three independent methods; that is, E_c from the angular distribution with the CASTAGNOLI method, (7.4), E_{ch} from the emission angles of produced secondary particles (7.6), and ΣE_γ from the size measurement of the associated cascade shower. The size measurement of the associated cascade shower is made by constructing the transition curve of the number of shower particles within a certain distance from the center, and comparing the result with the theoretical calculation of the lateral structure function, called the Nishimura-Kamata function[3]. Although the actual cascade shower is multi-photon

---

[1] The inelasticity in the mirror system, K_M, can be expressed in the following way. In the mirror system the incident energy is γM, while the sum of the energies of secondary particles is $\gamma(E_i - p_i \cos \vartheta_i)$, where γ is the Lorentz factor for the transformation between the laboratory and the mirror systems. Hence K_M is given by

$$K_M = \frac{\Sigma_i (E_i - p_i \cos \vartheta_i)}{M}, \tag{8.1}$$

which is obtained without reference to the Lorentz factor.

[2] See, ref. (13). The ICEF project was initiated by the late Professor M. SCHEIN of the niversity of Chicago, and carried out by M. KOSHIBA. Participants in this project were as many as one hundred from fourteen countries.

[3] See J. NISHIMURA: Handbuch der Physik, this volume, Chapter I.

Sect. 9. Large emulsion stacks. 133

initiated while the calculation is valid for a single-photon primary shower, the integration over the transition curve, which is called the track length, is known to give results almost independent of the initial condition of the shower.

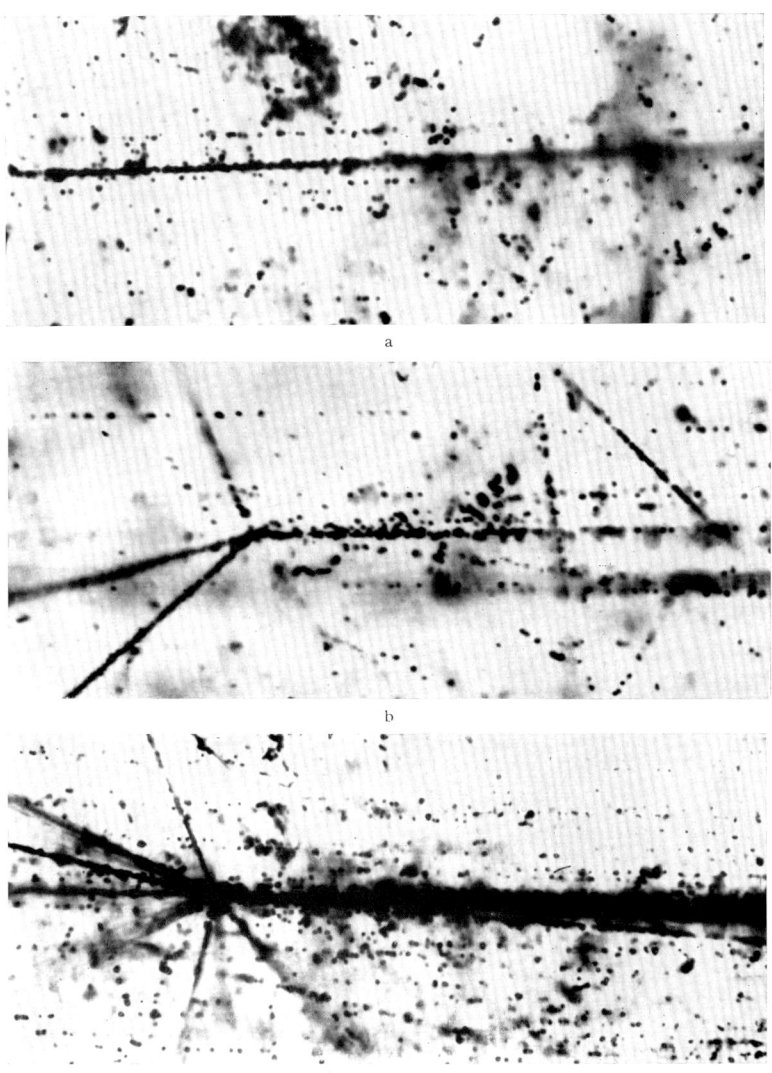

Fig. 8a—c. Successive nuclear interactions by heavy primary fragments observed in the ICEF stack (event ICEF Nr. 1).

Now we describe the relation among the energies estimated with the three different methods and the inelasticity thus obtained, in further reference to the Brawly stack (47) exposed in 1962. In Fig. 9 are plotted the values of E_{ch} and ΣE_γ obtained for all jets observed. The detection bias against events of small ΣE_γ can be removed by assuming a linear relation between E_{ch} and ΣE_γ. With this correction one obtains,

$$\Sigma E_\gamma = 2 E_{ch}, \tag{9.1}$$

which is consistent with the result given in II.5.

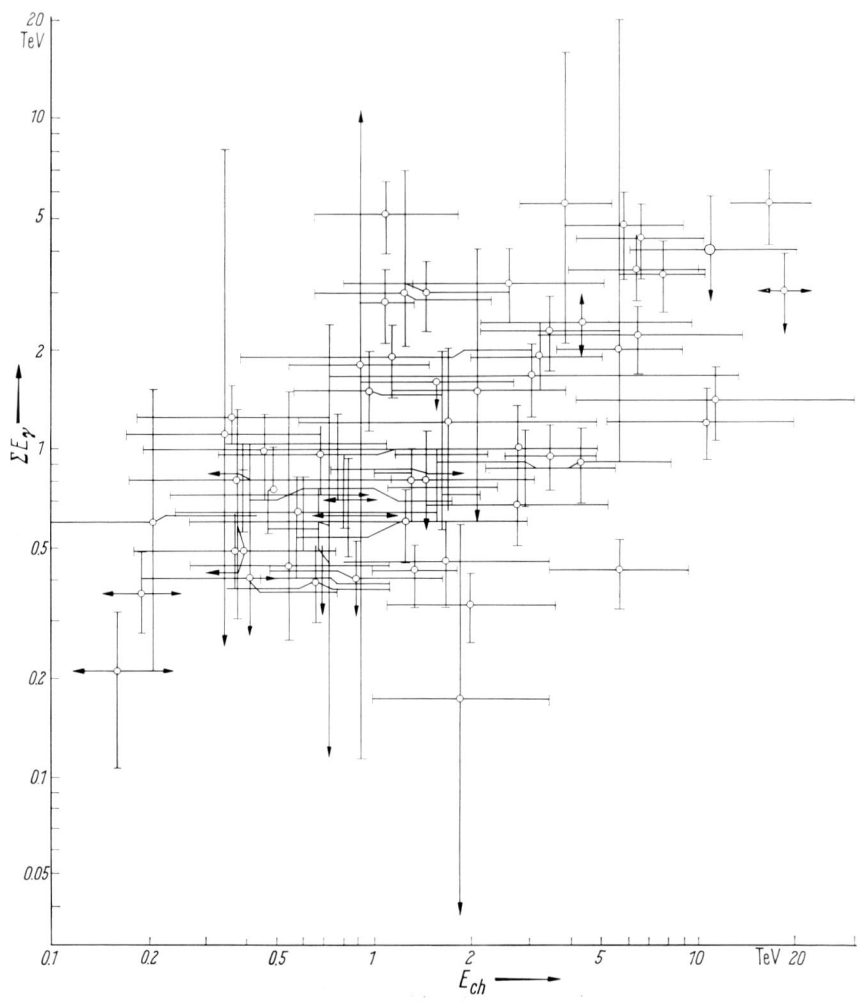

Fig. 9. Correlation between the energies transferred to charged secondary particles, E_{ch}, and γ-rays, ΣE_γ, observed with the I.C.E.F. and Brawly stacks.

Table 5. *Estimated energies (E_c and E_{ch}) of interactions of break-up products of a heavy primary particle.*

Event*	E_c TeV	E_{ch} TeV	K_{ch} ($E_{ch}/\langle E_c \rangle$)	Event*	E_c TeV	E_{ch} TeV	K_{ch} ($E_{ch}/\langle E_c \rangle$)
(3+5+H) H	2.32	0.26	—	(3+45) n	0.21	0.43	0.28
(11+44+H) H	2.44	2.7	—	(0+15) n	1.43	0.51	0.33
(7+8) n	0.044	0.037	0.024	(0+44) Li	5.3	4.5	—
(0+3+H) H	21.8	1.04	—	(9+40) p	0.98	0.52	0.34
(0+51+Li+He) H	5.04	5.53	—	(2+4) n	7.06	0.22	0.14
(8+6) p	4.92	1.03	0.68	(3+24) p	0.92	0.90	0.59
(1+4) He	18.0	0.44	—	(16+23) n	0.76	0.42	0.27
(2+1) p	0.105	0.003	0.002	(14+15) p	0.139	0.35	0.22

$\langle E_c \rangle = 1.53$ TeV.

* The events are classified in the customary notation: ($N_h + n_s$ + heavy fragments) incident particle. — The symbol H stands for an unidentified heavy nucleus.

The relation between E_{ch} and E_c can be plotted in a similar way, but here fluctuations are so large that one can hardly derive any conclusions.

The relation between E_{ch} and E_c can be investigated in a better way by use of the jets produced by the break-up products of a heavy primary particle, because the break-up products form a nucleon beam in which all particles have, within our accuracy, the same energy. Table 5 shows an example of the data on the interactions of such breakup products.

As is seen from Table 5, the fluctuation in E_c is enormous, a factor of about ten either way, although all the events are expected to have the same primary energy per nucleon. In order to reduce the fluctuation a composite angular distribution may be constructed by adding the distributions of all jets[1]. The primary energy is estimated by applying the CASTAGNOLI method on the composite distribution, and it is expressed as $\langle E_c \rangle$ in Table 5.

Now the inelasticity is obtained by comparing E_{ch} with $\langle E_c \rangle$ obtained as above. Including another break-up event of $\langle E_c \rangle = 20$ TeV, the resulting value of K_{ch} is

$$K_{ch} = 0.31 \pm 0.06. \tag{9.2}$$

Once K_{ch} is determined, the primary energy of each event can be estimated with reasonable reliability as

$$E_0 = E_{ch}/K_{ch}. \tag{9.3}$$

With this value of E_0, K_γ can be calculated:

$$K_\gamma = \Sigma E_\gamma / E_0 = 0.16. \tag{9.4}$$

Adding the contribution of neutral secondary particles, K_n, we now obtain the total inelasticity

$$K = K_{ch} + K_\gamma + K_n = 0.50 \pm 0.07. \tag{9.5}$$

An inelasticity of about 0.5 is, however, found also at lower energies with the calorimeter method as well as in accelerator experiments. Hence this may be regarded as a quantity nearly independent of energy, at least up to 10 TeV. The decrease of K with increasing energy found in several previous experiments might be erroneous, due to the use of the Castagnoli method for estimating the primary energy of individual events which could well introduce a systematic error.

Table 6. *Multiplicities in the TeV region.*

Energy TeV	Multiplicity $\langle n_s \rangle$	Remark
$0.47^{+0.09}_{-0.06}$	12.7 ± 0.9	secondary jets
1.5	18 ± 1.5	break-up products
12.3	24 ± 4	break-up products

Finally, we briefly mention the multiplicity observed with large stacks. Since essentially monenergetic beams are available in the break-up of heavy primaries, the results summarized in Table 6 give more reliable information on the multiplicity-energy relation than was available before.

10. Interactions of muons. It is known that a muon interacts with a nucleus through electromagnetic interaction. However, it has been suspected that some novel interactions of muons might appear in the high-energy region, because the muon has a much larger mass compared to the electron, and this mass could originate from an unknown process.

---

[1] The fluctuation is much larger than the one expected from a single statistical origin. It is found that asymmetry in forward-backward emission is the main cause of the fluctuation in E_c.

For experiments on muons either of cosmic rays or of accelerators, it is essential to reduce contamination of particles of other kinds as much as possible. Since the expected cross-section of muons is very small, even a small admixture of nuclear-active particles will affect the results in a serious way.

Now the best method for obtaining pure cosmic-ray muons is to perform experiments (48) underground. Particles of other kind incident from the atmosphere will be absorbed completely in thick layers of rocks. Furthermore, the average energy of the muons increases with the depth underground.

Even if one goes underground to observe nuclear interactions of muons, one cannot eliminate contamination from the nuclear-active particles produced locally by the muons themselves. This situation remains unchanged even at the largest depths. The following simple calculation makes the situation clear.

Let the cross section for the nuclear interaction of a muon be σ_μ, and the average number of nuclear-active particles thus produced by m. Since the attenuation length of nuclear-active particles is much smaller than the range of the muons, nuclear-active particles are in equilibrium with muons. The flux intensity of the former, J_N, is given, relative to the flux intensity of the latter, J_μ, by

$$J_N = \text{const}\,(m\sigma_\mu/a\sigma_N)\,J_\mu, \tag{10.1}$$

where σ_N is the nuclear interaction cross section, and $a\sigma_N$ is the cross section corresponding to the attenuation length. The constant factor is of the order of unity, and depends on the details of the interactions such as the inelasticity and the energy spectrum of the muons.

For comparison of the rates of nuclear interactions due to these two components, one has

$$\sigma_N J_N/\sigma_\mu J_\mu = \text{const}\,(m/a). \tag{10.2}$$

One now sees that the contribution of nuclear active particles is comparable to that of muon, since m and a are both of the order of unity.

Let us now describe the experiment of Osaka City University Group (49) as a typical example of studies on muon interactions. This experiment has clearly demonstrated the contribution of pion interactions among the nuclear events underground. Fig. 10 shows the experimental apparatus which consists of two large multi-plate cloud chambers and several trays of counters, operated at underground depths of 50 and 250 m water equivalent.

The identification of muons and nuclear-active particles was made by assuming that the latter are found in a nuclear cascade shower. This means that the events with multiple incidence are assumed to be due to interactions by nuclear-active particles, while those with single incidence are due to muons. The apparatus is set as close as possible to the ceiling of the tunnel, so that the loss of accompanying particles due to the angular divergence is small.

With such a selection criterion, the numbers of penetrating showers produced by nuclear-active particles and muons are compared in Table 7. By the triggering condition two or more penetrating particles are required for the interaction to be registered. — A typical photograph is reproduced in Fig. 11.

Table 7. *Characteristics of interactions in cloud chambers underground.*

Position of cloud chamber	Producer	Incidence	Multiplicity $\langle n_s \rangle$	Multiple/single frequency ratio
Upper	Pb	multiple	3.7 ± 0.6	0.26 ± 0.11
		single	3.6 ± 0.3 *	
Lower	Fe	multiple	3.6 ± 0.8	0.35 ± 0.20
		single	4.0 ± 0.3 *	

* Including a surviving muon.

Sect. 10. Interactions of muons. 137

Assuming that all single primary events are due to muons, the nuclear cross sections of muons for Pb and Fe are obtained:

$$\left.\begin{array}{l}\sigma_\mu(\text{Pb}) = (2.6 \pm 0.4) \times \\ \qquad \times 10^{-31}\ \text{cm}^2/\text{nucleon}, \\ \sigma_\mu(\text{Fe}) = (3.6 \pm 0.8) \times \\ \qquad \times 10^{-31}\ \text{cm}^2/\text{nucleon}.\end{array}\right\} \quad (10.3)$$

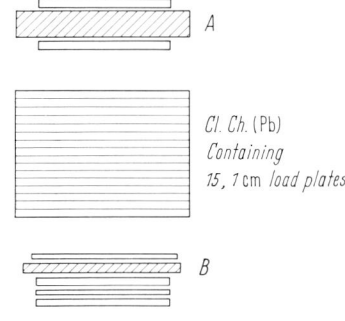

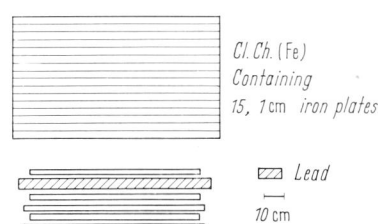

Fig. 10. Apparatus for the study of muon interactions, set up in a tunnel by the Osaka City University group.

Considering that the interaction is caused by the virtual photons which make up the muon field, we are able to deduce the photo-nuclear cross section σ_γ through

$$\sigma_\mu(E) = \int \sigma_\gamma(\varepsilon)\, n(E, \varepsilon)\, d\varepsilon, \quad (10.4)$$

where $n(E, \varepsilon)\, d\varepsilon$ is the number of virtual photons with energy between ε and $\varepsilon + d\varepsilon$ associated with a muon of energy E. Since monenergetic muons are not

Fig. 11a and b. Cloud chamber photograph of a nuclear interaction induced by an underground μ-meson (Osaka City University group), a in top chamber and b in bottom chamber.

available, the integral of (10.4) over the muon spectrum is performed. Referring to the muon spectrum given in III.14, using (50)

$$n(E, \varepsilon) = \frac{2}{\pi} \frac{1}{\varepsilon} \left\{ \ln\left(\frac{E}{\varepsilon}\right) - 0.38 \right\} \qquad (10.5)$$

and assuming energy independence of σ_γ, we obtain

$$\sigma_\gamma = (1.4 \pm 0.3) \times 10^{-28} \text{ cm}^2/\text{nucleon} \qquad (10.6)$$

for $\varepsilon \geq 5$ GeV.

The photonuclear cross section given in (10.6) is consistent with that obtained with accelerators. It does not seem to increase appreciably with energy, so that the nuclear absorption of muons plays an insignificant role in the energy loss of those muons which penetrate to extremely great depths, as will be discussed in III.14.

III. Behaviour of high-energy cosmic rays.

11. Genetic relation of the various components. As described in II, direct observations of high-energy interactions are not feasible at energies beyond 10 TeV, mainly because the flux of cosmic rays of such high energies is too small to obtain a sufficient number of events even with the largest detector available at present. On the other hand, the features of interactions above 10 TeV are considered as very important, because most of the models of multiple production thus far proposed can be distinguished from each other only in this high-energy region. Accordingly it is our urgent need to obtain every possible information on extremely energetic interactions, even if it is indirect.

Extensive air showers have met this need, and still provide the only means feasible for the investigation of interactions at energies higher than 10^{16} eV[1]. An extensive air shower is initiated by an ultra-high energy particle coming from outer space. When it impinges on the atmosphere, its collisions with air nuclei cause multiple production of particles. Among them charged pions as well as surviving nucleons form the N-component which develops itself into a large N-cascade composed of nuclear-active particles. These nuclear-active particles produce neutral pions at every nuclear collision, the latter decay into γ-rays which subsequently develop into cascade showers. Some of the pions and kaons produced in the nucleonic events decay into muons and neutrinos. Thus a great number of particles of various kinds form a huge shower spreading over a wide area whose dimension is as large as 1 km. This means that the area of an air shower detector can be as large as 1 km², and tells us why a small flux of particles of ultra-high energies can still be detected.

In the energy region between those of air showers and jets different methods are available. They are not as indirect as the observation of air showers, but are partly direct in the sense that the particles generated by high-energy interactions or their decay products are observed before further processes complicate the picture. Observations of γ-rays and muons of high energies are typical examples. The information thus obtained may be used not only for checking that obtained from direct observations, but also for finding indications of as yet unknown features of interactions at higher energy.

The intensity of muons had been measured to a depth as large as 1500 m water equivalent earlier than World War II, and was extended during the War to 3000 m water equivalent (51). The intensity-depth curve shows a well-known change in

---

[1] See review articles on extensive air showers, for example, G. COCCONI, Handbuch der Physik, Bd. XLVI/1, S. 215. Berlin-Göttingen-Heidelberg: Springer 1961.

its slope at several hundreds of m water equivalent. Formerly this was considered as an indication of an anomaly in high-energy interactions, but now has been recognized as due to the $\pi-\mu$ decay effect (52), (53). Experiments of underground muons have been continued to still greater depths, and now the deepest point reached in measurements of vertical particles is at 8400 m water equivalent (54).

The observation of atmospheric γ-rays was stimulated by the development of the emulsion chamber technique for studies of nuclear interactions. With large emulsion chambers exposed either at mountain altitudes or in the stratosphere, one now has information on γ-rays with energies up to 10 TeV. Furthermore one is able to obtain data on the nucleon spectrum, with the same apparatus, through the measurement of ΣE_γ.

As is well known, the main source of γ-rays and muons is pion decay through $\pi^0 \to 2\gamma$ and $\pi^\pm \to \mu^\pm + \nu$. By observing γ-rays and muons we are, therefore, able to learn about the features of pion production. Or to put it more conservatively, we are able to check the consistency of the features of multiple production found by means of direct observations of jets, by comparing them with the results obtained on γ-rays and muons. For example, the contribution of kaons to muons can be examined by such a comparison. Furthermore, in the gamut of cosmic-ray interactions the behaviour of γ-rays, muons, and N-component is incorporated as well as the properties of primary cosmic rays.

The genetic relations we are interested in are schematically shown below:

$$\text{primary cosmic rays} \to N\text{-component} \begin{array}{l} \to \pi \begin{pmatrix} \pi^0 \to \gamma, e \\ \pi^\pm \to \mu^\pm \end{pmatrix} \\ \to K (K^\pm \to \mu^\pm). \end{array}$$

12. Observation on high-energy γ-rays. In the course of the observation of γ-rays with an emulsion chamber, as described in II.6, some of the γ-rays are found to come not from the shower producer at the top but from the atmosphere above. They are considered as due to neutral pions produced by jets in the atmosphere.

In order to observe such γ-rays of higher energies, the detecting area of an emulsion chamber has to be increased. The first of such chambers exposed at Mt. Norikura had an area of 3 m² (55). The area was too large for microscopic scanning. Scanning was, therefore, made by searching by naked eye for black spots on highly sensitive X-ray films stacked together with the emulsion plates. In coarse-grained X-ray films the track of a charged particle cannot be identified, but a group of tracks form a clear black spot because the sensitivity of the film is higher than that of nuclear emulsions. The black spot is visible near the shower maximum of a cascade with energy greater than $0.3 \sim 0.5$ TeV, depending on the background. From the location of a black spot one finds the corresponding shower tracks under a microscope in the attached nuclear emulsion plate, and the angles of azimuth and dip of the shower are measured. Then the shower is followed successively to all other layers of X-ray films and emulsion plates in the chamber. The construction of the emulsion chamber of this type is shown in Fig. 12[1].

The energy of a γ-ray is measured by constructing the transition curve of the size of a cascade shower developed in the chamber. Let the number of electrons at depth t in a circle of radius R surrounding the axis of a shower initiated by a

---

[1] The emulsion chamber detects γ-rays as well as nuclear interactions produced in the chamber itself. The latter contribution (Pb jets) is estimated from the measurement of the N-component through an analysis of γ-ray families, and also from that of showers produced under thick absorbers.

γ-ray of energy E_0 be $N(E_0, R, t)$. Details of the function $N(E_0, R, t)$ are worked out by means of the three-dimensional cascade theory[1]. Near the shower axis an approximate relation holds stating that N is a function of the product $(E_0 R)$:

$$N(E_0, R, t) = N(E_0 R, t), \tag{12.1}$$

so that similarity of the shower structure is found for $E_0 R = \text{const}$. The primary energy E_0 can, therefore, be obtained by measuring the number of tracks in the circle as a function of depth, and comparing the result with the theoretical calculations of $N(E_0, R, t)$.

The accuracy of the energy measurement can be checked by using the kinematical relation in the decay of a neutral pion into two γ-rays. This is done by selecting γ-ray pairs to which a parent neutral pion can be assigned, and measuring their opening angles. Of course the selection of such pairs is not completely free from ambiguity in individual cases[2]. One finds that the absolute values of the energies of γ-rays measured with this method have an error of 30% or less.

For energy determination, a photometric measurement of the blackness of the spots either in the X-ray films or in the nuclear emulsion plates can be used instead of the method of counting the number of shower tracks. But one has to be careful in the calibration of the photometry measurement, because the photometry signal is dependent on various measuring conditions, and in particular on the developing condition of the photographic material.

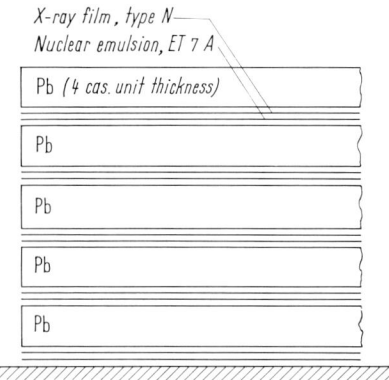

Fig. 12. Side view of emulsion chambers exposed at Mt. Norikura and at Mt. Chacaltaya.

There are several observations of γ-rays which make use of detectors of other kind than nuclear emulsions, such as scintillators and ionization chambers. These electronic detectors are convenient for studying coincidences with other events, particularly with extensive air showers. On the other hand, the spatial resolution of these devices is usually poorer by several orders of magnitude compared to that of nuclear emulsions. Since atmospheric γ-rays do not always arrive singly but sometimes in a group with small distances between them[3], fine spatial resolution is of essential importance for the study of the γ-ray spectrum. Furthermore a calibration of the energy measurement with the track-counting method is not easy for an apparatus other than nuclear emulsions.

The intensities and the energy spectra of γ-rays and electrons were measured at various altitudes, with use of emulsion chambers as well as with other methods. The experiments we are going to refer to are listed in Table 8.

These observations of γ-rays were made with different techniques, and both the absolute flux intensities and the energy calibrations have to be subjected to critical examination. Let us first compare the absolute intensities at given energies and various altitudes. This can be carried out with the help of the theoretical consideration that γ-rays are generated by nuclear-active particles through the decay of neutral pions, and the intensity of the γ-rays is determined by the

---

[1] J. NISHIMURA, Handbuch der Physik, this volume, Chapter I.
[2] Calibration with neutral pion decays is performed in an emulsion chamber with producer. See discussion in II.6, "Emulsion chambers and transverse momentum".
[3] See discussion in III.13, "γ-ray families".

combined effect of the N-cascade and the electron-photon cascade. Since the intensity of the N-component decreases exponentially and its attenuation length is larger than the radiation length, the γ-rays are considered to be in equilibrium

Table 8. *Observations on the γ-Ray Energy Spectrum.*

Atmospheric depth g/cm^2	Apparatus	Energy range TeV	Exponent of the power law	Reference
9 (26)*	balloon emulsion stack	0.1 — 2	$1.9^{+0.3}_{-0.2}$	Chicago[1]
22 (37)	balloon emulsion chamber	0.3 — 5	1.75 ± 0.20	Bristol[2], Bombay[3]
30 (57)	balloon emulsion chamber	0.1 — 2	2.0 ± 0.2	Japan[4,5]
197	airplane ion chamber	0.03 — 2	1.76 ± 0.11	Moscow[6]
220	airplane emulsion stack	0.3 — 2 2 — 10	2.3 ± 0.2 2.8 ± 0.3	Bristol[7]
310	airplane ion chamber	0.03 — 2	1.83 ± 0.13	Moscow[6]
550	Chacaltaya emulsion chamber	0.5 — 10	2.2 ± 0.15	Japan-Brazil[8]
730	Norikura emulsion chamber	0.15 — 1 1 — 10	2.0 ± 0.2 2.3 ± 0.15	Japan[8,9]

* The value in brackets represents the effective depth.

[1] F. ABRAHAM, J. KIDD, M. KOSHIBA, R. LEVI SETTI, C. H. TSAO e W. WALTER: Nuovo Cim. **28**, 221 (1963). — J. KIDD: Nuovo Cim. **27**, 57 (1962).
[2] J. BOULT, M. G. BOWLER, P. H. FOWLER, H. L. HACKFORTH, K. KEEREETAVEEP, V. M. MAYES, and S. N. TOVEY: Internat. Conference on Cosmic Rays at Jaipur, 1963, Vol. 5, 182.
[3] P. K. MALHOTRA, P. G. SHUKLA, S. A. STEPHENS, and B. VIJAYALAKSHMI: Internat. Conference on Cosmic Rays at Jaipur, 1963, Vol. 5, 232.
[4] O. MINAKAWA, Y. NISHIMURA, M. TSUZUKI, H. YAMANOUCHI, H. AIZU, H. HASEGAWA, Y. ISHII, S. TOKUNAGA, Y. FUJIMOTO, S. HASEGAWA, J. NISHIMURA, K. NIU, K. NISHIKAWA, K. IMAEDA and M. KAZUNO: Nuovo Cim., Suppl. **8**, 761 (1958).
[5] Y. FUJIMOTO, S. HASEGAWA, M. KAZUNO, J. NISHIMURA, K. NIU, and N. OGITA: Proc. Moscow Conference **1**, 41 (1960).
[6] L. T. BARADZEI, V. I. RUBTSOV, Y. A. SMORODIN, M. V. SOLOVYEV, and B. V. TOLSKACHEV: J. Phys. Soc. Japan **17**, Suppl. A III, 433 (1962), and Internat. Conference on Cosmic Rays at Jaipur, 1963, Vol. 5, 283.
[7] J. DUTHIE, P. H. FOWLER, A. KADDOURA, D. H. PERKINS e K. PINKAU: Nuovo Cim. **24**, 122 (1962).
[8] C. M. G. LATTES, C. Q. ORSINI, I. G. PACCA, M. T. CRUZ, E. OKUNO, and S. HASEGAWA, M. AKASHI, K. SHIMIZU, Z. WATANABE, J. NISHIMURA, K. NIU, N. OGITA, Y. TSUNEOKA; T. TAIRA, T. OGATA, A. MISAKI, I. MITO, Y. OYAMA, S. TOKUNAGA, A. NISHIO, S. DAKE, K. YOKOI, Y. FUJIMOTO, and T. SUZUKI: Internat. Conference on Cosic Rays at Jaipur, 1963, Vol. 5, 326.
[9] M. AKASHI, K. SHIMIZU, Z. WATANABE, T. OGATA, N. OGITA, A. MISAKI, I. MITO, S. OYAMA, S. TOKUNAGA, M. TAMURA, Y. FUJIMOTO, S. HASEGAWA, J. NISHIMURA, K. NIU, and K. YOKOI: J. Phys. Soc. Japan **17**, Suppl. A III, 427 (1962).

with the nuclear-active particles at lower altitudes. Hence the attenuation of γ-rays will be parallel to that of the N-component.

For a quantitative argument a brief mathematical treatment may be helpful (*56*). Let the differential production spectrum of γ-rays with energy E in dE at

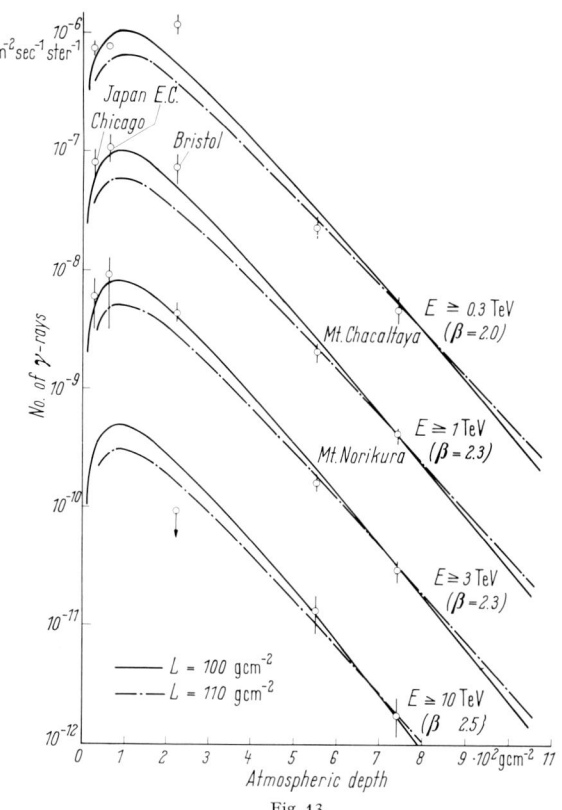

Fig. 13. Altitude variations of the γ-ray intensities in several energy ranges. Experimental values taken from five emulsion exposures are compared with those calculated with the source function given in (12.2). The full curves stand for the attenuation length of 100 gcm⁻², the dot-dashed curves for that of 110 gcm⁻².

Fig. 14. Energy spectra of γ-rays obtained with emulsion chambers exposed at Mt. Chacaltaya and Mt. Norikura. The data are taken from references 8 and 9 in Table 8.

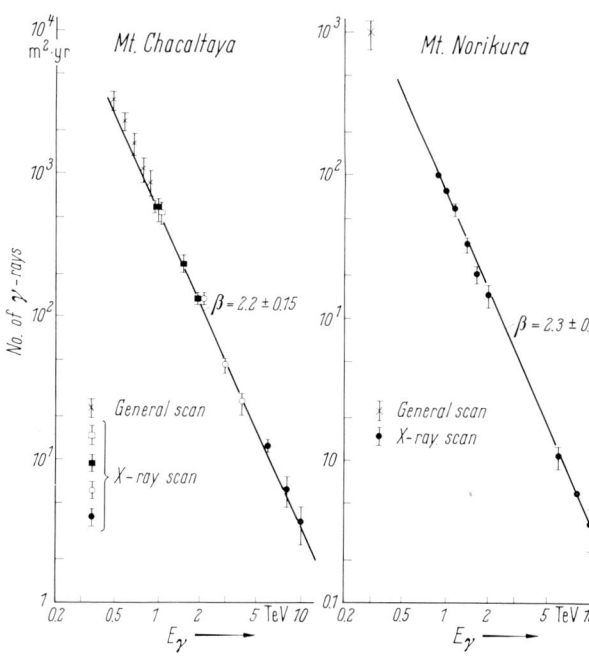

Fig. 14.

depth x in dx be

$$g_\gamma(E, x)\, dE\, dx \\ = g_\gamma(0)\, (E_0/E)^\beta \times \\ \times (dE/E)\, e^{-x/L}\, dx, \quad (12.2)$$

where L is the attenuation length of the N-component. A single power spectrum with exponent β is assumed. The altitude dependence of γ-rays is calculated by the cascade theory with the assumption of no primary γ-ray. The value of L is adjusted so as to give the best fit to the experimental data. The result of such an analysis for several energy regions is shown in Fig. 13. The curves are those calculated for

$$\beta = 2.3 \\ \text{and} \\ L = 100\ \text{gcm}^{-2} \\ \text{or} \\ 110\ \text{gcm}^{-2}, \quad (12.3)$$

The agreement is fair with the parameters given in (12.3), in spite of the variety of experimental methods. $L = 100$ gcm⁻² seems to give a better fit to the intensity variation between balloon altitudes and mountain altitudes, whereas the altitude dependence between Mt. Chacaltaya and Mt. Norikura is better reproduced with $L = 110$ gcm⁻².

Now let us examine the values of the exponent of the energy spectra measured by various authors. In the energy region below 1 TeV, most of the experiments give spectral indices

$$\beta = 1.8 \sim 2.0, \tag{12.4}$$

as seen from Table 8.

In the higher-energy region of $1 \sim 10$ TeV, there are two series of experiments. At mountain altitudes, the work of the Japanese Emulsion Group at Mt. Norikura (730 g/cm²), and of the Japan-Brazil collaboration at Mt. Chacaltaya (550 g/cm²) give a consistent result of

$$\beta = 2.2 \sim 2.3, \tag{12.5}$$

thus showing a gradual steepening of the spectrum. Fig. 14 presents the result of these mountain experiments. At stratosphere altitudes, data are available from experiments of the Bristol group in an airplane (220 g/cm²), and of the Bristol-India collaboration from a balloon flight (20 g/cm²). Unfortunately the results from the two observations — $\beta = 2.8$ in one case and $\beta = 1.75$ in the other — are in serious disagreement, so that further experiments will be needed before a final conclusion can be drawn[1].

The steepening of the γ-ray spectrum raises an interesting question. If this is attributed to a change in the mechanism of pion production, the break point of the spectrum should be more or less independent of altitude. If, on the other hand, this is due to the steepening in the spectrum of the primary particles, the break point would shift towards higher energy with increasing altitude. Experimental data available at present are not sufficient to distinguish between these alternatives, but there seem to be some indications in favour of the latter case.

13. γ-ray families. Emulsion chambers exposed to atmospheric γ-rays frequently register a group of γ-rays entering the apparatus together. This can be recognized by a determination of the arrival direction of the γ-rays, which in some cases demonstrates closely parallel incidence. Those are called a γ-ray family.

Like the initial stage of an extensive air shower, a γ-ray family is, in general, the composite product of nuclear interactions and cascade processes which occurred in the atmosphere above the detector. It consists not only of γ-rays directly generated by the nuclear interactions, but also of those produced in the course of their cascade development. If the interaction takes place far above the detector, the latter effect is so dominant that the reconstruction of the original nuclear interaction from the family observed is subject to considerable ambiguity. If, on the other hand, the interaction occurs nearby and the effect of cascade degradation is not appreciable, the family of γ-rays directly represents the features of a jet. In Fig. 15 these characteristic cases are schematically shown.

In most of the cases it is not difficult to identify an air cascade. After having passed several cascade units of air, the lateral spread of γ-rays in the atmospheric cascade reaches a nearly constant value, and is given for a γ-ray of energy E by

$$R \approx a(E_c/E)\, r_M, \tag{13.1}$$

---

[1] The errors in the exponent of the energy spectrum is not confined to the simple statistical one. There are groups of γ-rays arriving together, so that individual γ-rays observed should not be regarded as statistically independent of each other. The correction for this effect is important. Detailed arguments on the correlation effect of γ-ray families will be discussed by Japanese Emulsion Group; Prog. Theor. Phys. Suppl., to be published. In spite of this, there are several indications suggesting that the previous Bristol value of the exponent is too large. The spectrum in the stratosphere could extend to 10 TeV with an index of about 2.0.

where r_M is the Molière unit, E_c is the critical energy in air, and a is a constant of the order of 0.1 depending on the shower age. For a γ-ray of 1 TeV at Mt. Chacaltaya, for example, R is as small as 2 mm, and accordingly an air cascade can be identified without difficulty by its characteristic spatial distribution.

For a more accurate discussion we refer to the Monte Carlo calculations of three-dimensional cascade showers[1]. Its results not only confirm the qualitative

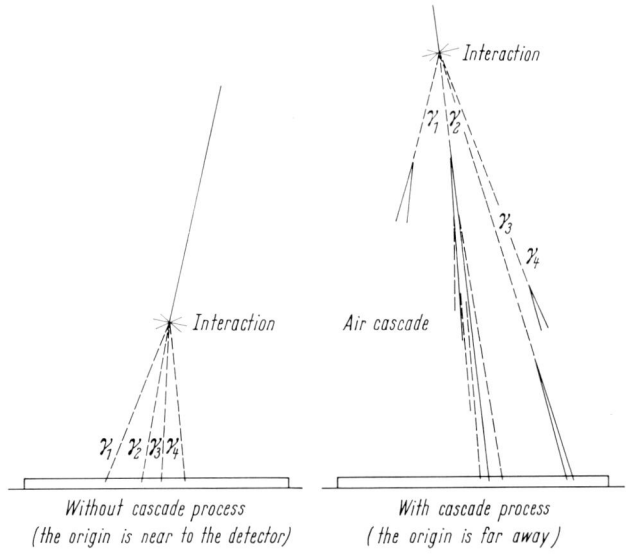

Fig. 15. Schematic representation of the γ-ray families to be observed with a large emulsion chamber.

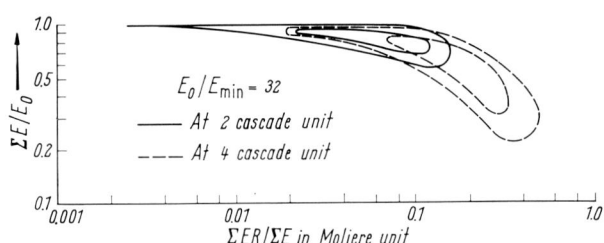

Fig. 16. Development of cascade showers in air calculated with the Monte Carlo method. The ordinate represents the fraction of energy carried by γ-rays of energies greater than $E_0/32$, where E_0 is the initial energy, and the abscissa represents the average lateral spread in Molière units corresponding to $E_0/32$. The contour lines indicate loci of equal probabilites at 2 cascade units (full line) and at 4 cascade units (dashed line).

relation, (13.1), for developed showers but also demonstrate that the lateral spread depends on the shower age. In Fig. 16 an example of the correlation diagram between the fraction of surviving energy, $\sum_i E_i/E_0$, and the energy-weighted average of lateral spread

$$\langle R \rangle_E = \Sigma E_i R_i / \Sigma E_i. \tag{13.2}$$

is represented. — The Monte Carlo calculation permits us to determine the shower age, and with it the energies of the γ-rays at production, E_0.

Once the energy of the initiating γ-rays is estimated, we can apply exactly the same procedure of analysis as in the case of the emulsion chamber with

---

[1] For details see Japanese Emulsion Group, Progr. Theor. Phys. Suppl. (to be published), and J. NISHIMURA: Handbuch der Physik, this volume, Chapter I.

producer[1]. The only difference lies in the fact that because of the remoteness of their production point, we are not able to measure the angle of convergence between the air cascades, so that the location of the original atmospheric nuclear interaction has to be obtained from the kinematical relation for the γ-decay of a neutral pion. The procedure of selecting a pair of γ-rays and identifying them as the decay products of a specific neutral pion is often called the "marriage of a pair". Some ambiguity remains in the marriage procedure of γ-rays into neutral pions, but this can be checked by the consistency of the shower ages determined from several pairs. Thus, this method allows an investigation of nuclear interactions occurring at altitudes higher than the observation level, and of very high energies. In Fig. 17 the ΣE_γ and the location of those atmospheric interactions are plotted which were identified by this method. One notices that events of $\Sigma E_\gamma = 10-100$ TeV could be studied[2,3]. — An example of a large γ-ray family is shown in Fig. 18.

A further analysis was carried out for those atmospheric interactions which originated within a certain elevation from the observation plane. This criterion was introduced in order to omit from the sample all events of ambiguous interpretation, and to avoid a possible selection bias in the data.

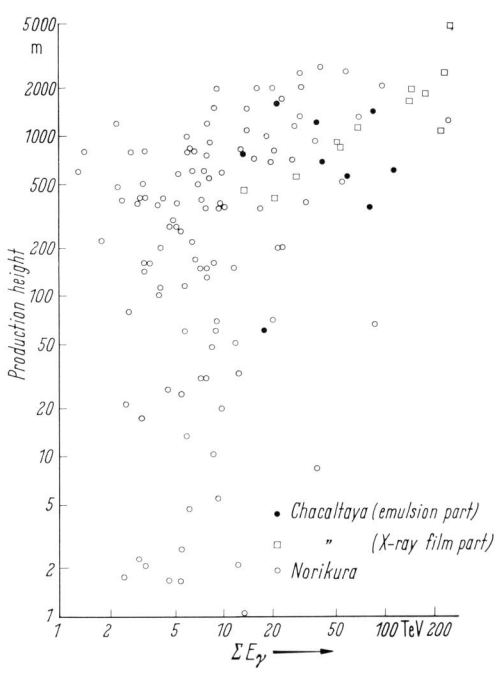

Fig. 17. Distribution of the production heights of γ-ray families, obtained with emulsion chambers exposed at Mt. Chacaltaya and Mt. Norikura. Open squares represent γ-ray families detected with X-ray films alone (no emulsions were inserted for some chambers).

Among the various results thus obtained, the distribution of the transverse momenta P_T was already presented in II.6. The energy spectrum of the γ-rays, and consequently that of the neutral pions, will now be given. They are expressed in fractional form, that is in terms of $E_\gamma/\Sigma E_\gamma$ and $E_{\pi^0}/\Sigma E_{\pi^0}$ respectively. This histogram of $E_{\pi^0}/\Sigma E_{\pi^0}$ is shown in Fig. 19 for various ranges of ΣE_γ. It will be noticed that in some of the most energetic events a single neutral pion carries off most of the total energy. Actually, in this energy region ($\Sigma E_\gamma \geqq 20$ TeV) one seems to observe two kinds of events: In one type most of the energy ΣE_γ is transferred to a single pion, while in the other many pions of comparable energies

---

[1] See also the discussion in II.6: "Emulsion chambers and transverse momentum".

[2] The results of the analysis of γ-ray families presented here are based on data from the emulsion chambers exposed at Mt. Norikura and Mt. Chacaltaya. For reference, see the reports of the Japanese Emulsion Group and of the Brasilian Emulsion Group, Intern. Conf. on Cosmic Rays, Jaipur (1963), Vol. 5, 326.

[3] As discussed in II.6, "Emulsion chambers and transverse momentum", the analysis of a γ-ray family becomes ambiguous for events originating at heights of two cascade units or more above the detector. This is due to uncertainties in the identification of the individual γ-rays resulting from the overlapping of the air cascades. Moreover, the energy correction factor $E_0/\Sigma E_\gamma$ becomes large in those cases. For events produced at lower altitudes there is no serious ambiguity.

Handbuch der Physik, Bd. XLVI/2.

are emitted. Further investigations will be needed to clarify the significance of this tentative classification into two groups.

The "family" analysis directly yields the production spectrum of γ-rays, which can be compared with the energy spectrum of γ-rays discussed in III.12.

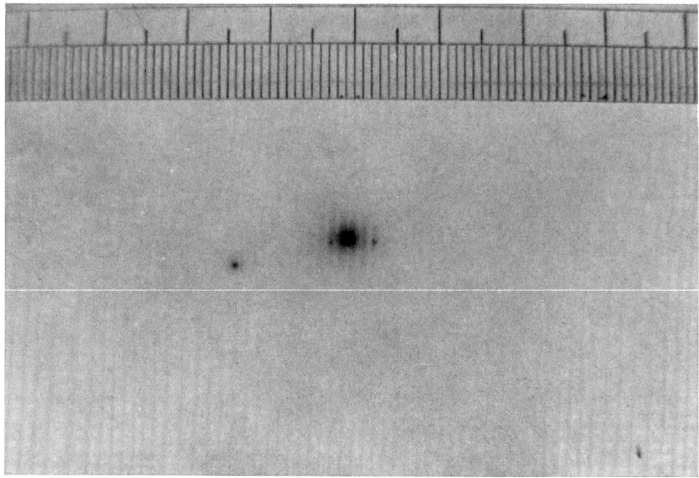

Fig. 18a and b. A large γ-ray family found in an emulsion chamber exposed at Mt. Norikura. a in X-ray film, Sakura N-type, and b in nuclear emulsion, Fuji ET 7 A. The event has $\Sigma E_\gamma \sim 80$ TeV. A central dot in the X-ray film is composed of two high-energy showers shown in b (nuclear emulsion), which probably come from a single π^0-meson carrying most of the total energy. (Japanese Emulsion Group).

Furthermore it offers the spectrum of ΣE_γ, which is related to the energy spectrum of nuclear-active particles. Table 9 shows the results of observations on ΣE_γ-spectra by various authors.

The spectral indices of the ΣE_γ-spectra are presented relative to those of the E_γ spectra, because most of the experiments observe both spectra at the same time, and the difference between these two spectral indices is regarded as more reliable than their individual values.

Table 9. ΣE_γ-Spectrum.

Atmospheric depth g/cm²	Energy range TeV	Difference of exponent between $E\gamma$ and $\Sigma E\gamma$	Reference*
22 (37)**	0.3—10	+0.29± 0.22	Bristol-Bombay [2,3]
197	0.1— 5	+0.16± 0.14	Moscow [6]
220	0.3— 2	+0.2± 0.2	Bristol [7]
	2—10	+0.7± 0.3	
310	0.1— 5	+0.3± 0.17	Moscow [6]
730	2—100	+0.2± 0.3	Japan [9]

* References are the same as in Table 8.
** The value in brackets represents the effective depth.

It is interesting to see that the exponent of the ΣE_γ spectrum is slightly smaller than that of single γ-rays. This may be due to the gradual increase of the multiplicity with energy. It is difficult at present to distinguish between the $E^{\frac{1}{4}}$ and $\log E$ laws of the multiplicity, but the energy dependence does not appear to be steeper than $\propto E^{\frac{1}{4}}$. The exponent of the ΣE_γ-spectrum should be equal to that found for nuclear-active particles, provided that K_γ is energy-independent. This seems to be the case, although the experimental values are too widely scattered to draw a definite conclusion.

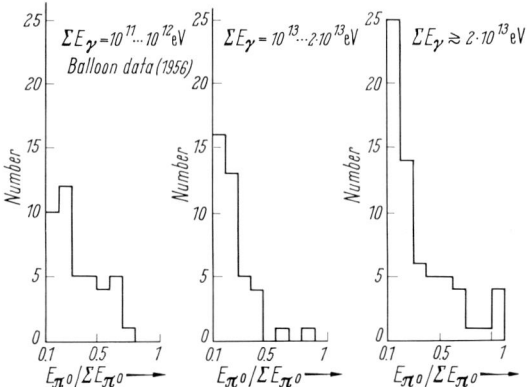

Fig. 19. Fractional differential energy spectra of neutral pions observed with emulsion chambers exposed in balloons and at Norikura and Chacaltaya.

14. High-energy muons. The energy spectrum of muons has been determined with three different methods: momentum measurement with a magnetic spectrometer, measurement of the size distribution of bursts, and measurement of muon intensities at various depths underground.

The first method is the most direct one, and consequently is regarded to give more accurate results than the other two for energies below 100 GeV (57). This energy region is lower than that of our interest, but the spectrum thus obtained can be used for the calibration of the other two methods, and also as a standard one to be extended to higher energies. Above 100 GeV the association of electron showers with a muon is the most disturbing source of experimental bias; the probability of the association of electron showers is as high as 20% at 100 GeV (58) and increases rather rapidly with increasing energy. Within the uncertainty in this correlation, the spectrum has been obtained up to 1 TeV.

From 100 GeV to 10 TeV the burst size spectrum can be used to determine the muon energy spectrum (59), (60). High-energy muons produce γ-rays by bremsstrahlung in a thick layer of matter surrounding an ionization chamber. These γ-rays develop into large cascade showers which give rise to large ionization pulses in the chamber. However, the size of the burst produced by a muons of given energy fluctuates so much that the muon spectrum cannot be related to the burst size spectrum in a straightforward way. For our present purpose it may suffice merely to mention that the burst size distribution gives a muon spectrum not inconsistent with that obtained by other methods.

The intensity-depth relation gives the muon spectrum covering the whole energy range, in part of which the other two methods are applicable. The flux of muons as a function of the depth — which is customarily measured in units of meter water equivalent (m. w. e.) — has been measured by many authors, as summarized in Fig. 20 (*61*). Down to 1800 m.w.e. many experimental points are available, and a smooth curve can be drawn rather easily through them. At about 3000 m.w.e. observations in a tunnel (*62*) and under water (*63*) give considerably different intensities. The intensities derived from measurement on inclined muons are closer to the former (*64*). A recent measurement of the vertical intensities down to 8400 m.w.e. has given values lying in between (*54*). It is worth while to note that no count was obtained in a period of 2800 hrs at the deepest point.

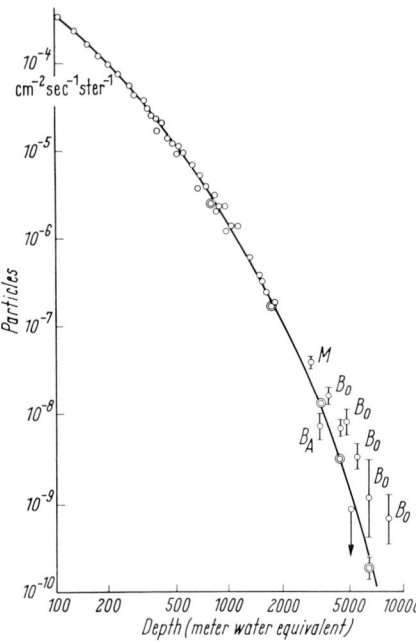

Fig. 20. Intensity-depth relation of muons, compiled in ref. (*54*). Double circles represent the results obtained by these authors, and the points indicated by M, B_A and B_0 are those of ref. (*62*) to (*64*).

Looking at the whole intensity-depth curve, one notices a steepening of the slope with increasing depth. The change of slope at about 1000 m.w.e. is attributed to the $\pi-\mu$ decay effect. The further steepening is due to the fact that the rate of energy loss increases gradually as the muon energy increases. Below 0.5 TeV the energy loss is due essentially to the ionization process, according to which the loss rate consists of a constant term plus a logarithmic term. At 1 TeV the ionization loss competes with other processes of energy dissipation, such as bremsstrahlung and direct pair creation. Since the loss rate due to the latter processes is known to be proportional to the muon energy, we express the energy loss rate in the standard earth as

$$-\frac{dE}{dx} = a + c \ln\left(\frac{E_{max}}{m_\mu c^2}\right) + bE \quad \text{MeV/gcm}^{-2}, \qquad (14.1)$$

where

$$\left.\begin{aligned} E_{max} &= \text{maximum transferable energy,} \\ m_\mu c^2 &= \text{muon rest energy,} \\ a &= 1.84 \text{ MeV/g cm}^{-2}, \\ c &= 0.076 \text{ MeV/g cm}^{-2}, \\ b &= 3.9 \times 10^{-6} \text{ g}^{-1} \text{ cm}^2; \end{aligned}\right\} \qquad (14.2)$$

the values of a, b and c here hold for the experiment at Kolar Gold Fields (*54*).

According to theoretical calculations, the value of b is made up of the bremsstrahlung loss, 2.2×10^{-6}, and the pair creation loss, 1.7×10^{-6}. However, the accuracy of the theoretical values, especially for the direct pair creation, is not better than about 20%. Furthermore, the energy loss rate due to nuclear interactions may give an additional contribution of the order 10^{-6} to b. It is, therefore, possible to put the plausible limits of b as

$$3.2 \times 10^{-6} < b < 4.8 \times 10^{-6}. \qquad (14.3)$$

Even if the value of b is as uncertain as (14.3) indicates, its contribution is so small below 0.5 TeV that the choice of parameters in (14.2) results in a good agreement of the energy spectrum deduced from the intensity-depth curve with that obtained with the magnetic spectrometer.

For deducing the energy spectrum above 1 TeV, not only the absolute value of b, but also the fluctuations in the bremsstrahlung loss are found to be factors of primary importance. The fluctuation problem has been worked out by the Monte Carlo method as well as by an analytic method similar to the cascade shower theory[1]. Here we make use of the latter calculation.

Taking into account these considerations, we are now able to construct the energy spectrum of muons up to 10 TeV, as shown in Fig. 21 in III.15. The following points may be worth noting. Below 1 TeV the spectra obtained with the magnetic spectrometer and from the intensity-depth curve are in good agreement with each other; a difference by a factor of 1.5 is unessential in view of difficulties in determining the absolute flux intensity. The exponent of the spectrum increases gradually with increasing energy; for the integral spectrum of parent pions and kaons its value changes from 1.7 at several hundreds of GeV, to 2.5 around 10 TeV.

The results above imply the following important consequences. Firstly, the energy loss due to nuclear interactions contributes to b at most 0.9×10^{-6} g^{-1} cm^2, so that an upper limit of the photonuclear cross-section is given by

$$\sigma_{\gamma N} \lesssim 3 \times 10^{-28} \text{ cm}^2 \quad \text{per nucleon.} \tag{14.4}$$

Secondly, a neutrino experiment is feasible at depths greater than 8000 m.w.e., at which the contribution of muons is negligible. The muon intensity at these large depths decreases exponentially with the depth.

15. Interrelationship between γ-rays and muons.

As discussed in III.11, both γ-rays and muons have the same genetic origin, pions. They should, therefore, be related to each other.

The production spectrum of γ-rays given in (12.2) is related to that of neutral pions through the kinematical relation of the 2γ decay of a neutral pion as

$$g_\gamma(E_\gamma) = 2 \int_{E_\gamma}^\infty g_{\pi^0}(E_{\pi^0}) \frac{dE_{\pi^0}}{E_{\pi^0}} = \frac{2}{\beta+1} g_{\pi^0}(E_\gamma), \tag{15.1}$$

where the last relation is derived by assuming a power spectrum. Doubling the production spectrum of neutral pions, on account of the charge independence, we obtain that of charged pions. The energy spectrum of muons can then be calculated by taking account of the $\pi-\mu$ decay effect[2].

Fig. 21 shows the flux intensity of muons thus calculated from the γ-ray intensities measured by various authors. The agreement with the observed muon spectrum is, in general, satisfactory. For detail, however, the following possible difference may be noticed. Below 1 TeV, the muon spectrum calculated from the γ-ray spectrum appears to be slightly steeper than that directly observed, whereas the absolute flux values agree well with each other. Above 1 TeV, on the contrary,

---

[1] Mathematical details of the analytic method are described by J. Nishimura, Handbuch der Physik, this volume, p. 108. — See also ref. (65).
[2] This procedure is discussed in G. N. Fowler and A. W. Wolfendale, ref. [7]. Here we use a slightly different method with better accuracy; see S. Hayakawa, J. Nishimura and Y. Yamamoto (56).

the calculated absolute flux appears to be a little smaller than that directly observed, whereas the slopes of these two are in very good agreement.

Considering that the relative measurements are more reliable than the absolute ones, and using only the data from the Norikura and Chacaltaya experiments, one finds a better fit with the muon data in the slope of the spectrum. A slight difference in the absolute value can be an indication of the contribution of kaons decaying into muons. The results are not inconsistent with the K/π-ratio found by the direct observations shown in II.5.

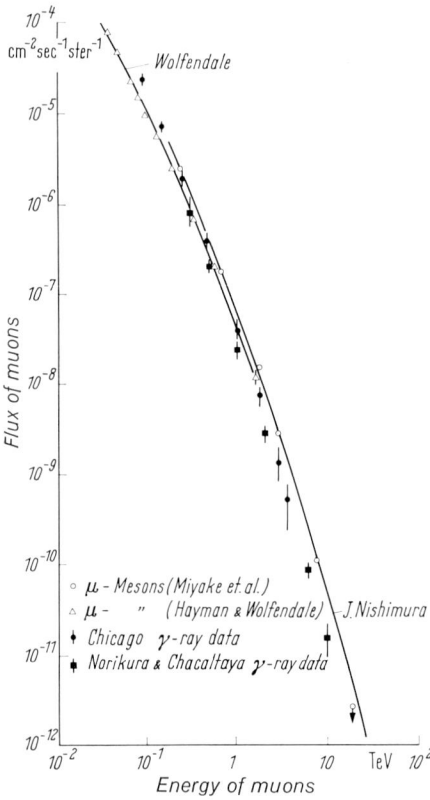

Fig. 21. Integral energy spectrum of muons observed and calculated from the γ-ray spectra under the assumption that all muons are the decay products of charged pions. ▲ the muon intensities measured with a magnetic spectrograph. ○ the muon intensities converted by means of the range-energy relation calculated by Nishimura from underground measurements. ◆, ■ the muon intensities calculated from the γ-ray intensities.

16. Extensive air showers. The highest energy of primary cosmic rays observed to date reaches 10^{20} eV, and will extend to still higher energies if new techniques are developed. Extensive air showers (EAS) produced by primary cosmic rays of such high energies provide information on nuclear interactions at the highest energy known. However, the structure of an extensive air shower is so complex that the investigation of nuclear interactions by means of EAS is comparable to scratching an itchy foot with shoes on. The record of an event observed in the Tokyo Air Shower Project, reproduced in Fig. 22, may serve as an example.

EAS are ordinarily observed with a number of detectors distributed over a large area. Each detector permits us to measure the number of particles incident on it. Since most of the charged particles in EAS are electrons, these detectors reveal the spatial distribution of the electron density. Integrating the density over the entire area, one can obtain the total number of particles, called the size of the EAS. — A typical cloud chamber photograph of a medium-sized shower with double-core structure is shown in Fig. 23.

The lateral distribution of the density is found to be independent of the shower size, and is expressed by the Nishimura-Kamata function. A simple analytic approximation to the Nishimura-Kamata function, obtained by Greisen,

$$f(r) = c(s) r^{s-2} (r+1)^{s-4.5} \tag{16.1}$$

is often referred to as the N-K-G function[1]. Here r is the distance from the shower axis in Molière units, and $c(s)$ is a normalization factor. s is the age parameter which is of primary importance in the analysis of EAS.

The value of s is obtained for each shower by fitting (16.1) to the observed lateral distribution for r between several meters and a hundred meters or so; at

---

[1] See, J. Nishimura, Handbuch der Physik, this volume, and also K. Greisen, Progress in Cosmic Ray Physics, III. 1 (1956).

very small r the lateral distribution fluctuates very much from one shower to another, and at very large r the effect of single scattering and the contribution of muons make the applicability of (16.1) inadequate (*66*). In this way the value of s is found to be on the average about 1.2. The fact that the value of s is practically independent of the size N proves that an EAS is not a pure electronic cascade,

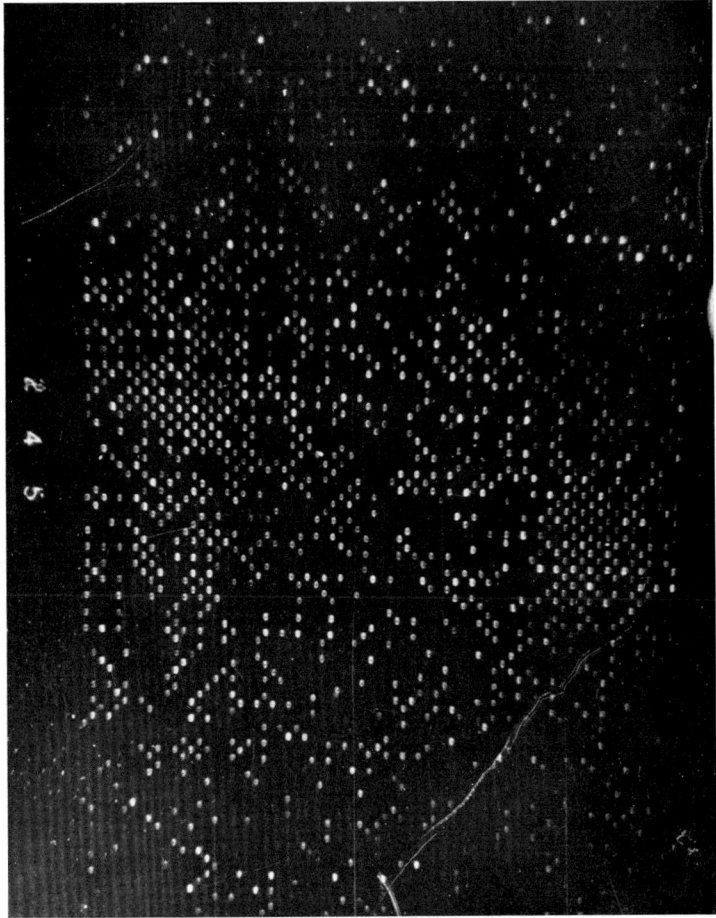

Fig. 22. A neon-hodoscope photograph showing the double-core structure of an EAS. Each neon tube has a diameter of 2 cm. (Tokyo Air Shower Project). — See also discussion in Sect. 6, p. 128.

but a highly complex mixture of nuclear and electronic cascades. In order to get direct information on nuclear interactions, therefore, some knowledge of nuclear cascade processes is indispensable.

The structure of the nuclear cascade depends essentially on the following three quantities: mean free path, inelasticity, and multiplicity. The energy distribution of the secondaries produced in an interaction is, of course, an important factor, but this can be accounted for by using "effective" values of inelasticity and multiplicity, taking into consideration that only high-energy secondary particles participate in the further cascade development.

If all of these fundamental quantities were independent of energy, the depth at which the shower size reaches a maximum would increase, and the attenuation length of the shower below this depth would become larger with increasing primary energy. Actually, however, both the depth of the maximum size and the attenuation length are found experimentally to vary little with size and consequently with primary energy. The size independence of the depth of the shower maximum, of the attenuation length, and of the age parameter demonstrate an energy dependence of one or some of these fundamental quantities acting in such a way that energy is more rapidly shared among the numerous particles as the primary energy increases.

Usually this effect is attributed to an increase in the multiplicity with energy. This serves to suppress the production of those γ-rays of enormous energies which

Fig. 23. A photograph of a double-core EAS observed in a large multiplate cloud chamber ($200 \times 130 \times 65$ cm³ in sensitive volume). The shower size is estimated as $6 \cdot 10^5$. (Mt. Norikura Air Shower Project.) See also discussion in Sect. 6, p. 128.

develop into large electronic cascades. It also prevents the production of very energetic charged pions which otherwise would initiate young nuclear cascades at any given level. However, this is not enough to eliminate all young nuclear cascades, because there would still be surviving nucleons capable of carrying them deep into the atmosphere. Therefore, the inelasticity is presumed to increase with the energy so as to reduce the fraction of energy transferred to the surviving nucleon[1]. The relation between the attenuation length and these fundamental quantities is shown in Table 10, on the basis of a model calculation (67).

Table 10. *Attenuation length of extensive air showers at sea level in gcm⁻².*

Primary energy E_0 in eV	Inelasticity K	
	1	$\frac{1}{2}$
10^{14}	—	144
10^{15}	112	172
10^{16}	123	202
10^{17}	145	242

Thus far we have quoted a number of indications that the characteristics of nuclear interactions may change when the energy increases beyond

---

[1] This has been suggested by G. TANAHASHI, J. Phys. Soc. Japan **20**, 883 (1965), by measuring the energy flow of the nuclear-active and the electronic components, as well as the energy spectrum of the nuclear-active particles.

10^{15} eV. However, as the following remark will show, a clear-cut conclusion should be reserved until further investigations both on EAS and from direct observation will have clarified the picture.

This remark concerns the effect of fluctuations on the development of an EAS, following an argument of MIYAKE (68). It is generally assumed that the primary energy of an EAS will be proportional to the shower size N_e, which is essentially equal to the total number of electrons in the shower. MIYAKE showed that in certain cases this assumption — which is tantamount to neglecting fluctuations — will introduce a serious error.

Suppose a primary particle of a given energy produces an EAS. The size of this EAS at a certain atmospheric depth can vary over a wide range because of the fluctuations in the multiplication processes, among which a variation of the depth of the first nuclear interaction produces the most significant effect. For example, primary particles with constant energy colliding at different levels give rise to EAS with a power-law size spectrum, the exponent of which is approximately equal to L_a/L_{int}, the ratio of the attenuation length L_a near the observation level and the interaction mean free path L_{int}. Therefore, the observed EAS spectrum is compatible with the assumption of primary particles confined to a narrow energy range if the primary spectrum is very steep. Consequently, all arguments derived from an analysis of EAS and based on the averaging procedure have to be re-examined very carefully, in particular since the primary spectrum does show indications of deviating from a simple power law. Fig. 24 presents a summary of the presently available information on the primary spectrum (69).

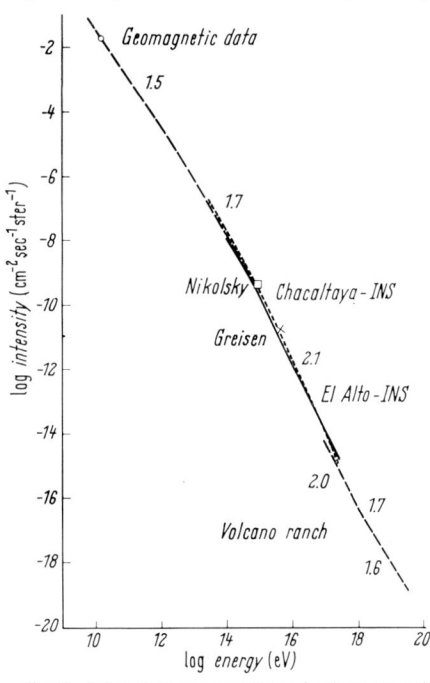

Fig. 24. Integral energy spectrum of primary cosmic rays. The figures attached are the spectral indices in the energy ranges concerned [S. J. NIKOLSKY, Proc. of the Fifth Interamerican Seminar on Cosmic Rays, vol. II (1962); K. GREISEN, Ann. Rev. Nuclear Sci. **10**, 63 (1960)], and the geomagnetic data give the absolute flux. The spectrum between 10^{11} eV and 10^{15} eV is derived on the basis of the energy spectra of γ-rays, muons and protons by Miyake and by the Durham group. The solid line is obtained from air shower data at Mt. Chacaltaya and at El Alto by the Bolivian Air Shower Joint Experiment and by the M.I.T. Group. The dotted line is obtained from air shower data at the Institute for Nuclear Study, Tokyo. The dashed line in the highest energy region is obtained from air shower data at Volcano Ranch by J. Linsley, Jaipur Conference (1963).

IV. Models of multiple production.

17. Thermodynamical and hydrodynamical considerations. As an introduction to the discussion on the models of multiple production, let us describe briefly the thermodynamical and hydrodynamical considerations prevalent at the period at which the existence of multiple production was established.

As described in I.1 "Early theoretical attempts," the theoretical treatment of multiple production was essentially based on the following argument. Writing A for the square of a matrix element for one-pion production, the probability of emitting n pions is given by

$$W_n \propto \int (A^n/n!) \Pi_i \, d\mathbf{p}_i/E_i \simeq (A\bar{p}^2)^n/n!. \tag{17.1}$$

In the second expression $d\mathbf{p}_i/E_i$ is the invariant phase volume for the i-th pion of momentum p_i and energy E_i, in C.M.S., and in the last expression $\bar{p}$ is the average momentum of these pions with relativistic energies. The most probable multiplicity is approximately obtained as

$$\bar{n} \approx (A\bar{p})^{\frac{3}{2}} \propto E_0^{*\frac{3}{2}} \propto E_0^{\frac{3}{4}}. \tag{17.2}$$

The second relation is obtained on account of $p \propto E_0^*$, which is derived under the assumption of a constant matrix element A. E_0^* is the incident energy in C.M.S., whereas E_0 is that in L.S. The energy dependence in (17.2) is steep, and is hard to reconcile with the experimental evidence. Furthermore, the transverse momentum of a pion produced turns out to be $\sim p$, because pions are supposed to be emitted isotropically. This again contradicts the experimental information.

In this procedure it is assumed that the pions thus emitted are uncorrelated. If, on the contrary, the correlation is so strong that there is a phase relation between these pions, the particle picture of pions does no longer hold but the whole system of pions may be described in terms of a classical wave. The number of particles with momentum $\mathbf{p}$ in $d\mathbf{p}$ is then obtained from the Fourier component of such a wave, $f(\mathbf{p})$, in the final stage as

$$n(\mathbf{p}) d\mathbf{p} = f(\mathbf{p})^2 d\mathbf{p}/E, \tag{17.3}$$

Before reaching the final stage, the wave form may be subject to a temporal change essentially governed by the non-linearity in the equation of motion. The non-linearity could cause the energy contained in high-frequency components to be transferred to low-frequency components, as in the case of hydrodynamical turbulence. In the particle picture this is equivalent to the creation of a large number of pions from a high-energy component. Thus a great part of the energy is concentrated into low-frequency components, and, as a result, the energies of the pions emitted are low. This result is in agreement with the experimental evidence showing a small and constant transverse momentum. But the predicted multiplicity increases rapidly with energy, essentially proportional to E_0^* (70).

This large multiplicity could be avoided by introducing a quasi-elasticity of the collision. However, a rapid decrease of the inelasticity with increasing energy would be required, which does not seem to be borne out by the experimental data. Moreover, a partial inelasticity can be introduced merely as an *ad hoc* assumption which has nothing to do with the wave picture.

A thermodynamical model was proposed to solve this difficulty of the energy dependence of the multiplicity (71). Let us now consider the system of the created particles confined in a volume Ω, which is in thermodynamical equilibrium due to the strong interactions between them. If the total energy in C.M.S., $2E_0^*$, contained in the volume Ω is distributed over all possible modes according to the condition of thermal equilibrium, then the energy density is given by the Stefan-Boltzmann law, and the temperature of this system, T, in units of the Boltzmann constant, is given by

$$E_0^*/\Omega \propto T^4. \tag{17.4}$$

Hence the most probable number of pions produced is

$$n = E_0^*/T \propto \Omega T^3 \propto E_0^{*\frac{3}{4}} \Omega^{\frac{1}{4}}. \tag{17.5}$$

Now Fermi assumes the volume Ω to be the Lorentz-contracted collision volume, $\Omega = (M/E_0^*)\Omega_0$, where M is the nucleon mass. Then we obtain the multiplicity law

$$n \propto E_0^{*\frac{1}{2}} \Omega_0^{\frac{1}{4}} \propto E_0^{\frac{1}{4}}. \tag{17.6}$$

Sect. 17.   Thermodynamical and hydrodynamical considerations.   155

This $\frac{1}{4}$-power law of multiplicity is in agreement with the experimental evidence, but the energy of a pion emitted in C.M.S. is so high as to give too large a value of the transverse momentum if the emission is isotropic in that system. The forward-backward peaking of the angular distribution can be derived from a consideration of the conservation of angular momentum (72), but this effect is not sufficient to reduce the theoretical transverse momentum to a value as small as the observed one.

The forward-backward peaking in the angular distribution may be accounted for, if the available energy is concentrated not in one collision complex but is shared between two bodies, each having a colliding nucleon as a core (73), (74). In each of the bodies which may be called fireballs, thermodynamical equilibrium is assumed, and since the fireball is considered as spherical in its rest system, the volume Ω is energy independent. Accordingly, the multiplicity depends on energy as

$$n \propto E_0^{*\frac{3}{4}} \propto E_0^{\frac{3}{8}}, \qquad (17.7)$$

in the limit of a totally inelastic collision. If part of the energy is shared with the translational motion of the fireball, the energy dependence of the multiplicity could be weaker, and the transverse momenta of particles emitted could be reduced.

The development of FERMI's theory of thermal equilibrium began in another direction with the following criticism. In the initial stages of the collision the particle number should not be well defined because of the strong coherence of the waves, while the energy density could be a well defined quantity. Now the macroscopic quantities can be introduced in the spatial region with linear dimension as large as the coherence extends, so that a hydrodynamical description of the whole system may be possible (75). That is to say, the macroscopic quantities such as the energy density ε and the local pressure p are given as functions of the space and time coordinates in the system. They are related to each other by the equation of state

$$p = \frac{\varepsilon}{3}, \qquad (17.8)$$

The local temperature T and the entropy density s can also be introduced, and they are related by the black body radiation as

$$\varepsilon \propto T^4, \quad s \propto T^3. \qquad (17.9)$$

The strong interaction in the system results in a short transport mean free path, and consequently in the essential conservation of entropy. The body of high temperature assumed in FERMI's thermodynamical model does not immediately radiate pions but expands in the direction of the collision, keeping the entropy density at a constant value. The process of expansion is described in terms of ordinary hydrodynamics, and it proceeds until the density decreases to so small a value that the mean collision length of the individual waves is as large as the dimension of the whole system. Then the individual parts are allowed to fly apart as free particles, whereby one can define the number of particles. Owing to the assumption of the isentropic expansion, the particle number is equal to the total entropy in the initial stage, so that the multiplicity law is identical with FERMI's:

$$n \propto S = s\Omega \propto \varepsilon_0^{\frac{3}{4}} \Omega = E_0^{*\frac{3}{4}} \Omega^{\frac{1}{4}} \propto E_0^{*\frac{1}{2}}. \qquad (17.10)$$

In the third relation the initial energy density ε_0 is related to the entropy density through (17.9)

In the final stage of expansion the temperature may decrease to a value as small as the rest energy of the pion, and the expansion in the transverse direction is small. Consequently, the transverse momentum of a particle due to the thermal motion and the transverse fluid velocity will be small and insensitive to the primary energy. Another important consequence of the low final temperature is the preferential emission of lighter particles (76). Hence the K/π-ratio is expected to be small, in agreement with observations, and the creation of anti-nucleons is even less likely to occur.

A further refinement of this hydrodynamical model of LANDAU was achieved by several authors (77), (78), showing the results to be applicable also to nucleon-nucleus collisions. The validity of this model was investigated on the basis of the statistical mechanics of irreversible processes, and it was demonstrated that local equilibrium would hold in most parts of the fluid except its very front (79). Local equilibrium is shown to hold within a space-time region whose spatial dimension is larger than the correlation length, and whose time scale is larger than the correlation time, the space and time scales being connected with each other in the relativistic fluid. The correlation time is approximately given as

$$\tau_c \approx \frac{\hbar}{T_0}, \qquad (17.11)$$

where T_0 is the initial temperature, provided that the pion-nucleon coupling is not of strong singularity. This is compared with the time scale for the fluid to disintegrate into pions,

$$\tau_d \simeq \frac{\hbar}{T}. \qquad (17.12)$$

Since $\tau_d \gg \tau_c$, the condition for local equilibrium is satisfied. At the front of the fluid, however, the spatial dimension within which the energy density changes by a considerable amount is shorter than the correlation length.

The hydrodynamical description does not hold at the very beginning of the collision either. The time scale needed for the exchange of energies between two colliding particles, the impact time, may be estimated as

$$\tau_i \approx r/c\gamma_c, \qquad (17.13)$$

where r is the effective force range. τ_i is greater than τ_c only if $\gamma_c \lesssim (Mcr/\hbar)^2$, since $T_0 \sim M c^2 \gamma_c^{\frac{1}{2}}$. If r were as large as the pion Compton wave length, the hydrodynamical description would be applicable even at the initial stage of the collision up to an incident energy as high as a few TeV. However, such a large value of r would result in too small a value of the momentum transfer compared with the experimental data, as will be shown in subsequent sections. It is, therefore, plausible that the force range effective to multiple production is as short as the radius of the nucleon core introduced in the theory of nuclear forces[1]. In consequence the hydrodynamical theory is not adequate to describe the production of the meson fluid. This failure is revealed in the fact that no room is left for introducing an inelasticity. Moreover, if one goes too far into details, a number of results is found which often are not in agreement with the experimental evidence. Such discrepancies are a common feature of all dynamical models, such as HEISENBERG's and FERMI's. It is, therefore, recommended to stop one step before reaching a dynamical theory, and to spend some thought on a phenomenological model, so that the characteristic features which any dynamical theory should take into account are investigated in full detail first.

---

[1] For further detail on space-time scales important for high energy interaction, the readers are referred to references [2] and [9].

18. Fireballs. Recent investigations of the multiple production of shower secondaries, reviewed in the previous sections in Part II and III, suggest that this process takes place through an intermediate stage which may be described by a complex, similar to the collision complex in low-energy nuclear reactions. The complex may be called a fireball, which makes us imagine a spherical entity from which pions are emitted isotropically. But in the following we shall use the term "fireball" to express a collision complex in general without specifying its physical properties or geometrical shape. The reason why fireball and compound nucleus are analogous may be understood if one considers that in either system many particles are concentrated within a small spatial region, and are strongly interacting with each other. It is, however, to be remembered that the particles in the fireball necessarily cannot be well-defined entities, whereas the concept of nucleons in the compound nucleus has a definite meaning.

In low-energy nuclear reactions, the energy brought in by an impinging particle is dissipated quickly into various modes of motion inside the nucleus, so that the whole nucleus reaches a state of thermal equilibrium. Thus in nuclear reactions at low energies, the properties of the final state are mainly responsible for the principal features of the reactions. Multiple production at extremely high energies is analogous to the nuclear reactions described above, and the process of pion emission from the fireball can be attributed essentially to a low-energy pion phenomenon, thus being independent of the incident energy. This is the very basis on which the assumption of an intermediate state called "fireball" rests.

The characteristic features of the fireball are experimentally revealed by the following facts. Firstly, the constancy of the transverse momentum holds over a wide range of energies, that is, from the accelerator region of several GeV up to the air shower regions of $10^{16}-10^{17}$ eV. Secondly, the K/π-ratio of about $\frac{1}{4}$ stays nearly constant from 10 GeV to 10 TeV at least.

These two properties are closely correlated with each other, and can be understood by thermodynamical arguments concerning the fireball. Both the magnitude of the transverse momentum and the K/π-ratio are determined by the temperature of the fireball, and their experimental values indicate a temperature of the order of the pion rest energy. Hence the pion emission process from the fireball will be accounted for in terms of low-energy pion physics. On the other hand, the production of the fireball itself belongs to the problems of high-energy interaction.

The fireball is identified from the density maximum in the log tan ϑ plot of the angular distribution (see II.7, "Angular distribution"). Thanks to the Lorentz invariance of the distribution of particles expressed in the log tan ϑ scale, the properties of the fireball can be analysed without accurate knowledge of the primary energy and of the translational energy of the fireball. Several typical examples of the log tan ϑ plots are shown in Fig. 25.

In Fig. 25a, we observe one bunch in the log tan ϑ plot, indicating that particles are isotropically emitted from one center. This is characteristic for most of the events in the primary energy region around 100 GeV or lower, as discussed in II.8 in connection with the calorimeter experiments.

As the energy increases, the bunch begins to split into two, thus indicating the presence of two fireballs, as seen in Fig. 25b. In some cases four bunches may be identified as in Fig. 25c, thus suggesting that the multiplicity of fireballs may not be restricted to one or two.

In all these cases it is evident that the distribution of particles in a log tan ϑ scale around the center of a bunch is essentially the same. This represents the distribution of longitudinal momenta of particles emitted from a fireball. The distribution thus obtained is similar to that of the transverse momenta, in agreement with the assumption of isotropic emission, and furthermore it is found to be consistent with PLANCK's law. All these facts justify the introduction of the fireball concept.

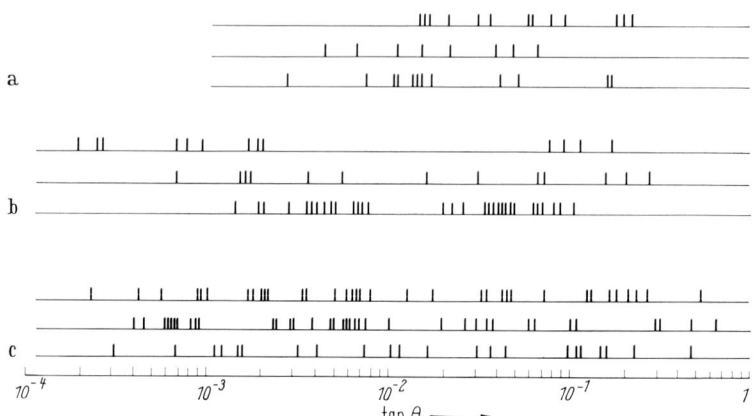

Fig. 25. Typical examples of the log tan ϑ plots.

Other quantities characteristic for the fireball model are the multiplicity of particles emitted from a fireball and its velocity. The former can be measured easily by counting the number of particles belonging to a bunch in the log tan ϑ plot. The latter quantity is obtained by measuring the distance in the log tan ϑ plot between the center of all particles and the center of a bunch, but the center of the log tan θ distribution does not always represent the center-of-mass system, as often emphasized in II in connection with the estimates of the primary energy.

Another important quantity to be measured is the four-momentum transfer associated with the emission of a fireball. The method of deriving it from observable quantities, and the physical implications of the momentum transfer will be discussed later for specific models.

19. Emission of one fireball in low-energy jets.

The characteristic properties of nuclear interactions with primary energy $\lesssim 100$ GeV are well interpreted by assuming the emission of one fireball in the collision, as is described in II.8 "Calorimeter method and asymmetry". It was noted that the usual symmetry criteria of the system do not always hold even for proton-proton collisions. This means that the fireball created in the interaction does not always stay at rest in C.M.S., but sometimes moves either in forward or in backward direction.

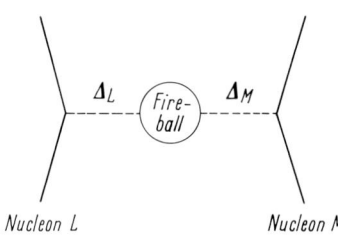

Fig. 26. Diagram of one-fireball emission.

Now the whole collision process can be graphically represented by the schematic drawing of Fig. 26. The suffices, L and M, refer to the incident particle in the laboratory and the mirror systems, respectively. The four-momentum transfer with respect to the incident particle is defined by

$$\Delta_L^2 \equiv (P_L - P_L')^2 - (E_L - E_L')^2, \tag{19.1}$$

where symbols with and without prime refer to the parameters after and before the collision, respectively. In the extremely relativistic case this is approximated by

$$\Delta_L^2 = \frac{1}{1-K_L}(K_L^2 M^2 + P_T'^2), \qquad (19.2)$$

where K_L is the inelasticity in L.S. defined as

$$K_L \equiv \frac{E_L - E_L'}{E_L} = \frac{1}{E_L}\Sigma E_i, \qquad (19.3)$$

and ΣE_i is the sum of the energies of all secondary particles in L.S. The same relation holds for the target particle, if the laboratory system is replaced by the mirror system. For example, the mirror inelasticity, K_M, is

$$K_M = \frac{1}{E_M}\Sigma_i(E_i - P_i \cos\vartheta_i), \qquad (19.4)$$

where ϑ_i is the emission angle of the i-th particles in L.S.

The values of K_L and K_M are not always equal in individual events, but their average values should be equal in nucleon-nucleon collisions, as discussed in II.8. A difference between them demonstrates existence and direction of an asymmetry. The importance of knowing both K_L and K_M is, therefore, evident. — Another method of estimating Δ_M is based on observations on the recoil proton in the laboratory system, that is, on the relation

$$\Delta_M^2 = P_M'^2 - (M - E_M')^2 = 2W_M' M, \qquad (19.5)$$

where W_M' is the kinetic energy of the recoil proton.

Now the mass of the created fireball is related to the four-momenta, Δ_L and Δ_M, or the inelasticities, K_L and K_M, as follows,

$$M_F = |\Delta_L + \Delta_M|^2 = 2\gamma_c\sqrt{K_L K_M} M, \qquad (19.6)$$

and its Lorentz factor in C.M.S. is

$$\gamma_F = \gamma_c M(K_L + K_M)/M_F = (K_L + K_M)/2\sqrt{K_L K_M}. \qquad (19.7)$$

Therefore the number of secondary particles emitted from the fireball is given by

$$n = M_F/T = 2\gamma_c\sqrt{K_L K_M} M/T, \qquad (19.8)$$

where T is the final temperature of the fireball.

The assumption of emission of one fireball can be checked experimentally through the relations (19.7) and (19.8). Agreement with these relations was indeed observed by the calorimeter experiment of the Soviet group (46). Furthermore, the multiplicity law, $n \propto K E_L^{\frac{1}{4}}$, given in (19.8), was first confirmed by KANEKO and OKAZAKI through the analysis of emulsion data (80). The validity of this model is, however, limited to primary energies up to 100 GeV or so.

It is interesting to compare nuclear interactions produced by muons with those produced by protons. In II.10 "Muon interaction" an underground experiment was described in which interactions induced by muons of energies of several tens of GeV were studied. The composite angular distributions in the log tan ϑ scale constructed for the events observed by the Osaka City University experiment (49), is reproduced in Fig. 27. It is found that the distribution shows the characteristic feature of one-fireball emission, similar to the low-energy proton interactions discussed above.

Under the assumption of one-fireball emission, the analysis of muon interactions can be carried out in the same way as for proton interactions. The resulting diagram for the interactions is represented in Fig. 28.

In analogy to the relation for Δ_L, the four-momentum transfer of a muon, Δ_μ, is expressed as follows

$$\left.\begin{aligned}\Delta_\mu^2 &= 4 E_0 (E_0 - E_t) \sin^2(\vartheta_\mu/2) \\ &= (E_t + P_T \operatorname{cosec} \vartheta_\mu) P_t \tan(\vartheta_\mu/2).\end{aligned}\right\} \qquad (19.9)$$

E_0 and ϑ_μ are the primary energy and the scattering angle of the muon, and E_t is the energy transferred to all secondary particles. If one has information on the P_T of the muons, it is possible to estimate Δ_μ without knowledge of the primary energy E_0.

There is some experimental evidence that the P_T for muons are nearly the same as those observed in

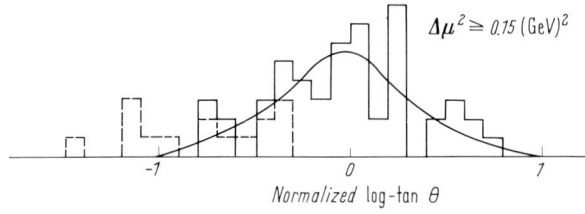

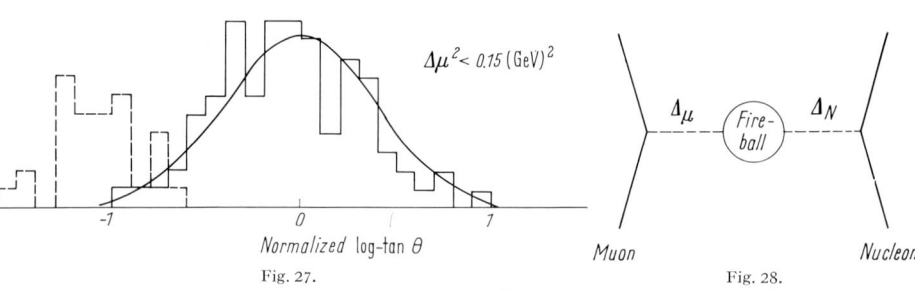

Fig. 27.

Fig. 28.

Fig. 27. Log-tan ϑ histograms of μ-interactions, the upper one for $\Delta_\mu^2 \geq 0.15$ (GeV)², the lower one for $\Delta_\mu^2 < 0.15$ (GeV)². The solid curves represent the Gaussian distributions with $\sigma = 0.39$, the isotropic distribution. The dotted parts represent surviving muons.

Fig. 28. Diagram of nuclear interaction of muons.

ordinary nuclear interactions. Assuming, therefore, $P_T = 400$ MeV/c, one can calculate the Δ_μ for every observed event. Their average value is

$$\langle \Delta_\mu^2 \rangle^{\frac{1}{2}} = 0.52 \text{ GeV}, \qquad (19.10)$$

with a dispersion of $\sigma = 0.35$ GeV. This value of Δ_μ is a little smaller than that for the nucleon-nucleon collision, which gives $\langle \Delta_L^2 \rangle^{\frac{1}{2}} \sim 1$ GeV.

The four-momentum transfer of a target nucleon, Δ_N, is obtained from the formula for the mirror system, (19.5). The average value is

$$\langle \Delta_N^2 \rangle^{\frac{1}{2}} = 0.3 \text{ GeV}, \qquad (19.11)$$

which is essentially equal to $\langle \Delta_\mu^2 \rangle^{\frac{1}{2}}$. The mass of the fireball is given by

$$M_F^2 = (\Delta_N + \Delta_\mu)^2 = \Delta_N^2 + \Delta_\mu^2 + 2 \Delta_N E_t \qquad (19.12)$$

This allows us to deduce the average mass of the fireball as

$$M_F = 1.8 \text{ GeV}. \qquad (19.13)$$

This is not inconsistent with the average observed multiplicity of about four, if neutral particles are taken into account, and the average energy of particles emitted in the fireball system is 0.4 GeV.

Thus the muon interaction appears analogous to the nucleon interaction in many respects. Although the absolute values of Δ_μ^2 and Δ_N^2 are smaller than the corresponding ones in nucleon interactions, this may be considered as due mainly to the smallness of the interaction energy, $E_t \sim$ several GeV. In this energy region the value of Δ_N^2 found in nucleon interactions is as small as that in (19.10) or (19.11).

20. Niu model of two-fireballs emission. When the primary energy increases up to about 1 TeV or more, the angular distribution of the secondary particles deviates appreciably from that of isotropic emission from one center (one-fireball model), and one often finds a pronounced forward and backward collimation. This may be interpreted as due either to collimated emission from one fireball or to the formation of two fireballs, each of which acts as an isotropic emitter.

An example of the former interpretation is the hydrodynamical model of LANDAU, details of which are described in IV.17. The LANDAU model is found successful in accounting for the experimental information on P_T and the K/π-ratio, and its predictions both on the multiplicity and on the forward-backward collimation in the angular distribution appear to be in agreement with the averaged experimental data. However, it fails to explain the fluctuations of the various quantities observed; the jet multiplicity fluctuates more than the thermodynamic law would lead one to expect, and the degree of collimation varies considerably from event to event. These observations suggest that the whole process of multiple production cannot be described by the hydrodynamical theory alone. Moreover, the theory has no room to introduce a physical quantity which corresponds to the inelasticity.

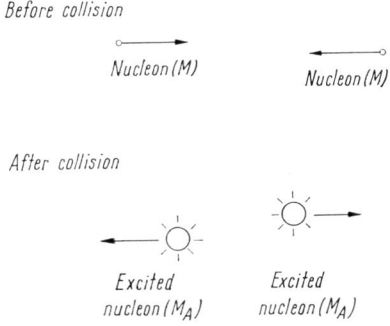

Fig. 29. Diagram of the TAKAGI model. Two nucleon of mass M collide and two fireballs of mass M_F are produced. The masses of these two may be different.

Therefore, the empirical evidence seems to favour the alternative interpretation, particularly for a phenomenological analysis. TAKAGI proposed a model assuming that after the collision the two nucleons are in highly excited states, each of which subsequently decays into a nucleon by emitting several pions (73). Fig. 29 shows a schematic diagram of TAKAGI's model.

Writing M_F for the mass of the excited nucleon or fireball, and M for the nucleon mass, kinematic relations for the TAKAGI model can be derived from the following considerations (81): The energy of the nucleon after de-excitation, that is, the elasticity of the event, is obtained from

$$1 - K = E_L'^*/E_L^* = M/M_F. \quad (20.1)$$

On the other hand, the inelasticity K and the multiplicity of pions are related by

$$K = n_F T/M_F \quad (20.2)$$

assuming an average energy T of the pions. From (20.1) and (20.2) one obtains

$$K/(1-K) = n_F T/M. \quad (20.3)$$

This relation between the multiplicity n_F and the inelasticity K does not seem to be supported by the experimental evidence, since for the right-hand side values appreciably greater than unity are found, whereas the left-hand side is of the order of unity.

In order to account for the considerably elastic nature of the collision, and at the same time to obtain a multiplicity as large as that experimentally observed, the fireball and the outgoing nucleon have to be assumed to move separately. As shown in Fig. 30, it is likely that two fireballs are created in a collision, one going forward in C.M.S. and the other backward, with velocities smaller than that of the outgoing nucleons. This model of emission from two fireballs separated from the two outgoing nucleons was first proposed by NIU (82).

Now let us derive kinematical relations based on NIU's two-fireballs model. Suppose two fireballs of masses M_F and M_B are produced in a collision, and move with the Lorentz factors γ_F^* and γ_B^* in C.M.S., respectively. Then the inelasticity coefficients are

$$K_L = \frac{\gamma_F^* M_F}{(\gamma_c - 1) M}, \qquad K_M = \frac{\gamma_B^* M_B^*}{(\gamma_c - 1) M}. \tag{20.4}$$

The relations of the inelasticities, K_L and K_M, to the four-momentum transfer, Δ_L and Δ_M, are the same as the one given in IV.19. There is one new quantity in this model, which represents the four-momentum transfer between two fireballs, Δ. The value of Δ can be evaluated from the relation

$$\begin{aligned}\Delta^2 &= (P_f - P_b)^2 - (E_f - E_b)^2 \\ &= \{(P_f - E_f) - (P_b - E_b)\} \times \\ &\quad \times \{(P_f + E_f) - (P_b + E_b)\},\end{aligned} \tag{20.5}$$

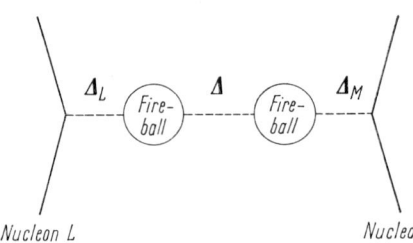

Fig. 30. Diagram of the NIU model. Two nucleons collide, and two surviving nucleons and two fireballs emerge.

where the suffices f and b refer to the forward and the backward groups respectively. Taking into account the energy-momentum conservation and neglecting the terms of higher order in the transverse momenta and masses in the relativistic approximation, one obtains the following useful formula for the estimate of Δ (83):

$$\Delta^2 = \langle P_T^2 \rangle \Sigma_F \tan \vartheta_i \Sigma_B \cot \vartheta_j \tag{20.6}$$

where the sums in the right-hand side are taken over particles emitted from the forward and backward fireballs respectively.

The expression for estimating the four-momentum Δ was originally proposed on the basis of the NIU model, but it was found to be applicable for many other cases as well. One may divide the secondary particles of a jet into two groups, and ask for the magnitude of the momentum transfer Δ between these two. For instance, the result of the above procedure of estimating Δ between the forward and backward groups depends, of course, on the way in which the secondary particles are assigned to the two groups. In some cases a division from inspection of the log tan ϑ plot is ambiguous. It may be remarked that the particular division for which the transfer Δ has its minimum value can be regarded as physically significant.

Analysing the empirical data on jets in accordance with the two-fireballs model, NIU obtained, in most of the cases, for the $\sqrt{\Delta^2}$ a magnitude of $\sim 1-2$ GeV. The Japanese Emulsion Group for ICEF constructed the $\sqrt{\Delta^2}$ distribution from the data of ICEF collaboration (84). In this work, ambiguities in defining the forward and backward groups were carefully examined by adopting various conventions. These two studies give essentially the same results; the distribution of $\sqrt{\Delta^2}$ has a peak at about 2 GeV. Fig. 31 shows the result of the ICEF work.

Now let us derive the relation between the momentum transfer, Δ, and the mass and the velocity of a fireball. In the case of symmetric emission in the forward and backward directions, we have $\gamma_F^* = \gamma_B^*$, $M_F = M_B$, the time component of Δ vanishes in C.M.S., and one has the relation

$$M_F = 2\gamma_F^* \Delta. \qquad (20.7)$$

From (20.4) and (20.7), one obtains for the Lorentz factor of the fireball

$$\gamma_F^* = [K(\gamma_c - 1) M / 2\Delta]^{\frac{1}{4}} \qquad (20.8)$$

and for the mass of the fireball

$$M_F = [K(\gamma_c - 1) M \cdot 2\Delta]^{\frac{1}{2}}. \qquad (20.9)$$

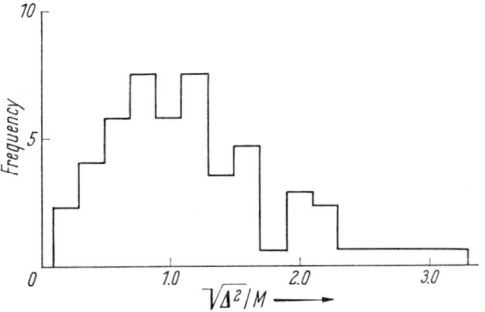

Fig. 31. Distribution of four-momentum transfers between two fireballs.

If both K and Δ are independent of the energy, γ_F^* and M_F are proportional to $\gamma_c^{\frac{1}{4}}$. Since the multiplicity is proportional to the mass of the fireball, (20.9) gives a $\frac{1}{4}$-power law for the multiplicity. Furthermore, the relation (20.9) indicates that, for a given primary energy, the multiplicity variations will be essentially determined by the factor $\sqrt{K\Delta}$.

It may be worth while to add a remark on the relation between the two-fireballs model and the one-fireball emission discussed in the previous section, IV.19. According to the original proposal of NIU, multiple production always takes place through the formation of two fireballs. But the two fireballs will not be resolved and will appear as one fireball when the Lorentz factor in C.M.S. γ_F^* is not large enough. To illustrate this point, let us assume $K = \frac{1}{2}$ and $\Delta = 2M$. Then we have from (20.8)

$$\gamma_F^* = [(\gamma_c - 1)/8]^{\frac{1}{4}}.$$

In order is to have $\gamma_F^* > 1$, we require

$$\gamma_c > 9, \qquad (20.10)$$

or a primary energy higher than 150 GeV. This shows that two fireballs created in a collision of 100 GeV or less will not be separated from each other, so that the collision will be well described in terms of the one-fireball model.

A quantitative study of the conditions under which the two fireballs can be distinguished was made by the Polish group (85). For this purpose, the evaluation of the dispersion of the log tan ϑ values,

$$\sigma = \langle (\log \tan \vartheta_i)^2 - \langle \log \tan \vartheta \rangle^2 \rangle^{\frac{1}{2}}, \qquad (20.11)$$

is useful. For an isotropic distribution, we should have

$$\sigma = 0.39. \qquad (20.12)$$

They found that for jets of primary energies around 1 TeV about three quarter of the events have σ greater than 0.6. On the other hand, the angular distributions in the forward and backward fireball systems give, respectively,

$$\sigma_F = 0.40 \pm 0.05, \quad \sigma_B = 0.43 \pm 0.05 \tag{20.13}$$

which are consistent with isotropy in the fireball system.

21. Excited baryons. So far, most of the discussion has centered on problems of the pion production process, and problems of the surviving nucleons were left aside. It is difficult in experiments to identify the outgoing nucleons among the jet particles, and to measure their physical properties. Nevertheless, increasing attention is now being paid to the surviving nucleons because of their important role in the interaction process (*86*).

In this section we shall mainly discuss the problem of a possible excitation of the outgoing nucleons. They may not always remain in the ground state, but may emerge from the collision as hyperons or excited baryons. Indeed, accelerator experiments on the quasi-elastic scattering of protons suggest that the probability of the outgoing nucleon being in excited states will not be negligible. For protons of momenta between 3.6 and 11.8 GeV/c, three excited nucleon states of energies 1238 MeV, 1512 MeV and 1690 MeV were resolved, and the quasi-elastic scattering cross-section of forming them was measured to be at least 1 mb. Furthermore, the probability of producing hyperons in 22 GeV proton collisions was estimated as 10—30% from the K^+/π^+- and K^+/K^--ratios (*87*).

Let us discuss at first what consequences of a possible existence of excited baryons we might except to find in cosmic-ray phenomena. To begin with, pions originating from the decay of excited baryons have velocities comparable to that of the outgoing nucleon, so that their energies will be higher than those of pions from fireballs. Although the number of pions from excited baryons will be smaller than that from fireballs, their contribution to the atmospheric γ-rays and underground muons may be significant, as illustrated by the following estimate. Let the probability of producing an excited baryon be f, the energy of its decay pion E_e and assume a fireball decay into n pions of energy E_f. Then the ratio of the contributions from these two sources is $f(E_e/E_f)^\alpha/n$, α being the exponent of the integral energy spectrum. If we put, for example, numerical values $\alpha=2$ and $n=10$, we find that the contribution from excited baryons is significant for the case $E_e/E_f \gtrsim 3/\sqrt{f}$.

In the case of the Niu model of two-fireballs emission, the distinction between pions from excited baryons and from fireballs becomes clear above a certain energy. Approximate values of E_e and E_f in this case are

$$E_e = K \gamma_c^2 E_X, \tag{21.1}$$

$$E_f = \sqrt{\tfrac{1}{4}K} \gamma_c^{\frac{3}{2}} \mu. \tag{21.2}$$

μ and E_X being the pion mass and the excitation energy of a baryon. Assuming $E_X \sim 3\mu$, one has

$$\frac{E_e}{E_f} \approx 4 \gamma_c^{\frac{1}{2}}. \tag{21.3}$$

A similar argument can be made for the angular distribution. One finds the following relation in the log tan ϑ scale for these two kinds of pions:

$$\log \vartheta_f - \log \vartheta_e = (\tfrac{1}{2})(\log \gamma_c + \log K) + \log 2. \tag{21.4}$$

On the other hand, the effect of excited baryons is not clearly resolved in the case of the multi-fireball model discussed in the next section, because in some cases these fast-moving fireballs have velocities comparable to that of the outgoing nucleon.

Information from cosmic ray experiments on this problem of surviving baryons is still poor. An analysis of the angular distribution of jets of several TeV gives no evidence yet for pions from excited baryons. The probability of excitation f can, therefore, not be as large as unity in this energy region (47). An analysis of γ-ray families (88), which covers a higher energy range, shows the existence of energetic pions among the secondaries in several examples. They may be due either to decay pions from the excited baryons, or to those from fast-moving fireballs.

22. Hasegawa model and multi-fireball emission. In the Niu model described in the previous section, IV.20, the mass of the fireball created in a collision is permitted, in principle, to take any value under the restriction of energy and momentum conservation. There is another possible approach, in which the fireball is considered to be of particle nature, and it is assumed to have a definite mass value like other known elementary particles or their excited states. A fireball of this nature was proposed by Hasegawa (89), and called the "H-quantum". In compensation for the definite mass value, the number of fireballs created is not restricted to two but is considered to be variable. Indeed, it increases with increasing interaction energy.

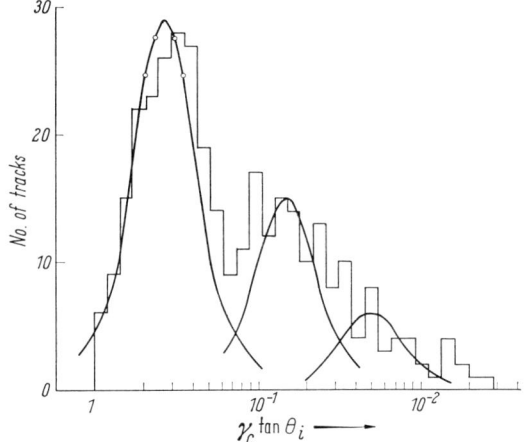

Fig. 32. Composite angular distribution of secondary particles from 20 jets with $\gamma_c > 100$. The three peaked distributions are those expected from the Hasegawa model.

Hasegawa's proposal was motivated by the observations on γ-ray families found in emulsion chambers exposed at Mt. Norikura (see III.13, "γ-ray families"). There a number of nuclear interactions of extremely high energies were observed in which several pions have energies much larger than the rest. The existence of such energetic pions among the secondary particles is hardly understandable in terms of the conventional two-fireballs model. Rather, it suggests that the number of fireballs emitted in the collision is higher than two, and that some of them have large translational energies. An indication of the multi-fireball emission is found also in the log tan ϑ plots of jet showers. As illustrated in Fig. 25, it appears that an appreciable fraction of the observed jets show an angular distribution which is hardly compatible with the two-fireballs model, and moreover in some of them one can find four-group structures.

The multi-bunch structure can be checked by constructing, in a log tan ϑ plot, the composite angular distribution in C.M.S. This is demonstrated in Fig. 32, in which one finds a minimum in the distribution between the first and the second H-quantum.

Incidentally, the bunch-structure found in the log-tan ϑ plot above appears to be more pronounced than one would expect from the hypothesis. This led Hasegawa to conjecture that the emission of pions from an H-quantum is not isotropic but of dipole type, that is, proportional to $\sin^2 \vartheta \, d\Omega$. The anisotropic emission of pions from an H-quantum could be interpreted as an indication that the H-quantum has a high spin value. A more natural explanation of the anisotropy can, however, be given in the following way (90). Since the magnitude of the four-momentum transfer $\sqrt{\Delta^2}$ is about 1 GeV, as shown in IV.20, the transverse momentum of the H-quantum, $P_T^{(H)}$, could be as large as

$$P_T^{(H)} \sim \sqrt{\Delta^2} \sim 1 \text{ Gev.} \tag{22.1}$$

If this transverse motion of the H-quantum exists, the angular distribution of pions will be affected, and exhibit an apparent anisotropy as observed.

Once the assumption of multi-fireball emission is made, the next step is to look for the properties of the unit fireball, called H-quantum. The distribution of the number of particles emitted

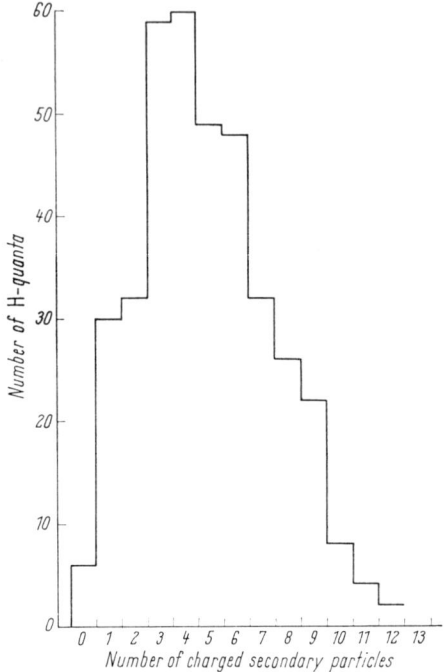

Fig. 33. Distribution of the numbers of charged particles emitted from H-quanta.

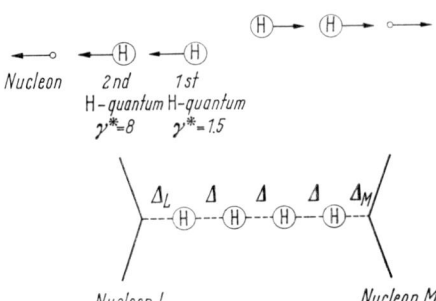

Fig. 34. Diagram of the HASEGAWA model. H represents an H-quantum. The case of the emission of four H-quanta is illustrated.

from an H-quantum, shown in Fig. 33, has a rather narrow width peaked at $n_s \approx 4$, and is similar to that in the antiproton-proton annihilation events. Thus one may postulate that the mass of the H-quantum equals two nucleon masses:

$$M_H = 2M. \tag{22.2}$$

This is consistent with the fact that the average multiplicity of shower particles from their disintegration is four, and the average energy, of the particles emitted in the H-quantum system is about twice the pion rest energy. This provides the basis for postulating that the fireball is of particle nature. It may not be unreasonable to list the H-quantum among the "fundamental" particles whose number is increasing rapidly with time. In view of its mass value close to that of a nucleon pair, one may suspect that the H-quantum is one of the particles responsible for the strong interaction.

Another important consequence of the H-quantum hypothesis is a regularity of the velocities of the H-quanta in C.M.S. HASEGAWA found that γ_H^*, the Lorentz factor of an H-quantum in C.M.S., does not vary continuously but appears to take only a series of discrete values, namely

$$\gamma_H^* = 1.5, 8, 45, \ldots. \tag{22.3}$$

The values (22.3) form a geometrical series, and express that the relative velocity between the two successive H-quanta is constant and the same in all cases. Fig. 34 illustrates a model of multiple production according to the H-quantum hypothesis. Together with the illustration, we present the diagram which expresses the H-quantum production. In this diagram a line connecting two adjacent H-quanta represents a four-momentum Δ. This is an extension of the momentum transfer between two fireballs introduced in IV.20 in connection with the NIU model. The definite mass of the H-quantum (22.2), and the definte relative velocity (22.3), result in a unique value of the magnitude of the momentum transfer, Δ, between two successive H-quanta, that is,

$$\sqrt{\Delta^2} \geq M_H / \sqrt{\gamma_{H,r}^{*2} - 1} \approx 0.9 \text{ GeV}, \qquad (22.4)$$

where in the second expression $\gamma_{H,r}^* = 2.5$ is the relative Lorentz factor between two adjacent H-quanta, and the transverse momenta are neglected. In fact, the average value of $\sqrt{\Delta^2}$ given above is verified by applying the HASEGAWA-YOKOI method to most of the jet showers observed. Therefore, instead of assuming (22.2) and (22.3) as the fundamental postulates, there is an alternative to consider (22.2) and (22.4) as fundamental and (22.3) as subsiduary.

A consequence of this model is a decrease of the inelasticity with increasing primary energy until the threshold for the further production of new H-quanta is reached. For example, in the energy region where only two H-quanta are produced, the inelasticity depends on the primary energy as

$$K^* = 2\gamma_H^* M_H / 2(E_0^* - M). \qquad (22.5)$$

Since the numerator is constant, the value of K^* is inversely proportional to the kinetic energy of the incident particle in C.M.S.

The second important consequence is the multiplicity law. Since the multiplicity is proportional to the number of H-quanta emitted, whereas the threshold energy for the new H-quantum production increases according to a geometrical series, the multiplicity law is expressed as

$$n_s \propto \ln E_0. \qquad (22.6)$$

This differs insignificantly from the $E_0^{\frac{1}{4}}$-law up to $E_0 = 1$ TeV, and deviates from the latter as E_0 exceeds 10 TeV.

In its original form, HASEGAWA's model may be too much idealized in many respects. One can relax a number of restrictions, and still obtain modified multi-fireball models. One such modification is that of NISHIMURA and OGATA who assigned a transverse momentum to the fireballs. In a modification proposed by FRAUTSCHI (91), it is assumed that two fireballs are emitted in the first stage, and successively each of them splits into two in a symmetrically way in the rest system of the parent fireball. In the case of four-fireball emission, one fireball has a constant Lorentz factor as small as $\gamma_H^* = 1.5$, whereas the other has a large but energy dependent Lorentz factor, $\gamma_H^* \sim \gamma_c^{\frac{3}{4}}$. Owing to this energy dependence, the inelasticity becomes independent of the primary energy.

The asymmetric emission of H-quanta forward and backward in C.M.S. was taken into account by HASEGAWA and YAZIMA (92). Then the identification of groups in the log tan ϑ plot with the fireballs becomes less clear than before, but is still possible in the majority of events. The threshold effect is smeared out in such a way that in the energy region where four H-quanta are emitted in the original model, events of two and four H-quanta emissions co-exist. Accordingly the velocities of H-quanta thus defined are distributed almost continuously over a range which is rather wide, though definitely narrower than the kinematically permissible range. This naturally results, on account of (22.5), in a continuous distribution of inelasticities. These analyses demonstrate that the models originally proposed by HASEGAWA and the modification of FRAUTSCHI correspond to the two extreme cases of the multi-fireball picture.

These considerations reveal the importance of fluctuations in the multi-fireball model. This is indeed an essential point of the fireball model, as discussed in VI.18. The effect of fluctuations has been emphasized by KOBAYASHI, NAMIKI, OHBA and ORITO (93), who studied the relation between the multiplicity and the momentum transfer, first found in the Japanese contribution to ICEF (84). According to a more exact expression of (22.6)

$$n = \frac{1}{\cosh^{-1}(1 + M_H^2/2\Delta^2)} \times \left[\ln\frac{2E_0}{M} - 2\sinh^{-1}\left(\frac{\sqrt{\Delta^2}}{2M}\right)\right], \quad (22.7)$$

the multiplicity n is a function of the momentum transfer. This relation is found to hold for ICEF events as shown in Fig. 35.

A model which takes account of fluctuations can be derived by extending the multiperipheral model proposed by AMATI et al. (94); the latter theory is known to be invalid at high energies where it predicts too small a value of the momentum transfer, and too small a value of the inelastic cross section (95). In the linked-cluster model by KOBAYASHI et al., there is a knot which connects the forward and backward links.

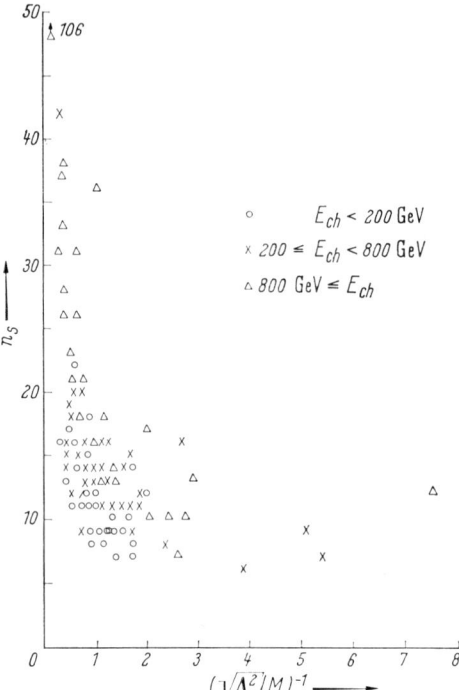

Fig. 35. Correlation between multiplicity n_s and momentum transfer Δ between the adjacent "lumps".

The assymmetry effect is accounted for by the position of the knot which may be located in any part in the whole link. The cross section is expressed as a function of the momentum transfer, so that all other quantities depend on Δ.

If fluctuations are taken into account in an appropriate way, the two-fireballs model can also reproduce most of the essential features of multiple production. One should, however, remember that the one-fireball model and the two-fireballs model give essentially the same prediction in the energy region of 100 GeV or below, and their difference becomes sufficiently clear for experimental checks only beyond ~1 TeV. In a similar way, the difference between the two-fireballs model and the multi-fireball models will not be appreciable in the energy region of ~1 TeV, and experimental checks on the difference have to be made in the energy region of 10 TeV or higher. At present experimental information on these extremely energetic events is meager, and a detailed discussion of the effects expected for multi-fireball emission in comparison with the experimental data has to be left to the future. In this respect the H-quantum hypothesis and its development described in this section will provide a useful guide for future investigations.

V. Summary of cosmic-ray information.

23. Characteristic features of high-energy interactions. The features of high-energy interactions discussed in Parts II, III and IV may be described in terms of at least two kinds of mechanisms. The first is more or less thermodynamical, and reveals itself with a characteristic energy of several hundreds of MeV, whereas the second is dynamical, and has a characteristic energy of a few GeV. We shall call them mechanisms A and B respectively. Furthermore, some indications at ultra-high energies suggest the presence of a third mechanism which may have a characteristic energy of the order of 10 GeV, and will be called mechanism C.

The experimental evidence which provides support for this point of view is summarized as follows.

Mechanism A. The transverse momentum P_T, the K/π-ratio, and the average multiplicity $\bar{n}$ are quantities characteristic of mechanism A. Their observed features are accounted for in terms of the thermal emission from a fireball of a temperature nearly equal to the pion rest energy. The decay process of the fireball into a number of mesons, that is the final state of the fireball, is determined in such a way that the fireball loses its identity as its dimension becomes comparable to the mean free path of pions in the fireball. The mean free path is a function of the multiplicity and the pion-pion interaction cross-section. The former quantity is governed by thermodynamics, and the latter by low-energy pion physics. Indeed, the observed final temperature of the fireball is consistent with that expected from the pion-pion cross-section estimated from various low-energy reactions.

Mechanism B. A remarkable feature of high-energy interactions is the rather large magnitude of the momentum transfer, $\sqrt{\Delta^2}$, which is about 2 GeV or of the order of the rest energy of a nucleon pair. Because of the success achieved by low-energy pion physics in the accelerator energy region, one might think by extrapolation that the collision would be overwhelmingly peripheral even in the high-energy region, so that the one-pion exchange process would be mainly responsible for interactions of multiple production. Some attempts have been made along this philosophy on multiple peripheral interactions (*94*). However, they predict too small a value of the momentum transfer and a cross section far smaller than that observed in these events.

A small value of momentum transfer, as expected from the exchange of a single pion, is found only for incident energies not higher than 10 GeV. This means that the reaction is mostly governed by mechanism A, and mechanism B is not in operation in this low-energy region. The magnitude of the momentum transfer increases to 1 GeV at a few tens of GeV, and remains constant, about 2 GeV, at least up to about several TeV. Incidentally, the Δ^2-distribution for 30 GeV protons is found to be essentially the same as that of cosmic-ray jets (*96*). Thus this is a quantity essentially independent of the energy over a wide energy region. Presumably this quantity is characteristic for the creation process of the fireball itself, and thus not of thermodynamical nature. Furthermore, so high a value of the momentum transfer cannot be accounted for by the exchange of a pion, but possibly by the exchange of a nucleon pair.

The average value of the multiplicity is, as stated, accounted for by thermodynamical arguments. However, the observed fluctuations of the multiplicity are of the order of $\Delta n \sim \bar{n}$, much greater than those expected from thermodynamics, $\Delta n \sim \bar{n}^{\frac{1}{2}}$.

It has been found that correlations exist between the mass of the fireball or the multiplicity, and the magnitude of the momentum transfer (*93*). Both the average value and the dispersion of the momentum transfer are found to increase gradually with the multiplicity. In other words, multiplicity fluctuations and momentum transfer derive from the same mechanism. Since the multiplicity fluctuations are greater than those of thermodynamical nature, they must be attributed to the production mechanism of fireballs, for example, to fluctuations in the mass or the number of fireballs.

Now we turn to the question of the transverse momenta of the surviving nucleon and the fireball. Although they have an important bearing on the mechanism, information from experimental data is not conclusive at present. If we suppose that there is no correlation between the pions emitted in multiple production, then a simple consideration proves that the transverse momentum of the surviving nucleon will be about $\sqrt{n}\, P_T(\pi)$. There are some indications that in the energy region up to several TeV, the transverse momentum of a nucleon is as small as that of a pion. If this is true, the transverse momenta of the pions would have to balance themselves, again in accordance with the model of fireballs formed separately from the surviving nucleons. Furthermore, the energy-momentum exchange between the nucleon and the fireball would have to take place mostly in the longitudinal component, leaving only a small fraction to the transverse component.

However, there is an indication mentioned before that the transverse momentum of the nucleon and of the fireball could be as high as 1 GeV or more. The evidence is stronger in the region of higher energies, and particularly in extensive air showers one observes some examples of transverse momenta as large as 10 GeV. Therefore, it is not unlikely that the distribution of transverse momenta has two components, one falling off very rapidly towards high transverse momenta, whereas the other has a rather long tail. The former could be thermodynamical, but not the latter, and this component may become more dominant as the energy increases.

If this is the case, the average transverse momentum would increase beyond 1 GeV with increasing energy, and the inelasticity would approach unity accordingly. Thus we suspect the existence of a third "mechanism C", of a characteristic energy of the order of 10 GeV. Although it must be left for future investigation to decide whether mechanism C really does exist, we shall keep this possibility in mind.

The situation as discussed here may be summarized in Table 11:

Table 11. *Characteristic features of high-energy interactions*

	Mechanism		
	A	B	C
Characteristic energy	2μ	$2M$	10 GeV
Incident energy (eV)	$\lesssim 10^{10}$	$10^{10} \sim 10^{14}$	$\gtrsim 10^{14}$
Interaction	$\pi\pi$	$N\overline{N}$	?
Characteristic quantities	P_T, K/π, $\bar{n}$	Δ, K, Δn	large P_T
Elastic scattering			
angular distribution	steep fall	slow decrease	
energy dependence	weak	strong	

In terms of the fireball model, the difference between the various mechanisms can be expressed as follows: mechanism A is revealed in the decay process of a fireball and is thermodynamical in nature, whereas mechanism B takes part in the formation of a fireball and is dynamical in nature. They may also be interpreted in terms of the composite model (97) of elementary particles. A pair of a nucleon and an antinucleon which form a pion behave in different ways depending on the characteristic time of the process concerned. Mechanism B appears in the collision process where the impact time is so short that the nucleon and the antinucleon act in an incoherent way, whereas only mechanism A plays a role in the decay process because the characteristic time is so long that they act in a coherent way.

One may also express the same situation in terms of the core-cloud model of nucleon scattering. In mechanism A the hard core of the nuclear force is inert, and only the pion cloud participates in the reaction. On the other hand, in mechanism B the core is resolved, so that it is no longer a hard core which repels particles or waves but plays a major part in the inelastic processes.

Whatever interpretation of mechanisms A and B may be chosen, it is manifest that these two mechanisms co-exist in a wide range of energies contrary to the philosophy fashionable some years ago that even in multiple production at high energies mechanism A is so dominant that no essential information can be gained from cosmic-ray studies which is not available also from low-energy pion physics.

24. Nucleon elastic scattering. The elastic scattering of nucleons is closely related to the mechanism of high-energy interactions. It is of particular interest to study this phenomenon in the accelelator energy region, with reference to the mechanisms A and B discussed in the previous section (98).

In the energy region of several GeV in which we are interested elastic scattering at small angles in a nucleon-nucleon collision is mainly due to diffraction, and the cross-section is found to vary as a Gaussian function of the angle ϑ,

$$\frac{d\sigma_{el}}{d\Omega} \propto \exp(-\vartheta^2/\gamma^2), \qquad (24.1)$$

where γ is the width of the diffraction peak. Its energy dependence is given by

$$\gamma = 1/kR. \qquad (24.2)$$

Here k is the momentum in C.M.S., and R may be interpreted, as the effective range of the interaction which is found to depend weakly on the energy. $1/\gamma$ is regarded as the angular momentum effective to the diffraction scattering. The diffraction angular distribution (24.1) results from the scattering amplitude of the l-th partial wave

$$\eta_l = \eta_0 \exp(-\gamma^2 l^2/2), \qquad (24.3)$$

Here η_0 represents the forward amplitude. Thus the elastic, inelastic and total cross-sections are expressed as

$$\sigma_{el} = \pi |\eta_0|^2/k^2\gamma^2 = \pi |\eta_0|^2 R^2, \qquad (24.4)$$

$$\sigma_{in} = \pi \frac{4 \operatorname{Im} \eta_0 - |\eta|^2}{k^2 \gamma^2} = \pi (4 \operatorname{Im} \eta_0 - |\eta_0|^2) R^2, \qquad (24.5)$$

$$\sigma_{tot} = 4\pi \operatorname{Im} \eta_0/k^2\gamma^2 = 4\pi \operatorname{Im} \eta_0 R^2. \qquad (24.6)$$

The experimental values of these cross-sections for $5 \sim 30$ GeV are approximately 10 mb, 30 mb, and 40 mb, respectively. This would suggest

$$\eta_0 = i, \qquad R = 0.57 \times 10^{-13} \text{ cm}, \qquad (24.7)$$

and indicate that the interaction responsible for the nucleon-nucleon collision in the energy region between 5 and 30 GeV is mainly absorptive, and the real amplitude relative to the imaginary one is not greater than $\frac{1}{5}$. Expressed in terms of a potential it states that the nucleon-nucleon potential is strongly absorptive with the effective radius R.

This information is quite different from that obtained in low-energy scattering experiments, which suggests the existence of a hard core near the origin. One may interpret these two pieces of information, for example, by assuming that the core in the nuclear potential is energy dependent; the core is strongly repulsive at low energies but turns itself into an absorptive core as the energy increases, somewhat similar to the variation observed in nucleus-nucleus scattering (99).

It is interesting to look for a possible variation of the effective interaction range R with the energy. The shrinkage of the diffraction peak observed in $p-p$ scattering suggests an increase of R with the energy. On the other hand, the magnitudes of the cross-section are found essentially constant in the energy range where the shrinkage is found. This would indicate a decrease of $\mathrm{Im}\,\eta_0$ with increasing energy, whereas $\mathrm{Re}\,\eta_0$ may increase to reach a finite value. If the real part is still small, the slow increase of R with energy leads to an increase of the inelastic cross-section relative to the elastic one. These trends could provide a solution consistent with the experiments.

(24.1) represents a Gaussian distribution of the transverse momenta, with the average momentum of

$$P_T = \frac{\sqrt{\pi}}{2} \frac{1}{R} \sim 310 \text{ MeV/c}. \tag{24.8}$$

This is not inconsistent with the observed value. However, the experimental results seem to be better expressed by an exponential distribution, especially for large P_T (100).

The superposition of a Gaussian and an exponential distribution is a common feature expected from the fluctuation-dissipation theorem and the theory of stochastic processes (101). Thus one may expect that the P_T distribution is Gaussian at small P_T and exponential at large P_T. The average value of P_T is related to the multiplicity n by

$$\langle P_T \rangle \sim \sqrt{n}\, P_T^0 \tag{24.9}$$

P_T^0 being the transverse momentum in interactions with a pion.

This general theory, however, predicts a decrease of the effective range R with increasing multiplicity, which is apparently in contradiction to the shrinkage of the diffraction peak of $p-p$ scattering, and to the energy independent diffraction pattern of $\pi-p$ scattering. Therefore it is suggested that strong correlation exists between the secondary particles, and the simple random walk argument cannot be applied here. This is what we discussed in V.23, namely, that pions are not produced in a random way but themselves form a strongly correlated system. This system is called a fireball, and plays an essential part in high-energy interactions.

The production process of a fireball is associated with a momentum transfer larger than that responsible for the diffraction scattering at small angles. The angular distribution of nucleon elastic scattering tends to become flat as the scattering angle increases beyond the diffraction region. The large-angle elastic cross-section was found to be highly energy dependent (102). In the flat portion the large angle elastic scattering cross-section may be expressed as (103)

$$\frac{d\sigma}{d\Omega} \approx \frac{\sigma_c}{2\pi N}. \tag{24.10}$$

This formula is based on the assumption that the nucleon collision forms a compound system with cross-section σ_c, and subsequently decays through N channels, one of which is the elastic channel. The number of possible channels is determined by the entropy S and the temperature T of the compound system:

$$\left. \begin{array}{c} N = \exp(S), \\ \dfrac{dS}{dw} = \dfrac{1}{T} \end{array} \right\} \quad (24.11)$$

the dependence of the entropy S on the available energy w is given by

$$S = aw, \quad (24.12)$$

which implies energy independence of the temperature. The choice of $1/a \approx 0.3$ GeV and $\sigma_c \sim 1$ mb gives a reasonable fit to the elastic cross-section, offering one of the possible interpretations of the phenomena.

25. Cosmic-ray neutrino experiments. In considering future prospects, investigations on neutrino interactions at high energies are as important as the studies of the strong interactions so far discussed. As is well known, neutrino interactions are very difficult to observe, because the cross-section is by far smaller than that of strong interactions. For illustration, let us estimate the frequency of cosmic-ray neutrino interactions expected to be observed with a conventional method.

Let the flux intensity of neutrinos be J_ν, and the cross-section per nucleon for the interaction be σ_ν. Suppose one has a huge scintillator weighing W, set up so that all interactions taking place in it are detectable. Then the counting rate of neutrino interactions is approximately

$$C = J_\nu \sigma_\nu N W, \quad (25.1)$$

where N is number of nucleons in a gram. For neutrinos of several GeV, we expect roughly

$$J_\nu \sim 10^{-2} \text{ cm}^{-2} \text{ sec}^{-1},$$
$$\sigma_\nu \sim 10^{-38} \text{ cm}^2.$$

In order to get one count per month, the weight of the detector W has to be

$$W \sim 5 \times 10^3 \text{ tons}.$$

This neutrino counting rate may be compared with a hypothetical mass annihilation rate. As an upper limit, suppose that any particle annihilates and transforms itself into energy with a mean lifetime equal to the cosmic age τ. Then the annihilation rate would be NW/τ. One knows that $1/\tau \approx 3 \times 10^{-18}$ sec^{-1} is much larger than $J_\nu \sigma_\nu \approx 10^{-40}$ sec^{-1}, so that the neutrino interactions would be completely masked by the hypothetical mass annihilation. Although such a mass annihilation process is not at all acceptable, the discussion above demonstrates not only that a detector for neutrino interactions should be huge, but also that any small background has to be eliminated.

Experimental sites with small background can be found when one goes deep underground. In fact, at the depth of 8400 meter water equivalent in the Kolar Gold Fields in India, no count was detected with a large scintillation detector during two months, as described in II.14 (*54*). This result has given us an optimistic view that neutrino experiments with cosmic rays are feasible at such great depth (*104*). An important source of harmful background is γ-radiation from radioactivity in the earth. This may be avoided by using a coincidence method, and at the same time discriminating scintillation pulse heights against low pulses which are produced by γ-rays. The effects of muons which may still remain can be reduced by employing a horizontal telescope, in which the intensity of muons will decrease by a large amount due to the great thickness of earth traversed, whereas the intensity of neutrinos will increase due to the increasing decay probability. It might be worth while to observe neutrinos coming upwards through the earth, because neutrinos penetrate through the whole earth without appreciable absorption, whereas muons are perfectly absorbed.

The intensity of neutrinos can be estimated from the decay of pions and muons,

and
$$\pi^{\pm} \to \mu^{\pm} + \nu_\mu (\bar{\nu}_\mu) \tag{25.2}$$

$$\mu^{\pm} \to e^{\pm} + \nu_e (\bar{\nu}_e) + \bar{\nu}_\mu (\nu_\mu). \tag{25.3}$$

Here we distinguish the two kinds of neutrinos, one being associated with muons and the other with electrons, by the symbols ν_μ and ν_e.[1] The energy spectra of vertical neutrinos from pions and muons are expressed respectively by (105),

$$J_{\pi\nu}(E_\nu)\, dE_\nu = \begin{cases} 1.85 \times 10^{-2} (0.08 + E_\nu)^{-2.80}, & 1 \leq E_\nu \leq 10 \text{ GeV}, \\ 6.65 \times 10^{-2} (1.1 + E_\nu)^{-3.22}, & 10 \leq E_\nu \leq 300 \text{ GeV}, \end{cases} \tag{25.4}$$

$$J_{\mu\nu}(E_\nu)\, dE_\nu = \begin{cases} 7.65 \times 10^{-2} (0.37 + E_\nu)^{-3.75}, & 1 \leq E_\nu \leq 10 \text{ GeV}, \\ 1.48 (3.5 + E_\nu)^{-4.51}, & 10 \leq E_\nu \leq 100 \text{ GeV}. \end{cases} \tag{25.5}$$

It should be noted that the latter is steeper than the former, because the probability of muon decay is appreciable only below 10 GeV. For the same reason the intensity increases with the zenith angle more rapidly for the neutrinos from muons than for those from pions.

In Fig. 36 we represent the energy spectra of $\nu_\mu (\bar{\nu}_\mu)$ and $\nu_e (\bar{\nu}_e)$ according to (25.2)—(25.5), separately in the vertical and horizontal directions. The angular distributions are given for the two kinds of neutrinos at three representative energies in the same figure. Adding the neutrinos from kaons, we find a large intensity of neutrinos, about one tenth of the intensity of primary cosmic rays at the same energy.

The detection of ν_μ is possible by observing muons produced in the processes

$$\nu_\mu + n \to p + \mu^-, \tag{25.6a}$$

$$\bar{\nu}_\mu + p \to n + \mu^+. \tag{25.6b}$$

The cross-sections for these processes, calculated under the assumption of point interaction, are

$$\sigma_\nu \approx 1.5 \times 10^{-38} E_\nu \text{ cm}^2, \tag{25.7a}$$

$$\sigma_{\bar{\nu}} \approx 0.5 \times 10^{-38} E_{\bar{\nu}} \text{ cm}^2 \tag{25.7b}$$

with E_ν in GeV. This may be regarded as maximum estimates, because the linear increase with energy is suppressed if the interaction is subject to some form factor (106).

If we adopt (25.7), the effective thickness of the target is approximately proportional to the neutrino energy, since the energy of a muon produced is

---

[1] The possible existence of two kinds of neutrinos was suggested, in connection with the two-meson hypothesis, by SAKATA, INOUE and TANIKAWA [see ref. (11)]. The experimental evidence for this was presented by G. DANBY, J. M. GAILLARD, K. GOULIANOS, L. M. LEDERMAN, N. MISTRY, M. SCHWARTZ and J. STEINBERGER: Phys. Rev. Letters 9, 36 (1962).

nearly equal to E_ν, and the target thickness is determined by the range of the muons. If we observe horizontal muons with a detector of an effective area of 10 m², and take into account the contribution of kaons which could be greater than

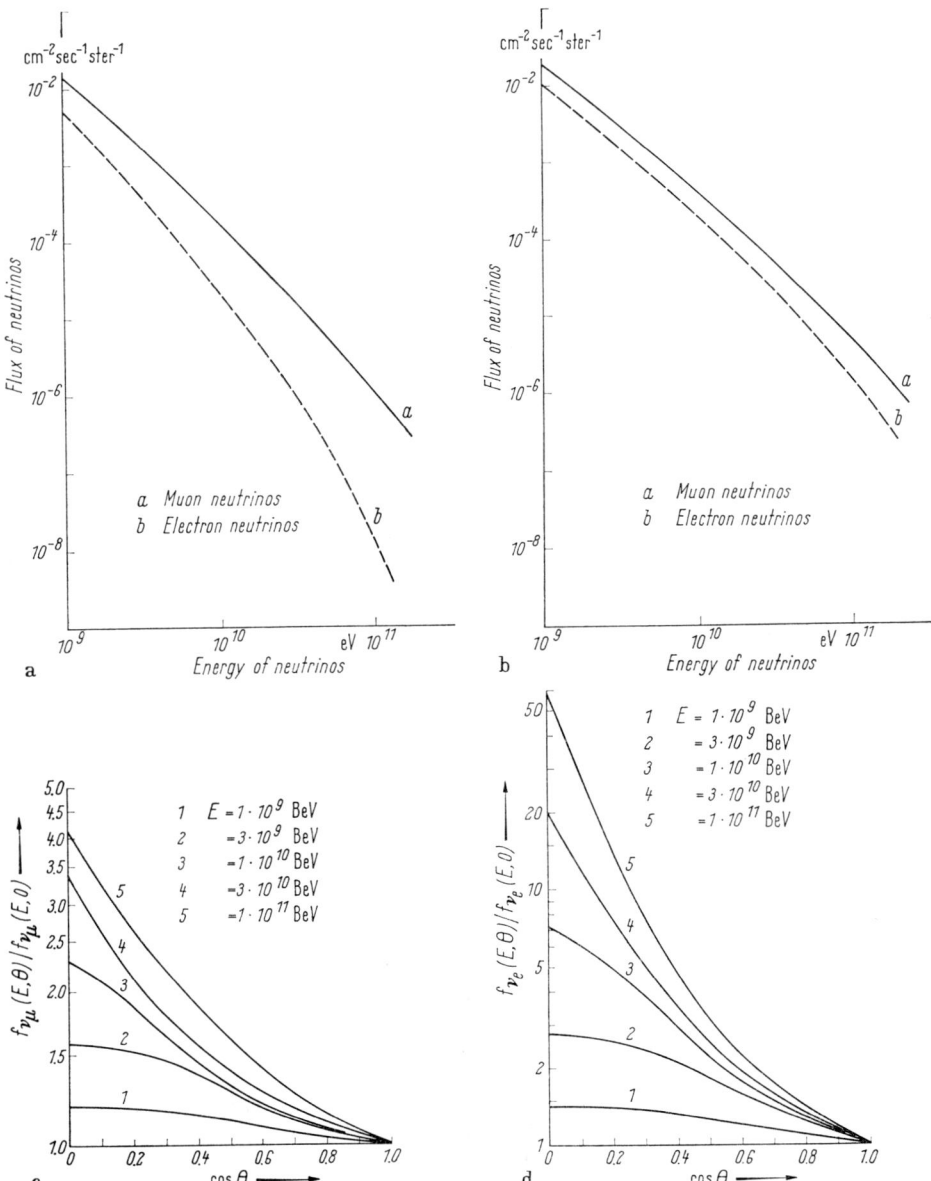

Fig. 36a—d. Intensities of neutrinos. (a) Integral energy spectra of vertical neutrinos. (b) Integral energy spectra of horizontal neutrinos. (c) Zenith angle distributions of muon neutrinos. (d) Zenith angle distributions of electron neutrinos.

that of pions, a detailed calculation gives us the following counting rate:

$$C(\nu_\mu) \sim 0.3 \text{ month}^{-1}. \tag{25.8}$$

If the linear increase of the cross-section is suppressed, the counting rate would be smaller.

Neutrinos of the other kind, ν_e, produce interactions

$$\nu_e + n \to p + e, \tag{25.9a}$$

$$\bar{\nu}_e + p \to n + e^+. \tag{25.9b}$$

If self-current interactions do exist, there would also be electrons from neutrino-electron collisions. But this component is expected to be much smaller than that of (25.9), because the energy available in C.M.S. is small in a $\nu - e$ collision. The cross-sections for (25.9a) and (25.9b) are practically equal to (25.7a) and (25.7b) in the relativistic limit. But the detection of those electrons will be more difficult, because the electrons dissipate their energy quickly by cascade processes, and their effective range is shorter than that of muons. Thus the electron intensity is expected to be smaller than that of the muons.

The physical implications of a neutrino experiment may be summarized as follows (107). Since the neutrino beams available in existing accelerators have energies not exceeding 10 GeV, an advantage of using cosmic rays lies in the fact that we are able to study neutrino interactions at higher energies. Particular emphasis must, therefore, be paid to a number of characteristic features which could reveal themselves only at energies higher than several GeV:

(i) The energy dependence of the neutrino cross-section can be measured. In particular, it will be interesting to find whether the cross-section continues to increase with energy as expressed in (25.7) or tends to level off forwards a constant value.

(ii) Information on the intermediate boson will be obtained. The idea of the intermediate vector boson, motivated by the conserved vector current in the universal weak interaction, has been considered seriously as a revival of the original idea of YUKAWA, according to which the interaction responsible for β-decay is mediated by a meson interacting with both the nucleon and the lepton. The production of muons is then associated with the creation of a boson as well as with its decay. The intensity of muons thus expected is comparable to that based on the energy-independent cross section in (i), provided that the mass of the intermediate boson is equal to the nucleon mass.

(iii) If the intermediate boson mediates an interaction between leptons, muons may be produced through $\bar{\nu} + e^- \to W \to \mu^- + \bar{\nu}$. Since this is a resonance reaction, the intensity of muons expected would be considerable. The present experiment at the deepest station (54) rules out the existence of such a boson that has a mass smaller than twice the nucleon mass.

Nearly the same mass limit of the intermediate boson has been obtained also with accelerator experiments (108), on the basis of the absence of a muon pair expected from

$$\bar{\nu}_\mu + A \to W^- + \mu^+ + A' \to \bar{\nu}_\mu + \mu^- + \mu^+ + A',$$

where W^- represent the negative intermediate boson. By expanding cosmic ray experiments at great depths, it does not seem to be impossible to push the mass limit to several times the nucleon mass within a few years. At the same time question (i) will also be answered.

26. Future prospects. By clarifying the process of fireball emission in multiple production, cosmic-ray investigations have proven the existence of mechanism B in strong interactions. It is interesting to notice that indications of mechanism B are beginning to be observable also in recent accelerator experiments with a proton beam of energy 10 GeV or more. The process which appears directly connected to mechanism B is the large-angle elastic scattering of protons by protons, but it has a cross-section smaller by several orders of magnitude than the total cross-section. The experiment is possible only with the use of intensive, monenergetic and collimated beams of accelerated protons, for otherwise a large-angle elastic scattering would not be observed.

If, however, giant accelerators will come into being, the situation will be greatly different. As soon as beams of protons with energy of ~ 100 GeV will be available, it will become possible to study the full details of mechanism B in a quantitative way. In the cosmic ray experiments, one-fireball production is found dominant at this energy range, so that one can hardly hope to investigate the features of two- and multi-fireball emission. Accelerators, however, will make it comparatively easy to study such rare events. If the beam energy can be increased to 1 TeV, one expects that even studies on mechanism C would become feasible. It is the purpose of the considerations offered here to suggest, from the experience of cosmic-ray investigations, guide lines for future accelerator experiments.

Though one can reasonably expect present and future accelerators to reveal details of mechanism B, questions of mechanism C will remain, for the time being, mostly the subject of cosmic-ray experiments. This is because mechanism C would play a main part in interactions of 10^{15} eV or more, but would occupy only a minor part in interactions of about 1 TeV. To investigate its effects it will, therefore, be necessary either to perform experiments on ~ 1 TeV interactions with good accuracy and large statistics, or direct observations on interactions of $\sim 10^{14-15}$ eV. Since the energy spectrum of cosmic rays decreases steeply with increasing energy, experiments of both kinds will have difficulties of the same order of magnitude.

In practice, it may be that the selection of few exceptional events from a large number of lower-energy interactions is more difficult than to observe common features from a smaller number of higher-energy events. The selection will be feasible if the fraction is greater than, say, 10^{-3}. Thus, mechanism C may be detectable preferably by accurate experiments at several TeV, if the characteristic energy of mechanism C is a few GeV. If, however, the characteristic energy is about 10 GeV or more, observations on events with higher energies will be more suitable. As far as present cosmic-ray evidence can be relied upon, the latter value appears more plausible. Observations on events of energies greater than 100 TeV are, therefore, of particular importance for the progress of our knowledge of the structure of elementary particles.

References

General references.

[1] ROZENTAL, I. L., i D. S. CHERNAVSKY: Uspekhi Fiz. Nauk 52, 185 (1954). — Fortschr. Phys. 4, 560 (1956). — A critical review of theoretical and experimental aspects of high-energy interactions in favour of LANDAU's hydrodynamical model.
[2] KOBA, Z., u. S. TAKAGI: Fortschr. Phys. 7, 1 (1959). — A review of various theoretical attemps on a standpoint as unbiased as possible.
[3] POWELL, C. F., P. H. FOWLER, and D. H. PERKINS: The Study of Elementary Particles by the Photographic Method. London: Pergamon 1959.
[4] PERKINS, D. H.: Progress in Elementary Particle and Cosmic Ray Physics, vol. V, p. 257. Amsterdam: North-Holland Publ. Co. 1960. — A summary of experimental information on jets.
[5] SITTE, K.: Encyclopedia of Physics, vol. XLVI/1, p. 157. Berlin-Göttingen-Heidelberg: Springer 1961. — A review of experimental and theoretical information on high-energy nuclear interactions.
[6] COCCONI, G.: Encyclopedia of Physics, vol. XLVI/1, p. 215. Berlin-Göttingen-Heidelberg: Springer 1961. — Extensive air showers.
[7] FOWLER, G. N., and A. W. WOLFENDALE: Encyclopedia of Physics, vol. XLVI/1, p. 272. Berlin-Göttingen-Heidelberg: Springer 1961. — Behaviour of muons in cosmic rays.

[8] FOWLER, G. N., and A. W. WOLFENDALE: Progress in Elementary Particle and Cosmic Ray Physics, vol. IV, p. 105. Amsterdam: North-Holland Publ. Co. 1958. — Interactions of muons with matter.
[9] HAYAKAWA, S.: Theoretical Physics, p. 485, Vienna, IAEA (1963). — Fireball models.
[10] FEINBERG, E. L., i. D. S. CHERNAVSKY: Uspekhi Fiz. Nauk 82, 3 (1964). — A theoretical review of high-energy nuclear interactions.
[11] J. Phys. Soc. Japan 17, Suppl. A-III-4, 5, 6, 7 (1962). Proceedings of the International Conference on Cosmic Ray and Earth Storms, Kyoto, 1961.
[12] Proceedings on 1963 Internat. Conference on Cosmic Rays, Jaipur, India.
[13] Nuovo Cim., Suppl. 1 (1963). — Contributions of the International Cooperative Emulsion Flight (ICEF).

Literature cited.

(1) HEISENBERG, W.: Z. Physik **101**, 533 (1936).
(2) HEISENBERG, W.: Z. Physik **113**, 61 (1939).
(3) HAMILTON, J., W. HEITLER, and H. W. PENG: Phys. Rev. **64**, 78 (1943). — HEITLER, W., and P. WALSH: Rev. Mod. Phys. **17**, 252 (1945).
(4) JÁNOSSY, L.: Phys. Rev. **64**, 345 (1943).
(5) HEITLER, W., and L. JÁNOSSY: Proc. Roy. Soc. Lond. A **62**, 374 (1949).
(6) MIYAZIMA, T., and S. TOMONAGA: Sci. Pap. I.P.C.R. **40**, 21 (1942).
(7) LEWIS, H. W., J. R. OPPENHEIMER, and S. A. WOUTHUYSEN: Phys. Rev. **73**, 127 (1948).
(8) SAKATA, S., and Y. TANIKAWA: Phys. Rev. **57**, 548 (1940).
(9) TAKETANI, M.: Reports of the Meson Symposium 1943. — Progr. Theor. Phys. **3**, 349 (1948).
(10) HAYAKAWA, S., and S. TOMONAGA: J. Sci. Res. Inst. **43**, 67 (1948).
(11) SAKATA, S., and Y. TANIKAWA: Reports of the Meson Symposium 1943. — SAKATA, S., and I. INOUE: Progr. Theor. Phys. **1**, 134 (1946). — TANIKAWA, Y.: Progr. Theor. Phys. **2**, 220 (1947).
(12) LATTES, C. M. G., G. P. S. OCCHIALINI, and C. F. POWELL: Nature, Lond. **160**, 543, 486 (1947).
(13) For details on this technique, see for instance ref. [3].
(14) LEPRINCE-RINGUET, L., T. F. HOANG, F. BOURSSER, L. JANNEAU, and D. MORELLET: Phys. Rev. **76**, 1273 (1949). — C. R. Acad. Sci. Paris **229**, 163 (1949).
(15) BRADT, H. L., M. F. KAPLON, and B. PETERS: Phys. Rev. **76**, 1735 (1949).
(16) LORD, J. J., J. FAINBERG, and M. SCHEIN: Phys. Rev. **80**, 970 (1950).
(17) HEITLER, W., and C. TERREAUX: Helv. Phys. Acta **24**, 551 (1951). — Proc. Phys. Soc. Lond. A **66**, 929 (1953).
(18) MESSEL, H., R. B. POTTS, and C. B. A. McCUSKER: Phil. Mag. **43**, 889 (1952).
(19) KUSUMOTO, C., S. MIYAKE, K. SUGA, and Y. WATASE: Phys. Rev. **90**, 998 (1953).
(20) FOWLER, W. B., R. P. SHUTT, A. M. THORNDYKE, and W. L. WHITTENMORE: Phys. Rev. **90**, 758 (1953); **95**, 1026 (1954).
(21) ZATSEPIN, G. T.: Dokl. Akad. Nauk. USSR **67**, 993 (1949).
(22) FUJIMOTO, Y., S. HAYAKAWA, and Y. YAMAGUCHI: Progr. Theor. Phys. **5**, 197 (1950).
(23) ROZENTAL, I. L., i D. S. CHERNAVSKY: Uspekhi Fiz. Nauk **52**, 185 (1954).
(24) KOSHIBA, M., and M. F. KAPLON: Phys. Rev. **97**, 193 (1955); LOHRMANN, E., M. W. TEUCHER, and M. SCHEIN: Phys. Rev. **122**, 672 (1961). — See also ref. (16).
(25) HANSEN, L. F., and W. B. FRETTER: Phys. Rev. **118**, 812 (1960).
(26) CHIKOVANI, G. E., V. A. MIKHAILOV, and V. N. ROINISHVILI: Phys. Letters **6**, 254 (1963).
(27) The conversion length was measured for γ-rays of several hundreds of GeV by K. PINKAU, Phil. Mag. **2**, 1389 (1957); see also M. KOSHIBA, ref. [13]. — The result of 3.4 ± 0.5 cm agrees well with the theoretical calculation.
(28) BRISBOUT, F. A., C. DAHANAYAKE, A. ENGLER, Y. FUJIMOTO, and D. H. PERKINS: Phil. Mag. **1**, 605 (1956).
(29) EDWARDS, B., J. LOSTY, K. PINKAU, D. H. PERKINS, and J. REYNOLDS: Phil. Mag. **3**, 237 (1958).
(30) LOHRMANN, E., and M. W. TEUCHER: Phys. Rev. **112**, 587 (1958). — See also ref. [13].
(31) KOBA, Z.: Progr. Theor. Phys. **15**, 461 (1956). — See also discussion in IV.7.
(32) COOL, R. C.: Proc. 1961 Intern. Conf. on High-Energy Accelerators, Brookhaven National Laboratory, p. 15 (1961).
(33) KOSHIBA, M.: I. C. E. F. results as reported in ref. [13].
(34) KIM, C. O.: Phys. Rev. **136** B, 515 (1964).
(35) CHUDAKOV, A. E.: Izv. Akad. Nauk. USSR **19**, 650 (1955).

(36) MINAKAWA, O., Y. NISHIMURA, M. TSUZUKI, H. YAMANOUICHI, H. AIZU, H. HASEGAWA, Y. ISHII, S. TOKUNAGA, Y. FUJIMOTO, S. HASEGAWA, J. NISHIMURA, K. NIU, K. NISHIKAWA, K. IMAEDA and M. KAZUNO: Nuovo Cim., Suppl. **8**, 761 (1958).
(37) A. BIGI, S. BRANDT, R. CARRARA, W. A. COOPER, A. DE MACOR, G. R. MCLEOD, CH. PEYROU, R. SOSNOWSKI, and A. WRABLEWSKI: Proc. 1962 Intern. Conf. on High-Energy Physics, CERN, p. 247 (1962).
(38) DAKE, S., K. KAWAMURA, I. OTA, and K. YOKOI: Private communication.
(39) HASEGAWA, S., and N. YAJIMA: Private communication.
(40) NISHIMURA, J.: Soryuchiron Kenkyu **12**, 24 (1956).
(41) MIURA, I., and Y. TANAKA: Proc. 1962 Intern. Conf. on High-Energy Physics, CERN 1962.
(42) MIYAKE, S., K. HINOTANI, K. KANEKO, and N. ITO: J. Phys. Soc. Japan **18**, 592 (1963).
(43) OREAR, J.: Phys. Rev. Letters **12**, 112 (1964).
(44) COCCONI, G., L. J. KOESTER, and D. H. PERKINS: UCRL 10022, p. 167.
(45) CASTAGNOLI, C., M. CORTINI, C. FRANZINETTI and L. MORERO: Nuovo Cim. **10**, 1537 (1953).
(46) GUSEWA, V. V., N. A. DOBROTIN, N. G. ZELEVINSKAYA, K. A. KOSTELNIKOV, A. M. LEBEDEV, and S. A. SLAVATINSKY: J. Phys. Soc. Japan **17**, Suppl. III-A, 375 (1962). — The "calorimeter method" is an extension of the technique first employed by F. E. FROEHLICH, E. M. HARTH, and K. SITTE, Phys. Rev. **87**, 504 (1952).
(47) KOSHIBA, M.: Intern. Conf. on Cosmic Rays, Jaipur 1963.
(48) Early results are summarized by E. P. GEORGE, Progr. in Cosmic Ray Physics I, p. 393 (1952), and ref. [8].
(49) WATASE, Y., S. HIGASHI, T. KITAMURA, Y. MITANI, S. MIYAMOTO, T. OSHIO, H. SHIBATA, and K. WATANABE, Proc. Intern. Conf. on Cosmic Rays and Earth Storms, Kyoto, III-A, 362 (1962).
(50) HEITLER, W.: Quantum Theory of Radiation, p. 265. Oxford: at the Clarendon Press 1954.
(51) MIYAZAKI, Y.: Phys. Rev. **76**, 1723 (1949).
(52) GREISEN, K. I.: Phys. Rev. **73**, 521 (1948).
(53) HAYAKAWA, S.: Progr. Theor. Phys. **3**, 199 (1948).
(54) MIYAKE, S., V. S. NARASINHAN, and P. V. RAMANA MURTHY: Proc. Intern. Conf. on Cosmic Rays, Jaipur 1963; ref. [12].
(55) FUJIMOTO, Y., S. HASEGAWA, M. KAZUNO, N. OGITA, J. NISHIMURA, and K. NIU: Proc. Intern. Conf. on Cosmic Rays, Moscow, I, 41 (1960).
(56) HAYAKAWA, S., J. NISHIMURA, and Y. YAMAMOTO: Prog. Theor. Phys. Suppl., in press.
(57) HAYMAN, P. J., N. S. PALMER, and A. W. WOLFENDALE: Proc. Roy. Soc. Lond. A **275**, 391 (1963).
(58) GIJSBERS, K., P. J. HAYMAN, Y. KAMIYA, and N. S. PALMER: Intern. Conf. on Cosmic Rays, Jaipur 1963; ref. [12].
(59) KRASILNIKOV, D. D.: Intern. Conf. on Cosmic Rays, Jaipur 1963; ref. [12].
(60) HIGASHI, S., T. KITAMURA, M. ODA, Y. TANAKA and Y. WATASE: Nuovo Chim. **32**, 1 (1964).
(61) RAMANA MURTHY, P. V.: Intern. Conf. on Cosmic Rays, Jaipur 1963; ref. [12].
(62) MIYAZAKI, Y.: Phys. Rev. **76**, 1733 (1949).
(63) BARTON, J. C.: Phil. Mag. **6**, 1271 (1961).
(64) BARRETT, P. H., L. M. BOLLINGER, G. COCCONI, Y. EISENBERG, and K. GREISEN: Rev. Mod. Phys. **24**, 133 (1952).
(65) J. NISHIMURA: Proc. 1963 Internat. Conference on Cosmic Rays, Jaipur, India, Vol.
(66) For details on the fluctuations near the shower core, see S. FUKUI, H. HASEGAWA, T. MATANO, I. MIURA, M. ODA, K. SUGA, G. TANAHASHI, and Y. TANAKA: Progr. Theor. Phys., Suppl. **16**, 1 (1961); and S. MIYAKE, ref. [11], p. 291.
(67) UEDA, A., and N. OGITA: Progr. Theor. Phys. **18**, 269 (1957). — OGITA, N.: Progr. Theor. Phys. **27**, 105 (1962).
(68) MIYAKE, S.: Progr. Theor. Phys. **20**, 844 (1958).
(69) HASEGAWA, H.: Progr. Theoret. Phys. Suppl. No. 30 (1964).
(70) HEISENBERG, W.: Z. Physik **133**, 65 (1952).
(71) FERMI, E.: Progr. Theor. Phys. **5**, 570 (1950).
(72) FERMI, E.: Phys. Rev. **81**, 683 (1951); **92**, 452 (1953); **93**, 1434 (1954).
(73) TAKAGI, S.: Progr. Theor. Phys. **7**, 123 (1952).
(74) KRAUSHAAR, W. L., and J. L. MARKS: Phys. Rev. **93**, 326 (1954).
(75) LANDAU, L. D.: Izv. Akad. Nauk. USSR **17**, 57 (1953).
(76) KOBA, Z.: Progr. Theor. Phys. **15**, 461 (1956).
(77) BELENSKIJ, S. Z., i. L. D. LANDAU: Uspekhi Fiz. Nauk USSR **56**, 309 (1955).
(78) AMAI, S., H. FUKUDA, C. ISO, and M. SATO: Progr. Theor. Phys. **17**, 241 (1957).
(79) NAMIKI, M., and C. ISO: Progr. Theor. Phys. **18**, 591 (1957).

(80) KANEKO, S., and S. OKAZAKI: Nuovo Cim. **8**, 521 (1958).
(81) SALZMANN, F., and G. SALZMANN: Phys. Rev. **125**, 1703 (1962).
(82) NIU, K.: Nuovo Cim. **10**, 944 (1958). — Similar ideas were proposed independently by P. CIOK, T. COGHEN, J. GIERULA, R. HOLYNSKI, A. JURAK, M. MIESOWITZ, T. SANIEWSKA and O. STANISZ: Nuovo Cim. **8**, 166 (1958); **10**, 741 (1958); and by G. COCCONI: Phys. Rev. **111**, 1699 (1958).
(83) HASEGAWA, S., and K. YOKOI: Proc. Intern. Conf. on High-Energy Physics, CERN 1962.
(84) FUJIOKA, G., Y. MAEDA, O. MINAKAWA, M. MIYAZAKI, I. MITO, K. KOBAYAKAWA, H. SHIMOIDA, N. NAKATANI, O. KUSUMOTO, K. NIU and K. NISHIKAWA: ref. [*13*].
(85) GIERULA, J.: Proc. Intern. Conf. on High-Energy Physics, Rochester, p. 816 (1960).
(86) PETERS, B.: Nuovo Cim. **23**, 88 (1962). — See, also, ref. [*11*] and [*12*].
(87) COCCONI, G.: Proc. Intern. Conf. Sienna 1963. — G. DAMGAARD and K. HANSEN, loc. cit.
(88) Japanese-Brazilian Collaboration Project. Intern. Conf. on Cosmic Rays, Jaipur 1963; ref. [*12*]. p. 326.
(89) HASEGAWA, S.: Progr. Theor. Phys. **26**, 150 (1961); **29**, 128 (1963).
(90) NISHIMURA, J.: Private communication; and OGATA, T., Progr. Theor. Phys. **30**, 924 (1963).
(91) FRAUTSCHI, S.: Nuovo Cim. **28**, 409 (1963).
(92) HASEGAWA, S., and N. YAZIMA: Private communication.
(93) KOBAYASHI, T., M. NAMIKI, I. OHBA, and S. ORITO: Progr. Theor. Phys. **31**, 840 (1964); **32**, 738 (1964).
(94) AMATI, D., S. FUBINI and A. STANGHELINI: Nuovo Cim. **26**, 896 (1962).
(95) KOBA, Z., and A. KRYZWICKI: Nuclear Phys. **46**, 471 and 485 (1963).
(96) TERANAKA, K.: Thesis, Osaka University 1962.
(97) SAKATA, S.: Progr. Theor. Phys. **16**, 686 (1956).
(98) HOVE, L. VAN: Nuovo Cim. **28**, 798 (1963). — YAMAGUCHI, Y.: Proc. Stanford Conf. 1963.
(99) OTSUKI, S., R. TAMAGAKI, and M. WADA: Progr. Theor. Phys. **32**, 220 (1964).
(100) OREAR, S.: Phys. Rev. Letters **12**, 112 (1964).
(101) NAMIKI, M.: Private communication.
(102) COCCONI, G., V. T. COCCONI, A. D. KRISH, J. OREAR, R. RUBINSTEIN, B. D. SEARL, W. F. BAKER, E. W. JENKINS, and A. L. READ: Phys. Rev. Letters **11**, 499 (1963); **12**, 132 (1964).
(103) COCCONI, G.: Nuovo Cim. **33**, 643 (1964).
(104) MENON, M. G. K., P. V. RAMANA MURTHY, B. V. SREEKANTAN, and S. MIYAKE: Phys. Letters **5**, 272 (1963). — Intern. Conf. on Cosmic Rays, Jaipur 1963; ref. [*12*].
(105) ZATSEPIN, G. T., i. V. L. KUZMIN: Soviet Phys. JEPT **14**, 1294 (1962).
(106) YAMAGUCHI, Y.: Progr. Theor. Phys. **23**, 1117 (1960).
(107) On High-Energy Neutrino Physics, D-577, Dubna 1960; — MARKOV, M. A., and I. M. ZHELEZNYKH: Nuclear Phys. **27**, 385 (1961). — MENON, M. G. K.: Intern. Conf. on Cosmic Rays, Jaipur 1963; ref. [*12*].
(108) BLOCK, M. M., H. BURMEISTER, D. C. CUNDY, B. EIBEN, C. FRANZINETTI, J. KEREN, R. MØLLERUD, G. MYATT, A. OORKIN-LECOURTOIS, M. PATY, D. PERKINS, C. A. RAMM, K. SCHULTZE, H. SLETTEN, K. SOOP, R. STUMP, M. VENUS, and H. YOSHIKI: Physics Letters **12**, 281 (1964). — BERNARDINI, G., J. K. BIENLEIN, G. VON DARDEL, H. FAISSNER, F. FERRERO, J. M. GAILLARD, H. J. GERBER, B. HAHN, V. KAFTANOV, F. FRIENEN, C. MANFREDOTTI, M. REINHARZ, and R. A. SALMERON: Physics Letters **13**, 86 (1964).

The Spectrum and Charge Composition of the Primary Cosmic Radiation.

By

W. R. WEBBER.

With 49 Figures.

I. Introduction.

1. We shall try to present, in what follows, the current status (as of mid-1964) and some of the outstanding problems relating to the study of the primary cosmic radiation. This writing is made at a time when satellite and space probe observations are beginning to contribute greatly to this knowledge and at a crucial time near sunspot minimum when the primary spectrum, particularly at the low energy end, is undergoing extensive study. The term primary radiation will be used to denote that corpuscular radiation which arrives from outside of the solar system (also called galactic radiation). Thus, we specifically exclude the very interesting field of solar cosmic rays as well as the exciting new fields of γ-ray and neutrino astronomy. We shall be mainly interested in the following features of the data relating to the primary radiation:

1. The charge and isotopic composition;
2. The energy or rigidity spectra of the various components.

These data will be presented in detail, together with an examination of their significance in relation to theories of the origin and propagation of this radiation to the earth.

In order to interpret the measurements we are about to discuss, it is useful to define a number of quantities.

dj/dP or dj/dE represent the uni-directional intensity of particles per second passing through a unit area placed normally to the direction of incidence (particles/unit area-solid angle-unit rigidity or energy interval-sec).

$$\frac{dj}{dP} = \frac{dN}{d\Omega \, dA \, dP \, dt}. \tag{1}$$

The integral uni-directional intensity above a certain rigidity P_0 or energy E_0 is defined as

$$j(>P_0, E_0) = \int_{P_0, E_0}^{\infty} \frac{dj}{dP} \, dP; \quad \frac{dj}{dE} \, dE. \tag{2}$$

The density of particles is given by

$$\varrho = \int \frac{j}{v} \, d\Omega, \tag{3a}$$

which for an isotropic distribution becomes

$$\varrho = \frac{4\pi j}{v}. \tag{3b}$$

In many instances, it is necessary to consider radiation incident upon a surface from all directions. In this case, if the surface is a sphere and considering the uni-directional particle intensity isotropic over one hemisphere and zero over the other, we may define an omni-directional particle intensity

$$J\,dA = \int_{\text{sphere}} j\,d\Omega\,dA, \tag{4}$$

and we obtain $J_0 = 2\pi j$ with similar relations for the differential intensities. If the surface is flat then $J_0 = \pi j$.

A most important aspect of the motion of cosmic ray particles is their motion in a magnetic field. In the case of circular motion, we have

$$B\varrho = \frac{pc}{Ze}, \tag{5}$$

where ϱ is the particles radius of curvature or LARMOR radius, B is the magnetic field strength and p is the particle momentum $= \gamma m v$. $B\varrho$ is frequently known as the magnetic rigidity P of the particle and is measured in gauss-cm in esu.

We may write

$$P = \frac{pc}{Ze}. \tag{6}$$

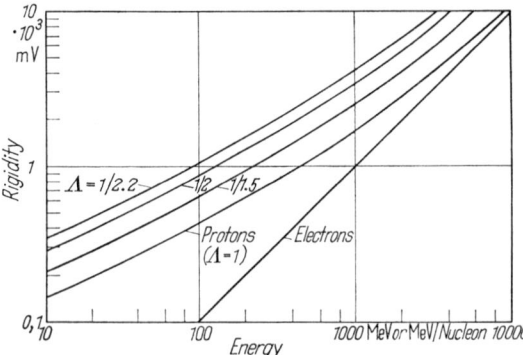

Fig. 1. The energy or energy/nucleon versus rigidity for particles of various Λ.

It is most convenient to express energies of cosmic ray particles in electron volts (eV). Since dimensionally $p = E/c$, the units of momentum are expressed in eV/c. Finally then, it is useful to express the rigidity, P, in comparable units to p and E. In this case, $P = p/Z$ (momentum/unit charge) in units of volts.

In these units, the radius of curvature of the particle (in cm) is simply

$$\varrho = \frac{1}{300}\frac{P}{B} \quad (B \text{ in gauss}). \tag{7}$$

The details of the above relationships may be obtained through the relativistic equation

$$T = c[p^2 + m_0^2 c^2]^{\frac{1}{2}}, \tag{8a}$$

where T is the total energy (per nucleon) $= E + m_0 c^2$, E is the kinetic energy per nucleon, and $m_0 c^2$ is the rest mass energy of a nucleon. Thus,

$$T = [P^2 + (m_0 c^2)^2]^{\frac{1}{2}}, \tag{8b}$$

for singly charged particles or

$$T = [\Lambda^2 P^2 + (m_0 c^2)^2]^{\frac{1}{2}}, \tag{8c}$$

for heavier particles where $\Lambda = Z/A$, the charge to mass ratio. Conversely,

$$P^2 = \frac{1}{\Lambda^2}[T^2 - (m_0 c^2)^2]. \tag{9}$$

The variation of rigidity with kinetic energy or kinetic energy per nucleon is shown in Fig. 1 for various values of Λ.

In order to interpret cosmic ray experiments performed within or outside the earth's atmosphere, we need to specifically know the motion of cosmic ray particles in the earth's magnetic field (assumed to be a dipole). In particular, we find from STÖRMER's work [1] that particles below a certain rigidity P_c are not allowed to arrive at the earth's surface. This rigidity is a function of the geomagnetic latitude and direction of incidence of the particle and may be written for a dipole field

$$P_c = 59.6 \frac{\cos^4 \lambda}{\{1 + (1 - \sin \varepsilon \cos \varphi \cos^3 \lambda)^{\frac{1}{2}}\}^2}, \quad (10)$$

where λ is the geomagnetic latitude, ε the zenith angle and φ the azimuth angle measured from the east. For vertically arriving particles, this expression reduces to

$$P_c = 14.9 \cos^4 \lambda \text{ (GV)}. \quad (11)$$

It is also found that in certain cases particles with rigidities greater than P_c cannot arrive at the earth, primarily at latitudes $<45°$. This theory of the main cone or penumbral effects has been examined in numerical detail extensively by SCHWARTZ [2], and the results in Fig. 2 illustrate the corrections necessary for vertically arriving particles.

In addition, it is found that higher order terms in the earth's magnetic potential, which averaged over the surface of the earth comprise ~ 20 percent of the dipole term, may also effect the so-called cut-off rigidity P_c. The effect of these terms may be taken into account rather simply by introducing a new "effective" λ in Eqs. (10) and (11). These effective latitudes are, in general, within a few degrees of those determined by the dipole approximation, and extensive tabulations of the cut-offs revised for these effects may be found in the work of QUENBY and WENK [3] and SAUER [4].

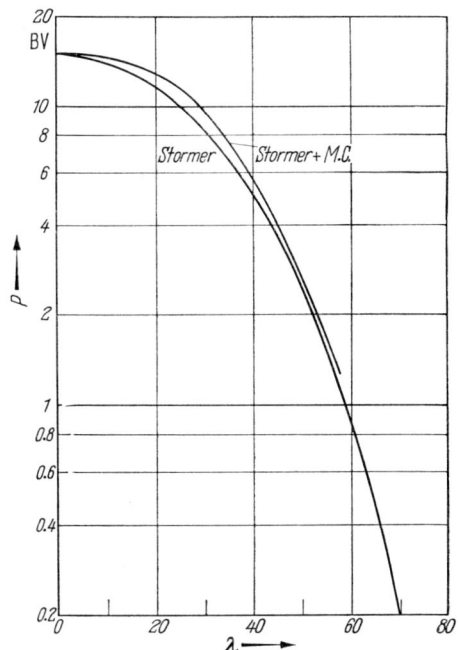

Fig. 2. Vertical cut-off rigidities as a function of geomagnetic latitude in the earth's dipole field. STÖRMER cut-offs are shown as well as average corrections due to penumbral effects (after QUENBY and WENK [3]).

In effect, the arrival of a particle at a particular location is determined only by its rigidity. Thus, we may, and certainly do, use the earth's field as a magnetic rigidity analyzer of the primary particles. Since most detection methods essentially measure the energy of the particles, it is often necessary to convert from one to the other and to understand clearly how this conversion is affected by different values of Λ.

The conversion from a measured differential energy spectrum dj/dE (or dj/dT as well) to a differential rigidity spectrum dj/dP is given by

$$\frac{dj}{dE} = \frac{1}{\Lambda^2} \frac{T}{P} \frac{dj}{dP} = \frac{1}{\Lambda \beta} \frac{dj}{dP}. \quad (12)$$

Frequently, one obtains a set of integral spectra for various charges and wishes to compare the ratios of intensities above a given energy/nucleon or rigidity. If we call R_{ij} the ratio of the i-th component to the j-th component, then

$$R_{ij}(P) = R_{ij}(E) \left[\frac{(m_0 c^2)^2 + P^2 \Lambda_j^2}{(m_0 c^2)^2 + P^2 \Lambda_i^2}\right]^{\gamma/2}, \quad (13a)$$

where γ is the exponent of the spectrum — assumed to be the same and constant with energy for all charges. At high energies this becomes

$$R_{ij}(P) = \left(\frac{A_j}{A_i}\right)^\gamma R_{ij}(E). \tag{13b}$$

At low energies

$$R_{ij}(P) = R_{ij}(E). \tag{13c}$$

also at high energies

$$R_{ij}(T) = \left(\frac{A_j}{A_i}\right)^\gamma R_{ij}(E). \tag{13d}$$

If, for example, we find $R_{pHe}(P) \approx$ const and if this is to be reconciled instead with an integral energy spectrum, then it must be assumed that the spectral indices of the integral spectrum for protons or alphas must be a function of energy with, in fact, a particularly rapid change in the region energies $\approx m_0 c^2$.

Finally, we should mention the application of LIOUVILLE's theorem to cosmic ray studies. Basically, this theorem requires that

$$D = \frac{\delta N}{\delta x\, \delta y\, \delta z\, \delta p_x\, \delta p_y\, \delta p_z} = \text{const}. \tag{14}$$

In other words, the density of particles in phase space remains constant with time (if the Jacobian of the canonical and real momenta $=1$). This has at least two important applications.

1. If the particle energy or momentum changes with time along the trajectory of a particle, then the differential intensity will change as well. Under these conditions, for a differential momentum spectrum,

$$\frac{\frac{dj}{dp}\, m}{p^3} = \text{const} \tag{15a}$$

or for a differential energy spectrum,

$$\frac{\frac{dj}{dE}}{p^2} = \text{const}. \tag{15b}$$

If one then calculates $dj/dE\,(E-\Delta E)$ in terms of $dj/dE\,(E)$ the result is

$$\frac{dj}{dE}(E - \Delta E) = \frac{dj}{dE}(E)\left\{1 - \frac{\Delta E}{E}\left[\frac{2E + 2m_0 c^2 - \Delta E}{E + 2m_0 c^2}\right]\right\}. \tag{16}$$

2. If the energy is a constant and the particles move in a constant magnetic field such that their magnetic moment remains constant, then dj/dp is a constant along the trajectory; and if we assume the cosmic radiation is isotropic at large distances from the earth, then it will also be isotropic at the earth. If, however, the magnetic field varies, then LIOUVILLE's theorem requires that in going from a region where the field is B_0 to one where the field is $B\left(\frac{B}{B_0} \ll 1\right)$, the omni-directional intensity is reduced by a factor $J/J_0 = \frac{1}{2}(B/B_0)$.

II. Solar influences on the spectrum.

2. Since our stated objective is to present and discuss the features of that cosmic radiation which reaches us from outside of the solar system, it is of utmost importance to have some way of estimating the magnitude and energy or rigidity dependence of the solar-induced variation. This solar cycle (or 11-year) variation

is in anticorrelation with sunspot activity, and some appreciation of its magnitude and temporal characteristics may be obtained from Fig. 3, which shows the variation observed by a neutron monitor sensitive to a mean energy ∼15 GeV (Mt. Washington) and an ion chamber sensitive to a mean energy ∼3 GeV and on an inverse scale, the monthly average Zürich sunspot numbers. The amplitude of this sunspot cycle variation is ∼20 percent in the neutron monitor and ∼45 percent in the ion chamber. A hysteresis effect is also noticeable in which the change

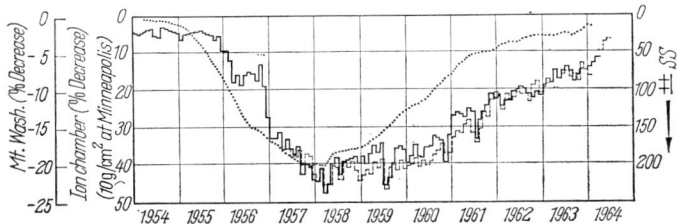

Fig. 3. Examples of the solar cycle or 11-year variations of (a) Monthly average neutron monitor intensity at Mt. Washington (solid curve); (b) Monthly average Zürich sunspot numbers (dotted curve — plotted on an inverse scale); (c) adjusted monthly average ion chamber intensity at 10 g/cm² at Minneapolis (dashed curve). (Courtesy Dr. J. R. WINCKLER.)

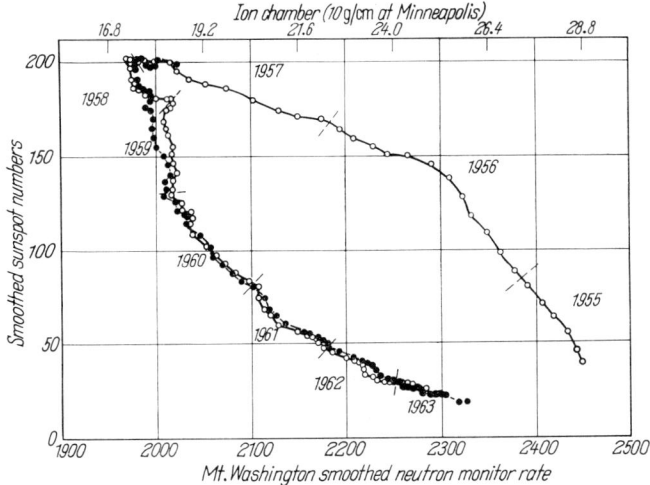

Fig. 4. Regression curves of smoothed monthly average neutron monitor (open circles) and ion chamber intensity at 10⁹/cm² (solid circles) rates versus the Zürich sunspot number. (Courtesy of Dr. C. J. WADDINGTON.) Notice the identical regression curves for the neutron monitor and ion-chamber indicating a very small hysterisis effect between these components.

of cosmic ray intensity clearly lags behind the change in sunspot number by roughly 12 months. There is some evidence that this lag is less (∼9 months) during the period of increasing solar activity than during the period of decreasing solar activity (∼18 months). This hysteresis effect is presumably related to the time constants of the buildup and decay of the solar modulation process.

The correlation between the temporal variations of the neutron monitor and ion chamber shows up more clearly in Fig. 4 where we have plotted this data directly against the sunspot number. The similarity of the hysterisis for both ion chamber and neutron monitor are clearly shown. These figures rather convincingly show that at all energies $\lesssim 30$ GeV and particularly at the low energies to which the ion chambers are sensitive, the solar modulation severely affects the incident extra-solar or galactic radiation. During the periods near sunspot maximum, the solar modulation is so great that the galactic spectrum at low energies is altered completely out of recognition. With our present incomplete knowledge of the energy and charge dependence of this modulation (particularly its effects at low energies), it is necessary to restrict ourselves to measurements of the primary spectrum made at times when the neutron monitor intensity at Mt. Washington is within 10 percent of its sunspot minimum value (1954). Only in exceptional cases, when no other data is available, will we accept other measurements for a determination of the primary spectrum. This effectively eliminates a great body of measurements made during the years from 1957—1961 which, although very valuable, serve only to delineate mainly the features of the solar modulation process rather than the features of the primary galactic spectrum. Even with this restriction, the solar modulation effects are large and important, and we must thus now proceed to devote an appreciable fraction of this review to delineating these effects (for which, of course, we shall make use of data obtained during the 1957—1961 period).

We begin our discussion of the solar modulation process by presenting some results obtained using ion chambers which, although they are not capable of determining changes in the individual charge components, do have the advantage of being a high precision measurement of great statistical accuracy of the changes taking place in the total radiation. We shall then compare these changes with those measured for the various charge components.

a) Solar produced changes in the total radiation.

3. A most comprehensive series of balloon measurements using carefully inter-calibrated ion chambers and extending over the period from the last sunspot minimum to the present time has been made by NEHER and co-workers at Thule, near the geomagnetic pole, and other locations [5]. A similar set of balloon ion chamber measurements has been made by WINCKLER and co-workers at Minneapolis during the period 1956 to the present time [6]. It is possible to directly compare the intensities reported by NEHER and WINCKLER as a result of simultaneous balloon flights made in the summer fo 1958. These suggest that 1 count/sec, according to Minnesota ion chambers, $=13.1$ ion pairs/cm^3-sec-atm according to NEHER. In Fig. 5, we show a regression curve of the ionization at 10 g/cm^2 at Minneapolis and Thule (or Churchill) versus the Mt. Washington neutron monitor for the period 1954—1964. The general features of the solar modulation can be inferred by inspecting this figure as well as Figs. 3 and 4.

Measurements made during the period of increasing solar activity (1954—1958) and decreasing activity (1958—1964) lie on the same curve in both cases, indicating there is no hysteresis or lag in the spectrum of the changes. That is to say, the spectrum of the changes occurring depends only on the ambient level of intensity, and high and low energies as represented by the neutron monitor and the ion chamber decrease and recover according to a unique relationship. This is a very important feature of the solar modulation. It should not be confused with the time lag hysteresis which is evident in Figs. 3 or 4.

The difference between Thule and Minneapolis is shown as excess in Fig. 5. This presumably represents the variation of particles with rigidities between the air cut-off at Thule (0.5 GV) and the geomagnetic cut-off at Minneapolis (1.3 GV). Note that the fractional change of this excess is much greater than the fractional change of the integral intensity at Minneapolis. This indicates how strongly the solar modulation affects the low energy particles.

Another approach to the problem of determining the features of the solar modulation is to examine the counting rates of the ion chambers as they rise effects for helium nuclei seem compatible with those for the total radiation. In fact, there is no observable difference between the modulation of the integral intensives above 1.3 and 4.5 GV of the protons and of the helium nuclei. This is strongly suggestive that, in fact, the modulation is primarily rigidity dependent since if the modulation were dependent on energy or some other parameter such as velocity, a difference would be expected to be observed. We shall discuss this point further in the next section. Meanwhile, to examine this question more definitively, we present in Fig. 11 the variations of the differential helium nuclei intensity in three intervals: $2.0 \leq P \leq 3.0$ GV ($430 \leq E \leq 830$ MeV/Nuc), $1.3 \leq P \leq 2.0$ GV ($200 \leq E \leq 430$ MeV/Nuc) and finally, $0.9 \leq P \leq 1.3$ GV ($100 \leq E \leq 200$ MeV/Nuc).

The data from all observers appears to be consistent with single regression curves although some variations outside of those to be expected from statistical fluctuations seem to occur. In particular, the behavior of the regression curve or the data points obtained near sunspot minimum is not clearly defined. This of course, increases the difficulty of deriving an effective sunspot minimum spectrum for these nuclei. Qualitatively, the modulation of these low energy helium nuclei appears to be similar to that for protons and the total radiation. We shall examine this point in more detail presently.

d) Solar produced changes in the S nuclei ($Z \geq 6$).

5. The limitation on the study of the modulation of nuclei heavier than helium is usually one of statistics. Because of this and other problems connected with the resolution of these nuclei, we shall discuss at this point only the variation of the so-called S nuclei ($Z \geq 6$). In Fig. 12, we display all of the available results on the integral intensities of these nuclei above 1.3 and 4.5 GV again plotted as a function of the Mt. Washington neutron monitor rate. Also shown is the variation of the ion chamber rate at a depth of 10 g/cm² at Minneapolis obtained from Fig. 5. There is again no observable difference between the modulation of the integral intensities of these nuclei above 1.3 and 4.5 GV and the protons, helium nuclei and the total radiation. The simularity of the behavior of these nuclei and the helium nuclei is not unexpected since both of these components have $\Lambda \approx \frac{1}{2}$.

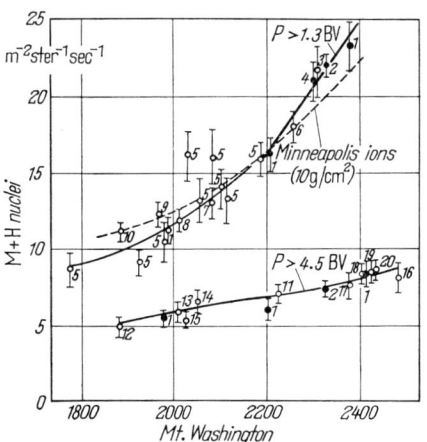

Fig. 12. Integral intensities of $M+H$ nuclei >1.3 and 4.5 GV at the top of the atmosphere as a function of the Mt. Washington neutron monitor rate. The numbers beside the points refer to the following references: *1* [23], *2* [14], *3* [42], *4* [43], *5* [35], *6* [44], *7* [39], *8* [56], *9* [36], *10* [46], *11* [38], *12* [47], *13* [48], *14* [25], *15* [40], *16* [49], *17* [50], *18* [51], *19* [52], *20* [53]. Measurements of various observers corrected for an L/S ratio of 0.25 when a more detailed charge breakdown is not reported.

Results exist on the solar modulation of the differential spectrum of the S nuclei at low energies, generally in the range $1.3-2.5$ GV ($200-700$ MeV/Nuc). These results are much less definitive than those for the helium nuclei; however, qualitatively, the two exhibit the same degree of modulation. We shall defer a discussion of the differential spectra of the S nuclei to later sections.

e) From a spectrum of the solar modulation.

6. We shall now utilize the previously presented data to try and understand the characteristics of the solar modulation. The objective is to determine whether this modulation is primary a function of rigidity or energy, or of velocity, or

c) Solar produced changes in the helium component.

Although the helium component is only ~10 percent as abundant as protons, a great deal of information is available on its spectrum, particularly at rigidities <5 GV. This is because the problem of secondaries and albedo is not as important as for protons and also because instrumental background and resolution problems are not as severe for this component. Since the helium nuclei have a different A than protons, a comparison of the changes in the two components should provide evidence on the rigidity or energy dependence of the solar induced changes.

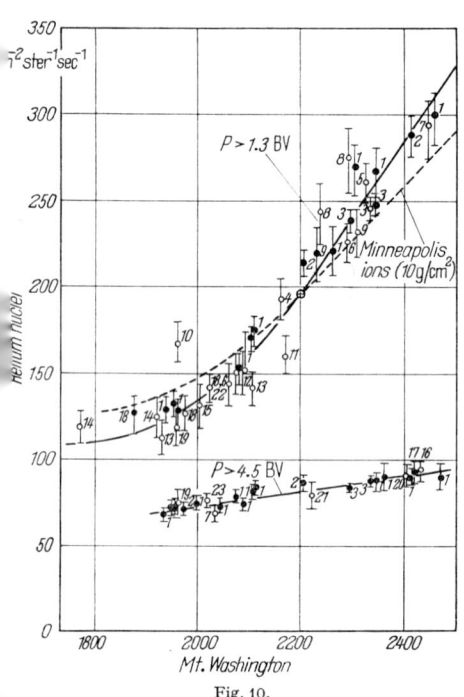

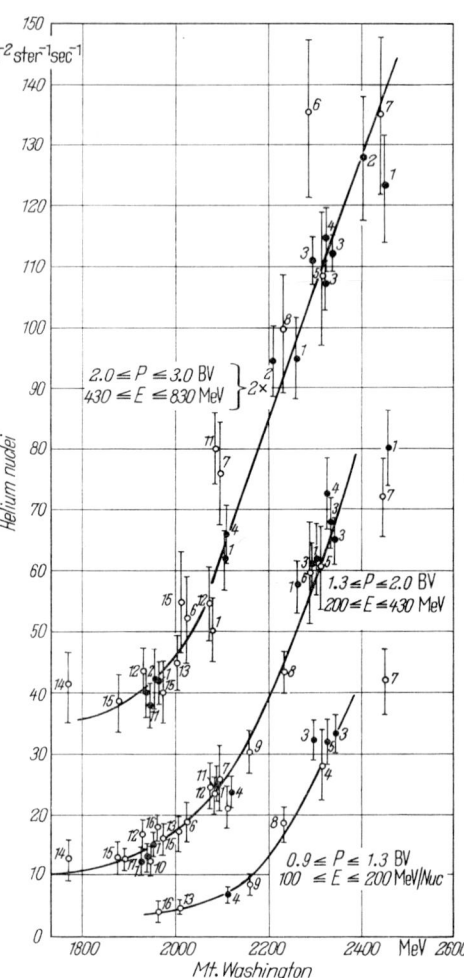

Fig. 10.

Fig. 11.

Fig. 10. Integral intensities of helium nuclei >1.3 and 4.5 GV at the top of the atmosphere as a function of the Mt. Washington neutron monitor rate. The numbers beside the points refer to the following references: *1* [*10*], *2* [*23*], *3* [*14*], *4* [*20*], *5* [*18*], *6* [*24*], *7* [*25*], *8* [*26*], *9* [*19*], *10* [*27*], *11* [*28*], *12* [*29*], *13* [*30*], *14* [*31*], *15* [*32*], *16* [*33*], *17* [*34*], *18* [*35a*], *19* [*36*], *20* [*37*], *21* [*38*], *22* [*39*], *23* [*40*].

Fig. 11. Differential intensities of helium nuclei in the intervals $0.9 \leq P \leq 1.3$ GV (100–200 MEV/Nuc), $1.3 \leq P \leq 2.0$ GV (200–430 MeV/Nuc) and $2.0 \leq P \leq 3.0$ GV (430–830 MeV/Nuc) as a function of Mt. Washington neutron monitor rate. The numbers beside the points refer to the following references: *1* [*10*], *2* [*23*], *3* [*14*], *4* [*41*], *5* [*18*], *6* [*24*], *7* [*25*], *8* [*19*], *9* [*20*], *10* [*27*], *11* [*29*], *12* [*30*], *13* [*32*], *14* [*31*], *15* [*35a*], *16* [*36*].

In Fig. 10 we display all of the available results on the integral intensities of helium nuclei above 1.3 and 4.5 GV plotted as a function of Mt. Washington neutron monitor Also shown is the variation of the ion chamber at a depth of 10 g/cm² at Minneapolis obtained from Fig. 5. As with the protons, we see that the variation of helium nuclei >1.3 GV is somewhat greater; however, if the ion chamber changes are extrapolated to the top of the atmosphere, the modulation observed by neutron monitors or expected on the basis of the changes observed in the total radiation. The variations of the primary protons in the ranges 0.45—0.65 GV (100—200 MeV) and 0.65—1.0 GV (200—450 MeV) as deduced from the measurements of a number of observers are shown in Fig. 9. The picture in these intervals is complicated by the fact that the observations of MEYER and VOGT [11], [12], [13] appear to lie along different regression curves than those of other observers. We shall discuss this discrepancy in more detail later when we present some additional evidence to suggest that the instrument of MEYER and VOGT appears to locally generate low-energy protons which are somehow not corrected in the analysis,

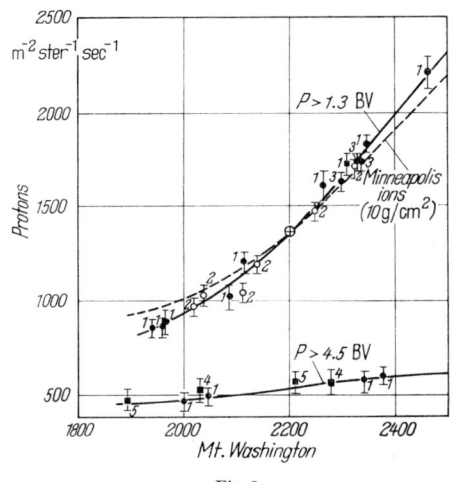
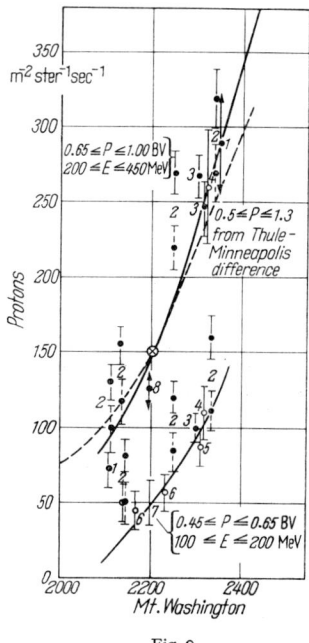

Fig. 8.   Fig. 9.

Fig. 8. Integral intensities of primary protons >1.3 and 4.5 GV at the top of the atmosphere as a function of the Mt. Washington neutron monitor rate. The numbers beside the points refer to the following references: 1 [10], 2 [11], [12],[13], 3 [14], 4 [15], 5 [16], 6 [17]. The use of circles for counter measurements and squares for emulsion measurements is followed in this figure and Figs. 9 through 12.

Fig. 9. Differential intensities of primary protons in the intervals $0.45 \leq P \leq 0.65$ GV (100—200 MEV) and $0.65 \leq P \leq 1.00$ GV (200—450 MeV) as a function of Mt. Washington neutron monitor rate. The numbers beside the points refer to the following references: 1 [10], 2 [11], [12], [13], 3 [14], 4 [18], 5 [19], 6 [20], 7 [21], 8 [22]. The circles with dotted error bars represent the MEYER and VOGT intensities reduced according to the discussion in the text.

thus tending to make their intensities too high. Meanwhile, if we take a constant 50 protons/m²-ster-sec away from their intensities in each of the above intervals, then the data of all observers is more or less consistent with single regression curves. The regression curve for the 0.65—1.0 GV interval is somewhat steeper than that obtained for the difference between Thule-Minneapolis ions. This is reasonable, however, since the ion chamber data refers to a somewhat higher rigidity interval. The curve for the 0.45—0.65 GV interval shows a still greater fractional change in accordance with the expected rigidity dependence of the modulation.

In general, the data on the protons alone is in very good agreement with that for the total radiation in regard to the solar variations.

The data presented in Figs. 8 and 9 as well as Figs. 5 and 6 thus provides a basis on which to derive the proton spectrum and spectrum of the total radiation existing at sunspot minimum as will be done in a later section.

0.5 and 1.0. This value does not appear to depend strongly on the level of modulation. We see here what appears to be a general similarity in the behavior of the low and high rigidity particles, the fractional modulation of the low rigidity particles showing, in effect, a behavior compatible with that of the high rigidity particles. The data on the changes in the low energy protons and helium nuclei seems to substantiate this type of behavior, as we presently shall see, and brings into fuller focus the characteristics connected with the modulation of the low rigidity particles.

Examples of the rigidity dependence of the fractional modulation at two different periods in the solar cycle are shown in Fig. 7. In each case a change ~5 percent is observed in the Mt. Washington neutron minor rate. The magnitude and rigidity dependence of the modulation as determined utilizing high energy data and the excess curve in Fig. 5 is shown by the solid lines, whereas the actually observed changes in the low energy protons are shown by the circles.

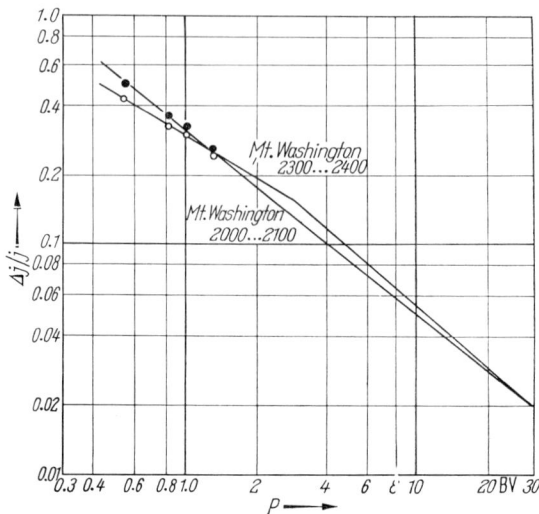

Fig. 7. Fractional modulation as a function of rigidity for two periods in the solar cycle. In each case the variation in Mt. Washington neutrons is ~5 percent. For a counting rate change from 2300—2400, the high energy modulation is given by $\gamma=1.0$ and the low energy data (to be presented in the following section), by open circles ($\gamma=0.7$); for a counting rate change from 2000—2100, the high energy modulation by $\gamma=0.7$ and the low energy data by open circles ($\gamma=0.8$).

b) Solar produced changes in the proton component.

4. Since the proton component is the most abundant in the primary radiation, its changes should closely reflect those of the ion chambers just presented. The rapid buildup of secondaries in the atmosphere, as well as upward moving (splash) and reentrant albedo, complicate the measurement and interpretation of the primary proton spectrum, however, particularly at low energies. Most of the experimental results have been obtained by a variant of the Čerenkov-scintillation counter or by emulsions. The results do not have the statistical accuracy of the ion chamber results, but they are able to provide direct differential measurements of the intensity of a single component. In Fig. 8, we have collected all of the available results on the integral intensities of protons above 1.3 and 4.5 GV and plotted them against the Mt. Washington neutron monitor (these are the only two rigidities at which sufficient data exist to determine a regression curve). The data from various experimenters and different techniques are consistent, and since the Gm cut-off at Minneapolis is 1.3 GV, we may directly compare the ion chamber changes there with the changes in proton intensity above 1.3 GV. The modulation for protons is seen to be slightly greater; however, the ion chamber data is at 10 g/cm² depth. If the ion chamber changes are extrapolated to the top of the atmosphere as are the proton intensities, the modulation effects for protons alone seem to be compatible with those of the total radiation.

A more definitive test of the form of the modulation is to compare changes in proton intensities measured in certain selected rigidity intervals with those has a profound effect on the low energy particles. This possibility, along with the uncertainties in the modulation at low energies, as we shall see, severely limits our ability to determine the correct spectrum of cosmic rays existing outside of the solar system.

Conversely the modulation which prevails near sunspot maximum indicates that large changes are taking place at this time even in the higher energy region (10—30 GV). It is believed that γ must be ~ 1 or greater at energies in excess

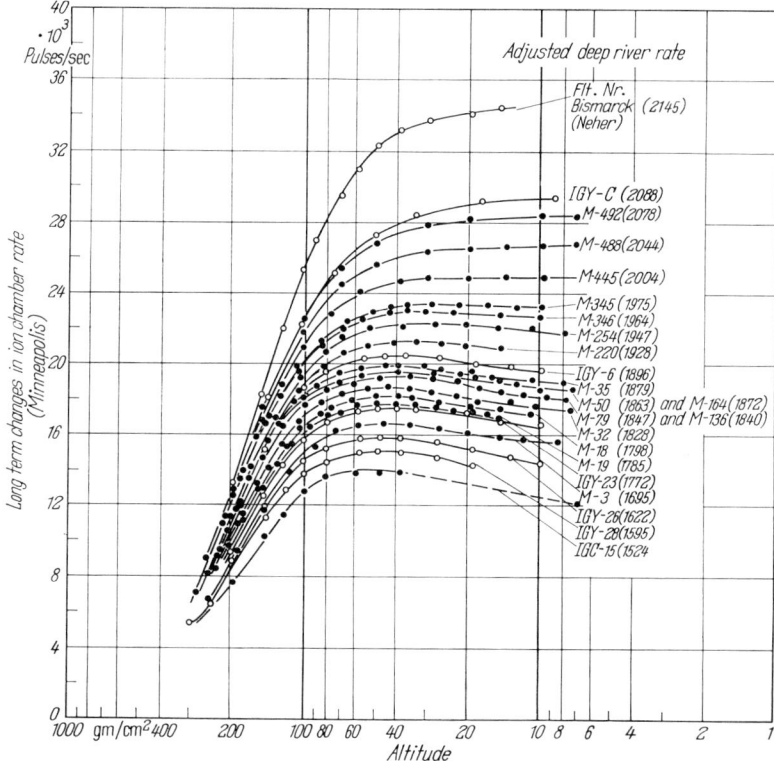

Fig. 6. Summary of ion chamber counting rates as a function of atmospheric depth at Minneapolis during the recent solar cycle after WINCKLER [6]. Flight identification numbers and normalized Deep River neutron monitor hourly counting rates at the flight times are shown. Open circles refer to the period of increasing solar activity between 1954—1959, dots to the period of decreasing activity between 1959—1963.

of 100 GeV [9]. Thus, in this instance, γ varies only slightly with energy. It may be that such a variation is of great importance in determining the spectral changes occurring at lower energies, however.

The effect of the ambient level of modulation on γ as discussed above has been derived mainly from the ion chamber ascent data and represents some sort of average value appropriate to the integral intensity >1.5 GV. It is also possible to derive an effective γ by comparing the difference in Thule and Minneapolis counting rate as a function of neutron monitor rate, the excess shown in Fig. 5, and the integral rate curve also shown in this figure. Assuming that the mean rigidity of the difference interval changes slowly from ~ 1.1 to 0.9 GV in accordance with the spectral changes, and that the fraction of albedo contributing to the ion chamber rate does not vary greatly over the sunspot cycle, the instantaneous values of γ required to reproduce the changes in the excess curve and the 10 g/cm^2 integral counting rate at Minneapolis (mean rigidity ~ 3 GV) lies between through the atmosphere on their way to altitude. Roughly, the mean response of the instrument shifts toward lower energies as the balloon rises in the atmosphere, thus scanning a part of the primary spectrum in a manner somewhat equivalent to conducting a latitude survey. In Fig. 6, we show the data from the ascents of over 50 balloon flights at Minneapolis covering the period from 1954—1964 (essentially from sunspot minimum to sunspot maximum and back again; courtesy of Professor J. A. WINCKLER). Here we see even more strikingly the similarity of the modulation during both the increasing and decreasing periods of solar activity and the increasing degree of modulation at the lower rigidities as represented by the greater changes at higher altitudes.

The intensity at different altitudes is related to the primary spectrum in a complicated manner. It is possible to express this relation in terms of yield functions or coupling coefficients. A detailed study of the yield functions for ion chambers in the atmosphere has been made by WEBBER and NERURKAR [7]. This allows us to translate the variations presented in Figs. 5 and 6 into spectral changes in the total cosmic radiation at the top of the atmosphere.

A number of studies indicate that an expression of the form

$$\left. \begin{array}{l} \dfrac{dj}{dP} = \left(\dfrac{dj}{dP}\right)_0 \\ \times \left[1 - \left(\dfrac{P_0(t)}{P}\right)^{\gamma(t)}\right], \end{array} \right\} \quad (17)$$

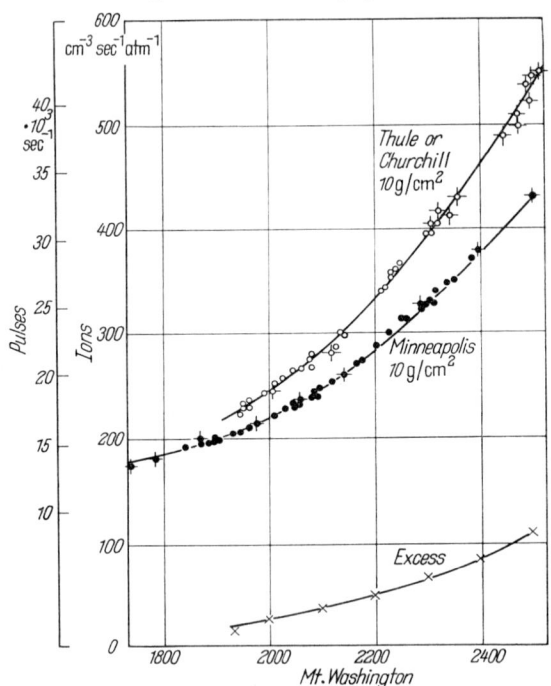

Fig. 5. Individual measurements of the ionization at 10 g/cm² at Minneapolis and Thule (or Churchill) versus the Mt. Washington neutron monitor rate (covering the period 1954—1964) after NEHER [5] and WINCKLER [6]. Crossed circles refer to measurements during period of increasing solar activity, circles to period of decreasing activity. Difference between the Minneapolis and Churchill curves is shown as excess.

is useful for representing changes in the primary cosmic ray spectrum due to the 11-year modulation [8]. Here P_0 and γ are both functions of time. The exponent γ is assumed to be independent of energy although later we shall see this is probably not strictly true.

We have here considered a rigidity dependence for the modulation, although the ion chamber results themselves cannot distingush between a rigidity or an energy dependence. A comparison of the modulation of the various charge components to be presented later will indicate that rigidity may be the more appropriate quantity.

The studies of WEBBER and NERURKAR appropriate to rigidities >1.5 GV show that the average value of γ remains almost constant or varies at most from a value ~1.2 near sunspot minimum when the solar modulation is small to a value ~0.8 near sunspot maximum when the modulation is greatest. Thus, as the modulation increases, the spectrum of the modulation may harden very slightly but on the whole remains remarkably constant. The value of $\gamma \gtrsim 1$ which prevails near sunspot minimum suggests that a small modulation effect at high energies perhaps whether the relevant features of the modulation are, in fact, changing as a function of the particle rigidity and, if possible, what the spectral form of this modulation is. Of crucial importance in these arguments is a comparison of the relative behavior of the proton and helium components.

Fig. 13 shows the ratios of the integral intensities of protons and helium nuclei above 1.3 and 4.5 GV as a function of the degree of modulation as represented by the Mt. Washington neutron monitor. If the modulation were only a function of rigidity, we should expect these ratios to remain constant. Actually, there is a slight tendency for the ratios to increase with increasing modulation, this tendency

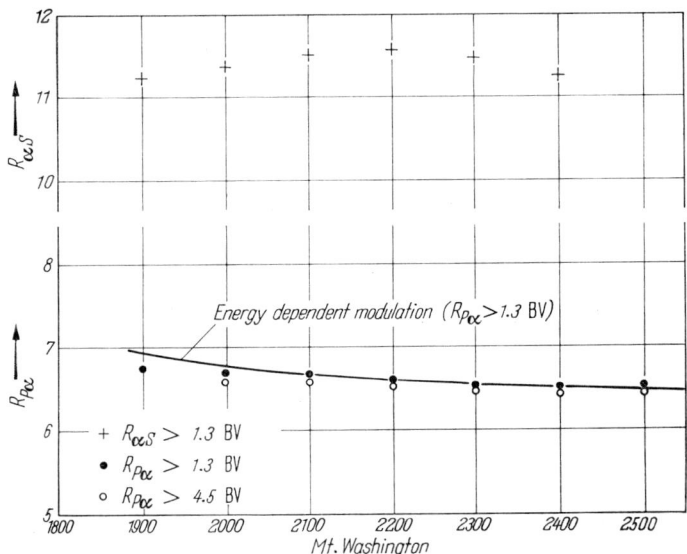

Fig. 13. Ratios of the integral intensities of protons and helium nuclei >1.3 and 4.5 GV as a function of the Mt. Washington neutron monitor rate. Also shown is the ratio of helium to S nuclei >1.3 GV. The variation of the proton to helium nuclei ratio above 1.3 GV to be expected if the modulation were energy dependent is shown as a solid curve.

being slightly greater for the >1.3 GV data. Whether or not this increase is real appears marginal, however, and depends upon the accuracy with which the best fit curves can be drawn through the data in Figs. 8 and 10. An increasing ratio with increasing modulation means that the helium nuclei are changing more rapidly than the protons at a given rigidity. This would be expected, for example, if the modulation were energy or velocity dependent. Following this line of thought, we also show in Fig. 13 a curve of R_{pHe} (>1.3 GV) to be expected if the modulation were strictly energy dependent. The expected increase in this ratio is ~ 10 percent over the solar cycle or somewhat larger than the observed change. One must conclude that using integral intensities, the differences in an energy or rigidity dependent modulation on protons and helium nuclei is small indeed and probably not clearly detectable with present techniques.

We note a slight but not significant variation in R_{HeS} (>1.3 GV) with the degree of solar modulation. The average value of this ratio is 11.4 ± 0.2.

Turning our attention now to the differential measurements at low energies, we find that only recently has sufficient data become available to define the modulation of protons and helium nuclei in overlapping energy/nucleon intervals and nearly overlapping rigidity intervals.

In Fig. 14 we show R_{PHe} as obtained from the best fit curves in Figs. 9 and 11 for two overlapping energy/nucleon intervals; 100—200 MeV/Nuc and 200—450 MeV/Nuc, and for rigidity intervals 0.65—1.00 GV for protons and 0.9—1.3 GV for helium nuclei, the closest one can come, at the present time, to overlapping rigidity intervals for these two nuclei.

Considering the energy intervals first, we see that R_{PHe} (200—450 MeV/Nuc) decreases by about 30 percent during the modulation covering an appreciable part of the solar cycle; and R_{PHe} (100—200 MeV/Nuc) while slightly lower in absolute value appears to decrease by about the same amount. If the modulation were rigidity dependent, we should expect these ratios to decrease with increasing modulation. This amount of decrease will depend on the spectral form of the modulation, but using reasonable values of this dependence derived from Fig. 7, the estimated range of variation of R_{PHe} for these two intervals is covered by the barred region in Fig. 14. The uncertainties in the data, particularly that for protons in both intervals and the helium nuclei in the 100—200 MeV interval, are large enough so that any differences in behavior of the two intervals are probably not significant at the present time. It is seen that the observations are not compatible with either a simple rigidity or energy dependence of the modulation but seem to suggest a more complicated form, perhaps a combination of the two such as a vP dependence, for example.

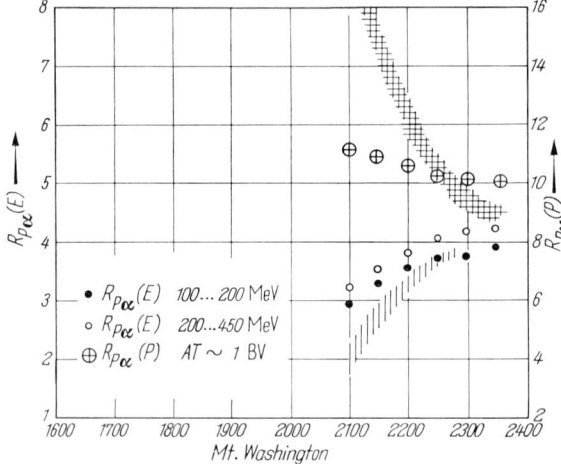

Fig. 14. Ratios of the differential intensities of protons and helium nuclei in the overlapping energy ranges 100—200 MeV/Nuc and 200—450 MeV/Nuc and in the comparative rigidity intervals 0.65—1.0 GV for protons and 0.9—1.3 GV for helium nuclei. Expected range of variation of this ratio in the overlapping energy intervals for a rigidity dependent modulation is shown by barred region. Expected range of variation of this ratio in the rigidity intervals for an energy dependent modulation is shown by cross-hatched region.

Turning now to the comparison of R_{PHe} on a rigidity basis, we note that for the intervals taken, this ratio increases by ~10 percent from its value near 10 at sunspot minimum during the course of the increasing solar modulation. The helium nuclei are thus undergoing slightly more rapid modulation than the protons. The mean rigidity of the proton interval is ~0.85 GV as opposed to 1.1 GV for the helium interval, and if the data is adjusted for exactly equivalent rigidity intervals using a reasonable dependence for the modulation, then this increase in R_{PHe} is enhanced slightly. Such an increasing R_{PHe} with increasing modulation in equivalent rigidity intervals is what is to be expected if the modulation is energy or velocity dependent. The change in this ratio will depend on the exact spectral dependence of the modulation. If we select a dependence which is equivalent to the assumed rigidity dependence of the low rigidity particles shown in Fig. 7, then the change in R_{PHe} is given approximately by the cross-hatched area in Fig. 14, much greater than the observed variation.

It appears then the features of the solar modulation at low energies cannot simply be described in terms of the quantities energy or rigidity alone. Such a conclusion can only be reached as a result of recent and improved measurements on low energy protons and helium nuclei in overlapping energy or rigidity intervals and is still subject to rather large uncertainties in the manner in which the low energy protons, in particular, vary. Improved measurements are certain to clarify this situation shortly. It is possible that these measurements will ultimately show that a single energy or rigidity dependence is sufficient for describing the solar modulation at low energies. However, the tentative estimate of a more complicated form, quite similar to a vP dependence at low energies, must be regarded as the most reasonable interpretation of the present data.

While the above discussion suggests the features which may be of importance in the solar modulation, it does not yet give us a basis for describing the rigidity or energy dependence of this modulation. The rigidity dependence or equivalent energy dependence implied by Eq. (17) and Fig. 7, while quite artifical and subject to large uncertainties at low energies, seems to be the most reasonable approach that can be made at the present time. Attempts to determine a sunspot minimum spectrum or to determine the ambient primary cosmic ray spectrum in the region outside of the solar system by extrapolation will then be subject to these limitations.

III. Spectrum of the total radiation.

a) Primary spectrum derived from ion chamber and counter measurements in the latitude sensitive region.

7. Previous attempts to derive a spectrum for the total radiation in the latitude sensitive region have required the collection and interpretation of many measurements made at different locations with diverse detectors [10], [54]. Recently, comprehensive data has become available for single detectors as a result of the latitude survey of NEHER [55] using ion chambers, and the satellite measurements of LIN and VAN ALLEN [56] using counters. We shall attempt to interpret these and other measurements in terms of the spectrum of galactic radiation existing at times near sunspot minimum (and maximum). The relation between such spectra and that for the main individual charge components, protons and helium nuclei, may be quite complex as we shall presently see. However, because these spectra can be obtained in detail over a wide energy range in contrast to the limited measurements available on the individual components, it seems worth while to attempt such a derivation.

The most direct way of obtaining these spectra is to use measurements in the latitude sensitive region made by counters or ion chambers at or near the top of the atmosphere. As such, the measurements are dependent on our knowledge of the correct geomagnetic cut-off as a function of latitude. Although the above detectors are essentially omni-directional, it is usually customary to relate the measured intensity with the vertical cut-off at the point of measurement. A second order calculation would consider the changing geomagnetic cut-off as a function of azimuth and zenith angles and the influence of the primary spectrum in order to determine the intensity as a function of arrival direction. Such a calculation leads to an effective cut-off which is ~ 10 percent less than the vertical one, nearly independent of latitude, provided the entire contribution to the counting rate is due to primary particles. In view of uncertainties in the cut-offs themselves and, in particular, the important role the essentially isotropic albedo plays, we shall neglect such effects and relate the measurements to the vertical cut-off only.

Detailed measurements of the latitude effect at the top of the atmosphere at a time near sunspot maximum are available for ion chambers [55] and for a single counter [56] and are shown in Fig. 15. The counter measurements were made in a satellite at ~1000 km altitude and refer to an effective solid angle for primary radiation of 9.5±1 steradian as opposed to ~2π steradians for the ion chamber (the effective solid angle for albedo is the same at both locations). The counter data has been adjusted to the top of the atmosphere for the changing effective solid angle for primary radiation by utilizing the fraction of the counting rate due to albedo as obtained from Fig. 16.

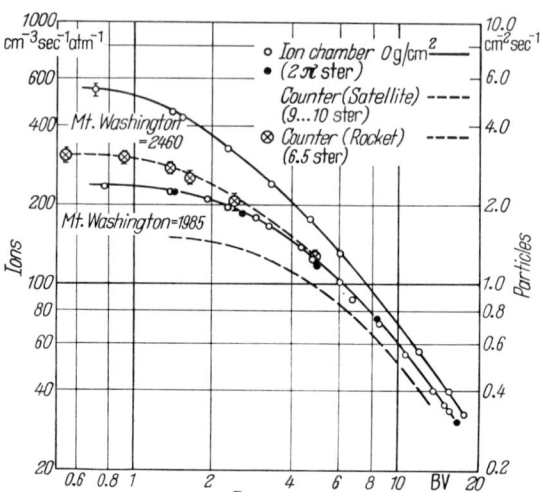

Fig. 15. The latitude effect as measured by an ion chamber or a single Geiger counter at the top of the atmosphere at times of sunspot minimum and maximum. Ion chamber data is shown by solid curve; open circles are measurements of NEHER [5], [55] solid circles those of WINCKLER [6], counter data is shown by dotted curve with crossed circles referring to the Iowa rocket measurements [57] (all data normalized to the top of the atmosphere-2π steradians).

Measurements of the latitude effect at times near sunspot minimum are not as comprehensive; however, using ion chamber measurements summarized by NEHER [5], it is possible to construct the latitude curve shown in Fig. 15. It is also possible to construct a partial latitude curve for a counter from data of the Iowa workers [57] and this is shown in Fig. 15. These counter measurements were made in rockets and thus are directly comparable with the ion chamber measurements extapolated to 0 g/cm².

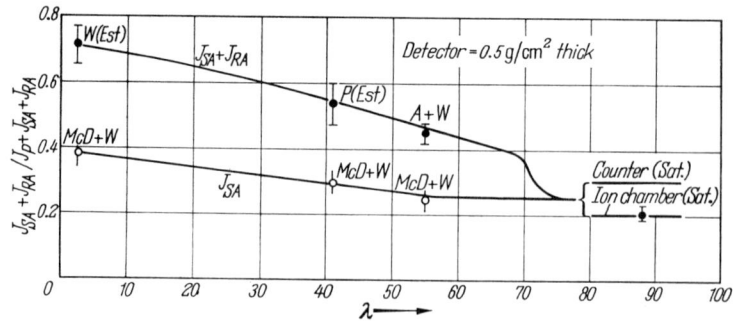

Fig. 16. The fraction of albedo counts as a function of latitude contributing to an omni-directional detector 0.5 g/cm² thick at the top of the atmosphere. J_{SA} points are derived from the measurements of McDONALD and WEBBER [58]. The $J_{SA}+J_{RA}$ are derived from the measurements of PERLOW, et al. [59], and a direct comparison of intensity measurements at the top of the atmosphere and measurements in space away from the earth [60], [61], [62].

From the curves in Fig. 15, it is immediately possible to obtain a relation between particle intensity and ion chamber rate and so, in effect, to calibrate the ion chamber results. This relation will depend on latitude and sunspot number; for example, at a cut-off of 4.5 GV, 1 particle/cm²-sec = 128 ions/cm³-sec-atm at SS max and 133 ions/cm³-sec-atm at SS min whereas at 1.5 GV, the values are

Sect. 7. Primary spectrum derived from ion chamber and counter measurements. 197

149 ions/cm³-sec-atm and 165 ions/cm³-sec-atm respectively. An intensity of 1 particle/cm²-sec of minimum ionizing particles would produce 66 ions/cm³-sec-atm.

The latitude curves for either the counter or ion chamber may then be used to obtain the primary spectrum of the total radiation provided that we know the contribution of splash and reentrant albedo to the counting rates. Fig. 16 summarizes our best estimates of the fraction of albedo as a function of latitude for an omni-directional detector 0.5 g/cm² thick. Following the early work of TREIMAN [63] and the knowledge that most of the albedo is relatively low energy electrons [58], [59], we have taken $J_{RA}=J_{SA}$ up to a latitude $\sim 65°$. Above that latitude, the fraction of reentrant albedo falls off sharply. It is seen that the reentrant albedo contributes in excess of 0.5 of all the counts obtained in such a detector at latitudes $\sim 55°$, increasing to ~ 0.7 at the equator. Thus, the accuracy of the derived primary integral spectra shown in Fig. 17 is mainly governed by the accuracy with which the very large albedo correction can be made.

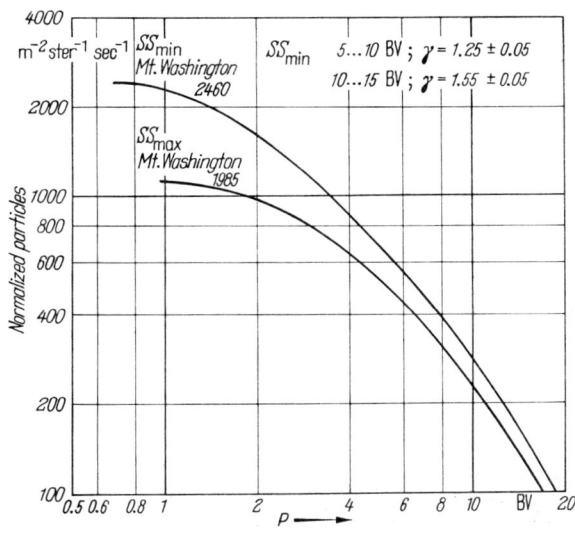

Fig. 17. Integral spectrum of the total radiation existing at sunspot minimum and sunspot maximum as derived from the latitude curves of ion chambers and counters and corrected for effects of albedo.

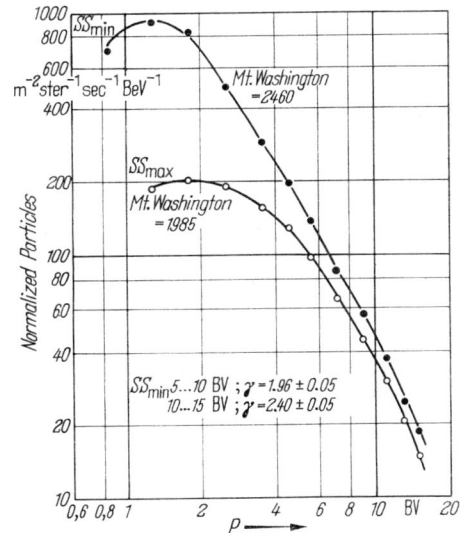

Fig. 18. Differential spectrum of the total radiation existing at sunspot minimum and sunspot maximum as derived from the integral spectrum in Fig. 17.

The integral spectrum for the total radiation derived after subtracting the albedo contribution (sunspot minimum) has an average exponent (on a rigidity basis) of 1.25 ± 0.05 between 5—10 GV, 1.53 ± 0.05 between 10—15 GV and a value of 1.60 ± 0.05 at 18 GV. If we differentiate these integral spectra, we obtain the differential spectra of the total radiation existing at SS minimum and maximum as shown in Fig. 18. We shall later compare these spectra with those obtained for the individual charged components. The average exponent on the differential spectrum at sunspot minimum has a value 1.96 ± 0.05 between 5—10 GV, 2.40 ± 0.05 between 10—15 GV and 2.50 ± 0.05 at 18 GV. The exponents of the differential and integral spectra are consistent.

Since these spectra are based on counter data and since on a rigidity basis we have seen the proton-helium ratio is ~ 7.0, these spectra predominately reflect the behavior of the proton component (decreasing all curves by the 18 percent contribution of nuclei with $Z>2$ directly gives the proton spectra). At rigidities >5 GV, the rigidity, total energy or energy spectra of protons are practically identical so that the above quoted exponents apply to these types of spectra as well.

b) Primary spectrum at energies of 100—1000 GeV derived from the spectrum of γ-rays at high altitudes.

8. A number of measurements are available on the production of electromagnetic cascades at balloon altitudes [64], [65]. Since it is believed that these cascades arise mainly from secondary γ-rays produced in the atmosphere, a measurement of the energy spectrum of these cascades gives a direct measure of the (neutral) pion production spectrum. Over the range 100—2000 GeV, the integral γ-ray spectrum at balloon altitudes has an exponent $=1.9\pm0.2$ constant with energy. A measurement of the γ-ray spectrum alone cannot differentiate between the contributions of the primary spectrum and the π^0 production spectrum without independent data such as, for example, the π^0 meson multiplicity. However, consideration of a model for the nucleon-air nucleus interaction allows MIYAKE [66] to derive an integral spectrum of primaries from the γ-ray data. This spectrum has an exponent of -1.63, constant with energy between 500—1000 GeV, and with the value 0.35 particles/m²-ster-sec >500 GeV.

c) Primary spectrum at energies from 100—10000 GeV derived from the spectra of muons and protons at sea level.

9. A number of detailed measurements are available on the spectrum of muons in the vertical direction at sea level [67], [68]. These cover the energy range 20—10000 GeV. The sea level proton spectrum also has been measured over the range 0.1—100 GeV [69]. Using these measurements, it is possible to draw quantitative conclusions about the primary spectrum if a model is assumed for the interaction between a primary nucleon and an air nucleus. Essentially, the proton spectrum at sea level gives information about the energy retained by the nucleons and the muon spectrum indicates what fraction of the energy lost by the nucleons is given to the parents of the muons. An essential step in the computation is the derivation of the pion production spectrum. Thus, this approach is closely related to that using the spectrum of γ-rays obtained at high altitude. The integral spectrum of primaries derived by WOLFENDALE [68] from this data has an exponent of -1.58 constant with energy between $10-10^4$ GeV and a value $j(>E)=229$ particles/m²-ster-sec at 10 GeV.

Thus, all of the available evidence to date suggests that the spectrum of the total radiation can be represented as a power law with integral exponent $=-1.60$ and differential exponent $=-2.60$ essentially constant from 10 GeV to at least 10^4 GeV. Below 10 GeV, the spectrum becomes much flatter as evidenced in Figs. 17 and 18.

IV. The spectrum of the singly charged component.

Both numerically and energetically, these nuclei represent the most important single component of the primary radiation. Nevertheless, for various experimental reasons, the available information about these nuclei is somewhat less detailed and precise than that about the helium or heavier nuclei. One of the limitations is that the energy of these nuclei can uniquely be determined only up to about 1 GeV (1.7 GV) with the presently used experimental techniques. This means that all measurements at higher energies must be measurements of the integral intensity above some rigidity determined by the geomagnetic cut-off at the location of measurement.

a) Integral spectrum in the latitude sensitive region.

10. To date, these integral measurements have essentially been restricted to three locations for which we can determine a reasonably accurate sunspot minimum intensity corrected for solar modulation effects. These are the 1.3 and 4.5 GV points discussed earlier and another point near the equator with average cut-off rigidity of 16.5 GV. The available data for protons above this latter rigidity is shown in Fig. 19 as a function of the Mt. Washington neutron monitor state.

This data has been corrected where necessary for reentrant albedo equal to the upward albedo measured by McDonald and Webber [58], and measurements at slightly different cut-off rigidities have been adjusted to 16.5 GV. From this data, it is not possible to determine a slope for the regression curve at this rigidi-

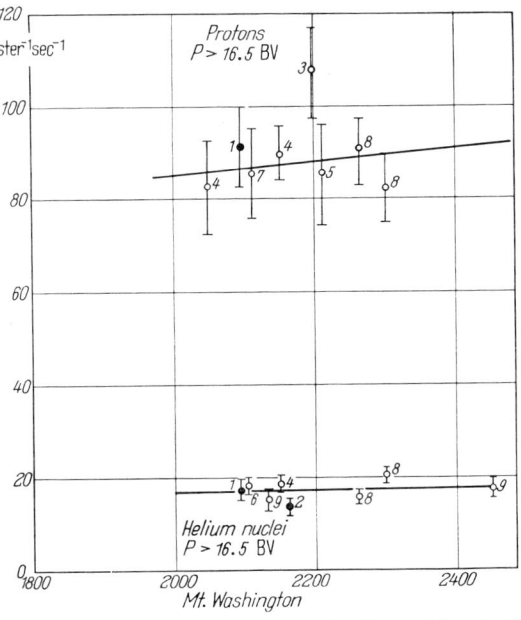
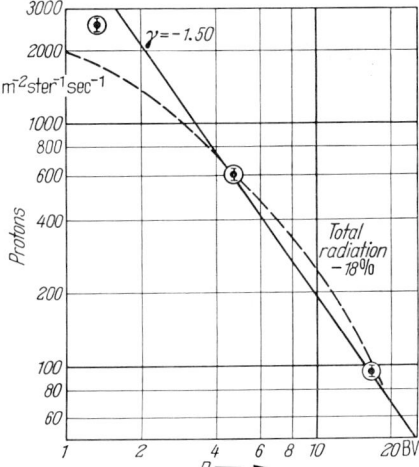

Fig. 19. Integral intensities of protons and helium nuclei >16.5 GV as a function of the Mt. Washington neutron monitor rate. The numbers beside the points refer to the following references: 1 [58], 2 [70], 3 [16], 4 [71], 5 [17], 6 [72], 7 [73], 8 [74], 9 [75].

Fig. 20. Integral spectrum of primary protons at sunspot minimum. Circled points refer to average of direct measurements; dotted curve is the integral spectrum for the total radiation given in Fig. 17 minus 18 percent.

ty although it is clear from other data on the solar modulation effects that a small positive slope should exist. A regression curve giving the best fit to the data and with a slope determined from the features of the solar modulation on the total radiation gives an intensity of 96 ± 5 protons/m²-ster-sec>16.5 GV at sunspot minimum. The corresponding values obtained from Fig. 8 are 2500 ± 60 proton/m²-ster-sec>1.3 GV and 610 ± 30 protons/m²-ster-sec >4.5 GV. These values serve to define the integral spectrum of protons existing at sunspot minimum and are shown in Fig. 20. Also shown in this figure is the spectrum of the total radiation reduced by 18 percent and a spectrum with an exponent $=-1.5$. The agreement between the spectrum of the total radiation and that implied by the direct proton measurements is satisfactory considering the approximations entering into the derivation of the total radiation spectrum. The proton data gives no indication as to how the spectral form is changing with energy or rigidity, however, and the points at 4.5 and 16.5 GV are consistent with a simple power law spectrum in rigidity with exponent $=-1.5$. More likely, the exponent is changing slowly over this range as is evidenced by the spectrum of the total radiation. Below 4.5 GV the evidence is clear that the exponent is decreasing, however. Because of the fact that the directly measured proton intensity >1.3 GV is higher than that implied by the spectrum of the total radiation, we should expect the measured differential proton spectrum at low energies then to exceed that derived for the total radiation as presented in Fig. 18.

b) Differential spectrum at low energies.

11. We shall now turn our attention to the features of the proton spectrum at energies less than 1 GeV where it is possible to measure a differential spectrum. Only recently has this been possible and, in fact, during the summer of 1963, at least five different measurements of the proton differential spectrum at low energies are available. At the time of these measurements, the high latitude neutron monitor intensity was still ~ 7 percent below that existing at sunspot minimum; thus the solar modulation was still very appreciable at this time. We shall first present this data and derive a best spectrum existing at this time. Then, using the regression curves for different rigidities presented in Fig. 9 and features of the solar modulation described in the previous section, we shall attempt to derive the proton spectrum existing at sunspot minimum.

We note, however, that most of the above measurements of the proton spectrum (as well as all of the previous ones) have been made in the earth's atmosphere Here the primary protons have to be distinguished from a relatively copious background of secondaries produced in the atmosphere above and perhaps in the detectors themselves, as well as both upward (splash) and downward (reentrant) albedo protons. The separation of these components is not always clear, and, in fact, is the most serious source of the rather marked disagreement that exists between experimenters on the shape and interpretation of the low energy proton spectrum. We have the work of VOGT [*11*] and MEYER and VOGT [*12*], [*13*] which suggests a differential spectrum (of primaries) steeply rising with decreasing energy below 150 MeV in the summers of 1960 and 1961 and still flat or rising slowly in 1962 and 1963 (it is falling in the range 150—350 MeV in all of these years). On the other hand, the work of WEBBER and McDONALD [*41*] suggests a steeply falling primary spectrum below 300 MeV in 1959 as does that of FICHTEL, et al. [*19*], [*20*], BRYANT, et al. [*21*], [*22*], and FREIER [*76*] in the summers of 1960 and 1961; and FICHTEL, et al. [*19*], ORMES and WEBBER [*14*], FREIER and WADDINGTON [*18*], and BALASUBRAMANIAN and McDONALD [*44*] in the summers of 1962 and 1963. In view of this discrepency and its effect on the interpretation of the low energy protons, we would like to briefly discuss this problem in more detail.

12. Secondary protons in the atmosphere. The intensity of low energy protons measured at high altitudes at high latitudes may be thought to be composed of the following components: j_P, the intensity of primary protons which may or may not be allowed in by the earth's field; j_S, the intensity of secondary protons produced by nuclear interactions of higher energy particles; j_{SP}, the intensity of upward moving secondary protons, the so-called splash albedo; and j_{RA}, the fraction of the splash albedo that is trapped by the earth's field and is again incident on the top of the atmosphere.

Most proton measurements have been made either near Minneapolis (cut-off = 1.3 GV) or Churchill (cut-off = 0.2 GV). Let us consider measurements of the proton spectrum at energies less than ~ 450 MeV (1 GV). Then conditions at these two locations are quite different. At Churchill, j_{RA} does not exist since the splash albedo is not trapped by the earth's field. At Minneapolis, j_{RA} does exist but on the other hand, j_P does not since these particles are not allowed in by the earth's field.

Thus, at Churchill

$$j = j_P + j_S + j_{SA},$$

while at Minneapolis

$$j = j_S + j_{SA} + j_{RA}.$$

A number of measurements of j exist at locations near Minneapolis at *different* depths in the atmosphere. These are shown in Fig. 21. The measurements of McDonald and Webber [58] using Čerenkov-scintillators are at a depth of 6.5 g/cm² and include $j_S + j_{SA} + j_{RA}$. The measurements of Freier [76] and Hasegawa [77] are using emulsions and at depths of 10 g/cm² and 2 g/cm² respectively and include $j_S + j_{RA}$. Finally, the measurement of Bryant, et al. (22) is using a $dE/dx - E$ detector at a depth of 4 g/cm² and includes $j_S + j_{RA}$. These measurements may be adjusted to a common level of primary intensity assuming a linear relationship between j_S, j_{RA} and j_{SA} and the integral intensity of protons > 1.3 GV (Fig. 8). This adjustment is shown by the dotted curves in Fig. 21. Starting with the information contained in this figure, a number of approaches may be used to separate out j_S, j_{RA} and j_{SA}. All of these lead to a consistent

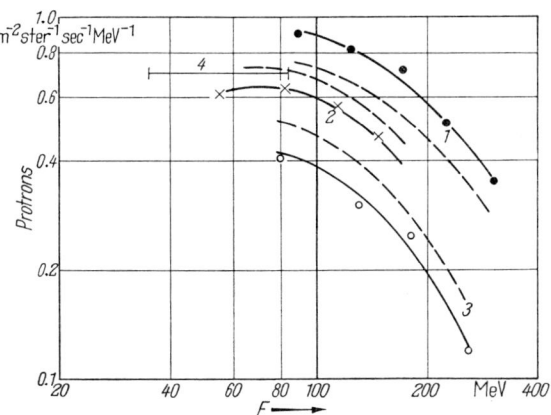

Fig. 21. Measured differential spectra of all protons at Minneapolis at depths between 2—5 g/cm². The numbers beside the points refer to the following references: *1* [58], *2* [76], *3* [77], *4* [22]. The correction to a constant intensity level is shown by the appropriate dashed curves.

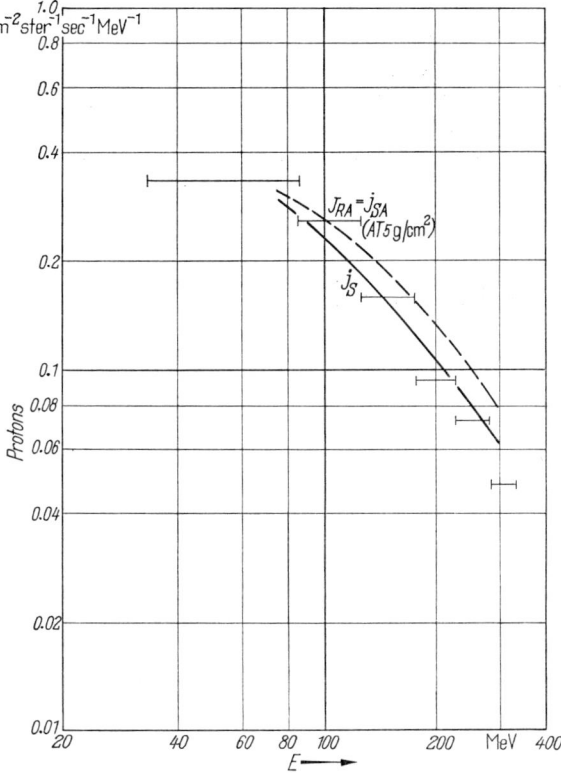

Fig. 22. Derived differential spectra for secondary protons (solid curve), splash and reentrant albedo protons (dashed curves) at a depth of 5 g/cm² at Minneapolis. Barred points are those calculated by Meyer and Vogt [11], [12] for secondary protons at 5 g/cm².

picture for the intensity of these components as a function of energy, and this is shown in Fig. 22. This suggests that $j_{RA} \approx j_{SA}$ at 5 g/cm² with an absolute differential flux of 0.3 protons/m²-ster-sec-MeV at 80 MeV, a factor of almost five less than calculated in an approximate manner by RAY [78]. The intensity of these components is just slightly greater than that of the secondary protons at 5 g/cm². We note also that this deduced secondary spectrum is in approximate agreement with that calculated by VOGT [11] using empirical star data. It thus appears that MEYER and VOGT's correction for atmospheric secondaries is valid.

Wherein then does the discrepancy in quoted primary proton intensities at low energy lie? Some insight into this problem can be obtained from a set of results obtained at Ft. Churchill during a one-month period in the summer of 1961 by three different measurements. In each case, we show in Fig. 23 the actual proton intensities measured at the balloon altitude which, in all flights, was ~ 3 g/cm².

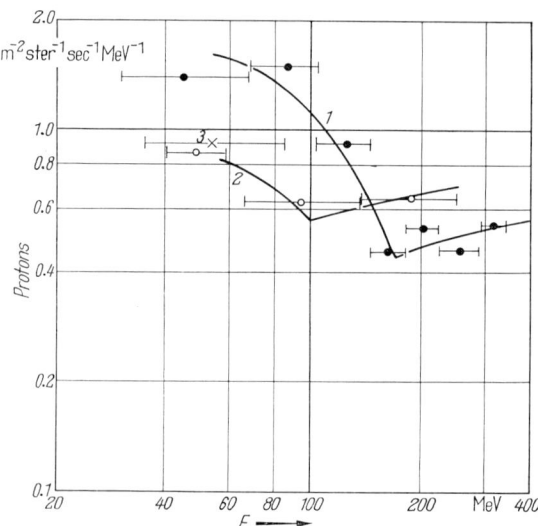

Fig. 23. Measured differential spectra of all protons at a depth ~ 3 g/cm² at Ft. Churchill during a one-month period in 1961. The numbers beside the points refer to the following references: *1* [*11*] Aug. 1, *2* [*20*] July 8, *3* [*21*] July 31.

The results of MEYER and VOGT show consistently higher intensities than the others at energies <150 MeV even though above this energy, their intensities are less than those of FICHTEL, et al. [20] in accordance with the slightly lower neutron monitor intensities prevailing at the time of the MEYER and VOGT flight on August 1. That this difference at the lowest energies is not simply a residual of the solar cosmic ray events occurring in July of that year is evidenced by the measurement of BRYANT, et al. [21], which was made on July 31, only one day before the MEYER and VOGT flight. This measurement gives an intensity consistent with the measurement of FICHTEL, et al., earlier in the month. It appears then that the instrument of MEYER and VOGT is measuring an extra intensity of low energy protons not observed by other detectors. A possible explanation is that these particles are somehow manufactured in the detector itself and are carried along as a "background" amounting to ~ 0.5 particles/m²-ster-sec-MeV at 100 MeV. Such a background would invalidate the methods of correction for the low energy particles used by VOGT [11]. FREIER [76] has reached essentially the same conclusion regarding background in the MEYER and VOGT instrumentation based on a detailed analysis of secondary protons produced in emulsions. We shall see that again in 1963, this excess intensity at low energies appears to exist in the work of MEYER and VOGT.

c) The proton spectrum in 1963 and its extrapolation to sunspot minimum.

13. In Fig. 24 we present the data of five different measurements of the differential primary proton spectrum at low energies during the summer and fall of 1963. These measurements all refer to a Mt. Washington neutron level of 2320 ± 1 percent. Thus, time variations are not expected to play a role in any

Sect. 13. The proton spectrum in 1963 and its extrapolation to sunspot minimum. 203

differences observed. Good agreement is seen between four of the observations, and, in particular, we see that the results of McDonald and co-workers [79] at energies from 20—80 MeV made on board the Imp satellite fit very nicely onto the measurements made in the atmosphere at higher energies by four different observers.

If a best fit curve is drawn through this data, then the intensities in the rigidity intervals between 0.45 and 1.6 GV may be extrapolated to those that would be expected at sunspot minimum (Mt. Washington = 2450—2500) using the regression curves in Fig. 9 and the modulation discussed earlier. Thus, we find that to obtain the spectrum existing at sunspot minimum, we must multiply the 1963 data between 1.0 and 1.6 GV by 1.8, between 0.65—1.0 GV by 2.2 and 0.45—0.65 by 3.0. The accuracy of these extrapolations is, of course, essentially dependent on how well we know the regression curves in Fig. 9, and it is clear that for the lower rigidities, it is very uncertain.

Below 0.45 GV, we have no way of estimating the sunspot minimum proton spectrum except by assuming a form of the modulation at low rigidities. As we know from our earlier as well as the regression curves, a rigidity dependence of the discussion fractional modulation $\sim P^{-1.5}$ is certainly

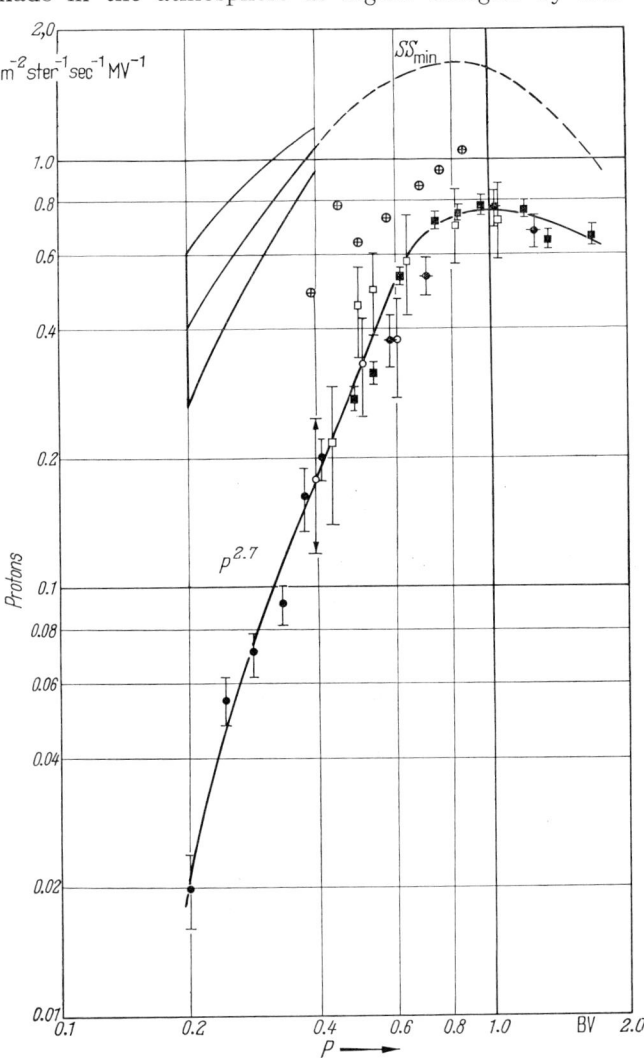

Fig. 24. Measured differential intensities of low energy primary protons in 1963. The various symbols refer to the following references: ■ [14], ● [79], ◆ [44], □ [18], ⊕ [13], ○ [19]. Dotted curve shows estimated sunspot minimum spectrum above 0.5 GV; solid curve shows the estimated limits on this spectrum below 0.5 GV, based on extrapolation from 1963 spectrum (lower solid curve).

an upper limit appropriate for this time in the solar cycle. An extension of this form of modulation to lower rigidities gives a sunspot minimum spectrum shown by the solid curve in Fig. 24.

Despite these very large uncertainties, certain basic features of the data should be noted: 1. The differential spectrum in 1963 has a peak at 0.9 ± 0.1 GV. 2. Below this, the spectrum of protons falls sharply and $dj/dP \sim P^{2.7}$ between 0.2—0.7 GV. 3. Protons exist in the primary radiation down to < 20 MeV.

Because of the steeply decreasing proton spectrum noted above, we can say that unless the solar modulation process has a rigidity dependence $> P^{-2.7}$, the ambient galactic differential spectrum must have a maximum near the earth.

We shall discuss more fully the implications of this proton spectrum and the relation between the sunspot minimum spectrum and the primary spectrum in the galaxy in a later section.

d) Presence of deuterium in the primary radiation.

14. At the present time, there is no positive experimental evidence regarding the magnitude of the deuterium component in the primary radiation. It may just be possible to set a crude limit on the d/p ratio at low energies, however. At present, four measurements exist which pertain to this problem, three having been made in the upper atmosphere and thus faced with the problem of secondary deuterons and one made in free space. HASEGAWA, et al. [77], have measured a d/p ratio ≈ 0.12 between 50—200 MeV/Nuc at a depth of 2 g/cm² at Sioux Falls. Since the normal geomagnetic cut-off at the location of this measurement is ~ 1.9 GV and since the rigidity range for the deuterons above is from 0.65—1.27 GV, it is likely that these particles are secondaries (or reentrant albedo) even though the cut-off was somewhat reduced from its normal value at this time. Both BRYANT, et al. [21], and ORMES and WEBBER [14] have measured the d/p ratio at two locations, Minneapolis or Sioux Falls and Churchill. BRYANT, et al., obtain a ratio = 0.11 at 4 g/cm² in the energy range 25—85 MeV/Nuc at Sioux Falls. However, this ratio at Churchill is <0.08. ORMES and WEBBER find a ratio = 0.25 at 8.5 g/cm² at Minneapolis and 0.15 at 4 g/cm² at Churchill in the range 30—45 MeV/Nuc at the detector. The errors are large in these measurements and it is difficult to compare the results in different energy intervals and to evaluate the effect of reentrant albedo in the measurements at lower latitudes. If one assumes the intensity of reentrant albedo to be comparable to the intensity of secondaries at these altitudes as is roughly the case for protons, then an upper limit of ~ 0.05 is obtained for the d/p ratio at the top of the atmosphere in the range 25—85 MEV. In any case, it does not appear to be larger than 0.1.

These results agree with the satellite observations of McDONALD and co-workers [79] which place an upper limit of 0.06 on this ratio in the range 25—80 MeV/Nuc in a measurement free of the complications of atmospheric background.

V. The spectrum of the helium component.

The form of the energy spectrum of the helium nuclei is of special interest since it can be compared with the proton energy spectrum to provide tests for theories of the origin and propagation of the primary cosmic rays. The primary helium nuclei have been investigated in detail by a number of groups using both emulsions and counters. Since there is no albedo and these nuclei are easily identifiable in the above detectors, the intensity of these nuclei is known with greater precision than the proton component. In addition, the energy of these nuclei can be determined up to ~ 1.5 GeV/Nuc (4.5 GV rigidity) in both emulsions and counters and to even higher energies in emulsions on occasion. This means that the differential spectrum of helium nuclei can be obtained up to a rigidity which permits a useful comparison with the integral measurements made at selected latitudes.

a) Integral measurements in the latitude sensitive region.

15. As with the proton component, the integral measurements of helium nuclei have essentially been restricted to three locations. Since these are the same for both components, we may compare ratios of intensities of the two components without introducing the uncertainties of the geomagnetic cut-offs. These locations are the 1.3 and 4.5 GV points discussed earlier and another point near the equator with a rigidity of 16.5 GV. The available data for helium nuclei at this higher rigidity is shown in Fig. 19, plotted as a function of the Mt. Washington neutron monitor rate. As with the protons, it is not possible to determine accurately the slope of the regression curve from the data in this figure although other data on the solar modulation suggest that a small positive slope should exist. The point at sunspot minimum is not too dependent on such an interpretation, however, since the data for helium nuclei covers a most useful part of the solar cycle. The sunspot minimum value derived from this curve is 17.5 ± 1.0 helium nuclei/m²-ster-sec >16.5 GV. The corresponding values obtained from the regression curves of the integral intensity >1.3 and 4.5 GV in Fig. 10 are 348 ± 12 helium nuclei/m²-ster-sec and 94 ± 4 helium nuclei/m²-ster-sec respectively. These values serve to define the integral spectrum of helium nuclei existing at sunspot minimum and are shown in Fig. 25. Also shown in this figure is a power law spectrum with an exponent $=-1.5$. The data between 4.5 and 16.5 GV are suggestive of a spectrum which has an average exponent slightly less than -1.5. However, it gives no indication on how this spectral index is changing with energy or rigidity. Most likely, this index is changing slowly over this range and to examine this point in more detail, we shall now investigate the differential spectrum of helium nuclei.

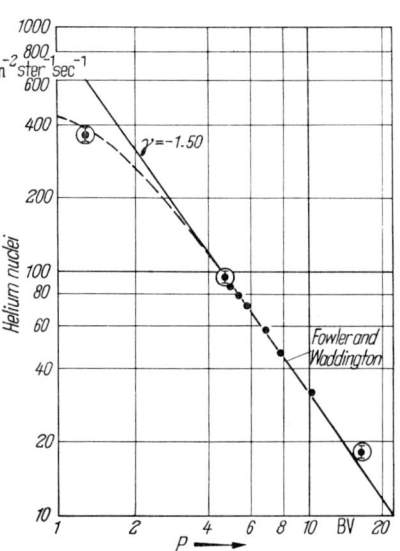

Fig. 25. Integral spectrum of helium nuclei at sunspot minimum. Circled points refer to average of direct measurements; dashed curve represents the integral spectrum deduced from the differential spectrum presented in Fig. 27. Results of FOWLER and WADDINGTON [52] at higher energies shown by solid circles.

b) Differential spectrum of helium nuclei at low energies.

16. Since we are interested in differential measurements that can be most simply related to the spectrum of helium nuclei existing at sunspot minimum, we shall concentrate upon and compare those made in the period 1954—56 and again in 1963 when the solar modulation effects were not large. In addition, when examining this spectrum at low rigidities, we must be sure that geomagnetic effects do not influence our interpretation of the spectrum.

Considering the 1954—56 period first, we see there are emulsion measurements available [24], [25], [26] on three occasions and counter measurements on four occasions [10], [41]. During the times of these seven measurements, the neutron monitor rate at Mt. Washington varied from 2310 to 2400 with a mean of 2360, about four percent below the sunspot minimum value. In view of the quite large statistical fluctuations on the individual measurements, we have averaged the respective emulsion and counter data above 1.5 GV rigidity. Thus, at the expence of some fuzziness in the appropriate neutron monitor rate, we have determined more accurate spectra in the range 1.5—4.5 GV. Below 1.5 GV only one measurement each by emulsions [25], [26] and counters [10], [41] is appropriate and this is shown individually in Fig. 26. The data in this figure indicates that the emulsion and counter measurements of the spectrum are consistent over the range 1.2—4.0 GV. Between 2.5 and 4.5 GV the slope of the differential spectrum is given by an average exponent of 1.95 ± 0.05. This compares with an exponent of 1.96 ± 0.05 at sunspot minimum for the total radiation between 5 and 10 GV and suggests that no inconsistency arises in interpreting the helium nuclei and protons as having similar rigidity spectra over this rigidity range at least.

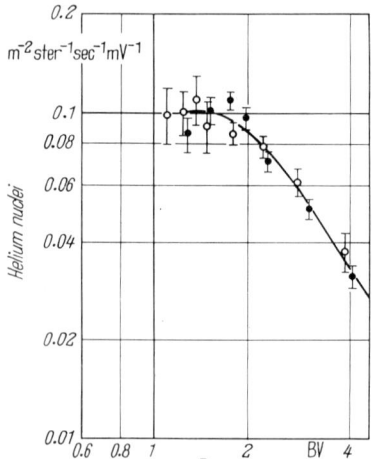

Fig. 26. Differential spectrum of helium nuclei observed in 1954—56 period. Open circles refer to averages of three emulsion measurements; solid circles to averages of four counter measurements; except below 1.5 GV where circles refer to one measurement only.

The differential spectrum of helium nuclei in this rigidity range is also consistent with the integral intensities given in Fig. 25.

Much more extensive data is available on the low energy end of the helium spectrum from data obtained in 1963, however. Here measurements are available from two emulsion experiments [18], [19] and two counter experiments [14], [44], [80] on balloons and a satellite measurement at the very low energy end [81]. All of these measurements are appropriate to a Mt. Washington neutron monitor rate of 2320 ± 1 percent or about 2.5 percent less than the mean of the 1954—56 measurements. The results of each measurement are shown in Fig. 27. The best fit curve through the 1963 data appears to lie 10—20 percent lower than that for the 1954—56 data in the range 2—4 GV. The 2.5 percent lower neutron monitor rate would imply a rate from 6—12 percent lower in 1963 in this range of the primary spectrum. Thus, it is quite likely that the lower intensities in 1963 are real although it is not useful to attempt to define the rigidity dependence of this change by comparing the two curves.

Below 2 GV, the differential intensity measured by BALASUBRAMANIAN and MCDONALD [44] appears to be less than that seen by the other observers. We know of no reason for this difference; however, it appears unlikely that it is due to a time variation. The main feature of the helium spectrum below 1 GV is the extension to very low energies by the satellite measurements of FAN, GLOECKLER and SIMPSON [81]. The spectrum obtained by these observers seems to fit onto the best fit balloon measurements at Minnesota [14], [18], [80] at higher energies only if we neglect the lowest energy balloon points. It is questionable whether we are justified in doing this, however, since although this is near the lower energy limit attainable, two completely independent balloon measurements suggest, in effect, a much more steeply falling spectrum than would be implied by the satellite measurements. The spectrum obtained by BALASUBRAMANIAN and MCDONALD is also not compatible with the satellite measurements unless we assume a sharp change in the helium spectrum at about 0.9 GV.

Since we do not know how this part of the spectrum is varying with time, it is possible that a time variation may explain this apparent discrepancy. In any case, it is clear that measurements currently being made will help to clear up this problem.

As a best fit working spectrum in this low energy region, we shall use the solid line in Fig. 27. From this figure we observe that at rigidities less than 2 GV, the differential helium spectrum begins to turn over and appears to reach a maximum at ∼1.4 GV (considerably higher than the protons at ∼0.9 GV).

If we extrapolate this best fit spectrum for 1963 to sunspot minimum conditions using the regression curves given in Fig. 11, we obtain the spectrum shown as a dotted line in Fig. 27. This extrapolation necessitates increasing the intensity in the range 2—3 GV by 1.6, in the range 1.3—2 GV by 2.0, and in the range 0.9—1.3 GV by 2.6. The accuracy of these extrapolations is determined by how well we know the regression curves in the region of extrapolation, and it is clear from Fig. 11 that these are very poorly known. In fact, the extrapolation depends importantly on the measurement of FREIER, et al. [25], at Saskatoon in 1954.

Despite these uncertainties, certain features of the helium spectrum, as measured in 1963, stand out. These are: 1. The differential spectrum has a peak at 1.35 ± 0.1 GV. 2. Below this, the spectrum falls sharply and $dj/dP \sim P^{3.3}$ between 0.55 and 1.0 GV. 3. Helium nuclei exist in the primary radiation in measurable quantities down to < 20 MeV/Nuc.

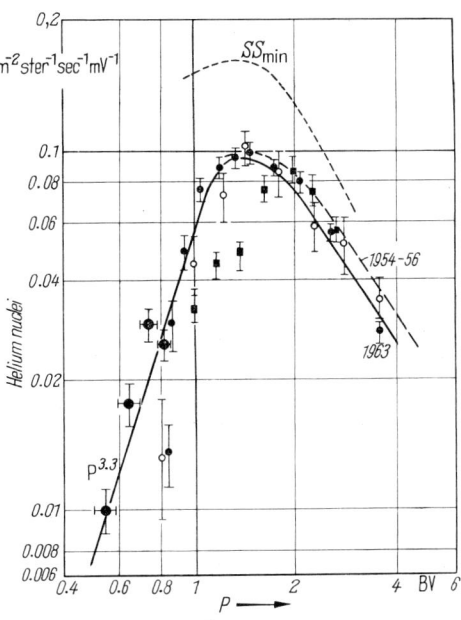

Fig. 27. Measured differential intensities of low energy helium nuclei in 1963. The various symbols refer to the following references: ● [14], ○ [18], ■ [44], ● [81]. 1954—56 spectrum shown by dashed curve. Estimated sunspot minimum spectrum by dotted curve.

The comparison between the spectra of the proton and helium components is obvious from this presentation, and we shall discuss this more fully in a later section.

c) Differential spectrum of helium nuclei at higher energies.

17. A number of attempts have been made using emulsion techniques to obtain the differential spectrum of helium nuclei at energies greater than 1.5 GeV/Nuc (4.5 GV). The first of these relates to a sunspot minimum spectrum measured from 1.5—6.0 GeV/Nuc in a stack exposed over Northern Italy on September 14, 1954 [52]. The integral spectrum measured at this time is shown as a series of circles in Fig. 25. This spectrum is consistent with an integral rigidity spectrum with exponent $= -1.4 \pm 0.1$ or an integral total energy spectrum with exponent $= -1.5 \pm 0.1$.

Recently workers at Washington University at St. Louis [82], [83], [84] have attempted to measure the helium spectrum above 1.5 GeV/Nuc. These results pertain to periods of large solar modulation and as a result, are not useful for defining the sunspot minimum spectrum and perhaps mainly illustrate what can be done with considerable effort in emulsions. The exponent of the integral total energy spectrum is given as -1.45 ± 0.1 between 1.5 and 7 GeV/Nuc on May 14, 1962 [84].

d) Isotopic composition of helium nuclei.

18. Since the measurement of APPA RAO (85), which first indicated that the helium nuclei in the primary radiation did not consist of a single isotope He^4 but contained a measurable fraction of He^3, the measurement of the relative abundance of He^3 and He^4 has received considerable attention. The identification of the He^3 and He^4 nuclei presents a formidable problem in emulsions and even with painstaking analysis of two parameters such as scattering and ionization loss, the resolution of these two isotopes is marginal. Because of these difficulties in analysis, the measurements have been restricted to a single value of $R=$ He^3/He^3+ He^4 in the energy range from 160—360 MeV/Nuc. The results of the presently available emulsion determinations of this ratio are given in Table 1 and range from 0.06 to 0.40 on an energy/nucleon basis. It is possible that the values of $R=0.06$ and 0.10 reported in references 87 and 89 may not represent

Table 1. *Measurement of helium isotopes in the primary cosmic radiation.*

Reference	Location	Lat.	Date	Alt. (g/cm²)	Energy Range $\left(\dfrac{\text{MeV}}{\text{Nuc}}\right)$	$\dfrac{\text{He}^3}{\text{He}^3+\text{He}^4}$ $\left(\dfrac{\text{Energy}}{\text{Nucleon}}\right)$
[86]	Minneapolis	55°	6. 30. 57	8.5	200—400	>[0.38± 0.09]
[86]	Neepawa	61°	8. 3. 58	3.8	160—355	0.31± 0.08
[87]	Minneapolis	55°	4. 21. 61	3.8	250—360	>[0.06± 0.03]
[88]	Neepawa	61°	8. 3. 58	3.8	160—350	0.15± 0.05
[89]	Sioux Falls	53.5°	9. 4. 59	2.0	155—320	>[0.10± 0.05]
[90]	Churchill	70°	8. 4. 62	—	160—370	0.20± 0.06

the ratio of primary He^3 and He^4 outside of the earth's field in equivalent energy/nucleon intervals. These measurements were made at $\lambda=53.5°$ and 55° and because of the rigidity intervals over which R was obtained, the He^3 nuclei would be partially cut off by the geomagnetic field. The remaining measurements give values of R ranging from 0.15 to 0.38, and all results seem to be more or less consistent with a value of 0.20—0.30 in this energy range. In view of the above results, there is no evidence that this ratio varies with solar activity, although if the modulation were rigidity dependent instead of energy dependent, such a variation should be expected if the particles are compared on the basis of their energy/nucleon.

The above results give little evidence as to the energy or rigidity dependence of the ratio R, or in fact, whether energy or rigidity is a more suitable parameter for interpreting this ratio. Recently, a measurement of R has been made using counters which may shed some light on this problem [91]. Using the earth's field as a magnetic analyzer, these workers have compared the geomagnetic cutoff effects of the primary singly charged component (assumed to be protons) with that of the helium component. They conclude that the differences observed can be explained in terms of a value of $R=0.3\pm0.1$ essentially constant with energy. However, on a rigidity basis even a value of $R(P)\sim1$ at low rigidities is insufficient to explain their results. This is because of the steeply falling spectrum of helium nuclei at low rigidities. If the value of R is essentially constant with energy then R should still be ~0.3 at the energies observed in satellite experiments; however, if R were essentially constant with rigidity, all of the helium in the energy range observable in satellite experiments should be He^4.

Attempts to measure the He^3/He^4 ratio at higher energies have so far given no definitive results [92].

VI. Nuclei heavier than helium.

19. The total intensity of all nuclei heavier than helium is only ~ 2 percent of the total primary radiation above a given rigidity. There are also large differences in intensities of different and sometimes adjacent charge components. Also, these nuclei have very short interaction mean free paths in the atmosphere. When they interact, they produce secondary nuclei belonging to a lesser charge group which are indistinguishable from primaries. In order to circumvent these problems a number of specialized approaches are made when dealing with these nuclei. First of all, in order to permit the measurement of intensity values having meaningful statistical weight, it has become conventional to separate these nuclei into groups based on their charge. These groups are usually defined as:

1. Lithium, Beryllium and Boron nuclei; $3 \leq Z \leq 5$, the so-called L nuclei,

2. Carbon, Nitrogen, Oxygen and Fluorine nuclei; $6 \leq Z \leq 9$, the so-called M nuclei,

3. Neon and heavier nuclei; $Z \geq 10$, the so-called H nuclei.

Recently, it has been found necessary to further sub-divide the H nuclei into the H_1 nuclei with $10 \leq Z \leq 14$, H_2 nuclei with $15 \leq Z \leq 19$ and the H_3 nuclei with $Z \geq 20$.

This particular breakdown mainly results from the astrophysical evidence that CNO are cosmically the most abundant heavy elements, whereas there is almost a total lack of Li, Be, B on a cosmological scale. The breakdown of the H nuclei is one of experimental convenience because of the relative abundance of these nuclei and the manner in which fragmentation changes the charge distribution.

Because of the importance of the fragmentation of these nuclei into nuclei of other charges, it is necessary to study this fragmentation in detail in order to interpret measurements made in the atmosphere in terms of actual intensities present at the top of the atmosphere as well as in terms of propagation of the particles through the interstellar medium. Historically, the magnitude of these corrections has represented an important source of controversy on the estimate of the L nuclei flux at the top of the atmosphere and many problems in connection with the interpretation of the abundances of the various H nuclei as well.

In order to apply a correction for the absorption of nuclei in the overlying atmosphere and to allow for production of secondary nuclei by the fragmentation of heavier nuclei, it has been customary to use a one-dimensional diffusion extrapolation due to NOON and KAPLON [93]. The applicability of this approach to propagation in the interstellar medium may be questioned and even in the atmosphere its application may be limited; but, nevertheless, it provides a relatively simple method for evaluating the effects of this fragmentation which, in some instances, has been verified by direct absorption measurements in the atmosphere. If $N_j(x)$ is the measured number of j type nuclei at depth x, then

$$\frac{dN_j(x)}{dx} = -\frac{1}{\lambda_j} N_j(x) + \sum_{i \geq j} \frac{1}{\lambda_i} P_{ij} N_i(x), \tag{18}$$

where λ_j is the interaction mean free path of j type nuclei and P_{ij} is the fragmentation parameter or actually the ratio of the number of secondary j nuclei produced by i type primary nuclei to the total number of interactions of i nuclei. In this equation it is assumed that a parallel beam of nuclei is passing through a medium and ionization loss is neglected (in most instances, it is corrected for separately). Further, it is assumed that the fragmentation products of the interactions maintain the same direction and energy per nucleon as the primary nuclei and that the P_{ij} are energy independent.

Some features of the energy dependence of the fragmentation parameters have been examined by BADHWAR, et al. [94], and BADHWAR and DANIEL [95], mainly in connection with the propagation of the L nuclei through interstellar material, and we shall discuss the implications of these results in a later section.

Meanwhile, to apply the above equation to the propagation of cosmic rays, we must know in detail the fragmentation parameters and mean free paths for the different nuclei in various materials.

a) Determination of fragmentation parameters.

20. Determinations of the fragmentation parameters in nuclear emulsion and other materials have been made by a number of workers; unfortunately, similar direct determinations of these parameters in air or hydrogen have not been made. Thus, in order to calculate the corrections introduced by the overlying atmosphere or in propagation through interstellar hydrogen, it is necessary to deduce these parameters from those observed in other media. The validity of these deduced values has been the subject of much controversy. The best values available for air were presented by WADDINGTON [96]. Since that time, two papers have appeared which enlarge upon and generally confirm the parameters presented by WADDINGTON. FRIEDLANDER, et al. [97], have studied the fragmentation parameters in polyethylene and teflon. DANIEL and DURGAPRASAD [98] have studied the parameters in celluloid. The results of these experiments are more directly related to air and hydrogen than are those in emulsion, enabling one to calculate the appropriate parameters as discussed by FRIEDLANDER, et al. [97]. In Table 2 are shown the parameters in air as deduced by FRIEDLANDER, et al., along with those of WADDINGTON [96] and of DANIEL and DURGAPRASAD [98]. The work of DANIEL and DURGAPRASAD on the absorption of the various H nuclei groups in the atmosphere suggests, as we shall see, that the parameters for air in the H subgroups given in Table 2 may be underestimated. In order to obtain more detailed conclusions on the P_{ij} parameters, particularly for the H nuclei, and

Table 2. *Fragmentation parameters in air* (after FRIEDLANDER, et al. [97]).

Primary	Secondary					
	H_3	H_{23}	H	M	L	α
H_3	(0.26) 0.25*	(0.36) 0.42*	(0.62) 0.17*			
H_{23}		(0.15) 0.13*	(0.28)			
H			0.17 (0.31) 0.27*	0.29 (0.33) 0.34*	0.26 (0.14) 0.16*	1.34 (1.30) 1.16*
M				0.11 (0.16) 0.04*	0.24 (0.21) 0.26*	1.00 (0.93) 0.78*
L					0.15 (0.13)	0.51 (0.81)

( ) WADDINGTON [96]. * DANIEL and DURGAPRASAD [98].

Table 3. *Fragmentation parameters in hydrogen.*

Primary	Secondary		
	H	M	L
H	0.41 ± 0.15 (0.39)	-0.02 ± 0.13 (0.28)	0.50 ± 0.18 (0.28)
M		0.04 ± 0.07 (0.25)	0.41 ± 0.12 (0.48)
L			0.21 ± 0.17

to examine the applicability of the simple diffusion equation, considerably greater statistical accuracy will be required than has heretofore been obtained, or else direct absorption measurements of the various components in air will have to be made.

The fragmentation parameters in hydrogen, as deduced by FRIEDLANDER, et al. [97], are shown in Table 3.

Also shown in this Table are those calculated by BADHWAR, et al. [94], on the basis of theoretical considerations (uncorrected for radioactive decay).

It can be seen that while the fragmentation parameter P_{iL} are in reasonable agreement, a discrepency does seem to exist for the P_{iM} values. This discrepency must be settled before we can interpret clearly the effects of propagation through interstellar hydrogen on the composition of the radiation.

b) Interaction mean free paths.

21. The interaction mean free path λ is defined as $\lambda = 1/n\sigma$ where n is the number of target nuclei/cm³ and σ is the cross section between a cosmic ray nucleus and a target nucleus. For the cross section, one usually writes

$$\sigma = \pi (r_T + r_{CR})^2, \quad (19)$$

where the effective radii of the target r_T, and the cosmic ray nucleus r_{CR}, are given by

$$r = r_0 A^{\frac{1}{3}} = 1.17 \times 10^{-13} A^{\frac{1}{3}},$$

[DANIEL and DURGAPRASAD [98]; GINZBURG and SYROVATSKY [99] use a value of $r_0 = 1.24 \times 10^{-13}$), except when protons are one of the interacting particles. Then r is somewhat less, and data is available from the measurements of CHEN et al. [100], which deals with the cross sections for protons interacting with different target nuclei.

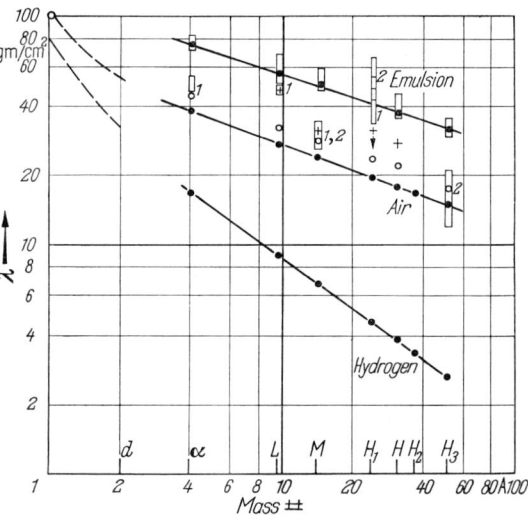

Fig. 28. Mean free paths of cosmic ray particles of different mass in emulsion, air and hydrogen. Calculated interaction M.F.P. using Eq. (19) are shown as a solid curve -●- and solid circles. Barred regions are the measured range of values for the interaction M.F.P. in emulsions. Calculated absorption M.F.P. as defined by Eq. (20) for various charge groups in air are shown by open circles, ○. Calculated attenuation M.F.P. of various charge groups in air is denoted by crosses, +. Range of measured attenuation M.F.P. in air is shown by barred regions where (1) refers to reference [101], (2) to reference [98].

Experimentally, the interaction mean free paths have been measured in emulsions for the various groups of nuclei, and this work is summarized by WADDINGTON [96]. Recently, these measurements have been extended to poly-ethylene and teflon by FRIEDLANDER, et al. [97], and celluloid by DANIEL and DURGAPRASAD [98]. The measured interaction MFP are in essential agreement with those calculated using the above model with the value of r_0 given by DANIEL and DURGAPRASAD and give no cause for belief that either can be in serious error. The interaction mean free paths calculated in emulsion, air and hydrogen for various charge groups are shown in Fig. 28. Also shown in this figure are the measured range of values for the interaction mean free paths in emulsion.

The interaction mean free path is related to the so-called absorption mean free path by

$$\Lambda_i = \frac{\lambda_i}{(1-P_{ii})}. \tag{20}$$

In case where the production of the i-th component by the j-th component is small, the absorption mean free path may represent the actual attenuation of the i-th component. This is typically the case for protons and helium nuclei and is a fairly close approximation for the M and H_3 nuclei as well. For the L, H_1 and H_2 of H components, the absorption is strongly influenced by fragmentation of nuclei from other charge groups. The calculated attenuation lengths for these nuclei as well as the above defined absorption lengths for p, He, M and H nuclei are shown in Fig. 28 for air only. Note that the attenuation length for L nuclei is ~ 48 g/cm² or almost twice the interaction mean free path. The same large difference occurs for the H_1 of H nuclei.

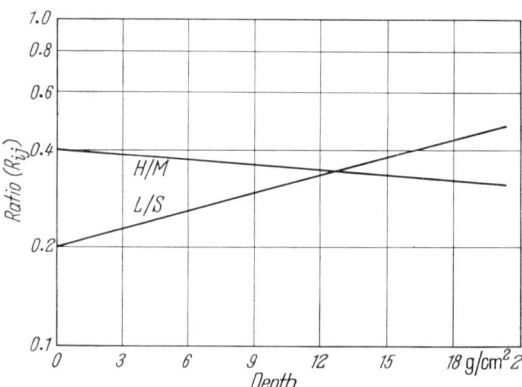

Fig. 29. Calculated changes in the L/S and H/M ratios as a function of atmospheric depth using fragmentation parameters in Table 2.

Recently, in some counter [101] and emulsion [98] experiments, it has been possible to directly measure the attenuation of some of these components in the atmosphere. The results of WEBBER and ORMES [101] for helium nuclei, L, M and H_1 nuclei are labeled 1 in Fig. 28. The results of DANIEL and DURGAPRASAD [98] for M, H_1 and H_3 nuclei are labeled 2. These attenuation lengths are compatible with the earlier measurements of DANIELSON, et al. [102], on M and H nuclei. The measured attenuation lengths in air for helium, L, M and H_3 nuclei are in close agreement with those calculated using the accepted fragmentation parameters and interaction mean free paths. For the H_1 or H nuclei, the measured values are considerably in excess of those calculated. This discrepancy appears to mainly be in the fragmentation parameters for these nuclei and will be discussed in more detail when the intensities of the specific charges are examined.

A convenient way of comparing the intensities and attenuation of the various charge groups is to express the intensities in terms of ratios. The most commonly used ratios are the L/S ratio, a ratio of L nuclei to all nuclei with $Z \geq 6$, the H/M ratio, and He/M ratio. In Fig. 29, we present the attenuation of the various charge components in terms of the calculated change of the various ratios with depth in the atmosphere. This rate of change of the ratios will depend both on the ratio at the top of the atmosphere and the difference in the attenuation lengths. The L/S value is shown normalized to 0.2 at 0 g/cm². It increases at a rate ~ 0.10 per 10 g/cm² near the top of the atmosphere (slightly less at greater depths), according to the diffusion-fragmentation analysis which gives an attenuation length for L nuclei larger by ~ 16 to 18 g/cm² than that for the S nuclei as a whole. The H/M ratio decreases very slightly at a rate ≈ 0.04 per 10 g/cm² from its normalized value of 0.4 at 0 g/cm². The difference in attenuation lengths for these components is only ~ 3 g/cm².

VII. The spectrum of the L nuclei.

a) Integral intensities.

22. We shall attempt to obtain the integral intensities of L nuclei appropriate to sunspot minimum conditions at the same cut-off rigidities used in the proton and helium nuclei analysis. Since we must correct for both time variations and atmospheric secondaries, as well as deal with a component for which a great deal of uncertainty still exists, particularly at the higher energies, it seems more appropriate to approach this problem in terms of the L/S ratio.

The correction of the $S(M+H)$ intensities for solar induced variations can be made using Fig. 12; and for the reasonable assumption that the L/S ratio is constant for the solar modulation, we have only to determine this ratio at the top of the atmosphere for the various cut-off rigidities to obtain the integral spectrum of L nuclei. By far the greatest amount of data exists to enable an accurate extrapolation to the top of the atmosphere to be made for the integral L/S ratio greater than about 3.5—4.5 GV. The actual L/S ratios obtained at this energy at various depths have been summarized up to 1961 by WADDINGTON [15]. In Fig. 30, we show a reproduction of his figure along with the additional more recent measurements of DANIEL and DURGAPRASAD [98], McDONALD and WEBBER [23], WEBBER and ORMES [14], and BALASUBRAMANIAM and McDONALD [44]. These results indicate no serious discrepancy between emulsion and counter measurements of the intensities of these nuclei. They further suggest that we may use these results to obtain an extrapolation of this ratio to the top of the atmosphere independent of the fragmentation approach. If this is done, the best fit curve through the data gives an L/S ratio of 0.23 ± 0.03 at the top of the atmosphere and a rate of change of 0.11 per 10 g/cm² appropriate to the integral intensity greater than about 4 GV. This is virtually indistinguishable from the rate of change of this ratio with altitude as determined from the fragmentation approach.

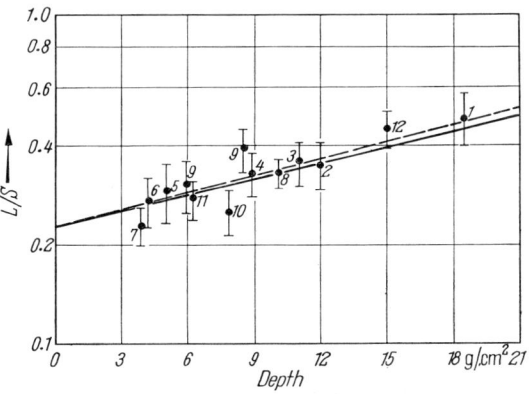

Fig. 30. The L/S ratio measured by various observers plotted at the depth in g/cm² of the observations. Dotted line shows diffusion extrapolation, solid line the best fit to the data for a L/S ratio at 0 g/cm² of 0.23. The numbers beside the points refer to the following references: 1 [103], 2 [104], 3 [105], 4 [48], 5 [106], 6 [47], 7 [107], 8 [98], 9 [23], 10 [101], 11 [44], 12 [108].

Table 4. *Summary of various recent integral L/S determinations above ~1.3 GV.*

Reference	Alt. (g/cm²)	Mt. Wash.	L/S at 0 g/cm²
[36]	6.2	1960	0.23 ± 0.04
[42]	2.5	1890	0.31 ± 0.04
[23a]	5.5	1975	0.29 ± 0.04
[23b]	6.5	2305	
[43]	6.8	2307	0.20 ± 0.04
[101]	6.5	2330	0.30 ± 0.03
[44]	5.0	2325	0.28 ± 0.06

We shall use this extrapolation then for measurements at other latitudes, thus assuming that all quantities of importance in the fragmentation calculation are energy independent. For the low energy point, comparable to the proton and helium nuclei integral intensities >1.3 GV, the measured values of L/S as extrapolated to the top of the atmosphere are shown in Table 4. The original ratios quoted by the various observers have been adjusted slightly, using a common set of the latest fragmentation parameters and H/M ratio. The average of all counter measurements is 0.27 ± 0.03 and for emulsion measurements, 0.25 ± 0.03 with a mean of 0.26 ± 0.03 for the integral L/S ratio above approximately 1.3 GV rigidity.

Within experimental error, this is the same as the L/S ratio just derived above about 4 GV rigidity. However, as we shall see later, the integral ratio is not very sensitive to changes in this ratio on a differential basis.

For the L/S ratio above 16.5 GV, we have only a limited number of observations. The measurements of KERLEE, et al. [109], giving $L/S=0.26\pm0.06$ and those of WEBBER [70], giving $L/S=0.35\pm0.09$, appear to be the only ones reported at high energies despite the fact that the S nuclei have been measured on at least ten occasions at comparable energies. The average of these measurements is 0.30 ± 0.05 and although of considerably less statistical weight than the measurements at lower energies, suggest that no marked change in the integral L/S ratio occurs over the range 4—16.5 GV.

The integral intensities of L nuclei at sunspot minimum may be determined from the data on the solar modulation of S nuclei given in Fig. 12 and the L/S ratios we have just derived. This procedure gives 7.2 ± 1.0 L nuclei/m²-ster-sec >1.3 GV, 2.1 ± 0.2 L nuclei/m²-ster-sec >4.5 GV and from the intensity of S nuclei derived from the results in Table 5, 0.38 ± 0.08 L nuclei/m²-ster-sec >16.5 GV.

These values are shown in Fig. 31 along with a power law integral rigidity spectrum with exponent $=-1.5$.

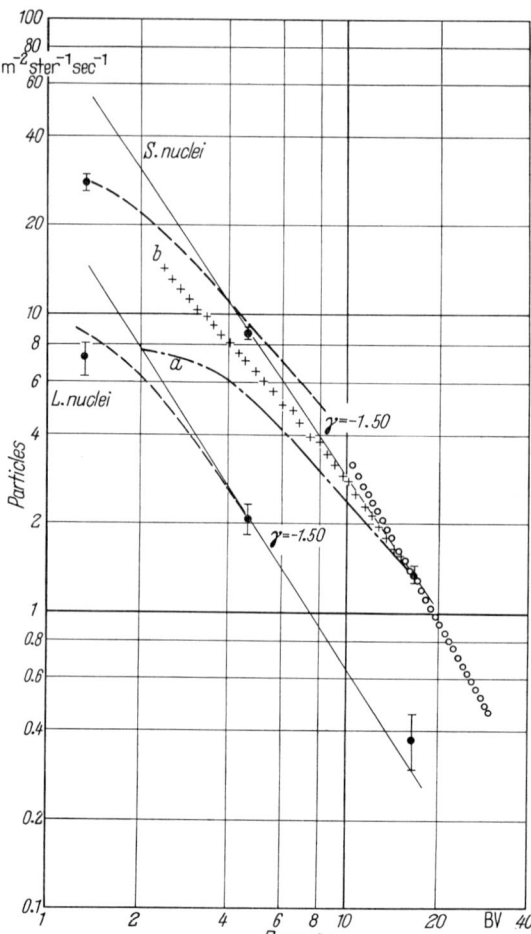

Fig. 31. Integral spectra of L and S nuclei at sunspot minimum derived in the latitude sensitive region. Integral L and S nuclei intensities derived in the text are shown by solid circles with error bars. The derived integral spectrum of L nuclei from the differential spectrum shown in Fig. 32 is given by a dashed line. Integral S nuclei spectrum below 4 GV derived from the sunspot minimum differential spectrum shown in Fig. 33 is shown as a dashed line. Above 4 GV this line refers to the measurements of EVANS [43]. The open circles between 10 and 30 GV refer to the spectrum of S nuclei deduced from the work of DANIELSON [110]. Curves a and b are explained in the text (p. 216).

b) Differential spectrum of L nuclei.

A number of measurements exist on the differential spectrum of L nuclei in the range 1.2—4 GV. Of the six independent observations reported to date, three are by counters and three are by emulsions. We may group three of these measurements as appropriate to sunspot maximum conditions (Mt. Washington $=1950\pm50$) and three appropriate to a period near sunspot minimum (Mt. Washington $=2320\pm20$). All of these measurements are shown in Fig. 32. From this figure, we again conclude that counter and emulsion data give essentially the same results. Although there is perhaps some tendency for the counter measurements to give somewhat higher intensities near sunspot minimum than the emulsion results of EVANS [43], the data is good enough so that we can observe clearly the effect of solar modulation on the spectrum of L nuclei. Best fit curves are drawn through the different measurements for the appropriate neutron monitor intensity levels, and we may extrapolate the spectrum to sunspot minimum conditions using the regression curves for the helium nuclei in Fig. 11. This spectrum is shown as the dashed curve in Fig. 32.

From this differential spectrum, we may construct the appropriate integral spectrum, and this is shown in Fig. 31, normalized at 4 GV. No large inconsistency appears to exist in the integral and differential measurements of the L nuclei, although the differential intensities are slightly higher than those inferred from the integral spectrum. This, of course, reflects in a higher L/S ratio at low energies when differential measurements are compared instead of integral ones. Thus, although we have seen that there is no evidence from the integral measurements that the L/S ratio does vary with energy, this is not necessarily so when the differential measurements are compared in more detail. We shall return to this point in a later section after we have presented data on the differential spectrum of the S nuclei.

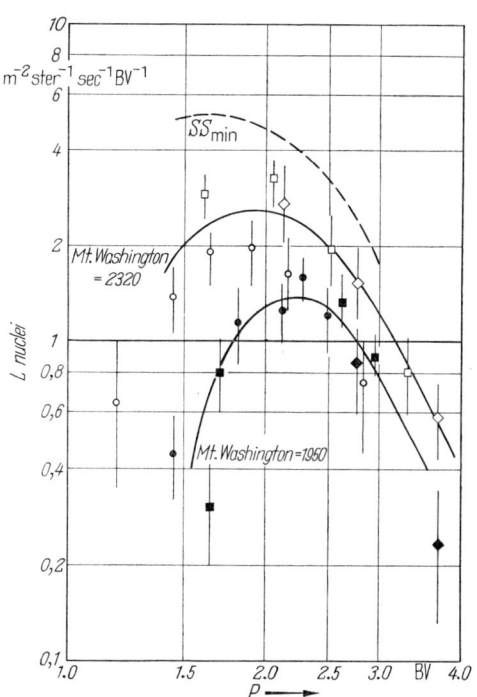

Fig. 32. Measurements of the differential spectrum of L nuclei at low energies by various observers. Measurements appropiate to a Mt. Washington neutron monitor rate = 1950 include: ◆ [23], ■ [36], ● [42]; those appropriate to a rate = 2320, ◇ [23], ○ [43], □ [101]. Estimated sunspot minimum spectrum shown as a dased curve.

Summarizing the features of the measured L nuclei spectrum appropriate to near sunspot minimum conditions we observe: 1. the differential spectrum has a peak at 1.8 ± 0.2 GV, a higher rigidity than either the protons or helium nuclei at this same level of the solar cycle. 2. below this rigidity, the spectrum appears to fall off rapidly, although, at present, the slope of this fall-off is very poorly known.

VIII. The spectrum of the $S(M+H)$ nuclei.

Because of the way the data has been presented in the literature, we shall, at first, present the data of this group as a whole and then evaluate separately the features of the M and H spectra, utilizing the measurements of the H/M ratio as a function of energy. This will enable us to obtain a spectrum for nuclei heavier than helium with a statistical accuracy comparable to that for the helium nuclei. It is possible with these nuclei to measure the differential spectrum with useful accuracy up to ~ 10 GeV/Nuc in emulsions, and this will provide a valuable comparison with the integral spectra derived using measurements made at different geomagnetic cut-offs.

a) Integral measurements in the latitude sensitive region.

23. As with the protons and helium nuclei, the measurements of the integral intensity of S nuclei have essentially been restricted to three locations. The measurements at 1.3 and 4.5 GV allow us to obtain a regression curve of the integral S nuclei intensity as a function of neutron monitor intensity (Fig. 12) and give the following values at sunspot minimum:

$$28 \pm 1.5 \ S \ \text{nuclei/m}^2\text{-ster-sec} > 1.3 \ \text{GV}$$

and

$$8.7 \pm 0.3 \ S \ \text{nuclei/m}^2\text{-ster-sec} > 4.5 \ \text{GV}.$$

Recent measurements of the intensity of S nuclei at a cut-off ∼16.5 GV are shown in Table 5. The best value for this intensity corrected to sunspot minimum conditions is 1.35 ± 0.08. S nuclei/m²-ster-sec > 16.5 GV.

Table 5. *Summary of recent measurements of intensities of S nuclei.*

Reference	Mt. Wash.	H/M	$j(s)$ at 0 g/cm² (Particles/m²-ster-sec)
[109]	2155	0.54 ± 0.14	1.03 ± 0.14
[111]	2150	0.37 ± 0.09	1.30 ± 0.15
[70]	2160	0.50 ± 0.21	1.54 ± 0.30
[112]	2150	0.37 ± 0.10	1.22 ± 0.17
[113]	2225		1.25 ± 0.10
Avg.	2170	0.42 ± 0.06	1.26 ± 0.08

The data between 4.5 and 16.5 GV are suggestive of an integral spectrum for S nuclei with an average exponent $= -1.5$. The details of this spectrum may be filled in more closely by examining the measured differential spectrum for these nuclei. Before we do this, we should mention two counter experiments performed on satellites in which the spectrum of S nuclei was measured using, in effect, the earth's field as a magnetic analyzer, as we have done to obtain the discrete integral points. These experiments were carried out by POMERANTZ et al. [*114*], using a pulse ion chamber and DURNEY et al. [*115*], using a Čerenkov detector. The measurements are appropriate to sunspot maximum conditions, but since they have a statistical accuracy far in excess of that previously obtained, we show them in Fig. 31 normalized to a point 10 percent below the calculated sunspot minimum spectrum at 15 GV. The POMERANTZ et al. measurements (curve a in Fig. 31) corresponds to an average Mt. Washington neutron monitor intensity of 2050, whereas those of DURNEY, et al. (curve b in Fig. 31), to an intensity ∼2250; thus these two measurements which seem to be directly comparable, even though flux values are lacking, give a very powerful method for evaluating the energy dependence of the solar modulation on the S nuclei. In particular, they are compatible with the integral spectrum we have derived for sunspot minimum, even though inaccuracies in these spectra are probably such that they do not permit a comparison precise enough for a determination of the spectral form of the modulation referenced to sunspot minimum.

b) The differential spectrum of the S nuclei at low energies.

24. Here we shall first consider the combined spectra of the M and H nuclei. As with the L nuclei, it is possible to group the available data into two periods, one corresponding to sunspot maximum conditions (Mt. Washington $= 1950 \pm 50$), the other to conditions usefully close to those existing at sunspot minimum (Mt. Washington $= 2320 \pm 30$). The data is summarized in Fig. 33.

The measurements of McDONALD and WEBBER have been adjusted assuming an H/M ratio of 0.35, constant with energy. This time the agreement between counter and emulsion measurements at low energies near sunspot minimum is good; however, a difference is to be noted in the various emulsion measurements at low energies near sunspot maximum. Some of this difference may be due to the fact that the neutron intensity at the time of the measurement of Aizu et al. (36) was slightly higher (~5 percent) than the other two measurements.

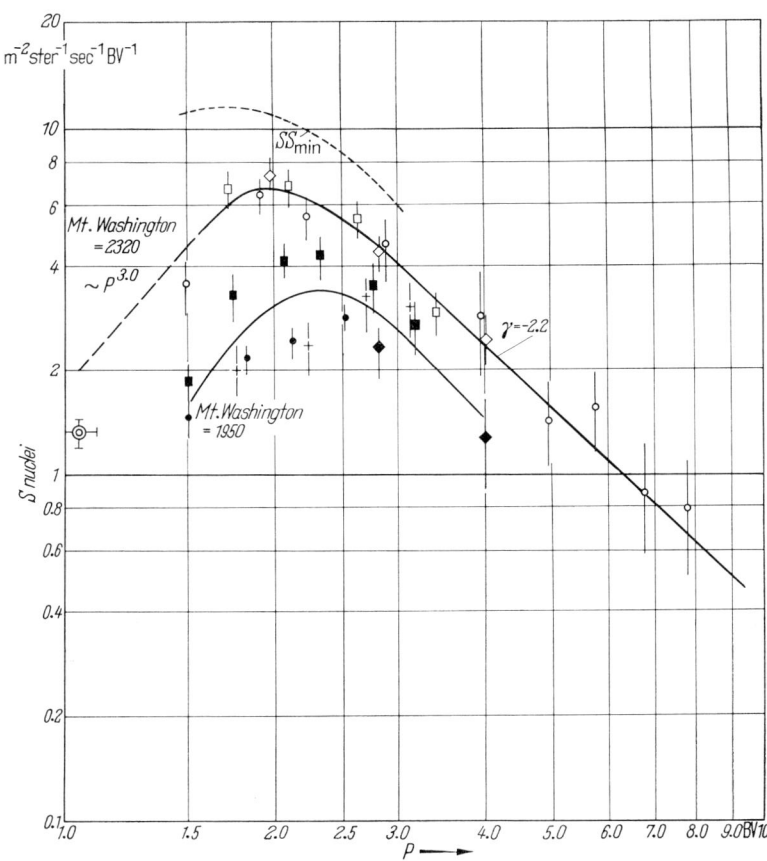

Fig. 33. Measurements of the differential spectrum of S nuclei at low energies by various observers. Measurements appropriate to a Mt. Washington neutron monitor rate=1950 include: ■ [36], ● [42], ◆ [23], + [39]: those appropriate to a rate=2320, ◇ [23], ○ [43], □ [101]. The measurement of Yagoda and Fukui [116] on a satellite is shown thusly, ⊙ Estimated sunspot minimum spectrum shown as a dashed curve.

Despite this difference, the solar cycle changes in the differential spectrum of the S nuclei stand out clearly. The spectrum may be extrapolated to absolute sunspot minimum conditions using the regression curves for helium nuclei in Fig. 11. In Fig. 31, we show the integral spectrum constructed from this differential spectrum. At rigidities <4 GV where the differential spectrum is best known, the spectra are consistent. Above 4 GV, the measurement of Evans [43] is the only one available. This work implies an exponent of -2.2 for a differential rigidity spectrum between 2.5 and 8 GV or as originally quoted by Evans, an exponent $=-2.5$ in a power law total energy/nucleon spectrum. At any rate, the integral spectrum resulting from this differential spectrum seems a bit too flat in the region 4—8 GV as compared with that determined using the latitude effect. Since the actual exponent is probably changing slowly with rigidity, there is probably no serious inconsistency in the two spectra.

The lowest rigidity measurements presently available on the S nuclei extend to only ~ 1.5 GV except for a very important measurement made at ~ 1.1 GV in a satellite (YAGODA and FUKUI [116]). This measurement was made in 1962 and corresponds to a Mt. Washington rate of 2230, and is most nearly comparable with the near sunspot minimum balloon data. It suggests that the differential spectrum of S nuclei is indeed falling at low rigidities.

In summary, the following features of the spectrum of S nuclei at a period near sunspot minimum seem to be important:

1. The differential spectrum has a peak at 1.9 ± 0.1 GV. 2. Below this, the spectrum falls sharply and $dj/dP \sim P^{3.0}$ between $1.0 - 1.5$ GV would be consistent with the data.

c) Differential spectrum of S nuclei at higher energies.

25. From time to time, attempts have been made to measure the differential spectrum of S nuclei directly in the range $10-30$ GV. One of these measurements made at the geomagnetic equator attempts to deduce the spectrum from a study of the azimuthal variation of intensity and a knowledge of the variation of geomagnetic cut-off as a function of azimuth (DANIELSON [110]). In the range 10 to 30 GV Danielson finds an exponent on an integral power law energy spectrum of -1.82 ± 0.19. The value at sunspot minimum would not be expected to be much different from this as would the exponent on an integral power law rigidity spectrum. This data, normalized at 15 GV, is shown in Fig. 31.

Measurements in emulsions on the M and H nuclei separately in the range $2-18$ GV give integral power law energy/nucleon spectra with exponents of -1.65 ± 0.27 and -1.82 ± 0.59 respectively [45]; and for the H nuclei alone, -1.7 ± 0.30 between $4.5-20$ GV [117]. Although these measurements are appropriate to sunspot maximum conditions, the uncertainties are large enough and the modulation effects small enough to let them be considered appropriate to sunspot minimum conditions as well. All of the direct measurements suggest a somewhat steeper spectrum than suggested by the integral points in Fig. 31. At present, it is not possible to say whether this difference is real or not, and it would appear that the integral spectrum defined from the latitude effect is, despite its short-comings, still the most accurate available.

d) The H/M ratio as a function of rigidity.

The most detailed information on the integral H/M ratio is available for cut-off rigidities of 4.5 GV and 1.5 GV. This is particularly true for the altitude dependence of this ratio. The integral values of this ratio measured in the atmosphere at both rigidities are shown in Fig. 34. The results form a very consistent pattern and do not suggest any clear differences in the integral H/M ratio at either 1.5 and 4.5 GV. Furthermore, the general trend of the data is for an H/M ratio increasing slightly with depth at both locations. The best value of this ratio is 0.30 ± 0.02 extrapolated to the top of the atmosphere, and the slope of the best fit curve through the data is $+0.044$ per 10 g/cm². The additional data upon which our summary is based has thus changed only slightly the values of 0.32 ± 0.01 and 0.014 per 10 g/cm² respectively, reported earlier by WADDINGTON [15]. The increasing H/M ratio as a function of depth is in direct contradiction to the predictions of diffusion-fragmentation theory which, as we have seen, gives a negative slope $= -0.03$ per 10 g/cm². A number of individual emulsion measurements have confirmed the increase in H/M with increasing depth from angular distribution studies (APPA RAO, et al. [118]; DANIEL and DURGAPRASAD [98]).

Also, this ratio has been found to increase with increasing depth in the atmosphere during the ascent of a counter experiment (WEBBER and ORMES [101]). The rate of increase of H/M with depth is 0.05 per 10 g/cm² between 10 and 60 g/cm² in the latter two experiments. It seems clear that the diffusion-fragmentation approach inadequately explains the behavior of the H nuclei in the atmosphere. The breakdown of the H group into three subgroups, and the study of the variation of these subgroups with altitude by DANIEL and DURGAPRASAD, and WEBBER and ORMES demonstrates that fragmentation into the H_1 and H_2 groups by particles in the H_3 group is the dominant cause for this behavior. The fragmentation parameters for this process must, therefore, be underestimated.

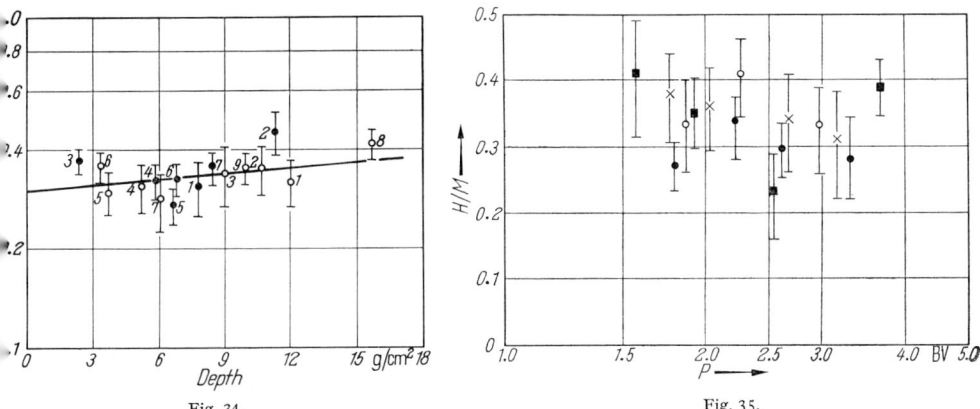

Fig. 34. Measurements of the integral H/M ratio at various depths in the atmosphere. Integral data >1.3 GV shown thusly, ●, with numbers corresponding to the following references: 1 [39], 2 [45], 3 [42], 4 [36], 5 [23], 6 [101]. Integral data >4.5 GV shown thusly, ○, with numbers corresponding to the following references: 1 [104], 2 [105], 3 [48], 4 [106], 5 [47], 6 [107], 7 [36], 8 [108], 9 [98].

Fig. 35. The differential H/M ratio between 1.5 and 4.0 GV as determined by various observers. The symbols refer to the following references: ● [101], ○ [43], □ [42], × [36].

The integral measurements of the H/M ratio at 1.5, 4.5 and 16.5 GV give no evidence that this ratio is varying with energy. The original average value of 0.42 given for 16.5 GV in Table 5 was based on the diffusion-fragmentation extrapolation, and if one uses instead the same extrapolation as for the data at 1.5 and 4.5 GV, the resulting H/M ratio is 0.32 ± 0.06, evidently in good agreement with the value at the lower rigidities. To examine this behavior in more detail, we show in Fig. 35 the results of four measurements of the differential H/M ratio between 1.5 and 4 GV. Because of relatively poor statistics there is wide scatter in the data; however, there is certainly no evidence, at present, that this ratio changes substantially over this rigidity range although a small increase as one goes to lower rigidities cannot be ruled out.

IX. Intensity and spectrum of the electronic component.

26. Up to now, we have discussed only the features of the nucleonic component in the primary radiation. A number of other components are thought to exist and, of these, only the electronic component has been clearly identified and measured to date. The features of the acceleration and propagation of the electronic component may be very different from those of the nucleons and because of this, they are of extreme interest, even though the intensity is small and the spectrum is very uncertain.

Only very recently have electrons been identified positively in the primary radiation, two observations appearing almost simultaneously in the literature (EARL (119); MEYER and VOGT [120]). The two measurements were made using quite different techniques; EARL used a randomly expanded multiplate cloud chamber and observed the electrons and determined their energies by the showers they produced, whereas MEYER and VOGT determined the presence of electrons by a measurement of dE/dx and the range of the particles. The integral intensities of these experimenters presented in Fig. 36 are more or less in agreement for the flux of electrons of energy greater than a few hundred MeV, but the implications of the high intensities and steeply rising MEYER and VOGT spectrum at lower energies are of great importance.

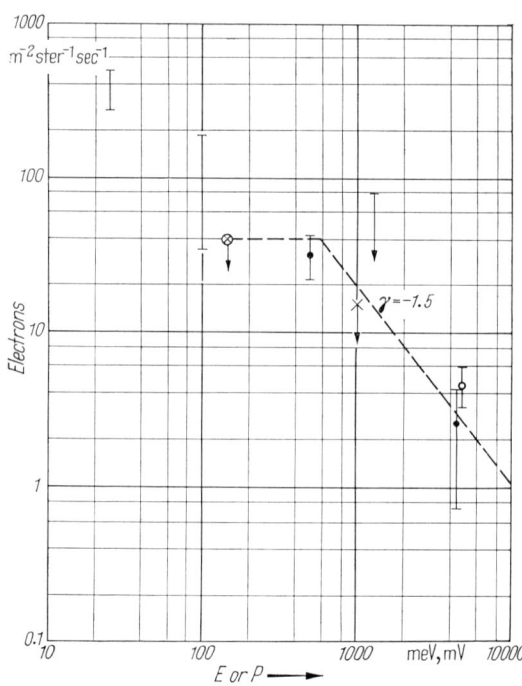

Fig. 36. Integral intensities of primary electrons as deduced by various observers. I refers to reference [120], ♦ to references [119] and [122], ○ to reference [124], × to reference [125] and ⊗ to reference [121]. The spectrum with exponent = −1.5 is shown for illustration only.

Of utmost importance in interpreting these measurements is the presence of background counts and of the production of secondary electrons in the atmosphere above the detectors. EARL's measurements infer that secondaries above 500 MeV contribute less than 10 percent of the observed intensity of electrons above that energy, and since reentrant albedo is not a problem at these energies at Minneapolis, the observed electrons must be primaries. The interpretation of the MEYER and VOGT results at low energies is somewhat more difficult and depends to some extent on the geomagnetic cut-off at Churchill, the location of their flight. The calculated cut-off based on the internal field is ≈100—150 MeV [3]; however, it is known that the cut-off is reduced well below this by external currents even at quiet times. The most reasonable estimate is that the effective cut-off at Churchill normally is certainly less than 50 MeV and probably less than 25 MeV. In this case, the low energy electrons observed by MEYER and VOGT must either be primaries or atmospheric secondaries. MEYER and VOGT compute that the intensity of secondary electrons greater than 25 MeV is ∼10—20 particles/m²-ster-sec at a depth of 4 g/cm² and thus conclude that their intensity of ∼400 electrons/m²-ster-sec > 25 MeV consists essentially of primary electrons.

Recently, a number of new measurements have been reported which begin to define the features of the spectrum and intensity of the electronic component more clearly. FREIER and WADDINGTON [121] have studied the minimum ionizing particles in an emulsion exposed in Churchill in 1963. In addition to obtaining the primary proton intensity and spectrum from measurements on over 200 protons, they observe only four downward moving electrons between 150 MeV — 1 GeV. Assuming these are all primaries gives an intensity upper limit of only 40 particles/m²-ster-sec in this energy range.

EARL has also continued his studies using cloud chambers with a single lead plate as well as multiplate chambers in six flights in the summers of 1961—63 at Churchill, Minneapolis, and in Texas [122], [123]. From an intercomparison of the intensities observed at the various locations, he is able to estimate the relative contributions of primaries, secondaries, and reentrant albedo. Of particular interest are the results of the magnetic cloud chamber which can resolve electrons in the range 40—150 MeV. An intensity of 38 ± 19 electrons/m²-ster-sec is observed in this energy range at Churchill which, when corrected for the estimated secondary intensity of 50 electrons/cm²-ster-sec at 5 g/cm², is consistent with the absence of primary electrons in this range.

A measurement which confirms the intensity above 4.5 GV given by EARL has been reported by AGRINIER, et al. [124]. These workers, identifying the electron showers in spark chambers, report an intensity of 5 ± 1.5 electrons/m²-ster-sec, a value corrected for secondaries and of considerably greater statistical weight (~ 20 events) than EARL's measurement based on two events which gives an intensity of 2.5 ± 1.8 electrons/m²-ster-sec.

A tentative integral spectrum more or less consistent with the present experimental data is shown as a dashed line in Fig. 36. The spectrum has an exponent of -1.5 above 1 GeV, and perhaps more significantly, there appears to be a break in this spectrum

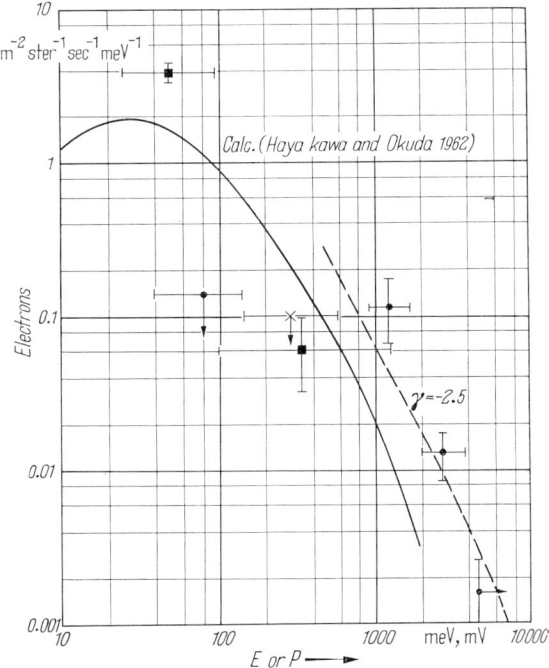

Fig. 37. Differential intensities of primary electrons as measured by various observers. ■ refers to reference [120], ● to references [119], [122], and [123], ⊗ to reference [121]. Differential spectrum shown by dotted line is differential of integral spectrum presented in Fig. 36. Solid curve is differential spectrum of secondary electrons according to HAYAKAWA and OKUDA [172].

at some not very clearly defined energy less than 1 GeV, perhaps not unlike that of the proton spectrum. To examine this important point more closely, we have attempted to construct a differential spectrum for the electrons, and this is shown in Fig. 37. The dotted curve is simply the differential of the integral spectrum presented in Fig. 36 with the point at 4.5 GV representing the differential of the measured integral intensity above 4.5 GV. The measurements of three separate observers are in essential agreement between 100—1000 MeV and taken along with the spectrum above 1 GeV, indicate a break in the differential spectrum more clearly than do the integral measurements. It seems difficult, however, to reconcile the MEYER and VOGT measurement between 25—100 MeV with those of the other observers and, in particular, to that of SCHMOKER [123]. The MEYER and VOGT measurements are, in fact, quite consistent with a differential spectrum of exponent near -2.5, extending down to ~ 25 MeV. The other results are clearly incompatible with this and although of low statistical weight, are unambiguous in their identification of electrons.

Recently, McDonald and co-workers [79] have reported an intensity of electrons ~150 particles/m²-ster-sec between 3—10 MeV as observed in interplanetary space. The differential power law spectrum of these electrons appears to have an index ~ —2.5, however, and an extension of this spectrum above 20 MeV is insufficient to explain the Meyer and Vogt point between 25—100 MeV by an order of magnitude. It is, however, not incompatible with the measurements of Schmoker.

Thus, while it may be extremely tempting to claim the spectrum of electrons rises continuously from energies of a few GeV all the way down to a few MeV with a differential exponent between —2.0 and —2.5, we prefer to conclude that what is being observed is two components, a high energy component with an ill-defined peak in the range 300—1000 MeV and a low energy component which becomes dominant in the range 10—20 MeV.

We should point out that no attempt has been made to disentangle the effects of solar modulation on this measured electron spectrum. These effects, particularly at energies less than 1 GeV, might be even larger than the uncertainties in the electron spectrum itself and will certainly reflect in any comparison of the measured electron spectrum with the observed galactic radio spectrum as will be done in a later section.

Finally, we report the very important measurements of Deshong, Hildebrand and Meyer [126]. These workers have measured the positron to electron ratio = $R_e \equiv e^+/e^+ + e^-$ of the electronic component at an altitude of ~5 g/cm² at Ft. Churchill. They used an elaborate instrument consisting essentially of two spark chambers to obtain the trajectory of the particle, a bending magnetic of ~6000 gauss, another set of spark chambers to again determine the trajectory of the particle and finally, a shower spark chamber to identify the particles as electrons. With this instrument, they could cover the range 50—1000 MeV and since the geomagnetic cut-off at Churchill is believed to be less than 50 MeV, the observed particles must be either primaries or secondaries produced in the atmosphere above the detector or in the detector itself. The authors quote the following values for R_e from an average of two flights at slightly different altitudes; between 25—100 MeV = 0.31 ± 0.12, 100—300 MeV = 0.38 ± 0.07 and 300—1000 MeV = 0.16 ± 0.04. Thus, R_e appears to decrease with increasing energy. Since the authors regard the atmospheric and local secondary contributions as negligible, these values are interpreted in terms of both cosmic primary and cosmic secondary origins of these particles. The authors are unable to quote intensities, however, due to detection efficiencies changing with energy, and this is extremely unfortunate in view of the discrepency that exists in the interpretation of electron intensities below 300 MeV. If we accept the lower spectra for electrons presented in Figs. 36 and 37 as the one most verified at the present time, then, indeed, the secondary flux of electrons produced in the atmosphere will be an important fraction of total number of observed electrons, particularly at the lower energies. Since R_e for this secondary component is > 0.5 (average of 0.68 between 50—1000 MeV according to the above authors), the observed values of R_e in this experiment could be explained by ~40 percent atmospheric secondaries and the remainder consisting of primary electrons only. The measurements at different altitudes (3 and 5 g/cm² respectively) are apparently not capable of ruling out this possibility. Needless to say, the difference in interpretation of these results has important astrophysical implications. If one believes the lower spectrum for electrons and in the absence of any actual flux values obtainable in this experiment, then it seems that the latter interpretation of these results is more nearly correct even though it is at variance with that of the authors. We shall discuss this point further in a later section.

X. Charge composition and spectrum of the cosmic radiation between 30 and 10 000 GeV.

27. In Sect. III, the spectrum of the total radiation has been estimated in the range up to 3×10^{13} eV. This is about the highest energy at which the individual charge components can be measured as well, the intensities becoming too small above this energy to obtain meaningful intensity values. The measurement of individual charges in the range 10^{11} — 10^{13} eV has been made almost exclusively by studying interactions occurring in emulsions. As such, the energy estimate of the particles producing the interactions may be uncertain, perhaps by a factor as great as two. The measurements of the intensities for protons, helium nuclei and S nuclei in this energy range are shown in Fig. 38, superimposed upon the curve for the total radiation derived from the previous data. The intensities for protons are about a factor of three higher at a given energy than those deduced for the total radiation. This obvious discrepancy could be explained if the proton energies were over-estimated by a factor of two. At present, this seems like a plausible explanation since, although the intensities for the total radiation are derived in an indirect manner, the parameters that enter into this derivation are believed to be known better than the uncertainties in estimating the energies of the high energy particles that interact. Then, too, as we shall see later, the intensities evaluated from extensive air shower data (EAS) seem to be more compatible with the total radiation spectrum.

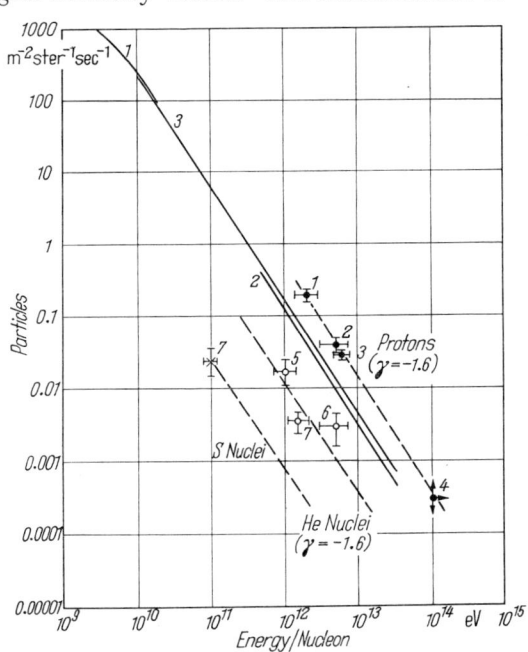

Fig. 38. Integral intensities of protons, helium nuclei and S nuclei in the energy range 10^{11}—10^{14} eV/Nuc. References corresponding to the proton points are as follows: *1* [127], *2* [128], *4* [130]. References corresponding to the Helium and S nuclei are as follows: *5* [131], *6* [132], *7* [112]. The spectrum of the total radiation are those presented in Sect. 3. [*1*] refers to the spectrum in Fig. 15, [*2*] is the spectrum deduced by MIYAKE [66], [*3*] is the spectrum deduced by WOLFENDALE [68].

The average helium nuclei spectrum, evaluated between 10^{12} and 10^{13} eV/Nuc is a factor ~ 20 less than that measured for protons, although only a factor ~ 5 less than the spectrum for the total radiation. The measured values represent energy per nucleon, however, and if the spectra are compared on a rigidity basis, the P/He ratio is ~ 7. Both of these ratios are compatible with those derived from the latitude sensitive data at $< 1.6 \times 10^{10}$ eV and do not indicate an appreciable change in this ratio over at least three orders of magnitude in energy.

This picture is again based on the interpretation that the helium nuclei energies are likewise overestimated by a factor of two as with the protons.

With regard to the S nuclei, there is only one measurement in the relevant energy range, that of JAIN, et al. [112], at 100 GeV/Nuc. This measurement is compatible with an He/S ratio that does not change with energy, at least up to this energy.

In view of the importance of determining the charge composition and energy spectrum in the range 10^{11}—10^{15} eV, it would seem worthwhile that a concerted effort be made in this direction. This is particularly so for a comparison with EAS measurements at higher energies. The ultimate in emulsion technique and one requiring much concentrated effort in analysis would seem to be ∼1 m² area exposed at <100 g/cm² depth, perhaps by an airplane, for ∼2000 hours. This would represent an exposure of 7×10^6 m² sec. Allowing for the attenuation in the air and assuming the probability for interacting in the stack is 1/3, then ∼3000 proton interactions should be expected $>10^{13}$ eV with ∼100 $>10^{14}$ eV. This would permit a statistical analysis of the interactions and their fluctuations and lead to a useful proton spectrum up to ∼5×10^{14} eV. In a like manner, the helium nuclei spectrum could be determined accurately up to 10^{14} eV and that of the S nuclei up to 3×10^{13} eV.

XI. The energy spectrum of the total radiation at energies $>10^{14}$ eV/Nuc.

28. At these energies we must exclusively rely on measurements of the size of extensive air showers (EAS) made at or near sea level. Most of the arrays that are used for these measurements sample the density of particles at various locations in the shower and attempt to determine the location of the shower axis or core (point of impact of the primary particles). Then, using a suitable lateral distribution function for the particles, either one theoretically or empirically derived, or better yet, one derived directly from the measurement itself, the total number of shower particles is obtained. In this sense, the differential or integral frequency-size distribution is the measurement of basic interest. The interpretation of this in terms of an energy spectrum of particles at the top of the atmosphere is a separate and complex problem.

Considering first the frequency-size distribution as obtained near sea level, we present in Fig. 39 a summary of the most recent results of the various groups working in the field. The numbers associated with the curves run from the earlier to the latest measurements. As our knowledge of the features of the EAS has improved, it is presumably possible to improve the estimates of N, the total number of particles in a shower, and the later measurements are corrected for many effects left out in earlier analyses. Nevertheless, a number of important uncertainties still exist in determining the absolute rate of EAS with number of particles $>N$. These are connected with 1. a systematic variation in lateral distribution with N, 2. fluctuations in shower development, and 3. geometrical size of the detecting area, as well as other lesser effects.

A casual glance at Fig. 39 suggests that no inconsistency arises if we assume that the integral number (size) spectrum of cosmic ray air showers is given by a power law spectrum with a single exponent $=-1.85 \pm 0.05$ over the range $N = 2 \times 10^5 - 10^9$. Thus

$$F(>N) = 3.6 \left(\frac{N}{10^6}\right)^{-1.85 \pm 0.05}. \tag{21}$$

Furthermore, an examination of the differential number-frequency spectra, out of which these integral spectra are constructed, reveals that it is almost impossible to interpret any fine structure in this region of shower sizes. The limit of our present ability would seem to be interpret any distinct change in slope of this size-frequency spectrum. In fact, a number of such interpretations have been made in the past. The Japanese workers (MURA and HASEGAWA [138]) have reported a break at precisely the point where the spectra in Fig. 39 start, that is $N \sim 1.3 \times 10^5$ particles. Below this number, the exponent on their size-frequency

Sect. 28. The energy spectrum of the total radiation at energies $<10^{14}$ eV/Nuc. 225

spectrum is -1.3. Also, the Moscow group (e.g., S. N. VERNOV, et al. [140]) have reported a break between $10^5 \leq N \leq 10^6$ with an exponent $= -1.4$ below $N = 10^5$. The place where this break is supposed to occur is in a region which is in transition from the very small air shower arrays to the very large ones designed to look at only the larger showers. New arrays in operation in a number of countries should help to define the characteristics of the spectrum in this range more closely, and, in fact, results from the Bolivian experiment (CLARK, et al. [142]; CLARK [143])

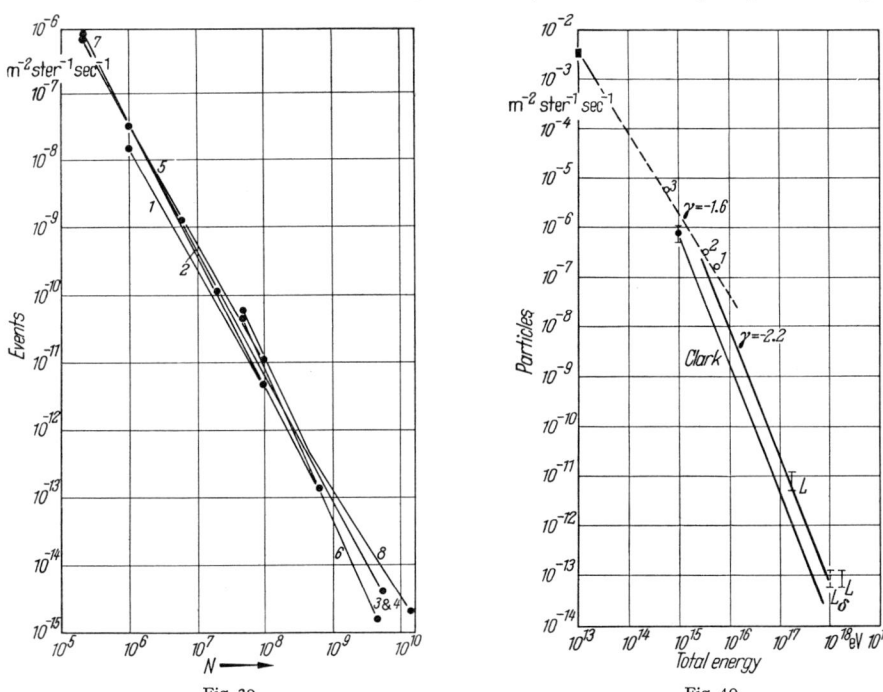

Fig. 39. The frequency of occurrence of EAS with total number of particles $>N$ where $10^5 \leq N \leq 10^{10}$ as measured by different groups. The numbers on these integral spectra refer to the following references: 1 [134], 2 [135], 3 [136], 4 [137], 5 [138], 6 [139], 7 [140], 8 [141].

Fig. 40. Energy spectrum of primary cosmic rays in the range $10^{13}-10^{19}$ eV; total energy as derived from the measured number-frequency distribution at sea level. The dashed curve represents the extension of the spectrum of the total radiation measured at energies $<10^{14}$ eV and having an exponent $= -1.60$. The three circles represent points derived from detailed energy loss studies of EAS in the atmosphere by 1 [136], 2 [145], 3 [146]. The solid curve corresponding to a primary spectrum with exponent $= -2.20$ is derived from the number frequency spectrum with $\delta = 1.15$. The points labelled L are from the number-frequency distribution assuming $E_0 = 2 \times 10^9 N$, constant with energy ($\delta = 1.0$).

again seem to confirm this break in more detail in the range $N = 10^5 - 10^6$. The question of whether such a break is a property of a change in the energy spectrum of primaries or due to some change in the properties of the fundamental nuclear interactions at high energies will be discussed presently when we attempt to derive the energy spectrum of the primaries.

If one now attempts to derive a spectrum of primary particles from the number frequency spectrum, one must introduce the additional uncertainties relating to the development and propagation of the shower through the atmosphere. A number of excellent reviews on EAS have covered the problems connected with this derivation thoroughly (e.g., COCCONI [144]), and here we shall present only a very simple approach to the problem.

For showers with $N \sim 10^5$, a point can be obtained on the integral total energy spectrum by integrating the energy lost in an EAS over the whole atmosphere, a

Handbuch der Physik, Bd. XLVI/2. 15

detailed balance energy consideration. Such points have been calculated by at least three workers for showers in the range $5 \times 10^4 \leq N \leq 5 \times 10^5$, and the corresponding total energies associated with the average frequency size spectrum just presented in Fig. 39 are shown for these points in Fig. 40. At these sizes, all workers arrive at an average primary energy per shower particle at sea level of from $1-1.5 \times 10^{10}$ eV. An extension of the spectrum for the total radiation deduced at lower energies is also shown in Fig. 40, and it is seen that there is no inconsistency if the spectrum of the primary particles is represented by a power law of single exponent $= -1.60 \pm 0.03$ over the entire range $5 \times 10^9 - 5 \times 10^{15}$ ev — 6 orders of magnitude. Thus, we have between these limits

$$J(>E) = 230 \left(\frac{E}{10^{10}}\right)^{-1.60 \pm 0.03}. \tag{22}$$

The spectrum above this energy can also be determined from the observed size spectrum provided we assume some model for the development and propagation of the EAS in the atmosphere. The arguments enabling this derivation to be made clearly indicate that the average primary energy per particle is greatest in smaller showers. This is equivalent to saying that the exponent on the integral primary spectrum is greater than the exponent on the integral number spectrum, e.g.,

$$\delta = \frac{\partial \ln N}{\partial \ln E_0} > 1. \tag{23}$$

Most analyses place the value of δ between 1.0 and 1.3. A number of calculations have been made on this problem. OLBERT [147], using various models for the high energy nuclear interactions, finds that $1.10 \leq \delta \leq 1.16$; CHUDAKOV et al. [148], have derived that $\delta \approx 1.15$; GREISEN [136], from an analysis of the way showers are absorbed in the air, gives $\delta \approx 1.21$; ODA [149] obtains the expression, $\delta \approx 1.52 - 0.045 \log N$, which for $N \approx 10^5 - 10^6$ gives $\delta \approx 1.25$. In a more direct experimental approach, comparing the number frequency distributions for the total particles and the μ-meson component, MURA and HASEGAWA [138] find $\delta \approx 1.3 - 1.4$ in the range of shower sizes $N = 10^5 - 10^7$.

An interesting observation relating to the value of δ is obtained by comparing the deduced primary spectrum of Fig. 40 with exponent $= -1.6$ in the range $10^{13} - 5 \times 10^{15}$ eV, and the reported measurements of the number frequency spectrum in the range $N = 10^3 - 3 \times 10^5$ which give an exponent $\approx 1.3 - 1.4$. The value of δ so obtained by relating these two numbers is ≈ 1.20.

Assuming that δ remains constant with energy and has a value of 1.15, we may transform the frequency size distribution into a spectrum of primary particles, normalizing at a value of $N = 2 \times 10^5$ corresponding to a primary energy of 3×10^{15} eV. This primary spectrum then has an exponent $= -2.2$ and is shown in Fig. 40. The exponent is almost exactly the same as the value of 2.17 given by CLARK, et al. [135]; however, the intensities are a factor of 3—4 higher as a result of the slightly different number-frequency spectrum used.

Another approach toward deriving the primary energy spectrum above $\sim 5 \times 10^{15}$ eV may be made by considering the fact that LINSLEY [141] argues that for $5 \times 10^7 \leq N \leq 10^{10}$, the relation $E_0 = 2 \times 10^9 N$ may be used to convert the number spectrum to an energy spectrum. This is a factor of six less than the primary energy E_0 per particle for a shower of size $N \approx 10^5$ particles and is, in fact, compatible with an average value of $\delta \approx 1.20$ throughout the range $N = 10^5 - 10^8$ particles. Applying LINSLEY's criterion to showers with $N =$ both 10^8 and 10^9 particles (e.g., assuming $\delta \approx 1.0$ at these sizes) along with the frequencies for these shower sizes given in Fig. 39 gives the limits in Fig. 40 marked L. If we assume δ is still ≈ 1.15 at these sizes and $E_0 = 2 \times 10^9 N$ at $N = 10^8$, we obtain the point marked L_δ in Fig. 40.

A comparison of the spectrum at high energies with that at lower energies suggests that there is a change in slope of the primary spectrum at $\sim 3 \times 10^{15}$ eV. Thus, this change in slope is not only implied by the change in slope of the number frequency spectrum as measured by a number of observers but also in what is effectively an independent manner by attempting to fit the fairly accurately known primary spectrum in the energy range $<10^{13}$ eV with the accurately known number frequency spectrum corresponding to energies of $10^{17}-10^{18}$ eV. Undoubtedly, experiments presently in progress will define this break more clearly, but meanwhile, it appears that the change in exponent is about 0.65 ± 0.1 at this energy.

At shower sizes $>10^9$ particles, corresponding to energies $>10^{18}$ eV, the only definitive measurements have been those reported by LINSLEY [*141*] which suggest that the number-frequency distribution begins to flatten out again. Between $N = 10^9 - 10^{10}$, the exponent on this spectrum is about -1.5. The only reasonable interpretation of this seems to be that the primary spectrum also flattens out, and between 10^{18} eV and the highest energy yet measured (corresponding to $N \approx 5 \times 10^{10}$) of $3 - 10 \times 10^{19}$ eV, the exponent on this spectrum lies between -1.5 and 2.0. The interpretation of the inflection favored by LINSLEY is that the inflection marks a transition between galactic and metagalactic cosmic rays. Obviously, there is need for continued detailed measurements of the number spectrum, particularly in the range $N = 10^5 - 10^7$ and again above 10^9.

XII. Detailed charge features of the radiation.

29. We have already discussed in some detail the breakdown of the primary radiation into various charge groups and have derived the following ratios appropriate to the top of the atmosphere:

$$L/S = 0.23, \quad H/M = 0.30.$$

Thus, it follows that $H/L = 1.00$. We further have seen that the variation of these ratios with energy is presently undetectable except perhaps at energies $\lesssim 1$ GeV/Nuc. We now seek to obtain a more complete breakdown of the charge composition and in order to increase the accuracy of this data, we shall consider measurements at a number of energies. Before we do this, we should briefly discuss the ability to resolve adjacent charges, however, since this is probably the limiting factor in obtaining reasonable values of intensities of the charges.

In emulsions with present-day techniques, it seems possible, with careful attention to detail, to resolve adjacent charges out to the range $Z = 14 - 16$ if the charges are relativistic. As the charge increases and as one approaches this limit, it becomes increasingly difficult to separate the adjacent charges and, indeed, a comparison of experimental results shows that some ambiguities apparently still exist in the $Z = 5$, 6, 7 and 8 region. The problem is enhanced when adjacent charges have widely different intensities, e.g., the odd-even atomic numbers above $Z = 8$. Above $Z \approx 16$, about all one can do is attempt to tell whether all of the charges are either even or odd and get some estimation of such a ratio. In the region of Fe ($Z = 26$), the difference in ionization of adjacent charges with the same velocity is ≈ 2.5 percent and the main objective is to decide whether the particle belongs to the Fe group or is perhaps some heavier nucleus with $Z \geq 30$.

Counter measurements of individual charges are not as comprehensive as those of emulsions, but present indications are that they can give comparable resolution in the relativistic range for charges with $Z \geq 4$. In fact, the counter measurements may be somewhat better than the emulsions for $Z \geq 16$ simply because the measure of ionization loss, which is the main identifying parameter, is made over a longer range interval than is usually practical in emulsions.

In Table 6, we present a list of the more definitive measurements of the fractional abundances of various charge components with Z between 3 and 10 and the total intensity of particles with $Z \geq 10$ as extrapolated to the top of the atmosphere. The results are presented in terms of the total number of nuclei with $Z \geq 3$; thus, according to our previous evaluations, the L nuclei should comprize 19 percent,

Table 6. *Fraction of different nuclei in cosmic rays as % of total nuclei $Z \geq 3$.*

Reference	Li	Be	B	C	N	O	F	$Z \geq 10$
a) Emulsions								
[104]	7.8	6.7	13.1	23.8	14.1	13.1	4.1	17.3
[48]	5.7	7.5	12.5	23.4	15.0	15.4	1.9	18.6
[47]	4.1	3.4	15.0	19.1	19.9	12.8	5.4	20.3
[107]	5.9	2.9	9.6	29.4	9.5	19.0	2.3	21.4
[36]	8.8	6.0	10.9	29.2	14.8	14.4		18.0
[39]	7.4	5.7	9.0	27.1	15.3	14.4		21.7
[45]			17.8	20.9	16.6	8.9		31.4
[108]		4.1	16.6	20.2	8.1	23.1	1.6	23.8
[42]	6.9	4.4	10.1	25.1	13.0	17.7	1.5	21.3
[98]			12.7	28.1	13.0	37.6	1.5	20.7
Mean of above	6.6	4.7	12.8	24.7	13.9	15.6	2.2	21.5
Mean according to Waddington [96]	5.2	4.3	11.9	25.1	14.9	14.5	4.0	20.1
b) Counters								
[23]		6.7	10.1	28.6	13.3	17.9		16.6
[101]	8.1	4.4	14.1	25.2	11.0	16.5	3.6	19.1
[44]	8.8	6.4	8.1	31.4	9.2	16.6		19.5
Mean of above	8.4	5.6	11.0	28.0	10.8	16.8	3.6	18.8
Mean of all measurements	7.0	4.9	11.9	26.3	12.3	16.0	2.9	20.2
			23.8		57.5			

the M nuclei, 62 percent and the H nuclei, 19 percent. In some cases in this table, different extrapolations to the top of the atmosphere were used for the different charge groups, and this will possibly change the percentages somewhat from those given above. This effect is clearly smaller than the other systematic errors in the analysis, however, and is neglected in this summary.

In the charge range $3 \leq Z \leq 10$, we see the results of the recent experiments are in close agreement with the summary of earlier data presented by Waddington [96]. Furthermore, there is no essential disagreement between emulsion and counter measurements of the charge composition. No individual measurement is significantly different than the average of all measurements given in the last row.

The main features of the charge distribution in this range of Z are as follows:
1. Boron is the most abundant and Beryllium the least abundant of the L nuclei (the mean of all measurements gives 0.41 for this ratio, appropriate to energies $\lesssim 1.5$ GeV/Nuc). 2. The abundances of CNO are approximately in the ratio 4/2/3 with the abundance of F being an order of magnitude less and probably just marginally detectable.

The picture regarding the H nuclei as shown in Table 7 is much less consistent. The breakdown into H_1, H_2 and H_3 nuclei gives more or less consistent results in

Sect. 29. Detailed charge features of the radiation.

that H_1 nuclei predominate, and an appreciable fraction of H_3 nuclei exist. The fraction of H_2 nuclei measured ranges from 3—10 percent of the total H nuclei however, with the more recent measurements giving an almost complete absence of these nuclei. The individual charge measurements have an even less comforting consistency. It is generally agreed that Ne and Mg are the most abundant, but major differences exist in the fraction of these nuclei reported. Similar differences exist for Na, which is one of the most abundant odd nuclei. Most measurements confirm a substantial abundance of Si, but the measurements of S are extremely variable. The charge range from 15—19 is certainly not very abundant, and the most definitive measurements, i.e., those of DANIEL and DURGAPRASAD [98] and KRISTIANSSON, et al. [108], are consistent with an almost complete absence of these nuclei. Above $Z=20$, the particles appear to be concentrated in the even nuclei with Fe contributing ~50 percent of all these nuclei. The abundance of Ni is ~10 percent of that of Fe, but the abundance of nuclei with $Z \geq 30$ must be ~1 percent of that of Fe with not more than one or two of these nuclei having been observed in emulsions in the last 10—15 years. A study of all reported counter and emulsion measurements of nuclei with $Z \geq 3$ since they were first discovered in emulsions reveals that ~1500 L nuclei have been identified, ~4000 M nuclei and ~1300 H nuclei. These numbers are of some interest for comparison with Russian measurements made using Čerenkov detectors carried on interplanetary probes and earth satellites, summarized by GINZBURG et al. [150]. These results, while not as definitive as those just discussed, have the benefit of much greater statistics (in all, ~20000 nuclei with $Z \geq 3$ have been recorded). The intensity of nuclei with $Z \geq 15$ observed by the Russians is ~5 percent of the total number of nuclei with $Z \geq 3$. From our analysis, we would predict 5.2 percent. A measurement of nuclei in the $Z \geq 24$ range by the Russians also fixes an upper limit for these nuclei of ~5 percent of all nuclei with $Z \geq 3$, indirectly suggesting an

Table 7. *Fraction of heavy nuclei as % of total nuclei with $Z \geq 10$.*

Reference	H_1	H_2	H_3	Ne	Na	Mg	Al	Si	P	S	$16 \leq Z \leq 19$	Ca	Ti	Cr	Fe	Odd $Z > 20$
Rochester*	68	9	23	19.2	8.6	20.7	4.6	12.5	2.0	5.0	9.3	6.0	1.3	0.7	9.3	3.0
Bristol*	73	8	19	19.1	15.0	24.0	6.0	7.2	2.4	3.0	7.8	3.0	3.0	5.4	4.2	0.7
[108]	70	4	26	24.0	4.0	21.3	1.3	18.7	0.7	4.7	4.4	0.7	2.7	6.7	8.0	
Minnesota*	66	9	25	36.0	2.0	9.5	2.5	15.0	1.0	4.5	9.0	1.5	2.0	2.0	16.0	
[117]	73		31	17.1	8.1	25.2	8.1	12.6	2.7	9.0						
[98]	65	<4	24	21.5	13.5	23.5	4.5	8.0	1.0	0	<4.0	4.5				
[36]	73	3	(27)								3.0					
[42]	75		(25)	28.0	16.5	7.1	1.4	15.0								
[101]	65		(35)													
Mean according to WADDINGTON [96]	71	3	24	21.4	13.4	23.2	4.5	8.0	0.9	0	3.1	4.7	3.0	5.0	9.9	1.6
Mean of all measurements	68	3—6	26	23.5	10.0	18.2	4.0	13.6	1.3	4.4	3—8	3.3	2.2	3.7	9.5	1.6

* The Rochester, Bristol and Minnesota entires represent a summary of all published data from these groups.

absence of nuclei in the $15 \leq Z \leq 24$ range. An estimate by GINZBURG et al. [150], of an upper limit of between 0.1 and 0.01 percent of all nuclei with $Z \geq 3$ is obtained for nuclei with $Z \geq 34$. This is $\sim 0.5-5$ percent of the intensity of Fe nuclei.

Among the most interesting features of the heavy nuclei are the odd-even abundances. For the H_1 nuclei, this ratio is 0.27, for the H_2 nuclei, the ratio is not clearly defined, while for the H_3 nuclei, the ratio is ~ 0.08. Thus, in fact, the nuclei with $Z \geq 20$ are almost entirely even nuclei, and a case might well be made for the complete absence of the odd nuclei.

An appreciation of the full significance of the abundances of the various nuclei in the primary cosmic rays can be derived by attempting to compare these

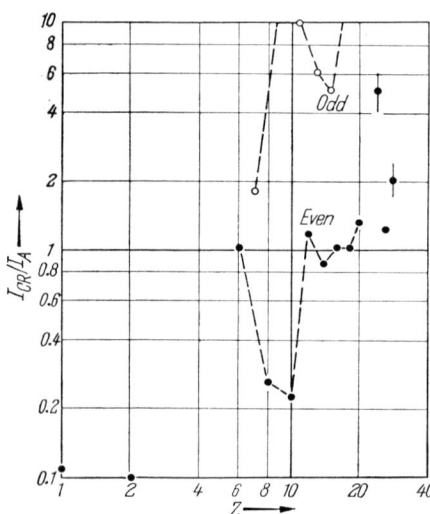

Fig. 41. Ratio of abundances of various nuclei measured in cosmic rays to the cosmical abundances given in Table 8. The abundance of Carbon is taken equal to 1, and the ratios are shown separately for both the even and odd nuclei.

Table 8. *Abundance table of various charges with the abundance of carbon = 1.*

Element	Solar (ALLER 1961)	Cosmical (Avg. 4 Refs.)	Primary Cosmic Rays
He	?	400	38
Li	≪0.001	≪0.001	0.27
Be	≪0.001	≪0.001	0.19
B	≪0.001	≪0.001	0.43
C	1.0	1.0	1.0
N	0.16	0.27	0.46
O	1.7	2.3	0.61
F	≪0.001	≪0.001	0.09
Ne	?	0.80	0.18
Na	0.004	0.006	0.08
Mg	0.050	0.12	0.15
Al	0.004	0.005	0.03
Si	0.063	0.13	0.11
P		0.002	0.01
$16 \leq Z \leq 19$	0.050	0.020	0.02—0.05
Ca		0.02	0.026
Ti		<0.001	0.017
Cr		0.006	0.030
Fe		0.06	0.080
Ni		0.008	0.015
Odd $Z > 20$		<0.001	0.01
$Z \geq 30$		~0.001	<0.004

abundance with so-called "cosmic" abundances. The later represent an attempt to determine the average abundance of matter in the galaxy, as derived from observations of the matter, generally in stellar atmospheres, available for our inspection. Specific regions or stars in the galaxy may and do have abundances which are appreciably different from the "cosmic" abundances. At least four attempts have been made to derive such cosmic abundances (SUESS and UREY [151], ALLER [152], CAMERON [153] and ALLER [154]). The derivations differ sometimes by a factor of 3—4 for the abundance of a particular charge, and we can surmise that uncertainties here are comparable to those in the cosmic ray measurements. More accurately known are the solar abundances of the various charges. These are shown in Table 8 along with the average cosmical abundances and those measured for cosmic rays.

A comparison of these abundances reveals a number of interesting features which may be used as guidelines for further discussion in a later section. These are 1. the well-known striking overabundance of the L nuclei in the primary radiation, believed to be one of the clearest indicators of the passage of the cosmic rays through a non-negligible amount of material enroute to the earth from their source, 2. a relative abundance of heavy nuclei in the cosmic radiation that greatly exceeds that found in cosmic abundances. Or perhaps it might be better to express that cosmic rays contain relatively less hydrogen and helium as compared to the heavier elements. In actual fact, as can be seen from Fig. 41, neither of these statements is strictly true. Considering, at first, only the even nuclei, which are both the most abundant and most accurately measured, we see that, based on a C abundance ratio of 1, all even nuclei with $Z \geq 12$ appear to have comparable abundance ratios. Any tendency for the ratio to increase with increasing Z appears to be small, indeed. From this viewpoint, it is O and Ne among the heavier nuclei that are deficient. On the other hand, it might be argued that C (and perhaps Mg) are overabundant, and some systematic increase in the ratio with increasing Z does exist for the other even nuclei. What is certainly true is that the even heavy nuclei in cosmic rays have a relative abundance 10—20 times that for hydrogen and helium as compared with the cosmical abundance measurements.

3. A relative abundance of odd nuclei in the cosmic radiation that greatly exceeds that obtained in cosmological abundance measurements, and, in fact, considerably exceeds the abundance excess of even nuclei. It may be significant that the two most abundant, and possibly the only two clearly identified odd nuclei with $Z \geq 6$, namely N and Na, are less overabundant in cosmic rays than the other odd nuclei. This odd-even nuclei behavior may simply be one of insufficient resolution between adjacent charges, or it may be a result of propagation of the particles in which the very weak source abundance of the odd nuclei is enhanced by fragmentation; or it may represent a real favoritism for odd nuclei in the cosmic ray sources. These points will be discussed more in a later section.

a) Isotopic composition of the primary radiation.

30. Outside of the work done on the isotopes of hydrogen and helium, discussed earlier, very little in a definitive sense has been achieved experimentally on the isotopes of heavier nuclei. A number of interesting problems exist, however, which are now just becoming amenable to experimental observations. One of these relates to the relative abundance of C^{12} and C^{13}. As HASEGAWA and ITO [155] have pointed out, this ratio should vary greatly depending upon the energy processes occuring in the sources of cosmic rays, varying from $C^{13}/C^{12} \sim 0.2$ for helium burning processes such as might occur in red giants, or in the CN cycle processes occuring in stars similar to the sun, to ~ 1 in sources in which the rapid CNO cycle occurs (e.g., late stars such as supernova). The above authors have presented some very preliminary evidence based on interaction characteristics to show that this ratio may be ≈ 1 in the primary radiation.

More recently, ALVIAL and co-workers [156] have attempted to differentiate between these isotopes directly in emulsion by comparing their range with detailed measurements of ionization loss along the track of the particle. The above methods require a great deal of painstaking analysis and seem to be pushing the emulsion techniques near to their limit. Very preliminary work from these workers also suggests a ratio of $C^{13}/C^{12} \approx 1$.

Another isotopic problem of great interest concerns Be^{10} with a half life of 2.5×10^6 years, somewhat comparable to the lifetimes of cosmic rays in the galaxy as deduced by various other methods. If one assumes that the L nuclei are all created as secondaries, then following the work of BADHAWAR, DANIEL and VIJA-YALAKSHMI [157], the relative abundance of Be and B can be calculated assuming all of the radioactive isotopes produced by fragmentation, decay. This, of course, should be different than the observed ratio of production (in emulsion, for example) since many of the short-lived isotopes will eventually decay during the subsequent propagation of the particles. The reaction $Be^{10} \to B^{10} + \beta^-$, however, has a long enough half life so that particularly at high energies Be^{10} should contribute along with stable Be^9 to the measured intensity of these nuclei. The actual "age" of the primary radiation depends on the assumed average density of interstellar matter and the amount of material the particles have gone through and is believed to lie in the interval $10^7 - 3 \times 10^8$ years according to the best estimates of these quantities. The variation of the Be/B ratio as a function of energy might just be observable then and would give extremely valuable information on the propagation of the radiation. Based on fairly simple assumptions regarding the cross sections for the production of various Be and B isotopes, the above authors [157] calculate Be/B ≈ 0.33 at energies of a few hundred MeV/Nuc, increasing to ≈ 0.52 at energies where the decay of Be^{10} is negligible. WEBBER and ORMES in a series of flights have studied the variation of this ratio with energy [158]. At energies <1 GeV/Nuc, they find Be/B $= 0.28 \pm 0.02$ and at energies > 5 GeV/Nuc, Be/B $= 0.30 \pm 0.08$. In view of the small effect observed it appears that the lifetime of these nuclei must be $\gtrsim 2 \times 10^7$ years.

XIII. A comparison of the spectra of the different charge groups.

31. We shall now attempt to compare at all energies the spectra of the various charge components as have been derived in the previous sections and inferred by other workers. Such a comparison has important implications relating to almost all of the cosmic ray problems from acceleration of the radiation to its propagation through interstellar space.

As we have seen in Sect. X, in the range 3×10^{10} to $\approx 10^{14}$ eV, the experimental results, while falling considerably short of the accuracy that one would like, particularly in the case of the L nuclei, and the sub-groups of the S nuclei, give no evidence that there are any significant differences in the spectra of the different

Table 9. *Ratios of the various charge groups in the primary radiation in the energy range* $3 \times 10^9 - 10^{14}$ eV.

	Rigidity	Energy/Nucleon	Total Energy/Nucleus	Nucleons $> E$
P/He	6.4 ± 0.3	17.6 ± 1.2	4.4 ± 0.3	4.4 ± 0.3
He/S	11.6 ± 0.2	11.6 ± 0.2	1.5 ± 0.03	1.5 ± 0.03
L/S	0.25 ± 0.02	0.25 ± 0.02	0.08 ± 0.01	0.08 ± 0.01
H/M	0.30 ± 0.02	0.30 ± 0.02	0.66 ± 0.04	0.66 ± 0.04
$P/Z \geq 2$	5.8 ± 0.3	16.0 ± 1.2	2.4 ± 0.2	2.4 ± 0.2

charge components. In the range $3-30$ GeV, much more accurate measurements are available, and the results summarized earlier again give no reason to believe that there are any significant differences in the spectra of the different charge components (when compared as a function of rigidity). Almost all workers who have made measurements on the intensity or spectrum of a charge component in this range also attempt to compare the spectra of the different charge groups and without citing each of these numerous references individually, we can say that the general consensus is that the spectra of the different groups are the same in this energy range in accordance with our findings. Individual measurements may frequently be accompanied by relatively large experimental errors which may be overcome in summarizing many similar measurements, and in this connection, it is well to note that the reviews of WADDINGTON [15], [96] arrive at essentially the same conclusion.

This gives a range from $\sim 3 \times 10^9 - 10^{14}$ eV, nearly five orders of magnitude in energy in which there are no observable differences in the rigidity (and energy) spectra of the various charge components. The ratios of the various charge groups as a function of rigidity, energy/nucleon, total energy/nucleus and nucleons above a given energy in this range are shown in Table 9.

At both the high and low energy ends of this range, apparent differences in the charge composition do exist, and it is these we wish to discuss here at some length, beginning with the high energy end.

a) Charge composition at energies $> 10^{14}$ eV.

32. The first experimental evidence of a possible change in the composition of the primary cosmic radiation at high energies was reported at the International Conference on Cosmic Rays held in Japan in 1961. The results were of a preliminary nature only and based on the features of EAS in the range $10^5 - 10^7$ particles. The MIT group [139] noted a lack of fluctuation in the proportion of penetrating particles (μ-mesons) in showers with $N > 10^6$ particles. According to these workers, this observation is most likely reconciled with a charge composition in which all the primary particles are either lighter than or heavier than carbon (more recent interpretations of these results have suggested that for the very largest showers these particles are most likely protons [141]). From very similar measurements covering the range $10^5 \leq N \leq 10^7$ the Tokyo group [159] concluded, however, that the frequency of heavy primary initiated showers is $\approx$ a few percent of the total, a result more or less consistent with the known spectrum at lower energies. At this same conference, the Russian workers [140] suggested that a change in composition occurs at energies corresponding to sizes $> 5 \times 10^6$ particles as a result of their measurements that the exponent of the size-frequency spectrum of EAS changes at this point. It is evident that the appearance of the shower size spectrum will be influenced not only be changes of the primary cosmic ray energy spectrum but also by changes in its composition. As we have seen in our earlier discussion, the experimental evidence regarding this change in exponent is now very strong.

Turning to more recent results relating to features other than the change in the exponent of the size-frequency spectrum of EAS, we see that although considerable controversy still exists, a number of lines of evidence point to a change in character of EAS at a size of about 10^6 particles ($\approx 3 \times 10^{15}$ eV total), just the size at which the change in the spectrum occurs. Although it is possible to interpret some of these measurements in terms of a change in the characteristics of the nuclear interactions at these high energies, a possible explanation is that we are seeing the manifestations of a change in composition of the primaries themselves.

Very detailed measurements of shower structure by the Sydney group [160] show a change from mainly (≈ 60 percent) single cored showers at sizes $< 10^6$ particles to multiple cored or flat topped showers above this size (21 multiple cored; 2 single cored). The multiple cored showers are interpreted as being due to heavy nuclei.

The interpretation of the Bombay group [162] on the fluctuations of the penetrating particles (e.g., μ/N ratio) in showers seems to support the above conclusions; however, the work of the Tokyo group [161] apparently does not admit of such a definite interpretation.

Some Russian work [146] on the relative fluctuations of Čerenkov light intensity (a measure of total ionization) in showers with $10^5 \leq N \leq 10^7$ also gives no evidence for a pronounced change of composition in this range.

b) Spectra at energies $< 3 \times 10^9$ eV.

33. Turning now to the low energy end of the primary spectrum, we find quite definite and clear-cut differences in the spectra of the different charge components and hence in the charge composition. Here one of the most important aspects of the discussion is the relative spectra of protons and helium nuclei and the significance of energy/nucleon or rigidity as a parameter for describing the spectrum. In the high energy region that we have just been discussing this difference is unrecognizable, except perhaps in the features relating to the change of the spectrum (or composition) at the highest energies; however, at low energies it becomes of crucial importance.

Since we have been discussing the different charge components in terms of rigidity spectra, let us first present the data in this form. In Fig. 42 we show the rigidity spectra of the protons, helium nuclei $\times 6.5$ and the S nuclei $\times 75$ in the range 4 GV to the lowest energy yet measured (as determined in 1963). The spectral differences between these components are striking, and the difference between the helium and S nuclei would be identical on an energy plot as well since both groups have $A \approx 1/2$. Slight differences would arise because of the presence of He^3, but these could explain only a small fraction of that observed. The relative proton and helium nuclei spectra would, of course, be much different on an energy/nucleon basis, and we shall come to this point shortly.

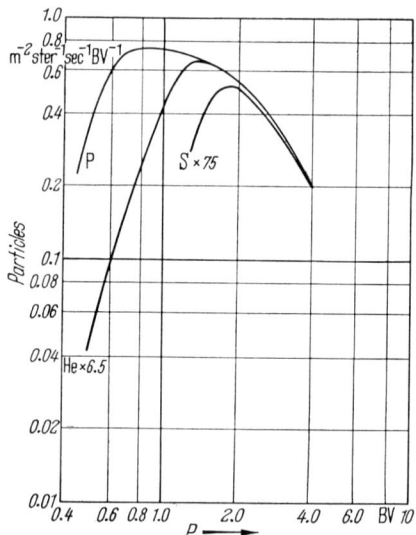

Fig. 42. Differential rigidity spectra below 4 GV of protons, helium nuclei and S nuclei as measured during 1963. Spectra are normalized to protons in the 3—4 GV region using a factor of 6.5 for helium and 75 for S nuclei.

Meanwhile, in Fig. 43, we show the P/He and He/S ratios as a function of rigidity. The most striking feature of this figure is the abruptness with which these two ratios change, the He/S ratio changing at ≈ 2 GV (460 MeV/Nuc), the P/He ratio changing at ≈ 1.3 GV. A number of possible causes of a change in these ratios at low energies exist but a cause of such an abrupt change is not immediately apparent. The effect of energy loss by ionization in the interstellar hydrogen is illustrated in the figure, and while this does produce marked changes in these ratios at low energies, these changes are not as abrupt as those actually observed, at least for a path length constant with energy.

The discovery of the change in P/He ratio at low energies is a prime example of the improvement that has been achieved by balloon and satellite techniques without a really substantial improvement in instrumentation. The earlier work of McDonald and Webber [10], [41], which is the only data existing before

1962 comparing these components, did not extend the helium spectrum below ≈ 1.2 GV or the proton spectrum above ≈ 1 GV. As a result, these workers were unable to determine that, in fact, the spectra of the two components were different.

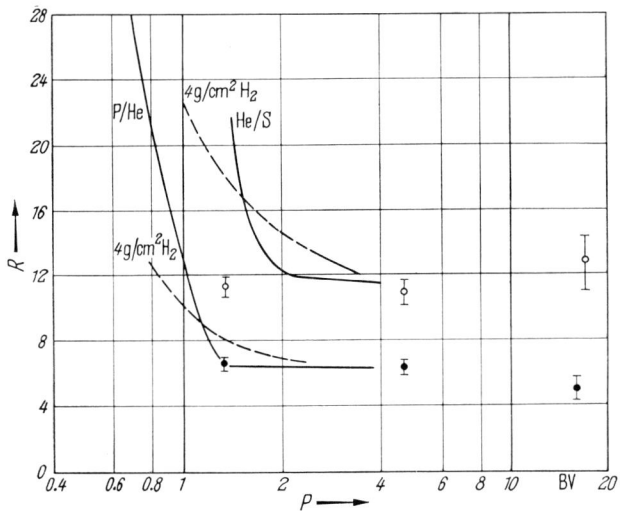

Fig. 43. The P/He, and He/S differential rigidity ratios as a function of rigidity, Solid circles refer to integral P/He ratios above a given rigidity, open circles to integral He/S ratios, solid lines are derived from the differential spectra. (Note insensitivity of lowest integral He/S point to the change in differential ratio at slightly higher rigidities). Dotted curve represents how a ratio, constant with rigidity at the source, would change if the measured spectrum is reproduced after passage through 4 g/cm² of interstellar hydrogen.

It was FICHTEL, et al. [19], [20], who, on the basis of measurements made in 1961 and 1962, first suggested a difference in the two spectra of the character we now observe. Measurements in the summer of 1963 by many observers, as presented in Figs. 24 and 27, clearly elucidated these differences.

If we further subdivide the nuclei heavier than helium, we recall from Fig. 35 that a summary of four measurements between 1.5 and 4 GV shows that the H/M ratio has a slight tendency to increase with decreasing rigidity, although certainly within the rather large experimental errors the evidence at present suggests that this ratio remains sensibly constant down to ≈ 1.5 GV. This is in general agreement with the conclusions of AIZU, et al. [36], KOSHIBA, et al. [42] and WEBBER and ORMES [101], although the work of EVANS [43] seems to be more suggestive of an increase in this ratio as one goes to lower energies.

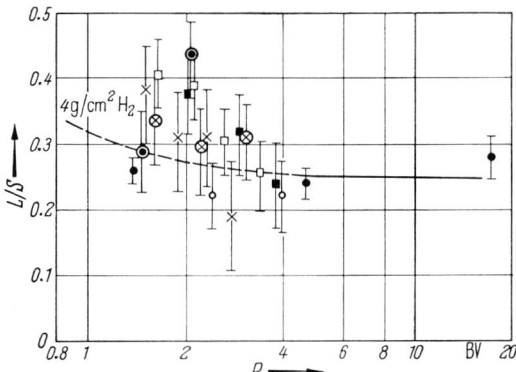

Fig. 44. The differential L/S ratio between 1.3 and 4 GV as measured by various observers. Solid circles and solid line refer to the integral ratios determined in Sect. VII. The various other symbols refer to the following references: ■ [23], ⊗ [36], ○ [42], × [43], □ [101], ○ [44]. Dotted curve represents how a ratio, constant with rigidity at the source would change if the measured spectrum is reproduced after passage through 4 g/cm² of interstellar hydrogen.

Turning to the L/S ratio, we exhibit in Fig. 44 the results of six separate measurements in various differential intervals between 1.3—4 5 GV. Here, in spite of rather large errors and differences of opinion from different observers, it appears that a definite increase in this ratio occurs as one goes to lower rigidities. This agrees with the specific conclusions of Aizu, et al. [36], Koshiba, et al. [42], Evans [43] and Webber and Ormes [101], although McDonald and Webber [23] and Balasubramanyan and McDonald [43] conclude that their results are consistent with an essentially constant ratio. This disagreement seems to be purely

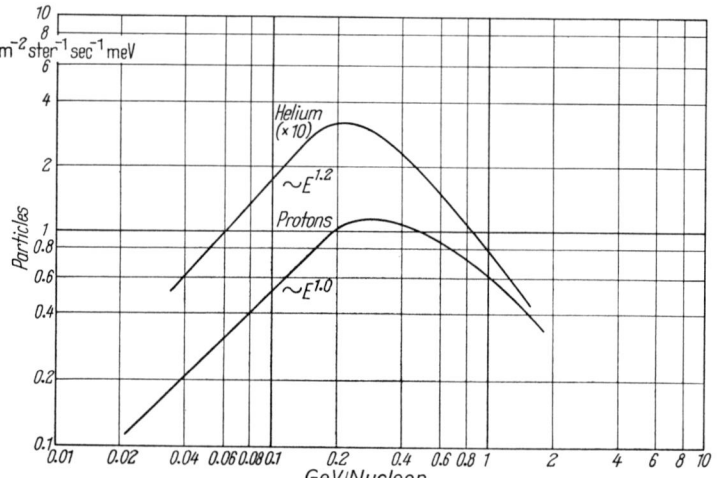

Fig. 45. Differential energy spectrum of protons and helium nuclei (× 10). Note the behavior at energies above the peak, the relative location of the peaks, and the fall off at low energies (based on 1963 proton and helium nuclei spectra presented in Figs. 24 and 27).

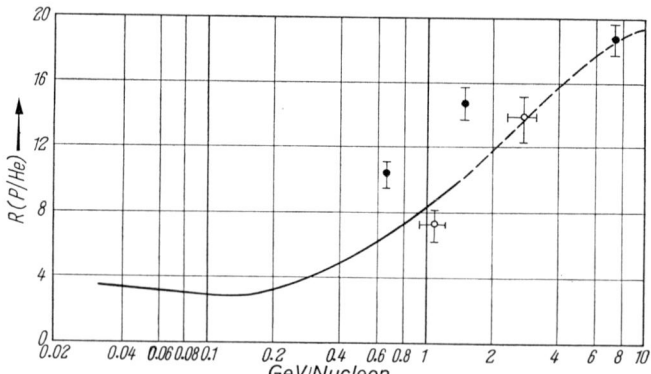

Fig. 46. Ratio of differential intensities of protons and helium nuclei an on energy/nucleon basis. Solid curve is based on the differential spectra. Solid circles are based on integral intensity measurements. Open circles and dotted curve are based on differential of integral spectra between 1.5—7.2 GeV/Nuc.

one relating to the statistical accuracy of the data, and improvements in counting rates of factors of 10 that should be soon forthcoming will settle this discrepency as well as defining more clearly the possible differences in the spectra of L nuclei at low energies.

We should point out that the above conclusions will also hold if the data of the heavier nuclei are compared in terms of energy/nucleon. For the M and H nuclei the average A is very closely $1/2$; however, for the L nuclei, it is $\approx 1/2.2$. The observed difference in the L nuclei spectrum cannot be completely resolved by comparing them on an energy/nucleon basis, although such a comparison reduces the differences.

Sect. 34. Effects of solar modulation on the low energy end of the spectrum. 237

If the differential spectra of protons and helium nuclei are now compared on an energy/nucleon basis, or what is equivalent for differential spectra a total energy/nucleon basis, we find a completely different behavior than in the case rigidity is used as a parameter (Figs. 45 and 46). Here there is evidence that the $P/$He ratio begins to change from its high energy value of ≈ 18 at perhaps 5 GeV/Nuc, decreasing steadily to a value ≈ 3 at 200 MeV/Nuc and remaining very nearly constant at this value down to the lowest energies yet measured (30 MeV/Nuc). The peak in the proton differential energy spectrum occurs at a slightly higher energy (≈ 300 MeV) than does that of the helium nuclei (≈ 200 MeV/Nuc). At energies below the peak, however, both spectra fall off very closely as $E^{1.0}$ thus producing the constant ratio at low energies. The slightly more rapidly fall-off of the spectrum for helium nuclei cannot at present be regarded as significant, although if it could be firmly established, it would be of extreme importance relating to the propagation of these two components.

In this study, the helium component is treated as being entirely He^4. If ≈ 20 percent is He^3 instead, the spectrum will be altered somewhat but none of the effects just discussed can be explained by this composition. Instead, we must regard the behavior of the $P/$He ratio as a function of energy/nucleon as very strange indeed, particularly the change in this ratio beginning at energies as high as ≈ 5 GeV/Nuc.

In lieu of a satisfactory explanation of this behavior in terms of acceleration and/or propagation effects (see next section), it seems that a comparison of this ratio in terms of rigidity spectra for the two charge components leads to behavior more in keeping with our ideas concerning how these two effects will modify the injection spectrum at low rigidities. If so, and if we can believe that rigidity is the parameter characterizing the primary spectrum at low energies, then we have a most important clue as to the acceleration process.

In addition, it would be extremely valuable to compare in some detail the spectra of protons and of the electronic component; however, owing to the large uncertainties in the latter about all one can say is that over the range from ≈ 50 MeV to a few GeV in energy the differential energy ratio P/e apparently increases with increasing energy from values perhaps ≈ 50 at the low energy and to values nearer 200 at the high energy end.

c) Effects of solar modulation on the low energy end of the spectrum.

34. The discussion in this section has been based on spectra obtained during 1963 at a time when the solar modulation still decreased the counting rates of high latitude neutron monitors by ≈ 6 percent below their sunspot minimum values. As we have seen in Sect. II, the modulation at low energies may be considerably larger than this, and its energy or rigidity dependence is not well understood. Nevertheless, a few useful comments pertinent to the preceeding discussion can be made.

First of all, for all of the heavier components with $\varLambda \approx \frac{1}{2}$, the modulation, whether it is energy (velocity) dependent or rigidity dependent or a combination of the two, will not effect the ratios of the different components. Thus, the differences between the spectra of the helium and S components and the L and S components are in all likelihood preserved and are a property of the primary radiation outside of the solar system. We cannot, of course, make such a simple statement for the proton and helium components, although we can consider two simple possibilities.

Let us assume that the modulation at low energies ($\lesssim 1$ GeV/Nuc) is essentially energy (velocity) dependent. Then, if we compare differential energy/nucleon spectra of these two components, the solar modulation will not change the ratio of the two spectra. (If the modulation becomes rigidity dependent at higher energies, the ratio may change; but since the modulation is believed to be small above a few GeV, it cannot produce the observed large change in this ratio between a few hundred MeV and ≈ 5 GeV). Thus, on this picture the differences in the two spectra are a property of the primary radiation outside of the solar system.

If, instead, we compare the differential rigidity spectra of the two components, then since the energy dependent solar modulation will change the helium nuclei more than the protons at a given rigidity, for a decreasing percentage of modulation the ratio P/He will decrease and tend to offset the observed rapid increase in this ratio at low rigidities. Just how effective such a solar modulation effect would be and whether it also disturbs the apparent constancy of this ratio above 3 GV depends on the spectrum and the amount of modulation at low rigidities, and also if the energy/nucleon or velocity type of modulation extends to higher rigidities. A precise unravelling of these effects is not possible at present and will have to await more precise measurements of the modulation near sunspot minimum.

Comparing the behavior of the electron component with that of the protons, we see that, unlike the case of nuclei, there is an important observational difference between an energy and a velocity dependent modulation. If the modulation depends on velocity (as in the case of a diffusion picture), then the change in the electronic component will be negligible except at the very lowest energies, and decreasing modulation should increase the P/e ratio. If, on the other hand, the modulation depends on energy (as in the case where FERMI deceleration dominates), protons and electrons should vary in a like manner, resulting in a constant P/e ratio with changing modulation. This important differnce may be of great value in determining the relative importance of these different effects in the solar modulation process.

Finally, we should mention the existence of a peak in the differential spectrum for all of the charge components. To discuss how this peak will be modified by the solar modulation process requires a knowledge of the energy or rigidity dependence of the modulation. This is very poorly known at low energies near sunspot minimum; however, the best evidence as summarized in Sect. II is that if we consider a modulation $\approx P^{-\gamma}$, or $E^{-\gamma}$, then γ is surely > 0 but most likely less than one (high energy data would suggest that it might be steeper than this, however). This being the case, since the observed spectra of protons and helium (and possibly S nuclei, as well) in 1963 decreases as $\approx P^{3.0}$ or $E^{1.0}$ at low energies, the solar modulation will not remove the peak in either the differential rigidity or energy spectrum for these components. Under these circumstances, the peaks will be a property of the primary radiation outside of the solar system. Although they will occur at a lower energy (or rigidity), the relative position will not change for the heavier nuclei. For the protons relative to the helium nuclei, assuming an energy dependent modulation, the separation of the peaks will be enhanced if we take rigidity spectra and remain essentially the same if we take energy spectra for the primary particles.

XIV. Astrophysical consequences of the above results in relation to the acceleration and propagation of cosmic rays.

35. In this section, we shall discuss some of the astrophysical consequences of the experimental summary presented earlier. These results shall be examined in the light of the origin and propagation of the cosmic radiation. Recently, a number of excellent summary papers on the origin of cosmic rays have appeared [99], [163], [164]. These works have been more or less oriented toward the theoretical aspects relating to the origin, and it is our intention here to approach the subject in a somewhat different manner. We shall use the experimental data as a basis and investigate how this restricts and guides our ideas of the origin and propagation of the radiation. The limitation of our experimental knowledge has, in the past, allowed a wide scope in the types of sources postulated for the primary cosmic radiation as well as the ideas regarding the acceleration and propagation of these particles to the earth. It is hoped that the experimental progress as summarized in this article as well as the progress of our general knowledge of the related astrophysics leads us now to more clearly defined ideas regarding these problems.

In the discussion to follow we shall consider the primary cosmic radiation to be essentially produced and to propagate in our galaxy. It would seem that cosmic rays of metagalactic origin cannot play an important role except perhaps at the highest energies ($>10^{17-18}$ eV).

For convenience in discussing the experimental results, we shall consider separately the following aspects or stages relating to the origin and propagation of the radiation to the earth:

1. The propagation of the cosmic rays from the source region to the earth,

2. The process of acceleration and escape from the source region, and an aspect which depends on the interpretation of the above two stages, namely:

3. The source region in which cosmic rays are accelerated.

We shall discuss these aspects beginning with the known features of the radiation at the earth (corrected as well as possible for solar modulation effects) and attempting to summarize the experimental data which is relevant to a particular stage.

a) Propagation of the primary radiation in the galaxy.

36. The observed cosmic ray beam is nearly isotropic. In spite of our very limited knowledge of the magnitude and character of the magnetic fields in the spiral arms and halo of our galaxy, we believe that the interaction of the particles with these fields has caused the particle directions to be randomized so that statistically they strike the earth from all directions even though the galactic position of the earth is by no means central. This stirring process is most usefully described in terms of a diffusion picture, the characteristics of which are importantly related to a large body of the experimental results presented in this article. This stirring may be, in effect, the acceleration process itself. We are thus led to consider the possibility that particles of different charge and different energy take differing times (and hence have different path lengths) to arrive to us. This time of storage or mean delay time between injection and detection is clearly an important parameter relating to the propagation of the radiation. In addition, cosmic ray particles can be lost from the system either by a catastrophic process such as a nuclear collision in which a heavier nucleus can be broken up into other nuclei, as well as eventually electrons and other components; by energy loss in one of many processes which move the particles to a new point in momentum space; or by direct escape from the galaxy.

These are the essential ingredients of any model that seeks to describe the propagation of the radiation. Let us now examine this propagation more closely from the point of view of the experimental observations. Perhaps the most important observational quantity related to the propagational characteristics is the galactic magnetic field, its magnitude, degree of fluctuation and spatial orderlines. Recently, measurements have been carried out using the ZEEMAN splitting of the 21 cm interstellar absorption lines as well as polarization observations of galactic and extra-galactic radio emissions which give some indication of the characteristics of these fields. The average values obtained by different observers looking in the same direction differ by factors of 3 or 4, however, and values from $\approx 2 \times 10^{-5}$ G in some directions to $\approx 2 \times 10^{-6}$ G in other directions have been reported [165]. There is also some evidence that the product $B_\perp n_e$ decreases with increasing distance from the galactic plane [166]. These polarization and FARADAY rotation measurements of extra-galactic radio sources are consistent with fields decreasing from $\approx 5 \times 10^{-6}$ G near the galactic equatorial plane to perhaps $\approx 2 \times 10^{-6}$ G in the halo. The corresponding electron densities would be $10^{-1}/cm^3$ and $10^{-3}/cm^3$ respectively. Available data on the degree of fluctuation of the magnetic field magnitude and spatial orderliness is difficult to interpret, although there is some evidence of the existence of a more ordered field along the spiral arms. In particular the details and character of the halo field are not well understood.

Cosmic ray observations themselves may be interpreted in terms of the characteristics of the galactic magnetic field. The energy density contained in the estimated galactic cosmic ray spectrum at sunspot minimum is ≈ 1.2 eV/cm³. It is possible that there is a concentration of cosmic rays near the sun itself, but it is quite likely that this is a characteristic energy density associated with the spiral arms. In this case if the magnetic field is to control the motion of the cosmic rays and produce isotropicity, then $B \gtrsim 7 \times 10^{-6}$ G. If, in addition, cosmic rays are to be assumed to move essentially freely from the spiral arms to the halo as is required by a number of other observations, and if stirring is maintained in the halo as well, B_H must also be $\gtrsim 7 \times 10^{-6}$ G.

This rather high halo field requirement can be alleviated somewhat if the conservation of the adiabatic invariant $\sin^2 \theta/B$ is applicable to some extent for particles in going from the disk to the weaker halo fields. Then as we have seen, $J/J_0 \approx \frac{1}{2}(B/B_0)$. Thus, if the average field is decreased by a factor of 5 between the disk and the halo, B_H need only be $\gtrsim 4 \times 10^{-6}$ G to maintain an equivalent stirring.

37. Implications of the measurement of the electronic component on propagation conditions. The measurement of the spectrum of the electronic component in the primary radiation and comparison with the spectrum of non-thermal radio emission from the galaxy (halo) has extremely important implications in connection with the interpretation of the magnetic fields in the galaxy. The association of this non-thermal radio emission with the synchrotron radiation from the cosmic ray electrons moving in the galactic magnetic fields is generally well accepted, following the pioneering work of SHKLOVSKI [167] and GINZBURG [168].

If we have a differential electron spectrum (constant along the path) given by

$$\frac{dj_e}{dE} = \frac{K}{E^\gamma},$$

then, for monochromatic synchrotron emission at the peak frequency given by $\nu_m = 1.4 \times 10^6 \, B_\perp (E/m_e c^2)^2$, (B in gauss), the index of the radio spectrum α is independent of B and is determined only the exponent γ such that the radio intensity $I_\nu \sim \nu^\alpha$ where $\alpha = \frac{\gamma - 1}{2}$ (for $\nu > \nu_m$). In effect, if the electron spectrum is steep enough, the contribution of higher energy electrons to the observed peak of the synchrotron emission spectrum is small, and electrons near the maximum in the differential spectrum will dominate the synchrotron emission spectrum as well. The intensity of the radio emission does depend on the magnetic field, however, and

$$I_\nu \sim B^{\frac{(\gamma+1)}{2}}.$$

The concensus of measurements indicates that both the halo and spiral arms have on the average very similar non-thermal emission spectra even though the intensity in the spiral arm is ~ 10 times greater per unit frequency interval than in the halo [165]. The results summarized by the Michigan group [169] suggest that the halo radio spectrum can be approximated quite well by $\alpha \sim 0.7$ essentially constant between 10 and 1000 Mc/s, with a peak in the spectrum at from $3-8$ Mc/s and a decrease roughly proportioned to ν at lower frequencies. It is tempting to associate this new data on the features of the radio spectrum with the cosmic ray data which also suggest the possibility of a peak in the differential spectrum of electrons occurring in the energy range $300-800$ MeV. Directly relating these two spectral peaks through the equation for ν_m gives $\bar{B}_\perp = 3.5 \times 10^{-6}$ G, a value uncertain by a factor ≈ 3 but subject to rapid improvement as experimental measurements of these two features improve and, in actual fact, probably as reasonable an approach to the estimate of the halo field as is presently available. Using this value of the field and the observed differential spectrum of

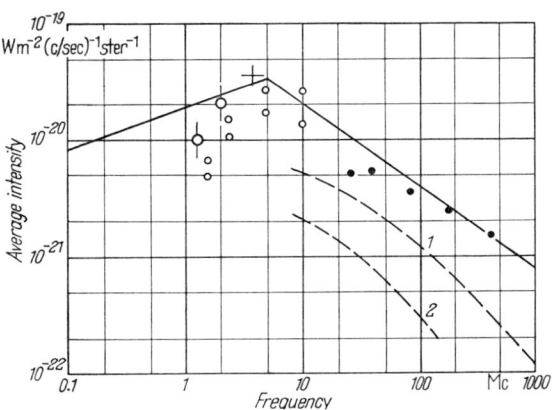

Fig. 47. Measured and calculated intensities of non-thermal ratio emission from the galactic halo. Measured intensities denoted by various symbols are from reference [169]. Solid line with slope $\nu^{-0.75}$ above 5 Mc/s is determined from differential electron spectrum presented in Fig. 37 and magnetic field $\bar{B}_\perp = 3.5 \times 10^{-6}$ G. Solid line below 5 Mc/s is calculated for a spectrum of electrons $dj/dE = $ const $= 0.1$ electron/M²-ster-sec-MeV below 500 MeV. Dashed curves (1) and (2) are the expected non-thermal ratio emission spectrum from secondary electrons only as calculated by HAYAKAWA [173] and GINZBURG and SYROVATSKY [171] respectively.

electrons near the earth ($\gamma \approx -2.5$), (with the intensity decreased by a factor of 2 to allow for a reduction in going to the weaker halo fields), we can estimate the halo radio emission spectrum and intensity employing the formulation for the power radiated in a magnetic field as given by OORT [170]. The results are shown in Fig. 47 which is modified from the work of SCHMOKER [123].

We are thus led to a self-consistent picture between the magnetic field and the intensity and spectrum of both the electrons and radio emission.

Originally, the Michigan workers attributed the decrease in intensity of the radio emission at lower frequencies to absorption by the interstellar gas. Viewed in the light of a differential electron spectrum which also shows a maximum, however, such a decrease could in part be related directly to the electron spectrum. The exact character of this decrease depends on the location of the peak in the electron spectrum as well as on its behavior at low energies (and possible absorption by interstellar gas); however, even in the limiting case of a complete absence of electrons below the peak in the differential spectrum, the fall-off in the radio spectrum cannot be greater than $I(\nu) \approx (\nu/\nu_m)^{\frac{1}{3}}$, which is the limiting condition for the synchrotron emission spectrum at low frequencies ($\nu \ll \nu_m$). We illustrate in Fig. 47 the radio fall-off to be expected if the spectrum of electrons is given by $dj/dE = $ const $= 0.1$ electron/M^2-ster-sec-MeV below 500 MeV.

The above treatment neglects the effects of solar modulation on the low energy electrons. Most papers seem to assume that these effects will be similar to those of protons (reference [171], [172] for example); however, this is not necessarily the case. If the solar modulation is predominantly velocity dependent as opposed to energy dependent (as would be the case in a diffusion picture), then it is quite possible that the variations in the electron component are negligible above ≈ 50—100 MeV.

The above discussion is intended to be mainly illustrative rather than definitive in view of the large uncertainties associated with the electron measurements but clearly points the way to a future great improvement in our understanding of the interrelation between electronic component, non-thermal radio emission and magnetic fields in our galaxy.

Before completing our discussion of the electronic component, let us turn to the ideas regarding the primary or secondary nature of this component as this will lead us directly into a discussion of interstellar particle density in the galaxy, the other important quantity related to the propagational characteristics of cosmic rays. Recall that in our discussion in Sect. IX it was argued that the positron-electron measurements of DESHONG, et al. [126], could be interpreted in terms of a primary component that was almost completely electrons (eg > 80%), a conclusion which is, incidentally, at some variance with that of the above workers. As has been pointed out in the literature [99], a measurement of the positron to electron ratio is of crucial importance in determining what fraction of the electronic component is secondaries, that is, are produced in $p-p$ or $p-n$ collisions between cosmic ray primaries and interstellar hydrogen. This secondary process has been studied in detail by HAWAKAWA and OKUDA [172] and by JONES [174] and yields a prediction for the spectrum of the electronic component as well as of the ratio of positrons to electrons. This e^+/e^- ratio has a value ≈ 2 at 100 MeV, decreasing towards 1 with increasing energy.

On the other hand, if the primary electronic component is accelerated in supernovae as proposed by GINZBURG [164] and others, then one is expected to observe predominantly electrons. An observation that the primaries are mainly electrons suggests then that the secondary contribution, which surely must occur, is small.

Recently, HAYAKAWA [173] and GINZBURG and SYROVATSKY [171] have approached this problem in some detail. They have calculated the total power and spectrum of radiation emitted by secondary electrons and compared it with the observed power and spectrum of galactic non-thermal radio emission. The calculation of the total power emitted by the secondary electrons depends on 1. the rate with which energy is transformed to the electronic component from the primary radiation which, in turn, depends on the interstellar hydrogen density and cosmic ray intensity, and 2. the lifetime of the electrons in the galaxy (and the average value of $B_\perp$). The calculation of both authors for the power emitted and spectrum for $B_\perp = 3 \times 10^{-6}$ G and a lifetime $T_e = 1.5 \times 10^8$ years is shown in Fig. 47. At the electron energies corresponding to these frequencies and for this value of the field, the degradation of the electron spectrum by synchrotron losses is negligible; and the spectrum is essentially that of production or injection. Although there seems to be a difference in the two calculations by a factor of ≈ 3 for identical fields and lifetimes, both calculations predict that the total power emitted by secondary electrons is inadequate to explain the non-thermal halo radio emission (by a factor of from 3—10). This is in accordance with our earlier statement that most of the electronic component apparently consists of primary electrons.

The observed radio spectrum could be reproduced from the secondary electrons, however, by a halo field $\gtrsim 10^{-5}$ G and/or an increased lifetime. Such a large field would be more or less inconsistent with other observations; let us see briefly what freedom there is regarding a lifetime. The lifetime of 1.5×10^8 years is compatible with passage through ≈ 4 g/cm^2 of interstellar matter of average density $\approx 2 \times 10^{-2}$ particles/cm^3. The value of ≈ 4 g/cm^2 is derived from a great wealth of other evidence to be discussed shortly and certainly is not likely to be changed by more than a factor of 3. The work of GINZBURG and SYROVATSKY [175] has shown that for a given age the radio emission is almost independent of the interstellar matter density. Thus, in order to explain the radio emission in terms of secondary electrons, we may increase T_e to $\approx 3 \times 10^8$ years at the very most, which along with a field $B \approx 1 \times 10^{-5}$ G, would be adequate.

If we accept that not more than 20 per cent of the electrons are secondaries, however, we are led by essentially the same arguments to an upper limit for $T_e \approx 1 \times 10^8$ years. Pursuing this approach further recall that the measurements of the Be/B ratio as a function of energy suggest $T_L \gtrsim 3 \times 10^7$ years, thus assuming the lifetime of these two types of particles (electrons and nuclei) are directly comparable and assuming that the average path length is, in fact, ≈ 4 g/cm^2, we have a very indirect estimate bracketing the average interstellar hydrogen density between 0.2—0.02/cm^3. This, of course, fits in quite well with the values surmised from other radio astronomical observations of the type discussed earlier (see also summary of observations of n_H in reference [176]).

A much more unambiguous determination of n_H would be to measure the directional γ-ray intensity $> 10^8$ ev (assumed to come from the neutral pions produced in the collisions of the primaries with interstellar hydrogen) as has been pointed out in a number of papers [173], [175], [176]. The intensity sensitivity $I_\gamma (> 10^8 \text{eV}) \approx 10^{-5}$ cm^{-2} sec^{-1} st^{-1} presumably required for such a measurement has still to be achieved, however.

The preceding discussion, despite its obvious limitations, makes one point exceedingly clear. Namely that unless we have almost perfectly reflecting boundaries to the galaxy, cosmic ray particles cannot move in straight line paths through the galaxy as a whole if we are to reconcile the interstellar hydrogen densities with the total amount of material the particles have passed through. Straight line paths would imply a total of only 10^{-1}—10^{-3} g/cm^2 of material per traverse. This must be enlarged by a factor of 10^2—10^4 to account for the observed amount of material. The most obvious way that this can take place is if the particles random walk or diffuse their way through the interstellar magnetic fields. In such a picture, effects at the boundary of the diffusing region (galaxy) are almost certain to be of great importance, and it may be necessary to "confine" the particles or to allow a certain probability for them to escape. It is quite likely then that this effective lifetime will be energy or rigidity dependent over some range of energies at least.

38. Measurements indicating total amount of material traversed in the galaxy.

Let us now then turn to the measurements that serve to indicate the total amount of material the particles have passed through since acceleration. These measurements refer to the intensity of L nuclei or the L/S ratio, the He^3 nuclei or the He^3/He^3+He^4 ratio, and the deuteron intensity or the d/P ratio, among others. There are strong grounds for believing that all three of these components are absent in the source regions of cosmic rays, and their presence in the radiation near the earth indicates their production in the interstellar medium and so gives us a measure of the total amount of material the cosmic rays have passed through.

The presence of any one of these components in the source region is possible, however, and in this sense the value for the amount of material passed through must be regarded as an upper limit.

This calculation depends critically on the accuracy of the fragmentation parameters in hydrogen (e.g. Table 3) as well as on the model for diffusion or propagation that one assumes. Since the calculation depends rather importantly on the diffusion approach that is used, it is well to pause here and discuss the basis on which this model can be applied to the cosmic ray propagation.

In a random walk in three dimensions $d^2 = 2N\lambda^2$ where d^2 is the mean square distance reached after N magnetic scatterings, the mean free path between such scatterings being λ. It is usually customary to discuss the diffusion in terms of a coefficient $D = v\lambda/3$ where v is the particle velocity. The net distance travelled in time τ is thus $d \approx \sqrt{2D\tau}$ and the velocity with which the particle progresses is $V_D \approx \sqrt{2D/\tau}$. If we have a particle diffusing in a region whose smallest dimension is R, it will cross this region and thus tend to "leak out" in a time $\tau = R^2/2D$. The description of particle motion is thus reduced to an interpretation of the quantity D or λ. The mean free path in such a magnetic diffusion has been discussed at some length by PARKER [178] (see also the review by MORRISON [163]). For our purposes we need to recognize two limiting cases. (a) When the radius of curvature of the particle r_c (conveniently given by $P/300B$, P = rigidity in GV, B = field in gauss) is $\ll Bl$, where l is the characteristic size of a scattering region, then

$$\lambda_t = \frac{1}{\pi l^2 n}, \tag{24}$$

n being the density of such "clouds" per unit volume; and (b) where $r_c \gg Bl$, then

$$\lambda_t = \frac{1}{\pi l^2 n} \left(\frac{r_c}{l}\right)^2, \tag{25}$$

a rigidity dependent mean free path. What sort of estimates can be made of λ_t or D that would be appropriate to interstellar space?

A similar diffusion picture which can seemingly be applied quite well to interplanetary propagation of solar cosmic ray particles [178], [179] leads to $D \approx 10^{22}$ to 10^{23} cm²/sec and to a value of $P_c \approx 5$ GV separating the two types of scattering description. In interstellar space, the scales of these quantities must certainly be much greater, even though the typical value of B characterizing a scattering center is not greatly reduced (perhaps from 10^{-4} to 10^{-5} G). GINZBURG [164] and GINZBURG and SYROVATSKY [99] have taken values of $D \approx 10^{29} - 10^{30}$ cm²/sec based on their interpretation of astronomical observations. If one assumes that the inhomogeneities in the magnetic field that are acting as scattering centers may be associated with the large clouds of partially ionized gas which circulate through the galaxy, then according to ALLEN [180], these clouds are estimated to have diameters $l \approx 5 \times 10^{19}$ cm, and their density within the galaxy, n, is estimated as $\approx 2 \times 10^{-60}$/cm³. Thus, we could have $\lambda_t = 1/\pi l^2 n \approx 6 \times 10^{19}$ cm giving $D \approx 10^{30}$ cm²/sec. Associating a field of $\approx 10^{-5}$ G with such centers gives $P_c \approx 5 \times 10^{16}$ eV. Thus, practically speaking, all of the low energy particles would have free paths which are independent of energy.

This approach, while necessarily crude and geometrical, whereas the actual scattering must be considerably more complex, serves to produce some of the guidelines for the diffusion picture and in particular argues strongly for the fact that the diffusion mean free path must be sensibly constant with energy to quite high energies at least. Thus, whereas the actual value for the effective diffusion coefficient used in this way can only crudely be estimated from astro-

Sect. 38. Measurements indicating total amount of material traversed in the galaxy.

physical data, and is probably best obtained by fitting the cosmic ray data itself, the observation that the scattering mean free path is relatively constant with energy up to $\approx 10^{15}-10^{16}$ eV depends only on the ratio of the magnetic field strengths to the size of the scattering centers. Very great changes in our ideas of these quantities will not alter the fact that over a very wide range of energies one should expect such a constant mean free path.

The lifetimes of cosmic ray particles under such circumstances may be calculated and are: (a) spiral arms; $R \approx 10^{21}$ cm, $D \approx 10^{29}$ cm^2/sec; $T_{esc} \approx 3 \times 10^5$ years. (b) halo; $R \approx 5 \times 10^{22}$ cm, $D \approx 10^{30}$ cm^2/sec; $T_{esc} \approx 5 \times 10^7$ years. These values are illustrated in Fig. 48 along with the lifetime for nuclear interaction, T_{nuc}, and for

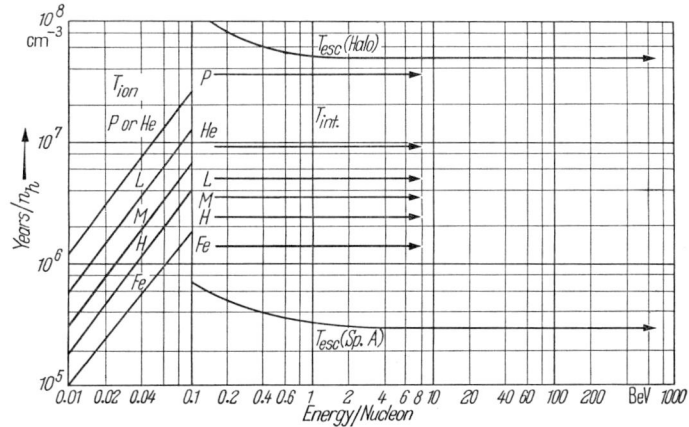

Fig. 48. The lifetimes, T, of cosmic ray particles of different charge, or charge group in the galaxy. The ionization and interaction lifetimes, T_{ion}, T_{int} are calculated on the basis of interstellar hydrogen only and are expressed in terms of years over the interstellar hydrogen density n_H. (Interaction cross sections assumed independent of energy). The escape lifetimes in the halo and spiral arms are calculated on the basis of diffusion theory as discussed in the text and are expressed terms of years.

ionization loss, T_{ion}. The relative importance of these effects in determining the spectra and ratios of the charge components will be discussed later. To confine particles in the spiral arm for the lifetime suggested from earlier discussions, that is $1 \times 10^7 - 10^8$ years, thus would seem to require a "closed" picture in which reflection from the boundaries must play an important role. However, if one allows the particles to move more or less freely from the arms to the halo, then because of the increased lifetimes resulting from the dependence of τ on R^2 the lifetimes are more compatible with an "open" model in which the cosmic rays diffuse relatively easily into metagalactic space. Such a model puts a greater drain on the power required from the sources of the cosmic rays, however, and we shall discuss this point in a later section (see ref. [99] for similar arguments).

We should make note of two effects which can change D at lower energies. First of all, the low energy particles are more likely to spiral around the lines of force under such conditions that the adiabatic invariant $\sin^2\theta/B = $ const is conserved. Under such circumstances, the encounters with the magnetic irregularities may be greatly modified, and instead of a simple diffusion picture, it may be necessary to consider a more ordered flow of particles along field lines, still preserving a high degree of isotropy (see, for example, discussions of Davis [180]). Secondly, even if the diffusion picture does apply, D will begin to change at low energies simply because the particles velocity v does. As a result, even if λ remains a constant the lifetime of a particle at 100 MeV/Nuc will be $\approx$ twice that for the same particle above 1 GeV/Nuc.

Returning now to the question of the change in chemical composition of the cosmic rays in interstellar space, it is necessary to specify a definite model for the propagation. Two models are usually considered: 1. a diffusion picture which in the steady state reduces to:

$$\nabla(D_i \nabla N_i) - \frac{\partial}{\partial E}(b_i N_i) \frac{N_i}{T_i} + \sum_{j \geq i} P_{ji} \frac{N_j}{T_j} + Q_i(r) = 0, \qquad (26)$$

where $N_i(r)$ represents the concentration of nuclei of type i, $b_i(E)$ is the rate of energy gain or loss, T_i is the lifetime for catastropic loss of nuclei of type i, and Q_i is the source density. 2. a "regular" diffusion picture, corresponding essentially to motion along a definite path as for example along magnetic force lines and given by

$$\frac{dN_i}{dt} = -\frac{N_i}{T_i} + \sum_{j \geq i} P_{ji} \frac{N_j}{T_j} \qquad (27)$$

(for a point source at center of the galaxy). This is essentially Eq. (18) of Sect. VI.

It is assumed in picture 2. that all the particles along the path from the source to the point of observation pass through one and the same thickness of matter X (in g/cm^2). Thus, in this picture, if one assumes an absence of a particular component in the source or at injection, the solution in terms of the measured intensity at the earth gives the amount of material X that the radiation has passed through in order to produce this measured intensity. Picture 1. gives essentially the same information except in terms of some other equivalent parameter of the model, for example the ratio of the coefficient of diffusion D to the average concentration of the interstellar gas, together with the known average distance to the sources. This model then gives information on the source distribution in addition to providing a more elaborate description of the propagation of the particles. In discussing this model it is useful to employ the parameter $\varepsilon = \frac{D}{n_H}$. This also may be written $\varepsilon = \frac{m_H v R_s^2}{X}$ where R_s is a characteristic source distance, and X is an "average" thickness of matter through which particles of a given energy have traversed. Since $X = n_H m_H v \tau$, $\varepsilon = \frac{R_s^2}{n_H \tau}$ or simply $\tau = \frac{R^2}{D}$ as noted before. In effect then, the amount of material passed through as determined from picture 2. must be related to the average material in picture 1. as expressed in terms of the other parameters of this model.

α) *Propagation based on the regular diffusion model.* Although model (2) corresponds to an orderliness of magnetic field not known to exist in the galaxy except perhaps in certain regions of the spiral arms, this model admits a very simple solution and is most often discussed in the literature; therefore, we shall consider it first. If the thickness of matter is not large as is the case, the effect of secondary or higher order collisions can be neglected, and the solution of the equation becomes for the L/S ratio:

$$\frac{N_L(x)}{N_S(x)} = \left\{ \frac{P_{HL}}{\lambda_H} \frac{N_H(x)}{N_S(x)} + \frac{P_{ML}}{\lambda_M} \frac{N_M(x)}{N_S(x)} \right\} X + \frac{N_L(0)}{N_S(0)}, \qquad (28)$$

with other equally simple forms for other ratios.

A large number of attempts have been made in the past to deduce X from the observed L/S ratio at the earth. These have differed according to the different values of the L/S ratio used and particularly according to the different fragmentation parameters that have been used but have generally been in the range 2 to 8 g/cm^2. We illustrate in Fig. 49 the growth curves of L/S ratio as a function of X

as calculated by a number of workers. These curves depend on the fragmentation parameters into the L group and we find that curves (1) and (3) are based on P_{HL} and P_{ML} values in hydrogen most nearly like those in Table 3. In particular, a P_{HL} value ≈ 0.4 is most consistent with present day experimental evidence. This particular fragmentation parameter has undergone rather marked changes in past presentations and according to current belief, calculations of X made with $P_{HL} \approx 0.1$ [88], [99] will overestimate this distance. Taking the most recent fragmentation parameters along with the measured L/S value of 0.25 appropriate above ≈ 2 GV rigidity gives a total path length of $3.5-4$ g/cm² for these particles.

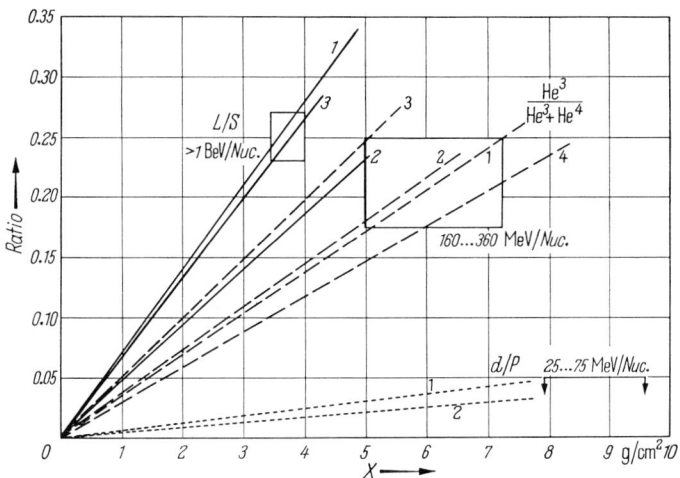

Fig. 49. Growth curves of L nuclei, He³ and d in interstellar hydrogen (Model 2). The numbers on the curves for L nuclei (solid lines) refer the the following references: 1 [94], 2 [88], 3 [36], for the He³ nuclei (dashed lines), 1 [88], 2 [86], 3 [181] 4 [95], and for deuterons (dotted lines), 1 [95], 2 [181].

The growth curves for the He³ nuclei may be somewhat less accurately known than those for the L nuclei with curves (1), (2) and (4) probably representing our best current estimates. (Most of the He³ nuclei come from fragmentation of He⁴ either directly to He³ or first to H³). These curves coupled with the measured He³/He³+ He⁴ ratio lead us to a total path length for production of these nuclei of $5-7$ g/cm².

The growth curves for deuterium clearly indicate that for reasonable path lengths this ratio is small indeed. From the assumed upper limit of 0.06 for the d/p ratio the total path length for production of deuterium is found to be <8 to 12 g/cm².

Before discussing the possible energy dependence of the amount of material through which the cosmic rays have passed, and the implications of the diffusion model for the propagation of the radiation, let us briefly review the other experimental measurements that reflect on this problem. Two features of the charge distribution which clearly reflect on the passage of the radiation through an appreciable amount of material are the odd-even ratios in different charge groups, and the relative absence of the H_2 nuclei. Both of these features are limited in their interpretation by a lack of knowledge of the corresponding partial fragmentation probabilities. However, a number of useful approaches to these problems have been made in the literature [36], [95], [177] and we shall follow these guidelines.

First of all, recall that the odd-even ratio for the H_3 nuclei was determined to be 0.08. If we suppose that the odd-even ratio at the source is negligibly small

($\ll 0.01$) in accordance with cosmical abundance measurements, then essentially all of the odd nuclei with $Z>20$ must be produced by fragmentation of the even H_3 nuclei. Both Aizu et al. [36] and Badhwar et al. [95] conclude that $P_{H_3 H_2} \approx 0.4$ and if we follow these workers and assume that odd and even nuclei will occur in the fragmentation products in proportion to the number of odd and even stable isotopes, then since this ratio is $\approx 1/3$, $P_{H_3 H_2 \text{(odd)}} \approx 0.13$. Passage of these nuclei through $\approx$ one M. F. P. (≈ 2.5 g/cm²) would thus be sufficient to produce the observed odd-even ratio in the H_3 group.

The above workers also estimate that $P_{H_3 H_2} \approx 0.15$; thus the observed relative absence of H_2 nuclei (H_2/H_3 ratio $\approx 0.1-0.2$) could also be rather simply explained by passage of the H_3 group through ≈ 1 M. F. P. of material, the H_2 group being essentially absent in the source regions.

Fragmentation into the H_1 group depends essentially on $P_{H_3 H_1}$ and $P_{H_1 H_1}$, and if we take $P_{H_3 H_1} \approx 0.15$ and $P_{H_1 H_1} \approx 0.25$ and allow the particles to pass through 4 g/cm² of material, then almost half of the H_1 group at the earth represents products of fragmentation; $\approx 20\%$ from the H_3 nuclei and another 20% from the H_1 nuclei themselves. Under these circumstances, it is possible, in the absence of any appreciable source abundance of odd H_1 nuclei, to attribute an odd-even ratio ≈ 0.15 at the earth to the H_1 nuclei after passage through ≈ 4 g/cm² of interstellar hydrogen. The observed ratio is 0.27 which in view of the possible existence of a non-negligible source abundance of Na as well as the uncertainties in the fragmentation parameters themselves is not incompatible with a fragmentation origin of these odd nuclei as well.

In fact, all of the charge features and in particular the abundance of the odd nuclei are suggestive of the passage of the primary radiation through a relatively small amount of matter after injection from the source regions. The value of 4 g/cm² is consistent with our rather crude knowledge of the fragmentation parameters and ideas of the relative absence of odd nuclei in the source regions. Path lengths $\gtrsim 10$ g/cm² for the "regular" diffusion model seem to be incompatible with the presently measured charge distribution.

Such conclusions using very similar arguments have also been reached by Kristiansson et al. [108], Aizu et al. [36] and Badhwar et al. [94].

These limitations make it extremely unlikely that any model requiring the passage of particles through $\gtrsim 10$ g/cm² at energies 5—100 GeV/Nuc can be seriously considered as realistic today. Thus, the suggestions of Singer [182] or Korchak and Syrovatsky [183] that the source of cosmic rays contains primarily H_3 nuclei and the observed charge spectrum is due mainly to fragmentation of these particles are not very realistic at the present time.

Theories in which cosmic rays are produced at some earlier stage of evolution of the galaxy (generally $>10^9$ years ago) and are retained in it (for example, Piddington [184], Bierman and Davis [185], Burbidge [186]) also meet with great difficulty if such a restriction is placed on the total amount of matter passed through, the value of n_H required becoming $<10^{-3}-10^{-4}$/cm³, a value which is apparently not consistent with other measurements (see reference [99]).

Let us now turn our attention to the variations of the L/S ratio as a function of energy (Fig. 44) in terms of a path length of material which is also a function of energy. Thus for particles of energy ≈ 250 MeV/Nuc, assuming the fragmentation parameters are energy independent down to at least this energy, since $L/S=0.4$ we obtain $X=6.5$ g/cm². We also note from Fig. 47 that the $\frac{He^3}{He^3+He^4}$ ratio at this energy/nucleon also may be interpreted in terms of a path length of 6 ± 1 g/cm². At this energy (velocity ≈ 0.6 c) both the L nuclei and He³ nuclei measurements are thus consistent with a path length $\approx 1.6\times$ that at relativistic energies. The deuteron measurements at lower energies (≈ 50 MeV/nuc, corresponding to a velocity ≈ 0.33 c) can at best place an upper limit of 12 g/cm² for the amount of material passed through by particles at this energy. Thus, although in principle an energy dependent change in fragmentation parameters could produce the change in ratios that is observed, the fact that both L nuclei and He³ nuclei measurements imply the same change in the fragmentation parameter make this possibility unlikely. The work of BADHWAR and DANIEL [95] summarizing data on these parameters as a function of energy shows that there is very likely a change in the cross section occurring at low energies in the case of both the L and He³ nuclei. However, this change occurs at lower energies than those we are considering. For example, in the production of the He³ nuclei in the He⁴$+p$ reaction an increase in cross section (by a factor ~ 2) takes place at energies < 200 MeV/Nuc. The data on the energy dependence of the fragmentation parameters for the L nuclei is much more limited, with the best qualitative estimates of the above authors, suggesting that no significant changes occur above 100 MeV/Nuc.

If this is the case, we must attribute the changes in path lengths occurring at low energies to a real increase in lifetime of these cosmic ray particles in the galaxy.

β) *Propagation based on the diffusion picture.* The most pertinent characteristic of the solutions to the diffusion equation (no energy change), relative to the regular diffusion is the much slower growth of the charge components believed absent in the source regions. Using the same fragmentation parameters a path length of 8 g/cm² is needed to reproduce the observed L/S ratio at higher energies- with correspondingly increased values needed to reproduce the observed L/S and He³ measurements at lower energies. In the diffusion picture, recall that our information is expressed in terms of a parameter $\varepsilon = \dfrac{D}{n_H}$ which is also related to the path length X through the characteristics of the source distribution and the conditions at the boundary of the galaxy. In fact, taking $X=8$ g/cm² and boundary conditions corresponding to free motion into intergalactic space along with a source distribution effectively equivalent to a source near the center of the galaxy, we directly obtain $\varepsilon = 2\times10^{30}$ cm⁵/sec ($n_H\, T = 4\times 10^9$ years/cm³) for relativistic particles. Thus, if we assume $n_H \sim 5\times 10^{-2}$/cm³ then $D=1\times 10^{29}$, a value appropriate to the galaxy as a whole. This value of D may be compared with that estimated in the previous section on the basis of other (including astrophysical) evidence.

Other variations of the boundary conditions and source distribution in the diffusion picture have been studied by DAVIS [181] and GINZBURG and SYROVATSKY [99]. Their conclusions, which are not essentially altered by the refinements in the fragmentation parameters used here are:

(i) The diffusion model in the presence of strong reflection from the boundaries of the galactic halo cannot yield the composition observed at the earth no matter what the composition of the source may be. If, however, we forego the idea of a stationary process such as implied above and assume that cosmic rays have been produced during an earlier stage of the evolution of the galaxy and are retained in it [184], [185], [186] (e.g. lifetimes of 10^9-10^{10} years) then no difficulties need arise with the composition. However, the limitations on the path length, X, and the lifetime, T, as inferred from other measurements correspond to an unacceptably low value of n_H as we have noted before.

(ii) Diffusion models in which a uniform source distribution throughout the galaxy is assumed require a considerable leakage of particles from the galaxy as well as leading to a rather unrealistic lifetime [99].

39. Propagational effects on the spectrum at low energies. Let us now discuss the rapid changes in ratios of the different charge components at low energies, in terms of propagational effects. A possible interpretation is in terms of energy loss by ionization which becomes particularly important for the long path lengths implied by the diffusion model; and/or a path length in interstellar hydrogen increasing with decreasing energy. Such effects might explain the L/S ratio variation with energy and the He^3 measurements. Now $X = n_H m_H v \tau$ which for a diffusion picture $\approx n_H m_H \dfrac{R_s^2}{\lambda}$ if the lifetime is determined by escape. Thus if X is to increase either the effective R_s must increase for low energy particles or else λ must decrease. A study of the characteristic times associated with escape, interaction and ionization loss in Fig. 48 reveals that for densities $<0.1/\text{cm}^3$ the limiting lifetimes are those of escape at high energies and ionization loss at low energies (except for the very heaviest nuclei). Under these circumstances the lifetime at low energies will decrease, however, in apparent contradiction to the observations.

Despite this difficulty the energy at which these two lifetimes becomes comparable is an extremely important parameter of the propagation and should be reflected in an important change in the charge composition at low energies — an effect that presumably is already evident in the data. If for example $\bar{n}_H$ is $\sim 0.1/\text{cm}^3$ and $T_{\text{esc}} \sim 5 \times 10^7$ years, then for protons or helium nuclei the lifetimes T_{ion} and T_{esc} become comparable to this at 70 MeV/Nuc whereas for the M nuclei this occurs at 180 MeV/Nuc. If we relate the observed peaks in the differential spectra to this effect then recalling that these peaks occur at ~ 400 MeV/Nuc for M nuclei and ~ 200 MeV/Nuc for both protons and helium and using the following approximate relation: $E_p \text{(MeV/Nuc)} = 10^{-2} \left(\dfrac{Z^2}{A} n_H T_{\text{esc}} \right)^{\frac{1}{2}}$; A = mass number, and T = lifetime for escape in years, we find $n_H T = 4 \times 10^7$. Previous estimates of this product based on other considerations range from $1 \times 10^6 - 4 \times 10^7$.

If the peaks in the differential spectra are moved to lower energies by the solar modulation, as expected, the values of this product will be reduced somewhat and be more in line with the upper extreme of the other estimates. In any case the features of ionization loss should be of great importance at low energies and will not only modify the charge distribution in accordance with the lifetimes, T_{ion}, but will also greatly modify the injection spectrum itself.

40. Propagational effects on the spectrum at high energies. One of the most significant characteristics of the primary radiation is the constancy of exponent of the spectrum over such a wide range of energies. In addition, over this same range of energies the charge composition also apparently remains constant. A change in this situation occurs at $\sim 10^{15} - 10^{16}$ ev, at which point the spectrum becomes steeper and possible changes in the charge composition occur.

What are the effects that could cause such a change?

We must recognize the possibility that such a change would be brought about by limitations on the acceleration mechanism itself occurring at approximately this energy. Here, however, we are discussing propagational effects. In this picture the constancy of the spectral exponent and the charge composition over many orders of magnitude can be regarded as a reflection of the fact that the ages of particles of different charge and energy (presumably determined by the diffusion coefficient and escape from the galaxy) are the same. In this case the injection spectrum is preserved. The changes that occur at $\sim 10^{15}-10^{16}$ ev can thus be interpreted in terms of a change in the age of the particles, similar to that which may be occurring at low energies. The age may either increase [187] or decrease.

Let us consider the type of diffusion picture presented earlier relative to the changes that are occurring at the high energies. We here shall follow an approach similar to that discussed by BYAKOV [188] and also utilized by LASTER [189]. Recall that the limiting rigidity for a scattering mean free path independent of energy occurs when $Br_c \sim Bl$. On the rather crude astrophysical evidence on possible scattering "centers" presented earlier, $Bl \sim 2 \times 10^{14}$ G cm so that this limiting rigidity might be $\sim 5 \times 10^{16}$ V. At some rigidity less than this, perhaps by a factor 10, the scattering mean free path would then become energy dependent — increasing with increasing energy. At rigidities where $r_c \sim l$ the energy dependence of λ_t becomes

$$\lambda_t \sim \frac{1}{\pi l^2 n} \left(\frac{r_c}{l}\right)^2, \tag{29}$$

according to the arguments of DORMAN [190]. This increasing diffusion coefficient with increasing energy means that the age of the particles will decrease $\sim \frac{1}{P}$ since the particles will leak out of the galaxy more rapidly. Thus even if the exponent of the spectrum of injection or escape from the source regions were maintained through this energy range, the observed exponent would begin to change — increasing to a value $\sim \gamma+1$ if γ is the exponent at lower energies.

Since this type of breakdown depends on rigidity and since the number of particles in an air shower is more closely related to the total energy of the primary particle, heavy nuclei would be expected to make a relatively greater contribution at shower sizes corresponding to primary total energies above this limiting energy. The details of the change in charge composition arising from a rigidity dependent limitation on the spectrum at high energies have been investigated by PETERS [191].

On this picture, both an increase in the exponent of the spectrum and a change in the charge composition (as a function of total energy) of the types that are seemingly observed should be expected.

FICHTEL [187], on the other hand, has pointed out that a change in composition may also arise because of an increase in path length at high energies. He suggests that the trapping of the highest energy particles takes place essentially in the halo and calculates the amount of material through which they pass on the basis that the amount of time spent in the spiral arms and halo is proportioned to their relative volume.

Assuming an age of 10^{10} years for the high energy particles, along with $n_H = 10^{-2}/cm^3$ gives $X=150$ g/cm². After passage through such a great amount of material the high energy particles should consist almost exclusively of protons with a steeper energy spectrum.

Another process whereby the intensity of heavy nuclei above $\sim 10^{17}$ ev might be reduced relative to that of protons has been suggested by MORRISON [163] and by GERASIMOVA and ZATSEPIN [192]. These authors have pointed out that, at these energies, interactions with photons Doppler shifted to energies above 10 MeV will preferentially remove the heavy nuclei as a result of their larger cross section for photo decay.

Since the data on the charge composition in the range $10^{15}-10^{18}$ ev is still uncertain, a selection of the possibilities considered above cannot yet conclusively be made. It may be helpful in this connection to examine other effects attendant with such large path lengths at high energies, however, such as the production of high energy secondary electrons and γ-rays.

41. Acceleration in the interstellar medium. So far our discussion of propagational effects has assumed that the main energy of the primary cosmic rays is supplied by the local source regions or the injection mechanism, and that propagation results in an essential degradation of this injection spectrum. The possibility that the main acceleration and energy comes from the interstellar medium itself via processes such as those envisaged by FERMI [193] or FAN [194], has been investigated by a number of workers (see e.g. references [49], [163], [164], [177]). Their general conclusion is that statistical acceleration in interstellar space cannot play an important role in modifying the energy spectrum of the primary nuclear radiation.

Consider for example the diffusion picture for the propagation of the radiation. In this picture "statistical acceleration" of the FERMI type is an inevitable consequence of the motion of the particles through the random (moving) fields and leads to the following rate of increase of the mean particle energy (omitting the effects of general expansion terms which may contribute a deceleration) [99].

$$\frac{dE}{dt} = \alpha E = \frac{U^2 v}{c^2 \lambda} E, \tag{30}$$

where λ is the effective mean free path between collisions and is essentially the same as used before, U is the random velocity of the scattering irregularities and v is the particle velocity. We have already seen that $\lambda \sim 6 \times 10^{19}$ cm, and U is usually taken to be $\lesssim 10^7$ cm/sec. Thus for $v \sim c$ we have $\alpha \lesssim 6 \times 10^{-17}$ sec^{-1}. The lifetime for retaining cosmic rays in the galaxy (no reflection from the boundaries) is $\sim 2 \times 10^{15}$ sec using the same parameters according to the diffusion model. Now since $E = E_0 \, e^{\alpha T}$ the increase in energy will be negligible unless α is in fact much greater, or significant reflection occurs thus increasing the "age" of the particles.

Another limitation on the value of α can be obtained from data on the energy spectrum of the primary radiation. For protons and nuclei of energy great enough for ionization losses to be negligible the differential energy spectrum in the presence of statistical acceleration is given by

$$\frac{dj}{dE} = \frac{K}{E^\gamma}, \tag{31}$$

where $\gamma = 1 + \frac{1}{\alpha T}$. From the observed fact that $\gamma = 2.6$ for all charges at these high energies we require $\alpha T \sim 0.6$, independent of charge. For this to be so it is likely that T is not determined by nuclear interactions even for the heaviest charges, but is determined by an escape process which is independent of charge as we have noted before. This means that $\alpha \sim 3 \times 10^{-16}$ sec^{-1}, if $T \sim 2 \times 10^{15}$ sec. This value of α is a factor ~ 50 larger than that suggested in the previous paragraph. To make the previous value of α consistant, we require that U be increased by this amount since changing λ will not cause the product αT to change (unless very effective reflection from the boundary of the galaxy occurs). Such a large value of U seems unlikely. As a result acceleration by more than one order of magnitude in energy and particularly to the very highest energies observed is very difficult to envision.

If the statistical acceleration process is important in the interstellar medium, then the competition between this and ionization loss will determine the spectrum at low energies. In terms of energy per nucleon we have for the gain $\frac{dE}{dt} \sim \frac{Z}{A} E$, and for the loss due to ionization $-\frac{dE}{dt} \sim \frac{n_H Z^2}{A v}$.

Equating the two we find the critical energy/nucleon at which the rates of energy gain and loss are equal.

$$E_p = \text{const } (n_H Z)^{\frac{2}{3}}. \qquad (32)$$

This indicates how the peak in the differential spectrum might vary for different charges and should be compared with the relationship we have presented earlier in Sect. XIV a-3, which results from the competition of escape and ionization loss in a diffusion picture for the propagation of the particles. The two relations give a different behavior for protons and helium nuclei. In the case that escape competes with ionization loss the peaks occur at the same energy/nucleon since the escape lifetime depends only on energy/nucleon for both charges. When energy gain competes with ionization loss the rate of increase of energy for helium nuclei is only $\frac{1}{2}$ that of protons for the same energy/nucleon, thus the peak for helium energy occurs at higher energies. The observation that the peak for helium nuclei occurs at the same or perhaps even a smaller kinetic energy/nucleon than for protons would thus seem to argue against the importance of acceleration in determining the features of the spectrum at low energies.

A similar type of argument can be used to explain the characteristics of the primary electron spectrum in the energy range where statistical acceleration might be expected to compete with magnetic bremsstrahlung losses. In this case

$$\frac{dE}{dt} = \alpha E - \beta E^2 = \alpha E - 4 \times 10^{-15} B_\perp^2 E^2 \quad (E \text{ in ev}). \qquad (33)$$

If a peak in the electron spectrum at ~ 500 MEV is to be produced on this basis (assuming now a primary origin as opposed to a secondary one), then with $B_\perp = 3 \times 10^{-6}$ G, we require $\alpha = 4 \times 10^{-17}$ sec^{-1}. This is one to two orders of magnitude less than required to reproduce the spectrum of the nucleonic component and in fact is quite close to that calculated on the basis of astrophysical evidence [Eq. (30)].

On these grounds it is possible to visualize a kink in the electron spectrum at ~ 500 MeV as resulting from a competition between statistical acceleration and synchrotron energy loss on an ambient spectrum of electrons injected from primary sources. Above this energy the tendency will be to slow down the ambient electrons, below this energy to accelerate them. If we believe that the lifetime for escape of the electrons is the same as for nuclei and is $\sim 2 \times 10^{15}$ sec (certainly no larger than 10^{16} sec), then we have a further competition between escape and energy loss processes occurring at much lower energies. For example, the lifetime of electrons for ionization loss becomes comparable to the lifetime for escape at ~ 50 MeV and at this energy both of these lifetimes are much less than that for synchrotron loss. Conditions will be appropriate to the formation of a kink in the differential spectrum at this energy as well. As a result a complex behavior of the electron spectrum can arise because of the competition of two energy loss processes and energy gain and escape and the relationships between B, n_H, and α. A study of this spectrum and in particular the features of a kink at $\lesssim 500$ MeV offers the possibility of greatly enhancing our understanding of the relative magnitude of these parameters and the importance of the different processes relating to the propagation of the electrons.

Other circumstances have been discussed in the literature under which more efficient acceleration can occur. One such process has been discussed by FAN [194], in which a particle may be trapped between two interstellar clouds and accelerated by the combined action of statistical processes and betatron acceleration. In this "push-pull" process values of $\alpha \sim 10^{-14}$ sec^{-1} can be obtained at least over certain regions of the galaxy, and for certain energy ranges.

b) The process of acceleration and escape from the source region.

42. In the light of our previous arguments the acceleration and escape of the primary radiation from its source regions is a step at least twice removed from our actual observations on these particles at the top of the atmosphere. Thus in order to isolate the process by which the galactic cosmic rays are accelerated to high energies we are forced into a framework of assumptions regarding the subsequent steps in the lifetime of the particles.

We have seem that it is very likely that the interstellar medium acts in such a way that we can regard acceleration in this medium as negligible. This greatly simplifies and isolates this aspect in the evolution of the primary cosmic rays. Even so we must separately recognize the features of the acceleration process itself, and the subsequent escape of the particles from the source region.

In order to approach this problem systematically let us make the following division of these processes: 1. Escape or post acceleration. This includes a) modifications of the acceleration spectrum by magnetic or electric field processes, presumably in much the same manner as the solar cosmic ray spectrum is modified in its passage through and eventual escape from the solar system, b) modifications of the spectrum by passage through material in the local source region. 2. Type of acceleration. Here we shall make the division into a) fast processes, those occuring on a time scale $\sim$ a few years or perhaps much less, b) slow processes, those occurring on time scales $>10^2$ years.

Before discussing these aspects in some detail let us make one final comment on the features of the propagation as they relate to the low energy portion of the spectrum. Recall that the observations on that part of the spectrum for protons and helium nuclei below the peak in differential spectrum reveals an intensity falling off approximately as $E^{1.0}$ in 1963. The residual solar modulation at this time will be in such a direction as to reduce this fall-off but it was argued in Sect. XIV, that this would most likely be insufficient to remove the peaks in the differential spectrum and we should regard these as a property of the propagation of the particles. In fact the character of the peak and the fall-off at low energies have been directly related to the fact that the lifetime for ionization loss in the interstellar medium controls the particle intensities at low energies. If this is indeed the case then $T \approx T_{\text{ion}} \approx E^{1.5}$. Thus for a given source differential spectrum we would expect it to be changed by at least this factor in arriving at the earth. These arguments imply that if ionization loss is important the differential primary spectrum at the sources must have a negative exponent. In other words, present measurements give no evidence of the existence of a peak in the source differential spectrum of the cosmic ray particles. It is quite possible, and in fact likely, that the exponent on this source spectrum is less at low energies than its value of -2.6 at energies above a few GeV, however. (A more exact calculation of the spectrum at low energies uses the diffusion equation (26) with the instantaneous particle energy loss. Some calculations of this type have been performed by HAYAKAWA et al. [195], WEBBER [10] and APPA RAO and KAPLON [129].

43. Escape of cosmic ray particles from the source region. Useful guidelines for a discussion of the escape of particles from the source regions are already available. Both LASTER [189] and KAPLON and SKADRON [196] have suggested that the amount of material that the primary radiation passes through in the localized source region is comparable or larger than that in interstellar space. KAPLON and

SKADRON have additionally obtained a value of X in the source region that is rigidity dependent by requiring that the reflectivity from the magnetic boundary of the source region be rigidity dependent. In this manner they can account for the differing spectra of the charge components at low energies quite apart from our earlier discussion of interstellar propagation effects.

The parameters entering the diffusion calculation (n_H, T, and D for example) are uncertain enough so that it is quite possible that the particles have passed through only a few tenths of a g/cm² in interstellar space. At the present time it is not possible to differentiate between material passed through in a localized source or in the extended interstellar medium since the data gives us essentially (average) values of X for each energy. A number of possibilities exist, however, that might enable limits to be put on the amount of material passed through the localized sources. These depend in part on the time scale of the local trapping process. In the picture of KAPLON and SKADRON this time is 10^3-10^4 years for intermediate energies, a time small compared with the lifetime of the particles in the galaxy but long enough so that effects from such recent possible sources as the Crab Nebula should still be present. One effect from such a source region in which nuclei were still trapped would be γ-rays from the nuclear interactions. Another approach to the (time) distribution of matter makes use of the comparatively short lifetime in the sources relative to that in the galaxy. From measurements on the presence of the radioactive Be^{10} (half life 2×10^6 years) in the primary radiation and on the variation of its intensity with energy using the relativistic time dilation effects it might be possible to discriminate a source of Be^{10} "local" in time or extended in time.

HAYAKAWA, KOSHIBA, and TERASHIMA [*195*] have also discussed in some detail the effect of local escape processes on the low energy part of the primary spectrum. They argue that the cooperative relationship between the acceleration and escape is all-important in shaping the low energy part of the injection spectrum and illustrate this by a number of models for acceleration of the particles along with an escape picture much the same as that to be expected from the solar system on the basis of PARKER'S [*197*] early disordered field model. Their discussion clearly emphasizes the complexity of the situation. It also emphasizes the possible importance of our earlier conclusion that the present spectral data at low energies is best interpreted in terms of rigidity spectra which are similar for the different charge components, rather than similar energy spectra.

It is perhaps worthwhile to reflect on what may be a very similar situation occurring in the solar system. Here we regard the acceleration as a separate and distinct process and can observe at first hand the influence of the interplanetary medium on both the energy spectra and the charge composition of the solar cosmic rays. Despite a great wealth of information regarding these processes [*178*], [*179*] we still cannot calculate with any degree of accuracy the spectrum and composition of the escaping particles.

44. The process of acceleration of particles to high energies. Let us first consider the acceleration process as distinct from the escape process (as is the case for solar cosmic rays) and examine some possible differences between slow or fast acceleration processes as defined earlier.

The most plausible slow processes might arise in the envelopes of supernovae, where α for the statistical acceleration might be as high as $10^{-12}-10^{-13}$ sec^{-1} [*177*]. For such values the time scale of acceleration might range from 10^4 years for the lowest energy particles to $>10^8$ years for the highest energy particles, depending upon the initial injection of particles and the value of n_H in the accelerating regions. A number of limitations to this type of process can be advanced on the basis of experimental results.

One of these, suggested by BADHWAR et al. [198], is based on the calculated relative production cross sections of the various L nuclei in the energy region 20—100 MeV/Nuc from spallation of heavy nuclei and the corresponding figures observed experimentally at $E>1.5$ GeV/Nuc in the primary radiation. These authors conclude that the amount of matter traversed by the radiation during the time of its acceleration from 20 to 100 MeV/Nuc is $\ll 2.5$ g/cm^2 of hydrogen. Thus a process in which a slow acceleration occurs and the low energy end of the spectrum is determined by ionization loss at the same time can presumably be ruled out.

Another limitation is related to the fact that it is possible to explain the behavior of the primary spectrum in the $10^{15}-10^{17}$ ev range in terms of propagational effects alone. This means that there is no compelling reason for limiting the acceleration process at these energies, indeed it is possible that the features of the spectrum of accelerated particles extend to much higher energies. If this is the case then it becomes increasingly difficult to explain the acceleration in terms of a relatively slow statistical process, which if occurring in the envelopes of supernovae for example, seems limited to $\sim 10^{14}-10^{15}$ ev, (much less than the limitations imposed by propagation in the galaxy). Indeed there is a very real question whether acceleration will occur at all in an expanding supernova envelope or whether deceleration or other energy loss effects will be dominant to degrade an already present spectrum.

Regarding the possibility of fast processes, we presumably need to consider catastrophic processes such as those directly related to the energetics of a supernova explosion itself [203] (literally shock wave processes) or statistical processes with $\alpha \sim 10^{-5}-10^{-7}$ sec^{-1}. In view of the complexity of stages that may occur between the source and the earth we know of no present information that will help select between the various types of fast processes that can occur. As in the case of our understanding of the solar cosmic ray acceleration, the best information would seem to be obtainable only when an actual acceleration event occurs. On the basis of our earlier arguments it seems that the fast processes should be favored, however.

Before completing our discussion of acceleration mechanisms we should take note of a number of papers which have related the features of the accelerated particles to the quite general characteristics of the accelerating regions. In this picture the processes of acceleration and escape are essentially combined and act cooperatively. This is, as we know, what effectively happens in a statistical acceleration process where a particle may escape from some region via the same type of statistical process in which it is accelerated.

According to the original suggestion of SYROVATSKY [199] the spectrum of cosmic ray particles ejected from a particular source is obtained in the form of a power law by assuming the equipartition of source energy among the magnetic field, turbulence and the high energy particles. Independently of the character of the acceleration process and the spectrum of cosmic rays inside the region, particles leaving the region can be shown to have a spectrum with an exponent $=-2.5$ for a volume constant process. Unfortunately the dynamics whereby such equilibrium is set up has been little studied so far; however, it now seems possible to evaluate the escape spectrum from the solar system (or the escape spectrum from the accelerating regions themselves on the sun which may be more pertinent to such a model) and so to determine the validity and limitations of such a general picture.

c) Source regions for primary cosmic rays.

The attempt to isolate the source regions of cosmic rays depends quite closely on the view point taken with regard to the relative importance of propagation and source effects on the radiation. Certain observational features are of great importance in this process of induction, however, and let us consider them within the framework of our previous discussions regarding the propagation of the radiation in the galaxy.

45. Chemical composition of the sources. The most important of these observational features is the chemical composition of the radiation. We now recognize that heavy elements can be accelerated on such stars as the sun and that their relative abundance is closely that in the source itself [200]. Thus whether or not we require a mechanism of preferential acceleration of the heavier nuclei, we have one example of an acceleration process in which the chemical composition of the cosmic rays apparently represents that of the source region.

In Table 10 we have attempted to estimate the abundances of the various nuclei in the galactic radiation at the source assuming a path length from injection to earth of 4 g/cm² (following the work of AIZU, et al. [36] and BADHWAR, et al. [95]). This is compared with the measured abundances of solar cosmic rays reported by BISWAS and FICHTEL [200]. It is clearly evident that the galactic radiation has at its source an excess of heavier nuclei, and possibly a deficiency of helium nuclei relative to the solar cosmic rays. In view of the similarity between actual solar abundances and solar cosmic ray abundances we may interpret this as a real anomalous abundance of the source region of galactic cosmic rays relative to stars similar to the sun. This enhancement of the heavier nuclei in the source regions has been discussed in countless papers, but its significance takes on a new dimension in the light of the recent solar cosmic ray measurements.

Our ideas of nuclear synthesis in stars are sufficiently well developed [153], [201], [202] to lead us to believe that the most likely candidates for such anomalous abundances are those stars in an advanced stage of development — the so called Extreme Population I stars — distributed throughout the spiral arms of the galaxy. Only in these stars has the helium burning proceeded far enough to form the heavier alpha particle nuclei C, O, Ne and Mg, and the temperatures are high enough to synthesize the Fe group. This group of stars includes the supernovae which GINZBURG [164], SHKLOVSKY [167], HAYAKAWA et al. [177], and others have suggested are the sources of the primary radiation. The recent measurements on solar cosmic rays, and measurements of the galactic radiation as presented in this article have only increased the ascendency of this picture, we believe. Moreover the more specific calculations relating to the characteristics of supernova explosions by COLGATE and colleagues [203], are also compatible with the observed charge characteristics of the primary radiation.

A number of interesting aspects of the observed charge distribution are not yet to be fully understood in terms of the supernova theory, however. These include the relative abundance of C and O, and also of Ne and Mg.

For the case of C and O, HAYAKAWA et al. [177] have pointed out that in the rapid cycle expected to take place in supernovae, at temperatures $>10^9$ °K, O is expected to be more abundant than C. The observed cosmic ray abundances are expected at a temperature of about 4×10^8 °K. This is too low to synthesize the heavier alpha particle nuclei and the features of the Fe peak.

In the case of Ne and Mg the observed comparable abundances seems to require synthesis at temperatures $\sim 2 \times 10^9$ °K and in a region of low density, perhaps in the envelope of the supernova. It may be that these features of the charge distribution will lead us to an increased understanding of the radial features of the source regions (e.g., in supernovae) or require us to consider other specialized sources for these charge components (e.g., reference [36], [181]).

Table 10. *Comparative abundance of galactic and solar cosmic rays.*

	Cosmic Rays		Solar Cosmic Rays	R (Gal./Sol.)
	At Earth	At Source (4 g/cm² mat)		
He	38	33	175	0.19
Li	0.27	⎫		
Be	0.19	⎬ ~0	<0.03	—
B	0.43	⎭		
C	1.0	1.0	1.0	1.0
N	0.46	0.49	0.30	1.6
O	0.61	0.70	1.7	0.4
Fl	0.09	<0.03	<0.05	—
Ne	0.18	0.18 ⎫	0.21 ⎫	0.9 ⎫
Na	0.08	<0.03 ⎬ 0.78	— ⎬ 0.33	— ⎬ 2.2
Mg	0.15	0.33 ⎪	0.07 ⎪	4.7 ⎪
Si	0.11	0.26 ⎭	0.05 ⎭	4.0 ⎭
H_2	0.03—0.06	~0	0.9	—
H_3	0.16	0.60	<0.03	≳20

46. Energetics and spatial distribution of the sources. The energy density of cosmic ray particles near the earth is ~ 1 eV/cm³. If we assume that, averaged over the entire galaxy including halo, it is about $\frac{1}{2}$ of this, then since best estimates put the volume of the halo at $\sim 10^{68}$ cm³ we have a total energy of cosmic rays in the galaxy of 5×10^{67} eV contained in $\sim 2 \times 10^{58}$ particles. (If we assume that the lifetime of these particles is between $1-2 \times 10^8$ years then we arrive at a source power of not less than 10^{40} ergs/sec or 10^{43} particles/sec.)

This is an extremely large amount of energy — comparable to that contained in starlight, in interstellar magnetic fields, and in turbulent gas motions in the galaxy. Because of this it is customary to look for sources that are energetically very important, possibly individual stars or major features related to the energetics of the galaxy itself such as the origin of the halo [186].

Let us consider individual stars first and start with the most energetic of these, the various classes of supernovae. Estimates of the total energy emitted in a supernova explosion range from $\sim 10^{50} - 10^{52}$ ergs which taken with a rate of occurrence of one every 100 years leads to a total power output of $\sim 10^{41} - 10^{43}$ ergs/sec. The uncertainties in this number are large but not as great as the uncertainties as to what fraction of this energy might be expected to appear in the form of high energy nuclei. The best estimates of various authors, based on arguments mainly related to the observed synchrotron emission from the energetic electrons known to be trapped in a few supernova remnants, give a total power output in energetic nuclei of $10^{37} - 10^{40}$ ergs/sec. Thus energetically speaking supernovae may be regarded as a likely candidate but only if the uncertainties in our present-day astrophysical knowledge are stretched to the limit. Perhaps another way of viewing the problem is to say that the efficiency for production of cosmic rays in a supernova explosion must be $\sim 10^{-1} - 10^{-3}$ according to our present knowledge if they are to be considered as primary sources.

Even if we regard these supernovae, or any other stars as merely injectors to a main interstellar process of acceleration, then with the parameters for this acceleration that apparently exist we still require the injector to supply 0.1—1 per cent or more of the energy [164].

Another type of star for which we have more direct evidence regarding cosmic ray production concerns the main disk population of which the sun is one. This class of main sequence stars is the most abundant in the galaxy countaining ~90 per cent of all stars, and perhaps 50 per cent of all of the mass of the galaxy contained in stars.

The total flux of solar cosmic ray particles above a few MeV reaching the earth during a solar cycle can now be evaluated with a reasonable accuracy and turns out to be $\sim 10^{11}/cm^2$ [179]. The calculation of that fraction that actually escape the solar system (or for that matter the spectrum of the escaping particles) is somewhat more difficult and depends upon the model assumed for propagation through the interplanetary medium. Considering that $\sim \frac{1}{3}$ of all of the particles observed at the earth eventually escape leads to an average solar injection of 10^{28} cosmic ray particles/sec, corresponding to $\sim 10^{24}$ ergs/sec. Since there are $\sim 10^{11}$ stars of this type in the galaxy, the total expected cosmic ray production would be $\sim 10^{39}$ particles/sec corresponding to $\sim 10^{35}$ ergs/sec, too small by many orders of magnitude on both counts to be considered as a source of cosmic rays. (It is well to note that in specific events on the sun in which cosmic rays are produced, the fraction of the total energy emitted that can be estimated as going into energetic particles is $\sim 10^{-3} - 10^{-4}$.)

A number of other types or classes of stars have from time to time been considered as possible sources for cosmic rays [181]. Even less can be surmised about energetic particle production on these objects, however, than in the case of supernova or solar type stars. About the only candidates that can even be considered from the energetic point of view are the Population II stars including the so-called supergiant and red giant stars. Although helium burning may have started in the interiors of these stars, they are believed to have a relatively low abundance of elements with $Z \geq 10$ in contradiction with the comparatively rich abundance of these elements found in the primary cosmic radiation.

In connection with the isolation of specific source regions for the production of cosmic rays our measurements are not yet quite sensitive enough to tell us anything definite about the distribution of sources in the galaxy. (Although measurements of the different charge components at low energies where the lifetimes are dominated by ionization loss and thus reflect the source characteristics of much smaller regions of space may soon provide some valuable progress in this direction.) In particular we might seek to differentiate between a central source such as might be expected in the explosive injection theories [185], [186] or in the case of an origin in Population II type stars; or a disklike source, associated with the spiral arms and an origin in Extreme Population I stars (e.g., supernovae).

Before completing our discussion of the possible sources of galactic cosmic rays we should perhaps remark on the hierarchical theories for the origin of the radiation (see e.g., reference [164]). It seems quite likely that almost all of the stellar objects are, over a wide range of their lifetime, capable of producing energetic particles which must certainly populate the galaxy to some extent. If what is meant by these theories is that the features of the primary radiation presently being observed represent some combination of a number of different types of sources, then according to our present knowledge these different sources (as opposed to a number of similar sources such as supernovae, for example) must a) provide very closely the same spectral index at injection to the interstellar medium (certainly a possibility), b) provide very closely the same charge distribution at injection to the interstellar medium (a much less likely possibility), c) have very closely the same source power (very unlikely). Thus while there is evidence at the very highest energies that we may be seeing an extra-galactic component of cosmic rays, it appears that our measurements are not yet sensitive enough to say that we are seeing a truly hierarchical radiation at low energies — indeed it seems as though one homogeneous component must dominate.

References.

[1] STÖRMER, C.: The Polar Aurora. London: Oxford Univ. Press 1955.
[2] SCHWARTZ, M.: Nuovo Cim. 11, Suppl. I, 27 (1959).
[3] QUENBY, J. J., and C. J. WENK: Phil. Mag. 7, 1457 (1962).
[4] SAUER, H. H.: J. Geophys. Res. 68, 957 (1963).
[5] NEHER, H. V.: Phys. Rev. 103, 228 (1956) and NEHER, H. V., and H. R. ANDERSON: J. Geophys. Res. 67, 1309 (1962) and references contained therein.
[6] WINCKLER, J. R.: J. Geophys. Res. 65, 1331 (1960) and MASLEY, A. J., T. C. MAY, and J. R. WINCKLER: J. Geophys. Res. 67, 3243 (1962) and references contained therein.
[7] WEBBER, W. R., and N. NERURKAR: J. Geophys. Res. (to be published).
[8] E.g., see MCDONALD, F. B., and W. R. WEBBER: Phys. Rev. 115, 194 (1959).
[9] MATHEWS, T.: Phil. Mag. 8, 387 (1963).
[10] Work of MCDONALD and WEBBER, summarized in W. R. WEBBER, Progr. in Cosmic Ray Physics 6, 75 (1963).
[11] VOGT, R.: Phys. Rev. 125, 366 (1962).
[12] MEYER, P., and R. VOGT: Phys. Rev. 129, 2275 (1963).
[13] MEYER, P., and R. VOGT: Proc. Int. Conf. on Cosmic Rays, Jaipur 3, 49 (1964).
[14] ORMES, J., and W. R. WEBBER: Proc. Int. Conf. on Cosmic Rays, Jaipur 3, 3 (1964).
[15] WADDINGTON, C. J.: Proc. Int. School of Physics, Varenna, p. 135. New York: Academic Press 1963.
[16] ALY, H. H.: Canad. J. Phys. 40, 1049 (1962).
[17] DANIEL, R. R., and N. SREENIVASAN: Proc. Int. Conf. on Cosmic Rays, Jaipur 3, 60 (1964).
[18] FREIER, P. S., and C. J. WADDINGTON: Phys. Rev. Letters 13, 108 (1964).
[19] FICHTEL, C. E., D. E. GUSS, D. A. KNIFFEN, and N. A. NEELAKANTAN: J. Geophys. Res. 69, 3293 (1964).
[20] FICHTEL, C. E., D. E. GUSS, G. R. STEVENSON, and C. J. WADDINGTON: Phys. Rev. 133, 3818 (1964).
[21] BRYANT, D. A., G. H. LUDWIG, and F. B. MCDONALD: V. Interamerican Seminar on Cosmic Rays 1, XII (1962).
[22] BRYANT, D. A., T. L. CLINE, V. D. DESAI, and F. B. MCDONALD: J. Geophys. Res. 67, 4983 (1962).
[23] MCDONALD, F. B., and W. R. WEBBER: J. Geophys. Res. 67, 2119 (1962).
[24] DUKE, P. J.: Phil. Mag. 5, 1151 (1960).
[25] FREIER, P. S., E. P. NEY, and C. J. WADDINGTON: Phys. Rev. 114, 365 (1959).
[26] FOWLER, P. H., P. S. FREIER, and E. P. NEY: Nature, Lond. 181, 1319 (1958).
[27] STEVENSON, G. R., and C. J. WADDINGTON: Phil. Mag. 6, 571 (1961).
[28] STEVENSON, G. R.: Nuovo Cim. 24, 557 (1962).
[29] ENGLER, A., M. F. KAPLON, J. KLARMANN, A. KEENAN, C. FICHTEL, and M. W. FRIEDLANDER: Nuovo Cim. 19, 1090 (1961).
[30] GREER, G.: M. S. Thesis, Univ. of Minnesota 1964.
[31] FREIER, P. S., E. P. NEY, and J. R. WINCKLER: J. Geophys. Res. 64, 685 (1959).
[32] ENGLER, A., F. FOSTER, T. L. GREEN, and J. MULVEY: Nuovo Cim. 20, 1157 (1961).
[33] FOWLER, P. H., P. S. FREIER, and E. P. NEY: Nuovo Cim. 8, Suppl. 492 (1958).
[34] ENGLER, A., M. F. KAPLON, and J. KLARMANN: Phys. Rev. 112, 597 (1958).
[35] FREIER, P. S., and C. J. WADDINGTON: Phys. Rev. 135, B 724 (1964).
[35a] FREIER, P. S., G. D. GREER, J. VALDEZ, and C. J. WADDINGTON: Phys. Rev. (to be published).
[36] AIZU, H., Y. FUJIMOTO, S. HASEGAWA, M. KOSHIBA, I. MITO, J. NISHIMURA, and K. YOKOI: Progr. Theoret. Phys., Suppl. 16, 54 (1960).
[37] WADDINGTON, C. J.: Nuovo Cim. 6, 748 (1957).
[38] NEELAKANTAN, K. A., and S. BISWAS: Bull. Amer. Phys. Soc. II, 8, 293 (1963).
[39] FICHTEL, C. E.: Nuovo Cim. 19, 1100 (1961).
[40] STANTIC, S., and D. E. GUSS (Reported at Midwest Cosmic Ray Conf. by F. Foster, Denver, 1964).
[41] WEBBER, W. R., and F. B. MCDONALD: J. Geophys. Res. 69, 3097 (1964).
[42] KOSHIBA, M., E. LOHRMANN, H. AIZU, and E. TAMAI: Phys. Rev. 131, 2692 (1963).

References.

[43] EVANS, D. E.: Nuovo Cim. **27**, 394 (1963).
[44] BALASUBRAHMANYAN, V. K., and F. B. MCDONALD: J. Geophys. Res. **69**, 3289 (1964).
[45] BISWAS, S., P. J. LAVAKARE, K. A. NEELAKANTAN, and P. G. SHUKLA: Nuovo Cim. **16**, 644 (1960).
[46] FOSTER, F., and A. DEBENEDETTI: Nuovo Cim. **28**, 1190 (1963).
[47] VAN HEERDEN, I. J., and B. JUDEK: Canad. J. Phys. **38**, 964 (1960).
[48] GARELLI, C. M., B. QUASSIATI, and M. VIGONE: Nuovo Cim. **15**, 121 (1960).
[49] YAGODA, H.: Geophys. Res., Paper No. 60, AFCRC (1958).
[50] YAGODA, H.: Canad. J. Phys. **34**, 122 (1956).
[51] APPA RAO, M. V. K., S. BISWAS, R. R. DANIEL, K. A. NEELAKANTAN, and B. PETERS: Phys. Rev. **110**, 75 (1958).
[52] FOWLER, P. H., and C. J. WADDINGTON: Nuovo Cim. **1**, 637 (1956).
[53] DANIEL, R. R., and N. SREENIVASAN: Proc. Int. Conf. on Cosmic Rays, Jaipur **3**, 60 (1964).
[54] SINGER, S. F.: Progress in Cosmic Ray Physics, vol. 4, p. 205. Amsterdam: North-Holland Publishing Co. 1958.
[55] NEHER, H. V.: J. Geophys. Res. **66**, 4007 (1961).
[56] LIN, W. C., D. VENKATESAN, and J. A. VAN ALLEN: J. Geophys. Res. **68**, 4885 (1963).
[57] MERIDITH, L. H., J. A. VAN ALLEN, and M. B. GOTTLIEB: Phys. Rev. **99**, 198 (1955).
[58] MCDONALD, F. B., and W. R. WEBBER: Phys. Rev. **115**, 194 (1959).
[59] PERLOW, G. J., L. R. DAVIS, C. W. KISSINGER, and J. D. SHIPMAN: Phys. Rev. **88**, 321 (1952).
[60] VAN ALLEN, J. A., and L. A. FRANK: Nature, Lond. **184**, 219 (1959).
[61] NEHER, H. V.: Proc. Int. Conf. on Cosmic Rays, Jaipur **2**, 139 (1964).
[62] ARNOLDY, R. L., J. R. WINCKLER, and R. A. HOFFMAN: J. Geophys. Res. **69**, 1679 (1964).
[63] TREIMAN, S. B.: Phys. Rev. **89**, 130 (1953).
[64] DUTHIE, J., P. H. FOWLER, A. KADDOURA, D. H. PERKINS, and K. PINKAU: Nuovo Cim. **24**, 122 (1962) and references therein.
[65] KIDD, J. M.: Nuovo Cim. **27**, 60 (1963).
[66] MIYAKE, S.: J. Phys. Soc. Japan **18**, 1226 (1963).
[67] HAYMAN, P. J., N. S. PALMER, and A. W. WOLFENDALE: Proc. Roy. Soc. Lond. **275**, 391 (1963).
[68] WOLFENDALE, A. W.: Proc. Int. Conf. on Cosmic Rays, Jaipur **6**, 3 (1964).
[69] BROOKE, G., and A. W. WOLFENDALE: Proc. Roy. Soc. Lond. (to be published).
[70] WEBBER, W. R.: Nuovo Cim. **8**, 532 (1958).
[71] BALASUBRAHMANYAN, V. K., S. V. DAMLE, G. S. GOKHALE, M. G. K. MENON, and S. K. ROY: Proc. Int. Cosmic Ray Conf., Jaipur **3**, 110 (1964).
[72] KAJAREKAR, P. J.: Proc. Int. Cosmic Ray Conf., Jaipur (1964).
[73] BHATT, V. L., and R. R. DANIEL: Proc. Int. Cosmic Ray Conf., Jaipur **3**, 58 (1964).
[74] DAYTON, B., and F. K. KUNTE: Proc. Int. Cosmic Ray Conf., Jaipur **2**, 9 (1964).
[75] HILDEBRAND, B., F. W. O'DELL, M. M. SHAPIRO, and B. STILLER: Proc. Moscow Cosmic Ray Conf. **3**, 115 (1960).
[76] FREIER, P. S.: Private Communication 1964.
[77] HASEGAWA, H., S. NAKAGAWA, and E. TAMAI: Proc. Int. Conf. on Cosmic Rays, Jaipur **3**, 83 (1964).
[78] RAY, E. C.: J. Geophys. Res. **67**, 3289 (1962).
[79] BALASUBRAMANYAN, V. K., T. L. CLINE, G. H. LUDWIG, and F. B. MCDONALD: Proc. NASA Imp. Symposium, Nov. 1963.
[80] ORMES, J., and W. R. WEBBER: Phys. Rev. Letters **13**, 106 (1964).
[81] FAN, C. Y., G. GLOECKLER, and J. A. SIMPSON: Proc. NASA Imp Symposium, Nov. 1963.
[82] FICHTEL, C., and M. W. FRIEDLANDER: Nuovo Cim. **18**, 825 (1960).
[83] FRIEDLANDER, M. W., e C. T. SPRING: Nuovo Cim. **26**, 1292 (1962).
[84] FOSTER, F.: Reported at Mid-West Cosmic Ray Conf., Denver, 1964.
[85] APPA RAO, M. V. K.: Phys. Rev. **123**, 295 (1961) and references therein.
[86] APPA RAO, M. V. K.: J. Geophys. Res. **67**, 1289 (1962).
[87] SHAPIRO, M. M., B. HILDEBRAND, F. W. O'DELL, R. SILBERBERG, and B. STILLER: Proc. 3rd Int. Space Science Symposium, Washington, D. C. 1962. Amsterdam: North Holland Publishing Co.
[88] FOSTER, F., and J. H. MULVEY: Nuovo Cim. **27**, 93 (1963).
[89] AIZU, H.: Proc. Int. Cosmic Ray Conf., Jaipur **3**, 90 (1964).
[90] APPA RAO, M. V. K., K. DAHANAYAKE, M. F. KAPLON, and P. J. LAVAKARE: Proc. Int. Cosmic Ray Conf., Jaipur **3**, 95 (1964).

[91] WEBBER, W. R., and J. F. ORMES: Phys. Rev. (to be published).
[92] BALASUBRAMANIAN, V. K., S. V. DAMLE, G. S. GOKHALE, M. G. K. MENON, and S. K. ROY: Proc. Int. Cosmic Ray Conf., Jaipur 3, 110 (1964).
[93] NOON, J. H., and M. F. KAPLON: Phys. Rev. 97, 769 (1955).
[94] BADHWAR, G. D., R. R. DANIEL, and B. VIJAYALAKSHMI: Progr. Theoret. Phys. Japan 28, 607 (1962).
[95] BADHWAR, G. D., and R. R. DANIEL: Progr. Theoret. Phys. Japan 30, 615 (1963).
[96] WADDINGTON, C. J.: Progr. Nuclear Phys. 8, 1 (1960).
[97] FRIEDLANDER, M. W., K. A. NEELAKANTAN, S. TOKUNGA, G. R. STEVENSON, and C. J. WADDINGTON: Phil. Mag. 8, 1691 (1963).
[98] DANIEL, R. R., and N. DURGAPRASAD: Nuovo Cim. 23, Suppl., 82 (1962).
[99] GINZBURG, V. L., and S. I. SYROVATSKY: Progr. Theoret. Phys. Japan, Suppl., 20, 1 (1961).
[100] CHEN, F. F., C. P. LEAVITT, and A. M. SHAPIRO: Phys. Rev. 99, 1857 (1955).
[101] WEBBER, W. R., and J. F. ORMES: Proc. Int. Cosmic Ray Conf., Jaipur 3, 69 (1964).
[102] DANIELSON, R. E., P. S. FREIER, J. E. NAUGLE, and E. P. NEY: Phys. Rev. 103, 1075 (1956).
[103] WEBBER, W. R.: Nuovo Cim. 4, 1285 (1956).
[104] WADDINGTON, C. J.: Phil. Mag. 2, 1059 (1957).
[105] KOSHIBA, M., G. SCHULTZ, and M. SCHEIN: Nuovo Cim. 9, 1 (1958).
[106] FREIER, P. S., E. P. NEY, and C. J. WADDINGTON: Phys. Rev. 113, 921 (1959).
[107] SHAPIRO, M. M., B. HILDEBRAND, F. W. O'DELL, R. SILBERBERG, and B. STILLER: Space Research 3, 1097 (1962).
[108] KRISTIANSSON, K., O. MATHIESEN, and B. WALDESKOG: Ark. Fysik 23, 479 (1963).
[109] KERLEE, D. D., O. K. KRIENKE, J. J. LORD, and M. E. NELSON: Phys. Rev. 118, 828 (1960).
[110] DANIELSON, R. E.: Phys. Rev. 113, 1311 (1959).
[111] WADDINGTON, C. J.: Nuovo Cim. 14, 1205 (1959).
[112] JAIN, P. L., E. LOHRMANN, and M. W. TEUCHER: Phys. Rev. 115, 654 (1959).
[113] BADHWAR, G. D., S. BISWAS, R. R. DANIEL, and N. DURGAPRASAD: Proc. Int. Cosmic Ray Conf., Jaipur 3, 38 (1964).
[114] POMERANTZ, M. A., S. P. DUGGAL, and L. WHITTEN: Space Research 4, 9722 (1964).
[115] DURNEY, A. C., H. ELLIOT, R. J. HYNDS, and J. J. QUENBY: Proc. Roy. Soc. Lond., Series A, 281, 553 (1964).
[116] YAGODA, H., and K. FUKUI: Proc. Int. Cosmic Ray Conf., Jaipur 3, 24 (1964).
[117] NEELAKANTAN, K. A., and P. G. SHULKA: J. Phys. Soc. Japan 17, Suppl. A-3 20 (1962).
[118] APPA RAO, M. V. K., S. BISWAS, R. R. DANIEL, K. A. NEELAKANTON, and B. PETERS: Phys. Rev. 110, 751 (1958).
[119] EARL, J. A.: Phys. Rev. Letters 6, 125 (1961).
[120] MEYER, P., and R. VOGT: Phys. Rev. Letters 6, 193 (1961).
[121] FREIER, P. S., and C. J. WADDINGTON (to be published).
[122] EARL, J.: Trans. Amer. Geophys. Union 44, 72 (1963).
[123] SCHMOKER, J. H.: M. S. Thesis, Univ. of Minn. 1964.
[124] AGRINIER, B., Y. KOECHLIN, B. PARLIER, G. BOELLA, G. DEGLI ANTONI, C. DILWORTH, L. SCARSI, and G. SIRONE: Phys. Rev. Letters 13, 377 (1964).
[125] CRITCHFIELD, C. I., E. P. NEY, and S. OLEKSA: Phys. Rev. 85, 461 (1952).
[126] DESHONG jr., J. A., R. H. HILDEBRAND, and P. MEYER: Phys. Rev. Letters 12, 3 (1964).
[127] LAL, D.: Proc. Ind. Acad. Sci. 38, 93 (1953).
[128] KAPLON, M. F., B. PETERS, H. L. REYNOLDS, and D. M. RITSON: Phys. Rev. 85, 295 (1952).
[129] APPA RAO, M. V. K., and M. F. KAPLON: Nuovo Cim. 27, 700 (1963).
[130] TEUCHER, M. W.: Proc. Moscow Conf. on Cosmic Rays 1, 26 (1960).
[131] FOWLER, P. H., and C. J. WADDINGTON: Phil. Mag. 1, 637 (1956).
[132] LOHRMANN. E., and M. W. TEUCHER: Phys. Rev. 115, 638 (1959).
[134] CRANSHAW, T. E., J. F. DEBEER, W. GELBRAITH, and N. A. PORTER: Phil. Mag. 3, 377 (1958).
[135] CLARK, G., J. EARL, W. KRAUSHAAR, J. LINSLEY, B. ROSSI, F. SCHERB, and D. SCOTT: Phys. Rev. 122, 637 (1961).
[136] GRIESEN, K.: Ann. Rev. Nuclear Sci. 10, 63 (1960).
[137] DELVAILLE, J., F. KENDZIORSKI, and K. GREISEN: J. Phys. Soc. Japan 17, Suppl. III, 76 (1962).
[138] MURA, I., and H. HASEGAWA: J. Phys. Soc. Japan 17, Suppl. A-3, 84 (1962).
[139] LINSLEY, J., L. SCARSI, and B. ROSSI: J. Phys. Soc. Japan 17, Suppl. A-3, 91 (1962).

References.

[140] VERNOV, S. N., G. B. KHRISTIANSEN, V. I. ATRASHKEVICH, V. A. DMITRIEV, YU. FOMIN, B. A. KHRENOV, G. V. KULIKOV, YU. A. NECHIN, and V. I. SOLOVIYEVA: J. Phys. Soc. Japan **17**, Suppl. A-III, 118 (1962).
[141] LINSLEY, J.: Proc. Int. Cosmic Ray Conf., Jaipur **4**, 77 (1964).
[142] CLARK, G., I. ESCOBAR, K. MURAKAMI, and K. SUGA: Proc. V. Interamerican Seminar on Cosmic Rays **2**, No. 37 (1962).
[143] CLARK, G., H. BRADT, M. LA POINTE, V. DOMINGO, I. ESCOBAR, K. MURAKAMI, K. SUGA, Y. TOYODA, and J. HERSIL: Proc. Int. Cosmic Ray Conf., Jaipur **4**, 65 (1964).
[144] COCCONI, G., Handbuch der Physik, Bd. 46, S. 215. Berlin-Göttingen-Heidelberg: Springer 1961.
[145] COCCONI, G.: Nuovo Cim. **8**, Suppl., 560 (1958).
[146] ZATSEPIN, G. T., S. I. NIKOLSKY, and G. B. KHRISTIANSEN: Proc. Int. Cosmic Ray Conf., Jaipur **4**, 100 (1964).
[147] OLBERT, S.: Quoted by G. CLARK, J. EARL, W. KRAUSHAAR, J. LINSLEY, B. ROSSI, and F. SCHERB, Nature, Lond. **180**, 353 (1957).
[148] CHUDAKOV, A. E., M. M. NESTROVA, V. I. ZATSEPIN, and E. I. TUKISH: Proc. Moscow Conf. **2**, 50 (1959).
[149] ODA, M.: Nuovo Cim. **5**, 615 (1957).
[150] GINZBURG, V. L., L. V. KURNOSOVA, L. A. RAZORENOR, and M. L. FRADKIN: Space Sci. Reviews **2**, 778 (1963).
[151] SUESS, H. E., and H. C. UREY: Rev. Mod. Phys. **28**, 53 (1956).
[152] ALLER, L. H.: Astrophys. J. **125**, 84 (1957).
[153] CAMERON, A. G. W.: Astrophys. J. **129**, 676 (1959).
[154] ALLER, L. H.: Abundances of the Elements. New York: Interscience 1961.
[155] HASEGAWA, H., and K. ITO: J. Phys. Soc. Japan **17**, Suppl. III, 53 (1962).
[156] ALVIAL, G.: Proc. Int. Conf. on Cosmic Rays, Jaipur **3**, 116 (1964).
[157] BADHWAR, G. D., R. R. DANIEL, and B. VIJAYALAKSKI: Progr. Theoret. Phys. **28**, 607 (1962).
[158] WEBBER, W. R., and J. ORMES: Phys. Rev. (to be published).
[159] HASEGAWA, H., T. MATANO, I. MIURA, M. ODA, S. SHIBATA, G. TANAHASHI, and Y. TANAKA: J. Phys. Soc. Japan **17**, Suppl. III, 86 (1962).
[160] MCCUSKER, C. B. A.: Proc. Int. Conf. on Cosmic Rays, Jaipur **4**, 35 (1964).
[161] MATANO, T., I. MIURA, M. NAGANO, M. ODA, S. SHIBATA, Y. TANAKA, G. TANAHASHI and H. HASEGAWA: Proc. Int. Conf. on Cosmic Rays, Jaipur **4**, 129 (1964).
[162] SREEKANTAN, B. V.: Proc. Inf. Conf. on Cosmic Rays, Jaipur **4**, 143 (1964).
[163] MORRISON, P.: Handbuch der Physik, Bd. 46, S 1. Berlin-Göttingen-Heidelberg: Springer 1961.
[164] GINZBURG, V. L.: Progr. in Cosmic Ray Physics **4**, 337 (1958).
[165] See summary by G. Swarup, Proc. Int. Conf. on Cosmic Rays, Jaipur **1**, 3 (1964).
[166] GARDNER, F. F., and J. B. WHITEOAK: Nature, Lond. **197**, 1162 (1963).
[167] SHKLOVSKI, I. S.: Cosmic Radio Waves. Cambridge, Harvard University Press 1960; and references therein.
[168] GINZBURG, V. L.: Ups. Fiz. Nauk **51**, 343 (1953).
[169] WALSH, D., F. T. HADDOCK, and H. F. SCHULTE: Proc. 4th Internat. Space Science Symposium, Warsaw 1963, p. 935.
[170] OORT, J. H.: Handbuch der Physik, Bd. 53, S. 100. Berlin-Göttingen-Heidelberg: Springer 1959.
[171] GINZBURG, V. L., and S. I. SYROVATSKY: Soviet Astronomy **8** (1964).
[172] HAYAKAWA, S., and H. OKUDA: Progr. Theoret. Phys. **28**, 517 (1962).
[173] HAYAKAWA, S.: Proc. Int. Conf. on Cosmic Rays, Jaipur **3**, 125 (1964).
[174] JONES, F. C.: J. Geophys. Res. **68**, 4399 (1963).
[175] GINZBURG, V. L., and S. I. SYROVATSKY: Proc. Int. Conf. on Cosmic Rays, Jaipur **3**, 301 (1964).
[176] POLLACK, J. B., and G. G. FAZIO: Phys. Rev. **131**, 2684 (1963).
[177] HAYAKAWA, S., K. ITO, and Y. TEROSHIMA: Progr. Theoret. Phys. Japan **6**, Suppl. (1958).
[178] PARKER, E. N.: Interplanetary Dynamical Processes. New York: Interscience 1963; and references contained within.
[179] WEBBER, W. R.: Proc. NASA Symposium on Solar Flares. NASA Document SP-50 (1964).
[180] ALLEN, C. W.: Astrophysical Quantities. London: The Althone Press 1963.
[181] DAVIS, L.: Proc. Moscow Cosmic Ray Conf. **3**, 220 (1960).
[182] SINGER, S. F.: Nuovo Cim. **8**, Suppl., 549 (1958).
[183] KORSHAK, A. A., and S. I. SYROVATSKY: Proc. Moscow Cosmic Ray Conf. **3**, 211 (1960).

[184] Piddington, J. H.: Austral. J. Phys. **10**, 515 (1957).
[185] Bierman, L., u. L. Davis: Z. Naturforsch. **13**a, 909 (1958); also Proc. Moscow Cosmic Ray Conf. **3**, 228 (1960).
[186] Burbidge, G.: Proc. Int. Conf. on Cosmic Rays, Jaipur **3**, 229 (1964).
[187] Fichtel, C. E.: Phys. Rev. Letters **11**, 172 (1963).
[188] Byakov, V. H.: Soviet Astronomy **7**, 480 (1964).
[189] Laster, H.: Technical Report No. 373, University of Maryland 1964.
[190] Dorman, L. I.: Proc. Moscow Cosmic Ray Conf. **4**, 320 (1960).
[191] Peters, B.: Proc. Moscow Cosmic Ray Conf. **3**, 157 (1960).
[192] Gerasimova, and G. T. Zatsepin: J. Exp. Theoret. Phys. **40**, 1245 (1961).
[193] Fermi, E. Astrophys. J. **119**, 1 (1954).
[194] Fan, C. Y.: Phys. Rev. **101**, 314 (1956).
[195] Hayakawa, S., M. Koshiba, and Y. Terashima: Proc. Moscow Cosmic Ray Conf. **3**, 181 (1960).
[196] Kaplon, M. F., and G. Skadron: Nuovo Cim. (to be published).
[197] Parker, E. N.: Phys. Rev. **109**, 1734 (1958).
[198] Badhwar, G. D., R. R. Daniel, and B. Vijayalakski: Proc. Int. Conf. on Cosmic Rays, Jaipur **3**, 390 (1964).
[199] Syrovatsky, S. I.: J. Exp. Theoret. Phys. **40**, 1788 (1961).
[200] Biswas, S., and C. E. Fichtel: Astrophys. J. **139**, 941 (1964).
[201] Burbidge, E. M., G. R. Burbidge, W. A. Fowler, and F. Hoyle: Rev. Mod. Phys. **29**, 547 (1957).
[202] Burbidge, G. R.: Ann. Rev. Nuclear Sci. **12**, 507 (1962).
[203] Colgate, S. A., and R. H. White: UCRL Report, 7551 (1963) and references therein.

High Energy Photons and Neutrinos from Cosmic Sources*.

By

R. J. GOULD and G. R. BURBIDGE.

With 11 Figures.

I. Introduction.

1. Cosmic rays were first detected more than 50 years ago and after a terrestrial origin was disproved they were first taken to be γ-rays. Only when they were found to be deflected in the earth's magnetic field was it realized that they are charged particles. Until a few years ago there was no direct evidence for the presence of hard radiation from the cosmos. In 1957 x-rays were first detected from the sun by the NRL group (1), but only in the last two or three years have x-ray sources which lie outside the solar system been detected and γ-ray observations been attempted.

From the theoretical standpoint it is clear that since a flux of charged particles is known to be present in the cosmos, fluxes of energetic quanta must always be present since there are many mechanisms which give rise to photons as secondary quanta. At present, many of the observations of high energy photons are very preliminary and, in some cases, contradictory. However, it is clear that the interpretation of these observations can provide significant information on a large number of astronomical problems. Of special interest are questions of cosmology and the early observational data was soon employed as a means of testing cosmological theories. Actually, further interpretation of the data already available may provide additional answers to cosmological questions. The data itself is difficult to come by, because of the necessity of carrying photon detectors above the earth's absorbing atmosphere by means of balloons, rockets, or satellites. In this article we shall not attempt to describe the ingenious techniques developed for carrying out such observations, and shall concentrate on the problem of the *interpretation* of the results. However, on occasion we shall comment on the question of whether certain experimental results are suspect [1].

There are essentially three general sources of uncertainty involved in the interpretation of the observations on energetic photons: 1) experimental uncertainties or errors, 2) uncertainties in the calculations of the basic physical processes which produce the photons, and 3) uncertainties in the astronomical parameters employed in calculating the photon flux. Usually this last source of uncertainty is the most serious; for example, the mean gas density, (low energy) stellar photon density, magnetic field, and cosmic ray (proton) intensity in the Galaxy and intergalactic medium are known only approximately and in some cases may be different by

---

* The authors originally prepared an article on this topic for presentation at the I.A.U. Symposium on Astronomical Observations from Space Vehicles which was held at Liège in August, 1964. At the request of Dr. SITTE this was revised for inclusion here.

[1] A number of other reviews of x-ray and γ-ray astronomy are listed in the Bibliography.

several orders of magnitude. Moreover, the photon flux received from sources at great distances depends on the detailed structure of the universe.

The basic physical processes responsible for photon production may be summarized as follows: 1) Bremsstrahlung is emitted in the interaction of charged particles with matter. It results from $e-p$ COULOMB scattering at non-relativistic energies and from both $e-p$ and $e-e$ scatterings at relativistic energies. 2) The Compton scattering of a low energy thermal photon by a high energy electron produces a high energy scattered photon, the energy being transferred from the electron. This process was first discussed by FEENBERG and PRIMAKOFF (2). FERMI pointed out that this was probably the mechanism by which electrons were removed from the primary cosmic ray flux. 3) Electrons moving in magnetic fields emit synchrotron radiation; this is the primary mechanism for radio emission in galaxies. Very high electron energies are required to produce high energy photons by this process; cosmic synchrotron spectra probably extend at most to photon energies in the keV-MeV range. 4) Gamma rays result from the decay of π^0-mesons ($\pi^0 \to 2\gamma$) following the production of mesons in collisions between primary cosmic ray particles and nuclei of the interstellar and intergalactic gas. Cosmic ray nuclear collisions are also a source of high energy electrons via charged pion production and ($\pi \to \mu \to e$) decay, as was proposed by BURBIDGE and GINZBURG in the early attempts to understand radio sources. A recent discussion applying to galactic radiation has been given by POLLACK and FAZIO (3) and by GINZBURG and SYROVATSKY (4). π^0-gammas are also produced following meson production in matter — anti-matter annihilation. Some processes which produce line radiation are: 5) Characteristic x-rays are produced following the ejection of an atomic inner shell electron by, for example, a high energy particle or photon flux. The resulting cascade transitions give rise to the emission of K, L, etc.-series x-rays. 6) Gammy rays are produced in the annihilation of electrons and positrons ($e^+ + e^- \to 2\gamma$). Energetic positrons in the interstellar (but not intergalactic) medium come essentially to rest by various energy loss processes (see Sect. II) before annihilating, and the resulting γ-rays are essentially mono-energetic at about 0.51 MeV. 7) The formation of deuterium via $n+p \to d+\gamma$ (the inverse of photodisintegration) produces a photon of energy 2.23 MeV. This is the only low energy nuclear reaction we have listed here which gives rise directly to γ-radiation. There are many low energy reactions which give rise to γ-rays either directly or indirectly, but in general they will occur in stellar interiors so that the γ-rays do not escape. However, there are some indications that nuclear reactions sometimes take place in stellar surfaces, so that these γ-rays may be observable. Both the 0.51 and 2.23 MeV lines were mentioned in an early paper by MORRISON (5) on the subject of gamma ray astronomy. 8) Finally, we mention the more general process called *inner bremsstrahlung* which really includes some of the processes mentioned above. If an electron is suddenly accelerated from rest to a velocity βc by *any* mechanism, the probability that in the acceleration process an additional *soft* photon of energy within $\hbar\, d\omega$ is emitted is given by the simple expression

$$dw = \frac{\alpha}{\pi} \left(\frac{1}{\beta} \ln \frac{1+\beta}{1-\beta} - 2 \right) \frac{d\omega}{\omega} \to \frac{2\alpha}{3\pi} \beta^2 \frac{d\omega}{\omega}, \quad \text{if } \beta \ll 1, \tag{1.1}$$

where α is the fine structure constant (cf. reference 14).

Most of the photon-producing processes are treated in Part II of this article where the effects of the high energy electrons produced in cosmic ray nuclear collisions are considered. Part III is devoted to the problem of discrete sources of x-rays and the possibility of observing extragalactic sources of high energy photons. High energy neutrino astronomy is intimately related to high energy photon astronomy, since in the production of a shower of pions from a nuclear collision neutrinos result from the charged pion and muon decays and photons result directly from neutral pion decays. Neutrino sources are discussed in Part IV. As we have already emphasized, the presently available data on both the photon fluxes and astronomical parameters are very rough; for this reason, we feel that in attempting interpretation no elaborate calculations are warranted. We have tried to give as simple a treatment of the physical processes as is possible while still doing justice to the data.

II. Production in the interstellar gas, the galactic halo, and the intergalactic medium.

2. In this section we consider the general background flux of cosmic photons produced in electromagnetic interactions involving *non-thermal* particles. A source of high energy particles is provided by the ordinary cosmic rays, in particular the cosmic ray protons, whose energy spectrum is known and extends up to $\sim 10^{20}$ ev. The protons themselves are not efficient at producing photons in direct electromagnetic interactions, due to their large mass. However, high energy electrons can result from nuclear collisions of cosmic rays in which a shower of pions is produced; the charged pions then decay into electrons via $\pi \to \mu \to e$. The energetic "secondary" electrons which result can produce high energy photons by a number of processes, and these will be considered later in this section. The photon spectrum produced by a specific process is determined (among other things) by the electron spectrum, which in turn is determined by the cosmic ray proton spectrum. We shall assume a *universal* cosmic ray spectrum, that is, except near local sources of cosmic rays, the cosmic ray flux at any place in the universe is assumed to be the same as that measured at the earth. There is some difference of opinion as to whether the primary cosmic rays are predominantly of galactic or extragalactic origin. For a discussion of two extreme schools of thought on this point the reader is referred to the work of GINZBURG and SYROVATSKY (*6*) and BURBIDGE and HOYLE (*7*). However, to make the calculations described here we have made the assumption that a universal cosmic ray flux with the same energy density inside and outside galaxies is present. We take no position on the validity of this hypothesis in this article, as this has been done simply to facilitate the computations. The results are easily adjusted for other assumptions. GINZBURG and SYROVATSKY have argued against a universal cosmic ray flux and estimate that the intergalactic cosmic ray density is smaller than the local (galactic) value by a factor $\sim 10^{-3}$. However, their reasoning is based on equipartition arguments and is, in our opinion, not convincing. Of course, it may be that there exists a "primary" cosmic ray electron component, where by primary electrons we mean those which may have been accelerated by the same process and in the same sources that produced the cosmic ray protons. This question is open. Recent experiments by DE SHONG, HILDEBRAND, and MEYER (*8*) measuring the electron/positron ratio in the local cosmic ray flux are certainly relevant to this problem, but the experiments still do not allow a definite conclusion regarding the primary of secondary origin of these electrons and positrons. We shall consider only the contribution from secondary electrons. It might be remarked that the acceleration of protons without an accompanying acceleration of electrons can be envisaged easily, since the electrons, with their smaller mass, lose energy by electromagnetic processes more readily.

We shall take a universal differential cosmic ray flux given by

$$dJ_p = K_p \gamma_p^{-\Gamma_p} d\gamma_p, \tag{2.1}$$

where dJ_p is the number of incident protons per cm² per second having LORENTZ factors $\gamma_p(=E_p/m_p c^2)$ within $d\gamma_p$ (centered at γ_p); here K_p and Γ_p are constants. By appropriate choice of K_p and Γ_p the power law (2.1) can be used to describe the observed flux for any range of γ_p. The choice $\Gamma_p = 2.6$, $K_p = 100$ cm⁻² sec⁻¹ fits the observations (9) over many orders of magnitude of γ_p in the *high* energy range. At lower energies the actual flux is smaller than that described by this choice of Γ_p, K_p. The extrapolation from high energies is too large by a factor ~ 2 at $\gamma_p \sim 100$ and by a factor ~ 4 at $\gamma_p = 10$. Since we are interested in the effects of the high energy cosmic rays we shall adopt the above values for the parameters Γ_p, K_p in the calculations outlined in this section. Given the astronomical parameters (gas density, magnetic field, etc.), the cosmic photon fluxes from various processes (synchrotron radiation, bremsstrahlung, COMPTON effect, etc.) are essentially determined by the cosmic ray spectrum. However, due to uncertainties in our knowledge of the physics of certain processes, in particular that of meson production in high energy nuclear collisions, the calculated photon fluxes must be considered at best only order of magnitude estimates. Uncertainties in the astronomical parameters further complicate the interpretation of the results. In view of this, a number of simplifying assumptions and approximations are made in the calculation of the physical processes.

After discussing meson production in cosmic ray collisions (part a) the electron production spectrum is derived in (b). Electron energy losses in the galaxy and intergalactic medium are treated in (c) and (d) and the resulting electron spectra are derived in (e). The photon fluxes are calculated in (f) and a discussion and comparison with the observational results follows. Some cosmological considerations of photon production in the intergalactic medium are given in (h).

a) Meson production in cosmic ray nuclear collisions.

All of the laboratory results on meson production are for incident proton energies less than 10 Gev at which it is possible energetically to produce only a few relatively low energy pions per inelastic collision. Our knowledge of meson production by high energy protons is based primarily on theory, and the theories of meson production are very crude; of course, an accurate theoretical treatment of the problem would be extremely difficult, probably beyond our present knowledge of elementary particle interactions. The simplest theory of meson production in high energy nuclear collisions is that of FERMI (10) and is outlined briefly below. The theory predicts the correct shape for the spectrum of high energy γ-rays resulting from π^0's produced in cosmic ray collisions.

3. Fermi theory of meson production. Consider the collision of a proton of (lab) energy $\gamma_p m_p c^2$ incident on a proton at rest. In the center of mass (c.m.) system the total energy of the two protons is $2\bar{\gamma}_p m_p c^2 = [2(\gamma_p+1)]^{\frac{1}{2}} m_p c^2$, where $\bar{\gamma}_p$ is the LORENTZ factor of the protons in the c.m. system. Each proton carries a cloud of virtual pions; in the proton's rest frame the radius of this cloud is approximately $\Lambda_\pi = \hbar/m_\pi c$, where m_π is the pion mass. The interaction cross section is then $\sigma \sim \pi \Lambda_\pi^2$. In the c.m. system each cloud is contracted in the direction of motion by a factor $\bar{\gamma}_p$, and when the protons collide the maximum common volume of the meson clouds (which, presumably, is where the interaction is strongest) is

$$\Delta V = \frac{4\pi}{3} \Lambda_\pi^3 \frac{1}{\bar{\gamma}_p}. \qquad (3.1)$$

For high proton energies it is possible energetically to produce many pions in an inelastic collision and FERMI made the assumption that the interaction in the volume (3.1) was strong enough to produce a distribution of pion energies corresponding to *thermal equilibrium* with most of the initial proton kinetic energy having been fed into the pion gas. Also, the pions are predominantly highly relativistic and thus have a PLANCKian distribution. The "temperature" for this distribution is easily shown to be $kT \approx \gamma_p^{\frac{1}{2}} m_\pi c^2$, so that in the c.m. system the mean pion energy corresponds to

$$\langle \bar{\gamma}_\pi \rangle \approx \gamma_p^{\frac{1}{2}}, \tag{3.2}$$

and in the lab system (where one of the protons is initially at rest)

$$\langle \gamma_\pi \rangle \approx \bar{\gamma}_p \gamma_p^{\frac{1}{2}} \approx \gamma_p^{\frac{3}{2}}. \tag{3.3}$$

FERMI assumed that the distribution arising when the pion clouds of the colliding protons overlap is "frozen in", so that Eq. (3.3) would apply to the pions produced in the collision. Eq. (3.3) also implies that the *multiplicity* of pions produced is proportional to (and is, in fact, roughly given by) $\gamma_p^{\frac{1}{2}}$.

A number of attempts have been made to improve the FERMI theory and some authors have taken a quite different approach to the problem. However, these alternative theories usually predict a pion production spectrum not radically different from that of the FERMI theory. The assumption of thermal equilibrium in the FERMI theory has been questioned by LANDAU (11), who has developed his own theory of meson production. Another defect in the simple FERMI theory is that the effects of the production of other unstable particles (for example, K-mesons), which eventually decay into pions, has not been taken into account. Nevertheless, for our purposes essentially the only result which need be specified is the relation between multiplicity (and mean pion energy) and γ_p. The detailed shape of the pion energy spectrum produced by an incident proton of given energy need not concern us.

4. Pion production spectrum. The number of pions produced per second per cm³ within the energy range $d\gamma_\pi$ in $p-p$ collisions would be computed from

$$q_\pi(\gamma_\pi) \, d\gamma_\pi = \int dJ_p \, n_H \sigma \, f(\gamma_\pi; \gamma_p) \, d\gamma_\pi, \tag{4.1}$$

where dJ_p is the differential incident cosmic ray proton flux, n_H the local density of hydrogen nuclei, $\sigma (\approx \pi \Lambda_\pi^2)$ the total (excluding the multiplicity factor) cross section for the event, and $f(\gamma_\pi; \gamma_p)$ the distribution function for the pion production spectrum. We approximate the spectrum $f(\gamma_\pi; \gamma_p)$ by a product of the multiplicity $(\approx \gamma_p^{\frac{1}{2}})$ and a δ-function at the mean energy $(\approx \gamma_p^{\frac{3}{2}})$ of the pion spectrum for given γ_p:

$$f(\gamma_\pi; \gamma_p) \approx \gamma_p^{\frac{1}{2}} \delta(\gamma_\pi - \gamma_p^{\frac{3}{2}}). \tag{4.2}$$

With a cosmic ray spectrum given by the power law (2.1) we then obtain

$$q_\pi(\gamma_\pi) \approx (4\pi/3) \Lambda_\pi^2 K_p n_H \gamma_\pi^{-\Gamma_\pi}, \quad \Gamma_\pi = \tfrac{4}{3}(\Gamma_p - \tfrac{1}{2}). \tag{4.3}$$

The δ-function approximation (4.2) does not introduce an appreciable error. For example, if one computes $q_\pi(\gamma_\pi; \gamma_p)$, using the WIEN approximation to the PLANCK thermal distribution, one obtains a slowly varying function of γ_π times γ_π to the power $-\tfrac{4}{3}(\Gamma_p - \tfrac{1}{2})$, that is, essentially the same result as Eq. (4.3). Moreover, the exponent in the spectrum (4.3) will be the same for the case where the mass of the incident cosmic ray particle is different from that of the "target" nucleus. In such a case the analysis follows analogously, since the LORENTZ factors in the c.m. system are still proportional to $\gamma^{\frac{1}{2}}$ (when γ is large), where γ is the LORENTZ factor of the incident particle in the rest frame of the target particle.

5. An experimental test for $q_\pi(\gamma_\pi)$. For nuclear collisions at high energy the number of π^+, π^-, and π^0 mesons produced are the same, as is their energy distribution. The π^0 decays via $\pi^0 \to 2\gamma$, with the mean (lab) γ-ray energy being roughly $E_\pi/2$. Thus, a measurement of the γ-ray spectrum from π^0-mesons produced in primary cosmic ray events would give the pion source spectrum $q_\pi(\gamma_\pi)$. Recently, KIDD (12) has measured the spectrum of high energy γ's from π^0-mesons produced by cosmic rays at the top of the atmosphere. By performing the experiment at high altitudes he was able to observe γ's from π^0's produced predominantly in primary jets. KIDD found for the differential energy spectrum of the γ-ray flux a power law with exponent $\Gamma_0 = 2.9^{-0.2}_{+0.3}$. The γ-ray energy range observed by KIDD was 0.7×10^{11} eV $< E_0 < 10^{12}$ eV, corresponding to $10^3 < \gamma_\pi < 10^4$ and $10^4 < \gamma_p < 2 \times 10^5$. At these proton energies the cosmic ray spectrum is described by the high energy fit with $\Gamma_p = 2.6$. The corresponding Γ_π from Eq. (4.3) is 2.8 and is consistent with the value (Γ_0) measured by KIDD. We should like to emphasize that KIDD's experiment confirms the *results* of the FERMI theory, but not the fundamentals of the theory itself.

b) The electron production spectrum.

6. In the charged pion decay $(\pi^\pm \to \mu^\pm + \nu)$ most of the center of mass kinetic energy released to the products μ, ν is carried away by the neutrino whose energy is small compared with $m_\pi c^2$. The resulting lab energy of the muon is then approximately $(m_\mu/m_\pi) E_\pi$, where E_π is the lab energy of the pion before decay. The electron resulting from the muon decay $(\mu^\pm \to e^\pm + 2\nu)$ is highly relativistic and behaves kinematically like the two neutrinos in the decay products. Thus, the mean energy in the spectrum of electron energies is about $\frac{1}{3} m_\mu c^2$ in the rest frame of the μ, and the mean lab energy $\langle E_e \rangle$ of the electron resulting from the $\pi \to \mu \to e$ decay is roughly $\frac{1}{3}(m_\mu/m_\pi) E_\pi \approx \frac{1}{4} E_\pi$; thus, $\langle \gamma_e \rangle \approx \frac{1}{4}(m_\pi/m_e)\langle \gamma_\pi \rangle$. Approximating the electron spectrum $f(\gamma_e; \gamma_\pi)$ by a δ-function at this energy we get for the electron source spectrum,

$$q_e(\gamma_e) d\gamma_e \approx \frac{2}{3} \int q_\pi(\gamma_\pi) d\gamma_\pi \, \delta\left(\gamma_e - \frac{m_\pi}{4 m_e} \gamma_\pi\right) d\gamma_e = \frac{8 m_e}{3 m_\pi} q_\pi\left(\frac{4 m_e}{m_\pi} \gamma_e\right) d\gamma_e; \quad (6.1)$$

a factor $\frac{2}{3}$ has been introduced because only charged pions decay into electrons.

We shall consider production and energy losses of electrons with $10^2 \leq \gamma_e \leq 10^{10}$ corresponding to $1 \leq \gamma_\pi \leq 10^8$ and to $1 \leq \gamma_p \leq 10^{11}$.

c) Electron energy losses in the galaxy.

Here we consider the various processes tending to decrease the energy of high energy electrons in the galaxy. We calculate the *average* rate of energy loss in the galaxy which we consider as the region within the galactic halo of radius $R_h \sim 5 \times 10^{22}$ cm. Actually, energy losses involving interactions with the galactic gas occur predominantly near the plane of the galaxy where most of the gas lies and where the gas is predominantly unionized. The volume of this disk of galactic interstellar gas is $\sim 10^{-2}$ of the volume of the galactic halo.

7. Ionization losses. The energy loss due to ionization and excitation of the interstellar gas may be computed from BETHE's formula for the stopping power. For high energy electrons this formula is

$$-\left(\frac{dE_e}{dx}\right)_I = \frac{2\pi n e^4}{m_e c^2} \ln \frac{\gamma_e^3 m_e^2 c^4}{2 I_0^2} \quad (7.1)$$

where I_0 is the mean excitation energy of the stopping material (hydrogen), and n is the number density of atoms of the material. The argument of the logarithm in Eq. (7.1) is very large and I_0 may be set equal to the RYDBERG energy $\frac{1}{2}\alpha^2 m_e c^2 (\alpha^{-1} \approx 137)$. We then have for the ionization loss in a hydrogen gas of mean density $\langle n \rangle$:

$$-\left\langle \frac{d\gamma_e}{dt} \right\rangle_I = 2\pi c r_0^2 \langle n \rangle \ln\left(\frac{2\gamma_e^3}{\alpha^4}\right). \tag{7.2}$$

Here $r_0 (= e^2/m_e c^2)$ is the classical electron radius. The energy loss computed from Eq. (7.2) is shown as a function of γ_e in Fig. 1 for a mean gas density $\langle n \rangle = 0.03$ cm^{-3}. This mean galactic gas density corresponds to a mean density near the plane of the galaxy of 3 cm^{-3}. This value (3 cm^{-3}) is about there times the observed density of atomic hydrogen. The higher value may be more appropriate if there is a high abundance of interstellar molecular hydrogen (13).

8. **Bremsstrahlung.** The energy loss rate by bremsstrahlung emission would be computed from

$$-\left(\frac{dE}{dt}\right)_B = nc \int \hbar\omega \, d\sigma_B, \tag{8.1}$$

where n is the density of hydrogen nuclei and $d\sigma_B$ is the differential cross section for the emission of a bremsstrahlung photon of energy within $\hbar\, d\omega$; in Eq. (8.1) the integral is over ω from 0 to $\gamma_e m_e c^2/\hbar$. For $d\sigma_B$ we take the approximate simplified expression (14) $d\sigma_B \approx 4\alpha r_0^2 \omega^{-1} d\omega \ln 2\gamma_e$ and calculate the mean bremsstrahlung loss rate:

$$-\left\langle \frac{d\gamma_e}{dt} \right\rangle_B \approx 4c\alpha r_0^2 \langle n \rangle \gamma_e \ln 2\gamma_e. \tag{8.2}$$

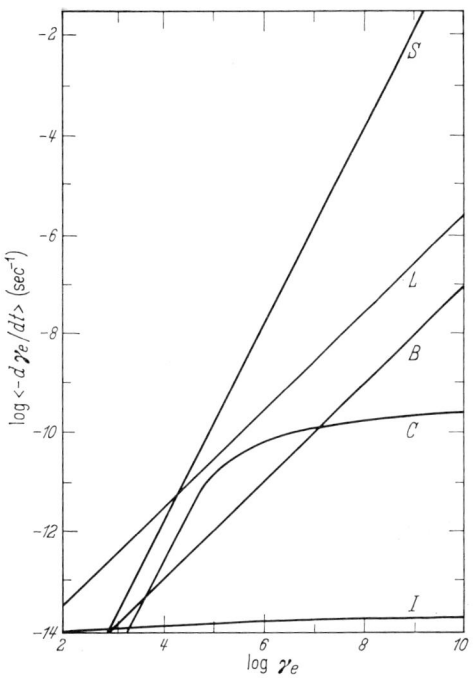

Fig. 1. Electron energy loss rate in the Galaxy by synchrotron emission (S), leakage out of the halo (L), bremsstrahlung (B), COMPTON scattering (C), and ionization (I).

This is the bremsstrahlung loss rate for interaction of electrons with *protons* and would be appropriate for calculating the energy loss in regions of ionized hydrogen. Actually, most of the galactic bremsstrahlung is likely to be produced near the galactic plane where the gas is predominantly atomic or molecular, and a correction for the associated shielding effects of the atomic electrons must be made. In fact, for the electron energies of interest the *strong shielding* expression would be more appropriate. In this case the argument of the logarithm in Eq. (8.2) should be replaced by ~ 137 [see HEITLER (15)]. Using this corrected expression the bremsstrahlung loss rate was computed for $\langle n \rangle = 0.03$ cm^{-3} and is shown in Fig. 1.

9. **Synchrotron losses.** It is well known that a highly relativistic electron of energy E_e in a magnetic field H moves in a circle with a LARMOR radius $r_L = E_e/eH$ and radiates energy by the synchrotron process at a rate

$$-\left\langle \frac{dE_e}{dt} \right\rangle_S = \frac{2}{3} c r_0^2 \langle H^2 \rangle \gamma_e^2. \tag{9.1}$$

The frequency spectrum of the radiation consists of a continuum with a maximum around $\nu_L \gamma_e^2$, $\nu_L (= eH/2\pi m_e c)$ being the LARMOR frequency. The loss rate $-\langle d\gamma_e/dt \rangle_S$ is shown in Fig. 1 for a magnetic field $H = 3 \times 10^{-6}$ gauss corresponding to the galactic halo.

10. Compton scattering by stellar photons. The COMPTON process, whereby a high energy electron makes an elastic collision with a thermal stellar photon, and transfers some of its kinetic energy to the photon, has been considered in some detail by FEENBERG and PRIMAKOFF (2) and by DONAHUE (16). More recently, FELTEN and MORRISON (17) have suggested this process as a mechanism for producing energetic photons. Consider the collision between an electron of energy $\gamma_e m_e c^2$ and a thermal photon of the galactic radiation field of initial energy ε_r. Let ε_r' denote the photon energy after scattering; let ε_r^* denote the initial energy of the photon in the rest frame of the electron; $\varepsilon_r^* \approx \gamma_e \varepsilon_r$. For $\varepsilon_r^* \ll m_e c^2$ the cross section for the scattering process is given by the THOMPSON limit:

$$\sigma_I \to \frac{8\pi}{3} r_0^2, \qquad (10.1)$$

while the mean energy loss per scattering may easily be shown, by the kinematics of the problem, to be

$$(\bar{\varepsilon}_r')_I \approx \gamma_e^2 \varepsilon_r. \qquad (10.2)$$

For collisions with very high energy electrons in which $\varepsilon_r^* \gg m_e c^2$, the KLEIN-NISHINA formula must be used to compute the scattering cross section. For high energies this formula approaches

$$\sigma_{II} \to \pi r_0^2 \frac{m_e c^2}{\varepsilon_r^*} \ln \frac{2\varepsilon_r^*}{m_e c^2}, \qquad (10.3)$$

while the mean energy loss per scattering is now comparable to the initial energy of the electron:

$$(\bar{\varepsilon}_r')_{II} \approx \gamma_e m_e c^2. \qquad (10.4)$$

The electron energy loss is computed from

$$-\left\langle \frac{dE_e}{dt} \right\rangle_C = c \left\langle \int \sigma n_r(\varepsilon_r) \bar{\varepsilon}_r' d\varepsilon_r \right\rangle, \qquad (10.5)$$

where $n_r(\varepsilon_r) d\varepsilon_r$ is the number density of photons of energy within $d\varepsilon_r$ in the radiation field. We shall lump the stellar radiation field into one mean photon energy $\bar{\varepsilon}_r$. Then $\int n_r(\varepsilon_r) d\varepsilon_r \equiv n_r \to \varrho_r/\bar{\varepsilon}_r$, where ϱ_r is the radiation energy density and n_r the number density of photons. For a thermal (black body) radiation field $\bar{\varepsilon}_r$ is approximately $3 k T_0$, where T_0 is the temperature of the thermal distribution. By employing the expressions for σ and $\bar{\varepsilon}_r'$ for the low energy region (I) where $\gamma_e \ll m_e c^2/\bar{\varepsilon}_r$ and the high energy region (II) where $\gamma_e \gg m_e c^2/\bar{\varepsilon}_r$, we get for the energy loss rates:

$$-\left\langle \frac{d\gamma_e}{dt} \right\rangle_{CI} = \frac{8\pi}{3} \frac{r_0^2}{m_e c} \langle \varrho_r \rangle \gamma_e^2, \qquad (10.6\text{ I})$$

$$-\left\langle \frac{d\gamma_e}{dt} \right\rangle_{CII} = \pi r_0^2 m_e c^3 \frac{\langle \varrho_r \rangle}{\bar{\varepsilon}_r'} \ln \frac{2\gamma_e \bar{\varepsilon}_r}{m_e c^2}. \qquad (10.6\text{ II})$$

It is interesting to note that at low energies the energy loss rate is proportional to the radiation energy density $\langle \varrho_r \rangle$ while at high energies it is essentially proportional to $\langle n_r \rangle / \bar{\varepsilon}_r$. Most of the contribution to the radiation field in the galactic halo comes from the relatively cool stars in the nuclear region of the galaxy. We shall take $\bar{\varepsilon}_r = 3$ eV and $\langle \varrho_r \rangle = 10^{-13}$ erg/cm³ as representative values for the radiation field in the halo. The corresponding energy loss rate is shown in Fig. 1. The curves for regions I and II were joined smoothly.

11. Leakage out of the galactic halo. Even for electron energies as high as $\gamma_e \sim 10^{10}$ the LARMOR radii are only ~ 1 pc which, presumably, is much less than the scale of "magnetic field condensations" in the halo. For this reason the high energy electrons moving in the halo are likely to penetrate only the outer edges of the magnetic field regions, and the paths of the electrons would resemble that of *Brownian motion*. The mean free path would correspond to motion between magnetic field condensations and, because of the smallness of the electrons' LARMOR radii, would be independent of energy if the magnetic field between the condensations were very small. The mean leakage time τ_L for escape from the halo would be roughly

$$\tau_L \sim R_h^2/\lambda c, \tag{11.1}$$

where $R_h (\approx 5 \times 10^{22}$ cm) is the radius of the halo and λ is the mean free path for the Brownian motion. The appropriate value of λ to be used to calculate τ_L is very uncertain. In the galactic disk the mean distance between gas clouds is ~ 100 pc; λ for the halo is probably larger than this. Taking $\lambda = 1$ kpc we calculate $\tau_L \sim 3 \times 10^{15}$ sec.

In a leakage process the energy of the electron is not lost gradually; instead essentially the *total* energy of the particle is lost (to the intergalactic medium) instantaneously. The *equivalent* loss is rate then

$$-\left\langle\frac{d\gamma_e}{dt}\right\rangle_L = \frac{\gamma_e}{\tau_L}, \tag{11.2}$$

and this quantity is plotted in Fig. 1 for $\tau_L = 3 \times 10^{15}$ sec.

d) Electron production and energy losses in the intergalactic medium.

12. The calculation of processes in the intergalactic medium is made difficult by our lack of knowledge of the astronomical parameters such as the gas density and magnetic field. Here we shall present results for *assumed* values of the parameters. The calculated production rates and energy losses are simply related to the parameters and can be easily revised when better astronomical data are available. Actually, it may be that some additional knowledge of these poorly known data may be gained from further interpretation of the high energy cosmic photon experiments.

As mentioned earlier, we assume a universal cosmic ray flux. The pion and electron production rates are then proportional to the intergalactic gas density and this gas is very likely to be predominantly hydrogen. Observationally, the upper limit to the intergalactic density of *atomic* hydrogen is (*18*) $\sim 10^{-5}$ cm^{-3}; the amount of *ionized* hydrogen is unknown. The usually *assumed* total density of intergalactic hydrogen is $\langle n_H \rangle \sim 10^{-5}$ cm^{-3}; this is the so-called cosmological[1] value and is the figure which we shall adopt. Also, we shall assume that the intergalactic hydrogen is fully ionized. We adopt 10^{-7} gauss for the mean intergalactic magnetic field. Certainly the intergalactic medium must have some, if only random,

---

[1] Several cosmological theories, including HOYLE's formulation of the steady-state theory, lead to values of this order for the mean density in the universe. One can arrive at this result by simply setting $E_0 + V = 0$, where $E_0 (= M c^2$; M is the "mass of the universe") is the rest energy of the universe, and $V (\sim - G M^2/R$; R is the "radius of the universe" or HUBBLE radius) is the gravitational energy. The resulting mean density is about two orders of magnitude greater than the observed smeared out density ($\sim 3 \times 10^{-31}$ gm/cm^3) from galaxies. The bulk of the matter in the universe is then attributed to the uncondensed intergalactic gas.

Handbuch der Physik, Bd. XLVI/2.

magnetic field. The intergalactic radiation field can be estimated with some reliability. The contribution from all galaxies in the universe results in a radiation field similar to the galactic (halo) field but diluted by about a factor of ten. Thus we take $\langle \varrho_r \rangle = 10^{-14}$ erg/cm^3 and, again, $\bar{\varepsilon}_r = 3$ eV.

Assuming the above values for the gas density, magnetic field, and radiation density in the intergalactic medium the various processes can be calculated readily by employing the relations given in part (c) of this section for galactic processes. However, for the bremsstrahlung contribution one must include the effects of electron-electron bremsstrahlung B_{ee} [14] as well as the contribution from B_{ep}. Since the cross section for high energy B_{ee} is approximately equal to that for B_{ep}, and since $n_e = n_p$ for the assumed fully ionized intergalactic medium, the total bremsstrahlung loss $-\langle d\gamma_e/dt \rangle_B$ is given by simply twice the expression (8.2) with $\langle n \rangle = 10^{-5}$ cm^{-3} [1]. The "ionization losses" for the fully ionized intergalactic medium actually correspond to production of plasma oscillations. The associated expression for the electron energy loss at high energies reduces to (19)

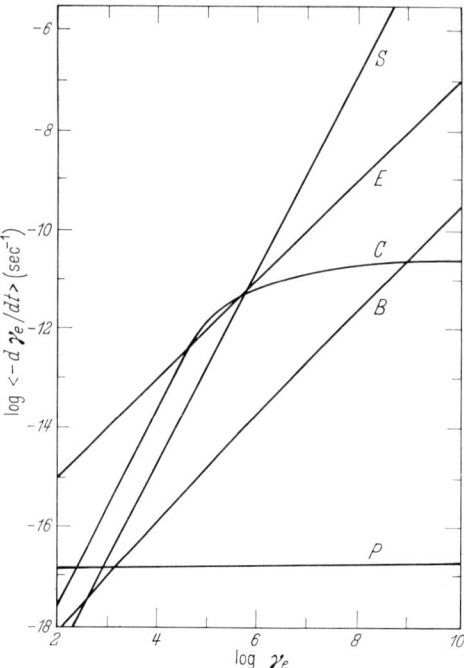

Fig. 2. Electron energy loss rate in the intergalactic medium by synchrotron emission (S), cosmic expansion (E), Compton scattering (C), bremsstrahlung (B), and excitation of plasma oscillations (P).

$$-\left\langle \frac{d\gamma_e}{dt} \right\rangle_P = 4\pi r_0^2 c \langle n \rangle \ln \frac{2m_e c^2 \gamma_e}{\hbar \omega_p}, \quad (12.1)$$

where $\omega_p (=[4\pi e^2 \langle n \rangle/m_e]^{\frac{1}{2}})$ is the plasma frequency. The result is plotted in Fig. 2.

For the intergalactic medium one should consider another "effective" energy loss process. The expansion of the universe results in an effective energy loss for the electrons in a given volume of

$$-\left\langle \frac{d\gamma_e}{dt} \right\rangle = \frac{\gamma_e}{\tau_E}, \quad (12.2)$$

where τ_E is the characteristic expansion time given by $\frac{1}{3} H^{-1} \sim 10^{17}$ sec ($H =$ HUBBLE constant). The factor $\frac{1}{3}$ takes into account the fact that the expansion is three dimensional, that is, H^{-1} is the characteristic time for the one dimensional expansion. The effective energy loss due to expansion is plotted in Fig. 2 for $\tau_E = 10^{17}$ sec, along with the energy losses due to bremsstrahlung, synchrotron radiation, and COMPTON scattering.

e) The electron energy spectrum in the halo and intergalactic medium.

13. Here we consider the electron spectrum which results from production (via $\pi - \mu - e$ decay) in nuclear collisions of cosmic rays and from the various loss processes. Let $n_e(\gamma_e) d\gamma_e$ denote the number of electrons per cm^3 with energies within $m_e c^2 d\gamma_e$. The spectral electron density $n_e(\gamma_e)$ satisfies a continuity equation

---

[1] Although $B_{ee} \approx B_{ep}$ for highly *relativistic* electrons, for *non-relativistic* electrons $B_{ee} \ll B_{ep}$. Essentially, this is because the photon emission by the non-relativistic system results from the dipole moment formed by the $e-p$ system.

in γ_e (energy) space:

$$\frac{\partial n_e(\gamma_e)}{\partial t} + \frac{\partial}{\partial \gamma_e}\left(n_e(\gamma_e)\frac{d\gamma_e}{dt}\right) = \sum_i q_i(\gamma_e). \tag{13.1}$$

In Eq. (13.1) the terms on the right hand side (r.h.s.) represent *sources* and *sinks* of high energy electrons corresponding to production, annihilation, and to processes leading to a sudden loss of a large fraction of the energy of the electron; terms representing leakage out of the halo or the expansion of the universe would also be included on the r.h.s. The factor $d\gamma_e/dt$ represents the *total* gradual energy loss from processes described earlier. We shall consider steady state conditions, so that $\partial n_e(\gamma_e)/\partial t = 0$.

14. Electron spectrum in the galactic halo.
From Fig. 1 we see that for $\gamma_e \leq 10^4$ (region I) the effective energy loss is primarily by leakage from the halo and the continuity equation reduces to

$$0 = q_e(\gamma_e) - n_e(\gamma_e)/\tau_L, \tag{14.1}$$

where $q_e(\gamma_e)$ is the production spectrum given by Eq. (6.1) and is of the form $k_e \gamma_e^{-\Gamma_\pi}$, and τ_L is the leakage time. Thus, for $\gamma_e \leq 10^4$, $n_e(\gamma_e)$ is of the form

$$\left.\begin{array}{l} n_e^{(I)}(\gamma_e) = K_e^{(I)} \gamma_e^{-\Gamma_\pi}, \\ K_e^{(I)} = \tau_L k_e. \end{array}\right\} \tag{14.2}$$

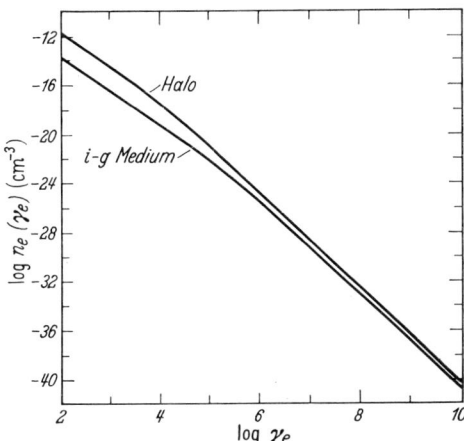

Fig. 3. Calculated energy spectrum of relativistic electrons in the galactic halo and in the intergalactic medium.

The electron spectrum in region I is essentially the same as the production spectrum, that is, the electrons escape from the galaxy without losing an appreciable amount of their original production energy.

For $\gamma_e \geq 10^5$ (region II) the electrons lose their energy primarily by synchrotron radiation for which $d\gamma_e/dt = -b\gamma_e^2$, and the continuity equation reduces to

$$-b\frac{\partial}{\partial \gamma_e}\left(\gamma_e^2 n_e(\gamma_e)\right) = q_e(\gamma_e) = k_e \gamma_e^{-\Gamma_\pi}. \tag{14.3}$$

The solution is then

$$n_e^{(II)}(\gamma_e) = K_e^{(II)} \gamma_e^{-(\Gamma_\pi + 1)}, \quad K_e^{(II)} = \frac{k_e}{b(\Gamma_\pi - 1)}. \tag{14.4}$$

With the assumed values for the parameters and with k_e computed from Eqs. (4.3) and (6.1) the calculated spectral electron density is shown in Fig. 3. The solutions for $n_e(\gamma_e)$ in regions I and II were joined smoothly.

15. Electron spectrum in the intergalactic medium.
The approximate spectrum of the intergalactic electrons is calculated by similar procedures. We approximate the effective energy loss for $\gamma_e \leq 10^5$ (region I, see Fig. 1) by the expansion loss and for $\gamma_e \geq 10^6$ (region II) by synchrotron losses. The electron spectrum in the two regions is then given by expressions similar to Eqs. (14.1) and (14.2) for the halo, essentially with τ_L replaced by τ_E. The calculated spectrum, with the curves for the two regions joined smoothly, is shown in Fig. 3 for the previously stated assumed values of the astronomical parameters.

18*

f) High energy photon flux from various processes.

16. Absorption of high energy photons. The *absorption* of cosmic photons is important in certain energy ranges. For x-ray photons traversing matter in the plane of the Galaxy absorption by the photoelectric effect in various elements is appreciable at the longer wavelengths. In Fig. 4 we give the optical thickness as a function of wavelength for a path of 1 kpc in neutral atomic gaseous matter of "cosmic" composition with $n(H)=1$ cm^{-3}. The curve with the total contribution from all elements is given as well as that including only hydrogen and helium. The discontinuity in the total at 23 Å is due to the onset of K-shell photoionization of oxygen. The curves have a slope of approximately 3 due to the (approximate) λ^3 dependence of photoelectric absorption, and are taken from the results of STROM and STROM (20).

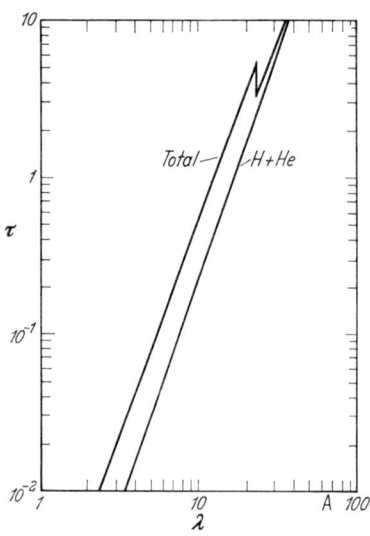

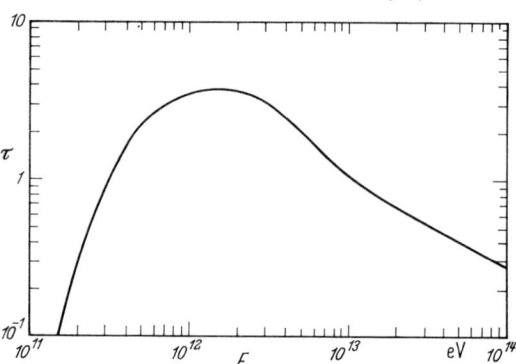

Fig. 4. Optical thickness τ as a function of wavelength λ in the x-ray range for photons traversing a distance of 1 kpc in which the matter is gaseous and atomic at a cosmic abundance with $n(H)=1$ cm^{-3}.

Fig. 5. Optical thickness τ as a function of photon energy E at very high energies for photons traversing 5×10^{27} cm of intergalactic matter in which the radiation field is PLANCKian with $kT = 0.5$ eV with a total energy density of 0.01 eV/cm^3.

Except for the pronounced edge due to oxygen, we have smoothed over the data which shows several other small jumps due to the onset of photoionization edges. Since the distance to the galactic center is ~ 10 kpc, and over this distance $\langle n(H) \rangle \sim 1$ cm^{-3}, we see that $\tau > 1$ for $\lambda \geq 5$ Å. For photons traversing intergalactic matter the path length is $\sim 5 \times 10^{27}$ cm, so that $\tau > 1$ for $\lambda > 10$ Å if $\langle n(H) \rangle \gg 10^{-6}$ cm^{-3} (here a similar "cosmic" abundance has been assumed). A density of $\langle n(H) \rangle \sim 10^{-5}$ cm^{-3} is a reasonable value to assume for intergalactic space; however, this material is also likely to the composed essentially of pure hydrogen, or perhaps hydrogen and helium; material with this composition would be ionized if the temperature of intergalactic matter were $\geq 1.5 \times 10^4$ °K (hydrogen), and $\geq 8 \times 10^4$ °K (helium). The absorption by the ionized matter would be negligible.

The only other instance where absorption of high energy cosmic photons is appreciable is when very high energy photons travel distances comparable with the classical radius of the universe. As NIKISHOV (21) has shown, absorption by pair production in photon-photon collisions ($\gamma + \gamma \rightarrow e^+ + e^-$) prevents us from seeing to the "outer edge" of the universe in photons of energy $\sim 10^{12}-10^{13}$ eV. For photons of this energy the cross section for pair production in collisions with the thermal stellar photons in the intergalactic radiation field has a maximum. The intergalactic radiation field in the thermal stellar range ($\sim$eV) is due to emission from galaxies. NIKISHOV calculated the optical thickness for a black body radiation field of temperature $kT = 0.5$ eV and total energy density 0.1 eV/cm³ out to the distance of Cygnus A ($R_C = 6.6 \times 10^{26}$ cm). We feel that the energy density which he employed may be on the high side and shall give the results for a radiation field one-tenth as large but for a distance of $R = 5 \times 10^{27}$ cm (the "cosmological cut-off"). The associated optical thickness is shown in Fig. 5 as a function of photon energy. We should like to emphasize again the uncertainty in the intergalactic radiation field and thus in the magnitude of the effect. Because of this and other uncertainties we shall ignore absorption in the remainder of this article; however, it should still be kept in mind that it could be appreciable for certain photon energies and could effect the high energy cosmic photon spectrum.

17. Photon spectra. The photon production spectrum by a given process may be computed from the electron (energy) spectrum $n_e(\gamma_e)$ and the expression for the photon emission spectrum by this process as a function of γ_e. Denote the photon energy by ε. The energy loss by an electron of energy E_e in time dt due to the emission of dN photons of energy within $d\varepsilon$ is

$$-dE_e \equiv \varepsilon \, dN = f(E_e, \varepsilon) \, d\varepsilon \, dt, \tag{17.1}$$

where $f(E_e, \varepsilon)$ is the emission spectrum. The number of photons emitted per cm³ per second per interval of ε by an electron spectrum $n_e(\gamma_e)$ would then be

$$\frac{dn}{dt \, d\varepsilon} \equiv \frac{dn(\varepsilon)}{dt} = \int d\gamma_e \, n_e(\gamma_e) \frac{dN}{d\varepsilon \, dt}. \tag{17.2}$$

We now approximate the emission spectrum by a δ-function at the characteristic photon energy ε_c:

$$\frac{dN}{d\varepsilon \, dt} = \frac{1}{\varepsilon} f(E_e, \varepsilon) \to -\frac{m_e c^2}{\varepsilon} \frac{d\gamma_e}{dt} \delta(\varepsilon - \varepsilon_c), \tag{17.3}$$

where $\varepsilon_c = \varepsilon_c(\gamma_e)$. The photon spectral *flux* due to emission along a line of sight of path $\int ds = R$ would be

$$j(\varepsilon) = \frac{dJ}{d\varepsilon} = \int \frac{dn(\varepsilon)}{dt} \, ds = \left\langle \frac{dn(\varepsilon)}{dt} \right\rangle R. \tag{17.4}$$

The incident photon spectra from both the galaxy and the intergalactic medium are readily calculated from the Eqs. (17.2), (17.3), and (17.4) using the derived electron spectra $n_e(\gamma_e)$ and the expressions for the energy losses $-\langle d\gamma_e/dt \rangle$. For synchrotron emission $\varepsilon_c \approx \hbar \omega_L \gamma_e^2$; for bremsstrahlung $\varepsilon_c \approx m_e c^2 \gamma_e$; for the COMPTON process from electrons with $\gamma_e \ll m_e c^2/\bar{\varepsilon}_r$, $\varepsilon_c \approx \bar{\varepsilon}_r \gamma_e^2$; for the COMPTON process from electrons with $\gamma_e \gg m_e c^2/\bar{\varepsilon}_r$, $\varepsilon_c \approx m_e c^2 \gamma_e$.

Taking a path length $R = 5 \times 10^{22}$ cm (the radius of the galactic halo) for the Galaxy and a path length $R = 5 \times 10^{27}$ cm (half the HUBBLE radius) for the intergalactic medium, the resulting photon spectra are shown in Fig. 6. The photon energy η is in units of $m_e c^2$, that is, $\eta = \varepsilon/m_e c^2$, $j(\eta) = dJ/d\eta$. The spectra are for synchrotron radiation, bremsstrahlung, COMPTON scattering, and π^0-decay. The spectra from π^0-decay are calculated directly from the pion production spectrum which is of the form $q_\pi(\gamma_\pi) = k_\pi \gamma_\pi^{-\Gamma_\pi}$ [Eq. (4.3)]. One-third of the pions produced are π^0's, and each π^0 gives two photons of mean energy $\frac{1}{2} \gamma_\pi m_\pi c^2$. The π^0-decay photon production spectrum is then approximately

$$\frac{dn^0}{d\eta \, dt} \approx \frac{2}{3} k_\pi \left(\frac{2 m_e}{m_\pi} \right)^{-(\Gamma_\pi - 1)} \eta^{-\Gamma_\pi}. \tag{17.5}$$

We have not included the effects of *photopion production* and decay by collisions of cosmic ray protons with the low energy photons in the intergalactic mediun (*102*). This process appears to be important only for the production of ultra-high energy photons ($E \gtrsim 10^{16}$ eV).

The galactic photon fluxes plotted in Fig. 6 are averaged over all directions; the average magnitude of the flux *per steradian* is $1/4\pi$ times the flux in Fig. 6.

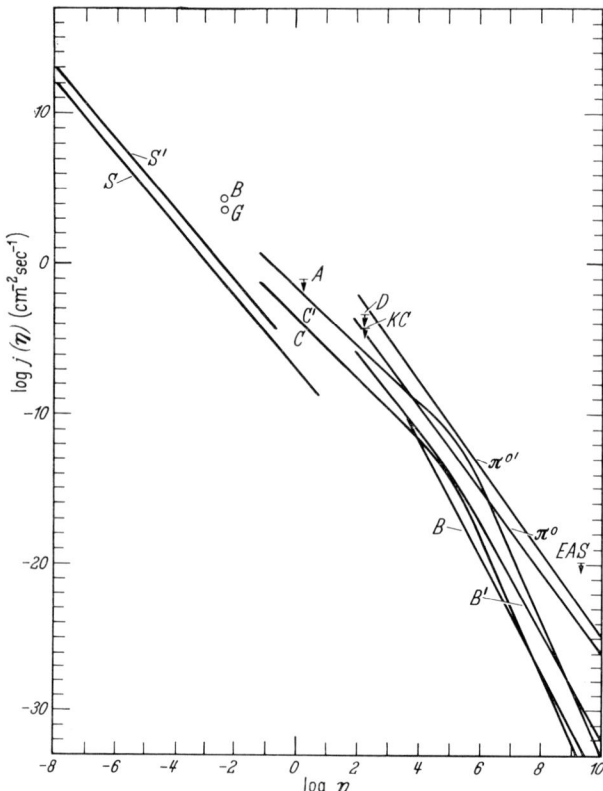

Fig. 6. Calculated high energy photon background fluxes from synchrotron radiation, COMPTON scattering, bremsstrahlung, and π^0-decay. The unprimed-designated spectra represent the galactic contributions and the primed denote the spectra from the intergalactic medium. Observational points are denoted by circles and arrows (limits). The letters next to the points refer to the observers (see Table 1).

Actually, the photon flux per steradian from bremsstrahlung and π^0 decay would be greatest in the direction along the galactic plane where the production and interaction with the gas takes place. The synchrotron radiation and COMPTON photons would also show a moderate anisotropy due, at least, to our off-center position in the Galaxy. We have not computed the spectrum from positron annihilation. The cross section for direct positron (energy: $\gamma_e m_e c^2$) annihilation with an electron at rest is, at high energies (*14*),

$$\sigma_a \approx \pi r_0^2 \frac{\ln 2\gamma_e}{\gamma_e} \quad (17.6)$$

so that the bremsstrahlung spectrum dominates the annihilation spectrum by a factor $\sim \alpha \gamma_e$ for $\gamma_e \gtrsim 10^2$. At lower positron energies ($\gamma_e < 10^2$) ionization losses are dominant (see Fig. 1) and the positron comes essentially to rest before annihilating, giving two photons each of energy $\eta \approx 1$.

To calculate the photon flux from the intergalactic medium we have taken essentially a static Euclidian universe cut off at $R = 5 \times 10^{27}$ cm. It is natural to inquire into the effects of the expansion (differential red shift) and detailed structure of the universe on the resulting photon spectra. It can be shown that only if the photon production spectrum is a *power law*, will the observed flux show the same shape spectrum (power law with the same index), independent of the structure (including expansion) of the universe. This results essentially because the Doppler-shifted photon energy is *proportional* to the unshifted energy. As a result, our calculated spectra, which are of the power law type in different energy regions, depend on the detailed structure of the universe only as far as the energy at which the spectra change their slope is concerned (at $\eta \sim 10^5$ for B and C, Fig. 6).

However, the shift in this critical energy is likely to be less than an order of magnitude.

We should like to emphasize again that the calculated photon fluxes are only approximate, and this should be kept in mind when we attempt possible interpretations of the observations. In particular, our treatment of meson production in cosmic ray collisions is very rough, especially at low energies where the FERMI theory should be invalid. Moreover, as mentioned earlier, our assumed cosmic ray spectrum is too large at the low energy end; this effect alone would produce a bend in the calculated photon fluxes at low energies such that the low energy ends of the curves in Fig. 6 should be reduced by about an order of magnitude.

g) Comparison with observations.

18. General discussion. The experimental points exhibited in Fig. 6 correspond to the observed cosmic background photon fluxes as summarized in Table 1 below[1]. The observations are in essentially four energy regions and are over ranges such

Table 1. *The observed high energy cosmic photon spectrum.*

Author	Designation in Fig. 6	Energy	$\bar{\eta}$	$j(\bar{\eta})$ (cm^{-2} sec^{-1})
GIACCONI et al. (22)	G	~2—3 kev	5×10^{-3}	4×10^3
BOWYER et al. (23)	B	~2—3 kev	5×10^{-3}	2×10^4
ARNOLD et al. (24)	A	~1 Mev	2	≤ 0.08
KRAUSHAAR and CLARK (25)	K-C	~100 Mev	200	$\leq 4 \times 10^{-5}$
DUTHIE et al. (26)	D	~100 Mev	200	$\leq 3 \times 10^{-4}$
FIROWSKI et al. (27), SUGA et al. (28)	EAS	~10^{15} ev	2×10^9	$\leq 10^{-20}$

that $\Delta \eta / \bar{\eta} \sim 1$. There is, of course, another range of energies where cosmic photons are observed, namely, the radio range. The radio spectrum is represented fairly well by the low energy range (not included in Fig. 6) of the calculated synchrotron radiation spectrum. We shall return to this question of the radio spectrum shortly.

We now consider the possibilities of interpreting any of the observed photon fluxes in terms of the various calculated spectra represented in Fig. 6. First consider the x-ray observations. The flux $j(\eta)$ for point B (Fig. 6) is five orders of magnitude above the curve S' and six orders of magnitude above S. This discrepancy is, in our opinion, sufficient to rule out the interpretation of the point B as due to synchrotron radiation, at least if the high energy electrons are of a secondary origin. The curves C and C' do not extend to lower energies because we have considered electrons with $\gamma_e \geq 10^2$, and in our approximate calculations have assumed that $\langle \varepsilon \rangle_C = \gamma_e^2 \bar{\varepsilon}_r$, giving $\langle \varepsilon \rangle_C \geq 30$ keV. However, due to the distribution of thermal photon energies there is, of course, a *distribution* of COMPTON photon energies which can be produced by an electron of given energy. Moreover, for a pion decaying at rest there is still an appreciable probability for a low energy (say, $\gamma_e \sim 30$) electron being produced. Therefore, the COMPTON spectra C and C' certainly do extend to the x-ray region. In spite of this, we do not believe that the x-ray point can be due to the COMPTON process, if the electrons responsible for the scattering have a secondary origin. For, as previously mentioned, the actual cosmic ray spectrum which produces the low energy pions and finally

---

[1] In this discussion we have taken the observational values given in Table 1 at their face value. However, it appears now that, while the background x-ray fluxes have been detected at the levels quoted, the γ-ray results are more uncertain and should all be treated as upper limits to the fluxes which may be present. That we are, therefore, only discussing possible explanations of hypothetical γ-ray fluxes in this section is to be emphasized.

electrons is smaller by about a factor of 10 than the power law spectrum used to compute the curves in Fig. 6. A realistic extrapolation of, for example, the curve C' to the x-ray region would still fall about three order of magnitude below the observational point B.

FELTEN and MORRISON (17) suggested that not only the x-ray flux, but also the photon fluxes at ~ 1 Mev and ~ 100 Mev (see Table 1), are due to the COMPTON process in the intergalactic medium. They suggested that the sources of the high energy intergalactic electrons are the strong radio sources. We can see from the curve C' in Fig. 6 that the intergalactic spectral density $n_e(\gamma_e)$ required to explain the results is about 20—30 times as large as the density which we estimated to result from secondary production in intereglactic space. The COMPTON spectrum must, of course, extrapolate to the x-ray region and this precludes a secondary origin for the electrons, unless they are produced by a cosmic ray spectrum which has a much higher intensity at low energies than that for cosmic rays observed at the earth. We cannot rule out the FELTEN-MORRISON hypothesis; in fact, elementary considerations of the necessary number of sources (radio galaxies) of high energy electrons in the universe suggest that the hypothesis is reasonable quantitatively. As we have shown, for our Galaxy this relatively low energy part of the electron spectrum, that is, the radio electrons, does escape from the Galaxy into the intergalactic medium before losing an appreciable amount of its initial energy. We shall show presently that, if the FELTEN-MORRISON idea is correct, the amount of synchrotron radiation which these electrons would produce places an upper limit to the intergalactic magnetic field.

Regarding the possible interpretation of the observations at 1 Mev (A, Fig. 6), we see that the observed flux is about an order of magnitude above the calculated curve C'. In view of the inaccuracies involved this "agreement" within an order of magnitude indicates that COMPTON scattering by secondary-produced intergalactic electrons provides a possible explanation for the observed photon flux at 1 Mev. Of the calculated processes represented in Fig. 6 this appears to be the only possible association with the observations at 1 Mev. The spectrum from π^0 decay certainly does not extend below $\log \eta = 2$ ($E \approx 50$ Mev), and the bremsstrahlung spectra B and B' must be less steep below $\log \eta = 2$ since, although the energetic secondary electrons can emit a bremsstrahlung spectrum extending to the lower energies, the corresponding bremsstrahlung photon would then carry away only a small fraction of the electron's energy, and the photon production process would be less efficient.

It would appear from Fig. 6 that the ~ 100 Mev photon flux which KRAUSHAAR and CLARK (25) first reported could be accounted for by π^0's produced in the Galaxy or in the intergalactic medium. However, our calculated π^0 spectrum, based on the Fermi theory, is very unsatisfactory at the low energy end. For low energy $p-p$ collisions it is primarily π^+ mesons that are produced and a more accurate treatment of meson production than our extrapolation of the Fermi theory must be employed. Now, in the Kraushaar-Clark observation the photon flux observed included essentially the whole spectrum from decays of π^0's of all energies, and most of the π^0's produced are of low energy. By employing the available data on meson production by incident protons of energy less than 10 Bev and the observed low energy cosmic ray spectrum, POLLACK and FAZIO (3) have computed the rate of production of pions by $p-p$, $p-\alpha$, and $\alpha-p$ collisions per hydrogen nucleus as the rate of production of π^0 decay and positron annihilation (after $\pi^+ \to \mu^+ \to e^+$ decay) photons:

π^0-decay: $q^0 \approx 1 \times 10^{-26}$ photons/sec-ster
positron annihilation: $q^+ \approx 2 \times 10^{-26}$ photons/sec-ster

The π^0-decay photons have energies above about 70 Mev and the galactic positron annihilation photons have energies of about 0.5 Mev, since the positrons come essentially to rest before annihilating. The π^0-decay photon flux from a region of density $\langle n_H \rangle$ of extend R would then be $4\pi q^0 \langle n_H \rangle R$ and in this manner we estimate fluxes of 2×10^{-4} photons/cm²-sec and 6×10^{-3} photons/cm²-sec from the Galaxy and intergalactic medium respectively; the galactic flux is a directional average. The KRAUSHAAR-CLARK flux is roughly the same as the calculated contribution from the intergalactic medium while the flux observed by DUTHIE et al. (26) is an order of magnitude larger. The origin of the discrepancy between the KRAUSHAAR-CLARK and DUTHIE et al. observations may lie in the latter's extrapolation of their balloon observations to zero atmospheric depth. At any rate, it is clear that an upper limit to essentially the product of the intergalactic cosmic ray flux and gas density is established by these observations. The calculated intensity of the positron annihilation line using POLLACK and FAZIO's value of q^+ and again the "standard" intergalactic gas density (10^{-5} cm^{-3}) is 1×10^{-2} photons/cm²-sec which is just below the upper limit of 1.5×10^{-2} photons/cm²-sec established by ARNOLD et al. However, *intergalactic* relativistic positrons do not slow down before annihilating (see Fig. 2) and would not produce a 0.51 MeV line but rather an annihilation *continuum* extending to higher energies.

The point denoted by EAS in Fig. 6 results from observations of *Extensive Air Showers* (27), (28) in which an abnormally low number of muons was observed, indicating possibly that the shower was initiated by electromagnetic processes rather than by a nuclear collision. If these showers result from primary photons the flux of these photons would be $\sim 10^{-3}$ times the flux of cosmic ray protons at the same energy. The results of these experiments are questionable and may only represent an *upper limit* to the primary cosmic photon flux at these high energies. In Fig. 6 we see that the EAS point lies 2 or 3 orders of magnitude above the curve corresponding to the decay of high energy secondary-produced π^0-mesons in the intergalactic medium.

As was mentioned in the footnote at the beginning of this section, it is necessary to emphasize the preliminary nature of all of these observations of high energy photons. While the existence of cosmic x-ray sources seems well established, the existence of positive fluxes at higher energies (the $\sim$MeV, 100 MeV, and 10^{15} eV observations) is *not* established. The fluxes given for these higher energy photons probably should all be taken as upper limits until the observational situation is clarified. For example, the $\sim$MeV observations may be plagued by radioactivity induced in the crystal of the scintillation detector (29).

19. X-rays from external galaxies. We should like to mention another possible explanation for the observed background flux of x-rays, which we proposed earlier (30). About 10 discrete sources of x-rays have been observed, and because of their apparent concentration toward the plane of the Galaxy, are assumed to be *galactic* and presumably at a galactic distance $R_g \sim 10$ kpc (see Sect. III). Since our Galaxy is believed to be a normal "average" galaxy, one would expect that this is a general characteristic of galaxies, so that external galaxies have x-ray luminosities L_x not too different from that of our own Galaxy. If the average x-ray luminosity for galaxies is $\langle L_x \rangle_g$, the isotropic background flux per steradian observed at the earth would be roughly

$$f_x \approx \langle L_x \rangle_g n_g R_c / 4\pi, \tag{19.1}$$

where $n_g (\sim 3 \times 10^{-75}$ cm^{-3}) is the number density of galaxies, and $R_c (\sim 5 \times 10^{27}$ cm) is a cosmological cut-off distance. Neglecting absorption, the total flux received from sources in our Galaxy is

$$F_x = \sum_i L_x^{(i)}/4\pi r_i^2 \approx L_x/4\pi R_g^2, \tag{19.2}$$

where $L_x \left(=\sum_i L_x^{(i)}\right)$ is the total x-ray luminosity of the Galaxy. Assuming $\langle L_x \rangle_g \approx L_x$, we find

$$f_x \approx F_x n_g R_g^2 R_c. \tag{19.3}$$

The total flux from all galactic sources is (see Sect. III) about $F_x \approx 33$ photons/cm²-sec, so that we estimate from Eq. (19.3) $f_x \approx 0.5$ photons/cm²-sec-ster, which is about an order of magnitude smaller than the observed flux. In view of the uncertainties and the assumptions involved, agreement within an order of magnitude must be regarded as satisfactory. Clearly, we make no assumption as to the production mechanism of these x-rays, but only that our Galaxy is "typical".

20. Radio emission. We conclude our discussion here with a few remarks about the observed cosmic *radio* spectrum from the galactic halo, which is undoubtedly due to synchrotron radiation by relativistic electrons. We attempt to answer the question as to whether the electron spectrum can be accounted for by secondary production by cosmic rays. This problem has been considered by a number of authors in a manner similar to our treatment. However, our view differs somewhat in that we consider leakage from the halo as the primary loss process for the radio electrons[1].

If the energy radiated per second per interval of frequency by an electron of energy $\gamma_e m_e c^2$ is $P(\nu, \gamma_e)$, the spectral intensity (erg/sec-cm²-ster-frequency interval) of radiation received from a direction $\boldsymbol{r}$ is

$$I_\nu = dE/dt\, dA\, d\Omega\, d\nu = (4\pi)^{-1} \iint n_e(\gamma_e) P(\nu, \gamma_e)\, d\gamma_e\, dr, \tag{20.1}$$

where $n_e(\gamma_e)\, d\gamma_e$ is the differential electron density. For an electron spectrum $n_e(\gamma_e) = K_e \gamma_e^{-\Gamma_e}$ the intensity I_ν may be computed approximately by taking $P(\nu, \gamma_e)$ to be equal to the expression (9.1) for dE_e/dt times a δ-function $\delta(\nu - \nu_L \gamma^2)$ at the frequency where $P(\nu, \gamma_e)$ is a maximum. Assuming a constant magnetic field H and a path length $\int dr = R_h$, the halo radius, we obtain a familiar result:

$$I_\nu \approx (12\pi)^{-1}\, c r_0^2\, K_e R_h H^2 \nu_L^{(\Gamma_e - 3)/2} \nu^{(\Gamma_e - 1)/2}; \tag{20.2}$$

a power law spectrum with exponent $\alpha \equiv (\Gamma_e - 1)/2$ is also obtained using the exact expression for $P(\nu, \gamma_e)$. The constant K_e may be determined by the observed value (500° K) of the radio brightness temperature $T_b = I_\nu \lambda^2/2k$ at 100 Mc/sec in the direction of the galactic pole. Employing Eq. (20.2) with $H = 3 \times 10^{-6}$ gauss, $R_h = 5 \times 10^{22}$ cm, $\Gamma_e = 2.8$ ($\alpha = 0.9$), we obtain $K_e = 1 \times 10^{-6}$ cm^{-3}. This number is to be compared with the value calculated from the production and loss processes. By Eqs. (4.3), (6.1), and (14.1) we get for the calculated K_e:

$$K_e \approx \left(\frac{8\pi}{9}\right) \Lambda_\pi^2 \left(\frac{m_\pi}{4 m_e}\right)^{\Gamma_e - 1} K_p \langle n_H \rangle \tau. \tag{20.3}$$

---

[1] Our conclusions also differ. We conclude that the spectrum of radio electrons in the galactic halo can be accounted for by the production (via $\pi - \mu - e$ decay) by cosmic-ray nuclear collisions in the galactic plane and subsequent diffusion to the halo. However, GINZBURG and SYROVATSKY (*31*) conclude, by a similar analysis, that the halo radio electrons *cannot* be explained in this manner, and that the expected secondary electron spectrum is smaller by one or two orders of magnitude than the value derived from radio observations. We feel that the astronomical data are not known sufficiently accurately to *expect* agreement within an order of magnitude. Moreover, we do not understand the results given in Figs. 5 and 6 of the paper by GINZBURG and SYROVATSKY; it appears to us that the electron spectra computed for the higher assumed mean galactic gas density n should be proportionally *larger*, offsetting the small effect of increased collisional energy loss.

Using the previously assumed values $K_p = 100$ cm^2-sec^{-1}, $\langle n_H \rangle = 0.03$ cm^{-3}, $\tau_L = 3 \times 10^{15}$ sec we calculate $K_e = 1 \times 10^{-6}$ cm^{-3}; the agreement with the radio value is fortuitous. Acutally, the observed radio spectrum has an index $\alpha \approx 0.7$ to 0.8, and we have adopted the "theoretical" value 0.9. This discrepancy may not be serious; the observed slightly flatter spectrum could be accounted for by a slight variation with γ_e of the effective value of τ_L. For example, if τ_L were slightly shorter for the low energy electrons (caused, perhaps by another, energy loss process at low energy) the smaller value of α and Γ_e could be understood. A more accurate treatment of the production spectrum could also indicate a smaller value for α and Γ_e. Further, we might mention that with our assumed values of the parameters (density, magnetic field, etc.) for the intergalactic medium, the calculated synchrotron intensity in the radio region from the intergalactic medium is comparable to that from the halo, while, as is seen from Fig. 6, the calculated intergalactic synchrotron radiation is actually greater by ~ 10 at the high energy end. Admittedly, our calculations are based on many assumptions, but these assumptions may well be valid, and much of the observed non-thermal radio background radiation may be coming from the intergalactic medium[1].

It is of interest to consider the requirements on the intergalactic magnetic field if the FELTEN-MORRISON idea is correct. From Fig. 6 we see that for the curve C' to pass near the points X, A, and K-C, the value of K_e must be larger by a factor ~ 30, or must be $\approx 3 \times 10^{-7}$ cm^{-3}. One can then compute the intergalactic magnetic field, by Eq. (20.2) with $R_h \to 5 \times 10^{27}$ cm, $\frac{1}{2}$ the HUBBLE radius, necessary to produce a brightness temperature of 500° K at 100 Mc/sec. One then finds 1×10^{-8} gauss for this magnetic field. Thus, if the FELTEN-MORRISON idea is correct, the intergalactic magnetic field must be less than 1×10^{-8} gauss.

Finally, we should like to mention some further checks on the calculated spectrum of the halo electrons. Recently the French-Italian group [AGRINIER et al. (8)] has reported the measurement of a primary cosmic ray electron flux of 6.6×10^{-4} particles/cm^2-sec-ster for $E_e > 4.5$ BeV, corresponding also to an electron/proton cosmic ray ratio of 1×10^{-2}. This measurement of the primary electron flux at fairly high energies is probably more reliable than results of measurements at lower energies which are influenced by solar activity. The measured flux is to be compared with that from the calculated spectra above 4.5 BeV ($\gamma_e > \gamma_0 = 4.5$ BeV/$m_e c^2$). One finds, with $K_e = 1 \times 10^{-6}$ cm^{-3}, $\Gamma_e = 2.8$, a flux

$$f_e = (4\pi)^{-1} \int_{\gamma_0}^{\infty} c K_e \gamma_e^{-\Gamma_e} d\gamma_e \approx 1 \times 10^{-4} \text{ particles/cm}^2\text{-sec-ster.} \qquad (20.4)$$

This flux is somewhat smaller than the observed one, but in view of the uncertainties involved in the calculations, agreement within an order of magnitude is all that one could hope for.

Another check on whether the observed cosmic ray electrons result from secondary production can be made by a measurement of the positron/electron ratio. This has been done by the group at the University of Chicago (8), who conclude that their measurements are inconsistent with the assumption that the bulk of the electron and positron spectrum is a result of secondary production. If indeed electrons and positrons of galactic origin are being observed, this would settle the question. Again, we take the conservative view that the question is still open, since the measured ratio is only off by a factor ~ 2 from the ratio expected on the basis of secondary production.

---

[1] Recently this view has also been expressed by some radio astronomers, for example, J. E. BALDWIN at the Second Texas Symposium on Relativistic Astrophysics (proceedings to published by the University of Chicago Press).

h) Tests of cosmological theories.

21. The hot universe model — bremsstrahlung from the intergalactic medium.
GOLD and HOYLE (*32*) have suggested a cosmological model in which the intergalactic medium is at a very high temperature ($\sim 10^9$ °K). The high temperature is supposed to arise from the ~ 1 Mev electrons which would result after the decay of spontaneously created neutrons as envisioned by the steady-state theory. Galaxy formation within the framework of this model was considered by BURBIDGE, BURBIDGE, and HOYLE (*33*). An observational test of this model can be made, since such a hot intergalactic medium would emit thermal bremsstrahlung photons in the x-ray region where observations have been made (*23*). For a mean thermal electron energy $\langle E_e \rangle = 50$ keV, and a density $n_e = n_p = 1.2 \times 10^{-5}$ cm^{-3} the rate of production of bremsstrahlung photons within the energy range of the observations is about $r_b = 1.17 \times 10^{-25}$ photons/cm^3-sec (*30*). Taking a cut-off radius $R = 5 \times 10^{27}$ cm for the universe, one calculates a flux $f_b = r_b R/4\pi \sim 50$ photons/cm^2-sec-ster to be expected at the earth. This flux is ~ 10 times the observed x-ray background flux and is evidence against the hot universe model (and the steady-state theory with spontaneous creation of *neutrons*). Actually, if the appropriate intergalactic density to be used is four times the usually adopted 2×10^{-29} g/cm^3, as suggested by SCIAMA (*34*), the disagreement with observations is even more violent. In any case it appears that the x-ray observations have established an upper limit of 10^7 °K for the temperature of the intergalactic medium.

22. Matter and anti-matter and the steady state cosmological theory. The attractive feature of the steady state theory is its simplicity. The unique feature is a spontaneous creation rate of "new" matter $dn/dt \sim 3 Hn$, where $n \sim 10^{-5}$ cm^{-3} is the mean matter density in the universe (taken to be the mean hydrogen density in the intergalactic medium) and H is the HUBBLE constant ($3H \sim 10^{-17}$ sec^{-1}). One might expect that in the spontaneous creation process, to conserve baryon and lepton number, particles and anti-particles are created. Since the expansion rate constant $3H$ is about two orders of magnitude greater than the annihilation rate (see below), BURBIDGE and HOYLE (*35*) suggested the possibility of an appreciable abundance of anti-matter in the universe. This idea can be put to a test, since the end products of matter and anti-matter annihilation are observable high energy γ-rays.

Let us suppose that (p, e^-) and $(\bar{p}, e^+)$ are spontaneously produced and have a steady state mean number density $n = 10^{-5}$ cm^{-3} and $\bar{\alpha}n$ respectively, where $\bar{\alpha}$ denotes the mean ratio of anti-matter to matter (or vice-versa). The electron-positron annihilation cross section at non-relativistic energies is (*14*) $\sigma_a = \pi r_0^2/\beta$, where r_0 is the classical electron radius and $\beta = v/c$. The annihilation rate is then $dn_a/dt = \bar{\alpha} n^2 \pi r_0^2 c \sim \bar{\alpha} \times 10^{-24}$ cm^{-3} sec^{-1}, and the expected flux of 0.51 MeV photons from the intergalactic medium out to a distance $R \sim 5 \times 10^{27}$ cm is $2R \, dn_a/dt \sim \bar{\alpha} \times 10^4$ photons/cm^2-sec. This can be reconciled with the upper limit of 10^{-2} photons/cm^2-sec suggested by ARNOLD et al. (*24*) only if $\bar{\alpha} < 10^{-6}$. This means that if there is appreciable anti-matter in the universe, it must be *separated* from matter, so that it cannot annihilate and produce observable γ-radiation.

A limit on the amount of anti-matter in the universe can also be provided from an analysis of the γ-ray experiments at higher energies which can detect π^0-decay γ's. In the proton-antiproton annihilation ~ 5 pions are produced, some of which are π^0's which produce γ-rays of energy ~ 100 MeV in their decay. In each $p-\bar{p}$ annihilation about $m_\gamma = 4$ γ-rays are produced. The cross section for $p-\bar{p}$ annihilation is $\sigma'_a = \sigma_0/\beta$, where $\sigma_0 = 5 \times 10^{-26}$ and βc is the relative velocity

(*36*). Again setting $n\,(=10^{-5}\,\mathrm{cm^{-3}})$ and αn equal to the matter and anti-matter densities respectively, the mean number of γ's produced in the universe per cm³ is $m_\gamma\bar{\alpha}n^2\sigma_0 c \sim \bar{\alpha} \times 6 \times 10^{-25}$ cm⁻³ sec⁻¹. If this production occurred in the universe out to a distance $R = 5 \times 10^{27}$ cm, the resulting flux of γ-rays would be consistent with the Kraushaar-Clark experiment only if $\bar{\alpha} < 10^{-6}$ — the same limit established from the observed upper limit for the intensity of the cosmic positron annihilation line. Thus, it appears that in the steady state cosmology matter and anti-matter cannot be created in comparable amounts in the same region.

Finally, regarding cosmological tests, we should like to mention the recent discussion by Gould and Sciama (*37*). They indicate how the measurement of the shape of an emission line, smeared into a continuum by the cosmic differential red shift, would provide information about the structure of the universe at great distances.

III. Discrete sources of high energy photons.

It has now been demonstrated quite conclusively by the NRL (*23*), (*38*), (*39*), MIT-ASE (*22*), (*40*), (*41*), and Lockheed (*42*)—(*44*) groups that there exist discrete sources of cosmic x-rays. About 10 such sources have been found and their properties are described below. The problem of the types of astronomical object and the mechanisms of emission which give rise to these sources is as yet unsolved, although it appears that there are only a few possible explanations. While the basic mechanism by which the x-rays are produced is not known, the present indication is that the x-ray sources are *galactic* and, in fact, are supernova remnants. This viewpoint is advanced here where supernova remnants, as source of x- and γ-rays, are discussed in some detail; the Crab Nebula in particular receives considerable attention. We also discuss the x-ray source at the galactic center. However, before considering these specific objects, we give a general review of possible galactic sources and then discuss the possible physical mechanisms for high energy photon production in discrete sources.

a) General summary of the observations.

23. The positions and intensities of the discrete x-ray sources as they are known at present are given in Table 2 which is taken from the paper by Bowyer et al. (*38*). Results reported by the other experimental groups (MIT-ASE and Lockheed) are in essential agreement with these in regards to both the positions[1] and intensities of the sources. The uncertainty in the positions of the sources is given as 1.5°, while the Tau XR-1 source has been localized to within 1' of the optical center of the Crab Nebula. The observational results on the Crab source will be summarized in more detail later (Sect. 29). The discrete sources appear to have a spatial distribution showing a concentration toward the plane of the Galaxy, indicating that the sources are probably *galactic* and at characteristic distances of $\sim 1-10$ kpc. The most intense of the x-ray sources is that in the constellation Scorpius (Sco XR-1) from which an x-ray flux of about $F_x \sim 10^{-7}$ erg/cm²-sec is detected. The flux from most of the other sources is about one-tenth as large as that from Sco XR-1. With the exception of Tau XR-1, none of the sources have been identified with a reasonable degree of certainty with any radio or optical objects, although there have been some tentative identifications.

Attempts have been made (*45*) to identify the Scorpius x-ray source with the so-called spur of radio emission which some have argued is a comparatively

---

[1] Recent unpublished work by the Lockheed group has given a more accurate position for the Sco XR-1 source: $\alpha = 16^h\,14^m$, $\delta = -15°\,36'$, with an expected error of $\pm 20'$.

nearby supernova remnant. However, it has been pointed out (46) that this positional identification is very poor since even with the uncertainties quoted for the position the center of the radio source component is some 27° away. Thus the situation here is uncertain.

Of the other x-ray sources, it is of interest to note that Oph XR-1 is 1.1° away from the position of KEPLER's 1604 supernova[1] and that Sgr XR-1 is 2.3° from the galactic center.

The size of Sco XR-1 has been established to be less than about 0.2° by both the NRL and MIT-ASE groups. With the exception of Tau XR-1, all that is known about the sizes of the other sources is that they are less than about 10° in extent. Little is known at present about the spectra of the x-ray sources. From the

Table 2. X-ray sources. [After BOWYER et al. (38).]

Source	R.A.	(1950) Dec.	Flux		
			(counts/cm²-sec)	(10^{-8} ergs/cm²-sec)	
			(a)	(b)	(c)
Tau XR-1	$05^h\ 31.5^m$	22.0°	2.7	5.5	1.1
Sco XR-1	$16^h\ 15^m$	−15.2°	18.7	38	7.9
Sco XR-2	$17^h\ 8^m$	−36.4°	1.4	2.9	0.6
Sco XR-3	$17^h\ 23^m$	−44.3°	1.1	2.3	0.5
Oph XR-1	$17^h\ 32^m$	−20.7°	1.3	2.7	0.6
Sgr XR-1	$17^h\ 55^m$	−29.2°	1.6	3.3	0.7
Sgr XR-2	$18^h\ 10^m$	−17.1°	1.5	3.0	0.6
Ser XR-1	$18^h\ 45^m$	5.3°	0.7	1.5	0.3
Cyg XR-1	$19^h\ 53^m$	34.6°	3.6	7.3	1.5
Cyg XR-2	$21^h\ 43^m$	38.8°	0.8	1.7	0.4

(a) Uncorrected for atmospheric absorption. Measured with 1/4-mil Mylar window.
(b) Computed for 2×10^7 deg K black body, 1.5−8 Å.
(c) Computed for 5×10^6 deg K black body, 1.8−8 Å.

change in the counting rate from Sco XR-1 as the rocket was passing through the upper atmosphere (in which absorption is wavelength dependent) the NRL group (38) has concluded that $\frac{1}{3}$ of the observed flux the strong Scorpius source is emitted in the 1−6 Å band and $\frac{2}{3}$ of the flux is emitted in the 6−10 Å band. Such a spectrum is compatible with emission from a black body having a temperature of 2 or 3×10^6 degrees. However, these results on the spectra of x-ray sources are suspect; the LOCKHEED group (44) has found an effective black body temperature of $\sim 2 \times 10^{7\circ}$ K for the Scorpius source — an order of magnitude higher than that reported by the NRL group.

b) Possible galactic sources.

If an x-ray source at 10 kpc is to produce an x-ray flux of 10^{-8} erg/cm²-sec, its x-ray luminosity must be 10^{38} erg/sec. Clearly, no individual *normal* star could produce such a luminosity in x-rays. For, although there are stars which have a *total* luminosity this large, the atmospheric conditions in these stars are such that most of the radiation emitted is at much lower energies ($\sim$, say, 10 eV). The sun emits x-rays from its corona and from flares, but at a much smaller rate (10^{21} to 10^{26} erg/sec). Only the cumulative effect of very populous clusters or the integrated effect of the stars in the galactic bulge could possibly produce a significant x-ray

---
[1] Recent unpublished work seems to indicate that KEPLER's supernova is *not* an x-ray source, however.

flux. This possibility will be discussed later (Sect. 33); suffice it to say for the moment that these combined effects of stellar coronae appear likely to be unimportant. However, there is an *abnormal* type of star which, at least for part of its evolutionary phase, emits a spectrum peaked on the x-ray region; this is the *neutron star*. We prefer to discuss neutron stars after first considering supernovae, which are known sources of large amounts of energy and high luminosity.

24. Supernovae. Although much of the energy released in a supernova is emitted soon after the outburst, the remnants still possess a large amount of energy and could possibly maintain an x-ray luminosity of 10^{38} erg/sec for much longer times. The required characteristic loss time τ_l for x-ray emission can be determined approximately as follows. Let us suppose that the $N_x \sim 10$ x-ray sources are galactic and resulted from supernova outbursts. Since the rate of supernova outbursts and formation of x-ray sources in the Galaxy is $dN_s/dt \sim 1/100$ yr, we must have

$$\left(\frac{dN_s}{dt}\right) \tau_l = N_x, \qquad (24.1)$$

$$\tau_l \sim 1000 \text{ yr}.$$

Clearly, this result must hold for whatever type of mechanism is to produce the x-rays, as long as the origin of the x-ray sources is to be supernovae outbursts. Moreover, if a characteristic time for x-ray emission by some process is computed to be much shorter than 1000 yr, then without regeneration that process cannot account for the x-ray sources. It is of interest to compare τ_l with the expected lifetime of *total* emission from the Crab Nebula (see Sect. 29), for which $E_{\text{tot}} \sim 10^{48}$ erg and $L_{\text{tot}} \sim 10^{37}$ erg/sec. Then $E_{\text{tot}}/L_{\text{tot}} \sim 3000$ yr $\sim 3 \tau_l$.

Supernova remnants can emit high energy photons through a variety of processes and at very different power levels. After the initial outbursts, emission can occur by the synchrotron process, by bremsstrahlung in the high temperature gas produced by the expanding ejecta, and by the radioactivity produced. The initial outburst is more spectacular, however, and we shall now consider what can be expected during this very early, violent stage.

25. Early phases of supernova outbursts. At this phase two processes may be important. These are:

(a) Nuclear γ-rays emitted in the process of nucleosynthesis at the time of the outburst.

(b) γ-rays emitted through the early interaction of a cloud of relativistic particles with the magnetic field and material in the expanding shell.

If, in a supernova outburst, the inner part implodes and the outer part is suddenly heated so that hydrogen burning takes place very rapidly, we can suppose that the bulk of the energy released is degraded through its passage through the material, but some fraction, perhaps the energy released in burning $0.01\, M_\odot$ of hydrogen, will be emitted as γ-rays in the Mev range. Thus we might suppose that 10^{50} ergs is emitted in ~ 1000 seconds. For a galactic supernova at assumed distances of 1 and 10 kpc this gives fluxes at the earth of 10^3 and 10^5 erg/cm² sec⁻¹, fantastic rates. However, the appearance of a galactic supernova is highly improbable. From extragalactic supernovae at characteristic distances of 10 and 100 Mpc the fluxes would be 10^{-5} and 10^{-7} erg/cm² sec⁻¹ respectively. These rates are obviously uncertain by several powers of 10. It might also be expected that some part of the flux is degraded to the energies of a few kilovolts and is emitted as x-rays. As an upper limit we might suppose that this flux is of comparable intensity for a few days with the flux at maximum light from the supernova.

If we suppose that it reaches a value of $M_v = -18$ this corresponds to 10^{43} erg/sec and at distances of 1 kpc and 10 kpc (Galactic) and 10 Mpc and 100 Mpc (extragalactic) fluxes at earth of 10^{-1} and 10^{-3} erg/cm² sec⁻¹ (Galactic) and 10^{-9} and 10^{-11} erg/cm² sec⁻¹ (extragalactic) may be expected.

A large flux of relativistic electrons is currently present in many supernova remnants, and it is possible that this in part is the remnant of a much larger flux of relativistic particles which was produced at the time of the outburst. Let us suppose that some 10^{50} ergs of particles, largely protons, was generated in the explosion. If they are originally confined in an expanding shell containing a magnetic field (they are the relativistic plasma component), then because of the density in the shell in the first hours they will largely be destroyed, and their energy will be dissipated in the form of neutrinos, γ-ray, and electrons and positrons which radiate in the magnetic field. A large flux of high energy (≥ 100 Mev) γ-rays will thus be generated and we might expect fluxes to escape over this period at a rate of perhaps $10^{44}-10^{45}$ erg/sec. For reasonable magnetic field values the synchrotron radiation will not lie in the x-ray or γ-range. However, it is possible that some part of the electron-positron flux will be dissipated by COMPTON collisions with thermal photons in which γ-rays are emitted. It is obvious that these suggestions are highly speculative. However, it is clear that detection of a supernova explosion by x-ray and γ-ray telescopes would give much information on the conditions at the early phases. For example, if there are no high energy γ-rays emitted this might be interpreted as meaning that there was no early generation of a large flux of relativistic protons.

26. Hard radiation emitted through radioactivity. It has been suggested that in a supernova outburst considerable nucleosynthesis takes place (*47*). In this a large flux of neutrons is added very rapidly to seed nuclei (*r*-process) building up to nuclei with $A \simeq 270$ and giving rise to large numbers of neutron-rich nuclei which subsequently decay. It is still not clear what fraction of the supernovae goes through this process but in connection with the possibility of checking this theory CLAYTON and CRADDOCK (*48*) have made calculations of the fluxes of γ-rays to be expected following such a process. The γ-ray line spectrum is calculated from the production curve for the *r*-process isotopes (*49*). Using these abundances the best estimates are made of the prompt γ-ray spectrum using the nuclear energy levels. Also, an estimate has been made of the γ-ray flux which is emitted in spontaneous fission in such isotopes as Cf^{252}. The fluxes to be expected for a supernova remnant at the distance of the Crab (~ 1000 pc) are shown in Fig. 7 taken from the calculations of CLAYTON and CRADDOCK. The strongest line (390 kev line from Cf^{249}) radiates at a rate 10^{39} photon/sec at the source. The calculations have been normalized for the assumption that in a supernova remnant $1.5 \times 10^{-4} M_\odot$ (3×10^{29} gm) of Cf^{254} are produced. This is adequate to explain the light curve of a Type 1 supernova on the assumption (*47*) that this is due to Cf^{254}. Detection of such a flux is being attempted at the time of writing. This will give a direct observational test of this hypothesis of *r*-process isotope synthesis in Type I supernovae.

27. Neutron stars. It has been pointed out by CHIU (*50*) and FINZI (*51*) that, since it is possible that neutron configurations may be reached as an end phase of stellar evolution by processes which leave the star extremely hot, such configurations may, for rather a short period, be thermal x-ray emitters. However, from the theoretical standpoint it must be conceded that at the present time we cannot demonstrate conclusively that stable neutron configurations are ever formed or can exist if formed. The presumption of these authors is that the neutron configurations are formed during a supernova outburst, as was first proposed by BAADE and ZWICKY (52) many years ago. There are many theoretical uncertainties associated with neutron configurations which we mention briefly.

It is well known that there is a critical mass for a degenerate neutron configuration above which no stable equilibrium is possible. This result was first derived by LANDAU (53) and calculations by OPPENHEIMER and VOLKOFF (54) gave a value of about 0.7 $M_\odot$ for this observable mass limit. While in later calculations this mass limit has been slightly revised, it is clear that the mass limit lies near 1 $M_\odot$. Even the doubtful assumption of a hard-core nuclear potential, which is known to be incorrect from relativistic considerations, only extends the maximum

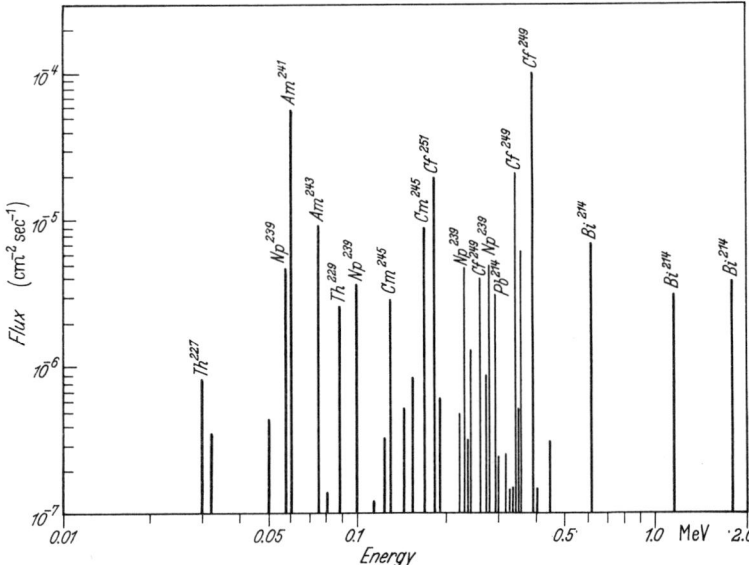

Fig. 7. The spectrum of the line fluxes anticipated from the Crab Nebula as calculated by CLAYTON and CRADDOCK (48).

mass to about 3 $M_\odot$. In fact it is clear from the earliest considerations (53) that the maximum mass is very insensitive to the equation of state at nuclear densities and above. For masses above the critical mass it appears that implosion must occur (55). For a modern review see HOYLE, FOWLER, BURBIDGE, and BURBIDGE (56). Thus, if neutron configurations which can exist long enough to be detected as sources of x-rays coming from supernova outbursts, it is required that in the supernova outburst sufficient mass is ejected so that the resulting configuration falls below the limit for support by a degenerate neutron configuration. None of the attempts to unravel the processes of supernova outbursts have yet given any real indication that such conditions can be achieved. The attempts by the California-Cambridge group (47), (57) have not been able to answer this question. Even the range of masses of stars which become supernovae is in doubt, but it appears highly probable that the Type II supernovae are stars of quite large mass $\sim 30\ M_\odot$. All of the discussion of the supernova outburst as it applies to the last phases of nucleosynthesis, and neutrino emission, etc. have been carried out by neglecting the effects of rotation. However, as has been shown by HOYLE et al. (56) this *may* have the effect of allowing a massive star to fragment, either into white dwarf, or into neutron configurations [cf. Eq. (45) of that paper (56)]. In the work of CHIU (50) no conclusion as to whether a degenerate neutron configuration with a mass below the critical mass is left has been reached.

The only attempt at a hydrodynamical calculation of the implosion of a supernova before relativistic effects become important is that by COLGATE and his colleagues (58). This calculation follows the collapse until nuclear densities are reached, but then it is supposed that a bounce occurs and the outer envelope is ejected. The calculation is not able to determine what fraction of the mass is left as a degenerate neutron configuration.

The only supernova remnant which can be studied in any detail is the Crab Nebula. While there are uncertainties in the mass of the nebula, analysis shows that it is only (59) $\sim 0.64\ M_\odot$ so that if the outburst originated from a star with a mass in excess of about $3\ M_\odot$ (and the type of supernova involved is still uncertain, as is the relation of type with mass) it must be concluded either that a large remnant has imploded or else that it is fragmented into a number of neutron stars.

Finally, there is some question about the stability of neutron configurations. The question of their dynamical stability has recently been considered for a range of models by MISNER and ZAPOLSKY (60) who have concluded that dynamically stable solutions exist for stars below the maximum mass for cold static equilibrium.

There is thus considerable uncertainty as to whether neutron stars are ever formed. If they are then detection of their x-rays emitted while the surfaces are still hot might provide the only direct observational test of their existence. Whether they are likely to be detected depends on the time that they may be expected to spend with their atmospheres hot enough to emit x-rays. The first calculations of the cooling rates (61), (62) suggested that such stars might emit for periods $\sim 10^3$ years. Thus if they were embedded in supernovae remnants such as the Crab which exploded in 1054 A.D. we might expect to detect them. The cooling is dominated by the neutrino production rate in the interior since the neutrinos escape from the stars with a negligible probability of being scattered or absorbed. A recent investigation by BAHCALL and WOLF (63) [see also FINZI (64)] takes into account the cooling reactions

$$n+n \to n'+p+e^-+\bar{\nu}_e$$

and

$$n+\pi^- \to n'+\mu^-+\bar{\nu}_\mu$$

and their inverses. If the first reaction alone is operating, the cooling rate is such that atmospheric temperatures only remain $\sim 2-3 \times 10^6$ degrees for about 10 years. There is still some doubt as to whether the second process operates, but if it does the cooling times are very much shorter than this. In any case, because of the argument previously given [see Eq. (24.1) and discussion], the short cooling time for neutron stars would rule them out as likely sources of x-rays.

From the observational side also there are very strong arguments against the neutron star hypothesis. The occulation observation of the NRL group (39) which shows that the source in the Crab has a diameter ~ 1 light year rules out its being a single neutron star and the existence of a cluster of such stars is improbable. Also, the observation of fluxes of 10—50 kev x-rays is not explicable in terms of a thermal source, since temperatures $\sim 10^8$ degrees are required, and these are far above that which the surface of a neutron star could attain for any significant time.

c) Mechanisms for x-ray production in discrete sources.

28. Apart from the mechanisms discussed earlier in this section, there are three possible mechanisms for the x-ray production: 1) COMPTON scattering, 2) bremsstrahlung, and 3) synchrotron emission. When x-ray sources were first discovered, the possibility that they were neutron stars was discussed at length, but as was shown in the previous sub-section this explanation now appears to be untenable. It has been suggested (65) that the x-rays from the Crab are due to COMPTON scattering of the radio-optical synchrotron photons by the associated synchrotron electrons; in this manner the synchrotron photon energy is amplified by a factor γ_e^2, where $\gamma_e m_e c^2$ is the energy of the (synchrotron) electron involved in the scattering. However, as has been emphasized recently (66), the intensity of this COMPTON-synchrotron radiation can be shown to be far too small to explain the observations. The effect is small essentially because the probability that a synchrotron photon undergoes such a COMPTON scattering before escaping from the nebula is very small. The Crab is one of the most intense galactic radio emitters and if the effect is small for it, one should not expect to observe the effect in other galactic objects. One might think that COMPTON scattering might produce a large x-ray flux from quasi-stellar radio sources in which the photon density and high energy electron density are large. Again, however, simple calculations indicate a completely negligible and unobservable x-ray flux from this process. Consequently, we are led to rule out COMPTON scattering as an x-ray production mechanism in discrete sources[1]. This leaves only the synchrotron and bremsstrahlung processes as possible x-ray sources.

First we consider the possibility that the x-rays from discrete sources are synchrotron radiation. We shall assume that an x-ray flux $F_x=10^{-8}$ erg/cm²-sec comes from a galactic source at a distance $r=10$ kpc; the x-ray luminosity of the source is then $L_x = 4\pi r^2 F_x = 1 \times 10^{38}$ erg/sec. Further, we assume for simplicity that the x-ray flux is at an effective wavelength 3 Å and frequency $\nu=10^{18}$ c/s, which is the characteristic synchrotron frequency $\nu_L \gamma_e^2$ emitted by electrons of energy $E_e = \gamma_e m_e c^2$. For a magnetic field $H=10^{-4}$ gauss (the assumed value in the Crab Nebula) the electron energy required is $E_e (\propto H^{-\frac{1}{2}}) = 3 \times 10^{13}$ eV. For such a high energy electron the lifetime against energy loss by synchrotron emission is only $E_e(dE_e/dt)^{-1} = \tau_e (\propto H^{-\frac{3}{2}}) = 30$ yr. The total energy in these electrons necessary to produce the flux F_x is $E_t (\propto F_x r^2 H^{-\frac{3}{2}}) = 1 \times 10^{47}$ erg. We note that: 1) The electron energies required to produce synchrotron x-ray are extremely high, 2) their lifetime is very short, and 3) the total energy involved is comparable to the energy released in a supernova outburst. Actually, the energy E_t quoted above is really the minimum energy of the highly relativistic electrons, since it includes only the synchrotron electrons producing x-rays. The contributions of the lower energy extension of the electron spectrum to the energy would increase the value of the total energy by an amount depending on the index of the spectrum and the low energy cutoff. For the case of the Crab Nebula (see Sect. 29) the extension of the x-ray spectrum (which has an index of about 1.1) to the visible leads to a total electron energy which is not excessively large ($\sim 10^{48}$ erg). However, it is very significant that the lifetime of the high energy electrons required to produce synchrotron x-rays is appreciably less than the age of the Crab Nebula and other supernova remnants, because it would mean that the electrons would have to be continuously or at least periodically produced. If they are spasmodically produced or accelerated, one might expect to observe variations in the x-ray intensity over time scales ≤ 10 yr.

---

[1] See, however, V. L. GINZBURG, L. M. OZERNOI, and S. I. SYROVATSKY (67): Dokl. Akad. Nauk SSSR **154**, 557 (1964), transl. Soviet Phys. — Doklady **9**, 3 (1964). They consider circumstances where one might be able to detect Compton-synchrotron photons of energy $\sim 10^7 - 10^8$ eV at a rate $\sim 10^{-5}$ photons/cm²-sec from the quasistellar object 3C 273-B which has a negative radio spectral index α. Because of the negative index, most of the Compton-synchrotron photon flux comes from the *high* energy end of the spectrum. The expected number of x-ray photons is smaller and unobservable.

Because of the difficulties associated with the hypothesis that the x-rays from discrete sources are produced by the synchrotron process, it is worthwhile considering an alternative model in which it is supposed that an outburst gives rise to a small very hot cloud which continues to emit x-rays as part of the thermal bremsstrahlung. We now discuss the properties associated with such a model. Earlier, we had suggested that the source at the galactic center resulted from bremsstrahlung. At the time we envisaged bremsstrahlung production by *non-thermal* electrons of energy greater than the energy of the x-ray photons. However, as was first pointed out by Rossi (68), about 10^5 times as much energy would be lost by these electrons in inelastic atomic collisions, so that if the x-ray luminosity of the source at the galactic center is 10^{38} erg/sec, about 10^{43} erg/sec must be supplied. This energy rate is excessively large on a galactic time scale (10^{10} yr), although perhaps it may be supplied during shorter times.

In spite of this difficulty the conditions whereby x-rays are produced by non-thermal particles may still exist. If so, there will also be production of *characteristic* x-rays, as was pointed out by us (30) and by Hayakawa and Matsuoka (69). These x-ray lines are produced in the radiative cascade following the ejection of K-shell electrons by the incident electrons or protons. Actually, most of the K-shell vacancies produced result in the emission of an Auger electron. The probability of x-ray emission by an element is given by the so-called *K-fluorescence yield* which is small for the light elements. The x-ray line emission, say of the K_α line, is approximately proportional to the product of the element abundance and K-fluorescence yield. One finds that the total intensity of the x-ray lines in the 2—8 Å region should be $\sim 10\%$ of the intensity of the bremsstrahlung continuum in the same wavelength range. This result holds essentially *independent* of the spectrum of the incident suprathermal particles and holds whether the incident particles are protons or electrons. As Hayakawa and Matsuoka have shown, the incident protons produce knock-on electrons and a radiation continuum by *inner bremsstrahlung* during the knock-on process. The observation of x-ray lines produced in this manner would be of great importance because, among other things (Cf. [37]), the abundances of the elements producing x-ray lines could be determined in this manner. In Table 3 we list the K_α wavelengths of the elements from carbon to iron along with their abundance and K-fluorescence yield. It appears that the most intense lines would be from Si (7.1 Å) and S (5.4 Å).

Energetically, bremsstrahlung x-ray production is more efficient in a high temperature ($T \sim 10^7$ °K) and low density gas where the bremsstrahlung is produced by *thermal* electrons and constitutes a major source of cooling and energy loss for the gas.

For the production of thermal bremsstrahlung the Gaunt approximation (70) to the bremsstrahlung cross section provides an adequate simplification. The differential cross section for the production of a bremsstrahlung photon of energy

Table 3. *Characteristic x-ray data.* [After Gould and Burbidge (30).]

Element	Logarithmic abundance	K-Fluorescence yield	K_α Wavelength (Å)
C	8.60	0.00126	45
N	8.05	.00223	31
O	8.95	.00397	24
F	6.0	.00634	18
Ne	8.70	.00963	15
Na	6.30	.0140	12
Mg	7.40	.0197	9.9
Al	6.22	.0269	8.3
Si	7.50	.0360	7.1
P	5.40	.0468	6.1
S	7.35	.0597	5.4
Cl	6.25	.0748	4.7
Ar	6.88	.0923	4.2
K	4.82	.112	3.7
Ca	6.19	.134	3.4
Sc	2.85	.158	3.0
Ti	4.89	.184	2.7
V	3.82	.212	2.5
Cr	5.38	.241	2.3
Mn	5.12	.272	2.1
Fe	6.57	0.304	1.9

within $\hbar\,d\omega$ by an electron of velocity βc incident on a nucleus of charge Ze is

$$d\sigma_B(\beta,\omega;Z) = \frac{16\pi}{3\sqrt{3}} Z^2 \alpha\, r_0^2 \frac{1}{\beta^2} \frac{d\omega}{\omega}, \tag{28.1}$$

where α is the fine structure constant and r_0 the classical electron radius. The bremsstrahlung energy spectrum emitted per unit volume by encounters with ions of charge Ze is then

$$\frac{dE_B(Z)}{dt\,dV\,d\omega} = n_e n_Z \int \frac{d\sigma_B(\beta,\omega;Z)}{d\omega} \hbar\omega\, v f(v)\, dv, \tag{28.2}$$

where $f(v)$ is the Maxwellian velocity distribution of the electrons; the integration in Eq. (28.2) is over v from $(2\hbar\omega/m)^{\frac{1}{2}}$ to ∞. One obtains

$$\frac{dE_B(Z)}{dt\,dV\,d\omega} = n_e n_Z 2^4\, 3^{-\frac{3}{2}} \alpha r_0^2\, \hbar c^2 Z^2 (2\pi m/kT_e)^{\frac{1}{2}} e^{-\hbar\omega/kT_e}, \tag{28.3}$$

and for the total emission between ω_1 and ω_2

$$\frac{dE_B(Z)}{dt\,dV} = n_e n_Z 2^4 3^{-\frac{3}{2}} \alpha r_0^2\, m c^2 Z^2 (2\pi kT_e/m)^{\frac{1}{2}} \cdot (e^{-\hbar\omega_1/kT_e} - e^{-\hbar\omega_2/kT_e}). \tag{28.4}$$

$(\omega_1 < \omega < \omega_2)$

The total energy emitted per unit volume over all frequencies $(\omega_1 \to 0, \omega_2 \to \infty)$ is

$$P_B = \sum_Z \frac{dE_B(Z)}{dt\,dV} = 1.43 \times 10^{-27} T_e^{\frac{1}{2}} n_e \sum_Z n_Z Z^2 \text{ c.g.s. units.} \tag{28.5}$$

It is noteworthy that the bremsstrahlung spectrum [Eq. (28.3)] is significantly different from the spectrum of a black body, so that from measurements of the

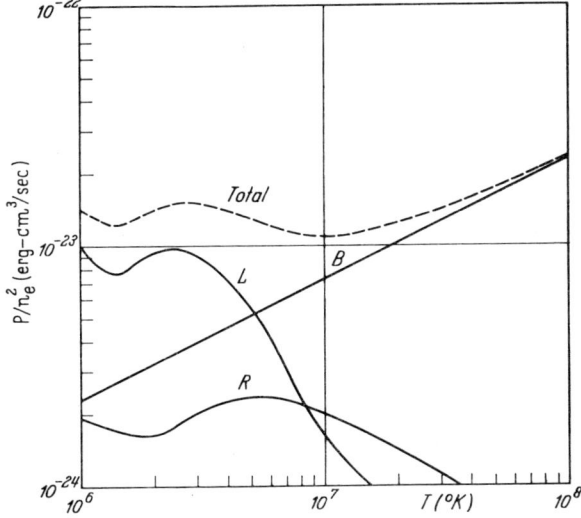

Fig. 8. Cooling rate as a function of temperature. P denotes the rate of energy emission per cm³. Cooling by bremsstrahlung (B), line emission following inelastic electron collisions (L), and recombination (R) is shown. The ions of the following elements have been included: (B) H+He, (L) He+C+N+O+Ne+Mg+Si+S, and (R) H+He+O+Ne.

x-ray spectrum it may be possible to establish that some sources are hot optically thin gases.

In addition to bremsstrahlung, there is also cooling and x-ray emission by electron-ion radiative recombination and by inelastic electron collisions with ions followed by radiative de-excitation (line emission). The line emission is due mainly to oxygen and neon in high stages of ionization and the calculation of the cooling and x-ray production involves a calculation of the ionization equilibrium. Equilibrium is established between ionization by electron collision and radiative and dielectronic recombination. Here we give only the results; the details will be published elsewhere. Preliminary results have already been published (66). Fig. 8 gives[1] the rate of loss P(erg/cm³-sec) of the free P (erg/cm³-sec) of the free electron kinetic energy density by various processes in the temperature range between 10^6 and 10^8 °K. It is seen that bremsstrahlung dominates the cooling at higher temperatures. In Fig. 9 we give the rate of production of x-rays ($p_x = P_x/n_e^2$) in the 1—10 keV range as a function of temperature. In the calculation of these processes a general cosmic abundance of the elements has been assumed. The cooling time ($\tau_c \approx 3\,kT_e/n_e\Lambda_e$), density ($n_e$), and mass ($M$) of a volume V of gas required to produce the observed x-ray fluxes are of prime interest. We assume the source to be at a distance $r = 10$ kpc and to produce an x-ray energy flux $F_x = 10^{-8}$ erg/cm²-sec on the range 1—10 keV. Further, we assume the gas to be at a temperature of 10^7 °K; parameters for other values of the temperature may be determined readily from Figs. 8 and 9. Since

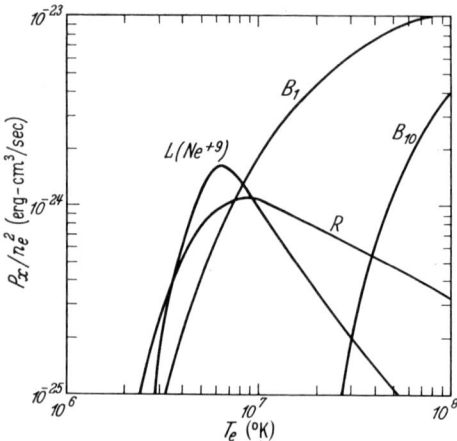

Fig. 9. X-ray production rates in the 1—10 keV range by bremsstrahlung (B_1), recombination radiation (R) and line emission (L). The bremsstrahlung rate (B_{10}) in the 10—20 keV range is also shown. Ions of the following elements have been included: (B_1, B_{10}) H+ He, (L) Ne, and (R) H + He + H + O + Ne. The line emission is due to the $1s-2p$ transition in Ne^{+9} ($E = 1.02$ keV). This is the strongest line emitted in 1—10 keV range.

$$F_x = p_x n_e^2 V/4\pi r^2, \qquad (28.6)$$

this choice of F_x, T, and r fixes the product $n_e^2 V$ at 4×10^{61} cm^{-3}. Then for a range $n_e = 0.1$ to 10^4 cm^{-3}, $\tau_c \sim 10^8$ to 10^3yr, $V \sim 10^8$ to 10^{-2} pc³, and $M \sim 4 \times 10^5$ to 4 solar masses. The associated *optical* bremsstrahlung intensity is of interest and depends only on the choice of T. One finds that this intensity corresponds to a 12th magnitude visual object which may be observable, depending on the extent of the source.

d) The Crab nebula.

29. The general observational data on the Crab are probably more extensive than for any other celestial object with the exception of the sun, although the general physical state of the Crab as derived from these observations is poorly known. Photon fluxes have been detected over a frequency range from 10^7 to 10^{19} c/s. In this section we shall consider what can be inferred from the more recent observations in the high energy end of the spectrum.

Observations of continuum radiation emitted by the Crab have been made in essentially three frequency ranges: the radio range (71), the optical range (59), and the x-ray range (23), (72); Fig. 10 summarizes the results. The radio spectral flux F_ν [watts/m²— (c/s)] is of the form $C_r \nu^{-\alpha}$, where $\alpha = 0.27$, and C_r can be determined by the value (71) [1.23×10^{-23} w/m²— (c/s)] of F_ν at $\nu = 400$ Mc/s. The synchrotron spectrum apparently retains this form up to a frequency $\nu_m =$

---

[1] The curve L in this figure differs from that given in (66) at lower temperatures. In our earlier work a rough estimate of the dielectronic recombination was used. More accurate calculations by W. Tucker at UCSD give the curve in Fig. 8.

10^{14} c/s at the beginning of the optical region. Designating this region $\nu<\nu_m$ as the radio range the radio *luminosity* L_r can then be computed from an assumed distance $d=1030$ pc to the Crab:

$$L_r = 4\pi d^2 \int_0^{\nu_m} C_r \nu^{-\alpha} d\nu = 4\pi d^2 C_r (1-\alpha)^{-1} \nu_m^{1-\alpha} \approx 7.4 \times 10^{36} \text{ erg/sec}. \quad (29.1)$$

The luminosity in, for example, the visible range ($\nu = 4-8 \times 10^{14}$ c/s) of the optical region is

$$L_v \approx 1.7 \times 10^{36} \text{ erg/sec}.$$

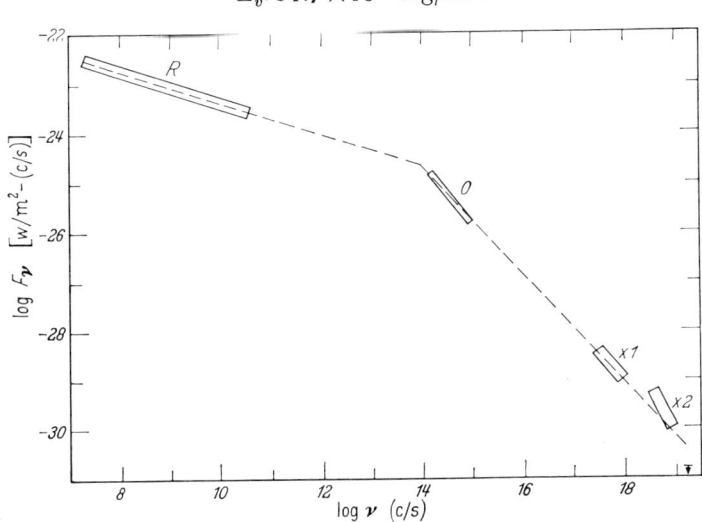

Fig. 10. The observed emission continuum of the Crab Nebula in the radio (R), optical (O) and x-ray regions ($X1$, $X2$). The observational data is shown somewhat schematically, the rectangles showing errors in the observed fluxes. The point at the highest energy represents an upper limit (72).

Assuming a spectrum $F_\nu \propto \nu^{-\alpha'}$ with $\alpha'=1.1$ for the low energy part of the x-ray region, the observations of the NRL group (23) in the range $3 \times 10^{17} < \nu < 10^{18}$ c/s indicate an x-ray luminosity

$$L_{x1} = 1.6 \times 10^{36} \text{ erg/sec}.$$

The observations of CLARK (72) in the higher energy x-ray region between $5 \times 10^{18} < \nu < 10^{19}$ c/s suggest an index $\alpha''=2$ and an x-ray luminosity in this range of

$$L_{x2} = 1.6 \times 10^{36} \text{ erg/sec}.$$

It should be noted that in the x-ray spectrum at the higher energies there is an apparent cutoff (see Fig. 10) or at least another change in slope.

When it was discovered that the Crab was an x-ray source, the suggestion was made that the x-rays were coming from a neutron star formed during the initial supernova outburst (23), (73). An observation designed to test this hypothesis was carried out by the NRL group in July of 1964 during the lunar occultation of the Crab (39). Since a neutron star would be essentially a point source of x-rays, as it would be occulted by the limb of the moon the observed counting rate would drop abruptly to zero. Had this effect been observed, it could have been taken as strong evidence for the existence of a neutron star. It was not observed. When the RNL group sent up a rocket during the time of the occultation (which lasted only a few minutes) with a detector system designed to look at the Crab, they found that the x-ray counting rate changed *continuously* during the occultation. This meant that the x-rays were coming from an *extended* source. The angular diameter of the x-ray source was found to be about 1', compared with an optical diameter of 2' and a radio diameter of 5'.

Assuming the radio spectrum $C_\nu \nu^{-\alpha}$ is due to synchrotron emission by relativistic electrons, an energy spectrum $n_e(\gamma_e) = K_e \gamma_e^{-\Gamma_e}$ with $\Gamma_e = 1 + 2\alpha = 1.54$ is implied. If the mean magnetic field in the Crab is $H = 10^{-4}$ gauss (59) the LARMOR frequency is $\nu_L = 280$ c/s, and the frequency $\nu_m = 10^{14}$ c/s would be emitted primarily by electrons with $(\gamma_e)_m = (\nu_m/\nu_L)^{\frac{1}{2}} \approx 6.0 \times 10^5$. The optical radiation from the Crab would be emitted by slightly higher energy electrons. If the radio emission originates from a volume $\int dV = V_0$, the radio flux F_ν is related to V_0, d, K_e, H by [see Eq. (20.1)].

$$F_\nu \approx (12\pi d^2)^{-1} V_0 K_e c r_0^2 H^2 \nu_L^{(\Gamma_e - 3)/2} \nu^{-(\Gamma_e - 1)/2}. \tag{29.2}$$

From the value of the product $V_0 K_e$ determined from the radio brightness we can compute the total energy of the radio electrons in the Crab:

$$\begin{aligned} E_r &= \int dV K_e m_e c^2 \int d\gamma_e \gamma_e^{-(\Gamma_e - 1)} = V_0 K_e m_e c^2 (2-\Gamma_e)^{-1} (\gamma_e)_m^{2-\Gamma_e} \\ &\approx 1.0 \times 10^{48} \text{ erg.} \end{aligned} \tag{29.3}$$

The age τ of the Crab is 910 years and we see that $E_r/\tau = 3.5 \times 10^{37}$ erg/sec $\gg L_r$, L_v, L_x.

For the assumed magnetic field 10^{-4} gauss the electrons lose energy at a rate [Eq. (9.1)] $-\gamma_e^{-1}(d\gamma_e/dt) \approx \gamma_e \times 1.94 \times 10^{-17}$ sec^{-1}; for $\gamma_e \leq 1.8 \times 10^6$ $(\nu = \nu_L \gamma_e^2 \leq 9.0 \times 10^{14}$ c/s), $-\gamma_e^{-1}(d\gamma_e/dt) < \tau^{-1}$. Thus, for the radio and optical electrons the characteristic time for energy loss by synchrotron emission is greater than the age of the nebula. The rough coincidence of the critical electron energy and synchrotron emission frequency with the value (Fig. 10) above which the spectrum is apparently reduced or perhaps cut off may be interpreted as an indication that the relativistic electrons in the Crab were produced in the initial supernova outburst. The absence of any continuous production of high energy electrons would preclude any interpretation of the x-ray point in Fig. 10 as being due to electron synchrotron emission, since the lifetime against energy loss through synchrotron emission by the energetic electrons necessary to produce this synchrotron frequency is about 30 years $\ll \tau$. An important parameter in this discussion is the strength of the magnetic field in which the electrons radiate. We have chosen a value of $H = 10^{-4}$ gauss, for which the lifetimes of the radio and optical electrons are longer than 10^3 years. However, if the assumed value of H is increased perhaps to 5×10^{-4} gauss, the lifetimes of the electrons emitting the same synchrotron frequencies are decreased by a factor of $(5)^{\frac{3}{2}} (=11.2)$, and the optical electrons have lifetimes less than the age of the nebula so that continuous injection of such electrons is required to explain the optical radiation. Since the magnetic field strength is uncertain we shall consider the possibility of continuous injection of electrons in what follows. We also might mention that as GINZBURG, PIKELNER, and SHKLOVSKY (74) have shown, there might exist in the Crab an energy loss by scattering by magnetic field condensations in the expanding nebula. These scatterings lead to a FERMI-type statistical *deceleration* of the electrons. The corresponding energy loss is approximately $-d\gamma_e/dt \approx \gamma_e V/r$, where V is the expansion velocity of the nebula and r its size; thus $r/V \approx \tau$, the age of the nebula. This energy loss process, if it is operative, dominates synchrotron losses for the radio and optical electrons but is negligible for higher energy electrons. With only this type of energy loss ($\propto \gamma_e$) the electron spectrum $n_e(\gamma_e)$ retains the power law shape of its production spectrum $q_e(\gamma_e)$.

Consider the case where the radio electrons of the Crab are produced continuously and, for simplicity, at a constant rate since the origin of the nebula. Neglecting energy losses[1] the continuity Eq. (13.1) reduces to

$$\partial n_e(\gamma_e)/\partial t = q_e(\gamma_e) = k_e \gamma_e^{-\Gamma_e}, \tag{29.4}$$

and so the electron spectrum at the present time would have become

$$n_e(\gamma_e) = \tau q_e(\gamma_e) = K_e \gamma_e^{-\Gamma_e}. \tag{29.5}$$

If the continuous production is via meson production in nuclear collisions, as was proposed (75) by one of us, there will also be continuous production of π^0-decay photons, and it is of interest to compute the resulting π^0-photon flux. For a pion production spectrum $q_\pi(\gamma_\pi) = k_\pi \gamma_\pi^{-\Gamma_\pi}$ the π^0-decay photon production spectrum is approximately [Eq. (17.5)]

$$d n^0/d\eta \, dt \approx \tfrac{2}{3} k_\pi (2 m_e/m_\pi)^{-(\Gamma_\pi - 1)} \eta^{-\Gamma_\pi}. \tag{29.6}$$

The observed spectral flux of π^0-photons would then be

$$j^0(\eta) = d J/d\eta = (4\pi d^2)^{-1} \int dV \, (d n^0/d\eta \, dt). \tag{29.7}$$

Employing the relation (6.1) between k_π and k_e and Eqs. (29.2) and (29.5) to determine k_π from the radio spectrum we find

$$j^0(\eta) = 1.0 \times 10^{-4} \times \eta^{-1.54}. \tag{29.8}$$

Table 4. *Upper limits to the high energy photon flux from various sources*
[After FRUIN et al (76).]

Source	Photon Flux (photons/cm²-sec)
Crab Nebula	1×10^{-10}
3 C 147	1×10^{-10}
3 C 196	5×10^{-11}
3 C 273	3×10^{-10}

For photons of energy around $\eta = 200$ ($E \sim 100$ MeV) the integrated spectrum with $\Delta\eta/\eta \sim 1$ gives $\int j^0(\eta) \, d\eta \sim 10^{-4} \eta^{-0.54} \sim 5.7 \times 10^{-6}$ photons/cm²-sec. This photon flux is almost four orders of magnitude smaller than the upper limit established by KRAUSHAAR and CLARK (25).

One can also calculate the high energy proton flux required to produce the pion production rate necessary to account for the secondary electron density and the radio spectrum. From this proton flux one can then compute the amount of K-series and inner bremsstrahlung x-rays in the wavelength range of the observations of BOWYER et al. (23). The calculated x-ray flux, for a low energy proton cut-off $\gamma_p = 1$ is about 6 orders of magnitude smaller than the observed flux.

A more definite conclusion regarding secondary electron production in the Crab Nebula may be provided by an analysis of the observations of FRUIN et al. (76). By employing ČERENKOV light detectors to observe light pulses from showers in the atmosphere they were able to set upper limits for the high energy photon flux from the Crab Nebula and also from the quasi-stellar radio sources 3 C 147, 3 C 196, and 3 C 273. The threshold energy for their detection system was 5×10^{12} eV ($\eta \approx 10^7$). The established upper limits to the photon fluxes are listed in Table 4.

If photons of energy $\eta = 10^7$ result from the decay of π^0's produced in nuclear collisions, the corresponding synchrotron emission frequency in the Crab's magnetic field by electrons resulting from the decay of charged pions of the same energy is about $\nu = 10^{16}$ c/s for $H = 10^{-4}$ gauss. This frequency is about midway (on the logarithmic scale) between the optical and x-ray frequencies at which the Crab has been observed (see Fig. 10). It is of interest to compute the π^0-photon flux at $\eta = 10^7$ from the Crab on the assumption that the optical — x-ray flux (if it exists) from the Crab is due to synchrotron emission by secondary-produced electrons.

---

[1] A similar result would be obtained if the Fermi-type statistical deceleration were operative since the characteristic loss time for this process is approximately τ, the age of the nebula.

In the region around $\nu=10^{16}$ c/s the apparent index of the synchrotron spectrum is (Fig. 10) $\alpha=1.1$, so that the electron spectrum in this region is of the form $n_e(\gamma_e)=K_e\gamma_e^{-\Gamma_e}$ with $\Gamma_e=3.2$. Moreover, for these high energy electrons the dominant energy loss process is synchrotron emission and K_e is related to the electron production spectrum $q_e(\gamma_e)=k_e\gamma_e^{-\Gamma_\pi}(\Gamma_\pi=\Gamma_e-1)$ by Eq. (14.4)

$$K_e=k_e/b(\Gamma_\pi-1), \tag{29.9}$$

with k_e related to k_π by Eq. (6.1). Calculating the π^0-photon flux as in Eqs. (29.5) and (29.6) and again determining the parameter $(4\pi d^2)^{-1} V_0 k_\pi$ from the supposed synchrotron emission rate [$F_\nu \approx 1.4_9 \times 10^{-27}$ w/m²— (c/s) at $\nu=10^{16}$ c/s] one calculates a π^0-photon spectrum given by

$$j^0(\eta)=2.2_1\eta^{-2.2}. \tag{29.10}$$

For photons of energy $\eta=10^7$ we find the integrated spectrum with $\Delta\eta/\eta\sim 1$ gives $\int j^0(\eta)\,d\eta\sim 2.2_1\,\eta^{-1.2}\sim 8\times 10^{-9}$ photons/cm²-sec. This calculated photon flux is almost two orders of magnitude *above* the observational upper limit (Table 4). Thus, the present preliminary observations are inconsistent with the interpretation of the x-ray emission from the Crab as synchrotron radiation *if* the necessary continuous production of high energy electrons is through secondary production via $\pi-\mu-e$ decay. If electrons are produced by secondary processes at a lower energy and then accelerated by Fermi processes to energies at which they will radiate synchrotron x-rays, it may be possible to explain the observed x-ray flux without coming into conflict with the results of Fruin et al.

In summary, regarding synchrotron radiation and the relativistic electrons in the Crab, provided that the magnetic field is as weak as 10^{-4} gauss, the view that the energetic electrons responsible for the radio and optical radiation in the Crab were produced in the initial supernova outburst is quite consistent. In fact, the apparent reduction below the extrapolated radio spectrum $F_\nu=C_\nu\nu^{-\alpha}$ in the optical region may possibly be interpretated as a result of energy losses by the more energetic electrons; that is, higher energy electrons would have already decayed in energy since the birth of the nebula. On the other hand, the electrons required to produce synchrotron radiation in the x-ray region would have to be continuously produced.

It should be pointed out that x-rays can also be emitted by the synchrotron process if electrons which are normally radiating in the optical range spiral into regions of much higher magnetic field. Since the critical frequency is proportional to H this means the field must be increased by a factor of $(\nu_x/\nu_0)\sim 10^3$. Thus this would imply that there are regions in the Crab with magnetic field strengths as high as 10^{-1} gauss. There are many difficulties associated with such a model partly because it would require continuous production of particles which move into regions of high field, since the lifetimes are proportional to $(H)^{-2}$. Also, the mechanism by which such concentrations of magnetic flux can be maintained is difficult to understand.

Regarding the possibility that the x-rays from the Crab are from the bremsstrahlung process, we must emphasize again the difficulties of the energy requirement if the bremsstrahlung is by non-thermal electrons. On the other hand, as Clark (72) has emphasized, to explain his observations at 50 keV a temperature of about 2×10^8 °K would be required to produce such energetic thermal bremsstrahlung. This temperature is about an order of magnitude larger than the values predicted from theories of the heating of the gas by the shock front resulting from the expanding ejecta. In view of these difficulties, which seem very great, it would seems that the "least objectionable" explanation for the x-ray production in the

Crab is that the synchrotron process is responsible. Such an explanation has also been suggested by SHKLOVSKY and WOLTJER (77). This problem of the Crab x-ray source has not as yet received a thorough theoretical treatment, and present conclusions must be regarded as tentative.

e) The galactic center.

30. The x-ray source Sgr XR-1 at (or near) the galactic center is of special interest if it is indeed connected with processes in the nucleus of the galaxy. The first discrete x-ray source discovered (22) was identified with the galactic enter, although apparently most of the observed counting rate was actually due to the stronger Scorpius source which, with the poor resolution, could not be distinguished from the galactic center. On the assumption that the x-ray source was the galactic center, we attempted to connect the effect with phenomena observed in the nuclei of external galaxies and with the radio observations of the galactic center (30). As we mentioned earlier (Sect. 28), our initial hypothesis of production by bremsstrahlung by non-thermal electrons meets with difficulties of energy requirements. A more plausible explanation is that the x-rays are due to thermal bremsstrahlung, in which case the characteristics (density, mass, etc.) of the source would correspond to those enumerated at the end of Sect. 28.

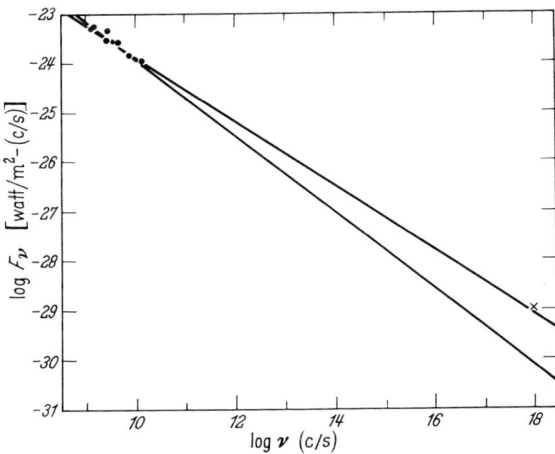

Fig. 11. The observed radiation spectrum from the galactic center. Dots denote the radio observations; an x denotes the x-ray point, determined from an energy flux 10^{-8} erg/cm²-sec and bandwidth $\Delta \nu/\nu = 1$ at $\nu = 10^{18}$ c/s.

Alternatively, the x-rays from the galactic center could be explained as synchrotron radiation. The energies of the synchrotron electrons would then have to be very large and their lifetime very short. However, it is interesting to plot (66) the x-ray observations of the galactic center along with the radio observations (78) of the non-thermal source, as in Fig. 11. The lines are the extension of the power law spectra derived for indices within limits (-0.72 ± 0.05) such as to fit the radio data. It is seen that the x-ray point lies within the limits defined by the extrapolated curves, although the extrapolation is over a factor 10^{10} in frequency. While this might be taken to mean that a single mechanism is responsible for both the radio and x-radiation, it must be remembered that the ratio of the lifetimes τ_r/τ_x of the electrons giving rise to synchrotron radiation in the two spectral regions ν_r and ν_x is $(\nu_r/\nu_x)^{\frac{1}{2}}$. Since the lifetimes of the x-ray synchrotron electrons must be very short ($\sim$30 yr, see Sect. 28), and there is apparently no change in spectral index over the radio to x-ray frequency range, this would mean that the radio synchrotron source was also formed recently.

f) Solar system sources.

31. The sun. The solar corona is well known to be a source of x-rays. A detailed review of solar x-ray astronomy has been given by FRIEDMAN [79] and our discussion will be extremely brief. Since the corona has a temperature $\sim 10^6$ degrees, the x-ray spectrum will consist of lines and continuum emitted in a variety of atomic processes; the first detailed computations of the spectrum which is emitted were made by ELWERT (80). Early observations were made by the NRL group and since then there have been many observations from rockets and satellites. References are given by FRIEDMAN. At times of solar activity the x-ray emission is very greatly increased. The quiet sun is emitting in the keV range about 10^{21} erg/sec while at the time of a class 3 flare the emission can increase to $\sim 10^{26}$ erg/sec.

PETERSON and WINCKLER (81), using balloon techniques, detected hard radiation in a short burst lasting only a few seconds during a 2^+ flare. They deduced that the quanta had energies near 0.5 Mev. Later other observers (82) also detected radiation in short bursts at the times of flares; these have energies in the range 20—80 kev and 20—150 kev. A theoretical discussion of the γ-rays which may be emitted from the sun has been given by DOLAN and FAZIO (83). It is well known that a burst of high energy charged particles is accelerated in a solar flare, and thus a flux of hard radiation is to be expected along with radio emission and enhancement of the visible light. SHKLOVSKY (84) proposed that these hard photons are produced by the COMPTON effect. However, it has been shown by ACTON (85) that the flux of electrons required in SHKLOVSKY's model ($\sim 10^{30}$ ergs of ~ 50 Mev electrons) will produce bremsstrahlung fluxes greater than those detected by ANDERSON and WINCKLER (82). His computations show that a similar situation to that described for the Crab (Sect. 29) exists, i.e., a flux of relativistic electrons moving in any other than an exceedingly intense radiation field will emit far more quanta by bremsstrahlung than through the COMPTON effect. Thus the hard photons emitted in flares are most likely of bremsstrahlung or synchrotron origin. The latter mechanism is entirely probable since the magnetic fields are high and high energy electrons are known to be present.

32. The planets and the interplanetary medium. Fluxes of hard quanta may be produced whenever charged particles are present. Thus Jupiter and the Earth both of which contain trapped fluxes of charged particles must emit some x-rays and γ-rays. Discussion of the hard photons associated with the VAN ALLEN belts lies outside the scope of this article. In the case of Jupiter the total flux of radio emission which is believed to be of synchrotron origin is $\sim 10^{16}$ erg/sec. We may suppose that some fraction of the electron energy is dissipated in the upper atmosphere of Jupiter by bremsstrahlung processes which give rise to some hard photons. An attempt to measure x-rays from Jupiter in the 4—8 kev range by FISHER et al. (86) set a limit of < 2.4 photons/cm² sec. This corresponds to setting a limit to the energy emitted in x-rays of $< 10^{19}$ erg/sec. This is far from being a meaningful limit, since the flux of x-rays is likely to be $\leq 10^{16}$ erg/sec.

HAYAKAWA and MATSUOKA (87) have attempted to estimate the amounts of hard radiation which are produced through the collision of cosmic ray primaries with the surfaces of the moon and planets and the interplanetary medium. Information can be obtained on the compositions of lunar and planetary material by detecting the characteristic x-rays which may be emitted from some elements.

g) Hard radiation from stellar coronae.

33. Since the sun is the only star whose corona is directly detectible, all theories concerning the origin and conditions in a corona have stemmed from it. The first question that arises is therefore whether it is plausible to suppose that other stars have coronae similar to that of the sun. To answer that question it is necessary to consider the probable origin and source of heating of the solar corona.

The theory of the expanding solar corona (88) is based on the concept that the convection below the photosphere generates wave motions (both acoustic and hydromagnetic waves have been discussed) which propagate upward and dissipate, and it is the dissipative heating which leads to coronal expansion. It therefore may be supposed that all stars which have extensive outer convection zones will maintain expanding coronae. This would imply that all main sequence stars below about $F2$ ($M \leq 1.5\, M_\odot$) would have extensive coronae and these stars comprise a considerable fraction of the mass of a galaxy. Also, all stars in the giant stage of their evolution would have coronae. The critical question next is to estimate the average temperature of such hypothetical coronae.

PARKER has pointed out that coronae heated at their bases will have temperatures given approximately by the relation $GMm_H/RkT \lesssim 4$, or $T \lesssim 5.8 \times 10^6$ (M/R) °K with (M/R) measured in solar units. For stars on the main sequence M/R is of the order of unity so that coronal temperatures in the range $10^6 - 10^7$ degrees are to be expected. For giant stars M/R is ≤ 0.1 and for supergiants it is ≤ 0.01. Thus the temperatures of the hypothetical coronae of giants are expected to be $\leq 10^6$ degrees, while for supergiants they are $\leq 10^5$ degrees, and it would appear that only main sequence stars are likely to have hot enough coronae to emit x-rays. PARKER has given various arguments for supposing that more massive main-sequence stars also may have coronae. However, the spectroscopic evidence for extended atmospheres in those stars suggests that the gas has temperatures only $\sim 10^4$ degrees (the heating is by dilute stellar radiation). Thus it is highly improbable that they have hot coronae.

In a previous discussion (89) we attempted to estimate the x-ray flux from the coronae of all of the stars in the galactic disk. There are a large number of uncertainties involved in making this estimate since many assumptions have to be made about the luminosity function of the stars, etc. However, from the early work of TUCKER (90) we estimated that the background flux from stars in the galactic bulge with "quiet" coronae would amount to about 4×10^{-11} erg/cm²-sec in the 2—8 Å region at the earth with an uncertainty of about an order of magnitude. More detailed work on this aspect of the problem is being carried out by TUCKER but the possible flux levels still appear likely to lie in the range $10^{-10} - 10^{-12}$ erg/cm²-sec. Though many stars may be continuously flaring, their integrated contribution is not likely to affect this estimate appreciably.

h) Extragalactic discrete sources.

34. There is still the possibility that some of the sources are extragalactic, and we consider in particular the Scorpius source from which an energy flux of $J_E \sim 10^{-7}$ erg/cm²-sec is observed in the x-ray region. If the Scorpius source were at a typical galactic distance (that is, within our own Galaxy) $d_g \sim 10$ kpc, its x-ray *luminosity* would be $L_g = 4\pi d_g^2 J_E \sim 10^{39}$ erg/sec. If it were at a typical *intergalactic distance* (the distance to a nearby galaxy) $d_{i-g} \sim 1$ Mpc, its luminosity would be $L_{i-g} \sim 10^{43}$ erg/sec, while if it were at a cosmological distance (to a distant galaxy) $d_c \sim 1000$ Mpc, its luminosity would be $L_c \sim 10^{49}$ erg/sec. We now make several observations concerning the energetics of the problem of establishing the distance to and nature of the Scorpius source. On a cosmic time scale[1] $\tau_c \sim 10^{10}$ yr. the energy $L_{i-g}\tau \sim 3 \times 10^{60}$ erg is small compared with the optical energy radiated by a normal galaxy ($\sim 10^{62}$ erg), but a normal galaxy would be expected to radiate a very much smaller amount of energy in x-rays. No unusual external galaxies are observed in the direction of the Scorpius source which is about 20°

---

[1] This time is also roughly the characteristic time for the evolution of a galaxy.

off the galactic center, although interstellar extinction of our own Galaxy prevents observations at lower galactic latitudes (say $\leq 10°$). However, there are no strong radio sources in the direction of Scorpius. Regarding the possibility that the Scorpius source is a distant galaxy, we note that $L_c \tau_c \sim 3 \times 10^{66}$ erg, much greater even than the rest mass energy $M_g c^2$ of a galaxy. Moreover, in the matter-anti-matter annihilation of a galactic mass which we might conceive took place in a time $\ll \tau_c$, the photon energies would be ≥ 0.5 MeV, not x-ray (keV) energies. On the other hand, the size of a small radio source, for example a quasi-stellar object, is $s \sim 10$ kpc, and the time τ_s for a light signal to propagate this distance is $s/c \sim 10^{12}$ sec. The product $L_c \tau_s$ is then $\sim 10^{61}$ erg, roughly the energy E_r of strong radio sources which may be stored in the relativistic particles.

In summary, it appears that *normal* distant galaxies (including radio galaxies) are incapable of producing the observed energy flux J_E corresponding to the Scorpius source over evolutionary time scales $\sim 10^{10}$ yr. However, an outburst over a shorter time might be capable energetically of producing the required x-ray luminosity. Let us consider further such a hypothetical outburst in a galaxy at a distance d involving the release of an amount E of energy, of which a fraction f_γ is emitted in high energy photons of mean energy $\overline{E}_\gamma$. If the outburst occurs during a time τ, the observed resulting photon flux would be

$$J_\gamma = \frac{f_\gamma E/\overline{E}_\gamma}{4\pi d^2 \tau}. \tag{34.1}$$

For $E = 10^{60}$ erg, $d = 1000$ Mpc, and with $\overline{E}_\gamma$ in MeV and τ in years we have

$$J_\gamma \approx 170 \, f_\gamma/\overline{E}_\gamma \, \tau \text{ photons/cm}^2\text{-sec},$$

and for $\overline{E}_\gamma \sim 100$ MeV (mean energy from π^0-decay) and $\tau \sim 1000$ yr (time scale for outburst), $J_\gamma \sim 10^{-3} f_\gamma$ photons/cm²-sec. Unless f_γ is very small, a flux of this magnitude could be observable. The detection of such a discrete source of γ-rays (or x-rays) might then possibly be interpreted as the observation of the birth of a strong radio source. Finally, we might mention that DUTHIE et al. (26) report a possible (~ 100 MeV) γ-ray flux of ~ 0.002 photons/cm²-sec from Cygnus A which is at a distance ~ 100 Mpc.

IV. Neutrino sources.

35. Any review of the fluxes of hard radiation which may be present in the universe would not be complete without mention of neutrinos. In principle detection of neutrino fluxes would give valuable direct evidence concerning conditions in stellar interiors, and also if high energy neutrinos could be observed in formation on the high energy particle flux could be obtained. Moreover, evidence of the energy density of neutrinos in the universe may have cosmological significance. The subject of neutrino astronomy has been discussed and reviewed ad nauseam in the last two or three years following developments in the theory of weak interactions and the realization that neutrino emission processes will become the dominant energy loss mechanism in the final stages of stellar evolution. There are a number of recent papers and reviews which have given some account of these processes and their repercussions on stellar evolution, nucleosynthesis, supernovae and cosmology; full references can be obtained in papers by PONTECORVO (*91*), FOWLER and HOYLE (*92*), BURBIDGE (*93*), WEINBERG (*94*), FODOR, KÖRVESSY, and MARX (*95*), CHIU (*50*), and BAHCALL (*96*). We only give a very brief summary here.

While the energy density in the flux of neutrinos is very considerable, so that, for example, for a normal galaxy it will be some 4% of the total luminous flux or about 4×10^{42} erg/sec, the very small interaction cross-sections ($\sim 10^{-44} \, \eta_\nu^2$ cm^2), (unless resonances are present, cf. below) obviously make the fluxes very difficult of detection. Moreover, no method of detecting low energy neutrinos with energies below those necessary to induce inverse beta decays is known. We illustrate the problems by discussing the work on solar neutrinos and then consider fluxes from more distant stars and galaxies.

Neutrinos (ν_e) are emitted in the normal hydrogen burning processes in stars. About 2% of the energy released in the $p-p$ chain and about 6% in the CNO cycle is emitted as neutrinos. Undoubtedly the sun is likely to be the strongest apparent source of neutrinos and direct detection of them is of the greatest importance. Following an early suggestion of PONTECORVO, BAHCALL (97) and DAVIS (98) have considered in detail the possibility of the detection of neutrinos emitted in Be^7 (e^-, ν) Li^7 and B^8 (e^+, ν) Be^8 through their absorption Cl^{37} (ν, e^-) Ar^{37}; the activity of Ar^{37} is then measured. On the basis of the best solar models (99) BAHCALL (97) has estimated that the fluxes at the earth's surface will be 1.2×10^{10} neutrinos/cm^2/sec and 2.5×10^7/cm^2 sec from the decay of Be^7 and B^8 respectively. From BAHCALL's analysis of the cross sections for Cl^{37} (ν,e^-) Ar^{37} DAVIS has concluded that the expected neutrino captures in 10^5 gallons of C$_2$Cl$_4$ in a mine would be about 4—11 a day which would be an order of magnitude above the background produced by the production of Ar^{37} by cosmic rays underground through Cl^{37} (p, n) Ar^{37}. The flux of detectable neutrinos from the central bulge of the galaxy will be less than that from the sun by a factor 10^7-10^8 while the flux to be expected from a nearby galaxy such as M 31 would be less than the sun by a factor $\sim 10^{11}$. While neutrinos are emitted in the normal energy producing cycles in the stars, neutrinos and anti-neutrinos are emitted with positrons and electrons respectively by beta unstable nuclei in the processes of energy generation and element synthesis beyond hydrogen. However, for a galaxy in a steady state it is easily shown that the fluxes to be expected are small compared with those emitted in hydrogen burning.

In the high temperature phase of stellar evolution (for core temperatures $\geq 5 \times 10^8$ degrees) neutrino pair emission becomes the dominant mechanism of energy loss. They arise by a variety of reactions in all of which they replace photon emission. An important process is

$$e^+ + e^- \rightarrow \nu + \bar{\nu}. \tag{35.1}$$

While this mechanism of energy loss is important from the point of view of the evolutionary process, an individual object (perhaps the immediate forewarning of a supernova) would be very difficult to detect even if a mechanism of detecting low energy neutrinos were found, because of the very short time scale associated with such evolutionary phases. Thus, for example, FOWLER and HOYLE (92) have calculated that if one solar mass in the center of a massive star reaches a temperature of 3.5×10^9 degrees the neutrino flux will amount to $\sim 10^{47}$ erg/sec. However, this phase will only last a few seconds. At a later stage, after a star has exploded and if a neutron configuration remains, the initial neutrino flux for a core temperature of 10^9 degrees will be much less [see (61), (63) and (64)].

We turn finally from the low energy neutrinos emitted in stellar evolution to consider the possibility as to whether high energy neutrinos ($E_\nu \geq 100$ MeV) are emitted in supernova outbursts and from radio sources in which large fluxes of high energy particles are present. Neutrinos are produced whenever a flux of high energy nuclei interacts with the nuclei of the local gas atoms to produce pions. In the $\pi \rightarrow \mu \rightarrow e$ decay of the charged pions both neutrinos and anti-neutrinos of the electron and muon type result. That is, in the pion decay

$$\left.\begin{array}{l}\pi^+ \to \mu^+ + \nu_\mu, \\ \pi^- \to \mu^- + \bar{\nu}_\mu,\end{array}\right\} \quad (35.2)$$

while in the muon decay

$$\left.\begin{array}{l}\mu^+ \to e^+ + \nu_e + \bar{\nu}_\mu, \\ \mu^- \to e^- + \bar{\nu}_e + \nu_\mu.\end{array}\right\} \quad (35.3)$$

Thus, a single charged pion pair π^+, π^- results in $2(\nu_\mu + \bar{\nu}_\mu) + (\nu_e + \bar{\nu}_e)$; twice as many μ-neutrinos as e-neutrinos are produced. In the pion decay the muon is essentially non-relativistic in the rest frame of the pion and most of the energy is carried away by the neutrino; here $E_\nu \approx (m_\pi - \mu_\mu) c^2 \approx \frac{1}{4} m_\pi c^2$. In the muon decay the mean neutrino energy is about $\frac{1}{3} m_\mu c^2 \approx \frac{1}{4} m_\pi c^2$ in the rest frame of the muon and pion (see Sect. IIb). Thus, the mean lab energy of the neutrinos in both decays is about $\frac{1}{4} \gamma_\pi m_\pi c^2 = \frac{1}{4} E_\pi$.

The (anti) neutrino production spectrum is readily computed from the pion production spectrum and is of the form similar to that for the production spectrum of π^0-decay photons [see Eq. (17.5)], that is, $dn_\nu/d\eta_\nu \, dt \propto \eta_\nu^{-\Gamma_\pi}$, where Γ_π is the index of the pion production spectrum. The ratio of the (anti)neutrino production spectrum (or of the spectral) flux) to the π^0-decay photon spectrum at the same η is, assuming equal number of π^+, π^-, π^0 produced, roughly

$$\left.\begin{array}{l}\left(\dfrac{\nu_\mu}{\gamma}\right)_{\text{same } \eta} \approx 2^{-(\Gamma_\pi - 2)} \\ \left(\dfrac{\nu_e}{\gamma}\right)_{\text{same } \eta} \approx 2^{-(\Gamma_\pi - 1)}\end{array}\right\} \quad (35.4)$$

BAHCALL and FRAUTSCHI (100) have discussed the detection of high energy neutrinos and have considered the possibility of observing a neutrino flux from the Crab Nebula and other radio sources. They assume neutrino production via $\pi - \mu$ decay, which implies also continuous production of pions and π^0-decay photons. Assuming a continuous constant production of high energy radio electrons through $\pi - \mu$ decay in the Crab since its birth, the associated π^0-decay photon flux was calculated in Sect. III [Eq. (29.8)]. The corresponding neutrino flux is of the same form

$$j_\nu(\eta_\nu) = k\eta_\nu^{-\Gamma_\pi}, \quad (35.5)$$

with $\Gamma_\pi = 1.54$. For μ-neutrinos $k_\mu \approx 2^{0.46} \times 1.0 \times 10^{-4}$ cm^{-2} sec^{-1}, while for e-neutrinos $k_e \approx \frac{1}{2} k_\mu$. This neutrino spectrum and also the π^0-decay photon spectrum is associated with (if there is continuous production) the radio synchrotron spectrum for 10^7 c/s $<\nu< 10^{14}$ c/s and with electron energies $200 < \gamma_e < 6 \times 10^5$. The range of η_ν over which Eq. (35.5) should represent the neutrino spectrum is the same as the range of γ_e, that is for 100 MeV $\leq E_\nu \leq$ 300 BeV. A neutrino spectrum similar to Eq. (35.5) was derived by BAHCALL and FRAUTSCHI. However, we *doubt* that such a neutrino flux will ever be observed from the Crab. In Sect. III we showed that there is evidence against continuous production via $\pi - \mu$ decay of very high energy electrons which would produce synchrotron radiation in the optical — x-ray range; moreover, the lifetime against synchrotron losses for the radio and optical electrons in the Crab is longer than the age of the nebula. We therefore feel that probably there is little or no continuous production of radio electrons in the Crab and no associated neutrino of π^0-decay photon production.

Regarding possible neutrino production in other radio sources, in particular in extragalactic objects, similar considerations apply. If there does exist continuous production of radio electrons via $\pi-\mu$ decay, and a steady state exists, then the energy radiated in neutrinos would be comparable to the total energy emitted in synchrotron radiation by the relativistic electrons produced with the neutrinos. For "normal" radio galaxies with steep spectra (index $\alpha \approx 0.8$) most of the neutrinos produced would have fairly low energies ($E_\nu \sim 100$ MeV), while sources with flat radio spectra (e.g., Crab Nebula, M 82) might be expected to emit predominantly higher energy neutrinos (say, $E_\nu \sim 100$ GeV). The strong extragalactic radio sources and quasi-stellar objects would be emitting lower energy neutrinos with $E_\nu \sim 1$ GeV at power levels of $10^{44}-10^{45}$ erg/sec.

However, it appears probable now that such steady state conditions are not present in these sources, so that even if large proton fluxes are present, the neutrino fluxes will be much lower than this (*101*). On the other hand it is possible that at an early phase when a violent outburst in a galaxy gives rise to some 10^{60} ergs high of energy particles (perhaps over a period of 1000 years), a large fraction of which may be protons, the collisions of some part of these with the interstellar gas before they escape into regions of very low density might give rise to a flux of high energy neutrinos several orders of magnitude greater than the values corresponding to steady state conditions. Thus one might expect to observe both neutrinos and π^0-decay photons from a violent outburst in a galaxy (see Sect. III).

The possibility of detecting such fluxes of high energy neutrinos has been considered by BAHCALL and FRAUTSCHI (*100*). They have pointed out that the very small cross sections for the interaction of neutrinos with matter mean that from very strong radio sources with a dominant proton flux only one neutrino-induced event per day would be experienced in a 10^5 ton absorber. However, as BAHCALL and FRAUTSCHI have proposed, the possibility exists that resonances in neutrino interaction processes are present. As they suggest, the reaction

$$\bar{\nu}_e + e^- \to \bar{\nu}_\mu + \mu^- \tag{35.6}$$

may have a resonance and may be detected by neutrino interactions with material in the earth's crust. Clearly a great deal of information might be gained from observations of neutrinos from extragalactic objects. Thus the most pressing requirement is to devise a neutrino telescope which has good angular resolution. BAHCALL and FRAUTSCHI have suggested that the muons ejected in (35.6) may enable this to be achieved.

V. Conclusion.

36. We have tried to summarize those mechanisms which may give rise to hard radiation in the universe. At present, apart from observations of the sun, there is little observational evidence which can be used in conjunction with the theoretical estimates. The brilliant work of the NRL, MIT-ASE, and LOCKHEED groups has shown that there are sources of x-rays at flux levels which are detectable with present techniques. Moreover, the absence of a large isotropic flux of x-rays has enabled us to set limits on the temperature of the intergalactic medium. As far as γ-rays are concerned it is not yet clear whether high energy γ-rays are present at the flux levels calculated in Sect. II. The detection of high energy neutrino fluxes would be very exciting but the preliminary results must be viewed with caution.

What are the possibilities for further investigations in this field? To us the parallel of this field of research with that of the early days in radio astronomy is strong. There is one major difference, however, and this concerns the theoretical expectations in the field.

The discovery of significant fluxes of radio emission from the cosmos was totally unexpected, and in the first decade after the war theoreticians only gradually came to understand that the process by which the non-thermal sources radiate is the synchrotron mechanism. Of course the process of thermal emission was well understood but could not explain the strength or the spectral characteristics of the bulk of the radiation. During this period there was much confusion because of the unexpected nature of the discoveries and it was the interplay between the theory and *optical* observation which led to the elucidation of the mechanism by which the sources radiate. The theoretical problem then evolved into that of understanding how the vast fluxes of relativistic particles and magnetic fields originate.

As far as the hard radiation is concerned, the physical mechanisms by which such radiation can be emitted are well known and the level at which fluxes have been detected (or not detected) suggests that no objects with the unexpected character of the radio sources are likely to be found by observational techniques in this energy range. If hard radiation is emitted by hot bodies such as neutron stars, then they must have very hot surfaces and they will cool very rapidly and soon cease to emit hard radiation. If very hot low density regions are generated in supernova outbursts they may well have much longer lifetimes so that it is possible that they can be detected (cf. Sect. III). Otherwise the mechanics by which hard quanta are emitted all stem from the interaction of fast charged particles with matter, radiation, or magnetic fields. Knowledge gained through cosmic-ray and radio astronomical discoveries enables predictions to be made of the fluxes of hard radiation to be expected with a range of parameters associated with the present uncertainties in these quantities. Thus detection and even non-detection of hard radiation will be most valuable in determining the state of matter and radiation in the universe.

The parallel between the developments in radio astronomy and x-ray, γ-ray, and neutrino astronomy is very close when we consider the problem of the discrete sources. In the early days in radio astronomy resolution was very poor and at least one of the strongest sources was put in the wrong constellation by one notable group of investigators. All of the major developments in the study of discrete radio sources have come in step with the increase in precision with which positions of sources could be determined. This has enabled the objects to be observed optically with large telescopes. With optical identification has come measurement of distance and with this a beginning of quantitative study of the physical conditions in the sources. It is the absence of a method of determining the distance of an extragalactic source which has required the cooperation of optical and radio telescopes[1]. The same situation appears to apply in γ-ray and x-ray astronomy, since the flux emitted in lines will in general be small. A considerable improvement in resolution is required in order to determine better positions for the x-ray sources so far detected. The lunar occultation observation of the Crab by the NRL group is a first step in this direction and at the time of writing a more accurate position for the Scorpius source is being obtained. Already the combination of observational arguments concerning the angular diameter and spectral characteristics of the Crab source together with theoretical discussions of the cooling rate for neutron stars leads to the conclusion that this source is almost certainly not a neutron star, and it is very doubtful whether any of the sources discovered so far are neutron configurations.

---

[1] In principle the 21 cm line is a powerful tool for determining distance by redshift measurements, but in practice it cannot be used since to detect the feature in galaxies at only very modest distances (≤ 20 Mpc) is beyond the capability of present day radio telescopes.

It is clear that the various theoretical estimates of fluxes which we have given in this paper suggest that a great increase in sensitivity of detectors as well as good resolution will be needed to exploit this field to the utmost. Finally, it is not out of place to remark that the x-ray observations have already shown that the universe is not very hot, and it may in fact be rather cool. In this case, apart from the neutrino flux which is part of the general cosmological thermal radiation field, the flux of hard radiation may be rather weak.

We are indebted to many friends and colleagues who have provided much material prior to publication. We wish to thank Mrs. JEAN FOX for typing a difficult manuscript with expedition and efficiency. This research has been supported in part by the National Science Foundation and in part by NASA through contract NsG-357.

References.

A. Review articles on x-ray and γ-ray astronomy, and on techniques for the detection of hard quanta.

[1] GINZBURG, V. L., i S. I. SYROVATSKY: Uspekhi Fiz. Nauk SSSR **84**, 201 (1964); translation in Space Sci. Rev. **4**, 267 (1965).
[2] HAYAKAWA, S., and M. MATSUOKA: Progr. Theoret. Phys., Suppl. **30**, 204 (1964).
[3] HAYAKAWA, S., H. OKUDA, Y. TANAKA, and Y. YAMAMOTO: Progr. Theoret. Phys., Suppl. **30**, 153 (1964).
[4] GARMIRE, G., and W. L. KRAUSHAAR: Space Sci. Rev. **4**, 123 (1965).
[5] GIACCONI, R., and H. GURSKY: Space Sci. Rev. **4**, 151 (1965).
[6] FAZIO, G. G.: Ann. Rev. Astron. and Astrophys. (to be published).

B. References cited.

(1) CHUBB, T. A., H. FRIEDMAN, R. W. KREPLIN, and J. E. KUPERIAN jr.: J. Geophys. Res. **62**, 389 (1957); see also H. FRIEDMAN, Rep. Progr. Phys. **25**, 163 (1962).
(2) FEENBERG, E., and H. PRIMAKOFF: Phys. Rev. **73**, 449 (1948).
(3) POLLACK, J. B., and G. G. FAZIO: Phys. Rev. **131**, 2684 (1963). — Astrophys. J. **141**, 1161 (1965).
(4) GINZBURG, V. L., i S. I. SYROVATSKY: Zhur. Eksp. Teor. Fiz. **45**, 353 (1963); **46**, 1865 (1964). — Soviet Phys. **18**, 245 (1964); **19**, 1255 (1964).
(5) MORRISON, P. M.: Nuovo Cim. **7**, 858 (1958); see also M. P. SAVEDOFF, Nuovo Cim. **7**, 1584 (1959).
(6) GINZBURG, V. L., i S. I. SYROVATSKY: Astron. Zhur. **40**, 466 (1963). — Soviet Astron. **7**, 356 (1963).
(7) BURBIDGE, G. R.: Progr. Theoret. Phys. **27**, 999 (1962). — BURBIDGE, G. R., and F. HOYLE: Proc. Phys. Soc. Lond. **84**, 141 (1964).
(8) DE SHONG, J. A., R. H. HILDEBRAND, and P. MEYER: Phys. Rev. Letters **12**, 3 (1964). — ARGINIER, B., Y. KOECHLIN, B. PARLIER, G. BOELLA, G. DEGLI ANTONI, C. DILWORTH, L. SCARSI, and G. SIRONI: Phys. Rev. Letters **13**, 377 (1964).
(9) GINZBURG, V. L., and S. I. SYROVATSKY: Origin of Cosmic Rays. London: Pergamon Press 1964.
(10) See, for example, R. MARSHAK: Meson Physics. New York: McGraw Hill Book Co. 1952.
(11) LANDAU, L.: Izv. Akad. Nauk SSSR **17**, 51 (1953).
(12) KIDD, J. M.: Nuovo Cim. **27**, 57 (1963).
(13) GOULD, R. J., T. GOLD, and E. E. SALPETER: Astrophys. J. **138**, 408 (1963); DORSCHNER, J., J. GÜRTLER, u. K.-H. SCHMIDT: Astron. Nachr. **288**, 149 (1965).
(14) JAUCH, J. M., and F. ROHRLICH: Theory of Photons and Electrons. Cambridge: Addison-Wesley Publ. Co. 1955.
(15) HEITLER, W.: The Quantum Theory of Radiation. London: Oxford University Press 1954.
(16) DONAHUE, T. M.: Phys. Rev. **84**, 972 (1951).
(17) FELTEN, J. E., and P. MORRISON: Phys. Rev. Letters **10**, 453 (1963).
(18) FIELD, G.: Astrophys. J. **135**, 684 (1962). — DAVIES, R. D., and R. C. JENNISON: Monthly Notices Roy. Astron. Soc. **128**, 123 (1964). — DAVIES, R. D.: Monthly Notices Roy. Astron. Soc. **128**, 133 (1964).
(19) HAYAKAWA, S., and K. KITAO: Progr. Theoret. Phys. **16**, 139 (1956).

(20) STROM, S. E., and K. M. STROM: Publ. Astron. Soc. Pacific **73**, 43 (1961).
(21) NIKISHOV, A. I.: Zhur. Eksp. Teor. Fiz. **41**, 549 (1961). — Soviet Phys. **14**, 393 (1962).
(22) GIACCONI, R., H. GURSKY, F. R. PAOLINI, and B. ROSSI: Phys. Rev. Letters **9**, 439 (1962). — GURSKY, H., R. GIACCONI, and F. R. PAOLINI: Phys. Rev. Letters **11**, 530 (1963).
(23) BOWYER, S., E. T. BRYAM, T. A. CHUBB, and H. FRIEDMAN: Nature, Lond. **201**, 1307 (1964).
(24) ARNOLD, J. R., A. E. METZGER, E. C. ANDERSON, and M. A. VAN DILLA: J. Geophys. Res. **67**, 4878 (1962). — METZGER, A. E., E. C. ANDERSON, M. A. VAN DILLA, and J. R. ARNOLD: Nature, Lond. **204**, 766 (1962).
(25) KRAUSHAAR, W. L., and G. W. CLARK: Phys. Rev. Letters **8**, 106 (1962); see also A. BRACESI e M. CECCARELLI: Nuovo Cim. **17**, 691 (1960).
(26) DUTHIE, J. G., E. M. HAFNER, M. F. KAPLON, and G. G. FAZIO: Phys. Rev. Letters **10**, 364 (1963).
(27) FIRKOWSKI, A., J. GAWIN, R. MAZE, and A. ZAWADSKI: J. Phys. Soc. Japan **17**, Suppl. A-III, 123 (1962).
(28) SUGA, K., I. ESCOBAR, G. W. CLARK, W. HAZEN, A. HENDEL, and K. MURAKAMI: J. Phys. Soc. Japan **17**, Suppl. A—III, 128 (1962).
(29) PETERSON, L. E.: J. Geophys. Res. **70**, 1762 (1965).
(30) GOULD, R. J., and G. R. BURBIDGE: Astrophys. J. **138**, 969 (1963); see also G. B. FIELD and R. C. HENRY, Astrophys. J. **140**, 1002 (1964).
(31) GINZBURG, V. L., i S. I. SYROVATSKY: Astron. Zhur. **41**, 430 (1964). — Soviet Astron. **8**, 342 (1964).
(32) GOLD, T., and F. HOYLE: Paris Smyposium on Radio Astronomy, ed. by R. N. BRACEWELL. Stanford: Stanford University Press 1958.
(33) BURBIDGE, E. M., G. E. BURBIDGE, and F. HOYLE: Astrophys. J. **138**, 873 (1963).
(34) SCIAMA, D. W.: Quart. J. R. A. S. **5**, 196 (1964).
(35) BURBIDGE, G. R., e F. HOYLE: Nuovo Cim. **4**, 1 (1956).
(36) COMBES, C. A., B. CORK, W. GALBRAITH, C. R. LAMBERTSON, and W. A. WENZEL: Phys. Rev. **112**, 1303 (1958).
(37) GOULD, R. J., and D. W. SCIAMA: Astrophys. J. **140**, 1634 (1964); regarding cosmological considerations, see also J. G. C. MCVITTIE, Phys. Rev. **128**, 2871 (1962).
(38) BOWYER, S., E. T. BRYAM, T. A. CHUBB, and H. FRIEDMAN: Science **147**, 394 (1965).
(39) BOWYER, S., E. T. BRYAM, T. A. CHUBB, and H. FRIEDMAN: Science **146**, 912 (1964).
(40) GURSKY, H., R. GIACCONI, and F. R. PAOLINI: Phys. Rev. Letters **11**, 530 (1963).
(41) ODA, M., G. CLARK, G. GARMIRE, M. WADA, R. GIACCONI, H. GURSKY, and J. WATERS: Nature, Lond. **205**, 554 (1965). — GIACCONI, R., H. GURSKY, J. WATERS, G. CLARK, and B. ROSSI: Nature, Lond. **204**, 981 (1964).
(42) FISHER, P. C., and A. J. MEYEROTT: Astrophys. J. **139**, 123 (1964); **140**, 821 (1964).
(43) FISHER, P. C., D. B. CLARK, A. J. MEYEROTT, et K. L. SMITH: Ann. d'Astrophys. **27**, 809 (1964).
(44) FISHER, P. C., D. B. CLARK, A. J. MEYEROTT, and K. L. SMITH: Private communication.
(45) SHKLOVSKY, I. S.: Private communication 1964.
(46) QUIGLEY, M. I., and C. G. HASLAM: Nature, Lond. **203**, 1272 (1964).
(47) BURBIDGE, E. M., G. R. BURBIDGE, W. A. FOWLER, and F. HOYLE: Rev. Mod. Phys. **29**, 547 (1957).
(48) CLAYTON, D. D., and D. D. CRADDOCK: Astrophys. J. **142**, 189 (1965).
(49) CLAYTON, D. D., W. A. FOWLER, and P. SEEGER: Astrophys. J., Suppl. (to be published).
(50) CHIU, H. Y.: Ann. Phys. **26**, 364 (1964).
(51) FINZI, A.: Astrophys. J. **139**, 774 (1964).
(52) BAADE, W., and F. ZWICKY: Astrophys. J. **88**, 411 (1938). — ZWICKY, F.: Astrophys. J. **88**, 522 (1938).
(53) LANDAU, L.: Phys. Z. Sowj. **1**, 285 (1932).
(54) OPPENHEIMER, J. R., and G. M. VOLKOFF: Phys. Rev. **55**, 374 (1939).
(55) DATT, B.: Z. Astrophys. **108**, 314 (1938). — OPPENHEIMER, J. R., and H. SNYDER: Phys. Rev. **56**, 455 (1939).
(56) HOYLE, F., W. A. FOWLER, G. R. BURBIDGE, and E. M. BURBIDGE: Astrophys. J. **139**, 909 (1964).
(57) FOWLER, W. A., and F. HOYLE: Ann. Phys. **10**, 280 (1960). — Astrophys. J., Suppl. **9**, 201 (1964).
(58) COLGATE, S.: Proc. Intern. Conf. on Cosmic Rays, Jaipur 1963.
(59) O'DELL, C. R.: Astrophys. J. **136**, 809 (1962).
(60) MISNER, C. W., and H. S. ZAPOLSKY: Phys. Rev. Letters **12**, 635 (1964).
(61) CHIU, H. Y., and E. E. SALPETER: Phys. Rev. Letters **12**, 413 (1964).
(62) MORTON, D. C.: Astrophys. J. **140**, 460 (1964).

(63) BAHCALL, J. N., and R. A. WOLF: Phys. Rev. Letters **14**, 343 (1965).
(64) FINZI, A.: Phys. Rev. **137**, B 472 (1965); see also ELLIS, D. G.: Phys. Rev. (to be published).
(65) MORRISON, P.: Proc. Second Texas Symposium on Relativistic Astrophysics Chicago: Chicago University of Chicago Press (in press).
(66) BURBIDGE, G. R., R. J. GOULD, and W. H. TUCKER: Phys. Rev. Letters **14**, 289 (1965).
(67) GINZBURG, V. L., L. M. OZERNOI i S. I. SYROVATSKY: Dokl. Akad. Nauk. SSSR **154**, 557 (1964); transl. Sovient Phys. - Doklady **9**, 3 (1964).
(68) ROSSI, B.: Private communication.
(69) HAYAKAWA, S., and M. MATSUOKA: Progr. Theoret. Phys. **29**, 612 (1963).
(70) See, for example, H. A. BETHE and E. E. SALPETER: Quantum Mechanics of One- and Two-Electron Atoms. New York: Academic Press 1957.
(71) CONWAY, R. G., K. J. KELLERMANN, and R. J. LONG: Monthly Notices Roy. Astron. Soc. **125**, 261 (1963).
(72) CLARK, G. W.: Phys. Rev. Letters **14**, 91 (1965).
(73) MORTON, D. C.: Nature, Lond. **201**, 1308 (1964).
(74) GINZBURG, V. L., S. B. PIKELNER i I. S. SHKLOVSKY: Astron. Zhur. **32**, 503 (1955).
(75) BURBIDGE, G. R.: Nuovo Cim., Suppl. **8**, 403 (1958).
(76) FRUIN, J. H., J. V. JELLEY, C. D. LONG, N. A. PORTER, and T. C. WEEKES: Phys. Letters **10**, 176 (1964).
(77) SHKLOVSKY, I. S.: Private communication 1964. — WOLTJER, L.: Astrophys. J. **140**, 1309 (1964).
(78) MAXWELL, A., and B. DOWNS: Nature, Lond. **204**, 865 (1964).
(79) FRIEDMAN, H.: Ann. Rev. Astron. and Astrophys. **1**, 59 (1963).
(80) ELWERT, G.: Z. Naturforsch. 7a, 202, 432 (1952); 9a, 637 (1954). — J. Geophys. Res. **66**, 391 (1961).
(81) PETERSON, L. E., and J. R. WINCKLER: J. Geophys. Res. **64**, 697 (1959).
(82) ANDERSON, K. A., and J. R. WINCKLER: J. Geophys. Res. **67**, 4103 (1962).
(83) DOLAN, J. F., and G. G. FAZIO: Rev. Geophys. **3**, 319 (1965).
(84) SHKLOVSKY, I. S.: Nature, Lond. **202**, 275 (1964).
(85) ACTON, L. W.: Nature, Lond. **204**, 64 (1964).
(86) FISHER, P. C., D. B. CLARK, A. J. MEYEROTT, and K. L. SMITH: Nature, Lond. **204**, 982 (1964).
(87) HAYAKAWA, S., and M. MATSUOKA: Rep. Ion. Space Res., Japan (to be published).
(88) PARKER, E. N.: Interplanetary Dynamical Processes. New York: Interscience 1963.
(89) GOULD, R. J., et G. R. BURBIDGE: Ann. d'Astrophys. **28**, 171 (1965).
(90) TUCKER, W. H.: private communication (see Ref. [89]).
(91) PONTECORVO, B.: Soviet Phys. — Uspekhi **6**, 1 (1963).
(92) FOWLER, W. A., and F. HOYLE: Astrophys. J., Suppl. **9**, 201 (1964).
(93) BURBIDGE, G. R.: Ann. Rev. Nuclear Sci. **12**, 507 (1962).
(94) WEINBERG, S.: Phys. Rev. **128**, 1457 (1962).
(95) FODOR, L., A. KÖRVESSY, and G. MARX: Acta Phys. Acad. Sci. Hung. **17**, 171 (1964).
(96) BAHCALL, J. N.: Science **147**, 115 (1965).
(97) BAHCALL, J. N.: Phys. Rev. Letters **12**, 300 (1964).
(98) DAVIS jr., R.: Phys. Rev. Letters **12**, 303 (1964).
(99) SEARS, R. L.: Astrophys. J. **140**, 477 (1964).
(100) BAHCALL, J. N., and S. C. FRAUTSCHI: Phys. Rev. **135**, B 788 (1964).
(101) BURBIDGE, G. R., E. M. BURBIDGE, and A. R. SANDAGE: Rev. Mod. Phys. **35**, 947 (1963).
(102) HAYAKAWA, S.: Phys. Letters **1**, 234 (1962).

Some additional references added in proof January 1966:

(a) FELTEN, J. E.: Inverse Compton Radiation from Intergalactic Electrons and Blackbody Photons. Phys. Rev. Letters **15**, 1003 (1965).
(b) CHODIL, G., R. C. JOPSON, H. MARK, F. D. SEWARD, and C. D. SWIFT: X-Ray Spectrum from Scorpius (SCO-XR-1), etc. Phys. Rev. Letters **15**, 605 (1965).
(c) TUCKER, W. H., and R. J. GOULD: Radiation from a Low Density Plasma at $10^6 - 10^8$ degrees. Astrophys. J. **143** (in press).
(d) PETERSON, L. E., A. S. JACOBSON, and R. M. PELLING: The Spectrum of Crab Nebula X-Rays to 120 KeV (submitted to Phys. Rev. Letters).
(e) GOULD, R. J.: High-energy Photons from the Compton-synchrotron Process in the Crab Nebula. Phys. Rev. Letters **15**, 577 (1965).
(f) GOULD, R. J., and G. SCHRÉDER: The Opacity of the Universe to High Energy Photons (submitted to Phys. Rev. Letters).

The Time Variations of the Cosmic Ray Intensity.

By

J. J. QUENBY.

With 37 Figures.

I. Introduction.

1. Time variations in the intensity of cosmic radiation arriving at the Earth's surface have been extensively studied over the last thirty years. Some of these variations are of atmospheric origin, but others are due to changing conditions in interplanetary space. It is reasonable to regard the solar system as being bombarded by an almost isotropic flux of energetic nuclei with energies ranging from 10^8 electron volts to 10^{19} electron volts. Particles in the range 10^8 eV to 10^{11} eV are modulated by the electromagnetic fields existing in interplanetary space so that a reduced intensity arrives at the Earth.

At times, the Sun produces energetic particles with energies up to 10^{10} eV and these propagate through interplanetary space to arrive at the Earth and cause an increase in the cosmic ray intensity. Present work in the field of time variations is directed towards understanding the details of the intensity changes in terms of postulated models for the interplanetary electromagnetic conditions. Both the galactic cosmic rays and the solar-accelerated particles provide evidence about this interplanetary field. Other experimental evidence comes from space probes, as such Pioneer V and Mariner II, which directly sample the magnetic fields and plasma densities and velocities in space.

It is the purpose of this article to select the experimental data which are most useful in giving information about the interplanetary fields and then to compare this evidence with various theoretical models. Variations of atmospheric origin will not be discussed here but these are dealt with by SCOTT E. FORBUSH [1] in Bd. 49 of this series. Intensity variations which can arise from the interaction of interplanetary plasma and the Earth's magnetic field will be mentioned, however.

Apart from the review by FORBUSH, three other articles are available which together cover most of this field of study. WEBBER [2] has carefully considered a great body of experimental data on the energy dependence of various types of time variations obtained from ground level and balloon-borne equipment. He also discusses current theoretical models for the long-term and Forbush variations. DORMAN [3] similarly compares experimental data with various theories and investigates in especial detail the onset of the sudden Forbush decreases in intensity. PARKER [4], in a monograph devoted to the dynamics of the interplanetary medium, discusses particle motion in his own model for the interplanetary field. Of these sources of information, PARKER gives the most comprehensive theoretical treatment, WEBBER shows the greatest appreciation for experimental data, while DORMAN lies somewhere in between the two extremes and gives the most detailed review of this field of study. Much of the theory of charged particle motion in electromagnetic fields is given by MORRISON in a previous Handbuch article [5] and will not be repeated here.

Since many observations consist of the measurement of secondary particles at ground level, it is necessary to understand how to relate these particles first to the primary cosmic rays at the top of the atmosphere and then, secondly, to be able to follow the primary particles along their trajectories through the geomagnetic field out into interplanetary space. The following section is therefore devoted to a study of geomagnetic effects and cosmic ray specific yield functions.

II. Geophysical effects of cosmic radiation.

a) Geomagnetic threshold rigidities.

2. Definition of a threshold rigidity. At any point on the Earth's surface, one can define a threshold rigidity for cosmic rays arriving at a particular zenith and azimuth angle. Below this rigidity, particles are excluded by the action of the geomagnetic field. The magnetic rigidity of a particle is the momentum-to-charge ratio and has the dimensions of voltage. It is usually written in units of GV, equal to 10^9 volts. Threshold rigidities are high near the equator but reduce towards zero at high latitudes and thus the cosmic ray intensity increases at high latitudes.

3. The dipole approximation. From a spherical harmonic analysis of the surface geomagnetic field elements, it is found that the chief contribution to the field comes from harmonic terms which can be represented by a dipole situated at the centre of the Earth [CHAPMAN and BARTELS [6]]. According to the 1955 magnetic survey, this dipole has a magnetic moment of 8.1×10^{25} gauss cm^3 and is tilted so that the magnetic axis intersects the Earths surface at a geographic latitude of 78.2° N and a geographic longitude of 69.0° W. A better representation of the geomagnetic field as a whole is obtained by using instead an eccentric dipole which takes into account some of the $n=2$ terms of the spherical harmonic analysis. These terms have been neglected in the centre dipole approximation. In 1955, the eccentric dipole was situated 433 km from the geographic centre of the Earth at 15.3° N latitude and 151.1° E longitude (*1*). The magnetic moment and angle of tilt were the same as from the centre dipole. STÖRMER [7] has made an extensive study of the motion of charged particles in a dipole field in connection with work on auroral particles, and this has been applied to cosmic ray motion in the geomagnetic field. To arrive at the threshold rigidity, STÖRMER found the following analytical integral of the equation of motion which describes particle trajectories in the meridian plane containing the moving particle and the dipole axis:

$$2\gamma = R \cos \lambda \sin \theta + \frac{\cos^2 \lambda}{R} \qquad (3.1)$$

γ is a constant proportional to the particle impact parameter about the dipole axis when at infinity, λ is the geomagnetic latitude, and θ is the angle between the velocity vector of the particle and the meridian plane and is positive if the particle crosses from east to west. R, the radial distance from the dipole, is measured in Störmer units, where one STÖRMER is equal to $\sqrt{\frac{M}{P}}$. M is the dipole moment and $P = \frac{pc}{Ze}$ and is the momentum-to-charge ratio or particle magnetic rigidity.

At a particular value of γ, areas in the meridian plane where $|\sin \theta| > 1$ are forbidden to the particle, while the boundary between allowed and forbidden regions is obtained by putting $|\sin\theta|=1$ in Eq. (3.1). In general, there is an inner and an outer allowed region and particles close to the threshold must penetrate the narrow pass between these two regions in order to reach the Earth (Fig. 1). The shape of the allowed regions depends on γ and the critical value where the pass point just closes is $\gamma=1$. Using this value of γ in Eq. (3.1) and solving for particles arriving vertically at the Earth's surface, it is found that

$$P = \frac{M}{4 r_e^2} \cos^4 \lambda = 14.9 \cos^4 \lambda \text{ GV} \qquad (3.2)$$

gives the vertical threshold rigidity where r_e is the Earth's radius.

According to Störmer theory, all rigidities above the threshold are allowed and we can easily compute the particle intensity if we assume an isotropic particle flux, at all rigidities, very far from the dipole. LIOUVILLE's theorem for an electromagnetic field states that the particle directional intensity I is given by

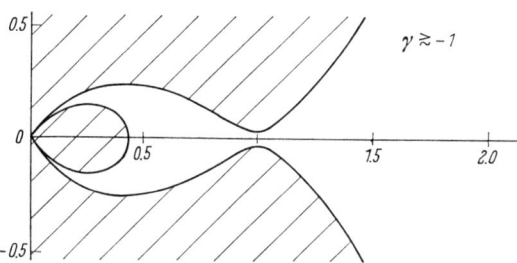

Fig. 1. Allowed regions (unshaded) and forbidden regions (shaded) for particle motion in the meridian plane under a dipole field. Distances are given in Störmer units.

$$I = \frac{p^3}{m} D \qquad (3.3)$$

where p and m are respectively the relativistic momentum and mass of a typical representative particle in a small group in phase space and D is a constant. I is typically measured in units of particles per cm² sec ster GV. Since the geomagnetic field is static, I remains constant along particle trajectories, and the directional intensity above the threshold rigidity is equal to the directional intensity at infinity.

The simple theory outlined so far takes no account of the effect of the impenetrable Earth. Detailed integration of trajectories of particles just above the threshold by VALLARTA and his colleagues shows that some trajectories intersect the Earth elsewhere before arriving in a direction allowed by Störmer theory [8]. These penumbral trajectories contain loops representing a turning away from and a turning towards the dipole. At higher rigidities, the main cone threshold is reached, and above this value all trajectories come from infinity without loops and can not be obstructed by the Earth in this way. The penumbral region between the STÖRMER and main cone thresholds is completely forbidden for vertical arrival between the equator and 20° geomagnetic latitude, but at higher latitudes alternate bands of allowed and forbidden rigidities are found and the penumbra becomes increasingly trans-

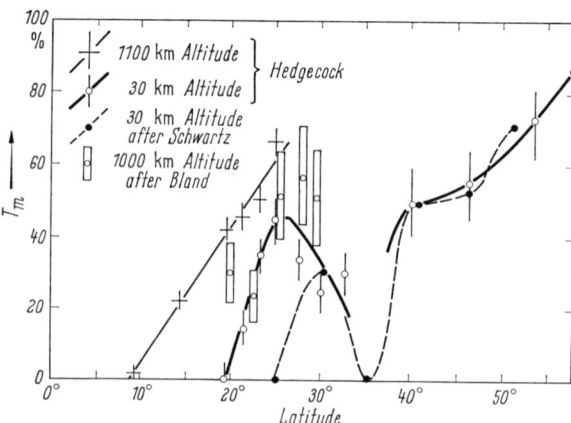

Fig. 2. Percentage transparency of the penumbra as a function of geomagnetic latitude. After HEDGECOCK (4).

parent. Schwartz (2) has extended Vallarta's transparency calculations to high latitudes, and Bland (3) and Hedgecock (4) have determined the transparency using a model of the dipole field in a vacuum tank with an electron beam to represent the cosmic rays. Fig. 2 shows these results in a plot of transparency against geomagnetic latitude. There is good agreement between Hedgecock's results applicable to an altitude of 30 km and the combined calculations of Vallarta and Schwartz, except around 25° where few or no numerical computations have been made. Both Bland and Hedgecock find an increase in transparency with altitude above the surface of the Earth. Knowing the penumbral width, an effective correction to the Störmer threshold can be estimated taking into account the percentage transparency, and this is shown in Fig. 3.

An additional effect of the solid Earth is that even uncomplicated trajectories, arriving at zenith angles close to the horizon, may be intercepted. This effect gives rise to the simple shadow correction and has been investigated by Kasper (5) for both sea-level and satellite altitudes. In general, it is a small correction which raises the horizon for cosmic rays above the geographical.

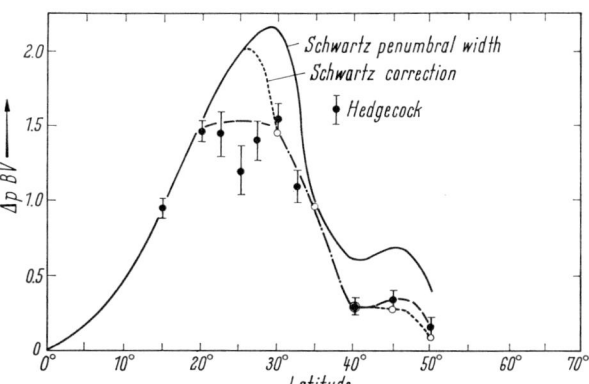

Fig. 3. Effective penumbral correction to the Störmer threshold rigidity. After Hedgecock (4).

So far the discussion has assumed a centre dipole approximation to the geomagnetic field, but the results can be applied to an eccentric dipole by making geometrical corrections for the shift of the dipole centre. Kodama et al. (6), for example, calculate the Störmer thresholds appropriate to the eccentric dipole.

4. Corrections to the threshold rigidities due to the non-dipole part of the field.

Discrepancies between the measured cosmic ray intensity distribution and that expected from the dipole representation of the geomagnetic field were first noticed by Johnson [9] and have been very clearly demonstrated by the neutron monitor latitude surveys of Simpson et al. (7). At a longitude of 30° W, the minimum in the cosmic ray latitude curve was found to be 9° north of its expected position on the geomagnetic equator. Other experiments have confirmed the world-wide nature of these discrepancies[1]. It was originally suggested (7) that the discrepancies arose as a result of the interaction between the geomagnetic field and the interplanetary plasma distorting the field. Although it is now known that interaction between solar plasma streams and the field is responsible for some changes in the threshold rigidities, the most important perturbation to cosmic ray trajectories arriving at low latitudes was shown by Rothwell and Quenby (8) to be due to the non-dipole part of the internal field.

The magnetic potential, V, of the Earth may be represented by a series of spherical harmonics. That part of V due to fields of internal origin may be written as

$$V = \sum_{n=1}^{\infty} r_e \left(\frac{r_e}{r}\right)^{n+1} \sum_{m=0}^{n} P_n^m(\theta) \left(g_n^m \cos m\Phi + h_n^m \sin m\Phi\right), \tag{4.1}$$

---
[1] For a comprehensive list of references, see [11].

where r is the distance from the centre of the Earth, θ is the geographic co-latitude, Φ the geographic longitude, g_n^m and h_n^m are the Gauss coefficients, and $P_n^m(\theta)$ are the partly normalized spherical harmonics introduced by SCHMIDT [9]. The term $n=1$ in this series corresponds to the centre dipole, tilted relative to the geographic axis. Using the values of g_n^m and h_n^m appropriate to the 1955 field survey [2], the root mean square values of V_n, averaged over the whole spherical surface, can be calculated. These higher order contributions to V are found as percentages of the $n=1$ or dipole term and are then tabulated (Table 1) for different distances r.

Table 1. *Relative importance of various spherical harmonic terms (% of $|V_1|$)*

Distance	$|V_2|$	$|V_3|$	$|V_4|$	$|V_5|$	$|V_6|$	$\sum_{n=2}^{6}|V_n|$
$1.0\, r_e$	10.4	5.9	2.8	0.9	0.4	20.4
$1.2\, r_e$	8.7	4.1	1.6	0.4	0.2	15.0
$1.5\, r_e$	6.8	2.6	0.8	0.2	0.1	10.5
$2.0\, r_e$	5.2	1.5	0.3	0.1	<0.1	7.0
$3.0\, r_e$	3.5	0.7	0.1	<0.1	<0.1	4.2

An approximate correction to the Störmer threshold rigidities which takes into account the non-dipole terms from $n=2$ to $n=6$ has been given by QUENBY and WEBBER [10]. Using these corrected thresholds, a much improved fit is obtained to both the measured cosmic ray equator (Fig. 4) and the intensity distribution resulting from the energetic solar particle event of 23 February 1956. At the equator the correction was found by following STÖRMER'S method for integrating the equation of motion but with the inclusion of the magnetic potentials due to the higher order terms. It was necessary to neglect the θ and Φ dependence of these terms and take into account only the fall-off with radial distance. For high latitudes, it was assumed that particles with rigidities close to the threshold followed lines of force into the Earth. Far out, the particle followed a dipole field line (EP' Fig. 5) but close in the non-dipole terms caused the field line, and hence the particle, to deviate along a path EP.

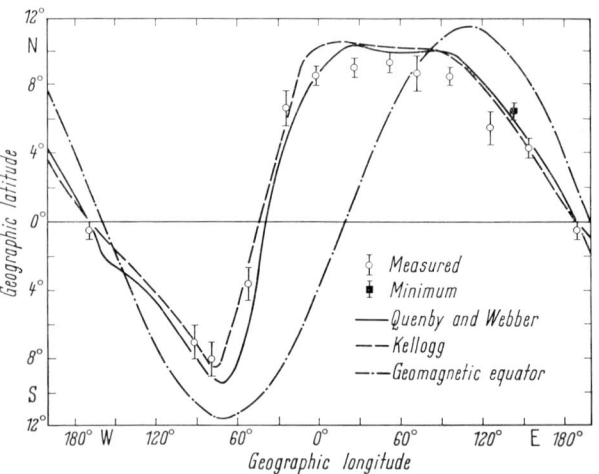

Fig. 4. Positions of minimum cosmic ray intensity measured at aeroplane altitudes (references given in [8]) compared with predicted cosmic ray equator.

The latitude $\bar{\lambda}$ of P', the arrival point of the unperturbed dipole line, is defined as the effective latitude of the actual point of arrival, P. The threshold rigidity, P_T, at P is then said to be

$$P_T = \frac{M}{4 r_e^2} \cos^4 \bar{\lambda} \tag{4.2}$$

since the particles start out along path EP as if they were going to arrive at a latitude $\bar{\lambda}$ in a dipole field.

QUENBY and WEBBER estimated $\bar{\lambda}$ from the values of the surface field elements, but QUENBY and WENK [11] were able to improve the approximation employing computations of the paths of magnetic field lines. These latter authors also demonstrated that a penumbral correction was important at low latitudes in the real

Sect. 4. Corrections to the threshold rigidities due to the non-dipole part of the field. 315

field of the Earth — a result confirmed by the model experiment of BLAND (3). The penumbra had been neglected in the previous approximation. Using the field line computations at high latitudes and the intensity distribution of the nucleonic component measured around the equator by KATZ, MEYER and SIMPSON (11), QUENBY and WENK constructed a new table of thresholds. The successive refinements in the approximation are illustrated in Table 2 for the flare event of 23 February 1956. This table gives the root mean square deviations, in percentage of rigidity, of the individual points from the curve of best fit to a graph of intensity increase versus threshold rigidity. Observations were taken at a time when the spatial distribution of solar protons was thought to be isotropic.

RAY (12) has developed an approximate theory which leads to the expression

$$P = \frac{M}{4L^2} \quad (4.3)$$

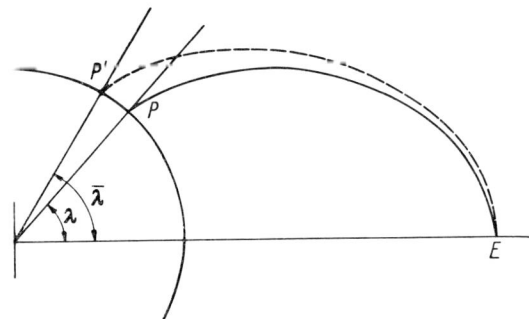

Fig. 5. Unperturbed dipole field line, EP, between the equatorial plane and the Earth, and a perturbed field line, EP'.

valid for $L > 2r_e$, for the threshold rigidity where L is the McIlwain (12) parameter describing the motion of trapped radiation. This expression provides a reasonable representation of satellite data on the cosmic ray intensity distribution (14). If we recall McILWAIN's definition of L, we can see a marked similarity in expression (4.3) and expression (4.2) involving the QUENBY-WEBBER effective latitude. For a dipole field, McILWAIN showed that

$$\frac{R_0^3 B}{M} = F\left(\frac{I^3 B}{M}\right), \quad (4.4)$$

where B is the magnitude of the field at a point A, R_0 is the distance at which the line of force through A cuts the equatorial plane, and I is the second adiabatic invariant given by

$$I = p^{-1} \oint_A^{A'} p_\parallel \, ds. \quad (4.5)$$

Table 2. *Root mean square deviation in percentage of threshold rigidity for solar flare of 23 February 1956.*

Threshold approximation		Neutron stations (14)	Ion chamber stations (11)
P_E	Eccentric dipole	12.6	14.3
	Eccentric dipole + dipole penumbra	6.6	11.7
P_M	Quenby-Webber	7.5	5.3
	Quenby-Webber + dipole penumbra	4.7	5.3
P_T	Quenby-Wenk	3.5	4.4

The integral over all elements of path length ds is taken between the mirror points A, A' of a particle with total momentum p and with a component of momentum $p_\parallel$ along the field line. For the real field, L is defined by

$$\frac{L^3 B}{M} = F\left(\frac{I^3 B}{M}\right), \quad (4.6)$$

where I is determined by numerical integration. In practice, L is found to be approximately constant along field lines. At high L values and large radial distances, the real and dipole fields approach each other (see Table 1) if external current systems are neglected. Thus, as can be seen by comparing Eqs. (4.4) and

(4.6), $L \sim R_0$ and it is also the distance between the dipole and point E in Fig. 5. Thus computation of L for some point close to the Earth, P, has identified where the unperturbed, dipole line of the Quenby-Webber formulation cuts the equatorial plane.

The equation for a dipole field line is

$$r = R_0 \cos^2 \lambda. \tag{4.7}$$

Thus with $r = r_e$, $R_0 = L$

$$r_e = L \cos^2 \bar{\lambda} \tag{4.8}$$

where $\bar{\lambda}$ is effective latitude. Substituting this last equation in Eq. (4.2) we get

$$P = \frac{M}{4 L^2}, \tag{4.9}$$

which is the expression given by RAY (4.3). The inherent identity between QUENBY and WEBBER's work on cosmic ray thresholds and McILWAIN's formulation of the motion of trapped particles was demonstrated by ELLIOT [12]. During the course of Project Argus, three small fission bombs were exploded at high altitude, so that electrons from the beta decay of fission fragments were injected into the magnetic field, forming a narrow shell of trapped radiation. The points in Fig. 6 represent the projections of the Argus III shell along the magnetic field lines onto the Earth's surface, while the line drawn through them is the 3.2 GV threshold rigidity contour of QUENBY and WENK [11]. The good agreement between the line and points, taken together with the fact that McILWAIN has shown L to represent well the Argus III shell, demonstrates the identity of the two different approaches.

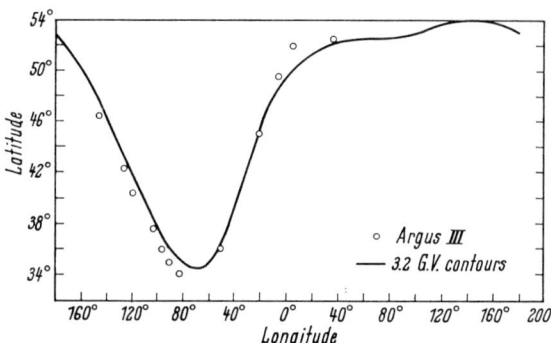

Fig. 6. Comparison of Argus III shell and the 3.2 GV contour of vertical threshold rigidity [8]. After ELLIOT [12].

Although the approximate modifications to Störmer theory discussed above have achieved considerable success in representing cosmic ray data, some discrepancies remain. POMERANTZ and AGARWAL (15) have measured the nucleonic intensity at an atmospheric level of 680 g/cm² over many parts of the world. They find that the QUENBY-WENK threshold rigidities can be as much as 2 GV too low over the north Atlantic near the Canary Islands. It is, of course, possible to improve the theoretical calculations by actually integrating particle trajectories. The technique is to follow the motion outwards of a negative particle of the same rigidity as that of the required incoming positive particle. When the rigidity is increased to a point where the particle escapes, the threshold has been reached. Further increase of rigidity will reveal any penumbral structure. A disadvantage of this method is that several trajectories must be integrated for each position and direction of arrival.

KELLOGG (16) has found the position of the cosmic ray equator by computation (Fig. 5), while McCRACKEN (17) has developed a widely used programme for trajectory integration. Using this programme, KONDO, KODAMA and MAKINO (18) have obtained much better agreement between the nucleonic intensity and the thresholds than had previously been achieved for measurements on a route between Japan and the Antarctic south of Africa. Some remaining discrepancy between the intensities north and south of the equator may be due to inadequate knowledge of the higher order field terms.

Threshold rigidities in the region 1 GV to 2.5 GV may be determined experimentally by measuring the primary proton or alpha spectrum directly with detectors carried in balloons to a few g/cm^2 of the residual atmosphere. Although reasonable agreement with the theoretical thresholds is found above 2 GV, anomalies exist in the region 1.5 GV or less (WEBBER [2]). During one flight at 45° N, 95° W, McDONALD and WEBBER (19) measured particles down to 0.7 GV, while the QUENBY and WENK threshold at the point is 1.39 GV. Further, in confirmation of the discrepancy, FOSTER and KLARMANN (20) have computed trajectories at 44° N, 94° E and find that below 1.3 GV particles are probably excluded. The region between 1.3 GV and 1.5 GV was indeterminate but higher rigidities are allowed. QUENBY and WEBBER (WEBBER [2]) have suggested that the true experimental threshold represented by the experimental results is 1.25 GV, and various experimental errors contribute to an apparent spread below this value. This interpretation is uncertain since the true primary spectrum existing outside the geomagnetic field was not well-known at the time. Clearly, a simultaneous measurement of the primary spectrum at two different geomagnetic latitudes is required in order to obtain the threshold at the lower latitude; otherwise, one cannot determine whether the lowest rigidity particles arriving represent the full external flux or the tail of some distribution of experimental measurements. FRIEDLANDER and SPRING (21) have made one such simultaneous measurement. Another similar series of measurements made during solar proton events will be mentioned in the following section.

b) Effect of magnetic fields due to external current systems.

5. An alternative explanation of the discrepancy between the measured and the calculated threshold rigidities for values <1.5 GV is that the thresholds are permanently lowered due to the action of external current systems. It is now established that during some magnetic storms, at least, the thresholds can be lowered to let in particles below the quiet-time threshold, and we shall proceed to discuss the occurrence of, and reasons for, these events.

6. Observations of threshold decreases. Experimental studies suggesting the lowering of geomagnetic threshold rigidities were made on the solar proton event of 11 May 1959 by NEY, WINCKLER and FRIER (22) and REID and LEINBACH (23). The latter authors studied the results of cosmic noise absorption measurements made by several riometer stations. During the event, enhanced ionisation over the polar cap was observed characteristic of the influx of solar protons in the 20 MeV to 200 MeV energy range. At effective latitudes greater than 70°, the absorption built up to a maximum early on 12 May and then decayed slowly away on 13 May. Stations at latitudes < 66° exhibited a rapid increase of absorption at about 0200 to 0300 U.T. on 12 May in association with the onset of the main phase of a magnetic storm. The absorption decerased equally rapidly towards the end of 12 May.

During the period 0400 to 1600 U.T. on 12 May, Ney, WINCKLER and FRIER observed an anomalous increase in particles, measured at balloon altitude, over Minneapolis. Their results suggested that the threshold at Minneapolis had been reduced from the quiet time value of 1.2 GV [FRIER (24)] to less than 0.5 GV. Such a reduction would explain the simultaneous increase of cosmic noise absorption at relatively low latitudes.

Further evidence for the relationship between the magnetic field changes during the main phase of a magnetic storm and the reduction in threshold rigidities is given by WINCKLER, BHAVSAR and PETERSON (25) for the events of July 1959. Cosmic noise absorption measurements at College, Alaska, indicated the continued presence of solar protons from three separate flares during the period 10 to 18 July, but at Minneapolis only short-lived bursts of particles were detected at balloon altitude. One such event is illustrated in Fig. 7, where we see a sudden increase above the galactic cosmic ray rate for an ion chamber flown over Minneapolis on 15 July. The increase, lasting about 5 hours, started 30 minutes after the sudden commencement of the storm but at the same time as the beginning of the main phase decrease in horizontal field intensity. An additional observation by these authors is of several half-hour period oscillations in the intensity at Minneapolis, which may indicate periodic changes in the geomagnetic field or may be characteristic of particle motion in the near interplanetary field. Similar oscillations, with a period of 75 minutes, were noticed by MATHEWS, THAMBYAHPILLAI and WEBBER (26) during the later stages of the 12 November 1960 event while the geomagnetic thresholds were lowered.

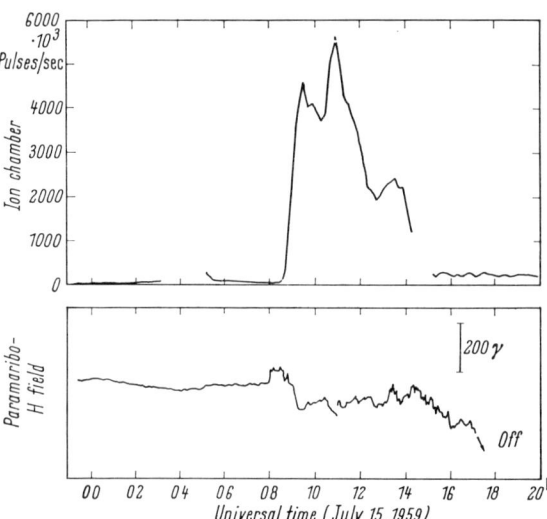

Fig. 7. Comparison of the counting rate of an ion-chamber carried by a balloon over Minneapolis and sea-level variations in the horizontal component of the magnetic field. After WINCKLER et al. (25).

Simultaneous measurements of the solar proton spectra, using nuclear emulsions, made on balloon flights at Minneapolis and at Churchill, Canada, during 11 and 12 July 1959 enabled FRIER (24) to determine the shape of the threshold at Minneapolis as a function of time. The results may be fitted by an exponential dependence of particle intensity, I, on the rigidity difference, $P_c - P$, from the rigidity, P_c, at which 100% intensity, I_0, is seen.

$$I = I_0 \exp\left(-\frac{2.3(P_c - P)}{0.15}\right), \qquad (6.1)$$

is given by FRIER for Minneapolis. It is suggessted that details of the penumbra between the main cone threshold and the Störmer threshold produce this shape, rather than experimental errors such as balloon drift to the north or south. The quiet-time value of P_c is ~1.2 GV. Since this would correspond to the main cone threshold while the predicted Störmer threshold at Minneapolis is 1.34 GV, FRIER's explanation would suggest a quiet-time lowering of the Minneapolis threshold.

Satellite measurements have also revealed storm-associated decreases in threshold rigidities. Three such events in July 1961 are analysed by PIEPER, ZMUDA, BOSTROM and O'BRIEN (27). During this period results were obtained from a silicon junction detector carried in the satellite Injun I, which had an orbital inclination of 67°. The $p-n$ detector was sensitive to protons in the energy range 1 MeV to 15 MeV, and thus positions where the flux of particles of rigidity

0.04 GV to 0.16 GV become zero could be determined provided a flux of solar protons in this range was present in space close to the Earth. Fig. 8 shows the flux of solar protons as a function of L for the Injun I observations during the magnetic storm of 13 July 1961. From 1717 U.T. 12 July to 0222 U.T. 13 July the solar protons only began to appear within a threshold region of $4.8 < L < 6.1$, while the expected Störmer thresholds at these L values are 0.62 GV to 0.40 GV respectively. Thus even before the storm the thresholds were reduced. After the sudden commencement at 1115 U.T. 13 July, the threshold L value dropped to $L=4.4$, and during the main phase a bigger reduction to $L=3.0$ to 3.5 occurred. During the later part of the main phase, the thresholds recovered to their prestorm value. Similar results were obtained during three other storm events. There was insufficient data to determine whether any significant local time dependence of the pre-storm threshold reductions occurs.

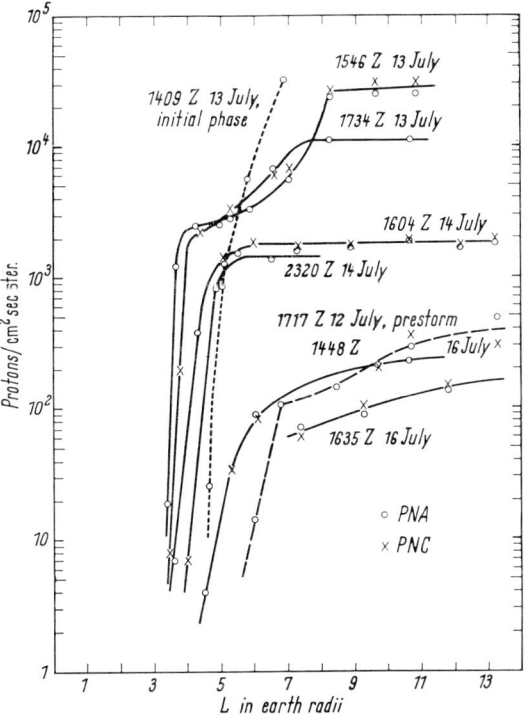

Fig. 8. Solar proton fluxes measured by the $p-n$ junction detector on Injun I during the magnetic storm of 13 July 1961. After PIEPER et al. (27).

Reductions in threshold rigidities during solar proton events were observed by Explorer VII, which carried Geiger counters sensitive to protons of 18 MeV and 30 MeV minimum energy respectively. AKASOFU, LIN and VAN ALLEN (28) show on the basis of these results that during magnetically quiet times the 0.5 GV threshold value at $L=5.5$ is reduced to 0.24 GV, while during magnetic storms particles of 0.24 GV can enter as far as $L=3.5$, corresponding to a threshold due to the internal field alone of 1.2 GV. Fig. 9 relates the minimum L value of arrival for 0.24 GV protons to the planetary three-hourly value of the change

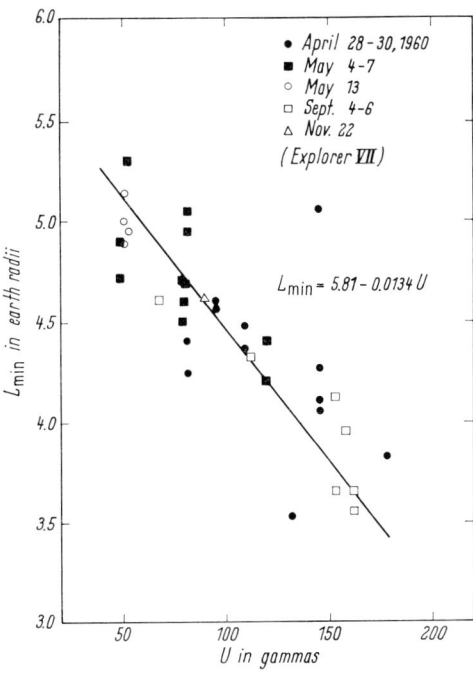

Fig. 9. Minimum L value for the arrival of 0.24 GV protons plotted against the planetary 3-hourly value, U, of the change in the horizontal field component. After AKASOFU et al. (28).

Fig. 9.

in the horizontal component of the Earth's surface field U. Although both the Injun I and the Explorer VII detectors were sensitive to alpha-particles of a higher magnetic rigidity, the 5% of alpha-particles in the solar radiation is unlikely to have affected noticeably the interpretation of the results on the minimum L value of arrival.

Recent satellite results obtained by STONE (28a) during a magnetically quiet time showed that solar protons of 1.5 MeV could reach the earth down to an effective latitude of 65° on the night side and 67° on the day side, the corresponding L values being 5.6 and 6.6 respectively. At higher latitudes throughout the polar cap, the thresholds were always found to be lower than 1.5 MeV.

Threshold rigidity changes have been detected at ground level during three events. KONDO, NAGASHIMA, YOSHIDA and WADA (29) have found during the Forbush decreases of 11 September 1957 and 9 February 1958 increases in intensity lasting a few hours and occurring soon after the initial large decrease. These increases correlate in time with the maximum depression of the horizontal component of the geomagnetic field. DORMAN (3) has analysed these events to show that the increase is longitude- as well as latidude-dependent, the greatest effect occurring at a threshold of about 6 GV on the day side of the Earth. The interpretation of these events was complicated by a large storm-time daily variation in the cosmic ray intensity, shifted in phase towards morning hours.

Analysis of the 12 November 1960 solar proton event by HATTON and MARSDEN (30) has revealed an asymmetric depression of the thresholds during the initial stages of the main phase of the magnetic storm which conicided with the event. For stations with a normal threshold of around 2 GV, the maximum reduction occurred on the night side of the Earth, the average reduction being about 0.5 GV.

7. The distant geomagnetic field.

STÖRMER [7] long ago postulated the existence of an equatorial ring current flowing in the westward direction far out to explain the position of the auroral zone. The discovery of charged particles trapped in the geomagnetic field meant that some contribution to the field, even at geomagnetically quiet times, must come from such an external current system. It is found that, out of all the observed trapped particles, protons in the energy range 150 KeV to 4.5 MeV provide the greatest contribution to the external field. Based on the proton measurements of DAVIS and WILLIAMSON (31), AKASOFU, CAIN and CHAPMAN (32) have calculated the distortion of the field due to the ring current distribution. They showed that the Earth's field was reduced inside 6.7 Earth radii and enhanced outside this distance, and that at the Earth's surface the equatorial field was reduced by 38γ. Far out, the effect was equivalent to a dipole 0.17 of the strength of the internal field moment. These calculations neglected the effect of the field changes on the motion of the trapped particles themselves.

Magnetic field measurements by Pioneer V on the day side and by Explorer VI on the evening side of the Earth are consistent with a ring current due to a particle distribution peaked at 6 Earth radii (33), (34), compared to the peak in the measured proton distribution at 3 to 4 Earth radii. However, Explorer X (35) and Explorer XIV (36) on the night side and Explorer XII (37) on the day side failed to find a ring current field at 6 Earth raii. The Explorer X results did indicate a possible weak ring current at less than 3 Earth radii. But all these last three Explorer measurements were consistent with the model of the distorted geomagnetic cavity in which the solar wind compresses the geomagnetic field on the day side while pulling out the field lines on the night side into an extended tail [e.g. JOHNSON (38)]. FRANK and VAN ALLEN (39) have shown that low energy trapped electrons exhibit a day-night asymmetry in their spacial distribution in general accordance with this model (Fig. 10). During quiet times the boundary of the geomagnetic cavity is located at about 10 Earth radii in the equatorial plane in the Earth-Sun direction.

Satellite data on the storm-time geomagnetic field is scanty. There are indications that the low energy proton flux increases by up to a factor of three during a storm, and changes in the field strength which correlate with the average surface field changes have been seen by Explorer VI. These field decreases are in qualitative agreement with the field changes calculated for a threefold increase in ring current intensity (40).

From this brief discussion of magnetometer evidence for the distant geomagnetic field configuration, we conclude that a bounded cavity, asymmetrically shaped with respect to the Earth-Sun line, together with a ring current located at 3 to 4 Earth radii, is a possible model for the quiet-time field. During magnetic storms, the ring current intensity is enhanced. An alternative model suggested for the main phase of the magnetic storm is that the Earth is enveloped by a large-scale uniform magnetic field, directed so as to oppose the equatorial geomagnetic field and brought into the vicinity of the Earth by solar plasma (29). Interaction of the solar wind and the geomagnetic field is neglected in this model.

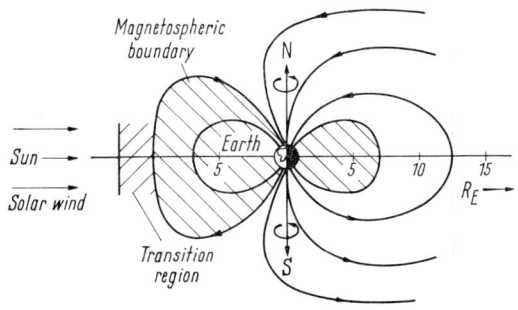

Fig. 10. Regions in the magnetosphere where 40 KeV electrons are trapped. After FRANK and VAN ALLEN (39).

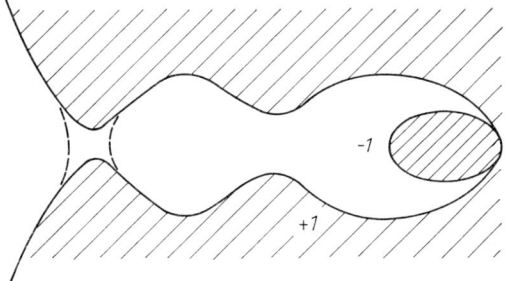

Fig. 11. Allowed and forbidden (shaded) regions in the meridian plane for particle motion under a dipole plus ring current field.

8. Calculation of threshold rigidity changes.
Simple solutions for the changes in the threshold rigidities can only be obtained for models involving an external perturbation to the geomagnetic field which is symmetrical about the dipole axis. In these cases, the external field may be represented by a magnetic potential, A'_Φ, which has a component only in the Φ direction, Φ being geomagnetic longitude Then STÖRMER's integral of the equation of motion may be written

$$b = r \cos \lambda \sin \theta + \frac{(A_\Phi + A'_\Phi)}{P} r \cos \lambda, \tag{8.1}$$

b being a constant, A_Φ being the magnetic potential of the dipole $\left(A_\Phi = \frac{M \cos \lambda}{r^2}\right)$, and the other symbols having the same meaning as in (3.1).

α) *Ring currents.* RAY (41) has considered the effect of a westward filamentary ring current, located in the equatorial plane, on the thresholds. He demonstrates that the meridian plane is divided up into allowed and forbidden regions by putting $\sin \theta = \pm 1$ in Eq. (8.1) as shown in Fig. 11. As in Störmer theory, the shape of this diagram is a function of b and the threshold is determined by the condition that the inner and outer allowed regions are just separated by the closure of a pass point. For low rigidity particles the critical value of b which just closes the outer pass point between forbidden regions must be found, while at higher rigidities it is the inner pass point which must be closed. Substitution of this critical value of b back in Eq. (8.1) then yields the threshold rigidity. Fig. 12 shows the results of RAY's calculation for a ring current located at 7.5 Earth radii which produces a change $\Delta H = -150\,\gamma$ in the surface geomagnetic field at the equator. Here, the percentage change in threshold from the dipole value is plotted as a function of latitude. HEDGECOCK (42) has used RAY's method to compute the threshold drop using the ring current configuration suggested by TREIMAN (43), and these results also are shown in Fig. 12. TREIMAN's current flowed on a spherical surface, the current density being proportional

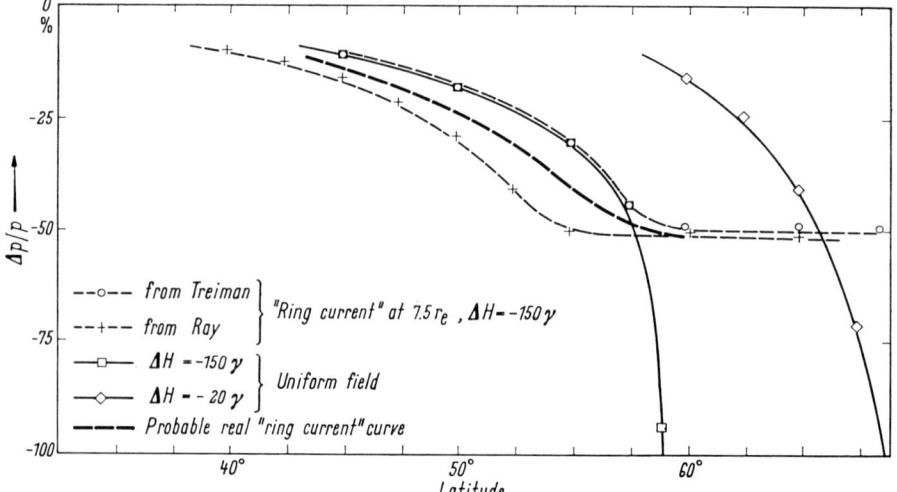

Fig. 12. Reduction of Störmer threshold rigidities due to various types of external fields. After HEDGECOCK (42).

to $\cos \lambda$. In this way, a uniform field opposing the Earth's field is produced inside the ring and a dipole field outside the ring $\left(A'_\Phi = \dfrac{M' \sin \theta}{r^2},\ r > \text{ring radius}\right)$. HEDGECOCK suggests that a current flowing in the manner calculated by AKASOFU, CAIN and CHAPMAN would produce threshold changes intermediate between the Ray and Treiman models. KELLOGG and WINCKLER (44) also have considered the effects of ring currents and show that at very high latitudes the Störmer thresholds P_0 are reduced to a value P where

$$P = \frac{P_0}{1 + \dfrac{M'}{M}}, \qquad (8.2)$$

M is the dipole moment and M' the ring current moment. This expression corresponds to the flat portion of the curves in Fig. 12. AKASOFU and LIN (45) have computed the magnetic moments of the ring currents corresponding to several models of the trapped particle distribution, and in no case do they find the ratio M'/M exceeding unity. Indeed, it seems physically implausible that the field due to trapped particles should exceed the trapping field at large distances from the Earth, as would be the case if $M'/M > 1$. This probable limit to M' has led AKASOFU, LIN and VAN ALLEN (28) to exclude ring currents as being the sole cause of threshold reductions. From Eq. (8.2) the maximum change of P_0 at high latitudes can only be 50% when $M'=M$, whereas storm-time reductions in P_0 down to 20% of the initial value have been cited in the experimental evidence discussed previously. Although magnetically-quiet time thresholds have been observed at up to 50% of the unperturbed value, the quiet-time ring current is estimated to correspond to a moment of $M'=0.17 M$ and not $M'=M$ which would be required for complete consistency.

β) *Uniform external field.* The author has considered the case of a field due to the sum of the dipole field and a large-scale and uniform external field orientated parallel or anti-parallel to the dipole axis. A vector potential $A'_\Phi = -\dfrac{H_z r \cos \lambda}{2}$ is used where H_z is the northward-pointing external field. The vertical threshold under these combined fields may be evaluated using a method similar to RAY's yielding the following expression for the modified value when the change is small:

$$P = \frac{M}{4 r_e^2}\left[1 + \frac{M^{\frac{1}{2}} H_z}{2 P^{\frac{3}{2}}}\left(1 - \frac{\cos^6 \lambda}{4}\right)\right]\cos^4 \lambda. \tag{8.3}$$

If H_z is southward-pointing, the threshold goes to zero at a latitude given by

$$\cos^2 \lambda = \frac{\tfrac{3}{2} M^{\frac{1}{2}} H_z^{\frac{1}{2}}}{M^{\frac{1}{2}} r_e^{-1} + \tfrac{1}{2} H_z r_e^2 M^{-\frac{1}{2}}}. \tag{8.4}$$

At latitudes where the threshold change is large, numerical methods must be used to find the value of the integration constant, b, which just closes the jaws of the forbidden regions. Fig. 12 shows the reductions calculated by the author for a 20 γ southward-pointing external field and by HEDGECOCK (*42*) for a similar field of 150 γ strength. HEDGECOCK has also partially checked these calculations with a model experiment. Results which are very similar have been obtained by OBAYASHI (*46*), who has superimposed external fields which make part of the dipole field vanish at a fixed distance, R_0, from the Earth. The two conditions used to determine the external magnetic potential were that the dipole horizontal field or, alternatively, the vertical field vanished at R_0 at all latitudes. The first case is similar to making H_z positive in the above calculation and the second corresponds to making H_z negative.

Although a uniform external field which reduced the surface equatorial field by 50 γ would explain the quiet-time reduction in the thresholds seen by Injun I, satellite and space probe measurements give no indication that such a field exists over a long period. Further, the interaction of the solar wind and the magnetosphere should compress the field, producing an effect more akin to a superposition which enhances the dipole field and hence the threshold rigidities. It has been suggested that an opposing external field of 300 γ arriving at the time of a magnetic storm could produce measurable increases in the sea-level neutron monitors in the course of a Forbush decrease. However, Fig. 12 shows that a ring current, which causes the same change in the surface field as does the external uniform field, will produce similar changes in low latitude thresholds and therefore similar increases in counting rate. While there is positive satellite evidence for the enhancement of the ring current during magnetic storms from Explorer VI field measurements and Explorer XII particle measurements, there is no indication yet of the presence of the uniform field during these events. Even if solar plasma did carry a suitably oriented field to the vicinity of the Earth, it has still to be demonstrated that such a field could penetrate the plasma trapped in the internal field.

γ) *Combination of a ring current and a bounded magnetosphere.* So far it has been shown that although a ring current probably exists in the magnetosphere it is unable to cause all the observed threshold changes. AKASOFU, LIN and VAN ALLEN (28) demonstrate that a field configuration combining the effect of a ring current and the termination of the geomagnetic cavity by the solar wind will produce roughly the required decreases. Plasma probe results from Mariner II (47) indicate the continued presence of conducting gas streaming from the Sun, while Explorer XII field measurements show a definite boundary to the geomagnetic field at times both of geomagnetic quiet and of geomagnetic activity. As a first approximation to the effect of the currents set up in the interface, the CHAPMAN and FERRARO (48) image dipole model is used. If the magnetospheric boundary on the Sun side of the Earth is located at a distance r_B from the centre of the Earth, the image dipole is situated at $2r_B$ in the equatorial plane. The dipole strength of the image is $f(M+M')$ where M' is the ring current moment and $f=1$ in a two-dimensional approximation to the interface interaction. This image causes the horizontal equatorial field just inside the interface to double in accord with the boundary condition given by DUNGEY [13]. However, outside the boundary, where the interplanetary field is dominant, it can give a completely wrong field representation. The two parallel dipoles produce neutral points north and south of the equator in the N-S plane through the point r_B in the equatorial plane. Thus there is a high latitude position where particles of zero rigidity can enter freely and reach the Earth. However, at lower latitudes the effect of the image is similar to that of a northward-pointing external field, and so the thresholds tend to be increased by the image field and reduced by the ring current field.

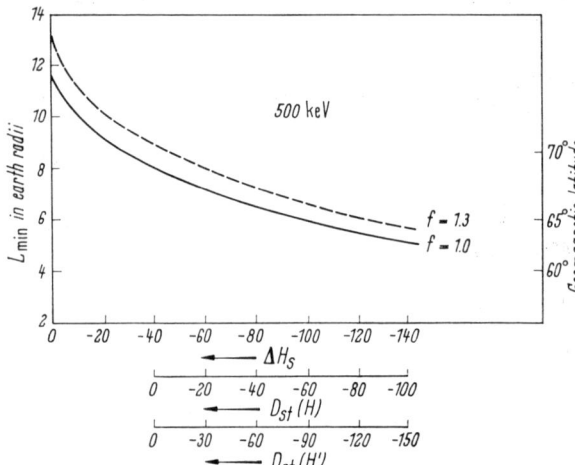

Fig. 13. Minimum L value at the Earth's surface for the arrival of 500 KeV protons as a function of the disturbance to the horizontal component of the field at the equator. After AKASOFU et al. (28).

AKASOFU, LIN and VAN ALLEN obtain approximate values for the reduced thresholds under this model by neglecting the asymmetries of the magnetosphere, and they therefore assume that the field configuration in the Sun-Earth plane applies at all longitudes. An isotropic particle flux is allowed to impinge on the magnetosphere boundary. Very low energy particles (~500 KeV) just penetrate the boundary and attach onto a line of force which guides them to the Earth at a high latitude. AKASOFU (49) has calculated the deformation of field lines as a function of the equatorial horizontal field change, ΔH_S, produced by a ring current plus image dipole field. Thus the minimum latitude of arrival of 500 KeV protons can be derived as a function of ΔH_S from his results (Fig. 13). In this figure, $L_{\min}$ is acutally plotted where $L_{\min} = \dfrac{1}{\cos^2\lambda}$. The quiet time ring current yields a value $\Delta H_S = -40\,\gamma$; thus storm-time changes $D_{St}(H)$ are measured from this zero. Also plotted are $D_{St}(H')$ values, these being the observed field change when Earth-induced currents are taken into account. For $D_{St}(H') =$

$-150\,\gamma$, the figure gives a value of L_{min} for 500 KeV protons equal to 5.0, provided that the boundary is at $r_B = 8$ Earth radii. MAEHLUM and O'BRIEN (*50*) report that during the storm of 13 July 1961 500 KeV protons arrived on the magnetic shell $L = 5.5$ when $D_{St}(H') \cong -150\,\gamma$.

The model just discussed provides a promising approach to the problem at very low rigidities, but becomes less valid at higher values of the thresholds. An alternative attack at high rigidities is the model experiment of HEDGECOCK (*42*), which enables the threshold reductions to be measured when the field configuration due to an asymmetric ring current is added to the dipole field. He finds that the thresholds are reduced most at longitudes opposite the point where the ring current is closest to the Earth.

c) Asymptotic directions and the analysis of anisotropies in the incident particle flux.

9. To determine the rigidity and spatial dependence of anisotropies in the flux of primary particles incident on the Earth, it is necessary to know in detail the asymptotic directions, far out, of trajectories arriving at the surface after deflection in the geomagnetic field. Initially, these asymptotic directions were determined by model experiments [MALMFORS (*51*), BRUNBERG and DATTNER (*52*)], but with the advent of improved digital computers many trajectories, especially at low rigidities, have been integrated on machines [e.g. JORY (*53*)], and recently non-dipole terms up to $n = 6$ have been included in the geomagnetic field representation used [MCCRACKEN (*17*)]. Methods of analysis for two classes of phenomena will be considered, namely solar proton events due to flares on the Sun and anisotropies in the flux of galactic radiation. Since the first class involves particles of 1 GV to 10 GV while the second is a modulation in the 1 GV to 100 GV range, the methods used will be considered separately.

10. Analysis of solar proton events. Assuming a point source of particles on the Sun and a field-free interplanetary medium, FIROR (*54*) calculated the positions of impact on the Earth's surface for flare particles of 1 GV to 10 GV rigidity. Further assuming a flat particle spectrum, FIROR found that the primary zone of impact was located around 0900 hrs local time, although appreciable intensity also occurred at 0400 hrs local time and a third, less intense zone, was located at 2000 hrs local time. Even higher order impact zones seemed to occur for particles very near the threshold rigidity at a particular latitude, which probably gave rise to a low intensity background at most longitudes. FIROR found rough agreement between his calculations and sea-level observations of three solar proton events, provided that the source had an angular width of 30°. More detailed analysis of the source, especially on occasions when the action of the interplanetary magnetic field on the particles may appreciably alter its apparent position, is complicated by the presence of several different impact zones. MCCRACKEN has introduced an important simplification into the problem of deducing the source-position and angular dependence from a set of world-wide observations made by neutron monitors. He points out that only particles of rigidity greater than 1.1 GV for neutron monitors and 4.0 GV for meson telescopes, and arriving at zenith angles less than 32° to the vertical, make an appreciable contribution to the couning rate of sea-level detectors. Furthermore, solar particles of the highest rigidity arrive in the 0900 hrs L.T. impact zone, while lower rigidity particles arrive in the higher order zones. Above a certain effective latitude, particles of $P > 1.1$ GV only populate the 0900 hrs L.T. impact zone and, therefore, by restricting the observations to sufficiently high latitude neutron monitors only one impact zone need be considered. For a particular monitor, an asymptotic cone of acceptance is defined which corresponds to the directions far out of all trajectories arriving within 32° of the vertical and with 1.1 GV$<P<$5.7 GV. Fig. 14 shows these asymptotic cones for eleven stations, the co-ordinates being the geographic latitude and longitude of the asymptotic directions. McCracken included non-dipole terms up to $n=6$ in the preparation of this figure and further showed that an external ring current made little practical difference.

The small size of the asymptotic cones of acceptance in the flare particle rigidity range indicates that all particles detected by a particular station have come from roughly the same direction in space. Thus it is possible to define a mean asymptotic direction of viewing, A_m, for each detector so that the counting rate increase is given by

$$\Delta N = \int_{P_a}^{\infty} J(P, A_m, t) S(P) \, dP, \tag{10.1}$$

where J is the differential flare spectrum as a function of direction, P_a is the atmospheric threshold, and $S(P)$ is the specific yield function defined by

$$J_0(P, t) S(P) = \frac{\partial N}{\partial P}.$$

J_0 is the differential galactic spectrum and $\partial N/\partial P$ is the change in counting rate of a monitor with vertical threshold rigidity for galactic particles. Changes in the atomic composition of the particles between the galactic and solar spectra have been neglected, and any zenith angle dependence of the counting rate due to atmospheric effects is assumed to be independent of rigidity.

The mean direction of viewing may be defined by

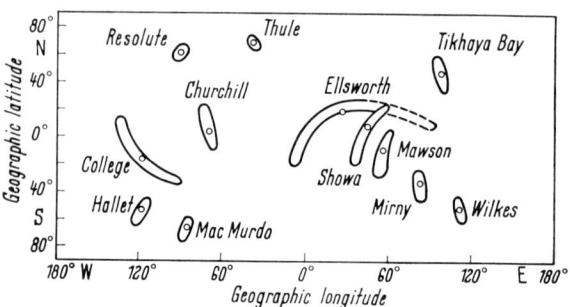

Fig. 14 Asymptotic cones of acceptance and mean directions of viewing appropriate to a solar energetic particle event observed by sea-level neutron monitors. After McCracken (17).

$$A_m = \frac{\int_{P_a}^{\infty} P^{-\gamma} S(P) A(P) \, dP}{\int_{P_a}^{\infty} P^{-\gamma} S(P) \, dP}, \tag{10.2}$$

where the flare spectrum has been represented by a negative power law of exponent γ. Mean asymptotic directions appropriate to a flare-type spectrum are shown in Fig. 14. If γ is independent of direction, the relative flare-increases at the different stations can be marked on Fig. 14 against the mean directions to provide a contour map of the source distribution.

11. Analysis of anisotropies in the galactic flux. The concept of a mean direction of viewing has been used in the past to determine the direction of anisotropy in space corresponding to the daily variation in the galactic cosmic ray intensity [e.g. Brunberg and Dattner (52)]. Analysis of the daily variation for a particular detector averaged over many days yields a certain local time when the intensity reaches maximum amplitude. It is then possible to find the asymptotic directions from which particles of a given rigidity must have originated in order to arrive at that particular local time and at the position and direction of view of the detector. By plotting these asymptotic directions as a function of rigidity for results from several different detectors, an attempt is made to find a unique rigidity and direction which satisfy all the curves, thus identifying the source-direction of the daily variation, together with the mean rigidity of the particles most affected. This method suffers from the disadvantage that we have little previous knowledge about the rigidity dependence of the modulation and, in principle, particles from at least 1 GV to 100 GV can be affected. The asymptotic cone of acceptance for an equatorial station in this rigidity range spreads over 180° of longitude, and thus the use of a mean direction may lead to large errors if the modulation is only slowly dependent on rigidity.

RAO, MCCRACKEN and VENKATESAN (55) are among the workers who have attempted a more rigorous analysis of the daily variation. They define a variational coefficient, $\nu(\Omega_i, \beta)$, given by

$$\nu(\Omega_i, \beta) = \int_{P_T}^{\infty} \frac{\partial N}{\partial P} \frac{1}{N} P^\beta \frac{Y(\Omega_i, P)\,dP}{Y(4\pi, P)}. \tag{11.1}$$

Here it is assumed that the fractional change in primary intensity in the solid angle Ω_i at rigidity P is

$$\frac{\Delta J_i}{J} = A P^\beta. \tag{11.2}$$

where A is a function of asymptotic direction and β is a constant. It is further assumed that the effect of the atmosphere is a separable function of direction

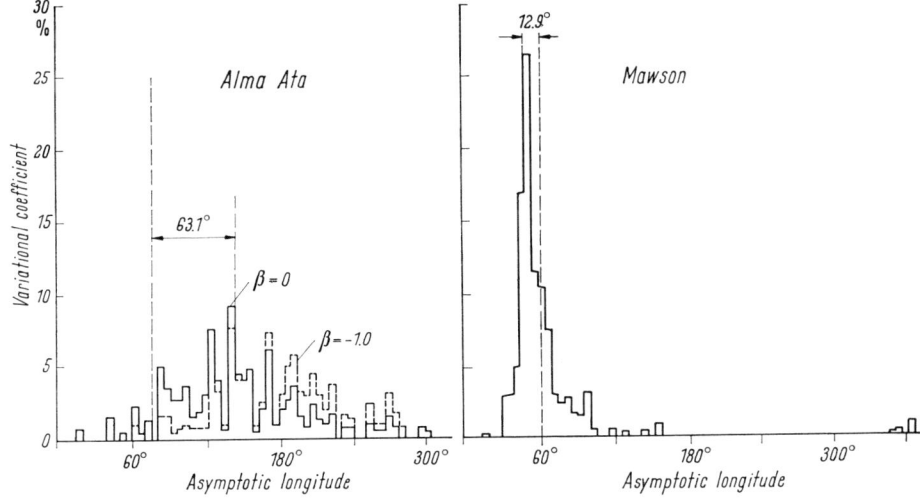

Fig. 15. Variational coefficients as a function of asymptotic longitude. After RAO et al. (55).

and rigidity. A suitable approximation for the atmospheric effect is shown to be that the observed detector rate per unit primary particle arriving at any zenith angle between 0° and 40° is constant per unit zenith angle. At zenith angles > 40°, the response of the detectors is zero. Then $Y(\Omega_i, P)$ is the number of unit zenith and azimuth intervals accessible from Ω_i while $Y(4\pi, P)$ is the number accessible from any asymptotic direction whatsoever. Fig. 15 shows variational coefficients for a high and a low latitude station when $\beta = 0$ and when the anisotropy has a cosine dependence on asymptotic latitude. The total fractional change in counting rate is obtained by summing the product $A \cdot \nu(\Omega_i, \beta)$ over all primary solid angles. Fig. 16 has been computed for two primary anisotropies with a square wave dependence on local time, and under the above assumptions we notice that only the observed variation at the high latitude station Mawson reflects the true time-dependence of the anisotropy and that, in the bottom curve, even though the anisotropy is independent of rigidity ($\beta=0$) the low latitude variation (Alma Ata) is much reduced due to the cone of acceptance of this station being of approximately the same width as the primary anisotropy. These calculations were performed using asymptotic trajectories computed in the $n=1$ to $n=6$ expansion of the geomagnetic field potential. A general difficulty in the analysis of the daily variation by methods such as the one just outlined is that several assumptions as to the form of the primary anisotropy must first be made and only then can the calculations be compared to experimental data.

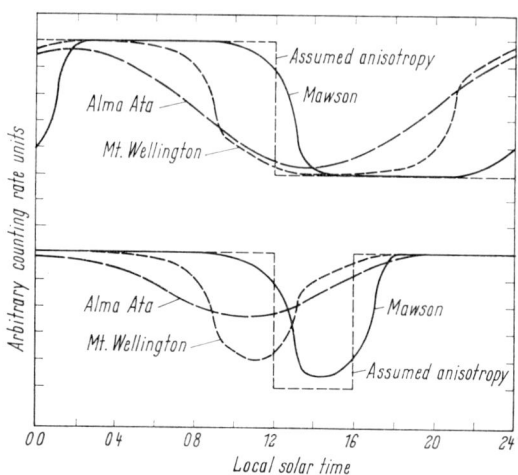

Fig. 16. Predicted counting rate dependence on local time for two types of assumed primary anisotropy. After Rao et al. (55).

d) Specific yield functions.

12. It has already been demonstrated in connection with the study of anisotropies that it is necessary to know the counting rate yield of a sea-level detector per incident primary particle of a given magnetic rigidity if changes in the primary spectrum are to be investigated. The specific yield function $S(P, x)$ of a cosmic ray detector of a particular type located at a certain atmospheric depth, x gm/cm², can be defined by

$$J(P) S(P, x) = \frac{\partial N(P, x)}{\partial P}. \tag{12.1}$$

Here $J(P)$ is the differential primary spectrum as a function of rigidity and $N(P, x)$ is the counting rate of the detector located at a vertical threshold rigidity, P. Although primaries are incident on the top of the atmosphere from all directions in the upper 2π solid angle, most of the detector response comes from particles arriving near the vertical, and therefore the vertical threshold is used to represent an approximate lower limit to the rigidity of any particle producing detector counts. $\partial N/\partial P$, the differential response function, may be derived from the latitude curve for the particular detector and then, if the primary spectrum is known for the time when the latitude curve was measured, $S(P, x)$ can be calculated. No attempt is made here to separate out the contributions to $S(P, x)$ due to the different species of nuclei in the primary flux for the following reasons. Theories of cosmic ray modulation at relativistic energies generally show that the particle mean free path in a magnetic field is the parameter which determines the relative behaviour of different particles. Because mean free path depends on rigidity, it is sufficient to know the total J, S and $\partial N/\partial P$ for all particles of the same rigidity since their expected modulation will be similar. Analysis of experimental data by Webber [2] suggests that the modulation of relativistic particles is rigidity-dependent. In the region where particle velocities are becoming measurably less than the velocity of light, theoretical predictions often do not give a pure rigidity-dependence, but this does not matter here since $S(P, x) \to 0$

Sect. 12. Specific yield functions. 329

at these energies. A method is given by WEBBER and QUENBY (56) whereby $S(P, x)$ can be split up into the contributions from different primary nuclear species, if necessary.

Yield functions have been given by NEHER [14] and by several authors since [see WEBBER and QUENBY (56)]. However, the shape of the curves obtained depends very much on the value of the threshold rigidities used and so we give here only the latest results based on the currently available threshold calculations. MATHEWS (57) has found $\partial N/\partial P$ at sea level for neutron monitors and μ-mesons at both sunspot minimum and sunspot maximum, and these results are tabulated below. For the sunspot minimum neutrons, latitude surveys by ROSE, FENTON, KATZMAN and SIMPSON (58) were plotted using QUENBY and WENK [11] thresholds, while at sunspot maximum the neutron survey of KONDO, KODAMA and MAKINO (18), plotted against machine-calculated thresholds, was employed, this latter result corresponding to an epoch when the Mt. Washington neutron monitor was 18% below the 1954/55 minimum value. The results are normalized so that the high latitude minimum intensity equals 100. At $P \geq 15$ GV, the minimum curve has been extrapolated according to

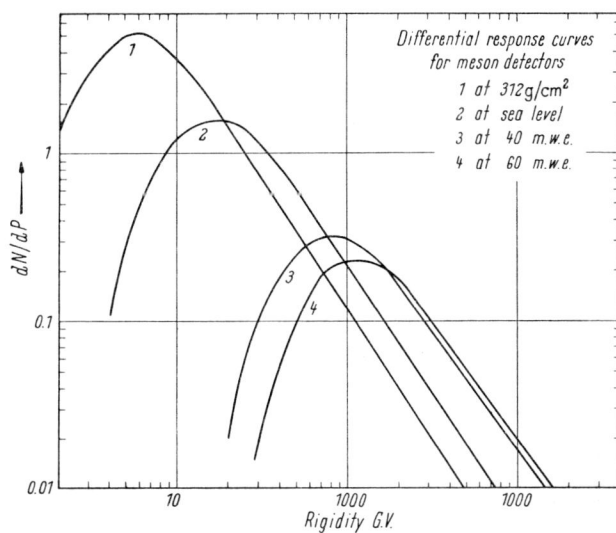

Fig. 17. Differential response curves for meson detectors at different depths. After MATHEWS (59).

$$\frac{\partial N}{\partial P} = 116 \cdot 8 \, P^{-1.517} \,. \qquad (12.2)$$

Table 3. *Sea-level detectors.*

P(GV)	Neutrons				μ-Mesons			
	Sunspot N	Minimum $\partial N/\partial P$	Sunspot N	Maximum $\partial N/\partial P$	Sunspot N	Minimum $\partial N/\partial P$	Sunspot N	Maximum $\partial N/\partial P$
15	55.70	1.87	51.70	1.70	88.77	1.62	85.51	1.41
14	57.70	2.13	53.48	1.86	90.36	1.56	86.89	1.35
13	59.95	2.40	55.42	2.06	91.98	1.51	88.21	1.29
12	62.51	2.76	57.60	2.29	93.39	1.46	89.47	1.22
11	65.39	3.05	59.99	2.52	94.80	1.36	90.65	1.12
10	68.60	3.39	62.64	2.78	96.12	1.24	91.71	0.94
9	72.16	3.70	65.55	3.01	97.27	1.05	92.53	0.73
8	76.00	3.98	68.67	3.21	98.22	0.85	93.17	0.56
7	80.11	4.23	71.96	3.34	98.97	0.63	93.65	0.40
6	84.47	4.40	75.35	3.35	99.49	0.43	93.97	0.26
5	88.90	4.39	78.67	3.10	99.82	0.25	94.16	0.09
4	93.26	4.06	81.54	2.39	100.00		94.16	
3	97.02	3.02	83.45	1.28				
2	99.50	1.49	84.10	0.32				
1	100.00		84.10					

The μ-meson results are based on the work of WEBBER and QUENBY, renormalized and corrected for the solar cycle. They are shown in Fig. 17, together with the curves for underground μ-meson telescopes derived by MATHEWS (59) following an empirical method given by DORMAN (60).

If at any two times in the solar cycle the shape of the latitude curve is known, the change in the primary spectrum can be found from

$$\frac{\frac{\partial N_1}{\partial P}}{\frac{\partial N_2}{\partial P}} = \frac{J_1(P)}{J_2(P)}.$$ (12.3)

According to WADDINGTON [15], $J(P)$ at sunspot minimum is a negative power law in rigidity with an exponent of between 2.4 and 2.5.

III. Experimental evidence on the modulation mechanism.

a) Cosmic ray time variation.

13. Introduction to time variations. Using the ideas developed in Chapter II concerning geomagnetic effects and those in the account of meteorological effects given by FORBUSH [1], we suppose that it is possible to correct sea-level cosmic ray data for all geophysical influences and thus arrive at a knowledge of the primary intensity variations outside the magnetosphere. A great body of experimental evidence has been assembled on time variations and it is only possible for the author to attempt to select some of the more significant results. These will be mentioned in this chapter, together with measurements made by space probes on interplanetary plasma and magnetic fields, which are also very relevant in determining the cause of cosmic ray modulation.

A variation in solar activity with a period of about 11 years has been established by cosmic ray observations over two solar cycles which clearly exhibits a negative correlation between the mu-meson rate at sea-level and sunspot number. Between the sunspot minimum of 1954 and the maximum of 1958, the ion chamber intensity at Huancayo fell by 6%, the high latitude, mountain altitude neutron intensity fell by 20%, the Geiger counter rate at 15 mb altitude changed by a factor of 2, and the ionisation recorded over the north pole changed by a factor of 4. Emulsion and counter observations at balloon altitudes show that alpha-particles and heavy nuclei participate in the 11-year variation as well as primary protons.

Forbush decreases in the cosmic ray intensity usually occur in association with magnetic storms on the Earth, but there is no one-to-one correlation between the amplitude of the decrease and the size of the magnetic field change. The decrease may be as much as 12% in a high latitude neutron monitor, taking place with an onset time varying between 2 hours and 2 days. Recovery of the intensity to the pre-storm level can last from 2 days to 2 weeks or more. Alpha-particles seem to be affected as well as primary protons. A tendency for small decreases to recur with a 27-day period, in good correlation with the 27-day cycle of magnetic activity, suggests that some small Forbush events are repeated each time an active period on the Sun rotates past the Earth.

A variation in intensity with a period of 1 day and an amplitude of 0.1% to 0.2% for both neutron monitors and mu-meson detectors exists through the solar cycle. The average local time of maximum is around 1300 to 1500 hrs, although on some days it is shifted to morning hours. There is evidence for a 22-year cycle in the value of this mean. During the first few days of some Forbush decreases, the daily variation may increase to 0.5% amplitude and exhibit an earlier time of maximum. The phase and amplitude of the 12-hour wave in the primary radiation have not been established with any certainty.

Apart from the continuous production of kilovolt particles to form the solar wind, the Sun is capable of accelerating protons to energies between 10^6 eV and 10^{10} eV. Acceleration occurs at the time of large solar flare outbursts in the region of a sunspot group. Usually a flare producing these energetic particles simultaneously emits radio noise in the 1000 to 10000 Mc region, known as a Type IV outburst. During the period 1956—1961, of the order of 50 events were detected but only 10 included sufficiently energetic particles to cause measurable changes in the intensity recorded by neutron monitors. Quite often, Forbush decreases follow 1 to 2 days after these solar proton events. It is the propagation characteristics of the solar protons through interplanetary space which are of chief interest in this article, rather than the acceleration mechanism operative on the Sun. Alpha-particles and heavy nuclei are also accelerated by the Sun.

Thus there are three major classes of time variations of the galactic cosmic ray intensity and an additional class resulting from the acceleration of solar particles. Further details are given by FORBUSH [1], DORMAN [3], and WEBBER [2].

14. Scale-size of the Forbush decrease mechanism. Clearly, it is of interest to decide whether the modulating mechanism for the three classes of time variations just mentioned acts on a scale comparable to the size of the Earth's magnetosphere or to that of the solar system. It has already been stated that all the variations mentioned have been corrected for meteorological effects, but there remains the possibility that geomagnetic field changes, especially at the time of storms, could contribute to the time variations. However, if we go to a sufficiently high latitude, where the threshold is less than 1 GV, primary particles close to the threshold can produce no measurable ground-level effects because the specific yield functions at these rigidities tend to zero. Thus, here, threshold changes due to field changes cannot cause any variations in intensity. All time variations are observed at stations where the threshold <1 GV. Therefore, apart from small effects such as the increase apparently taking place during the initial stages of some Forbush decreases (Sect. 6) we have to look outside the magnetosphere to find the modulation mechanism. This argument was developed by SIMPSON (*61*).

Simultaneous observation of variations at the Earth and in space some distance away also indicates the scale-size of the modulation mechanism. Such measurements have been possible for the Forbush effect. Pioneer V recorded a decrease in the counting rate of a telescope responding to protons >75 MeV when the space probe was 0.04 astronomical units away from the Earth but on the Earth-Sun line (*62*). A decrease was recorded by the Climax neutron monitor which was known to coincide to within a few hours with the decrease in deep space. Using a relationship between Forbush changes at Climax and above the atmosphere, established with an identical telescope on Explorer VI, it was found that the change at Pioneer V was 1.3 times as large as at the Earth. This figure is only valid if the energy-dependence of the events recorded by Pioneer V and Explorer VI were the same.

Correlation between short-term variations measured simultaneously by an ion-chamber carried in Mariner II and the Deep River neutron monitor was studied by NEHER (*63*). The two instruments showed very similar variations while the probe was within 0.1 A.U. of the Earth, but further away the correlation became less marked. However, a Forbush decrease recorded when the probe was 0.25 A.U. closer to the Sun was seen 2 days later on the Earth, thus giving a velocity for the phenomena causing the decrease of the order of 200 km/sec. This velocity may be compared to the solar plasma velocity deduced from the Mariner II measurements of 300 km/sec to 700 km/sec. In this connection, it may be mentioned that studies of the correlation between disturbances on the Sun and the onset of geomagnetic fluctuations yield a velocity for the solar corpuscular streams of 200 km/sec for small storms and 1000 km/sec for large storms (*64*). Forbush decreases often occur within 2 hours of the time of the sudden commencement of a magnetic storm.

The simultaneous measurements made so far only show conclusively that the scale-size of the Forbush decrease is $\gtrsim 0.1$ A.U. However, if we take 200 km/sec as a lower limit to the speed of the edge of the region in space where the intensity is depressed, we can increase the size estimate. The Mariner II event just mentioned lasted for about 2 days, thus implying that this region in space was 0.25 A.U. across in a direction along the solar radius vector. Some FORBUSH events last at least four times as long, thus suggesting that the scale is $\gtrsim 1$ A.U.

15. Anisotropies in the primary particle modulation. To distinguish between models of the modulation mechanism involving ordered fields on the one hand and disordered fields on the other, it is important to know the directional dependence of primary intensity variations. In general, one might expect more ordered fields to produce greater anisotropies than disordered fields.

Since the long-term variation exhibits an 11-year cycle in intensity changes, the size of daily variation averaged over a long period is a measure of the anisotropy associated with this modulation. Even if the cause of the daily variation is not connected with that of the 11-year cycle, the magnitude of the daily variation will provide a reasonable upper limit to the anisotropy. There is general agreement that the amplitude of the anisotropy represents a change in primary flux of 0.5% or less at rigidities greater than a few GV, and that the direction of maximum intensity is roughly 90° east of the Earth-Sun line: but there is no consensus of opinion on the rigidity-dependence of the variation. For example, using counter-telescope data from MAWSON corrected for the temperature effect, QUENBY and THAMBYAHPILLAI (*65*) suggested that maximum intensity arrives from a direction 70° east of the Earth-Sun line and that the mean rigidity of the particles causing the variation is between 10 and 20 GV. However, their method could only give an average rigidity and would not yield a complete spectrum. DORMAN [*16*] assumes an energy-dependence of the fractional change in intensity $\delta J(E)/J(E)$ of the form

$$\begin{aligned}\frac{\delta J(E)}{J(E)} &= 0, & E < E_0, \\ \frac{\delta J(E)}{J(E)} &= K E^{-\gamma}, & E > E_0.\end{aligned} \quad (15.1)$$

E being the total energy of the particles. He obtains a best fit to experimental results with $\gamma = 1$ and $E_0 = 6.6$ GeV, the direction of anisotropy lying 82° east of the Earth-Sun line. Some of the data used by DORMAN included mu-meson telescope results and were thus susceptible to errors introduced by an imperfect knowledge of the correction necessary for the 24-hour variation in atmospheric temperature [e.g. QUENBY and THAMBYAHPILLAI (*65*)]. DUGGAL, NAGASHIMA and POMERANTZ (*66*) and RAO, McCRACKEN and VENKATESAN (*57*) have analysed data from all available neutron monitors working during the IGY. Thus they avoid difficulties due to temperature effects in the atmosphere, since a neutron monitor is very insensitive to these changes. The latter authors used a geomagnetic field model with terms up to $n=6$ to obtain the asymptotic direction, whereas the former authors used a dipole field. DUGGAL et al. obtain a variation of the form

$$\frac{\delta J(P)}{J(P)} = 0, \qquad P < 6 \text{ GV},$$
$$\frac{\delta J(P)}{J(P)} = K P^{-0.5}, \qquad P > 6 \text{ GV},$$
(15.2)

with a direction of anisotropy 69° east of the Earth-Sun line, while RAO et al. obtain, for particles in the ecliptic plane, $\frac{\delta J(P)}{J(P)} = 410^{-3}$, $1 \text{ GV} \leq P \leq 200 \text{ GV}$, the angle being 85° east of the Earth-Sun line. Both sets of authors find that the magnitude of the anisotropy falls off as the cosine of the angle made with the plane of the ecliptic. The upper limit of 200 GV in the last results depends on underground measurements where the yield functions are poorly known. None of these attempts to establish the spectral dependence of the variation has allowed for the possibility of a high rigidity peak in the distribution. One might imagine a modulation mechanism which is very effective at low rigidities, therefore producing almost complete isotropy but becoming less efficient at higher rigidities, and thus probably more anisotropic in its effects.

Although there are days on which the phase of the daily variation is completely reversed, the long-term average variation changes little from year to year. HYNDS (67) has considered the variation in the mean phase, averaged over a year, for low latitude ion chambers which have been corrected for atmospheric temperature variations according to the empirical method given by QUENBY and THAMBYAHPILLAI. Between 1936 and 1959, the time of maximum ranged from 1400 to 1630 hrs L.T.

Turning now to the Forbush decrease, we still find that the modulation is an essentially isotropic phenomenon. To illustrate this we show in Fig. 18 the bi-hourly intensities of a world-wide distribution of neutron monitors recording the decrease of 28—29 August 1957. LOCKWOOD and RAZDAN (68) have investigated the onset times, as a function of mean asymptotic longitude, for 13 large decreases in the period 1957—1961. In only 6 events could differences in onset time of 2 hours or more be distinguished. Early onset occurred almost always for stations whose asymptotic directions were west of the Earth-Sun line. The position and angular width of the early onset zone varied from event to event. During one event at least (22 October 1957) there is evidence that onset came earlier, at a given asymptotic longitude, for stations looking at high asymptotic latitudes, but this effect is not distinguishable in all events. During the Forbush decrease of 20 August 1956, FENTON, MCCRACKEN, ROSE and WILSON (69) showed that the intensity 90° to the west of the Earth-Sun line was depressed relative to that 90° to the east for a day before the effect became world-wide. Another decrease they studied where the anisotropy lasted more than 12 hours is shown in Fig. 19.

Days of high geomagnetic activity, characterised by K-figures ~ 9, are correlated with days when the daily variation of the mu-meson component increases to an amplitude $\sim 0.5\%$ and the direction of maximum anisotropy moves round to correspond to particles arriving from the solar direction (70). Because of the association between Forbush decreases and high geomagnetic activity, we can therefore expect some enhancement in the daily variation during these cosmic ray events.

Variations recorded at sea level are very roughly one half of the size of primary variations. Thus we obtain the following picture for a large decrease producing

334   J. J. Quenby: The Time Variations of the Cosmic Ray Intensity.   Sect. 15.

a 10% neutron monitor change at high latitudes. The intensity of primaries of a few GV decreases by 20% in a period ~1 day or more, the change 90° west of the Earth-Sun line probably leading that at 90° east by 2 hours. During the first

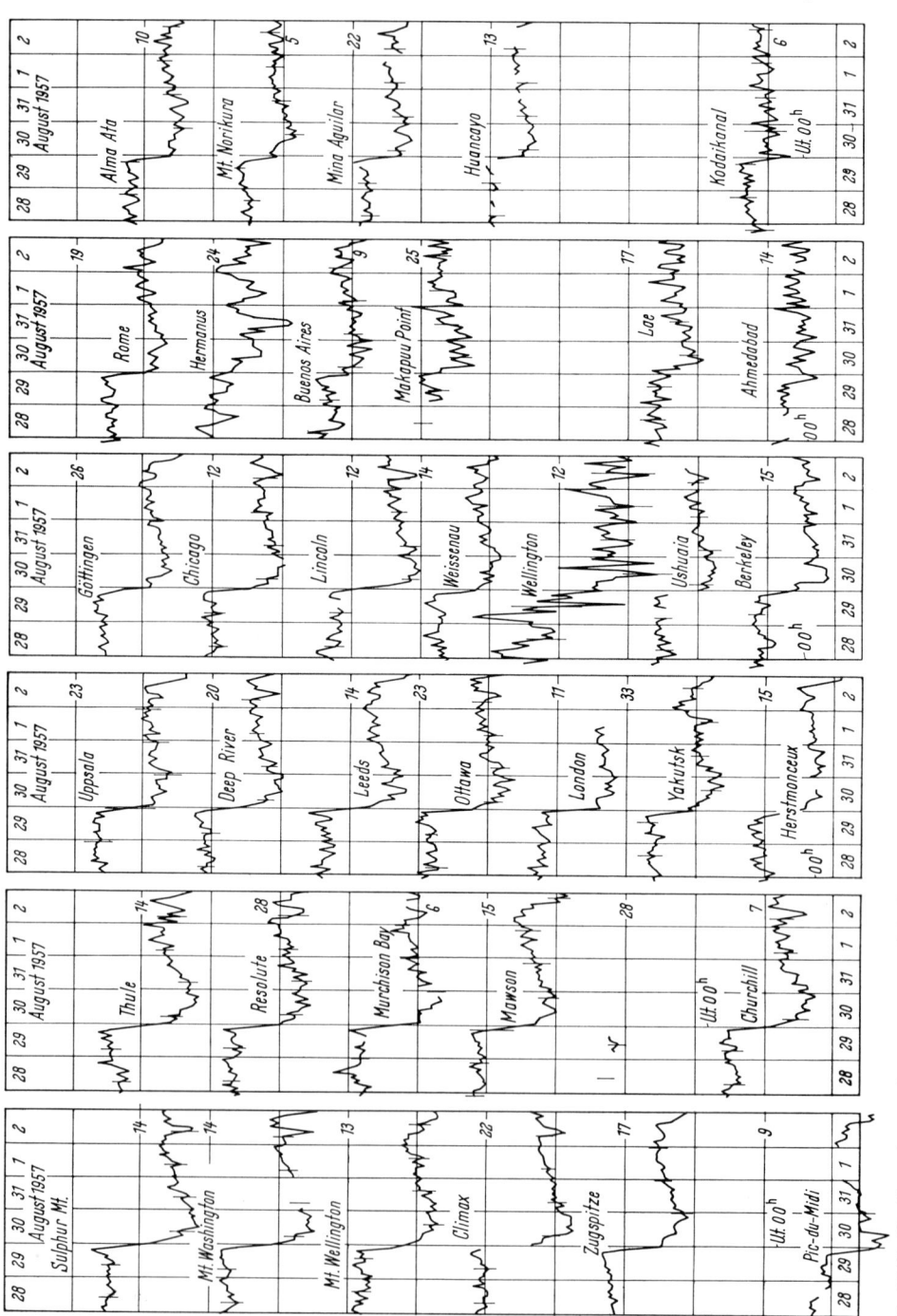

Fig. 18. World-wide neutron monitor measurement of the Forbush decrease of 28—29 August 1957. Data received at IGY World Data Centre C2 and compiled by Dr. Y. Miyazaki

few days of the event, the intensity in the anti-solar direction may be depressed 1% more than in the solar direction. Later on, the intensity may or may not recover to the pre-storm value.

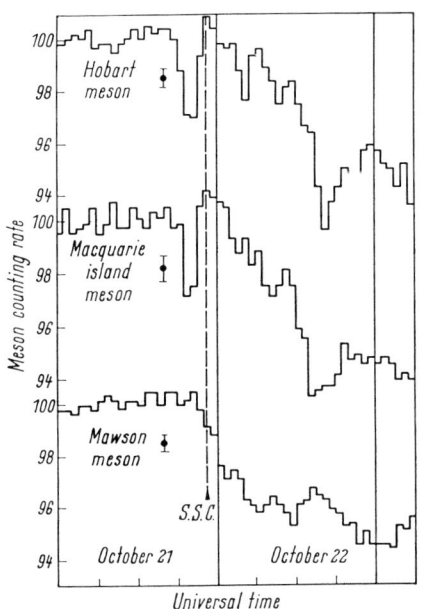

Fig. 19. Forbush decrease of 21 October 1957. At the time of the initial decrease centred on 2000 U.T. at Hobart and Macquarie Island, these two stations were sampling radiation from a direction 45° west of the Earth-Sun line. After FENTON et al. (69).

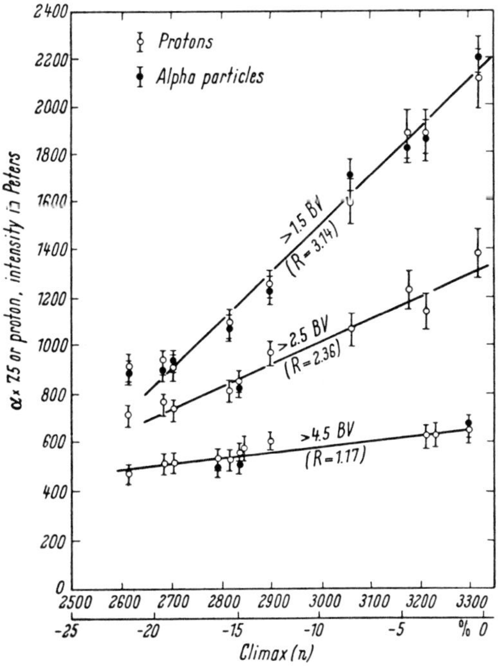

Fig. 20. Integral intensity above a rigidity R of primary protons and alpha-particles plotted against the climax neutron monitor rate, measured as a percentage of the sunspot minimum value. After McDONALD and WEBBER (71).

16. Relative variations of protons and alpha-particles. Measurement of the relative time variations of the different primary nuclei can help to distinguish between various modulation mechanisms. In particular, at the low end of the spectrum, where the energy per nucleon is ∼1 GeV, large differences in relative behaviour are predicted by some theories. McDONALD and WEBBER (71) have been largely responsible for the experimental evidence at present available on this question, and an account of their results is given by WEBBER [2].

Fig. 20 shows the integral intensities of protons and alpha-particles above certain rigidity thresholds, obtained on several flights at different stages in the solar cycle, plotted against the corresponding rate of the Climax neutron monitor. All alpha fluxes have been multiplied by a factor 6.8. To within an accuracy of ±5%, the relative integral intensity variations above 1.5 GV are similar. Differential spectra data above 1.5 GV are rather scanty. Nor is there any very conclusive evidence available on the relative time variations during a Forbush decrease.

In an accompanying Handbuch article, WEBBER[1] has reviewed recent balloon and satellite data and suggests that the ratio of protons to alpha particles as a function of time simply depends on the energy per nucleon or velocity of the particles. This result is valid in the energy per nucleon range from 100 MeV to 450 MeV, equivalent to a proton rigidity range between 0.5 GV and 1.0 GV.

---

[1] W. R. WEBBER: The Spectrum and Charge Composition of the Primary Cosmic Radiation, p. 181.

17. Rigidity-dependence of the 11-year and Forbush variation.

From the many measurements of intensity as a function of time at ground level and at balloon and satellite altitudes, it is possible to obtain an estimate of the energy-dependence of the larger time variations. Analysis is made convenient if we assume all primary nuclei of the same magnetic rigidity behave in a similar fashion. The previous section indicated that this assumption is reasonable for $P > 1.5$ GV when the 11-year cycle is considered, but that it must remain as a hypothesis in the case of the Forbush decrease. Here, we will not

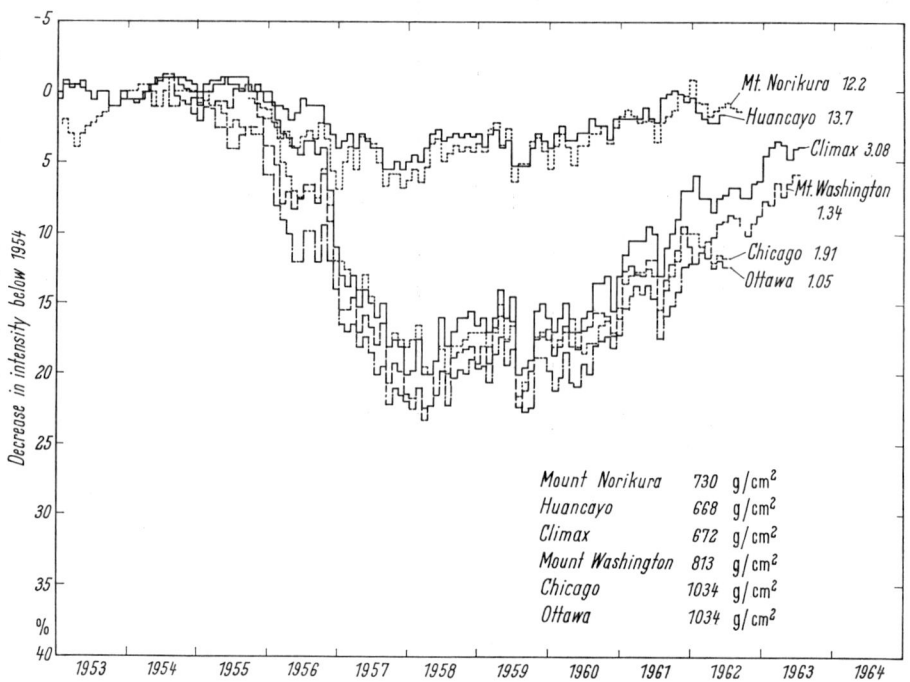

Fig. 21. Monthly averages of the counting rates for six neutron monitors, measured as percentage differences from the sunspot minimum value.

be concerned with variations at $P \simeq 2$ GV. Fig. 21 shows the monthly average intensities for 6 neutron monitors extending over nearly a whole solar cycle. Threshold rigidities and atmospheric depths appropriate to the stations are given. It is rather difficult to deduce the change in the primary differential spectrum δJ from the few ground results available taken in conjunction with some isolated direct flux determinations at the top of the atmosphere. The method used is to assume a form for δJ and then compute $\delta N/N$, the fractional change in the observed ground-level couting rate, from

$$\frac{\delta N}{N} = \frac{\int_{P_T}^{\infty} \frac{\delta J(P)}{J(P)} \frac{\partial N}{\partial P} dP}{N(P_T)}, \qquad (17.1)$$

where P_T is the station threshold.

The form of δJ giving a best fit between the observed and computed values of $\delta N/N$ at all stations is adopted. Using this method, Elliot, Hynds, Quenby and Wenk (72) and Webber [2] have estimated the ratio of the sunspot maximum

(1957/58) to sunspot minimum (1954/55) spectra, and their results are plotted in Fig. 22. The work of WEBBER is based on more data than that of ELLIOT et al. A better method, involving no initial assumptions, is to know the ratio of the differential response functions at all rigidities in the latitude-sensitive region at both stages in the solar cycle. Then

$$\frac{J(P)_{\max}}{J(P)_{\min}} = \frac{\left(\frac{\partial N}{\partial P}\right)_{\max}}{\left(\frac{\partial N}{\partial P}\right)_{\min}}. \qquad (17.2)$$

MATHEWS has obtained this ratio using the response functions given in Sect. 12 and his results also are shown in Fig. 22. Although the sunspot maximum curve

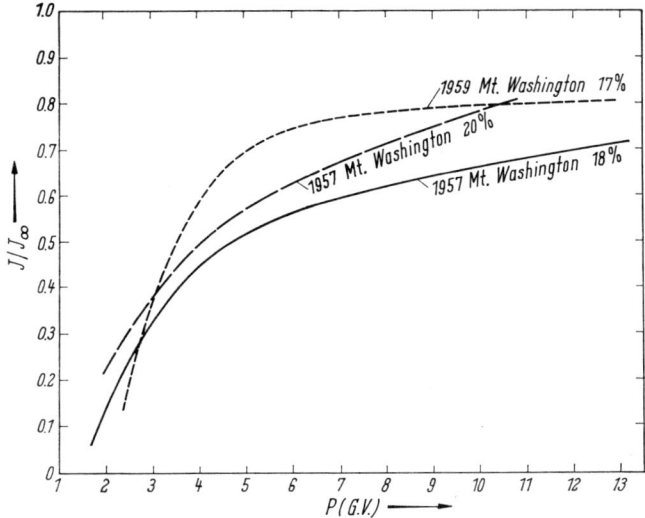

Fig. 22. Ratio of the differential primary spectrum near sunspot maximum (1957/58) to the sunspot minimum (1954/55) spectrum. Full curve estimated by ELLIOT et al. (72), dashed curve estimated by WEBBER [2], and dotted curve estimated by MATHEWS (Sect. 12). The year of observation and the value of the Mt Washington neutron monitor rate are also given.

is based on computed threshold rigidities along the course of a voyage between Japan and the Antarctic, it could be in error due to imperfect knowledge of the geomagnetic field in the region of the South African anomaly. All these estimates agree in ascribing a greater than 50% modulation at $P<3$ GV and a 10% to 20% modulation at $P\sim15$ GV as a result of the long-term variation between 1954 and 1958.

More recently, two sets of satellite results giving information on the long-term variation have become available. DURNEY, ELLIOT, HYNDS and QUENBY (73) have measured the rigidity spectrum of $Z \geq 6$ nuclei with a Čerenkov detector, utilising the latitude effect to obtain the particle rigidities. Above 8.5 GV the integral spectrum in May and June 1962 followed a $P^{-1.5}$ power law in agreement with the spectrum appropriate to the previous sunspot minimum. Below 8.5 GV the spectrum flattened to give a $P^{-1.2}$ rigidity dependence. The shape of the modulation below 8.5 GV, obtained by assuming that $P^{-1.5}$ is the correct unmodulated spectrum, is given in Fig. 23. Using this modulation function and suitable specific yield functions, it is possible to compute the expected 11-year decrease for the stations shown in Fig. 21. This comparison is given in Table 4. The Ariel results imply that in May—June 1962 no modulation occurs >8.5 GV.

To check this suggestion, we show in Fig. 24 the monthly mean counting rates of 7 equatorial neutron monitors for the years 1961, 1962, and 1963. Within the rather wide experimental errors, no long-term increase is intensity is discernible for these stations, all of which have thresholds >12 GV.

From data obtained with an ionisation chamber carried in Explorer VII, POMERANTZ, DUGGALL and WITTEN (74) found the average integral spectrum of $Z \leqq 6$ nuclei between October 1959 and May 1960 could be represented by

$$N(>P) = K \exp\left(-\frac{P}{P_0}\right), \qquad (17.3)$$

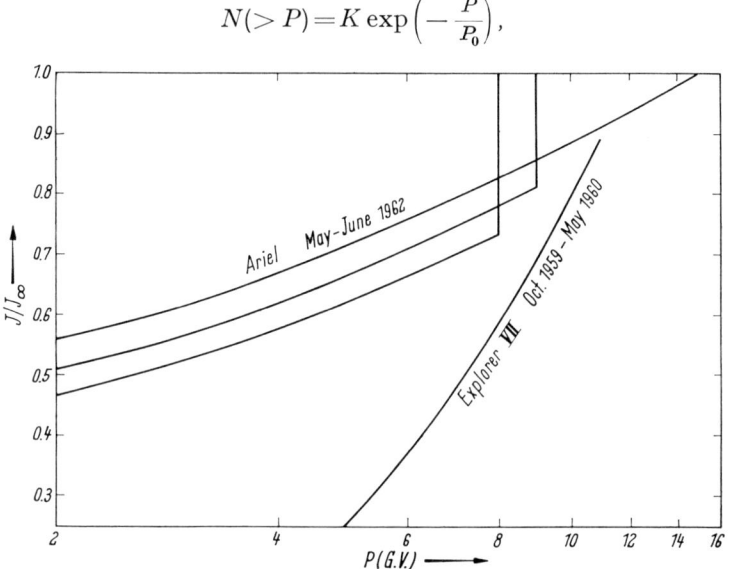

Fig. 23. Long-term modulation obtained by DURNEY et al. (73) from Ariel I data, and by using a similar method for Explorer VII data given by POMERANTZ et al. (74).

where $P_0 = 7.9$ GV. Assuming a $P^{-1.5}$ form for the sunspot minimum spectrum, the modulation function derived from this result is also shown in Fig. 23. It was necessary to know the absolute size of modulation at one rigidity to construct this graph, and so we took a figure of 20% modulation at 10 GV, based on comparable results shown in Fig. 22.

Table 4.

Threshold rigidity	Station	Decrease estimated on basis of Ariel spectrum %	Observed decrease %
1.05 GV	Ottawa	8	(11±1)
1.91	Chicago	8	12±1
1.34	Mt Washington	11	9±1
3.08	Climax	11	7±1
12.2	Mt Norikura	0	0 to 2
13.7	Huancayo	0	0 to 2

Forbush decreases have been recorded simultaneously at many different stations, especially during the IGY, and it is therefore relatively easy to obtain the latitude curve for the effect and so use Eq. (17.2) to find the fractional change in primary spectrum. R. P. KANE (75) has made a very comprehensive survey of the short-term changes which took place during the IGY and has kindly sent the author his raw results. These take the form of the total magnitude of particular changes for each station after bad data has been excluded. The events we show are, firstly, initial decrease phases of Forbush events whose amplitude exceeded 6% and, secondly, recovery phase of Forbush events where again the total change

Sect. 17. Rigidity-dependence of the 11-year and Forbush variation. 339

exceeded 6%. When applying Eq. (17.2) to these events, we have used the differential response function appropriate to sunspot maximum (Sect. 12). Ratios of the differential primary spectra before and during the events are shown in Figs. 25 and

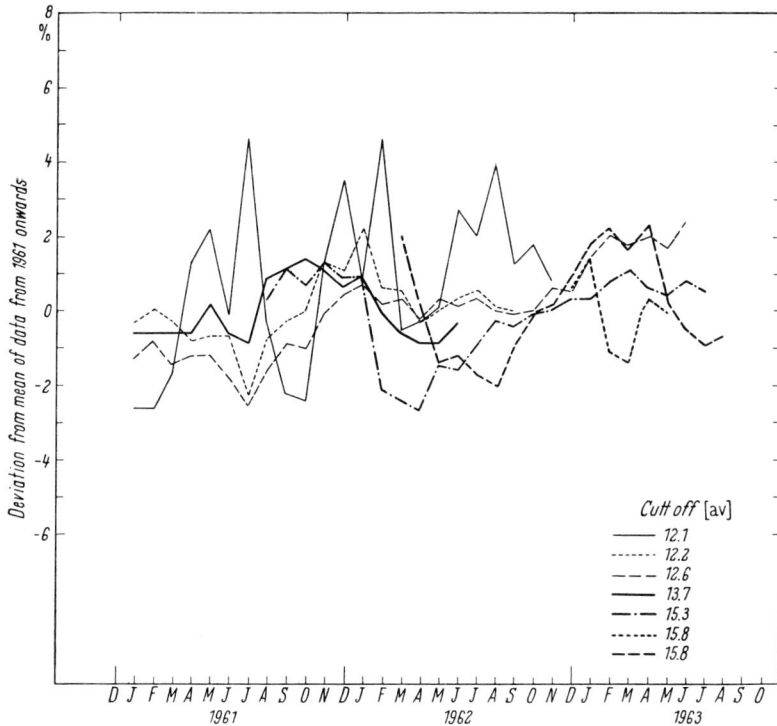

Fig. 24. Monthly averages, as percentage deviations from the mean, for seven equatorial neutron monitors working during 1961—1963.

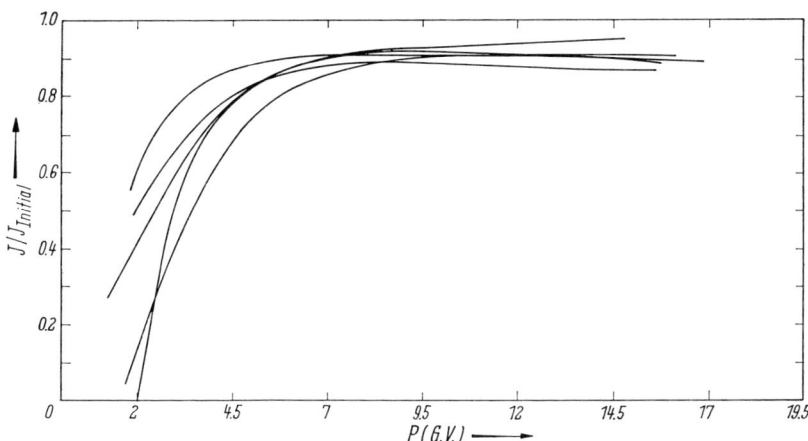

Fig. 25. Ratio of differential primary spectrum during a Forbush decrease to the pre-decrease spectrum. Five events exceeding 6% in high latitude sea-level neutron monitors during the IGY are shown.

26. Only sea-level neutron monitors have been included in this analysis. We notice that the modulation is remarkably flat between about 5 GV and 15 GV, in reasonable agreement with the experimental results given by DORMAN [3] for the decreases

22*

of 11 May 1959 and July 1959. Both phases of the decreases are rather similar, on average, while most of the variability between individual events occurs at $P < 4$ GV. Around 2 GV these results become unreliable, since the neutron monitor yield function becomes small here. WEBBER [2], who has considered balloon altitude results in his analysis of Forbush decreases, shows that the modulation is rarely greater than 50% in this rigidity region.

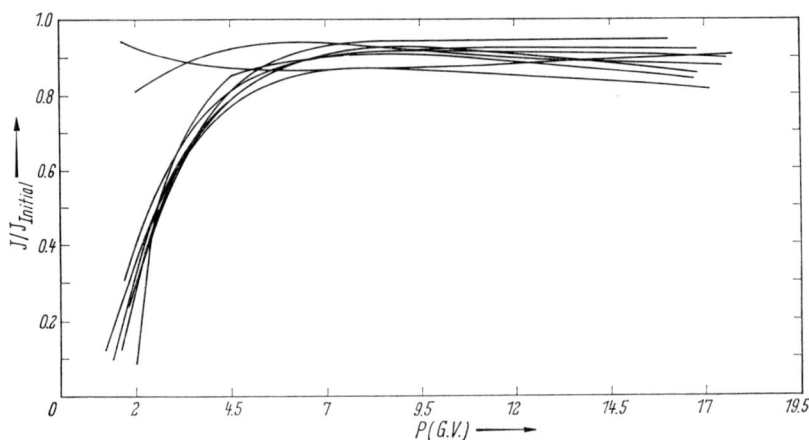

Fig. 26. Ratio of differential primary spectrum after recovery from a Forbush decrease to the spectrum during the decrease. Eight events exceeding 6% in high latitude sea-level neutron monitors during the IGY are shown.

18. The gradient in the interplanetary flux. Closely related to the estimation of the scale size of the modulation mechanism and the associated anisotropies is a measurement of the gradient of the cosmic ray flux in interplanetary space. Three experiments which provide some idea of this gradient in the ecliptic plane have been performed.

SIMPSON, FAN and MEYER (76) obtained data with a proportional counter telescope sensitive to protons of energy > 75 MeV during the flight of Pioneer V in March and April 1960. Data were received until the space probe was 0.1 A.U. closer in to the Sun. To find the gradient in intensity a correction must be made for time variations during the flight. This correction is attempted by assuming that the percentage variation of primaries at the Earth is equal to the percentage variation at the space probe. Correcting the data, these authors first claimed they found a gradient of $15 \pm 20\%$ per A.U. Later they suggested that the apparent gradient was due to the removal of particles in the 75 MeV to 1 GeV energy range during the Forbush decrease of 31 March 1960. These particles did not return during April and their removal could not be detected by the ground level neutron monitor used in correction the Pioneer V data. Thus they found the flux gradient to be $0 \pm 20\%$ per A.U. However, ARNOLDY and WINCKLER (77) do not find a corresponding permanent removal of particles by the FORBUSH event when analysing their ion chamber data from Pioneer V, but believe that the low energy particle flux returns to within 1% of the prestorm value.

Some information on the gradient away from the Sun comes from a shielded Geiger counter carried in the Russian probe Mars I (78). Data obtained between 11 November 1962 and 1 January 1963, extending 0.24 A.U. further out than the Earth, are given in Fig. 27. These data have been corrected for time variations using Geiger counter rates measured over MURMANSK within a few gm/cm² of the top of the atmosphere. It has been assumed that intensity changes in space recorded by a Geiger counter are 2.5 times as large as those recorded by a similar counter at balloon altitudes (WEBBER [2]).

Mariner II ionisation chamber data (63) also are plotted in Fig. 27. Both sets of data have been normalized to 100% close to the Earth. The Mariner II data extend about 0.25 A.U. closer in towards the Sun and were obtained between 1 September 1962 and mid-December 1962. They have been corrected using the

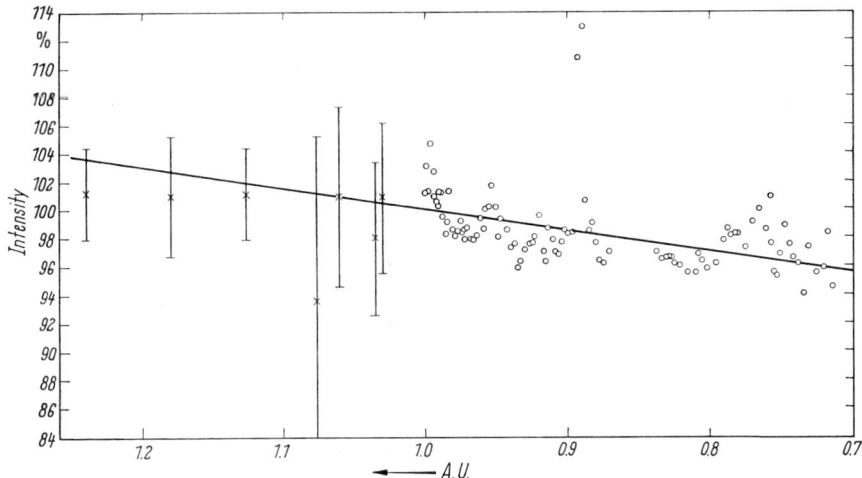

Fig. 27. Interplanetary energetic particle flux as a function of distance from the Sun. Crosses-Mars I Geiger counter data; circles-Mariner II ion chamber data.

fact that the short-term variations of the Mariner II ion-chamber are observed to be 3 times as large as those seen by the Deep River neutron monitor. No account is taken of delays in the onset of these variations between Mariner II and the Earth, or of any fluctuations of individual events about the mean 3:1 ratio. The peak at about 0.9 A.U. is thought to be due to solar protons. Assuming that our corrections and normalizations are valid, the combined data yield a gradient in the flux of low energy cosmic rays of $14\pm7\%$ per A.U.

19. Direct measurement of the interplanetary magnetic field. Great difficulty is experienced in measuring directly the interplanetary magnetic field with magnetometers carried in space probes since the field to be measured — a few gammas in strength (1 gamma $=10^{-5}$ oersted) — is often smaller than the residual field due to the spacecraft itself. Valuable results have been obtained, however, from Pioneer V (79), Mariner II (80) and IMP-I (80a).

Pioneer V carried a spinning coil magnetometer which gave the component of the field perpendicular to the spin axis free from errors due to spacecraft fields. It is believed that the spin axis always pointed within 27° of the radius vector between the Sun and the vehicle and thus a measurement of the field was made, without any knowledge of phase, in the plane normal to the direction to the Sun. This direction lay nearly in the ecliptic plane. An average perpendicular field component of 2.7 gamma was measured in March and April 1960, but on 1 April field strengths of up to 50 gamma were observed. As a result of a flare lasting from 1455 to 1858 hrs U.T. on 30 March, solar plasma is believed to have arrived at Pioneer V at 0720 hrs U.T. on 31 March, causing the onset of a Forbush decrease together with an increase in magnetic field (81) (Fig. 28). A flare of importance 3 at 0845 to 1222 hrs. U.T. on 1 April gave rise to a solar proton event observed at Pioneer V (Fig. 28) and at the Earth, commencing at 0935 hrs

U.T. (82). It is clear from this figure that solar protons could arrive at the Earth while fields of 10 to 20 gamma or more existed in interplanetary space.

From the time lapse between the occurrence of the flare on 30 March and the onset of geomagnetic disturbances at the Earth, a velocity of 2000 km/sec has been deduced for the solar plasma causing the event. Now detailed magnetometer data during the Pioneer V storm show times of relatively constant field for periods of 20 minutes or more (83), suggesting that the scale size of magnetic inhomogeneities must be $\gtrsim 2\times 10^{11}$ cm. However, an examination of the figures published in this reference reveals times when the field changed by $\sim 25\%$ in a $\frac{1}{2}$-minute interval, showing that some inhomogeneities are $\sim 10^{10}$ cm in scale.

At the time of writing, a final analysis of Mariner II data has not been completed and we can only mention here some preliminary results (84). Three fluxgate magnetometers mounted in the stabilised vehicle gave all three components of the field, although only the component perpendicular to the ecliptic plane was reasonably easily obtained free from error due to residual vehicle fields. For this last component, field values ranging between ± 5 gamma were seen, the average field for a day or more being close to zero. In the ecliptic plane the results may be variously interpreted at present as giving a field at 45° to the Earth-Sun line and ranging between $\pm(5-10)$ gamma on average, or as giving a field more nearly radial in direction but varying between 0 and 10 gamma in amplitude.

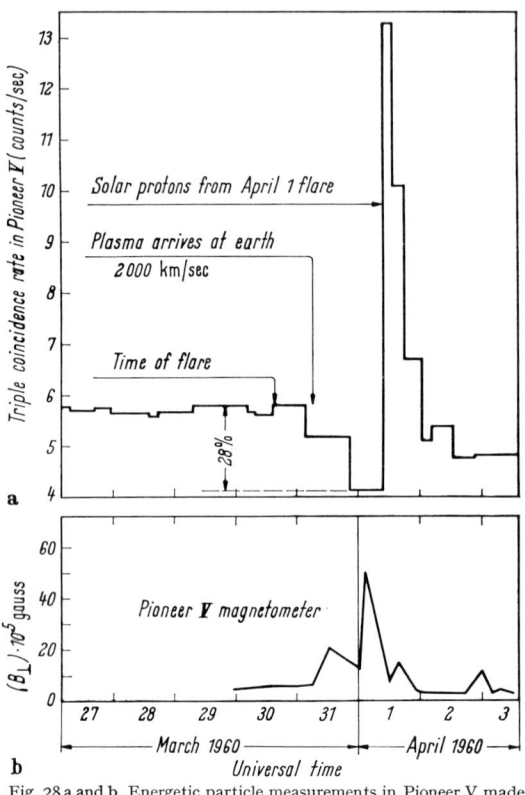

Fig. 28 a and b. Energetic particle measurements in Pioneer V made with a counter telescope sensitive to protons >75 MeV and compared to Pioneer V magnetometer data during the 1 April 1960 solar flare event. After FAN et al. (81).

IMP-I, launched in November 1963 with an apogee of 31 earth radii carried fluxgate magnetometers to provide vector field measurements. The results (80a) distinguished between the boundary of the magnetosphere at 10 earth-radii and the shock front, due to the solar wind interacting with the geomagnetic field, at 13 earth-radii on the sun-earth line. Field measurements outside the shock front gave average values of the interplanetary field between 4 and 7 gamma. Zero field was rarely observed. The average direction was 10° to 20° below the ecliptic plane and the field component in the ecliptic plane was at an angle of either 30°—50° to the earth-sun line or exactly opposed to these directions. However almost every conceivable direction was seen at some stage. Comparison of the fluxgate data with rubidium vapour magnetometer measurements on the same satellite, together with the absence of spin modulation of the rubidium data, suggested that the field values observed were accurate to ± 0.3 gamma.

20. Plasma measurements in interplanetary space. The existence of plasma, emitted from the Sun and propagated through interplanetary space, has been inferred from indirect evidence such as the orientation of comet-tails and the correlation of solar and geomagnetic disturbances. Now, some direct measurements of the plasma have been achieved with recent space probes and eccentric orbit satellites.

Explorer X, carrying a multi-grid Faraday cup, obtained plasma data on 25—27 March 1961 during the initial stages of its orbital life as it moved out to a distance of 43 Earth-radii (85). Out to about 20 Earth-radii, no plasma reached the satellite because it was still inside the geomagnetic cavity. Further out, protons were observed coming from a direction within at least 10° of the Sun.

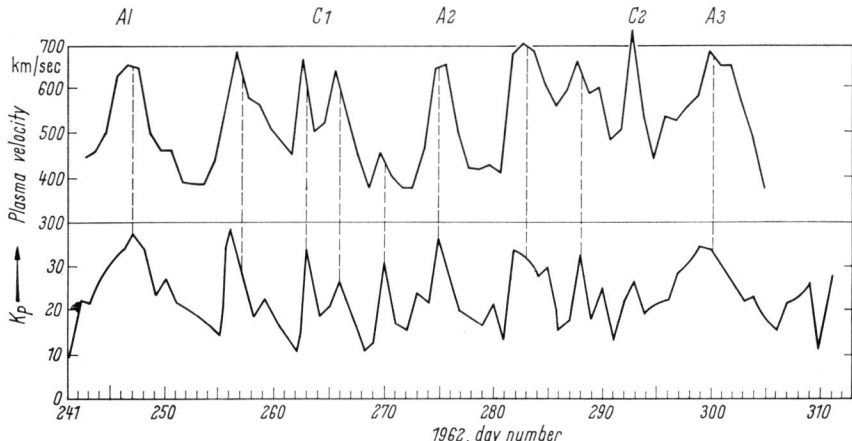

Fig. 29. Solar plasma velocity deduced from the Mariner II data and compared with the geomagnetic activity index, K_p. After SNYDER et al. (86).

From the mean measured proton velocity of $3 \cdot 10^7$ cm/sec and the mean flux of $3 \cdot 10^{-8}$/cm²/sec, the average density of plasma protons was found to be 10/cm³. Using the spread in measured proton energies, it was estimated that the temperature of plasma protons is $\sim 10^5$ °K.

Mariner II carried a cylindrical electrostatic analyser which gave the energy per unit charge of positively charged ions in the solar plasma. The aperture of the detector ($\pm 8°$ roughly in angular width) was kept pointing at the Sun. At all times during the Mariner II flight, plasma was seen coming from the Sun (86) and two ionic components were distinguished, presumably corresponding to protons and alpha-particles of the same velocity. The proton velocities lay between 360 and 700 km/sec, the temperatures between $6 \cdot 10^4$ and $5 \cdot 10^5$ °K, the number densities, n_p, between 0.3 and 10 protons/cm³, and the energy densities between $2 \cdot 10^{-7}$ and $2 \cdot 10^{-8}$ erg/cm³. Since the magnetic field energy density, $\left(\frac{H^2}{8\pi}\right)$, in the plasma is $\sim 10^{-9}$ to 10^{-10} erg/cm³, the plasma energy density is dominant and the plasma velocity exceeds the Alfvén velocity, V_A, given by

$$\frac{1}{2} n_p V_A^2 = \frac{H^2}{8\pi}, \tag{20.1}$$

thus enabling hydromagnetic shock phenomena to occur.

Fig. 29 (86) shows the daily average of the plasma velocity compared to the planetary magnetic activity index, K_P. Good correlation between these two parameters is shown, together with the tendency for the peaks to recur after 27-day intervals. Thus the Mariner II results are not consistent with the idea of a steady, symmetrical solar wind, but rather with the existence of long-lived local regions in the corona with have higher than average activity. The steady velocity component must be limited to 400 km/sec according to these data. Little correlation is found between the Mariner II measurements of cosmic ray intensity and the plasma velocity, and Elliot (87) has shown that the amplitude of the daily variation of cosmic ray intensity measurement at Deep River, is also uncorrelated with the wind velocity. A theory of Ahluwalia and Dessler (88), attributing the cause of the anisotropy to the effects of a regular spiral magnetic field generated by the solar wind, predicts an increased amplitude with increased wind speed.

Although both Explorer X and Mariner II demonstrate that abrupt changes in the plasma correlate in time with abrupt interplanetary magnetic field changes, no pattern for these variations has yet emerged. To establish certain properties of the interplanetary plasma, the following parameters have been assumed: the particle density $=10/cm^3$, the temperature $=10^5$ °K, and the plasma velocity $=300$ km/sec. These not unreasonable values yield [e.g. Lüst (89)] an electrical conductivity for the plasma of $1\cdot 8.10^{14}$ sec^{-1} near the Earth and an Ohmic dissipation time of 10^{13} years, assuming a characteristic length of 1 A.U. Therefore, magnetic fields present in the interplanetary space should be frozen into the moving plasma.

b) Solar energetic particles.

A great amount of experimental data has been amassed recently on the arrival from the Sun of particles in the energy range 10^6 eV to 10^{10} eV. This evidence can be conveniently divided into two categories. First, there is the determination of the arrival direction of the particles which travel most quickly, this measurement giving some indication of the large-scale regularities in the interplanetary magnetic field. Second, a study of the time variations in intensity for as long as solar particles remain in the vicinity of the Earth indicates the ability of the interplanetary field to trap or scatter these particles.

Table 5.

Event	Arrival direction		Remarks
	°E or W of Earth-Sun line	Declination	
17 July 1959 . . .	isotropic		small event
4 May 1960	55° W	$+10°$	well-established
12 November 1960	40° W	$-20°$	well-established
15 November 1960	50° W		well-established
18 July 1961 . . .	isotropic		small event
20 July 1961 . . .	60° E		uncertain
28 September 1961	55° W or 125° E	0°	for 100 MeV protons
	90° E to 125° E	0°	for 2 to 15 MeV protons

21. Arrival characteristics. Definite impact zones have been noticed on the Earth during the initial stages of energetic solar particle events. Use of the techniques mentioned in Sect. 10 enabled Firor (54) to find the asymptotic directions of the solar particles for three events seen at sea level during the early days of cosmic ray recording, namely those occurring on 28 February 1942, 7 March 1942, and 19 November 1949. The particles had come from within roughly 30° of the Earth-Sun line. A similar analysis shows that during the large event of 23 February 1956 particles also came initially from within 30° to 50° of the solar direction. More precise measurements are available for the recent sea-level events, since several very high latitude neutron monitors are in operation and MCCRACKEN's method (Sect. 10) which depends on these stations can be used. Table 5 summarizes these results.

CARMICHAEL (90) gives a very useful review from which these results were taken. For some events, the dependence of intensity on the angular departure, δ, from the mean asymptotic direction has been found, and the results for the flare of 4 May 1960 are shown in Fig. 30. Directional observations (91) in a very eccentric-orbit satellite yielded the data on the event of 28 September 1961. The 2 to 15 MeV proton directions are rendered uncertain by possible corrections

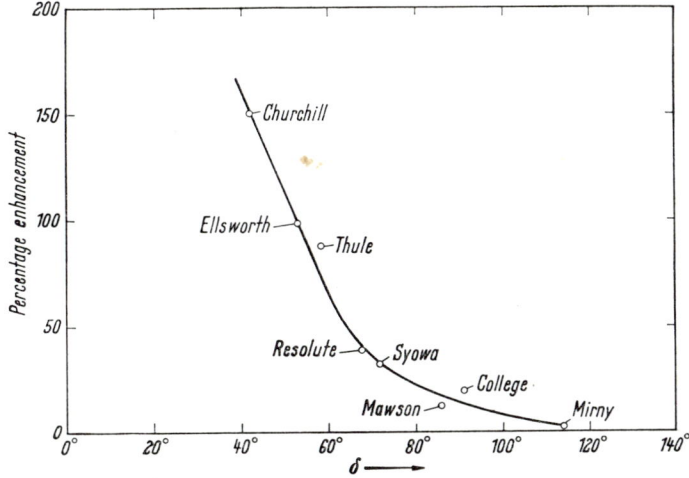

Fig. 30. Intensity-dependence on the angular departure, δ, from the mean asymptotic direction for the arrival of flare particles during the 4 May 1960 event. After K. G. MCCRACKEN, J. Geophys. Res. **67**, 435 (1962).

for the geomagnetic field. Whereas it is certain that in some events solar particles come from a direction 40° to 55° W of the Earth-Sun line, and probable that other events exhibit and isotropic onset for 1 GV particles, the existence of other arrival directions — especially to the east — is not established or denied on the basis of present analysis.

Events with well marked impact zones tend to originate from solar flares on the west limb of the Sun. Further, the time taken for the flare particle intensity to reach a maximum at the Earth is often shorter for these west limb events. Table 6, given by MCCRACKEN [17] illustrates these points.

REID and LEINBACH (23) have demonstrated the tendency in low energy ($\sim$100 MeV) events for the delay time between particle acceleration and particle arrival at the Earth to be greater for east limb flares.

To explain the various results just mentioned, MCCRACKEN proposes that a roughly spiral shaped interplanetary magnetic field is set up in the ecliptic plane, due to the constant dragging out of magnetic field lines by plasma emitted from active regions on the Sun. The combination of the east-to-west solar rotation, as seen by the Earth, and the radial outward plasma flow produces the spiral shape which should guide solar particles to the Earth to arrive from 40° to 55° W. On this model, solar flares occurring on the west limb are more likely to result in impact zones and fast arrival. One version of this model demands that each time impact zone effects are seen a previous cloud of plasma has recently passed the Earth and arranged the interplanetary field in the spiral configuration.

All the events listed in Table 5 took place during a Forbush decrease, except those of 12 November 1960 and 28 September 1961. However, a sudden-commencement type of geomagnetic disturbance occurred at the beginning of the 12 November event. Thus, in almost all cases there is indirect evidence for the necessary plasma cloud. Despite this favourable circumstance, and also the location of the flare source on the west limb, two of the listed events (17 July 1959 and 18 July 1961) showed an isotropic particle increase.

Further complexity is introduced when we study the way in which the anisotropic flux distribution becomes isotropic. McCracken (92) cites evidence sug-

Tabelle 6.

Event	Time Scale	Position on Solar Disk	Impact Zones
4 May 1960	1	90° W	Very marked*
23 February 1956..	4	80° W	Marked*
15 November 1960 .	6	45° W	Very marked*
19 November 1949 .	6	70° W	Very marked*
7 March 1942. ...	7	90° W	Marked
28 February 1942 .	13	4° E	Poorly defined
25 July 1946 ...	17	15° E	Not noticeable
12 November 1960 .	20	10° W	Poorly defined*
17 July 1959	36	30° W	None*
3 September 1960. .	50	90° E	None

* Event for which a large number of observations are available and on which the greatest reliance can be placed.

gesting that during the 4 May 1960 event the radiation remained anisotropic for at least 9 hours although the magnitude of the anisotropy became much smaller as time progressed. Isotropy set in suddenly on 12 November 1960 with the onset of a Forbush decrease; but the plasma responsible for this cosmic ray modulation also appeared to be able to trap solar particles, since the flux of these latter increased significantly at this time (93). Later still, on 13 November, a further plasma disturbance — from the flare which caused the acceleration — reached the Earth and appeared to remove all remaining solar particles from the neighbourhood (94). In other events, the radiation became isotropic without the possible assistance of a Forbush decrease. For about 5 minutes after the onset of the event of 23 February 1956, no radiation was seen outside the impact zones but the flux became completely isotropic after $1\frac{1}{2}$ hours. During the first 30 minutes of the event of 15 November 1960, there was a similar absence in some directions and the radiation took 2 hours to become isotropic although the initial increase outside the impact zone was sudden. These time delays for particle trajectories apparently not connected directly to the Sun led Meyer, Parker and Simpson (95) to postulate the existence of a scattering barrier at 1.4 A.U. enclosing a field-free region, and McCracken [17] to suggest that the two types of particle arrived by two separate routes differing in length. Bryant et al. (96) found that although on 28 September 1961 the anisotropy at 100 MeV lasted only about $\frac{1}{4}$ hour, at 2 to 15 MeV it was detectable for 6 hours. They also found low energy (9 to 30 MeV) particles brought along by the Forbush decrease cloud 2 days later in the final stages of decay of this event.

Carmichael (90) has estimated delay times between the production of particles on the Sun and their incidence on the Earth, based on the spiral model of the interplanetary magnetic field which gives a 55° inclination to the Earth-Sun line at the Earth. For four impact zone events, the calculated transit times were around 12 minutes but the observed times were about twice as long.

One solar particle event is known to be associated with a flare on the back side of the Sun. This occurred on 20 November 1960 and was due to a flare at least 20° in longitude behind the west limb (*90*). The increase at sea level was only ∼5% and no impact zone effects were detected.

The spectrum of solar particles varies from event to event, and tends to steepen with time during individual events. For particles >1 GV a differential spectrum which is a negative power law in rigidity seems applicable. Some values of this exponent are listed in the following table.

Table 7.

Event	Time	Exponent	Reference	Remarks
23 February 1956	0350	−4.0	Pfotzer (*96*)	Anisotropic phase
	0600	−7.0	Meyer et al.	Isotropic phase
4 May 1960	1045	−5.0	McCracken (*92*)	Anisotropic phase
	1100	−6.0	McCracken (*92*)	Anisotropic phase
12 November 1960	1600	−7.0	Mathews et al. (*26*)	Anisotropic phase
13 November 1960	0200	−8.5	Mathews et al. (*26*)	Isotropy + Forbush decrease
15 November 1960	0500	−7.2	Mathews et al. (*26*)	Isotropic phase

In the rigidity range 0.05 GV to 1 GV, the spectrum is best represented by an exponential form, $J(P) = J_0 \exp(-P/P_0)$, where P_0 is a function of time during an event. Frier and Webber (*97*) have demonstrated this form to fit the results for 16 events between 1956 and 1961. There is a suggestion that in this rigidity range the spectrum is steeper if the flare event occurs during a Forbush decrease (*98*).

22. Decay characteristics. Several investigators have noticed that once the flare particle flux has become isotropic the rise of intensity to its maximum and the initial stages of its decay may be represented by the law for the diffusion of particles injected at $t=0$ in an infinite and static diffusion medium with an isotropic mean free path, λ. If D is diffusion coefficient for the medium, the density of particles n at distance r from the point of injection is

$$n = \frac{N}{8(\pi D t)^{\frac{3}{2}}} \exp\left[-\frac{r^2}{4Dt}\right] \tag{22.1}$$

where N is the total number of particles injected and $D = \lambda v/3$ where v is the particle velocity.

Dorman [*3*] finds some agreement between this expression and the time-dependence of the first five sea-level events recorded up to 23 February 1956. A mean free path ∼10^{12} cm best fits the results for particles ∼1 to 2 GV. However, since an examination of Dorman's diagram ([*3*], Fig. 85) shows that the actual intensity curves exhibit sharper peaks than the predicted curves, we must regard this diffusion approach as a preliminary approximation. More recent events, investigated using this approximation, are listed in Table 8 together with the value of λ obtained and the time from the start of the event during which the approximation held.

During the final stages of decay of solar particle events, lasting up to 6 days from the time of injection, the time-dependence is found to be roughly exponential.

One — but not unique — explanation for this form is that the particles are diffusing in a medium which is spherical but surrounded by free space. On reaching the boundary, particles are lost catastrophically. PARKER [*4*] has given an expression for the particle density in a medium of radius r_0, based on the theory of heat flow which is mathematically similar [e.g. (*101*)]. After a time such that $Dt \gg r_0^2$, PARKER finds

$$n = \frac{N \sin(\pi r/r_0)}{2 \cdot r_0^2 \cdot r} \exp\left(-\pi^2 \frac{D \cdot t}{r_0^2}\right) \tag{22.2}$$

where the particles are released at $r=0$, $t=0$.

By determining experimentally the relationship between n and t, the constant term in the exponential may be found and, assuming the value of D obtained from the earlier stages of diffusion, r_0 can be calculated. Values or r_0 also are given in Table 8 for four events which fitted infinite diffusion theory early on and later

Table 8.

Event	Mean free path (A.U.)	Barrier distance (A.U.)	Rigidity of particles	Time applicable	Reference
3 September 1960	0.02	2.2	~0.5 GV	5 hrs	WINCKLER (*99*)
18 July 1961	0.025	1.8	~0.5 GV	1 hr	HOFMANN (*100*)
20 July 1961	0.04	1.7	~0.5 GV	2 hrs	HOFMANN (*100*)
28 September 1961	0.04	3.1 / 1.8	0.7 GV / 1.8 GV	Between 7 and 20 hrs	BRYANT (*91*)

exhibited an exponential particle decay. WEBBER [*2*] illustrates the decay of at least five other solar particle events which show a roughly exponential time-dependence.

We can conclude that the isotropic phases of solar proton events can be very crudely represented by diffusion in a spherical medium which has a mean free path $\sim 3 \cdot 10^{11} - 10^{12}$ cm and is about 2 A.U. in radius. Note that the results obtained for infinite medium diffusion in the earlier stages cannot be much modified by the presence of a boundary at 2 A.U. Most of the results apply to an intermediate stage in the sunspot cycle and none correspond to the extraordinary peak in activity around 1957/58. Detailed analysis of the September 28, 1961 event by BRYANT, CLIVE, DESAI and McDONALD (*101*a) suggests that the diffusion mean free path is independent of magnetic rigidity between 0.06 GV and 1 GV. They found that the relative intensities of protons of different energies followed the same curve when plotted as a function of the distance travelled by the particles.

One other class of observation which can yield data on the propagation of solar particles is the study of the charge composition. BISWAS, FICHTEL, GUSS and WADDINGTON (*102*) have obtained emulsion measurements of the solar particle flux at six different times during the later stages of three events. The relative abundances of protons, alpha-particles and medium nuclei (C, N, O, F) seem similar during all flights when measured either as a function of kinetic energy per nucleon, which is simply a function of particle velocity, or as a function of the product of particle velocity and particle rigidity. There is some suggestion that at early times during an event it is velocity, and later on $v \times P$, which is the best parameter; but the result has poor statistical weight. Thus a theory giving a diffusion coefficient proportional to $v \cdot P$ or v would be consistent with these results.

IV. Modulation mechanisms.

a) Introductory survey.

23. It is generally believed that the time variations in the cosmic ray intensity arise from single-particle interaction with electric or magnetic fields in interplanetary space. The density of matter here is too low to affect the intensity by Coulomb collisions. Both static and time-varying electric or magnetic fields have been considered as causes of modulation. NAGASHIMA (*103*) postulated the existence of an electric potential on the Earth, positive relative to the galaxy, and EHMERT (*104*) suggested that a similar potential was centred on the Sun. JANOSSY (*105*) discussed the effect of the solar dipole moment assuming an empty interplanetary space, and ELLIOT (*106*) considered a dipole field set up by current systems flowing in the solar corona. The possibility that solar plasma pushes back the galactic magnetic field to form a cavity around the solar system 10 to 100 A.U. in radius is discussed by DAVIS (*107*) and BEISER (*108*). The former author suggests a field-free cavity, while the latter postulates a diamagnetic cavity with a field $\sim 10^{-4}$ gamma inside.

This class of models, involving static fields, is rendered unlikely by the present knowledge of interplanetary conditions. The high conductivity of interplanetary space (Sect. 20) makes the existence of the necessary electric potentials, $\sim 10^9$ volts, implausible. Measurement of the possible dipole moment of the Sun yields figures which would cause negligible effects on cosmic rays at the Earth. Since existing interplanetary magnetic fields must be frozen in to the interplanetary plasma (Sect. 20), the presence of a dipole-like field with the moment orientated along the axis of solar rotation and set up by coronal currents seems incompatible with the observed continual outflow of solar plasma forming the solar wind. Direct field measurements are also in contradiction to this model. A field-free or diamagnetic cavity in the galactic field cannot exist since it is known to be filled with fields ~ 1 to 10 gamma out to at least the Earth's orbit. Further, the degree of perfection in the magnetic walls of the cavity necessary to prevent particles leaking in faster than the time constant of the 11-year variation would seem to be impossible to achieve in practice.

Turning to the time-varying class of models, we must clearly have in each case both a varying magnetic field and a varying electric field. The distribution, between the models lies in the relative importance given to the two effects. Suppose in a system of co-ordinates at rest with respect to the solar system we measure electric and magnetic fields E and H in an element of plasma. Then in the reference frame moving with the plasma velocity v the fields are E' and H' given by

$$E' = \left(E + \frac{v}{c} \times H\right) \frac{1}{\sqrt{1 - v^2/c^2}}, \tag{23.1}$$

$$H' = \left(H - \frac{v}{c} \times E\right) \frac{1}{\sqrt{1 - v^2/c^2}}. \tag{23.2}$$

for space where $\varrho = \varepsilon = 1$. Because of the high conductivity of plasmas in interplanetary space, $E' \approx 0$ and, since $v \ll c$,

$$E \approx -\frac{v}{c} \times H. \tag{23.3}$$

Also

$$H' \approx H. \tag{23.4}$$

If the field in the plasma frame $\boldsymbol{H'}$ is time-varying, for example due to the steady expansion of the plasma as each individual particle flows radially out from the Sun, then there is a contribution to $\boldsymbol{E'}$ from

$$\operatorname{curl} \boldsymbol{E'} = -\frac{\partial \boldsymbol{H'}}{\partial t}. \tag{23.5}$$

FÄLTHAMMAR (109) discusses one simple case of radial flow where $\boldsymbol{E'}$ is of first order in small quantities while curl $\boldsymbol{E'}$ is of zero order.

ALFVÉN [18] was the first to suggest that cosmic ray variations were due to deceleration in the electric field $\boldsymbol{E}$ associated with outward moving streams carrying rather regular frozen in magnetic fields. A transverse field of 10 gamma and a plasma velocity $v = 3 \cdot 10^7$ cm/sec yields a field of $3 \cdot 10^{-5}$ volt/cm. However, this field can only act for as long as the particle remains within the plasma cloud, and this is in turn determined by the magnetic field $\boldsymbol{H'}$. A quite general approach to the problem is due to FERMI (110), who considers the particle approaching a plasma and transforms momentum-energy to plasma co-ordinates where $\boldsymbol{E'} \approx 0$ so that there can be no energy change. The plasma field only alters the particle direction, and a final transformation back to the rest frame with this new velocity vector then yields an energy change. Maximum energy loss is obtained in a tail-on collision where the velocity vector is reversed and

$$\frac{\Delta U}{U} = -\frac{2v}{c} \text{ of } \omega \approx c \tag{23.6}$$

where U is the total energy and ω is the particle velocity. Thus in practice $\frac{\Delta U}{U} \sim -2 \cdot 10^{-3}$ and clearly a single collision cannot account for the large spectrum changes associated with the 11-year and Forbush variations. DORMAN [16] has considered this effect as a cause of the small daily variation, however.

Two possibilities exist for making ALFVEN's mechanism more effective. The particle may make a large number of collisions if many solar gas clouds inhabit interplanetary space at one time. Then statistically the particle may lose much more energy. SINGER, LASTER and LENCHEK (111) have studied this effect and have also included the deceleration associated with curl $\boldsymbol{E'}$ in Eq. (23.5) for each contact of a particle with an expanding gas cloud. Alternatively, one can emphasise the screening effect of the regular magnetic field associated with the solar streams, as does DORMAN [16]. This magnetic field configuration has been given more reality by COCCONI, GOLD, GREISEN, HAYAKAWA and MORRISON (112), who suggest that the loops of magnetic field observed in the inner corona are dragged out to form magnetic tongues which can exclude cosmic rays.

These extensions to ALFVÉN's mechanism will be considered in more detail, as will the models which depend only on the diffusion of cosmic rays through a cloud of magnetic scattering centres. MORRISON (113) studied the filling of an originally empty diffusing region, while PARKER [4] demonstrated the modulation effected by outward-streaming scattering centres. Finally, we must look at the class of mechanisms postulated by PARKER [4] involving a rather regular spiral interplanetary field into which various kinks or constrictions are introduced.

Since the difficulty is to find a model which gives a sufficiently large modulation to explain the observed cosmic ray effects, we shall examine first those depending on regular field configurations. A regular arrangement of magnetic field is much more economical in producing a given amount of screening than is a disordered field.

b) Modulation by regular magnetic fields.

24. Magnetic tongues. DORMAN [16] has attributed the Forbush decrease to the enveloping of the Earth by a solar stream carrying with it a homogeneous magnetic field. To compute the decrease in intensity for such a stream two effects must be considered. First, particles moving at right angles to the magnetic field can only arrive if their rigidity, P, is greater than

$$P_{\min} = 300\, H\, \frac{d}{2}, \qquad (24.1)$$

where d is the minimum distance between the Earth and the edge of the stream. Particles just above $P_{\min}$ will only be able to arrive from certain directions in space. Second, to find the intensity of particles moving parallel to the field direction, the configuration of the field lines in interplanetary space must be known in order to determine whether or not galactic cosmic rays could ever get onto these lines of force. This second question is only answered in the magnetic tongue model which has been developed further by GOLD (114). Fig. 31 illustrates the solar field lines, present originally above some active groups, which are frozen into plasma ejected by the activity and are therefore carried out to form an extended dipole enveloping the Earth. Later still, the field lines may break off from their anchoring positions on the Sun. Thus the Gold model completes the field line topology which was not present in the Dorman model. If the field lines were very regular, low energy cosmic rays would be completely excluded from inside the tongue since at no stage could they enter and hook onto the field lines. To obtain some idea of what happens at higher rigidities, let us approximate the tongue by a dipole moment located on the Sun. Using Störmer theory and remembering that the Earth is in the dipole equatorial plane, we find a minimum and maximum rigidity of arrival given by

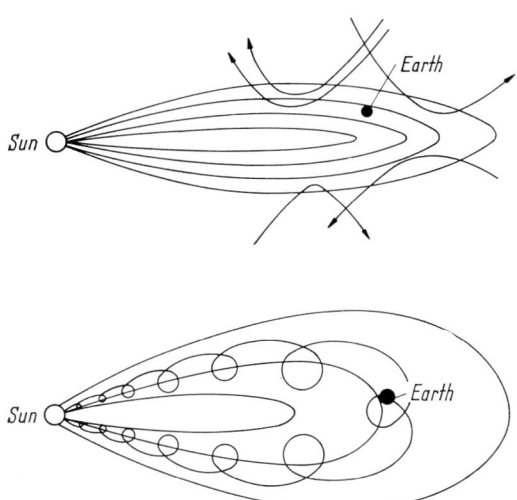

Fig. 31. Gold tongue model of the interplanetary field associated with the passage of plasma emitted by an active region. The deflection of galactic particles and the trapping of solar particles are shown. After T. GOLD, J. Geophys. Res. **64**, 1665 (1959).

$$P_{\max} = \frac{300\, M}{r_e^2}; \quad P_{\min} = \frac{1}{(\sqrt{2}+1)^2}\, \frac{300\, M}{r_e^2}. \qquad (24.2)$$

The directions of $P_{\max}$ and $P_{\min}$ would be 180° apart but at right angles to the Earth-Sun line. Thus a dipole moment of $2.3 \cdot 10^{34}$ gauss cm³ would exclude all particles of $P < 5$ GV and produce a daily variation in particles with $5\ \mathrm{GV} < P < 30\ \mathrm{GV}$.

However, we have seen that the daily variation in a Forbush decrease is probably no greater than one-tenth of the total modulation (Sect. 15) and that the percentage of screened particles shows little change from 3 GV up to at least 15 GV (Sect. 17). Therefore, regular fields are unlikely to explain the observed characteristics of the Forbush effect.

It might be argued that small inhomogeneities inside the tongue can cause enough scattering largely to remove the anisotropies. In order to assess this possibility, it is necessary to estimate the spatial extent of the tongue. Forbush decreases can last for at least a week and, in some cases, up to two weeks or more. Thus, for all this time, the Earth must be immersed in the tongue. Fig. 32 shows the expected field line configuration in the equatorial plane in the most favourable case for a long decrease. We suppose that some field lines have very quickly reached the Earth while others are just starting out from the Sun. The solar rotation gives the spiral twist to the field lines. Now even the slowest field lines must pass the Earth after a transit time given by the quiet-time solar wind velocity, that is in a maximum time of $1.5 \cdot 10^{13}$ cm $\div 3.5 \cdot 10^7$ cm/sec, which is $4 \cdot 10^5$ sec. Thus on this model the decrease can only last 5 days.

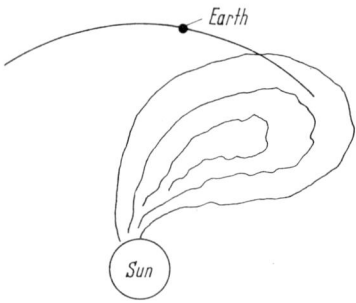

Fig. 32. Gold tongue in the ecliptic plane.

Although the tongue model does not seem to give quite the required time-scale, let us pursue it further. We shall assume a well ordered field barrier around the periphery of the tongue to exclude particles with greater disorder inside to bring about near anisotropy in the directions of particles which slowly leak in from the exterior. It is necessary to limit this leakage time to a minimum of $4 \cdot 10^5$ sec, otherwise there can be no decrease. Now the most likely source of leakage is particle drift perpendicular to a magnetic field gradient. The drift velocity is given by

$$\omega_D = \frac{1}{2} \omega_\perp R \frac{|\mathrm{grad}_\perp B|}{|B|}, \qquad (24.3)$$

where R is the Larmor radius and $\omega_\perp$ is the velocity perpendicular to the field direction. The drift must take place over ~ 1 A.U. For 10 GV particles in a mean field of 10 gamma, we find $L \sim 6 \cdot 10^{13}$ cm where $\frac{|\mathrm{grad}\, B|}{|B|} \sim \frac{1}{L}$. Since the total dimension of the tongue is $\sim L$, it is unlikely that the necessary lack of field gradients actually exists.

25. Shock wave production of a magnetic kink. A model for the Forbush effect which similarly relies on regular fields has been given by PARKER [4]. He is responsible for showing that, because of the high temperature of the corona near the Sun, static equilibrium is impossible and the only adequate way in which the observed coronal energy could be dissipated is by the continual outward flow of solar material. A coronal temperature of $6 \cdot 10^5$ °K led, according to PARKER, to a 300 km/sec wind velocity. Near the base of the corona, the gas pressure exceeds the magnetic pressure of the general, 1 gauss, solar field, and so this field is carried out by the wind. Since the field lines are rigidly fixed in the Sun at their "ends" but are carried out radially by the wind, the resulting field configuration is like a spiral where

$$\left. \begin{aligned} B_r(r, \theta, \Phi) &= B_r\left(r_1, \theta, \Phi - \frac{r\Omega}{v}\right)\left(\frac{r_1}{r}\right)^2, \\ B_\theta(r, \theta, \Phi) &= 0, \\ B_\Phi(r, \theta, \Phi) &= B_r\left(r_1, \theta, \Phi - \frac{r\Omega}{v}\right) \frac{r_1 \Omega}{v} \frac{r_1}{r} \sin\theta. \end{aligned} \right\} \qquad (25.1)$$

Polar co-ordinates are used with Φ in the equatorial plane, Ω is the solar angular velocity, v is the wind velocity, and the suffix 1 indicates values close to the solar surface. With $v=300$ km/sec, the field lines make an angle of 56° at the Earth with the radial direction. This idealized picture must be disturbed by plasma instabilities, and PARKER estimates that irregularities with a scale-size $\sim 5 \cdot 10^{11}$ cm can be present at 1 A.U.

It is onto the quiet-time field that the disturbances due to solar outbreaks are superimposed. GOLD (19) was the first to suggest that magnetic storm Sudden Commencements are due to shock waves in interplanetary space, and PARKER

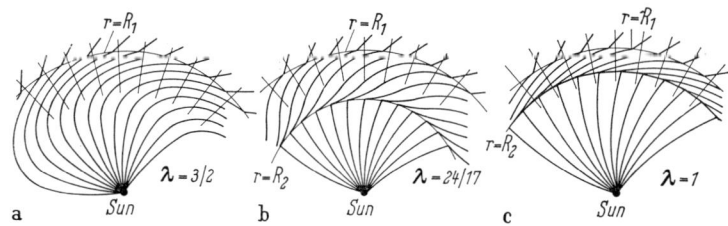

Fig. 33 a—c. Various types of kinks formed in the interplanetary field due to the passage of a shock wave. After PARKER [4].

has used this idea in connection with the cosmic ray effects. A solar flare leads to the ejection of solar material at velocities ~ 1000 km/sec — in excess of the velocity of sound in the interplanetary medium, which is ~ 100 km/sec. Thus a shock front propagates ahead of the solar material. This front causes various types of kinks in the spiral interplanetary field (Fig. 33) and once the shock front has passed the Earth some reduction in cosmic ray intensity may occur due to the restricted passage of particles through the kink.

To understand the intensity reduction occurring we will consider the motion of charged particles through the kink. Fig. 34a represents the equatorial plane with a weak field region B_1 separated by a shock front from the strong field region B_2. ψ_1 and ψ_2 are the angles which the respective field lines make with the normal to the shock front. The Earth is in the field region B_1. Particles perform helical motion in each field region and, after crossing the shock boundary, the instantaneous velocity vector from the previous region is preserved. Consider negatively

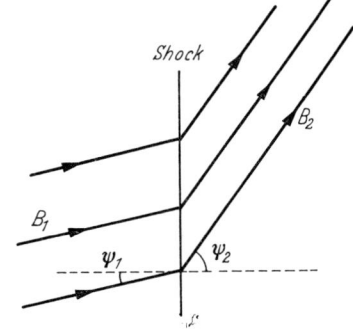

Fig. 34a. Geometry of Magnetic Shock.

charged particles moving from B_2 to B_1. Some particles with large pitch angles will cross the boundary and then recross back into B_2 again. However, since they will always have a smaller pitch angle in B_2 than previously, they will eventually stay permanently in B_1. Thus all particles reaching the boundary from B_2 are transmitted. This theorem has been proved in detail by WENTZEL (114a). Now reverse all particle trajectories by changing the sign of the charge. A positive particle distribution in B_1, which is isotopic over the 2π solid angle centred on the $\boldsymbol{B_1}$ direction towards the boundary, produces a similar isotropic distribution over the 2π solid angle centred along the $\boldsymbol{B_2}$ direction away from the shock.

Consider now negative particles approaching the boundary from B_1. Some particles with large pitch angles go back towards the boundary on entering B_2 and are thus ultimately reflected. Fig. 34b, computed by WENTZEL, shows regions of transmitted and reflected particles as a function of pitch angle θ and phase angle α in B_1 after interaction with the shock for two values of S where

$S = \cos\psi_1/\cos\psi_2$. Hence an isotropic distribution of positive particles along $\boldsymbol{B}_2$ towards the boundary will result in a distribution which is only isotropic in B_1 in the transmission region of velocity space and many large pitch angles in B_1 are unfilled.

A completely isotropic particle distribution on both sides of the boundary for all time is one distribution satisfying LIOUVILLE's theorem. Here, the directions in B_1 left vacant by particles transmitted from B_2 are filled by particles reflected back into B_1. If the shock front is moved slowly in space so that electric field effects are unimportant, no change in particle density can result on either side of the boundary. If a particle density gradient already exists with B_1 at the lower density and the shock is moving so as to expand the B_1 region, then there is a time lag while particles come through from the high density region B_2 to fill up the newly created B_1 space to its former level. A temporary reduction in intensity can thus occur. In this region of depressed intensity the directions in space with small pitch angles which are connected through the boundary to B_2 by transmitted trajectories at all phase angles will still see the original particle intensity. All other directions will eventually be refilled by transmitted particles which reflect or scatter deeper in B_1.

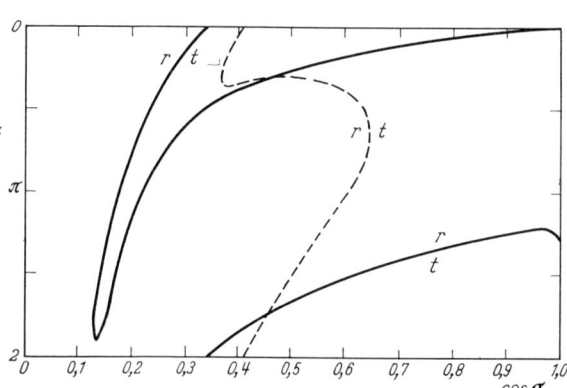

Fig. 34b. Boundaries in velocity space between trajectories transmitted (t) and reflected (r) into the weak-field side for a moderate shock with $S = 2^{\frac{1}{2}}$, $\psi_1 = 45°$ (dashed curve) and a strong shock with $S = 3$, $\psi_1 = 9°$ (solid curve). After WENTZEL (114a).

Two points emerge from this discussion. First, the mechanism requires a previously existing particle density gradient. Second, the intensity reduction would be seen as a large daily variation with some directions recording no decrease at all, rather than the almost isotropic effect observed. To produce both the original density gradient and the observed isotropy, we probably require scattering centres in the interplanetary field and it is perhaps best to class the shock front as a special type of centre to be considered in a discussion of modulation mechanisms caused by diffusion processes.

A second objection to PARKER's mechanism for the Forbush decrease lies in the time-scale of the effect. To obtain a decrease lasting 2 weeks with a shock wave travelling at 1000 km/sec required the inner field region B_2 to extend to a radial distance of 8 A.U. At this distance the angle between the quiet-time spiral field and the radial direction is 84°. It is a little doubtful whether the regular field can be maintained out to this distance.

A further difficulty arises from the leakage into the inner field region from the sides. Doing the same calculation as for the magnetic tongue model we again require the average field gradient, $\frac{|\text{grad}_\perp B|}{|B|}$, to be $\sim \frac{1}{L}$, where $L = 6 \cdot 10^{13}$ cm for a recovery time of 5 days. This seems difficult to achieve in practice. Further, if particle drift determines the recovery time, this time is $\propto \frac{1}{\omega_P} \propto \frac{1}{P}$, where P is particle rigidity. But we have shown that in the latitude-sensitive region the recovery seems to be independent of rigidity (Sect. 17) and it is only at underground depths that Forbush decreases exhibit a shorter time scale (*59*), corresponding to primary rigidities > 30 GV.

Although we have found difficulty in using the magnetic tongue or shock wave models to explain all features of a Forbush decrease, we must nevertheless admit the existence of strong evidence in solar particle events for the presence of this type of field configuration, at some times at least, between the Sun and the Earth. Before mentioning this evidence, we note that McCracken [*17*] has suggested that the early onset of some Forbush decreases from directions west of the Earth-Sun line (Sect. 15) can be qualitatively explained on either of these two models. With the Earth located just ahead of the advancing tongue or shock kink, the quiet-time spiral field connects detectors on the Earth looking to the west of the Earth-Sun line with regions of depressed cosmic ray intensity inside one or other of the two types of trap. Thus these detectors record the decrease first.

26. Regular fields and solar particle propagation. It has already been stated that in several energetic solar particle events beams of particles arrive initially at the Earth collimated to within $\sim \pm 45°$ of the mean direction. In the three most well attested cases, this mean direction lay 40° to 55° west of the Earth-Sun line. Further, delay times from west limb events are much shorter on average than from east limb events. McCracken [*17*] points out that the simplest way to explain these facts is to use the spiral model of the interplanetary magnetic field. Clearly, the mean free path for large-angle scattering is of the order of 1 A.U. for this type of propagation. Since it seems certain that the interplanetary field is greater than 1 gamma on the average, the Larmor radius of a 1 GV particle is less than 0.02 A.U.

However, easy passage from the Sun can be achieved if the particle spirals around a connecting line of force. Because of the solar rotation, such a field line is more likely to connect with an active region on the west limb. The field lines are formed either by the quiet-day solar wind or by the passage of a large plasma disturbance previous to a flare event. Evidence presented in Sect. 21 does not distinguish between these possibilities. Another means of easy passage from the Sun is along a neutral line. The interplanetary field must consist of field lines directed both towards and away from the Sun, because the net flux into the solar surface should be zero. Thus, separating groups of opposed field lines, there must run neutral lines. A particle starting off down such a line is reflected from the walls in a manner such as always to return to the zero field line, and thus the particle is guided along this line.

Without detailed knowledge concerning the region in which flare acceleration occurs, it is only possible to assume that the particles are more likely to find a strong field line than a neutral line. During their motion, the first adiabatic invariant is conserved if the field is regular. Now the field varies roughly as $1/r^2$ [Eq. (25.1)] so even if the particles were injected isotropically at 0.1 A.U. from the Sun, where $B = B_0$, use of the conservation law $\sin^2 \theta = \frac{B_E}{B_0}$ gives the maximum pitch angle, θ, at the Earth ($B = B_E$) to be only 6°. Hence, to explain the 45° beam width, McCracken suggested the presence of scattering centres in the field with dimensions of the order of the particle Larmor radius. Thirty-six scattering centres located on the connecting field line, and each capable of changing the pitch angle by 15°, would give the right order of scattering. If the field changed direction by 15° in a distance comparable to that covered by the particle in one Larmor revolution, then the pitch angle would change by roughly this amount. Alternatively, if the particle distribution were isotropic 0.3 A.U. from the Earth, the field variation over the remaining distance would provide the required amount of collimation provided that scattering gave a total pitch angle change less than 10°.

At high rigidities, where the Larmor radius is greater than the scale size of the field inhomogeneity, the mean free path is expected to vary as P^{-1}. Thus the time for travel through scattering centres is greater as P decreases. Only the event of 23 February 1956 has yielded detailed data on transit time as a function of rigidity. LÜST and SIMPSON (115) showed that it varied between 19 minutes at 11 GV and 28 minutes at 2 GV. Scattering throughout 1 A.U. cannot explain this difference with a P^{-1} dependence for the mean free path, but some fit to the results is obtained if the scattering region is restricted to within 0.3 A.U. of the Sun (115). To explain the particle collimation it seems necessary to require that only part of the space between the Earth and the Sun is filled with regular fields and the existence of a diffusion cloud close around the Sun is allowed. Such a cloud may explain the transit time anomalies noticed in three other events (Sect. 21).

In the type of model we have been discussing, particles from the east limb of the Sun have to cross lines of force to reach the Earth and therefore have a much slower arrival time than the west limb events. The event of 17 July 1959 was a west limb event which did not exhibit impact zones, and particles of all energies took 12 hours to reach their maximum intensity[1]. Thus there are times when lines of force do not connect the Earth with the Sun in a simple manner.

The spiral field configuration affords an explanation of the 4 April 1960 solar particle event observed on Pioneer V (Sect. 19). Here it was found that particles arrived promptly at the Earth (WEBBER [2]) while fields of 10 to 20 gamma existed between the Sun and the Earth. It may be assumed that the field measured by Pioneer V was the component in the azimuthal or easterly direction of a field lying mainly in the ecliptic plane and extending over a volume ~ 1 A.U. in scale size.

A magnetic tongue or travelling shock will also explain the delayed arrival of many solar particles during the event of 12 November 1960 (Sect. 21). While the flare at 1323 U.T. caused the prompt arrival of some particles, the intensity was almost doubled after 1900 U.T. when the plasma cloud or shock wave from the flare of 0305 U.T. on 11 November reached the Earth. A magnetic tongue would certainly trap many particles, but a shock kink may also fit the results since the transmission coefficient through the kink is not 100% and some particles can be confined. Let us assume that escape from the magnetic trap is predominantly due to particle drifts caused by an average gradient of magnetic inhomogeneity in the azimuthal direction. Then the drift velocity should not exceed the velocity given by the travel time of the trap to the Earth if appreciable numbers of particles are to be retained.

Thus

$$\left. \begin{array}{l} \omega_D \lesssim 1.2 \cdot 10^8 \text{ cm/sec} \\ \dfrac{1}{2} \omega_\perp R_\perp \dfrac{1}{L} \sim \omega_D \end{array} \right\} \tag{26.1}$$

where $\dfrac{1}{L} \sim \dfrac{|\text{grad}_\perp \boldsymbol{B}|}{|\boldsymbol{B}|}$ in the azimuthal direction. Taking the field to be 10^{-4} gauss and considering 1 GV particles, we find $L \gtrsim 2 \cdot 10^{12}$ cm ~ 0.1 A.U. Thus the field must be reasonably homogeneous over distances of this order. We note that this estimate for L is more than an order of magnitude less than that required to explain Forbush decreases on the regular field model.

---

[1] See WEBBER [2], page 140.

McCracken [17] has pointed out that the 30-minute delay in the arrival of particles from the anti-Sun direction during the event of 15 November 1960 and the rapidity of the subsequent increase suggest that either the prompt and late particles travelled by two different routes or a sharp reflecting barrier existed at 1.5 A.U. The former possibility is consistent with the tongue model and the latter with the model involving a kink in the field line. A diffuse barrier would be expected to cause too slow an onset. One positive statement favouring one of these two possibilities is made by Carmichael (90). He has evidence — though it is not conclusive — that the direction of anisotropy during the flare of 20 July 1961 was 60° east of the Earth-Sun line. The shock wave model can only give directions of anisotropy to the west.

Summarizing this discussion, it appears that studies of solar particle propagation yield evidence for the existence of homogeneous field regions with sizes probably at least ~ 0.3 A.U. but perhaps up to ~ 0.7 A.U. between the Sun and the Earth. Unperturbed, spiral field lines extending the whole distance are improbable, however. The magnetic traps formed by these homogeneous field regions may play some part in causing Forbush decreases but it is difficult to explain the whole event on this model.

c) Modulation by disordered magnetic fields.

27. Diffusion and convection. The remarkable lack of anisotropy in the long-term modulation and the rather small anisotropy present in the Forbush decreases imply that particle diffusion over large distances is a dominant process. Further, the little that is known about the average scale size of the inhomogeneities in the interplanetary field from both direct measurement and plasma theory suggests a value of 10^{10} to 10^{11} cm. Morrison (113) was the first to examine a modulation mechanism associated with diffusion. He supposed that turbulent magnetised plasma ejected from active regions on the Sun in a large cloud were responsible for Forbush decreases. The cloud is initially empty of cosmic rays; but as it expands and moves out into interplanetary space it becomes filled with particles according to

$$\frac{\delta J(P)}{J(P)} = -\exp\left(-\frac{\pi^2 \lambda \omega t}{3 r_0^2}\right), \qquad (27.1)$$

where $\delta J/J$ is the fractional depression of intensity at rigidity, P, λ is the mean free path at P, ω is the particle velocity, and r_0 is the radius of the cloud. This expression is obtained by solving the diffusion equation for a spherical, homogeneous region with fixed scattering centres, and yields the same time constant as Eq. (22.2) which describes the leakage of flare particles from a similar region. The most favourable situation is when the centre of the cloud just reaches the Earth so $r_0 \sim 1$ A.U. $=1.5 \cdot 10^{13}$ cm. Putting $t=10^5$ sec as the time during which the cloud was filling as it travelled to the Earth, Morrison found $\lambda = 6 \cdot 10^{10}$ cm for a 5% reduction in intensity. Since 10 GV particles are modulated by this amount and the Larmor radius cannot exceed the mean free path, the field strength H in the scattering centres is given by $H \equiv \dfrac{10^{10}}{300\,\lambda} = 50$ gamma. This value of H already appears to be larger than the observed fields and is still inadequate to explain Forbush decreases lasting longer than the order of a day.

Parker [4] has pointed out that an efficient modulation mechanism is obtained when the motion of the scattering centres is taken into account. Even regular fields swept out by the solar wind will break into inhomogeneities as various forms of plasma instability set in outside the Earth's orbit. The long-term modulation arises because, as particles diffuse in through this region of instability, some particles are swept outwards again after unfavourable collisions with scattering centres. Enhancement of the local wind velocity and field strength may help to explain the Forbush effect. PARKER's diffusion equation and method of solution under these circumstances is as follows.

Let $n(x, y, z, P)$ represent the number of particles per unit volume of rigidity P and of a particular charge Z. $\lambda(x, y, z, P)$ is the mean free path for large-angle scattering and $\Phi'(x, y, z, P)$ is the flux of particles per unit area in a given sense due to a gradient in $n(x, y, z, P)$. Kinetic theory gives

$$\Phi' = -\tfrac{1}{3} \lambda \operatorname{grad}[\omega \cdot n], \tag{27.2}$$

where ω is particle velocity. This is true in a frame of reference moving with the local scattering centres but in a fixed frame

$$\Phi = v n - \tfrac{1}{3} \lambda \operatorname{grad}(\omega \cdot n). \tag{27.3}$$

where v is the general, outward drift of the scattering centres and $v \cdot n$ is the local flux due to this drift. If there is no acceleration or absorption of particles, the equation of continuity is

$$\frac{\partial n}{\partial t} = -\operatorname{div} \Phi = -\operatorname{div}(v \cdot n) + \frac{1}{3} \operatorname{div}(\lambda \operatorname{grad} \omega \cdot n). \tag{27.4}$$

In the steady state with no sinks for galactic particles, no solar particles present and for a spherically symmetrical system, (27.4) reduces to

$$n(r, P) v = \frac{1}{3} \lambda \omega \frac{\partial n(r, p)}{\partial r}. \tag{27.5}$$

This simply expresses the fact that the outward flux due to the general sweeping caused by the motion of the scattering centres exactly balances the inward flux due to density gradient. Making the assumption that λ is not a function of position, integration gives

$$n = A \exp\left(\frac{3 v r}{\lambda \omega}\right). \tag{27.6}$$

Suppose that scattering is only effective out to $r = r_0$ and beyond this point $\lambda \to \infty$. Then

$$n_\infty = A \exp\left(\frac{3 v r_0}{\lambda \omega}\right) \tag{27.7}$$

and is the galactic cosmic ray density. Thus

$$n(r, P) = n_\infty(r_0, P) \exp\left(-\frac{3 v (r_0 - r)}{\lambda \omega}\right). \tag{27.8}$$

The depression in intensity caused by moving scattering centres is more effective than a static system where absorption by the Sun is the only particle sink. PARKER [4] shows that solar absorption only changes the intensity significantly within a few solar radii of the Sun if the barrier to galactic particles is static.

Before discussing the values to be given to the parameters in Eq. (27.8) and comparing the results with experiment, we will consider the modulation in intensity which may come about due to changes when particles encounter scattering centres.

28. Particle deceleration. Since the plasma energy is dominant, the interplanetary magnetic field participates in the outward motion of the plasma. This motion is chiefly radial. Thus each field inhomogeneity will gradually move away from the others as the plasma moves out. Further, the element of plasma holding any one inhomogeneity will expand in volume and thus the field will decrease as measured in a reference frame moving so as to be rigidly attached to one point in this element of plasma. Two sources of energy change are possible. The net velocity of recession will cause a deceleration by the Fermi mechanism. Also, while particles are in contact with expanding magnetic field an inverse betatron mechanism can occur.

To investigate the energy changes we shall, basically, follow the work of SINGER, LASTER and LENCHEK (*111*), but we shall use a cruder and simpler approximation which should, however, bring out the basic physical principles involved.

Let us first consider the Fermi effect. Each scattering centre is considered to be a rigid, hard sphere, separated from the next one by a distance equal to the mean free path, λ. Fig. 35 shows a scattering centre, distance r from the Sun,

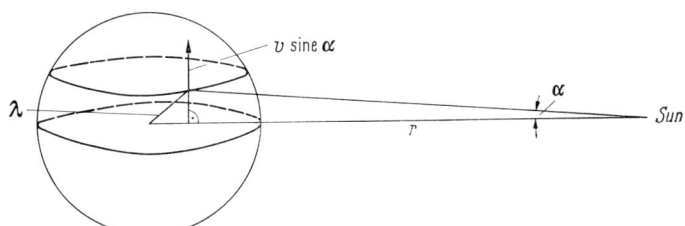

Fig. 35. Diagram illustrating the relative motion of two scattering centres — one at the centre of the sphere and one on the surface, a distance λ away. The vector $v \sin \alpha$ is at right angles to r.

with which a particle has just collided. It will then meet another centre somewhere on the surface of a sphere, radius λ. The relative velocity of recession between the centres is only in the direction at right angles to the centre-Sun line and is $\sim v \sin \alpha$ where α is the angle subtended at the Sun by the two centres considered. It is assumed that on collision with the second centre the particle direction of motion is reversed. Averaging for collisions over the whole sphere, and taking into account the dependence of $\delta P/P$ on the cosine of the angle between the particle and recession velocities, we find

$$\frac{\delta P}{P} \sim -\frac{2v\lambda\omega}{c^2 r} \frac{2}{3}, \qquad (28.1)$$

for the mean fractional momentum change per collision of a relativistic particle. The number of collisions per unit time is ω/λ so

$$\frac{\delta P}{P} = -\frac{4vdt}{3r}, \qquad (28.2)$$

since $\omega \sim c$.

To obtain an upper limit to the betatron deceleration, let us assume that the first adiabatic invariant is conserved for the whole of the motion inside the distance r_0. We start with the particle approaching the solar system in a galactic frame of reference. Next, its motion is transformed to a system moving with one point of the expanding solar plasma. The particle is followed through the solar system from moving co-ordinate system but on arrival at the Earth quantities are transformed back into the galactic frame. The energy change due to collision with the solar wind is given by the transformations between co-ordinates, and in the discussion of the Alfvén mechanism this was seen to be small. More important is the deceleration due to a finite curl $\boldsymbol{E}'$ [Eq. (23.5)] in the moving frame.

At distances greater than 1 A.U., the angle which Parker's quiet-time spiral field makes with the radial direction is greater than 45° and thus the field is mainly azimuthal. We suppose that the field, B, causing the particle deceleration, depends on distance in the same way as this azimuthal field, that is

$$B \cdot r = c = \text{constant}. \tag{28.3}$$

Then

$$\frac{dB}{B} = -\frac{dr}{r} \tag{28.4}$$

and, since $\dfrac{dr}{dt} = v$

$$\frac{dB}{B} = -\frac{v}{r} dt. \tag{28.5}$$

Thus, as the plasma moves out a distance dr, a time variation in the field of magnitude dB is seen by a particle moving in the plasma reference frame. If the adiabatic invariant is conserved

$$\frac{P_\perp^2}{B} = \text{const}, \tag{28.6}$$

where $P_\perp$ is the momentum component perpendicular to the field direction. Then, since

$$P^2 = P_\perp^2 + P_\parallel^2, \tag{28.7}$$

where $P_\parallel$ is the component of momentum parallel to the field, we find

$$dP = \frac{1}{2P} \cdot \frac{P_\perp^2}{B} dB. \tag{28.8}$$

For an isotropic pitch angle distribution, the average value

$$\overline{\frac{P_\perp^2}{B}} = \frac{2}{3} \frac{P^2}{B}. \tag{28.9}$$

so

$$\overline{dP} = \frac{P}{3B} dB, \tag{28.10}$$

and from (28.5)

$$\frac{\overline{dP}}{P} = -\frac{v}{3} \cdot \frac{dt}{r}. \tag{28.11}$$

The change in momentum of a particle due to interaction with an electromagnetic field affects the particle intensity in two ways. First, according to Liouville's theorem, the particle intensity j per unit of area, time and solid angle is given by

$$j = \frac{P^3}{m} \cdot D = P^2 D', \tag{28.12}$$

if $\omega \approx c$, where m is the relativistic mass and D and D' are constants. Thus

$$\frac{\delta j}{j} = \frac{2\, dP}{P}. \tag{28.13}$$

This is the change in directional intensity following a group of particles from an initial momentum P to a final momentum $(P+dP)$. However, since the particles ending up at a momentum P have come from a point on the spectrum appropriate to a momentum $(P-dP)$, there is an additional intensity change given by

$$\delta j = -\frac{\delta j}{dP} \delta P, \tag{28.14}$$

where dj/dP is the slope of the differential momentum spectrum. Now the best estimate of the galactic spectrum is

$$j = K P^{-\gamma}, \tag{28.15}$$

where $\gamma = 2.5$, so

$$\frac{\delta j}{j} = 2.5 \frac{dP}{P}. \tag{28.16}$$

Summing both effects, the total change in j is

$$\frac{dj}{j} = (2+2.5) \frac{dP}{P}. \tag{28.17}$$

Thus, adding the Fermi deceleration given by (28.2) and the betatron deceleration given by (28.11),

$$\frac{dj}{j} = -\frac{7.5\, v\, dt}{r}. \tag{28.18}$$

During the time dt, the particle diffuses a distance dr by a random walk. If the adiabatic invariant is conserved, the distance between scattering centres is the distance in which the field direction changes by a large angle. The question to ask now is whether diffusion theory is appropriate in this case. If particles remain on one line of force throughout their motion, then only changes in pitch angle need be considered. However, we can show that drift motion perpendicular to the field gradient probably makes particles lose all memory of the field line they were following initially.

Suppose that the path length of the particle following a field line is 10^{14} cm (~ 6 A.U.). Then the transit time is $3 \cdot 10^3$ sec for a relativistic particle. Choosing the Larmor radius as 10^{11} cm, corresponding to a 3 GV particle in 10^{-4} gauss, and choosing both the mean free path and the average value of the reciprocal of the field gradient as 10^{12} cm, the drift velocity is

$$\omega_D = \frac{\omega_\perp}{2} R_L \frac{|\mathrm{grad}_\perp B|}{|B|} = 7.5 \cdot 10^8 \text{ cm/sec}, \tag{28.19}$$

where $\dfrac{|\mathrm{grad}_\perp B|}{|B|} \sim \dfrac{1}{10} \cdot 2 \text{ cm}^{-1}$.

Thus the particle drifts $2.3 \cdot 10^{12}$ cm during transit — a distance exceeding the mean free path. Therefore, before a particle can reach the Earth along one field line it can be transferred to another going in a completely different direction and inward progress is by means of a succession of random scatterings. Although, strictly, the mean free path is not isotropic, we use diffusion theory appropriate to an isotropic mean free path to give a crude approximation to what must actually occur.

After N collisions in an infinite, isotropic diffusing medium, the average particle moves a distance $\sim d$ away from its starting point where

$$d^2 = 2 N \lambda^2. \tag{28.20}$$

The particle actually travels a distance $N\lambda$. Hence we define an average velocity ω_e for diffusion over a distance d where

$$\omega_e = \omega \frac{d}{N\lambda} = \frac{2\lambda\omega}{d}. \tag{28.21}$$

This is only really true for stationary scattering centres, and also it conceals the fact that the particle will travel quicker over the first part of its journey.

Returning to Eq. (28.18), we then suppose that the particle spends a time

$$dt = -\frac{dr}{\omega_e}, \qquad (28.22)$$

diffusing a distance dr. Hence

$$\frac{dj}{j} = \frac{7.5v}{r}\frac{dr}{\omega_e}. \qquad (28.23)$$

Integrating and putting $j=j_\infty$ at $r=r_0$,

$$\frac{j}{j_\infty} = \left(\frac{r}{r_0}\right)^{\frac{7.5v}{\omega_e}}. \qquad (28.24)$$

But $d=r_0-r$, so

$$\frac{j}{j_\infty} = \left(\frac{r}{r_0}\right)^{\left(\frac{3.25v(r_0-r)}{\lambda\omega}\right)}. \qquad (28.25)$$

This expression for modulation by a combination of Fermi and betatron deceleration may be compared to expression (27.8) for the modulation caused by outward convection. To obtain the relative magnitude of the two effects we substitute two different sets of parameters in the expressions. If, first, $\lambda = 3\cdot 10^{11}$ cm, $r_0 = 2$ A.U., and $v = 4\cdot 10^7$ cm/sec, it is found that deceleration yields a 22% reduction in intensity and convection gives an 18% drop. Alternatively, putting $\lambda = 1.2\cdot 10^{11}$ cm, $r_0 = 4$ A.U., and $v = 4\cdot 10^7$ cm/sec, the deceleration effect causes a 91% modulation while a 78% modulation results from convection. Thus both sources of modulation cause roughly the same size of intensity change. This statement applies for two alternative models of the interplanetary magnetic field. If the scattering centres are separated by field-free space, then the betatron term has been overestimated but the Fermi term, which is the larger, is still correct. However, if the adiabatic invariant is preserved through the motion, the two expressions, which both depend on diffusion theory, can only apply if particle drifts are sufficiently large and there is a sufficient number of magnetic constrictions present in the field lines. Then a typical particle will drift off the field line to which it was originally attached and suffer several reflections at the constrictions before reaching the Earth. There is probably insufficient evidence on the detailed field structure from magnetometer measurements to distinguish between the two models. The only firm limitation on the field provided by the measurements is to require the field strengths to be generally less than 10^{-4} gauss. At this stage, it is better to deduce the diffusion mean free path by fitting the theoretical modulation functions to cosmic ray data, provided the scale size and field strengths are not unreasonable.

It is noteworthy that PARKER [4] has considered deceleration when computing his Forbush decrease model involving an outward-mowing kink in the field lines. Energy loss arises from tail-on collisions with the moving barrier at the shock front. PARKER also finds that deceleration produces a modulation comparable in magnitude to the diffusion term, although here the former is smaller.

For the following reasons we shall only attempt to compare the convective term with experimental results on modulation. Data on both particles and fields are rather crude anyway and many quantities are not known to much better than a factor of two. Further, the rigidity-dependence of both modulation effects is given by the rigidity-dependence of λ, which appears in a similar position in both expressions and should cause similar changes of both functions. Then, for large modulations, our approach does not tell us how to combine the two effects. Finally, the derivation of the convective effect is more rigorous than that for the

29. Comparison of diffusive modulation with experimental data.

At low rigidities, particles will either follow field lines or suffer large deflections when encountering a magnetic scattering centre because their radii of curvature are small compared to the scale size of field variations. Thus in this region the mean free path does not vary with rigidity. At higher rigidities where the radius of curvature, R, exceeds the scale size, l, of the scattering centres, small angle scattering will occur, the deflection, θ, per encounter being given by

$$\theta \sim \frac{l}{R}. \tag{29.1}$$

After n encounters, the mean scattered angle is $\Delta \sim n^{\frac{1}{2}} \cdot \theta$ so the particle loses any memory of its initial direction after $\sim \frac{1}{\theta^2} = \frac{R^2}{l^2} = n$ collisions. If a typical particle encounters ν scattering centres on its way in from galactic space over a distance $(r_0 - r)$ then

$$\varrho (r_0 - r) l^2 = \nu \tag{29.2}$$

where ϱ is the density of scattering centres and $\nu \leq \frac{r_0 - r}{l}$. Thus the mean free path, λ, for large-angle scattering is

$$\lambda = \frac{n}{l^2 \varrho} = \frac{R^2 (r_0 - r)}{l^2 \nu}. \tag{29.3}$$

The mean free path becomes independent of rigidity when $R = R_M = l$ at a rigidity P_M so we can write expression (27.8) as

$$j = j_\infty \exp\left[-\frac{3v\nu}{\omega} \left(\frac{P_M}{P}\right)^2\right], \tag{29.4}$$

at high rigidities and

$$j = j_\infty \exp\left[-\frac{3v\nu}{\omega}\right], \tag{29.5}$$

at low rigidities less than P_M.

A rough fit to Ariel I data (Sect. 17) can be obtained with the Parker mechanism (116) using reasonable parameters. We assume the solar wind to be effective out to either 4 A.U. or 2 A.U. so $(r_0 - r) = 4.5 \cdot 10^{13}$ cm or $1.5 \cdot 10^{13}$ cm, the solar wind velocity, $v = 4 \cdot 10^7$ cm/sec, $l = 1.8 \cdot 10^{11}$ cm, the field in the scattering centres, $H = 9.3 \cdot 10^{-5}$ gauss, and the free space between scattering centres to be l across, so $\nu = \frac{r_0 - r}{2l}$, $\omega = 3 \cdot 10^{10}$ cm/sec. Using these parameters, Fig. 36 has been constructed. The mean experimental modulation, j/j_∞, is given together with an estimate of the error limits. Hard sphere scattering occurs up to 5 GV. Note that if $\nu \to \frac{r_0 - r}{l}$, the adiabatic invariant will always be conserved at $P \ll P_M$ and we must rely on particle drifts to make the diffusion approximation valid. Remembering that deceleration has been neglected, the curve corresponding to $(r_0 - r) = 2$ A.U. is probably a reasonable fit.

Explanation of the long-term modulation, observed at sunspot maximum, on the Parker model is more difficult. Elliot, Hynds, Quenby and Wenk (72) show that the experimental modulation, j/j_∞, exhibits a flatter rigidity-dependence at high rigidities than does the theoretical curve, which tends to a P^{-2} form when the modulation is small (Fig. 37). The modulation at 10 GV is at least 15% to 20% (Sect. 17). To obtain some rough fit to the experimental data we assume two

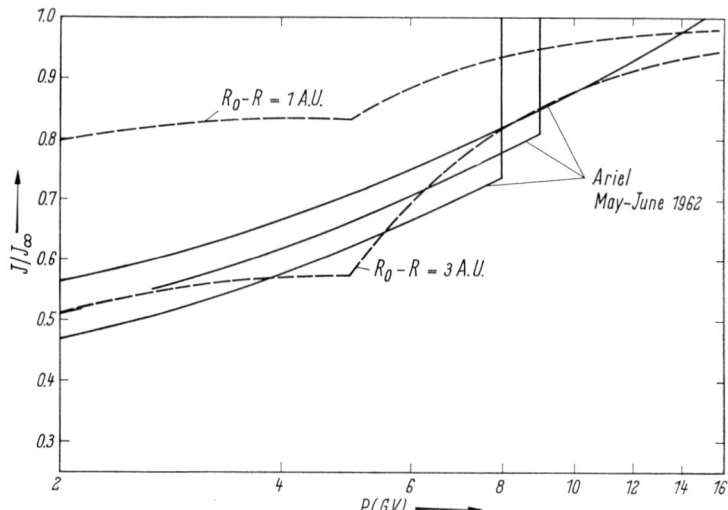

Fig. 36. Long-term modulation deduced from Ariel I measurements compared to the modulation produced by the Parker convection process.

sizes of scattering centres. Their properties are summarized in Table 9. In practice, of course, a whole variety of scattering centres may be present, but this idealization will serve to illustrate the problem.

We put the quantities $(r_0 - r) = 3$ A.U., and v and ω are as before. Hard sphere scattering occurs below 2 GV for the small centres and below 10 GV for the large centres. 10 GV particles suffer little deflection when traversing the small centres, while 2 GV particles pass fairly freely through the large centres due to adiabatic invariant conservation. These parameters produce a 78% modulation at 2 GV and a 9% modulation at 10 GV, in reasonable agreement with observation when we remember that deceleration has still to be included. However, fields 17 gamma in magnitude are required every 0.1 A.U., while Pioneer V saw fields of this magnitude only at times of terrestrial magnetic storms. This space probe was launched in 1960 — a time when the differential intensity at 10 GV was probably down by 5% to 10% on the sunspot minimum value.

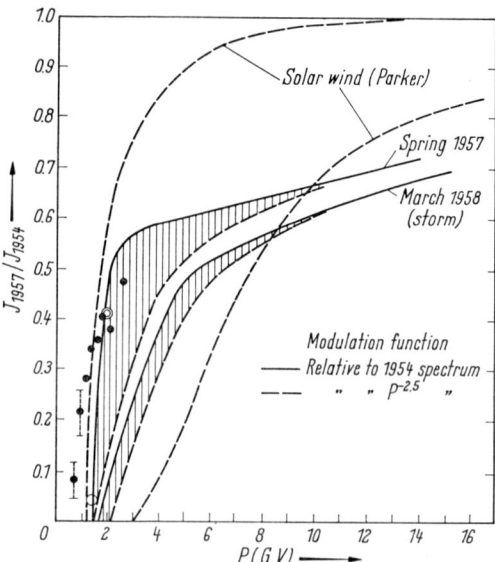

Fig. 37. Long-term modulation deduced from ground measurements in Spring 1957 compared to the modulation produced by the Parker convection process. After ELLIOT et al. (72).

An interesting test of the theory under discussion is a measurement of the relative modulation of protons and alpha-particles of the same rigidity since the effect also depends on the particle velocity, ω. WEBBER's analysis[1] suggests that the modulation only depends on ω in the range 100 meV/nucleon to 450 meV/nucleon.

---

[1] W. R. WEBBER, This Handbook, p. 181.

Increases in the solar wind velocity and the field strength associated with the scattering centres at the time of solar outbursts will explain the Forbush decreases on the Parker diffusion mechanism. It is assumed that the energy produced in the outburst is spread over a wide solid angle and that it is produced for a time comparable with the length of Forbush decrease. l, which is a measure of the size of instabilities in the plasma, may also change at this time, so no one-to-one correlation of cosmic ray intensity and wind velocity is necessarily expected.

Large Forbush decreases show 10% intensity changes at 10 GV and exhibit a very flat spectrum in most of the latitude-sensitive region (Sect. 17). Thus, if the parameters applicable to hard centres in Table 9 were changed so that the exponents in Eqs. (29.4) and (29.5) were doubled, we would have parameters applicable to a decrease at sunspot maximum. Since 20 gamma to 50 gamma fields were seen by Pioneer V during disturbed times, the explanation seems promising. Strictly, Eqs. (29.4) and (29.5) refer to equilibrium conditions, but as the time-scale of Forbush events is several days we may take them to estimate the final magnitude of the decrease. PARKER [4] considers the time-varying case and shows that various types of onset characteristics may be fitted with mixtures of solar streams with different parameters.

Table 9.

	l (cm)	H (gauss)	Distance between centres (cm)
Small centres	$1.2 \cdot 10^{11}$	$5.6 \cdot 10^{-5}$	0
Large centres	$2.0 \cdot 10^{11}$	$1.7 \cdot 10^{-4}$	$1.8 \cdot 10^{12}$

Some decreases are observed underground where the effective primary energy is ~ 100 GV. To produce the observed 1% change on the diffusion model, a mean free path ~ 1 A.U. with scattering fields ~ 50 gamma is required. With $\lambda \sim (r_0 - r)$ diffusion theory seems inapplicable and, anyway, the fields required seem too high when it is remembered that they must exist out to 4 A.U. all over the solar system. It is more likely that a Gold or Parker type arrangement of regular fields is the dominant mechanism causing modulation at these energies. The large daily variations and short time-scales associated with these theories appear to be present for underground detectors. Moreover, studies of solar proton events have suggested the existence of homogeneous field regions ~ 1 A.U. in scale size. We must suppose that lower rigidity particles diffuse or drift reasonably freely through these large-scale fields due to the action of small inhomogeneities which affect the 100 GV particles very little. Thus it is still correct to apply diffusion theory at the low end of the spectrum.

Assuming that convection by outward-sweeping scattering centres is the cause of the long-term modulation, we can obtain an estimate of the mean free path from measurements of the gradient in the interplanetary flux. Differentiation of Eq. (27.8) gives the fractional intensity change per unit distance as

$$\frac{1}{n}\frac{dn}{dr} = \frac{3v}{\lambda \omega}. \tag{29.6}$$

Using the experimental value of 0.14 (Sect. 18) on the left-hand side and the usual values for v and ω, we find $\lambda = 0.021$ A.U. This applies to particles ~ 1 GV rigidity and is in reasonable agreement with the value of λ used to fit the Ariel I data obtained in the same year (1962) as the gradient was measured. For Ariel I, we had $\lambda = 2l = 0.024$ A.U.

Other estimates of the mean free path have already been made in Sect. 22. Here work was reported which represented the initial, isotropic stages of solar particle propagation by diffusion theory for a point source in an infinite medium.

Outward motion of the scattering centres was neglected. Since most flares produce terrestrial disturbances 1 to 2 days later, this period is an estimate of the transit time for the scattering centres during solar particle events. For the first seven hours, where infinite diffusion theory seems to fit, it might be expected that the additional sweeping out by the scattering centres is not too serious and hence the mean free path estimates are reasonable. Table 8 shows that two events in 1960—1961 yielded mean free paths ~ 0.02 A.U., in agreement with the above estimates for 1962, while two other events gave $\lambda \sim 0.04$ A.U.

Use of diffusion theory for a finite, spherical medium for the later, isotropic stages of solar events gave values for the radius of the outer boundary of the medium ~ 2 A.U. (Table 8). However, the long-term modulation at sunspot maximum has been fitted using a barrier radius at 4 A.U. The flare results just mentioned apply to a time when the modulation was close to the sunspot maximum level. Decreasing the barrier radius by a factor of two leads to a factor of three decrease in the mean free path and a corresponding increase in the value of the magnetic field necessary in the scattering centers, which is an unsatisfactory conclusion. It is doubtful, however, whether the diffusion theory employed in the solar particle analysis is valid since, once again, static scattering centres are assumed. The observations of intensity decay are usually spread over 2 or more days — a period comparable to the Sun-Earth transit times for the scattering centres. Qualitatively, additional removal by convection gives the same sign of effect as does reduction in the barrier radius, so an underestimate of this radius is not unexpected.

30. Deceleration of flare particles. Inverse Fermi plus adiabatic deceleration of flare particles can also remove flux at a given rigidity faster than is predicted by the simple diffusion theory. Again it will be assumed that flare particle propagation is represented by diffusion through static scattering centres, and the extreme case will be considered where the first adiabatic invariant is conserved throughout the motion. For relativistic particles, the previous analysis for galactic cosmic rays holds with some minor modifications. Eq. (28.5) becomes

$$\frac{dB}{B} = -\frac{2v}{r} dt, \qquad (30.1)$$

since the field between the Sun and the Earth is better represented by $B \propto 1/r^2$. Integration is now carried out between $r = R_0$, where R_0 is the radius of a cavity surrounding the Sun and outside of which diffusion theory is applicable and $r = 1$ A.U. What happens inside R_0 is determined by the acceleration mechanism. Then, proceeding as before,

$$j = j_0 \left(\frac{R_0}{r}\right)^{\frac{\alpha v (r - R_0)}{2 \lambda \omega}}, \qquad (30.2)$$

where $\alpha = 2(2+\gamma)$ and γ is the exponent of the initial rigidity spectrum, j_0, inside R_0.

For non-relativistic particles, a similar calculation may be made using

$$\frac{\delta W}{m_0 c^2} \sim -\frac{2vw}{c^2} \cdot \frac{\lambda}{r} \cdot \frac{2}{3}, \qquad (30.3)$$

to give the average change in kinetic energy, δW, per collision and

$$\frac{\frac{1}{2} m_0 (\omega \sin \theta)^2}{B} = \text{const.}, \qquad (30.4)$$

as the adiabatic invariant, where θ is the pitch angle.

Then the relation between the fractional change in the spectrum and time is

$$\frac{dj}{j} = -\left(\frac{3}{2} + \gamma_1\right) \cdot \frac{4v}{r} dt, \tag{30.5}$$

and finally we find

$$j(w) = j_0(w) \left(\frac{R_0}{r}\right)^{\frac{\alpha_1 v (r - R_0)}{2 \lambda \omega}}, \tag{30.6}$$

for $W \ll m_0 c^2$ where $\alpha_1 = 4(\gamma_1 + \frac{3}{2})$ and where the initial differential spectrum, $j_0(w)$, is represented by a negative power law in kinetic energy with γ_1 as exponent. Expressions (30.2) and (30.6) refer only to the time when the intensity at the Earth due to an instantaneous injection of particles is a maximum.

Choice of a value for R_0 is difficult but, since it must be greater than λ for the theory to have any validity, the minimum possible value is probably obtained by putting $R_0 = 0.1$ A.U.

Table 10 has been calculated using $\lambda = 3.8 \cdot 10^{11}$ cm, $v = 4 \cdot 10^7$ cm/sec and for $R_0 = 0.1$ A.U. and $R_0 = 0.3$ A.U. The second value of R_0 would correspond to a rather homogeneous field region close to the Sun where field instabilities had not yet been able to grow.

Table 10.

Kinetic energy	j/j_0 ($R_0 = 0.1$ A.U.)	j/j_0 ($R_0 = 0.3$ A.U.)
>1 GeV	0.47	0.73
100 MeV	0.035	0.26
50 MeV	0.015	0.18
10 MeV	$6 \cdot 10^{-5}$	0.019

The figures for j/j_0 refer to the time when the intensity at a particular energy is a maximum and hence do not refer to the same lapse of time after the flare. To a first approximation, the intensity reduction at any instant is independent of energy. The large reduction seen at 10 MeV, corresponding to only 10 hours after the event, makes it difficult to understand how any particles are seen 2 days or more later. If, however, diffusion theory does not apply between the Sun and the Earth at low energies but the particles are stored outside 1 A.U. in a thick shell, then the deceleration is much reduced, as can be seen from the r^{-1} dependence of Eq. (30.5). It may be that 10 MeV particles move fairly freely through the field inhomogeneities, conserving the adiabatic invariant, while 100 MeV particles are scattered by these field irregularities between the Sun and the Earth. Moreover, it is known that the arrival at the Earth of the main bulk of 10 MeV particles is frequently controlled by the presence of a solar stream arriving at the same time and producing a large geomagnetic disturbance. Thus the motion of the particles is determined by the plasma motion, for as long as they remain trapped, and not by diffusion theory. This trapping may last for only a fraction of the plasma transit time and the remainder of the Sun-Earth journey may be through regular fields. It is also necessary to point out that PARKER [4] has considered particle diffusion in a geometry consisting of a circular cone of injection from the Sun attached to a thick diffusing shell. He finds a time dependence similar to that for a finite diffusing region, namely a $t^{-\frac{3}{2}}$ law going over into an exponential decay. Thus the observed time-dependence does not necessarily require scattering centres throughout the space enclosed by 1 A.U. However, as HOFFMANN and WINCKLER (100) point out, the inner boundary of the shell must be far inside the orbit of the Earth to explain the slow rise-time of some events. It seems then that low energy particles must be stored between the Sun and the Earth in some events, and if this time exceeds a few hours our approximate calculation has shown that significant reductions in intensity can occur due to deceleration.

d) Summary.

31. Summarizing the discussion on cosmic ray modulation and flare particle propagation, it would seem that we are presented with a variety of models, none of which is capable in itself of explaining all the facts. Particle convection and deceleration in a diffusing medium with a mean free path of 0.02 A.U. and with field strengths limited to 10^{-4} gauss is capable of explaining the modulation of galactic particles of less than 10 GV rigidity. Some aspects of solar proton events also fit this model. Long-term modulation above 10 GV is less satisfactorily explained by diffusion, requiring scattering centres with fields of $2 \cdot 10^{-4}$ gauss located every 0.1 A.U. at sunspot maximum. Forbush decreases, during which particles of 100 GV are affected, probably require the presence of homogeneous magnetic fields $\sim$1 A.U. in dimension, although decreases caused by such fields give large anisotropies and do not last for very long. The impact zone effects noticed in some solar proton events are consistent with regular fields inside the Earth's orbit where the particle mean free path is $\sim$1 A.U. The existence of low energy solar particles for several days after a flare also suggests that the mean free path inside 1 A.U. is long while trapping chiefly takes place farther out.

Probably we are only able to make general statements as a result of our limited understanding of cosmic ray time variations. Within a space limited to 4 A.U. of the Sun, the interplanetary field is such as to yield particle mean free paths varying between 0.01 A.U. and 1 A.U. in length, the shorter mean free paths occurring more frequently for a given particle rigidity; larger mean free paths are more likely to be found at distances less than 1 A.U. from the Sun. The overall field configuration is roughly spiral shaped, as predicted by Parker's solar wind theory and the original theory of solar streams given by Chapman [6].

References.

General references.

[1] Forbush, S. E.: Handbuch der Physik, Bd. XLIX, S. 159. Berlin-Heidelberg-New York: Springer 1966.
[2] Webber, W. R.: Progress in Elementary Particle and Cosmic Ray Physics VI. Amsterdam: North-Holland Publ. Co. 1962.
[3] Dorman, L. I.: Progress in Elementary Particle and Cosmic Ray Physics VII. Amsterdam: North-Holland Publ. Co. 1963.
[4] Parker, E. N.: Interplanetary Dynamical Processes. New York: Interscience Publishers, John Wiley & Son 1963.
[5] Morrison, P.: Handbuch der Physik, Bd. XLVI/1, 1—87. Berlin-Göttingen-Heidelberg: Springer 1961.
[6] Chapman, S., and J. Bartels: Geomagnetism, II. Oxford: Clarendon Press 1940.
[7] Störmer, C.: The Polar Aurora. Oxford: Clarendon Press 1955.
[8] Vallarta, M. S.: Handbuch der Physik, Bd. XLVI/1, S. 88. Berlin-Göttingen-Heidelberg: Springer 1961.
[9] Johnson, T. S.: Rev. Mod. Phys. **10**, 193 (1938).
[10] Quenby, J. J., and W. R. Webber: Phil. Mag. **4**, 90 (1959).
[11] Quenby, J. J., and G. J. Wenk: Phil. Mag. **7**, 1457 (1962).
[12] Elliot, H.: Rep. Progr. Phys. **26**, 145 (1963).
[13] Dungey, J. W.: Cosmic Electrodynamics. Cambridge: Cambridge University Press 1958.
[14] Neher, H. V.: Progress in Cosmic Ray Physics I. Amsterdam: North-Holland Publ. Co. 1952.
[15] Waddington, C. J.: Progress in Nuclear Physics, vol. 8. Oxford: Pergamon Press 1960.
[16] Dorman, L. I.: Cosmic Ray Variations. Moscow: State Publishing House 1957.
[17] McCracken, K. G.: J. Geophys. Res. **67**, 447 (1962).
[18] Alfven, H.: Cosmical Electrodynamics. Oxford: Clarendon Press 1950.
[19] Gold, T.: Gas Dynamics of Cosmic Clouds, ed. H. C. van de Hulst and J. M. Burgers. Amsterdam: North-Holland Publ. Co. 1955.

References cited.

(1) WEBBER, W. R.: Nuovo Cim. (8), Suppl. 11, 532 (1958).
(2) SCHWARTZ, M.: Nuovo Cim. 11, Suppl. 1, 27 (1959).
(3) BLAND, C. J.: Phil. Mag. 7, 1487 (1962).
(4) HEDGECOCK, P. C.: Communication by H. ELLIOT, Intern. Conf. on Cosmic Rays, Jaipur 1963.
(5) KASPER, J. E.: J. Geophys. Res. 65, 39 (1960).
(6) KODAMA, M., I. KONDO, and M. WADA: J. Sci. Res. Inst. Tokyo 51, 138 (1957).
(7) SIMPSON, J. A., K. B. FENTON, J. KATZMAN, and D. C. ROSE: Phys. Rev. 102, 1648 (1956).
(8) ROTHWELL, P., e J. J. QUENBY: Nuovo Cim. 8, Suppl. 2, 249 (1958).
(9) SCHMIDT, A.: Beitr. angew. Geophys. 41, 346 (1934).
(10) FINCH, H. P., and B. R. FENTON: Monthly Not. Roy. Astronom. Soc. Geophys. Suppl. 1, 314 (1957).
(11) KATZ, L., P. MEYER, e J. A. SIMPSON: Nuovo Cim. 8, Suppl. 2, 277 (1958).
(12) RAY, E. C.: Ann. of Phys. 24, 1 (1963).
(13) MCILWAIN, C. E.: J. Geophys. Res. 66, 3681 (1961).
(14) LIN, W. C., D. VENKATESAN, and J. A. VAN ALLEN: Geophys. Res. 68, 4885 (1963).
(15) POMERANTZ, M. A., and S. P. AGARWAL: Phil. Mag. 7, 1503 (1962).
(16) KELLOGG, P. J.: J. Geophys. Res. 65, 2701 (1960).
(17) MCCRACKEN, K. G.: J. Geophys. Res. 67, 423 (1962).
(18) KONDA, I., M. KODAMA, and T. MAKINO: Int. Conf. on Cosmic Rays, Jaipur 1963.
(19) MCDONALD, F. B., and W. R. WEBBER: Phys. Rev. 115, 194 (1959).
(20) FOSTER, F., and J. KLARMANN: Private communication 1964.
(21) FRIEDLANDER, M. W., e C. T. SPRING: Nuovo Cim. (X) 26, 1292 (1962).
(22) NEY, E. P., J. R. WINCKLER, and P. S. FRIER: Phys. Rev. Letters 3, 183 (1959).
(23) REID, G. C., and H. LEINBACH: J. Geophys. Res. 64, 1801 (1959).
(24) FRIER, P. S.: J. Geophys. Res. 67, 2617 (1962).
(25) WINCKLER, J. R., P. D. BHAVSAR, and L. PETERSON: J. Geophys. Res. 66, 995 (1961).
(26) MATHEWS, T., T. THAMBYAHPILLAI, and W. R. WEBBER: Monthly Not. Roy. Astronom. Soc. 123, 97 (1961).
(27) PIEPER, I. G. F., A. J. ZMUDA, C. A. BOSTROM, and B. J. O'BRIEN: J. Geophys. Res. 67, 4959 (1962).
(28) AKASOFU, S. I., W. C. LIN, and J. A. VAN ALLEN: J. Geophys. Res. 68, 5327 (1963).
(29) KONDO, I., K. NAGASHIMA, S. YOSHIDA, and M. WADA: Proc. IUPAP Moscow Conf. (Moscow) 4, 210 (1960).
(30) HATTON, C. J., and P. L. MARSDEN: Phil. Mag. 7, 1145 (1962).
(31) DAVIS, L. R., and J. M. WILLIAMSON: 3rd Int. Space Sci. Symp., May 1962.
(32) AKASOFU, S. I., J. C. CAIN, and S. CHAPMAN: J. Geophys. Res. 67, 2645 (1962).
(33) AKASOFU, S. I., and S. CHAPMAN: J. Geophys. Res. 66, 321 (1961).
(34) APEL, J. R., S. F. SINGER, and R. C. WENTWORTH: Advances in Geophysics, ed. by H. E. LANDSBERG, and J. V. MIEGHAN. New York: Academic Press 1962.
(35) HEPPNER, J. P., N. F. NESS, C. S. SCEARCE, and T. A. SKILLMAN: J. Geophys. Res. 68, 1 (1963).
(36) CAHILL, L. J.: COSPAR Meeting, Warswa, June 1963.
(37) CAHILL, L. J., and P. G. AMAZEEN: J. Geophys. Res. 68, 1835 (1963).
(38) JOHNSON, F. S.: J. Geophys. Res. 65, 3049 (1960).
(39) FRANK, L. A., and J. A. VAN ALLEN: Symp. on the Results of the IGY-IGC, Los Angeles, August 1963.
(40) SMITH, E. J., C. P. SONETT, and J. W. DUNGEY: Preprint (1963).
(41) RAY, E. C.: Phys. Rev. 101, 1142 (1956).
(42) HEDGECOCK, P. C.: Thesis, London University, 1964.
(43) TREIMAN, S. B.: Phys. Rev. 86, 917 (1952).
(44) KELLOGG, P. J., and J. R. WINCKLER: J. Geophys. Res. 66, 399 (1961).
(45) AKASOFU, S. I., and W. C. LIN: J. Geophys. Res. 68, 973 (1963).
(46) OBAYASHI, T.: Ark. Geofys. 3, Nr. 21, 507 (1961).
(47) NEUGEBAUER, M., and C. W. SNYDER: Science 138, 1095 (1962).
(48) CHAPMAN, S., and V. C. A. FERRARO: Terr. Magn. Atmosph. Electr. 36, 77 (1931).
(49) AKASOFU, S. I.: J. Geophys. Res. 68, 15 (1963).
(50) MACHLUM, B., and B. J. O'BRIEN: J. Geophys. Res. 68, 997 (1963).
(51) MALMFORS, K. G.: Ark. Mat. Astronom. Fys. 32A (8) (1945).
(52) BRUNBERG, E. A., and A. DATTNER: Tellus 5, 269 (1953).
(53) JORY, F. S.: Phys. Rev. 103, 1068 (1956).

(54) FIROR, J.: Phys. Rev. **94**, 1017 (1954).
(55) RAO, U. R., K G. MCCRACKEN, and D. VENKATESAN: J. Geophys. Res. **68**, 345 (1963).
(56) WEBBER, W. R., and J. J. QUENBY: Phil. Mag. **4**, 654 (1959).
(57) MATHEWS, T., and M. KODAMO: Preprint (1964). — MATHEWS, T.: Thesis, London University, 1962.
(58) ROSE, D. C., K. B. FENTON, J. KATZMAN, and J. A. SIMPSON: Canad. J. Phys. **34**, 968 (1956).
(59) MATHEWS, T.: Phil. Mag. **8**, 387 (1963).
(60) DORMAN, L. I.: Section VII, IGY Programme (CR) Nr. 1 (Publ. House Acad. Sci. USSR, 1959).
(61) SIMPSON, J. A.: Phys. Rev. **94**, 426 (1954).
(62) FAN, C. Y., P. MEYER, and J. A. SIMPSON: Phys. Rev. Letters **5**, 272 (1960).
(63) NEHER, H. V.: Int. Conf. on Cosmic Rays, Jaipur 1963.
(64) MUSTEL, E. N.: Soviet Astron. **5**, 19 (1961).
(65) QUENBY, J. J., and T. THAMBYAHPILLAI: Phil. Mag. **5**, 585 (1960).
(66) DUGGAL, S. P., K. NAGASHIMA, and M. A. POMERANTZ: J. Geophys. Res. **66**, 1970 (1961).
(67) HYNDS, R. J.: Thesis, London University, 1961.
(68) LOCKWOOD, J. A., and H. RAZDAN: J. Geophys. Res. **68**, 1581 (1963).
(69) FENTON, A. G., K. G. MCCRACKEN, D. C. ROSE, and B. G. WILSON: Canad. J. Phys. **37**, 970 (1959).
(70) SANDSTRÖM, A. E.: Tellus **7**, 204 (1955).
(71) MCDONALD, F. B., and J. R. WEBBER: J. Phys. Soc. Japan **17**, Suppl. A II, 428 (1962).
(72) ELLIOT, H., R. J. HYNDS, J. QUENBY, and G. WENK: Proc. IUPAP Conf. Moscow, IV, 319 (1960).
(73) DURNEY, A. C., H. ELLIOT, R. J. HYNDS, and J. J. QUENBY: To be published in Proc. Roy. Soc. 1964.
(74) POMERANTZ, M. A., S. P. DUGGAL, and L. WITTEN: COSPAR Meeting, Warsaw 1963.
(75) KANE, R. P.: Nuovo Cim. (X) **27**, 441 (1963).
(76) SIMPSON, J. A., C. Y. FAN, and P. MEYER: J. Phys. Soc. Japan **17**, Suppl. A II, 506 (1962).
(77) ARNOLDY, R. L., and J. R. WINCKLER: Int. Con. on Cosmic Rays, Jaipur 1963.
(78) VAKULOV, P. V., S. N. VERNOV, E. B. GORCHAKOV, YU. I. LOGACHEV, A. N. CHARAKHCHYAN, T. N. CHARAKHCHYAN, and A. E. CHUDAKOV: COSPAR Meeting, Warsaw 1963.
(79) COLEMAN, P. J., L. DAVIS, and C. P. SONETT: Phys. Rev. Lettres **5**, 43 (1960).
(80) COLEMAN, P. J., L. DAVIS, E. J. SMITH, and C. P. SONETT: Science **138**, 1095 (1962).
(81) FAN, C. Y., P. MEYER, and J. A. SIMPSON: Phys. Rev. Lettres **5**, 269 (1960).
(82) ALLEN, J. A. VAN, and W. C. LIN: J. Geophys. Res. **65**, 2998 (1960).
(83) COLEMAN, P. J., C. P. SONETT, and L. DAVIS: J. Geophys. Res. **66**, 2045 (1961).
(84) SMITH, E. J.: Private communication 1963.
(85) BONETTI, A., H. A. BRIDGE, A. J. LAZARUS, E. F. LYON, B. ROSSI, and F. SCHERB: COSPAR Symposium, Washington, D.C. (to be published in Space Research III, North-Holand Publ. Co., Amsterdam).
(86) SNYDER, Cl W., and M. NEUGEBAUER: COSPAR Meeting, Warsaw 1963.
(87) ELLIOT, H.: Int. Conf. on Cosmic Rays, Jaipur 1963.
(88) AHLUWALIA, H. S., and A. J. DESSLER: Planet. Space Sci. **9**, 195 (1962).
(89) LÜST, R.: Space Sci. Rev. **1**, 480 (1962/63).
(90) CARMICHAEL, H.: Space Sci. Rev. **1**, 28 (1962).
(91) BRYANT, D. A., T. L. CLINE, U. D. DEASI, and F. B. MCDONALD: J. Geophys. Res. **67**, 4983 (1963).
(92) MCCRACKEN, K. G.: J. Geophys. Res. **67**, 435 (1962).
(93) STELJES, J. F., H. CARMICHEL, and K. G. MCCRACKEN: J. Geophys. Res. **66**, 1363 (1961).
(94) ROEDERER, J. G., J. R. MANZANO, O. R. SANTOCCHI, N. NERURKAR, O. TRONCOSO, R. A. R. PALMEIRA, and G. SCHWACHHEIM: J. Geophys. Res. **66**, 1603 (1961).
(95) MEYER, P., E. N. PARKER, and J. A. SIMPSON: Phys. Rev. **104**, 768 (1956).
(96) PFOTZER, G.: Nuovo Cim. **8**, Suppl. II, 180 (1958).
(97) FRIER, P. S., and W. R. WEBBER: J. Geophys. Res. **68**, 1605 (1963).
(98) CHARAKHCHYAN, A. N., V. F. TULINOV i. T. N. CHARAKHCHYAN: Zh. Eksper. Teor. Fiz. **41**, 735 (1961).
(99) WINCKLER, J. R., and P. D. BHAVSAR: J. Geophys. Res. **68**, 2099 (1963).
(100) HOFMANN, D. J., and J. R. WINCKLER: J. Geophys. Res. **68**, 2067 (1963).
(101) e.g. JOST, W.: Diffusion in Solids, Liquids, Gases. New York: Academic Press 1952. — CARSHAW, H. S., and J. C. JAEGER: Conduction of Heat in Solids. London: University Press 1947.

(102) BISWAS, S., C. E. FICHTEL, D. E. GUSS, and C. J. WADDINGTON: Hydrogen, Helium, and Heavy Nuclei from the Solar Event Nov. 15, 1960, X-611-62-235, Goddard Space Flight Center, Greenbelt, Md., U.S.A.
(103) NAGASHIMA, K.: J. Geomagn. and Geoelectr. **3**, 100 (1951).
(104) EHMERT, A.: Space Research. Proc. 1st Space Science Symp., Nice; ed. H. KALLMANN-BIJL. Amsterdam: North-Holland Publ. Co. 1960.
(105) JÁNOSSY, L.: Z. Physik **104**, 430 (1937).
(106) ELLIOT, H.: Phil. Mag. **5**, 60 (1960).
(107) DAVIS, L.: Phys. Rev. **100**, 1440 (1955).
(108) BEISER, A.: J. Geophys. Res. **63**, 1 (1958).
(109) FÄLTHAMMER, C. G.: J. Geophys. Res. **67**, 1791 (1962).
(110) FERMI, E.: Phys. Rev. **75**, 1169 (1949).
(111) SINGER, S. F., H. LASTER, and A. M. LENCHEK: J. Phys. Soc. Japan **17**, Suppl. A II, 583 (1962).
(112) COCCONI, G., T. GOLD, K. GREISEN, S. HAYAKAWA e P. MORRISON: Nuovo Cim. Suppl. VIII, Nr. 2, 161 (1958).
(113) MORRISON, P.: Phys. Rev. **101**, 1397 (1956).
(114) GOLD, T.: Astrophys. J., Suppl. **4**, 406 (1960).
(115) LÜST, R., and J. A. SIMPSON: Phys. Rev. **108**, 1563 (1957).
(116) e.g. ELLIOT, H.: Cosmic Ray Work at Imperial College in Relation to Solar Modulation and Geomagnetic Effects. Survey Paper presented at IUPAP Int. Conf. on Cosmic Rays, Jaipur 1963.

Nukleonen in der Atmosphäre[1].

Von

E. SCHOPPER

unter Mitarbeit von E. LOHRMANN und G. MAUCK.

Mit 88 Figuren.

A. Einleitung.

Mit ihrem Eintritt in die Erdatmosphäre ändert die einfallende primäre kosmische Strahlung, bestehend aus Protonen und Atomkernen, ihre Zusammensetzung durch Bildung sekundärer Strahlung infolge der Wechselwirkung mit den Atomkernen der Luft.

Die primären Kerne fragmentieren bereits in den oberen Atmosphärenschichten und gehen entweder in ein Büschel hochenergetischer Nukleonen oder in Zwischenschritten in leichtere Kernbruchstücke[2] über. Wir benützen die Kenntnis dieses Prozesses, um die innerhalb der Atmosphäre zwischen 4 bis 10 g/cm² Tiefe bei Ballonmessungen beobachteten Kernbruchstücke[3] auf Fluß und Massenspektrum der Primären am Gipfel der Atmosphäre zurückzurechnen.

Der Strom der primären Protonen wie der sekundären Nukleonen löst seinerseits Wechselwirkungen mit den Kernen der Luft aus.

Bei Wechselwirkungen mit einer Primärenergie bereits oberhalb einiger 10^9 eV kommt es dabei zur Auslösung lokaler durchdringender Schauer mit der Emission von Anstoßnukleonen und von Mesonen, meist Pionen, aus dem Kern. Die starken Wechselwirkungen bestimmen den Ablauf dieser für die Erzeugung sekundärer Nukleonen dominanten Prozesse. Ihr Mechanismus ist im Beitrag von SITTE im Bd. 46/1 dieses Handbuches im Detail behandelt.

Verfügt das stoßende Nukleon nach dem ersten Stoß noch über ausreichende Energie, so kann es, ebenso wie energiereiche sekundäre Nukleonen, weitere Kernstöße machen. Vom Gipfel der Atmosphäre aus wiederholen sich solche Einzelprozesse mit einer gewissen mittleren Stoßlänge und führen zur Ausbildung einer Nukleonenkaskade in der Atmosphäre. Die lokalen durchdringenden Schauer müssen wir als Träger der Nukleonenkaskade durch die Atmosphäre hindurch ansehen. Zu ihrem Stamm tragen Nukleonen mit Energien oberhalb 10 GeV bei. Die höchsten gemessenen Primärenergien liegen bei 10^{19} eV.

Bei hohen Energien (>100 GeV) wird wegen der relativistischen Zeitdilatation auch die geladene Komponente der Pionen für Kernwechselwirkungen wirksam und trägt zur Nukleonenkaskade bei. Strange particles wie K-Mesonen und Hyperonen, sowie Teilchen-Antiteilchenpaare scheinen — auch bei sehr hohen Energien — vergleichsweise seltener gebildet zu werden; man schätzt ihren Anteil auf 15%;

---

[1] Die Kapitel C und D dieses Beitrags sind von einer früheren Fassung (LOHRMANN, SCHOPPER, 1958) übernommen, mit Ergänzungen in Kapitel C und D in den Ziffern 5—15. Die Kapitel A, B, C (Ziffer 4) und E wurden neu verfaßt (MAUCK, SCHOPPER).

[2] Die Produktion von Radionukliden durch die kosmische Strahlung wird in dem Beitrag von LAL und PETERS "Cosmic Ray Produced Radioactivity on the Earth" in diesem Band behandelt.

[3] Die Fragmentation wird ausführlich im Beitrag von WEBBER "Spectrum and Charge Composition of the Primary Cosmic Radiation" behandelt.

hingegen wird bei hohen Energien offenbar das Auftreten und die Beteiligung isobarischer Nukleonenzustände an der Kaskade wahrscheinlich (vgl. Ziff. 4).

Die neutralen Pionen zerfallen über die elektromagnetische Wechselwirkung in Photonen
$$\pi^0 \to 2\gamma,$$
die geladenen Pionen führen in bekannter Weise über schwache Wechselwirkungen durch die Zerfallsprozesse
$$\pi^\pm \to \mu^\pm + \nu_\mu; \quad \mu^\pm \to e^\pm + \nu_\mu + \nu_\beta$$

— und über analoge Zerfallsprozesse die Kaonen und Hyperonen — zu den Teilchen der Myon- und der Elektron-Photon-Komponente, die auf dem Wege durch die Atmosphäre durch die Nukleonenkaskade gespeist werden.

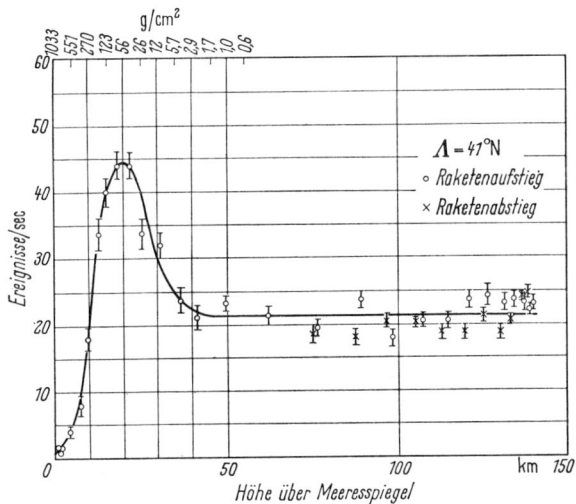

Fig. 1. Gesamtintensität geladener Teilchen in der Atmosphäre (Raketenmessung) nach van Allen und Singer [Al 50 b].

Der um die Anstoßnukleonen verarmte Restkern eines Schauervorgangs ist meist hoch angeregt und kühlt sich durch verzögerte Emission von Verdampfungsnukleonen ab; solche Verdampfungsprozesse werden auch durch Kernstöße energiearmer Nukleonen ausgelöst. Sie stellen in der Atmosphäre die lokale Quelle für die Produktion niederenergetischer Sekundärnukleonen ($E \lessgtr 10$ MeV) und häufig das jeweilige Ende eines nuklearen Kaskadenastes dar.

Die Vervielfachung der Intensität der geladenen Teilchen durch Sekundärenbildung in der Atmosphäre ist experimentell schon seit langem durch Ballonmessungen als Pfotzer-Maximum [Pf 36] bekannt und durch Raketenmessungen von van Allen und Singer [Al 50b] eindeutig bestätigt (Fig. 1). Die oberhalb der Atmosphäre gemessene Gesamtintensität erweist sich bis 150 km Höhe als konstant. Sie setzt sich aus der einfallenden Primärintensität und dem Albedo-Anteil, d.h. einer Rückstreuung geladener Teilchen aus der Atmosphäre zusammen.

In einer Höhe zwischen 40 und 50 km — oft als Gipfel der Atmosphäre bezeichnet — beginnt als Folge der Wechselwirkung der Primären mit der Atmosphäre die Ausbildung der Sekundären, die sich nach Fig. 1 in einem Intensitätsanstieg der geladenen Komponenten äußert und über das Pfotzer-Maximum bei etwa 20 km Höhe hinter 60—80 g/cm² Luftschicht führt.

Das Intensitätsverhalten der sekundär in der Atmosphäre erzeugten und seit 1933 nachgewiesenen Neutronen diskutierten Bethe, Korff und Placzek [Be 40], und FLÜGGE [Fl 43a] sagte 1943 Lage und Gestalt eines Neutronenmaximums voraus. YUAN [Yu 48, 49, 51] und SIMPSON [Si 51b] gelang dessen Nachweis. Für die Neutronen besteht kein Zustrom von außen, dagegen ein gewisser Verluststrom in die Exosphäre. Dieser Neutronenstrom kann ungehindert vom Erdmagnetfeld in den Außenraum wegfließen. Diese Verlustneutronen werden überwiegend radioaktiv zerfallen: $n \to p^+ + e^- + \nu_\beta$. Ein Teil ihrer Zerfallsprotonen und Zerfallselektronen soll nach SINGER [Si 58a] und HESS [He 59b] vom Magnetfeld der Erde in den inneren van-Allen-Gürtel eingefangen werden und zu dessen teilweiser Speisung dienen. Der Diffusionsfluß bzw. Transportfluß der Neutronen in der Atmosphäre kann neuerdings durch den Einsatz großer Rechenmaschinen über die früheren analytischen Abschätzungen hinaus so genau errechnet werden,

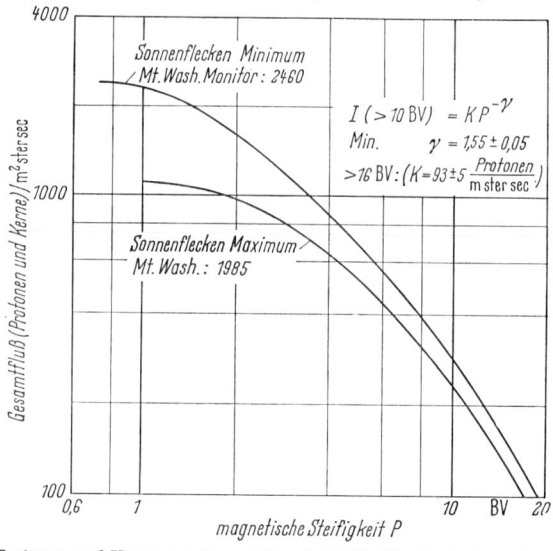

Fig. 2. Gesamtfluß der Protonen und Kerne am Atmosphärengipfel (Steifigkeits-Spektrum) im Sonnenfleckenmaximum und -minimum.

daß es möglich ist, etwa den Neutronenverluststrom in Abhängigkeit einer solaren Modulation oder einer solar flare-Störung anzugeben. Die Übereinstimmung zwischen Experiment und Theorie ist befriedigend.

Im Nukleonenhaushalt in der Erdatmosphäre spiegelt sich die Intensität der einfallenden primären kosmischen Strahlung wider. Im sog. Breiteneffekt zeigt sich der Einfluß des erdmagnetischen Feldes auf Fluß und Spektrum der Primären, die den Atmosphärengipfel an einem bestimmten Ort erreichen. Er wird im niederenergetischen Anteil, dem geomagnetischen oder breitenabhängigen Bereich ihres Spektrums (vgl. Ziff. 2) naturgemäß besonders wirksam. In analoger Weise wirkt sich hier, im *Steifigkeitsbereich* bis etwa 30 GV, die solare Modulation im Elfjahreszyklus der Sonnenflecken aus. Sowohl der Absolutwert des Teilchen- und des Energieflusses, wie auch die relativen Änderungen zwischen Sonnenfleckenmaximum und -minimum sind in diesem Bereich am größten und wirken auf den Nukleonenhaushalt in der Erdatmosphäre zurück.

Im Detail wird hierüber im Artikel von WEBBER[1] in diesem Bande berichtet. Wir bringen hier nur einige orientierende Daten.

In Fig. 2 ist das *Steifigkeits*spektrum des Gesamtflusses der Protonen und Kerne für das Sonnenfleckenmaximum und -minimum dargestellt, wobei die

---

[1] W. R. WEBBER: The Spectrum and Charge Composition of the Primary Cosmic Radiation, Ch. III, S. 181.

integrale Intensität $I(>P)$ gegen die magnetische Steifigkeit P aufgetragen ist. Man erhält es, wenn man die Erde im Verein mit der Magnetosphäre als magnetisches Multipolfeld-Spektrometer betrachtet, und vom Pol (Abschneidesteifigkeit $P_A = 0{,}2$ GV) zum Äquator ($P_A = 16{,}5$ GV) mißt.

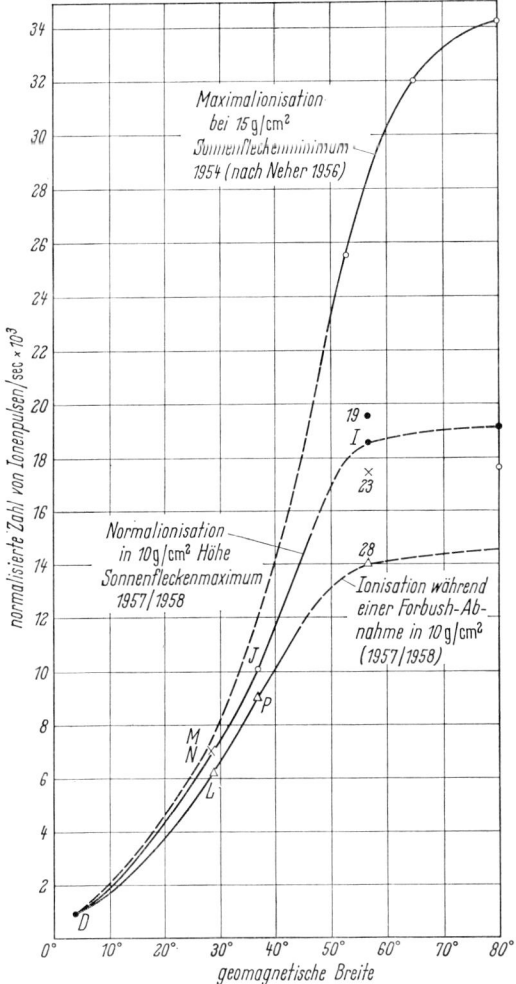

Fig. 3. Breiteneffekt der geladenen Teilchen in der Atmosphäre in Abhängigkeit von der solaren Modulation nach WINKLER [Wi 60c] — dort auch Angabe der Meßpunkte. Nach NEHER [Ne 66] stimmt die Maximalionisation 1964/65 mit derjenigen von 1953/54 bis auf einige Prozent überein.

Fig. 3 zeigt die Intensität geladener Teilchen nach Ionisationsmessungen in einer Höhe von etwa 30 km in ihrer Breitenabhängigkeit für verschiedene Fälle der Sonnenaktivität. Das in der Phase erhöhter Sonnenaktivität vorhandene „Knie" bei etwa 60° geomagnetischer Breite — ein Abschneideeffekt solaren Einflusses am niederenergetischen Ende des Primärspektrums — ist im Sonnenfleckenminimum nahezu verschwunden.

Ähnlich intensiv und den Nukleonenhaushalt der Erdatmosphäre kurzfristig modulierend wirken solar flares mit heftigem Zustrom solarer Teilchen magnetischer Steifigkeiten von 0,2—2 GV; in extrem starken Ausbrüchen werden Störungen bis zum Bereich von 20 GV beobachtet.

Absolute Intensitätsangaben bedürfen daher eines Bezuges auf die Phase der Sonnenaktivität, für den man die langfristigen Mittelwerte der ruhigen kosmischen Strahlung wählt (vgl. Ziff. 1 und 2).

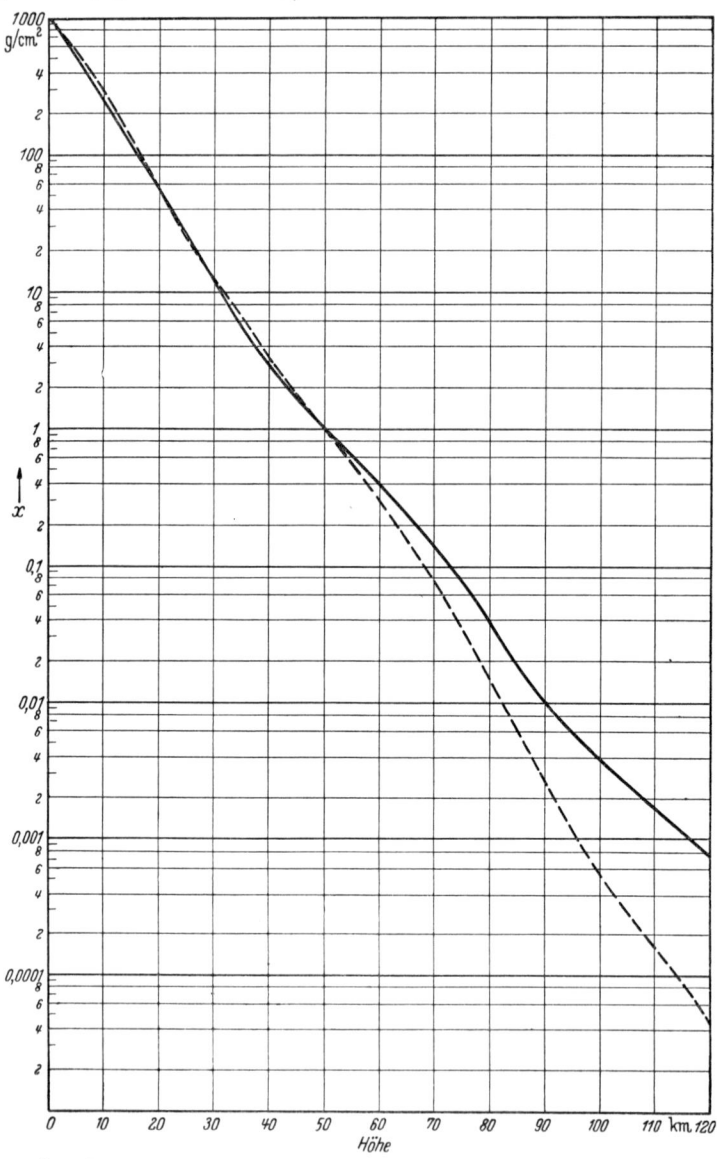

Fig. 4. Massenverteilung der Erdatmosphäre. ——— NACA-Modell; — — — Daten aus Raketenmessungen. Den Druck p der darüberliegenden Luftsäule in [mb] erhält man durch Multiplikation von x mit 0,981.

Im Kapitel D werden wir eine Übersicht über die Protonenkomponente in der Atmosphäre, in Kapitel E über die Neutronenkomponente geben.

Im Internationalen Geophysikalischen Jahr (IGY) erwies sich die Notwendigkeit einer Standardtabelle der Atmosphäre mit Daten für Druck, Dichte und Temperatur. Die bis dahin für wissenschaftliche Zwecke international angenommene Standardatmosphäre war die I.C.A.O.- (International Civil Aviation

Organisation-) Atmosphäre, deren Tabellen aber nur bis 22 km Höhe reichen. Oft wurden inoffizielle amerikanische Tabellen benutzt, die Daten bis 800 km Höhe enthalten. COSPAR (Comittee on Space Research) hat im Sommer 1961 eine etwa 200 Seiten umfassende Tabelle veröffentlicht [Ci 61], die alle bis 1961 verfügbaren atmosphärischen Daten berücksichtigt. Die U.S.-Standardatmosphäre [US 62] vervollständigt teilweise die COSPAR-Daten oder diskutiert diese kritisch. Zum Zwecke der Übersicht und für eine orientierende Information ist in Fig. 4 die Massenverteilung in der Atmosphäre nach dem NACA-Modell [Pe 52a] dargestellt, ergänzt durch Messungen aus Raketenaufstiegen [Ha 52, Wh 52].

B. Die Primärkomponente der kosmischen Strahlung[1].

1. Vorbemerkung. Direkte Beobachtungen und Messungen der primär auf die Atmosphäre der Erde auffallenden Höhenstrahlung sind seit etwa 25 Jahren möglich. Einen ersten wichtigen Schritt bedeutete die Ende der dreißiger Jahre von E. REGENER begonnene Verwendung großer Kunststoffballone, die es heute erlauben, in Höhen bis 40 km ü.M. mit etwa 3 g/cm² darüberliegender Atmosphärenschicht, d.h. praktisch bis nahe zum „Atmosphärengipfel", vorzudringen. Einen gleich wichtigen Schritt bedeutet die Verwendung von Raketen zur Messung der kosmischen Strahlung oberhalb der Atmosphäre. Am 16. April 1946 wurde eine erbeutete V2-Rakete mit VAN ALLEN's Geigerzähler gestartet; bis 1958 waren 68 V2-Raketen, 63 Aerobees und 7 Viking-Raketen zu Forschungszwecken in die obere Atmosphäre geschossen worden. Eine Übersicht hierüber findet sich bei LUDWIG [Lu 63].

Die primäre kosmische Strahlung scheint nach den heutigen Meßergebnissen isotrop oder nahezu isotrop in den Raum um die Erde einzufallen; nach möglichen Anisotropien wird gesucht. Eine Aussage über das zeitliche Verhalten der Intensität ist schwer zu machen: allgemein wird als einfachste Annahme zeitliche Konstanz gewählt. SANDSTRÖM [Sa 65] nimmt eine Gleichverteilung[2]) im Zeitmittel, um eine als Mittelwert der „ruhigen kosmischen Strahlung (quiescent cosmic rays)" definierte Größe an. Die Definition dieses Mittelwertes als „Intensität isotroper Primärstrahlung nach Abzug aller Überlagerungsphänomene und aller Modulationseffekte" erscheint uns jedoch nicht ganz unproblematisch. Einerseits kennen wir die Quellstärken der solaren, galaktischen und eventuell extragalaktischen Primärstrahlung noch nicht, zum andern ist die Gesamtheit der Modulationsmechanismen der solaren und galaktischen Strahlung heute nicht genügend bekannt.

Manche Autoren identifizieren diese „ruhige kosmische Strahlung" mit der galaktischen kosmischen Strahlung. Die Tatsache, daß das Energiespektrum des primären Gesamtflusses ebenso wie die Energiespektren der primären Einzelkomponenten im breitenempfindlichen Bereich bei einer magnetischen Steifigkeit von etwa 1—2 GV ein Maximum aufweisen, dessen Lage zwischen Sonnenfleckenminimum und -maximum sogar etwas variabel ist, deutet jedoch auf einen Übergangseffekt im interplanetaren Raum (Coronabereich) hin. Es ist nicht ohne weiteres einzusehen, daß die ruhige kosmische Strahlung zur Zeit der Sonnenfleckenminima identisch mit der tatsächlichen galaktischen Strahlung sein soll.

---
[1] In der ersten Fassung dieses Beitrags, war eine ausführliche Darstellung der Primärkomponente enthalten. Inzwischen konnte das vielfältige Datenmaterial des Primärflusses und der Energiespektren der Gesamtintensität sowie einzelner Komponenten durch solare Modulation der Primärstrahlung im breitenabhängigen Energiebereich gedeutet werden. Eine ausführliche Darstellung der Primärdaten gibt WEBBER in seinem Beitrag in diesem Band. Wir beschränken uns hier auf die Wiedergabe einiger für spätere Kapitel erforderlicher Daten.
[2] "The Time Variations of the Cosmic Ray Intensity" behandelt J. J. QUENBY in diesem Band; ferner sei auf den Beitrag von S. E. FORBUSH "Time Variations of Cosmic Rays" im Handbuch der Physik, Bd. 49/1, hingewiesen.

Die Satelliten- und Raumsondenmessungen in den letzten Jahren haben gezeigt, daß die solare Modulation sich in einer Variation der Primärstrahlung nicht nur in engster Erdnähe äußert, sondern zu einem gewissen Teil weit in den interplanetarischen Raum übergreift (vgl. FAN et al. [Fa 60]; ROTHWELL et al. [Ro 60b]). Man wird daher den niederenergetischen Anteil der tatsächlichen galaktischen Primärintensität erst nach Erforschung der Corona, der Stoßwelle des solaren Winds, der Magnetopause und der Magnetosphäre im interplanetarischen Raum erschließen können.

Zusammenfassende Darstellungen über die Primärkomponente finden sich z.B. bei TEUCHER [Te 53d], ROSSI [Ro 55] und neueren Datums im Beitrag von WEBBER in diesem Bande.

Einen ausführlichen zusammenfassenden Bericht über ältere Arbeiten über schwere Kerne in der Primärkomponente hat PETERS [Pe 52b] gegeben.

2. Fluß und Energiespektrum der Primärteilchen im breitenempfindlichen Bereich (0.2—30 GV). Die Daten der Flußwerte und die Spektren bei niedriger Energie stellten noch vor wenigen Jahren — auch bei Messungen am gleichen geomagnetischen Ort zu verschiedenen Zeiten — einen unverstandenen Komplex schwankender Zahlwerte bzw. abweichender Spektralverteilungen dar.

Die Vermutungen von FORBUSH [Fo 54] und MEYER und SIMPSON [Me 55a] sowie von NEHER [Ne 56], daß diese Schwankungen mit dem elfjährigen Zyklus der Sonnenfleckentätigkeit zusammenhängen müßten, haben sich inzwischen bestätigt. Nach MEYER und SIMPSON [Me 57] soll die solare Modulation des Primärspektrums bis zu Energiewerten von etwa 30 GeV heraufreichen.

Im Internationalen Geophysikalischen Jahr (IGY 57) wurde eine quantitative Neubestimmung des Erdmagnetfeldes vorgenommen. Dies war eine wesentliche Voraussetzung für die neueren Messungen. (Vergleiche dazu die Zusammenfassung mit ausführlicher Literatur von QUENBY und WEBBER im Teilbeitrag "Cosmic Rays and Earth Magnetic Field" [We 62]; weitere Daten und Literatur findet man in MCCRACKEN, RAO, FOWLER, SHEA und SMART [McCr 65] und [Sh 65].

WEBBER[1] hat den verwertbaren Teil der früheren Messungen des Primärflusses einschließlich der neueren Meßdaten unter Umrechnung von Energien auf magnetische Steifigkeiten über der ideal geeigneten Zeitachse, nämlich den mittleren zeitlichen Schwankungen der Neutronen-Monitorwerte vom Mt. Washington, aufgetragen. Auf diese Weise ordnen sich die früher so uneinheitlich anmutenden Meßdaten gleicher oder benachbarter Werte der *Steifigkeit* sowohl für die Protonen und die α-Teilchen, als auch für die schweren Kerne zwanglos auf Regressionskurven innerhalb des Sonnenfleckenzyklus. Durch Extrapolation zum Monitorstand beim Sonnenfleckenminimum kann für einige *Steifigkeits*werte auf den Fluß beim Sonnenfleckenminimum geschlossen werden (vgl. dazu die Fig. 9—13 und 20 bei WEBBER). Die extrapolierten Werte sind in guter Übereinstimmung mit den Meßwerten des Breiteneffektes (Fig. 3), die kurz nach dem letzten Sonnenfleckenminimum von NEHER et al. [Ne 62] und WINCKLER [Wi 60c] ermittelt und von WEBBER auf Albedo-Einflüsse korrigiert worden sind. Mit Hilfe dieser Regression verfügt man nun über Flußwerte für die Gesamtintensität (nach Fig. 2); insbesondere ist damit ein Verständnis für den Modulationsmechanismus eingeleitet und das Spektrum charakteristisch angebbar. Wir übernehmen die durch Regression und Extrapolation gewonnenen Flußwerte für die Einzelkomponenten der Protonen, He-Kerne und der Kerne der L, $S=M+H$-Gruppe im breitenabhängigen Bereich.

---

[1] Siehe insbesondere seinen Beitrag in diesem Band, wo in Fortführung zu [We 62] die Meßdaten des Flusses noch auf Albedoeinflüsse korrigiert sind.

Solare Protonen wurden mit Energien bis etwa 20 MeV herab gemessen. Die Spektren zeigen im breitenabhängigen Bereich alle sehr ähnlichen Verlauf: sie haben bei höheren Energien (Steifigkeit $> 16{,}5$ GV) für alle Primären einen negativen Exponenten des integralen Potenzspektrums mit $\gamma = 1{,}50$, durchlaufen bei 1—2 GV ein Maximum, um dann zu niederen Energien steil abzufallen.

Wir stellen für die verschiedenen Teilchenarten die Daten der breitenabhängigen Spektren in einigen Zahlenwerten zusammen (vgl. auch Fig. 5).

a) *Protonen* $(z=1)$:

$$J(>16{,}5 \text{ GV}) = 93 \pm 5 \frac{\text{Protonen}}{\text{m}^2 \text{ sterad, sec}}, \quad \gamma = 1{,}5,$$

$$J(>4{,}5 \text{ GV}) = 610 \pm 30 \frac{\text{Protonen}}{\text{m}^2 \text{ sterad, sec}},$$

$$J(>1{,}3 \text{ GV}) = 2140 \pm 40 \frac{\text{Protonen}}{\text{m}^2 \text{ ster. sec}}.$$

b) *He-Kerne* $(z=2)$:

$$J(>16{,}5 \text{ GV}) = 17{,}5 \pm 1{,}0 \frac{\text{Heliumkerne}}{\text{m}^2 \text{ ster. sec}}, \quad \gamma = 1{,}5,$$

$$J(>4{,}5 \text{ GV}) = 94 \pm 4 \frac{\text{Heliumkerne}}{\text{m}^2 \text{ ster. sec}},$$

$$J(>1{,}3 \text{ GV}) = 328 \pm 12 \frac{\text{Heliumkerne}}{\text{m}^2 \text{ ster. sec}}.$$

c) *L-Kerne* $(z=3 \text{ bis } 5)$:

$$J_L(>16{,}5 \text{ GV}) = 0{,}38 \pm 0{,}08 \frac{L\text{-Kerne}}{\text{m}^2 \text{ ster. sec}}, \quad \gamma = 1{,}5,$$

$$J_L(>4{,}5 \text{ GV}) = 2{,}1 \pm 0{,}2 \frac{L\text{-Kerne}}{\text{m}^2 \text{ ster. sec}},$$

$$J_L(>1{,}3 \text{ GV}) = 7{,}2 \pm 1{,}0 \frac{L\text{-Kerne}}{\text{m}^2 \text{ ster. sec}}.$$

d) $S=(M+H)$-*Kerne* $(z>5)$:

$$J_S(>16{,}5 \text{ GV}) = 1{,}55 \pm 0{,}08 \frac{S\text{-Kerne}}{\text{m}^2 \text{ ster. sec}}, \quad \gamma = 1{,}5,$$

$$J_S(>4{,}5 \text{ GV}) = 8{,}7 \pm 0{,}3 \frac{S\text{-Kerne}}{\text{m}^2 \text{ ster. sec}},$$

$$J_S(>1{,}3 \text{ GV}) = 28 \pm 1{,}5 \frac{S\text{-Kerne}}{\text{m}^2 \text{ ster. sec}}.$$

3. Fluß und Spektrum der Primären bei sehr hohen Energien. Im Bereich niedriger Energien, dem sogenannten breitenempfindlichen Bereich (vgl. Fig. 5 und Ziff. 2) ist Fluß und Spektrum der primären Protonen bzw. der verschiedenen Gruppen primärer Kerne direkt meßbar; im Gegensatz dazu müssen Fluß und Spektrum der Primären hoher Energie indirekt über verschiedene Phänomene ihrer Wechselwirkung in der Atmosphäre bzw. innerhalb geeigneter Meßapparaturen erschlossen werden. Eine Übersicht über die in verschiedenen Energiebereichen angewandten Methoden und über den daraus erschlossenen, integralen Fluß der primären Protonen vermittelt Fig. 5. Soweit sie bekannt sind, sind die Daten für primäre He-Kerne bzw. schwere S-Kerne miteingetragen. Auf die übrigen Kurven der Fig. 5, die den Intensitätsverlauf der Nukleonen bzw. der nukleoaktiven Komponente in verschiedenen Tiefen in der Atmosphäre darstellen gehen wir weiter unten ein.

Im Bereich der Energien zwischen 10^{11} und einigen 10^{13} eV sind die Informationen aus Messungen an lokalen, durchdringenden Schauern (jets) mit Kern-

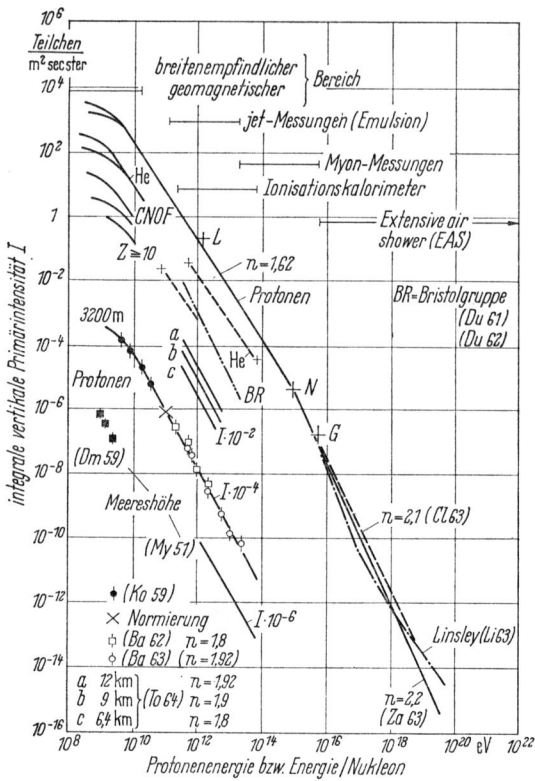

Fig. 5. Primärspektrum der Protonen mit experimentellen Normierungspunkten nach LAL [*La 53*], N = NIKOLSKY [*Ni 62*], G = GREISEN [*Gr 60c*]. Die Primärspektren der Kerne sind im geomagnetischen bzw. im jet-Bereich bekannt (Details und Literatur s. Fig. 6). Mitaufgenommen (teilweise mit verschobenem Ordinatenmaßstab) sind gemessene Spektren der Protonen (Nukleonen) und der nukleoaktiven Komponente in verschiedenen Tiefen in der Atmosphäre. (Details s. Ziff. 4, S. 383 ff.). Charakteristisch sind die verschiedenen Steigungen n der Spektren. Die Primärspektren spiegeln sich also nicht homogen, d.h. mit konstanter Steigung n in den Spektren der Sekundären wider. — Der Spektralverlauf bis 10^{15} eV kann nach COCCONI [*Co 61a*] analytisch beschrieben werden; hier ist der etwas flachere Verlauf nach ZATSEPIN et al. [*Za 61*] mit n = 1,62 eingetragen. Über 10^{15} eV kann man die Spektren mit n = 2,2 [*Za 63*] oder mit n = 2,1 nach CLARK et al. [*Cl 63*] bzw. nach LINSLEY [*Li 63*] zugrundelegen. (Zitate: [*Ba 62*] lies [*Ba 62a*]; [*Ba 63*] lies [*Ba 63a*]).

spuremulsionen gewonnen. Etwa im gleichen Bereich kann man mit Ionisationskalorimetern Intensität und Energieumsatz der nukleoaktiven Komponente messen. Daran schließen sich die Daten aus Intensitätsmessungen an energiereichen μ-Mesonen an. Im Bereich oberhalb 10^{15} eV stellen die ausgedehnten Luftschauer (Extensive Air Showers, EAS) gegenwärtig die einzige Informationsquelle über die auslösenden Primären dar. Hier ist der Rückschluß auf die Masse der auslösenden Primären nicht ganz eindeutig: Besonders im Bereich der höchsten Energien bleibt es offen, ob die Gesamtenergie des Schauers *einem* Proton oder den Nukleonen eines Kerns zuzuordnen ist. Die Charakteristiken des Luftschauers, Teilchenzahl und Aufbau, lassen eine Unterscheidung zwischen Einzel- und Mehrfachstößen bei sehr hohen Energien zur Zeit nicht treffen. Im Rahmen unseres Berichts soll darauf aber nicht eingegangen werden.

Ziff. 3. Fluß und Spektrum der Primären bei sehr hohen Energien. 381

Von GREISEN [*Gr 56b*] ist das Problem der großen Luftschauer in einem zusammenfassenden Bericht bis zum Stand des Jahres 1956, von COCCONI bis 1957 behandelt worden (Bd. 46/1 dieses Handbuches; vgl. auch GREISEN [*Gr 60c*]).

In den letzten Jahren hat eine stürmische Entwicklung auf diesem Gebiet begonnen. Ihr Ergebnis ist ziemlich vollständig bis zum Jahre 1965 den umfang-

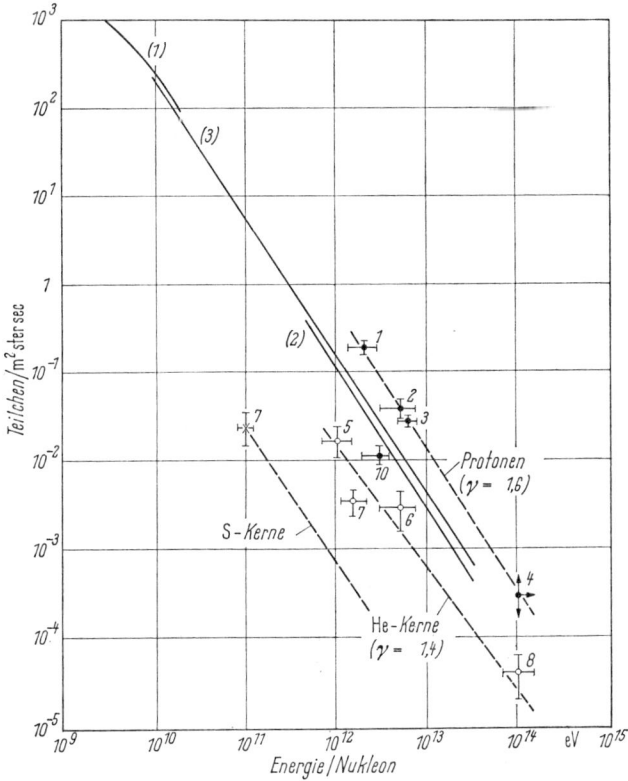

Fig. 6. Integrales Spektrum und Fluß der Gesamt-Intensität der Primären und der einzelnen Komponenten: P-, He- und S-Kerne. Die an die Meßpunkte angeschriebenen Zahlen weisen auf den Autor.

Protonen.
1 D. LAL: Proc. Ind. Acad. Sci. **38**, 93 (1953).
2 M. F. KAPLON, B. PETERS, H. L. REYNOLDS u. D. M. RITSON: Phys. Rev. **85**, 295 (1950). — M. F. KAPLON u. D. M. RITSON: Phys. Rev. **88**, 386 (1952).
3 P. H. BARRETT, L..M. BOLLINGER, G. COCCONI, Y. EISENBERG u. K. GREISEN: Rev. Mod. Phys. **24**, 133 (1952).
4 M. W. TEUCHER, E. LOHRMANN et al.: Proc. Mosc. Conf. Cosmic Rays **1**, 26 (1960).
10 C. B. A. MCCUSKER u. L. S. PEAK: Nuov. Cim. **31**, 525 (1964).

Helium- und S-Kerne.
5 P. H. FOWLER u. C. J. WADDINGTON: Phil. Mag. **1**, 637 (1956).
6 E. LOHRMANN u. M. W. TEUCHER: Phys. Rev. **115**, 638 (1959).
7 P. L. JAIN, E. LOHRMANN u. M. W. TEUCHER: Phys. Rev. **115**, 654 (1959).
8 P. H. FOWLER, P. S. FREIER, C. LATTES u. E. P. NEY: Nuov. Cim. **8**, Suppl. 725 (1959).

Gesamtspektrum
(1) Konstruiert nach WEBBER aus Daten von H. V. NEHER: Phys. Rev. **103**, 228 (1956). — J. Geophys. Res. **66**, 4007 (1961); **67**, 1309 (1962) u. Ref. dort. — J. R. WINKLER: J. Geophys. Res. **65**, 1331 (1960).
(2) MIYAKE, S.: J. Phys. Soc. Japan **18**, 1226 (1963).
(3) A. W. WOLFENDALE et al.: Proc. Phys. Soc. Lond. **83**, 853 (1964).

reichen Konferenzberichten der IUPAP über *Kosmische Strahlung* zu entnehmen: Moskau 1959 (gedruckt 1960); Kyoto 1961 [(in Journ. Phys. Soc. Japan. Vol. 17, Suppl. AI, AII, AIII (1962)]; Jaipur 1963 (Tata-Institut of Physics, Bombay 1964) und London 1965 (Institute of Physics and Physical Society, London 1966). Wir haben uns für die in Fig. 5 und 6 wiedergegebenen Daten auf die erwähnten

Quellen sowie auf einige frühere Arbeiten, z.B. die MIT-Agassiz-Messungen von CLARK et al. [Cl 57, 58a, 61] und die Volcano-Ranch-Messungen [Cl 63] gestützt. Aus diesen Unterlagen kennt man das Spektrum bis etwa 10^{19} eV.

In Fig. 6 erscheint der aus den Emulsionsmessungen 1—4 entnommene Fluß der Protonen höher als der Fluß der Gesamtprimären, im Gegensatz zu dem von McCUSKER und PEAK aus Emulsionsmessungen gewonnenen, jedoch auf einen Energiemittelwert $\bar{E}_c = 2{,}8 \cdot 10^{12}$ eV bezogenen Fluß (Meßpunkt 10). Die Bildung eines Mittelwertes E_C aus den nach CASTAGNOLI et al. [Ca 53] bestimmten Primärenergien $E_{c,i}$ der einzelnen ausgewerteten jets wird nach ALI et al. [Al 63][1] auf Grund eines Maschinenexperiments nahegelegt: Protonen bekannter Energie von 30 GeV wurden in Kernemulsion geschossen und die jets nach CASTAGNOLI ausgewertet, wobei sich für die Primärenergien „Schätzwerte" von 2 GeV bis zu 200 GeV ergaben. Deren Mittelwert $\bar{E}_C = 1/n_i \sum_{\varepsilon=1}^{n} E_{C,i} = 29{,}8$ MeV stimmt jedoch erstaunlich gut mit der Einschußenergie überein.

Die Spektren primärer Kerne sind offensichtlich besser bestimmbar. LOHRMANN, TEUCHER und SCHEIN [Lo 61] haben die Gültigkeit der Castagnoli-Formel [Ca 53] überprüft. Sie benutzten Fragmentationen schwerer Primärteilchen, die in Bündeln nahezu monoenergetischer Nukleonen zerfallen, deren Wechselwirkungen sie in Emulsion verfolgen und miteinander vergleichen konnten. Die Castagnoli-Formel

$$\left. \begin{array}{l} -\ln \gamma_c = \dfrac{1}{n_s} \ln \operatorname{tg} \Theta_i + C \\[4pt] \dfrac{\beta_c}{\beta_i^*} = 1 \quad \text{ergibt} \quad C = 0; \quad (\beta = v/c) \end{array} \right\} \tag{3.1}$$

gestattet den Lorentzfaktor γ_c des CMS (Schwerpunktsystems) unter der Annahme zu bestimmen, daß das CMS mit demjenigen System übereinstimmt, in dem die Winkelverteilung mit den Winkeln Θ_i der n_s geladenen Schauerteilchen symmetrisch ist. Gilt ferner $\beta_c/\beta_i^* = 1$, so verschwindet die Konstante C (vgl. wegen weiterer Details [Pe 60]).

Die Untersuchungen von LOHRMANN et al. ergaben, daß eine starke Abhängigkeit des Ergebnisses von der Zahl der schwarzen Spuren n_h vorliegt. Für $n_h \leq 5$ ist die mit Hilfe der Castagnoli-Formel bestimmte Primärenergie um einen Faktor 1,3 höher als die wirkliche Primärenergie. Für Sterne mit $n_h > 5$ ist die Castagnoli-Energie um einen Faktor 1,8 zu klein. Zudem treten starke Schwankungen der Einzelergebnisse auf. Die Abweichung um einen Faktor 1,3 bei $n_h \leq 5$ wird darauf zurückgeführt, daß viele Mesonen im CMS so niedrige Energien haben, daß die Bedingung $\beta_c/\beta_i = 1$ und damit $C = 0$ nicht erfüllt ist. Für $n_h > 5$ führt man die Abweichung darauf zurück, daß im getroffenen Kern mehrere Wechselwirkungen stattfinden, bei denen die Winkelverteilung verbreitert wird.

Der Knick des Primärspektrums der Protonen bei Energien oberhalb von etwa 10^{15} eV in Fig. 5 resultiert aus Messungen an ausgedehnten Luftschauern (EAS). Angelehnt an die experimentell gewonnenen Meßpunkte $L = $ [La 53], $N = $ NIKOLSKY [Ni 62] und $G = $ GREISEN [Gr 60c] kann man nach ZATSEPIN et al. [Za 63] das Spektrum bis zu Energien von etwa 10^{15} eV fortführen; die Meßdaten nach CLARK et al. [Cl 63] mit einer Steigung von $n = 2{,}1$ und das noch zweifach geknickte Spektrum nach LINSLEY [Li 63] deuten an, daß die endgültige Form des gesamten primären Spektrums noch unsicher ist.

Wie erwähnt, sind auch im mittleren Energiebereich von $10-10^5$ GeV Betrag und Steigung des Flusses nur indirekt erschlossen; eine *direkte* Messung — z.B. nach

---

[1] Vgl. auch die ICEF-Collaboration (ICEF 63), Seiten: 1047, 1084, 1116f.

dem Vorschlag von SITTE [*Si 65*] — für Energien von 10^{12} bis 10^{13} eV wäre sehr wünschenswert.

Abgesehen vom breitenempfindlichen Bereich und von der Unsicherheit im Bereich des Knicks bei ca. 10^{15} eV kann man die integrale spektrale Verteilung grob durch zwei Potenzgesetze annähern.

$$N(\geq E) = A E^{-n} \quad \text{(Teilchen/m}^2 \cdot \text{sec} \cdot \text{sterad)} \tag{3.2}$$

mit den Werten:

nach WOLFENDALE [*Br 64b*]

$$A_W = \left(0{,}87 \, {}^{+0{,}53}_{-0{,}3}\right) \cdot 10^4; \quad n = 1{,}58;$$

E in GeV für 10^{10} eV $\leq E \leq 3 \cdot 10^{13}$ eV

nach CLARK et al. [*Cl 63*]

$$A_{Cl} = (3{,}2 \pm 0{,}5) \cdot 10^{-6}; \quad n = 2{,}1;$$

E in 10^{15} eV für 10^{15} eV $\leq E \leq 10^{17}$ eV.

Entsprechend gelten für diese Potenzverteilungen differentielle Spektren

$$N(E) \, dE = A' \cdot E^{-(n+1)} \, dE \quad \text{(Teilchen/cm}^2 \cdot \text{sec} \cdot \text{sterad} \cdot \text{GeV)} \tag{3.3}$$

Der Festlegung des obigen Spektrums nach WOLFENDALE et al. [*Br 64a*] wurde ein differentielles „Versuchs"-Spektrum nach LINSLEY, SCARSI und ROSSI ([*Li 62*, Kyoto-Konferenz]) zugrunde gelegt mit den Zahlenwerten (vgl. S. 387).

$$A' = 9{,}43; \quad n = 2{,}7;$$

und dem Gültigkeitsbereich von 10^{12} bis 10^{14} eV.

C. Kernwechselwirkungen der Nukleonen-Komponente in der Atmosphäre

4. Einzelheiten zur Nukleonenkaskade in der Atmosphäre.

α) *Protonen und die Nukleoaktive Komponente*: Messungen der nukleoaktiven Komponente (d.h. an Nukleonen und Pionen) liegen in Meereshöhe, in verschiedenen Gebirgshöhen, verschiedenen Flughöhen und in Ballonhöhen vor.

KOCHARYAN, SAAKYAN und KIRAKOSYAN [*Ko 59*] haben in Gebirgshöhe (3 200 m, 700 g/cm²) mit einem magnetischen Impulsspektrometer das Protonen-Impulsspektrum aufgenommen (vgl. S. 413). Die Meßwerte — umgerechnet in eine Energieverteilung — können durch das integrale Spektrum

$$N(>E) \cdot = 3{,}56 \cdot 10^{-3} [2 + E(\text{GeV})]^{-1{,}8} \frac{\text{Protonen (Teilchen)}}{\text{m}^2 \cdot \text{sec} \cdot \text{sterad}} \tag{4.1}$$

gültig für $(10^{10}$ eV $\leq E \leq 10^{11}$ eV$)$

wiedergegeben werden (vgl. Kurve ♦ [*Ko 59*] in Fig. 5 bis zum Normierungspunkt).

In derselben Gebirgshöhe wurde die nukleoaktive Komponente mit Hilfe zweier anderer, sehr wichtiger Meßverfahren untersucht: a) in einer großen Ionisationskammer (10 m²×3 m) nach GRIGOROV, SHESTOPEROV et al. [*Gr 60a, b*] in einem Ionisationskalorimeter nach GRIGOROV, MURZIN und RAPOPORT [*Gr 58, 60b*]. Es sei wegen der apparativen und experimentellen Details auf die Originalarbeiten verwiesen. Diese Meßverfahren wurden in erheblich größeren Meßanlagen weiterentwickelt, die im Bau oder schon in Betrieb sind und zum Teil durch

Funkenkammern, Emulsionskammern, Blasenkammern ergänzt werden (vgl. z. B. den Bericht von DOBROTIN über die Tienshan-Meßstation bei GUSEVA et al. [Gu 66]).

GRIGOROV hat mit seinen Mitarbeitern die nukleoaktive Komponente sowohl mit einer Ionisationskammer (BABAYAN et al. [Ba 62a]), als auch mit einem Ionisationskalorimeter (GRIGOROV et al. [Ba 61] und [Gr 63] gemessen. Für eindeutige Einzelereignisse nukleoaktiver Teilchen, die die Kammer oder das Kalorimeter auslösen, finden sie eine gute Übereinstimmung der gemessenen Potenzspektren: Mit beiden Verfahren gemessene Spektren haben gleiche Steigung im Energiebereich von $3 \cdot 10^{11}$ bis 10^{13} eV mit $n=1,8$ (bzw. $n=1,92$) [Gr 64a]. In Fig. 5 schließen diese Meßwerte direkt an die Messung von KOCHARYAN an und lassen das Spektrum (4,1) gut fortsetzen. Die eingezeichnete Intensität ist um einen Faktor 10^{-4} erniedrigt, um eine übersichtliche Darstellung zu gewährleisten. Man vergewissert sich leicht, daß die Intensität der nukleoaktiven Komponente in 3200 m noch niedriger ist als diejenige der primären S-Kerne. Auffällig ist jedoch, daß für die Sekundärteilchen, die die Atmosphäre in etwa 8—9 nuklearen Stoßlängen durchlaufen haben, der Exponent des Spektrums größer ist, als der für die auslösenden Primären; dabei sei vom Übergangsgebiet des Knicks und den höchstenergetischen Protonen zunächst abgesehen.

Bei früheren Messungen mit großen Ionisationskammern (vgl. [Fa 57], [Za 58], [Mu 59], [De 61] und [Ba 62a]) an intensiven Ionisationsbursts wurden, rückschließend auf die nukleoaktive Komponente, Potenzspektren mit flacherem Verlauf gefunden, mit Exponenten $1,35 \leq n \leq 1,7$. Sehr frühe Messungen an Ionisationsbursts von LAPP [La 46] und STINCHCOMB [St 51b] mit kleindimensionierten Ionisationskammern ergaben $n=2,0$.

Detaillierte Untersuchungen durch die Grigorov-Gruppe lassen diese Diskrepanz erklären: BOYADZHYAN et al. [Ba 63a] zeigen, daß ein Burst, ausgelöst durch *mehrere* nukleoaktive Teilchen, die als Schauer- bzw. Core-Teilchen noch eng gebündelt sein können, sich anders verhält als ein „Einzel"-Burst, ausgelöst durch ein *einzelnes* nukleoaktives Teilchen. Schließlich zeigen sie, daß in derselben Ionisationskammer, bei jeweils einem und demselben Ereignis, die Kopplung einzelner Zählrohre zu einem großen Zählrohr mit 10 m² Seitenfläche zu einer Steigung $n=1,37$ führt, bei Zählerflächen von 1 m² zu $n=1,58$ und bei Einzelzählrohrschaltung mit jeweils 15×15 cm² Stirnfläche der 3 m langen Zählrohre zu $n=1,92 \pm 0,03$.

GRIGOROV et al. [Gr 63] [Ba 61] [Ba 62a] haben in einem Ionisationskalorimeter in Verbindung mit einer Nebelkammer einige hundert Ereignisse auf Mehrfacheinfall und Einzeleinfall untersucht und finden für die einzeln auslösenden Nukleoaktiven bei Energien von $2-20 \cdot 10^{11}$ eV die Steigung $n=1,92$.

Völlig unabhängig von diesen Messungen findet man bei den Untersuchungen von TOLKACHEW [To 64] in verschiedenen Flughöhen mit einer Zählkammer, ebenfalls steilere Spektren als die der auslösenden Primären. Nach Messungen mit Emulsionen durch die Bristolgruppe von PERKINS u. Mitarb. [Du 61] und [Du 62], in Flughöhen von 200 g/cm² verläuft das Nukleonenspektrum ebenfalls steiler. GRIGOROV [Gr 64a, b] hat einen Vergleich seiner Meßergebnisse in Gebirgshöhe mit denjenigen der Bristolgruppe bzw. den Messungen von AKASHIY, SHIMIZU, WATANABE und NISHIMURA (mit $n=2,3$) versucht. Wegen der verschiedenen Meßhöhen bereitet jedoch die Normierung Schwierigkeiten. Wir dürfen hier auf die Originalarbeit verweisen (vgl. auch [Ba 62b]).

In Meereshöhe verfügen wir über die frühen Messungen von MYLROI und WILSON [My 51]. Die Flußwerte sind in Fig. 5 um den Faktor 10^{-6} verschoben eingezeichnet. Sie liegen knapp unterhalb der Flußwerte in Gebirgshöhe.

Weitere Messungen stammen von DIMITREV et al. [*Di 59*] und BABECKI et al. [*Ba 61*]. Der Wert des Exponenten in Meereshöhe wird durch die neuesten Messungen in Leeds in einem Ionisationskalorimeter durch KELLERMANN und SILK [*Ke 66*] bestätigt. Sie finden einen Wert von n zwischen 1,8 und 1,9.

β) *Ausbreitung der Nukleonen und der nuklearen Komponente in der Atmosphäre — Einfachste Modelle:* Die eigentlichen Träger der Kaskade der kosmischen Strahlung in der Atmosphäre sind hochenergetische Nukleonen. Diese tragen die hohen Energien E_p primärer Protonen bzw. der Nukleonen primärer Kerne im Verlauf einiger Stöße tief in die Atmosphäre, dabei verbleibt jedem „Trägernukleon" (persisting nucleon) bei relativ hoher, mittlerer Inelastizität K_t über die „Elastizität" η

$$\eta = (1 - K_t) \tag{4.2}$$

nach dem ersten Stoß die Energie $E_1 = \eta E_p$; nach j-Stößen verfügt es noch über eine Energie

$$E_j = \eta^j E_p \equiv (1 - K_t)^j E_p. \tag{4.3}$$

Gemäß einer mittleren, totalen Inelastizität $K_t (\approx \text{const})$ geht jedem Trägernukleon bei jedem Stoß ein Anteil an Energie

$$E_t = K_t E_p \tag{4.4}$$

dissipativ verloren. Aus diesen Teilbeträgen E_t der Energie werden die weiteren Transportprozesse der kosmischen Strahlung gespeist; für unsere Betrachtung soll der relativ kleine Beitrag primärer Elektronen, primärer Photonen und primärer Neutrinos außer Betracht bleiben. Der weitaus größte Bruchteil dieser Energie E_t kommt der Pion-Komponente zugute. Kennzeichnet man diesen Anteil mit K_π, der Inelastizität für Pionproduktion, so steht pro Stoß im Mittel ein Energiebetrag

$$E_\pi = K_\pi \cdot E_p \tag{4.5}$$

für geladene Pionen (nukleoaktive Komponente und daraus folgende Zerfallsmyonen) und für neutrale Pionen (und folgende Photonen sowie Elektronen) zur Verfügung. Wegen der beträchtlichen Multiplizität n_s der Erzeugung geladener Pionen bei hochenergetischen Wechselwirkungen (vgl. PERKINS [*Pe 60*] und MALHOTRA [*Ma 63a, b*]) siehe auch: v. LINDERN [*Li 61a*]; COCCONI et al. [*Co 61b*]; GUSEVA et al. [*Gu 62*]; LAL et al. [*La 62a* und *62b*]; LOHRMANN et al. [*Lo 61*]

$$n_s = a \left(\frac{E_p}{1 \text{ GeV}}\right)^\alpha; \quad \text{mit} \quad a \equiv \begin{Bmatrix} a_{\text{Perkins}} = 2 \\ a_{\text{Brooke}} = 1{,}8 \end{Bmatrix}; \quad \text{und} \quad \alpha = \tfrac{1}{4} \tag{4.6}$$

verteilt sich diese Energie E_π auf eine größere Zahl vergleichsweise „energieärmerer" Pionen, die sich normalerweise im CM-System als zwei scharfe Bündel in Vorwärts- und Rückwärtsrichtung ausbreiten, bzw. die im Laborsystem in zwei Emissionskegeln in Vorwärtsrichtung unter verschiedener Energie emittiert werden (vgl. PERKINS [*Pe 60*], GUSEVA et al. [*Gu 62*]; wegen asymmetrischer Schauer und doppelter Maxima sei auf ICEF, S. 1119ff, 1156ff, verwiesen, dort auch weitere Literatur).

Für diese starke Pionvervielfachung hat sich inzwischen nach COCCONI der Sprachgebrauch „Pionisation" eingebürgert. Zur theoretischen Beschreibung der Pionisation sind verschiedene Modelle vorgeschlagen worden, wobei der oder die im CM-System entstehenden „Feuerbälle" entweder aus peripheren Ein-Mesonaustauschprozessen, peripheren Zwei-Mesonaustauschprozessen, zentralen Stoßprozessen, Regge-Pol-Modellen oder über Diffraktionsstreuung entstehen sollen. Einen Überblick, dem weitere Literatur entnommen werden kann, gab FEINBERG in Jaipur (1963, Bd. 5); s. auch FERRARI et al. [*Fe 62*] und Cern-Report 61-22, p. 253—330.

Entscheidende Bedeutung gewann die Tatsache, daß der mittlere Transversalimpuls der Pionisationsprodukte $p_t \approx 0{,}4$ GeV/c ist, worauf Koba [Ko 56] und Milechin und Rozenthal [Mi 57] zuerst hingewiesen haben, und was inzwischen von vielen Seiten bestätigt worden ist.

Die Winkelverteilung der Pionisationsprozesse ist sowohl in jets in Kernemulsionen (vgl. ICEF-Collaboration [ICEF 63]), als auch im Ionisationskalorimeter von Dobrotin et al. [Gu 62] gemessen worden. Eine endgültige Modellbeschreibung des Nukleonentransports in der Atmosphäre muß nicht nur symmetrische Schauer im CM-System beschreiben, sondern auch asymmetrische Verteilungen, bei denen noch zwischen Laborsystem und Spiegelsystem unterschieden werden muß.

Die Annahme eines Potenzspektrums für die Energieverteilung primärer Nukleonen ist eine Näherung, die nur in beschränkten Energiebereichen gilt [vgl. (3.2), bzw. Fig. 5]. Für unseren Überblick ist es ausreichend, an dieser Näherung festzuhalten. Mit der vereinfachten Modellannahme einer konstanten mittleren Inelastizität K_t (4.4) kann eine Verknüpfung zwischen der Primärintensität und der Sekundärintensität in der atmosphärischen Tiefe x hergestellt werden. Nukleonen derselben Sekundärenergie $E = E_j$, die aus Stoßprozessen mit verschiedener Stoßzahl j in der Atmosphäre hervorgegangen sind, stammen wegen (4.3)

$$\frac{E}{(1-K_t)^j} = \frac{E_j}{(1-K_t)^j} = E_p \tag{4.7}$$

aus verschiedenen Energiebereichen E_p des primären Flusses.

Will man den gemessenen differentiellen Fluß (etwa in Meereshöhe $x_0 = 1030$ g/cm²) mit einem angenommenen Primärspektrum (3.3) in Beziehung setzen, so muß man die Anzahl der Stöße j mit in Rechnung setzen. Erreicht das Nukleon in einem Zeitintervall τ die Tiefe x, so ist — bei einer mittleren freien Stoßlänge λ_i — die Wahrscheinlichkeit dafür, daß es j-mal gestoßen hat, durch eine Poissonverteilung gegeben:

$$p(j) = e^{-x/\lambda_i} \frac{(x/\lambda_i)^j}{j!}. \tag{4.8}$$

Der differentielle Fluß $N_{\text{sek}}(x, E)\, dE$ setzt sich aus Beiträgen $n_j(x, E)\, dE$ mit verschiedenen Stoßzahlen j zusammen und wird so vom primären Fluß aus verschiedenen Primärenergien $E(1-K_t)^{-j}$ gespeist:

$$n_j(x_0, E)\, dE = p(j)\, N\!\left(0;\, \frac{E}{(1-K_t)^j}\right) dE_p. \tag{4.9}$$

Ersetzt man die Energieintervalle $dE_p = (1-K_t)^{-j}\, dE$, so kann man bei einem Potenzspektrum für den integralen Sekundärfluß schreiben

$$\begin{aligned}
\sum_{j=0}^{\infty} \int_E^{\infty} n_j(x, E)\, dE &= \sum_{j=0}^{\infty} e^{-x/\lambda_i} \frac{(x/\lambda_i)^j}{j!} \cdot \frac{1}{(1-K_t)^j} \cdot \int_E^{\infty} \frac{S_0(1-K_t)^{j(n+1)}}{E^{n+1}}\, dE \\
N_{\text{sec}}(x;\, \geq E) &= \frac{S_0}{n} \cdot \frac{1}{E^n} e^{-x/\lambda_i} \sum_{j=0}^{\infty} \frac{[(1-K_t)^n (x/\lambda_i)]^j}{j!} \\
&= N(x=0;\, \geq E)\, e^{-x/\lambda_i} \cdot e^{[(1-K_t)^n \cdot x/\lambda_i]} \\
&= N(x=0;\, \geq E) \cdot e^{-x/\lambda_a}.
\end{aligned} \tag{4.11}$$

Als Absorptionslänge λ_a (attenuation length) wird hierbei ein Ausdruck eingeführt, der aus der Zusammenfassung der beiden Exponentialfunktionen hervorgeht und der die experimentell beobachtete Schwächung des Flusses darstellt: (Tabelle 1)

$$\lambda_a = \frac{\lambda_i}{1-(1-K_t)^n} \equiv \frac{\lambda_i}{1-\eta^n}. \tag{4.12}$$

Direkte Rückschlüsse aus den gemessenen Spektren auf die mittlere totale Inelastizität K_t haben WOLFENDALE, BROOKE, HAYMAN und KAMIYA [*Br 64b*]

Tabelle 1. *Mittlere freie Weglängen für Stoß λ_i und Absorption λ_a in Luft.*

Stoßlänge λ_i (g/cm²)	Absorptionslänge λ_a (g/cm²)	E (GeV)	Autor
—	123 ± 6	30	*Az 51*
81 ± 5	119 ± 1		*Bo 62*
—	114 ± 10		*Ge 50*
—	$118 \pm 1{,}1$		*Ho 52c*
—	114 ± 10		*Ry 55*
—	125 ± 5		*Schu 54*
—	118 ± 2		*Ti 48*
95 ± 8	121 ± 3	300	*To 64*
—	120 ± 4	600	*To 64*
—	112 ± 8	1200	*To 64*
—	125 ± 22	2400	*To 64*
81 ± 5	—		*Wa 50a*
Theoretische Werte			
83	—	24	*Al 61; Co 61a*
			Wi 60; Co 61a
80	—	—	*Ud 62*

gezogen. Sie legen das von BROOKE und WOLFENDALE [*Br 64a*] sehr genau gemessene differentielle Protonensekundärspektrum in Meereshöhe zugrunde (vgl. S. 414, Tabelle 8 und S. 463, Fig. 52). Zusammen mit dem weniger genau bekannten Primärspektrum nach LINSLEY, SCARSI und ROSSI [*Li 62*] gemäß (3.3), errechnen sie zunächst für verschiedene Energien von 10 GeV—100 GeV und für verschiedene Stoßlängen gemäß (4.9) K_t-Werte; wählt man für die Stoßlänge den wahrscheinlichsten Wert $\lambda_i = 80$ g/cm², so ergeben sich über dieses Vergleichsspektrum die Werte K_t (10 GeV) = 0,575 und K_t (100 GeV) = 0,54. Legen sie dagegen die zugehörigen integralen Spektren zugrunde, wobei zu berücksichtigen ist, daß die Steigung n im Energiebereich 10 GeV—100 GeV sich erheblich verändert (vgl. Fig. 52), so folgt aus (4.12) die Beziehung

$$K_t = 1 - \left(\frac{\lambda_a - \lambda_i}{\lambda_i}\right)^{1/n}, \qquad (4.13)$$

in der sich jedoch die Energieabhängigkeit von $\lambda_a(E)$ und $n(E)$ im angegebenen Energiebereich fast völlig kompensieren, so daß sich praktisch ein nahezu konstanter Wert $K_t = 0{,}505$ mit maximaler Abweichung $\pm 0{,}005$ einstellt; das integrale Spektrum ist daher ungeeignet für Abschätzungen von K_t.

Das eigentliche Anliegen von WOLFENDALE et al. ist, eine verbesserte Abschätzung des Primärspektrums zu erreichen. Neben dem Protonspektrum legen sie das von HAYMAN und WOLFENDALE [*Ha 62*] gemessene Myonspektrum $N_\mu(E)$, das nach großen Energien nach den Meßergebnissen von DUTHIE et al. [*Du 62*] ergänzt ist, zugrunde, um daraus nach der bekannten Formel von BARRET et al. [*Ba 52*] bzw. PINE et al. [*Pi 59*] ein „experimentelles" Pionproduktionsspektrum $N_\pi(E)$ in Meereshöhe zu errechnen:

$$N(E_\pi) \, dE_\pi = N_\mu\left(\frac{E}{\gamma}\right)\left(1 + \frac{E}{\gamma B}\right)\frac{1}{\gamma} D(E) \, dE. \qquad (4.14)$$

Dabei ist $\gamma = \frac{m_\pi}{m_\mu} = 1{,}32$; $B = 90$ MeV; $D(E)$ ist ein Faktor, der den Verlust an Pionen durch μ-e-Zerfall und Energieverluste durch Ionisation in der Atmosphäre berücksichtigen läßt. Nun haben COCCONI, KOESTER und PERKINS [*Co 61b*] eine empirische Beziehung, das CKP-Modell, aufgestellt, das erlaubt das beobachtete Energiespektrum der Pionen bei Wechselwirkungen von Protonen mit leichten Elementen bei Beschleunigerenergien (≤ 30 GeV) gut wiederzugeben. Die Beziehung gibt die Zahl der Pionen $N(E_\pi)$ *eines* Ladungsvorzeichens an, die im CM-System in Vorwärtsrichtung emittiert werden.

$$N(E_\pi)\, dE_\pi = \frac{A}{T_\pi} \exp\left(-\frac{E_\pi}{T_\pi}\right) dE_\pi. \qquad (4.15)$$

E_π stellt hierbei die Pion-Energie im Labor-System dar. A kennzeichnet die mittlere Multiplizität der Pionen eines Ladungsvorzeichens in Vorwärtsrichtung im CM-System; T_π ist die mittlere Pion-Energie. Die verfügbare Pion-Energie für ein Ladungsvorzeichen ist nach (4.5) $1/3\, K_\pi \cdot E_p$, wobei E_p wieder die Primärenergie des jeweils stoßenden Protons (Nukleons) darstellt, diese verteilt sich auf die Teilchen in Vorwärtsrichtung gemäß

$$A \cdot T_\pi = \tfrac{1}{3} K_\pi \cdot E_p. \qquad (4.16)$$

Wählt man für die mittlere Multiplizität der Pionen wieder die Fermi-Gleichung (4.6), wobei jetzt n die gesamte Zahl der Sekundären sei, d.h. geladene und neutrale Pionen erfaßt, so gilt

$$n = 6 \cdot A = C \cdot E_p^\alpha. \qquad (4.17)$$

Der Faktor 6 berücksichtigt, daß drei Ladungszustände existieren und daß 50% der geladenen Pionen in Rückwärtsrichtung fliegen; von diesen jedoch wird angenommen, daß sie im L-System eine vernachlässigbar „kleine" Energie haben. Man kann nach (4.6) für die geladenen Pionen die Zahl n_s mit $n_s = 1{,}8 \cdot E_p^{\frac{1}{4}}$ gut approximieren; dies führt nach (4.17) auf

$$A = \frac{3}{2} \cdot \frac{n_s}{6} = a \cdot E_p^\alpha \qquad (4.18)$$

mit

$$a = 0{,}45;\ \alpha = \tfrac{1}{4}.$$

Es gibt aber keinen begründeten Anlaß, bei hohen Energien das CKP-Modell zur Beschreibung des Pion-Spektrums anzuwenden. Faltet man das primäre, differentielle Protonenspektrum (3.3) mit dem CKP-Produktionsspektrum (4.15), so findet man für die geladenen Pionen in Vorwärtsrichtung eine Pion-Verteilung, die von der *Form* des Pion-Spektrums (4.15) nicht stark abhängt:

$$N(E_\pi)\, dE_\pi = \frac{2\, dE_\pi}{1-(1-K_t)^{n-1}} \cdot \frac{3 a^2 \cdot B}{K_\pi} \int_{3E_\pi}^{\infty} E_p^{-n-1+2\alpha} \exp\left(\frac{-3 a E_\pi}{K_\pi E_p^{1-\alpha}}\right) dE_p \qquad (4.19)$$

mit $n = 2{,}70$; $B = 9{,}43$; $\alpha = \tfrac{1}{4}$; $a = 0{,}45$ nach (4.18) und (4.5). Der Faktor $\frac{1}{1-\eta^{n-1}} = \frac{1}{1-(1-K_t)^{n-1}}$ gibt gerade die Summe der Pionbeiträge von jeder Generation wieder, d.h. er entspricht der Summe von Pionen, erzeugt durch die primären, sekundären, tertiären Nukleonen, die einfach, zweifach ... $j$fach zur Wechselwirkung kamen.

WOLFENDALE et al. bestimmen über (4.19) die Inelastizität K_π; wobei sie das „experimentelle" Pion-Spektrum (4.14) mitbenutzen.

Schließlich bilden sie die Differenz $K_t - K_\pi$. Wird berücksichtigt, daß (vgl. PERKINS [Pe 61] und POWELL [Po 62]) 75% der sekundären Teilchen Pionen sind, die etwa 30—45% der Stoßenergie E_p aufnehmen und daß ferner, etwa 50% der Energie vom Trägernukleon mitgenommen wird, so verteilt sich der Rest der

Energie beim Stoßprozeß gemäß den Grenzen $0{,}05 \leq K_t - K_\pi \leq 0{,}2$ auf Nichtpionen. Als beste Abschätzung für diese Differenz wurde $K_t - K_\pi = 0{,}12 \pm 0{,}05$ gewählt. Damit wurde die mittlere Inelastizität $K_t = 0{,}47$ und $K_\pi = 0{,}35$ geschätzt. Mit diesen Zahlwerten und nach detaillierter Behandlung der beträchtlichen Fluktuationen von K_t und K_π (wozu auf die Originalarbeit verwiesen sei) wird auf ein verbessertes integrales Primärspektrum umgerechnet, der Form

$$N(0, \geq E) = \left(\left(0{,}87 \begin{array}{c} +0{,}52 \\ -0{,}31 \end{array} \right) E^{-1{,}58} \frac{\text{Protonen}}{\text{cm}^2 \text{sec. sterad.}} \right) \quad (4.20)$$
$$\text{für } 10^{10} \text{ eV} \leq E \leq 3 \cdot 10^{13} \text{ eV}.$$

Daraus kann man folgern, daß die früher bestimmten Intensitäten der primären Protonen erheblich überschätzt wurden; für höhere Energien von $3 \cdot 10^{13}$ eV bis 10^{15} eV wird das Wolfendale-Spektrum steiler als (4.20) angibt. Das in Fig. 5 angegebene Protonenspektrum stimmt in weitem Bereich mit (4.20) überein. Die Berechnung eines inhomogenen Mesonquellspektrums nach (14.18) mit $\alpha \neq 0$ bringt es mit sich, daß die Steigung der Spektren der Folgeteilchen geändert wird, wie man im Integral (4.19) leicht erkennt. Zugleich wird aber durch ein solches inhomogenes Produktionsspektrum die Gestalt des Pion-Spektrums geändert, da das Argument der Exponentialfunktion unter dem Integral von α abhängt und kein homogenes Polynom 0. Grades hinsichtlich der Energie ist.

PINKAU [Pi 64] hat die Nukleonenkaskade[1] und die elektromagnetische Kaskade im Sinne von BUDINI und MOLIÈRE [Bu 53] unter Berücksichtigung der heutigen Vorstellungen über Wechselwirkung, Elastizität und Multiplizität erschöpfend behandelt. Speziell der Einfluß homogener und inhomogener Produktionsspektren auf die Form der Sekundärspektren der Folgegenerationen ist dort ausführlich untersucht. Das Cocconi-Köster-Perkins-Modell (4.15) liefert auch hier — beispielsweise für die Quellstärkenspektren der Gammaquanten, die von DUTHIE et al. [Du 62] gemessen sind — sehr gute Übereinstimmung, falls PINKAU noch eine interessante Zusatzannahme macht. Das CKP-Spektrum (4.15) enthält über T_π in der Exponentialfunktion [vgl. etwa (4.16)] noch den Faktor K_π der Pion-Inelastizität (4.5). Damit unterliegt die Gestalt des CKP-Spektrums noch den Fluktuationen von K_π. Durch den speziellen Ansatz für die Pion-Multiplizität

$$A = A' \cdot K_\pi, \quad (4.21)$$

wird es möglich, die mittlere Pionenergie T_π von den Fluktuationen von K_π zu befreien, und damit ein CKP-Spektrum zu formulieren, das den (π_0, γ)-Prozeß besser wiedergibt.

Ferner sei angemerkt, daß dort eine ausführliche Zusammenfassung kosmischer Strahlungsdaten bis Ende 1963 vorgenommen ist, die die Daten von PERKINS [Pe 61], COCCONI [Co 61a], POWELL [Po 62] weitgehend ergänzt.

γ) *Allgemeines Isobaren-Modell:* Die Entdeckung vieler Baryonzustände — besonders von Baryonresonanzen — ließ etwa ab 1959/60 vermuten, daß bei hochenergetischen Nukleon-Targetnukleon-Stößen nicht nur Nukleon-Endzustände sondern auch Baryon-Endzustände zu erwarten sind.

PERKINS [Pe 61] erwähnt die möglichen Prozesse; ZATSEPIN [Za 62] analysiert die experimentellen Daten hochenergetischer Stöße und zieht bei der theoretischen Interpretation auch angeregte Nukleonenisobare in Betracht. Die Hauptschwierigkeit war, die Pionisation über einen Feuerball-Mechanismus mit den Isobareneigenschaften in Einklang zu bringen.

---

[1] NISHIMURA behandelt die Kaskade in diesem Band.

PETERS [1961, Kyoto] hat bei seinem ersten Ansatz Hyperonen in Betracht gezogen. Seit der CERN-Konferenz 1962 [Pe 62] zeichnet sich jedoch das durchsichtige und elementare Modell angeregter Baryonzustände für die „Träger-Baryonen" (persisting baryons) der nuklearen Kaskade ab. PETERS und PAL [Pa 63a; Pe 63b; Pa 63b] haben dieses verallgemeinerte Isobaren-Modell gemeinsam ausgearbeitet. Der Grundgedanke des Modells ist einfach: die Pion-Produktion erfolgt über zwei verschiedene Prozesse oder Kanäle:

a) Beim Nukleon-Targetnukleon-Stoß entstehende Nukleon-Resonanzen, bzw. Baryon-Resonanzen sollen überwiegend in hochenergetische Pionen zerfallen, und zwar jeweils über Zweiteilchen-Zerfall-Prozesse der angeregten Baryonen-Resonanzen, wobei Pionen oder Isobosonen (mit nachfolgendem Pionzerfall) entstehen.

b) Der Pionisationsprozeß über einen Feuerballmechanismus verläuft wie bisher. Als Vereinfachung kommt hinzu, daß die Pion-Erzeugungsspektren den Anteil einiger hochenergetischer Pionen [Resonanzzerfallpionen nach a)] nicht mehr beschreiben müssen.

Die Voraussetzung für die Anwendbarkeit des Modells ist, daß isobare Zustände bei hoher und höchster Energie ähnlich häufig sind wie bei Maschinenenergien. Die Gültigkeit des Modells diskutieren PAL und PETERS mit Rücksicht auf den Knick im Primärspektrum nur bis $\approx 10^{14}$ eV.

Zählt man den Nukleon-Zustand mit einer Ordnungszahl $r=0$, und ordnet man die Baryonzustände nach wachsenden Baryonmassen M_r aufsteigend mit $r=1, 2, 3, \ldots$, so soll ein Baryon-Zustand r entsprechend einer Baryon-Elastizität $\eta_r = (1-K_{t,r})$ einen Bruchteil der Energie des stoßenden Nukleons übernehmen. Diesen übernommenen Energiebetrag, der wahrscheinlich wieder etwa die Hälfte der Primär-Energie sein wird, teilt die Resonanz bei ihren folgenden Zweiteilchenzerfällen auf die Pionen und Isobosonen auf, bis zuletzt ein hochenergetisches Sekundär-Nukleon übrigbleibt und gelegentlich weiterstößt. Die Absorptionslänge Λ_a dieses Nukleons ist durch den erreichten isobaren Zustand r und dessen Zerfallswahrscheinlichkeiten δ in Pionen mitbestimmt. Berücksichtigt man dies, so kann man in völliger Analogie zur Absorptionslänge λ_a des Nukleonen-Zustands ($r=0$) in Gl. (4.12) über die Poisson-Fluktuationen der Nukleonen-Stoßfolge eine Absorptionslänge Λ_a im Rahmen des Isobarenmodells errechnen. Diese gehorcht der Gl. (4.12) formal analogen Beziehung:

$$\frac{1}{\Lambda_a} = \frac{1}{\lambda_i}(1-\langle\eta^n\rangle). \qquad (4.22)$$

Der Mittelwert über die $n$fache Potenz der Elastizität $\langle\eta^n\rangle$, der in (4.22) eingeht, berücksichtigt die Stoßlängen $\lambda_{i,r}$, bzw. Elastizitäten η_r des vollständigen Systems isobarer Zustände:

$$\langle\eta^n\rangle = \sum_{r=1}^{\infty} \frac{\lambda_i}{\lambda_{i,r}} \eta_r^n, \qquad (4.23)$$

$$\frac{1}{\lambda_i} = \sum_{r=1}^{\infty} \frac{1}{\lambda_{i,r}}. \qquad (4.24)$$

Zur Bestimmung von $\langle\eta^n\rangle$ in Abhängigkeit von den Massen und den Zerfallsarten der Isobare sei auf die Diskussion in der Originalarbeit verwiesen.

Das verallgemeinerte Isobar-Modell verlangt eine abgeänderte Multiplizitätsbeziehung

$$n_t = 2 \cdot s \cdot n_B + a \cdot E^\alpha. \qquad (4.25)$$

Im ersten Glied von (4.25) gibt s die Wahrscheinlichkeit an, mit der eine Baryonanregung erfolgt; im Falle des Nukleonstoßes folgt mit $s=0$ nur der übliche Pionisationsprozeß, etwa mit der Fermimultiplizität (4.5). PAL und PETERS haben aus Eigenschaften der sekundären kosmischen Strahlung näherungsweise auf die Eigenschaften der anregbaren Baryonzustände geschlossen. Sie finden für isobare Massen $M_B \geq 2300$ MeV eine Anregungswahrscheinlichkeit von $s = 0{,}7 \pm 0{,}07$. Die mittlere Zahl n_B der beim Isobarzerfall emittierten Pionen beträgt $n_B = 3{,}5 \pm 0{,}5$ und der mittlere Ladungsüberschuß unter den Zerfallspionen ist $|\mu^+ - \pi^-| = 0{,}35 \pm 0{,}15$. Schließlich geben sie noch das Verhältnis der Hyperonen zu Nukleonen mit $\frac{Y}{N} = (7 \pm 7)\%$ an.

Die Isobaren-Hypothese kann das Verhältnis $\mu^+/\mu^- = \frac{5}{4}$ in Meereshöhe mit dem erwähnten Pionladungsüberschuß in Zusammenhang bringen und gibt eine einfache Voraussage: die Myonen in Meereshöhe gehen durch Zerfall aus den (im Mittel) $3 - 3{,}5$ Pionen der Isobarzerfallsprozesse hervor.

Das in Emulsionen gefundene Verhältnis von K^+/K^--Mesonen ist ein weiteres unabhängiges Anzeichen für die Baryonanregung bei hochenergetischen Stoßprozessen mit $K^+/K^- = 20$. Dieser Beobachtung entspricht im Spiegel-System des Nukleon-Nukleon-Stoßes, daß das Kaon mehr als 25% der Energie des stoßenden Teilchens übernimmt. Ein so hoher Energieanteil ist normal, falls ein Kaon beim Zerfall des Vorwärts-Isobars entsteht. Der große positive Exzeß folgt direkt aus der Annahme, daß die meisten angeregten Isobare die Strangeness $S = 0$ haben, also überwiegend angeregte Nukleon-Zustände sind.

Mit diesen wenigen Beispielen sei angedeutet, daß das verallgemeinerte Baryonmodell für die komplexen Probleme der sekundären kosmischen Strahlung eine einfache und durchsichtige Erklärung anbietet. Zur Frage der Ausbreitung der kosmischen Strahlung durch die Atmosphäre im Sinne des Isobaren-Modells über die Nukleon-, Pion-, Myon-, Photon-, Elektron- und Neutrino-Komponente sei auf die Originalarbeit [Pa 63b] verwiesen.

5. Sterne in Kernspurplatten. Die ersten Beobachtungen von Kernwechselwirkungen von Teilchen der kosmischen Strahlung in Kernspurplatten wurden von BLAU und WAMBACHER [Bl 37a, 37b] publiziert, die beim Durchmustern von Platten häufig mehrere von einem Punkt ausgehende Spuren sahen. STETTER und WAMBACHER [St 39] fanden weiterhin, daß die Häufigkeit solcher Ereignisse mit wachsender Höhe zunahm; auch waren die gefundenen Spuren länger als diejenigen von α-Teilchen aus radioaktiven Verunreinigungen. Man schloß daraus, daß es sich um Zertrümmerungen von Kernen der Emulsion durch die Teilchen der kosmischen Strahlung handeln mußte. Diese Ereignisse werden als „Sterne" bezeichnet. Die Teilchen aus solchen Sternen wurden als Protonen identifiziert; SCHOPPER [Scho 37] und IDANOV und Mitarbeiter [Fi 39] zeigten weiterhin, daß ein Teil der Spuren auch von α-Teilchen und gelegentlich von schweren Kerntrümmern herrührte. Eine Übersicht über frühere Arbeiten auf diesem Gebiet findet man bei SCHOPPER [Sch 39] und SHAPIRO [Sh 41]. Später hat TEUCHER [Te 53c] einen ausführlichen Überblick über die grundlegenden Arbeiten gegeben.

Die photographische Kernspuremulsion als Nachweisinstrument hat seit dem Beginn des vergangenen Jahrzehntes entscheidende Beiträge zu unserer Kenntnis in der Elementarteilchenphysik und der kosmischen Strahlung geliefert. Nachdem es gelungen war, ihre Empfindlichkeit so zu steigern, daß Bahnspuren aller geladenen Teilchen, auch im Minimum ihrer Ionisation nachgewiesen wurden, und nachdem man das Verfahren beherrschte, große Emulsionsvolumina der Größenordnung von 100 dm³ in Form von „Stacks" aus vielen Emulsionsblättern zu verarbeiten und auszuwerten, war sie ein ideales Instrument für Ballonmessungen in der Stratosphäre und langdauernde Messungen in den unteren Atmosphärenschichten geworden.

Wir berichten in diesem Abschnitt vorwiegend über die mit der Kernspurmethode gewonnenen Informationen über die Kernwechselwirkungen der Nukleonenkomponente. Bei der Übertragung der Wechselwirkungsvorgänge in der Kernspuremulsion auf die entsprechenden Vorgänge in der Atmosphäre ist zu beachten, daß diese neben O- und N-Kernen einen hohen Prozentsatz schwerer Ag- und Br-Kerne enthält, und die Art des jeweils getroffenen Kernes der Emulsion meist nicht leicht zu ermitteln ist. Man ist auf statistische Methoden angewiesen.

In der Literatur haben sich bezüglich der Meß- und Auswertetechnik folgende Begriffe eingebürgert, die hier kurz erwähnt seien: Bei der Beurteilung der Ionisationsdichte einer Spur geht man aus von der Spur eines Teilchens der Ladung 1, das sich praktisch mit Lichtgeschwindigkeit bewege. Es war früher üblich, die photographische Korndichte G der Spuren solcher Teilchen als unabhängig von der Energie anzunehmen und sie als Minimum-Korndichte zu bezeichnen. Spätere Untersuchungen haben jedoch gezeigt [*St 53*], daß auch in der photographischen Emulsion ein Wiederanstieg der Korndichte bei sehr großer Energie ($E > 10\ Mc^2$) um etwa 10% auf einen Plateauwert auftritt. Wir beziehen uns im folgenden immer auf diese sog. Plateaukorndichte, die wir mit G_{Pl} bezeichnen. (Für eine grobe Einteilung der Spuren, die wir im folgenden geben werden, ist allerdings der geringe Unterschied zwischen G_{Pl} und der Minimumkorndichte belanglos.) G_{Pl} beträgt bei den zur Zeit gängigen elektronenempfindlichen Schichten im Mittel 20 bis 30 Körner pro 100 μ, abhängig von der entsprechenden Emulsion und der Entwicklung. Man unterscheidet „dünne" (thin) (Korndichte $G < 1,4\ G_{Pl}$) und „starke" (heavy) ($G > 1,4\ G_{Pl}$) Spuren. Die starken Spuren werden wieder in „graue" und „schwarze" eingeteilt. Die Grenze zwischen beiden liegt zwischen 4 und 8 G_{Pl}. Die einzelnen Gruppen dieser Einteilung entsprechen ungefähr den weiter oben besprochenen Arten von Teilchen, die bei dem komplexen Mechanismus eines Kerntreffers entstehen. Die dünnen Spuren, die auch als Schauerteilchen bezeichnet werden, werden hauptsächlich von π-Mesonen und sehr energiereichen Anstoßnukleonen gebildet, die grauen Spuren von Anstoßnukleonen mittlerer Energie (bis zu einigen 100 MeV bei Protonen) mit einigen wenigen langsamen π-Mesonen. Die schwarzen Spuren kommen in der Hauptsache von Teilchen her, die bei der Kernverdampfung entstehen; das energiereichere Ende ihres Spektrums enthält einen geringen Anteil von Anstoßnukleonen.

Nach einer von der Physikergruppe in Bristol eingeführten Bezeichnungsweise kennzeichnet man einen Stern durch die Zahl seiner starken Spuren N_h und die Zahl seiner dünnen Spuren N_S. Ein neutrales auslösendes Teilchen wird durch Anfügen von n, ein einfach geladenes Teilchen durch Anfügen von p bezeichnet. Beispielsweise ist die Benennung für einen Stern mit drei starken und zehn dünnen Spuren, der von einem einfach geladenen Teilchen ausgelöst ist: $3 + 10p$. Bei der statistischen Untersuchung sehr vieler Sterne ist es in der Regel nicht möglich, die Art des Primärteilchens mit Sicherheit zu ermitteln. Geht eine dünne Spur in den oberen Halbraum — bezogen auf die Lage der Platte bei der Exposition — so wird sie als die Primäre angesehen; bei Vorhandensein von mehreren solcher Spuren nimmt man diejenige mit dem kleinsten Winkel zur Vertikalen. Diese Übereinkunft ist vernünftig, da die überwiegende Zahl schneller Teilchen von oben und, vor allem in großen Tiefen der Atmosphäre, bevorzugt aus der Zenitrichtung kommt. Diese Festlegung bedeutet außerdem eine untere Grenze von 300 bis 500 MeV für die Energie des auslösenden Teilchens, wenn es ein Proton ist. Ein Proton kleinerer Energie erzeugt eine graue Spur und wird nicht mehr als Primärteilchen angesprochen. In der Stratosphäre kommen als auslösende Teilchen auch schwere Primärkerne in Frage. Eine sichere Bestimmung des auslösenden Teilchens ist bei gut gebündelten Mesonenschauern möglich, da in diesem Fall das Primärteilchen in der rückwärtigen Verlängerung der Schauerachse gelaufen sein muß.

6. Höhenabhängigkeit der Sterne. Die Abnahme der Intensität der sternerzeugenden Komponente macht sich in einer Abnahme der Sternhäufigkeit in der Emulsion bemerkbar. Da sich die mittlere Teilchen-Energie mit der Tiefe in der Atmosphäre nicht sehr rasch ändert und außerdem die Wirkungsquerschnitte nicht stark energieabhängig sind, wird der Fluß sternerzeugungsfähiger Teilchen ungefähr der Sternhäufigkeit in der Emulsion proportional sein. Unterhalb des Übergangsmaximums in der Stratosphäre beobachtet man eine exponentielle Abnahme der Sternhäufigkeit mit zunehmender Tiefe.

Die Sternhäufigkeit in der Höhe x ist also gegeben durch die Beziehung

$$I = I_0 \, e^{-x/L}. \tag{6.1}$$

L wird als Absorptionslänge bzw. Absorptionsschicht bezeichnet und in cm bzw. g/cm^2 gemessen. Sie wird im allgemeinen länger sein als die „Stoßschicht", die mittlere freie Weglänge zwischen zwei Stößen eines Teilchens, die ungefähr einem geometrischen Wirkungsquerschnitt entspricht. Dies hat seinen Grund darin, daß die Teilchen im Kern meist nicht katastrophisch absorbiert werden, sondern auch nach einem Stoß entweder selbst oder über Anstoßnukleonen weitere Sterne erzeugen können.

Die Ergebnisse verschiedener Autoren hinsichtlich der Höhenabhängigkeit der Sterne stimmen im einzelnen nicht immer überein; dies gilt vor allem für die Werte der Absoluthäufigkeit der Sterne. Dies mag durch die Verschiedenheit der Auswerteverfahren bedingt sein. Ältere Untersuchungen wurden oft unter Verwendung von Ilford C2-Platten gemacht, die nicht elektronenempfindlich sind und Protonen nur bis ungefähr 50 MeV nachweisen. Auf diese Weise können Sterne verloren gehen. Ebenso kann die Definition eines Sternes Anlaß zu Diskrepanzen geben. Meist zählt man nur Ereignisse mit drei oder mehr Spuren, von denen mindestens eine länger als 60 μ sein muß. Damit werden Sterne von radioaktiven Verunreinigungen getrennt. Ebenfalls sind dadurch Anstoßprotonen und Einzelteilchen mit großen Winkelablenkungen ausgeschlossen. Trotz dieser Ausscheidung unsicherer Ereignisse hat man oft Schwierigkeiten bei der Definition sehr kurzer Spuren, und damit bei der Bestimmung der Zahl der Spuren eines Sterns. Weitere Unsicherheitsfaktoren sind das Übersehen von Sternen bei der Durchmusterung unter dem Mikroskop, das allmähliche Verschwinden des latenten Bildes bei langdauernder Exposition, sowie ungenaue Kenntnis der Dicke der verwendeten Platten. Aus all diesen Gründen wird man für die Absolutwerte der Sternhäufigkeit keine große Genauigkeit erwarten dürfen, während man für die Absorptionslänge L, die ja meist durch Vergleich von relativen Häufigkeiten bestimmt wird, zuverlässigere Werte erhalten sollte. Eine Fehlerquelle bei der Bestimmung von L besteht in der Rückstreuung, vor allem von Neutronen, in umgebendem Material. Sie kann einen merklichen Einfluß auf die Absorptionsschicht haben [*Te 53a*].

Tabelle 2 gibt eine Übersicht über die Werte der Absorptionslänge L, die von verschiedenen Autoren erhalten wurden. Sie liegen zwischen 130 und 170 g/cm^2.

Ältere, mit C2-Emulsionen durchgeführte Messungen sind mit einem Stern (*) bezeichnet. Messungen in niedrigen geomagnetischen Breiten [Ro 52a, Di 55]

Tabelle 2. *Absorptionslänge für Sternerzeugung in der Atmosphäre.*

Autor	geomagnetische Breite λ	Meßbereich (m über NN)	L in g/cm^2
Pe 47	47°—50°	0— 4300	150 (*)
Ge 49	47°—50°	0— 3500	150 (*)
Be 49	45°	3500—Ballon	135 ± 4 (*)
Lo 49	51°—56°	0—30000	148 (*)
Pe 49	47°—50°	3500—25000	140 (*)
Ya 49	48°—50°	0—27000	143 (*)
Lo 51	54°	3500—30000	145 (p-Sterne)
			170 (n-Sterne)
Te 52	47°—50°	150— 3774	127 ± 4
Te 53a	47°—50°	550— 4550	141
Ca 49	47°—50°	200— 3500	130
Ro 52a	21° S	2630— 5350	149 ± 2
Di 55	3° S	1680— 5960	149
Br 54	47°—50°	0—Gebirge	132 ± 4 (Nebelkammer)
Ka 53		Gebirge—Ballon	120 (Energie $>10^{12}$ eV)

scheinen etwas größere Werte für L zu ergeben. Dasselbe scheint für die Messungen in C2-Emulsion zu gelten, vermutlich wegen der kleineren mittleren Energie der registrierten Ereignisse. Dazu würde auch der höhere Wert von LORD [Lo 51] für die neutral-ausgelösten Sterne (n-Sterne) gegenüber den geladen-ausgelösten (p-Sterne) passen, und ebenso das Ergebnis von LATTIMORE [La 49], der die Höhenabhängigkeit sehr energiearmer Prozesse, nämlich von Einzelspuren, zwischen 0 und 3500 m untersucht hat. Er findet dabei für L den sehr großen Wert von 190 g/cm^2. Ganz allgemein scheint L mit zunehmender Energie der Primären abzunehmen.

Die Werte können weiterhin verglichen werden mit den Ergebnissen für die Absorptionslänge der Neutronenkomponente in der Atmosphäre (s. Ziff. 17 und 21). Sie beträgt für dieselbe geomagnetische Breite von 47 bis 50° rund 160 g/cm^2, ist also deutlich größer als die meisten Werte für Sterne. Eine der dafür in Frage kommenden Ursachen ist wohl die, daß an der Neutronenerzeugung relativ energiearme Prozesse stark beteiligt sind, während für die Sterne die untere Grenzenergie bei ungefähr 150 bis 200 MeV liegt [Br 49, Ti 52]. Eine ausgeprägte Abhängigkeit der Absorptionslänge von der Sternenergie findet ROEDERER [Ro 52a] auch direkt bei Messungen unter 20° geomagnetischer Breite. Die Zunahme von L mit abnehmender Energie paßt nach seinen Angaben gut zu der unter 20° Breite gemessenen Absorptionslänge von 205 g/cm^2 für die Neutronenkomponente. (Für weitere Ausführungen zu dieser Frage s. Ziff. 19 und 21.)

Ein brauchbarer Mittelwert der Absorptionslänge für Sternerzeugung unter rund 50° geomagnetischer Breite dürfte

$$L = 138 \text{ g/cm}^2$$

sein.

Über die Intensität der sternerzeugenden Komponente in der Stratosphäre ist verhältnismäßig wenig bekannt. Messungen mit photographischen Emulsionen wurden von LORD [Lo 51] durchgeführt. Daneben besitzen wir Beobachtungen mit Ionisationskammern. Durch passende Diskriminierung lassen sich hier Ereignisse aussondern, die zum größten Teil auf Sterne und schwere Kerne der Primärstrahlung zurückzuführen sind. Der von schweren Kernen herrührende Anteil kann auf Grund ihrer ungefähr bekannten Absorptionslängen ermittelt und von der Gesamtzählrate abgezogen werden. Ältere Messungen dieser Art stammen von McClure und Pomerantz [Cl 50] und von Whyte [Wh 51]. Die Messungen von McClure und Pomerantz zeigten für die Sterne einen Übergangseffekt in der Atmosphäre, gekennzeichnet durch ein Anwachsen der Sternhäufigkeit beim Eindringen in die Atmosphäre, die in einer Tiefe von etwa 49 g/cm² ein Maximum erreicht und in größerer Tiefe in der Atmosphäre wieder abfällt (Fig. 7).

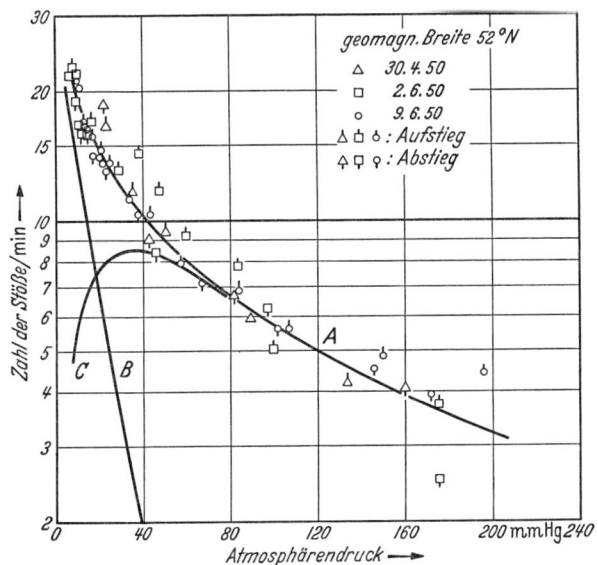

Fig. 7. Stoßzahlen in der Ionisationskammer nach [Cl 50]. A Gesamtstoßzahl, B Beitrag der schweren Kerne der Primärstrahlung (berechnet), C Differenz zwischen A und B.

Diese Messung wurde unter 52° geomagnetischer Breite gemacht. In verschiedenen geomagnetischen Breiten wurde der Übergangseffekt von Whyte untersucht. Das Maximum der Sternhäufigkeit verschiebt sich nach seinen Messungen mit zunehmender geomagnetischer Breite nach kleineren Tiefen in der Atmosphäre hin, ein Effekt, der auch für den Übergangseffekt der Neutronenkomponente beobachtet wird und in derselben Weise zu deuten ist (s. Ziff. 19 und 21). Im einzelnen fand er das Maximum bei 100, 70, 60 und 50 g/cm² unter der geomagnetischen Breite von 0, 29, 35 und 52°. Das Maximum in 52° Breite ist aber nur undeutlich ausgeprägt; dies ist unter anderem auf die experimentelle Schwierigkeit zurückzuführen, daß die Ansprechwahrscheinlichkeit seiner Ionisationskammer für die schweren Primärteilchen sehr groß ist und diese somit einen erheblichen Anteil an der Gesamtzählrate ausmachen. Diese Schwierigkeit umgeht man bei Messungen mit der Kernemulsion, die außerdem eine getrennte Untersuchung nach Sterngrößen, d.h. nach Sternenergien, erlaubt. Fig. 8 und 9 zeigen nach Messungen von Lord [Lo 51] die gefundene Höhenabhängigkeit der Sterne in zwei verschiedenen geomagnetischen Breiten Λ. Für $\Lambda = 52$ bis 56° wächst die Sternhäufigkeit zunächst exponentiell mit der Höhe bis zu ungefähr 80 mb. Die weitere Zunahme erfolgt dann wesentlich langsamer; in der höchsten untersuchten Höhe (15 mb) bleibt die Sternhäufigkeit konstant. Demgegenüber zeigen die großen energiereichen Sterne bis zu den höchsten erreichten Höhen eine exponentielle Zunahme, die einer Absorptionslänge von 120 g/cm² entspricht. Die Messungen in der obersten Schicht der Atmosphäre unter 28° geomagnetischer Breite ergeben für die größten Sterne ebenfalls eine Absorptionslänge derselben Größenordnung, nämlich 130 g/cm². Der Unterschied gegenüber den Messungen von WHYTE, der wenigstens für niedrige Breiten ausgeprägte Maxima findet, mag in einer Verschiedenheit der registrierten mittleren Energie begründet sein. Auf jeden Fall läßt sich oberhalb von etwa 100 mb für Sterne kein Wert der Absorptionslänge mehr angeben, mit Ausnahme der sehr großen Sterne.

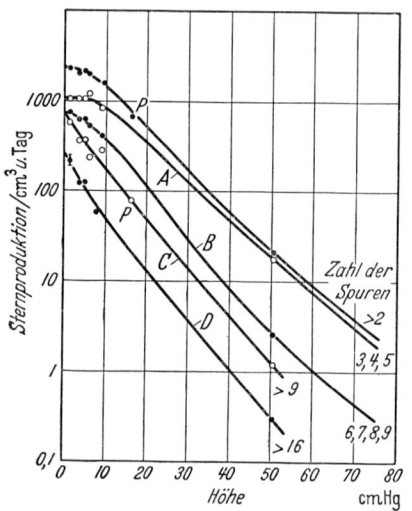
Fig. 8. Sternproduktion in der Stratosphäre als Funktion der Höhe für eine geomagnetische Breite von 52 bis 56° nach [Lo 51]

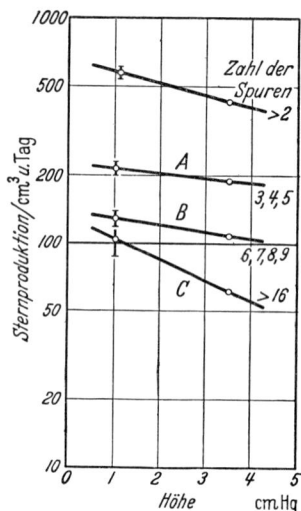
Fig. 9. Sternproduktion in der Stratosphäre als Funktion der Höhe für eine geomagnetische Breite von 28° nach [Lo 51]

Der oben erwähnte Wert von 120 g/cm² wurde auch bei anderen Messungen an der sehr energiereichen Nukleonkomponente in der Atmosphäre gefunden.

Eine Zusammenstellung älterer Werte für die absolute Häufigkeit von Sternen in der Kernemulsion findet man bei PUPPI und DALLAPORTA [Pu 52]. Inzwischen sind zahlreiche weitere Messungen gemacht worden. Tabelle 3 gibt eine Übersicht über Messungen in 47 bis 50° nördlicher geomagnetischer Breite. Messungen in niederen geomagnetischen Breiten sind in Tabelle 4 zusammengestellt. PUPPI

Tabelle 3. *Absolutwerte für die Sternhäufigkeit (mittlere geomagnetische Breiten)*.

Autor	Höhe mb	Sterne/cm³ · Tag	Autor	Höhe mb	Sterne/cm³ · Tag
Ge 49	1030	1,46	Te 53a	590	38,5 ± 1,5
Be 50b	680	14,2		1000	2,8 ± 0,2
La 49	670	16,4	Ba 52b	680	17 ± 0,4
Ad 50b	15	1960	Rö 52	720	13,3 ± 0,4
Sa 50	12	2000		970	2,5 ± 0,2
Lo 51	15	2390	Te 52	1030	1,34 ± 0,15
	48	2030		930	3,1 ± 0,7
	64	2150		850	5,9 ± 0,7
	82	2040		730	13,2 ± 0,5
	121	1610		650	26,6 ± 0,9
	680	22			

und DALLAPORTA [Pu 52] haben eine Reihe solcher Messungen auf 680 mb reduziert; es zeigen sich dann recht erhebliche Unterschiede in den Absolutwerten bis zum Faktor 2. Da man infolge der anfangs erwähnten systematischen Fehlerquellen erwarten muß, daß bei der Analyse unter Umständen Sterne verloren gehen, sind die hohen Werte als zuverlässiger zu betrachten.

Tabelle 4. *Absolutwerte für die Sternhäufigkeit.* (niedere geomagnetische Breiten)

Autor	Höhe mb	Sterne cm^{-3}d^{-1}	geomagnetische Breite
Ro 52a	760	6,2 ± 0,5	20° S
	650	11,5 ± 0,7	
	580	18,3 ± 0,7	
	530	25,6 ± 0,8	
Di 55	811	2,9	3° S
	565	15,1	
	475	27,6	
Jo 57	560	10,1 ± 0,9	0°
Lo 51	15	575	28° N
	48	425	

7. Statistische Auswertung der Sterne.
Ausgangspunkt für die meisten statistischen Untersuchungen der Sternerzeugung durch die Nukleonenkomponente bildet eine Tabelle, in der die Häufigkeit der Sterne als Funktion der Sterngröße eingetragen ist. Die Aufteilung einer solchen Tabelle geschieht in der Regel nach der Zahl der schweren Spuren N_h, der Zahl der dünnen N_s, sowie nach der Art des auslösenden Teilchens [geladen (p) oder ungeladen (n)]. Die von CAMERINI und Mitarbeiter [Ca 51] angegebene Tabelle findet sich im Artikel von E. RAY in Bd. 46/1[1]. Dort sind auch weitere Angaben über die Eigenschaften der emittierten Teilchen sowie ihre Häufigkeit und ihr Energiespektrum besprochen, so daß wir

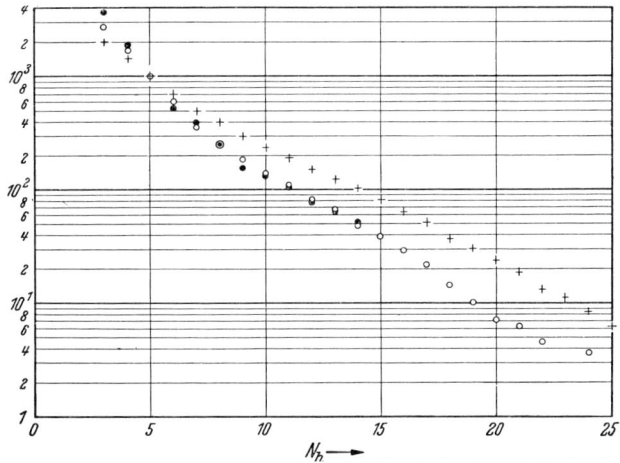

Fig. 10. Sterngrößenverteilung in verschiedenen Höhen. Ordinate: Normierte Gesamtzahl der Sterne mit Anzahl von schweren Spuren ≥ N_h. Statistik des Max-Planck-Instituts für Physik, Göttingen: ● 150 m (287 Sterne); ○ 3774 m (6358 Sterne); + etwa 20 km (5434 Sterne). (Nach TEUCHER [Te 53c]).

uns hier kurz fassen können. Es sei auf die von TEUCHER [Te 53b] aufgestellten Sternstatistiken für Höhen von 15 bis 25 km, 3770 m und 150 m hingewiesen, die einen unmittelbaren Vergleich der Sterngrößen in verschiedenen Tiefen in der Atmospäre gestatten. Ältere statistische Angaben über Sterne in 3600 m Höhe findet man bei BROWN und Mitarbeiter [Br 49].

Einen raschen Überblick gestatten die sog. Sterngrößen-Diagramme. Sie geben die Häufigkeit von Sternen mit mehr als N Spuren als Funktion von N. Fig. 10 zeigt diese Sterngrößenverteilung in verschiedenen Höhen (nach [Te 53c]).

---
[1] E. RAY: Experimental Results of Flights in the Stratosphere. Handbuch der Physik, Editor S. FLÜGGE, Bd. 46, 1.

Trägt man den Logarithmus der integralen Sternhäufigkeit gegen die Zahl N_H der starken Spuren auf, so kann man den Zusammenhang ungefähr durch zwei Gerade verschiedener Neigung darstellen, die sich bei $N_h \sim 7$ schneiden. Die Erklärung für diesen Knick ist wohl folgende: die photographische Emulsion enthält zwei Gruppen verschieden schwerer Kerne (C, N, O, H und Ag, Br). Der Beitrag der leichten Kerne reicht natürlich nur bis $N_h \leq 8$; ihm überlagert sich der Beitrag der schweren Kerne. Für diese Deutung sprechen vor allem die Versuche von BIRNBAUM und Mitarbeiter [Bi 52] mit „verdünnten" Emulsionen von geringerem Br- und Ag-Gehalt. Auch sie fanden den Knick unabhängig von der Höhe ungefähr bei $N_h = 7$; der Vergleich ihres Sterngrößendiagramms in verdünnten Schichten von Kernemulsionen mit denen in normalen erklärt den Knick zwanglos durch die beiden Gruppen von verschiedenen Kernen. TEUCHER [Te 53b] hat allerdings darauf hingewiesen, daß auch die Energie der auslösenden Teilchen eine Rolle spielen muß. Wenn er für das Sterngrößendiagramm nur die energiereichen Sterne berücksichtigt, erhält er keinen Knick, sondern eine Steigung der Geraden, die dem „schweren" Ast des normalen Sterngrößendiagramms ($N_h > 7$) entspricht (s. Fig. 11). Die andere Steigung für $N_h < 7$ könnte demnach auch von der maßgeblichen Beteiligung energiearmer Teilchen (größtenteils Neutronen) an der Auslösung kleiner Sterne verursacht sein. Dazu würde auch passen, daß der Knick in Stratosphärenhöhe weniger deutlich ausgeprägt ist, wo noch kein Gleichgewicht zwischen den Teilchen verschiedener Energie besteht.

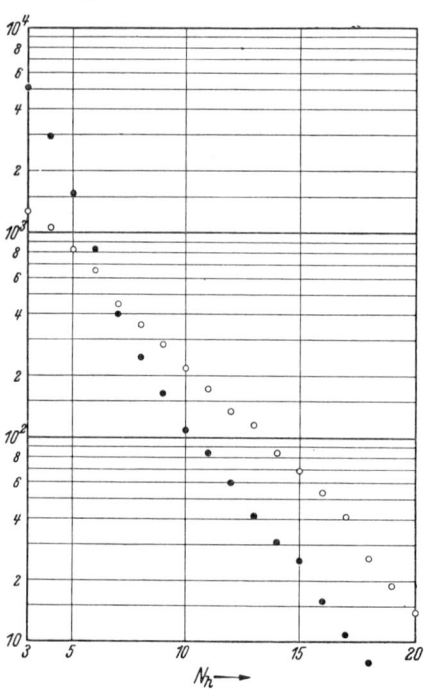

Fig. 11. Sterngrößenverteilung in 3774 m, getrennt für energiearme (0_n)-Sterne (●) und energiereiche ($0_p+1_n+1_p+2_n+2_p+\cdots$)-Sterne (○). Ordinate: Gesamtzahl der Sterne mit Anzahl von schweren Spuren $\geq N_h$. Statistik des Max-Planck-Instituts für Physik, Göttingen (6300 Sterne). (Nach TEUCHER [Te 53c].)

Die Aufstellung von sog. Schauergrößendiagrammen, d.h. der Häufigkeit von Sternen mit mehr als N_s dünnen Spuren als Funktion von N_s, erfordert umfangreiches statistisches Material. Fig. 12 zeigt die von der Bristolgruppe für Ballon- und Bergeshöhen gefundene Schauergrößenverteilung für geladene und für neutrale Primärteilchen (nach der Darstellung von TEUCHER [Te 53c]). Für $N_s \geq 4$ unterscheiden sich die beiden Verteilungen kaum; für $N_s < 4$ ist die relative Zahl der Schauer in Bergeshöhen größer. Mit wachsender Schauergröße nimmt das Verhältnis der Zahl der geladenen zu den neutral ausgelösten Schauern zu.

Für die Ableitung der Energie des auslösenden Teilchens aus der Sterngröße läßt sich keine befriedigende Vorschrift angeben. Alle Formeln geben nur Anhaltspunkte für den Mittelwert der Energie aus einer großen Zahl von Sternen.

Für Energien >1 GeV ist es zweckmäßig, für die Energieabschätzung von der Zahl der Schauerteilchen auszugehen, doch ist die hieraus abgeleitete Energie für den Einzelfall in der Regel nicht sinnvoll, da man die Inelastizität (s. Ziff. 4) nur im Mittel kennt. Im einzelnen sind folgende Formeln angegeben worden:

Für relativ energiearme Sterne ohne Schauerteilchen

$$E = (37 N_h + 4 N_h^2) \text{ MeV} \tag{7.1}$$

nach [Br 49a].

Dieser Wert soll jedoch für die 3er Sterne ($N_h = 3$) nach TIDMAN und HODGSON [Ti 52] um 35% zu niedrig sein.

Für energiereiche Sterne gaben CAMERINI und Mitarbeiter [Ca 51] die Formel

$$E = (155 N_h - 100) \text{ MeV}. \tag{7.2}$$

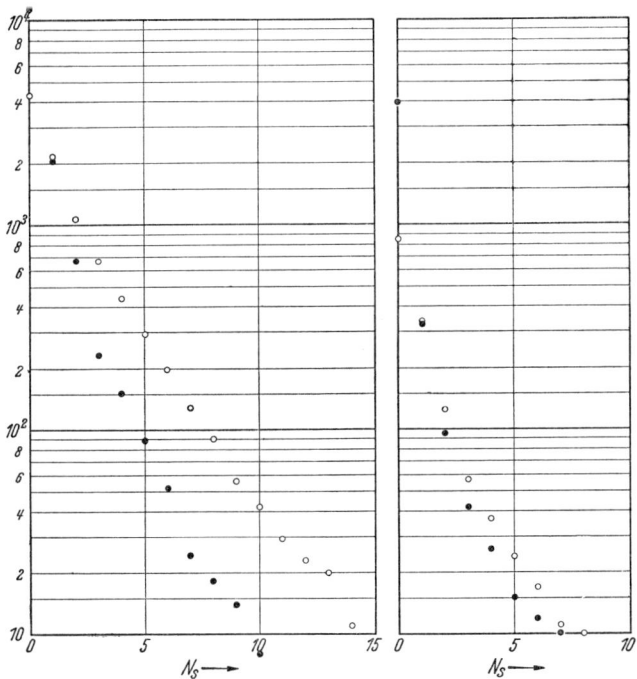

Fig. 12. Schauergrößendiagramme in Ballon- und Bergeshöhen. Ordinate: Gesamtzahl der Sterne mit Anzahl von Schauerspuren $\geq N_s$. a) ● Ohne Primärteilchen; o mit Primärteilchen; Bristol-Statistik 22000 m (15000 Sterne). b) ● Ohne Primärteilchen; o mit Primärteilchen; Bristol-Statistik 3500 m (5000 Sterne). (Nach TEUCHER [Te 53c].)

Hierbei mußte noch die in den Schauerteilchen steckende Energie hinzugezählt werden, die größenordnungsmäßig 1 GeV pro Schauerteilchen beträgt. Von ROEDERER stammte der Ansatz

$$E = (1500 N_s + 155 N_h - 100) \text{ MeV}, \tag{7.3}$$

der jedoch für größere Werte von N_s nicht mehr galt.

Neuere Arbeiten mit künstlich beschleunigten Protonen [Jo 55] und mit der kosmischen Strahlung [Hä 56] liefern für E jedoch Werte, die rund um den Faktor 2 höher liegen.

Diese Arbeiten haben weiterhin gezeigt, daß die mittlere Zahl der aus einem Stern emittierten Teilchen sehr großen Schwankungen unterworfen ist. Schon deshalb hatten die Formeln (7.1) bis (7.3) nur eine sehr beschränkte praktische Anwendung. Daß sie für höhere Energien ($\gtrsim 10$ GeV) nicht mehr gültig sind, geht auch aus den Ausführungen von Ziff. 4 deutlich hervor.

Ein Großteil der Untersuchungen mit Kernemulsion diente früher dem Zweck, nähere Einzelheiten über die Einzel- und Kaskadenprozesse, Kernverdampfung,

Wirkungsquerschnitte usw. zu gewinnen; diese Fragen gehören indessen eher zur Kernphysik als zur kosmischen Strahlung. Mit* den großen Beschleunigern hat man heute auch die Möglichkeit, viele dieser Fragen unter Bedingungen zu studieren, die eindeutiger und besser reproduzierbar sind als die in der kosmischen Strahlung herrschenden. Wir wollen deshalb hier nur kurz darauf eingehen und nur die Hauptergebnisse früherer Untersuchungen mit der Höhenstrahlung erwähnen, die wir den zusammenfassenden Darstellungen von TEUCHER [Te 53c] SYMANZIK [Sy 53] und GOTTSTEIN [Go 53b] entnehmen. Das allgemeine Bild der Sterne läßt sich gut durch den eingangs erwähnten Kaskadenmechanismus beschreiben. Die Stöße im Kern kann man hierbei mit der sog. „Stoßnäherung" behandeln, bei der man annimmt, daß die Nachbarnukleonen des gestoßenen Nukleons nur Energien von der Größenordnung der Bindungsenergie aufnehmen können. Für hohe Energien (>100 MeV) kann man also die Nukleonen im Kern als frei annehmen, insbesondere sind die Wirkungsquerschnitte ungefähr dieselben wie für die freien Teilchen. Nach etwa 10^{-22} sec haben alle schnellen Teilchen den Kern verlassen, der in der Regel hoch angeregt zurückbleibt. Die Anregungsenergie wird in etwa 10^{-18} sec durch Abdampfen von Nukleonen und leichten Kerntrümmern wieder abgegeben. Für den Ablauf der Kaskade ist der experimentelle Befund in Übereinstimmung mit theoretischen Kaskadenrechnungen, unter anderen von SYMANZIK [Sy 53]. BERNARDINI und Mitarbeiter [Be 52a, b] haben Monte Carlo-Rechnungen für 340 MeV-Protonen durchgeführt, die mit den beobachteten Sternen in Kernspurplatten gut übereinstimmen. Besonders für die grauen Spuren (meist Protonen von etwa 500 MeV) bestätigen sich die theoretischen Vorstellungen gut. Diese Teilchen entstammen praktisch alle dem Kaskadenmechanismus im Kern und zeigen somit nur eine schwache Abhängigkeit ihrer Zahl und mittleren Energie von der Primärenergie. Unter den Schauerteilchen befinden sich etwa 80% π-Mesonen [Fo 50, Ca 50]. Die mittlere Energie der Schauerteilchen scheint nach CAMERINI und Mitarbeiter [Ca 52] nur langsam mit ihrer Multiplizität (d.h. der Primärenergie) anzusteigen; man findet in Sternen mit zwei bis vier Schauerteilchen z.B. eine mittlere Energie je Teilchen von 850 MeV. Dagegen nimmt die mittlere Energie der Schauerteilchen mit wachsender Zahl der schwarzen Spuren etwas ab. Das Energiespektrum der identifizierbaren π-Mesonen läßt sich im Bereich von 200 bis 1100 MeV darstellen durch [Ca 52]

$$n_\pi(E)\, dE \sim E^{-1,4}\, dE. \tag{7.4}$$

WHITTEMORE und SHUTT [Wh 52] finden in der Nebelkammer für die differentielle Impulsverteilung schneller Mesonen ebenfalls ein Potenzspektrum, doch erhalten sie den Exponenten $-2,82\pm0,07$. Für die π^0-Mesonen findet KIM [Ki 56] in der Nebelkammer ein differentielles Energiespektrum, dessen Exponent von $-1,5\pm0,2$ bei 400 bis 900 MeV auf $-2,7\pm0,5$ für Energien von 900 bis 2000 MeV abnimmt. Ausführlichere Angaben über die Erzeugung von π-Mesonen und hochenergetische Nukleon-Nukleonstöße finden sich im Artikel von K. SITTE in Band 46/1 [Si 61].

In diesem Zusammenhang sind noch drei neuere Arbeiten zu erwähnen. GRIGOROV und Mitarbeiter [Gr 59] haben Sterne in Pb und in Luft in 9 km und 20 km Höhe untersucht. ALIKHANOV [Al 59a] und Mitarbeiter haben das Massen- und Impulsspektrum von Sekundärteilchen aus Wechselwirkungen hoher Energie in Pb in 3250 m Höhe mit Hilfe eines magnetischen Spektrometers und mit Proportionszählern in mehrfachen Lagen gemessen. Untersuchungen über den Fluß von Deuteronen in der Atmosphäre in 3200 m Höhe wurden von BADALIAN

---
* 2. bis 25. Zeile von oben siehe Ergänzung am Schluß des Bandes (S. 633).

[*Ba 59b*] berichtet. Seine Meßapparatur, bestehend aus einem Spektrometer und zwei Nebelkammern, gestattete eine Massenbestimmung der Teilchen. Die Zahl der Deuteronen, die von außen auf die Apparatur auffallen, beträgt 8,6±1,0% der entsprechenden Zahl von Protonen im Impulsintervall von 1,2 bis 1,39 GeV/c. Alle Messungen sind verträglich mit der Annahme, daß die Deuteronen sekundär in Kernreaktionen in der Luft erzeugt wurden. Für Impulse von der Größenordnung 1 GeV/c ergibt sich die Produktionsrate in Luft:

$$n(p)\, dp = (7{,}85 \pm 1{,}48) \times 10^{-7}\, p^{-3{,}14 \pm 0{,}44}\, dp/\text{g sec ster. GeV/c}. \tag{7.5}$$

Dieses Spektrum ist etwas steiler als das Spektrum der Protonen in der Atmosphäre. Dies bedeutet, daß das Verhältnis Deuteronen/Protonen mit wachsender Energie abnimmt.

Energiespektrum und Winkelverteilung der langsamen Teilchen aus Sternen werden im allgemeinen durch die Verdampfungstheorie gut wiedergegeben. Für das Energiespektrum der Protonen erhält man lediglich am oberen und unteren Ende geringe Abweichungen vom Verdampfungsspektrum; die Abweichungen kommen wohl bei den energiearmen Teilchen durch den Tunneleffekt zustande. Unter den energiereichen Protonen (von etwa 30 MeV) wird man schon einen merklichen Beitrag von Anstoßprotonen erwarten, die nicht nur die erwähnte Abweichung der Energieverteilung verursachen, sondern auch eine Abweichung der Winkelverteilung von der Isotropie.

Nicht so gut sind die Voraussagen der Verdampfungstheorie bei den α-Teilchen erfüllt. Während für kleine Anregungsenergien das Energiespektrum etwa den Erwartungen entspricht, ist das Energiespektrum bei großen Sternen verbreitert und zu kleineren Energien hin verschoben. Es läßt sich verstehen, wenn man annimmt, daß ein Teil der α-Teilchen aus hochangeregten Kerntrümmern kommt, die vom Kern emittiert werden und außerhalb zerfallen. Dazu würde auch passen, daß man für α-Teilchen keine ganz isotrope Winkelverteilung erhält. Man findet außerdem Winkelkorrelationen zwischen den α-Teilchen, allerdings auch für Protonen [*Ce 53*]. Angaben über die Winkelverteilung von Rückstoßkernen finden sich bei GRILLI und VITALE [*Gr 53b*].

Ein ziemlich großer Anteil der Spuren wird von α-Teilchen gebildet. Nach PAGE [*Pa 50*] sind bei den schweren Kernen (27 ± 1)% der Verdampfungsspuren α-Teilchen, bei den leichten Kernen dagegen 54%. PERKINS [*Pe 50b*] hat den Anteil der α-Teilchen an den Sternspuren als Funktion der Sterngrößen bestimmt und findet bei kleinen Sternen zunächst ein Anwachsen des Verhältnisses und ein breites Maximum bei Sternen mit etwa 10 Spuren. HODGSON [*Ho 51a*] erhielt in Pb einen Anteil von (34 ± 7)%, gemittelt über alle Sterngrößen. Nach PAGE [*Pa 50*] sind 25%, nach HARDING und Mitarbeiter [*Ha 49b*] 30% aller einfach geladenen Teilchen Deuteronen oder Tritonen. Dies ist in guter Übereinstimmung mit dem theoretischen Wert [*Le 50*] von 28%. Seit längerem bekannt ist auch, daß gelegentlich Kernsplitter höherer Ladung als 2 emittiert werden. Die ersten Beobachtungen stammen von HEITLER u. a. [*He 39*], SCHOPPER [*Scho 39, Scho 47*], BONETTI und Mitarbeiter [*Bo 49*] und HODGSON und Mitarbeiter [*Ho 49*]. Eine eingehendere Untersuchung dieser Splitter hat gezeigt ([*Sö 49, Sö 51, Pe 50c, Cr 52*]), daß eine Reihe feinerer Einzelheiten in der Winkelverteilung, der relativen Häufigkeit und dem Energiespektrum nicht allein durch die Verdampfungstheorie beschrieben werden können.

8. Sterne in definiertem Material. Eine Schwierigkeit der genaueren Analyse von Sternen in Kernemulsion besteht in dem schon mehrfach erwähnten Vorkommen zweier Gruppen sehr verschieden schwerer Kerne in der Emulsion. Wenn man mit dem geometrischen Wirkungsquerschnitt rechnet, beträgt der Anteil für

Reaktionen an den Kernen der schweren Gruppe (Ag, Br) ~65%, der leichten Gruppe (C, N, O) ~30% und an Wasserstoff rund 5%. Diese Zahlen sind in guter Übereinstimmung mit den auf experimentellem Weg erhaltenen. Der Prozentsatz von Sternen der leichten Gruppe ist nach PAGE [*Pa 50*] 27%, nach HARDING [*Ha 49c*] (36±4,5)%.

Es ist in der Regel nicht möglich im Einzelfall den getroffenen Kern zu identifizieren. Die Aussagen der Sternanalyse gelten deshalb für einen „mittleren" Kern, wobei die Art der Mittelung unsicher ist wegen der Unsicherheit in der Abhängigkeit der einzelnen Meßgrößen von der Massenzahl. Insbesondere kann man z.B. die Messungen an den Sternen in Kernspurplatten nicht ohne weiteres für die Analyse der Nukleonenkaskade in der Atmosphäre heranziehen, deren

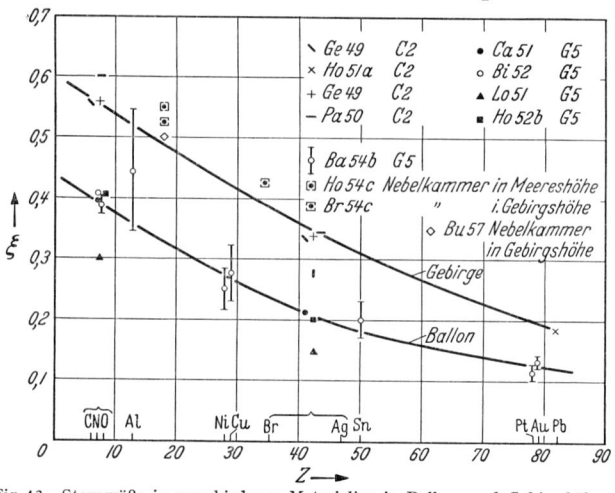

Fig. 13. Sterngröße in verschiedenen Materialien in Ballon- und Gebirgshöhen.

Kerne zur „leichten Gruppe" gehören. Zur Untersuchung der Eigenschaften von Sternen, die in verschieden schweren Kernen erzeugt worden sind, sind verschiedene Wege beschritten worden, z.B. Beobachtungen in Hochdrucknebelkammern, sowie Untersuchungen in verdünnten Emulsionen, d.h. Emulsionen mit kleinerem Bromsilbergehalt; durch einen Vergleich mit Sternen in normalen Emulsionen kann man auf statistischem Wege Aussagen über Sterne in der Gelatine erhalten. Ein anderer Weg, der direkte Aussagen liefert, wurde von BARBOUR [*Ba 54b*] beschritten. Er legte dünne Schichten verschiedener Metalle zwischen je zwei Emulsionsschichten und beobachtete Wechselwirkungen in den Metallfolien an Hand der in die angrenzenden Emulsionen laufenden Spuren. Die Folien müssen dünn sein, um den Verlust von Spuren möglichst klein zu halten. BARBOUR verwendete Folien aus Au, Pt, Sn, Cu, Ni und Al von 25 bis 50 μ Dicke, die von der Emulsion noch durch 10 μ starke Zellophanfolien getrennt waren. Die Platten wurden in rund 30 km Höhe exponiert. Wegen der Schwierigkeit, Teilchen mit Minimumionisation solchen Sternen zuzuordnen, gibt die Arbeit hauptsächlich Daten über die Anstoß- und Verdampfungsspuren. Die Methode macht eine Reihe von Korrekturen für Spuren notwendig, die bei der Analyse übersehen werden, sowie für das Übersehen sehr kleiner Sterne. Für das Sterngrößendiagramm findet BARBOUR eine Beziehung von der Form

$$M(m) = M_0 e^{-\xi \cdot m} \tag{8.1}$$

(m Zahl der Spuren eines Sterns; M Zahl der Sterne mit $\geq m$ Spuren).

Für die Sterne in den Folien findet er keinen Knick bei $m = 7$, wohl aber für die Sterne in der Emulsion. Die Deutung, daß der Knick in der Sterngrößenverteilung (s. Ziff. 6) hauptsächlich durch das Vorhandensein zweier verschieden schwerer Kerngruppen in der Emulsion verursacht sei, ist also in guter Übereinstimmung mit diesen Experimenten, die eine starke Abhängigkeit des Exponenten ξ von der Massenzahl A ergeben. Fig. 13 enthält eine Zusammenstellung verschiedener auch mit anderen Methoden gefundener Werte von ξ (entnommen [Ba 54b]). Sie sind in zwei Gruppen, nach der Höhe der Exposition, aufgetragen. Außerdem ist die verwendete Emulsionsart angegeben. Die Trennung der Punkte für die in verschiedener Höhe erfolgte Exposition zeigt die Abhängigkeit von ξ von der mittleren Energie der sternauslösenden Teilchen. Zu einer Zunahme von ξ

Tabelle 5. *Zahl der grauen Spuren.*

Beobachtete Spurenzahl Wahre Spurenzahl	3—4	3—5 5—7	6—9 8—13	10—13 14—19	14—20 20—30	alle Sterne
Emulsion		1,2 ± 0,2	3,1 ± 0,6	3,8 ± 0,7	6,8 ± 0,8	2,0 ± 0,2
Pt	1,2 ± 0,3	1,2 ± 0,2	3,4 ± 0,6	4,9 ± 0,8	6,2 ± 0,7	2,9 ± 0,2
Ag und Br			3,1 ± 0,6	3,8 ± 0,7	6,8 ± 0,8	(2,1 ± 0,3)
Cu		1,9 ± 0,4	2,5 ± 0,5	6,4 ± 1,6		1,9 ± 0,3
Ni		1,7 ± 0,4	3,0 ± 0,7	5,7 ± 1,7	7,1 ± 3,2	2,0 ± 0,3
Al		1,2 ± 0,5	3,5 ± 2,5			1,4 ± 0,4
C, N, O	(1,2)	(1,3)				(1,2)
Mittelwert	(1,2)	1,3	2,9	4,6	6,4	

mit abnehmender Primärenergie passen auch die Werte von Maschinenexperimenten; sie ergeben für 385 MeV-Protonen $\xi \sim 0,8$ [Be 50a] für Emulsionssterne; für kleinere Teilchenenergien findet man noch höhere Werte [Bo 51].

BARBOUR hat weiterhin die Häufigkeit von Sternen mit mehr als vier Spuren in den verschiedenen Materialien bestimmt. Hieraus erhält er eine Abhängigkeit des Wirkungsquerschnittes σ von der Massenzahl A in der Form:

$$\sigma = c A^k \quad \text{mit} \quad k = 0,70 \pm 0,12, \tag{8.2}$$

also eine gute Übereinstimmung mit einem $A^{\frac{2}{3}}$-Gesetz.

Statistische Untersuchungen an grauen Spuren, die von Anstoßprotonen einer Energie zwischen 25 MeV und 100 MeV erzeugt sind, sind in Tabelle 5 zusammengefaßt. Die Zahl der grauen Spuren wächst mit der Sterngröße. Dies ist zu erwarten, da eine große Zahl von Anstoßnukleonen mit einer höheren Kernanregung verknüpft ist. Dagegen ist für Sterne einer gegebenen Größe keine deutliche Zunahme der Zahl der grauen Spuren mit der Massenzahl festzustellen; die Zunahme der Gesamtzahl der grauen Spuren mit der Massenzahl scheint eher mit der allgemeinen Zunahme der Sterngröße für schwerere Kerne zusammenzuhängen. Auf Grund von qualitativen Vorstellungen über den Mechanismus der Kaskade im Kern wäre zu erwarten, daß die Zahl der grauen Spuren rascher als mit $A^{\frac{1}{3}}$ ansteigt. Dies ist nach den Beobachtungen aber nicht der Fall, braucht jedoch noch nicht in Widerspruch mit dem Modell einer Kaskade im Kern zu stehen.

Eine Analyse der dünnen Spuren mit Hilfe dieser Methode erscheint nicht möglich. Dagegen kann man bei Verwendung einer unempfindlichen Emulsion bei den Verdampfungsspuren noch zwischen einfach und doppelt geladenen

Teilchen unterscheiden. ZANGGER und ROSSEL [Za 56] haben solche Untersuchungen an α-Teilchen durchgeführt mit Hilfe von Kodak NT 2a-Emulsionen, die eine sehr gute Diskriminierung der α-Teilchen liefern, allerdings für Protonen nur bis 25 MeV empfindlich sind. Sie haben damit das Verdampfungsspektrum von α-Teilchen an Sternen in Ag- und Pb-Folien studiert, die von der kosmischen Strahlung in Gebirgshöhe (3600 m) ausgelöst wurden.

9. Sterne hinter dichtem Material. Die Intensität der sternerzeugenden Komponente der kosmischen Strahlung in bzw. hinter Absorbern aus dichtem Material ist mit Kernspurplatten sowie mit elektronischen Nachweismethoden untersucht worden. Man interessierte sich einerseits für die Absorptionslänge in Funktion der Massenzahl, andererseits wurde nach Übergangseffekten gesucht, die beim Übergang der Nukleonenkaskade von Luft in dichtes Material in Analogie zum Rossi-Maximum der weichen Komponente erwartet wurden.

α) *Absorptionslänge.* Tabelle 6 enthält eine Zusammenstellung der von verschiedenen Autoren gemessenen Absorptionslängen L in einigen Materialien.

Tabelle 6. *Absorptionslänge L in verschiedenen Materialien.*

Autoren	Material	L in g/cm²	Meßmethode
Tabelle 2 (Mittelwert)	Luft	138	Sternerzeugung
Ba 51	Kohlenstoff	166± 8	Kernspurplatte
		143± 10	ohne Anteil der Mesonen
Cl 53	Kohlenstoff	260± 35	Szintillationszähler, kleine Energien
Ch 59	Kohlenstoff	182± 30	Zählrohre
Ha 49c	Eis	200± 10	Kernspurplatte
La 49	Eis	196	Kernspurplatte, Einzelspuren
Be 49	Aluminium	200	Kernspurplatte
Ge 49	Blei	310± 20	Kernspurplatte
Be 49	Blei	300± 20	Kernspurplatte
Ma 50, Scho 51a, b .	Blei	320	Kernspurplatte
Ro 51	Blei	305± 7	Kernspurplatte
Cl 53	Blei	350± 15	Szintillationszähler, kleine Energien
Ba 52b	Blei	350± 15	Kernspurplatte
Li 54	Blei	316± 50	Ionisationskammer
Ma 58	Blei	300	

Zwischen den einzelnen Werten für dasselbe Material bestehen teilweise erhebliche Differenzen, z.B. für Kohlenstoff.

Wie in Luft scheint auch in dichtem Material die Absorptionslänge mit wachsender Energie der Prozesse abzunehmen.

Die Abhängigkeit der Absorptionslänge von der Energie bzw. Sterngröße wurde von BARFORD und DAVIS [Ba 52b] in Pb untersucht. Hiernach liegt L für sehr kleine Sterne ($N_h \leq 3$) mit 340 g/cm² nur sehr wenig über dem für alle Sterne gefundenen Wert von 330 g/cm². Mit zunehmender Sterngröße nimmt L ab und erreicht für die größten Sterne ($N_h \geq 12$) den Wert (210± 70) g/cm².

MATHUR und GILL [Ma 58] haben in Gebirgshöhen die Absorptionslänge für Kernprozesse in Pb mit und ohne vorgeschaltetes Pb-Filter von 200 g/cm² untersucht. Sie finden ohne Pb-Filter $\lambda = 300$ g/cm², mit Pb-Filter $\lambda > 350$ g/cm². GRIGOROV und Mitarbeiter [Gr 59] haben die Häufigkeit von Sternen und Einzelteilchen in 9 und 20 km Höhe hinter verschiedenen Absorbern im Hinblick auf Übergangseffekte gemessen.

Ein Unterschied der Nukleonenkaskade in festen Materialien gegenüber derjenigen in Luft besteht offensichtlich in der Rolle der π-Mesonen. In der Atmosphäre zerfallen die meisten π-Mesonen und machen deshalb keine Kerntreffer, sofern ihre Energie nicht sehr groß ist. In festen Materialien dagegen beteiligen sich fast alle π-Mesonen durch Kerntreffer an der Nukleonenkaskade. BARTON und Mitarbeiter [Ba 51] haben den Anteil der π-Mesonen an der Sternerzeugung im Kohlenstoff abgeschätzt. Der Abzug dieses Anteils führt nach ihren Überlegungen zu einer Verkürzung der gemessenen Absorptionslänge um 14%. Der Einfluß der π-Mesonen beruht also einfach auf der Dichteänderung beim Übergang in den Absorber und sollte den Verlauf der Kaskade ändern, auch wenn die Massenzahlen von Luft und Absorber nicht wesentlich verschieden sind, wie z.B. im Fall von Kohlenstoffabsorbern. Wenn sich beim Übergang in den Absorber die Massenzahl stark ändert, werden sich auch die mittlere Multiplizität, Winkelverteilung und Energieverteilung der erzeugten Teilchen, die die Kaskade fortsetzen, sowie die Wirkungsquerschnitte für Sternerzeugung ändern (Ziff. 12). Alle diese Einflüsse werden sich in einer Abweichung von der rein exponentiellen Abnahme der Sternhäufigkeit mit der Tiefe äußern.

Tabelle 7. *Übersicht über die Übergangseffekte, Lage der Maxima.*

Autor	Absorber-material	1. Maximum (cm)	2. Maximum (cm)	Bemerkungen
Rö 54a, b	C	16	> 25	Kernspurplatte (3000 m Höhe und NN), $\cos^2$-Absorber
Rö 54a, b	C	9	22, fraglich	Kernspurplatte, ebener Absorber
Ku 54a, b	C	9	22	Szintillationszähler, ebener Absorber
Li 54c	C	—	20—25	Ionisationskammer
Va 60	C	10—15	—	ebener Absorber
Ma 50	Al	2	—	Kernspurplatte
Be 51	Al	2	—	Kernspurplatte
Rö 54a, b	Fe	4,5	20	Kernspurplatte, $\cos^2$-Absorber
Ku 54a, b	Fe		22	Szintillationszähler, $\cos^2$-Absorber
Ku 54a, b	Fe		18	Szintillationszähler, ebener Absorber
Bu 56	Fe	3,2	17, fraglich	Szintillationszähler, ebener Absorber
Rö 54a, b	Sn	4—5	20	Kernspurplatte, $\cos^2$-Absorber
Be 48a, b	Pb	2		Kernspurplatte, $\cos^2$-Absorber
Ma 50	Pb	1		Kernspurplatte, $\cos^2$-Absorber
Be 51	Pb	0,5		Kernspurplatte, $\cos^2$-Absorber
Scho 51a, b	Pb	1—2	20	Kernspurplatte, $\cos^2$-Absorber
Ku 54a, b	Pb		18	Szintillationszähler, $\cos^2$-Absorber
Ku 54a, b	Pb		16	Szintillationszähler, ebener Absorber
Li 54	Pb	1—2, fraglich		Ionisationskammer
Co 56	Pb	2	20	Szintillationszähler
Sh 55	Pb	1,4		Kernspurplatte, 5000 m Höhe
Va 58, 60	Pb	0,5		Kernspurplatte, 3000 m Höhe

Analog zum Übergangseffekt der weichen Komponente können also auch bei der Nukleonenkomponente Übergangseffekte mit Maxima der Sternhäufigkeit auftreten, die ein Zeichen für eine Störung des Gleichgewichts der Nukleonenkaskade sind.

β) *Übergangseffekte.* Hinter Pb-Absorbern beobachteten in Gebirgshöhen BERNARDINI und Mitarbeiter [Be 49] Übergangseffekte der Sternerzeugung in Kernspurplatten, ebenso MALASPINA und Mitarbeiter [Ma 50] hinter Pb und Al. MATHUR und GILL [Ma 58] finden in Gebirgshöhe ein Übergangsmaximum in Pb bei etwa 2 cm.

Von LORD und SCHEIN [Lo 49] wurde ein sehr ausgeprägter Übergangseffekt hinter Pb in 18 km Höhe gefunden. Ebenfalls in der Stratosphäre haben TABIKAJEW [Ta 56] sowie SHAPIRO und Mitarbeiter [Sh 51a, b] Übergangseffekte gemessen, letztere mit einer geometrischen Anordnung des Pb-Absorbers, mit der sie eine komplizierte, nicht leicht zu deutende Übergangskurve der Sternhäufigkeit erhielten.

Eingehende Untersuchungen mit Absorbern verschiedener Massenzahl haben SCHOPPER, KUHN, HÖCKER und RÖSSLE in Gebirgshöhen und in der Nähe des Meeresniveaus mit Kernspurplatten durchgeführt.

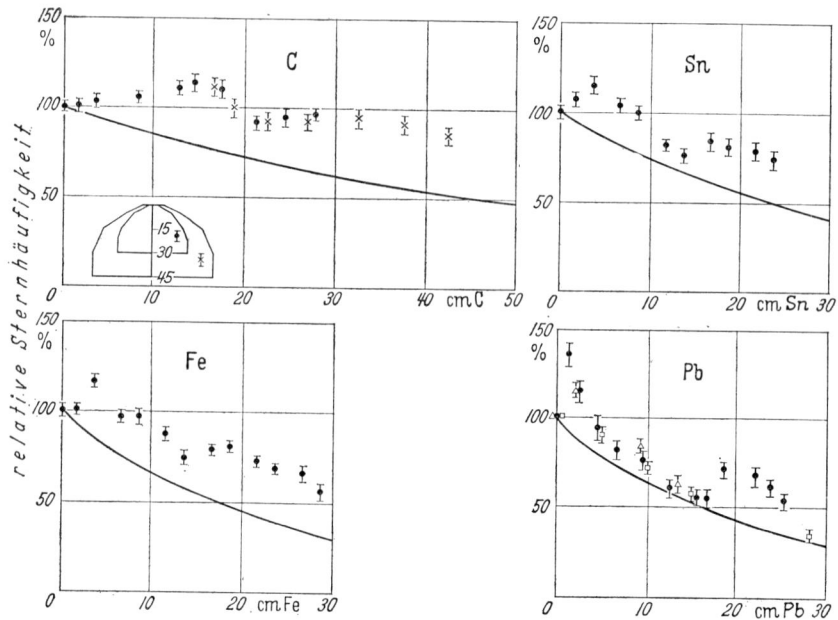

Fig. 14. Übergangseffekte der Sternhäufigkeit in C-, Fe-, Sn- und Pb-Absorbern (nach RÖSSLE u. SCHOPPER). Absorberform siehe Teilfigur C. Ausgezogene Kurve entspricht expontieller Schwächung im jeweiligen Material Pb-Absorber: ● Messungen von RÖSSLE u. SCHOPPER [Rö 54a], △ Messungen von BERNARDINI [Be 48], □ Messungen von GEORGE u. JASON [Ge 49].

Fig. 14 zeigt Übergangskurven der Sternhäufigkeit in verschiedenen Materialien (C, Sn, Fe, Pb) nach einer Arbeit von RÖSSLE und SCHOPPER [Rö 54a]. Die Figur zeigt außerdem die verwendete Absorberform; sie war so gewählt, daß die in bestimmten Tiefen der Absorberachse ausgelegten Kernspurplatten von dem aus der Luft einfallenden Teilchen mit jeweils annähernd gleichen Absorberweglängen erreicht wurden, wenn für sie eine $\cos^2\vartheta$-Zenitwinkel-Abhängigkeit zugrunde gelegt wurde.

Neben dem in schweren Materialien nach dem Vorangehenden zu erwartenden Übergangseffekt deutet sich in größerer Tiefe ein zweites Maximum an, das bei den leichteren Absorbern allerdings an der Grenze statistischer Signifikanz liegt. Auffallend ist der von RÖSSLE und SCHOPPER in Kohlenstoff beobachtete Übergangseffekt.

Für sein Zustandekommen könnten zunächst π-Mesonen verantwortlich sein. Die Autoren zeigen jedoch an Hand der unterschiedlichen Sterngrößenverteilungen für geladen- und neutralausgelöste Sterne in verschiedenen Absorbertiefen, daß die Maxima der Sternhäufigkeit durch energiearme, neutralausgelöste Sterne gebildet werden. Damit scheiden π-Mesonen für die direkte Deutung des Effektes aus. Ähnliche Ergebnisse wurden von anderen Autoren gefunden. Einen Überblick über die beobachteten Maxima gibt Tabelle 7.

Ein genauer Vergleich der Ergebnisse ist kaum möglich, da die einzelnen Autoren verschiedene Absorber-Anordnungen und zum Teil verschiedene Nachweismethoden verwendet haben.

Messungen mit Szintillationszählern z.B., enthalten Beiträge von Einzelteilchen und — besonders in Absorbern höherer Ordnungszahl — auch Beiträge von Elektronenkaskaden, wenn nicht besondere Vorkehrungen zu ihrer Elimination getroffen sind.

Über negative Ergebnisse bei der Suche nach einem Übergangsmaximum in Pb berichten HAUSER et al. [Ha 60c], die in Gebirgshöhe mit Kernspurplatten maßen. Ihre statistische Genauigkeit, sowie die Absorbergeometrie lassen, nach Ansicht des Referenten, eine signifikant negative Aussage jedoch nicht zu.

LINDENBERGER und MEYER [Li 54c] haben eine Hochdruck-Ionisationskammer mit Antikoinzidenz verwendet, die neutral im Füllgas ausgelöste Sterne registrierte. Vor einigen Jahren wurden von VARSIMASHVILI [Va 60] Messungen mit Kernspurplatten in Pb und C wiederholt. Sie bestätigen ein Übergangsmaximum in Pb bei 0,5 bis 1 cm Tiefe, das in 3 cm Tiefe praktisch abgeklungen ist. Sie können weiterhin zeigen, daß der Übergangseffekt in Kohlenstoff auch bzw. wieder auftritt, wenn der Kohlenstoffabsorber allseitig durch 2—3 cm Pb abgeschirmt war, während er verschwindet, wenn die Kernspurplatten *im* Kohlenstoffabsorber jeweils mit 3 cm Pb umgeben waren (Fig. 15).

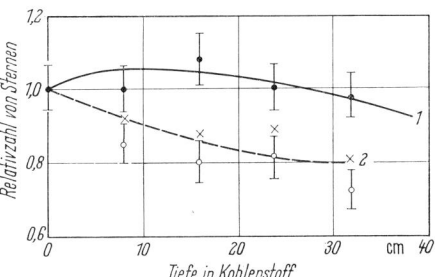

Fig. 15. Zahl der Sterne in verschiedenen Tiefen eines ebenen ausgedehnten Kohlenstoffabsorbers (nach [Va 60]). Kurve 1: Mit 3 cm Pb allseitig abgeschirmter C-Absorber. Kurve 2: Unabgeschirmter C-Absorber, Kernspurplatten jeweils mit 3 cm Pb abgeschirmt; exponentieller Abfall der Sternhäufigkeit.

Beim Vergleich mit Fig. 14 ist zu beachten, daß ein ebener ausgedehnter Absorber wegen der für allseitigen Einfall sehr unterschiedlichen Absorberweglängen die Lage des Maximums verschmiert.

Die Effekte in Gebirgshöhe und in der Nähe des Meeresniveaus [Rö 52, Li 54, Bu 56] sind völlig ähnlich. Die am Maximum beteiligte „Komponente" steht offensichtlich im Gleichgewicht mit der Nukleonenkomponente.

Neben der direkten Bestimmung der Sternhäufigkeit wurde auch der Übergangseffekt der Neutronenproduktion in schwerem Material mit Hilfe von BF_3-Zählern gemessen. Nach dem in Ziff. 6, S. 394 Gesagten sollte die Neutronendichte ein Maß für die Häufigkeit energiearmer Kernzertrümmerungen sein. Messungen in Pb von HOGREBE [Ho 52a], TREIMANN und FONGER [Tr 52], REICH [Re 55] sowie von SIMPSON [Si 53c] ergaben eine im wesentlichen exponentielle Abnahme der neutronenerzeugenden Strahlung mit der Absorberdicke. Die gemessenen Absorptionslängen von (310 ± 35) g/cm² [Ho 52a], 320 g [Tr 52] und (320 ± 32) g/cm² [Re 55] stimmen gut untereinander und mit den für die Sterne gefundenen Werten überein. Dasselbe gilt für die Messung von SIMPSON und URETZ [Si 53c] in 10000 m Höhe, die eine Absorptionslänge von (350 ± 40) g/cm² lieferte. Von SIMPSON und Mitarbeiter, TREIMANN und Mitarbeiter, sowie HOGREBE wurden aber auch hier Übergangseffekte beobachtet. SIMPSON sowie TREIMANN fanden ein Maximum bei etwa 1 cm Pb; für größere Absorbertiefen lagen die Meßpunkte so weit auseinander, daß keine Aussage über eine weitergehende Struktur der Absorptionskurve möglich ist. Letzteres gilt auch für einige ältere Messungen der Sternerzeugung mit Kernspurplatten [Ge 49], bei denen keine Übergangseffekte gefunden wurden. In 20 cm Tiefe erhielt HOGREBE [Ho 52a] in Pb eine Abweichung vom Exponentialverlauf, die eine Ähnlichkeit mit dem für die Sterne mit Kernspurplatten gefundenen zweiten Maximum aufweist. Neben Untersuchungen in Pb existieren auch Messungen in Kohlenstoff [We 55b] und Aluminium [Re 55]. Dabei findet WEISS [We 55b] in Kohlenstoff ein Maximum bei 9 cm, REICH [Re 55] in Aluminium bei 8 cm.

Die Messungen von WEISS wurden von CHAUDHURI und PFOTZER wiederholt [Ch 59]. Ihre Anordnung zeigt Fig. 16: Die hinter dem C-Absorber variabler Dicke auftretende nukleoaktive Komponente erzeugt in der 2 cm Pb-Schicht Neutronen, die in Paraffin thermalisiert mit BF_3-Zählern registriert werden. Ihre unkorrigierte Zählrate weist wieder ein Maximum hinter 10—15 cm Kohlenstoff auf. Nach Anbringung von Korrekturen für Rückstreueffekte der Neutronen in der Pb-Schicht verbleibt dann aber ein exponentieller Verlauf mit einer

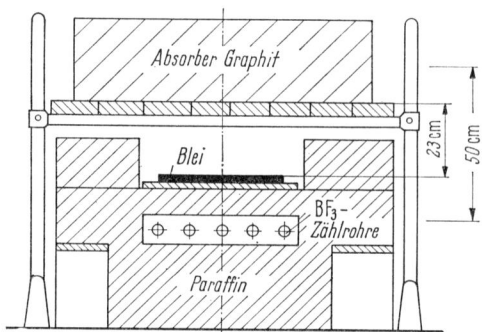

Fig. 16. Anordnung zur Messung der Neutronenproduktion in einer Pb-Schicht hinter einem Kohlenstoff-Absorber. (Nach [Ch 59].)

Absorptionslänge für Kohlenstoff von 260 g/cm². Die Autoren sehen sich daher zu dem Schluß berechtigt, daß die in Kohlenstoff-Absorbern beobachteten Übergangseffekte allgemein auf Reflexionseffekte von Verdampfungsneutronen im Absorber zurückzuführen seien.

Im Gegensatz zu ihren Ergebnissen mit C-Absorbern stehen ihre Messungen hinter Pb-Absorbern mit dem Ergebnis von RÖSSLE und SCHOPPER [Rö 54a] im Einklang und sprechen für die dort beobachteten Übergangsmaxima. In der Meßanordnung ist dabei der C-Absorber durch variable Schichten von Pb ersetzt unter Wegfall der 2 cm-Pb-Erzeugerschicht. Sie messen die integrale Neutronenproduktion.

Die Diskrepanz im Falle des C-Absorbers wird verständlich, wenn man die Verschiedenartigkeit der Nachweismethode berücksichtigt: bei RÖSSLE und SCHOPPER werden die hinter bestimmten Schichtdicken des Kohlenstoffabsorbers in Kernspurplatten erzeugten Sterne als Maß für den Fluß der erzeugenden Komponente beobachtet. Dabei wird das Übergangsmaximum von teilchen- und energiearmen Sternen mit zwei bis vier sichtbaren Spuren gebildet.

CHAUDHURY und PFOTZER messen den Fluß der sternerzeugenden Komponente über deren Neutronenproduktion in Verdampfungssternen der 2 cm-Pb-Schicht. Die Nachweiswahrscheinlichkeit für diese Neutronen ist aber der Multiplizität ihrer Erzeugung in Pb-Sternen proportional. Wir haben also die mittlere Multiplizität $\bar{\nu}=15$ in „normalen" Pb-Sternen zu vergleichen mit einem Wert von $\nu_m \sim 4$ bei den das Übergangsmaximum bildenden Sternen. Der Anteil der letzteren wird mit einer um den Faktor $\nu_m/\bar{\nu} = \frac{4}{15}$ reduzierten Wahrscheinlichkeit registriert und sinkt damit beim Übergangseffekt in Kohlenstoff unter die Nachweisbarkeit.

CHAUDHURY und PFOTZER haben somit nur die exponentielle Schwächung des „Normalanteils" der sternerzeugenden Komponente in Kohlenstoff bestätigt.

HAUSER [Ha 60b] hat die Neutronenproduktion in Pb-Schichten bis 2,4 cm Dicke mit B-beladenen Kernspuremulsionen in einem Moderator untersucht. Sie findet ebenfalls einen exponentiellen Verlauf ohne Übergangsmaximum. Für ihr Ergebnis gilt das oben Gesagte analog.

BAGGE und SKORKA [Ba 58] haben den Übergangseffekt der Neutronenkomponente in Wasser ebenfalls mit Hilfe von BF_3-Zählern untersucht. Diese Messungen werden in Ziff. 17 beschrieben.

Die Existenz der Übergangseffekte kann in Anbetracht ihrer Bestätigung durch zahlreiche unabhängige Messungen kaum bezweifelt werden. Ihre Deutung durch einen bestimmten Mechanismus ist bislang nicht gelungen: Die Existenz eines Übergangseffektes in Kohlenstoff, vor allem aber ein zweites Maximum in Pb, lassen sich bereits qualitativ schwer verstehen. Quantitative Überlegungen, um wenigstens das erste Maximum zu erklären, stammen von DALLAPORTA und Mitarbeiter [Da 50 b], KUHN [Ku 54 b], HÖCKER und Mitarbeitern [Hö 55] sowie von RÖSSLE [Rö 52], RÖSSLE und SCHOPPER [Rö 54a]. Eine ausführliche Theorie der Nukleonenkaskade in großen Absorbertiefen wurde von URBANIK und LOPUSZANSKI [Ur 55] gegeben.

Da die Sterne, die die Maxima verursachen, verhältnismäßig kleine Energien haben und neutral ausgelöst sind, liegt es zunächst nahe, für ihre Erklärung die Neutronenkomponente heranzuziehen. Eine Erklärung durch andere Prozesse (z.B. Neutronen des μ-Mesoneneinfangs oder durch γ-Quanten) scheint wegen der geringen Intensität dieser Komponenten bzw. wegen ihres geringen Wirkungsquerschnittes für Sternerzeugung auszuscheiden. RÖSSLE und SCHOPPER [54a, b] hat versucht, unter vereinfachenden Annahmen die Kaskade energiearmer Neutronen in Absorbern zu berechnen und damit die beobachteten Übergangseffekte zu deuten. Die Rechnung gibt die beobachteten und erwarteten schwachen Übergangseffekte der geladen-ausgelösten und der energiereichen Sterne gut wieder, doch ist keine befriedigende Übereinstimmung für das Maximum der energiearmen neutralausgelösten Sterne zu erreichen. Eine *gute* Übereinstimmung wäre in diesem Fall allerdings kaum zu erwarten, da die Rechnung nicht nur einige Näherungen benützt, sondern auch die verwendeten experimentellen Ausgangsdaten, wie z.B. Winkelverteilung und Multiplizität nicht ausreichend genau sind. Außerdem macht es Schwierigkeiten, die Form der Absorber genügend genau zu berücksichtigen. RÖSSLE hat allerdings den Einfluß der Absorberform und -größe experimentell untersucht und festgestellt, daß hierbei keine prinzipiellen Änderungen der Übergangseffekte auftreten. Dies geht schon aus Tabelle 7 hervor, die eine größenordnungsmäßige Übereinstimmung der meisten Daten zeigt, obwohl sie mit sehr verschiedenen geometrischen Anordnungen gemessen wurden.

Die *räumliche Verteilung* der Sterne in der Kernspur-Emulsion wurde von einer größeren Anzahl von Autoren gemessen; insbesondere wurde nach einer möglichen Korrelation der Lage der Sterne in der Emulsion bei kleinen Abständen bis zu einigen Millimetern gesucht. Genetisch zusammengehörende Sterne haben nach dem üblichen Bild, das man von der Nukleonen-Kaskade hat, in festen Materialien mittlere Abstände von der Größenordnung 30 cm. Es ist also mit den üblichen Mechanismen keine nennenswerte Korrelation bei Entfernungen von einigen Millimetern zu erwarten. Trotz dem wurde von einer ganzen Reihe von Autoren zum Teil erhebliche Effekte gefunden.

Man bestimmt dazu in der Regel die Zahl der Paare von Sternen, deren Entfernung voneinander kleiner als ein vorgegebener Betrag ist und die nicht durch ein ionisierendes Teilchen miteinander verbunden sind. Diese Zahl wird mit der bei statistisch regelloser Verteilung der Sterne theoretisch zu erwartenden verglichen. Bestimmt man lediglich die in die Plattenebene projizierte Entfernung r_p der Sterne, so erhält man für die Zahl $N(r_p)$ der zufälligen Sternpaare mit einer Entfernung $\leq r_p$

$$N(r_p) = \sum_{i=1}^{n} \frac{N_i(N_i-1)}{2} \frac{\pi r_p^2}{a_i b_i} \left[1 - \frac{4(a_i+b_i)r_p}{3\pi a_i b_i}\right] \tag{9.1}$$

(nach Li [Li 50]). Dabei bedeutet N_i die Zahl der Sterne in der i-ten rechteckig gedachten Probefläche mit den Seitenlängen a_i und b_i. Es ist die Voraussetzung $r_p \ll a_i, b_i$ gemacht. [In der eckigen Klammer von (8.1) ist ein Glied der Ordnung $r_p^2/(a_i b_i)$ weggelassen.] Um Korrelationen bei sehr kleinen Entfernungen zu untersuchen, benutzt man besser die wahre räumliche Entfernung r der Sterne. Die Zahl der zufälligen Sternpaare $M(r)$ ist in diesem Fall (unter Vernachlässigung der Randeffekte Gl. (13.1)

$$M(r) = \sum_{i=1}^{n} \frac{N_i(N_i-1)}{2 a_i b_i} \begin{cases} \frac{4\pi}{3d} r^3 \left(1 - \frac{3}{8}\frac{r}{d}\right) & \text{für } r \leq d \\ \pi r^2 \left(1 - \frac{1}{6}\left(\frac{d}{r}\right)^2\right) & \text{für } r \geq d, \end{cases} \tag{9.2}$$

d Dicke der Platte (nach [Lo 55c]).

Verschiedene Autoren haben erheblich mehr Sternpaare gefunden als es Gl. (8.1) und (8.2) entspricht. Dies läßt sich schwer erklären; lediglich bei [Co 54c] scheint eine Deutung durch sekundäre Neutronen möglich. Andererseits wurde von Barbanti und Mitarbeiter [Ba 52c] und Lohrmann [Lo 55c] in sorgfältig durchgemusterten Platten bis zu Abständen von 0,5 bzw. 1 mm kein Effekt gefunden. Dies und die Beobachtung von Davis [Da 52a], daß der von ihm beobachtete Effekt durch sehr kleine Sterne verursacht ist, deutet auf einen subjektiven Einfluß beim Durchmustern der Platten hin, indem kleine Sterne in der Nähe von großen weniger leicht übersehen werden. Gramenitskij und Mitarbeiter [Gr 56a] haben außerdem gezeigt, daß die Sterne die den beobachteten Effekt verursachen, nicht gleichzeitig erzeugt worden sind, indem sie zwei Kernspurplatten durch ein Uhrwerk langsam aneinander vorbeibewegten und die Zeit an Hand der Verschiebung der in der nächsten Platte weiterlaufenden Spuren bestimmten. Die früher berichteten Anomalien dürften also auf einen subjektiven Einfluß zurückzuführen sein.

D. Protonen in der Atmosphäre.

10. Allgemeine Bemerkungen zum Energiespektrum. Die wichtigsten Daten zur Beschreibung der Protonenkomponente in der Atmosphäre sind das Energiespektrum und die Gesamtintensität als Funktion der Tiefe in der Atmosphäre. Um diese Größen zu messen, ist es zunächst notwendig, die Protonen von den anderen Teilchen der kosmischen Strahlung zu unterscheiden.

Experimentelle Schwierigkeiten bestehen hierbei in der geringen Intensität der Protonen gegenüber der viel größeren Häufigkeit der μ-Mesonen. Um ein praktisches Beispiel zu geben, erwähnen wir, daß in einer von Filthuth [Fi 55] beschriebenen Apparatur großer „Lichtstärke" je nach den eingestellten Energieintervallen zwischen 28 und 2 Protonen am Tag in Meereshöhe registriert wurden. Der Fluß der μ-Mesonen ist dabei im Mittel 100mal größer. Die Verhältnisse sind etwas günstiger in Gebirgshöhe, aber auch dort machen die Protonen nur 10 bis 20% der gesamten durchdringenden Komponente aus. Dies sowie die geringe Absolutintensität bringt große Anforderungen an die Zeitkonstante und Stabilität der Meßapparatur mit sich; ebenso muß die Unterscheidung zwischen μ-Mesonen und Protonen mit einiger Sicherheit möglich sein. Das Energiespektrum der Protonen in der Atmosphäre ist deshalb nur in einem begrenzten Energiegebiet näher bekannt. Im Gebiet sehr hoher Energien ist die Intensität sehr gering; man ist dort in der Regel auf indirekte Methoden angewiesen. Die untere Grenze, bis zu der das Protonenspektrum meist gemessen wird, liegt bei 100 MeV. Unterhalb dieser Energie sind nur wenige Daten verfügbar; auf Meereshöhe ist das Spektrum bis auf 18 MeV herab von FILTHUTH [Fi 55] gemessen worden. LATTIMORE [La 49] hat versucht, Angaben bei niedrigen Energien durch Untersuchung von einzelnen Spuren in der Kernspurplatte zu gewinnen.

Zur Trennung der Protonen von der übrigen ionisierenden Strahlung (meist μ-Mesonen und Elektronen) kann man die starke Kernwechselwirkung der Protonen benutzen. Dieses Vorgehen bringt jedoch einige Schwierigkeiten mit sich. Einmal läßt sich die Energie des Teilchens nicht sehr zuverlässig ermitteln. Der Zusammenhang zwischen Schauergröße bzw. Sterngröße (bei Kernspurplattenmessungen) und Primärenergie ist nur statistisch bestimmt und nicht sehr genau. Eine direkte Energiebestimmung ist möglich durch Messung der Krümmung der Spuren im Magnetfeld oder durch Messung der Coulombschen Vielfachstreuung, doch versagen diese Methoden oberhalb einiger GeV. Bei den auf diese Weise gewonnenen Spektren hat man weiterhin zu beachten, daß vor allem bei hohen Energien auch π-Mesonen zum Fluß beitragen. Außerdem setzt die Methode die Kenntnis des Wirkungsquerschnittes für Kernwechselwirkungen als Funktion der Energie voraus. Eine Methode, die diese Schwierigkeiten vermeidet, besteht in einer Identifizierung der Protonen durch eine Massenbestimmung; hierbei erhält man ihre Energie automatisch. In den letzten Jahren sind eine ganze Reihe solcher Arbeiten durchgeführt worden, so daß wir wenigstens im Energiegebiet zwischen 100 MeV und einigen GeV das Energiespektrum mit einiger Genauigkeit und Zuverlässigkeit kennen.

11. Überblick über experimentelle Methoden. Es seien im folgenden die in Ziff. 10 erwähnten Methoden an Hand einiger (älterer) Arbeiten besprochen: Ein ausführliches Beispiel einer Messung mit Hilfe der Kernspurplattenmethode findet man in einer der grundlegenden Arbeiten der Bristolgruppe [Ca 51]. Die Untersuchungen wurden an Kernzertrümmerungen durchgeführt, die bei einem Ballonflug in 20 km Höhe registriert wurden. Die Verfasser wählten 200 Spuren aus Sternen aus, die nach den Kriterien von Ziff. 4 als Primärspur des Sternes aufgefaßt werden können und an denen Messungen der Vielfachstreuung möglich sind. Gleichzeitig wurde die Korndichte gemessen. Hieraus lassen sich Masse und Energie des Teilchens bestimmen. Bei hoher Energie läßt sich die Unterscheidung zwischen π-Mesonen und Protonen nicht mehr durchführen; in diesem Fall gibt die Messung der Vielfachstreuung allein die Energie des Teilchens. Die Verfasser schätzen, daß bei einer Energie > 1 GeV etwa 6% aller Primärteilchen π-Mesonen sind. Eine höhere Energie als 10 GeV kann mit dieser Methode kaum gemessen werden; spätere Untersuchungen haben darüber hinaus gezeigt, daß schon im Bereich einiger GeV systematische Fehler durch das sog. „spurious scattering" [Bi 55, Fa 55, Lo 56c, Br 56] auftreten können. TEUCHER [Te 53c] hat versucht, aus dem Sterngrößendiagramm von Sternen in Kernemulsionen ein Energiespektrum zu gewinnen, doch kommt er zu dem Schluß, daß die Methode nicht sehr zuverlässig ist. In Gebirgshöhen haben DILWORTH und Mitarbeiter [Di 53b] das Energiespektrum von sternerzeugenden Teilchen in der Kernemulsion durch Streumessungen bestimmt.

Messungen in rund 9000 m Höhe wurden unter anderem von CONVERSI [Co 50] in einem Flugzeug mit Hilfe einer Zählrohr-Apparatur durchgeführt.

Der Verfasser registrierte verzögerte Koinzidenzen und Anti-Koinzidenzen von Teilchen, die in einem Absorber zur Ruhe kommen. Aus der Zahl der verzögerten Koinzidenzen konnte er durch Extrapolation den Anteil der μ-Mesonen unter diesen Teilchen bestimmen und hieraus den Fluß der Protonen berechnen.

Untersuchungen, bei denen die Protonen direkt durch eine Massenbestimmung nachgewiesen wurden, sind z.B. von ROSEN [Ro 54] und MILLER und

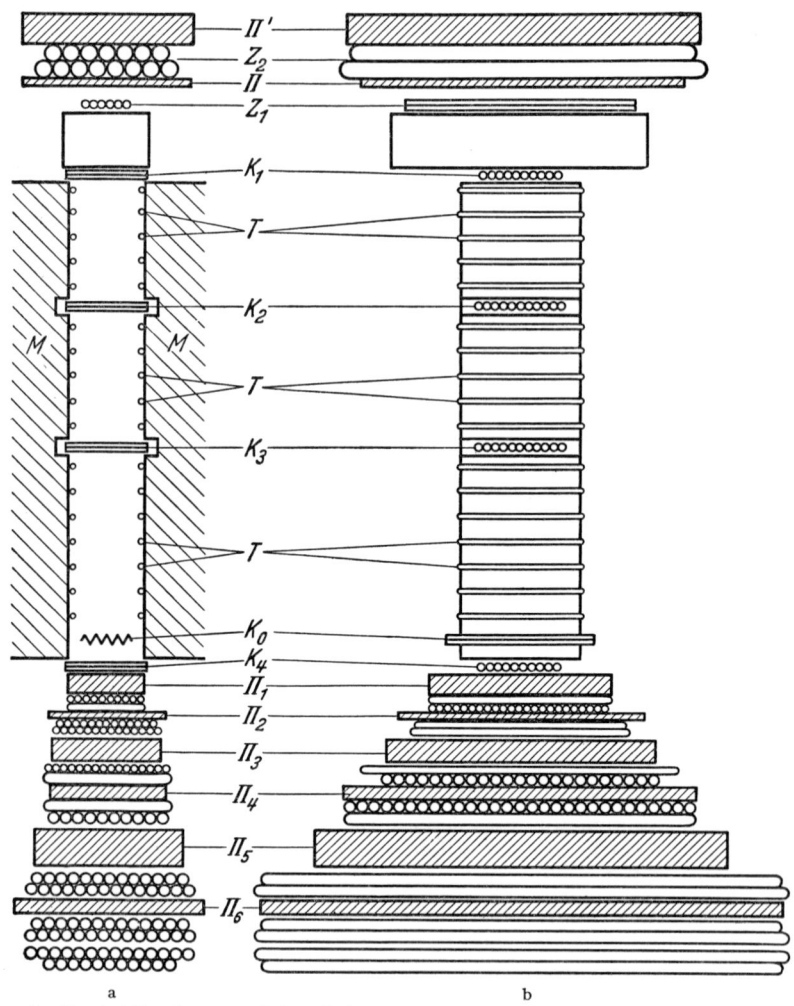

Fig. 17a u. b. Konstruktion des magnetischen Spektrometers nach [Ko 55]. a in einer Vertikalebene parallel zum Magnetfeld, b in einer Vertikalebene senkrecht zum Magnetfeld.

Mitarbeiter [Mi 54a, Mi 54b] in der Nebelkammer gemacht worden. Die Apparatur von ROSEN besteht aus drei übereinanderliegenden Nebelkammern. Die Auslösung erfolgt durch Zählrohr-Koinzidenzen. Mit Hilfe eines Magnetfeldes, das sich zwischen der oberen und mittleren Kammer befindet, wird das Teilchen abgelenkt und sein Impuls aus der Bahn in der mittleren Kammer bestimmt. Die Reichweite wird in der anschließenden dritten Kammer gemessen, die 25 Kupferplatten von insgesamt 86,6 g/cm² enthält. Aus Impuls und Reichweite läßt sich die Masse des Teilchens errechnen (dies ist auch das Prinzip der im folgenden beschriebenen Apparatur [Ko 55]). Ein Pb-Absorber über der Apparatur gestattet eine Erweiterung des Meßbereiches, doch sind die Ergebnisse wegen der Erzeugung sekundärer Teilchen im Absorber dann viel schwerer zu deuten.

Ein magnetisches Spektrometer großer Genauigkeit wird von KOCHARIAN [Ko 55] beschrieben. Die Ablenkung der Teilchen wird in einem Magnetfeld M (Abmessungen $80 \times 20 \times 10$ cm, Stärke 5800 Oersted) mit Hilfe von 4 Lagen von Zählrohren spezieller Konstruktion K_1, K_2, K_3, K_4 bestimmt (Fig. 17). Die Zähler T scheiden Teilchen aus, die an den Wänden des Magneten gestreut worden sind. Zur Bestimmung der Reichweite dienen die Absorber $\pi_1, \ldots \pi_6$, zwischen denen sich Zählrohrlagen befinden.

Eine andere Methode, die eine Massenbestimmung gestattet, besteht in der Messung von Ionisation und Reichweite des Teilchens. Eine solche Apparatur mit der ebenfalls sehr genaue Ergebnisse erzielt wurden, wurde von MESHKOVSKII und SOKOLOV verwendet [Me 58a]. Zur Ionisationsmessung benutzten sie fünf Szintillationszähler spezieller Konstruktion; die Reichweite wurde wie üblich mit Hilfe von Pb-Absorbern bestimmt, die von Zählrohrlagen umgeben waren. Die Apparatur ist bei MESHKOVSKII und Mitarbeiter [Me 56a, b] beschrieben.

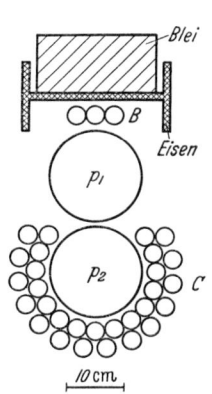

Fig. 18. Zählrohranordnung zum Nachweis von Protonen nach [Fi 55]. In Lage A und B darf ein und nur ein Zählrohr ansprechen. $P2$ und $P1$ sind Proportionalzählrohre.

Messungen bis zu sehr kleinen Energien herab (18 MeV) wurden von FILTHUTH [Fi 55] in 150 m Höhe durchgeführt. Fig. 18 zeigt die Anordnung der Apparatur. Ein wesentlicher Bestandteil sind zwei Proportionalzählrohre P_1 und P_2 von 15 cm Durchmesser und 1 m Länge. Es wurden alle Teilchen gezählt, die eine vierfache Koinzidenz der Zählrohre $A B P_1 P_2$ auslösen. Man mißt die Energieverluste E_1 und E_2 des Teilchens beim Durchgang durch die beiden Zählrohre P_1 und P_2. Da das Teilchen zwischen P_1 und P_2 eine definierte Materiemenge durchqueren muß, erhält man für jede Teilchenart einen anderen charakteristischen Zusammenhang zwischen E_1 und E_2. Der Verfasser konnte mit dieser Methode eine gute Separation zwischen Protonen und Mesonen erzielen. Die Zählrohre C sind in Antikoinzidenz geschaltet, um μ-Mesonen auszuschließen, die von Anstoßelektronen in P_1 und P_2 begleitet sind und ein Proton vortäuschen könnten. Die registrierte Protonenenergie wurde durch entsprechende Wahl des Pb-Absorbers eingestellt. Eine Messung großer Genauigkeit in einem engen Energieintervall wurde von BACCALIN und Mitarbeiter [Ba 55] mit Hilfe eines Zählrohrteleskops und eines Čerenkov-Zählers in Meeres- und Gebirgshöhe gemacht. Als weitere Methode sei noch eine Arbeit von CHOU [Ch 56] erwähnt, der die Geschwindigkeit der Teilchen durch Bestimmung ihrer Flugzeit gemessen hat.

12. Das differentielle Energiespektrum der Protonen. Eine Übersicht über das differentielle Impulsspektrum auf Grund von älteren Arbeiten findet man bei PUPPI und DALLAPORTA [Pu 52] sowie bei BUDINI und MOLIÈRE [Bu 53]. Später wurden vor allem in Gebirgshöhen weitere genaue Messungen durchgeführt [Ko 55, Me 58a]. Eine Übersicht über die Ergebnisse charakteristischer Arbeiten gibt Fig. 19. Sie zeigt das differentielle Energiespektrum der Protonen in verschiedenen Höhen. Aufgetragen ist die vertikale Intensität von Protonen pro cm², sec Einheit des Raumwinkels und MeV Energieintervall. 680 mb entsprechen einer Höhe von etwa 3400 m, 310 mb einer Höhe von etwa 9000 m. Messungen zum Energiespektrum in sehr großer Höhe, z.B. [Ca 51], haben wir nicht berücksichtigt. Die extrapolierte Kurve für 0 mb wurde lediglich aus Vergleichsgründen eingezeichnet, sie wurde von [Bu 53] übernommen. Für 310 mb

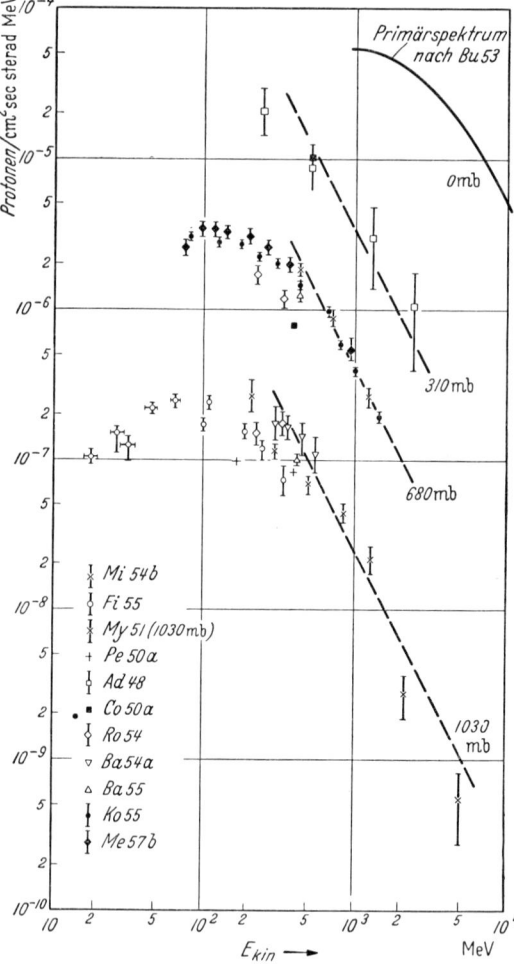

Fig. 19. Differentielles Energiespektrum der Protonen in verschiedenen Höhen. Lies: Me 57b ≡ Me 58a; Co 50a ≡ Co 50.

Tabelle 8. *Differentielle Flußwerte der Protonen (Meereshöhe).*

	GeV/c	Protonen cm² sec. steread (Gev/c)
MESH-KOVSKII et al. [Me 58a]	0,31−0,38	$(1,15\pm0,16)\,10^{-7}$
	0,38−0,44	$(1,45\pm0,18)\,10^{-7}$
	0,44−0,48	$(1,78\pm0,16)\,10^{-7}$
	0,48−0,55	$(1,84\pm0,16)\,10^{-7}$
WOLFENDALE, BROOKE [Br 64a] Vgl. auch Fig. 52	0,75	$(1,50\pm0,28)\,10^{-7}$
	1,03	$(8,40\pm0,75)\,10^{-8}$
	1,53	$(2,85\pm0,30)\,10^{-8}$
	2,12	$(2,26\pm0,20)\,10^{-8}$
	2,94	$(7,64\pm0,88)\,10^{-9}$
	4,60	$(3,40\pm0,36)\,10^{-9}$
	6,16	$(1,40\pm0,46)\,10^{-9}$
	10,3	$(4,6\ \pm1,2\ )\,10^{-10}$
	18,0	$(1,17\pm0,47)\,10^{-10}$
	27,5	$(2,95\pm0,94)\,10^{-11}$
	79,0	$(2,3\ \pm1,4\ )\,10^{-12}$

existieren nicht viele Daten. Saubere Untersuchungen in dieser Höhe lassen sich noch am besten mit Flugzeugen gewinnen und sind entsprechend schwierig.

Die mit [Ko 55] bezeichneten Punkte stellen Mittelwerte aus zwei Meßkurven dar, die gut zueinander stimmen. Weitere Daten über zum Teil ältere Messungen in Meereshöhe findet man bei [Me 56]. Die Spektren sind im Gebiet großer Energie verträglich mit der Annahme eines Potenzgesetzes für die Intensität von der Form

$$\Phi(E)\,dE = E^{-n}\,dE. \qquad (12.1)$$

Die eingezeichneten Geraden entsprechen einem Exponenten γ des differentiellen Energiespektrums von 1,95. Der Exponent scheint mit wachsender Tiefe in der Atmosphäre abzunehmen. Eine bessere Anpassung an die Meßpunkte würde man bei 1030 mb mit $n=2,2$ und bei 310 mb mit $n=1,3$ erhalten. Hierzu wären noch genauere Messungen bei höherer Energien zu wünschen. Solche Messungen sind aber recht schwierig, da eine Unterscheidung zwischen μ-Mesonen und Protonen auf Grund einer Massenbestimmung kaum mehr möglich ist, sobald die Energie einige GeV übersteigt und auch die Werte für die Energie unsicher werden. Aus diesen Gründen fehlen direkte Angaben bei höheren Energien.

Bei sehr hohen Energien besitzt man aber noch eine Angabe über den Fluß, die von KAPLON und Mitarbeiter [Ka 53] mit Hilfe einer ganz anderen Methode gewonnen wurde. Diese Autoren benutzten zwei „Emulsionskammern", die beide ungefähr gleich aufgebaut waren; davon exponierten sie die eine in der Stratosphäre, die andere in rund 3500 m Höhe. Daraufhin untersuchten sie beide Kammern auf sehr energiereiche Schauer mit einer Energie $> 10^{12}$ eV. Durch Vergleich der Intensitäten in beiden Höhen konnten sie nach einer entsprechenden Gross-Transformation Werte für den integralen Fluß in den beiden Höhen angeben. Demnach ist der integrale Fluß von Teilchen mit einer Energie $> 10^{12}$ eV:

$(0{,}22 \pm 0{,}075)/\text{m}^2$ sec sterad   am Gipfel der Atmosphäre

$(0{,}0009 \pm 0{,}0004)/\text{m}^2$ sec sterad   in 3500 m Höhe.

Dies entspricht einer Absorptionslänge in Luft von (129 ± 15) g/cm². Die Verfasser geben als untere Grenze 115 g/cm² an, falls man den Übergangseffekt in der Atmosphäre mitberücksichtigt. Der Wert am Gipfel der Atmosphäre ist in guter Übereinstimmung mit einer Formel für das integrale Energiespektrum der Primärstrahlung, das man aus der Beobachtung der großen Luftschauer gewinnt.

Man nimmt allgemein an [Ro 52b], daß sich das Energiespektrum in jeder Höhe für größere Energie als Potenzspektrum mit nur langsam veränderlichem Exponenten darstellen läßt, und kann unter dieser Voraussetzung die Messungen von KAPLON und Mitarbeiter dazu verwenden, einen mittleren Wert für den Exponenten γ des differentiellen Energiespektrums zwischen 10^{10} eV und 10^{12} eV zu gewinnen. Der Fluß am Gipfel der Atmosphäre liefert zusammen mit der in Fig. 19 eingezeichneten Kurve für 0 mb ein $n = 2{,}6$. Dies ist in guter Übereinstimmung mit dem von GREISEN [Gr 56b] gegebenen Wert für große Luftschauer. Für Gebirgshöhe erhält man einen Exponenten $n = 2{,}2$, wenn man von den Meßwerten der Fig. 19 ausgeht. Dies paßt ebenfalls gut zu den in Fig. 19 eingezeichneten Geraden, die ein Potenzspektrum mit $n = 1{,}95$ darstellen, da man erwartet, daß n mit wachsender Energie zunimmt.

Auf der Seite der kleinen Energien reichen die meisten Messungen des Spektrums wenig unter 100 MeV herab. Die meßtechnischen Schwierigkeiten liegen hier an der geringen Reichweite der Protonen, die eine Massenbestimmung in den in Ziff. 11 besprochenen Apparaturen schwierig macht. Aus Emulsionsexperimenten ist jedoch das Erzeugungsspektrum energiearmer Protonen bis zur Energien von der Größenordnung einiger MeV herab bekannt, aus dem sich das Energiespektrum der Protonen in der Atmosphäre bis zu sehr kleinen Energien im Prinzip berechnen läßt (s. Ziff. 13). Man sieht aus Fig. 19, daß sowohl in Meeres- als auch in Gebirgshöhe das differentielle Energiespektrum nach kleinen Energien hin wieder abnimmt. Das Maximum der Intensität liegt bei 100 MeV. Die Form des Spektrums am unteren Ende ist im wesentlichen auf den Energieverlust zurückzuführen, den die Protonen bei kleiner Energie in der Atmosphäre durch Ionisation erleiden.

Alle behandelten Messungen beziehen sich auf die vertikale Intensität der Strahlung. Die Strahlung, die einen Detektor insgesamt aus allen Richtungen trifft, läßt sich aus der Winkelverteilung berechnen (s. Ziff. 15). MILLER und Mitarbeiter [Mi 54a] haben das Impulsspektrum zwischen 0,4 GeV/c und 1 GeV/c auch für Einfallwinkel von 45° gegen die Vertikale gemessen, um Effekte einer möglichen Ost-West-Asymmetrie zu untersuchen.

BARADZEI und Mitarbeiter [Ba 59b] haben Messungen des Impulsspektrums der harten Komponente in 9000 m Höhe veröffentlicht. Der Anteil der Protonen in dieser Höhe ist $50 \pm 10\%$. Für das differentielle Impulsspektrum der Protonen finden sie zwischen 2 und 6 GeV/c ein Potenzgesetz mit dem Exponenten $-2{,}8 \pm 0{,}5$. Die totale Intensität der Protonen ist $(2{,}5 \pm 0{,}13) \times 10^{-2}/\text{cm}^2$ sec sterad. In Gebirgshöhe (3200 m) wurden neue Messungen von KOCHARIAN und Mitarbeitern [Ko 59] durchgeführt. Sie maßen das Energiespektrum der Protonen bis zu sehr hohen Energien (zwischen 3 und 100 GeV). Die experimentelle Schwierigkeit besteht in der Trennung der Protonen von dem Untergrund der μ-Mesonen und π-Mesonen. Die Trennung der Protonen und π-Mesonen von den μ-Mesonen erfolgte durch die Beobachtung der Kernwechselwirkung der Teilchen. Das Verhältnis Protonen/μ-Mesonen beträgt 2,4% bei einer Energie von 10 GeV, 2,7% bei 32 GeV und 4,3% bei einer Energie von 6 GeV. Die Bestimmung des Anteils der π-Mesonen unter den Teilchen mit starker Kernwechselwirkung erfolgte durch Beobachtung der Bahnkrümmung der Teilchen im Magnetfeld der Nebelkammer. Da man annehmen kann, daß die Zahl der negativen und der positiven π-Mesonen ungefähr gleich groß ist, läßt sich die Gesamtzahl der geladenen π-Mesonen aus der Zahl der negativen Teilchen berechnen. Dieses Verfahren läßt sich für Energien < 33 GeV durchführen. Für größere Energien kann die Krümmung nicht mehr beobachtet werden. Die Zahl der π-Mesonen in diesem Energiegebiet wurde rechnerisch ermittelt. Dies sollte zu keinen nennenswerten Fehlern Anlaß geben, da die Zahl der π-Mesonen verglichen mit den Protonen nicht sehr groß ist. Bei einer mittleren Energie von 5 GeV ist das Verhältnis der π-Mesonen/Protonen nach ihren Messungen ungefähr 25%. Messungen des Protonenspektrums sind in Fig. 5, 6, 19, 52 und Tabelle 8 dargestellt. Zwischen 3—100 GeV läßt sich das Energiespektrum der Protonen darstellen in der Form:

$$N(E)\, dE = 3{,}2 \times 10^{-3}(2+E)^{-2,8}\, dE \quad \text{pro } \text{cm}^2 \text{ sec sterad GeV} \tag{12.2}$$

E bedeutet die kinetische Energie in GeV.

Messungen in Meereshöhe wurden von MESHKOSKI und SOKOLOV [Me 58a] in dem Impulsintervall zwischen 0,30 und 0,55 GeV/c veröffentlicht. Fig. 21 in [Me 58] zeigt das Ergebnis dieser Messungen zusammen mit Messungen anderer Autoren [Me 56, My 51, Po 56]. Sie zeigen eine befriedigende Übereinstimmung untereinander, liegen jedoch höher als die Messungen von FILTHUTH [Fi 55]. Das Maximum des Impulsspektrums liegt bei ungefähr 0,6 GeV/c.

13. Berechnung des Energiespektrums bei kleineren Energien. Ältere Rechnungen zum Energiespektrum der Protonen stammen von BAGGE [Ba 39, 43]. ROSSI [Ro 52b] hat ausführlichere Rechnungen publiziert, die die Rolle des Ionisationsverlustes für die Form des Spektrums deutlich zeigen. ROSSI geht davon aus, daß alle Neutronen und fast alle Protonen in der Atmosphäre, vor allem bei kleineren Energien, sekundären Ursprungs sind, d.h. von Kernreaktionen stammen, die in einiger Entfernung vom Beobachtungspunkt stattgefunden haben. Falls man das Energiespektrum für die Erzeugung kennt, kann man hieraus das Spektrum am Beobachtungsort berechnen. Das differentielle Erzeugungsspektrum der Neutronen bzw. Protonen sei gegeben durch

$$q^n(E, x) \quad \text{bzw.} \quad q^p(E, x).$$

$q^n(E, x)\, dE\, dx$ ist z.B. die Zahl von Neutronen, die mit einer Energie zwischen E und $E\, dE$ in einer Luftschicht der Dicke dx in der Tiefe x pro cm², sec, sterad erzeugt werden. Die differentielle vertikale Intensität sei gegeben durch

$$\Phi^n(E, x)\, dE \quad \text{bzw.} \quad \Phi^p(E, x)\, dE;$$

dies ist die Zahl von Teilchen, die pro cm², sec, sterad auftreffen und die eine Energie zwischen E und $E + dE$ besitzen. Zwischen $\Phi^n(E, x)$ und $q^n(E, x)$ gilt

Ziff. 13.   Berechnung des Energiespektrums bei kleineren Energien.   417

folgende Beziehung:
$$\Phi^n(E, x) = \int_0^x q^n(E, x') \cdot \exp[-(x - x')/L_c] \, dx'. \tag{13.1}$$

Hierbei ist L_c die mittlere frei Weglänge für Stöße in der Atmosphäre. L_c ändert sich nur langsam als Funktion der Energie; wir nehmen sie im folgenden als konstant an.

Für Protonen muß der Einfluß des Energieverlustes durch Ionisation berücksichtigt werden. Mit $k(E)$ wird der differentielle Energieverlust bezeichnet:
$$k(E) = \frac{dE}{dR}, \tag{13.2}$$

wo R die Reichweite des Teilchens ist. Wir betrachten nun ein Volumenelement an der Stelle x', in dem Protonen der Energie E' erzeugt werden sollen. Ihre Zahl ist gegeben durch die Funktion $q^p(E', x') \, dE'$. An der Stelle x besitzen sie die Energie E, wobei zwischen E und E' folgende Beziehung besteht:
$$R(E) = R(E') + (x - x'). \tag{13.3}$$

$R(E)$ ist die Reichweite eines Teilchens mit der Energie E. Der vom Volumenelement an der Stelle x' herrührende Fluß an der Stelle x ist also
$$\Delta \Phi^p(E, x) \, dE = q^p(E', x') \, dE'. \tag{13.4}$$

Wir drücken dE und dE' noch vermöge Gl. (13.2) aus und erhalten:
$$k(E) \Delta \Phi^p(E, x) = k(E') q^p(E', x'). \tag{13.5}$$

Um den gesamten Fluß zu erhalten, muß man wieder über die ganze Erzeugungsschicht integrieren unter Berücksichtigung der Wahrscheinlichkeit, daß das Proton durch einen Kerntreffer ausscheidet. Man erhält somit
$$k(E) \Phi^p(E, x) = \int_0^x k(E') q(E', x') \cdot e^{-\frac{x-x'}{L_c}} dx'. \tag{13.6}$$

Zur Vereinfachung kann man näherungsweise den Ansatz machen:
$$q^p(E', x') = q^p(E', x) e^{-\frac{x-x'}{L}}, \tag{13.7}$$

wo L die mittlere freie Weglänge für Absorption ist. Man erhält dann:
$$k(E) \Phi^p(E, x) = \int_0^x k(E') q^p(E', x) e^{-(x-x')\left(\frac{1}{L_c} - \frac{1}{L}\right)} dx'. \tag{13.8}$$

Entferntere Luftschichten tragen nur wenig zu dem Integral Gl. (13.8) bei, da das Energiespektrum $q^p(E')$ mit wachsender Energie rasch abnimmt. Zur Durchrechnung eines Beispiels setzte Rossi deshalb in Gl. (13.8) den Exponenten der Exponentialfunktion ≈ 1. Für das Produktionsspektrum benutzte er die von Camerini [Ca 51] aus Sternen in Kernspuremulsion in der Stratosphäre bestimmten Daten, die es in der folgenden Form darstellen lassen:
$$q^p(E, x) = \frac{A \, e^{-x/L}}{(50 + E)^2}, \tag{13.9}$$

wo E in MeV gemessen wird. Man nimmt hierbei an, daß diese Form des Spektrums sich nicht ändert, wenn die Sterne in der Atmosphäre statt in der Emulsion entstehen. Rossi erhielt damit für die Protonen ein Gesetz von der Form
$$\Phi^p(E, x) = \frac{A \, e^{-x/L}}{k(E)} \cdot \frac{E_m - E}{(50 + E) \cdot (50 + E_m)}; \tag{13.10}$$

hierbei ist E_m gegeben durch
$$R(E_m) = R(E) + x.\qquad(13.11)$$
Für die Neutronen erhielt er unter Annahme derselben Form des Produktionsspektrums
$$\Phi^n(E, x) = \frac{A \cdot x \cdot e^{-x/L}}{(50+E)^2}.\qquad(13.12)$$
Für hohe Energien nähern sich die beiden Funktionen $\Phi^n(E, x)$ und $\Phi^p(E, x)$ einander an. Das differentielle Energiespektrum $\Phi^p(E, x)$, Gl. (13.10), hat ein Maximum bei 100 MeV und fällt nach kleineren Energien hin rasch ab in qualitativer, aber nicht genauer Übereinstimmung mit den Messungen Fig. 19 (s. hierzu [Fi 55]).

14. Die Gesamtintensität der Protonen. Die Gesamtintensität der Protonen läßt sich im Prinzip durch Integration der in Fig. 19 gegebenen Energieverteilung und durch Interpolation unter Annahme einer exponentiellen Absorption für jede Tiefe in der Atmosphäre gewinnen. Gewöhnlich finden sich in den Arbeiten über das Energiespektrum der Protonen auch Werte für die Intensität in einem größeren Energieintervall oder für die Gesamtintensität. Wegen der in Ziff. 13 besprochenen Schwächung der Protonenkomponente durch Ionisation ist der Beitrag zur Gesamtintensität bei kleinen Energien unter 50 bis 100 MeV nicht groß, so daß die Messungen einen guten Wert für die Gesamtzahl der Protonen geben dürften, auch wenn sie nur bis zu der oben angegebenen Energie herabreichen. In Tabelle 9 geben wir eine Zusammenstellung von Angaben über Vertikalintensitäten der Protonen in verschiedenen Höhen in der Atmosphäre.

Tabelle 9. *Vertikalintensität der Protonenkomponente.*

Autor	Energieintervall (MeV)	Impulsintervall (GeV/c)	Vertikalintensität (Teilchen/cm² sec sterad)	Höhe
Ko 55	80— 430	—	$(0{,}806 \pm 0{,}012) \times 10^{-3}$	3200 m
	430—1220	—	$(0{,}606 \pm 0{,}013) \times 10^{-3}$	(700 mb)
	>1220	—	$0{,}281 \times 10^{-3}$	
	>80	—	$(1{,}69 \pm 0{,}02) \times 10^{-3}$	
Wh 52	50—370	—	$0{,}96 \times 10^{-3}$	3400 m
				(680 mb)
Ha 49a	>10000	—	$1{,}1 \times 10^{-5}$	2800 m
				(730 mb)
Fi 55	18—350	—	$0{,}56 \times 10^{-4}$	Meereshöhe
My 51	>500	—	$0{,}55 \times 10^{-4}$	Meereshöhe
Yo 52	150—350	—	$(1{,}3 \pm 0{,}1) \times 10^{-4}$	Meereshöhe
Me 58	>50	—	$(1{,}88 \pm 0{,}05) \times 10^{-4}$	Meereshöhe
Me 58	—	0,37—1,04	$(12{,}15 \pm 0{,}30)\, 10^{-4}$	3250 m
Ro 54	—	0,59—0,77	$(9{,}6 \pm 1{,}2)\, 10^{-5}$	2700 m
		0,77—0,93	$(7{,}9 \pm 1{,}2)\, 10^{-5}$	
Me 58	—	0,37—1,04	$(1{,}12 \pm 0{,}03)\, 10^{-4}$	Meereshöhe
		$I(>1)$	$(1{,}88 \pm 0{,}05)\, 10^{-4}$	
Ro 54	—	0,59—0,77	$(9{,}1 \pm 1{,}9)\, 10^{-5}$	Meereshöhe
		0,77—0,93	$(12{,}1 \pm 2{,}3)\, 10^{-5}$	

Für Protonen einer Energie > 80 MeV erhält man in 3200 m Höhe aus den Messungen von [Ko 55] einen Wert von $1{,}69 \cdot 10^{-3}$ Teilchen/cm² sec sterad. Um den entsprechenden Wert in Meereshöhe zu erhalten, kann man die Angabe [Fi 55] und [My 51] und Fig. 19 benutzen. Man erhält für die Intensität bei einer Energie > 80 MeV einen Wert von $1{,}2 \cdot 10^{-4}$/cm² sec sterad. Der Vergleich der beiden Intensitätswerte bei 700 mb und 1030 mb führt auf eine Absorptionslänge von 120 g/cm², unter der Annahme einer exponentiellen Absorption.

Zusätzliche Angaben über die Gesamtintensität erhält man aus einigen weiteren Untersuchungen über die Häufigkeit der Protonen im Verhältnis zur gesamten durchdringenden Komponente.

Eine Übersicht über solche Messungen in verschiedenen Höhen gibt Tabelle 10. p bedeutet hierbei den relativen Anteil der Protonen an der gesamten durchdringenden Komponente, deren Intensität ROSSI [Ro 48] oder neueren Arbeiten entnommen

Tabelle 10. *Anteil p der Protonen an der gesamten durchdringenden Komponente.*
Geordnet nach Meßhöhe; geomagnetische Lage unberücksichtigt.

Autor	Energie (MeV)	Impuls p (GeV/c)	Myonexcess $K_\mu = \dfrac{N_{\mu+}}{N_{\mu-}}$	p Protonanteil (%)	Meßhöhe [m]
Al 64[1]	—	$0{,}5 \leq p \leq 50$	1,25	16 ± 3	2960
		$0{,}5 \leq p \leq 20$	1,25	$19{,}7 \pm 3$	
		$0{,}2 \leq p \leq 20$	1,25	$20{,}8 \pm 3$	
Gr 51a, b	> 400	—		7,9	3100
Ko 56	—	$0{,}4 \leq p \leq 14$	$f(p): 1—1{,}4$	12	3200
Va 57	—	$0{,}34 \leq p \leq 2{,}5$	1,25	14 ± 1	3250
Al 57	—	$9{,}3 \leq p \leq 17$	1,25	4,8	3250
Me 56a, b	—	$0{,}36 \leq p \leq 1{,}0$	$f(p)$	vgl. Tabellen	2770/3250
Wh 52	—	$\geq 0{,}3$	1,35	19 ± 2	3400
Mi 54	—	$0{,}7 \leq p \leq 2{,}0$	1,17	20 ± 2	3400
Lo 54	> 300	—		12	3500
Be 55	—	$> 0{,}31$		11 ± 1	3650
Al 65[1]	—	$0{,}5 \leq p \leq 50$	1,25	24 ± 3	5200
		$0{,}2 \leq p \leq 20$	1,25	?	
Ba 59	—	$1 \leq p \leq 5$	1,2—1,3	50 ± 10	9000

werden kann. Mit diesen Werten erhält man eine etwas größere Protonenintensität als die weiter oben angegebene. Dies mag an der Schwierigkeit liegen, die Protonen zuverlässig zu identifizieren. Auch zeigen die Messungen untereinander zum Teil keine Übereinstimmung. In Meereshöhe erhält CHOU [Ch 56] $p = (1{,}5 \pm 0{,}8)\%$ für Protonen von mehr als 280 MeV. Zusammen mit einer Intensität der harten Komponente in Meereshöhe von $0{,}83 \cdot 10^{-2}/\text{cm}^2$ sec sterad [Ro 48] erhält man für die Protonen hieraus $(1{,}3 \pm 0{,}6) \cdot 10^{-4}/\text{cm}^2$ sec sterad. Innerhalb der Fehlergrenzen ist dies in Übereinstimmung mit den weiter oben gemachten Angaben.

In der hohen Atmosphäre existieren Messungen über die Gesamtintensität der harten Komponente als Funktion der Höhe, die wenigstens einen größenordnungsmäßigen Aufschluß über die Häufigkeit der Protonen in sehr großer Höhe geben. Solche Untersuchungen wurden unter anderen durchgeführt von CLARK [Cl 52], VIDALE [Vi 52], RAO und Mitarbeiter [Ra 53], sowie von PULLAR und DYMOND [Pu 53].

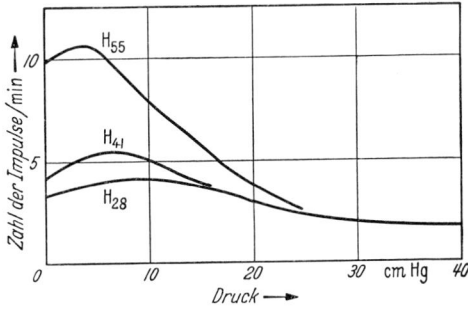

Fig. 20. Höhenabhängigkeit der harten Komponente, gemessen unter 12 cm Pb in einer geomagnetischen Breite $\Lambda = 55°$ (H_{55}), $\Lambda = 41°$ (H_{41}), und $\Lambda = 28°$ (H_{28}) nach [Vi 52].

Die registrierten Teilchen sind hauptsächlich μ-Mesonen und Protonen. Im folgenden sind einige Messungen über die Intensität der harten Komponente als

---
[1] Dr. ALLKOFER hat freundlicherweise den in [Al 65] nicht erwähnten Wert für p für $0{,}5 \leq p \leq 50$ GeV/c brieflich übermittelt; ebenso für [Al 64] (s. auch Fortschr. d. Physik: im Druck).

Funktion der Tiefe in der Atmosphäre bei verschiedenen geomagnetischen Breiten Λ zusammengestellt. Fig. 20 gibt Messungen von [Vi 52] bei $\Lambda = 55$, 41 und 28° unter 12 cm Pb wieder; diese Messungen werden ergänzt durch [Ra 53] bei $\Lambda = 19$ und 3° unter 10 cm Pb (Fig. 21). In niedrigen geomagnetischen Breiten findet man einen deutlichen Übergangseffekt der harten Komponente; dieser ist auf die im Vergleich zu polnahen Gebieten größere Mesonenproduktion zurückzuführen.

Eine Trennung von Protonen und μ-Mesonen in der harten Komponente kann auf Grund der Kernwechselwirkung der Protonen geschehen. Fig. 22 zeigt

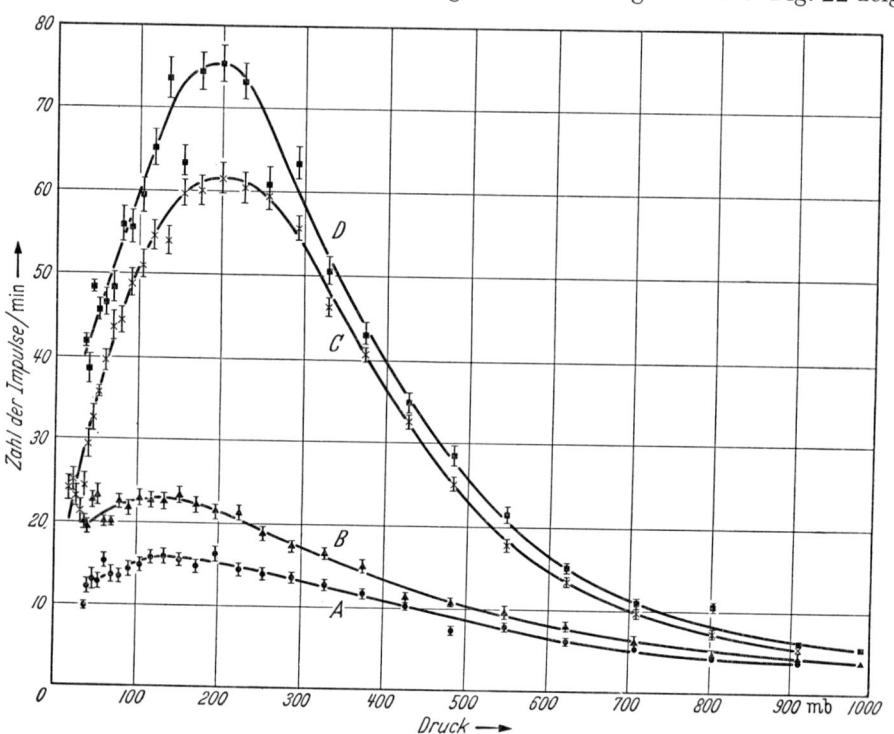

Fig. 21. Höhenabhängigkeit der harten Komponente, gemessen unter 10 cm Pb nach [Ra 53]. A geomagnetische Breite $\Lambda = 3°$ N. B $\Lambda = 19°$ N. Totale Vertikalintensität: C $\Lambda = 3°$ N, D $\Lambda = 19°$ N.

eine von Clark [Cl 52] verwendete Apparatur, mit der er unter 55° geomagnetischer Breite Messungen durchgeführt hat. Die Zählrohr-Koinzidenz AB wählt Teilchen aus, die 20 cm Pb durchdringen können. Wenn die Zählrohrlage C in Koinzidenz mit AB geschaltet ist, wird man vorzugsweise Teilchen messen, die in dem Pb-Absorber durch Wechselwirkung Sekundäre erzeugen und damit die Koinzidenz C auslösen. Um μ-Mesonen auszusondern, wird C in Antikoinzidenz geschaltet. Die gemessenen Zählraten $AB-C$ als Funktion der Tiefe in der Atmosphäre zeigten sich in guter Übereinstimmung mit theoretischen Abschätzungen über das zu erwartende Verhältnis Protonen/μ-Mesonen. Man kann hieraus schließen, daß die Koinzidenzen $AB+C$ hauptsächlich von Protonen ausgelöst sind. Der Verfasser schätzt, daß wegen der Art des Nachweises die Energie der registrierten Protonen >1 GeV ist. Er kann damit die Intensität solcher Protonen in der Atmosphäre zwischen 16 und 400 mb Höhe angeben. Sie zeigt eine exponentielle Abnahme mit wachsender Tiefe in der Atmosphäre mit einer Absorptionslänge $L = 110$ g/cm². Es ist kein Übergangseffekt zu erkennen. Durch Extrapolation auf den Gipfel der Atmosphäre erhält man einen primären Fluß von 0,19 Protonen/cm² sec sterad mit einer Energie >1 GeV. Ähnliche Messungen in sehr großer Höhe wurden von PULLAR und DYMOND [Pu 53] durchgeführt unter 55° geomagnetischer Breite. Sie finden für die Primärteilchen von durchdringenden Schauern ebenfalls eine exponentielle Absorption in der Atmosphäre mit einer Absorptionsschicht von 120 g/cm² Luft. Zusammenfassend wurden in Fig. 23 die Vertikal-Intensität der Protonen mit

Tabelle 11. *Vertikalintensität der harten Komponente am Gipfel der Atmosphäre.*

Autor	Vertikalintensität je cm² sec sterad ($\Lambda = 55°$)
Mi 55	0,21
Cl 52	0,19
Da 56b	0,19
Da 53	0,21
Vi 52	∼0,25
Pu 53	0,16

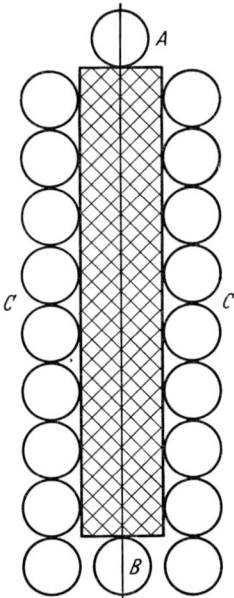

Fig. 22. Die Zählrohranordnung von CLARK [Cl 52]. Die Größe des Bleiabsorbers beträgt 20 cm × 3,5 cm × 15 cm.

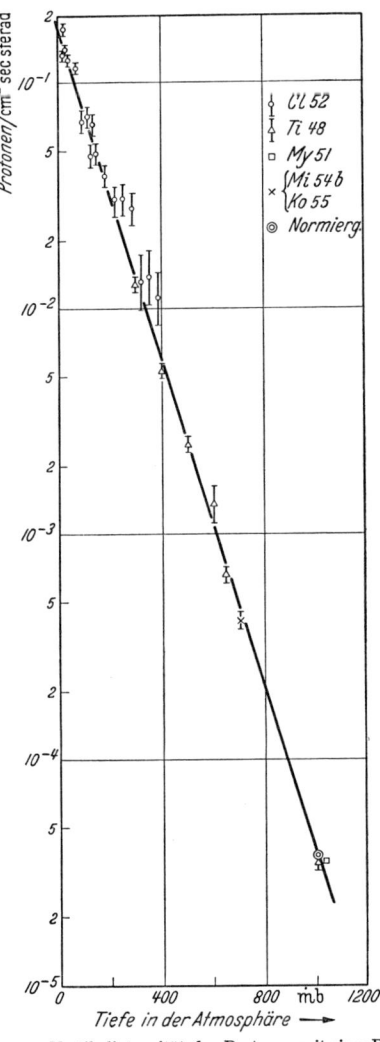

Fig. 23. Vertikalintensität der Protonen mit einer Energie >1 GeV in der Atmosphäre.

einer Energie >1 GeV in der Atmosphäre aufgetragen. Zwischen 16 und 400 mb wurden die Punkte von [Cl 52] benutzt. Als Wert am Gipfel der Atmosphäre haben wir 0,18 Protonen/cm² sec sterad angenommen. Dies ist verträglich mit den Messungen von CLARK selbst (0,19/cm² sec sterad) sowie mit dem von PETERS [Pe 52b] in einer zusammenfassenden Arbeit angegebenen Wert von 0,2/cm² sec sterad. Hiermit sind weiterhin neuere Messungen der primären Protonenintensität unter 55° geomagnetischer Breite zu vergleichen (Abschneideenergie etwa 800 MeV) (s. Tabelle 11). Bei 1030 und 700 mb haben wir die in

Tabelle 9 und Fig. 19 zusammengestellten Werte benutzt; sie stützen sich im wesentlichen auf die Arbeiten [My 51], [Mi 54b] und [Ko 55]. Die Verbindung zwischen den Messungen auf Gebirgshöhe und denen von CLARK kann man durch Messungen von TINLOT [Ti 48] erhalten. Dieser hat in einem Flugzeug mit einer Zählrohrkoinzidenzapparatur die Häufigkeit durchdringender Schauer registriert. Zwischen 300 und 1000 mb findet er eine exponentielle Abnahme mit einer Absorptionslänge von (118 ± 2) g/cm². Seine Meßpunkte in Gebirgs- und Meereshöhe wurden auf die oben erwähnten neueren Intensitätsmessungen normiert. Sie schließen sich dann nach oben hin gut an die von CLARK an. Alle Messungen lassen sich durch eine exponentielle Abnahme der Intensität mit einer einheitlichen Absorptionslänge von 118 g/cm² darstellen.

Die Gesamtintensität der Nukleonenkomponente in der Atmosphäre zeigt zeitliche Schwankungen. Sie sind in größeren Tiefen der Atmosphäre gering. SIMPSON und Mitarbeiter [Si 53] haben eine Standardanordnung aus Zählrohren und Absorbern angegeben („Simpson-pile"), mit der einheitlich an vielen Meßstellen auf der Erde die Intensität der Nukleonenkomponente überwacht wird. Auf die Ergebnisse dieser Messungen der zeitlichen Schwankung der Nukleonenkomponente soll hier nicht weiter eingegangen werden. Die Zählrohranordnung mißt im wesentlichen die Intensität der sternerzeugenden Komponente. Eine genauere Diskussion der Wirkungsweise und des von der Apparatur hauptsächlich erfaßten Energiebereiches findet sich in der eingangs erwähnten Arbeit [Si 53].

Die Gesamtintensität der Nukleonenkomponente hängt außerdem von der geomagnetischen Breite ab. Das Gebiet der Minimumintensität definiert eine Linie, die ungefähr mit dem geomagnetischen Äquator zusammenfällt, wenn man ihn um 40 bis 45° nach Westen dreht. Mit zunehmender geomagnetischer Breite steigt die Intensität zunächst an und bleibt von einer bestimmten geomagnetischen Breite Λ an konstant („Knie"). Wenn die Messungen mit dem „Simpson-pile" und in Meereshöhe vorgenommen werden, ist dieses Gebiet bei $\Lambda = 55$ bis 60°. Als Breiteneffekt wird das Verhältnis: Mittlere Intensität oberhalb des Knies/ Minimumintensität bezeichnet. Messungen des Breiteneffekt in Meereshöhe wurden von ROSE und Mitarbeitern [Ro 56] und von SKORKA [Sk 58] durchgeführt mit Hilfe eines etwas modifizierten „Simpson-piles". Die Ergebnisse der beiden Arbeiten zeigten befriedigende Übereinstimmung. ROSE und Mitarbeiter fanden einen Breiteneffekt von 1,77; SKORKA fand 1,78 im Atlantischen Ozean und 1,99 im Indischen Ozean (vgl. Anhang 2 und Ziff. 23).

15. Die Winkelverteilung der Protonen. Die Winkelverteilung der Protonen in der Atmosphäre ist mit verschiedenen Methoden bestimmt worden. Eine erste Gruppe von Autoren benutzte Kernemulsionen für die Untersuchungen der Winkelverteilung von geladenen sternauslösenden Teilchen [Br 49, Lo 50, Lo 51, Mo 52a, Ba 52b, Lo 55a]. Die Auswahl dieser Teilchen geschieht nach den Regeln, die in Ziff. 4 angegeben sind. Durch die damit hereingebrachte Willkür dürften jedoch wesentliche Fehler nicht entstehen. Die sternauslösenden Teilchen bestehen hauptsächlich aus Protonen und π-Mesonen. Nach Messungen der Bristolgruppe [Ca 51] ist der Anteil der π-Mesonen klein; dies gilt wenigstens für eine Höhe von 20 km, in der diese Messungen gemacht wurden, und für verhältnismäßig kleine Energien. Aber auch bei höheren Energien dürfte ein eventueller Fehler durch einen Beitrag der π-Mesonen wenig ins Gewicht fallen, da diese in der Regel unter sehr kleinen Winkeln zum auslösenden Teilchen emittiert werden und damit die Winkelverteilung der auslösenden Protonen und der π-Mesonen ähnlich sind. Die Kriterien für die Auswahl von „sternerzeugenden

Teilchen" schreiben vor, daß die Korndichte der Teilchenspuren kleiner sein muß als ein vorgegebener Wert (1,5 bis 1,7fache Minimum-Korndichte). Dies bedeutet, daß die ausgewählten Teilchen eine Mindestenergie von etwa 300 MeV haben, wenn sie Protonen sind.

Die Kernspurplatten-Methode, mit deren Hilfe Winkelverteilungen in der Stratosphäre, in Gebirgshöhe und in Meereshöhe angegeben worden sind, läßt zwar eine ziemlich saubere und direkte Bestimmung der Winkelverteilung zu. Sie ist jedoch mühsam. Eine hohe statistische Genauigkeit läßt sich mit dem Nachweis der Teilchen durch Zählrohre und Ionisationskammern erreichen, doch ist die Analyse hier nicht in so direkter Weise möglich. Solche Untersuchungen wurden von BASSI und Mitarbeiter [*Ba 52a*] sowie von CONVERSI und ROTHWELL [*Co 54*] durchgeführt. Die letzteren benutzten eine dünnwandige Ionisationskammer, die von einem Kranz von 36 Zählrohren umgeben war (ausführliche Beschreibung und Diskussion der Apparatur s. [*Co 53*]). MILLER und Mitarbeiter [*Mi 54*] haben die Winkelverteilung der Protonen mit Hilfe einer Nebelkammer mit Magnetfeld untersucht. BULLOCK [*Bu 57*] hat die Winkelverteilung von sternauslösenden Teilchen in einer Hochdruck-Nebelkammer gemessen.

Die hauptsächlichen qualitativen Ergebnisse aller dieser Messungen sind folgende: Mit zunehmender Tiefe in der Atmosphäre wird die Winkelverteilung „steiler", d.h. die Zenitrichtung wird immer mehr bevorzugt, schräg kommende Teilchen werden seltener. Teilchen mit höherer Energie haben eine steilere Winkelverteilung. Beides läßt sich leicht qualitativ einsehen: Es sei $P(\vartheta)\, d\Omega$ die Zahl von Teilchen, die unter dem Winkel ϑ zur Zenitrichtung in dem Raumwinkelelement $d\Omega$ beobachtet werden. Die Zahl von Teilchen im Zenitwinkelintervall von ϑ bis $\vartheta + d\vartheta$ ist dann bei azimutaler Isotropie gegeben durch $2\pi\, P(\vartheta) \sin\vartheta\, d\vartheta$.

Wir gehen nun von der Annahme aus, daß die Teilchen am Gipfel der Atmosphäre isotrop einfallen und exponentiell absorbiert werden. Man erhält dann in der Tiefe x der Atmosphäre

$$P(\vartheta) = e^{-\frac{x}{L}\left(\frac{1}{\cos\vartheta} - 1\right)}. \tag{15.1}$$

Hierbei ist L die Absorptionslänge in der Luft. Die Bevorzugung der Zenitrichtung nach Gl. (15.1) kommt also einfach dadurch zustande, daß schräg laufende Teilchen eine größere Schichtdicke durchdringen müssen. Dies wird für zunehmende Tiefe x in der Atmosphäre immer wichtiger; d.h. die Vertikalrichtung wird immer stärker ausgezeichnet. Hierbei ist allerdings die Voraussetzung gemacht, daß bei der Produktion die sekundären Teilchen die Richtung des Primärteilchens beibehalten. Diese Voraussetzung ist nur für hohe Energien näherungsweise erfüllt. Teilchen kleiner Energie dagegen werden meist größere Winkel mit der Primärrichtung bilden. Die Winkelverteilung wird deshalb vor allem für die energiearmen Teilchen nicht so stark im Zenitrichtung gebündelt erscheinen, wie es Gl. (15.1) angibt, sondern sich mehr der Isotropie annähern.

Wenn die Messungen auch hinsichtlich dieser qualitativen Aussagen dasselbe Bild geben, so stimmen sie doch im einzelnen nicht alle überein.

In der Stratosphäre haben LORD und SCHEIN [*Lo 50*] Messungen mit Kernemulsionen durchgeführt. In 30 km Höhe finden sie eine ungefähr isotrope Winkelverteilung für energiereiche Protonen, die mindestens zwei Schauerteilchen erzeugen. Erst für Zenitwinkel >60° macht sich die Absorption in der Atmosphäre geltend und führt zu einem Abfallen der Intensität. Für energieärmere Teilchen macht sich die azimutale Asymmetrie infolge des Abschneidens durch das erdmagnetische Feld sowie die stärkere Bremsung in der Atmosphäre bemerkbar. So fällt die Intensität unter 54° geomagnetischer Breite bei einem Winkel von rund 47° auf die Hälfte, in einer geomagnetischen Breite von 28° dagegen schon bei einem Winkel von 26°.

Die meisten Untersuchungen wurden jedoch in Gebirgshöhen durchgeführt. Die Winkelverteilungen lassen sich hier in der Regel mit befriedigender Näherung durch ein Gesetz der Form

$$P(\vartheta) = A \cdot \cos^m \vartheta \qquad (15.2)$$

darstellen. Die Messungen in Kernemulsion von BARFORD und DAVIS [Ba 52b] sowie von LOHRMANN [Lo 55b] zeigen gute Übereinstimmung. Die Messungen des letzteren [Lo 55b] wurden zum Teil unter Absorbern durchgeführt, doch wurde die Winkelverteilung der Teilchen wegen der speziellen Form der Absorber dadurch praktisch nicht beeinflußt. Nach [Ba 52b] ist in 3600 m Höhe $m = 5 \pm 1$; die Messung von [Lo 55b] in 3000 m Höhe gibt $m = 6$. In beiden Arbeiten wird eine leichte Zunahme des Exponenten m mit wachsender Energie der Teilchen bzw. mit der Sterngröße festgestellt. Dieselbe Abhängigkeit der Winkelverteilung von der Energie finden auch CONVERSI und ROTHWELL [Co 54a] in 3500 m Höhe mit dem auf S. 411 erwähnten Zählrohr-Hodoskop. Für den Wert von m selbst finden sie jedoch ein abweichendes Ergebnis. Man erhält mit ihrer Apparatur zunächst nur die Winkelverteilung $p(\vartheta')$ der Projektion der Spuren auf eine vertikale Ebene. Für die Umrechnung der projizierten Winkelverteilung $p(\vartheta')$ geben sie unter der Voraussetzung, daß die Intensität nicht vom Azimutwinkel abhängt, folgende Gleichung an (nach R. GATTO, erwähnt bei [Co 54a]):

$$P(\vartheta) = \frac{1}{\pi} \int_{\pi/2}^{\vartheta} \frac{d\vartheta'}{\sqrt{\cos^2 \vartheta - \cos^2 \vartheta'}} \frac{d}{d\vartheta'} \left(|\cos\vartheta'| p(\vartheta') \right). \qquad (15.3)$$

Die resultierende räumliche Winkelverteilung wird wieder durch ein $\cos^m \vartheta$-Gesetz angenähert mit dem Ergebnis $m = 2,1 \pm 0,3$ für eine mittlere Energie von 60 MeV und $m = 2,6 \pm 0,3$ für eine mittlere Energie von 700 MeV. Diese Winkelverteilung nähert sich mehr der Isotropie an als die von [Ba 52b] und [Lo 55b] gefundene. WALKER [Wa 50b] hat ebenfalls mit einer Zählrohrapparatur für die Winkelverteilung von Primären energiereicher Schauer $m = 7,3 \pm 1$ gemessen. Ungefähr in der Mitte zwischen beiden Gruppen von Ergebnissen liegt der Wert $m = 3,2$ von MILLER und Mitarbeiter [Mi 54a], die eine Nebelkammer in 3300 m Höhe benutzten. Eine weitere Nebelkammermessung von BULLOCK [Bu 57] in 2000 m Höhe ergibt $m = 7$, doch sind die statistischen Fehler hierbei groß. Es erscheint nicht ganz klar, woher die Unterschiede der einzelnen Messungen kommen.

In der Nähe des Meeresniveaus (420 m) wurde die Winkelverteilung von sternerzeugenden Teilchen in der Kernemulsion von LOHRMANN [Lo 55b] mit $m = 5,7$ bestimmt.

MIYAKE et al. haben in einer mit Wasserstoff gefüllten [Mi 57a] und in einer mit Stickstoff [Mi 57b] gefüllten Hochdruckhebelkammer die Winkelverteilung von Protonen und von geladenen Teilchen gemessen.

E. Neutronen in der Atmosphäre.

16. Ursprung der Neutronen in der Atmosphäre (und Modulation). Das Vorhandensein von Neutronen in der Erdatmosphäre wurde bald nach der Entdeckung des Neutrons (1932) nachgewiesen. Die Untersuchungen der Neutronenintensität wurde mit sehr verschiedenen experimentellen Techniken und bei verschiedenen Höhen ausgeführt: die ersten Messungen stammen von LOCHER

[*Lo 33, 34, 36, 37*] und RUMBAUGH und LOCHER [*Ru 36*]; gefolgt von FÜNFER [*Fü 37, 38*], SCHOPPER [*Scho 37, 39*], FROMAN und STEARNS [*Fro 38*]; von VON HALBAN, MAGART und KOVARSKY [*Ka 39*], MONTGOMERY [*Mo 39a, b*] und KORFF [*Ko 39a, 39b*]. KORFF [*Ko 39c*] faßte die Resultate dieser ersten Phase zusammen. Die experimentellen Befunde ergaben zwei wichtige Ergebnisse. Die Neutronenintensität zeigte einmal ein Anwachsen mit zunehmender Meßhöhe in der Atmosphäre; zum anderen eine Korrelation mit dem Auftreten von Schauern in der kosmischen Strahlung. Dies legte schon früh die Annahme nahe, daß die kosmische Strahlung für das Auftreten des gemessenen Neutronenflusses verantwortlich ist.

Mit solchen Überlegungen schließt an die erste Meßphase (bis 1939) eine erste kritische, theoretische Sichtung der Neutronenmessungen an. Die Neutrino-Hypothese des β-Zerfalls der Atomkerne (PAULI [*Pa 33*]), besonders aber die Formulierung des β-Zerfalls: $N \to P + e^- + \bar{\nu}$ durch FERMI [*Fe 34*] ließ erwarten, daß das Neutron selbst — im Gegensatz zum Proton — β-instabil sei. Energetisch war dieser Prozeß wegen der größeren Ruhmasse, $M_N > M_P + M_e$, möglich. Die theoretisch damals auf knapp eine Stunde geschätzte Halbwertszeit des Neutrons eröffnete die Möglichkeit, das Neutron hypothetisch als ein Sekundärteilchen der kosmischen Strahlung in der Atmosphäre einzuführen. BETHE, KORFF und PLACZEK [*Be 40*] nehmen für die (sekundären) Neutronen in der Atmosphäre eine Energieverteilung an, ohne spezielle Modelle der Neutronenproduktion zu diskutieren, und zeigen, daß für Neutronen mit Energien <100 keV theoretisch ein Diffusionsgleichgewicht in der Atmosphäre möglich ist, natürlich mit Ausnahme der Randbereiche von je etwa einem Meter Wasseräquivalent[1] am Gipfel der Atmosphäre, wie beim Übergang an der Erdoberfläche. Für die Geschwindigkeitsverteilung $N(v)\,dv$ der Neutronen wurde nach PLACZEK die Verteilung[2] vorgeschlagen:

$$N(v)\,dv = M \cdot q \cdot l(v) \frac{dv}{v^2} e^{-M \int_v^{v_1} \frac{\sigma_c(v')}{\sigma_s(v') + \sigma_c(v')} \cdot \frac{dv'}{v'}} \quad (16.1)$$

die wesentlich (von dem damals noch sehr ungenau bekannten) Einfangquerschnitt $\sigma_c(v)$ und dem Streuquerschnitt $\sigma_s(v)$ der Neutronen an Kernen der Masse M bestimmt wird. q stellt die Quellstärke der Neutronen (Zahl/g sec) dar, die eine geschwindigkeitsabhängige mittlere Weglänge $l(v)$ in Luft durchlaufen. Mit einer von BETHE entworfenen Verteilungsfunktion für die räumliche Verteilung der Neutronen zwischen Produktion und Absorption schätzen BETHE et al. einen mittleren Absorptionsweg $L = (76$ bis $103)$ cm Wasseräquivalent (oder g/cm²) in der Atmosphäre, was zur Folge hat, daß Neutronen in der Atmosphäre nur relativ kleine Wege zurücklegen, um dann überwiegend durch C^{14}-Bildung zu verschwinden (Vorschlag von MONTGOMERY [*Mo 39*]). Weiter erörtern BETHE, KORFF und PLACZEK ein mögliches Maximum der Neutronenintensität, wie es für sekundäre Teilchen zwangsläufig ist. Dessen wahrscheinliche Tiefe sollte zwischen 73 und 100 g/cm² liegen; die experimentellen Daten zeigten damals in jenem Bereich noch ein Ansteigen der Intensität.

---

[1] Normalerweise mißt man Längen der kosmischen Strahlung in g/cm² (atmosphärische Tiefe, Stoßlänge, Absorptionslänge etc.) 1 g/cm² ist dasselbe wie 1 cm Wasseräquivalent, eine oft angewandte Einheit in der kosmischen Strahlung.

[2] Die Neutronendiffusionstheorie in Form der „Alterstheorie der Neutronen", wird bei BETHE, KORFF und PLACZEK in ihren Ergebnissen und Folgerungen erstmals veröffentlicht. Erst in den declassified papers nach dem Krieg wird ein Teil der Originalarbeiten zugänglich gemacht; eine ausführliche Darstellung findet man bei R. E. MARSHAK, „Theory of slowing down of neutrons by elastic collision with atomic nuclei", Rev. Mod. Phys. 19, 185—238 (1947), besonders ab S. 212ff.; dort ist auch eine Darstellung der Grundlagen und Zusammenhänge für obige Arbeit zu finden.

Montgomery und Montgomery [Mo 39b] verglichen die Produktion der Neutronen mit derjenigen der Protonen, und es erschien in der Tat wahrscheinlich, daß diese beiden Teilchen im selben Prozeß erzeugt werden konnten. Bagge [Ba 43] diskutiert im Zusammenhang mit der Emission von schnellen Protonen in Kernzertrümmerungen (Sterne), daß auch schnelle Neutronen mitemittiert werden müssen. Bei großen Energien wird wegen der Symmetrie der Kernkräfte das Spektrum für Protonen und Neutronen etwa das gleiche sein, während die Verteilungen bei kleinen Energien wegen der Coulomb-Barriere der Protonen voneinander abweichen müssen.

Flügge [Fl 43a] behandelt die räumliche Verteilung der Neutronen in der Atmosphäre erstmals quantitativ. Wie Bethe et al. [Be 40] greift er auf die Diffusionsgleichung zurück, die er im Sinne der Alterungstheorie integriert. Weiter beschreibt er die Neutronenproduktion durch primäre Strahlung energetisch mit einem Quellspektrum $N(E)\,dE \sim E\,e^{-\alpha E}\,dE$ und wählt für die Höhenverteilung der Neutronenproduktion durch Primäre ein exponentielles Abklingen der Form $e^{-Z/L}$. Besonders für die Wahl der Kerntemperatur ($\alpha = 0{,}35$ MeV^{-1}) ist Flügge's Quellspektrum in befriedigender Übereinstimmung mit den heutigen Ansätzen für Verdampfungsneutronen; auch der Absorptionskoeffizient (d.h. eine Absorptionslänge von etwa $L = 145$ g/cm^2) war günstig gewählt. Die von Flügge errechnete theoretische Verteilung der langsamen Neutronen ließ ein Maximum bei ca. 150 g/cm^2 atmosphärischer Tiefe erwarten. Die Alterungstheorie erlaubte kein beliebiges Verschieben des Maximums nach oben, sondern setzte eine obere Grenze zwischen 110 bis 120 g/cm^2 atmosphärischer Tiefe, wie es später durch Yuan [Yu 48, 49] dann auch gefunden wurde.

Erst die Messungen des β-Zerfallspektrums an Reaktorneutronen ab 1948 durch Snell et al. [Sn 48, 50, 51] und Robson [Ro 51] bewiesen die β-Instabilität des freien Neutrons, und damit wurde diese erste spekulative Phase abgeschlossen. Inzwischen ist die Halbwertszeit des Neutrons in Präzisionsmessungen zu $\tau_{\frac{1}{2}} = 11{,}7 \pm 0{,}3$ min bestimmt worden, und zwar übereinstimmend in den USA im Argonne National Lab. [Bu 58], in Canada (Chalk River Laboratories of Atomic Energy of Canada) durch [Ro 55] und [Cl 58b] und schließlich in der Academy of Science der USSR durch [So 59] und [Tre 59]. Seit 1948 hat man daher geschlossen, daß in der primären kosmischen Strahlung galaktischen Ursprungs keine Neutronen (es seien denn verschwindend wenige mit extrem hohen Energien) vorhanden sein können. Primäre solare Neutronen [Bi 51] sind bei einer mittleren Lebensdauer von ca. 17 min jedoch durchaus denkbar. Bei Vorhandensein eines solaren Anteils primärer Neutronen muß auf Grund der elektrischen Neutralität ein halbtägiger Gang [Sw 52] erwartet werden. Fonger et al. [Fo 53] und Haymes [Ha 59] bzw. [Ha 64b] fanden inzwischen bei der Suche nach einem solchen Tagesgang keinen meßbaren Neutronenfluß von der Sonne. Man zog daraus die Konsequenz, daß *alle Neutronen in der Erdatmosphäre* (auch die „Albedo-Neutronen", die man besser als die Verlustflußneutronen in die Stratosphäre bzw. in den interplanetarischen Raum bezeichnet) *Sekundärprodukte der kosmischen Strahlung sind*. Ein möglicher Beitrag schneller und energiereicher solarer Neutronen wurde von Simpson [Si 63] vorgeschlagen; mit Hilfe von Messungen an Rückstoßprotonen in Kernemulsion — ausgelöst durch Neutronen von etwa 5—160 MeV — versuchen Apparao, Daniel, Vijayalakshmi und Bhatt [Ap 66] auf einen möglichen solaren Neutronenstrom zu schließen (s. auch [He 63]). Energiereiche solare Neutronen behandeln Lingenfelter, Flamm, Canfield und Kellman [Li 65] theoretisch.

Diese Hypothese wurde früh durch weitere Messungen der Neutronenintensität bekräftigt. Die Neutronen zeigen einmal — genau wie geladene Teilchen — einen

Breiteneffekt, d.h. eine starke Abhängigkeit von der geomagnetischen Breite. Für elektrisch neutrale Teilchen finden geomagnetische (bzw. magnetosphärische) Effekte nur dann eine einfache Erklärung, wenn die Produktion der Neutronen über geladene (primäre bzw. sekundäre) Teilchen (Protonen, α-Teilchen und schwerere Kerne oder Pionen) erfolgt. Sinngemäß spiegelt sich dann eine Abschneideenergie etwa in Form der magnetischen Steifigkeit für geladene Teilchen im Spektrum der Neutronen indirekt wieder. Ferner nimmt die Neutronen-Intensität mit der Höhe der Atmosphäre zu, erreicht etwa in 16 km über Meereshöhe (110 g/cm²) das schon erwähnte, von YUAN [$Yu\ 18$] erstmalig nachgewiesene Maximum, um danach zum Gipfel der Atmosphäre (40—50 km Höhe) wieder abzunehmen, und zwar auf die Werte des Neutronenverlustflusses

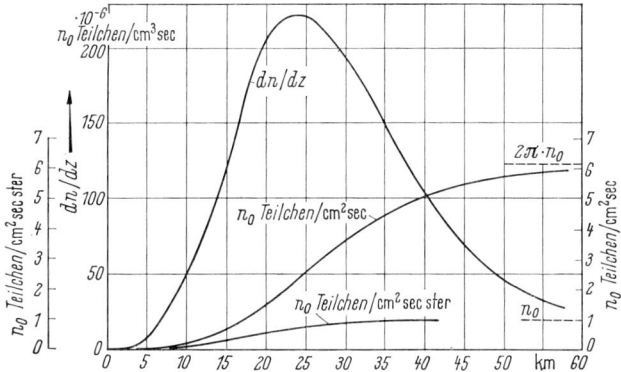

Fig. 24. Höhenabhängigkeit der inelastischen Nukleonenstöße in der Atmosphäre nach BRUNBERG [$Br\ 58$]. Kurve A ($= dn/dz$): Gesamtzahl der inelastischen Stöße pro Einheitsvolumen· Kurve C ($= n_0$(Teilchen/Sterad)): Gesamtzahl der Teilchen, die noch nicht inelastisch gestoßen haben. Kurve B ($= 2\pi n_0$): Primäre aus dem Zenithbereich, die noch nicht gestoßen haben.

(leakage flux). Dieses Intensitätsverhalten legt (im Verein mit der β-Instabilität) nahe, daß die Neutronen in der Atmosphäre als Sekundärerscheinung durch primäre geladene wie sekundäre) Teilchen der k. S. ausgelöst werden. Der exponentielle Abfall der Intensität weist darauf hin, daß sich die Neutronen im Gleichgewicht mit den Erzeugenden befinden. Experimentell wie theoretisch wurde dieses Gleichgewicht in den letzten Jahren intensiv untersucht.

BRUNBERG [$Br\ 58$] hat die inelastischen Stöße der Nukleonen in der Atmosphäre untersucht und gibt ihre Höhenverteilung innerhalb einer Normalatmosphäre bei einer mittleren Stoßlänge von 120 g/cm² an. Unter der Annahme einer konstanten mittleren Multiplizität konnte er die Gesamtzahl der inelastischen Stöße pro Einheitsvolumen in der Atmosphäre, vgl. Kurve A von Fig. 24, errechnen. Die Kurven B und C geben den Höhenverlauf der Primären an, die noch nicht inelastisch gestoßen haben, wobei die Teilchen in Kurve B aus der Zenitrichtung einlaufen. Man erkennt, daß in einer Tiefe von 7—10 km alle Primären gestoßen haben.

Ein Teil hochenergetischer Nukleonenstöße verläuft nahezu streifend durch obere Schichten der Atmosphäre. Neben einem Beitrag von Albedo-Protonen verlassen bei diesem Prozeß auch hochenergetische Neutronen in Form eines Verlustflusses die Atmosphäre. Fig. 25 soll dies schematisch wiedergeben, wobei die mögliche Winkelverteilung der energiereichen Verlustflußneutronen durch die Keulenform angedeutet ist.

Eine quantitative theoretische Beschreibung dieser Stoßprozesse besitzen wir bis heute noch nicht. Dies gilt sowohl für die Nukleonenkaskade[1], die durch hochenergetische primäre Protonen ausgelöst und durch sekundäre Nukleonen und Pionen in die Tiefe der Atmosphäre getragen wird, als auch für die komplexe nukleare Kaskade der Fragmentationsstöße der primären Kerne mit den Luftkernen. Aus diesen Gründen können wir nicht direkt aus den gemessenen Primärspektren der Protonen bzw. der Atomkerne die Höhenverteilung, Energieverteilung und den Breiteneffekt der sekundären Neutronen (bzw. Protonen) in der Atmosphäre quantitativ erschließen.

Die verschiedenen Möglichkeiten der Neutronenproduktion in der Atmosphäre wurden unter dem Gesichtspunkt der damaligen experimentellen Erfahrung von KORFF [Ko 48] diskutiert, und es wurde geschlossen, daß die größte Zahl der atmosphärischen Neutronen bei Kernexplosionen oder Kernzerlegungen (Sternen) emittiert werden müsse.

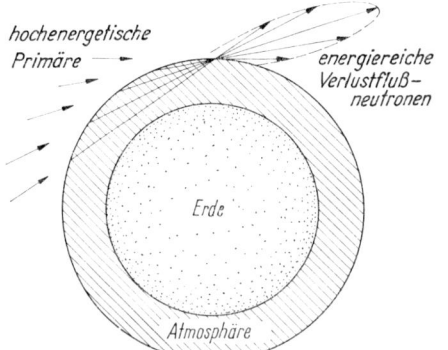

Fig. 25. Produktion hochenergetischer Anstoßneutronen bei streifendem Einfall primärer Protonen in den höheren Schichten der Atmosphäre (energiereiche Verlustflußneutronen) nach [He 59a].

Wir stützen uns auf die Tatsache, daß der hochenergetische wie der energiereiche Nukleonenstoß mit einem Atomkern nicht als zentraler Stoß mit der Gesamtheit der Nukleonen abläuft, sondern entweder in Form „peripherer Stöße" oder direkter Nukleon-Nukleon-Stöße. Das stoßende Nukleon gibt im Mittel etwa die Hälfte seiner Energie an das Anstoßnukleon (knock-on nucleon) und an die multipel dabei mitabgelösten Mesonen (Schauerteilchen) ab und verläßt selbst als relativ energiereiches Nukleon (persisting nucleon) den Kern.

Bei und ab welchen Energien das Anstoßnukleon innerhalb des Kerns plural noch eine intranukleare Kaskade auslöst, ist in Bd. 46/1 von SITTE [Si 61] ausführlich behandelt. Der aufgebrochene hochangeregte Zwischenkern geht unter Emission einiger Verdampfungs-Neutronen und -Protonen in einen stabilen Endkern über oder verbleibt als Radionuklid. Die bei diesen Kernzerlegungen oder Kernexplosionen freiwerdenden Anstoßneutronen (einschließlich der umgeladenen Protonen) und die dabei mitemittierten Verdampfungsneutronen sind die Quellen der Erzeugung von Neutronen in der Atmosphäre. SIMPSON [Si 51a, b] hat abgeschätzt, daß 90% der Neutronen der k. S. in Verdampfung Sternen durch primäre oder sekundäre kosmische Strahlungsnukleonen, die Energien von 300 MeV oder weniger haben, ausgelöst werden. POWELL und Mitarbeiter [Br 49] haben eine Beziehung zwischen Sterngröße in Emulsion und der mittleren Energie der sternerzeugenden Nukleonen hergeleitet, der wir entnehmen, daß einer Produktionsenergie von 300 MeV Sterne mit etwa 5–6 stark ionisierenden Spuren entsprechen.

In großen Höhen und am Gipfel der Atmosphäre kann die Neutronenzahl pro Stern im Mittel noch etwas modifiziert sein. Neben den durch Nukleonen ausgelösten Sternen findet man die komplexeren Stöße der Primärkerne mit Luftkernen. Bei der sukzessive sich wiederholenden Fragmentation — die entweder zu einem Bündel von Nukleonen führt oder gelegentlich zur Bildung von Radio-

---

[1] Die nukleare Kaskade wurde z. B. unter anderem von BUDINI und MOLIÈRE [Bu 53] behandelt, und von PINKAU [Pi 64] mit verbesserten Annahmen und Ansätzen neu dargestellt.

nukliden (s. Kapitel VI, LAL und PETERS) — muß eine größere Neutronenzahl frei werden, als beim Nukleonkernstoß. Rechnet man die Primärintensität der kosmischen Strahlung in Nukleonenzahlen um, so sind die primären Protonen mit ca. 60% beteiligt, die restlichen 40% an Nukleonen werden von den Primärkernen repräsentiert; etwa die Hälfte der letzteren entspricht gebundenen Neutronen. Diese müssen abgesehen von Umladung in Protonen und der Bildung von Radionukliden in großen Höhen bei Fragmentationsstößen frei werden. Messungen darüber scheinen nicht vorzuliegen. Ebenso reichte das vorliegende experimentelle Material nicht aus, um aus dem Primärspektrum über die Fragmentationsparameter ein Neutronenquellspektrum in großen Höhen zu errechnen oder etwa durch Verbesserung der vereinfachten Noon-Kaplonschen eindimensionalen Diffusion [No 55] zu erschließen.

Diese in solch komplexen Stoßprozessen produzierten Neutronen können durch ein Quellspektrum pauschal beschrieben werden. Dieses Neutronenquellspektrum enthält eine Verteilung hinsichtlich der Energie, ebenso der Winkel schließlich eine Verteilungsfunktion nach der Höhe und Breite des Erzeugungsorts.

17. Das Quellspektrum der „Verdampfungs"-Neutronen. α) *Energieabhängigkeit.* Die Zahl der Verdampfungsneutronen $N(E)\,dE$, die pro Sekunde im Energieintervall dE erzeugt wird, wird nach LE COUTEUR [Le 50, Le 52] oder HAGEDORN und MAKE [Ha 53] bei Kernexplosionen (Sternen) durch die Weisskopfsche Verdampfungsformel beschrieben:

$$N(E)\,dE = B \cdot E \cdot e^{-\beta E}\,dE; \text{ mit } B = \frac{A}{\Theta^2}. \quad (17.1)$$

Die Kerntemperatur Θ(MeV) $=1/\beta$ muß für Sternbildung in Luft geeignet gewählt werden (17.1) kann zur theoretischen Simulation der Quellung von Neutronen in der Atmosphäre dienen (vgl. Ziff. 35—38).

MIYAKE, HINOTANI und NUNOGAKI [Mi 57a] haben in einer Nebelkammer mit Wasserstoffüllung über Protonen das Energiespektrum der auslösenden „Anstoß"-Neutronen (s. wegen Direktreaktionen auch [Ci 66]) gemessen, das sie als Potenzspektrum [vgl. (21.3) S. 447] annähern. Diese Neutronendaten umgerechnet auf die Form des Weißkopf-Spektrums (17.1) — geben eine Kerntemperatur $\Theta_M =$ 2,6 MeV oder $\beta_M = 0,385$ MeV^{-1}, in schöner Übereinstimmung mit dem früher von FLÜGGE (vgl. S. 426) gewählten Wert. In Fig. 26 ist dieses „Verdampfungs"-Spektrum als nicht-normiertes differentielles Neutronen-Quellspektrum dargestellt; der differentielle Fluß im Maximum beträgt für Energien von 2—10 MeV etwa $2,3 \cdot 10^{-3}$ (Neutronen/cm² sec MeV); im Bereich von 10—20 MeV etwa das Doppelte[1]. Dieses Ergebnis ist auch im Einklang mit Messungen von MIYAKE et al. [Mi 57b] bei Neutronenreaktionen in einer Kammer mit Stickstoffüllung. NEWKIRK [Ne 63] überlagert diese Quellenspektren (vgl. auch S. 495).

GROSS [Gr 56c] hat Kohlenstoff mit Protonen von 190 MeV beschossen und mißt dabei u. a. ein Spektrum von Verdampfungsneutronen mit einer Kerntemperatur $\Theta_C = 1$ MeV. HESS et al. [He 61] errechnen damit ihr Quellspektrum der Verdampfungsneutronen.

β) Die Winkelverteilung der Verdampfungsneutronen ist isotrop. Die Anstoßneutronen zeigen besonders mit wachsender Energie eine anisotrope Streuung in Vorwärtsrichtung.

γ) *Höhenabhängigkeit der Verdampfungsneutronen.* (i) HESS et al. [He 61] vergleichen die Bremslänge (slowing down length) für Neutronen in der Atmosphäre (ca. 40 g/cm²) mit der Absorptionslänge L für Neutronen ($L=155$ g/cm²). Da die Abbremsung auf kurzen Wegen im Vergleich zur Absorption erfolgt, wählen

---
[1] Herrn Prof. MIYAKE sei für die freundliche briefliche Mitteilung und Diskussion gedankt.

sie für die Höhenabhängigkeit der Neutronenquellen die gleiche Verteilungsfunktion wie die von Hess, Patterson, Wallace und Chupp [He 59a] bei (geomagnetisch) $\Lambda = 44°$ N gemessene Höhenabhängigkeit des Gleichgewichtsspektrums

$$N(E, z)\, dE \cdot dz = e^{-z/155}\, dz \cdot N(E)\, dE. \tag{17.2}$$

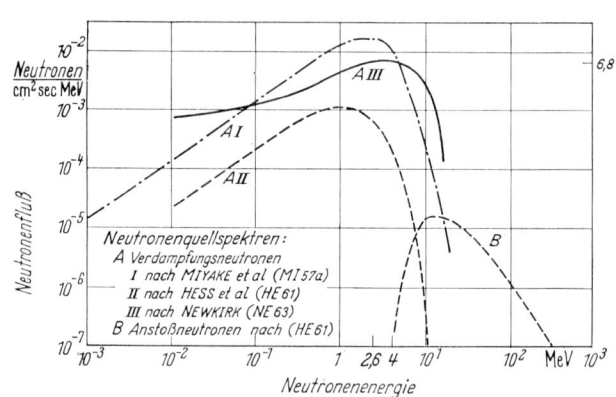

Fig. 26. Quellspektrum von „Verdampfungsneutronen" nach (Mi 57a) u. (He 61) und Quellspektrum von Anstoßneutronen nach [He 61]. Normierung bei Miyake willkürlich, vgl. Text S. 429.

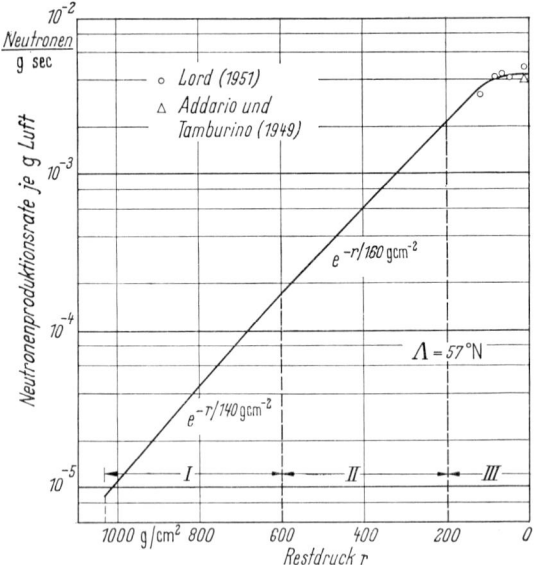

Fig. 27. Verteilungsfunktion der relativen Höhenabhängigkeit für Quellneutronen in der Atmosphäre nach Newkirk [Ne 63]. Die Fläche unter der Kurve ist auf 1 normiert.

Die Ähnlichkeit mit der von Flügge [Fl 43a] gewählten Verteilungsfunktion (vgl. S. 426) ist ersichtlich.

(ii) Newkirk [Ne 63] gliedert seine Höhenabhängigkeit für die geomagnetische Breite $\Lambda = 57°$ N in drei Teilbereiche (vgl. Fig. 27); auch den Gleichgewichtsbereich unterteilt er.

In größeren Tiefen der Atmosphäre (600 bis 1033 g/cm²) ergibt sich eine Absorptionslänge $L = 140$ g/cm², die nach Neutronenmessungen von Simpson und Fagot [Si 53a] (vgl. S. 437, 443) für alle geomagnetischen Breiten anwendbar ist.

In atmosphärischen Tiefen von 200 bis 600 g/cm² fanden SIMPSON et al. [Si 51c] bei Ionisationskammermessungen der Sterne der k. S. und SOBERMANN [So 56] bei Neutronenmessungen eine Breitenabhängigkeit der Absorptionslänge $L(\Lambda)$. Für $\Lambda = 57°$ N beträgt $L(57°) = 160$ g/cm².

Vom Gipfel der Atmosphäre bis zur Tiefe von 200 g/cm² werden die Häufigkeiten der durch kosmische Strahlung ausgelösten Sterne in Kernemulsion zugrundegelegt (vgl. Fig. 45). LORD [Lo 51] hat bei $\Lambda = 54°$ N und ADDARIO und TAMBURINO [Ad 49] haben bei $\Lambda = 55°$ N die Sternhäufigkeiten gemessen. NEWKIRK schätzt, daß diese verfeinerte Höhenabhängigkeit auf $\pm 20\%$ richtig ist; nur im Bereich von einigen g/cm² am Gipfel der Atmosphäre sind die experimentellen Daten unzureichend oder fehlerhaft.

Die Fläche unter der Höhenabhängigkeitskurve hat NEWKIRK auf 1 normiert. Damit kann man dieses höhenabhängige Quellspektrum auf eine absolute Neutronenproduktionsrate [Neutronen/cm² sec^{-1}] mit Hilfe direkter Neutronenmessungen normieren.

Der von NEWKIRK nach Messungen von SMITH et al. [Sm 62] zu 7.1 Neutronen/cm² sec^{-1} gewählte Normierungsfaktor ist mit Vorsicht zu betrachten. Er darf nicht als Normwert gewählt werden. Die Messungen von SMITH et al. über Bemidji ($\Lambda = 57°$ N geomagnetisch) am 19. Juli 1961 0130 UT (Einheitszeit-Greenwich) wurden wenige Stunden nach einer maximalen solaren (Protonen-)Störung im Gefolge einer gas-chromatischen Eruption (solar flare) der Größenordnung 3$^+$ ausgeführt. Die solaren Protonen im breitenempfindlichen Bereich des Primärspektrums modulieren auch den Neutronenhaushalt der Atmosphäre. Diesen gestörten Neutronenfluß haben SMITH et al. gemessen; wir werden bei der Diskussion der Newkirkschen Rechnungen auf die Auswirkungen dieser speziellen Normierung des Neutronenquellspektrums zurückkommen (vgl. S. 496 und 514f.).

Mit diesem Spezialfall haben wir die Auswirkung einer möglichen aperiodischen Modulation der primären Intensität der kosmischen Strahlung durch direkt einfallende Plasmawolken solarer Teilchen im Energiebereich von einigen MeV bis höchstens 20 GeV vor uns. Neben gelegentlichen — in manchen Jahren bis 15 oder mehr — solchen aperiodischen Modulationen unterliegt die einfallende Primärintensität der kosmischen Strahlung dauernd einer nachhaltigen periodischen Variation, die mit dem Sonnenfleckenzyklus korreliert ist. Der Modulationsmechanismus[1] des Sonnenfleckenzyklus ist noch nicht vollständig geklärt, man weiß, daß die Sonnenfleckenzahl zur kosmischen Strahlungsintensität antikorreliert ist, d.h. beim Sonnenfleckenminimum fällt maximale Intensität auf die Atmosphäre ein; der Modulationsbereich erstreckt sich auf den sog. breitenempfindlichen oder geomagnetischen Bereich des Primärspektrums bis ca. 30 GeV; Einflüsse bis 300 GeV konnten nachgewiesen werden (Fig. 2; 5). Dem Primärspektrum entnimmt man, daß der überwiegende Anteil der breitenempfindlichen Primären, die der solaren Modulation unterliegen, relativ niedere Energie besitzt. Sie haben auf Grund ihrer Kernstoßlänge von ca. 120 g/cm² nur geringe Chancen tief in die Atmosphäre einzudringen; so wirkt sich der Hauptteil der solar modulierten Teilchen — wie man leicht abschätzt — nur bis herab in atmosphärische Tiefen von etwa 300—400 g/cm² stark aus, bis herab zum Erdboden verliert sich die kennzeichnend nachweisbare Variation der Primärintensität in den Sekundärerscheinungen mehr und mehr.

---

[1] Wegen Einzelheiten sei auf den Beitrag von S. E. FORBUSH "Modulation of Cosmic Rays" im Handbuch der Physik, Bd. 49,2, verwiesen.

Die Sonnenfleckenminima (bzw. -maxima) wiederholen sich zyklisch in etwa elfjährigem zeitlichem Abstand; es scheint noch nicht endgültig geklärt, ob dieser Zeitraum die Periode oder die Halbperiode des Sonnenfleckenzyklus[1] darstellt.

Eine vollständige Beschreibung des Quellspektrums der Neutronen in der Atmosphäre muß nicht nur der energetischen und höhenabhängigen Verteilung, sondern auch der geomagnetischen (breitenempfindlichen) und zeitlichen Variation Rechnung tragen. Aus der Gesamtheit der Neutronendaten in und oberhalb der Atmosphäre, sowie den bisherigen Messungen der solaren Modulation hat LINGENFELTER [Li 63a] die Neutronenproduktionsraten (Quellstärken) in Abhängigkeit von der atmosphärischen Tiefe z, der geomagnetischen Breite Λ und der zeitlichen Variation bei Sonnenfleckenminimum bzw. -maximum angegeben (vgl. Ziff. 32).

18. Das Quellspektrum der schnellen Neutronen. Während für die Verdampfungsneutronen die charakteristische Energieverteilung nach dem Weißkopfschen Verdampfungsspektrum theoretisch gut verstanden und experimentell gut verifiziert ist, entbehren wir einer direkten Information aus dem Bereich der energiereichen Neutronen der kosmischen Strahlung. Man ist gezwungen ein Spektrum für die schnellen und energiereichen Neutronen im Energiebereich zwischen 10—500 MeV in der Atmosphäre indirekt zu erschließen. Anhaltspunkte dazu kann man für sehr schnelle Neutronen durch Analogieschlüsse mit schnellen Protonen gewinnen; andererseits können Maschinenexperimente weiterhelfen.

Neutronenspektren sind gemessen worden durch Beschuß verschiedener Elemente. GRAVES und ROSEN [Gr 53c] benutzten dazu 14 MeV-Neutronen, GROSS [Gr 56c] beschoß verschiedene Targets mit 190 MeV-Protonen. Die nachgewiesenen Neutronen zeigen bei Energien kleiner als 3 bis 4 MeV völlige Isotropie und können als Verdampfungsneutronen identifiziert werden. Ab Neutronenenergien $E_N > 4$ MeV findet man eine Neutronenausbeute, die größer als die von der Verdampfungstheorie vorausgesagte ist, und zudem findet man für solche Neutronen auch schon eine bevorzugte Emission in Vorwärtsrichtung. Man nimmt daher an, daß diese vorwärtsgestreuten, höherenergetischen Neutronen bei direkten Nukleon-Nukleon-Stößen aus dem Target herausgeschlagen wurden und bezeichnet sie — im Unterschied zu den Verdampfungsneutronen — als Anstoßneutronen. Bei sehr hohen Energien darf man voraussetzen, daß Anstoßneutronen und Anstoßprotonen mit gleichen Ausbeuten zustande kommen; bei niederen oder mittleren Energien ist diese Annahme nicht mehr erfüllt, denn die Protonenemissione wird durch die Coulomb-Barriere behindert, so daß die Neutronenausbeute überwiegt.

Zwei Fragestellungen sind von besonderem Interesse: α) Wie verteilen sich schnelle Neutronen von 10—30 MeV spektral, die inelastisch wechselwirken, aber keine Anstoßnukleonen mehr auslösen können? β) Wie stark tragen Neutronen von (30—200) MeV oder (200—300) MeV über Anstoßnukleonen und Verdampfungsneutronen zu einer intensiven, aber niederenergetischen spektralen Verteilung bei?

Wegen der Anisotropie des Neutronenflusses oberhalb 10 MeV ist die Diffusionstheorie eine unbrauchbare Näherung zur Beantwortung dieser Fragen.

Die Transporttheorie wäre in Grenzen anwendbar, jedoch sind die Neutronenwirkungsquerschnitte bis zu einigen 100 MeV für Absorption, elastische und inelastische Streuung am Sauerstoff und Stickstoff für Abschätzungen völlig unzureichend ([Hu 55, Hu 58]; und BNL-325-Supplement 1 und 2).

---

[1] Wegen Details — auch über solar flares — sei auf den Beitrag von DE JAGER "Structure and Dynamics of the Solar Atmosphere" im Handbuch der Physik, Bd. 52, verwiesen.

Ziff. 18. Das Quellspektrum der schnellen Neutronen. 433

Das Spektrum schneller Neutronen muß deshalb indirekt erschlossen und ergänzt werden. Oberhalb 10 MeV ist der Hauptmechanismus der Verringerung der Neutronenenergie der inelastische Stoß und die $(n, 2n)$- bzw. die (n, xn)-Reaktion (Fig. 67), während die Reaktionsraten aller anderen (n, x)-Reaktionen — ohne Neutronenemission — den Bruchteil des energiereichen Neutronenflusses angeben, der dem Neutronenhaushalt der Atmosphäre verloren geht. Nimmt man noch an, daß die Neutronenspektren solcher inelastischer Neutronenreaktionen denjenigen von Protonreaktionen ähnlich sind, so kann man aus inelastisch an Stickstoff oder Kohlenstoff gestreuten schnellen oder energiereichen Protonen versuchen, geeignete Neutronenanalogspektren zu konstruieren. HESS, CANFIELD und LINGENFELTER [He 61] haben zu diesem Zweck die Streudaten von 14 MeV Neutronen an Stickstoff nach SMITH [Sm 54], die inelastische Streuung von 96 MeV

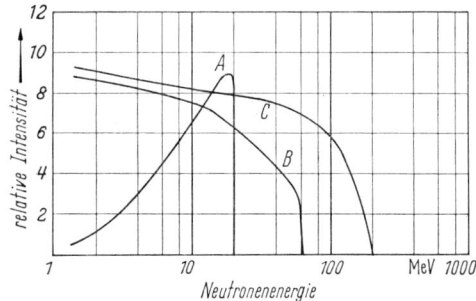

Fig. 28. Analogspektren für energiereich gestreute Neutronen nach [He 61]. Nähere Hinweise im Text.

Protonen an Kohlenstoff nach STRAUCH und TITUS [Str 56] und die inelastische Streuung an 300 MeV Protonen an Kohlenstoff durch HESS und MOYER [He 56] benutzt, um nach Mittelung über alle Streuwinkel auf die Analogspektren energiereicher und inelastisch gestreuter Neutronen zu schließen (vgl. Fig. 28). HESS et al. haben die Spektralkurven noch so modifiziert, daß Kurve A Neutronen von 10 bis 31,6 MeV; Kurve B Neutronen von 31,6 bis 100 MeV und Kurve C Neutronen von 100 bis 316 MeV entsprechen soll. Kurve A deuten HESS et al. so, daß Neutronen mit Energien ≤ 30 MeV selbst keine Anstoßneutronen mehr auslösen können; über wenige inelastische Stöße nimmt daher die Intensität dieser „energiearmen" schnellen Neutronen rasch ab. Völlig anders zeigten sich energiereiche schnelle Neutronen (Kurve B bzw. C): sie produzieren über einige Anstoßprozesse zugleich jeweils hinreichend viele energiearme Verdampfungsneutronen mit. Die relativ geringe Intensität weniger energiereicher Neutronen verschiebt sich zugunsten einer relativ hohen Intensität von Verdampfungsneutronen, die also gespeist werden, während die in den Energiebereich (10 bis 30) MeV fallenden Neutronen schließlich wieder aussterben. Diese Überlegungen dienen dazu, den Anteil an hochenergetischen energiereichen und schnellen Neutronen abzuschätzen, der mit einer gewissen Wahrscheinlichkeit aus dem Energiebereich > 10 MeV in den der Diffusionstheorie zugänglichen Energiebereich <10 MeV hineingestreut wird. Dieser Bruchteil der schnellen Neutronen kann dann dem Spektrum der Verdampfungsneutronen zugeschlagen werden.

Die Verteilung der energiereichen Anstoßneutronen kann man nach ROSSI mit Gl. (13.12) als „Quellspektrum" für Sekundärneutronen beschreiben. Will man diese Verteilung für Energien kleiner als etwa 8 bis 10 MeV abschneiden, so kann

man das Spektrum (13.12) nach HESS, CANFIELD und LINGENFELTER [*He 61*] mit
$$N(E)\, dE = K \cdot E^{-2} \cdot e^{-160 E^2} \cdot dE \tag{18.1}$$
annähern (vgl. Fig. 26).

Der „Gamow-Faktor" wurde eingeführt, um bei „Verdampfungsenergien" die direkte Neutronenemission zu unterdrücken. Dieses spezielle Anstoßneutronen-Spektrum (18.2) läßt den Anteil des energiereichen Neutronenflusses abschätzen, der in Energieintervalle der Transportnäherung oder der Diffusionsnäherung hineingestreut wird und so zum Neutronenhaushalt in der Atmosphäre beiträgt. Die Form dieses Spektrums ist erheblich unsicher, doch geht sie nur unempfindlich in die Rechnung ein.

19. Intensität langsamer Neutronen in der Atmosphäre. Langsame Neutronen wurden in Abhängigkeit von der Höhe in der Atmosphäre und in Funktion der geomagnetischen Breite sehr häufig gemessen.

Die Messungen wurden mit Hilfe von BF_3-Zählern durchgeführt, das bedeutet, daß der Hauptanteil der registrierten Neutronen aus dem Energiegebiet unterhalb einiger eV kommt. Die Zähler registrieren zusätzlich einen unvermeidlichen Untergrund, der von der Neutronenzählrate abgezogen werden muß. Der Untergrund wird von Sternen hervorgerufen, die in den Zählrohrwänden erzeugt werden oder von Rückstoßspuren im Zählgas oder durch weiche Schauer ausgelöst werden.

Die Elimination des Untergrundes geschieht mit Hilfe einer Differenzmessung, deren praktische Durchführung zu zwei verschiedenen Verfahren geführt hat: Man benutzt in beiden Verfahren gleiche Zählrohre; beim ersten Verfahren unterscheiden sich die Zählgase in ihrem Isotopengemisch B^{10}/B^{11}.

Ein Zählrohr wird mit natürlichem BF_3 gefüllt (80,4% B^{11} zu 19,6% B^{10}), während das andere Zählrohr mit stark angereichertem $B^{10}F_3$ gefüllt ist. Über die $B^{10}(n, \alpha)\,Li^7$-Reaktion wird durch die B^{10}-Anreicherung eine unterschiedliche Neutronenzählrate erreicht, während der Untergrund in beiden Zählrohren der gleiche bleibt und durch Differenzmessung eliminiert werden kann (DAVIS [*Da 50b*] und STAKER [*St 50a*]).

Das andere Verfahren besteht darin, daß man gleich angereichertes $B^{10}F_3$ verwendet, die beiden Zählrohre aber mit massengleichen Schichten aus Metall ummantelt, das eine mit Cd, das andere mit einem Metall, das Neutronen praktisch nicht absorbiert. Die Strahlung, die den Untergrund verursacht, wird in beiden Umhüllungen gleich geschwächt, dagegen zählt das mit Cd umgebene Zählrohr nur Neutronen mit einer Energie $>0{,}4$ eV. Durch Differenzbildung erhält man die Neutronen mit einer Energie $<0{,}4$ eV.

BETHE, KORFF und PLACZEK [*B 40*] haben ausführlich besprochen, daß bei einem $1/v$-Absorptionswirkungsquerschnitt (wie er beispielsweise bei B^{10} vorliegt) direkt die Dichte langsamer Neutronen gemessen wird. Bei den unabgeschirmten Zählern von DAVIS und STAKER trifft das zu. Im Gegensatz dazu kann bei dem Cd-Differenzverfahren nicht auf die Dichte geschlossen werden. So ergeben sich durch die beiden Meßverfahren Verschiedenheiten in der mittleren Energie der registrierten Neutronen. YUAN [*Yu 51*] und PFOTZER [*Pf 52a*] klären die Unterschiede beider Verfahren ausführlich und geben Korrekturen an, die es erlauben, die Festlegung von Absolutwerten auch bei Cd-Differenzmessungen genügend genau zu erreichen. Mit diesen beiden Arten von Zählern wurde der Fluß der langsamen Neutronen in der Atmosphäre gemessen. Als Träger der Meßgeräte dienten Flugzeuge bzw. Ballone.

Die Messungen vor 1948 sind bei KORFF [Ko 48] zusammengefaßt. Mit den neueren Messungen ab dem Sonnenfleckenmaximum 1948 verfügen wir über eine genaue Kenntnis der Höhen — bzw. der Breitenabhängigkeit der langsamen Neutronen in der Atmosphäre, die man im Einzelnen den Arbeiten von STAKER et al.

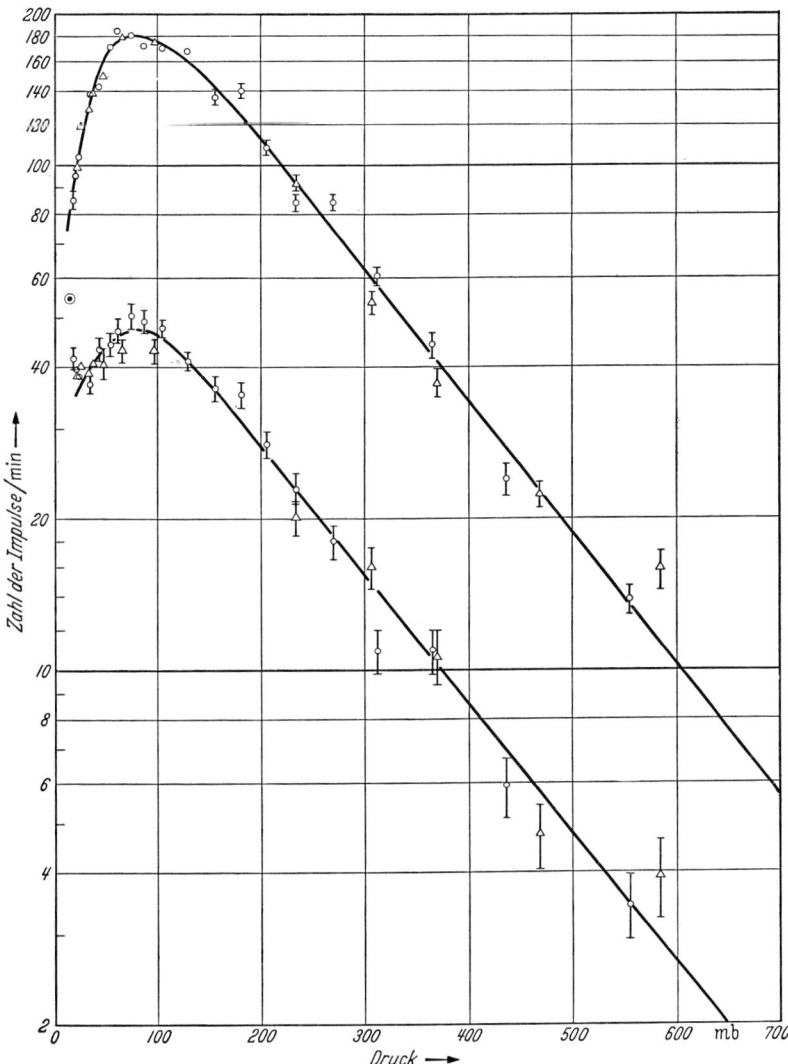

Fig. 29. Neutronenzählrate als Funktion des Atmosphärendrucks. Obere Kurve: BF$_3$-Zähler, Füllung angereichert an B^{10}. Untere Kurve: BF$_3$-Zähler, Füllung normales Bor. o Aufstieg, △ Abstieg. (Nach [So 56])

[St 50, St 51], DAVIS [Da 50], SIMPSON [Si 51a], YUAN [Yu 49, Yu 51], SWETNICK [Sw 54], der Korff-Gruppe bei SOBERMANN [So 55, So 56] sowie bei NEUBURG et al. [Ne 55a] entnehmen kann. Fig. 29 zeigt eine typische Flugkurve, die bei einem der Ballonflüge von SOBERMANN [So 56] gewonnen wurde, wegen Korrektur s. [Ko 58].

Die Elimination des Untergrundes wurde nach DAVIS und STAKER vollzogen. Vom Wert des Neutronenflusses, der natürlich erst am Gipfel der Atmosphäre erreicht werden könnte, steigt die Neutronenzählrate mit wachsender atmosphärischer Tiefe zum Flügge-Yuanschen Neutronenmaximum der Dichte an und fällt von 200 g/cm² exponentiell ab:

$$N = N_0\, e^{-p/L}, \tag{19.1}$$

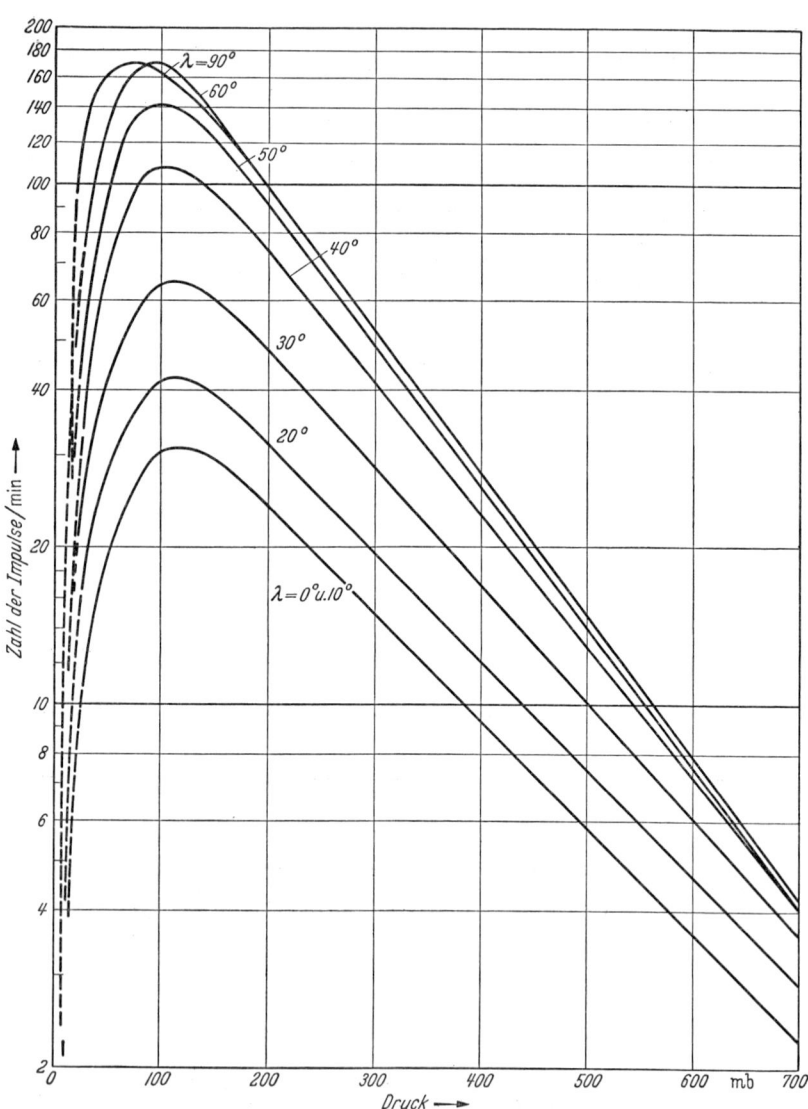

Fig. 30. Neutronenintensität als Funktion des Atmosphärendrucks für verschiedene geomagnetische Breiten Λ auf der Nordhalbkugel, Messung 1952—1954 durch [So 56].

wobei p entweder die Tiefe x in der Atmosphäre in (g/cm²) oder den Restdruck in (mb) darstellen kann und dementsprechend die Absorptionslänge in derselben Einheit gerechnet wird.

SIMPSON [Si 51c] hat festgestellt, daß unterhalb von 700 bis 800 mb Abweichungen von dem Exponentialgesetz (19.1) auftreten (vgl. S. 430, 447f.).

Messungen, die bei verschiedenen geomagnetischen Breiten von den erwähnten Autoren insbesondere von SOBERMANN, ausgeführt wurden, ergaben, daß die Lage des Übergangsmaximums und ferner die Absolutwerte der Neutronenintensität sowie die Absorptionslänge $L = L(\Lambda)$ in (19.1) von der geomagnetischen Breite Λ abhängen. Eine Übersicht über die Variation der Neutronenintensität mit der geomagnetischen Breite gibt Fig. 30 [*So 56*], s. auch [*Ko 58*].

Die meisten Messungen reichen bis in eine Tiefe von ca. 700 g/cm² herab. Messungen in größeren atmosphärischen Tiefen wurden von SIMPSON und FAGOT [*Si 53a*] durchgeführt. Mit Hilfe eines Blei-Paraffin-Pile (Neutronenmonitor) bestimmten sie die Abhängigkeit der Neutronenproduktion von der Höhe. Sie fanden unterhalb von 700 g/cm² Höhe ebenfalls eine exponentielle Abnahme der Intensität. Die Absorptionslänge beträgt in diesem Bereich 140 g/cm² und ist von der geomagnetischen Breite unabhängig. Dies deutet darauf hin, daß die primär einfallende kosmische Strahlung, die die Produktion der sekundären Neutronen in dieser Tiefe trägt, von der Abschneidesteifigkeit des Erdmagnetfeldes nicht mehr beeinflußt wird.

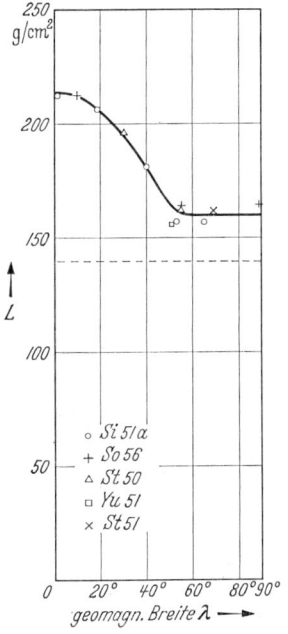

Fig. 31. Effektive Absorptionslänge L_{eff} in der Atmosphäre als Funktion der geomagnetischen Breite Λ. ——— Breitenabhängige Absorptionslänge L_{eff} für atmosphärische Tiefen 200 mb $\leq p \leq$ 700 mb. – – – – Breitenunabhängige Absorptionslänge $L = 140$ g/cm² für 700 mb $\leq p$.

Oberhalb von 700 g/cm² hängt die Absorptionslänge L von der geomagnetischen Breite ab und wird in diesem Bereich als effektive Absorptionslänge $L_{\text{eff}}(\Lambda)$ bezeichnet. Ihre Abhängigkeit von der geomagnetischen Breite Λ kann aus Fig. 31 abgelesen werden.

Die mitaufgeführten Werte für schnelle Neutronen[1] nach SIMPSON [*Si 51c*] stimmen gut mit den Daten für langsame Neutronen überein. Die Messungen von SOBERMANN fanden zwischen den Sommern 1952 und 1954 statt, d. h. sie fielen praktisch mit dem Sonnenfleckenminimum 1953/54 zusammen.

Die früheren Messungen von STAKER und SIMPSON liegen bei einer geomagnetischen Breite ($<50°$) unterhalb des Knies. In diesem Bereich sollte die Abhängigkeit von der Sonnenfleckenaktivität vernachlässigbar sein.

Bei niedrigen geomagnetischen Breiten wächst die effektive Absorptionslänge auf 130% des Wertes im Polgebiet (oberhalb des Knies) ab etwa $\Lambda = 50-60°$ geomagnetischer Breite an. Diesen Effekt kann man mit Hilfe des Schwellenwertes der magnetischen Steifigkeit von ca. 16 GV im Äquatorialbereich deuten. Die mittlere Energie der Primärteilchen, die in niedrigen geomagnetischen Breiten die nukleare Kaskade auslösen, ist höher als im Polgebiet. Dies hat jeweils eine größere Anzahl der pro Stoß erzeugten Neutronen zur Folge. Außerdem haben die Sekundärteilchen eine größere mittlere Energie, so daß sie weiterhin sternerzeugungsfähig sind; damit kann die Kaskade tiefer in die Atmosphäre eindringen.

Das gleiche Argument kann dazu dienen, die tiefere Lage des Neutronenmaximums bei niederen geomagnetischen Breiten (Äquatorbereich) gegenüber dem Polbereich zu erklären. Der kleine Schwellenwert der magnetischen Steifigkeit

---

[1] Zum weiteren Vergleich s. Tabelle 14, S. 449.

am Pol erlaubt es einem großen Strom energiearmer Primärer in die Atmosphäre einzudringen, so daß sich dort schon in geringer atmosphärischer Tiefe ein Neutronenmaximum ausbilden kann. Der von SOBERMANN gefundene Zusammenhang ist in Fig. 32 dargestellt.

Der Breiteneffekt der Neutronen der K. S., d.h. die Abhängigkeit der Neutronenzählrate von der geomagnetischen Breite wurde auch zur Zeit des Sonnenfleckenminimums gemessen, und zwar in drei verschiedenen Höhen:

ROSE et al. [Ro 56] haben in Meereshöhe bei einem Druck von 1030 g/cm² gemessen; SIMPSON and FAGOT [Si 53a] in einer Höhe von 680 g/cm². Dagegen

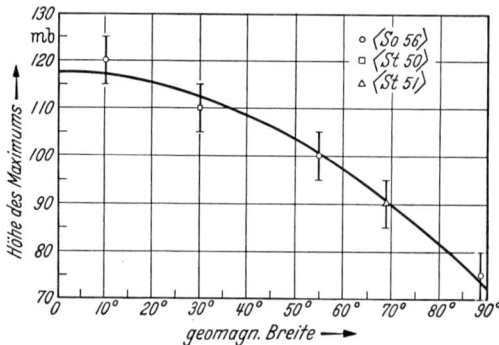
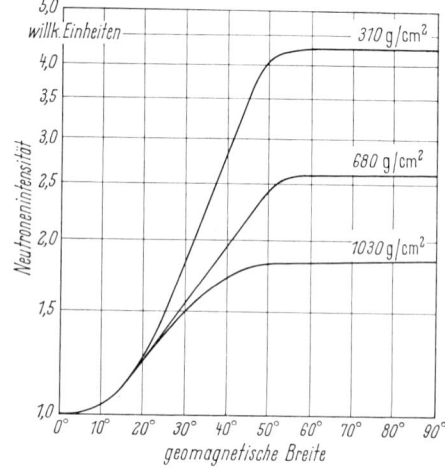

Fig. 32. Lage des Intensitätmaximums langsamer Neutronen als Funktion der geomagnetischen Breite.

Fig. 33. Der Breiteneffekt der Neutronenintensität während des Sonnenfleckenminimums in Abhängigkeit von der Höhe. Gemessen bei 1030 g/cm² durch [Ro 56], bei 680 g/cm² durch [Si 53a] und bei 310 g/cm² durch [Si 51b] im Jahre 1948. Diese letztere Messung wurde durch die Meßergebnisse von MEYER und SIMPSON [Me 55a] und SOBERMANN [So 56] auf 1953/54 korrigiert.

ist die Messung in 310 g/cm² Höhe aus dem Jahr 1948 (Sonnenfleckenmaximum) von SIMPSON [Si 51] mit Hilfe der Daten von MEYER und SIMPSON [Me 55a] und SOBERMANN [So 56] auf die Sonnenfleckenminimumswerte aus dem Jahr 1953/54 angepaßt worden. Der Verlauf des Breiteneffektes der Neutronen in der Atmosphäre für das Sonnenfleckenminimum 1954 ist in Fig. 33 dargestellt.

Die Messung des Breiteneffekts am Gipfel der Atmosphäre (0 g/cm²) ist aus leicht ersichtlichen Gründen nicht möglich. Eine Direktmessung scheidet aus; so kann nur indirekt über Messungen des Neutronenverlustflusses aus der Atmosphäre bzw. der in die Atmosphäre einfallenden primären Intensität und über die Ionisation durch die Primären bzw. durch deren Sternerzeugung die Höhen- und Breitenabhängigkeit der Neutronen oberhalb 200 g/cm² erschlossen werden. Wir kommen auf diese Methode in Abschnitt 20 zurück.

Die Messungen der Intensität der langsamen Neutronen wurden meist mit unabgeschirmten Borzählern ausgeführt. Wegen der $1/v$-Abhängigkeit des Bor-Wirkungsquerschnittes stellen diese Messungen — wie S. 434 erwähnt — die Neutronendichte dar. Die Lage des Maximums des langsamen Neutronenflusses fällt also mit dem Maximum der Dichte langsamer Neutronen zusammen.

Ein nächster Schritt war die Bestimmung der Lage des Übergangsmaximums bei wachsender Neutronenenergie. MEYER [Me 55b] hat durch Abschirmen der Zählrohre mit Cd bzw. mit Borkarbid im wesentlichen nur noch Neutronen mit einer Energie $>0,4$ eV bzw. >1 keV zählen können. Er führte seine Messungen in einer geomagnetischen Breite von 51 bis 52° N durch. Für Neutronenenergien >0,4 eV war das Maximum auf 68 mb hochgeschoben, bei Neutronenenergien >1 keV sogar bis zu einer Höhe von 48 mb, während das Maximum der thermischen Neutronen nach den Messungen von YUAN [Yu 51] vergleichsweise bei einer atmosphärischen Tiefe von 112 mb liegt. Die Höhenangaben von MEYER sind mit den heutigen Vorstellungen nicht verträglich.

HAYMES hat die Dichte thermischer bzw. nahezu thermischer Neutronen in der Atmosphäre bestimmt und zwar bei einer geomagnetischen Breite von 41° N im September 1958, d.h. zur Zeit des Sonnenfleckenmaximums.

MILES [Mi 64] hat eine neue Meßmethode mit einer Relativgenauigkeit von ±2% für die Messung langsamer Neutronen der K. S. entwickelt. Er benützte eine Ionisationskammer mit $B^{10}F_3$-Füllung. Die Entladungen über die $B^{10}(n, \alpha) Li^7$-

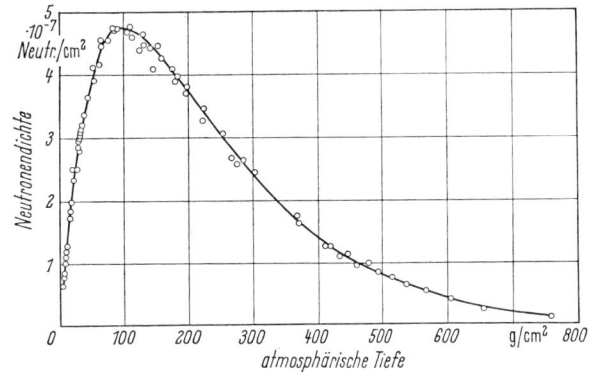

Fig. 34. Präzisionsmessung der Dichte der Neutronen in der Atmosphäre durch MILES [Mi 64].

Reaktionen zählte er mit Hilfe eines Quarzfiberelektrometers Neherscher Bauart. Eine zweite Ionisationskammer diente als Untergrund- und Differenzmesser.

Wegen der experimentellen Details wird auf die Originalarbeit verwiesen. Kalibrierungsunsicherheiten begrenzten die absolute Meßgenauigkeit schließlich auf ±20%. MILES wandte diese bisher genaueste Methode auf die Messung der Neutronendichte bei $\Lambda = 41°$ geomagnetisch N im Verlauf mehrerer Ballonflüge des Jahres 1962 an. Bei einer Höhe von 100 g/cm² (16,3 km) fand er das Maximum der langsamen Neutronen in der Erdatmosphäre mit einer Neutronendichte von $(4,7 \pm 1,2) \times 10^{-7}$ cm^{-3}. Oberhalb dieses Maximums nimmt die Dichte schnell bis auf kleine Werte ab. HAYMES und KORFF [Ha 60a] beobachteten eine Abnahme um einen Faktor 50 bei den höchsten Ballonhöhen von etwa 5 g/cm². HAYMES [Ha 64b] erwartet auf Grund der Messungen von MILES, daß der Neutronenverlustfluß überwiegend von schnellen Neutronen gebildet wird. Die Präzisionsmessung der Dichte der kosmischen Strahlungsneutronen in der Atmosphäre ist in Fig. 34 dargestellt. Weiterhin verglich MILES die Werte der Neutronendichte von HAYMES mit seinen eigenen, wobei er der zeitlichen Variation vom Jahre 1958 zum Jahre 1962 Rechnung getragen hat. Die mittlere Übereinstimmung beider Experimente ist sehr gut (vgl. Fig. 35).

Dagegen ergibt sich ein beträchtlicher Unterschied zu den Messungen von HESS [He 59a] et al., die bei einer geomagnetischen Breite von 44° N in einer Höhe von 200 g/cm² im Winter 1956/57 in einem Flugzeug vom Typ B-36 direkt gemessen wurden. Die von HESS et al. gemessenen Neutronendichten liegen um mehr als 60% höher als die Präzisionsmessungen von MILES (vgl. Fig. 36). Wir werden

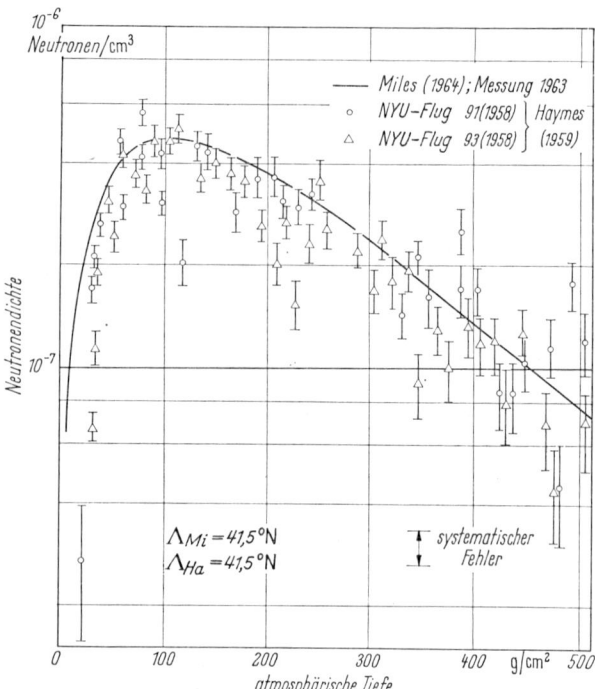

Fig. 35. Vergleich der Dichtemessungen von HAYMES und MILES [Mi 64].

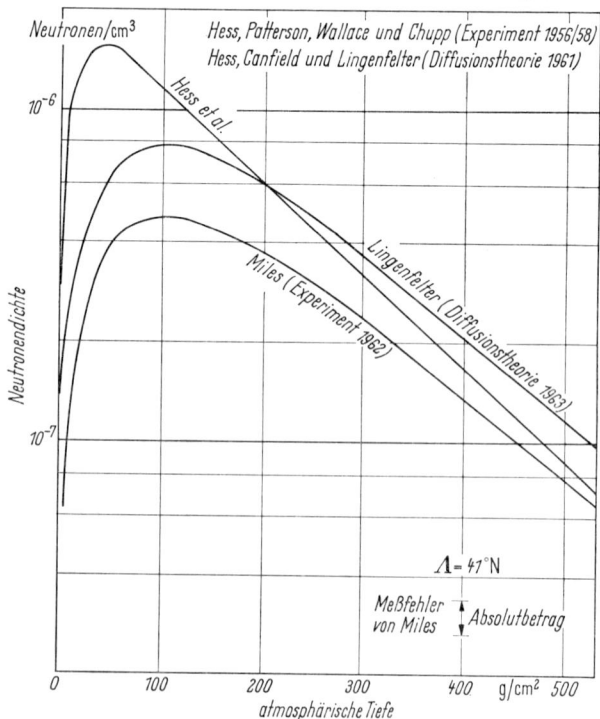

Fig. 36. Die Dichte der Neutronen in der Atmosphäre nach den Messungen von MILES [Mi 64] bzw. HESS et al. im Vergleich mit dem theoretischen Ergebnis von LINGENFELTER. Maximum und Dichte nach HESS et al. zu hoch; ebenso Dichte nach LINGENFELTER etwa um einen Faktor zwei überhöht.

noch darauf zurückkommen, daß dieser Unterschied nicht durch Berücksichtigung geomagnetischer Einflüsse bzw. durch die zeitliche Variation der kosmischen Strahlung erklärt werden kann.

Es ist unwahrscheinlich, daß ein Unterschied in der absoluten Kalibrierung als Erklärung herangezogen werden kann, da die Standardeichnormalen der beiden Laboratorien miteinander verglichen worden waren. Es ist möglich, daß die Abbremsung von Neutronen in dem Flugzeug für die höhere Neutronendichte, die HESS et al. beobachteten, verantwortlich gemacht werden kann.

Ein ähnlicher Effekt tritt voraussichtlich auch in den Körpern der Raketen und Satelliten auf, die zur Messung des Neutronenverlustflusses oberhalb der Atmosphäre benutzt worden sind (vgl. Ziff. 45).

BAGGE und SKORKA [Ba 58] haben die langsame Neutronenkomponente unmittelbar über einer Wasserfläche gemessen und den Übergangseffekt der langsamen Neutronen beim Übergang von der Atmosphäre in Wasser untersucht. Sie fanden einen Übergangseffekt in Wasser mit einem Maximum in $4{,}5 \pm 0{,}3$ cm Tiefe. In größerer Tiefe (> 50 cm) fanden sie eine exponentielle Abnahme des Flusses mit einer Absorptionslänge von 169^{+19}_{-16} g/cm². Die Messungen mit einem BF_3-Zählrohr mit 1,5 cm Durchmesser wurden bis zu einer Wassertiefe von 140 cm ausgedehnt. Oberhalb der Wasserfläche konnte ein Fluß thermischer Neutronen von $(0{,}90 \pm 0{,}02) \cdot 10^{-3}$ n/cm² sec als Cd-Differenz gemessen werden.

Messungen der Neutronendichte unter Wasser und Absolutmessungen der Produktion Neutronen der K. S. über den Zeitraum seit 1939 behandelt EDGE [Ed 59]. Er findet eine genaue neue Meßmethode notwendig, die er unter Gebrauch der Szillard-Chalmers-Reaktion entwickelt. HESS et al. [He 59] haben im Sommer 1958 die *Goldresonanz* bei 4,9 eV mit Cd-Abschirmung als weiteren Detektor für langsame Neutronen angewandt. Der Fluß betrug bei dieser Energie in einer Höhe von 700 g/cm² bei geomagnetischer Breite von 44° N $(5{,}0 \pm 1{,}7) \cdot 10^{-3}$ Neutronen/cm² sec.

20. Schnelle Neutronen in der Atmosphäre. Für schnelle Neutronen liegen Meßdaten vor, die zeitlich mit den Sonnenfleckenmaxima 1947/48 und 1957/58 zusammenfallen, bzw. kurz zuvor oder wenig danach — meist noch bei aktiver Sonne — durchgeführt wurden. Die erste systematische Untersuchung der schnellen Neutronen bei ruhiger Sonne (Minimum) ist 1964/65 erfolgt.

Wir geben in diesem Abschnitt eine Übersicht über Messungen schneller Neutronen. Zunächst besprechen wir Direktmessungen mit Hilfe von Zählern, dann gehen wir auf die Neutronendaten ein, die mit Hilfe der Kernemulsion erschlossen werden können. Danach geben wir einen kurzen Überblick über die Benutzung des Neutronenmonitors als Meßgerät für Neutronen (Nukleonen) in der Atmosphäre. Zum Abschluß gehen wir noch auf die großen Nukleonenteleskope ein, die im Verein mit einem Ionisationskalorimeter oder einem Neutronenmonitor zu detaillierten Untersuchungen der Nukleonenkomponente benutzt werden.

a) Direktmessungen.

21. Schnelle Neutronen konnten früher nur durch Abbremsung in wasserstoffhaltigem Material als langsame Neutronen gezählt werden. Erst seit wenigen Jahren kann man schnelle Neutronen in der Atmosphäre direkt messen.

α) *Messungen mit ummantelten Borzählern (bzw. Lithiumzählern).* Die schnellen Neutronen werden in geeigneten Ummantelungen aus Paraffin oder Polyäthylen (CH_2) moderiert und dann mit Zählern für langsame Neutronen unter Ausnutzung der $B^{10}(n, \alpha) Li^7$ bzw. $Li^6(n, \alpha) H^3$-Reaktionen gezählt Diese Zählgeräte haben Nachteile. Einmal werden auch geladene Teilchen mitgemessen, so daß mit Hilfe eines zweiten identischen Zählers mit einem relativ unempfindlichen Isotop, etwa B^{11}, der Untergrund bei diesen Experimenten beseitigt werden muß. Zum anderen werden durch Protonen in der Bremssubstanz Neutronen angestoßen, die mitgemessen werden.

Mit solchen ummantelten Bor- und Lithiumzählern wurden in den letzten Jahren zahlreiche Messungen schneller Neutronen in der Atmosphäre bzw. oberhalb der Atmosphäre (vgl. Ziff. 45) ausgeführt.

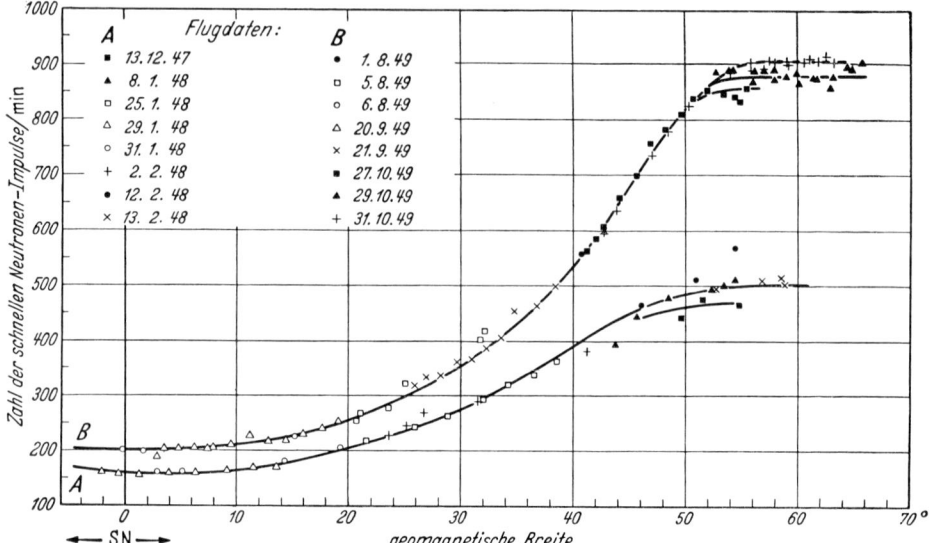

Fig. 37. Breitenabhängigkeit schneller Neutronen gemessen von SIMPSON [Si 51b] mit paraffinummantelten BF_3-Zähler im Zeitraum des Sonnenfleckenmaximum 1948. A in 8300 m Höhe; B in 9200 m Höhe.

Die ersten ausgedehnten Untersuchungen der Intensität schneller Neutronen bis zu einer Energie von einigen MeV hat SIMPSON [Si 51a, b] durchgeführt. Um langsame Neutronen der Atmosphäre bei seinen Messungen sicher auszuschließen, hat er die paraffinummantelten BF_3-Zähler außen noch mit einer Cd-Hülle abgeschirmt. Die Messungen erfolgten in einem Flugzeug bei einer Höhe von 300 mb. Besondere Sorgfalt wurde darauf verwendet, möglichst die Verhältnisse der freien Atmosphäre zu erhalten. Da die Messungen in verschiedenen geomagnetischen Breiten stattfanden, konnte SIMPSON zwischen 10° Süd und 60 bis 70° Nord etwa über den Zeitraum des vorletzten Sonnenfleckenmaximums 1947/48 eine Breitenabhängigkeit schneller Neutronen messen (vgl. Fig. 37). Die Meßkurven stellen ein Mittel aus sehr vielen Flügen dar. Während in den niederen und mittleren geomagnetischen Breiten die Meßpunkte gut zusammenfallen, treten bei hohen geomagnetischen Breiten signifikante Aufspaltungen auf. In diesen Änderungen des schnellen Neutronenflusses spiegelt sich indirekt über die Neutronen der Atmosphäre die Variation der primären kosmischen Strahlung in dem Zeitraum von etwa $1^3/_4$ Jahren um das Sonnenfleckenmaximum 1948 herum wider. Bei 56° geomagnetisch Nord und einer Meßhöhe von 10 km betragen diese Intensitätsschwankungen bis zu 30%. Die von SIMPSON [Si 51c] gefundenen Werte für die Absorptionslänge sind nicht in Übereinstimmung mit den Messungen der Sternhäufigkeit in Kernspurplatten. Schon in Ziff. 6 wurde bei der Besprechung der

Absorptionslänge für Sternerzeugung auf diese Diskrepanz hingewiesen. Um diese weiter zu klären, haben SIMPSON und FAGOT [Si 53a] schnelle Neutronen, die sie in Kohlenstoffabsorbern bzw. Bleiabsorbern erzeugten, gemessen. Die Produktion im Kohlenstoff entspricht dabei den Verhältnissen der Produktion in der Atmosphäre. Die Messungen wurden ebenfalls mit dem Flugzeug ausgeführt und später in größeren Tiefen der Atmosphäre wiederholt. SIMPSON und FAGOT fanden dabei eine Höhenabhängigkeit der Absorptionslänge in Luft, und zwar nimmt für die drei untersuchten geomagnetischen Breiten 0°, 41°, 52° die Absorptionslänge mit zunehmender Tiefe in der Atmosphäre ab. Unterhalb 600 mb nimmt sie bei allen geomagnetischen Breiten denselben Wert an, der etwas über 140 g/cm² liegt. Damit ist der Unterschied zu dem Wert der Absorptionslänge für Sternerzeugung, der in Gebirgs- und Meereshöhe (700 und 1000 mb) 130 bis 140 g/cm² beträgt, nicht mehr beträchtlich (vgl. auch S. 394 und 430).

HESS et al. [He 59a] haben mit ummantelten Borzählern am Erdboden und herauf bis zu Höhen von 12 km „schnelle" Neutronen von 10 keV bis 1 MeV gemessen. Mit ihren Meßdaten haben sie dann (vgl. Abschnitt 21) das Energiespektrum in diesem Bereich berechnet.

KORFF und HAYMES [Ko 60] haben in einem Ballon während des zeitlichen Ablaufs eines intensiven Nordlichts bei 40 km Höhe mit Hilfe paraffinummantelter Borzähler schnelle Neutronen gemessen. Sie diskutieren deren Erzeugung durch schnelle Protonen im Paraffin bzw. die tatsächliche Abbremsung schneller Neutronen durch das Paraffin. Der Differenzzähler zeigt auf jeden Fall auch das gleichzeitige Vorhandensein ionisierender Strahlung. Einige charakteristische und zeitlich aufeinanderfolgende steile Anstiege der Neutronenzählrate konnten KORFF und HAYMES bei ihrer Messung finden. Die gemessenen Neutronenproduktionsraten während des Ablaufs des Nordlichts waren einige hundertmal größer, als bei Experimenten in Ballonhöhen in ruhigen Zeiten. Der Neutronenfluß konnte nicht angegeben werden, da der $1/v$-Typ des Detektors nur die Neutronendichte unterhalb 1000 eV angibt; zur Bestimmung des Neutronenflusses wäre eine unabhängige Messung der Neutronengeschwindigkeit notwendig gewesen. Sieht man von den Intensitätsanomalien bei dieser Messung ab, so waren die Meßergebnisse für (abgebremste) schnelle Neutronen in Übereinstimmung mit den Rechnungen von HESS, CANFIELD und LINGENFELTER.

GAUGER [Ga 64] hat die Messungen von HESS et al. fortgeführt. Er mißt am Fluß der schnellen Neutronen den Breiteneffekt von 15° Süd bis 55° Nord von Mai bis Juli 1962. An zwei verschiedenen Meßtagen im Juli 1962 fand er erhebliche Unterschiede für die Neutronenzählrate von 30 bis 55° Nord. In Fig. 38 ist aus Übersichtsgründen der mittlere Verlauf für 30 bis 55° N nicht eingezeichnet sondern nur die charakteristische Verschiebung am 12. 7. nach größeren Breiten, am 21. 7. nach kleineren Breiten. Eine einfache Erklärung für die Verschiebung am 12. 7. wäre eine Forbush-Abnahme, am 21. 7. eine erhöhte Sonnenaktivität. Solche Störeffekte waren jedoch in diesem Zeitraum nicht nachweisbar. NEHER und ANDERSON [Ne 60a] fanden ebenfalls isolierte Bereiche in der Atmosphäre mit erhöhten Zählraten, die ebenfalls nicht mit solaren oder irdischen Störungen in Zusammenhang gebracht werden können. Als mögliche Erklärung erwähnt GAUGER Protonen, die zeitweilig in dem schlotartigen Bereich zwischen innerem und äußerem VAN ALLEN-Gürtel eingefangen sind und von dort gelegentlich in die Atmosphäre einbrechen können. Dieser Erklärungsversuch ist unbefriedigend, denn an allen diesen Tagen fand GAUGER auch für Neutronen über 60 MeV (die er mit der Wismut-Spaltungskammer, s. S. 469, gemessen hatte) eine analoge Schwankung. Mit Hilfe der BF_3-Zählraten hat GAUGER im Sinne von HESS et al. [He 59] das Neutronenspektrum im Bereich von 10 keV bis 1 MeV errechnet und findet Übereinstimmung. GAUGER [Ga 65] hat ferner im Bereich der brasilianischen Anomalie des Erdmagnetfeldes einen etwa 20 Minuten anhaltenden, erhöhten Fluß schneller Neutronen (0,1 bis 5 MeV) gemessen.

BOELLA, DEGLI ANTONI, DILWORTH, SCARSI et al. [Bo 63 und 65] haben bei einer geomagnetischen Abschneidesteifigkeit von 4,6 GV (New Mexico Holloman Air base) und Italien (Mailand) den Fluß von Neutronen von thermischen Energien bis 20 MeV mit nackten und ummantelten Zählern gemessen. Die Zählraten der schnellen Flußzähler wurden gegen örtliche Produktion korrigiert, trotzdem liegen diese aber etwas höher als vergleichbare Werte, die mit Rückstoßprotonenzählern gewonnen worden sind (vgl. Tabelle 15). Die mittlere Absorptionslänge L wurde in Tabelle 10 aufgenommen.

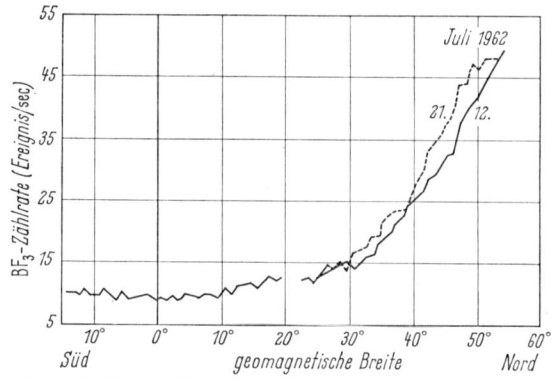

Fig. 38. Gemessene Schwankung des Breiteneffekts schneller Neutronen am 12. bzw. 21. Juli 1961 (nach [Ga 64]). Die Neutronenintensität war zudem von April bis Juli 1961 auf unerklärte Weise um etwa 50% niedriger als im Sonnenfleckenzyklus erwartet (s. Text).

GREENHILL, PHILLIPS, FENTON, FENTON und BOWTHORPE [Gr 65a][1] haben ebenfalls mit nackten und ummantelten Borzählern in den Monaten September bis Oktober 1964 und fünf Stationen auf der südlichen Halbkugel den Breiteneffekt der Dichte der langsamen und des Flusses der schnellen Neutronen gemessen, besonders von 400 mb bis 25 mb. Auch hier wurde die lokale Produktion eliminiert. Wegen der Daten muß auf die Originalarbeit verwiesen werden. Die ,,totale oder gesamte Breitenabhängigkeit", d.h. das Verhältnis der Zählraten an Pol und Äquator in der Höhe z wurde aus den Daten dieser Arbeit, den Daten von KORFF et al. [Ko 65] und den Daten für langsame Neutronen nach SOBERMANN [So 56] errechnet und in Tabelle 15 aufgenommen.

Der Fluß in 240 mb Höhe ist nach Modulationskorrektur in Übereinstimmung mit der Flußmessung nach GAUGER [Ga 64]. Die Neutronendichte wird von 500 g/cm² bis 25 g/cm² für die südliche Halbkugel dargestellt und mit der Messung von MILES [Mi 64] und den theoretischen Werten von LINGENFELTER verglichen (vgl. dazu S. 531 f.).

In Tabelle 12 sind die Absorptionslängen der schnellen Neutronen in verschiedenen geomagnetischen Breiten und Höhen nach den Daten von SIMPSON, HESS et al. und GAUGER zusammengefaßt.

Die Ausbeute all dieser moderierenden Zähler für schnelle Neutronenmessungen ist sehr gering. Ein typischer Wert ist nach HAYMES [Ha 64b] 0,01%; ebenso kann die Empfindlichkeit für ein beliebiges Spektrum und einen allseitig gerichteten Fluß nicht berechnet werden. Den größten Nachteil jedoch, den diese

---

[1] Den Herren Dr. FENTON und Dr. GREENHILL gilt besonderer Dank für die Einsicht in die unveröffentlichten Daten.

ummantelten Zähler aufweisen, muß man in der Tatsache sehen, daß einfallende Protonen über (p, n)-Reaktionen innerhalb des Bremsmaterials oder des Flugkörpers Neutronen erzeugen, die nach Abbremsung mitgezählt werden und die Neutronenintensität in nicht kontrollierbarer Weise verfälschen.

Es ist unmöglich, diese örtlich erzeugten Neutronen von der Gruppe der Neutronen der Atmosphäre, deren Messung wir anstreben, zu trennen. Der Neutronenfluß kann durch solche geladen ausgelösten Neutronen vielleicht bis zu einem Faktor 2 verfälscht werden; besonders ungünstig wird diese Situation, wenn bei einem speziellen Experiment ein großer Fluß energiereicher Protonen vorliegt, wie in dem oben erwähnten Nordlichtexperiment von KORFF und HAYMES bzw. bei solar flares oder den oben erwähnten Schwankungen.

Es sind viele Versuche unternommen worden, um diese Schwierigkeiten zu überwinden. ALBERT et al. [Al 62] zählen die Protonen in szintillationsfähigem Bremsmaterial und benutzen Antikoinzidenz zwischen dem Szintillator und dem langsamen Neutronenzähler. TRAINOR und LOCKWOOD [Tr 63] benutzen zum gleichen Zweck eine Lage von Proportionalzählern um den langsamen Neutronenzähler herum.

In beiden Techniken ist zunächst die Unempfindlichkeit gegenüber dem Nachweis schneller Neutronen erhalten geblieben, hinzu kommt die lange Dauer der Neutronenabbremsung in der Größenordnung von 100 μsec, welche schon bei mäßigen Flüssen geladener Teilchen den Szintillator überlastet.

β) Inzwischen wurden empfindliche *Zähler für schnelle Neutronen* entwickelt, die den Protonenrückstoß, den schnelle Neutronen auslösen, zur Zählung benützen.

(i) CH_2 (Polyäthylen-) ausgefütterte Proportionalzähler. MOYER [Mo 52] und THOMPSON [Th 55] haben einen Argon- und CO_2-Proportionalzähler entwickelt, in dem die Rückstoßprotonen, die von Neutronen bei Stößen in einer CH_2-Ausfütterung herausgeschlagen werden, gezählt werden. Für eine etwa 3 mm ($^1/_8''$) starke CH_2-Ausfütterung ist die Ausbeute dieses Proportionalzählers für Neutronen im Energiebereich von 50 keV bis 20 MeV nahezu energie-proportional. Aus diesem Grund mißt dieser Zähler tatsächlich den Energiefluß und nicht wie bei verschiedenen anderen Arten von Zählern den Teilchenfluß.

(ii) Schnelle Neutronenzähler mit flüssigem organischem Szintillator. Hier wird vom Protonenrückstoßmechanismus im Szintillator Gebrauch gemacht. Diese Zähler werden unter der Firmenbezeichnung Ne 213 von der Firma Nuclear Enterprises Ltd., Winnipeg, hergestellt. Eine ausführliche Beschreibung der physikalischen und technischen Daten findet man bei MENDELL [Me 63a] und MENDELL und KORFF [Me 63b] bzw. HAYMES [Ha 64].

(iii) Den Protonenrückstoß durch Neutronen bzw. die Spuren geladener Teilchen bei Kernreaktionen in einer Hochdruck-Nebelkammer nützen MIYAKE, HINOTANI und NUNOGAKI [Mi 57a] bei Wasserstoff-Füllung und MIYAKE, HINOTANI, KATSUMATA und KANEKO [Mi 57b] bei Stickstoff-Füllung. Die Nebelspuren der Protonen bzw. Kerne wurden stereographisch aufgenommen und vermessen, und damit die Neutronenverteilung, der Neutronenfluß und die Energieverteilung bestimmt.

γ) *Messungen mit Zählern für schnelle Neutronen.* Mit CH_2-ausgefütterten Proportionalzählern haben HESS, PATTERSON, WALLACE und CHUPP [He 59] am Erdboden und in verschiedenen Flughöhen bis zu 12 km Höhe den Energiefluß schneller Neutronen in einem Energiebereich von etwa 100 keV bis 25 MeV direkt gemessen. Diese Daten werden in Ziff. 31 bei der Bestimmung des Energiespektrums benutzt. Mit flüssigen Szintillatoren haben MENDELL und KORFF [Me 63b] bei einem Ballonaufstieg bis 27,4 km Höhe an einem geomagnetisch ruhigen Tag,

Tabelle 12. *Die Absorptionslängen schneller Neutronen für verschiedene geomagnetische Breiten und Höhen.* Die Zählraten sind mitangegeben (nach [*Ga 64*]).

Zähler-Charakteristiken			Hess et al. [*He 59a*]	Simpson [*Si 51b*]	Gauger [*Ga 64*]
Zählerwirkungsquerschnitt $\Sigma_{\text{eff}(n,\alpha)}$			2,6 cm²	8,8 cm²	18,8 cm²
Bremssubstanz, Paraffindicke			5 cm	4,4 cm	5 cm

Geomagnetische Breite Λ	Autor	Höhe g/cm²	Zähler	Absorptionslänge, mittlere () = Höhe g/cm²	Zählraten, bezogen auf 245 g/cm²	Meßdatum	rel. Zählraten. Effektive Flächen/cm² sec
0°	Simpson	306	200 cpm	212	254 cpm	Aug. 1949	0,48
	Simpson	272	250 cpm	212	284 cpm	Dez. 1947	0,54
	Simpson	245	265 cpm	212	265 cpm	Aug. 1949	0,50
	Simpson	231	270 cpm	212	253 cpm	Juni 1948	0,48
	Simpson	224	315 cpm	212	285 cpm	Dez. 1947	0,54
	Gauger	245	9,2 cps		9,2 cps	Mai—Juli 1962	0,50
19° N	Simpson	306	260 cpm	206	350 cpm	Juni 1948	0,86
	Simpson	272	314 cpm	206	358 cpm	Dez. 1947	0,88
	Gauger	245	12,5 cps		12,5 cps	Mai—Juli 1962	0,69
36° N	Hess et al.	193	300 cpm	150	212 cpm	Winter 1956/57	1,36
	Gauger	221	22,3 cps	130	19,3 cps	12. Juli 1962	1,04
	Gauger	245	22,3 cps		22,3 cps	21. Juli 1962	1,20
40° N	Simpson	306	530 cpm	181	741 cpm	April—Mai 1949	1,40
	Simpson	272	600 cpm	181	696 cpm	Dez. 1947	1,32
	Simpson	238	800 cpm	181	832 cpm	April—Mai 1949	1,57
	Simpson	224	800 cpm	181	898 cpm	Aug. 1949	1,70
	Simpson	306	520 cpm	181	727 cpm	Aug. 1949	1,37
	Hess et al.	590	31 cpm	150	332 cpm	Sommer 1956	2,14
	Hess et al.	412	90 cpm	150	272 cpm	Winter 1956/57	1,74
	Hess et al.	480	70 cpm	150	335 cpm	Winter 1956/57	2,15
	Gauger	221	28,6 cps	130	24,7 cps	12. Juli 1962	1,33
	Gauger	245	26,3 cps		26,3 cps	21. Juli 1962	1,42
42° N	Simpson	306	630 cpm	181	880 cpm	31. März 1949	1,67
	Gauger	202	36,3 cps	137 (von 202 bis 221) 127 (von 221 bis 245)	27,3 cps	12. Juli 1962	1,48
	Gauger	245	31,2 cps		31,2 cps	21. Juli 1962	1,68
	Boella et al.	200		180 ± 5 (200 bis 600)		Okt. 1963	
44° N	Hess et al.	860	5 cpm	150	300 cpm	Sommer 1956	1,92
	Hess et al.	720	12 cpm	150	282 cpm	Sommer 1956	1,81
	Hess et al.	638	26 cpm	150	359 cpm	Sommer 1956	2,30
	Hess et al.	388	110 cpm	150	287 cpm	Winter 1956/57	1,84
	Gauger	202	39,3 cpm	137 (von 202 bis 221) 127 (von 221 bis 245)	29,6 cps	12. Juli 1962	1,60
	Gauger	245	36,9[1]		36,9 cps	21. Juli 1962	1,99
46° N	Simpson	306	710 cpm	157	993 cpm	31. März 1949	1,88
	Gauger	202	44,0 cps	137 (von 202 bis 221) 127 (von 221 bis 245)	32,9 cps	12. Juli 1962	1,78
48° N	Simpson	306	760 cpm	157	1160 cpm	31. März 1949	2,12
	Hess et al.	324	200 cpm	150	339 cpm	Winter 1956/57	2,17
	Gauger	202	49,4 cps	137 (von 202 bis 221) 127 (von 221 bis 245)	37,3 cps	12. Juli 1962	2,02
	Gauger	245	43,8 cps		43,8 cps	21. Juli 1962	2,37
51° N	Hess et al.	590	40 cpm	150	428 cpm	Winter 1956/57	2,74
	Gauger	202	57,0 cps	132 (von 202 bis 221) 127 (von 221 bis 245)	42,6 cps	12. Juli 1962	2,31
	Gauger	245	47,2 cps		47,2 cps	21. Juli 1962	2,55
54° N	Simpson	306	830 cpm	157	1220 cpm	9. April 1949	2,31
	Gauger	184	70,8 cps	165 (von 184 bis 202) 132 (von 202 bis 221) 127 (von 221 bis 245)	47,6 cps	12. Juli 1962	2,57
	Gauger	245	47,6 cps		47,6 cps	21. Juli 1962	2,57

[1] Linear interpolation.   cps = counts per second.   cpm = counts per minute.

8. November 1962, in einer geomagnetischen Breite von 53° N den Fluß schneller Neutronen gemessen. Die Neutronenflußdaten wurden in zwei Energiebereichen gemessen. In Fig. 39 ist der Fluß schneller Neutronen für Protonenrückstoßenergien von 1 bis 10 MeV wiedergegeben.

Die mittlere Absorptionslänge für die gesamte Neutronenzählrate im Höhenbereich von 200 bis 600 g/cm² atmosphärischer Tiefe war $L = 145 \pm 6$ g/cm². In einer Höhe von 75 mb atmosphärischem Druck wurde ein breit ausgedehntes Neutronenmaximum gemessen.

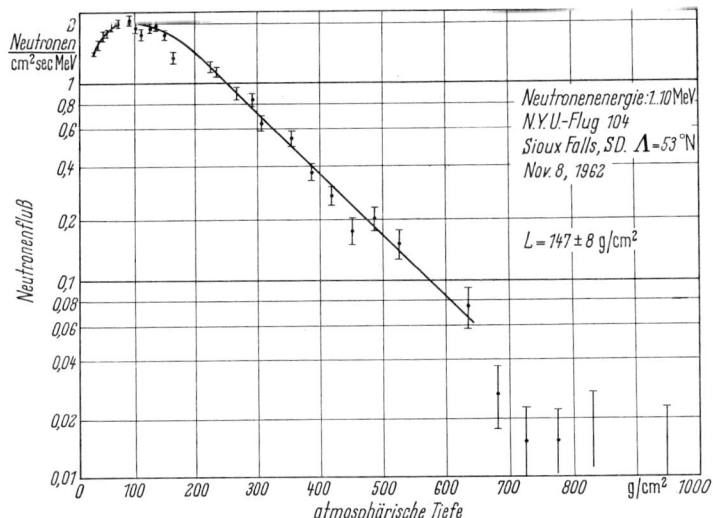

Fig. 39. Neutronenfluß bei Neutronenenergien von 1—10 MeV in Abhängigkeit von der atmosphärischen Tiefe in g/cm² nach MENDELL und KORFF [Me 63b].

Ein differentielles Energiespektrum zwischen 1 und 10 MeV wurde durch ein Potenzgesetz der Form

$$N(E)\, dE = N(1)\, E^{-n}\, dE \quad \text{Neutronen/cm}^2 \text{ sec MeV} \tag{21.1}$$

angenähert. Zwischen 200 und 700 mb wurde für den Exponenten n ein Wert von $1{,}16 \pm 0{,}2$ MeV gefunden, die Höhenabhängigkeit x (g/cm²) des differentiellen Flusses konnte im Bereich von 200 bis 600 g/cm² durch

$$N(E, x)\, dE = 2{,}6 \cdot E^{-1{,}16 \pm 0{,}2} \cdot e^{-0{,}0069\, x} \cdot dE \left(\frac{\text{Neutronen}}{\text{cm}^2 \text{ sec MeV}} \right) \tag{21.2}$$

angegeben werden.

MIYAKE, HINOTANI und NUNOGAKI [Mi 57a] nähern den gemessenen, differentiellen Fluß bei $\Lambda = 25°$ N in 760 g/cm² Tiefe mit dem Potenzspektrum

$$N(E)\, dE\, d\Omega = (1{,}2 \pm 0{,}48)\, 10^{-3} \cdot E^{-1{,}25 \pm 0{,}1}\, dE \cdot d\Omega \left(\frac{\text{Neutronen}}{\text{cm}^2 \text{ sec MeV sterad}} \right). \tag{21.3}$$

Bei isotropem, schnellem Fluß der Neutronen besteht hinreichende Übereinstimmung mit den Spektren (21.2) und (21.4). Somit wird ein Vergleich der Meßergebnisse des Flusses schneller Neutronen von MENDELL und KORFF in verschiedenen Höhen mit theoretischen Aussagen bedeutungsvoll. Verglichen wird mit den Rechnungen von NEWKIRK [Ne 63], HESS, CANFIELD und LINGENFELTER [He 61], deren grundlegende Arbeit auf den Messungen von HESS, PATTERSON, WALLACE und CHUPP [He 59a] aufbaut. In Fig. 40 wird diese Gegenüberstellung vorgenommen. Man erkennt, daß die gemessenen (und errechneten) Flußwerte bei

HESS et al. größer ausfallen als in den Messungen von MENDELL und KORFF, während befriedigende Übereinstimmung mit den theoretischen Kurven von NEWKIRK zu bestehen scheint, wo nur Abweichungen bis zu 30% vorliegen. NEWKIRK stützt sich auf

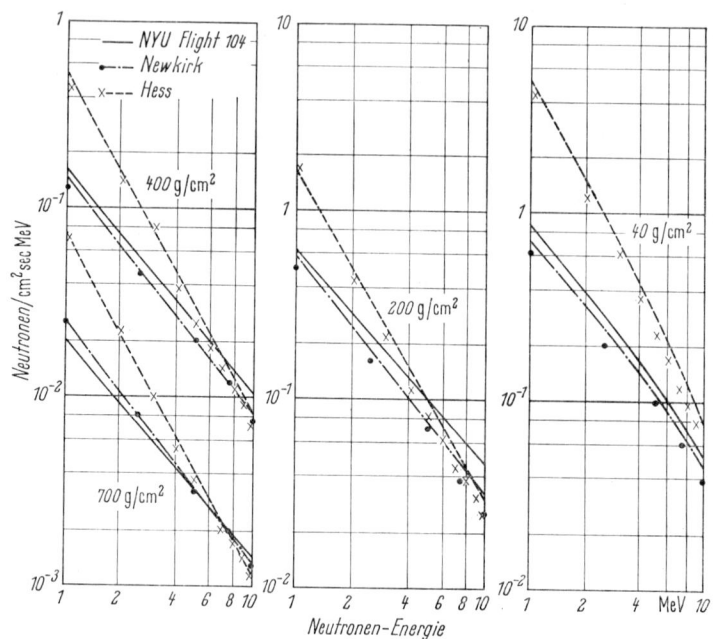

Fig. 40. Das gemessene differentielle Neutronenenergiespektrum von 1—10 MeV in vier verschiedenen Höhen (750, 400, 200 und 42 g/cm²) nach MENDELL und KORFF [Me 63b]. Verglichen wird mit den theoretischen Kurven von NEWKIRK [Ne 63] und HESS et al. [He 61].

Tabelle 13. *Vergleich der Exponenten der Energiespektren schneller Neutronen:* $N(E)\,dE = N_0(E_0)\,E^{-n}\,dE$; $E_0 = 1$ *MeV für ca. 3000 m über Meereshöhe.*

Autor	Exponent n	Λ geom. °N	Höhe g/cm²	$N_0(E_0) \times 10^{-2}$ Neutronen cm² sec MeV
MIYAKE et al. [Mi 57a]	$1{,}25 \pm 0{,}01$	29	760	$(1{,}5 \pm 0{,}6)$
HAYMES [Ha 64b]	$1{,}3 \pm 0{,}1$	41,5	700	$\left(5{,}6{}^{+3{,}1}_{-2{,}0}\right)$
HESS et al. [He 59a]	≈ 2	44	700	7,5
KASTNER et al. [Ka 63]	$\geq 2{,}1$	—	1030	(0,09)
MENDELL und KORFF [Me 63b]	$1{,}16 \pm 0{,}2$	53	700	2,0
KORFF, MENDELL und HOLT [Ko 65]	1,2	53	400	32,0
dieselben geben für $E \geq 10$ MeV	< 1	53	(700)	(5,0?)
		69	(700)	5,5
HESS et al. [He 61]	1,68	44	berechnet (Fig. 65 und 40)	
NEWKIRK [Ne 63]	1,13	57	berechnet (Fig. 66 und 40)	
LINGENFELTER [Li 63c]	1	beliebig wählbar	berechnet	

ein Neutronen-Quellspektrum nach MIYAKE et al. [Mi 57a, b], Ziff. 17. Zweckmäßig geben wir hier in Tabelle 13 für eine Höhe von rund 3000 m ü. M. (ca. 700 g/cm²) die von verschiedenen Autoren gefundenen Exponenten des Potenzspektrums des differentiellen Flusses an. Man sieht deutlich, daß die Werte bei den Flugzeugmessungen von HESS erheblich höher liegen als die Werte aller anderen Autoren. Die

Messungen in Meereshöhe durch KASTNER et al. [Ka 63] sind nahezu in Übereinstimmung mit dem steilen Spektrum von HESS. Es ist daher erforderlich, diese Werte noch durch eine größere Zahl von Messungen eindeutig zu klären. Nachdem die Unstimmigkeit im Bereich der Quellneutronen auftritt, erhebt sich die Fage, inwieweit zusätzliche Neutronenproduktion in der nächsten Umgebung der Meßgeräte Anlaß zu solchen Unterschieden gegeben haben kann. Dieselbe Fragestellung wird uns noch einmal begegnen, und zwar bei der Messung von Verlustflußneutronen, wo der Raketenkörper als Neutronenquelle störend den kleinen Verlustfluß überdecken kann. Eine Größe, die unabhängig von der Neutronenproduktion ist, sollte die mittlere Absorptionslänge in einem bremsenden Medium sein.

Für schnelle, direkt gemessene Neutronen liegen noch wenig Daten vor, um eine effektive Absorptionslänge L_{eff} in Abhängigkeit von der geomagnetischen Breite angeben zu können, wie dies in Fig. 31 für die Dichteverteilung der langsamen Neutronen im Gleichgewichtsbereich geschah. Zudem liegen sämtliche Messungen im Zeitraum des Sonnenfleckenmaximums und der nachfolgenden solar stark gestörten Periode. Stellt man sich aber die Werte der Absorptionslängen L (g/cm²) schneller Neutronen in Tabelle 14 zusammen, so erkennt man im Gang der Zahlwerte in Abhängigkeit von der geomagnetischen Breite Λ ein ähn-

Tabelle 14. *Absorptionslängen schneller Neutronen (direkte Phoswich-Messung).*
(Nachträglich ergänzt mit Daten vom Sonnenfleckenminimum 1964/1965.)

Autor	Absorptionslänge L g/cm²	Geomagnetische Breite °N	Jahr der Messung	Bemerkungen
KORFF, MENDELL und HOLT [Ko 65]	226	8	1965, März	
HAYMES [Ha 64 b]	169 ± 13	41,5	1963	
KORFF, MENDELL und HOLT [Ko 65]	204 ± 30	43	1964, Sept.	Meßhöhe: 200—400 mb
HESS [He 59 a]	155	44	1957/58	
MENDELL und KORFF [Me 63 b]	145 ± 6	53	1962	
KORFF, MENDELL und HOLT [Ko 65]	159 ± 7	53	1964, Sept.	von 335—550 mb "burst" Störungen; Daten nicht berücksichtigt
KORFF, MENDELL und HOLT [Ko 65]	159 ± 5	69	1965, Aug.	

liches Verhalten der schnellen Neutronen wie der langsamen Neutronen. Dies ist nicht verwunderlich, weil ja die Neutronen nach ihrer Erzeugung nur kurze Diffusions- bzw. Bremswege haben. Die schnellen Neutronen befinden sich ebenfalls nahe dem Entstehungsort und spiegeln so die geomagnetischen Einflüsse auf die Primären wider.

HAYMES [Ha 64b] hat 1963 eine Reihe von Ballonflügen bei einer geomagnetischen Breite $\Lambda = 41°$ Nord bis zu atmosphärischen Tiefen von 3,6 g/cm² durchgeführt. Neben γ-Messungen hat er mit dem oben erwähnten Phoswich-Szintillationszähler schnelle Neutronen von 1—14 MeV gemessen. Die gemessenen Flußwerte von HAYMES sind, temperaturkorrigiert, in Fig. 41 in einer Gesamtübersicht dargestellt. Wie üblich ist der Neutronenfluß über der atmosphärischen Tiefe aufgetragen. Um einen Vergleich mit anderen Arbeiten ziehen zu können, die einen linearen Höhenmaßstab benützen, hat HAYMES die Daten von Fig. 41 auf eine lineare Höhenskala übertragen (Fig. 43). Den Ergebnissen von HAYMES kann man entnehmen, daß sich der Fluß der schnellen Neutronen (bei niederen Höhen, im Gleichgewichtsbereich) sehr ähnlich verhält wie die Dichte der langsamen Neutronen. Der Fluß wächst exponentiell mit wachsender Höhe (im Druckbereich von

700 mb bis 200 mb) an. HAYMES hat versucht, die Höhenabhängigkeit der mittleren Absorptionslänge L zu überprüfen, die ja NEWKIRK [Ne 63] für die Höhenverteilung der Neutronenquellen (vgl. Fig. 41 und S. 430) zu $L=140$ g/cm² bzw. $L=160$ g/cm² vorgeschlagen hatte. Für die 5 Ballonflüge sind die Einzelwerte der

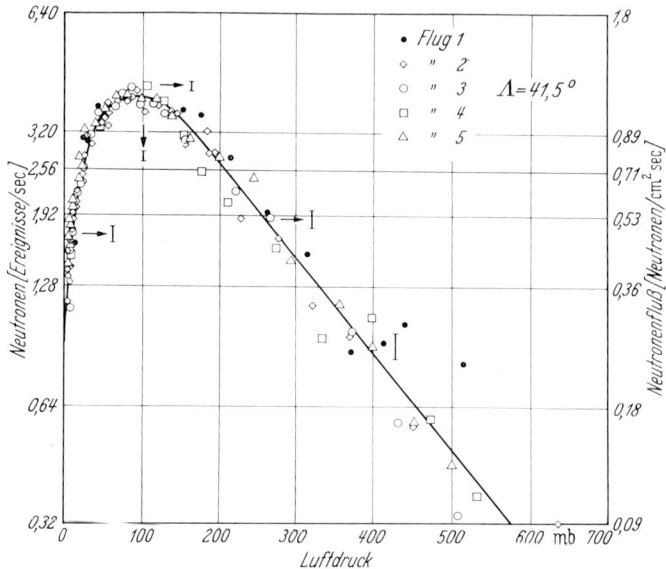

Fig. 41. Fluß schneller Neutronen nach HAYMES [Ha 64b]. Mittelwert über alle 5 Flüge im April und Mai 1963 bei $\Lambda_{\text{geom.}} = 41{,}5°$ N.

mittleren freien Absorptionslängen in Tabelle 15 zusammengestellt. Die Statistik ist nicht gut genug, um eine Stellungnahme zu NEWKIRKS Annahme zu geben; der Mittelwert, den HAYMES findet, ist 169 ± 13 g/cm². Vielleicht ist es zweckmäßig, noch darauf hinzuweisen, daß die effektive Absorptionslänge für langsame Neutronen bei 40° geomagnetisch Nord bei 180 g/cm² liegt, allerdings bezogen auf das Sonnenfleckenminimum.

Tabelle 15. *Mittlere freie Weglängen für Absorption in der tieferen Atmosphäre.*

Flug	Zeit 1963	L (g/cm²)
1	25. April	260 ± 79
2	1. Mai	179 ± 14
3	7. Mai	171 ± 12
4	11. Mai	183 ± 11
5	14. Mai	154 ± 14

Mittel $L = 169 \pm 13$ g/cm²

Das Übergangsmaximum für den Fluß schneller Neutronen wird bei ungefähr 90 g/cm² atmosphärischer Tiefe erreicht, das wiederum ein sehr ähnliches Verhalten zeigt wie die Dichte der langsamen Neutronen in diesen geomagnetischen Breiten. Im Maximum ist der schnelle Neutronenfluß 1,1 Neutronen/cm² sec. Man erwartet, daß die Lage des Maximums breitenabhängig ist, und zwar in der Weise, daß mit zunehmender geomagnetischer Breite die Lage des Flußmaximums schneller Neutronen in größere Höhen verschoben erscheint. Ebenso wahrscheinlich ist es, daß die Flußwerte der Übergangsmaxima zeitabhängig sind, denn die elfjährige Sonnenfleckenperiode wirkt sich über die Modulation der einfallenden, galaktischen Strahlung direkt im Fluß der Neutronen aus.

KORFF, MENDELL und HOLT [Ko 65][1] haben im Zeitraum des Sonnenfleckenminimums 1964/65 den schnellen Neutronenfluß (1—10 MeV) nach Breite

---

[1] Herrn Prof. Dr. KORFF, Frau Dr. MENDELL und Herrn Dr. HOLT sei an dieser Stelle nochmals warm für die Überlassung der Meßergebnisse vor der Publikation gedankt.

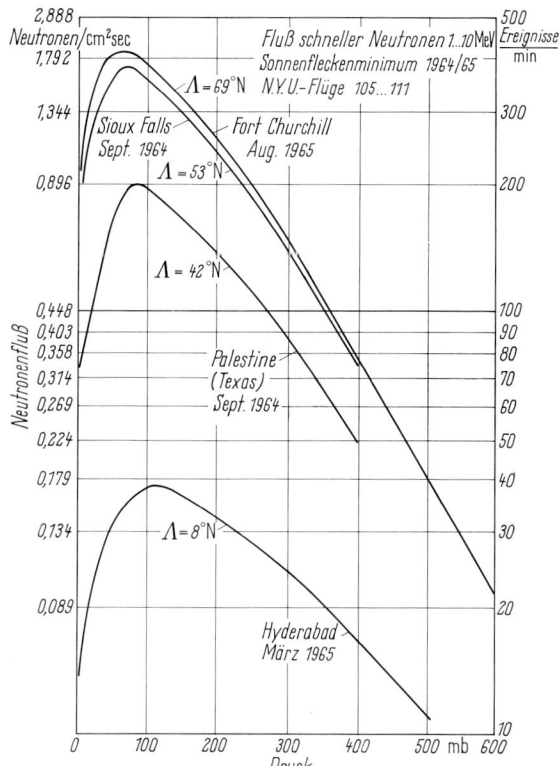

Fig. 42. Die Breitenabhängigkeit des Flusses schneller Neutronen (1—10 MeV) im Zeitraum der ruhigen Sonne 1964/65 nach [Ko 65].

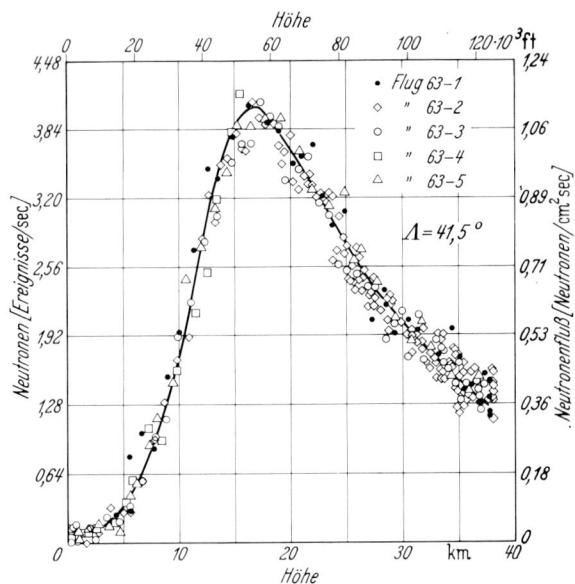

Fig. 43. Fluß schneller Neutronen in Abhängigkeit der Höhe (in linearem Maßstab) über der Erdoberfläche. Daten wie in Fig. 41. Die Beobachtung, daß der Fluß bei Tiefen von weniger als einer mittleren freien Weglänge abfällt, deutet auf einen wesentlichen Beitrag von großen Zenitwinkeln hin.

Tabelle 16. *Fluß schneller Neutronen im Zeitraum des Sonnenfleckenminimums 1964/1965 und kurz zuvor.*
Breitenabhängigkeit und extrapolierte Flußwerte für den Gipfel der Atmosphäre. Weiterhin Angabe der „gesamten Breitenabhängigkeit" = Zählratenverhältnis Pol/Äquator in Funktion der Höhe z für langsame und für schnelle Neutronen. Werte von KORFF et al. [Ko 65] für 1—10 MeV-Neutronen (Rückstoßprotonenzähler). Werte von BOELLA et al. [Bo 63, 65] für thermische bis 20 MeV Neutronen (ummantelte Zähler); entsprechendes gilt für GREENHILL et al. [Gr 65a].

Ort	Geomagnetische Breite oder magnetische Abschneidesteifigkeit in GV		Fluß schneller Neutronen Φ Neutronen/cm² sec			Autoren	Höhe z (mb)	Gesamte Breitenvariation = Zählratenverhältnisse (Pol/Äquator in der Höhe z) für		
	Nord	Süd	25 mb	4 mb	Φ_g (extrapoliert)			langsame Neutronen Minimum 1953/54 nach Daten von [So 56]	schnelle Neutronen nach Daten von	
									[Ko 65]	[Gr 65a]
Fort Churchill	69°		1,65	1,48	0,96	⎫				
Sioux Falls	53°		1,34	1,18	0,65	⎬ Ko 65	400	3,25	4,66	4,78
Palestine	42°		0,54	0,36	0,29		300	3,43	6,00	—
Holloman (New Mexico)	42°;	4,6	1,65[1]	0,56[1]	0,45[1]	Bo 65	240	3,9	7,6	5,36
	41,5°		0,71	0,40	0,24	Ha 64b	200	4,2	7,1	6,3
Hyderabad	8°		0,11	0,09	0,055	Ko 65	110	5,1	9,9	—
							100	5,58	10,35	9,68
Lae		15,8	66,2[3]	—	0,06	⎫	80	6,7	11,0	—
Brisbane		7,0	210[3]	—	0,17	⎬	65	7,1	12,35	—
Mildura		4,0	347[3]	—	0,36	⎬ Gr 65a[2]	50	8,5	13,5	11,55
Hobart		1,79	663[3]	—	0,68		25	11,3	15,4	13,65
Wilkes		<0,02	902[3]	—	0,96	⎭	15	(12—15)[4]	16,5	[5]
							4	(?)[4]	18,15	

(8°; 43°; 53°; 69°) und Höhe gemessen (vgl. Fig. 42 und Tabelle 16) und bestätigen diese Erwartungen (vgl. auch S. 436f).

In großen Höhen zeichnet sich ein Unterschied im Verhalten der schnellen Neutronen zu den langsamen ab. Ursache dafür ist der unterschiedliche Neutronenverlustfluß. HAYMES und KORF [Ha 60a] haben gezeigt, daß die Dichte der langsamen Neutronen vom Maximum bis zu den höchsten gemessenen Ballonhöhen (etwa 5 g/cm²) scharf abfällt, und zwar etwa um einen Faktor 50 (vgl. S. 439 und Fig. 34 und 35). Im Gegensatz dazu nimmt der Fluß schneller Neutronen grob nur um einen Faktor 3 im Vergleich zum Flußwert beim Neutronenmaximum ab (Fig. 43), wo diese geringfügige Abnahme direkt abgelesen werden kann.

HAYMES hat versucht, den in 3,6 g/cm² Höhe gemessenen Flußwert von 0,39 Neutronen/cm² sec zum Gipfel der Atmosphäre hin zu extrapolieren. Zu diesem Zweck stellt er seine Daten in Fig. 44 für die höchsten Höhen zusammen und versucht, zunächst den Mittelwert der Meßdaten exponentiell zum Gipfel der Atmosphäre zu extrapolieren. Dies führt auf Punkt A in der Fig. 44. Es gibt jedoch keinen Grund zur Annahme, daß die Exponentialkurve eine richtige Flußapproximation liefert. HAYMES stützt sich deshalb auf den von NEWKIRK [Ne 63] errechneten Fluß der Neutronen (vgl. Abschnitt 38). NEWKIRK errechnet über die Boltzmannsche Transportgleichung den Fluß mit der S_n-Näherung, die auch in

---
[1] Meßwerte für ummantelte Zähler (Polyäthylene) und Korrektur für lokal erzeugte Neutronen.
[2] [Gr 65a] extrapolieren den Fluß in 15 mb Höhe zum Verlustfluß und normieren mit dem Verlustflußwert von NEWKIRK [Ne 63].
[3] Zählraten der ummantelten Fluß-Zähler. Absolute Normierung auf Neutronen/cm² sec s. später erscheinende Originalarbeit.
[4] Extrapolierte Werte.
[5] Zählraten für 15 mb nicht angegeben. GREENHILL, FENTON et al. haben diese 15 mb Flußwerte mit Hilfe des theoretischen Verlustflusses von NEWKIRK als „experimentellen Verlustfluß" normiert; dafür ergibt sich eine Breitenvariation von 16,5.

Gebieten, in denen kein Diffusionsgleichgewicht mehr vorliegt, eine genaue Flußdarstellung ermöglicht. Im Gegensatz dazu kann man nicht erwarten, daß die Diffusionstheorie die Gestalt des Neutronenflusses innerhalb von zwei bis drei mittleren freien Weglängen am Rand der Atmosphäre wiedergeben wird. HAYMES hat mit einer Parallelkurve zu der Newkirkschen Flußkurve[1] seine Daten zum Gipfel der Atmosphäre hin extrapoliert und findet so Punkt B in Fig. 44. Er schließt damit auf einen Fluß von $0,24 \pm 0,02$ Neutronen/cm² sec als Mittelwert über alle Flüge. Es wird angenommen, daß dies eine genauere Abschätzung für den (isotropen?) schnellen Fluß der Neutronen am Gipfel der Atmosphäre ist. Der theoretische Verlustfluß beträgt davon etwa ein Viertel.

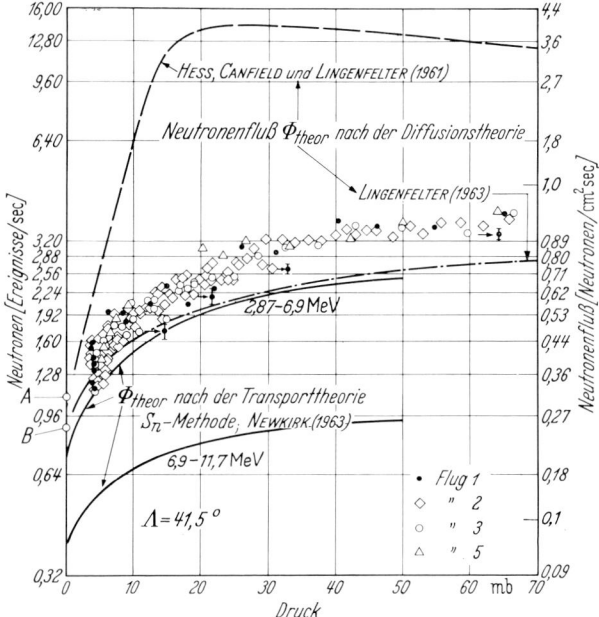

Fig. 44. Extrapolation der Flußwerte von HAYMES [Ha 64b] zum Gipfel der Atmosphäre mit Hilfe der S_n-Approximation nach NEWKIRK. Wegen der Diskussion der Punkte A und B wird auf den Text verwiesen. Alle Kurven sind auf 41,5° N korrigiert gemäß der Breitenabhängigkeit nach LINGENFELTER [Li 63a].

Aus ihrer Messung des Flusses schneller (1—10 MeV) Neutronen beim Sonnenfleckenminimum 1964/65 extrapolieren KORFF et al. [Ko 65] mit einer Exponentialkurve auf den Fluß Φ_g am Gipfel der Atmosphäre (vgl. Tabelle 16). Auch hier zeigt ein Vergleich mit Tabelle 36 (S. 520), daß der (korrigierte) theoretische Verlustfluß etwa $1/4\ \Phi_g$ beträgt.

HAYMES [Ha 64b] kann für den Gleichgewichtsbereich des Flusses schneller Neutronen im Höhenbereich ($150 \leq x \leq 700\ \text{g/cm}^2$) ein differentielles Flußspektrum angeben, das allerdings eine Abhängigkeit von der atmosphärischen Tiefe und eine ziemliche Ähnlichkeit mit dem von MENDELL und KORFF (21.2) aufweist:

$$N(E, x)\, dE = 2,5 \cdot e^{-(0{,}0059 \pm 0{,}0004) \cdot x}\, E^{-1{,}3 \pm 0{,}1} \cdot dE \left(\frac{\text{Neutronen}}{\text{cm}^2\ \text{sec MeV}} \right). \qquad (21.4)$$

b) Sterne in Kernemulsion.

22. Mehrere Autoren (GEORGES und EVANS [Ge 50b]; BERNARDINI, CORTINI und MANFREDINI [Be 50b]; LATTIMORE [La 49]; LORD [Lo 51]; BRIDGE [Br 47];

---
[1] Es handelt sich um die Flußkurve der Energiegruppe (2,87—6,9) MeV in Fig. 4 auf S. 1830 der Arbeit [Ne 63].

ADDARIO [Ad 49]) haben in Emulsion die Zahl der Sterne/cm³ Tag gemessen, die durch die nukleare Komponente in verschiedenen Höhen der Atmosphäre ausgelöst werden (vgl. Kap. C). In Fig. 45 ist diese Sternhäufigkeit in Kurve B dargestellt. Die Vergleichkurve A gibt die in Ziff. 19 erwähnte Messung von YUAN [Yu 51] wieder, die beweist, daß die Sternhäufigkeit als Produktionsmechanismus der Neutronen in der Atmosphäre und die Neutronendichte in der Atmosphäre gleichsinnig verlaufen. In diesem Zusammenhang haben wir dort auch auf die Messungen von SIMPSON [Si 51a] verwiesen.

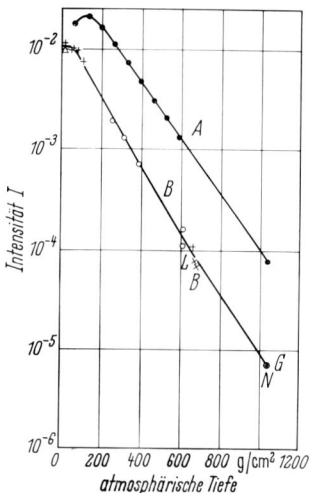

Fig. 45. Vergleich des Intensitätsverlaufs langsamer Neutronen (Kurve A) mit der Häufigkeit von Sternen und Kernexplosionen in Kernemulsion (Kurve B), jeweils aufgetragen über der atmosphärischen Tiefe (Neutroneneinfang und Sternintensität in Einheiten sec⁻¹g⁻¹ Luft) YUAN [Yu 51], ○ BRIDGE [Br 47], + LORD [Lo 51], × LATTIMORE [La 49], ■ BERNARDINI [Be 50], ● GEORGE [Ge 50], N Normalisationspunkt für Kernexplosionen, ADDARIO [Ad 49].

Die Intensität bzw. das Spektrum der Neutronen von einigen 100 MeV Energie soll über „Sterne" in der Kernspurplatte bestimmt werden. Wir beschränken uns hier auf energiearme Nukleonenprozesse mit schwarzen Spuren und vernachlässigen alle höherenergetischen Ereignisse mit grauen und dünnen Spuren, d.h. Sterne, bei denen gleichzeitig Pionen und relativistische Protonen mitbeteiligt sind. Die Gruppe der Sterne mit 3 Spuren kann man untergliedern in 3-Spuren-Sterne die von einem Neutron ausgelöst werden, und in „2-Spuren-Sterne", die durch Protonen ausgelöst werden, wobei das einfallende Proton durch eine der 3 Spuren mitgezählt ist.

Die mittlere freie Weglänge λ für die Sternproduktion kennzeichnen wir für Sterne, die zwei oder mehr geladene Spuren aufweisen mit λ_2, weiter sei λ_3 die mittlere freie Weglänge mit drei oder mehr geladenen Spuren. GERMAIN [Ge 51], BERNARDINI et al. [Be 52]; LOCK et al. [Lo 55]; JOHNSON [Jo 55] haben die mittleren freien Weglängen λ_2 und λ_3 und deren Energieabhängigkeit ausführlich untersucht. GERMAIN gibt direkt die Werte von λ_2 und λ_3. Die anderen Autoren geben nur die Häufigkeitsverteilung der Spuren in Abhängigkeit von der Energie an. HESS et al. [59a] rechnen diese in mittlere freie Weglängen um und benutzen dazu für die chemischen Bestandteile der Kernemulsion die jeweiligen geometrischen mittleren freien Weglängen für die Einzelnuklide, bei geeigneter Mittelbildung einen Wert von etwa 29 cm. Nach COCCONI [Co 60] beträgt die über Wirkungsquerschnitte errechnete mittlere freie Weglänge für Protonen in Emulsion $\lambda_{\mathrm{th}} = 35{,}3$ cm. Nach den Messungen kann man schließen, daß λ_n keine Abhängigkeit von der atmosphärischen Tiefe zeigt. Weiterhin nimmt man an, daß λ_n denselben Wert annimmt sowohl für neutronenausgelöste als auch für protonenausgelöste Ereignisse.

Die mittleren freien Weglängen benutzen wir zur Festlegung der Reaktionsraten Σ für Sternproduktion

$$\Sigma_n = \frac{1}{\lambda_n}\left(\frac{\text{Sterne mit} \geq n \text{ geladenen Spuren}}{\text{Neutronen (oder Protonen) cm}^3 \text{ Tag}}\right). \tag{22.1}$$

Eine Zusammenfassung der reziproken Werte $1/\lambda_n$ ist in Fig. 46[1] enthalten; dort sind die Reaktionsraten $\Sigma_2(E) = 1/\lambda_2$ und $\Sigma_3(E) = 1/\lambda_3$ in Abhängigkeit von der Energie zu verfolgen.

---

[1] Für $E \to \infty$ strebt $1/\lambda$ nicht wie Fig. 46 gegen den Grenzwert 0,033, sondern nach neueren Messungen gegen die Grenze $> 0{,}028$. Vgl. dazu etwa LOHRMANN, et. al. [Lo 61] u. a. etwa

Kennt man den sternauslösenden differentiellen Fluß Φ_p der Protonen (wobei jedes sternauslösende Proton in der Spurenzahl mitgezählt wird), so ist die differentielle Zählrate dZ_p protonenausgelöster Sterne mit $n=3$ oder mehr Spuren

$$dZ_p(n \geq 3) = \Phi_p(E) \cdot \Sigma_2(E) \cdot dE = \Phi_p(E) \frac{dE}{\lambda_2(E)}. \tag{22.2}$$

Entsprechend ist der Anteil durch den Neutronenfluß Φ_n neutral ausgelöster Dreispurensterne

$$dZ_n(n \geq 3) = \Phi_n(E) \Sigma_3(E) dE = \Phi_n(E) \frac{dE}{\lambda_3(E)}. \tag{22.3}$$

Die theoretischen Zählraten

$$Z(n \geq 3)\left[\frac{\text{Sterne}}{\text{cm}^3 \cdot \text{Tag}}\right] = \int_{E_0}^{\infty} \frac{dE}{\lambda_2(E)} \cdot \Phi_p(E) + \int_{E_0}^{\infty} \frac{dE}{\lambda_3(E)} \Phi_n \tag{22.4}$$

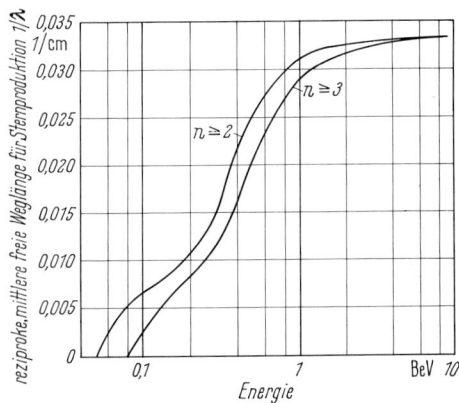

Fig. 46. Die reziproken mittleren freien Weglängen für die Erzeugung von 2-Spuren- bzw. 3-Spuren-Sternen in Kernemulsion in Abhängigkeit von der Energie. (Nach [He 59a].) Wegen Literaturangaben s. Text.

können mit den gemessenen Zählraten $Z_{\exp}(x)$ mit Sternen von $n \geq 3$ (für unseren Fall energiearmen) Spuren in der Höhe x (vgl. Tabelle 3, S. 396) in Zusammenhang gebracht werden. Auf diese Weise wird es möglich auf den Protonen- bzw. Neutronenfluß zu schließen, oder Zusammenhänge mit Flußwerten der Kaskadentheorie zu schaffen.

Gelingt es einen proportionalen Zusammenhang zwischen dem Protonenfluß Φ_p und dem Neutronenfluß Φ_n zu gewinnen

$$\Phi_p = K \cdot \Phi_n, \tag{22.5}$$

so kann man in (22.4) den Protonenfluß eliminieren:

$$Z(x) = \int_{E_0}^{\infty} dE \cdot \Phi_n(E) \cdot \left\{ \frac{K}{\lambda_2(E)} + \frac{1}{\lambda_3(E)} \right\}. \tag{22.6}$$

Eine vollständige Theorie der nuklearen Kaskade steht noch aus. Deshalb beschränken wir uns auf die vereinfachte, eindimensionale Kaskadenrechnung nach Rossi [Ro 52b] (vgl. Ziff. 13). Dort ist der Fluß der Sekundären spektral nach

[ICEF 63; p. 1184 ff. and 1095]. Die reziproken Weglängen gehen direkt in die Reaktionsausbeuten $Q(E, x)$ in Gl. (22.13) ein. Die oben erwähnte Abänderung macht sich im Bereich einiger GeV bis 30 GeV bemerkbar. Mit der Reaktionsausbeute $Q(E, x)$ wird nur im Bereich von 100 bis 500 MeV auf Intensität und Spektralform des Neutronenflusses geschlossen, der als Potenzgesetz wie etwa E^{-n} mit $n = 1,5$ bis 2,5 abfällt. Ein nennenswerter Einfluß ist nicht zu erwarten.

Energie und nach der Höhe für Neutronen $\Phi_n(E, x)$ in Gl. (13.12) und für die Protonen $\Phi_p(E, x)$ in (13.8) angegeben. Bildet man jeweils den integralen Fluß für Energien $E \geq 80$ MeV bei einer mittleren Absorptionslänge $L = 138$ g/cm², so ergibt sich für die Sekundärflüsse ein Verhältnis

$$R(x) = \frac{\Phi_n(E \geq 80 \text{ MeV}; x)}{\Phi_p(E \geq 80 \text{ MeV}; x)}, \qquad (22.7)$$

vgl. Fig. 47, ausgezogene Kurve. Für große atmosphärische Tiefen stimmt R wenigstens grob mit den Meßwerten $(N/G)_{\text{exp}} = $ Neutral/Geladen in Tabelle 17 überein. Für große Höhen ist die Übereinstimmung schlecht. Eine Erklärung dafür ist, daß im Verhältnis R nach (22.7) der Fluß primärer Protonen Φ'_p — dies ist eine

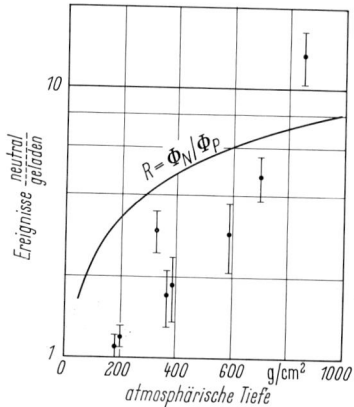
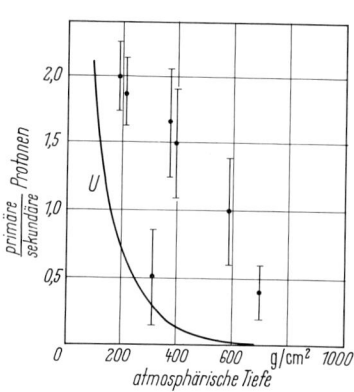

Fig. 47. Verhältnis der Neutronen zu den Protonen für Energien $E > 80$ MeV als Funktion der atmosphärischen Tiefe. Die ausgezogene Kurve wurde von Rossi berechnet und gilt nur für sekundäre Teilchen und ein Verhältnis Φ_n/Φ_p. (Nach [He 59a].)

Fig. 48. Verhältnis der primären Protonen zu sekundären Protonen als Funktion der atmosphärischen Tiefe. (Nach [He 59a].) Ausgezogene Kurve: gerechnet nach Rossi.

Eigenart des Rossischen Modells — nicht enthalten ist; mit anderen Worten der Nenner ist zu klein. Trägt man den primären Protonen Rechnung, so kann man ein geeigneteres Flußverhältnis definieren:

$$T(x) = \frac{\Phi_n(E \geq 80 \text{ MeV})}{\Phi'_p(E \geq 80 \text{ Mev}, x) + \Phi_p(E \geq 80 \text{ Mev}; x)} = \frac{\Phi_n/\Phi_p}{1 + \Phi'_p/\Phi_p}. \qquad (22.8)$$

Definiert man noch das Flußverhältnis der primären zu den sekundären Protonen mit

$$U = \frac{\Phi'_p}{\Phi_p}, \qquad (22.9)$$

so kann man das gemessene Verhältnis (Neutral/Geladen) mit T (22.8) direkt in Verbindung bringen:

$$(N/G)_{\text{exp}} = T = \frac{R}{1 + U}. \qquad (22.10)$$

Hess et al. [He 59a] benutzten den Zusammenhang Gl. (22.10), um die „Meßpunkte" in Fig. 48 über $U = (R - T)/T$ zu errechnen: in Fig. 47 wird R der Kurve entnommen; T stellt dort die Meßpunkte dar, die ohne Angabe der Autoren und weiterer Einzelheiten bei [He 59a] eingetragen sind. Eine detaillierte Diskussion ist daher erschwert. Der Meßpunkt bei etwa 850 g/cm² liegt sehr hoch; er würde in Fig. 48 sogar zu einem negativen Verhältnis U führen. Diese „Meßpunkte" stellen keinen Maßstab für Güte oder Richtigkeit des — ebenfalls über die Rossische Kaskade errechneten — Flußverhältnisses U dar. Mit diesem Verfahren kann man jedoch Hinweise gewinnen, dafür in einer Höhe von 190 g/cm² die Zahl der primären Protonen etwa doppelt so groß sei wie die der sekundären. Eine quantitative Überprüfung bleibt offen.

Die Proportionalität zwischen dem (primären und sekundären) Protonenfluß und dem Neutronenfluß, soll jetzt als Funktion der Energie und der atmosphärischen Tiefe angesetzt werden:

$$K'(E, x) = \frac{1}{T(E, X)} = \frac{\Phi'_p(E, x) + \Phi_p(E, x)}{\Phi_n(E, x)}, \qquad (22.11)$$

so erhält man mit Hilfe der Gl. (13.10) und (13.12)

$$K'(E, x) = \frac{(50 + E)^2}{x \cdot e^{-x/L} \cdot A} \left\{ \frac{A \cdot e^{-x/L}}{(50 + E)} \left(\frac{1}{(50 + E)} + \frac{(E_m - E)}{k(E)(50 + E_m)} \right) \right\}$$

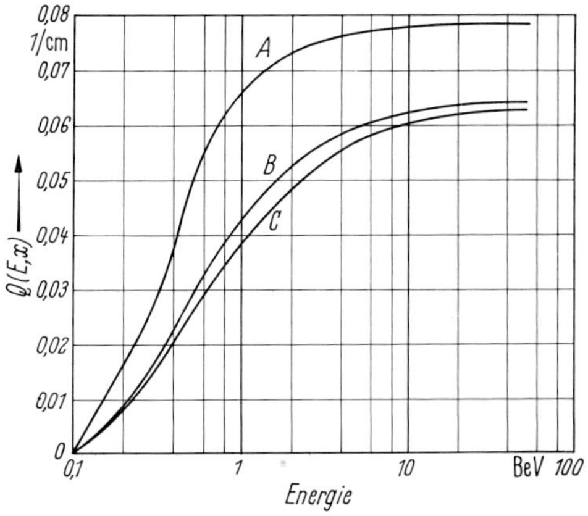

Fig. 49. Werte der Reaktionsausbeute $Q(E, x)$ (s. Text) als Funktion der Energie für verschiedene Tiefen in der Atmosphäre; Kurve A für 200 g/cm², Kurve B für 700 g/cm², Kurve C für 1000 g/cm². (Nach [He 59a].)

Tabelle 17. *Verhältnis neutral zu geladen ausgelöster Kernwechselwirkungen* $(N/G)_{\text{exp}}$.

Höhe m	Autor	Methode	N/G	Energie (GeV)
Meereshöhe	De 54	Nebelkammer	$0{,}4 \pm 0{,}06$	50
	Te 53b	Kernemulsion	1,33	>1
	Ba 54	Nebelkammer	5,9	>0,15
	Br 54b	Nebelkammer	$5{,}55 \pm 0{,}5$	>0,04
3 200	Gr 53a	Zählrohr	0,75	>100
3 000	De 54	Nebelkammer	$0{,}44 \pm 0{,}06$	50
2 800	Wa 55	Zählrohr	$0{,}72 \pm 0{,}2$	30
4 260	Wa 50a	Zählrohr	$0{,}83 \pm 0{,}1$	>10
3 600	Br 49a	Kernemulsion	1,0	>1
3 770	Te 53b	Kernemulsion	1,0	>1
3 500	Co 53	Zählrohr	$1{,}54 \pm 0{,}4$	>0,4
2 000	Ba 55	Zählrohr	2,8	0,4
2 040	Bu 57	Nebelkammer	$5{,}55 \pm 0{,}8$	>0,04
3 500	Br 54b	Nebelkammer	$3{,}7 \pm 0{,}4$	>0,04
22 000	Ca 52	Kernemulsion	0,95	>1
30 000	En 54b	Kernemulsion	0,1	>30
15 000—25 000	Te 53b	Kernemulsion	0,38	
36 000	Ba 60	Kernemulsion	0,077	>1000

Benutzt man weiter das von Rossi vorgeschlagene Primärspektrum (13.9) (wobei die Höhenabhängigkeit schon zuvor herausgezogen wurde), so bleibt:

$$K'(E, x) = \frac{1}{x}\left(1 + \frac{(E_m - E) \cdot (50 + E)}{k(E) \cdot (50 + E_m)}\right). \qquad (22.12)$$

Setzt man anstelle von K in Gl. (22.6) $K'(E, x)$ in die geschweifte Klammer ein, so stellt der Ausdruck der geschweiften Klammer die „Reaktionsausbeute $Q(E, x)$" für neutral und geladen ausgelöste Sterne mit drei und mehr Spuren dar:

$$Q(E, x) = \frac{K'(E, x)}{\lambda_2(E)} + \frac{1}{\lambda_3(E)}. \qquad (22.13)$$

In Fig. 49 sind die Reaktionsausbeuten $Q(E, x)$ für einige atmosphärische Tiefen x zu verfolgen; die Errechnung verwendet die λ_n-Werten nach Fig. 46 und Gl. (22.13). In Ziff. 31 wird die Reaktionsausbeute (22.13) zusammen mit den Zählraten $Z(x)$ (energiearmer) Sterne dazu dienen, mit Hilfe von Gl. (22.6) auf den Neutronenfluß $\Phi_n(E)$ zu schließen.

c) Neutronen-Monitor.

23. Eine kontinuierliche Messung der Intensität der kosmischen Primärstrahlung wäre wünschenswert, insbesondere um die Variation der Primärstrahlung genau verfolgen zu können. Eine Dauermessung über dem Gipfel der Atmosphäre wird in absehbarer Zeit mit Satelliten oder Raumstation verwirklicht werden. Zu einer Dauerkontrolle der einfallenden Intensität — zumindest für einen eingeschränkten Energiebereich — ist man daher vorläufig auf Meßmethoden in Bodenstationen angewiesen, die indirekt auf die Intensität der Primärstrahlung schließen lassen: sowohl auf den breitenabhängigen Anteil als auch einen Teil des breitenunabhängigen.

Verfügbar für solche Messungen sind Sekundärteilchen der Höhenstrahlung: einerseits die Myonen[1], die in geeigneten Teleskopen gemessen werden, zum anderen die Nukleonen in der Atmosphäre.

Die langsamen Protonen sind schwer zu messen; die langsamen Neutronen (Verdampfungsneutronen und thermische) unterliegen in hohem Grad den atmosphärischen Schwankungen. Dennoch dienten Neutronendaten aus direkten Messungen in großer Höhe früher zu Schlüssen auf die Variation der Primärintensität.

ADAMS [*Ad 50a*] und SIMPSON et al. [*Si 53b*] haben als geeignete Methode die indirekte Messung der Nukleonenintensität in Bodenstationen vorgeschlagen. Man läßt die relativ energiereichen Nukleonen (mit Energien $E > 500$ MeV) der sekundären Nukleonenkomponente der Atmosphäre mit Pb-Kernen wechselwirken. Bei diesen Nukleon-Kern-Stößen werden schnelle Neutronen emittiert, die nach geeigneter Abbremsung auf thermische Energien in großen $B^{10}F_3$-Zählrohren gemessen werden. Notwendigerweise müssen die schnellen wie thermischen Neutronen der Atmosphäre von der Zählung ausgeschlossen werden, was mit Hilfe einer dicken, protonenreichen Paraffinabschirmung hinreichend gelingt. Der schematische Aufbau eines solchen Simpson-Neutronen-Pile-Monitors, heute meist als „Neutronen-Monitor" gekennzeichnet, ist in Fig. 50 zu erkennen. SIMPSON [*Si 57*] hat für das Internationale Geophysikalische Jahr (IGY) diesen Monitor als Standardmeßgerät für das Netz der Meßstationen zur Messung der Intensität der kosmischen Strahlung und deren Schwankungen vorgeschlagen. Die meisten Bodenstationen (vgl. die Übersicht über Bodenstationen in Anhang I, S. 129) verfügen heute über Simpson piles in Form des „Standard-Monitors" oder mehr oder weniger abgeänderter Geräte.

Für das Jahr der ruhigen Sonne (IQSY) haben CARMICHEAL et al. [*Ha 64a*] den Supermonitor vorgeschlagen (Fig. 50 und 51). Wegen physikalischer, elektronischer

---
[1] Beitrag von FOWLER und WOLFENDALE "The Hard Component of μ-Mesons in the Atmosphere", Handbuch der Physik, Bd. 46/1.

und technischer Details wird auf Spezialliteratur verwiesen ([*Si 57*, *St 61a*, *Ca 64*] und das Buch von SANDSTRÖM [*Sa 65*]). Tabelle 18 gibt einen knappen

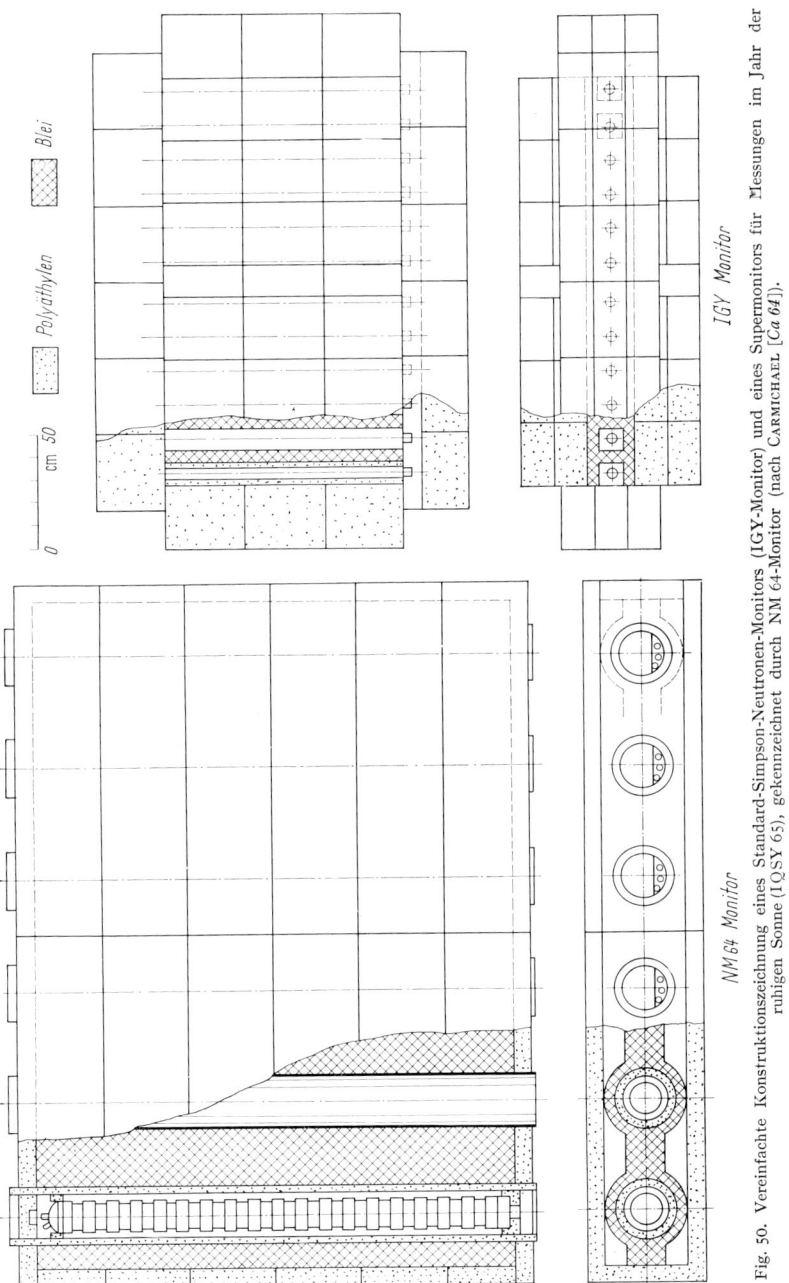

Fig. 50. Vereinfachte Konstruktionszeichnung eines Standard-Simpson-Neutronen-Monitors (IGY-Monitor) und eines Supermonitors für Messungen im Jahr der ruhigen Sonne (IQSY 65), gekennzeichnet durch NM 64-Monitor (nach CARMICHAEL [*Ca 64*]).

Einblick in den Aufbau und die Leistungsfähigkeit beider Monitortypen. Die Wirkungsweise der Neutronenvervielfachung sei weiter umrissen. Die Neutronen, die irgendwo im Bleikörper überwiegend durch hochenergetischen Nukleon-

Pb-Kern-Stoß als Anstoßneutronen und als Verdampfungsneutronen emittiert werden, entweichen als schnelle Neutronen ziemlich isotrop in alle Richtungen.

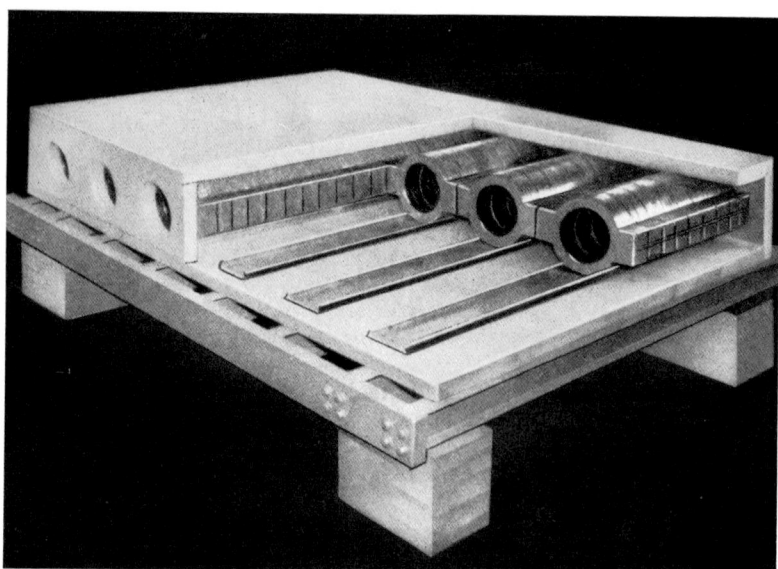

Fig. 51. Ansicht einer Einheit von 6 Zählern eines NM-64-Neutronenmonitors während des Aufbaus (nach [Ca 64]).

Tabelle 18. *Vergleich einer 6-Zähler-Einheit eines NH-64-Monitors mit einem 6-Zähler-Teil des IGY-Monitors.* (Nach CARMICHAEL et al. [Ha 64a].)

Parameter	NM-64	IGY
Counter		
Active length, cm	191	86.4
Internal diameter, cm	14.8	3.66
Gas	BF_3 96% ^{10}B	BF_3 96% ^{10}B
Pressure (at 0° C), cm	8.04	45
Active volume, liters	33	0.91
Volume BF_3 at NTP, liters	18.5	0.54
Diameter/m.f.p. of thermal neutrons	0.375	0.225
Moderator		
Thickness, cm	2.0	~3.7 (av,)
Section	24.1 cm dia.	10.2 cm sq.
Producer		
Mass of lead, kg	9650	1412
Length of lead, cm	207	102
Width of lead, cm	300	92
Area of lead, m²	6.21	0.94
Average depth of lead, cm	13.7	13.4
Reflector		
Overall length, cm	222	168
Overall width, cm	315	245
Overall depth, cm	52	79
Thickness, cm	7.5	30.5
Counting rate		
Counting rate (1963), hour^{-1}	~260000	~12000
Rate per counter, hour^{-1}	43000	2000
Rate per m² of lead, hour^{-1}	42000	12800

Über elastische und inelastische Stöße mit den Protonen im Paraffin werden sie langsam oder auf thermische Energien abgebremst und können dann beim Durchgang durch Bor in einem Zählrohr über die Bor-Reaktion das Zählrohr auslösen. Die äußere Neutronenabschirmung dient den schnellen Neutronen im Inneren als zweckmäßiger Neutronenreflektor, der die Wahrscheinlichkeit erhöht, daß ein Neutron im Verlauf seines Bremswegs, bzw. Diffusionswegs, in einem $B^{10}F_3$-Zähler wechselwirkt. Die Anreicherung von B^{10} beträgt ca. 93 bis 97%.

Ist die Multiplizität der Neutronen bei einem Proton-Pb-Kernstoß sehr groß, so können die aus einem Stoßereignis folgenden Neutronen durch die möglicherweise sehr verschiedenen Bremswege der einzelnen Neutronen im Monitor zu zeitlich schon erheblich verschobenen Neutronenregistrierungen führen.

DYRING [*Dy 62*] hat bei Monitordaten eine besondere Art eines systematischen Fehlers festgestellt, der auf eine Vielfachproduktion der Neutronen im Monitorbleikörper zurückgeführt werden kann. Man erkennt, daß eine zufällige Fluktuation der Zählrate in systematischer Weise die gewöhnlichen statistischen Fluktuationen überlagert. Numerisch kann man diesem zusätzlichen Fehler dadurch Rechnung tragen, daß man die Fehlergrenzen, gemäß einer Poisson-Verteilung mit Hilfe eines Faktors vergrößert. Dieser „Multiplizitätsfaktor" trägt dann den Neutronen Rechnung, die durch hohe Neutronenmultiplizität in ein und demselben Kernstoß erzeugt, lediglich durch die verschieden langen Bremswege die Zähler in Zeitintervallen auslösen, die lang im Vergleich mit den kurzen Totzeiten der Proportionalzähler sind. Solche Ereignisse könnten fehlerhaft als Neutronen gedeutet werden, die aus verschiedenen Kernstößen stammen und als zwei Primärereignisse gezählt werden. MEYER [*Me 61*] behandelt korrelierte Neutronen in einem SIMPSON-monitor in verschiedener Höhe (Zugspitze-München) und geomagnetischer Abschneidesteifigkeit. BACHELET, BALATA, DYRING und JUCCI [*Ba 64 b*] untersuchen für dieselben Parameter und die Sonnenfleckenzyklus-Variation den Vervielfachungsprozeß. Die Abschätzung des Standardfehlers der Monitordaten wird auch noch von FIELDEHOUSE, HUGHES und MARSDEN [*Fi 62*] diskutiert. Über das IQSY Committee, Sekretariat, 6 Cornwall Terrace, London NW 1, können IQSY Instruction Manuals — besonders auch über Neutronenmonitoren — bezogen werden.

BERCOVITCH et al. [*Be 60*] konnten zeigen, daß Sekundärnukleonen von Energien von 200 bis 300 MeV und mehr bei jeder Wechselwirkung im Mittel ungefähr 8 Verdampfungsneutronen mit einer mittleren Energie von ca. 2 MeV auslösen. Einige Anwendungen des Monitors beschreiben wir in Ziff. 25, S. 462f.

d) Neutronenmessungen (direkt und indirekt) mit speziellen kosmischen Strahlungsspektrographen.

24. Wismut-Spaltungskammer. HESS et al. [*He 59a*] haben eine Wismut-Spaltungsionisationskammer benutzt, um Neutronen im Energiebereich von ca. 80—500 MeV in der Atmosphäre direkt zu messen. Dem Meßvorgang liegt die Spaltungsreaktion des Bi^{209}-Kerns zugrunde, der bei Beschuß mit Neutronen oder Protonen bei Energien $E > 50$ MeV spaltet. Zur Unterscheidung der Bi^{209}-Spaltungsreaktion von den anderen Reaktionen und Ereignissen wurde eine identische wismutfreie Kammer zur Differenzmessung gebaut. Der Aufbau ist bei HESS, PATTERSON und WALLACE [*He 57*] beschrieben.

HESS, PATTERSON, WALLACE und CHUPP [*He 59a*] haben mit dieser Bi-Spaltungskammer die Zählrate der Neutronenereignisse bestimmt und daraus das Neutronenspektrum iterativ errechnet (vgl. Ziff. 31). Gemessen wurde im Sommer

1956 in ca. 3000 m Höhe; Winter 1956/57 im Flugzeug bis zu 12000 m (200 g/cm²; Bezugshöhe dieser Messungen 245 g/cm²; Umrechnung mit der Absorptionslänge $L=150$ g/cm²); 1958 in Meereshöhe. Die Messungen liegen kurz vor und innerhalb des Sonnenfleckenmaximums. GAUGER hat mit derselben Bi²⁰⁹-Spaltungskammer von HESS et al. [*He 59a*] gemessen (Meßzeit Mai bis Juni 1962, also 4 Jahre nach dem Sonnenfleckenmaximum bei gleicher geomagnetischer Breite, gleicher Höhe). Die Zählraten der Bi-Kammer und die der Neutronenereignisse wurden auf 45° N und eine Höhe von 245 g/cm² Restdruck umgerechnet (vgl. Tabelle 19).

Es zeigt sich ein erheblich steileres Spektrum. Es ist überraschend, daß der Fluß schneller Neutronen im Sommer 1962, d.h. 4 Jahre nach dem Sonnenfleckenmaximum, geringer ist als vor und während des Maximums. Solare Effekte oder Forbush-Abnahmen fanden nicht statt (vgl. auch S. 443f. und 531).

Die Werte von HESS et al. liegen für die Meßwerte um einen Faktor 7,2 und für das errechnete Spektrum um 5,4 höher gegenüber dem speziell gewählten Mittelwert $6,6 \pm 2,2$.

Weitere Messungen der Kernspaltung im Gebiet hoher Energie wurden von FLEROV und Mitarbeitern [*Fl 59*] durchgeführt.

Tabelle 19. *Neutronenzählrate N/h für Bi²⁰⁹-Spaltung zu verschiedenen Zeiträumen im solaren Zyklus.*

Autoren	Jahr (Monat) eventuell Tag	Neutronenzählrate N/h je Bi²⁰⁹-Spaltung		$N(E)\,dE$ $\sim E^{-n}dE$
		gemessen	über Spektrum berechnet	
HESS et al. [*He 59*]	1958 (Sommer) 1956/57 (Winter) 1958	48	36	$n \leq 2$
GAUGER [*Ga 64*]	1962 (Mai bis Juli)[1] 12. 7. 21. 7.	$6.6 \pm 1,4$ $4,86 \pm 0,96$ $16 \pm 2,2$	—	$n = 2,5$

25. Neutronen-Monitor in Kombination mit magnetischem Spektrometer. Der Effekt der Neutronenmultiplizität wurde auch von FIELDHOUSE, HUGHES und MARDSEN [*Fi 62*] untersucht. Es wurden einmal die Zählraten der Neutronenmultiplizität m_N für $1 \leq m_N \leq 9$ pro Tag gezählt, in einem weiteren Experiment wurde der Monitor in Koinzidenz mit einem magnetischen Spektrographen betrieben. Neben der mittleren Multiplizität $\langle m_N \rangle = 1,23$ für einen Standard-Simpson-pile wurden die relativen Häufigkeiten der den Monitor auslösenden Teilchen (Nukleonen bzw. Nichtnukleonen) bestimmt. Das Ergebnis zeigt Tabelle 20. Überwiegend fallen demnach sekundäre, hochenergetische Neutronen ein.

Tabelle 20. *Prozentuale Verteilung der Teilchenarten, die einen Neutronen-Monitor auslösen.*

77%	Neutronen
14,8%	Protonen
6,8%	Myon-Einfang-Ereignisse
1%	Myonen im Flug / Nukleonen in Schauern / Pionen

Die Neutronenproduktion in Blei durch Protonen der K. S. ist von HUGHES, MARSDEN, MEYER und WOLFENDALE [*Hu 62, 64*] gemessen worden. Protonen von 0,3 bis 150 GeV/c wurden im magnetischen Spektrometer und die Neutronenmultiplizität im Monitor gemessen. Die mittleren Neutronenmultiplizitäten (5—12) in Abhängigkeit von der Protonenenergie $0,3 \leq E \leq 2$ GeV sind in Übereinstimmung mit den Monte-

---

[1] Spezieller Mittelwert aus Mai bis Juli + 12. 7. + 21. 7.-Messung.

Carlo-Rechnungen von METROPOLIS et al. [Me 58]. Die gleiche Übereinstimmung finden bis 1 GeV BERCOVITCH et al. [Be 60]. Man vermutet, daß Pion-Erzeugung und nachfolgende Pion-Wechselwirkungen innerhalb des Kerns hauptsächlich für die große Zahl von Verdampfungsteilchen bei hochenergetischen Nukleon-Kernstößen verantwortlich sind. Für Energien über 2 GeV wächst die mittlere Multiplizität rasch auf über Hundert an, um dann abzufallen. Der experimentelle Kurvenverlauf und dessen Vergleich mit der Theorie der Mesonproduktion kann der Originalarbeit entnommen werden.

Pion/Proton- und Pion/Neutron-Verhältnis.

Das Experiment läßt das Pion-Neutron-Verhältnis nicht direkt bestimmen. Es besteht jedoch experimentell kein Grund anzunehmen, daß die Neutron-

Tabelle 21. *Relative Pion-Flußwerte bei Proton-Energien.* (Nach [Hu 62].)

E	1 GeV	10 GeV	100 GeV
π/Proton: rel. Fluß	0,05	0,10	0,10

produktion durch Pionen und Protonen verschieden ist, und deshalb hängt die relative Neutronenproduktion durch Pionen allein von deren relativem Fluß ab. Diese relativen Flußwerte sind für verschiedene Energien E beim Verhältnis π/p angegeben (Tabelle 21).

Tabelle 22. *Absorptionslänge des Neutrons bzw. Protons.*

GeV/c	$L(p)$ (g/cm²)
3	130 ± 5
100	100 ± 10

26. Messungen mit dem Spektrographen in Durham. WOLFENDALE et al. [Br 62, 64a] haben das Impulsspektrum von Protonen (und von Myonen) bestimmt. Der Vergleich mit dem Rossischen Primärspektrum [Ro 59, 60] wird gezogen. Man nimmt an, daß das Neutronenspektrum ober-

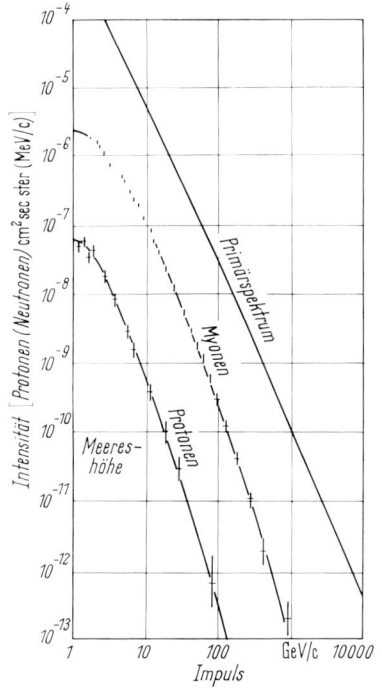

Fig. 52. Impulsspektrum von Protonen, gemessen von WOLFENDALE et al. [Br 62] in Meereshöhe. Ab 3 GeV/c entspricht dieses Spektrum auch dem Neutronenspektrum in Meereshöhe.

halb 3 GeV/c identisch mit demjenigen der Protonen ist (Fig. 52). Die mittlere Absorptionslänge ist nur langsam mit dem Impuls (vgl. Tabelle 22) veränderlich. Vgl. auch S. 389 Ziff. 4 und S. 413f., Ziff. 12.

e) Die Winkelverteilung schneller Neutronen.

27. Über die Zenitwinkelverteilung besitzt man sehr wenige Messungen, die alle indirekter Art sind. CONVERSI und ROTHWELL [Co 54] und LOHRMANN [Lo 55b] verglichen die Winkelverteilung von schnellen Teilchen, die von geladen ausgelösten und von neutral ausgelösten Wechselwirkungen emittiert wurden. Da diese schnellen Teilchen (wohl überwiegend Anstoßnukleonen bzw. umgeladene Protonen) ungefähr die Richtung des Primärteilchens beibehalten, lassen sich hieraus Rückschlüsse auf die Winkelverteilung des auslösenden Teilchens gewinnen. CONVERSI und ROTHWELL finden, daß die Winkelverteilungen aus geladen und neutral ausgelösten Sternen gleich sind und schließen hieraus, daß die Winkelverteilung schneller Neutronen ungefähr dieselbe ist, wie für schnelle Protonen. Falls man die Zenitwinkelverteilung in der Form

$$P(\vartheta) = \cos^m \vartheta$$

schreibt, erhält man für $m = 2,1$ bis $2,6$ (s. Ziff. 21). LOHRMANN fand ebenfalls in Kernemulsion mit dieser Methode $m = 2^{+2}_{-1}$. BARFORD und DAVIS benutzten ein anderes Verfahren [Ba 52b]; Sie exponieren Kernemulsion unter einem kegelförmigen Absorber, dessen Achse in die Zenitrichtung zeigt, so daß unter verschiedenen Zenitwinkeln einfallende Strahlung verschieden geschwächt wurde. Sie fanden für energiearme von Neutronen ausgelöste Sterne $m = 2,5 \pm 0,5$ mit einem Anzeichen, daß m für größere Sterne anwächst; für sehr große Sterne kann m den Wert 4 erreichen. Für höhere Neutronenenergien $E > 10$ GeV wird von FIDECARO und Mitarbeitern [Fi 58] mit einer Zählrohrapparatur eine Winkelverteilung $e^{-m/\cos \Theta}$ mit $m = 7,1 \pm 1,3$ gemessen.

MIYAKE et al. [Mi 57a und Mi 57b] haben mit einer Wasserstoff- bzw. Stickstoff-gefüllten Nebelkammer gemessen. Ihre Winkelverteilung geladener Teilchen wird von NEWKIRK [Ne 63] für die Neutronen übernommen.

Alle Messungen wurden in Gebirgshöhen durchgeführt und sind mit einem Wert des Exponenten m von 2 bis 3 verträglich. Die meisten Messungen für die geladene, sternerzeugungsfähige Komponente liefern einen etwas größeren Wert für m, d.h. diese zeigen eine etwas mehr in Zenitrichtung gebündelte Winkelverteilung.

28. Energiespektrum der Neutronen in der Atmosphäre. Ein Energiespektrum für langsame Neutronen, die sich im Diffusionsgleichgewicht in der Atmosphäre befinden, wurde erstmals von BETHE, KORFF u. PLACZEK [Be 40] angegeben. [Wegen Bezeichnungen vgl. (16.1') und Text.]

$$\Phi(E)\,dE = N(E)\,v(E)\,dE = \frac{M\,q\,l(E)}{2}\,\frac{dE}{E}\,\exp\left\{-\frac{M}{2}\int_E^{E_1} \frac{\sigma_c(E')}{\sigma_a(E')}\,\frac{dE'}{E'}\right\}. \quad (28.1)$$

Die Form dieses Spektrums ist Fig. 53 zu entnehmen. Dort findet man auch eine Darstellung der nachfolgenden Spektren. FERMI [Fe 50] errechnete für die langsamen Neutronen ein etwas anderes Spektrum

$$\Phi(E)\,dE = N(E)\,v\cdot dE = \frac{q}{\xi\,\Sigma_s(E)}\,\frac{dE}{E}\,\exp(-0{,}379/\sqrt{E}). \quad (28.2)$$

FREESE und MEYER [Fr 53] geben für die Energieverteilung der langsamen Neutronen die Darstellung:

$$\begin{aligned}\Phi(E)\,dE = N(E)\,v\,dE \sim \frac{dE}{E}\,\exp\Big\{&-15{,}52\left(\frac{E_c}{E}\right)^{\frac{1}{2}} + 5{,}49\left(\frac{E_c}{E}\right)^{\frac{2}{2}} -\\ &-3{,}09\left(\frac{E_c}{E}\right)^{\frac{3}{2}} + \cdots\Big\} \text{ mit } \sqrt{E_c} = 0{,}025 \text{ eV}^{\frac{1}{2}}.\end{aligned} \quad (28.3)$$

Das Spektrum der langsamen Neutronen in einem unendlich ausgedehnten Medium kann für einen gasförmigen, schweren und absorbierenden Moderator mit Hilfe der Differentialgleichung (vgl. COHEN und HURWITZ [Co 56b])

$$\frac{dq(E)}{dE} = \Sigma_a(E) \cdot \Phi(E) \quad (28.4)$$

errechnet werden. Die Bremsdichte $q(E)$ ist (vgl. POOLE, NELKIN und STONE [Po 58] Gl. (2.14)) mit dem Neutronenfluß Φ verknüpft:

$$q(E) - \int_0^E S(E') \cdot dE' = (\xi \Sigma_s) \left[E \cdot \Phi(E) + E \cdot kT \cdot \frac{d\Phi(E)}{dE} \right], \qquad (28.5)$$

(wobei $\xi = 2/A$; A = Atomgewicht), Σ makroskopischer Wirkungsquerschnitt; s = Streuung, a = Absorption). $S(E')$ kennzeichnet das Hereinstreuen von Neutronen der Energie E' ins Intervall E und $E + dE$.

COHEN und HURWITZ haben für eine $1/v$ Abhängigkeit von $\Sigma_a(E)$ diese Differentialgleichung numerisch gelöst, wobei das Verhältnis

$$\Delta = \frac{4 \Sigma_a(kT)}{\xi \Sigma_s} \qquad (28.6)$$

die Lösungen entscheidend bestimmt. Für Luft ergibt sich ein $\Delta \approx 3{,}9$, für das man in Fig. 1 bei POOLE et al. die Dichteverteilung der Neutronen entnehmen kann. Der zugehörige Fluß Φ ist in Fig. 53 unter der Kennzeichnung „POOLE, NELKIN und STONE" aufgenommen.

HURWITZ, NELKIN und HABETLER [Hu 56] haben die Verteilungsfunktion des Flusses Φ in einen Maxwellanteil $M(E)$ und einen Zusatzterm $H(E)$ aufgespalten, den sie mit Hilfe einer Störungsrechnung errechnen (vgl. Fig. 2 bei POOLE et al. [Po 58]).

Die meisten Autoren greifen heute auf das über (28.4) errechnete Spektrum für die langsamen Neutronen zurück. Ein experimenteller Entscheid über eines der 4 Spektren steht noch aus. Diese Spektren gelten bis zu Energien von einigen keV. Der weitere Spektralbereich war bis vor wenigen Jahren unbekannt. Bei sehr hohen Neutronenenergien kann man in Analogie von dem Protonspektrum der Kaskade auf das Neutronenspektrum schließen.

MIYAKE, HINOTANI und NUNOGAKI [Mi 57a] haben das Neutronen-Energiespektrum in der Atmosphäre im Energiebereich von 1 bis 15 MeV gemessen. Sie benutzten eine mit Wasserstoff gefüllte Nebelkammer in 2840 m Höhe (Mt. Norikura) bei $\Lambda = 25°$ N und untersuchten die Protonenstöße, welche durch Neutronen ausgelöst wurden. Sie vereinigen die Ergebnisse ihrer Messungen mit Berechnungen und experimentellen Daten anderer Autoren, um eine Neutronenflußverteilung im Bereich von 1 eV bis etwa 10 GeV zu bestimmen. Nach Integration über die Winkel erhält man den gesamten Neutronenfluß (differentielles Energiespektrum). Das Ergebnis von MIYAKE et al. ist in Fig. 55 aufgenommen worden (vgl. S. 447).

HESS, PATTERSON, WALLACE und CHUPP [He 59a] können mit ihren Meßdaten sowohl den Bereich der langsamen Neutronen wie auch den Bereich der schnellen Neutronen praktisch lückenlos bis zu Neutronenenergien von ungefähr 500 MeV überdecken. Nach höheren Energien schließen sie nach dem Vorbild von MIYAKE et al. an die Spektralform der Kaskadentheorie (MESSEL [Me 54]) an.

Die entscheidenden Ergebnisse von MIYAKE et al. und HESS et al. sind die Bestätigung des lange vermuteten Gleichgewichtsspektrums für den Neutronenfluß in der Atmosphäre in einem Energiebereich von 10 Zehnerpotenzen (10^{-2} bis $5 \cdot 10^8$ eV) und die Messung des Zwischenmaximums $\geq 0{,}1$ MeV, das HESS et al. als Quellspektrum für Verdampfungsneutronen, MIYAKE et al. als Potenzspektrum von Anstoß-neutronen deuten. MIYAKE et al. haben in einer fixen Höhe (2840 m) gemessen; die Messungen von HESS ergänzen von Meereshöhe bis 12000 m.

Die Gegenüberstellung der Flußwerte nach MIYAKE et al. und HESS et al. in Fig. 55 ergibt eine deutliche Diskrepanz; dasselbe gilt für die Spektralform im Bereich des flachen Zwischenmaximums (vgl. auch S. 448).

Wir legen der weiteren Darstellung die Ergebnisse von MIYAKE et al. [Mi 57a] und HESS et al. [He 59a] zugrunde.

Zuerst einmal unterteilt man das weite Intervall der Gleichgewichtsverteilung der Neutronenenergie in drei geeignete Energiebereiche:

Bereich I   von thermischen Energien bis 50 keV
Bereich II von 50 keV bis 1 MeV
Bereich III von 1 MeV bis ca. 500 MeV.

Die wesentlichen Schritte zur Festlegung der Spektralform und der Absolutwerte geben wir im folgenden.

Methodisch sei noch an den Zusammenhang zwischen dem differentiellen Neutronenfluß $\Phi(E)$ und der Zählrate C eines Neutronenmeßgeräts erinnert, der in bekannter Weise (vgl. etwa ROSSI [Ro 52b]; BECKURTS und WIRTZ [Be 64]) über die makroskopische Reaktionsausbeute $\Sigma(E)$ hergestellt wird:

$$C_i = \int_0^\infty \Phi(E) \cdot \Sigma_i(E)\, dE. \tag{28.7}$$

Damit kann über die gemessene Zählrate C_i und die bekannte (errechnete oder gemessene) Reaktionsausbeute $\Sigma_i(E)$ direkt oder iterativ auf $\Phi(E)$ geschlossen werden.

29. Bereich I: Von thermischen Energien bis 50 keV. HESS et al. benutzen als Hilfsmittel zur Berechnung dieses Teilspektrums folgende experimentelle Daten:

a) Die C_{14}-Produktionsrate in der Atmosphäre, wobei sie sich auf die ausführliche Darstellung der Daten von ANDERSON und LIBBY stützen [An 51, An 53, Si 51b] (vgl. S. 501).

b) Die Zählraten von nackten BF_3-Zählern; gezählt wurde ohne bzw. mit Cd-Mantel. Benutzt wurden die Arbeiten von YUAN [Yu 51], SIMPSON [Si 51b] und HESS et al. [He 59a] (vgl. Ziff. 19).

c) Die Zählrate von Cd-abgedeckten Goldfolien, die der kosmischen Strahlung ausgesetzt wurden und den Neutronenfluß bei 4,9 EV geben (HESS et al.) (vgl. S. 441).

Die Gestalt des Spektrums im Bereich I kann nicht einheitlich angegeben werden, sondern wird in zwei Teilbetrachtungen erschlossen.

α) Von 10 keV bis herab zu 1 keV zeigt das Neutronenspektrum roh eine $1/E$-Energieabhängigkeit. Dieses direkte Ergebnis der Neutronenbremstheorie darf auch auf Mischungen von Bremsmaterialien angewandt werden, vorausgesetzt, daß der fragliche Energiebereich hinreichend weit unterhalb der Energie der Neutronenproduktion liegt und daß im Bremsbereich keine Absorption stattfindet (vgl. [Be 64] oder [Gl 52]).

In einem unendlich ausgedehnten nicht absorbierenden Medium verläuft das Energiespektrum genau wie $1/E$, wenn der Streuquerschnitt energieunabhängig ist. Wenn aber dem Medium durch einen Verluststrom schnelle Neutronen verloren gehen oder aber in dem Medium eine Absorption stattfindet oder ein energieabhängiger Wirkungsquerschnitt vorliegt, dann wird die Form des Spektrums modifiziert und kann näherungsweise durch eine $1/E^\alpha$ Energieabhängigkeit wiedergegeben werden.

Der Verlustfluß an Neutronen aus der Atmosphäre hat im Gleichgewichtsbereich keinen Einfluß auf die Spektralform, d.h. α bleibt $=1$; nur am Gipfel der Atmosphäre muß der Einfluß des Neutronenverlustes auf die Spektralform mit geeignetem α berücksichtigt werden. Eine Veränderung der Spektralform am Gipfel der Atmosphäre sehen wir in Ziff. 37, 38.

Der Einfluß der Absorption über den (n, γ)-Prozeß ist sogar bei 50 keV nur unbedeutend. Dagegen ist die Energieabhängigkeit des Wirkungsquerschnittes ein entscheidender Faktor und bewirkt eine Abänderung der Form des Spektrums auf $\Phi(E) \sim 1/E^{0,88}$. Diese Energieabhängigkeit legen HESS et al. für den Verlauf des Spektrums von 1 eV bis 50 keV zugrunde.

β) Im 1 eV-Bereich wird die Absorption der Neutronen durch den Stickstoff der entscheidende Prozeß. Die Gestalt des Energiespektrums hängt von den Details des Absorptionsprozesses ab, der in Konkurrenz mit der thermischen

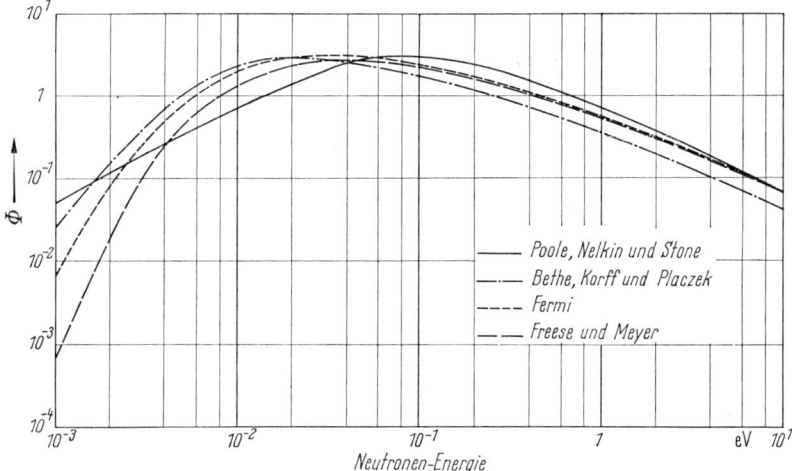

Fig. 53. Energiespektrum des Flusses Neutronen der k. S. bei niedrigen Energien. Alle 4 Kurven sind bei 100 eV normiert. Autoren: [Po 58], [Be 40], [Fe 50], [Fr 53].

Abbremsung der Neutronen steht. Wie besprochen ist die endgültige Form des Neutronenspektrums in Luft bis heute noch nicht entschieden; mögliche Ansätze wurden von verschiedenen Autoren ausgearbeitet und sind in Fig. 53 zusammengefaßt.

Die verschiedenen theoretischen Analysen ergeben keinen radikal verschiedenen Verlauf des Neutronenspektrums unterhalb von 1 eV, und es ist schwer, experimentell eine Entscheidung zu treffen. Eine mögliche Methode der Unterscheidung würde genauere Messungen des Cd-Verhältnisses erfordern; unglücklicherweise ist diese Methode sehr unempfindlich gegenüber den kleinen Unterschieden des Spektrums in Fig. 53, denn alle dort aufgeführten Spektren ergeben beispielsweise ein Cd-Verhältnis von ca. 2,2:1. Eine andere Möglichkeit der Festlegung des Spektrums bietet die Aktivierung verschiedener Materialien, die Neutronenresonanzen an Stellen haben, wo die vier zur Konkurrenz stehenden Spektren sich stark unterschieden.

Es scheint, daß die Frage der Spektralabhängigkeit durch POOLE, NELKIN und STONE [Po 58] am exaktesten behandelt wurde. Aus diesem Grund wurde ihre niederenergetische Spektralform für das Neutronenspektrum in der Atmosphäre benützt.

30. Bereich II: Von 50 keV bis 1 MeV. Das Energiespektrum in diesem Bereich war nicht bekannt. Unter der Annahme, daß man erst ab 50 keV sich den Energien nähert, bei denen Verdampfungsneutronen merklich produziert werden, deren Maximum bei 1 MeV liegen sollte, haben HESS et al. die spektrale Form des Bereichs I versuchsweise bis 500 keV fortgesetzt: $\sim E^{-0,88}$). Oberhalb

500 keV muß die Form des Spektrums durch das Quellspektrum und inelastische (n, γ)-Prozesse modifiziert erscheinen.

HESS et al. benutzen in diesem Energieintervall von 50 keV bis 1 MeV als geeigneten Detektor einen BF_3-Zähler. Diesen umgeben sie jeweils mit verschieden dicken Paraffin- bzw. Polyäthylen-(CH_2)-Ummantelungen. Je mehr Paraffin bzw. CH_2 hinzugefügt wird, desto mehr Neutronen höherer Energie können nachgewiesen werden. Für eine bestimmte atmosphärische Tiefe erhalten sie in Abhängigkeit von der Dicke der CH_2-Ummantelung verschiedene Zählraten.

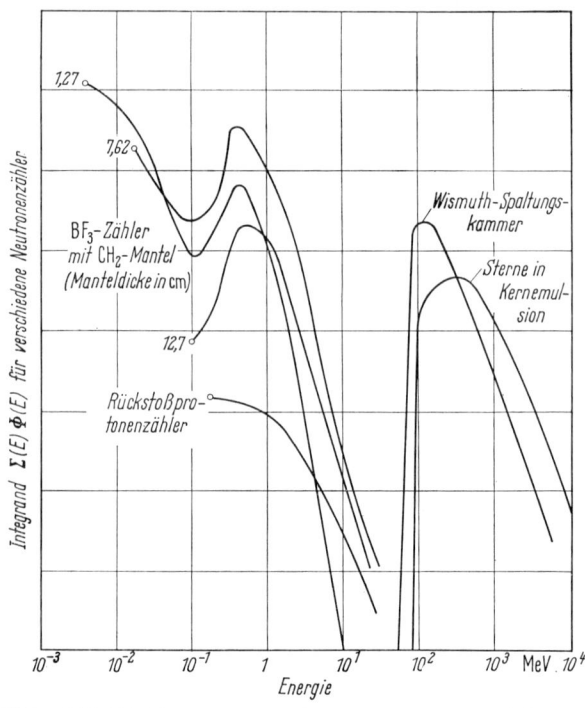

Fig. 54. Das Produkt $\Sigma(E) \times \Phi(E)$ (d.h. Detektorausbeute × Neutronenfluß) für verschieden mit CH_2-ummantelte BF_3-Zähler, für den Proportionalzähler, für die Wismuth-Spaltungskammer und für Sterne in Kernemulsion in Abhängigkeit von der Energie. Die Bereiche relativ starker Energieabhängigkeit sind deutlich zu erkennen. (Nach [He 59 a].)

Mit einem iterativen Verfahren errechnen sie das Spektrum im Bereich von 10 keV bis zu 1 MeV.

Man kann für jede CH_2-Ummantelung einer Dicke d die zugehörige Detektorausbeute $\Sigma_i(d, E)$ als Funktion der Energie bestimmen. Weiter mißt man für einen solchen Zähler der Dicke d die Zählrate $C_i(d)$ und kann so nach (28.5)

$$C_i(d) = \int_{\Delta E_d} \Phi(E) \Sigma_i(d, E) \, dE \tag{30.1}$$

den Mittelwert des unbekannten Neutronenflusses $\Phi(E)$ abschätzen, indem man über einen geeigneten Energiebereich ΔE_d aufsummiert.

In Fig. 54 sind für CH_2-ummantelte BF_3-Zähler mit Manteldicken von 1,27 cm (½″), 7,60 cm (3″) und 12,65 cm (5″) die Integranden von (30.1) also $\Phi(E) \cdot \Sigma_i(d, E)$ als Funktionen angegeben. Die Fläche unter jeder Kurve stellt als Integral über die Energie unmittelbar die Zählrate dar. Die untere Integrationsgrenze für den Integrationsbereich ΔE_d wird so festgelegt, daß die Fläche 98% der gemessenen Zählrate enthält. Bei einer Messung mit acht verschiedenen Ummantelungen konnten so zunächst acht Punkte für den wahrscheinlichen Verlauf des Neutronenflusses gefunden werden, was natürlich erst einen rohen Verlauf des Spektrums angibt. Durch weitere Iteration verfeinert man schließlich Schritt um Schritt diese Spektralmittelwerte, bis man im Bereich von 50 keV bis 1 MeV eine hinreichend genaue Energieabhängigkeit des Spektrums gewinnt.

Ab 500 keV zeichnet sich dabei ein Zwischenmaximum ab (vgl. Fig. 55).

HESS et al. interpretieren das Zwischenmaximum im Spektrum als die Auswirkung der Neutronenquelle, die bei der Sternbildung Verdampfungsneutronen bei einem Maximum von 1 MeV emittiert (vgl. Ziff. 16, aber auch S. 418). Die Autoren approximieren aus diesem Grund den Verlauf des Spektrums über 500 keV durch die Form des Verdampfungsspektrums in der Gegend des Maximums, Fig. 27.

Im Gegensatz dazu finden MIYAKE et al. nur ein flaches, kaum ausgeprägtes Zwischenmaximum. Mit dem Potenzspektrum (21.3) deuten sie diesen Sachverhalt als eine überwiegende Beteiligung von „Anstoß"-Neutronen, ohne nennenswerte Beiträge von Verdampfungsneutronen mit Maximum bei 2,6 MeV (vgl. Ziff. 17). Fig. 55 zeigt in den Kurven A und B diesen differentiellen Fluß (Neutronen/cm² sec MeVsterad) nach MIYAKE et al. Die schnellen Neutronen scheinen hinreichend isotrop, so folgt in einer näherungsweisen 4π-Geometrie die Vergleichskurve C. Dieses differentielle Energiespektrum weicht im fraglichen Quellbereich etwa um einen Faktor 6—9 vom gemessenen berechneten Fluß nach HESS et al. [He 59a; 61] ab. Diese Diskrepanz überschreitet den von HESS et al. angegebenen Fehler von $\pm 25\%$. Weitere Messungen (Ziff. 21) stützen die Daten von MIYAKE et al.

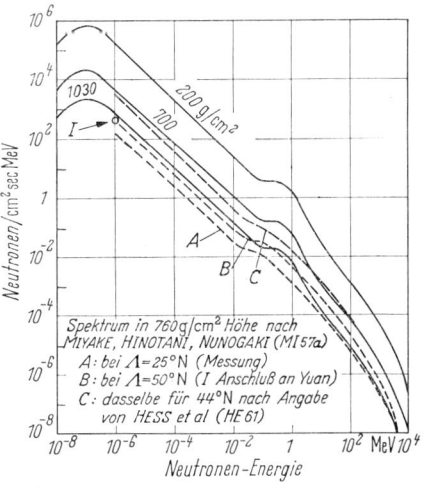

Fig. 55. Das Energiespektrum des Flusses kosmischer Strahlungsneutronen in Meereshöhe, 3000 und 12000 m Höhe nach [He 59]. Die Kurve A nach [Mi 57a] in einer Höhe von 2800 m. MIYAKE schließt in Punkt I an YUANS Werte an.

Im Energiebereich $< 0,04$ eV können Abweichungen um einen Faktor $\geqq 2$ im Vergleich zum Kurvenverlauf in Fig. 55 möglich sein.

31. Bereich III: Oberhalb 1 MeV. Oberhalb 1 MeV können — neben der Direkt-Messung durch MIYAKE et al. — folgende experimentelle Daten zur Berechnung des Spektralverlaufes herangezogen werden:

a) die Häufigkeit der durch Neutronen der k. S. in der Kernemulsion ausgelösten Sterne
b) die Zählrate der großen Wismuth-Spaltungs-Ionisationskammer
c) die Zählrate eines mit Polyäthylen ausgelegten Proportionalzählers (Zählung von Anstoßprotonen).

Die Gestalt des Spektrums ist in Bereich III ebenso wenig aus allgemeinen Gesichtspunkten anzugeben wie in Bereich II.

HESS et al. errechnen deshalb die Gestalt des Spektrums im Bereich III mit Hilfe der gemessenen Zählraten der oben erwähnten Experimente mit dem gleichen für Bereich II erwähnten iterativen Verfahren.

Die Produkte aus Fluß und Detektorausbeute $\Phi(E) \times \Sigma(E)$ führen nach Integration direkt auf die Zählrate.

In Fig. 54 ist dieses Produkt für verschiedene Detektoren aufgetragen. Die Integrale unter diesen Kurven ergeben dann sofort die Zählrate für diese Detektoren.

Die untere Grenze des Energieintervals wird so gewählt, daß die Fläche 98% der gemessenen Zählrate entspricht.

Berücksichtigt man bei der Zählrate der Emulsionssterne nur solche mit drei bis fünf schweren Spuren — verzichtet man also insbesondere auf Sterne mit einer oder zwei grauen Schauerspuren — so kann man als eine obere Grenze für die niederenergetisch produzierten Sterne 500 MeV ansetzen, d. h. wir erhalten etwa im Bereich von 100—500 MeV über die Sternproduktionsrate einen Beitrag zum Spektrum.

Der Spaltungsquerschnitt für Wismut ist ebenfalls nur bis etwa 500 MeV bekannt. Über die Zählrate der Wismut-Spaltungskammer wird das Neutronenspektrum nur von etwa 100—500 MeV bestimmbar.

Der mit Polyäthylen ausgelegte Proportionalzähler liefert Information im Bereich von etwa 100 keV bis 10 MeV. Damit ist eine Überschneidung mit Bereich II gegeben, so daß die Anpassung des Spektrums an die Bereiche I und II mit Hilfe des Proportionalzählers möglich wird.

Die Gestalt dieses zusammengesetzten Spektrums ist in Fig. 55 für verschiedene Höhen (1000, 700 und 200 g/cm²) wiedergegeben und gilt für eine geomagnetische Breite von $\Lambda = 44°$ N. Daten, die in anderen geomagnetischen Breiten gemessen und zur Festlegung dieses Neutronenspektrums mit verwendet worden sind, wurden unter Benutzung des von SIMPSON [Si 51b] gemessenen Breiteneffekts umgerechnet.

Die von HESS et al. benutzten geomagnetischen Daten sind in Anhang II zusammengestellt und können dort mit anderen geomagnetischen Daten verglichen werden.

Die Messungen zeigen, daß die Gestalt des Spektrums von einer atmosphärischen Tiefe von 200 g/cm² bis herab zu 1000 g/cm² (Meereshöhe) ungeändert ist.

Tabelle 23 vermittelt eine Zusammenfassung der über die verschiedenen Detektoren gewonnenen experimentellen Zählraten. Zugleich enthält die Tabelle die errechenbaren Zählraten, die aus dem konstruierten Spektrum in einer Höhe von 700 g/cm² folgen.

Tabelle 23. *Experimentelle Zählraten gegen errechnete Zählraten unter Zugrundelegung des Spektrums von Hess et al.*
Alle Daten bei einer atmosphärischen Tiefe von 700 g/cm² und einer geomagnetischen Breite von 44° N.

Detektor	Über das Spektrum errechnet	Experimentelle Werte
Goldfolie (K-Wert)	$4,9 \times 10^{-3}$ Neutronen/cm² sec	$(5,0 \pm 1,7) \times 10^{-3}$ Neutronen/cm² sec
BF_3 + 1,27 cm CH_2	0,097 Ereignis/sec	$0,084 \pm 0,005$ Ereignis/sec
BF_3 + 2,53 cm CH_2	0,141 Ereignis/sec	$0,150 \pm 0,006$ Ereignis/sec
BF_3 + 3,16 cm CH_2	0,168 Ereignis/sec	$0,173 \pm 0,006$ Ereignis/sec
BF_3 + 3,80 cm CH_2	0,212 Ereignis/sec	$0,192 \pm 0,006$ Ereignis/sec
BF_3 + 5,06 cm CH_2	0,250 Ereignis/sec	$0,208 \pm 0,005$ Ereignis/sec
BF_3 + 7,60 cm CH_2	0,221 Ereignis/sec	$0,200 \pm 0,008$ Ereignis/sec
BF_3 + 10,01 cm CH_2	0,136 Ereignis/sec	$0,133 \pm 0,006$ Ereignis/sec
BF_3 + 12,65 cm CH_2	0,098 Ereignis/sec	$0,105 \pm 0,006$ Ereignis/sec
Proportionalzähler	1,46 Ereignisse/min	$1,55 \pm 0,18$ Ereignisse/min
Bi-Spaltung	1,33 Ereignisse/hr	$1,6 \pm 0,3$ Ereignisse/hr
Sterne in Kernemulsionen	15 Sterne/cm³ Tag	15 Sterne/cm³ Tag
C^{14}-Atome/cm² sec	3,1 Atome/cm² sec	2,6 Atome/cm² sec
BF_3 + ohne Paraffin + Cd-Abschirmung	0,0022 Ereignis/cm²	$0,0026 \pm 0,0003$ Ereignis/cm²

Oberhalb von 200 g/cm² sind Veränderungen des Spektrums zu erwarten. Auf diese gehen wir in Ziff. 37 und 38 ein; unter anderem macht sich der Verlustfluß der Neutronen und zwar hauptsächlich der schnellen Neutronen aus der Atmosphäre heraus störend bemerkbar.

Das untere Ende der Atmosphäre begrenzt der Erdboden bzw. die Wasserfläche des Meeres. In beiden Medien werden weniger Neutronen absorbiert als in Luft, wie man den Messungen von BAGGE (vgl. S. 441) entnehmen kann. So kann man vermuten, daß das Spektrum am Grund der Atmosphäre reich an thermischen Neutronen ist, die in die Atmosphäre zurückgestreut werden.

Nach Angaben von MATHER (vgl. HESS et al. [He 59a]) tendiert die Erde dazu, einfallende Neutronen auf thermische Energien abzubremsen ohne viele von diesen einzufangen. Dann werden die meisten dieser langsamen Neutronen in die Luft reflektiert und gelegentlich durch N^{14}-Kerne eingefangen.

Neutronen mit Energien oberhalb 50 MeV können innerhalb der Atmosphäre nur als Anstoßneutronen bei Kernstoßprozessen emittiert oder durch Umladung energiereicher Protonen zustandegekommen sein. Ab Energien von einigen GeV verhalten sich Protonen und Neutronen bei Stoßprozessen so ähnlich, daß man mangels Kenntnis der spektralen

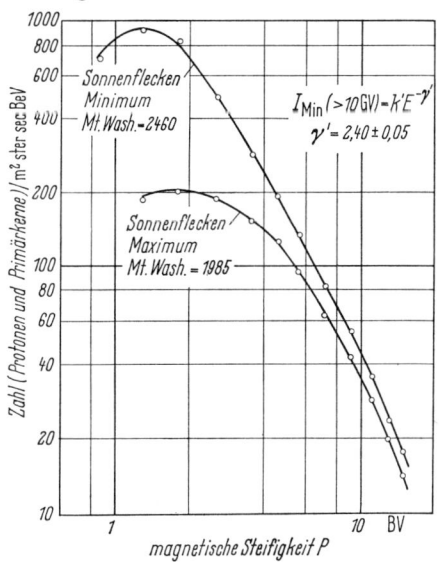

Fig. 56. Differentielles Spektrum der gesamten Primärstrahlung beim Sonnenfleckenminimum bzw. -maximum.

Verteilung dieser hochenergetischen Neutronen in Analogie das Spektrum der Protonenkaskade verwenden darf. Bei $\Lambda = 44°$ geomagnetisch Nord nehmen HESS et al., gültig für den Bereich ihrer Messungen, eine Abschneideenergie für Protonen von $E_c = 3$ GeV an.

Beschreibt man die primären Protonen der k. S. durch ein Potenzspektrum, dann muß gemäß der Messelschen Theorie der Nukleonenkaskade [Me 54] das differentielle Energiespektrum $\Phi_{\text{sec}} (>E_c)$ für sekundäre Nukleonen oberhalb der primären Abschneideenergie E_c dieselbe Form haben wie das Primärspektrum.

Für das differentielle Primärspektrum der Protonen benutzen HESS et al. das Primärspektrum nach SINGER [Si 58b] mit einem Exponenten von $\gamma = 2{,}15$ und führen mit diesem spektralen Verlauf ihr Spektrum vom Bereich III zu hohen Energien fort.

Das Primärspektrum der Protonen ist jedoch zeitlich nicht konstant. Im sog. breitempfindlichen Bereich des Primärspektrums, von ca. 400 MeV bis ca. 30 GeV, wirkt sich die solare Modulation zwischen Sonnenfleckenminimum (maximale Intensität) und Sonnenfleckenmaximum (minimale Intensität) (vgl. Fig. 56) sehr stark aus. Die zeitliche Variation der Neutronenproduktionsrate — als Funktion des variablen Primärflusses — findet ihren Ausdruck in einer zeitlichen Variation des Spektrums des Neutronenflusses in der Atmosphäre. Die quantitative Errechnung des Spektrums der schnellen und langsamen Neutronen mit oder ohne zeitliche Variation aus dem gemessenen Fluß der primären kosmischen Strahlung über eine *nukleare Kaskade* steht bekanntlich noch aus.

Anstelle von Kaskadenmodellen ist man zur quantitativen, theoretischen Beschreibung der Nukleonen in der Atmosphäre auf die Boltzmannsche Stoßgleichung angewiesen und benötigt dafür ein geeignetes Quellspektrum.

32. Neutronenproduktionsrate nach Breite, Höhe und Sonnenfleckenzyklus. Die Neutronen werden in der Atmosphäre nicht einheitlich produziert; man kann die Produktionsrate in einen energieabhängigen Anteil $N(E)\, dE$ aufspalten, sowie in einen räumlichen Anteil, der der Quellung der Neutronen in Abhängigkeit von der Höhe und der geomagnetischen Breite Rechnung trägt: $q(z, \Lambda)\, dz\, d\Lambda$. Die zeitliche Variation wirkt sich in einer periodischen bzw. aperiodischen Modulation der Intensität des Primärflusses aus. Die Energieabhängigkeit der Stoßprozesse und der Reaktionen wird davon nicht berührt; d.h. die zeitliche Variation des primären Flusses wirkt sich in einer zeitlichen Variation der *räumlichen* Verteilungsfunktion der Neutronenproduktionsrate aus.

Das zeitabhängige Quellspektrum der Neutronen in der Atmosphäre, die Produktionsrate, setzen wir an:

$$S = N(E, z, \Lambda, t) \cdot dE \cdot dz \cdot d\Lambda = N(E)\, dE \cdot \begin{Bmatrix} q_{\min}(z, \Lambda) \\ q_{\max}(z, \Lambda) \end{Bmatrix} dz \cdot d\Lambda, \qquad (32.1)$$

wobei $q_{\min}$ (bzw. $q_{\max}$) die räumliche Produktionsrate zum Zeitpunkt des Sonnenfleckenminimums (bzw. -maximums) kennzeichnen soll.

Die Energieverteilung des Quellspektrums (32.1) haben wir in Ziff. 17 nach den Ansätzen von FLÜGGE [*Fl 43a*], HESS, CANFIELD und LINGENFELTER [*He 61*] bzw. NEWKIRK [*Ne 63*] bzw. den Messungen von MIYAKE, HINOTANI und TAKETANI [*Mi 57a*] ausführlich dargestellt.

In Ziff. 17 ist die Verteilungsfunktion für die Höhenabhängigkeit der Neutronenquelle für einige Meßorte mit bestimmten geomagnetischen Breiten ebenfalls ausführlich diskutiert worden. FLÜGGE und HESS et al. rechnen mit der Verteilung Gl. (17.2); NEWKIRK hat diese Verteilung gemäß Fig. 27 verbessert angesetzt. LINGENFELTER [*Li 63a*] ergänzt diese Ansätze für alle geomagnetischen Breiten Λ durch Angabe der Verteilungsfunktion $q_{\min}$ für das Sonnenfleckenminimum bzw. für $q_{\max}$. Er stützt sich auf folgende Argumente:

Die zeitliche Variation der Neutronenproduktionsrate ist ursächlich mit der solaren Modulation der Primärintensität verknüpft und wirkt sich über den breitenempfindlichen bzw. geomagnetischen Teil des Primärspektrums aus. Im Zeitraum des Sonnenfleckenminimums fällt maximale Intensität der kosmischen Strahlung auf die Atmosphäre der Erde ein. Dementsprechend ist eine höhere Produktion von Neutronen in der Atmosphäre zu erwarten. Zum Zeitpunkt des Sonnenfleckenmaximums wird die Primärintensität (vgl. Fig. 2, 56) durch einen noch nicht vollständig erklärten solaren Modulationsmechanismus begrenzt, wobei insbesondere der primäre Fluß bei geringer magnetischer Steifigkeit von etwa 0,6 bis 3 GV um mehr als 50% reduziert wird. Bei hohen magnetischen Steifigkeiten, etwa der Abschneidesteifigkeit am Äquator, beträgt die relative Variation der Intensität nur noch etwa 5%. Dies wirkt sich in der Höhenabhängigkeit der Neutronenproduktion in der Atmosphäre stark aus. Beim Sonnenfleckenminimum dringen erheblich mehr niederenergetische Protonen und leichte Kerne auf Grund ihrer niedrigen magnetischen Steifigkeit in den polaren Bereichen, d.h. oberhalb des geomagnetischen Knies ($\Lambda > 60-70°$), in die Erdatmosphäre ein und bewirken eine höhere Neutronenproduktion in den höheren Schichten der Atmosphäre insbesondere in Polargebieten. Eine Messung der starken Schwankung der Neutronenproduktionsrate innerhalb des solaren Zyklus wird also zweckmäßig in der näheren Umgebung der geomagnetischen Pole durchgeführt. Wir befinden uns gegenwärtig wieder im Zeitraum eines Sonnenfleckenminimums. In internationaler Zusammenarbeit sollen dabei der Breiteneffekt langsamer und schneller Neutronen, die Neutronenproduktionsrate und viele andere Nukleonendaten gemessen werden, die im Zeitraum des letzten Sonnenfleckenminimums 1953/54 auf Grund des damals noch nicht erkannten solaren Modulationsmechanismus nicht gezielt in Angriff genommen werden konnten. FORBUSH [Fo 54] und MEYER und SIMPSON [Me 55a] hatten erstmals den inversen Zusammenhang zwischen Sonnenfleckenzahl und der Intensität der k. S. zur Diskussion gestellt. Das letzte Sonnenfleckenmaximum fiel in den Zeitraum 1957/58. Die Höhenabhängigkeit und der Breiteneffekt wurden im Zeitraum der Sonnenfleckenextrema umfassend gemessen. LINGENFELTER [Li 63a] hat einen Großteil dieser Ergebnisse benutzt, um eine Reihe von Kurven zu konstruieren, die die Produktion von Neutronen in der Atmosphäre bei jeder geomagnetischen Breite in ihrer Höhenabhängigkeit wiedergeben und zwar einmal für die Produktionsrate beim Sonnenfleckenminimum, zum anderen beim Sonnenfleckenmaximum. Er lehnt diese Kurvendarstellung an ähnliche Kurven von BENIOFF [Be 56] und LAL [La 58] an, die diese Autoren für die Höhen- und Breitenabhängigkeit der Sternproduktion in der Atmosphäre entwickelt haben, ohne jedoch eine zeitliche Variation zu berücksichtigen.

Zwei Randbedingungen kennen wir:

a) Der Gleichgewichtsbereich beginnt etwa bei 200 g/cm² in der Atmosphäre, und bis zu dieser Höhe ist der Breiteneffekt genau genug gemessen und bekannt.

b) Der Breiteneffekt der primär einfallenden Strahlung konnte durch Satelliten auf Polbahnen gemessen werden.

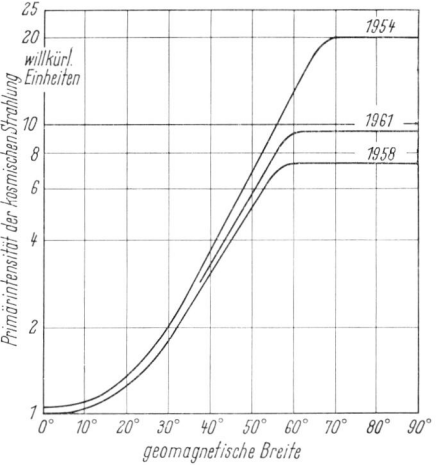

Fig. 57. Der Breiteneffekt der einfallenden kosmischen primären Strahlung am Gipfel der Atmosphäre (0 g/cm²), 1961 durch ALBERT et al. [Al 62] mit einem Polbahnsatelliten gemessen. Abschätzung der solaren Modulation für das Sonnenfleckenminimum (1954) bzw. Sonnenfleckenmaximum (1958) nach LINGENFELTER [Li 63a]. Die Polintensitäten wurden nach Ionisationsmessungen von NEHER [Ne 59]; ANDERSON [An 61] und NEHER u. ANDERSON [Ne 62] berechnet. Die Äquatorintensitäten sind nach LOCKWOOD [Lo 60] bestimmt. Nach NEHER [Ne 66] erreichen die Ionisationswerte 1964/65 innerhalb weniger Prozent diejenigen von 1953/54.

Es gilt für den Höhenbereich zwischen 0 und 200 g/cm² die Abhängigkeit der Neutronenproduktion bei höheren oder niedrigeren geomagnetischen Breiten, wo ja keine direkten Messungen vorliegen, abzuschätzen. Dazu wird es notwendig, die Breitenabhängigkeit der primär einfallenden Strahlung während des Sonnenfleckenzyklus zu bestimmen. NEHER [Ne 56] hat durch Ionisationsmessungen der niederenergetischen kosmischen Strahlung Werte gesammelt, die den allgemeinen Gang nicht nur der Breiten-, sondern auch der Höhenabhängigkeit über die Ionisation verfolgen lassen. Damit kann man sich ein Gefühl dafür schaffen, wie man die Extrapolation in diesem oberen Bereich für das Sonnenfleckenminimum zu vollziehen hat.

Zweckmäßig bezieht man sich bei der Festlegung der Breitenabhängigkeit auf die Werte während des Sonnenfleckenminimums. Durch Satellitenmessungen mit Polbahnen in einer Höhe von etwa 320 km wurde von ALBERT et al. im Jahre 1961 [Al 62] die Variation der primären Protonenintensität mit der Breite gemessen. Dieser gemessene Breiteneffekt ist in ausgezeichneter Übereinstimmung mit demjenigen, der für die gesamte primäre Strahlung durch SIMPSON, FONGER und

TREIMAN [Si 53b] von 0° bis 50° geomagnetischer Breite berechnet worden ist. Damit verfügt man über die schon erwähnte Randbedingung zur Bestimmung der Breitenabhängigkeit der primären Protonen im oberen Bereich der Atmosphäre.

Die Variation der einfallenden primären Intensität in Funktion der Sonnenfleckenaktivität wurde durch NEHER [Ne 59] und ANDERSON [An 61] bestimmt. Durch Ionisationsmessungen bei 88° geomagnetischer Breite errechnete NEHER [Ne 59] einen gesamten primären Fluß von 0,27 Teilchen/cm² sec ster während des Sonnenfleckenminimums 1954. Diesen Wert kann man mit dem Gesamtfluß von 0,10 während des Sonnenfleckenmaximums im Jahr 1958 [An 61] und

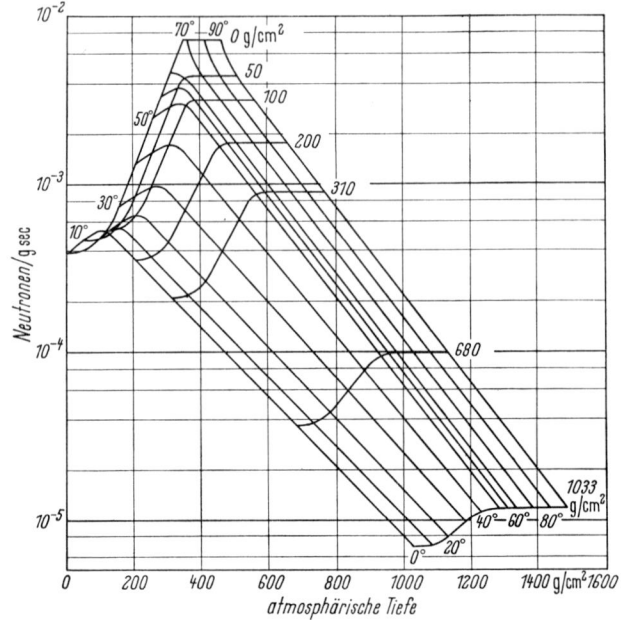

Fig. 58. Die Neutronenproduktionsrate $q_{min}(x, \Lambda)$ (Neutronen/g sec) als Kurvenschar der Parameter atmosphärische Tiefe x (g/cm²) und geomagnetische Breite Λ beim Sonnenfleckenminimum (nach LINGENFELTER [Li 63 a]); (willkürliche) Normierung der Gesamtproduktionsrate Q (Neutronen/cm² sec) auf ein Neutron in einer Luftsäule von 1 cm² Querschnitt über dem geomagnetischen Pol pro Sekunde. Wegen Absolutwerten s. Tabelle 32 (Seite 506) und weitere Diskussion.

einem Fluß von 0,13 im Jahr 1961 [Ne 62] vergleichen. Bei sehr niedrigen Breiten (Äquator) können wir annehmen, daß die Variation des einfallenden primären Flusses zwischen Sonnenfleckenminimum und -maximum roh dieselbe ist wie diejenige, die man mit abgeschirmten Ionisationskammern am geomagnetischen Äquator in einer Höhe (vom Druck 680 g/cm²) mißt, nämlich nach LOCKWOOD [Lo 60] eine Abnahme um ungefähr 5%. LINGENFELTER [Li 63a] bezieht das Verhältnis dieser Werte auf das Jahr 1961. Er konstruiert damit spekulativ einen Breiteneffekt sowohl für das Sonnenfleckenminimum als auch für das -maximum; dazu adjustiert er mit diesen unterschiedlichen Zahlverhältnissen im Polbereich (hohe Breiten) wie im Äquatorbereich den 1961 von ALBERT et al. gemessenen Breiteneffekt des primären Flusses der Protonen. Diese Breitenabhängigkeit ist in Fig. 57 zu sehen.

LINGENFELTER benutzt nun den konstruierten Breiteneffekt der einfallenden primären Strahlung für das Jahr 1954 des Sonnenfleckenminimums und normiert bei einer geomagnetischen Breite von 55° Nord und einer auf den Gipfel der Atmosphäre (0 g/cm²) extrapolierten Sternproduktionsrate diese Breiteneffekt-kurve. Er bezieht sich dafür auf die Werte von LORD [*Lo 51*]. Für den Höhenbereich von 200 bis 0 g/cm² extrapoliert LINGENFELTER die über den Breiteneffekt in der Atmosphäre vom Erdboden bis zu diesem fraglichen Höhenbereich schon bekannten Neutronenproduktionsraten (vgl. Fig. 58). Jetzt erscheint die Neutronenproduktionsrate als eine Raumfläche, die nach dem Parameternetz der atmosphärischen Tiefe x bzw. der geomagnetischen Breite Λ gegliedert ist. Die beschriebene Extrapolation zum Gipfel der Atmosphäre kann leicht verfolgt werden.

Für diese Parameterschar von Neutronenproduktionskurven wurde eine willkürliche Normierung vorgenommen und zwar derart, daß am geomagnetischen Pol $\Lambda_{geom} = 90°$ längs einer Luftsäule von 1 cm² Querschnitt die Neutronenproduktionsrate $q(x, \Lambda)$ (Neutronen/g sec) vom Gipfel der Atmosphäre von $x = 0$ g/cm² bis zur Meereshöhe $x_0 = 1033$ g/cm² aufsummiert wird:

$$Q(\Lambda = 90°) = \int_0^{x_0} q(x; \Lambda = 90°) \, dx = 1 \left(\frac{\text{Neutron}}{\text{cm}^2 \text{ sec}} \right). \tag{32.2}$$

33. Die Neutronenproduktion im Zeitraum des Sonnenfleckenmaximums. Für das Sonnenfleckenmaximum kann eine ähnliche Schar von Kurven für die Neutronenproduktionsrate konstruiert werden. MCDONALD [*McD 59*] hat ausgeführt, daß das primäre Spektrum im Einflußbereich des Erdmagnetfeldes hauptsächlich am niederenergetischen Ende des Spektrums modifiziert wird (vgl. Fig. 2,56). Die solare Modulation ist ausführlich im Beitrag von WEBBER[1] behandelt; als für uns entscheidendes Ergebnis fassen wir zusammen, daß die einfallende Strahlung am Gipfel der Atmosphäre für hohe geomagnetische Breiten stark reduziert wird, während bei niederen geomagnetischen Breiten, wo wegen der magnetischen Steifigkeit nur höherenergetische Primäre durch das Erdmagnetfeld einzudringen vermögen, eine relativ geringfügige Variation auftreten kann.

Experimentell wurden in jenem Zeitraum folgende Fakten gewonnen: Im Frühjahr 1958 — dem Zeitraum der höchsten Sonnenfleckenaktivität — fand LOCKWOOD [*Lo 60*], daß der Mittelwert der nuklearen Intensität am Mt. Washington, 55° geomagnetisch Nord bei einer atmosphärischen Tiefe 820 g/cm² ungefähr um 25% abgefallen war im Vergleich zu dem Mittelwert im Juli 1954, dem Zeitraum des Sonnenfleckenminimums. SIMPSON [*Si 58d*] fand dieselbe Abnahme der Neutronenintensität bei 48° geomagnetisch Nord bei einem Druck von 680 g/cm², und FENTON et al. [*Fe 58*] wies weiterhin nach, daß dieselbe Abnahme der nuklearen Komponente bei 83° geomagnetisch Nord auftrat wie am Mt. Washington. Zusätzlich konnte von LOCKWOOD [*Lo 60*] am Mt. Norikura (25,6° geomagnetisch Nord und 720 g/cm²) mit den dortigen Nukleonendetektoren und den abgeschirmten Ionisationskammern bei Huancayo (0,6° geomagnetisch Süd und 680 g/cm²) im Verlauf derselben Periode nur eine Abnahme von 5—6% festgestellt werden. (Wegen der experimentellen Ausstattung der Beobachtungsstationen wird auf Anhang I verwiesen, wegen der Umrechnung der dort aufgeführten geographischen in geomagnetische Koordinaten auf Anhang II.)

Für das Folgende ist es wichtig zu wissen, in welchen atmosphärischen Tiefen sich die Variation der kosmischen Strahlung in Abhängigkeit von der Energie der einfallenden Teilchen auswirkt. LINGENFELTER stützt sich dabei auf die Annahme, daß die effektive Absorptionslänge (vgl. Ziff. 19) in atmosphärischen Tiefen größer als 200 g/cm², also im Gleichgewichtsbereich der langsamen Neutronen, von

---

[1] W. R. WEBBER, dieser Band, Kap. III, S. 181.

der solaren Modulation nicht stark verändert wird. Wie wir bei der Behandlung der schnellen Neutronen gesehen haben, unterliegt der schnelle Fluß nach (21.2) und (21.3) einer Höhenabhängigkeit, die sich natürlich in der effektiven Absorptionslänge mitäußert. Wie stark sich allerdings diese Absorptionslänge L_{eff} zwischen Sonnenfleckenminimum und Sonnenfleckenmaximum unterscheidet, kann erst nach Auswertung der derzeitigen Messungen beim Minimum entschieden werden. Die Lingenfelterschen Überlegungen werden von experimentellen Ergebnissen von LOCKWOOD [Lo 60] und SIMPSON [Si 58d] unterstützt. Man sollte erwarten, daß im Gleichgewichtsbereich der Atmosphäre der größere Teil der Neutronenintensität durch Primäre von mittlerer und höherer

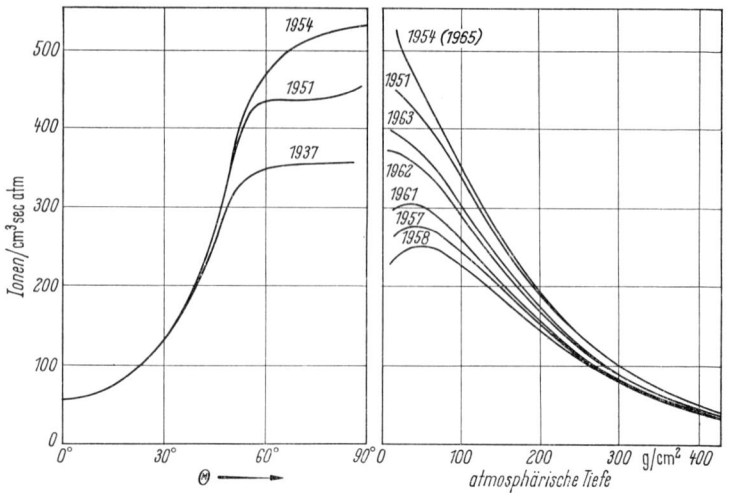

Fig. 59. Links: Breitenabhängigkeit der Ionisation einer Restatmosphäre von 15 g/cm². (NEHER [Ne 56]; SIMPSON [Si 58].] Im Jahr 1954 lag die Abschneidesteifigkeit unter 100 MeV. Rechts: Ionisation in Funktion der atmosphärischen Tiefe für verschiedene Jahre zwischen 1951 und 1963 nach NEHER [Ne 64]. Die Ionisationswerte 1964/65 zeigen praktisch Übereinstimmung mit den Werten von 1953/54 (s. [Ne 66]).

Energie erzeugt wird. Die oben erwähnten Messungen der Neutronenintensität sind in Bodenstationen ausgeführt worden. Die Intensitätsabnahme um 5% im Äquatorbereich bzw. um 25% im Polarbereich ist primären Teilchen von überwiegend mittleren bis höheren Energien des breitenempfindlichen Spektralbereichs zuzuordnen. Am Gipfel der Atmosphäre sind die Verhältnisse anders. Dort wirken sich die niederenergetischen Primären aus, die eine erheblich niedrigere Absorptionslänge haben, so daß sie schon ab einer atmosphärischen Tiefe ab 200 g/cm² keinen Einfluß mehr auf die Neutronenproduktion haben werden.

Diese starke Abnahme der Intensität niederenergetischer primärer Teilchen — und die damit korrelierte Neutronenproduktion — ist durch die Messungen von NEHER [Ne 59] und ANDERSON [An 61] klargestellt. Sie beträgt grob 63% und ist zur Errechnung der Intensitätsverhältnisse in Fig. 57 verwendet worden. Dies ist weiter in Übereinstimmung mit der Form der Variation des Primärspektrums, wie sie zwischen 1955 und 1958 durch McDONALD [McD 59] gefunden wurde. Sie wird in dem Beitrag von WEBBER ausführlich behandelt. In Fig. 59 kann die Variation der kosmischen Strahlung im Gipfelbereich der Atmosphäre im Verlauf mehrerer Jahre verfolgt werden.

Die Konstruktion der Schar der Neutronenproduktionskurven für den Zeitraum des Sonnenfleckenmaximums kann im wesentlichen wie für das Minimum ausgeführt werden. Für die Breitenabhängigkeit der Neutronenintensität wurde dazu auf die beim Sonnenfleckenmaximum 1948 durch SIMPSON [*Si 51b*] in einer Höhe von 310 g/cm² gemessene Werte zurückgegriffen. Diese Werte wurden über die prozentuale Abnahme nach LOCKWOOD [*Lo 60*] auf das Sonnenfleckenminimum normiert. Auf diese Weise sind wiederum vom Erdboden bis zu einer Höhe von 200 g/cm² Druck die Neutronenproduktionsraten als Parameterschar konstruierbar. Bis zum Gipfel der Atmosphäre werden sie unter Zugrundelegung des Breiteneffekts und der Lordschen Sterndaten nach Fig. 27 und 45 extrapoliert. Für niedere geomagnetische Breiten sind die Extrapolationskurven praktisch von gleicher Gestalt und sind nur in ihrem Betrag reduziert. Die Abweichungen im Polarbereich sind beträchtlich (vgl. Fig. 60).

Die experimentellen Daten, die der Konstruktion dieser Parameterschar von willkürlich normierten Neutronenproduktionsraten zugrunde liegen, sind nicht

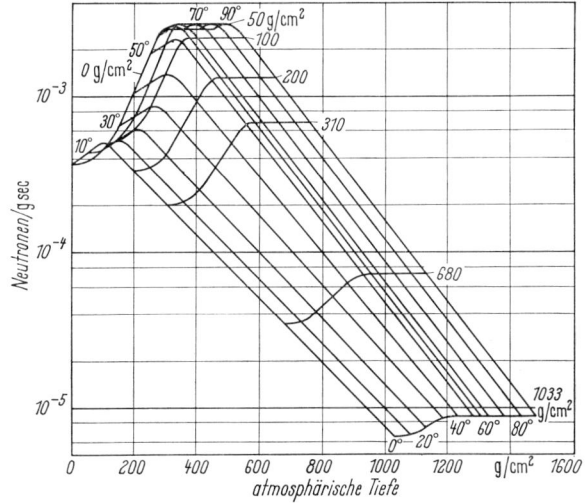

Fig. 60. Neutronenproduktionsrate $q_{max}(x, \Lambda_i)$ (Neutronen/gsec) in Funktion der atmosphärischen Tiefe g/cm² und der geomagnetischen Breite Λ während des Sonnenfleckenmaximums. Normierng wie in Fig. 58. (Bezugswert Sonnenfleckenminimum.) Wegen Absolutwerten s. Tabelle 32, S. 506 und weitere Diskussion.

statistisch verteilt über die ganze Erdoberfläche gewonnen worden, sondern fallen in einen relativ engen Sektor der nördlichen Hemisphäre, der von Längengraden 50° und 150°, jeweils West, begrenzt wird.

Mit den Darstellungen Fig. 58 und 60 verfügt man nach LINGENFELTER über willkürlich am geomagnetischen Pol normierte Neutronenproduktionsraten $q(x; \Lambda; t)$, deren Zahlwerte für jede Höhe x (in g/cm²) für die Parameterwerte der geomagnetischen Breite Λ und der zeitlichen Lage im Sonnenfleckenzyklus angebbar sind. Diese Produktionsraten gestatten es zusammen mit dem Energiespektrum der Verdampfungsneutronen (17.1) die Neutronenquellfunktion

$$S(x; \Lambda; t) = q(x; \Lambda; t) \cdot N(E) \cdot dE \cdot dx \qquad (33.1)$$

festzulegen.

Diese Produktionsraten stellen keine absoluten Werte dar. Man weiß, daß die Stärke der solaren Modulation schwankt. Wir benötigen also künftig eine Korrekturfunktion, die die effektiven Produktionsraten angeben läßt. Zum anderen bleibt zu prüfen, ob der von ALBERT et al. [*Al 62*] gemessene und von LINGENFELTER hier zugrundegelegte Breiteneffekt der Verlustflußneutronen mit einem Pol:Äquatorverhältnis von 10:1 richtig ist. Diese Frage ist besonders wichtig, da das von SIMPSON [Si 51b] in der Atmosphäre und das von TRAINOR und LOCKWOOD [Tr 64] über den Verlustfluß gemessene Breiteneffektverhältnis nur 4,5 : 1 beträgt (vgl. in diesem Zusammenhang auch die Diskussion in Ziff. 45).

Weiter wären noch mehr Sterndaten oder Neutronendaten zwischen 200 g/cm² und dem Gipfel der Atmosphäre wünschenswert.

Ob die verschiedenen geomagnetischen Anomalien, die in den letzten Jahren ausführlich gemessen worden sind, die von LINGENFELTER angenommene polare bzw. äquatoriale Symmetrie seiner Höhen- und Breitenverteilung in geomagnetischen Koordinaten nachhaltig stören können, ist gegenwärtig nicht bekannt. Unter Voraussetzung dieser polaren und äquatorialen Symmetrie sind die für die Erdatmosphäre gebildeten globalen Mittelwerte der Neutronenproduktionsrate bzw. der C^{14}-Produktionsrate (vgl. Ziff. 40) gebildet worden. Wir benutzen heute die von LINGENFELTER entwickelten Kurven der Neutronenproduktionsrate für das Quellspektrum der Neutronen in der Atmosphäre (33.1), das in die theoretische Behandlung sowohl der Diffusionsgleichung als auch der Transportgleichung eingeht.

34. Theorie der Neutronen in der Atmosphäre. Die Produktionsmechanismen liefern im wesentlichen sekundäre Neutronen in zwei praktisch getrennten Energiebereichen: energiereiche Anstoßneutronen (knock-on-neutrons) und Verdampfungsneutronen (evaporation neutrons) aus den hocherhitzten Atomkernen der Luft. Für die Theorie erübrigt sich eine detaillierte Behandlung dieser Stoßprozesse, denn wir können die Anstoßneutronen und die Verdampfungsneutronen durch geeignete Quellspektren beschreiben. Die Messungen haben ergeben, daß die Neutronen nicht nur im Bereich thermischer, langsamer und mittlerer Neutronen im Gleichgewicht stehen, sondern daß dieses Gleichgewicht bis zu Energien von etwa 10 GeV gesichert ist. Die mathematische Behandlung der Neutronen in der Atmosphäre muß dem Rechnung tragen und baut auf dem Gleichgewicht

Neutronenabsorption + Verlustfluß = Neutronenproduktion

auf. Im Falle eines stationären Gleichgewichts ist die Neutronenzahl N zeitlich konstant, d.h.

$$\frac{\partial N}{\partial t} = 0.$$

Diese Bedingung wird allerdings nach unserem heutigen Wissen innerhalb der Atmosphäre nicht voll erfüllt. Die primäreinfallende galaktische Strahlung wird durch den solaren Modulationsmechanismus in ihrer Intensität variiert. Dies macht sich in der elfjährigen Schwankung der auf die Atmosphäre einfallenden Primärintensität bemerkbar. Diese Modulation verläuft jedoch so langsam, daß wir immer ein quasistationäres Gleichgewicht vorliegen haben.

Neben dieser periodischen Modulation werden wir auch noch der aperiodischen Modulation begegnen, wenn solare Teilchen im Gefolge einer gas-chromatischen Eruption (solar flare) direkt in die Erdatmosphäre eindringen und einen beträchtlichen meist lokalen Intensitätszuwachs bewirken. Für diese kurzfristigen Störungen, die meist nur einige Stunden dauern, ist $\partial N/\partial t \neq 0$. Trotzdem kann auch in diesem Fall der Beitrag, den diese Störung auslöst, errechnet werden.

Nachdem wir einen Überblick über die experimentell bestimmten bzw. aus den Meßdaten entwickelten Verteilungsfunktionen der Neutronenproduktionsrate nach Höhe, Breite und zeitlicher Variation haben, und nachdem wir über das Energiespektrum der Neutronen im stationären Gleichgewicht verfügen, können wir zur Behandlung des Neutronenstoßes und des Neutronentransports in der Atmosphäre weitergehen.

Zur Behandlung solcher Stoß- und Neutronentransportprobleme dient allgemein die Boltzmannsche Stoßgleichung (Neutronentransportgleichung). Physikalisch läßt sich der Neutronenfluß in der Erdatmosphäre speziell durch die Lösung der stationären Neutronentransportgleichung unter den Rand- und Quellbedingungen für die Atmosphäre berechnen.

Das Prinzip der Erhaltung der Neutronenzahl wird durch die Neutronentransportgleichung gewährleistet. Die Produktion der Neutronen in der Atmosphäre wird darin einfach durch geeignete Quellspektren für die Anstoßneutronen und für Verdampfungsneutronen berücksichtigt. Die Verdampfungsneutronen haben eine isotrope Winkelverteilung. Für Anstoßneutronen trifft dies nicht zu. Diese werden bevorzugt anisotrop in Vorwärtsrichtung erzeugt und gestreut. Die Notwendigkeit, eine Anisotropie bei bestimmten Fragestellungen berücksichtigen zu müssen, hat Einfluß auf die Auswahl der Näherungen zur Lösung der Neutronentransportgleichung.

35. Näherungslösungen für die Neutronen-Transportgleichung.

Die Neutronen-Transportgleichung (vgl. BECKURTS-WIRTZ "Neutron Physics" oder DAVIDSON [*Da 57*] besonders p. 15ff.) ist für die meisten physikalischen Systeme so kompliziert und schwierig zu lösen, daß praktisch immer vereinfachende Lösungen notwendig werden. Unter der Vielzahl der vereinfachenden Lösungsmethoden scheinen die Alterungsgleichung, die Diffusionsnäherung und die S_n-Methode am zweckmäßigsten.

Die *Diffusionsnäherung* überführt die Transportgleichung in die einfachere Neutronendiffusionsgleichung (vgl. BECKURTS-WIRTZ "Neutron Physics" oder DAVIDSON [*Da 57*] p. 94ff. und 255ff.). Die Erdatmosphäre ist ein Medium, in dessen Inneren der Neutronendiffusionsfluß gut mit Hilfe der Diffusionsgleichung beschrieben werden kann (BETHE et al. [*Be 40*]). Die Gültigkeit der Diffusionslösungen in den Randzonen, hier also besonders am Gipfel der Atmosphäre, muß überprüft werden, am besten im Vergleich mit den direkten Näherungslösungen der Transportgleichung. Als weitere Voraussetzung muß die Diffusion isotrop bzw. nahezu isotrop erfolgen, d.h. man ist auf Neutronenenergien ≤ 10 MeV eingeschränkt.

Ein konkurrierendes numerisches direktes *Näherungsverfahren der Transportgleichung* ist die von CARLSON [*Ca 55, 58a, 59*] und CARLSON und BELL [*Ca 58b*] entwickelte S_n-Methode. Der Raumwinkel wird dabei in n-Segmente geteilt und die Neutronenpopulation in jedem Segment durch lineare Ausdrücke definiert und deren Wert für die extremen Richtungen innerhalb dieser Segmente errechnet. Die S_2-Methode ist meist schon eine Verbesserung gegenüber der Diffusionsnäherung; die S_4-Näherung ist den meisten praktischen Zwecken angemessen. Die S_n-Methode erlaubt anisotrope Quellen bzw. Streuungen zu berücksichtigen und gibt Lösungen, die bis zum Rand der Atmosphäre gültig sind. Die große Anzahl von Iterationsschritten macht Großrechenanlagen zur Lösung notwendig.

Fast alle Rechnungen über Neutronen in der Atmosphäre bedienen sich der Diffusionsnäherung. BETHE, KORFF und PLACZEK [*Be 40*] entwickelten bei der erstmaligen Behandlung der Diffusion der Neutronen in der Atmosphäre die erste Form der Bremstheorie (slowing down theory). FLÜGGE [*Fl 43a*] beschrieb die Neutronendiffusion in einem Orts- und Stoßraum, einem Formalismus, den wir heute als (Fermische) Alterungsgleichung bezeichnen [*Fl 43b* und *Ma 47*].

Diese letztere Methode wurde bis etwa 1953 mehrfach benutzt, um mit Hilfe analytischer Lösungen, in die oftmals noch weitere Näherungsannahmen eingingen, die Dichte oder den Fluß der Neutronen in der Atmosphäre zu berechnen. Wir bezeichnen diesen Zeitraum als Phase 1.

Die parallel gehende rapide Entwicklung der Neutronenphysik, Kernphysik und Reaktortheorie führte im Verlauf der letzten Jahre zu guten Meßdaten für Wirkungsquerschnitte, Reaktionsquerschnitte etc. einerseits, zum anderen wurden numerische Verfahren entwickelt, um mit Hilfe von Großrechenanlagen die Diffusion von Neutronen in Reaktoren berechnen zu können.

Damit setzte etwa ab 1960 die zweite Phase der theoretischen Behandlung der Neutronen in der Atmosphäre ein. Voraussetzung war die Messung des Energiespektrums der Neutronen im Gleichgewicht durch MIYAKE et al. [Mi 57a] und HESS et al. [He 59a] (vgl. Ziff. 19). HESS, CANFIELD und LINGENFELTER [He 61] haben in ihrer „Demographie der Neutronen der kosmischen Strahlung" erstmals eine vollständige numerische Lösung der Diffusionsgleichung durchgeführt. Die Behandlung desselben Problems durch NEWKIRK [Ne 63] unter Voraussetzung eines anderen Quellspektrums, insbesondere unter Benutzung der S_n-Methode, ergab eine befriedigende Übereinstimmung der Ergebnisse. In der Zwischenzeit hat LINGENFELTER [Li 63, 64a, 64b] unter Benutzung besserer Wirkungsquerschnitte im Bereich der Quellproduktion von Verdampfungsneutronen und darüber die ursprünglichen Rechnungen verbessert und mit seinen Verteilungsfunktionen der Neutronenproduktionsrate (vgl. Ziff. 32 u. 33) die Demographie der Neutronen und der C^{14}-Produktion in die Atmosphäre in eine übersichtliche Form gebracht. So verfügen wir heute über eine gute Kenntnis der Neutronen der Atmosphäre, die in mancher Hinsicht weiter entwickelt ist als die der Protonen in der Atmosphäre.

Phase 1. Behandlung der Neutronendiffusion mit geschlossenen analytischen Lösungen.

PLACZEK (in [Be 40] bzw. [Pl 46]) hat als erster die Wahrscheinlichkeit dafür angegeben, daß ein Neutron der Energie E, ohne dabei eingefangen zu werden, auf eine Energie E_2 herabgebremst wird,

$$w \sim e^{-\frac{M}{2}\int_{E_2}^{E_1}\frac{\sigma_c(E)}{\sigma_s(E)+\sigma_c(E)}\cdot\frac{dE}{E}}. \tag{35.1}$$

Für die Geschwindigkeitsverteilung haben wir die analoge Formel in (16.1) angegeben. Alle Überlegungen für die Neutronenabbremsung bzw. die Neutronendiffusion in der Atmosphäre sind mit dieser Beziehung verknüpft.

BETHE, KORFF und PLACZEK haben weiter gezeigt, daß für Neutronen in der Atmosphäre ein Diffusionsgleichgewicht besteht mit Ausnahme der Randbereiche. Für die Energieverteilung von Neutronen unter 100 keV diskutierten sie die Tatsache, daß wegen des $1/v$-Einfangs thermischer Neutronen sich kein thermisches Gleichgewicht einstellen kann, sondern eine Verteilung der Form

$$N(v)\,dv = \mathrm{Mql}(v)\cdot\frac{dv}{v^2}\,e^{\left\{-M\frac{\sigma_{c,\mathrm{th}}}{\sigma_s}\left(\frac{KT}{E}\right)^{\frac{1}{2}}\right\}} \tag{35.2}$$

zu erwarten ist. Zweckmäßig stellen wir Abbremslängen, Diffusionslängen wie zugehörige Zeiten für Neutronen in der Atmosphäre in Tabelle 24 zusammen.

In der gewöhnlichen Diffusionsgleichung beschreibt man die Neutronendichte bzw. den Neutronenfluß in einem Orts- und Zeitraum (vgl. auch FREESE u. MEYER [Fr 53]).

Tabelle 24. *Bremslängen und Abbremszeiten bzw. Diffusionslängen und -zeiten für langsame Neutronen in homogenen Atmosphären verschiedener Dichte (als Anhalt für verschiedene atmosphärische Tiefen).*

Höhe		Bremslänge (mittlere)	Diffusionslänge (mittlere)	Abbremszeit (mittlere)	Diffusionszeit (mittlere)
km	g/cm²	m	m	sec	sec
100	~4·10⁻³	3,8·10⁷	7,8·10⁶	3,7·10⁴	1,5·10⁴
50	1	7,3·10⁴	6·10⁴	71,5	112,8
31	10	2,9·10⁴	6·10³	28,6	11,3
21	200	1500	340	0,65	0,25
0	1033	330	75	0,15	0,06

Geht man zu dem mittleren logarithmischen Dekrement u der Energie E pro Neutronenstoß $u = \ln E_0 - \ln E = \ln E_0/E$ über, dann ist die Änderung Δu der Größe u nach ν Stößen im Mittel $\overline{(\Delta u)}_\nu = \nu \xi$. Dies gilt im Falle von kugelsymmetrischer Streuung im Schwerpunktsystem

$$\xi = 1 + \frac{1-\alpha}{\alpha} \ln(1-\alpha); \quad \text{mit} \quad \alpha = \frac{4A}{(A+1)^2} \tag{35.3}$$

(A = Atomgewicht). Setzt man sehr viele Stöße voraus, so kann man ein Differential bilden (wegen der Herleitung sei auf die Lehrbücher der Neutronenphysik verwiesen).

Mit diesen neuen Variablen ist es möglich, eine Differentialgleichung für die Neutronendichte als Funktion von Ort z und Energie E aufzustellen. Gesucht wird die Zahl der Neutronen

$$N(E_0, z, \nu) \, dE \cdot dz \cdot d\nu$$

in der Höhe Z, die ausgehend von einer Anfangsenergie E_0, ν-Stöße durchgemacht haben.

Man kann aus der Anschauung [Fe 43] etc. oder in Form einer Näherung der exakten Transportgleichung (MARSHAK [Ma 47]) eine der Diffusionsgleichung analoge Näherung entwickeln:

$$\frac{\partial N}{\partial t} = -\frac{\partial}{\partial z}(-D)\frac{\partial N}{\partial z} - \frac{\partial}{\partial \nu}\left(\frac{n v}{\lambda_s}\right) \tag{35.4}$$

(wobei λ_s = freie Weglänge für Streuung in cm; $D = v\lambda_s/3$ = Diffusionskonstante). Das erste Glied auf der rechten Seite entspricht einer Konvektion im Ortsraum, das zweite im „Stoßraum", denn nv/λ_s ist die Zahl der pro Sekunde im Intervall vorkommenden Stöße und jeder Stoß führt ein Neutron aus dem Intervall heraus. Um die Ortsabhängigkeit der freien Weglänge $\lambda_s = 1/N\sigma_s$ zu vermeiden, führt man zur Beschreibung der Neutronen in der Atmosphäre üblicherweise die atmosphärische Tiefe x ein.

$$dx = -\varrho \cdot dz; \quad l_s = \varrho \cdot \lambda_s = \frac{m}{\sigma_s}. \tag{35.5}$$

(In dieser Schreibweise ist l_s druckunabhängig.)

Um in der konventionellen Form zur Alterungsgleichung übergehen zu können, verwenden wir folgende Bezeichnungen

und
$$\left.\begin{array}{c}\dfrac{\xi \cdot N(z)}{\lambda_s} dz = -\dfrac{\xi \cdot N(x) \cdot v}{l_s} dx = \chi(E_0; x; \nu) \, dx \\[6pt] \dfrac{l_s^2}{3} d\nu = d\tau\end{array}\right\} \tag{35.6}$$

und erhalten für den stationären Fall

$$\frac{\partial^2 \chi}{\partial x^2} = \frac{\partial \chi}{\partial \tau}.$$
(35.7)

Wird die Anisotropie der Streuung im Laborsystem mitberücksichtigt, so ist [Ma 47b]

$$d\tau = \frac{l_s^2}{3(1-\overline{\cos\vartheta})}\, d\nu \quad \text{mit} \quad \overline{\cos}\,\vartheta = \frac{2}{3A}.$$
(35.8)

FLÜGGE [Fl 43a] hat diese Alterungsgleichung mit Hilfe einer Fourier-Entwicklung gelöst. Seine Neutronendaten entsprechen denen von BETHE et al. [Be 40]. Er wählt als

Anfangsbedingung:

$$\chi(E, x, \nu = 0) = \chi(E, x),$$
(35.9)

d.h. ein Neutron, das in einer Tiefe x mit einer Energie E erzeugt wird, hat noch nicht gestoßen, $\nu=0$. Dies entspricht der Energie- und Höhenverteilung des Quellspektrums.

Randbedingungen:
1.
$$\chi(E, x=0; \nu) = 0.$$
(35.10a)

Unabhängig von der Anzahl der Stöße ν verschwindet die Dichte der Neutronen am Gipfel der Atmosphäre $x=0$ g/cm².

2.
$$\chi(E, x=1033 \text{ g/cm}^2, \nu) = 0.$$
(35.10b)

Unabhängig von der Anzahl der Stöße ν verschwindet die Dichte der Neutronen in Meereshöhe oder am Erdboden $x_0 = 1033$ g/cm².

FLÜGGE stellt also den Verlustfluß der Neutronen nicht in Rechnung. Dazu hat FLÜGGE, wie in Ziff. 16 erwähnt, für langsame und schnelle Neutronen je ein Neutronenquellspektrum Weißkopfscher Art $\sim aE e^{-\alpha E} dE$ erstmalig benützt (Konstanten $a=0{,}178$, $b=6{,}16$; Exponenten: $\alpha=0{,}08$ MeV^{-1}; $\beta=0{,}35$ MeV^{-1}) und die Höhenabhängigkeit e^{-p/p_0}, mit einem Absorptionskoeffizienten $\mu=1/p_0 = 7$ atm^{-1}. DAVIS [Da 50b] hat mit verbesserten Neutronendaten die Flüggesche Dichteverteilung nachgerechnet und fand Übereinstimmung:

$$\varrho_0(p) = 2J_0 \sum_{n=1}^{\infty} \frac{n\pi}{\mu^2 + n^2\pi}\, 0{,}995\, n^2 \left\{ \frac{a(1-0{,}0166\, n^2)!}{\alpha^2 - 0{,}0166\, n^2} + \right. \\ \left. + \frac{b(1-0{,}0166\, n^2)!}{\beta^2 - 0{,}0166\, n^2} \right\} \sin n\mu p.$$
(35.11)

Die von FLÜGGE vorausgesagte Verteilung *langsamer* Neutronen ist später erstmals von YUAN [Yu 48, 49, 51] gemessen worden. Die Übereinstimmung zwischen Theorie und Experiment (vgl. Fig. 61) ist gut; wird — wie im folgenden — der Verlustfluß gemäß der Randbedingubg (35.14) berücksichtigt, so wird sie nahezu vollständig (vgl. auch Fig. 62). Wegen der Pyrex-Korrektur und wegen des Meßverfahrens s. DAVIS [Da 50a] und GABBE [Ga 58].

Die langsamen Neutronen in der kosmischen Strahlung wurden ebenfalls mit der Alterungstheorie gemäß (35.7) und (35.8) von FUJIMOTO und TAMURA [Fu 52] behandelt. Sie stellen für eine monochromatische ebene Neutronenquelle, die sie in einer atmosphärischen Tiefe x' annehmen, die Lösungen χ der Alterungsgleichung als Greensche Funktion $\chi(\tau, x', x)$ dar.

Anfangsbedingung:

$$\chi(\tau, x', x) \to \delta(x-x') \quad \text{für} \quad x'=x; \quad \text{für} \quad \tau=0,$$
(35.12)

d.h. ein Neutron, dessen Fermi-Alter $\tau = 0$ ist, d.h., das noch nicht gestoßen hat, wird am Ort x erzeugt.

Randbedingungen:
$$\chi(\tau, x, x') \to 0, \qquad (35.13)$$

falls $x \to \infty$ (d.h. 1033 g/cm) gültig für alle Fermi-Alter τ

$$\chi(\tau, x, x') = \frac{2}{3} \frac{l(u)}{(1 - \overline{\cos \vartheta})} \cdot \frac{\partial \chi(\tau, x, x')}{\partial x} = \frac{2}{3} \bar{l}(u) \cdot \frac{\partial \chi}{\partial x}, \qquad (35.14)$$

für den Gipfel der Atmosphäre $x = 0$, für alle Fermi-Alter τ.

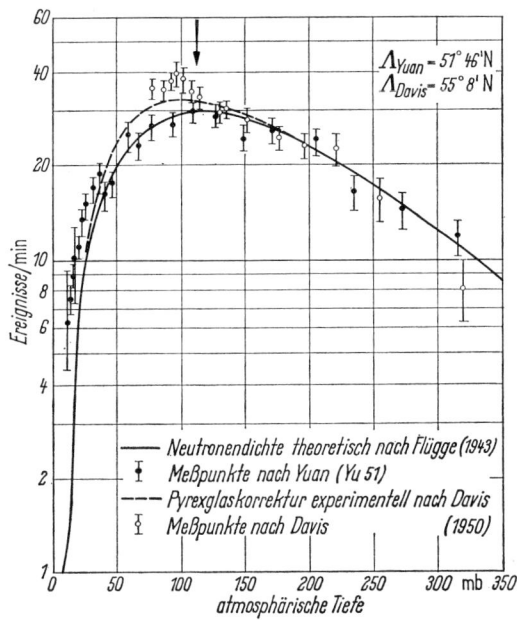

Fig. 61. Flügge-Yuansches Dichtemaximum thermischer und langsamer Neutronen (Pfeil). Theoretische Kurve nach Gl. (35.11) mit $J_0 = 1$. Die Meßpunkte ($^1/_3$ der Zählraten der Cd-Differenz ($\leq 0,4$ eV-Neutronen) der Messung vom 8. Januar 1949) nach YUAN geben das Dichtemaximum gut wieder. Die gemessene Dichte nimmt zum Gipfel der Atmosphäre hin langsamer ab, als die theoretische Voraussage. Dies hängt mit der Randbedingung (35.10a) zusammen; s. Text. — Die Verteilung von DAVIS enthält zusätzlich einen Anteil schneller Neutronen, und beruht zudem auf anderen Meßbedingungen (Bor in Pyrex-Zählerglas). Die Tendenz des Höherwanderns des Maximums ist zu erkennen. Zum Vergleich sei auf das „Simpsonsche Fluß-Maximum" schneller Neutronen (Fig. 39 und 42) hingewiesen.

Diese Autoren untersuchen den Einfluß, der durch die verschiedene Randbedingungen am Gipfel der Atmosphäre (35.14) im Vergleich mit der Flüggeschen (35.10a) in der Neutronenverteilung bemerkbar werden muß. Es läßt sich leicht zeigen, daß die Flüggesche Dichteverteilung (35.11) — eingeschränkt durch die Randbedingung (35.10) — nur vom Fermi-Alter, nicht aber von der mittleren freien Weglänge $l(u)$ abhängig ist. TAMURA und FUJIMOTO zeigen, daß für eine Flächenquelle (35.12) die Lösungen der Alterungsgleichung (35.7) mit der Randbedingung (35.10a) (dies entspricht (35.14) = 0) lauten

$$\chi_0(\tau, x, x') = \frac{1}{\sqrt{4\pi\tau}} \left\{ \exp\left(-\frac{(x-x')^2}{4\tau}\right) - \exp\left(-\frac{(x+x')^2}{4\tau}\right) \right\}. \qquad (35.15)$$

Eine Neutronenquelle mit einer Höhenabhängigkeit

$$S(x') = \begin{cases} J_0 \exp(-a/\lambda) & \text{für } x' < a \\ J_0 \exp(-x'/\lambda) & \text{für } x' > a \end{cases} \qquad (35.16)$$

gibt, die Häufigkeit der beobachteten Sterne in Emulsion (LORD [Lo 51]) gut wieder; die Lordschen Werte werden mit $a=100$ g/cm² und der mittleren freien Weglänge für Sternerzeugung (bzw. Neutronenabsorption) $\lambda=140$ g/cm² gut angenähert. Für diese Quelle folgt die Neutronendichte langsamer Neutronen

$$\begin{aligned}\Psi_0(x,\tau) &= \int dx'\, \chi(\tau, x, x')\, S(x') \\ &= \frac{J_0}{2} \exp(-a/\lambda)\{2\Phi(x/2\sqrt{\tau}) - \Phi((x-a)/2\sqrt{\tau}) - \\ &\quad -\Phi((x+a)/2\sqrt{\tau})\} + \\ &\quad + \frac{J_0}{2}\exp(\tau/\lambda^2)\left\{\exp(-x/\lambda)\left(1+\Phi\left(x-a-\frac{2\tau}{\lambda}\right)\big/2\sqrt{\tau}\right) - \\ &\quad -\exp(x/\lambda)\left(1-\Phi\left(x+a+\frac{2\tau}{\lambda}\right)\big/2\sqrt{\tau}\right)\right\}, \end{aligned} \quad (35.15\text{a})$$

wobei Φ das Gaußsche Fehlerintegral darstellt. Für große atmosphärische Tiefen x gibt es eine asymptotische Form:

$$\Psi_0(x,\tau) \simeq J_0 \exp(\tau/\lambda^2)\cdot e^{-x/\lambda}, \quad (35.15\text{b})$$

die nur vom Fermialter τ, nicht aber von der mittleren freien Diffusionsweglänge $l(u)$ abhängt; d.h., daß sich im Gleichgewichtsbereich die Höhenverteilung der Neutronenquelle in der Neutronendichte widerspiegelt. Auch hier ist ersichtlich, daß die Absorptionslänge λ der Neutronen die Verteilung bestimmt, nicht aber die Diffusionslänge $l(u)$. (Wir werden in Ziff. 41, bei den durch solaren Teilchen ausgelösten Neutronen auch andere Verhältnisse kennenlernen.)

Berücksichtigt man die Randbedingung (35.14) exakt, so kann man nach MARSHAK [Ma 47] die Alterungsgleichung geschlossen lösen mit

$$\chi(\tau,x,x') = \frac{1}{\sqrt{4\pi\tau}}\left\{\exp\left(-\frac{(x-x')^2}{4\tau}\right) - \exp\left(-\frac{(x+x')^2}{4\tau}\right) - \\ -2\sqrt{\pi\tau}\,B\exp(B^2\tau + B(x+x'))\left(1-\Phi\left(B\sqrt{\tau}+\frac{x+x'}{2\sqrt{\tau}}\right)\right)\right\}, \quad (35.17)$$

wobei

$$B = \frac{3}{2\bar{l}(u)} \quad \text{und} \quad \bar{l}(u) = \frac{l(u)}{1-\overline{\cos\vartheta}}.$$

(Auf eine Reihe von Druckfehlern in [Fu 52] sei aufmerksam gemacht.)

In den praktischen Anwendungen ist $B\tau$ viel größer als 1, so daß für das Gaußsche Fehlerintegral die asymptotische Entwicklung gilt. Damit erhält man

$$\chi = \frac{1}{\sqrt{4\pi\tau}}\left(\exp\left(-\frac{(x-x')^2}{4\tau}\right) - \frac{2B\tau - x - x'}{2B\tau + x + x'}\exp\left(-\frac{(x+x')^2}{4\tau}\right)\right),$$

Die spezielle Neutronenverteilung (35.15), die mit der Randbedingung (35.10a) verträglich ist, kann man mit dem Ansatz

$$\chi = \chi_0 + \delta\chi$$

abspalten. Der Einfluß der Randbedingungen (35.14) kann so im Anteil $\delta\chi$ verfolgt werden:

$$\delta\chi = \frac{1}{\sqrt{\pi\tau}}\frac{x+x'}{2B\tau+x+x'}\exp\left(-\frac{(x+x')^2}{4\tau}\right), \quad (35.17\text{a})$$

d.h. der Verlustfluß an Neutronen ist in $\delta\chi$ enthalten.

Die Neutronenverteilung $\Psi(x, \tau)$ in der Atmosphäre, die mit der Quellverteilung (35.16) errechnet wird, spalten wir auf in den schon in (35.15a) bekannten Anteil Ψ_0 und den Anteil

$$\begin{aligned}\delta\Psi(x,\tau) &= \int_0^\infty dx'\,\delta\chi(x,x',\tau)\,S(x') \\ &= J_0\exp(-a/\lambda)\,\frac{1}{\sqrt{\pi}}\,\frac{1}{B\sqrt{\tau}+x/2\sqrt{\tau}}\,\exp(-x^2/4\tau)- \\ &\quad -\frac{1}{B\sqrt{\tau}+\dfrac{a+x}{2\sqrt{\tau}}}\,\{\exp(-(x+a)^2/4\tau)\}+ \\ &\quad +J_0\,\lambda/B(\lambda B-1)\left\{\exp(-a/\lambda)\,\frac{1}{\sqrt{\pi}}\,\frac{1}{B\sqrt{\tau}+(a+x)/2\sqrt{\tau}}\times\right. \\ &\quad\left.\times\exp(-(x+a)^2/4\tau)\right\}- \\ &\quad -J_0\,\frac{1}{\lambda B-1}\,\exp(\tau/\lambda^2)\exp(x/\lambda)\times \\ &\quad\times\left(1-\Phi((a+x+2\tau/\lambda)/2\sqrt{\tau})\right).\end{aligned} \quad (35.18)$$

Die Abhängigkeit von der Diffusionslänge $\bar{l}(u)$ steckt implizit in B [vgl. (35.17)]. Wegen des berücksichtigten Verlustflusses ist die Neutronendichte am Gipfel der Atmosphäre etwas höher, was von FUJIMOTO und TAMURA ausführlich gezeigt wird. Zudem verschiebt sich das Maximum der Neutronendichte höher und etwas näher zum Gipfel der Atmosphäre hin (vgl. Fig. 62).

Wegen der weiteren Einzelheiten der Rechnung bzw. der graphischen Darstellung sei auf die Originalarbeit verwiesen. Ein Vergleich mit den Yuanschen Meßdaten ist dort ebenfalls ausgeführt.

FREESE und MEYER [Fr 53] haben die Höhenverteilung der langsamen Neutronen nach der Alterungsgleichung gerechnet und konnten zeigen, daß die Lösungsmethode von FLÜGGE [Fl 43a] nahezu identisch mit der direkten Lösung der Alterungsgleichung ist. Sie verwenden dieselbe Anfangsbedingung und Randbedingung wie FLÜGGE. Auch hier wird die Höhenverteilung in guter Übereinstimmung mit der Yuanschen Kurve gefunden. Wegen Details wird wieder auf die Originalarbeit verwiesen. Das Spektrum nach FREESE und MEYER findet sich in Fig. 53 und in Gl. (28.3).

GALLI [Ga 53] behandelt die Diffusion und Bremsung von Neutronen, die bei 4 MeV erzeugt werden. Er legt ebenfalls die Alterungsgleichung zugrunde und errechnet eine Albedo (d.h. Verlustfluß) von ungefähr 10%, die in guter Übereinstimmung mit den heutigen Messungen und Rechnungen ist. Der Vergleich

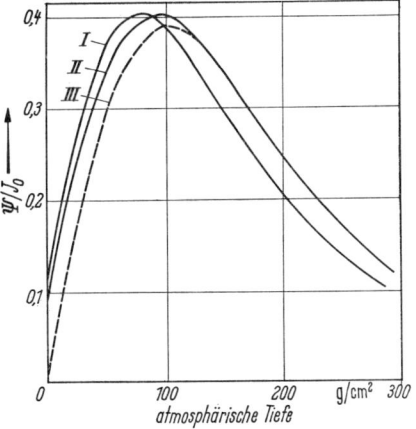

Fig. 62. Verhältnis der Dichte ψ langsamer Neutronen zur Quellstärke J_0 bei verschiedenen Annahmen über den Verlustfluß, bzw. bei unterschiedlicher Höhenabhängigkeit der Neutronenquelle $S(x)$ nach (35.16), charakterisiert durch den Parameter a. Mit $a=100$ g/cm², das etwa die von LORD gemessene Sternhäufigkeit wiedergibt, errechnet sich ohne Verlustfluß (d.h. FLÜGGES Randbedingung (35.10a) die Kurve III; mit Verlustfluß [Randbedingung (35.14)] stimmen die Kurven II und III unterhalb des Maximums überein, jedoch wird das Maximum angehoben, etwas zum Gipfel verschoben und die angehobene Neutronendichte ergibt am Gipfel den Verlustfluß langsamer Neutronen. — Die unphysikalische Annahme des exponentiellen Anstiegs der Neutronenquellung S bis zum Gipfel der Atmosphäre ($a=0$) senkt die Neutronendichte in der Atmosphäre und schiebt die Dichte, wie deren Maximum, zum Atmosphärengipfel hoch. (Man vergleiche in diesem Zusammenhang besonders die Fig. 36 und 65 und in Analogie für schnelle Neutronen Fig. 44.)

seiner Höhenverteilung für die Neutronendichte von 0,4 eV Neutronen stimmt sehr gut überein mit dem YUANschen Höhenmaximum. Die Berücksichtigung der Albedo bedeutet, daß er bei seiner Methode die Randbedingung (35.14) direkt oder indirekt mitbenutzt. Die Anhebung und die geringfügige Verschiebung der Neutronendichte zum Gipfel der Atmosphäre hin, verbessert die Übereinstimmung mit der YUANschen Kurve.

BAGGE und FINCKE [Ba 49] behandeln die Intensitätsverteilung der Höhenstrahlungsneutronen in der Atmosphäre, und zwar durch direkte Lösung der räumlichen Dichteverteilung der Neutronen in Art einer Multigruppenmethode, die sie geschlossen analytisch lösen können. Hinsichtlich des Neutronenquellspektrums wird auf ältere Überlegungen zurückgegriffen; dies ist nicht in Übereinstimmung mit der heutigen Erfahrung. Die YUANsche Höhenverteilung kann auch mit dieser Dichteverteilung wiedergegeben werden.

PFOTZER [Pf 52b] hat die Neutronenkomponente der kosmischen Strahlung einer kritischen Durchsicht unterzogen.

Als Resume dieser ersten analytischen Behandlung der Diffusionsgleichung zur Errechnung der Neutronendichte in der Atmosphäre können wir sagen:

Trotz mancher Vereinfachung, die zur geschlossenen analytischen Lösung der Diffusions- bzw. Alterungsgleichung gemacht werden mußten, stimmt die theoretische Voraussage der Höhenverteilung der langsamen Neutronen gut mit der gemessenen YUANschen Kurve überein. Dies gilt sowohl für die Randbedingung (35.10) für die die Neutronendichte 0 am Gipfel der Atmosphäre eingesetzt wird, wie auch für die Randbedingung (35.14), die dem Verlustfluß langsamer Neutronen durch geringfügige Anhebung des Maximums und durch unerhebliche Verschiebung der Neutronendichte zum Gipfel der Atmosphäre hin Rechnung trägt.

Auf Grund der genauen Messungen der Neutronendichte langsamer Neutronen nahezu am Gipfel der Atmosphäre durch HAYMES und KORFF [Ha 60a] wird dies verständlich. Die Dichte der langsamen Neutronen fällt vom Maximum zum Gipfel der Atmosphäre hin steil ab; in dem Experiment konnte eine Abnahme um den Faktor 50 angegeben werden. Wir wir in Ziff. 21, S. 452 ausgeführt haben, sind die Verhältnisse für schnelle Neutronen anders. Die Voraussagen der Diffusionstheorie im Rahmen der geschlossenen analytischen Lösungen waren also für die Erklärung der langsamen Neutronenereignisse ausreichend.

Phase 2. Numerische Lösungen der Diffusionsgleichung.

HESS, CANFIELD und LINGENFELTER [He 61] haben erstmals eine „Demographie" der kosmischen Strahlungsneutronen geben können. Ziel ihrer Arbeit war, den skalaren Neutronenfluß $\Phi(E, r)$ mit der Diffusionsgleichung zu errechnen.

Befinden sich in 1 cm³ der Atmosphäre, in einer Höhe r eine Zahl von $N(E, r)\, dE$ Neutronen, die dem Energieintervall E bis $E + dE$ angehören, bei einer skalaren Geschwindigkeit v, so ist

$$\Phi(E, r)\, dE = N(E, r)\, v\, dE \tag{35.18}$$

der skalare Fluß der Neutronen in Abhängigkeit von der Dichte der Neutronen.

Dieser skalare Gleichgewichtsneutronenfluß wird durch die eindimensionale, stationäre Diffusionsgleichung bestimmt.

$$\left.\begin{aligned} &-\nabla D(E, r)\, \nabla \Phi(E, r) + \Sigma(E, r)\, \Phi(E, r) \\ &-\int_0^\infty dE'\, \Sigma(E' \to E, r)\, \Phi(E', r) = S(E, r). \end{aligned}\right\} \tag{35.19}$$

Ziff. 35.   Näherungslösungen für die Neutronen-Transportgleichung.

Im einzelnen bedeuten die Symbole:

		Dimension
v	= die Neutronengeschwindigkeit	cm/sec
$n(E,r)\cdot dE$	= die Zahl der Neutronen pro cm³ in Energieintervall E und $E+dE$	Neutronen/cm² MeV
$\Phi(E,r)$	= $n(E,r)\cdot v\,dE$	Neutronen/cm² sec · MeV
$\Sigma(E,r)$	= Gesamter makroskopischer Wirkungsquerschnitt	cm⁻¹
$\Sigma(E'-E,r)$	= Makroskopischer Wirkungsquerschnitt für Streuung vom Energieintervall dE' ins Intervall dE	cm⁻¹/MeV
$D(E,r)$	= Diffusionskoeffizient = $\dfrac{1}{3\Sigma_{\text{Tr}}(E,r)}$	cm
$\Sigma_{\text{Tr}}(E,r)$	= makroskopischer Transportquerschnitt	cm⁻¹
$S(E,r)$	= Neutronenquellstärke = Zahl der Neutronen, die pro cm³ und sec pro Energieintervall ins System geliefert werden (Produktionsrate).	Neutronen/cm³ sec · MeV

Aus der Differentialgleichung (35.19) erhält man die übliche Multigruppengleichung für Neutronen in der i-ten Energiegruppe $E_{i-1}<E<E_i$, indem man alle Glieder der Gleichung mit dE multipliziert und gliedweise von E_{i-1} bis E_i integriert.

$$-\nabla D^i(r)\,\nabla\Phi^i(r)+\Sigma^i(r)\,\Phi^i(r)-(\text{Summe})_j\Sigma^{ij}(r)\,\Phi^j(r)=S^i(r). \qquad (35.20)$$

In dieser Multigruppengleichung bedeuten im einzelnen die Symbole:

Symbol = Bedeutung	Bezeichnung
$\Phi^i(r) = \int_{E_{i-1}}^{E_i} \Phi(E,r)\,dE$	skalarer Gruppenfluß
$S^i(r) = \int_{E_{i-1}}^{E_i} S(E,r)\,dE$	Gruppenquelle
$D^i(r) = \dfrac{1}{\Phi^i(r)}\int_{E_{i-1}}^{E_i} D(E,r)\,\Phi(E,r)\,dE$	Gruppendiffusionskoeffizient
$\Sigma_i(r) = \dfrac{1}{\Phi^i(r)}\int_{E_{i-1}}^{E_i} \Sigma(E,r)\,\Phi(E,r)\,dE$	gesamter Gruppenwirkungsquerschnitt
$\Sigma_{ij}(r) = \dfrac{1}{\Phi_j(r)}(\text{Summe})_j\int_{E_{i-1}}^{E_i} dE\,\dfrac{1}{\Phi_j(E,r)}\int_{E_{j-1}}^{E_j} dE'\,\Sigma'(E'\to E)\,\Phi_j(E',r)$	Gruppenwirkungsquerschnitt für Streuung von der Gruppe j nach der Gruppe i

Wir betrachten die Atmosphäre (auf deren Eigenschaften wir noch im einzelnen eingehen werden) als eine hinreichend dünne Schicht oberhalb der Erdoberfläche von etwa 40—50 km Dicke. Die mittlere freie Abbrems- bzw. Diffusionslänge eines Neutrons ist klein verglichen mit dieser Schichtdicke bzw. mit einer Bogensekunde der geomagnetischen Breite (vgl. Tabelle 21). Der Gleichgewichtsfluß bzw. die Dichteverteilung langsamer Neutronen ist bei irgendeiner geomagnetischen Breite nur mit der nächsten Nachbarschaft der Flußwerte (bzw. der Dichteverteilung) verkoppelt und wird deshalb nicht von irgendwelchen Veränderungen des Neutronenflusses in Abhängigkeit von der geomagnetischen Breite beeinflußt. Sowohl aus rechnerischen Gründen als auch aus Gründen der einfacheren physikalischen Interpretation ersetzt man die Kugelschalenatmosphäre durch eine ebene Atmosphäre, die als Schicht in zwei Dimensionen ins Unendliche sich erstreckt und eine Schichtdicke r annimmt. Für diese ebene Atmosphäre ist es zweckmäßig, so wie es schon bei der Alterungsgleichung (35.7) geschah, auf eine homogene Dichte von 1 g/cm² überzugehen. Die Masse x der darüberliegenden Luft kann dann als der Druck p gemessen werden, den eben diese Luftmasse auf eine Querschnittsfläche von 1 cm² in der Höhe r ausübt. Wir führen diese Transformation für die Multigruppengleichung (35.17) durch.

Im einzelnen erhalten wir für den Gradienten

$$V = \frac{d}{dr} = \frac{dr}{dx} \cdot \frac{d}{dx}. \tag{35.21}$$

Befinden sich in einem Intervall x bis $x+dx$ eine Zahl von $N(x)dx$ Neutronen bzw. $S(x)dx$ Quellneutronen, so definieren wir den Zusammenhang in den neuen Koordinaten

$$S(x)dx = S(r)dr \quad \text{und} \quad N(x)dx = N(r)dr. \tag{35.22}$$

Damit können wir für den Fluß sofort angeben

$$N(x) \cdot \frac{dx}{dt} = N(r) \frac{dr}{dt} \quad \text{oder} \quad \Phi(x) = \Phi(r). \tag{35.23}$$

Zuletzt benötigen wir noch die Transformation der Wirkungsquerschnitte

$$\Sigma(x) = \Sigma(r) \cdot \frac{dr}{dx}. \tag{35.24}$$

Mit den Ausdrücken (35.18) bis (35.21) führen wir die Transformation in der ursprünglichen Multigruppengleichung (35.17) auf die Einheitsmasse durch, so gewinnen wir

$$-\frac{d}{dx} D^i(x) \frac{d}{dx} \Phi^i(x) + \Sigma^i(x) \Phi^i(x) - (\text{Summe})_j \Sigma^{ji}(x) \Phi^j(x) = S^i(x). \tag{35.25}$$

Die Neutronendiffusion als Reaktorrechenproblem wurde vielfach programmiert; das spezielle Programm ZOOM von STUART, CANFIELD, DOUGHERTY und STONE [St 58] für IBM 704 und IBM 7090 Rechenanlagen wurde bis jetzt den Berechnungen der Neutronen in der Atmosphäre zugrundegelegt.

In dieser Schreibweise erscheint die Atmosphäre auf eine einheitliche Dichte von 1 g/cm³ komprimiert. Eine Säule von einer Höhe $x_0 = 1033$ cm von dieser Einheitsdichte entspricht einem Druck der Masse von 1033 g auf 1 cm² Fläche und das ist in diesem Maß gerade die Höhe der homogenen Atmosphäre. Man erhält in der homogen komprimierten Atmosphäre, wie in der diffusen Atmosphäre denselben Neutronenfluß, der durch 1 cm²/sec hindurchtritt. Somit wird weder die Normierung des Neutronenflusses noch die C^{14}-Produktionsrate beeinflußt. Aber es muß in diesem Zusammenhang erwähnt werden, daß in einer verdünnten Atmosphäre die mittlere freie Stoßlänge und damit die Abbrems- und Diffusionszeiten doch hinreichend lang sind, so daß einige der Neutronen im Fluge zerfallen können, bevor sie auf thermische Energien abgebremst oder eingefangen werden können. In Meereshöhe, wo die Abbremszeit von der Größenordnung von 0,1 sec ist (vgl. Tabelle 24), führt dies auf weniger als $10^{-2}\%$ Neutronenverlust; aber bei sehr großen Höhen, bei Drücken von der Größenordnung von 10 g/cm² wird dies ein kennzeichnender Verlustmechanismus, denn die Abbremszeit ist umgekehrt proportional zum Druck. Der gesamte Effekt des Verlustes durch β-Zerfall beträgt jedoch bei der Integration über die Höhe weniger als 0,5% des Neutronenverlustes.

Nachdem ein benutzbares Rechenprogramm zur Lösung der Multigruppengleichung (35.17) und (35.25) vorliegt, wird es notwendig, die verschiedenen Daten, die in dieses Programm eingehen, festzulegen. Die nuklearen Eigenschaften der Atomkerne der Luft (Stickstoff, Sauerstoff, eventuell Argon) werden mit Hilfe der gemessenen Wirkungsquerschnitte beschrieben. In der grundlegenden Arbeit von HESS, CANFIELD und LINGENFELTER [He 61] wurden dazu die Wirkungsquerschnitte nach den Tabellen von HUGHES und SCHWARZ [Hu 58] — des BNL-Berichts 325 — verwendet. Die Wirkungsquerschnitte für N und O wurden mit dem von HESS et al. [He 59a] gemessenen Neutronenspektrum der Neutronen in der Atmosphäre gewichtet. Die Übertragungskoeffizienten die die Streuung von einer Energiegruppe in die andere beschreiben, wurden auf der Annahme aufgebaut, daß die Atmosphäre eine Bremssubstanz darstellt, die wie ein Maxwellsches Gas isotrop im Schwerpunktsystem streut; wegen des Temperaturmittelwertes dieses Gases s. S. 492, Modellatmosphäre I bzw. II.

CANFIELD, STEWARD, FREIS und COLLINS [Ca 61] haben für die IBM 709/7090 ein Programm Sophist I entwickelt, mit dem die Multigruppenübertragungskoeffizienten $\Sigma_{ij}(x)$ für gasförmige Medien berechnet werden können. Dem Effekt der Anisotropie bei elastischer Streuung oberhalb von 1 MeV wird Rechnung getragen in Form einer linearen Korrektur mit dem Mittelwert $\mu = \overline{\cos\Theta}$ des Streuwinkels.

LINGENFELTER [Li 63c] legt für den niederenergetischen Bereich die Tabellen von HUGHES und SCHWARZ [Hu 58] zugrunde, für den hochenergetischen Bereich benutzt er zusätzlich die Tafeln von HOWERTON [Ho 58, 59], LUSTIG et al. [Lu 57, 58] und CHASE et al. [Ch 61]. Inzwischen ist noch ein weiteres Supplement zu BNL-325 erschienen, nämlich STEHN et al. [St 64].

Die Wirkungsquerschnitte im höherenergetischen Bereich sind nur unvollständig bekannt, so daß ihre Werte für die einzelnen Energiegruppen nicht vollständig tabuliert sind. Um trotzdem gute gewichtete Werte zu erhalten, werden für jede Energiegruppe diese Wirkungsquerschnitte iterativ über ein selbstkonsistentes Neutronenflußspektrum im Gleichgewicht gemittelt.

Die Diffusionskoeffizienten werden bestimmt mit Hilfe der Beziehung

$$D = \frac{\Sigma_a}{K^2}, \qquad (35.26)$$

wobei der Eigenwert K mit der asymptotischen Eingruppenlösung der Transportgleichung (vgl. GLASSTONE und EDLUND [Gl 52])[1] errechnet werden kann.

$$\frac{\Sigma_s}{2K} \ln \frac{\Sigma+K}{\Sigma-K} = \frac{K^2 + 3\Sigma_a \Sigma_s \bar{\mu}_0}{K^2 + 3\Sigma \Sigma_a \bar{\mu}_0}. \qquad (35.27)$$

Die makroskopischen Wirkungsquerschnitte Σ, Σ_a und Σ_s kennzeichnen den Gesamt-, den Absorptions- und den Streuquerschnitt; $\bar{\mu}_0$ ist der mittlere Wert des $\cos\Theta$ des Streuwinkels im Laborsystem. Bei einer direkten Multigruppenbehandlung zur Errechnung von K könnten sich die Diffusionskoeffizienten D um mehrere % ändern.

36. Die *Quellverteilung der Neutronen* wurde in Ziff. 32 und 33 hinsichtlich ihrer Höhenverteilung, Breitenabhängigkeit und ihrer zeitlichen Abhängigkeit innerhalb des Sonnenfleckenzyklus dargestellt (vgl. Fig. 58 und 60). Die Energie-

---
[1] Beachte Druckfehler in dieser Auflage.

verteilung der Quellneutronen (vgl. Fig. 26) wurde in Ziff. 17 sowohl für die Verdampfungsneutronen als auch für die Anstoßneutronen (18.2) angegeben. Nachdem die Diffusionsnäherung nur auf Neutronen von Energien $E \leq 10$ MeV angewandt werden kann, fallen die Anstoßneutronen praktisch außerhalb des berechenbaren Bereichs. Nachdem aber nur für einen gewissen Bruchteil der schnellen Neutronen eine Wahrscheinlichkeit dafür besteht, über den Verlustfluß aus der Atmosphäre entkommen zu können, und zwar bevorzugt bei streifendem Einfall hochenergetischer Primärer (vgl. Fig. 25) wird der Rest der schnellen Neutronen innerhalb der Atmosphäre durch inelastische Streuung Energie abgegeben. In der Sprache der Multigruppennäherung bedeutet dies, daß er in Gruppen tieferer Energie hineingestreut wird. HESS, CANFIELD und LINGENFELTER haben aus dem Analogspektrum für die Streuung schneller Neutronen (vgl. Fig. 28 und zugehörigen Text) Wahrscheinlichkeiten abgeschätzt, welcher Prozentsatz schneller Anstoßneutronen in den Energiebereich der Verdampfungsneutronen hineingestreut wird. Unter Vorwegnahme einiger Daten des Verlustflusses finden sie für 3—10 MeV-Neutronen eine kleine Wahrscheinlichkeit von 8% zu entkommen. Für die energiereichen Neutronen im GeV-Bereich schätzen sie auf Grund rein geometrischer Überlegungen 1%.

Unter Berücksichtigung des Satzes der Neutronenerhaltung geben HESS et al. für den in den Bereich der Verdampfungsneutronen hineingestreuten Anteil schneller Neutronen an: Von den Anstoßneutronen werden ungefähr 43% absorbiert, etwa 5% beteiligen sich am Verlustfluß mit Energien oberhalb 10 MeV. Die restlichen 52% werden in den Energiebereich unterhalb 10 MeV hineingestreut (44% fallen in die Energiegruppe zwischen 10 und 3,16 MeV, 6% zwischen 3,16 und 1 MeV; 2% unter 1 MeV).

Das Verhältnis der Quellstärken der Verdampfungsneutronen zu den Anstoßneutronen (vgl. Fig. 64)

$$\frac{N_V(E)\,dE}{N_A(E)\,dE} = 4{,}1 \tag{36.1}$$

mußte notwendig so gewählt werden, um das experimentell abgeschätzte Verhältnis des Flusses aus dem Anstoßneutronenbereich in den Verdampfungsneutronenbereich angeben zu können. Wir greifen zurück auf das Neutronenverdampfungsspektrum (16.2)

$$N_V(E)\,dE = K_V \cdot E \cdot e^{-E/\Theta} \cdot dE. \tag{36.2}$$

Die Quellstärke K_v des Verdampfungsspektrums berücksichtigt nur die Neutronen mit Energien zwischen 0 und 10 MeV. Der Beitrag, der von den Anstoßnukleonen in dieses Energieintervall hineingestreut wird, und bei dieser Diffusionsrechnung mitgerechnet und für das das gesamte Neutronenspektrum mit berücksichtigt werden muß, wird durch eine Anstoßneutronenquellstärke K_A für diesen Bereich ergänzt:

$$K_A = 0{,}52 \cdot \frac{K_V}{4{,}1}.$$

Die Quellstärke S für das Quellspektrum der Neutronen im Energiebereich $E \leq 10$ MeV wird dann

$$N(E)\,dE = S \cdot E \cdot e^{-E/\Theta}\,dE \tag{36.3a}$$

mit

$$S = K_V + K_A. \tag{36.3b}$$

Ziff. 36. Die Quellverteilung der Neutronen. 491

Die gesamte Quellverteilung (Produktionsrate/gsec) wurde von HESS et al. [He 61] mit (36.3a) und (17.2) angesetzt:

$$S(E, x)\, dE \cdot dx = e^{-x/L} \cdot N(E)\, dE\, dx. \tag{36.4}$$

LINGENFELTER [Li 63c] verwendet für die verbesserten Rechnungen ebenfalls (36.3a) und die in Ziff. 32 und 33 dargestellten Verteilungsfunktionen (33.1) für die Produktionsraten $q(x, \Lambda_i;, \tau_i)$, d.h. die Quellverteilungen, die die geomagnetische Breite Λ und zeitliche Lage im Sonnenfleckenzyklus berücksichtigen:

$$S(E, x, \Lambda_i, t_i)\, dE, dx = q(x, \Lambda_i, t_i) \cdot N(E)\, dE\, dx. \tag{36.5}$$

Die Erdatmosphäre stellt Quellort und Wanderungsweg der Neutronen dar. Die Eigenschaften der Erdatmosphäre sollen knapp zusammengestellt werden. Die

Tabelle 25. *Zusammensetzung der Luft in freier Atmosphäre.*

	N_2	O_2	A	CO_2	H_2	N_e	H_e	Kr	X
Volum-Prozente..	78,03	20,99	0,933	0,030	0,01	0,0018	0,0005	0,0001	$8 \cdot 10^{-6}$
Gewichts-Prozente	75,47	23,20	1,28	0,046	0,001	0,0012	0,00007	0,0003	$4 \cdot 10^{-5}$

Zusammensetzung der Luft ist nach LANDOLT-BÖRNSTEIN [La 52] in Gewichts- bzw. Volumenprozenten in Tabelle 25 zusammengestellt.

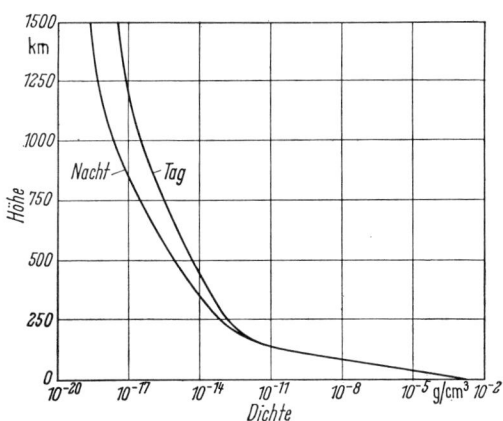

Fig. 63a. Dichte der Atmosphäre nach CIRA [Ci 61] und NICOLET [Ni 63]. Vgl. auch die NACA-Atmosphäre Fig. 4.

Das NACA-Modell der Atmosphäre nach PENNDORF [Pe 52a] Fig. 4 gibt den Zusammenhang zwischen atmosphärischer Tiefe x in (g/cm²) und der Höhe in km. Bis etwa 50 km sind die Unterschiede zu experimentell mit Raketen gefundenen Werten unerheblich. In großen Höhen zeigt die Luftdichte in der CIRA (Cospar-Atmosphäre) [Ci 61] oder nach NICOLET [Ni 63] (vgl. Fig. 63a) einen Tag- und Nachtgang. Ob sich dieser auf den Verlustfluß der langsamen Neutronen auswirkt ist nicht gesichert. Vielleicht kann der Rückeintritt der durch die Schwere auf die Erde zurückfallenden langsamen Verlustflußneutronen (nach Messungen von HAYMES [Ha 59], wurde nach Sonnenuntergang ein Anwachsen der langsamen Neutronendichte festgestellt) durch diese Luftdichte-Schwankung mitbedingt sein.

Das Temperaturfeld der Atmosphäre ist in Fig. 63b schematisch dargestellt. Gemessen wurde zum Zeitpunkt des Sonnenfleckenmaximums. Die Temperaturschichtung ist breitenabhängig. Beispielsweise beginnt die Stratosphäre am Äquator mit $-80°$ C etwa bei 18 km Höhe; in Mitteleuropa etwa bei 11 km. Danach nimmt die Temperatur ab 25—30 km (je nach Breite) wieder zu und erreicht bei 50 km Höhe, am Gipfel der Atmosphäre, einen Wert von $+30°$ C.

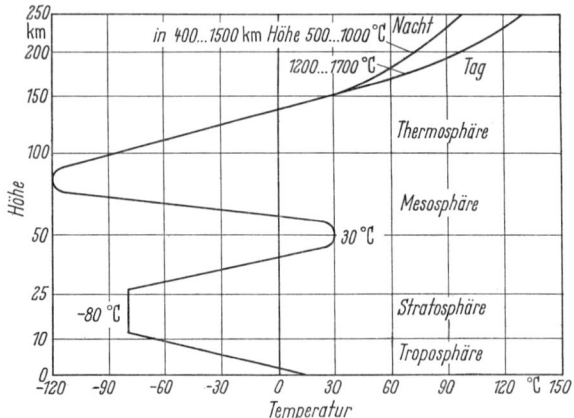

Fig. 63b. Der Temperaturaufbau der Atmosphäre zur Zeit des Sonnenfleckenminimums nach CIRA [Ci 61] und NICOLET [Ni 63].

In den bisherigen Rechnungen zur Lösung der Diffusionsgleichung bzw. der Transportgleichung wurden folgende Modellatmosphären benutzt.

Modellatmosphäre I (benutzt von HESS, CANFIELD, LINGENFELTER [He 61]).
A. Zusammensetzung: 80% Stickstoff,
    20% Sauerstoff.
B. Simulation der Erdoberfläche: 20 cm H_2O.
C. Temperaturen nach SHAPIRO [Sh 53] vgl. Tabelle 26. Zur Errechnung der Überführungskoeffizienten Σ_{ij} wurden diese Zonen ebenfalls benutzt.

Tabelle 26. *Temperaturzonen nach Shapiro.*

	Höhe in km	Atmosphärische Tiefe in (g/cm²)	Temperatur	
			°K	°C
Zone IV	9—50	301—0	219	-54
Zone III	5—9	545—301	234	-39
Zone II	2—5	789—455	260	-13
Zone I	0—2	1033—789	280	$+7$

Modellatmosphäre II (für C^{14}-Produktion LINGENFELTER [Li 63a]).
A. Zusammensetzung: 78,09% Stickstoff (nach KUIPER [Ku 52]),
    20,98% Sauerstoff,
    0,93% Argon.
B. Simulation der Erdoberfläche = 20 cm H_2O.
C. Temperaturfeld: Maximale C^{14}-Produktion in atmosphärischer Tiefe von 100—300 g/cm² Restdruck. Temperatur für diese Zone $T=215°$ K wurde auf die ganze Erdatmosphäre übernommen. Ebenso wurden die Überführungskoeffizienten Σ_{ij} der einzelnen Energiegruppen für diese Temperatur 215° K eines Maxwell-Gases errechnet.

Modellatmosphäre III (MAUCK [*Ma 66*]).

A. Zusammensetzung
B. Simulation der Erdoberfläche } wie bei Modellatmosphäre II.

C. Temperaturfeld: breitenabhängige Temperaturprofile (vgl. Beispiel Figur 63b). Berücksichtigung des Temperaturgangs pro Jahr und Sonnenfleckenzyklus. Außerdem Untersuchung relativ kurzzeitiger, aber großräumiger und starker Temperaturschwankungen der Stratosphäre.

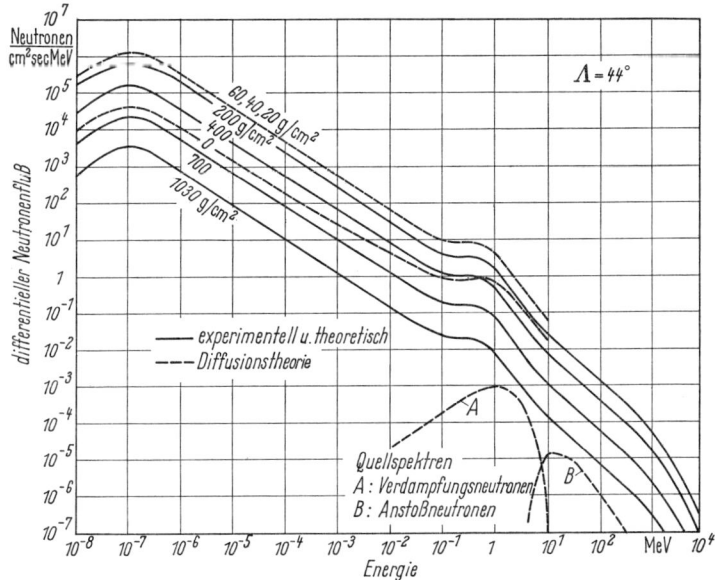

Fig. 64. Der Gleichgewichtsneutronenfluß aufgetragen über der Energie bei verschiedenen Tiefen in der Atmosphäre für $\Lambda_{geom}=44°$ Nord. Die mit „experimentell and theoretisch" gekennzeichneten Energiespektren sind experimentelle Werte (HESS, PATTERSON, WALLACE u. CHUPP [*He 59*]); mit diesen wurden die errechneten Kurven bei Abweichungen von ±25% zur Festlegung der gesamten Quellstärke normiert. Der Neutronenverlustfluß wurde berechnet. Das Verdampfungsquellspektrum und das Anstoßneutronenquellspektrum sind gestrichelt mit eingezeichnet. Oberhalb 500 MeV wurde der Neutronenfluß mit Hilfe der Messelschen Kaskadentheorie fortgesetzt (vgl. auch Fig. 65).

37. Ergebnisse der Diffusionsrechnung Mit den in Ziff. 36 zugrundegelegten Daten konnten HESS, CANFIELD und LINGENFELTER numerisch mit der Diffusionsgleichung einen Gleichgewichtsfluß errechnen. Dieser wurde auf das von HESS, PATTERSON, WALLACE und CHUPP [*He 59a*] gemessene differentielle Energiespektrum normiert ($\lambda_{geom}=44°$ Nord). Durch diese Normierung wird die gesamte Quellstärke festgelegt. Sie beträgt nach Gl. (36.3a) bzw. (36.2)

mit
$$S = K_A + K_V = 6{,}2 \text{ Neutronen/cm}^2 \text{ sec}$$
Verdampfungsneutronenquelle $K_V = 5{,}0$,
Anstoßneutronenquelle $K_A = 1{,}2$. (37.1)

Mit dieser Normierung stimmt der errechnete differentielle Gleichgewichtsfluß (vgl. Fig. 64) mit dem gemessenen Fluß innerhalb der Meßgenauigkeit von ±25% überein, und zwar über acht Größenordnungen der Energie, d.h. von thermischen Energien bis 10 MeV. Diese Übereinstimmung gilt von Meereshöhe bis etwa 200 g/cm² unterhalb des Gipfels der Atmosphäre, d.h. der oberen Grenze dieser Experimente. Einige weitere Kurven bis zum Gipfel der Atmosphäre sind errechnet und mit eingezeichnet. Bis 50 g/cm² atmosphärischer Tiefe ist die Gestalt des Flusses ähnlich wie bei 200 g/cm²; danach macht sich der Einfluß des Verlustflusses stark bemerkbar, was sich in der Kurve für 0 g/cm² ausdrückt.

In Tabelle 27 fassen wir die Daten der Rechnung nach HESS et al. [He 61] zusammen. Um den gesamten Mittelwert der Quellstärke Q (21.34) zu erhalten (vgl. ANDERSON [An 53]), multipliziert man mit 0,75 und erhält als Gesamtquellstärke $\bar{S} = 4{,}6$ Neutronen/cm² sec. Die Produktionsrate für die C_{14}-Bildung erhält

Tabelle 27. *Neutronen/cm² sec in der Atmosphäre. Nach* HESS *et al.* [He 61].

Neutronen	<1 MeV	1—10 MeV	>10 MeV	Gesamt
Eingefangen mit C¹⁴-Bildung . .	3,86	0,11	0	3,97
Durch andere Reaktionen . . .	0,18	0,49	0,53	1,20
Durch die Erde mit Einfang . .	0,00	0,00	0,00	<0,01
Verlustfluß aus der Atmosphäre	0,62	0,35	0,06	1,03
Gesamt	4,66	0,95	0,59	6,20

man dann anteilig aus dem Verhältnis 3,97/6,2 (vgl. Tabelle 27), welches einen Gesamtmittelwert der C¹⁴-Produktionsrate $Q = 2{,}9$ Neutronen/cm² sec, ebenfalls mit einer Genauigkeit von $\pm 25\%$ ergibt. Eine Diskussion dieses Wertes erfolgt in Ziff. 39.

Der gesamte Verlustfluß von 1,03 Neutronen/cm² sec bei $\Lambda_{\text{geom}} = 44°$ N (vgl. Tabelle 27) entspricht umgerechnet auf den Äquator einem Verlustfluß von 0,4 Neutronen/cm² sec und von ungefähr 1,8 Neutronen/cm² sec am Pol. Eine Diskussion dieses Wertes erfolgt in Ziff. 45.

Diese Untersuchung zeigt, daß der Gesamtmittelwert der Neutronenquellstärke $\bar{S} = 4{,}6$ Neutronen/cm² sec beträchtlich höher liegt als die errechnete C¹⁴-Produktionsrate $\bar{Q} = 2{,}9$ Neutronen/cm² sec, die mit dem Mittelwert von ANDERSON [An 53] $\bar{Q} = 2{,}6$ noch ziemlich gut übereinstimmt. Man muß daraus schließen, daß in der Atmosphäre erheblich mehr Neutronen produziert werden als man aus den C¹⁴-Daten ablesen kann, und diese über andere Reaktionsprozesse meist bei höherer Energie absorbiert werden.

Die Konkurrenzreaktionen (vgl. Ziff. 39) wirken sich auf Grund ihrer Wirkungsquerschnitte meist bei Energien $E > 1$ MeV aus. Beschränkt man sich auf den Bereich unterhalb 1 MeV, so bleibt die entscheidende Reaktion, $N^{14}(n, p) C^{14}$. Dieser Einfangprozeß des Neutrons zur Bildung von C¹⁴ bleibt nahezu unbeeinflußt von der Art wie die Energien der schnellen und hochenergetischen Neutronen — durch die zum großen Teil noch unbekannten Reaktionswirkungsquerschnitte — ein für allemal gefiltert oder verändert werden. In dem niederenergetischen Bereich unterhalb 1 MeV bleibt die Form des Spektrums erhalten; seinem Betrag nach kann es verändert werden.

Für langsame Neutronen vergleichen wir in Fig. 36 die Neutronendichte (Neutronen/cm³), die MILES [Mi 64] in seiner Präzisionsmessung bestimmt hat, mit dem (extrapolierten) experimentellen Ergebnis von HESS et al. [He 59a] (vgl. Fig. 64 und 65). Dort ist der differentielle Neutronenfluß $\Phi = N(E) \cdot v$ aufgetragen, der oberhalb von 200 g/cm² bis zum Gipfel der Atmosphäre nach HESS, CANFIELD und LINGENFELTER [He 61] mit Hilfe der Diffusionsnäherung festgelegt wird, da HESS et al. [He 59a] nur bis 12 km gemessen haben. Für die verbesserten Rechnungen hat LINGENFELTER [Li 63a, b] eine andere Höhenverteilung (Fig. 26) bzw. Breitenabhängigkeit (vgl. Ziff. 32 und 33) zugrunde gelegt und die nur wenig bekannten Neutronenwirkungsquerschnitte für $E > 1$ MeV berücksichtigt. Dieser Gleichgewichtsneutronenfluß ist wieder an den Messungen von HESS et al. [He 59a] normiert. Der Verlauf dieser Neutronendichte stimmt mit der von MILES gemessenen ziemlich überein; aber das Maximum liegt um etwa 1,7 höher und die Gesamtabweichung gegenüber MILES ist etwa 50%. Nachdem nach MILES [*Mi 64*] die Eichnormalen beider Institute ausgetauscht worden waren, muß dieser Unterschied andere Gründe haben; MILES vermutet als Ursache eine zusätzliche Neutronenproduktion im Trägerflugzeug der Hessschen Messung. LINGENFELTER [*Li 63*] legt die Modellatmosphäre II zugrunde und errechnet speziell die C^{14}-Produktionsrate und den Neutronenverlustfluß (vgl. detaillierte Diskussion seiner Ergebnisse in Ziff. 39).

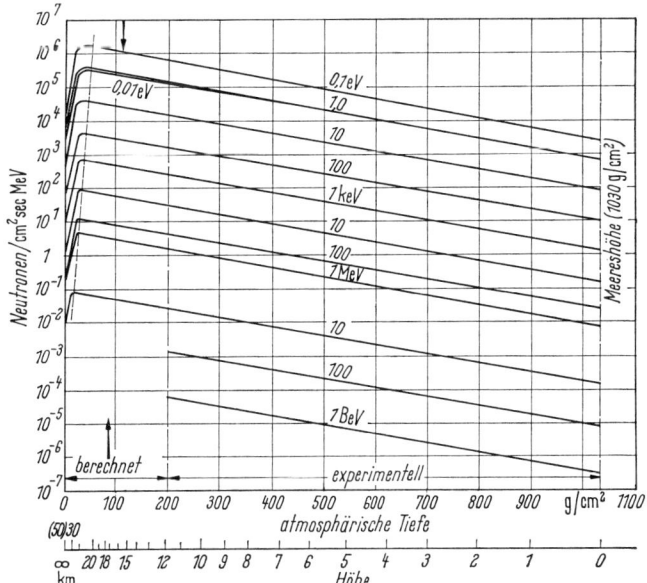

Fig. 65. Der Gleichgewichtsneutronenfluß aufgetragen über der Höhe nach [*He 61*]. Es liegen dieselben Daten zugrunde wie in Fig. 64. Der obere Pfeil weist auf die Lage des Flügge-Yuanschen Dichtemaximums, der untere Pfeil auf die Lage des Simpsonschen Flußmaximums hin. Die Lage der errechneten Maxima ist zu hoch.

38. Näherungsverfahren der Transportgleichung (S_n-Methode). Unter den verschiedenen Methoden zur genäherten, numerischen Lösung der Neutronentransportgleichung wurde die S_n-Methode von CARLSON [*Ca 55, 58a, 59*] und CARLSON und BELL [*Ca 58b*] entwickelt. Diese ist auf Differenzengleichungsverfahren aufgebaut und kann auf Großrechenanlagen durchgerechnet werden. Sie ist verschiedentlich auf Kernreaktorprobleme angewandt worden und lieferte eine erhebliche Genauigkeit. NEWKIRK [*Ne 63*] benutzte zur Behandlung des Neutronentransports in der Atmosphäre diese S_n-Methode.

Die Transportgleichung für Neutronen in der Atmosphäre kann in Analogie zu der zuvor beschriebenen Diffusionsgleichung ebenfalls in ebener Geometrie ausgeführt werden. Die zuvor besprochenen Diffusionsverhältnisse erlauben ein Neutron in ebener Geometrie durch seine atmosphärische Tiefe z und durch $\mu = \cos\vartheta$ — wobei ϑ den Winkel zwischen der skalaren Neutronengeschwindigkeit v und der z-Achse bedeutet — darzustellen. Die Neutronentransportgleichung für einen stationären Zustand, wie er in der Atmosphäre vorliegt, lautet dann

$$\mu \frac{d\mathbf{F}_g(z,\mu)}{dz} + \Sigma_g(z) \cdot \mathbf{F}_g(z,\mu) = S_g(z,\mu) \tag{38.1}$$

und gilt für jede Energiegruppe g.

$F_g = F_g(z, \mu)$ stellt den Gruppenfluß in der Energiegruppe g dar bei einer atmosphärischen Tiefe z pro Einheit μ, gemessen in Neutronen/cm² sec. $\Sigma_g(z)$ (cm⁻¹) ist der mittlere makroskopische gesamte Wirkungsquerschnitt für die Gruppe g. Er ist eine Funktion von $\varrho(z)$, der Dichte der Luft in g/cm³. $S_g(z, \mu)$ ist der Quellterm der Energiegruppe g in Einheiten Neutronen/cm³ sec. In diesem Quellterm sind die direkt in der Atmosphäre erzeugten Neutronen mitgezählt, wie auch diejenigen, welche in das Energieinterfall g hineingestreut worden sind. Dem Vorgang von HESS et al. [He 61] folgend, wurde auf die kondensierte Atmosphäre mit Einheitsdichte umgerechnet. Gl. (38.1) ist invariant gegen diese Transformation und kann einfacher integriert werden, weil nach der Transformation die Wirkungsquerschnitte unabhängig von den räumlichen Koordinaten werden. Wegen mathematischer Details sei auf die Originalarbeiten von CARLSON verwiesen. NEWKIRK hat für die S_4-Näherung der Transportgleichung (38.1) ein Fortran-Programm zur Benutzung auf einer IBM 7090 entwickelt. Es erlaubt, in 20 Energiegruppen und 100 räumliche Intervalle aufzugliedern. Weiter gestattet es, die Variation nach der Energie, nach dem Winkel und nach der Höhe zu berücksichtigen. NEWKIRK legt die Modellatmosphäre I, S. 492 zugrunde. Die Wirkungsquerschnitte wurden in der gleichen Weise über die Energie gemittelt unter Zugrundelegung des Flußspektrums der atmosphärischen Neutronen, wie wir es zuvor bei HESS beschrieben haben. Die Höhenverteilung des Quellspektrums der Neutronen, die NEWKIRK benutzt, haben wir in Ziff. 16 besprochen (vgl. Fig. 27). HAYMES [Ha 64b] hat bei der Messung schneller Neutronen versucht, die von NEWKIRK vorgeschlagene Höhenverteilung mit zwei Absorptionslängen, nämlich $L = 140$ g/cm² bzw. $L = 160$ g/cm² zu überprüfen. Die statistische Genauigkeit seiner Messung reichte nicht hin, eine Klärung zu bringen (vgl. Tabelle 12). Inzwischen liegen die Messungen von GAUGER [Ga 64] vor, der den Breiteneffekt schneller Neutronen von $\Lambda_{geom} = 15°$ Süd bis $54°$ Nord gemessen hat und seine Daten mit denen von SIMPSON und HESS (die ebenfalls mit ummantelten $B^{10}F_3$-Zählern gemessen haben), vergleicht (vgl. Tabelle 13). Nach diesen Daten ist eine Höhenabhängigkeit nicht auszuschließen, doch ist es sicher notwendig, solche Messungen mit direkten Neutronenzählern zu wiederholen.

Für die Energieverteilung benützt NEWKIRK im Gegensatz zu HESS et al. ein Quellspektrum nach den Messungen von MIYAKE, HINOTANI, KATSUMATA und KANEKO [Mi 57b], die in einer stickstoffgefüllten Nebelkammer geladene Teilchen (vorzugsweise Protonen) gemessen haben, welche durch kosmische Strahlung ausgelöst wurden. Er übernahm die gemessene Winkelverteilung für die geladenen Teilchen; die gemessene Energieverteilung für die Protonen korrigierte er auf Neutronen, indem er die Energiewerte um etwa 2 MeV verringerte; dies sollte den Unterschieden der Neutronen- und Protonenenergien Rechnung tragen, die durch die Coulomb-Barriere zustande kommen. Die Form dieses für alle Höhen angenommenen Energiespektrums kann aus Fig. 26, Kurve A III, entnommen werden.

NEWKIRK hat für die Normierung seines Neutronenquellspektrums auf eine Messung von SMITH et al. [Sm 61] zurückgegriffen, die unter anderem während eines solar flare [in Bemidji (Minnesota) $\Lambda_{geom} = 57°$ Nord] ausgeführt wurde. Die absolute Produktionsrate von $\bar{Q} = 7{,}1$ Neutronen/cm² sec liegt ziemlich hoch und entspricht keinem Normalmittel. Wir kommen bei der Besprechung der solar flare-indizierten C^{14}-Bildung in Ziff. 41 darauf zurück.

Die Ergebnisse von NEWKIRK mit der S_4-Methode sind in Fig. 66 dargestellt. Zum Vergleich sind die Kurven von HESS et al. [He 61] mit eingetragen. Diese wurden auf $\Lambda_{geom} = 57°$ Nord umgerechnet unter Benutzung der Breiteneffektmessungen von SIMPSON [Si 51b] für Höhen von 312 g/cm² Restdruck. Für Höhen von 680 g/cm² Restdruck wurden die Messungen von SIMPSON und FAGOT [Si 53a]

Ziff. 38. Näherungsverfahren der Transportgleichung (S_n-Methode). 497

zugrunde gelegt. Die Ergebnisse bei 40 g/cm² Restdruck wurden mit Hilfe der Messungen der Neutronenintensität bei 20 g/cm² von WHYTE [Wh 51] auf $\Lambda = 57°$ N übertragen. Für die Ergebnisse am Gipfel der Atmosphäre wurde keine Breitenkorrektur vorgenommen, da das Lingenfeltersche Verfahren damals noch nicht bekannt war. Das gemessene Spektrum von MIYAKE et al. [Mi 57] ist in Fig. 66 aufgenommen. Innerhalb der Fehlergrenze stimmen für 760 g/cm² die nach der S_n-Methode gewonnenen Kurven mit dem von MIYAKE gemessenen Spektrum überein; ebenso die Spektren nach HESS et al. — mit Ausnahme in einem Bereich von etwa 0,1 bis etwa 4 MeV. Dort sind die experimentellen Werte von HESS et al. um ungefähr einen Faktor 2 bis 3 größer als die mit der S_n-Methode er rechneten. Da eben in diesem Bereich die Verdampfungsneutronen sich als Quellneutronen in der Rechnung bemerkbar machen, hat NEWKIRK aus naheliegenden

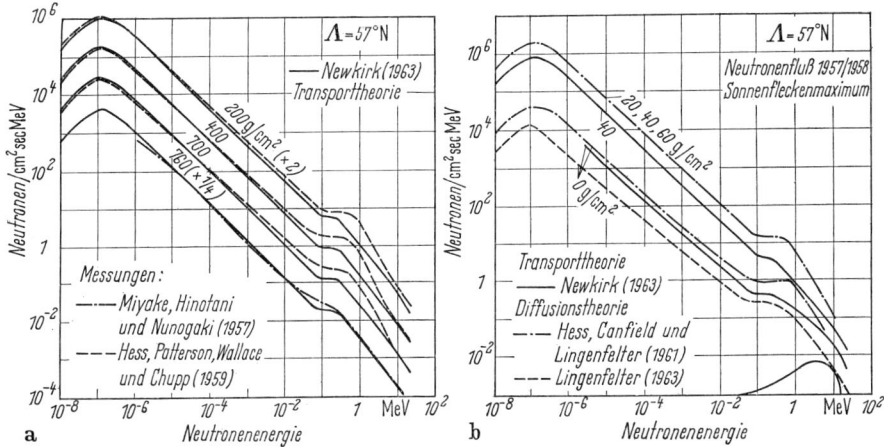

Fig. 66 a u. b. Neutronenfluß von NEWKIRK [Ne 63] mit S_n-Methode berechnet für $\Lambda_{geom} = 57°$ Nord. a Bei 760 g/cm² mit MIYAKE et al. [Mi 57], bei den weiteren Höhen mit HESS et al. [He 59]. Zwei Kurvenpaare (Faktor angegeben) wurden vertikal verschoben. Bei atmosphärischen Tiefen unter 760 g/cm² bleibt die Spektralform des Flusses erhalten und erscheint tiefer geschoben. Deshalb sind diese Kurven nicht mehr mitgezeichnet. b Vergleich der theoretischen Kurven von NEWKIRK mit denjenigen von HESS et al. [He 61] nach der Diffusionsnäherung. Die Form des von NEWKIRK benutzten Quellspektrums ist angedeutet. In a und b sind die Spektren auf gleiche geomagnetische Breite $\Lambda_{geom} = 57°$ Nord umgerechnet (vgl. Text) mit Ausnahme von 0 g/cm².

Gründen die Empfindlichkeit der S_n-Methode gegenüber dem zugrunde gelegten Quellspektrum überprüft. Zu diesem Zweck wurde das von HESS et al. verwendete Neutronenquellspektrum für die S_4-Rechnung übernommen; für den resultierenden Fluß ergab sich jedoch nur ein Unterschied von wenigen Prozent für diese verschiedenen Quellspektren. So kam keine Erklärung der Diskrepanz[1] zustande.

Der Fluß schneller Neutronen am Gipfel der Atmosphäre wurde von HAYMES [Ha 64a] gemessen. Mit Hilfe des von NEWKIRK mit der S_4-Methode bestimmten Neutronenflusses extrapoliert HAYMES den Fluß zum Gipfel der Atmosphäre (Fig. 44), denn in diesem Bereich der Atmosphäre, in dem kein Diffusionsgleichgewicht mehr besteht, ist die Transporttheorie die einzig angemessene Grundlage. Diese spezielle Extrapolation (Punkt B in Fig. 44) führt für $\Lambda_{geom} = 41,5°$ N auf einen extrapolierten Verlustfluß von 0,24 Neutronen/

---

[1] HAYMES [Ha 64a] führt aus, daß die Experimente die ursprünglichen Ergebnisse von HESS et al. [He 61] nicht bestätigen (vgl. dazu auch Fig. 41). Man nimmt an, daß die Unterschiede einmal von der unterschiedlich angenommenen Höhenverteilung herrühren, zum anderen hat R. E. LINGENFELTER HAYMES privat mitgeteilt, daß in der ursprünglichen Kurve ein Zeichenfehler vorliegen soll, vgl. aber auch Tabelle 23.

cm² sec. Wir schließen hier mit dem Ergebnis von NEWKIRK durch eine Zusammenfassung seiner Endwerte in Tabelle 28 ab. Wir betonen noch, daß die Ergebnisse

Tabelle 28. *Neutronenproduktion bzw. Reaktionen in der Atmosphäre* (nach NEWKIRK [*Ne 63*].) Normierung mit solar flare — Daten nach SMITH et al. [*Sm 62*].

Neutronen, welche auslösen:	Absolutwert Neutronen (Kerne) cm² sec	Relativverhältnis
C¹⁴-Produktionsrate	4,0	0,56
Tritium-Einfang	0,13	0,02
Einfang durch andere Prozesse .	2,2	0,31
Verlustfluß	0,8	0,11
Gesamt	7,1	1,00

der S_4-Methode und der Diffusionsrechnung — nach Einführung der Höhen- und Breitenabhängigkeit der Neutronenproduktionsrate für die Neutronenquelle bzw. bei Berücksichtigung der Wirkungsquerschnitte für $E > 1$ MeV nach [*Li 63*] — ziemlich gut übereinstimmen. Das erhöhte Zwischenmaximum im Bereich der Energien der Verdampfungsneutronen wird ausgeglättet und das Zwischenmaximum von etwa 800 nach 350 keV verschoben; so wird der Spektralverlauf dem Newkirkschen sehr ähnlich (vgl. Fig. 66b und dortige Diskussion; vgl. dazu auch den Spektralverlauf in Fig. 77).

Die genauesten theoretischen Daten, die heute vorliegen, haben inzwischen LINGENFELTER [*Li 63, 64*] und seine Mitarbeiter errechnet. Die Diskussion der C¹⁴-Produktionsrate erfolgt in Ziff. 40; die geophysikalische Bedeutung von Kohlenstoff-14 und Tritium in der Atmosphäre wird im Beitrag LAL und PETERS behandelt.

Der Verlustfluß der Neutronen aus der Atmosphäre (Ziff. 42) wird zuletzt zeigen, daß die bisherigen Neutronenmessungen *oberhalb* der Atmosphäre dagegen die Hess-, Canfield-, Lingenfeltersche Spektralverteilung zu begünstigen scheinen.

39. Neutroneneinfang (C¹⁴-Produktion). Absolute Normierung. In der Reaktortheorie gliedert man die Neutronen üblicherweise nach ihrer kinetischen Energie und teilt sie demgemäß in verschiedene Gruppen:

a) hochenergetische Neutronen mit $E > 2$ GeV, und energiereiche Neutronen mit $E \geq 20$ MeV. Dies sind überwiegend sternerzeugungsfähige Neutronen, die in der Reaktorphysik praktisch keine Rolle spielen.

b) schnelle Neutronen von $(10 \leq E \leq 20)$ MeV;

c) Verdampfungsneutronen 500 keV $\leq E \leq 10$ MeV;

d) Neutronen mittlerer Energie 1 keV $\leq E \leq 500$ keV;

e) langsame Neutronen $E < 1000$ eV; thermische Neutronen.

Für die Neutronen in der Atmosphäre haben wir uns an diese Terminologie gehalten. Bei der folgenden Behandlung der Absorptionsprozesse bzw. Einfangreaktionen ist es zweckmäßig, an dieser Gliederung festzuhalten.

Sieht man von den Reaktionen in Meßgeräten bzw. den Trägern von Meßgeräten (Rakete, Flugzeug, Ballon) ab, so haben wir es in der Atmosphäre mit Neutronenreaktionen am Stickstoff, Sauerstoff und den Edelgasen zu tun, wobei bei den letzteren nur das Argon entscheidend mitzählt. Auf die Kernreaktionen, bei denen Neutronen über die Fragmentation der einfallenden Primärkerne mitbeteiligt sind, brauchen wir nicht einzugehen, weil dies im Beitrag von LAL und PETERS behandelt wird.

Bei den Stößen der sekundären Neutronen in der Atmosphäre können folgende Reaktionen ablaufen:

Die hochenergetischen und energiereichen Neutronen und Protonen stoßen mit den Luftkernen unter Emission von Anstoßnukleonen, die sich bei hinreichender Energie weiter an der Kaskade beteiligen können; uns interessieren bei diesem

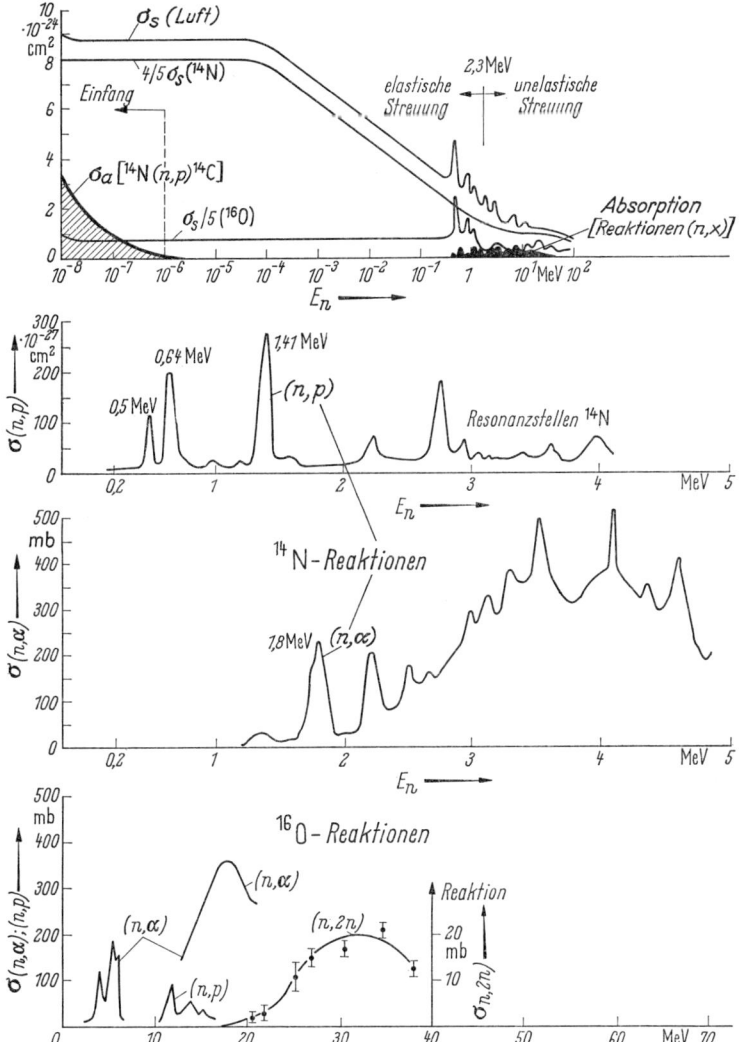

Fig. 67. Überblick über die verschiedenen Wirkungsquerschnitte der Luft als Gemisch von $\frac{4}{5}$ N^{14}; $\frac{1}{5}$ O^{16}. Soweit verfügbar wurden Reaktionswirkungsquerschnitte aus Tabellen schematisch für Stickstoff bzw. Sauerstoff übernommen.

Prozeß nur die dabei mitentstehenden schnellen bzw. Verdampfungs-Neutronen. Diese schnellen Neutronen beteiligen sich zunächst an inelastischen Stoßprozessen mit den Kernen der Luft (vgl. Fig. 67). Die Wirkungsquerschnitte für verschiedene Teilchenreaktionen (n, x) sind nur in geringem Umfang für die Energiebereiche von einigen MeV bis etwa 30 MeV bekannt. In Fig. 67 konnten deshalb nur sehr unvollständige Wirkungsquerschnitte der (n, p)- bzw. (n, α)-Reaktionen am Stickstoff und am Sauerstoff, ergänzt durch $(n, 2n)$-Prozesse, eingetragen werden.

Schnelle Neutronen haben wegen der zum Teil beträchtlichen Reaktionsquerschnitte in diesem Bereich eine Chance durch inelastischen Stoß geschwächt zu werden. Wegen der unvollkommenen Kenntnis der genauen Reaktionsquerschnitte kann man die Absorptionsrate der Neutronen in diesem Energiebereich trotz dem inzwischen hinreichend bekannten Neutronenspektrum theoretisch nur unvollkommen abschätzen.

Im Bereich von 0,5 bis zu 40 MeV treten auch inelastische $(n, \gamma\text{-})$ Reaktionen auf, die in der obersten Zeichnung der Fig. 67 als Absorption gekennzeichnet wurden.

Schnelle Neutronen, die durch inelastische Streuung in den Energiebereich der Verdampfungsneutronen bzw. in den Bereich mittlerer Neutronenenergien hereingestreut wurden, treten mit den Atomkernen der Luft überwiegend durch elastische Streuung in Wechselwirkung. Sind die Neutronen schließlich nach mehreren elastischen Stößen langsam geworden, erhöht sich die Wahrscheinlichkeit, daß sie durch die N^{14}-Einfangreaktion eingefangen werden. Diese Reaktion hat einen $1/v$-Wirkungsquerschnitt. Im Gegensatz dazu zeigt Sauerstoff einen praktisch nicht meßbaren Neutroneneinfangquerschnitt. Der relativ große thermische Neutroneneinfangquerschnitt von Argon mit 0,88 barn ist wegen der geringen prozentualen Häufigkeit von keiner nennenswerten Bedeutung.

Nach dieser Übersicht über mögliche Reaktionen der Neutronen in der Atmosphäre wenden wir uns der dominanten Einfangreaktion langsamer Neutronen durch N^{14}-Kerne zu. Diese Reaktion wurde von MONTGOMERY [Mo 39] erstmals diskutiert. KORFF [Ko 42] hat ausgeführt, daß der genaue Wert der Produktionsrate des β-aktiven C^{14} wünschenswert sei, um auf die je Zeiteinheit absorbierten Neutronen schließen zu können.

Die eigentliche Bedeutung des C^{14}-Produktionsprozesses wurde erkannt und gewürdigt, nachdem LIBBY [Li 46] die Suche nach dem „natürlichen" C^{14} im lebenden bzw. abgestorbenen Gewebe aufnahm und den C^{14}-Haushalt in der Atmosphäre untersuchte. Schon 1947 gelang ihm und ANDERSON [An 47] der Nachweis.

Innerhalb der Atmosphäre ist die $N^{14}(n, p)C^{14}$-Reaktion nicht der einzige Prozeß, der zur C^{14}-Produktion beiträgt. Die verschiedenen möglichen Kernreaktionen sind in Tabelle 29 zusammengestellt. Die Tabelle ist so aufgebaut, daß die relative atmosphärische Häufigkeit der verschiedenen Konkurrenzreaktionen

Tabelle 29. *Übersicht über C^{14}-Produktion in der Atmosphäre ausgelöst durch Neutronen-Reaktionen bzw. Spallation.*

Reaktionen mit C^{14}-Bildung	Wärmetönung der Reaktion Q ($Q>0$: exotherm)	Relative atmosphärische Isotopenhäufigkeit $N^{14} \equiv 1$	Relative Reaktionsrate pro Kern	Relative Reaktionsrate in der Atmosphäre
$C^{13}(n, \gamma)C^{14}$	$+ 8{,}17$	$0{,}23 \cdot 10^{-5}$	$5 \cdot 10^{-4}$	$1{,}1 \cdot 10^{-9}$
$N^{14}(n, p)C^{14}$	$+ 0{,}626$	1	1	1
$N^{15}(n, d)C^{14}$	$- 7{,}98$	$0{,}37 \cdot 10^{-2}$	$<1 \cdot 10^{-2}$	$<3{,}7 \cdot 10^{-5}$
$O^{16}(n, He^3)C^{14}$	$-14{,}6$	$0{,}269$	$<1 \cdot 10^{-2}$	$<2{,}7 \cdot 10^{-3}$
$O^{17}(n, \alpha)C^{14}$	$+ 1{,}82$	$0{,}99 \cdot 10^{-4}$	$2{,}3 \cdot 10^{-1}$	$2{,}3 \cdot 10^{-5}$
$Ne^{20, 21, 22}$ (Spallation)	vgl. LAL [La 58]	$0{,}12 \cdot 10^{-4}$	$<1 \cdot 10^{-2}$	$<1{,}2 \cdot 10^{-7}$
Werte	nach AJZENBERG-SELOVE [60], ASHBY-CATRON [As 59]	KUIPER [Ku 52], BAINBRIDGE u. NIER [Ba 50]	s. Text	

aus ihr entnommen werden kann. Spalte 1 kennzeichnet die Reaktion, Spalte 2 die Wärmetönung dieser Reaktion. Kennzeichnet man die relative atmosphärische Isotopenhäufigkeit von N^{14} mit 1, so ersieht man in Spalte 3 den Relativanteil der konkurrierenden Kerne, die jeweils mit der relativen Reaktionsrate pro Kern in Spalte 4 zur relativen Reaktionsrate in der Atmosphäre in Spalte 5 führt. Die einzige bedeutende Reaktion für C^{14}-Erzeugung bleibt der Neutroneneinfang im Stickstoff.

Zur Abschätzung der relativen Reaktionsrate

$$R = \int_0^\infty \Phi(E) \Sigma_{n,p}(E)\, dE \tag{39.1}$$

wurde ein Neutronenfluß $\Phi(E) \sim \frac{1}{E}$ angesetzt, während die Reaktionsausbeute $\Sigma(E)$ auf den Einzelkern bezogen ist. Die Reaktionsquerschnitte $\Sigma_{n,p}$ kann man für die exothermen Reaktionen den Tabellen von HUGHES und SCHWARZ [Hu 58] entnehmen. Für die endotherme (n, d)-Reaktion im Stickstoff bzw. die endotherme (n, He^3)-Reaktion im Sauerstoff kann man wie LINGENFELTER [Li 63] zur Abschätzung der Reaktionsquerschnitte die Arbeiten von CHASE [Ch 61] und LUSTIG et al. [Lu 48] heranziehen. Das durch Spallation von Neon gebildete C^{14} kann nach LAL [La 58] abgeschätzt werden.

Die Wahl des Neutronenflusses $\Phi \sim E^{-1}$ stellt eine obere Grenze für die C^{14}-Produktionsrate bei endothermen Reaktionen dar, denn man entnimmt Tabelle 11 und Fig. 54 für den Exponenten Werte zwischen $1{,}2 \leq n \leq 2$. Die geschätzten Reaktionsraten sind also erheblich überbewertet; wir dürfen deshalb schließen, daß innerhalb der Atmosphäre allein die $N^{14}(n, p) C^{14}$-Reaktion Bedeutung haben kann.

Die älteren Daten der gesamten C^{14}-Produktionsrate $\overline{Q}$, die auf eine Luftsäule von 1 cm^2 bezogen sind und in Einheiten $\frac{C^{14}\text{-Atome}}{cm^2 \text{ sec}}$ gezählt werden, haben wir in Tabelle 30 zusammengefaßt.

Tabelle 30. *Produktionsraten von C^{14} in der Atmosphäre in einer Luftsäule von 1 cm^2 Querschnitt.* Im Gleichgewicht entsprechen diese Produktionsraten den C^{14}-Zerfallsraten.

Autor Jahr	C^{14}-Produktionsrate $\left(\frac{C^{14}\text{-Atome}}{cm^2 \text{ sec}}\right)$		C^{14}-Zerfallsrate $\left(\frac{C^{14}\text{-Atome}}{cm^2 \text{ sec}}\right)$
LIBBY [Li 46]	0,8		—
ANDERSON-LIBBY [An 51]	2,6		2,2 ± 0,1
PFOTZER [Pf 52]	2,2 ± 0,3		—
LADENBURG [La 52a]	2,4		—
KOUTS und YUAN [Ko 52]	2,4		—
ANDERSON [An 53]	2,6 ± 0,4		2,3 ± 0,1
SOBERMANN [So 56] (korrigiert durch KORFF [Ko 58])	1,4		—
CRAIG [Cr 57]	—		1,8 ± 0,7
FERGUSSON [Fe 63]	—		1,9 ± 0,2
BIEN et al. [Bi 63]	—		2,0 ± 0,1
GREENHILL, FENTON et al. [Gr 65a]	1,85 ± 0,5		—
HESS, CANFIELD, LINGENFELTER [He 61][1]	2,9 ± 0,7		—
NEWKIRK [Ne 63][2]	2,1 ± 0,7		—
LINGENFELTER [Li 63a][1]	2,61 ± 0,5 Sonnenfleckenminimum	2,08 ± 0,4 Sonnenfleckenmaximum	—

[1] Theoretisch erschlossen über Diffusionsnäherung.
[2] Theoretisch erschlossen mit Transportgleichung (S_n-Methode).

Die absolute Neutronenzahl in der Atmosphäre wird über die C^{14}-Produktionsrate festgelegt. Man macht von folgendem Zusammenhang Gebrauch: Im $1/v$-Bereich verschwindet über die $N^{14}(n, p)\,C^{14}$-Reaktion die gleiche Anzahl langsamer Neutronen aus der Atmosphäre, wie C^{14}-Atome in der Atmosphäre gebildet werden. Mit einer anderen Reaktion, die ebenfalls einen $1/v$-abhängigen Wirkungsquerschnitt hat, der $B^{10}(n, \alpha)\,Li^7$-Indikatorreaktion, zählt man bei nacktem $B^{10}F_3$-Zähler atmosphärische Neutronen desselben Energiebereichs. YUAN [Yu 51] hat für jeden „Bor-Zähler" eine äquivalente Luftmasse in Gramm (g Luft) angegeben, die — unabhängig von der Energieverteilung — ebenso viele Neutronen absorbiert wie der Zähler „effektiv" zählt. Dabei muß allerdings der Störpegel aller Nicht-Neutronenereignisse von der Neutronen-Zählrate Z des $B^{10}F_3$-Zählers abgezogen werden.

Dieses Luftäquivalent erlaubt — unabhängig von der Energieverteilung der langsamen Neutronen in der Atmosphäre — proportional die BF_3-Neutronenzählrate Z (des nackten Zählers) in die Neutronenabsorptionsrate q umzurechnen.

Dies gilt für jede atmosphärische Tiefe x:

$$q(x)\left(\frac{\text{Zahl der Neutronen}}{\text{g sec}}\right) = K \cdot Z(x)\left(\frac{\text{Neutronenereignisse}}{\text{Minute}}\right). \qquad (39.2)$$

Die Proportionalitätskonstante K

$$K = \frac{\Sigma_{np}(C^{14})\,\text{cm}^2/\text{g}}{60 \cdot \Sigma_{n\alpha,\text{eff}}(B^{10} \cdot F_3)\,\text{cm}^2} \qquad (39.3)$$

verknüpft den effektiven Wirkungsquerschnitt $\Sigma_{n\alpha}$ des BF_3-Zählers, der nach YUAN [Yu 51] mit thermischen Neutronen kalibriert wird, mit dem Reaktionsquerschnitt $\Sigma_{n,p}$ des Neutroneneinfanges in Stickstoff. Dieser letztere wird nicht als makroskopischer Wirkungsquerschnitt in cm^{-1}, sondern in cm^2 pro g Luft berechnet. Für thermische Neutronenenergien

$$E_{\text{th}} = 0{,}025\text{ eV} \quad \text{ist} \quad \Sigma_{np}(\text{cm}^2/\text{g Luft}) = 5{,}85 \cdot 10^{-3}\text{ cm}^2/\text{g} \cdot \text{Luft}. \qquad (39.4)$$

Summiert man die Neutronenabsorptionsrate $q(x)$ vom Gipfel der Atmosphäre $x = 0$ bis zu Meereshöhe $x = 1033$ g/cm² (manche Autoren verwenden hier auch eine mittlere Tiefe bis zum Erdboden $x_0 = 1000$ g/cm²), so ergibt sich die Anzahl Q der Neutronen, die im $1/v$-Energiebereich in einer Luftsäule vom Querschnitt 1 cm² pro Sekunde absorbiert worden sind:

$$Q = \int_0^{x_0} q(x)\,dx = k \cdot \int_0^{x_0} Z(X)\,dX. \qquad (39.5)$$

Da für jedes langsame absorbierte Neutron in der Atmosphäre ein C^{14}-Atom entsteht, stellt die Zahl Q gleichzeitig die in derselben Luftsäule produzierten C^{14}-Atome dar.

PFOTZER [Pf 52b] vergleicht eine Reihe der mit nackten BF_3-Zählern bzw. mit Cd-ummantelten Bor-Zählern bestimmten C^{14}-Produktionsraten Q. Es gelingt ihm die Diskrepanz der damaligen Werte für Q zum großen Teil auf die Falschberechnung von $\Sigma_{n,p}$ [cm²/g] Luft zurückzuführen; andererseits legt er den Vorzug der Messung mit nacktem BF_3-Zähler gegenüber der Cd-Methode dar. Aus den Meßdaten von YUAN [Yu 50], DAVIS [Da 50a], STAKER [St 50], deren Daten wohl verteilt im Bereich von 100—300 g/cm² gemessen wurden, berechnet er $Q(\Lambda \geq 50°\text{ N}) = 3{,}8$ Neutronen/cm² sec. Die Mittelung dieses Wertes über alle geomagnetischen Werte nach den Messungen des Breiteneffektes durch [Ag 47] und [Si 51b] ergibt einen Reduktionsfaktor von 0,6, was zu einem Mittelwert von $Q = 2{,}3$ Neutronen/cm² sec führt.

Da ein Störpegel von 25% bei diesen Messungen nicht auszuschließen ist, ergibt sich eine Eingrenzung auf

$$Q_{\text{Gesamt}} = \begin{Bmatrix} 1,9 \\ 2,5 \end{Bmatrix} \frac{\text{Neutronen}}{\text{cm}^2 \text{ sec}} \quad \text{oder} \quad \frac{C^{14}\text{-Atome}}{\text{cm}^2 \text{ sec}}. \tag{39.6}$$

Die gesamte mittlere C^{14}-*Zerfallsrate* in dem Austauschreservoir der Erdatmosphäre ist ebenfalls auf der Grundlage der gemessenen spezifischen C^{14}-Aktivität in der Atmosphäre enthaltenen abgeschätzten Masse von C^{14} berechnet worden. Unter der Annahme, daß die Produktionsrate und die Zerfallsrate im Gleichgewicht sind, haben ANDERSON und LIBBY [*An 51*] die Zerfallsrate zu 2,2 ± 0,1 Zerfälle pro Luftsäule von 1 cm² Querschnitt vom Erdboden bis zum Gipfel der Atmosphäre je Sekunde errechnet. Weitere Werte finden sich ebenfalls in Tabelle 30, von denen die gegenwärtig besten diejenigen von CRAIG [*Cr 57*] und FERGUSSON [*Fe 63*] bzw. BIEN et al [*Bi 63*] sein dürften.

In Tabelle 30 sind (etwas abgegrenzt) Werte mit aufgenommen, die aus den Ergebnissen der Diffusionstheorie und der Transporttheorie in den letzten Jahren erschlossen worden sind. HESS, CANFIELD und LINGENFELTER [*He 61*] haben speziell für $\Lambda_{\text{geom}} = 44°$ den Neutronenfluß in der Atmosphäre berechnet, NEWKIRK bei $\Lambda_{\text{geom}} = 57°$ N. Inzwischen hat LINGENFELTER [*Li 63*] die Neutronenproduktion (und damit die C^{14}-Produktion) ziemlich allgemein für alle geomagnetischen Breiten und in Abhängigkeit von der Variation der Höhe und über den Sonnenfleckenzyklus abgehandelt (Ziff. 40). Seine Darstellung ist gegenwärtig die umfassendste. Zudem berücksichtigt er für Energien $E > 1$ MeV die inzwischen bekannt gewordenen Neutronenreaktionen.

Auf Grund der relativ kurzen Bremslängen und Diffusionslängen der (langsamen) Neutronen kann man erwarten, daß die Produktionsrate von C^{14} durch Neutronen eine ähnliche Höhenabhängigkeit und Breitenabhängigkeit zeigt, wie wir sie aus der Quellverteilung der Neutronen schon kennen (vgl. besonders Fig. 58 und 60).

Ein örtlicher bzw. zeitlicher Zusammenhang besteht auch zwischen den sekundären Neutronen in der Atmosphäre und den sie primär bzw. sekundär auslösenden Wechselwirkungen. Charakteristisch unterscheidet sich jedoch die Lage des Flügge-Yuanschen Höhenmaximums der Neutronen (dazu etwa Fig. 29), das nicht mit dem Pfotzerschen Maximum der einfallenden Gesamtintensität (vgl. Fig. 1) zusammenfällt. Dieser Unterschied ist bei allen geomagnetischen Breiten gefunden worden. Er hat seine Ursache in der Erniedrigung der Neutronendichte in großen atmosphärischen Höhen durch den beträchtlichen Verlustfluß (ca. 10% des Gesamtflusses) von Neutronen in die Stratosphäre, bzw. in den interplanetarischen Raum. Dieses tieferliegende Neutronenmaximum muß sich im Spektrum der C^{14}-Produktionsrate wiederfinden lassen.

Der Breiteneffekt der einfallenden primären kosmischen Strahlung spiegelt sich in bekannter Weise in der Neutronendichte wider. So ist zu erwarten, daß sich — wohl in etwas schwächerer Form — auch in der C^{14}-Produktionsrate ein Breiteneffekt abzeichnet.

Schließlich muß die zeitliche Variation der kosmischen Primärstrahlung ebenfalls ihren Niederschlag in der C^{14}-Produktion finden. Die Variation der einfallenden kosmischen Strahlung folgt nach bisherigem Wissen zwei Modulationsformen:

a) im 11jährigen bzw. 22jährigen periodischen Sonnenfleckenzyklus,
b) aperiodisch über den direkten Zustrom solarer Teilchen in die Erdatmosphäre bei gas-chromatischen Eruptionen (sog. solar flares) der Größenordnung 3 und 3⁺.

40. C^{14}-Produktion in Abhängigkeit von der periodischen Sonnenfleckenmodulation.
Wie wir gesehen haben (Fig. 2, 58 u. 60), wirkt sich die Modulation überwiegend auf den geomagnetischen Bereich der einfallenden Primärintensität aus, d.h. sie ist stark breitenempfindlich. In hohen geomagnetischen Breiten — auch oberhalb des Knies — wirkt sich die Intensitätsschwankung vom Sonnenfleckenmaximum am stärksten aus und damit muß auch in hohen geomagnetischen Breiten die C^{14}-Produktionsweise durch Neutronen besonders stark mitvariiert werden.

Die Bestimmung der C^{14}-Produktionsrate macht eine Berechnung des Neutronenflusses Φ_N notwendig. Dies geschieht über die numerische Integration der

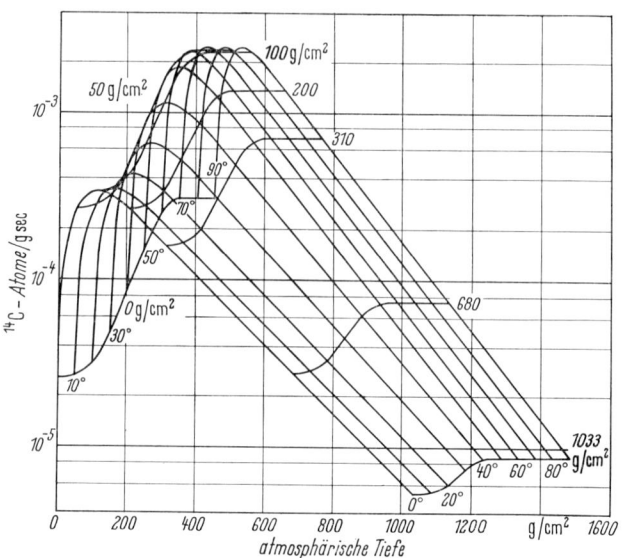

Fig. 68. Die Kohlenstoff 14-Produktionsrate P — ausgelöst durch Neutronen in der Atmosphäre — in (C^{14} Atome/g sec) beim Sonnenfleckenminimum. Die Produktionsrate hängt von 2 Parametern ab, der atmosphärischen Tiefe in g/cm² und der geomagnetischen Breite Λ. Die auslösende gesamte Neutronenintensität wurde so normiert, daß in einer Luftsäule von 1 cm² Querschnitt über dem geomagnetischen Pol 1 (Neutron/cm² sec) während des Sonnenfleckenminimums erzeugt wird. (Nach [Li 63a].)

Diffusionsgleichung (35.17). Als Quellspektrum der Neutronen

$$S(E; x, \Lambda; t) dE\, dx = q\left(x, \Lambda_i; /t^{\min}_{\max}\right) \cdot N(E)\, dE\, dx \qquad (40.1)$$

wählt man sich jeweils für einen fixen Parametersatz $(z, \Lambda_i; t_{\min})$ (aus Fig. 58) oder $(z, \Lambda_i; t_{\max})$ (aus Fig. 60) eine der Parameterkurven der Höhen- und Breitenverteilung der Lingenfelterschen Neutronenproduktionsraten q aus und geht damit in Gl. (40.1). Das Energiespektrum der Quellneutronen und dessen Normierung wurde schon in (35.25) besprochen. Auf diese Weise kann man nach LINGENFELTER [Li 63] numerisch für jeden Parametersatz den Neutronenfluß $\Phi_N = \Phi_N(E; z, \Lambda_i; t)$ errechnen. Mit dem bekannten makroskopischen Wirkungsquerschnitt Σ_{np} (cm²/g Luft) [vgl. 39.4] für die Reaktion $N^{14}(n, p) C^{14}$ findet man dann für die Parameterwerte die C^{14}-Produktionsraten

$$\left.\begin{array}{c} P(z, \Lambda_i; \tau)\left(\dfrac{C^{14}\text{-Atome}}{\text{g}\cdot\text{sec}}\right) = \displaystyle\int_0^{\infty} \Phi_N(E, z, \Lambda_i, \tau)\, \Sigma_{np}(E)\, dE \\ \text{entweder für} \quad \tau = \tau_{\min} \quad \text{oder} \quad \tau_{\max}. \end{array}\right\} \qquad (40.2)$$

Ziff. 40. C¹⁴-Produktion in Abhängigkeit von der periodischen Sonnenfleckenmodulation. 505

Die C¹⁴-Produktionsraten sind in Fig. 68 für das Sonnenfleckenminimum und in Fig. 69 für das Sonnenfleckenmaximum wiedergegeben. In beiden Darstellungen blieb die willkürliche Normierung (jeweils auf das Sonnenfleckenminimum bezogen) erhalten; sie beträgt am geomagnetischen Pol 1 Neutron/cm² sec (vgl. (32.1).

Diese Produktionsraten von C¹⁴-Atomen entsprechen zugleich der Zahl der langsamen Neutronen, die diese Reaktion auslösen. Die *absolute* Normierung der C¹⁴-Produktionsrate legt damit auch die Zahl der langsamen Neutronen in der Atmosphäre fest.

Wir haben die Vorzüge des Yuanschen Normierungsverfahrens schon besprochen, so daß wir nur noch geeignete Messungen langsamer Neutronen mit

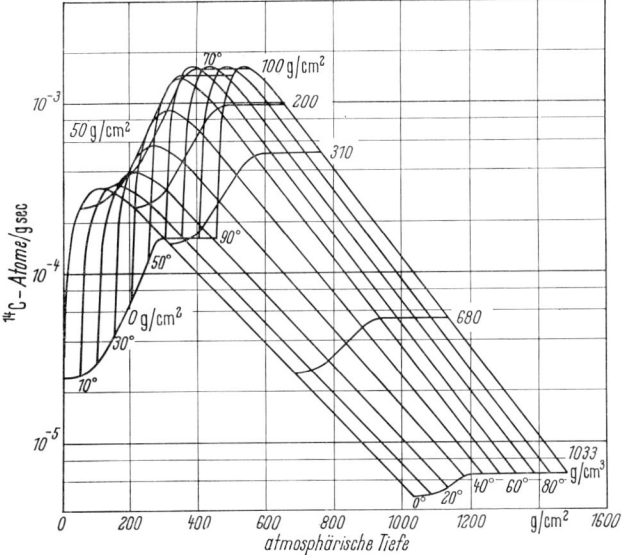

Fig. 69. Die Kohlenstoff 14-Produktionsrate P (C¹⁴ Atome/g sec) beim Sonnenfleckenmaximum als Funktion der Höhe (g/cm²) und der geomagnetischen Breite Λ bezogen auf eine Neutronenproduktionsrate von 1 (Neutron/cm² sec) für eine Luftsäule von 1 cm² Querschnitt am geomagnetischen Pol während des Sonnenfleckenminimums. (Nach [Li 63a].)

nackten Bor-Zählern auswählen müssen, um von den Neutronenzählraten auf die proportionale Zahl der erzeugten C¹⁴-Atome (und damit der langsamen Neutronen in der Atmosphäre) schließen zu können.

Nach (39.2) besteht Proportionalität zwischen der C¹⁴-Produktionsrate und der Zahl der Neutronenereignisse, die wir in Abhängigkeit von den Parametern der Tiefe, geometrischer Breite und zeitlicher Lage im Sonnenfleckenzyklus angeben wollen:

$$P_{C^{14}}(z, \Lambda_i, t_i)\left(\frac{C^{14}\text{-Atome}}{g \cdot sec}\right) = K\left(\frac{C^{14}\text{-Atome} \cdot min}{\text{Ereignisse} \cdot g \cdot sec}\right) \cdot Z(z, \Lambda_i, t_i)\left(\frac{\text{Ereignisse}}{min}\right). \quad (40.3)$$

Folgende Messungen liegen vor: Drei aus den Jahren 1948 und 1949 von YUAN [Yu 56] ($\Lambda = 52°$ geomagnetisch N); weiter sieben Messungen von 1952 bis 1954 ($\Lambda = 10°, 55°, 88°$ N) von SOBERMANN [So 56] und fünf Messungen im Jahre 1957 (bei $\Lambda = 36°, 48°, 58°$ N) von HESS et al. [He 59a] und eine Messung durch REIDY et al. [Re 62] bei $\Lambda = 49°$ N.

Zeitlich überdecken die Messungen einen Zeitraum von etwas mehr als einem Sonnenfleckenzyklus vom Maximum zum Maximum, örtlich sind sie hinreichend gestreut, und am besten bezieht man sich auf die Zählraten im Höhenbereich von etwa 100 bis 300 g/cm², in dem die überwiegende Masse von C^{14}-Atomen gebildet wird.

In die Proportionalitätskonstante K in (40.3) geht der „effektive" Wirkungsquerschnitt $\Sigma_{n\alpha,\text{eff}}(\text{cm}^2) \equiv A_{\text{eff}}$ des BF_3-Zählers ein. Die Festlegung von A_{eff} nach YUAN [Yu 51] kann im wesentlichen Tabelle 31 entnommen werden.

Tabelle 31. *Kalibrierungs-Werte des Yuanschen Zählers.*

Zähler-Durchmesser	5,0 cm
Zähler-Länge	20,0 cm
Zähler-Druck	50 cm Hg
Anreicherung B^{10}/B^{12} . . .	96%
Effektive Fläche	$(10.6 \pm 1,0)$ cm

Der effektive Wirkungsquerschnitt A_{eff} (cm²) eines unabgeschirmten $B^{10}F_3$-Zählers errechnet sich dann zu

$$A_{\text{eff}}(\text{cm}^2) = \frac{(\text{Volumen} \times \text{Druck}) \text{ für Zähler } y}{(\text{Volumen} \times \text{Druck}) \text{ für Zähler Yuan}} (10,6 \pm 1,0) \text{ cm}^2.$$

Diese absolute Normierung sei an Daten aus dem Flug II (8. Juli 1948, Sonnenfleckenmaximum) von YUAN [Yu 51] erläutert. In Tabelle 30 sind für die Breite $\Lambda = 52°$ N für verschiedene atmosphärische Tiefen z die Zählraten $Z = Z(z, 52°)$ angegeben. Gln. (39.3) und (40.3) ermöglichen damit die Berechnung der C^{14}-Produktionsraten $P_{C^{14}}$ (A_{eff} ist in Tabelle 32 angegeben). Spalte 1 enthält die Zahlwerte der theoretischen, willkürlich normierten C^{14}-Produktionsrate P gemäß Gl. (40.2), die in Fig. 69 für 50° und die Höhen z verfolgt werden können; Normierungsfaktoren N_C (vgl. Spalte 5) lassen die absolute Normierung der theoretischen Werte vollziehen: $P_{C^{14}} = N_C \cdot P$.

Für eine Höhe von 300—310 g/cm² beträgt der Normierungsfaktor $N_C = (8.7)$ — in dieser Spalte kennzeichnen eingeklammerte Zahlen den Ballonabstieg.

Tabelle 32. *Absolute Normierung der C^{14}-Produktionsrate mit Hilfe der B^{10}-F_3-Zählrate an Hand eines Fluges von YUAN [Yu 51].*
Die Normierungskonstante N_C ist auch innerhalb des Gleichgewichtsbereichs starken Schwankungen unterworfen. Das gleiche gilt für die Konstante N (s. Text).

C^{14}-Produktionsrate (theor.-willk. norm.) $P\left(\frac{^{14}C\text{-Kerne}}{g \cdot \text{sec}}\right) \times 10^{-3}$ (Fig. 69)	Atmosphärische Tiefe z (g/cm²)	Zählrate $Z=Z(z, 52°)$ YUAN, Flug II ( ) = extrapolierte Werte	C^{14}-Produktionsrate, experi.[1] $P_{C^{14}}\left(\frac{C^{14}\text{-Kerne}}{g\text{sec}}\right) \times 10^{-3}$ bzw. q_{exp} (Neutr./gsec)	Absolute Normierungskonstante N_C $P_{C^{14}} = N_C \cdot P$ ( ) = Ballonabstieg	Neutronen-produktionsrate $q(z, \Lambda = 50°) \times 10^{-3}$ Theorie (vgl. Fig. 60)	Absolute Normierungskonstante N ( ) = Ballonabstieg
	96	$149{,}0 \pm 5{,}5$	$12{,}8 \pm 0{,}475$		2,26	5,66
1,17	100	( )	12,45	10,6 (10,2)	2,22	5,6 (5,35)
	115,7	$128{,}2 \pm 5{,}2$	$11{,}05 \pm 0{,}46$		1,97	5,6
	190	$118{,}3 \pm 5{,}2$	$10{,}2 \pm 0{,}47$		1,36	7,5
0,96	200	( )	9,55	9,95 (9,6)	1,20	7,65 (7,35)
	252	$82{,}8 \pm 4{,}1$	$7{,}15 \pm 0{,}35$		0,846	8,45
0,35	300	( )	5,9	11,1 (8,15)	0,68	8,68 (6,35)
	316	$63{,}8 \pm 2{,}9$	$5{,}5 \pm 0{,}25$		0,622	8,8
	391	$31{,}2 \pm 2{,}5$	$2{,}7 \pm 0{,}21$		0,408	6,6
0,24	400	( )	2,42	10,2 (11,7)	0,37	6,54 (7,6)
	484	$18{,}2 \pm 2{,}0$	$1{,}57 \pm 0{,}17$		0,253	6,25
1,36	500	( )	1,46	(10,7 (11,4)	0,23	6,35 (6,75)
	578	$10{,}6 \pm 1{,}5$	$0{,}915 \pm 0{,}13$		0,14	6,5 (6,7)
8,12	600	( )	0,728	6,6 (8,65)	0,118	6,17 (7,6)

[1] A_{eff} für Zähler von Flug II = 11,3; damit $k = 8{,}63\ ^6\ 10^{-5}$ nach (39.3).

Ziff. 40. C¹⁴-Produktion in Abhängigkeit von der periodischen Sonnenfleckenmodulation. 507

Dieser Zahlwert stimmt nach Tabelle 33 mit dem dort von LINGENFELTER angegebenen Mittelwert für die Normierung der Yuanschen Meßdaten gut überein. Die anderen Normierungsfaktoren der Spalte 5 differieren sowohl für den Gleichgewichtsbereich, als auch für das Maximum erheblich von diesem Wert $8,8 \pm 1$. Bei Flug III ergeben sich ähnliche Daten. Bei Flug I fallen die Normierungsfaktoren für Höhen von $100-500$ g/cm² in den Bereich von $7,9-9,3$. Die Gesamtmittelwerte liegen nahe bei 10 oder darüber.

Analog zu den Normierungsfaktoren N_c kann man auch die Neutronen-Produktionsrate $q(z, \Lambda)$, Spalte 6, normieren, zunächst so, daß die willkürlich normierten Neutronenproduktionskurven Fig. 60 gerade der gemessenen C¹⁴-Produktion entsprechen. Diese Werte sind in Spalte 7 aufgenommen. Die Faktoren N sind kleiner als N_c. Für die einzelnen Yuanschen Flüge liegen die Mittelwerte von N bei 6,0; 6,2 und 7,8. Die tatsächliche Produktionsrate an Neutronen in der Atmosphäre liegt höher als mit N festgelegt, denn zu der (N¹⁴—C¹⁴)-Reaktion kommen die in Ziff. 39 erwähnten Konkurrenzreaktionen, sowie der Neutronenverlustfluß.

LINGENFELTER [Li 63] gibt nach Auswertung des gesamten Materials der Autoren die in Tabelle 33 aufgeführten Normierungsfaktoren an. Als Mittelwert der absoluten Normierung erhält er $N_c = (9,0 \pm 1,7)$.

Tabelle 33. *Absolute Normierung der C¹⁴-Produktionsrate [Atome/g sec].*

YUAN	$(8,8 \pm 1,0)$	(1948/1949)
SOBERMANN . . .	$(9,3 \pm 1,4)$	(1952/1954)
HESS et al. . . .	$(9,8 \pm 2,0)$	(1957/1958)
REIDY et al. . .	$(8,0 \pm 2,1)$	(1960)
Mittel .	$(9,0 \pm 1,7)$	

Mit diesem Zahlwert normiert er auch die Produktionsraten q für Neutronen. Diese Wahl führt allem Anschein nach zu einer zu großen Neutronenproduktion in der Atmosphäre, denn die Messungen der Neutronendichte im Verlauf des Sonnenfleckenminimums (Sept.—Okt. 1964) nach Höhe und Breite im Bereich der südlichen Halbkugel durch GREENHILL, PPILIPS, FENTON, FENTON und BOWTHORPE [Gr 65a] ergab, daß in Polnähe die theoretische Neutronendichte nach [Li 63] um einen Faktor 1,2, und in mittleren Breiten bzw. Äquatornähe um Faktoren 1,5 bis 1,7 gegenüber den Meßwerten zu groß ist. Eine Überprüfung dieses Sachverhalts ist im Gang.

Durch Integration dieser normierten Funktionen vom Gipfel der Atmosphäre $x=0$ bis zum Erdboden $x_0 = 1033$ g/cm² folgt aus (40.3) bzw. (40.1)

$$Q = \int_0^{x_0} P_{C^{14}}(z, \Lambda_i, t)\, dz \quad \text{bzw.} \quad S = \int_0^{x_0} q(z, \Lambda, t)\, dz; \qquad 40.4$$

damit ist die absolute Neutronenquellstärke $S(\Lambda_i, t_i)$ und die C¹⁴-Produktionsrate $Q(\Lambda_i, t_i)$ in einer Luftsäule von 1 cm² Querschnitt pro sec für den Zeitraum des Sonnenfleckenminimums $\tau_{\min}$ (1953/54) und das Sonnenfleckenmaximum (1957/58) bestimmt. Die Werte für alle 10° geomagnetischer Breite sind in Tabelle 34 angegeben.

HESS, CANFIELD und LINGENFELTER [He 61] haben als erste eine „Demographie" der Neutronen (vgl. Tabelle 27) in Angriff genommen. NEWKIRK [Ne 63] konnte diese Daten auf Grund seiner Benutzung der S_n-Methode hinsichtlich der Tritiumproduktion ergänzen (Tabelle 28). In beiden Arbeiten waren die Daten jeweils auf eine spezielle geomagnetische Breite bezogen, lagen die Messungen in irgend einem bestimmten Zeitraum des Sonnenfleckenzyklus ohne jedoch korreliert zu sein und ergaben sich hinsichtlich der Spektralform im Energiebereich der Verdampfungsneutronen Unterschiede.

LINGENFELTER [Li 63] hat diese Einschränkungen praktisch überwinden können. Im Rahmen der Diffusionstheorie berücksichtigt er über 1 MeV sämtliche bis heute bekannt gewordenen Reaktionswirkungsquerschnitte. Als nächstes führte er für die Neutronenproduktionsrate ein sehr realistisches Modell ein, das die Höhen- und die Breitenabhängigkeit berücksichtigt und zudem dem Zeitpunkt innerhalb des Sonnenfleckenzyklus Rechnung trägt. Damit hat er zum ersten Mal mit seinen Ergebnissen eine ziemlich allgemeine Demographie der Neutronen der Atmosphäre angeben können (vgl. Tabelle 34). Dort sind sowohl für das Sonnenfleckenminimum 1953/54 wie für das Sonnenfleckenmaximum 1957/58 von 10 zu 10° geomagnetischer Breite anwachsend die Neutronenquelle, die C^{14}-Produktion und der Neutronenverlustfluß mit Energien $E < 10$ MeV angegeben.

Tabelle 34. *Demographie der Neutronen* $\left[\dfrac{Teilchen}{cm^2\ sec}\right]$.

Geomagnetische Breite °	Sonnenflecken-Minimum, 1953/54			Sonnenflecken-Maximum, 1957/58		
	Neutronen-quelle	Kohlenstoff^{14}-Produktion	Neutronen-verlustfluß <10 MeV	Neutronen-quelle	Kohlenstoff^{14}-Produktion	Neutronen-verlustfluß <10 MeV
0	1,48	0,98	0,10	1,41	0,93	0,09
10	1,53	1,01	0,11	1,46	0,96	0,10
20	1,85	1,22	0,13	1,74	1,15	0,12
30	2,76	1,83	0,19	2,46	1,63	0,17
40	4,60	3,02	0,35	3,73	2,45	0,28
50	7,03	4,52	0,65	5,34	3,44	0,49
60	8,54	5,26	1,06	6,05	3,79	0,67
70—90	9,00	5,38	1,29	6,05	3,79	0,67
Gesamtmittelwert	4,10	2,61	0,41	3,22	2,08	0,29

In den Daten der Tabelle 34 zeichnet sich der Breiteneffekt für Produktion und Verlustfluß deutlich ab. Die Neutronenquelle ist in den Polargebieten während des Sonnenfleckenminimums ungefähr sechsfach so stark wie am Äquator. Dagegen ist sie beim Sonnenfleckenmaximum nur etwa viermal so stark. Ähnliches gilt für die C^{14}-Produktion, die beim Minimum zwischen Äquator und Pol um einen Faktor 5,5, aber beim Maximum etwa auf das Vierfache anwächst.

Der Neutronenverlustfluß für Neutronen der Energie $E < 10$ MeV steigt entsprechend beim Minimum nahezu um einen Faktor 13 an, beim Maximum knapp um 7,5.

LINGENFELTER gibt sodann die Gesamtmittelwerte. Man ersieht, daß im Zeitpunkt des Sonnenfleckenminimums 10% der in der Atmosphäre erzeugten Neutronen durch Verlustfluß verloren gehen. Andererseits werden von den in der Atmosphäre im Mittel vorhandenen Neutronen nur etwa 63,7% in C^{14} überführt, so daß rund 26% der Neutronen über andere Reaktionen innerhalb der Atmosphäre absorbiert oder eingefangen werden. Wie man sich vergewissert, bleiben diese Prozentualverhältnisse beim Sonnenfleckenmaximum nahezu erhalten.

Wie man sich weiterhin versichert, bleiben diese Prozentualverhältnisse für die einzelnen geomagnetischen Breiten grob erhalten.

Sehr gut kann man weiter den Einfluß der solaren Modulation aus der Tabelle 34 ersehen. Die Schwankung zwischen Minimum und Maximum ist im Bereich des geomagnetischen Äquators sowohl für die Neutronenquellstärke als auch für die C^{14}-Produktion und für den Neutronenverlustfluß kaum merklich.

Oberhalb des geomagnetischen Knies hingegen, in Polargegenden, sind die Unterschiede zwischen Minimum und Maximum erheblich; die Neutronenquellstärke sinkt auf etwa 67% des Minimumwertes ab, die C^{14}-Produktion etwa auf

70%, der Neutronenverlustfluß sogar auf 51%. Dies ist verständlich. Am Äquator beträgt die vertikale magnetische Abschneidesteifigkeit für die geladenen Primären etwa 16,5 GV. Dies bedeutet, daß nur energiereiche Primäre des breitenempfindlichen Spektralbereichs bei niederen geomagnetischen Breiten Neutronen in der Atmosphäre erzeugen können. Die solare Modulation wirkt sich auf Primärteilchen dieser Energie nur noch gering aus.

Anders liegen die Verhältnisse am geomagnetischen Pol; die magnetische Steifigkeit ist dort nur noch einige hundert MV, so daß nahezu die Gesamtheit der niederenergetischen Primären Gelegenheit hat, an den Polkappen in die Erdatmosphäre einzudringen. Dies führt in Höhen bis zu etwa 200 g/cm² zu einer erheblichen Produktion von Neutronen. Wir haben in Ziff. 32 darauf hingewiesen, daß NEHER [Ne 59] und ANDERSON [An 61] zwischen Minimum und Maximum eine Variation der Ionisation von etwa 63% fanden. Für die gesamte Luftsäule im Polgebiet findet man für die Neutronen eine ähnliche Variation von etwa 67%, wobei die überwiegende Mehrzahl dieser Neutronen ebenfalls in den Bereich vom Gipfel der Atmosphäre bis etwa 200 g/cm² fällt. Diese größere Produktion von Neutronen in den Polgebieten in großen Höhen hat natürlich auch einen nahezu doppelt so großen Verlustfluß (zwischen Minimum und Maximum) zur Folge. Integriert man die C^{14}-Produktionsrate über die geomagnetischen Breiten, so erhält man den Gesamtmittelwert für C^{14}-Produktion $\bar{Q}$ in der Erdatmosphäre. Innerhalb des Sonnenfleckenzyklus variieren diese errechneten Werte sehr stark, vgl. Tabelle 35.

Tabelle 35. *Gesamte mittlere C^{14}-Produktion $\bar{Q}$ im Verlauf des Sonnenfleckenzyklus.*

	Sonnenfleckenzahl
$\bar{Q}_{min} = 2{,}61 \pm 0{,}50 \; \dfrac{C^{14}\text{-Atome}}{cm^2 \, sec}$	9,1 (1953/54) Minimum
$\bar{Q}_{max} = 2{,}08 \pm 0{,}40 \; \dfrac{C^{14}\text{-Atome}}{cm^2 \, sec}$	187,5 (1957/58) Maximum

LINGENFELTER diskutiert einen linearen Zusammenhang zwischen C^{14}-Produktion und Sonnenfleckenzahl. Er nimmt an, daß die Abweichung der C^{14}-Produktionsrate von ihrem Maximalwert (während des Sonnenfleckenminimums) $\bar{Q}_{min} = 2{,}61$ zum Mittelwert der Sonnenfleckenzahl direkt proportional ist. Unter dieser Annahme gibt er die zeitlich veränderliche C^{14}-Produktionsrate, $\bar{Q}_T$, über die Sonnenfleckenperiode der Dauer T als Funktion der mittleren jährlichen Sonnenfleckenzahl $\bar{S}_T$ an:

$$\bar{Q}_T = 2{,}61 - \frac{0{,}53\,(\bar{S}_T - 9{,}1)}{178{,}4}. \tag{40.5}$$

Weitere Schlüsse kann man aus dem Mittelwert der Sonnenfleckenzahlen jeweils über einen 11jährigen Zyklus ziehen. Der Mittelwert zwischen den Jahren 1943 bis 1954 betrug $\bar{S}_T = 75{,}1$. Für diesen Zyklus ergibt sich eine mittlere C^{14}-Produktionsrate $\bar{Q}_T = 2{,}41 \pm 0{,}46$. Die mittlere Sonnenfleckenzahl[1] für diesen Zyklus (1954—1965) beträgt $\bar{S}_T = 80{,}5$, somit haben wir gegenwärtig eine extrapolierte C^{14}-Produktionsrate $\bar{Q}_T = 2{,}40 \pm 0{,}45$.

Noch signifikanter ist der Mittelwert über die insgesamt beobachteten 10 Sonnenfleckenzyklen von 1844—1954. In diesen 111 Jahren beträgt die ge-

---
[1] Vorläufiger Wert.

mittelte Sonnenfleckenzahl $\overline{S}_T = 47{,}7$. Es sollte also in diesem Zeitraum eine mittlere berechnete C^{14}-Produktionsrate $\overline{Q}_T = 2{,}50 \pm 0{,}50 \frac{\text{C}^{14}\,\text{Atome}}{\text{cm}^2\,\text{sec}}$ vorgelegen haben. Für diese Produktionsrate haben wir bis jetzt nur die periodischen Modulationen der Primärintensität berücksichtigt. Wir vergleichen das 110-Jahresmittel (bei periodischer Modulation) mit den neuesten Abschätzungen der C^{14}-Zerfallsrate (vgl. Tabelle 30), $1{,}8 \pm 0{,}2$ nach CRAIG [Cr 57]; $1{,}9 \pm 0{,}2$ nach FERGUSSON [Fe 64], und $2{,}0 \pm 0{,}1$ nach BIEN et al. [Bi 63]; es scheint trotz der großen Fehler, die noch eine Überdeckung der Meßergebnisse zulassen, doch ein Anzeichen dafür vorzuliegen, daß die gegenwärtige natürliche *Produktionsrate* die natürliche *Zerfallsrate* um etwa 25% übertrifft. Diese Ungleichheit kann sogar noch größer werden, wenn der Beitrag von solaren Teilchen aus gas-chromatischen Eruptionen (solar flare Teilchen) mit betrachtet wird. Allerdings ist nicht auszuschließen, daß die Zerfallsrate auch in den letzten Arbeiten erheblich zu nieder (etwa 10%) abgeschätzt wurde, da der C^{14}, der in pelagischen Sedimenten abgelagert ist, nicht mitgezählt wurde (vgl. für die geophysikalische Verteilung des C^{14} den Beitrag von LAL und PETERS).

Eine solche Ungleichheit muß — falls sie wirklich existiert — auf eine physikalische Ursache zurückzuführen sein. Zwei mögliche Modelle für eine Erklärung sind dafür bis jetzt vorgeschlagen worden: ELSASSER et al. [El 56] vermutet eine Variation des Magnetfeldes der Erde; dagegen denkt STUIVER [St 61b] an eine Variation der Sonnenaktivität mit mehrhundertjähriger Periode. ELSASSER et al. haben die zeitliche Variation der C^{14}-Produktionsrate berechnet, gemäß der Modulation des einfallenden Flusses der kosmischen Strahlung durch eine Variation des Erdmagnetfeldes. Sie nehmen auf der Grundlage der archäomagnetischen Messungen durch THELLIER und THELLIER [Th 46] eine exponentielle Abnahme des Erdmagnetfeldes in den letzten 2000 Jahren an.

Die Ungleichheit in der Produktions- und Zerfallsrate ist von derselben Größe und weist in dieselbe Richtung wie die Rechnung von LINGENFELTER dies ergibt, und die Änderung in der C^{14}-Konzentration ist nicht unverträglich mit der gemessenen C^{14}-Variation in den Jahresringen der Bäume. SCHOVE [Sch 55] hat eine langzeitige Variation der Sonnenaktivität vorgeschlagen. STUIVER hat die Variation in der atmosphärischen C^{14}-Konzentration gemäß dieser Modulation des primären kosmischen Strahlungsflusses berechnet. Er errechnet, daß 25%ige Variationen in der Produktionsrate bei über 100 Jahren währenden Perioden etwa 2%ige Änderungen in der atmosphärischen C^{14}-Konzentration ergeben.

f) Neutronenproduktion und C^{14}-Produktion durch solare Protonen.

41. Neben der periodischen Modulation der kosmischen Strahlung mißt man zeitlich aperiodisch überlagerte Störungen, die sich in der Form eines zusätzlichen solaren Teilchenflusses bis in die Erdatmosphäre hinein äußern. Im Gefolge gaschromatischer Eruptionen[1] (solar flares) der Größenordnung 3 und 3+ entstehen solche solare Teilchenströme bzw. Plasmawolken. Diese solaren Teilchen — meist Protonen — besitzen durchschnittlich geringe Energien, im Bereich von einigen MeV bis etwa 2 GeV; Höchstenergien bis 20 GeV wurden gelegentlich gemessen. Sie können aus diesem Grund nur an magnetisch begünstigten Stellen der Erde in die Atmosphäre eindringen, bevorzugt an den Polkappen (vgl. Fig. 70). SIMPSON

---

[1] Solar flares und solare Teilchen bzw. deren Transport- und Einbruchmechanismus in das Magnetfeld der Erde gehören nicht zu unserem Thema. Ausführliche Information darüber geben AAS-NASA 1963, WEBBER und FREIER [We 63] und WEBBER [We 62]; bzw. SANDSTRÖM [Sa 65]. Siehe auch BIERMANN, HAXEL und SCHLÜTER [Bi 51].

[Si 60] hat auf die Bedeutung der Einbrüche solarer Protonen in die Erdatmosphäre hingewiesen; sie sind an der Neutronenproduktion mit beteiligt und tragen so zum C^{14}-Haushalt der Erdatmosphäre bei. Eine direkte Messung der im Verlauf eines solaren Teilcheneinbruchs miterzeugten Neutronen bzw. der daraus entstandenen C^{14}-Atome ist experimentell sehr schwierig.

Die Bedeutung der Einbrüche solarer Teilchen für den Neutronenhaushalt der Erde ist jedoch so groß, daß LINGENFELTER und FLAMM [Li 63b, 64] sich entschlossen, die C^{14}-Produktionsrate durch solare Teilchen theoretisch zu bestimmen. Ihr Grundgedanke ist der folgende: es fallen nur Protonen in die Erdatmosphäre ein, die eine Energie $E \geqq E_C$ bzw. eine magnetische Steifigkeit $P \geqq P_C$ besitzen. An den

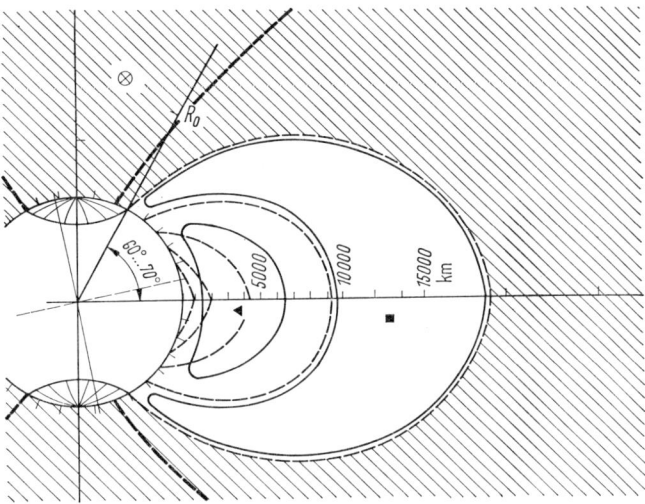

⊗ Einfallsbereich von solar flare Teilchen
▲ innerer Strahlungsgürtel
■ äußerer Strahlungsgürtel, ruhige Periode

Fig. 70. Schnittbild durch die Erdkugel sowie den inneren und äußeren van Allen-Gürtel. Aus dem schraffierten Außenraum können relativ niederenergetische (solare) Teilchen auf die Polarkappen eindringen (nach OGILVIE [Og 63]).

Polkappen liegt die magnetische Abschneidesteifigkeit P_C bei 0,1 GV und zum Teil niedriger, so daß auch energiearme Protonen in diesen Bereichen bis in die Atmosphäre eindringen können. Diese wechselwirken mit den Atomkernen der Luft und erzeugen in bekannter Weise Anstoß- und Verdampfungsneutronen.

Unter Zuhilfenahme neuer experimenteller Meßdaten bestimmen LINGENFELTER und FLAMM im Bereich von 0,03 bis 10 GeV die Anregungsfunktion für die Erzeugung von Verdampfungsneutronen durch Beschuß von Luftkernen mit Protonen. Unter Hinzunahme der Monte Carlo-Rechnungen von METROPOLIS [Me 58] können sie die dabei produzierten Anstoßneutronen mit erfassen. Diese konstruierte Anregungsfunktion für die gesamte Erzeugung von Neutronen in Luft überprüfen sie; sie errechnen mit einer dazu konstruierten Neutronenmultiplizitätsfunktion mit Hilfe dieser Anregungsfunktion die theoretisch zu erwartende Neutronenmultiplizität. SIMPSON [Si 51b] hat für Neutronen von einigen 100 MeV ein Verfahren angegeben, das es gestattet, die mittlere Neutronenmultiplizität aus experimentellen Werten zu errechnen. LINGENFELTER und FLAMM können zeigen, daß die errechnete Multiplizitätsfunktion mit den erwarteten Mittelwerten gut übereinstimmt. Diese Überprüfung zeigt, daß die vorgeschlagene Anregungsfunktion realistisch ist.

Das weitere Ziel der Autoren ist es, die Höhen- und Breitenabhängigkeit der in der Atmosphäre durch Protonen erzeugten sekundären Neutronen zu errechnen mit Hilfe der Gleichung:

$$\left.\begin{array}{l} N(x,\Lambda)\,dx = \int\limits_{E_c(\Lambda)}^{\infty} \left(\dfrac{dJ}{dE_0}\right) \cdot dE_0 \cdot \int\limits_{1}^{x/R(E_0)} e^{-x/\mu L(E')} \cdot \Sigma_{1n}(E') \, \dfrac{dx}{\mu} \cdot d\mu, \\[2ex] E'(E_0, x, \mu) = E_0 - \int\limits_{0}^{x/\mu} \dfrac{dE}{dr} \cdot dr. \end{array}\right\} \quad (41.1)$$

In der folgenden Aufstellung geben wir eine Übersicht über die verschiedenen mathematischen Größen.

Symbol	Bedeutung

$N(x, \Lambda)$ = der Neutronenproduktionsrate pro Gramm bei einer atmosphärischen Tiefe von x (g/cm²) und der geomagnetischen Breite Λ.

E_0 = der Energie des einfallenden Protons am Gipfel der Atmosphäre, d.h. für $x=0$.

$E_c(\Lambda)$ = der Energie, die der magnetischen Abschneidesteifigkeit für ein Proton bei der geomagnetischen Breite Λ entspricht.

dJ/dE_0 = dem differentiallen Energiespektrum der isotrop einfallenden solaren Protonen am Gipfel der Atmosphäre $x = 0$.

$L(E')$ = der Absorptionslänge des Protons der Energie E' in Luft, abgeschätzt vom totalen Wirkungsquerschnitt in Stickstoff.

$R(E_0)$ = der Reichweite eines Protons der Energie E_0 in Luft.

$\Sigma_{1n}(E')$ = dem makroskopischen Wirkungsquerschnitt für die Produktion eines Neutrons durch ein Proton der Energie E' in Luft.

$\mu = \cos \vartheta$ = wobei ϑ den Winkel zwischen der Geschwindigkeit des einfallenden Protons und der Normalen am Gipfel der Atmosphäre darstellt.

dE/dr = dem differentiellen Energieverlust von Protonen in Luft, nach den Tabellen von RICH und MADEY [Ri 54].

Die mittlere Absorptionslänge für Protonen in Luft wurde über den gesamten inelastischen Wirkungsquerschnitt in Stickstoff abgeschätzt. Für Protonenenergien oberhalb 1 MeV kann man grob mit einer konstanten Absorptionslänge L von 125/g cm² rechnen. Die Neutronenproduktion ist unempfindlich gegenüber der Absorptionslänge. Eine Änderung von L um 25% bewirkt eine Änderung der Neutronenproduktion um etwa 1%.

Solarer Protonenfluß. In Übereinstimmung mit der Analyse von FREIER und WEBBER [Fr 63] wurde das differentielle Spektrum der solaren Protonen durch eine Exponentialfunktion beschrieben:

$$\frac{dJ}{dP} = \left(\frac{dJ}{dP}\right)_0 e^{-P/P_0} \quad \text{(Protonen/cm}^2 \text{ sec)}. \quad (41.2)$$

Hierbei ist P die magnetische Steifigkeit des Protons und P_0 ist eine charakteristische Steifigkeit, die sich mit der Zeit ändert.

Im Zeitraum von 1956 bis 1961 haben 30 große bzw. größere solare Teilcheneinbrüche stattgefunden. Diese sind von WEBBER und FREIER [We 63] tabuliert. Da diese zeitliche Periode das letzte Sonnenfleckenmaximum einschließt, wurde der gesamte Teilchenfluß dieser 30 Ereignisse als der Gesamtfluß für die letzte Sonnenfleckenperiode gewertet. Der mittlere solare Teilchenfluß für die letzte Sonnenfleckenperiode ist dann

$$\frac{dJ}{dP} = 2 \times e^{-P/125 \text{ MV}} \quad \text{(Protonen/cm}^2 \text{ sec MV)} \quad (41.3)$$

Ziff. 41.   Neutronenproduktion und C^{14}-Produktion durch solare Protonen.   513

oder der integrale Fluß

$$J(>100 \text{ MeV}) = 4 \text{ Protonen/cm}^2 \text{ sec.} \tag{41.4}$$

Von den 30 größeren solaren Teilchenereignissen führten 5 Ereignisse integrale Flüsse mit Energien oberhalb 100 MeV von mindestens 10^8 Protonen/cm² sec. Der größte Fluß $3{,}5 \times 10^8$ Protonen/cm² sec oberhalb 100 MeV und $1{,}4 \times 10^9$ Protonen/cm² sec oberhalb 30 MeV, wurde im Gefolge eines $3+$ Solar flare am 12. November 1960 beobachtet.

So scheint es, daß der ganze jährliche solare Fluß von Protonen an einem oder zwei Tagen in die Erdatmosphäre einfällt und dabei scharfe Maxima der C^{14}-Produktion und des Verlustflusses an Neutronen auslöst. Die Rechnungen wurden mit

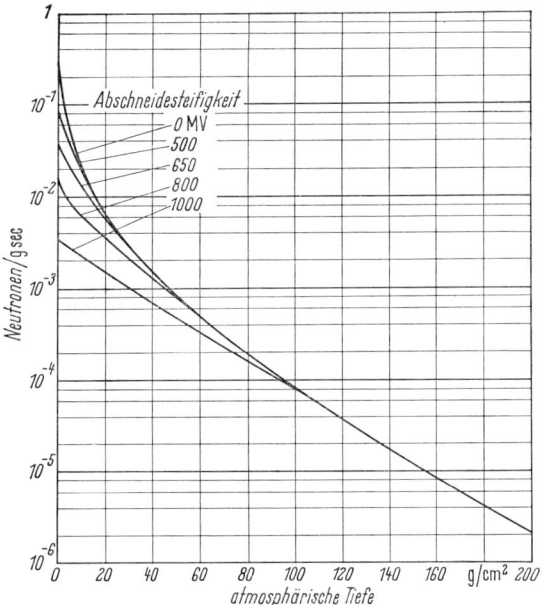

Fig. 71. Höhenabhängigkeit einer Neutronenquelle der Atmosphäre an den geomagnetischen Polen in Abhängigkeit von den erzeugenden solaren Protonen, errechnet mit dem exponentiellen Protonenflußspektrum $(dJ/dP) = (dJ/dP)_0 \cdot e^{-P/125 \text{ MV}}$ Protonen/cm² sec MV. Normierung gemäß Gl. (41,3).

einer charakteristischen Steifigkeit $P_0 = 165$ MV für dieses außergewöhnliche Ereignis und mit einem Wert $P_0 = 125$ MV für den Zeitmittelwert durchgeführt.

Die weitere Berechnung der Höhen- und Breitenabhängigkeit der Neutronenproduktionsrate nach (41.2) geschah in einer Reihe von numerischen Integrationen, deren Teilergebnisse jeweils in Form von Funktionsabbildungen in der Originalarbeit enthalten sind. Die Höhenabhängigkeit der durch solare Teilchen erzeugten Neutronen kann Fig. 71 entnommen werden. Als Quellspektrum der Neutronen wurde das Verdampfungsspektrum von Gl. (17,1) benutzt.

Zur Lösung der Multigruppen-Neutronendiffusionsgleichung (35.25) benutzte LINGENFELTER weiter die Daten der Modellatmosphäre II (S. 492). Die numerische Integration der Diffusionsgleichung wurde, wie in Ziff. 36 beschrieben, durchgeführt, wobei angenommen wurde, daß die solar ausgelösten Neutronen in der hohen Atmosphäre sich im Gleichgewicht befinden.

Wie man Fig. 71 entnimmt, fällt die Neutronenproduktionsrate für hohe geomagnetische Breiten (kleiner Wert der magnetischen Abschneidesteifigkeit) sehr steil ab. Falls die mittlere Bremslänge kleiner oder von der gleichen

Größenordnung wie die mittlere freie Diffusionslänge eines Neutrons ist kann sich die Verteilung der Neutronen in solchen Bereichen sehr stark von ihrer Quellverteilung der Neutronen unterscheiden. Der Gleichgewichtsneutronenfluß ist für verschiedene Energiegruppen in Fig. 72 wiedergegeben. Ein Vergleich beider Bilder zeigt deutlich, daß genau an den Polen keine Übereinstimmung von Quellverteilung und Gleichgewichtsverteilung vorliegt. Die Maximalintensität verschiebt sich zu etwas größeren Tiefen (30—40 g/cm^2) und niedrigeren Energien. Dies steht in deutlichem Gegensatz zu den durch galaktische kosmische Strahlung erzeugten Neutronen. Deren Quellverteilung zeigt eine effektive Absorptionslänge, die etwa um eine Größenordnung größer ist als die mittlere

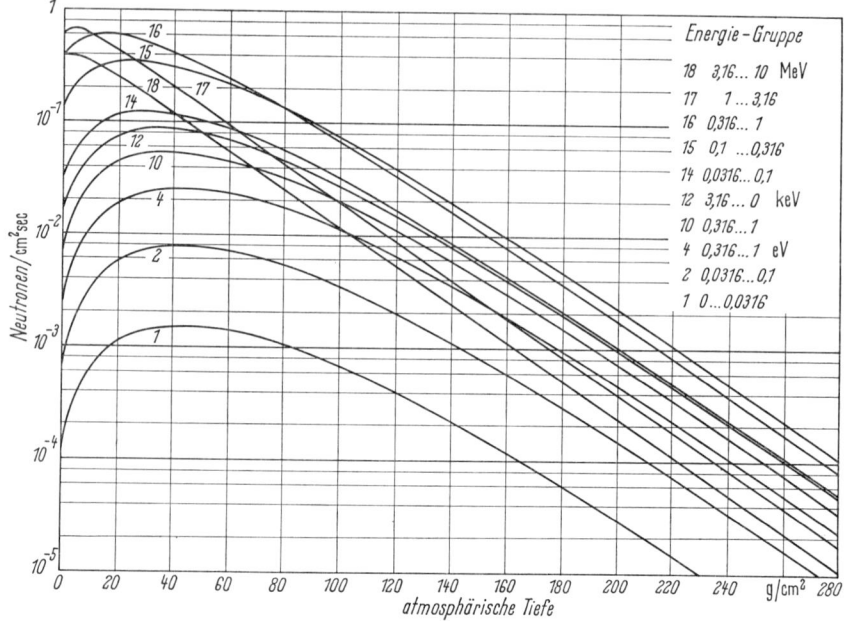

Fig. 72. Gleichgewichtsneutronenfluß in Funktion der atmosphärischen Tiefe für verschiedene Energiegruppen. (Nach [*Li 63 b*].)

freie Diffusionslänge und deren Gleichgewichtsverteilung deshalb einzig und allein durch die Quellverteilung bestimmt wird. Das Neutronenenergiespektrum bei verschiedenen Höhen ist in Fig. 73 dargestellt.

LINGENFELTER und FLAMM können mit dieser Methode der Berechnung eines durch solare Protonen ausgelösten Gleichgewichtsflusses zwei spezielle Erscheinungen diskutieren:

(i) den Überschußneutronenfluß bei solaren Teilcheneinbrüchen,

(ii) die C^{14}-Produktionsrate durch solare Teilchen.

(i) SMITH et al. [*Sm 62*] haben am 19. Juli 1961, wenige Stunden nach dem maximalen Protoneneinbruch im Gefolge eines 3+ solar flare, die Neutronenintensität mit einem nackten bzw. einem ummantelten Li$^6(n, \alpha)$H^3-Zähler mit zwei Ballonen in großer Höhe gemessen. (NEWKIRK hat, wie S. 496 erwähnt, diese Messungen zur Normierung seines Neutronenquellspektrums benutzt.) Wenige Tage nach diesem Ereignis, wieder bei ruhigen Strahlungsverhältnissen, haben SMITH et al. erneut gemessen. Die Differenz der beiden Flußmessungen ergibt den Überschuß der durch die solaren Teilchen ausgelösten Neutronen. Zur Berechnung müssen bekannt sein die Intensität, das Spektrum der solaren Protonen und die

Ziff. 41.   Neutronenproduktion und C^{14}-Produktion durch solare Protonen.

geomagnetische Abschneidesteifigkeit. Nach FREIER und WEBBER [*Fr 63*] ist das solare Protonenspektrum in der Nähe des fraglichen Zeitpunktes (19. Juli) und bei der geomagnetischen Breite $\Lambda = 57°$ N

$$\frac{dJ}{dP} = 4{,}0 \times e^{-P/125\,\text{MV}} \quad (\text{Protonen/cm}^2\,\text{sec sterad MV}). \tag{41.5}$$

Zum Zeitpunkt des Maximums der geomagnetischen Störung, d.h. ungefähr 3 Std vor dieser Messung von SMITH et al., wurde festgestellt, daß der Schwellwert

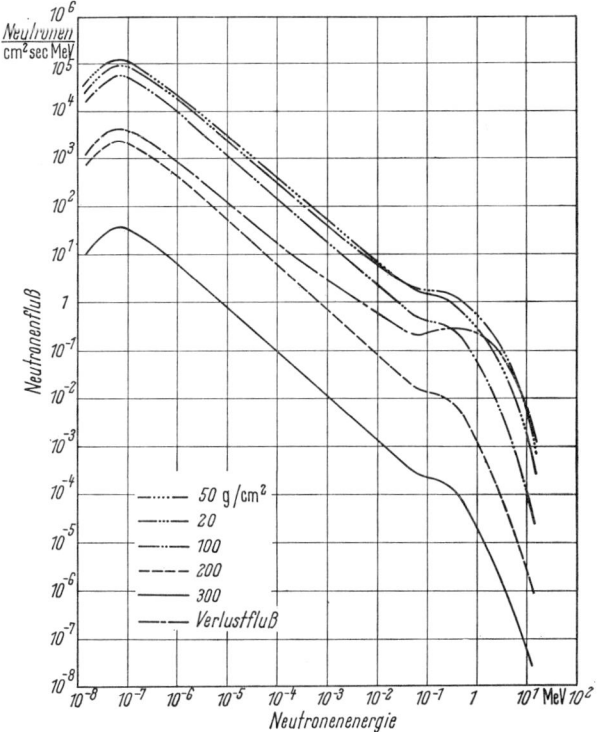

Fig. 73. Gleichgewichtsenergiespektrum für Neutronen bei verschiedenen Höhen berechnet mit Hilfe der höhenabhängigen Neutronenquelle in Fig. 71. Geomagnetische Abschneidesteifigkeit = 0. (Nach [*Li 63b*].)

für die geomagnetische Breite von Bemidji vom Normalwert = 1200 GV zu einer diffusen Grenze von 700 MV oder weniger führte. Durch Interpolation dieser Werte wurde für den Zeitpunkt der Smithschen Messung ein Abschneidewert von 900 MV geschätzt. Für eine scharfe Grenze von 900 MV wurde ein minimaler Gleichgewichtsneutronenfluß gemessen; ohne Schwelle ein maximaler. Diese Neutronenflußspektren wurden mit der Ausbeute des nicht abgeschirmten Neutronendetektors multipliziert und über die Energie integriert. In Fig. 74 sind diese errechneten Überschußzählraten als Kurven eingetragen. Die Meßwerte liegen für atmosphärische Tiefen zwischen 20 und 60 g/cm² gut bei der 900 MV-Abschneidekurve. In größeren Tiefen liegen sie befriedigend zwischen den Grenzkurven.

(ii) Die C^{14}-Produktion muß in einem bestimmten Zusammenhang mit den durch solare Protonen erzeugten Neutronen stehen. Wir haben schon erwähnt, daß in der Nähe des geomagnetischen Pols die Quellverteilung der Neutronen

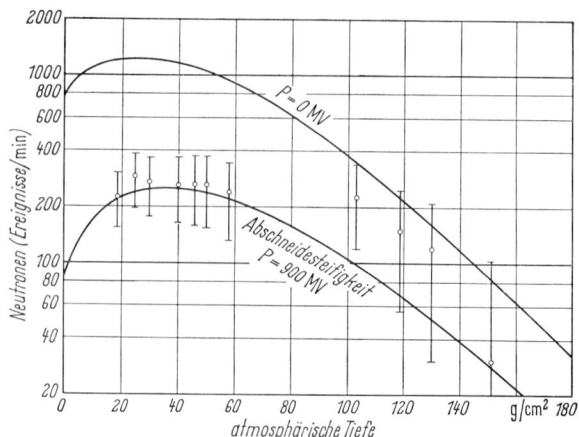

Fig. 74. Errechnete Überschußraten für nackten Li⁶ (n, α) H³-Detektor über Bemidji 013 UT (Weltzeit) am 19. Juli 1961 nach [Li 63b]. Obere Kurve: Geomagnetische Abschneidesteifigkeit = 0; untere Kurve: hohe geomagnetische Abschneidesteifigkeit bei 900 MV. Die experimentellen Meßpunkte und Fehler beziehen sich auf die Meßergebnisse von SMITH et al. [Sm 62].

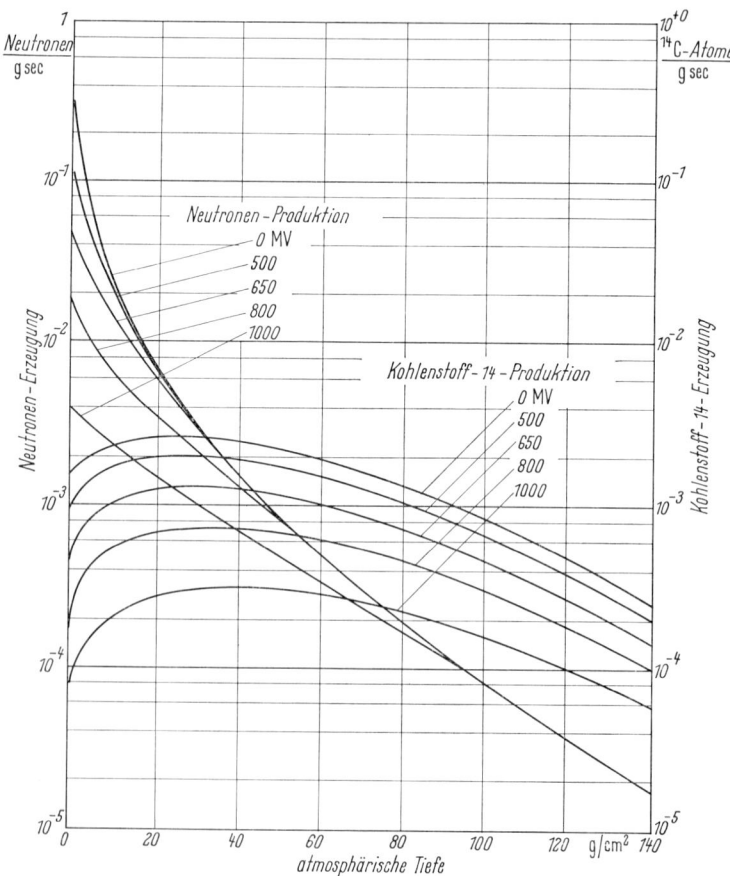

Fig. 75. Höhenabhängigkeit der Neutronen- und C¹⁴-Produktionsraten nach [Li 63b]; berechnet für einen solaren Protonenfluß $(dJ/dP)_0 \times e^{-P/125 \text{ MV}}$; normiert mit $(dJ/dP)_0 = 1$ Proton/cm² sec MV und bei verschiedenen geomagnetischen Steifigkeiten.

einen anderen Verlauf annimmt als die Verteilung der Neutronen im Diffusionsgleichgewicht. Dieser Unterschied muß sich auch in der C^{14}-Produktionsrate widerspiegeln (vgl. Fig. 75).

Integriert man die C^{14}-Produktionsrate in einer Luftsäule von 1 cm² Querschnitt über die Tiefe der Atmosphäre und normiert man auf den gesamten, mittleren solaren Protonenfluß während der Sonnenfleckenhalbperiode 1957—1962

$$\frac{dJ}{dP} = 2 \cdot e^{-P/125 \, \text{MV}} \quad (\text{Protonen/cm}^2 \, \text{sec}), \quad (41.2)$$

so erhält man eine C^{14}-Produktionsrate von 0,44 C^{14}-Atomen/cm² sec in den Polargebieten.

In der näheren geomagnetischen Umgebung des Pols, d.h. bis zu einer geomagnetischen Steifigkeit von 300 MV, sinkt der Polwert der C^{14}-Produktionsrate nur um 10% ab (vgl. Fig. 76). Mit abnehmender geomagnetischer Breite Λ fällt die C^{14}-Produktionsrate steil ab und erreicht etwa bei 1100 MV ein Zehntel des Polwertes.

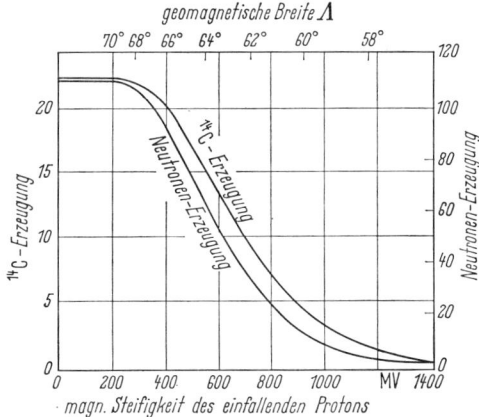

Fig. 76. Die Abhängigkeit der Neutronen- und C^{14}-Produktionsraten von der geomagnetischen Steifigkeit nach [Li 63b] bei solaren Einzelereignissen; Normierung wie in Fig. 75. Ordinatenmaßstab: Teilchen/cm² sec. Die geomagnetische Breite ist an der oberen Bildkante aufgezeichnet (nomaler Dipolwerte).

Um den Gesamtmittelwert der C^{14}-Produktionsrate festzulegen, muß eine Abschätzung vorgenommen werden über die geomagnetischen Abschneidesteifigkeiten, die zum Zeitpunkt des Eintreffens des solaren Teilchenflusses vorliegen. Die untere Grenze des Gesamtmittelwertes erhält man, wenn man die normalen Dipol-Abschneidewerte zugrundelegt, d.h. 11% des Polwertes oder eine C^{14}-Produktionsrate von 0,05 Atome/cm² sec, über den Verlauf des letzten Sonnenfleckenzyklus. Man erhält im wesentlichen denselben Zahlwert, wenn man nach SAUER [Sa 63] vertikal mit einem Feld 6. Ordnung abschneidet.

Inzwischen sind in Minneapolis mehrfach bei verschiedenen größeren solaren Teilchenereignissen Protonen bis herab zu 30 MeV beobachtet worden. Dies legt nahe, den normalen Dipolabschneidewert um einen Faktor 5 zu reduzieren. Der zugehörige gesamte Mittelwert der C^{14}-Produktionsrate für den letzten Sonnenfleckenzyklus würde dann 0,12 Atome/cm² sec betragen. Es ist bemerkenswert, daß auch dieser obere Grenzwert noch innerhalb des Fehlers $\pm 0,50$ Atome/cm² sec der galaktisch ausgelösten C^{14}-Produktion liegt. Die Sonne war in der letzten Sonnenfleckenperiode besonders aktiv. Der zeitliche Mittelwert der C^{14}-Produktionsrate kann über die mittlere Lebensdauer des C^{14}-Atoms noch kleiner ausfallen. Der Beitrag der solar erzeugten C^{14}-Atome scheint für die geochemische Untersuchung vernachlässigbar.

42. Der Verlustfluß der Neutronen aus der Atmosphäre. Die Demographie der Neutronen in der Atmosphäre wäre unvollständig, wenn nicht der Verlustfluß an Neutronen aus der Atmosphäre in die Darstellung eingeschlossen wäre. Der früher — und zum Teil heute noch — benutzte Ausdruck „Neutronen-Albedo" sollte zugunsten des Begriffs Neutronen-„Verlustfluß" (leakage flux) aufgegeben werden. Denn mit Ausnahme sehr langsamer Neutronen, die (vor ihrem Zerfall) auf Grund der Gravitation wieder zur Erde zurückfallen müssen (vgl. Fig. 82) und damit im Sinne einer Albedo zurück-„gestreut" werden, gehen die übrigen Neutronen des Verlustflusses der Atmosphäre real verloren, indem diese entkommen und zerfallen.

Der Neutronenverlustfluß ist nicht allein wegen der absoluten Normierung der gesamten Neutronenquelle in der Atmosphäre von Interesse, sondern insbesondere auch wegen seiner möglichen Bedeutung für die zumindest teilweise Speisung des inneren van Allenschen Strahlungsgürtels der Erde. Dieser kann durch direkt einfallende Protonen oder Elektronen nicht gespeist werden, wie man durch die Rechnungen von STÖRMER [*St 55*] und ALFVÉN [*Al 50a*] weiß. Nur unter bestimmten Streuprozessen und Bahnablenkungen können schnelle Protonen auf geschlossene Bahnen in einem Dipolfeld gebracht werden; normalerweise treffen die geladenen Teilchen entweder auf die Erdoberfläche oder werden im Feld magnetisch umgelenkt.

Als Mechanismus zur Speisung des inneren Van Allen-Gürtels [*Al 58*] schlugen SINGER [*Si 58a*] und VERNOV et al. [*Ve 59*] vor, daß ein Teil der Neutronen des Verlustflusses innerhalb der Magnetosphäre durch β-Zerfall in Protonen und Elektronen übergeht, die im inneren Strahlungsgürtel eingefangen werden. Zunächst wurden die Energiespektren dieser eingefangenen Protonen bzw. deren Höhenverteilung in der Äquatorebene durch SINGER [*Si 58*], HESS [*He 59b*], FREDEN u. WHITE [*Fr 60*], SHKLOVSKY et al. [*Sh 59*] untersucht. Im Jahr 1959 gaben FREDEN und WHITE [*Fr 59*] eine experimentelle Bestätigung der Existenz dieser Protonen. Die Physik der Strahlungsgürtel hat sich inzwischen als großes Sondergebiet entwickelt; es sei deshalb auf zusammenfassende Darstellungen[1] verwiesen, etwa SINGER [*Si 62*] oder HAERENDEL [*Hae 64*].

Wir beschäftigen uns hier nur mit dem tatsächlichen Neutronenverlustfluß. Das Interesse, das die Geophysik, die Plasmaphysik und die Physik der Weltraumerforschung den Strahlungsgürteln entgegenbringt, ist beträchtlich. Aus diesem Grunde wurden in den letzten Jahren ziemlich viele Neutronenmessungen in der hohen Stratosphäre mit Raketen oder mit Satelliten durchgeführt. Die Meßtechnik ist soweit entwickelt, daß man nicht nur nach der periodischen Variation des Verlustflusses mit dem Sonnenfleckenzyklus sucht, sondern auch nach einem stark aperiodischen Neutronenverlustfluß jeweils bei einem solaren Teilcheneinbruch in die Erdatmosphäre im Gefolge eines solar flare.

Die Übereinstimmung zwischen Experiment und theoretischer Voraussage ist noch nicht so gut wie bei den Neutronen *in* der Atmosphäre.

g) Verlustfluß im Sonnenfleckenzyklus. (Theoretische Ergebnisse.)

43. Ein erster Versuch den Fluß und das Energiespektrum hochenergetischer Verlustflußneutronen abzuleiten, beruht auf einer Untersuchung von WENTWORTH u. SINGER [*We 55a*]. Um die Protonen-Albedo festzulegen, griffen die Autoren auf die Daten von CAMERINI et al. [*Ca 50, 52*] zurück und versuchten aus Sternen in Emulsionen auf das Spektrum der Anstoßprotonen zu schließen. Von dem Protonen-Albedospektrum wurde in direkter Übertragung auf den Neutronen-

---
[1] Conference Report 1965 Summer School, Bergen, 1966.

Ziff. 43.   Verlustfluß im Sonnenfleckenzyklus. (Theoretische Ergebnisse.)   519

verlustfluß bzw. das Neutronenspektrum geschlossen: (43.1)

$$\Phi_N(E) \cdot dE = 2 \cdot E^{-1,8} \cdot dE,$$

für Energien $E > 50$ MeV schätzten sie einen Fluß von 0,4 Neutronen/cm² sec.

KELLOG [Ke 58] hat, auf Grund eines Vorschlags von ROTHWELL und GOLD, den Verlustfluß der Neutronen etwa gleichzeitig wie SINGER [Si 58] behandelt. Ausgehend von der Diffusionsgleichung nach BETHE et al. [Be 40] normiert er mit den unkorrigierten atmosphärischen Intensitätsmessungen von

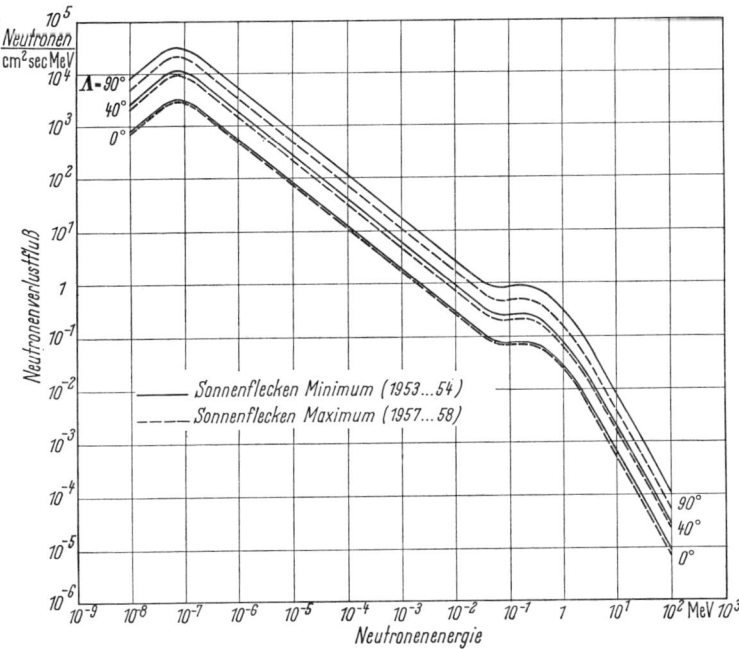

Fig. 77. Spektrum des Neutronenverlustflusses aus der Atmosphäre nach LINGENFELTER [Li 63c] für das Sonnenflecken-minimum 1953/54 und das Sonnenfleckenmaximum 1957/58 für drei geomagnetische Breiten $\Lambda = 0$, 40 und 90°. Die Gestalt des Spektrums oberhalb 10 MeV wurde von HESS et al. [He 59] übernommen.

SOBERMANN [So 56]. Die Kellogschen Werte betragen etwa ein Viertel des von LINGENFELTER [Li 63b] errechneten Verlustflusses. LINGENFELTER [Li 63a, 63b] hat unter den in Ziff. 36 und 37 ausführlich dargestellten Abänderungen gegenüber dem ursprünglichen Lösungsansatz von HESS, CANFIELD u. LINGENFELTER [He 61] die Diffusionsgleichung (35.23) mit der Modellatmosphäre I (S. 492) den Neutronenfluß in der Atmosphäre und den Verlustfluß am Gipfel der Atmosphäre nachberechnet. In Fig. 77 ist das Spektrum des Neutronenflusses ab 300 g/cm² Restdruck bis zum Gipfel der Atmosphäre (Verlustfluß) angegeben. Der Neutronenfluß ist für drei geomagnetische Breiten $\Lambda = 0$, 40 und 90° jeweils für das Sonnenfleckenminimum (1953/54) als auch für das Sonnenfleckenmaximum (1957/58) wiedergegeben.

Der Verlustfluß nach HESS et al. [He 61] ist uns schon in Fig. 64 bekannt geworden; er muß mit dem Verlustfluß bei LINGENFELTER im Sonnenfleckenmaximum verglichen werden. Man beachte beim Vergleich der beiden Darstellungen insbesondere auch die Abschwächung des Buckels des Zwischenmaximums im Energiebereich von 0,5 bis 10 MeV der Verdampfungsneutronen, die auf die Berücksichtigung inelastischer Prozesse durch LINGENFELTER zurückzuführen ist.

Die Intensitäten des Verlustflusses sind bei LINGENFELTER etwa um ein Drittel niedriger, als die von HESS et al. [He 61] bei $\Lambda = 44°$ N berechneten Werte. Dieser Unterschied findet eine Erklärung im Wechsel der Annahmen über die Höhenverteilung der Neutronenproduktionsrate. Diese verläuft bei HESS et al. mit konstanter Absorptionslänge nach (16.3) bis zum Gipfel der Atmosphäre. LINGENFELTER ändert von 200 g/cm² bis zum Gipfel gemäß der Lordschen Sternintensität ab [entsprechend wie NEWKIRK in (16.4)]. Diese Abänderung wirkt sich für den Neutronenfluß etwa zwischen 100 g/cm² bis zum Gipfel aus. Innerhalb der Atmosphäre sind die Neutronenflußrechnungen von LINGENFELTER [Li 63b] bis auf 20%, d.h. noch innerhalb der Meßfehler, in Übereinstimmung mit HESS et al.

Hinsichtlich der Breitenabhängigkeit unterscheiden sich die beiden Theorien ebenfalls. Dieser Unterschied kann den Figuren nicht leicht entnommen werden, da der Neutronenfluß von HESS et al. nur für 44° N dargestellt ist; Umrechnungen für andere geomagnetische Breiten müssen mit Hilfe des Breiteneffektes von SIMPSON [Si 51b] vorgenommen werden, den HESS et al. zugrunde legen. Der von SIMPSON gemessene Breiteneffekt ergibt ein Pol:Äquator-Verhältnis[1] von etwa 4:1. Die Flußwerte von HESS et al. am Pol müssen demnach um einen Faktor 11—12 höher liegen als die Äquatorwerte von LINGENFELTER. — Nun hat aber LINGENFELTER seine Breitenabhängigkeit (vgl. Ziff. 32) nach den Messungen von ALBERT, GILBERT u. HESS [Al 62] normiert. Deren Pol:Äquator-Verhältnis des gemessenen Breiteneffekts ist etwa 11:1. Die Lingenfelterschen Intensitäten sind im Polbereich etwa auf die gleichen Werte angehoben wie die von HESS et al.

Zwischen den Ergebnissen von NEWKIRK [Ne 63], die mit der S_4-Methode (Fig. 66) gewonnen worden sind, und der Lingenfelterschen Diffusionsrechnung besteht wegen der praktisch gleichen Höhenverteilung (s. Fig. 27 und Ziff. 26) quantitative Übereinstimmung. Die S_4-Rechung liefert für die gesamte Neutronenquelle (vgl. Tabelle 28) einen prozentualen Anteil des Verlustflusses von 11% der Quellneutronen. Das Rechenprogramm ZOOM führt nach LINGENFELTER auf denselben Wert.

Ein spezieller Vergleich zwischen beiden Rechnungen kann noch für die im Juli 1961 über Bemidji durch SMITH et al. [Sm 62] ausgeführte Messung erreicht werden. NEWKIRK normiert mit diesen Daten — in denen ein großer solarer Teilcheneinbruch mitgemessen wurde — und findet in einer Luftsäule von 1 cm²

Tabelle 36. *Verlustfluß[2] (Neutronen/cm² sec) aus der Atmosphäre in Abhängigkeit von der geomagnetischen Breite für verschiedene Energien im Verlauf des Sonnenzyklus.*

Geomagnetische Breite Λ	Sonnenflecken-Minimum (1953—1954)					Sonnenflecken-Maximum (1957—1958)				
	Neutronenquelle	Neutronenverlust				Neutronenquelle	Neutronenverlust			
		Gesamt	<1 MeV	1—10 MeV	>10 MeV		Gesamt	<1 MeV	1—10 MeV	>10 MeV
0	1,48	0,110	0,071	0,033	0,006	1,41	0,105	0,068	0,031	0,006
10	1,53	0,113	0,073	0,034	0,006	1,46	0,107	0,069	0,032	0,006
20	1,85	0,138	0,088	0,042	0,008	1,74	0,130	0,083	0,039	0,008
30	2,76	0,205	0,131	0,062	0,012	2,46	0,182	0,117	0,055	0,010
40	4,60	0,368	0,235	0,112	0,021	3,73	0,298	0,190	0,091	0,017
50	7,03	0,687	0,438	0,210	0,039	5,34	0,523	0,333	0,160	0,030
60	8,54	1,120	0,695	0,362	0,063	6,05	0,709	0,447	0,222	0,040
70—90	9,00	1,367	0,825	0,465	0,077	6,05	0,709	0,447	0,222	0,040
Gesamtmittelwert	4,10	0,432	0,276	0,131	0,025	3,22	0,302	0,192	0,092	0,018

[1] Neuere Daten vgl. Tabelle 16. S. 452.
[2] Nach einer privaten Mitteilung von LINGENFELTER an MARTIN [Ma 65] sind die hier wiedergegebenen Flußwerte um einen Faktor 2,13 zu groß.

Querschnitt bei $\Lambda = 57°$ N 0,8 Neutronen/cm² sec. LINGENFELTER errechnet für diesen Zeitpunkt und geomagnetischen Ort 0,85 Neutronen/cm² sec. Die geringfügige Differenz wird auf die spezielle Normierung bezogen.

LINGENFELTER hat inzwischen — wie für die C^{14}-Produktionsrate — auch für den Verlustfluß die Abhängigkeit von der solaren Modulation angegeben. In Tabelle 36 sind die Verlustflußwerte bei verschiedenen geomagnetischen Breiten (von 10° zu 10°) und für verschiedene Energieintervalle der die Atmosphäre verlassenden Neutronen sowohl im Sonnenfleckenminimum (1953/54) als auch im Sonnenfleckenmaximum (1957/58) errechnet.

Diese Flußwerte — für den Sonnenfleckenzyklus festgelegt — müssen durch Messungen des Verlustflusses der Neutronen noch bestätigt werden, vgl. S. 452f.

44. Die Winkelverteilung des Neutronenverlustflusses. Eine präzise Angabe der Winkelverteilung des Verlustflusses der Neutronen aus der Atmosphäre ist nur mit Hilfe des Transport-Vektorflusses möglich. Die benutzte Diffusionstheorie ist eine Näherung der Transportgleichung, in die von einer Kugelfunktionsentwicklung des Transportvektorflusses nur die beiden ersten Glieder (mit $l=0$ und $l=1$) eingehen. Durch die Randbedingung (35.14b) wird erreicht, daß der Anteil des Verlustflusses dem Betrag nach richtig errechnet wird. Der Verlustfluß verschwindet dann im Abstand $0{,}71\,\lambda_{\text{Tr.}} = 2{,}13\,D$ vom Ort der Randbedingung ($\lambda_{\text{Tr.}}$ = Transportweglänge, D = Diffusionskonstante). Im Gegensatz zum Betrag des Verlustflusses wird die Winkelverteilung nicht exakt wiedergegeben (vgl. Lehrbücher der Neutronentheorie, besonders GLASSTONE u. EDLUNDS [Gl 52], S. 403).

α) *Winkelverteilung für Neutronen mit Energien $E < 10$ MeV.* Der Vektorfluß $F(x, \mu)$ in der Diffusionsnäherung ist

$$F(x, \mu) = \sum_{l=0}^{1} \frac{2l+1}{2} \cdot P_l(\mu) \cdot F_l(x) = \frac{1}{2} \cdot F_0(x) + \frac{3}{2} \cdot \mu \cdot F_1(x). \quad (44.1)$$

$F_0(x)$ ist der mit Hilfe der Diffusionsgleichung (35.23) errechnete skalare Neutronenfluß $\Phi(x)$ in der Atmosphäre:

$$F_0(x) = \Phi(x). \quad (44.2)$$

$$F_1(x) = J(x) = -D \cdot \frac{d\Phi}{dx} \quad (44.3)$$

ist der Betrag der Neutronenstromdichte. Wir übertragen die Randbedingung (35.14b) auf unser Problem; d.h. wir finden

$$\left.\frac{d\Phi}{dx}\right|_{x=R} = \frac{\Phi(R+2{,}13 \cdot D) - \Phi(R)}{(R+2{,}13 D) - R} = -\frac{\Phi(R)}{2{,}13 D}. \quad (44.4)$$

Für den Vektorfluß $F(x, \mu)$ erhalten wir so genähert

$$\left.\begin{aligned}F(R, \mu) &= \frac{1}{2} \cdot \Phi(R) + \frac{3}{2} \mu \frac{\Phi(R)}{2{,}13}, \\ &= \frac{1}{2} \Phi(R) \cdot \left(1 + \frac{3\mu}{2{,}13}\right).\end{aligned}\right\} \quad (44.5)$$

Die Zahl der Neutronen pro cm³ und pro sec, die mit diesem Vektorfluß durch eine Einheitsfläche des Randes der Atmosphäre hindurchtreten, ist

$$\frac{\mu}{2} \cdot F(R, \mu) = \text{constant} \left(\mu + 3 \cdot \frac{\mu^2}{2{,}13}\right) = J(R, \mu) \quad (44.6)$$

mit $\mu = \cos\vartheta$, wobei $\vartheta =$ dem Zenitwinkel ist. Dann stellt $J(R, \mu)$ die Winkelverteilung der Verlustneutronen mit $E \leq 10$ MeV dar, wobei für $\vartheta = 0$ auf 1 normiert wurde (vgl. Fig. 78). Nimmt man nach HESS et al. [He 61] zur Abschätzung eine etwas einfachere Form für die Funktion

$$J(R, \mu) = f(R) \cdot \mu^2$$

und normiert man die Konstante $f(R)$ für den Erdradius R, so daß

$$J(R) = -D \cdot \frac{d\Phi}{dx}\bigg|_{x=R},$$

$$= \int_0^1 J(R, \mu) \cdot 2\pi \cdot d\mu = \frac{2\pi}{3} \cdot f(R)$$

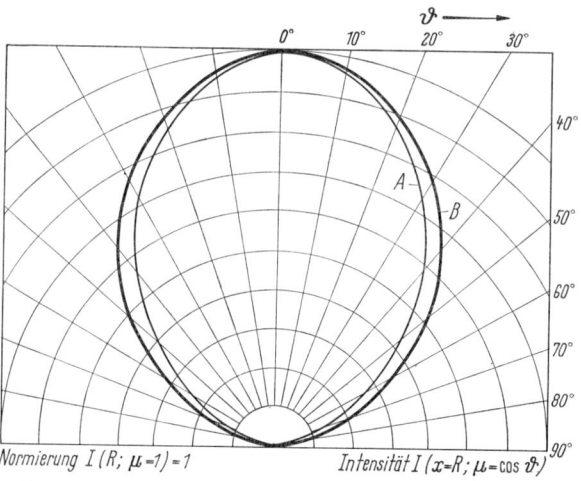

Fig. 78. Winkelverteilung langsamer Neutronen, die aus der Erdatmosphäre entweichen (Kurve B). Zum Vergleich zeigt Kurve A eine $\cos^2 \vartheta$-Verteilung. ϑ ist der Zenitwinkel. (Nach [He 61].)

ist, dann sind die vom Gipfel der Atmosphäre fortfliegenden Neutronen bestimmt durch

$$J(R, \mu) = \frac{3 \cdot J(R)}{2} \cdot \mu^2 \left(\frac{\text{Neutronen}}{\text{cm}^2 \text{ sec sterad MeV}}\right).$$

β) *Winkelverteilung des hochenergetischen Neutronenflusses für $E > 10$ MeV.* Hochenergetische Anstoßneutronen sind mit der bisherigen Winkelverteilung nicht erfaßt. Sie spielen aber sowohl im Verlauf der periodischen Variation der kosmischen Strahlung im Sonnenfleckenzyklus als auch bei aperiodischen Störungen durch solare Teilchen-Einbrüche eine Rolle.

Wir schließen uns der Abschätzung dieser Winkelverteilung hochenergetischer Neutronen an Hand rein geometrischer Überlegungen nach HESS et al. [He 61] an. Bei Maschinenexperimenten wurde gefunden, daß die Winkelverteilung stark in Vorwärtsrichtung weist. Der halbe Öffnungswinkel bei halbem Maximum betrug bei der Erzeugung von Neutronen durch 2 GeV-Protonen 6°, bei 5 GeV betrug er noch etwa 1°, und bei 10 GeV besitzt das erzeugte Neutron im wesentlichen die Richtung des einfallenden Nukleons.

Auf Grund dieser Winkelverteilung können nur noch höchstenergetische Neutronen, die von nahezu streifend in die Atmosphäre einfallenden Nukleonen erzeugt worden sind, gewissermaßen aus der Atmosphäre „herausgeblasen"

Ziff. 44. Die Winkelverteilung des Neutronenverlustflusses.

werden, wie dies schon in Ziff. 16, Fig. 25 schematisch dargestellt worden ist. Bei steilerem Einfall müssen diese sekundären Neutronen in die Tiefe der Atmosphäre weiterlaufen, und wechselwirken weiterhin auf ihrem Weg.

Für die Berechnung der Winkelverteilung wurde der aus einer Platte auslaufende Neutronenstrom errechnet, der durch eingeschossene Protonen erzeugt wird. Wegen Einzelheiten s. HESS et al. Das Ergebnis der numerischen Integration ist in Fig. 79 aufgetragen.

γ) *Der Neutronenstrom in den Raum.* HESS, CANFIELD u. LINGENFELTER errechnen diesen Neutronenstrom $\Phi(R, \Lambda, E)$, indem sie von einer Neutronenquelle Φ ausgehen, deren Emission durch die obige Winkelverteilung beschrieben wird, die von der geomagnetischen Breite Λ im Sinne des von SIMPSON [Si 51b] gemessenen Breiteneffekts abhängt und deren Höhenverteilung im Bereich von 200 g/cm² bis zum Gipfel der Sternintensität nach LORD entspricht. Integriert man diese Neutronenquelle $J(\mu, E, \Lambda)$ über die Oberfläche der Erde (besser dem

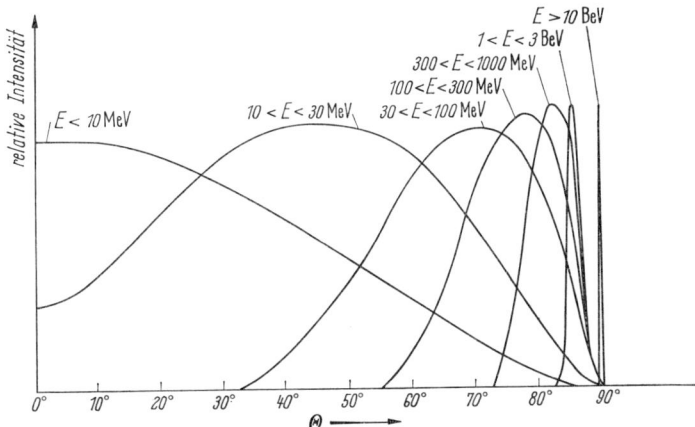

Fig. 79. Winkelverteilungen von Verlustflußneutronen bei verschiedenen Energien. Θ ist der Zenitwinkel. (Nach [He 61].)

Gipfel der Atmosphäre), so kann man den Neutronenfluß berechnen, der durch 1 cm² Oberfläche einer Kugel an irgend einem Ort im Raum strömt:

$$J(R, \Lambda, E) = \int_{\text{Erdoberfläche}} J(\mu, E, \Lambda) \cdot e^{-\frac{L}{v\tau}} \cdot \frac{1}{L^2} \cdot dF \quad \text{(Neutronen/cm}^2 \text{ sec MeV)}. \quad (44.7)$$

mit R = Abstand vom Erdmittelpunkt,
L = Abstand des Neutronen-Quellpunkts vom Beobachtungspunkt,
τ = mittlere Lebensdauer des Neutrons, ca. 17 min.
v = Geschwindigkeit des Neutrons.

Die Neutronenquelle kann noch vereinfacht angegeben werden:

$$J(\mu, E, \Lambda) = k \cdot F_1(\mu) \cdot F_2(E) \cdot F_3(\Lambda) \quad \text{(Neutronen/cm}^2 \text{ sec MeV sterad)}, \quad (44.8)$$

wobei wir im einzelnen wählen können: $F_1(\mu)$ = Winkelverteilung in Fig. 79 $F_2(E)$ ist das Energiespektrum (für feste Werte von R) nach Fig. 80 und $F_3(\Lambda)$ ist die Breitenabhängigkeit nach SIMPSON [Si 51b].

Die numerische Integration auf einer IBM 650 ergab für (40.7) ein Energiespektrum, das für den geomagnetischen Äquator ($\Lambda = 0$) für verschiedene Abstände R_e von der Erdoberfläche errechnet worden ist und in Fig. 80 dargestellt ist. In dieser Darstellung spiegelt sich natürlich das stärkere Zwischenmaximum

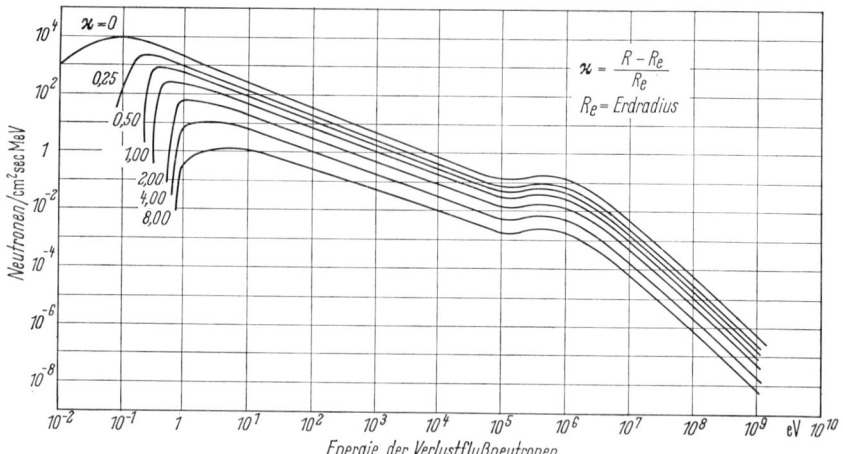

Fig. 80. Energiespektrum der Verlustneutronen im Raum bei verschiedenen Abständen von der Erde über dem geomagnetischen Äquator ($\Lambda = 0$ nach [He 61]). Die Abstandeinheit ist der Erdradius (R_e). Die Kurve $R_e = 0$ soll den Gipfel der Atmosphäre darstellen, der hier etwa 100 km hochliegt. Zwischenmaximum bei (10^5—10^6 eV wahrscheinlich überhöht.) (Nach [He 61].)

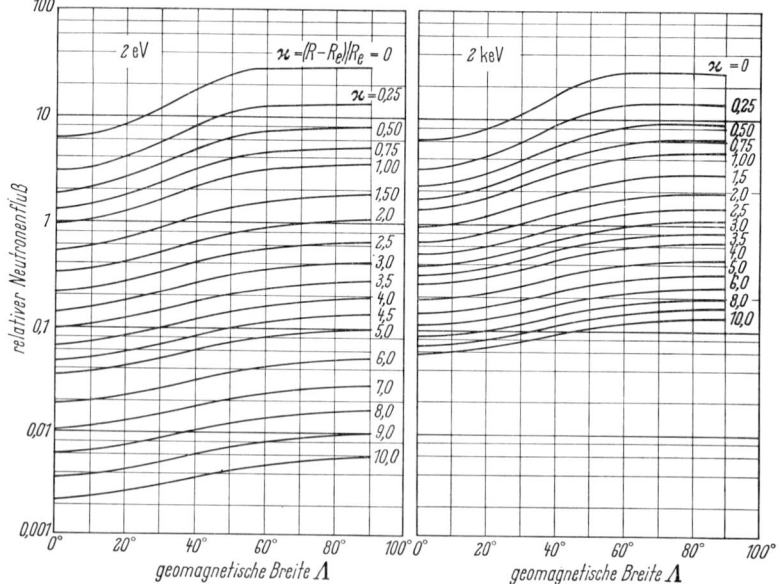

Fig. 81. Breiteneffekt für langsame Verlustflußneutronen von 2 eV und von 2 keV in verschiedenen Abständen $\varkappa = (R - R_e)/R_e$. (Nach [He 61].)

bei 1 MeV wider, weil für die Energieverteilung die Spektralfunktion von HESS et al., Fig. 64, eingesetzt wurde. In Fig. 81 ist dann für langsame Neutronen von 2 eV und 2 keV der Breiteneffekt oberhalb der Atmosphäre zu erkennen. Man entnimmt den beiden Figuren, daß für 2 eV Neutronen die Dichte bzw. der Neutronenfluß rascher abnimmt, da mehr von den langsamen Neutronen in einem gegebenen Abstand zerfallen.

δ) *Durch Schwerkraft eingefangene Neutronen* sind alle diejenigen langsamen Neutronen, die eine kleinere Energie als 0,66 eV haben. Diese können der Erdanziehung nicht entweichen und fallen zur Erde zurück. Ein Neutron von 0,35 eV

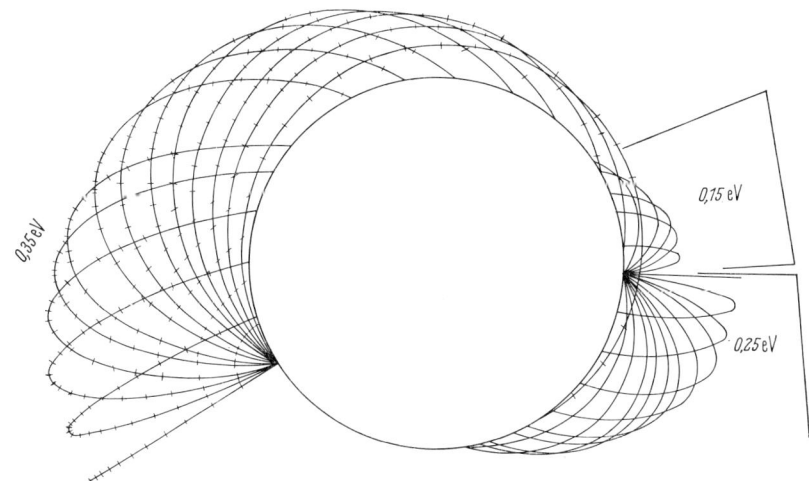

Fig. 82. Neutronen unter 0,66 eV können nicht aus dem Schwerefeld der Erde entweichen. Die Bahnkurven veranschaulichen (näherungsweise) (nach [*He 61*]) die Fallbewegungen von Verlustflußneutronen mit Energien $E_n = 0{,}35$, $0{,}25$ und $0{,}15$ eV. Die Punkte an den Bahnen der 0,36 eV Neutronen kennzeichnen jeweils eine Flugzeit von 200 sec; die mittlere Lebensdauer eines Neutrons beträgt ca. 1014 sec.

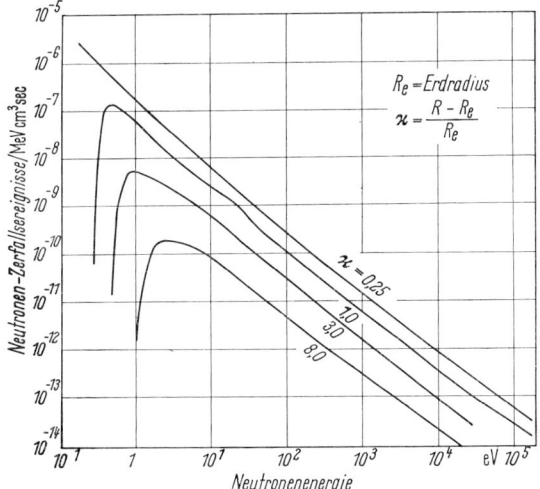

Fig. 83. Energiespektrum der Neutronen, die in verschiedenen Abständen von der Erdoberfläche zerfallen. Gültig für $\Lambda = 0$.

kann bei senkrechtem Start 80 min ansteigen, bis es zur Erde zurückfliegen würde; jedoch ist diese Aufstiegszeit schon die 7fache Halbwertszeit. Nahezu alle diese Neutronen zerfallen also bevor sie zur Erde zurückkehren. Typische elliptische Bahnen sind in Fig. 82 zu erkennen. Sie wurden von HESS et al. [*He 61*] genähert errechnet.

Der Fluß der durch die Schwerkraft eingefangenen Neutronen ist im Spektrum Fig. 80 einbezogen. Die Intensität dieser niederenergetischen Neutronen fällt mit der Höhe (Abstand von der Erdoberfläche) schneller ab als die Intensität der schnellen Neutronen, weil die langsamen schon nach einem kürzeren Weg zerfallen müssen.

Zum Schluß geben wir noch die *Neutronenzerfallsdichte* an, die notwendig ist um die mögliche Speisung des van Allen-Strahlungsgürtels (soweit es für unsere Zwecke erforderlich) zu verstehen. Wir versuchen die Neutronenzerfallsdichte dn/dV (zerfallende Neutronen pro cm³ und pro sec) in Abhängigkeit von der Energie E, dem Abstand von der Erdoberfläche R_e und der geomagnetischen Breite Λ anzugeben:

$$\frac{dn(\Lambda, E, R)}{D\,dV} = \frac{1}{v\tau} J(\Lambda, E, R), \qquad (44.9)$$

wo v wieder die Neutrongeschwindigkeit und τ die Zerfallzeit des Neutrons kennzeichnet. Das Spektrum der Neutronen, die in verschiedenen Abständen von der Erdoberfläche zerfallen ist in Fig. 83 dargestellt.

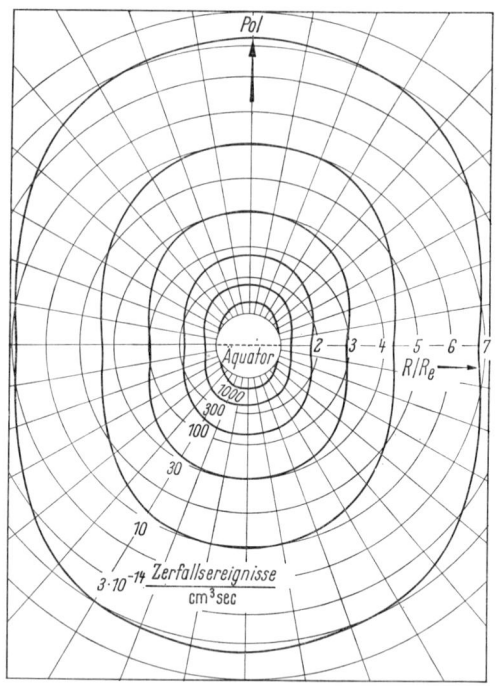

Fig. 84. Zerfallsdichte der Neutronen, aufsummiert über die Energie nach (44.10) bei verschiedenen Höhen (Erdabständen) und für alle geomagnetischen Breiten.

Wie Hess et al. ausführen, sollen die Elektronen bei allen Zerfällen schneller Neutronen im wesentlichen dieselbe Energie besitzen. Dies scheint bei den Untersuchungen der Strahlungsgürtel gefunden worden zu sein. Integriert man (44.9) über die Energie, so gewinnt man die Abhängigkeit der Neutronenzerfallsdichte von der geomagnetischen Breite und dem Abstand von der Erdoberfläche:

$$\frac{dn(\Lambda, R)}{dV} = \int_0^\infty \frac{d}{dV} n(\Lambda, R, E) \cdot dE. \qquad (44.10)$$

In Fig. 84 ist diese Breitenabhängigkeit der Neutronenzerfallsdichte dargestellt.

Aus diesen Darstellungen kann man ablesen, daß die durch die Schwerkraft eingefangenen Neutronen sehr wesentlich zu der Zerfallsdichte der Neutronen beitragen. Die langsamen Neutronen verbringen eine hinreichend lange Zeit im Raum nahe der Erde. In einem Abstand von 0,5 Erdradien von der Erdoberfläche beteigen sich beispielsweise die durch die Schwerkraft eingefangenen Neutronen zu 40% am Zerfall. Dies ist beträchtlich, denn sie tragen weniger als 1% zum Neutronenverlustfluß bei (vgl. dazu Tabelle 25).

Eine Abschätzung für sehr energiereiche Neutronen ist noch nützlich. Die Zerfallsrate (44.9) muß hier relativistisch korrigiert werden, die Breitenabhängigkeit wird von solch schnellen geladenen Teilchen nicht mehr wahrgenommen und kann unterdrückt werden:

$$\frac{d}{dV} n(E,r) \cdot dE = \frac{J(E,r)\, dE}{\beta \cdot c \cdot \tau \sqrt{1-\beta^2}}. \tag{44.9a}$$

Man errechnet leicht, daß ein energiereiches, relativistisches Verlustflußneutron im Verlauf seiner (dilatierten) Lebensdauer nahezu einen Weg von einer astronomischen Einheit zurücklegt. Neutronen von 10 bis etwa 1000 MeV dringen dann im Verlauf ihrer mittleren Lebensdauer ziemlich weit in den interplanetarischen Raum vor.

Die von KELLOG [Ke 59] errechnete Neutronenzerfallsdichte ist kleiner als die von HESS et al. [He 61] beschriebene. Die Gründe sind folgende: KELLOG berücksichtigt nur radial wegfliegende Neutronen, nicht aber die $\cos^2 \vartheta$-Verteilung. Ferner berücksichtigt er den Neutroneneinfang durch die Schwerkraft nicht und errechnet zuletzt einen Neutronenfluß, von nur $1/20$ dessen, den HESS et al. zugrunde legen. Diese raumzeitlichen Verhältnisse der Verlustneutronen können für die mögliche Speisung des inneren Strahlungsgürtels mit Protonen und Elektronen über den Neutron-β-Zerfall von Bedeutung werden.

h) Experimentelle Verlustfluß-Intensitäten.

45. Der Neutronenverlustfluß der Atmosphäre ist sowohl für langsame Neutronen ($B^{10}F_3$-Zähler) als auch für schnelle Neutronen (ummantelte $B^{10}F_3$-Zähler) gemessen worden. Messungen mit direkten Zählern für schnelle Neutronen sind bis jetzt noch nicht veröffentlicht worden.

Wir konnten auf S. 450 das unterschiedliche Verhalten der Dichte der langsamen Neutronen im Gegensatz zu dem schnellen Neutronenfluß am Gipfel der Atmosphäre darstellen. Zweckmäßig trennt man auch beim Verlustfluß in diese beiden Energiebereiche.

1. Langsame Verlustflußneutronen. Messungen liegen vor von REIDY, HAYMES und KORFF [Re 62] und MARTIN, WITTEN u. KATZ [Ma 63c]. Beide geben den Modul für die Neutronenzählraten N an, der sofort die Angabe der Neutronendichte ϱ erlaubt: $\varrho_{\text{Re}} = 7{,}40 \cdot 10^{-7} \cdot N$ (Neutronen/cm²) bzw. $\varrho_{\text{Ma}} = 3{,}40 \cdot 10^{-7} N$.

REIDY, HAYMES u. KORFF haben oberhalb des Gipfels der Atmosphäre eine Neutronendichte von $(3{,}7 \pm 0{,}4) \cdot 10^{-7}$ Neutronen/cm³ gemessen. Rechnet man den differentiellen Neutronenfluß von HESS, CANFIELD u. LINGENFELTER [He 61] (vgl. Fig. 64, wo er über der Energie aufgetragen ist) mit der jeweils zur Energie E gehörigen Neutronengeschwindigkeit v in eine differentielle Neutronendichte um, und integriert diese (graphisch oder numerisch) über die Energie, so gewinnt man für langsame Neutronen ($E \leq 10^3$ eV) die theoretisch zu erwartende Neutronendichte

$$\varrho_{\text{theor}} = 0{,}5 \cdot 10^{-7} \text{ (Neutronen/cm}^3\text{)} \quad \text{(vgl. Tabelle 37).}$$

Ein Vergleich der von HAYMES u. KORFF [*Ha 60a*] bei Ballonaufstiegen gemessenen Neutronendichten und dem im Höhenintervall von 9,5 bis 22 km beim Raketenaufstieg in der Atmosphäre gemessenen Wert (vgl. S. 439f.) zeigen gute Übereinstimmung. Auf Grund dieser Kontrolle gewinnt der hohe Meßwert von $3{,}7 \cdot 10^{-7}$ Neutronen/cm³ an Gewicht. Dieser Wert liegt etwa um einen Faktor 7 höher als der theoretisch vorausgesagte; gegenüber der Lingenfelterschen Voraussage liegt er sogar um einen Faktor 12 höher.

Als Erklärung für diese „hohe" Neutronendichte in 80—200 km Höhe geben REIDY, HAYMES u. KORFF in ihrer interessanten Analyse zwei Möglichkeiten an: a) die nicht ummantelten $B^{10}F_3$-Zähler konnten nicht hinreichend vom Raketenkörper entfernt oder getrennt montiert werden. Örtlich im Raketenkörper durch kosmische Primärstrahlung ausgelöste Neutronen wurden — nach deren Abbremsung — teilweise mitgezählt. Auch an die Bremsung schneller Verlustflußneutronen im Raketenkörper wurde gedacht, so daß der gemessene Wert nur eine obere Grenze darstellen kann. WILLIAMS u. BOSTROM [*Wi 64*] ergänzen mit dem Argument, daß die Verdampfungsneutronen im Raketenkörper ein Maximum bei ca. 7 MeV haben. Die langsamen Neutronendetektoren sind aber für Energien kleiner 1000 eV am empfindlichsten. Bei den Korrekturen der örtlich erzeugten schnellen Neutronen bleibt offen, welcher Bruchteil in der Rakete auf Energien für nachweisbare langsame Neutronen abgebremst wird. Es ist denkbar, daß ein Großteil schneller Neutronen entkommen kann. In diesem Fall würde eine hohe tatsächliche Neutronendichte gemessen. b) als zweiten Faktor betrachten REIDY, HAYMES u. KORFF die durch die Schwerkraft eingefangenen langsamen Neutronen als mögliche Ursache der Erhöhung der Neutronendichte oberhalb des Gipfels der Atmosphäre.

Wie schon erwähnt, können Neutronen mit einer Energie größer als 0,66 eV aus dem Feld der Erdanziehung entkommen. In einer Höhe von ca. 100 km in der Stratosphäre ist die Dichte des Restgases (bei beginnender Entmischung) 10^{13} Teilchen/cm³ (vgl. Fig. 63a). In dieser Höhe überschreitet die mittlere Absorptionslänge für ein thermisches Neutron ($E = 0{,}025$ eV) 10^5 km; dagegen ist die mittlere freie Weglänge gegen β-Zerfall von der Größenordnung 1700 km. Der in der Atmosphäre dominante Einfangprozess durch N^{14} spielt in dieser Höhe keine Rolle mehr. Die der Erdanziehung unterworfenen langsamen und thermischen Neutronen haben so im Gleichgewicht — vor ihrem Zerfall — die Chance beim Aufstieg auf radialen oder schrägen Bahnen bzw. beim Fall zum „Gipfel der Atmosphäre" ein relativ „dichtes" Neutronengas zu bilden (vgl. Fig. 82).

Die Ergebnisse von REIDY et al. und MARTIN et al. [*Ma 63c*] sind in Tabelle 37 aufgenommen. Die Korrektur der gemessenen Werte nach MARTIN et al. bedarf einer Erklärung. Der Grundgedanke ist einfach: die tatsächlich gemessene Neutronendichte ϱ_m kann für jede Rakete als Summe der tatsächlichen Neutronendichte ϱ und einem örtlich erzeugten Untergrund b im Raketenkörper angegeben werden:

$$\varrho_{m_1} = \varrho_1 + b_1 \quad \text{(Atlas-Rakete — MARTIN et al.)}$$

und

$$\varrho_{m_2} = \varrho_2 + b_2 \quad \text{(Aerobee-Rakete — REIDY et al.)}$$
$$= C_H \cdot C_A \cdot \varrho_1 + C_R \cdot B_1,$$

d.h. es soll mit Hilfe eines Höhenfaktors C_H und eines geomagnetischen Breitenfaktors C_A und eines Untergrundfaktors C_R von der Aerobee-Messung auf die Atlas-Messung umgerechnet werden. Mit $C_H = 1{,}17$; $C_A = 1{,}67$; $C_R = 3{,}94$ betragen die Korrekturen über 70% für die Atlas-Messung; über 80% für die

Experimentelle Verlustfluß-Intensitäten.

Aerobee-Messung. Für die bei größeren Höhen (800—1000 km) gemessenen Neutronendichten ist keine physikalisch sinnvolle Lösung nach diesem Verfahren angebbar. WILLIAMS u. BOSTROM [Wi 64] errechnen mit weiteren C_i-Korrekturfaktoren entweder kleinere Korrekturen, oder sogar negative Untergrundkorrekturen.

Zusammenfassend können wir sagen, daß die Messung der Dichte langsamer Verlustneutronen einen oberen Grenzwert von $3{,}7 \cdot 10^{-7}$ Neutronen/cm³ ergeben hat, und daß der wirkliche Wert wohl zwischen diesem Wert und den Voraussagen der Theorien nach HESS et al. bzw. LINGENFELTER liegen wird. Die neueren Messungen von MARTIN [Ma 65] sind in Tabelle 37 aufgenommen. Die Meßwerte — mit und ohne Korrektur — scheinen besser mit der Theorie von HESS et al. übereinzustimmen.

Tabelle 37. *Dichte langsamer Verlustflußneutronen.*

Messung	Jahr	Geomagnetische Breite Λ °	Höhe km	Neutronendichte ϱ					
				Experiment 10^{-7} n/cm³		Vergleich mit Theorie			
						HESS et al. [He 61]		LINGENFELTER [Li 63c]**	
				gemessen ϱ_{exp}	korrigiert* ϱ'_{exp}	ϱ_{theor} 10^{-7} n/cm³	$\varrho'_{exp}/\varrho_{theor}$	ϱ_{theor} 10^{-7} n/cm³	$\varrho'_{exp}/\varrho_{theor}$
MARTIN et al. [Ma 63c]	1961	35	480	$1{,}24 \pm 0{,}65$	$0{,}32 \pm 0{,}16$	0,32	$1{,}0 \pm 0{,}5$	$0{,}085 \pm 0{,}20$	3,8
MARTIN [Ma 65]	1962	35	671	$0{,}44 \pm 0{,}10$	—***	0,32	$1{,}38 \pm 0{,}31$	$0{,}085 \pm 0{,}20$	$5{,}18 \pm 1{,}2$
		49	209	$0{,}62 \pm 0{,}12$	—***	0,5	$1{,}94 \pm 0{,}38$	$0{,}20 \pm 0{,}40$	$3{,}1 \pm 0{,}6$
DY et al. [Re 62]	1960	49	84—200	$3{,}7 \pm 0{,}4$	$0{,}62 \pm 0{,}31$	0,5	$1{,}24 \pm 0{,}14$	$0{,}20 \pm 0{,}4$	7,3

* Korrektur nach MARTIN et al. [Ma 63c] im Körper der Atlas-Rakete beträgt 73 %; für die Aerobee-Rakete [Re 62] soll sie sogar 82 % betragen (vgl. auch Text).
** Vgl. auch Anmerkung bei Tabelle 36.
*** Eine geeignete Korrektur für lokale Neutronenproduktion kann nicht angegeben werden, und unterblieb.

2. *Messungen schneller Verlustflußneutronen.* Die ersten Messungen stammen von HESS u. STARNES [He 60]. Inzwischen sind weitere Messungen ausgeführt worden von BAME, DAVIS, GLORE u. BRUMLEY [Ba 60]; ALBERT, GILBERT u. HESS [Al 62]; BAME, CONNER, BRUMLEY, HOSTETLER u. GREEN [Ba 63b]; TRAINOR u. LOCKWOOD [Tr 63] und WILLIAMS u. BOSTROM [Wi 64].

In Tabelle 38 sind alle Messungen schneller Verlustflußneutronen nach Jahr, geomagnetischer Breite, Höhe (eventuell mittlerer Höhe) aufgenommen und mit den theoretischen Werten nach HESS et al. [He 61] bzw. LINGENFELTER [Li 63c] in Zusammenhang gebracht. Mehrere Meßdaten sind gegen örtliche Neutronenproduktion korrigiert; die Güte, selbst die Größenordnung dieser Korrekturen ist noch nicht gesichert. Mit oder ohne Korrektur zeichnen sich zwei Gruppen von Messungen ab: a) diejenigen, die besser mit den Voraussagen von HESS, CANFIELD und LINGENFELTER [He 61] zusammenpassen, b) diejenigen, die besser mit den Voraussagen von LINGENFELTER [Li 63c] übereinstimmen.

Zur Gruppe a) gehören die Messungen von ALBERT et al.; HESS et al.; WILLIAMS und BOSTROM. Die Meßdaten liegen — mit oder ohne Korrektur — noch über den Voraussagen von HESS et al. Gegenüber den Voraussagen von LINGENFELTER sind sie um Faktoren 2,5—9 zu groß.

Tabelle 38. *Verlustfluß[1] schneller Neutronen aus der Atmosphäre* nach [Mi 64].

Messung	Jahr	Geomagnetische Breite λ °	Höhe km	Gesamter Neutronenverlustfluß (Neutronen/cm² sec)		Verhältnis Messung/Theorie		Korrektur für Neutronenuntergrund
				Hess et al. [He 61]	Lingenfelter [Li 63]	Hess et al. [He 61]	Lingenfelter [Li 63c]	
Albert et al. [Al 62]	1961	0	275–575[2]	0,264	0,10	1,55±0,38	4,61±1,12	nein
Trainor und Lockwood [Tr 63]	1962	0	200–400[3]	0,298	0,10	0,34±0,10	1,0 ±0,3	ja
Bame et al. [Ba 63b]	1961	8,7	320	0,307	0,10	0,33±0,16	1,0 ±0,5	ja
Martin et al. [Ma 63c]	1961	~28–31	875–700	0,328–0,420	0,117–0,158	1,0 ±0,24[4]	2,6 ±0,4[4]	ja
Hess und Starnes [He 60]	~1959	31	1000	0,375	0,132	1,85±0,46	5,25±1,31	nein
Bame et al. [Ba 63b]	1961	36,5	650	0,545	0,220	0,40±0,15	1,0 ±0,36	ja
Albert et al. [Al 62]	1961	40	275–575[2]	0,681	0,297	2,45±0,59	5,62±1,35	nein
Trainor und Lockwood [Tr 63]	1962	40	200–400[3]	0,768	0,335	0,21±0,07	0,49±0,15	ja
Williams und Bostrom [Wi 64]	1961–1962	40	1000	0,538	0,324	2,01±0,52	0,55±2,64	ja
Hess und Starnes [He 60]	~1959	41	100	0,888	0,360	1,47±0,37	3,63±0,91	nein
Williams und Bostrom [Wi 64]	1961–1962	43	1000	0,656	0,296	1,88±0,59	8,86±2,79	ja
Bame et al. [Ba 60]	1959	44,5	120	1,06	0,40	0,36±0,08	0,95±0,20	ja
Trainor und Lockwood [Tr 63]	1962	80	200–400[2]	1,32	1,04	0,28±0,11	0,36±0,14	ja
Albert et al. [Al 62]	1961	80	275–575[3]	1,1	1,0	2,80±0,71	3,1 ±0,8	nein

[1] Beachte Anmerkung zur Tabelle 36; Korrektur hier nicht vollzogen.
[2] Mittlere Höhe von 425 km wurde zugrundegelegt.
[3] Mittlere Höhe von 300 km wurde zugrundegelegt.
[4] Mittelwert des Verhältnisses über den angegebenen Höhen- bzw. Breitenbereich.

TRAINOR und LOCKWOOD [Tr 63] diskutieren Gründe für die starken Abweichungen gegenüber den Lingenfelterschen Voraussagen. Im Experiment von ALBERT et al. war die Zählerkoinzidenzanordnung — wie S. 445 erwähnt — zu langsam, so daß auch örtlich produzierte Neutronen mitgemessen, und so die Zählrate der Verlustflußneutronen verfälscht werden konnte. Gegen das Experiment von WILLIAMS u. BOSTROM wenden sie ein, daß keine Diskriminierung gegen örtlich erzeugte Neutronen für den zentral im Raketenkörper angeordneten Neutronenzähler vorgesehen war, und daß bei dieser Anordnung eine Korrektur schwer möglich sei. Wie wir auf S. 445 darstellten, gilt aber für die Antikoinzidenzanordnung von TRAINOR u. LOCKWOOD ähnliches wie bei ALBERT et al.

Zur Gruppe b) gehören die Messungen von BAME et al. [Ba 60] und BAME et al. [Ba 63b] und TRAINOR u. LOCKWOOD. Die Messungen von BAME et al. [Ba 63b] ergeben (August und Dezember 1959; Mai 1960 und Dezember 1961) am geomagnetischen Äquator und bei 36,5° N einen Neutronenfluß, der um einen Faktor 3 niedriger liegt, als der von HESS, CANFIELD u. LINGENFELTER vorausgesagte, aber mit den neueren theoretischen Voraussagen von LINGENFELTER [Li 63a, 63b] und den unabhängig von NEWKIRK [Ne 63] gewonnenen, in guter Übereinstimmung zu sein scheint.

Die Messungen von TRAINOR und LOCKWOOD passen nur am Äquator gut zu der Lingenfelterschen Voraussage. Mit wachsender geomagnetischer Breite bis zum Pol hinken deren gemessene Neutronenflußwerte erheblich dem vorausgesagten Wert von LINGENFELTER nach. TRAINOR et al. messen ein Breiteneffektverhältnis von nur 1:4,5. Dieses ist erheblich niedriger als das von ALBERT et al. mit 1:10 gemessene, das LINGENFELTER seiner Bestimmung der Verteilungsfunktionen für die Neutronenproduktionsrate (vgl. Ziff. 32) zugrundelegt. Die Albertschen Werte (aus Gruppe a!) liegen — auch bei einer angemessenen Korrektur — weit über den Lingenfelterschen Voraussagen.

WILLIAMS u. BOSTROM [Wi 64] weisen nachdrücklich auf diese verschiedenen Verhältnisse für den gemessenen und in die Theorie aufgenommenen Breiteneffekt hin. Wie wir schon S. 520 ausgeführt haben, ist nach beiden Theorien am Pol der gleiche Wert des Verlustflusses zu erwarten. Auch hier ist über die Meßwerte die Diskrepanz nicht aufgelöst.

Der aus Messungen in der Atmosphäre extrapolierte Verlustfluß Φ_g nach Tabelle 16 kann mit den Ergebnissen der Satellitenmessungen, Tabelle 38, verglichen werden. Die gute Übereinstimmung der Flußwerte Φ_G nach KORFF et al. [Ko 65], HAYMES [Ha 64b], BOELLA et al. [Bo 65] und GREENHILL et al. [Gr 65a] mit den Satellitenwerten von BAME et al. [Ba 63b] kann jedoch darüber nicht hinwegtäuschen, daß durch die nachträgliche Korrektur der theoretischen Verlustflußwerte (vgl. Anmerkung bei Tabelle 36) die Klärung der Absolutwerte, sowie die der sehr differierenden Breitenabhängigkeit, noch offen steht.

Eine weitere, unverstandene Unsicherheit in unserer Kenntnis des schnellen Neutronenflusses liegt in den gemessenen aber nicht erklärten Schwankungen der Intensität sowohl innerhalb der Atmosphäre (GAUGER [Ga 64, 65], [Ko 65] und Ziff. 21) als auch oberhalb der Atmosphäre im Verlustflußgebiet (WILLIAMS u. BOSTROM [Wi 64]). Diese Autoren messen im Zeitraum April bis Juli 1961 einen etwa um 50% erniedrigten schnellen Fluß an Neutronen, mit Schwankungen, die anscheinend nicht mit solaren Effekten, mit Forbush-Effekten oder sonst bekannten Variationen in Zusammenhang gebracht werden können.

Die gemessenen Werte des Neutronenverlustflusses sind bis jetzt erst bis auf einen Faktor 3—5 sicher. Für den Breiteneffekt dürfte die Unsicherheit noch einen Faktor 2—3 betragen. Die gefundenen Schwankungen des schnellen Neutronenflusses bedürfen noch einer Deutung oder Erklärung.

Eine Nachmessung der Flußwerte schneller Neutronen mit direkten Neutronenzählern innerhalb und außerhalb der Atmosphäre ist notwendig. Das Jahr der ruhigen Sonne mit maximaler Intensität und geringsten solaren Störungen scheint dazu besonders geeignet. Ob sich daraus Korrekturen für die Theorie ergeben, bleibt abzuwarten.

i) **Aperiodisch emittierter Verlustfluß. (Solare Teilcheneinbrüche.)**

46. Eine anomale Komponente zusätzlich im inneren Strahlungsgürtel eingefangener Protonen wurde von Armstrong, Harrison, Heckman u. Rosen [*Ar 61*] und Naugle und Kniffen [*Na 61*] entdeckt. Man vermutete, daß dieser Überschuß von Protonen im Strahlungsgürtel von dem Zerfall von Verlustflußneutronen herrühren könnte, die in den Polarbereichen in der Atmosphäre durch Einbrüche solarer Protonen im Gefolge von solar flares produziert werden. (Vgl. wegen der Beschreibung des Gesamtvorgangs Ziff. 41 und Fig. 85.)

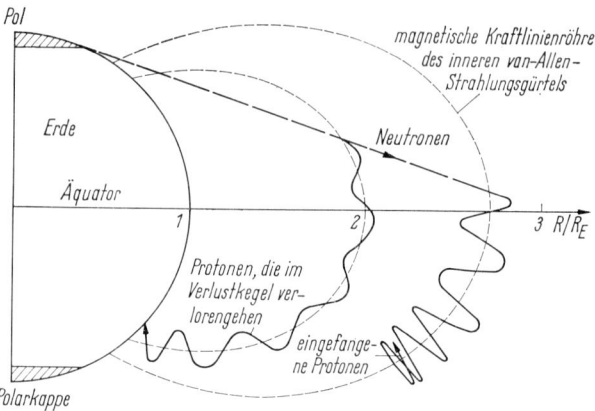

Fig. 85. Verlustneutronen aus den Polkappen speisen den inneren Strahlungsgürtel, nach [*Si 62*]. Man beachte, daß die meisten Zerfallsprotonen nicht in einer „magnetischen Flasche" eingefangen werden, sondern daß die meisten Protonen in den Verlustkegel injiziert werden und dem Strahlungsgürtel und dem Raum verloren gehen. Neben diesem möglichen Speiseprozeß der van Allen-Zone kommen wahrscheinlich noch Plasmateilchen des solaren Windes in Frage.

Diese Vorstellung wurde durch die direkte Beobachtung gefestigt, daß der mittlere Neutronenverlustfluß in mittleren Breiten nach dem größeren solar flare, vom 15. November 1960, um etwa das dreifache angewachsen ist. An jenem Tag konnten Hess, Curry u. Chupp [*He 64*] ab ungefähr 0605 UT in geomagnetischen Breiten von 31 bis 41° N in Höhen von 200 bis 1000 km einen Wert von 1,5 Neutronen/cm² sec messen, gegenüber dem Normalwert von 0,5 Neutronen/cm² sec.

Lenchek und Singer [*Le 62a*] schätzen eine untere Grenze für die Neutronenproduktion durch solare Protonen ab. Lenchek [*Le 62b*] hat diese Abschätzung erweitert, um den Neutronenverlustfluß für diese Ereignisse mitzuerfassen, von dem die Dichte der eingefangenen Protonen errechnet wurde. Der resultierende Fluß der eingefangenen Protonen war in roher Übereinstimmung mit der beobachteten Anomalie. Lenchek schließt weiter, daß bei adiabatischem Einfang die Lebensdauer solcher Protonen im Strahlungsgürtel — verglichen mit dem Zeitraum größerer Injektionsereignisse — sehr groß sein müsse (etwa 100 Jahre). Unter dieser Voraussetzung darf die anomale Komponente als Gleichgewichtsproblem behandelt werden.

LINGENFELTER und FLAMM [Li 64] haben die Behandlung solarer Teilchen im Rahmen der Diffusionstheorie (also bis 10 MeV), wie in Ziff. 41 beschrieben, auf den Verlustfluß ausgedehnt und geben für den oben erwähnten solar flare

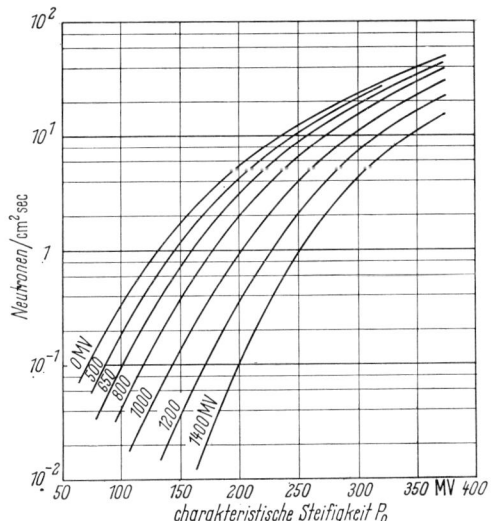

Fig. 86. Neutronenverlustfluß für verschiedene Abschneidesteifigkeiten in Funktion der charakteristischen Steifigkeit P_0, nach [Li 64]. Die theoretischen Zusammenhänge sind in Abschnitt 41 dargestellt. Das zugrundeliegende solare Protonenspektrum $dJ/dP = (dJ/dP)_0 \cdot \exp(-P/P_0)$ ist auf $(dJ/dP)_0 = 1$ Proton/cm² sec normiert. Für unser Beispiel gilt $(dJ/dP)_0 = 4{,}2$ als Faktor.

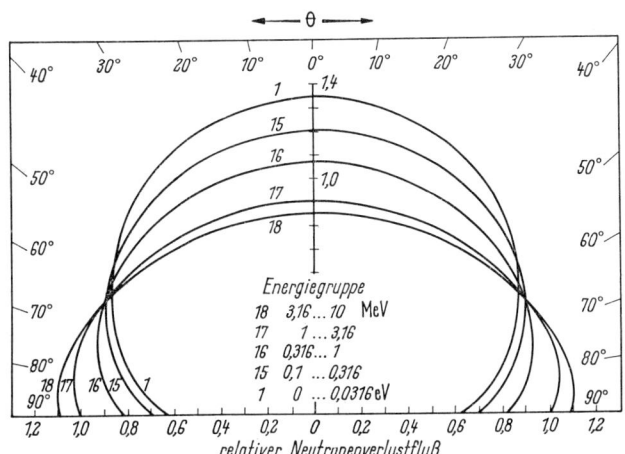

Fig. 87. Winkelverteilung des Neutronenverlustflusses für verschiedene ausgewählte Energiegruppen der Diffusionsgleichung. Die Rechnung und die zugrundeliegende Neutronenquelle nach [Li 64] sind in Abschnitt 41 erläutert. Der integrale Fluß innerhalb jeder Energiegruppe ist auf einen Gesamtverlust von 1 Neutron/cm² sec normiert.

vom 15. November 1960 mit Hilfe der Steifigkeits-Werte von FREIER und WEBBER [Fr 63] über das zugehörige exponentielle Spektrum für die solaren Protonen

$$\frac{dJ}{dP} = 4{,}2 \cdot e^{-P/375\,\text{MV}} \quad (\text{Protonen/cm}^2\,\text{sec})$$

charakteristische Steifigkeit $P_0 = 375$ MV,

einen Verlustfluß von 200 Neutronen/cm² sec an den Polen an, den man für die Abschneidesteifigkeit 0 MV (am Pol) direkt aus der errechneten Verlustflußkurve Fig. 86 ablesen kann. Auf Grund roher geometrischer Abschätzungen sollen im Bahnbereich der Rakete von HESS et al. aus dieser Neutronenproduktion im Polbereich 1—2 Neutronen/cm² sec als Verlustfluß folgen. Eine genauere Analyse ist angekündigt. Die Winkelverteilung der Überschußverlustneutronen kann in Fig. 87 verfolgt werden. Für langsame Neutronen liegt wieder eine cos² ϑ-Verteilung vor, der mittelenergetische Verlustfluß strömt nahezu isotrop und bei hohen Energien sind wieder die großen Winkel bevorzugt.

Einige Unterschiede der beiden Modelle von LENCHEK und LINGENFELTER sollen zum Abschluß noch erwähnt werden.

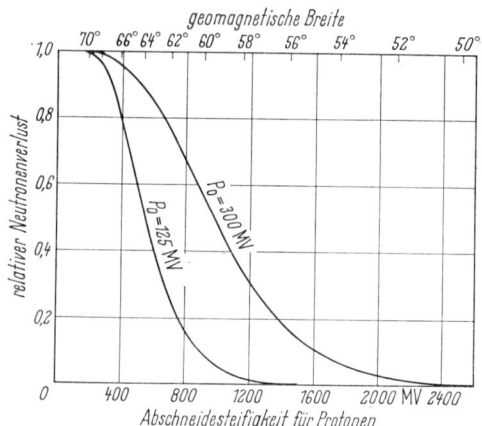

Fig. 88. Der Neutronenverlustfluß in Funktion der geomagnetischen Abschneidesteifigkeit für zwei charakteristische Steifigkeiten $P_0=125$ und 300 MV nach [Li 64]. Die Breitenskala am oberen Bildrand bezieht sich auf normale Dipolabschneidesteifigkeit.

Für zwei charakteristische Steifigkeiten $P_0=125$ MV und $P_0=300$ MV ist der Verlustfluß errechnet worden und in Fig. 88 dargestellt. Die Skala am oberen Rand der Figur gibt die geomagnetische Breite wieder, für die jeweils die normale vertikale Dipol-Abschneidesteifigkeit P_c ist. Die charakteristische Steifigkeit P_0 bewirkt je nach Wert einen steilen Abfall des Verlustflusses in verschiedenen geomagnetischen Breiten. Für $P_0=125$ MV liegt dieser Abfallbereich zwischen 67° bis 60° und für $P_0=300$ MV zwischen 64° bis 56°. Da aber die magnetischen Abschneidesteifigkeiten nach einer Forbush-Abnahme, die gewöhnlich ein solares Teilchenereignis begleitet, bemerkenswert diffus und reduziert werden, so erscheint der tatsächliche Abfall des Verlustflusses gewöhnlich bei viel niedrigeren geomagnetischen Breiten, als dies in Fig. 88 zu erkennen ist.

Fig. 88 kann man ferner entnehmen, daß etwa für die (zeitabhängige) charakteristische Steifigkeit $P_0=125$ MV der Verlustfluß nur um 10% abnimmt, wenn die magnetische Abschneidesteifigkeit vom Pol (0 MV) bis etwa 67° (350 MV) anwächst. Umgekehrt kann man daraus schließen, daß in diesem Fall mehr als 90% der erzeugten Verlustflußneutronen durch einfallende Protonen mit Energien größer als 100 MeV erzeugt worden sind.

Diese Überlegung muß man den Rechnungen LENCHEKs entgegenhalten, bei dem der Beitrag des hochenergetisch ausgelösten Verlustflusses durch Protonen mit Energien größer als 100 MeV vernachlässigt worden war.

Die Ergebnisse von LINGENFELTER und LENCHEK unterscheiden sich ferner darin, daß LINGENFELTER und FLAMM den Verlustfluß bis zu 10 MeV berechnen, während sich LENCHEK mit eingefangenen Protonen oberhalb 10 MeV, d.h. mit Verlustflußneutronen mit Energien über 10 MeV befaßt.

In den Größenordnungen der Neutronenproduktion liegen LINGENFELTER und FLAMM um einen Faktor 10 höher als LENCHEK. Diese Diskrepanz kann signifikant sein. Die höhere Produktionsrate von LINGENFELTER macht nur eine Einfangslebensdauer der Protonen von etwa einem Sonnenfleckenzyklus notwendig, während LENCHEK findet, daß die niederenergetische Komponente der anomal eingefangenen Protonen durch Verlustflußneutronen aus solaren Einbrüchen nur erklärt werden könne, wenn die eingefangenen Protonen eine Lebensdauer von etwa 100 Jahren aufweisen.

Eine Emulsionsmessung im Strahlungsgürtel durch FREDDEN und WHITE [Fr 62] zeigt, daß das anomale steile Maximum von ARMSTRONG et al. [Ar 61] nicht mehr nachweisbar ist. FREDDEN und WHITE geben für Energien von 15 bis 700 MeV differentielle Neutronenflußspektren, die aus Protonendaten erschlossen sind.

Anhang 1: Kosmische Strahlung — Registrierstationen[1].

SL = Meereshöhe, Sealevel
NM = Neutronen Pile Monitor
μT = Mesonen-Teleskop
ScT = Szintillations-Teleskop
I = Ionisationskammer
EAS = Extended Air shower Station

Schwellwerte der magnetischen Steifigkeit (Rigidity thresholds) nach QUENBY and WENK, Phil. Mag. 7, 2457 (1962). (Höhen in Meter über NN.)

Station	Höhe in m über Normal Null	Geographische Koordination Breite λ	Geographische Koordination Länge φ	Schwellwert der magnetischen Steifigkeit Threshold rigidity (GV)	Instrumente
Agassiz (USA)	100	N 35° 18'	W 111° 46'		EAS
Ahmedabad (India)	SL	N 23° 01'	E 72° 36'	15,8	NM, μT
Albuquerque (New Mexico)					EAS
Alma Ata (USSR)	806	N 43° 12'	E 76° 56'	6,6	NM
Amsterdam Holland)	SL	N 52° 23'	E 4° 54'	2,6	I
Baguéres (France)	550	N 43°	E		
Beirut (Libanon)	SL	N 38° 55'	E 35° 31'	8,2	NM
Berkeley (USA)	SL	N 37° 52'	W 122° 18'	4,6	NM, μT
Bergen (Norway)	SL	N 60° 24'	E 5° 24'	1,2	NM, μT
Bologna (Italy)	50	N 44° 30'	E 11° 21'	5,2	ScT, μT
Boulder (USA)	1600	N 40° 05'	W 105° 15'		
Brisbane (Australia)	SL	S 27° 04'	E 153°	7,0	
Buenos Aires (Argentine)	SL	S 34° 36'	W 58° 30'	11,4	NM, μT
Canberra (Australia)	800	S 35° 15'	E 149° 08'	11,8	NM, EAS
Cape Schmidt (USSR)	SL	N 68° 52'	W 179° 30'	0,6	NM, I
Chacaltaya (Bolivia)	5220	S 16° 19'	W 68° 10'	13,3	NM, μT, ScT
Chalais-Meudon (France)	100	N 48° 48'	E 2° 11'	3,4	NM, μT
Cape Town (South Africa)	SL	N 33° 55'	E 18° 22'	6,4	μT
Cheltenham[2] (England)	SL	N 38° 44'	W 76° 50'	2,2	I
Chicago (USA)	176	N 41° 50'	W 87° 40'	1,9	NM
Christchurch (New Zealand)	SL	S 43° 32'	W 172° 40'	2,8	μT, I
Churchill (Canada)	SL	N 58° 45'	W 94° 05'	0,2	NM, μT
Climax (USA)	3400	N 39° 22'	W 106° 11'	3,1	NM
College (USA)	SL	N 64° 05'	W 147° 50'	0,5	NM
Colombo (Ceylon)	SL	N 6° 56'	E 79° 58'	17,6	μT
Deep River (Canada)	145	N 46° 06'	W 70° 30'	1,0	NM, I
Darjeeling (India)	≃2000?	N 27° 03'	E 88° 30'	15,1	μT

[1] Erweiterte Zusammenstellung nach SANDSTRÖM [Sa 65].
[2] Die Ionisationskammermessungen in Cheltenham werden in Fredericksburg fortgesetzt.

Station	Höhe in m über Normal Null	Geographische Koordinaten Breite λ	Länge φ	Schwellenwert der magnetischen Steifigkeit Threshold rigidity (GV)	Instrumente
Ellsworth (Antarctica) . . .	SL	S 77° 42'	W 41° 6'	0,7	NM
Fredericksburg (South Africa)	SL	N 38° 18'	W 77° 29'	2,2	NM, I
Freiburg (Schauinslund) (Germ.)	1220	N 48° 0'	E 7° 50'	4,0	NM, I
Friedrichshafen (Germany) .	410	N 47° 39'	E 9° 29'	4,0	
Godhaven	SL	N 69° 20'	W 53° 30'	0,03	I
Gulmarg	2400	N 34° 03'	W 74° 21'	12,0	μT
Göttingen (Germany) . . .	273	N 51° 31'	E 9° 56'	2,9	NM
Hafelekar (Austria)	2290	N 47° 19'	E 11° 23'	4,2	μT, I
Haifa (Israel)	200	N 32° 48'	E 35° 00'	11,3	μT
Halle (Germany)	100	N 51° 30'	E 12° 00'	3,4	I
Heiss Island (USSR)	SL	N 80° 37'	E 58° 03'	<0,1	NM, μT
Hermanus (South Africa) . .	SL	S 34° 25'	E 19° 13'	6,4	NM, μT
Herstmanceaux	SL	N 50° 52'	E 0° 20'	2,8	NM, μT
Hobart (Tasmania) . . .	SL	S 42° 54'	E 147° 20'	1,8	μT
Hovra	SL	N 22° 37'	E 88° 27'	16,2	I
Huancayo (Peru)	3400	S 12° 02'	W 75° 20'	13,7	NM, I
Hyderabad (India)	29	N 25° 23'	E 68° 25'		
Invercargill (New Zealand) .	SL	S 46° 24'	E 158° 21'	1,9	NM, μT
Irkutsk (USSR)	433	N 52° 27'	E 104° 02'	3,5	NM, μT, I
Ithaca, N.Y. (USA)	260	N 42° 26'	W 76° 25'		EAS
Jungfraujoch (Switzerland) .	3550	N 46° 33'	E 7° 59'	4,8	NM
Kampala (Makerere) (Uganda)	1196	N 0° 20'	E 32° 34'	15,3	NM, μT
Kiel (Germany)	SL	N 54° 18'	E 10° 06'	2,3	NM, μT
Kiruna (Sweden)	390	N 67° 51'	E 20° 10'	0,5	μT
Kodaikanal (India) . . .	2343	N 10° 14'	E 77° 28'	17,5	NM, μT, EAS
Kühlungsborn (Germany) . .	SL	N 54° 07'	E 11° 46'	2,3	μT, I
Lae (New Guinea)	SL	S 6° 44'	E 147° 00'	15,8	NM, μT
Leeds (England)	100	N 53° 49'	W 1° 33'	2,1	NM
Lincoln (USA)	350	N 40° 49'	W 96° 41'	2,2	NM
Lindau (Germany)	140	N 51° 36'	E 10° 06'	2,9	NM, μT
Libon	SL	N 38° 42'	W 9° 10'	5,0	μT
Lomnicky Stit (CSR) . . .	2634	N 49° 11'	E 20° 13'	3,97	NM, μT
London (England)	SL	N 51° 32'	W 0° 06'	2,7	NM, μT
Makapuu Point	100	N 21° 18'	W 157° 39'	13,5	NM, μT
Macao	SL	N 21° 21'	E 133° 36'	16,0	μT
McMurdo Sound	SL	N 70° 00'	W 105° 00'	<0,1	NM, ScT
Macquari Island	SL	S 54° 29'	E 185° 58'	0,4	μT
Manchester (England) . . .	SL	N 53° 59'	W 2° 15'		
Mawson (Antare Fica) . . .	SL	S 67° 36'	E 62° 52'	0,2	NM, μT
Mexico City (Mexico) . . .	2200	N 19° 20'	W 99° 10'	10,0	I
Mina Aguilar (Argentine) . .	4000	S 23° 12'	W 65° 42'	12,6	NM
Minneapolis (USA)	255	N 44° 59'	W 93° 15'	1,34	
Mirny (Antarctica)	SL	S 66° 34'	E 92° 55'	<0,1	NM, I
Mt. Norikura (Japan) . . .	2770	N 36° 07'	E 137° 33'	12,2	NM, I
Mt. Washington (USA) . . .	1917	N 44° 20'	W 71° 40'	1,3	NM
Mt. Wellington (Australia) .	725	S 42° 55'	E 147° 14'	1,8	NM
Moskau (USSR)	200	N 55° 45'	E 37° 35'	2,3	NM, μT, I, EAS
München (Germany) . . .	500	N 48° 12'	E 11° 36'	4,0	NM
Murichson Bay (Norway) . .	SL	N 80° 03'	E 18° 18'	0,066	NM, μT
Murmansk (USSR)	SL	N 68° 57'	E 33° 10'	0,5	NM
Nederhorst den Berg	SL	N 52° 14'	E 5° 05'	2,6	NM
Nagoya (Japan)	SL	N 35° 10'	E 136° 58'	12,6	μT
Ottawa (Canada)	101	N 45° 24'	W 75° 54'	1,1	NM, μT
Oulu (Finnland)	SL	N 65° 01'	E 25° 24'	0,7	NM, T
Palestine (USA)		N 31° 48'	W 95° 43'		
Palo Alto (USA)		N 37° 25'	W 122° 15'	4,65	EAS.
Pic-du-Midi (France)	2860	N 42° 56'	E 0° 15'	5,6	NM, μT
Port aux France	SL	S 49° 21'	E 70° 13'	1,2	NM

Anhang 2. Das Erdmagnetfeld; geomagnetische Koordinaten; spezielle Wahl.

Station	Höhe in m über Normal Null	Geographische Koordinaten Breite λ	Länge φ	Schwellenwert der magnetischen Steifigkeit Threshold rigidity (GV)	Instrumente
Praha (Prague) (CSR) ...	187	N 50° 04'	E 14° 26'	3,4	NM, μT
Resolute Bay (Canada) ...	SL	N 74° 41'	W 94° 54'	0,1	NM, μT
Rio de Janeiro (Brazil) ...	SL	S 22° 54'	W 43° 14'	12,1	NM
Rome (Italy)	SL	N 41° 54'	E 12° 31'	6,4	NM, μT
Sacramento Peak (USA) ..	3000	N 27° 30'	E 70° 00'	5,1	NM
Sapporo (Japan)	SL	N 43° 01'	E 141° 21'	7,4	I
Simferopol (USSR)	570	N 44° 44'	E 34° 00'	5,8	μT
Stockholm (Sweden)	SL	N 59° 20'	E 18° 00'	1,4	μT
Sulphur Mountain (Canada) .	2283	N 51° 12'	W 115° 36'	1,1	NM, μT
Sverdlovsk (USSR)	SL	N 56° 14'	E 61° 04'	2,3	μT, I
Sydney (Australia)	SL	S 33° 53'	E 151° 10'	4,3	NM
Syowa-Base (Anarctica) ..	SL	S 69° 00'	E 39° 35'	0,4	NM, μT, I
Thule (Greenland)	SL	N 76° 33'	W 68° 50'	<0,1	NM, μT
Tbilisi (USSR)	SL	N 41° 43'	E 44° 49'	6,9	I, μT
Tixie Bay (USSR)	SL	N 71° 40'	E 128° 54'	<0,1	I
Tokyo, Itabashi (Japan) ..	SL	N 35° 45'	E 139° 43'	11,9	μT, I
Tokyo, Mabashi (Japan) ..	SL	N 35° 42'	E 139° 40'	11,9	ScT, μT, I
Trivandrum (India)	SL	N 08° 31'	E 77° 00'	17,5	μT
Tromsö (Norway)	SL	N 69° 42'	E 18° 59'	0,4	μT
Uppsala (Sweden)	SL	N 59° 51'	E 17° 35'	1,3	NM, μT, ScT
Ushuaia (Argentine)	SL	N 54° 48'	W 68° 18'	6,6	NM, μT
Volcano Ranch (USA) ...	100				EAS
Weissenau (Germany) ...	427	N 47° 48'	E 9° 30'	4,0	NM, μT
Wellington (Australia) ...	125	S 41° 17'	E 174° 46'	3,2	NM
Wilkes (Antarctica)	SL	S 66° 15'	E 110° 35'	<0,1	μT
Yakutsk (USSR)	105	N 62° 01'	E 129° 43'	1,4	NM, I
Zugspitze (Germany)	2960	N 47° 25'	E 10° 59'	4,5	NM, μT

Anhang 2: Das Erdmagnetfeld; geomagnetische Koordinaten; spezielle Wahl.

Das Erdmagnetfeld, und zwar das äußere Magnetfeld der Erde, kann in einer ersten Näherung durch einen Dipol dargestellt werden, der im Zentrum der Erde lokalisiert ist. Die räumliche Orientierung der magnetischen Dipolachse kann mit Hilfe einer Kugelfunktionsanalyse des (an der Erdoberfläche) gemessenen Erdmagnetfeldes gefunden werden. Der Dipol durchstößt bei $\lambda_0 = 78{,}5°$ N, $\varphi_0 = 69{,}0°$ W bzw. 78,5° S, 111,0° E (geographisch) die Erdkugel. (Die Angabe von Winkeln erfolgt entweder in ° ' '' oder in Dezimalen der Bogenminuten und Bogensekunden: $15{,}5° = 15° 30'$ oder $15{,}167° = 15° 7'$.) Man nennt diese Punkte die „*Geomagnetischen Pole*". Diese fallen *nicht* mit den Magnetpolen der Erde zusammen. (Magn. „Nord"pol bei den Melville-Inseln 70° 5' n.Br. 96° 46' w.L., magn. „Süd"pol auf dem antarktischen Kontinent 72° 25' s.Br. 154° ö.L.) Die Ebene des geomagnetischen Äquators ist um 11,5° gegen den geographischen Äquator geneigt und schneidet diesen bei 21° O und 159° W. Man nennt diese Magnetfeldnäherung das „zentrierte Dipol-Modell" (centred dipole model). Die geomagnetischen Koordinaten eines Punktes der Erde, Breite Λ und Länge Φ, können über die entsprechenden geographischen Koordinaten φ und λ errechnet werden. Die Verknüpfung geschieht nach CHAPMAN und BARTELS [Ch 40 bzw. Ch 51] — (Die Geomagnetiker verwenden die Symbole Φ (bzw. φ) für Breiten, Λ (bzw. λ) für Längen).

$$\operatorname{tg} \Phi = - \operatorname{tg}(\varphi - 291{,}0) \sin x \; \frac{1}{\cos(x + 78{,}5°)},$$
$$\operatorname{tg} \Lambda = - \cos \Phi \; \operatorname{tg}(x + 78{,}5°),$$
$$\operatorname{tg} x = \cos(\varphi - 291{,}0) \operatorname{ctg} \lambda.$$

Die geomagnetische Breite Λ des zentrierten Dipols kann auch (vgl. etwa [Ch 51], S. 646) durch die Beziehung

$$\sin \Lambda = \sin \lambda \cdot \sin \lambda_0 + \cos \lambda \cos \lambda_0 \cdot \cos (\varphi - \varphi_0)$$
$$= \sin \lambda \cdot \sin 78{,}5° + \cos \lambda \cdot \cos 78{,}5° \cdot \cos (\varphi - 69°)$$

wiedergegeben werden. HESS et al. [He 59a] und [He 61] — wie die späteren theoretischen Arbeiten — legen diese geomagnetische Breite zugrunde; man beachte Druckfehler bei [He 59a].

Nach BARTELS [Ba 36] stellt ein „exzentrisches" Dipolmodell eine bessere Näherung dar. Der Dipol ist im Abstand von 342 km vom Erdzentrum gelegen. Die geomagnetischen Pole liegen dafür bei 80,1° N; 82,7° W und bei 76,3° S und 121,2° O. SINGER (1958) hat darauf hingewiesen, daß der Ort des magnetischen Zentrums säkularen Schwankungen unterliegt. Aber nicht einmal das exzentrische Dipolmodell gibt das Magnetfeld an der Erdoberfläche einigermaßen gut wieder, was an den vielen Anomalien, die bei den geomagnetischen Vermessungen auftreten, sichtbar wird. Es ist möglich, Abweichungen von einem Dipolfeld durch Hinzufügen eines zweiten Dipols Rechnung zu tragen; man bezeichnet dann den ersten oft als den Haupt-Dipol.

Verwendet man den Begriff der „*magnetischen Koordinaten*", so bezieht man sich auf das *tatsächliche* magnetische Feld in einem Punkt; dabei kann der „magnetische Meridian" durch die Deklination definiert werden. Der Tangens der magnetischen Breite ist gleich dem Tangens der halben Inklination. Der Äquator der magnetischen Koordinaten wird üblicherweise als Dip-Äquator (dip equator) bezeichnet. Es ist dies der Ort der Punkte in denen die Inklination 0° ist. Diese Kurve sollte mit dem geomagnetischen Äquator übereinstimmen, falls das Erdmagnetfeld ein ideales Dipolfeld wäre. Tatsächlich folgt es einem irregulären Verlauf.

Die geladenen Teilchen der primären kosmischen Strahlung nehmen neben dem Dipolfeld des Erdmagneten sowohl das Multipolfeld der Erde, als auch die Magnetfeldeinflüsse der Magnetosphäre wahr. Das resultante Magnetfeld führt deshalb für den einfallenden Strom von Teilchen zu Abweichungen, die etwa anstelle des Dip-Äquators bzw. des geomagnetischen Äquators durch einen abweichenden „kosmischen Strahlungsäquator" meßbar sind (vgl. SIMPSON, FENTON, KATZMANN und ROSE [Si 56]). Dieses effektive Magnetfeld kennzeichnet man heute meist über die Schwellwerte der Abschneidesteifigkeit für die einfallenden primären Teilchen bzw. deren Penumbraeffekte. In Anhang I sind für einige wichtige Meßorte für kosmische Strahlungsexperimente neben den geographischen Koordinaten Abschneidesteifigkeiten angegeben.

Die Schwellenwerte der Abschneidesteifigkeiten sind inzwischen vielfach gemessen und theoretisch behandelt worden. Man vergleiche etwa: ROTHWELL [Ro 58b], QUENBY und WENK [Qu 62] und insbesondere MCCRACKEN u. Mitarb. [McCr 65] bzw. [Sh 65]. Diese Arbeiten enthalten auch ausgedehnte Literaturangaben und ergänzen die Darstellung bei SANDSTRÖM [Sa 65].

Literatur.

A. Grundlegende und allgemeinere Arbeiten.

[Ba 60] BARKOW, A., B. CHAMANY, D. M. HASKIN, P. L. JAIN, E. LOHRMANN, M. W. TEUCHER, and M. SCHEIN: Phys. Rev. **122**, 617 (1960).

[Ba 62a] BABAYAN, KH. B., J. S. BABECKI, N. G. BOYADZHYAN, Z. A. BUZA, N. L. GRIGOROV, E. S. LOSKIEWICZ, E. A. MAMIDZHANYAN, E. I. MASSALSKI, A. A. OLES, CH. A. TRET'YAKOVA i V. YA. SHESTOPEROV: Bull. Acad. Sci. USSR., **26**, No. 5 558—571 (1962). (Columbia Technical Translation.)

[Ba 63b] BABAYAN, KH. P., N. G. BOYADZHYAN, N. L. GRIGOROV, CH. A. TRET'YAKOWA i V. YA. SHESTOPEROV: Soviet Physics JETP **17**, 15—23 (1963); — J. Exp. Theoret. Phys. (USSR.) **44**, 22—34 (1963).

Literatur. 539

[Be 40] BETHE, H. A., S. A. KORFF, and G. PLACZEK: Phys. Rev. **57**, 573 (1940).
[Be 64] BECKURTS, K. H., and K. WIRTZ: Neutron Physics. Berlin-Göttingen-Heidelberg: Springer 1964.
[Br 64a] BROOKE, G., and A. W. WOLFENDALE: Proc. Phys. Soc. Lond. **83**, 843—851 (1964).
[Br 64b] BROOKE, G., P. J. HAYMANN, Y. KAMIYA and A. W. WOLFENDALE: Proc. Phys. Soc. Lond. **83**, 853—869 (1964).
[Br 64c] BROOKE, G., M. A. MEYER, and A. W. WOLFENDALE: Proc. Phys. Soc. Lond. **83**, 871—877 (1964).
[Co 61a] COCCONI, G.: Conf. on Theor. Aspects of very high energy phenomena. CERN 61—22, p. 128—150, Genf 1961.
[Co 61c] COCCONI, G.: Handbuch der Physik, Bd. 46/1, edit. by FLÜGGE.
[Da 50a] DAVIS, W. O.: Phys. Rev. **80**, 150—154 (1950).
[Du 62] DUTHIE, J., P. H. FOWLER, A. KADDOURA, D. H. PERKINS e K. PINKAU: Nuovo Cim. **24**, 122 (1962).
[Fl 43a] FLÜGGE, S.: Kap. 14, in: Vorträge über kosmische Strahlung, herausgeg. v. W. HEISENBERG. Berlin: Springer 1943 und engl. Ausgabe, edit. by W. HEISENBERG, Dover 1946.
[Fr 53] FREESE, E., u. P. MEYER: Kap. 14 in: Kosmische Strahlung, herausgeg. von W. HEISENBERG. Berlin-Göttingen-Heidelberg: Springer 1953.
[Gr 64a] GRIGOROV, N. L.: Soviet Physics JETP **18**, 1063—1071 (1964) [J. Exp. Theoret. Phys. (USSR.) **45**, 1544—1557 (1963)].
[Gr 64b] GRIGOROV, N. L.: J. Exp. Theoret. Phys. (USSR.) **45**, 1919—1925 (1963); — Soviet Physics JETP **18**, 1318—1322 (1964).
[Gu 62] GUSEVA, V. V., N. A. DOBROTIN, N. G. ZEVELINSKAYA, K. A. KOTEL'NIKOV, A. M. LEBEDEV i S. A. SLAVATINSKII: Bull. Acad. Sci. USSR. **26**, No. 5, 550—557 (1962). (Columbia Technical Translation.)
[Ha 64b] HAYMES, R. C.: J. Geophys. Res. **69**, 841—852, 853—859 (1964).
[He 59a] HESS, W. N., H. W. PATTERSON, R. WALLACE, and E. L. CHUPP: Phys. Rev. **116**, 445—457 (1959).
[He 61] HESS, W. N., E. H. CANFIELD, and R. E. LINGENFELTER: J. Geophys. Res. **66**, 665—677 (1961).
[ICEF 63] ICEF = International Co-Operative Emulsion Flight. Nuovo Cim., Suppl. **1**, No. 4, 1039—1229 (1963). Zitate entweder ICEF 63, mit Seitenangabe, oder nach Autoren.
[Li 63a] LINGENFELTER, R. E.: Rev. Geophys. **1**, 35—55 (1963).
[Li 63c] LINGENFELTER, R. E.: J. Geophys. Res. **68**, 5633—5639 (1963).
[Lo 61] LOHRMANN, E., M. W. TEUCHER, and M. SCHEIN: Phys. Rev. **122**, 672 (1961).
[Me 63b] MENDELL, R. B., and S. A. KORFF: J. Geophys. Res. **68**, 5487 (1963).
[Mi 57a] MIYAKE, S., K. HINOTANI, and K. NUNOGAKI: J. Phys. Soc. Japan **12**, 113—121 (1957).
[Mi 64] MILES, RALPH F.: J. Geophys. Res. **69**, 1277—1284 (1964).
[Ne 63] NEWKIRK, L. L.: J. Geophys. Res. **68**, 1825 (1963).
[Pa 63b] PAL, Y., u. B. PETERS: Mat. Fys. Medd. **33**, 1—55 (1964).
[Pe 60] PERKINS, D. H.: Progr. in Elementary Particle and Cosmic Ray Physic. vol. 5, Amsterdam: North-Holland Publ. Co. (1960).
[Pe 61] PERKINS, D. H.: CERN 61—22, Int. Conf. on Theoret. Aspects of very High Energy Phenomena 1961.
[Pf 52a] PFOTZER, G.: Naturwissenschaften **39**, 149—158 (1952).
[Pi 64] PINKAU, K.: Fortschr. Phys. **12**, 139—234 (1964).
[Re 62] REIDY, W. P., R. C. HAYMES, and S. A. KORFF: J. Geophys. Res. **67**, 459—465 (1962).
[Si 51b] SIMPSON, J. A.: Phys. Rev. **83**, 1175—1188 (1951).
[So 56] SOBERMANN, R. K.: Phys. Rev. **102**, 1399—1409 (1956).
[Tr 64] TRAINOR, J. H., and J. A. LOCKWOOD: J. Geophys. Res. **69**, 3115—3125 (1964).
[Wi 64] WILLIAMS, D. J., and C. O. BOSTROM: J. Geophys. **69**, 377—391 (1964).
[Yu 51] YUAN, L. C. L.: Phys. Rev. **81**, 175 (1951).

B. Literaturzitate.

[Ad 48] ADAMS, R. V., C. D. ANDERSON, P. E. LLOYD, R. R. RAU, and R. C. SAXENA: Revs. Mod. Phys. **20**, 334 (1948).
[Ad 49] ADDARIO, M., and S. TAMBURINO: Phys. Rev. **76**, 983—984 (1949).
[Ad 50a] ADAMS, N.: Phil. Mag. **41**, 503—505 (1950).
[Ad 50b] ADDARIO, M. M., and S. TAMBURINO: Phys. Rev. **80**, 749 (1950).
[Ag 47] AGNEW, H. M., W. C. BRIGHT, and D. FROMANN: Phys. Rev. **72**, 203 (1947).
[Aj 59] AJZENBERG-SELOVE, F., and T. LAURITSEN: Energy levels of light nuclei. Amsterdam: North-Holland Publ. Co. 1959. (Sonderdruck aus Nucl. Phys. **11**, 1—340 (1959).

[Ak 62]   AKASHI, SHIMIZU, WATANABE, and NISHIMURA: Kyoto Conference. J. Phys. Soc. Japan **17**, Suppl. A **3**, 427 (1962).
[Al 50a] ALFVÉN, H.: Cosmical Electrodynamics. Oxford: Clarendon Press 1950, 2nd edition 1963.
[Al 50b] ALLEN, J. A. VAN, and S. F. SINGER: Phys. Rev. **78**, 819; **80**, 116 (1950).
[Al 51]   ALIKHANOV, A. I., i G. P. ELISEEV: J. Exp. Theoret. Phys. **21**, 1009 (1951).
[Al 53]   ALIKHANOV, A. I., i G. P. ELISEEV: J. Exp. Theoret. Phys. **25**, 368 (1953).
[Al 57]   ALIKHANIAN, A. I., i A. O. VAISENBERG: J. Exp. Theoret. Phys. **32**, 413—416 (1957); — Soviet Physics JETP **5**, 349—351 (1957).
[Al 58]   ALLEN, J. A. VAN, G. H. LUDWIG, E. RAY, and C. E. McILWAIN: U.S. National Academy of Sciences, May 1 (1958).
[Al 59a] ALIKHANOV, A. I., G. P. ELISEEV, V. SH. KAMALYAN, V. A. LYUBIMOV, B. N. MOISEV i A. KHRIMIAN: Soviet Physics JETP **9**, 280 (1959).
[Al 59b] ALLEN, J. A. VAN, C. E. McILWAIN, and G. H. LUDWIG: J. Geophys. Res. **64**, 877 (1959).
[Al 60]   ALIKHANOV, A. I., A. KHRIMYAN, V. K. KOSMACHEVSKII, V. V. AVAKYAN, Y. u. V. GORODKOV, K. SH. EGIYAN i N. A. NALBANDYAN: Proc. Cos. Ray Conf. IUPAP, Moskau, 1960, vol. I, p. 183.
[Al 61]   ALEXANDER, i YEKUTIELI: HFCRL-1089 1961.
[Al 62]   ALBERT, R. D., F. GILBERT, and N. W. HESS: J. Geophys. Res. **67**, 3537 (1962).
[Al 63]   ALY, H. H.: Report to the Bristol Conference on Ultra High Energy Physics, Bristol 1963.
[Al 64]   ALLKOFER, O. C., u. J. TRÜMPER: Z. Naturforsch. **19**a, 1304 (1964).
[Al 65]   ALLKOFER, O. C., e E. KRAFT: Nuovo Cim. **39**, 1051—1056 (1965).
[An 47]   ANDERSON, E. C., W. F. LIBBY, J. WEINHOUSE, A. F. REID, A. D. KIRSHENBAUM, and A. V. GROSSE: Phys. Rev. **72**, 931—936 (1948).
[An 51]   ANDERSON, E. C., and W. F. LIBBY: Phys. Rev. **81**, 64—69 (1951).
[An 53]   ANDERSON, E. C.: Ann. Rev. Nuclear Sci. **2**, 63—78 (1953).
[An 61]   ANDERSON, H. R.: Thesis, Primary cosmic Radiation in 1958 and Variations. California Institute of Technology, 1961 (unpublished).
[An 62]   ANSARI, R.: Nuovo Cim. **23**, (Ser. X), 355—359 (1962).
[Ap 66]   APPARAO, M. V. KRISHNA, R. R. DANIEL, B. VIJAYALAKSHMI, and V. L. BHATT: TIFR-Report N. E. 66—1 (J. Geophys. Res. 1966).
[Ar 61]   ARMSTRONG, A. H., F. B. HARRISON, H. H. HECKMANN, and L. ROSEN: J. Geophys. Res. **66**, 351 (1961).
[As 59]   ASHBY, V. J., and H. C. CATRON: Lawrence Radiation Laboratory (Livermore, Calif.), Rept. UCRL-5419 (Feb. 10, 1959).
[Az 51]   AZIMOV, VISHNEVSKII i KHIL'KO: Dokl. Akad. Nauk S.S.S.R. **78**, 231 (1951).
[Ba 36]   BARTELS, J.: Terr. Magn. **41**, 225—250 (1936).
[Ba 39]   BAGGE, E.: Ann. phys., Lpz. **35**, 11 (1939); — [siehe auch ergänzend: Ann. phys., Lpz. **39**, 515 (1941), Phys. Z. **44**, 401 (1943) und Zusammenfassung in „Naturforschung und Medizin", Bd. 13, S. 29, herausgeg. von W. BOTHE u. S. FLÜGGE. Wiesbaden: Dietrich 1948].
[Ba 43]   BAGGE, E.: Kapitel 13 in: Kosmische Strahlung, herausgeg. von W. HEISENBERG. Berlin: Springer 1943.
[Ba 49]   BAGGE, E., u. K. FINKE: Ann. Physik **6**, 321—337 (1949).
[Ba 50]   BAINBRIDGE, K. T., u. A. O. NIER: Nat. Res. Council. Prelim. Rept. No. 9. Washington, D. C. (1950).
[Ba 51]   BARTON, H. M., E. P. GEORGE, and A. C. JASON: Proc. phys. Soc. Lond. A **64**, 175 (1951).
[Ba 52]   BARRETT, P. H., L. M. BOLLINGER, G. COCCONI, Y. EISENBERG, and K. GREISEN: Revs. Mod. Phys. **24**, 133 (1952).
[Ba 52a] BASSI, P. C. MANDUCHI e P. VERONESI: Nuovo Cim. **9**, 722 (1952).
[Ba 52b] BARFORD, N. C., and G. DAVIS: Proc. Roy. Soc. Lond. Ser. A **214**, 225 (1952).
[Ba 52c] BARBANTI SILVA, L., C. BONACINI, C. DEPIETRI, G. LOVERA e P. PERILLI FEDELI: Nuovo Cim. **9**, 630 (1952).
[Ba 53]   BAGGE, E., u. R. RÖHLER: Physik. Verhandl. **4**, 24 (1953).
[Ba 54a] BALLAM, J., and P. G. LICHTENSTEIN: Phys. Rev. **93**, 851 (1954).
[Ba 54b] BARBOUR, I. G.: Phys. Rev. **93**, 535 (1954).
[Ba 55]   BACCALIN, C., P. BASSI e C. MANDUCHI: Nuovo Cim. **1**, 657 (1955).
[Ba 58]   BAGGE, E., u. S. SKORKA: Z. Physik **152**, 34 (1958).
[Ba 59a] BADALIAN, G. V.: Soviet Physics JETP **8**, 209 (1959); — J. Exp. Theoret. Phys. (USSR.) **35**, 303—305 (1958).
[Ba 59b] BARADZEI, L. T., M. V. SOLOVEV, Z. I. TULINOVA i L. I. FILATOVA: Soviet Physics JETP **9**, 1151 (1959); — J. Exp. Theoret. Phys. (USSR.) **36**, 1617—1620 (1959).

Literatur. 541

[Ba 60] BAME, S. J., R. W. DAVIS, J. P. GLOVE, and F. B. BRUMLEY: Bull. Amer. Phys. Soc. (2) **5**, 360 (1960).
[Ba 61] BABECKI, J., Z. BUJA, N. L. GRIGOROV, J. LOSKIEWICZ, J. MASSALSKI, A. OLES i V. YA. SHESTOPEROV: J. Exp. Theoret. Phys. **40**, 1551 (1961); — Soviet Physics JETP **13**, 1089 (1961).
[Ba 62b] BARADZEI, L. T., V. I. RUBTSOV, YU. A. SMORODIN, M. V. SOLOV'EV i B. V. TOLKACHEV: Bull. Acad. Sci. USSR., **26**, No. 5 573—583 (1962) (Columbia Technical Translation).
[Ba 63a] BARADZEI, L. T., V. I. RUBTSOV, YU. A. SMORODIN, M. V. SOLOV'EV i B. V. TOLKACHEV, Veröffentl. (TRUDY) des P. N. LEBEDEV — Phys. Inst. **26** (1963) u. J. Phys. Soc. Japan **17**, Suppl. A III, 433—439 (1962).
[Ba 63b] BAME, S. J., J. P. CONNER, F. B. BRUMLEY, R. L. HOSTETLER, and A. C. GREEN: J. Coophys. Res. **68**, 1221—1228 (1963).
[Ba 64a] BABAYAN, KH. P., N. G. BOYADZHYAN, E. A. MAMIDZHANYAN, N. L. GRIGOROV, CH. A. TRET'YAKOVA i V. YA. SHESTOPEROV: J. Exp. Theoret. Phys. (USSR.) **46**, 110—122 (1964); — Soviet Physics JETP **19**, 80—88, 1032—1041 (1964).
[Ba 64b] BACHELET, F., P. BALATA, E. DYRING e N. JUCCI: Nuov. Cim. **31**, 1126—1130 (1964).
[Be 48a] BERNARDINI, G., G. CORTINI e A. MANFREDINI: Nuovo Cim. **5**, 511 (1948).
[Be 48b] BERNARDINI, G., G. CORTINI, and A. MANFREDINI: Phys. Rev. **74**, 845 (1948).
[Be 49] BERNARDINI, G., G. CORTINI, and A. MANFREDINI: Phys. Rev. **76**, 1792 (1949).
[Be 50a] BERNARDINI, G., E. T. BOOTH, and S. J. LINDENBAUM: Phys. Rev. **80**, 905 (1950).
[Be 50b] BERNARDINI, G., G. CORTINI, and A. MANFREDINI: Phys. Rev. **79**, 952 (1950).
[Be 51] BELLIBONI, G., L. FABRICHESI, L. DE MARCO e M. MERLIN: Nuovo Cim. **8**, 574 (1951).
[Be 52a] BERNARDINI, G., E. T. BOOTH, and S. J. LINDENBAUM: Phys. Rev. **85**, 826 (1952).
[Be 52b] BERNARDINI, G., E. T. BOOTH, and S. J. LINDENBAUM: Phys. Rev. **88**, 1017 (1952).
[Be 55] BECKER, J., D. P. CHANSON, E. NAGEOTTI et P. TREILLE: J. Phys. Radium **16**, 191 (1955).
[Be 56] BENIOFF, P. A.: Phys. Rev. **104**, 1122—1130 (1956).
[Be 60] BERCOVITCH, M., H. CARMICHAEL, G. C. HANNA, and E. P. HINCKS: Phys. Rev. **119**, 412—431 (1960).
[Bi 51] BIERMANN, L., O. HAXEL u. A. SCHLÜTER: Z. Naturforsch. **6a**, 47 (1951).
[Bi 52] BIRNBAUM, M., M. M. SHAPIRO, and B. STILLER: Phys. Rev. **86**, 86 (1952).
[Bi 55] BISWAS, S., B. PETERS, and RAMA: Proc. Ind. Acad. Sci. A **41**, 154 (1955).
[Bi 63] BIEN, G. S., N. W. RAKESTRAW, and H. E. SUESS: Radioactive dating, Vienna, IAEA, pp. 159—173 (1963).
[Bl 37a] BLAU, M., u. H. WAMBACHER: Sitzber. Akad. Wiss. Wien **146**, 469, 623 (1937).
[Bl 37b] BLAU, M., and H. WAMBACHER: Nature, Lond. **140**, 585 (1937).
[Bl 56] BLOKHINTSEV, D. I.: CERN-Symposium on High-Energy Accelerators and Pion Physics, vol. 2, Genf 1956.
[Bo 49] BONETTI, A., and C. DILWORTH: Phil. Mag. **40**, 585 (1949).
[Bo 51] BOGGILD, J., and F. TENNY: Phys. Rev. **84**, 1070 (1951).
[Bo 62a] BOZOKI, G., E. FENYVES, T. SANDOR, O. BALEA, M. BATAGUI, E. FRIEDLAENDER, B. BETEV, S. KAVLAKOV e L. MITRANI: Soviet Physics JETP **14**, 743—744 (1962); — J. Exp. Theoret. Phys. (USSR.) **41**, 1043—1045 (1961). — BOZOKI, G., E. FENYVES, and L. JANOSSY: Nuclear Phys. **33**, 236 (1962).
[Bo 62b] BOWLER, J., J. DUTHIE, P. H. FOWLER, A. KADDOURA, D. H. PERKINS, K. PINKAU, and W. WOLTER: J. Phys. Soc. Japan **17**, Suppl. A III, 424 (1962).
[Bo 63] BOELLA, G., G. DEGLI ANTONI, G. DILWORTH, G. CIANELLI, E. ROCCA, L. SCARSI e D. SHAPIRO: Nuovo Cim. **29**, 103—107 (1963).
[Bo 65] BOELLA, G., G. DEGLI ANTONI, C. DILWORTH, M. PANETTI, L. SCARSI, and D. S. INTRILIGATOR: J. Geophys. Res. **70**, 1019—1030 (1965).
[Br 47] BRIDGE, H., B. ROSSI, and R. WILLIAMS: Phys. Rev. **72**, 257 (1947).
[Br 49] BROWN, R. H., U. CAMERINI, P. H. FOWLER, H. HEITLER, D. T. KING, and C. F. POWELL: Phil. Mag. **40**, 862—881 (1949).
[Br 54] BROWN, W. W.: Phys. Rev. **93**, 528 (1954).
[Br 56] BRISBOUT, F. A., C. DAHANAYAKE, A. ENGLER, P. H. FOWLER e P. B. JONES: Nuovo Cim. **3**, 1400 (1956).
[Br 57] BROWN, R. R.: Nuovo Cim. **6**, 956 (1957).
[Br 58] BRUNBERG, E. Å.: Ark. Fysik **14**, 195 (1958).
[Br 62] BROOKE, G., P. J. HAYMAN, F. E. TAYLOR u. A. W. WOLFENDALE: J. Phys. Soc. Japan **17**, Suppl. A III-5-5; 311—312 (1962).
[Bu 53] BUDINI, P., u. G. MOLIERE: In: W. HEISENBERG, Kosmische Strahlung, S. 365. Berlin-Göttingen-Heidelberg: Springer 1953.
[Bu 56] BURCKHARDT, C.: Helv. phys. Acta **29**, 533 (1956).
[Bu 57] BULLOCK, F. W.: Proc. phys. Soc. Lond. A **70**, 134 (1957).

[Bu 58]   Burgy, M. T., V. E. Kohn, T. B. Novey, G. R. Ringo, and V. L. Telegdi: Phys. Rev. **110**, 1214 (1958); — Phys. Rev. Letters **1**, 324 (1958).
[Ca 49]   Camerini, N., T. Coor, J. H. Davies, P. H. Fowler, W. O. Lock, H. Muirhead, and N. Tobin: Phil. Mag. **40**, 1073 (1949).
[Ca 50]   Camerini, U., P. H. Fowler, W. O. Lock, and H. Muirhead: Phil. Mag. **41**, 413 (1950).
[Ca 51]   Camerini, N., J. H. Davies, P. H. Fowler, C. Franzinetti, H. Muirhead, W. O. Lock, D. H. Perkins, and G. Yekutieli: Phil. Mag. **42**, 1241 (1951).
[Ca 52]   Camerini, U., W. O. Lock, D. H. Perkins: In: Progress in Elementary Particles and Cosmic-Ray Physics, ed. by J. G. Wilson u. S. A. Wouthuysen, Vol. I, p. 1 ff. Amsterdam: North-Holland Publ. Co. 1952.
[Ca 53]   Castagnoli, C., G. Cortini, C. Franzinetti, A. Manfredini e D. Moreno: Nuovo Cim. **10**, 1539 (1953).
[Ca 55]   Carlson, B. G.: LA 1891, Los Alamos Sci. Lab., Los Alamos, New Mexico, 28 pp. (1955).
[Ca 58a] Carlson, B. G.: LAMS 2201, Los Alamos Sci. Lab., Los Alamos, New Mexico, 62 pp. (1958).
[Ca 58b] Carlson, B. G., and G. I. Bell: Proc. Intern. Conf. Peaceful Uses Atomic Energy; 2nd. Geneva, 1958.
[Ca 59]   Carlson, B. G.: LA 2260, Los Alamos Sci. Lab., Los Alamos, New Mexico, 30 pp. (1959).
[Ca 61]   Canfield, E. H., R. N. Stuart, R. P. Freis, and W. H. Collins: Sophist I — Lawrence Radiation Lab. (Livermore, Calif.) Rept. UCRL-5956, Oktober 19, 1961.
[Ca 64]   Carmichael, H.: IQSY Instruction manual No. 7. Cosmic Rays (published by the IQSY Secretariat, London).
[Ce 53]   Ceccarelli, M., e G. T. Zorn: Nuovo Cim. **10**, 540 (1953).
[Ch 40]   Chapmann, S., and I. Bartels: Geomagnetism. 2 Vols.. 1126 pp. Oxford: Oxford University press 1940, second ed. 1952.
[Ch 56]   Chou, C. N.: Phys. Rev. **102**, 848 (1956).
[Ch 59]   Sen Chaudhuri, P. K., u. G. Pfotzer: Z. Naturforsch. **14**A, 10 (1959).
[Ch 61]   Chase, L. F., G. R. Johnson, R. V. Smith, F. J. Vaughn, and M. Walt: Lockheed Aircraft Corp., Missiles and Space Division, Rept. LMSD-895076; Februar 1961.
[Ci 61]   CIRA (Cospar International Reference Atmosphere). Amsterdam: North-Holland Publ. Co. 1961; s. auch [US 62].
[Ci 66]   Cindro, N.: Revs. Mod. Phys. **38**, 391—446 (1966).
[Cl 50]   MaClure, G. W., and M. A. Pomerantz: Phys. Rev. **79**, 911 (1950).
[Cl 52]   Clark, M. A.: Phys. Rev. **87**, 687 (1952).
[Cl 53]   Clark, G. W.: Phys. Rev. **90**, 368 (1953).
[Cl 57]   Clark, G., J. Earl, W. Kraushaar, J. Linsley, B. Rossi, and F. Scherb: Nature **180**, 353 (1957).
[Cl 58a] Clark, G., J. Earl, W. Kraushaar, J. Linsley, B. Rossi, and F. Scherb: Nuovo Cim., Suppl. **8**, (Serie X), 623—652 (1958).
[Cl 58b] Clark, M. A., J. M. Robinson, and R. Nathans: Phys. Rev. Letters **1**, 100 (1958).
[Cl 61]   Clark, G. W., J. Earl, W. L. Kraushaar, J. Linsley, B. Rossi, F. Scherb, and D. W. Scott: Phys. Rev. **122**, 637 (1961), and Kyoto-Conf. J. Phys. Soc. Japan **17**, Suppl. A III (1962).
[Cl 63]   Clark, G., H. Bradt, M. La Pointe, V. Domingo, I. Escobar, K. Murakami, K. Suga, Y. Toyoda, and J. Hersil: Int. Conf. Cosmic Rays Jaipur 1963, vol. 4, p. 65—78.
[Co 50]   Conversi, M.: Phys. Rev. **79**, 749 (1950).
[Co 53]   Conversi, M., G. Martelli e P. Rothwell: Nuovo Cim. **10**, (Ser. IX), 898 (1953).
[Co 54]   Conversi, M., e P. Rothwell: Nuovo Cimento **12**, 191—210 (1954).
[Co 55]   Cohen, E. R.: Geneva Conference 1955, Paper P/611.
[Co 56]   McCormack, P. D.: Proc. phys. Soc. Lond. A **69**, 845 (1956).
[Co 60]   Cocconi, G.: Proc. 1960 Annual Intern. Conf. High Energy Physics, Rochester, p. 799. New York: Univ. Rochester/Interscience 1960.
[Co 61b] Cocconi, G., A. Koester, and D. H. Perkins: Lawrence Rad. Lab., UCID-1444 (1961).
[Cr 52]   Crussard, J.: Thèse, Faculté des Sciences de l'Université de Paris 1952.
[Cr 57]   Craig, H.: Tellus **9**, 1—17 (1957).
[Da 50b] Dallaporta, N., M. Merlin e G. Puppi: Nuovo Cim. **7**, 99 (1950).
[Da 52]   Davis, G.: Phil. Mag. **43**, 472 (1952).
[Da 53]   Davis, L. R., H. M. Caulk, and C. Y. Johnson: Phys. Rev. **91**, 431 (1933).
[Da 56a] Davis, L. R., H. M. Caulk, and C. Y. Johnson: Phys. Rev. **101**, 800 (1956).
[Da 56b] Danielson, R. E., P. S. Freier, J. E. Naugle, and E. P. Ney: Phys. Rev. **103**, 1075 (1956).

[Da 57]	DAVISON, B., and J. B. SYKES: Neutron Transport Theory. Oxford: Clarendon Press 1957.
[Da 64]	DAGAN, S., and G. YEKUTIELI: Proc. 8. Int. Conf. Cos. Rays 1963, Vol. 5, p. 460—462. Jaipur 1964.
[Da 64]	DANILOVA, V., E. V. DENISON i S. I. NIKOLS'KII: Soviet Physics JETP **19**, 1057—1066 (1964); — J. Exp. Theor. Phys. (USSR.) **46**, 1561—1577 (1964).
[De 54]	DEUTSCHMANN, M.: Z. Naturforsch. **9a**, 477 (1954).
[De 60a]	DEDENKO, L. G., and G.T. ZATSEPIN: Proc. Internat. Conf. Cosmic Rays, Moskau 1959.
[De 60b]	DEDENKO, L. G.: Soviet Physics JETP **13**, 439 (1961), aus J. Exp. Theoret. Phys. **40**, 630 (1960).
[De 61]	DENISOV, E. V., V. I. ZATSEPIN, S. I. NIKOLS'KII, A. A. POMANSKII, B. V. SUBBOTIN, E. I. TUKISH i V. A. YAKOVLEV: Soviet Physics JETP **13**, 287 (1961).
[Di 53a]	DILWORTH, C. C., S. J. GOLDSACK, T. F. HUANG e L. SCARSI: Nuovo Cim. **10**, 1261 (1953).
[Di 53b]	DILWORTH, C. C., e S. J. GOLDSACK: Nuovo Cim. **10**, 926 (1953).
[Di 55]	DIXIT, K. R.: Z. Naturforsch. **10a**, 339 (1955).
[Dm 59]	DMITRIEV, V. A., G. V. KULIKOV i G. B. KHRISTIANSON: Soviet Physics JETP **10**, 637 (1960), aus J. Exp. Theoret. Phys. **37**, 833 (1959).
[Du 58]	DUNGEY, J. W.: Cosmic Electrodynamics. Cambridge: Cambridge University press 1958.
[Du 61]	DUTHIE, J., C. FISHER, P. FOWLER, A. KADDOURA, D. H. PERKINS, K. PINKAU, and W. WOLTER: Phil. Mag. **6**, 89 (1961).
[Dy 62]	DYRING, E.: Tellus **14**, 33 (1962).
[Ed 59]	EDGE, R. D.: Nuclear Phys. **12**, 182—189 (1959).
[El 56]	ELSASSER,W., E. P. NEY, and J. R.WINCKLER: Nature, Lond. **178**, 1226—1227 (1956).
[En 54b]	ENGLER, A., U. HABER-SCHAIM, and W. WINKLER: Nuovo Cim. **12**, (Serie IX), 930 (1954).
[Fa 55]	FAY, H.: Z. Naturforsch. **10a**, 572 (1955).
[Fa 57]	FARROW, L. A.: Phys. Rev. **107**, 1687 (1957).
[Fa 60]	FAN, C. Y., P. MEYER, and J. A. SIMPSON: Phys. Rev. Letters **4**, 421 (1960); — **5**, 269 (1960).
[Fe 34]	FERMI, E.: Z. Physik **88**, 161 (1934).
[Fe 50]	FERMI, E.: Nuclear physics, p. 221. Chicago: Chicago University Press 1950.
[Fe 58]	FENTON, A. G., K. B. FENTON, and D. C. ROSE: Can. J. Phys. **36**, 824—839 (1958).
[Fe 62]	FERRARI, E., and F. SELLERI: Nuovo Cim., Suppl. **24** (Ser. X), 453—515 (1962).
[Fe 63]	FERGUSSON, G. J.: Geofis. Pura Appl. (1963).
[Fi 39]	FILIPOV, A., A. IDANOV i. I. GUREVICH: J. Exp. Theoret. Phys. **1**, 51 (1939).
[Fi 55]	FILTHUTH, H.: Z. Naturforsch. **10a**, 219 (1955).
[Fi 58]	FIDECARO, M. C., G. FIDECARO, G. MARINI e L. MEZZETTI: Nuovo Cim. **9**, 37 (1958).
[Fi 62]	FIELDHOUSE, P., E. B. HUGHES, and P. L. MARSDEN: J. Phys. Soc. Japan **17**, Suppl. A II, 518 (1962).
[Fl 43b]	FLÜGGE, S.: Z. Physik **121**, 298 (1943).
[Fl 59]	FLEROV, G. N., V. I. KALASHNIKOVA, A. V. PODGURSKAYA, E. D. VOROBEV i. G. A. STOLYAROV: Soviet Physics JETP **9**, 511—515 (1959).
[Fo 50]	FOWLER, P. H.: Phil. Mag. **41**, 169 (1950).
[Fo 53]	FONGER, W. H., J. W. FIROR, and J. A. SIMPSON: Phys. Rev. **89**, 891 (1953).
[Fo 54]	FORBUSH, S. E.: J. Geophys. Res. **59**, 525—542 (1954).
[Fr 59]	FREDEN, S. C., and R. ST. WHITE: Phys. Rev. Letters **3**, 9, 145 (1959).
[Fr 60]	FREDEN, S. C., and R. ST. WHITE: J. Geophys. Res. **65**, 1377 (1960).
[Fr 62]	FREDEN, S. C., and R. ST. WHITE: J. Geophys. Res. **67**, 25—29 (1962).
[Fr 63]	FREIER, P. S., and W. R. WEBBER: J. Geophys. Res. **68**, 1605 (1963).
[Fro 38]	FROMAN, D. K., and J. C. STEARNES: Phys. Rev. **54**, 969 (1938).
[Fü 37]	FÜNFER, E.: Naturwissenschaften **25**, 235 (1937).
[Fü 38]	FÜNFER, E.: Z. Physik **111**, 351 (1938).
[Fu 52]	FUJIMOTO, Y., and T. TAMURA: Prog. Theor. Phys. (Kyoto) **8**, 221—230 (1952).
[Ga 53]	GALLI, M.: Nuovo Cim. **10**, 1187—1195 (1953).
[Ga 58]	GABBE, J. D.: Phys. Rev. **112**, 497 (1958).
[Ga 64]	GAUGER, J.: J. Geophys. Res. **69**, 2209—2222 (1964).
[Ga 65]	GAUGER, J.: J. Geophys. Res. **70**, 3571—3574 (1965).
[Ge 49]	GEORGE, E. P., and A. C. JASON: Proc. Phys. Soc. Lond. A **62**, 243 (1949).
[Ge 50a]	GEORGE, E. P., and A. JASON: Proc. Phys. Soc. Lond. A **63**, 1081 (1950).
[Ge 50b]	GEORGE, E. P., and J. EVANS: Proc. Phys. Soc. Lond. A **63**, 1248 (1950) und A **64**, 193 (1950).
[Ge 51]	GERMAIN, L.: Phys. Rev. **82**, 596 (1951).
[Gl 52]	GLASSTONE, S., and M. C. EDLUND: The Elements of Nuclear Reactor Theory. Princeton: (N.J.): Princeton University Press 1952.

[Go 51] GOLDWASSER, E. L., and T. C. MERKLE: Phys. Rev. **83**, 43 (1951).
[Go 53] GOTTSTEIN, K.: In: W. HEISENBERG, Kosmische Strahlung, S. 180ff. Berlin-Göttingen-Heidelberg: Springer 1953.
[Gr 51a] GREGORY, B. P., and J. H. TINLOT: Phys. Rev. **81**, 667 (1951).
[Gr 51b] GREGORY, B. P., and J. H. TINLOT: Phys. Rev. **81**, 675 (1951).
[Gr 53a] GREISEN, K., and W. D. WALKER: Phys. Rev. **90**, 915 (1953).
[Gr 53b] GRILLI, M., e B. VITALE: Nuovo Cim. **10**, 1047 (1953).
[Gr 53c] GRAVES, E. R., and L. ROSEN: Phys. Rev. **89**, 343 (1953).
[Gr 54] GRAMENITSKII, I. M., G. S. EMELIANOV i M. I. PODGORETSKII: J. Exp. Theoret. Phys. **27**, 654 (1954).
[Gr 56a] GRAMENITSKII, I. M., M. I. PODGORETSKII i I. F. SHARAPOVA: Soviet Physics JETP **3**, 320 (1956), aus J. Exp. Theoret. Phys. **30**, 277 (1956).
[Gr 56b] GREISEN, K.: In: Progress in Cosmic Ray Physics, ed. by J. G. WILSON and S. A. WOUTHUYSEN, vol. III, p. 134. Amsterdam: North Holland Publ. Co. 1956.
[Gr 56c] GROSS, E.: Lawrence Radiation Lab. (Berkeley Calif.) Rept. UCRL-3330 (Feb. 29. 1956) (OTS).
[Gr 58] GRIGOROV, N. L., V. S. MURZIN i I. D. RAPOPORT: Soviet Physics JETP **7**, 348 (1958); — J. Exp. Theoret. Phys. **34**, 506 (1958).
[Gr 59] GRIGOROV, N. L., A. V. PODGURSKAYA, A. I. SAVELEVA i L. M. POPEREPOVA: Soviet Physics JETP **8**, 1 (1959).
[Gr 60a] GRIGOROV, N. L., M. A. KONDRAT'EVA, A. I. SAVEL'EVA, V. A. SOBINYAKOV, A. V. PODGURSKAYA i V. YA. SHESTOPEROV: Intern. Conf. on Cosmic Rays Moskau 1959, vol. 1, p. 122. Moskau 1960.
[Gr 60b] GRIGOROV, N. L., V. V. GUSEVA, N. A. DOBROTIN, K. A. KOTELNIKOV, V. S. MURZIN, S. V. RYABIKOV, S. A. SLAVATINSKY: Proc. Moscow Cos. Ray Conf. 1959, vol. I, p. 143, Moskau (1960).
[Gr 60] GREISEN, K.: Ann. Rev. Nucl. Sci. **10**, 63 (1960).
[Gr 63] GRIGOROV, N. L., I. N. EROFEEVA, V. S. MURZIN, L. G. MISHCHENKO, I. D. RAPOPORT, B. O. ROSTOMYAN, V. A. SOBINYAKOV i A. F. TITENKOV: Soviet Physics JETP **17**, 1213—1216 (1963).
[Gr 65a] GREENHILL, J. G., J. PHILLIPS, K. B. FENTON, A. G. FENTON, and M. BOWTHORPE: Preprint 9. Internat. Conf. Cosmic Rays London, 1965; Proceedings, London. (1966).
[Gr 65b] GRIGOROV, N. L., V. E. NESTEROV, I. D. RAPOPORT, I. A. SAVENKO, and G. A. SKURIDIN: Preprint 9. Internat. Cosmic Ray Conf. London 1965; Proceedings London 1966.
[Gu 65] GUSEVA, V. V., E. V. DENISOV, N. A. DOBROTIN, S. A. DUBROVINA, D. V. EMELJANOV, N. G. ZELEVINSKAYA, B. G. INATIJEVA, K. A. KOTELNIKOV, A. M. LEBEDEV, V. M. MAXIMENKO, E. A. MOROSOV, A. G. NOVIKOV, V. S. PUCHKOV, D. F. RAKITIN, O. F. OGURCHOV, I. N. FETISOV, N. E. CHROMICH, S. A. SLAVATINSKY, V. V. SOKOLOVSKY, and B. I. TITOV: Proc. 9. Internat. Conf. Cosmic Rays London 1965 (1966).
[Ha 39] HALBAN, H. v., M. MAGART et L. KOWARSKI: C. R. Acad. Sci. Paris **208**, 572 (1959).
[Ha 49a] HAZEN, W. E., C. A. RANDALL, and O. L. TIFFANY: Phys. Rev. **75**, 694 (1949).
[Ha 49b] HARDING, J. B., S. LATTIMORE, and D. H. PERKINS: Proc. roy Soc. Lond. A **196**, 325 (1949).
[Ha 49c] HARDING, J. B.: Nature, Lond. **163**, 440 (1949).
[Ha 52] HAVENS, R. J., R. T. KOLL, and H. E. LABON: J. Geophys. Res. **57**, 59 (1952).
[Ha 53] HAGEDORN, R., u. W. MACKE: In: Kosmische Strahlung, herausgeg. von W. HEISENBERG, S. 201ff. Berlin-Göttingen-Heidelberg: Springer 1953.
[Ha 59] HAYMES, R. C.: Phys. Rev. **116**, 1231—1237 (1959).
[Ha 60] HANSEN, L. F., and W. B. FRETTER: Phys. Rev. **118**, 812 (1960).
[Ha 60a] HAYMES, R. C., and S. A. KORFF: Phys. Rev. **120**, 1460—1462 (1960).
[Ha 60b] HAUSER, I.: Ann. Phys. **5**, 327—334 (1960).
[Ha 60c] HAUSER, I., P. LANDROCK, K. LANIUS, L. MITRANI u. A. PEEVA: Ann. Phys. **5**, 335—338 (1960).
[Ha 62] HAYMAN, P. J., and A. W. WOLFENDALE: Proc. Phys. Soc. Lond. **80**, 710 (1962).
[Ha 64a] HATTON, C. J., and H. CARMICHAEL: Canad. J. Phys. **42**, 2443 (1964).
[Hae 64] HAERENDEL, G.: Fortschr. Physik **12**, 273—344 (1964).
[He 39] HEITLER, W., C. F. POWELL, and G. E. F. FERTEL: Nature, Lond. **144**, 283 (1939).
[He 52] HEISENBERG, W.: Z. Physik **133**, 65 (1952).
[He 53] HEISENBERG, W.: Kosmische Strahlung, S. 148. Berlin-Göttingen-Heidelberg: Springer 1953.
[He 56] HESS, W. N., and MOYER: Phys. Rev. **101**, 337 (1956).
[He 57] HESS, W. N., H. W. PATTERSON, and R. WALLACE: Nucleonics **15**, 74 (1957).
[He 59] HESS, W. N.: Phys. Rev. Letters **3**, 11, 145 (1959).
[He 60] HESS, W. N., and A. K. STARNES: Phys. Rev. Letters **5**, 48—50 (1960).

[He 63]   Hess, W. N., and R. C. Kiefer: Trans. Am. Geophys. Union **44**, 83 (1963).
[He 64]   Hess, W. N., C. Curry, and E. L. Chupp: (to be published) (1964) preprint.
[Hö 55]   Höcker, K. H., H. Kuhn u. M. Ritzi: Z. Naturforsch. **10**a, 386 (1955).
[Ho 49]   Hodgson, P. E., and D. H. Perkins: Nature, Lond. **163**, 439 (1949).
[Ho 51a] Hodgson, P. E.: Phil. Mag. **42**, 82 (1951).
[Ho 52a] Hogrebe, H.: Z. Naturforsch. **7**a, 772 (1952).
[Ho 52b] Hornbostel, J., u. E. O. Salant: Zit. bei [Bi 52].
[Ho 52c] Hodson, L.: Proc. phys. Soc. London A **65**, 702 (1952).
[Ho 54c] Hones, E.: Phys. Rev. **83**, 1263 (1951) u. zit. bei [Ba 54b].
[Ho 58]   Howerton, R. I.: Lawrence Radiation Laboratory (Livermore Calif.) Rept. UCRL-5351, November 1958.
[Ho 59]   Howerton, R. I.: Lawrence Ladiation Lab. (Livermore, Calif.) Rept-UCRL-5226-revised October 1959.
[Ho 63]   Hove. L. van: Nuovo Cim. **28**, 798 (1963).
[Hu 55]   Hughes, D. I., and J. A. Harvey: Brookhaven Nat. Lab. Report BNL 325. Washington D.C.: U.S. Government Printing Office 1955.
[Hu 56]   Hurwitz jr., H. H., M. S. Nelkin, and G. J. Habetler: Nuclear Sci. Engng. **1**, 280 (1956).
[Hu 58]   Hughes, D. J., and R. B. Schwartz: Brookhaven Nat. Lab. Report BNL-325 (July 1, 1958). Suppl. I siehe: Hughes et al. [Hu 60]; Suppl. II siehe: Stehn et al. [St 64].
[Hu 60]   Hughes, D. J., B. A. Magurno, and M. K. Brüssel: BNL-325, 2nd edit., Suppl. I. Brookhaven Nat. Lab., Upton, N.Y. 1960.
[Hu 62]   Hughes, E. B., P. L. Marsden, M. R. Meyer, and A. W. Wolfendale: J. Phys. Soc. Japan, Suppl. A-II 516 und Proc. Phys. Soc. Lond. **83**, 253 (1962).
[Hu 64]   Hughes, E. B., P. L. Marsden, G. Brooke, M. A. Meyer, and A. W. Wolfendale: Proc. Phys. Soc. Lond. **83**, 239—251 (1964).
[Jo 55]   Johnson, R. W.: Thesis (unpublished). University of California Radiation Laboratory Repost UCRL-2979, May, 1955.
[Jo 57]   Jongejans, B., F. Schurink u. D. J. Holthuizen: Physica, Haag **23**, 164 (1957).
[Ka 53]   Kaplon, M. F., J. Z. Klose, D. M. Ritson, and W. D. Walker: Phys. Rev. **91**, 1573 (1953).
[Ka 63]   Kastner, J., B. G. Oltmann, and L. D. Marinelli: Intern. Symp. Nat. Radiation Environment, Rice University, Houston, Texas (April 1963).
[Ke 58]   Kellog, P. J.: Nuovo Cim. **11**, 48 (1959).
[Ke 66]   Kellermann, E. W., and M. G. Silk: Proc. 9. Cosm. Ray Conf. 1965, Vol. 2, p. 883, London 1966.
[Ki 56]   Kim, J. B.: Phys. Rev. **102**, 882 (1956).
[Ki 64]   Kim, C. O., Phys. Rev. **136**, B 515 (1964).
[Ko 39a] Korff, S. A.: Phys. Rev. **56**, 210 (1939).
[Ko 39b] Korff, S. A., and W. E. Danforth: Phys. Rev. **55**, 980 (1939).
[Ko 39c] Korff, S. A.: Rev. Mod. Phys. **11**, 211 (1939).
[Ko 42]   Korff, S. A.: Terr. Mag. **45**, 133 (1940).
[Ko 48]   Korff, S. A.: Rev. Mod. Phys. **20**, 327 (1948).
[Ko 52]   Kouts, H. J., and L. C. L. Yuan: Phys. Rev. **86**, 128—129 (1952).
[Ko 55]   Kocharyan, N. M.: Aus J. Exp. Theoret. Phys. **28**, 160 (1955) in Sov. Phys. JETP **1**, 128 (1955).
[Ko 56a] Koba, Z.: Proc. 6. Rochester. Conf. High Energy Nucl. Phys., Sess. IV, p. 46 (1956).
[Ko 56b] Kocharian, N. M., M. T. Aivazian, J. A. Kirakosian i A. S. Aleksanian: Aus J. Exp. Theoret. Phys. **30**, 243—247 (1956) in Sov. Phys. JETP **3**, 350 (1956).
[Ko 58]   Korff, S. A.: Nuovo Cim. **8**, 796—800 (1958).
[Ko 59]   Kocharian, N. M., G. S. Saakian i Z. A. Kirakosian: Soviet Physics JETP **8**, 933 (1959) aus J. Exp. Theoret. Phys. **35**, 1335—1349 (1958).
[Ko 60]   Korff, S. A., and R. C. Haymes: J. Geophys. **65**, 3163 (1960).
[Ko 64]   Korff, S. A.: Proc. Jaipur Conf. **5**, 255—260 (1964).
[Ko 65]   Korff, S. A., R. B. Mendell, and St. S. Holt: Proc. 9. Cos. Ray Conf. 1965, Vol. 1, p. 573, London 1966.
[Ku 52]   Kuiper, G. P. (Editor): The Atmosphere of the Earth and Planets, 2nd ed., p. 1. Chicago (Ill.): Chicago University Press 1952.
[Ku 54a] Kuhn, G., u. E. Schopper: Z. Naturforsch. **9**a, 851 (1954).
[Ku 54b] Kuhn, G.: Diss. Stuttgart 1954.
[La 46]   Lapp, R. E.: Phys. Rev. **69**, 321 (1946).
[La 49]   Lattimore, S.: Phil. Mag. **40**, 394 (1949).
[La 52a] Ladenburg, R.: Phys. Rev. **86**, 128 (1952).
[La 52b] Landolt-Björnstein: Bd. III, Astronomie und Geophysik, herausgeg. von J. Bartels und P. ten Grüggengate. Berlin-Göttingen-Heidelberg: Springer 1952.

[La 53]   LAL, D.: Proc. Ind. Acad. Sci. A **38**, 93 (1953).
[La 58]   LAL, D.: Investigations of nuclear interactions produced by cosmic rays. Bombay University, Bombay, India 1958 (unpublished thesis).
[La 62]   LAL, D., and B. PETERS: In: Progr. Cosmic Ray Phys., Vol. VI. Amsterdam: North-Holland Publ. 1962.
[La 62a] LAL, S., Y. PAL, and R. RAGHAVAN (Kyoto): J. Phys. Soc. Japan, Suppl. A III, **17**, 393 (1962a).
[La 62b] LAL, S., R. RAGHAVAN, B. V. SREEKANTAN, A. SUBRAHMANYAN, and S. D. VERMA (Kyoto): J. Phys. Soc. Japan, Suppl. A III, **17**, 390 (1962a).
[Le 50]   LeCOUTEUR, K. I.: Proc. Phys. Soc. Lond. A **63**, 259 (1950).
[Le 52]   LeCOUTEUR, K. I.: Proc. Phys. Soc. Lond. A **65**, 718 (1952).
[Le 62a] LENCHEK, A. M., and S. F. SINGER: J. Geophys. Res. **67**, 1263 (1962).
[Le 62b] LENCHEK, A. M.: J. Geophys. Res. **67**, 2145 (1962).
[Li 46]   LIBBY, W. F.: Phys. Rev. **69**, 671—672 (1946).
[Li 50]   LI, T. T.: Phil. Mag. **41**, 1152 (1950).
[Li 54c] LINDENBERGER, K. H., u. P. MEYER: Z. Physik **139**, 372 (1954).
[Li 61a] LINDERN, G. v.: Dissertation Ludwigs-Maximilian-Universität München 1961.
[Li 61b] LINDERN, G. v.: CERN-Report 1961.
[Li 62]   LINSLEY, J., L. SCARSI, and BR. B. ROSSI: Proc. Int. Conf. on Cosmic Rays and Earth Storm, Kyoto. J. Phys. Soc. Japan, Suppl. A III, **17**, 393 (1962).
[Li 63b] LINGENFELTER, R. E., and E. J. FLAMM: J. Atmosph. Sci. **21**, 134—140 (1963).
[Li 63d] LINGENFELTER, R. E.: Proc. A.P.L. John Hopkins Conf. on Albedo Neutrons, 1963, p. 15—31.
[Li 63e] LINSLEY, J.: Int. Conf. Cosmic Rays, Jaipur 1963, vol. 4, 77—86.
[Li 64]   LINGENFELTER, R. E., and E. J. FLAMM: J. Geophys. Res. **69**, 2199—2207 (1964).
[Li 65]   LINGENFELTER, R. E., E. J. FLAMM, E. H. CANFIELD, and S. KELLMAN: J. Geophys. Res. **70**, 4077—4086, 4087—4095 (1965).
[Lo 33]   LOCHER, G. L.: Phys. Rev. **44**, 774 (1933).
[Lo 34]   LOCHER, G. L.: Phys. Rev. **45**, 296 (1934).
[Lo 36]   LOCHER, G. L.: Phys. Rev. **50**, 394 (1936).
[Lo 37]   LOCHER, G. L.: J. Franklin Inst. **224**, 555 (1937).
[Lo 49]   LORD, J. J., and M. SCHEIN: Phys. Rev. **75**, 1956, 1957 (1949).
[Lo 50]   LORD, J. J., and M. SCHEIN: Phys. Rev. **77**, 19 (1950).
[Lo 51]   LORD, J. J.: Phys. Rev. **81**, 901 (1951).
[Lo 54]   LOVATI, A., A. MURA, C. SUCCI e G. TAGLIAFERRI: Nuovo Cim. **12**, 526 (1954).
[Lo 55a] LOCK, W. O., P. V. MARCH, and R. McKEAGUE: Proc. Roy. Soc. Lond. A **231**, 368—378 (1955).
[Lo 55b] LOHRMANN, E.: Nuovo Cim. **1**, 1126 (1955).
[Lo 55c] LOHRMANN, E.: Nuovo Cim. **1**, 1141 (1955).
[Lo 56c] LOHRMANN, E., e M. TEUCHER: Nuovo Cim. **3**, 59 (1956).
[Lo 60]   LOCKWOOD, J. A.: J. Geophys. Res. **65**, 19—25 (1960).
[Lo 61]   LOHRMANN, E., M. W. TEUCHER, and M. SCHEIN: Phys. Rev. **122**, 672 (1961).
[Lo 63]   LOCK, W. O.: Proc. 8 Int. Conf. Cosmic Rays, Jaipur 1963, vol. 5, p. 105.
[Lo 64]   LOCK, W. O.: Proc. 8. Cosm. Ray. Conf. 1963, vol. 5, p. 105—138. Jaipur 1964.
[Lu 57]   LUSTIG, H., H. GOLDSTEIN, and M. H. KALOS: Nuclear Development Corporation of America Rept. NDA-86-1, June 30, (1957).
[Lu 58]   LUSTIG, H., H. GOLDSTEIN, and M. H. KALOS: Nuclear Development Corporation of America Rept. NDA-086-2, January 31, 1958.
[Lu 63]   LUDWIG, G. H.: In: Cosmic Rays, Solar Particles ans Space Research, edit. by PETERS, Enrico Fermi School, vol. XIX, p. 320, 348. Varenna: Academic Press 1963.
[Ma 47]   MARSHAK, R. E.: Rev. Mod. Phys. **19**, 185—238 (1947).
[Ma 50]   MALASPINA, L., M. MERLIN, O. PIERUCCI e A. ROSTAGNI: Nuovo Cim. **7**, 145 (1950).
[Ma 58]   MATHUR, R. N., and P. S. GILL: Indian J. Phys. **32**, 19 (1958).
[Ma 63a] MALHOTRA, P. K.: Proc. Conf. on Cosmic Rays, Jaipur, 1963, vol. 5.
[Ma 63b] MALHOTRA, P. K.: Nuclear Phys. **46**, 559 (1963).
[Ma 63c] MARTIN, J. P., L. WITTEN, and L. KATZ: J. Geophys. Res. **68**, 2613—2618 (1963).
[Ma 65]   MARTIN, J. P.: J. Geophys. Res. **70**, 2057—2063 (1965).
[Ma 66]   MAUCK, G.: To be published.
[McCr 65] McCRACKEN, K. G., U. R. RAO, B. C. FOWLER, M. A. SHEA, and D. F. SMART: Cosmic Ray Tables (Asymptotic Directions, Variational Coefficients and Cut off Rigidities. IQSY Instruction Manual No. 10 (1965), London NW 1, 6 Cornwall Terrace.
[MCu 63] McCUSKER, C. B. A., and L. S. PEAK: Nuovo Cim. **31**, 525—540 (1963).
[McD 59] McDONALD, F. B.: Phys. Rev. **116**, 462—463 (1959).
[Me 50]   MERKLE, T. C., E. L. GOLDWASSER, and BRODE: Phys. Rev. **79**, 926 (1950).

[Me 54] MESSEL, H.: Progress in Elementary particles and Cosmic Ray Physics, edit. by JOHN G. WILSON and S. A. WOUTHUYSEN, Vol. 2, p. 176. Amsterdam: North-Holland Publ. Co. 1954.
[Me 55a] MEYER, P., and J. A. SIMPSON: Phys. Rev. 99, 1517—1523 (1955).
[Me 55b] MEYER, P.: Z. Physik 141, 28—32 (1955).
[Me 56a] MESHKOVSKI, A. G., i L. G. SOKOLOV: Soviet Physics JETP 3, 683—690 (1956), aus J. Exp. Theoret. Phys. 30, 840 (1956).
[Me 56b] MESHKOVSKI, A. G., i L. G. SOKOLOV: Soviet Physics JETP 4, 629 (1957), aus J. Exp. Theoret. Phys. 31, 752—755 (1956).
[Me 57] MEYER, P., and J. A. SIMPSON: Phys. Rev. 106, 568—571 (1957).
[Me 58a] MESHKOVSKII, A. G., i L. I. SOKOLOV: Soviet Physics JETP 6, 424 (1958); aus J. Exp. Theoret. Phys. 33, 542—544 (1957).
[Mo 58b] METROPOLIS, N., R. BIVINS, M. STORM, J. M. MILLER, and G. FRIEDLANDER: Phys. Rev. 110, 185, 204 (1958).
[Me 61] MEYER, B.: Thesis Maximilian-Universität München 1961.
[Me 63a] MENDELL, R. B.: Fast neutron flux in the atmosphere. Ph. D. Thesis: New York University 1963.
[Mi 54a] MILLER, C. E., J. E. HENDERSON, D. S. POTTER, J. TODD, W. M. SANDSTROM, G. R. GARRISON, W. R. DAVIS, and F. M. CHARBONNIER: Phys. Rev. 93, 590 (1954).
[Mi 54b] MILLER, C. E., J. E. HENDERSON, G. R. GARRISON, D. S. POTTER, W. M. SANDSTROM, and J. TODD: Phys. Rev. 94, 167 (1954).
[Mi 55] MITCHELL, E. N.: Phys. Rev. 98, 1163 (1955).
[Mi 57b] MIYAKE, S., K. HINOTANI, I. KATSUMATA, and T. KANEKO: J. Phys. Soc. Japan 12, 815—851 (1957).
[Mi 58] MILEKHIN, G. A., i I. L. ROZENTAL: Sovjet Phys. JETP 6, 154—156 (1958); — J. Exp. theoret. Fys. (USSR) 33, 197—199 (1957).
[Mo 39a] MONTGOMERY, C. G., and D. D. MONTGOMERY: Phys. Rev. 56, 10 (1939).
[Mo 39b] MONTGOMERY, C. G., and D. D. MONTGOMERY: Rev. Mod. Phys. 11, 255 (1939).
[Mo 52a] MORAND, M., et TSAI-CHU: C. R. Acad. Sci., Paris 235, 1502 (1952).
[Mo 52b] MOYER, B. J.: Nucleonics 15, No. 4, 14 (1952); 10, No. 5, 44 (1952).
[Mu 59] MURZINA, E. A., S. T. NIKOSLKII i V. I. YAKOVLEV: In Soviet Physics JETP 8, 906 (1959), aus J. Exp. Theoret. Phys. 35, 1298 (1958).
[My 51] MYLROI, M. G., and J. G. WILSON: Proc. Phys. Soc. Lond. A 64, 404 (1951).
[Na 61] NAUGLE, J. E., and D. A. KNIFFEN: Phys. Rev. Letters 7, 3 (1961).
[Ne 55a] NEHER, V.: Phys. Rev. 100, 959 (1955).
[Ne 56] NEHER, H. V.: Phys. Rev. 103, 228—236 (1956).
[Ne 59] NEHER, H. V.: Nature, Lond. 184, 423—435 (1959).
[Ne 60a] NEHER, H. V., and H. R. ANDERSON: Proc. Moscow Cosmic Ray Conf., Vol. 4, p. 101 (1960).
[Ne 60b] NEHER, H. V., and H. R. ANDERSON: J. Geophys. Res. 67, 1309—1316 (1962).
[Ne 62] NEHER, H. V., and H. R. ANDERSON: J. Geophys. Res. 67, 1309—1316 (1962).
[Ne 64] NEHER, H. V., and H. R. ANDERSON: J. Geophys. Res. 69, 807—814 (1964).
[Ne 66] NEHER, V.: Proc. Cosmic Ray Conf. 1965, Vol. 1, p. 153. London 1966.
[Ni 62] NIKOLSKY, S. I.: Proc. V. Interamerican Seminar on Cosmic Rays, Bolivio 1962, vol. II, No. 48.
[Ni 63] NICOLET, M.: General Aeronomy. In: Handbuch der Physik, Bd. 49, Teil 1. Berlin-Heidelberg-New York: Springer (im Druck).
[Ni 64] NIKOLSKY, S. I.: Soviet Physics Uspekhi 5, 849—877 (1963).
[No 55] NOON, J. H., and M. F. KAPLON: Phys. Rev. 97, 769 (1955).
[Og 55] OGILVIE, K. W.: Canad. J. Phys. 33, 746 (1955).
[Or 54] ORTEL, W. C. G.: Phys. Rev. 93, 561 (1954).
[Pa 33] PAULI, W.: Rep. on Solvay-Kongress. Inst. Solvay, 7me Conseil de Physique 1933, p. 324.
[Pa 50] PAGE, N.: Proc. Phys. Soc. Lond. A 63, 250 (1950).
[Pa 63a] PAL, Y.: Proc. Jaipur Conference 1963, vol. 5, p. 443—459.
[Pe 47] PERKINS, D. H.: Nature, Lond. 160, 707 (1947).
[Pe 49a] PEYROU, C.: Nuovo Cim. (Suppl.) 6, 408 (1949).
[Pe 49b] PEYROU, C., B. D'ESPAGNAT et L. LEPRINCE-RINGUET: C.R. Acad. Sci., Paris 228, 1777 (1949).
[Pe 50a] PEYROU, C., et A. LAGARRIGUE: J. Phys. Radium 11, 666 (1950).
[Pe 50b] PERKINS, D. H.: Phil. Mag. 41, 138 (1950).
[Pe 50c] PERKINS, D. H.: Proc. roy. Soc. Lond. A 203, 399 (1950).
[Pe 52a] PENNDORF, R.: Beitrag in LANDOLT-BJÖRNSTEIN, Bd. III, S. 328 — Meteorologie; S. 564—584; Astronomie und Meteorologie. Berlin-Göttingen-Heidelberg: Springer 1952. Darstellung der INA (= Internationalen Normal-) Atmosphäre. In den USA wurde die NACA (= National Advisory Committee for Aeronautics)-Atmosphäre

eingeführt, die mit der INA weitgehend übereinstimmt. Wegen neuerer Atmosphären siehe [Ci 61] und [Us 62].

[Pe 52b] PETERS, B.: In: Vol. I, p. 193. Progr. in Cosm.-Ray Phys. Amsterdam: North-Holland Publ. Co. 1952.
[Pe 62] PETERS, B.: Proc. Int. Conf. High Energy Phys. CERN 1962, p. 623.
[Pe 63a] PETERS, B.: Proc. Jaipur Conference. Vol. 5, 423—442 (1963).
[Pe 63b] PETERS, B.: In: Cosmic Rays, Solar Particles and Space Research, ed. by PETERS, Enrico Fermi Institute, Varenna 1963. New York and London: Academic Press.
[Pf 36] PFOTZER, G.: Z. Physik **102**, 23 (1936).
[Pf 52b] PFOTZER, G.: Z. Naturforsch. 7a, 145—149 (1952).
[Pi 59] PINE, J., R. J. DAVISON, and K. GREISEN: Nuovo Cim. **14**, (Ser. X) 1181 (1959).
[Po 56] POTAPOV, L. I., i N. V. SHOSTAKOVICH: Soviet Physics „Doklady" **1**, 85 (1956).
[Po 58] POOLE, M. J., M. S. NELKIN, and R. S. STONE: In: Progress in Nuclear Energy Series I: Physics, and Mathematics, Vol. 2, p. 91. (Editors: Hughes, SANDERS, HOROWITZ). London: Pergamon Press 1958.
[Po 62] POWELL, C. F.: KYOTO Conf., J. Phys. Soc. Japan **17**, Suppl. A3, 492 (1962).
[Pu 52] PUPPI, G., and N. DALLAPORTA: In: Progress in Cosmic Ray Physics, vol. I, p. 315—391, p. 378 and by J. G. WILSON and S. WORTHUGSEN. Amsterdam: North Holland Publ. Co. 1952.
[Pu 53] PULLAR, J. D., and E. G. DYMOND: Phil. Mag. **44**, 565 (1953).
[Pu 56] PUPPI, G.: In: J. G. WILSON, Progr. Cosmic Ray Physics **3**, 341 (1956).
[Qu 62] QUENBY, J. J., and G. J. WENK: Phil. Mag. **7**, 1457 (1962).
[Ra 53] RAO, A. S., V. K. BALASUBRAHMANYAM, G. S. GOKHALE, and A. W. PEREIRA: Phys. Rev. **91**, 764 (1953).
[Re 55] REICH, H.: Z. Naturforsch. 10a, 914 (1955).
[Ri 54] RICH, M., and R. MADEY: Range Energy Tables. Lawrence Radiation Lab., Berkeley, Calif. Rept. UCRL-2301 (1954).
[Rö 52] RÖSSLE, E.: Diplomarbeit Stuttgart 1952.
[Rö 54a] RÖSSLE, E., u. E. SCHOPPER: Z. Naturforsch. 9a, 836 (1954).
[Rö 54b] RÖSSLE, E.: Diss. Stuttgart 1954.
[Ro 48] ROSSI, BR. B.: Rev. Mod. Phys. **20**, 537—583 (1948).
[Ro 51] ROBSON, J. M.: Phys. Rev. **78**, 311 (1951).
[Ro 52a] ROEDERER, J. G.: Z. Naturforsch. 7a, 765 (1952).
[Ro 52b] ROSSI, BR. B.: High Energy Particles. Englewood Cliffs, New Jersey: Prentice Hall Inc. 1952.
[Ro 54] ROSEN, A. Z.: Phys. Rev. **93**, 211 (1954).
[Ro 55] ROBSON, J. M.: Phys. Rev. **100**, 943 (1955).
[Ro 55b] ROSSI, B.: Nuov. Cim. Suppl. **2** (Ser. X), 275 (1955).
[Ro 56] ROSE, D. C., K. B. FENTON, J. KATZMAN, and J. A. SIMPSON: Canad. J. Phys. **34**, 968—977 (1956).
[Ro 58a] ROBSON, J. M.: Canad. J. Phys. **36**, 1450 (1958).
[Ro 58b] ROTHWELL, P.: Phil. Mag. **3**, 961 (1958).
[Ro 59, 60a] ROSSI, B. B.: I.U.P.A.P. Cosmic Ray Conf., 1959 (Moskau) in Vol. II, p. 18 (1960).
[Ro 60b] ROTHWELL, P., and C. E. MCILWAIN: J. Geophys. Res. **65**, 799 (1960).
[Ru 36] RUMBAUGH, L. H., and G. L. LOCHER: Phys. Rev. **50**, 855 (1936).
[Ry 55] RYSHKOVA, K. P., i L. I. SARYCHEVA: J. Exp. Theoret. Phys. (USSR.) **28**, 618 (1955); — Soviet Physics JETP **1**, 572 (1955).
[Sa 50] SALANT, E. O., J. HORNBOSTEL, C. B. FISK, and J. E. SMITH: Phys. Rev. **79**, 184 (1950).
[Sa 63] SAUER, E. H.: J. Geophys. Res. **68**, 957 (1963).
[Sa 65] SANDSTRÖM, A. E.: Cosmic Ray Physics. Amsterdam: North-Holland Publ. Co. 1965.
[Sch 55] SCHOWE, D. J.: J. Geophys. Res. **60**, 127—146 (1955).
[Scho 37] SCHOPPER, E.: Naturwissenschaften **25**, 557 (1937).
[Scho 39] SCHOPPER, E. M., u. E. SCHOPPER: Phys. Z. **40**, 22 (1939).
[Scho 47] SCHOPPER, E.: Naturwissenschaften **34**, 118 (1947).
[Scho 51a] SCHOPPER, E., K. H. HÖCKER, and G. KUHN: Phys. Rev. **82**, 444 (1951).
[Scho 51b] SCHOPPER, E., K. H. HÖCKER u. E. RÖSSLE: Z. Naturforsch. 6a, 603 (1951).
[Schu 54] SCHULTZ, H.: Z. Naturforsch. **99**, 419 (1954).
[Se 64] SETTLES, R. D., and R. W. HUGGETT: Phys. Rev. **133**, B 1305—B 1317 (1964).
[Sh 41] SHAPIRO, M. M.: Revs. Mod. Phys. **13**, 58 (1941).
[Sh 51a] SHAPIRO, M. M., B. STILLER, M. BIRNBAUM, and F. W. O'DELL: Phys. Rev. **83**, 455 (1951).
[Sh 51b] SHAPIRO, M. M.: Phys. Rev. **83**, 456 (1951).
[Sh 53] SHAPIRO, A. H.: The Dynamics and Thermodynamics of Compressible Fluid flow, Vol. 1, app. B, pp. 612—613. New York: Ronald-Press 1953.

[Sh 55]	SHDANOW, A. P., i P. I. FEDETOW: Dokl. Akad. Nauk. S.S.S.R. **100**, 659 (1955).
[Sh 59]	SHKLOVSKY, I. S., V. I. KRASSOVSKY i G. I. GALPERIN: Izv. Acad. Nauk. SSSR., Ser. geophys. **12**, 1799 (1959).
[Sh 65]	SHEA, M. A., D. F. SMART, and G. MCCRACKEN: J. Geophys. Res. **70**, 4117—4130 (1965); dort auch weitere Literaturhinweise.
[Si 50]	SITTE, K.: Phys. Rev. **78**, 721 (1950).
[Si 51a]	SIMPSON, J. A.: Phys. Rev. **81**, 639, 895 (1951).
[Si 51c]	SIMPSON, J. A., H. W. BALDWIN, and R. B. URETZ: Phys. Rev. **84**, 332 (1951).
[Si 53a]	SIMPSON, J. A., and C. W. FAGOT: Phys. Rev. **90**, 1060—1072 (1953).
[Si 53b]	SIMPSON, J. A., W. FONGER, and S. B. FREIMANN: Phys. Rev. **90**, 934—950 (1953).
[Si 53c]	SIMPSON, J. A., and R. B. URETZ: Phys. Rev. **90**, 44 (1953).
[Si 56]	SIMPSON, J. A., K. B. FENTON, J. KATZMANN, and D. C. ROSE: Phys. Rev. **102**, 1648 (1956).
[Si 57]	SIMPSON, J. A.: Ann. Intern. Geophys. Y. **4**, 351 (1957).
[Si 58a]	SINGER, S. F.: Phys. Rev. Letters **1**, 171, 181 (1958).
[Si 58b]	SINGER, S. F.: Progress in Elementary Particles and Cosmic Ray Physics, (ed. J. G. WILSON and S. A. WOUTHUYSEN, Vol. 4, p. 203. Amsterdam: North-Holland Publ. Co. 1958.
[Si 58c]	SINGER, S. F.: Progress in Elementary Particles and Cosmic ray Physics, ed. J. G. WILSON und S. A. WOUTHUYSEN, Vol. IV, p. 278. Amsterdam: North-Holland Publ. Co. 1958.
[Si 58d]	SIMPSON, J. A.: Cosmic Ray Programm; first twelve month, compiled by S. E. FORBUSH. IGY Bull. **15**, 11 (1958).
[Si 60]	SIMPSON, J. A.: J. Geophys. Res. **65**, 1615—1616 (1960).
[Si 61]	SITTE, K.: Penetrating Showers. In: Handbuch der Physik, Bd. 46/1 ed. by S. FLÜGGE. Berlin-Göttingen-Heidelberg: Springer 1961.
[Si 62]	SINGER, S. F.: Progress in Elementary Particle and Cosmic Ray Physics, Vol. VI, ed. by J. G. WILSON und S. A. WOUTHUYSE. Amsterdam: North-Holland Publ. Co. 1962.
[Si 63]	SIMPSON, J. A.: Semaine d'Etude sur le problem du Rayonnement Cosmique dans l'Espace Interpanetaire p. 323—352. Vatican: Pontificia Academi Scientia 1963.
[Si 65]	SITTE, K.: Proc. 9. Int. Cos. Ray Conf. 1965. London 1966.
[Sk 58]	SKORKA, S.: Z. Physik **151**, 630 (1958).
[Sm 54]	SMITH, J. R.: Phys. Rev. **95**, 730 (1959).
[Sm 62]	SMITH, R. V., L. F. CHASE, W. L. IMHOF, J. B. REAGAN, and M. WALT: ARL-TDR-62-3, 6571 st, Aeromedical Res. Lab. Holloman AFB, New Mexico, 55 pp. (Study conducted by Lockhead Missiles and Space Company) 1962.
[Sn 48]	SNELL, A. H., and L. C. MILLER: Phys. Rev. **74**, 1217 (1948).
[Sn 50]	SNELL, A. H., F. PLEASONTON, and R. V. MCCORD: Phys. Rev. **78**, 310 (1950).
[Sö 49]	SÖRENSEN, S. O. C.: Phil. Mag. **40**, 947 (1949).
[Sö 51]	SÖRENSEN, S. O. C.: Phil. Mag. **42**, 188 (1951).
[So 55]	SOBERMANN, R. K., A. BEISER, and S. A. KORFF: Phys. Rev. **100**, 859—860 (1955).
[So 55]	SOBERMANN, R. K., A. BEISER, and S. A. KORFF: Phys. Rev. **99**, 608; **100**, 859 (1955).
[So 59]	SOSNOVSKII, A. N., P. E. SPIVAK, YU. A. PROFKOFÉEV, J. A. KULIKOV, and YU. DOBRYNIN: J. Exp. Theoret. Phys. (USSR.) **35**, 1059—1061 und **36**, 717 (1959); — Soviet. Physics JETP **8**, 737 und **9**, 717 (1959).
[St 39]	STETTER, G., u. H. WAMBACHER: Phys. Z. **40**, 702 (1939).
[St 50a]	STAKER, W. P.: Phys. Rev. **80**, 52 (1950).
[St 50b]	STINCHCOMB, T. G.: Phys. Rev. **78**, 321 (1951).
[St 51]	STAKER, W. P., M. PAWALOW, and S. A. KORFF: Phys. Rev. **81**, 889 (1951).
[St 53]	STILLER, B., and M. M. SHAPIRO: Phys. Rev. **92**, 735 (1953).
[St 55]	STÖRMER, C.: The Polar Aurora. Oxford: Clarendon Press 1955.
[St 56]	STRAUCH, K., and F. TITUS: Phys. Rev. **104**, 191 (1956).
[St 58]	STUART, R. N., E. H. CANFIELD, E. E. DOUGHERTY, and S. P. STONE: Law. Rad. Lab. (Livermoore, Calif.) Rept. UCRL-5293, November 19, 1958.
[St 61a]	STELJES, J. F., and H. CARMICHAEL: Solar Geographical Data, Part B. CRPL-F-204 und 205, Vb. (1961).
[St 61b]	STUIVER, M.: J. Geophys. Res. **66**, 273—276 (1961).
[St 64]	STEHN, J. R., M. D. GOLDBERG, B. A. MAGURNO, and R. WIENER-CHASMAN: Neutron Cross Sections. BNL 325, 2nd. edit. Suppl. 2 (1964).
[Str 56]	STRAUCH, K., and F. TITUS: Phys. Rev. **104**, 191 (1956).
[Sw 52]	SWETNICK, H. J., H. NEUBERG, and S. A. KORFF: Phys. Rev. **86**, 589 (A) (1952).
[Sw 54]	SWETNICK, H. J.: Phys. Rev. **95**, 793 (1954).
[Sy 53]	SYMANZIK, K.: In: W. HEISENBERG, Kosmische Strahlung, S. 164ff. Berlin-Göttingen-Heidelberg: Springer 1953.

[Ta 56] TAKIBAJEW, Z. S.: Soviet Physics JETP **3**, 559 (1956), aus J. Exp. Theoret. Phys. (USSR.) **30**, 713 (1956).
[Te 52] TEUCHER, M.: Z. Naturforsch. **7a**, 61 (1952).
[Te 53a] TEUCHER, M.: Helv. phys. Acta **26**, 434 (1953).
[Te 53b] TEUCHER, M.: Z. Naturforsch. **8a**, 127 (1953).
[Te 53c] TEUCHER, M.: In: W. HEISENBERG, Kosmische Strahlung, S. 69ff. Berlin-Göttingen-Heidelberg: Springer 1953.
[Te 53d] TEUCHER, M.: In: W. HEISENBERG, Kosmische Strahlung, S. 20ff. Berlin-Göttingen-Heidelberg: Springer 1953.
[Th 46] THELLIER, E., and O. THELLIER: C. R. Acad. Sci. Paris **222**, 905—907 (1946).
[Th 55] THOMPSON, B. W.: Nucleonics **13**, (3), 44 (1955).
[Ti 48] TINLOT, J.: Phys. Rev. **74**, 1197 (1948).
[Ti 52] TIDMAN, D. A., and P. E. HOGDSON: Phil. Mag. **43**, 992 (1952).
[To 64] TOLKACHEV, B. V.: Soviet Physics JETP **19**, 31—35 (1964).
[Tr 52] TREIMAN, S. B., and W. FONGER: Phys. Rev. **85**, 364 (1952).
[Tr 63] TRAINOR, J. H., and J. A. LOCKWOOD: Trans. Amer. Geophys. Union **44**, 73 (1963).
[Tre 59] TREBUKHOSKII, YU. K., V. V. KLADIMIRSKII, V. V. GRIGOREV i V. A. ERGAKOV: J. Expt. Theoret. Phys. (USSR.) **36**, 931 (1959); — Soviet Physics JETP **9**, 931—932 (1959).
[Ud 62] UDGAONKAR, B. M., and M. GELL-MANN: Phys. Rev. Letters **8**, 346 (1962).
[Ur 55] URBANIK, K., e J. LOPUSZANSKI: Nuovo Cim. (Suppl.) **2**, 1147, 1150, 1161 (1955).
[US 62] USA-1962-Standard-Atmosphäre, Washington, US-Printingoffice (December 1962).
a) die ICAO (International Civil Arration Organisation)-Atmosphäre bis 20 km, b) eine vorgeschlagene Ausdehnung dieser Atmosphäre bis 32 km Höhe, c) und versuchsweise und spekulative Tabellen bis 700 km. Die Unterschiede zur CIRA-Atmosphäre [Ci 61] werden besprochen. Wegen früherer Modellatmosphären vgl. PENNDORF [Pe 52a].
[Va 57] VAISENBERG, A. O.: Soviet Physics JETP **5**, 352—357 (1957), aus J. Exp. Theoretical. Phys. (USSR.) **32**, 416—422 (1956).
[Va 58] VARSIMASHVILI, T. V., i N. I. KOSTANASHVILI: Soviet Physics JETP **6**, 1183 (1958), aus J. Exp. Theoret. Phys. (USSR.) **33**, 1530—1531 (1957).
[Va 60] VARSIMASHVILI, T. V.: Soviet Physics JETP **11**, 231 (1960), aus J. Exp. Theoret. Phys. **38**, 319—332 (1960).
[Ve 59] VERNOV, S. N., N. L. GRIGOROV, I. P. IVANENKO, A. I. LEBEDINSKII, V. S. MURZIN i A. E. CHUDAKOV: Soviet Phys. Dokl. **4**, 154 (T) (1959).
[Vi 52] VIDALE, M. L.: Phys. Rev. **88**, 266 (1952).
[Wa 50a] WALKER, W. D., S. P. WALKER, and K. GREISEN: Phys. Rev. **80**, 546 (1950).
[Wa 50b] WALKER, W. D.: Phys. Rev. **77**, 686 (1950).
[Wa 55] WATASE, Y., K. SUGA, Y. TANAKA, and S. MITANI: Nuovo Cim. **2**, 1183 (1955).
[We 55a] WENTWORTH, R. C., and S. F. SINGER: Phys. Rev. **98**, 1546 (1955).
[We 55b] WEISS, H. M.: Z. Naturforsch. **10a**, 21 (1955).
[We 62] WEBBER, W. R.: In: Progress of Elementary Particle and Cosmic Ray Physics, Vol. VI, ed. by J. G. WILSON and S. A. WOUTHUYSEN: Amsterdam: North-Holland Publ. Co. 1962.
[We 63] WEBBER, W. R., and P. S. FREIER: Symposium on Protection Against Radiation Hazards in space, held in Getlinborg, Tenn. 1962. ORNL and NASA 12—32.
[Wh 51] WHYTE, G. N.: Phys. Rev. **82**, 204—208 (1951).
[Wh 52] WHITTEMORE, W. L., and R. P. SHUTT: Phys. Rev. **86**, 940 (1952).
[Wi 60a] WILLIAMS, R. W.: Nuovo Cim. **16**, 762 (1960).
[Wi 60b] WILLIS, E. H., H. TAUBER, and K. O. MÜNNICH: Amer. J. Sci. Radiocarbon, Suppl. **2**, 1—4 (1960).
[Wi 60c] WINKLER, J. R.: J. Geophys. Res. **65**, 1340 (1960).
[Ya 49] YAGODA, H., N. KAPLAN, and C. H. CONNER: Phys. Rev. **76**, 171 (1949).
[Yo 52a] YORK, C. M.: Proc. Phys. Soc. Lond. A **65**, 558 (1952).
[Yo 52b] YORK, C. M.: Phys. Rev. **85**, 998 (1952).
[Yu 48] YUAN, L. C. L.: Phys. Rev. **74**, 504 (1948).
[Yu 49] YUAN, L. C. L.: Phys. Rev. **76**, 165 (1949).
[Za 56] ZANGGER, C., u. J. ROSSEL: Helv. phys. Acta **29**, 507 (1956).
[Za 58] ZATSEPIN, G. T., KRUGOVYKH, E. A. MURZINA i S. I. NIKOLSKII: Soviet Physics JETP **7**, 207 (1958). — J. Exp. Theoret. Phys. (USSR.) **34**, 298—300 (1958).
[Za 62] ZATSEPIN, G. T.: Bull. Acad. Sci. USSR **26**, No. 5, 673—680 (1962); Columbia Technical Translation.
[Za 63] ZATSEPIN, G. T., S. I. NIKOLSKY, and G. B. KHRISTIANSEN: 8. Conf. on Cosmic Rays Jaipur 1963, vol. 4, p. 100—117.

Cosmic Ray Produced Radioactivity on the Earth.

By

D. LAL and B. PETERS.

With 24 Figures.

A. Introduction.

1. Matter which is exposed to cosmic radiation undergoes characteristic changes in its chemical and isotopic composition. Such irradiated material can be found in many places on earth and belongs to either of two categories:

terrestrial matter bombarded by cosmic ray particles incident on the earth, and
extraterrestrial matter bombarded in outer space and subsequently accreted by the earth.

The corpuscular radiation in the vicinity of the earth consists of protons and α-particles with a small admixture of heavier atomic nuclei and of electrons. Since this article will be concerned with isotopic changes, the term "cosmic radiation" will be used in a restricted sense to designate only those particles of the incident beam which are atomic nuclei and have sufficient energy to produce nuclear disintegrations before being brought to rest in the upper atmosphere by atomic collisions. One is therefore concerned only with primary nuclei with energy above a few tens of MeV/nucleon; a part of the radiation thus defined is in fact not "cosmic" but originates in the sun. Averaged over time, however, the major fraction of the particles responsible for nuclear reactions in the atmosphere comes from outside of the solar system.

α) *Irradiation of terrestrial material.* By far the largest number of nuclear transformations induced by cosmic rays takes place within the earth's atmosphere, since most of the cosmic ray energy is dissipated there. Almost all transformations are produced by nucleons; the contribution from mesons or from their decay products, i.e. muons, electrons and γ-rays, is small by comparison.

The outermost layer of the earth's crust constitutes a second region where nuclear transmutations occur. Down to a depth of one or two meters below the surface isotope production is dominated by nucleons and by the pions which they produce in the ground. The rate at which nuclear transformations occur in this region is several hundred times smaller than in the atmosphere, but since the earth's surface contains many elements which are not present in the atmosphere, the contribution may nevertheless be of importance for particular isotopes. For example, the long lived radioisotope Cl^{36} ($\tau_{\frac{1}{2}} = 3 \times 10^5$ years) is produced by spallation reactions in atmospheric Ar^{40}; but about 70% of the earth's inventory of Cl^{36} is due to another, more efficient, process which cannot occur in air, namely neutron capture by the stable nuclide Cl^{35} which is fairly abundant in the sea and in rocks [D2], [L0].

At greater depth below the earth's surface, strongly interacting particles have been eliminated and only the muon and neutrino components of the secondary cosmic radiation survive. High energy muons dominate interactions down to several thousand meters below the surface [G8], [P3]. Since they interact only by virtue of their electric charge, disintegrations in this part of the earth are due to these electromagnetic interactions of muons with nuclei and to nuclear interactions of the recoil nucleons and the pions to which they give rise.

Finally, deeper still in the earth's interior, one must expect isotopic changes to be induced by neutrinos of the secondary cosmic radiation as well as by neutrinos of stellar and solar origin [L8]. Although the rate at which such transformations occur is too low to be detected with presently available techniques, such effects, which are interesting particularly from the point of view of astrophysics, may become observable some day.

The corresponding reaction products in the earth's crust will, of course, have to be distinguished from those arising from natural radioactivity and spontaneous fission.

β) *Accretion of irradiated extraterrestrial material.* Much of the interplanetary matter accreted by the earth has been subjected to prolonged bombardment by cosmic radiation before entering the atmosphere and must be expected to show corresponding chemical and isotopic anomalies.

The galactic cosmic ray particles themselves are no exception; their original composition has been modified by nuclear collisions in the region where they were accelerated as well as by interactions with interstellar gas during transit. The presence in the primary radiation of comparatively high percentages of low abundance elements such as lithium, beryllium and boron and of rare isotopes like He^3 is attributed to such transformations [P4].

The earth collects also atoms which originate in the sun; these enter the earth's atmosphere as low energy particles. Possibly, solar material can also enter, when plasma clouds, ejected by the sun, impinge on the earth.

In addition there is considerable accretion by the earth of various types of "cosmic debris", such as meteorites, many of which are believed to represent matter from the asteroidal belt, and smaller objects such as meteoroids and interplanetary dust. Tektites, whose origin is uncertain as yet, may conceivably constitute debris from the surface of the moon [K4], [V1].

In terms of weight, the interplanetary dust contributes most of the incident extraterrestrial material, $\sim 10^4$ tons per day [A11]. The meteoroids which evaporate on entering the atmosphere and the meteorites which survive the traversal contribute less, each of the order of hundred tons per day [A11], [H6], [J2].

Until now, the study of transformations in extraterrestrial material has been confined to meteorites where the isotopic effects are large, and to tektites where the effects appear to be small or absent [V1], thereby putting in doubt the hypothesis of their extraterrestrial origin. Study of isotopic changes induced in the interstellar dust has commenced [K3], [W3]; if successful, it may throw light on the history of interplanetary irradiation and the accretion process in distant geological epochs.

γ) *History of cosmic radiation as preserved in meteorites.* Among the exposed materials which have been examined so far, meteorites show the largest cosmic ray induced changes because the reaction products have accumulated undisturbed for long periods preceding the arrival of the meteorites on earth. The presence of radioactive isotopes and abnormal abundances among stable isotopes can in principle be used to determine the intensity of cosmic radiation in the past, its energy spectrum and the accumulated radiation dose to which the meteorite was exposed during its life period as an independent object, i.e., since it became separated from a larger parent body.

A study [A 9] of the isotopic evidence in meteorites indicates that most stony meteorites have been exposed for periods of the order of 20—30 million years, whereas the iron meteorites have been exposed for much longer periods, i.e. 10^8—10^9 years. During these periods, the average intensity of the corpuscular radiation as deduced from concentration of radioactive and stable cosmogenic isotopes found in meteorites, seems to have remained constant within narrow limits and the energy spectrum also seems not to have undergone appreciable changes [A 4], [G 3]. The accuracy of these conclusions when applied to the billion year range is somewhat reduced by a lack of knowledge as to the rate at which surface material of meteorites is lost by space erosion and collisions [G 9], [V 2]. In this article, we will not discuss in detail the cosmic ray record found in meteorites; a review of the information obtained so far can be found in Chapter VIII by M. Honda and J. R. Arnold in this book.

Large cosmic ray induced changes in isotopic composition will most likely also be observed in interstellar dust [K 3], [W 3], and in material from the lunar surface when adequate samples become available.

δ) *Cosmic ray induced radioactivity on the earth.* Compared to radiation effects in extraterrestrial samples the effects of irradiation in terrestrial materials are weak. They are not preserved as well because the outer layers of the earth are in continuous motion; their atoms participate in various geochemical and geophysical cycles operating between the atmosphere, biosphere, hydrosphere, and lithosphere and, during these transfer processes, matter which has been irradiated becomes strongly diluted with non-irradiated matter. This dilution decreases the magnitude of observable effects. While the majority of fragments from cosmic ray induced nuclear reactions are lost in a vast mass of ordinary terrestrial material, two types of end products remain traceable:

radioactive nuclei, because they remain distinguishable from ordinary matter for some time, and

noble gas atoms, because, being unable to form compounds or attach themselves to dust particles, they do not participate effectively in the mixing cycles; most of their inventory is confined to the atmosphere.

The entire geosphere, i.e., the atmosphere and all parts of the earth which directly or indirectly exchange material with the atmosphere, contains cosmic ray produced radioactive isotopes. The activity has been found in the sea and in sediments on the ocean floor, in the biosphere and the top soil, in the polar ice cap, in surface rocks and in material eroded from surface rocks [L 0].

ε) *Important cosmic ray produced radioisotopes.* The radioisotopes created on earth differ appreciably from those created in meteorites. This is due to the fact that on earth the bulk of isotopic change occurs in interactions in the atmosphere where heavy target nuclei are rare and therefore practically only light fragments can be produced. Furthermore, in meteorites, a large number of cosmogenic stable nuclides can be detected in contrast to the terrestrial case where they get mixed with ordinary matter and become undetectable. For these reasons the variety of isotopes available for tracer work is larger in meteorites than in terrestrial materials. A list of isotopes produced by cosmic radiation in the atmosphere with half lives longer than one day, is given in Table 1 (Sect. C).

ζ) *Source and sink functions.* The radioactivity produced by cosmic radiation in terrestrial material, although weak, is nevertheless of considerable interest because of the many applications it has found in various sciences; it is particularly well suited as tracer for the study of a large variety of geophysical and geochemical processes both in the present and in the recent or more distant past.

The most essential condition, which needs to be fulfilled in order to make a substance useful for tracer work, is that its source and sink functions can be established in space and time, i.e., that one can obtain an accurate knowledge as to when, where, and at what rate the tracer substances are introduced and by what mechanisms they are removed from the region under consideration.

This condition is quite well satisfied in the case of cosmic ray produced isotopes. The geographical and the vertical distributions of nuclear interactions in the atmosphere and their rate of occurrence can be calculated from cosmic ray data, while the production cross sections for particular isotopes, where not measured directly, can usually be estimated from available data on related cross sections obtained in accelerator experiments. Thus, the source function, i.e., the manner in which radioactive isotopes of widely differing half lives are introduced into the atmosphere, may be considered to be adequately known [$L0$]. Cosmic ray data pertinent to isotope production in the atmosphere are discussed in Sect. B. The properties of source functions are discussed in detail in Sect. C. Isotopes of non-atmospheric origin are discussed in Sect. D.

The sink function, i.e., the manner in which cosmic ray produced isotopes are removed from the atmosphere, is also fairly well understood and reasonably simple. The nuclei are quickly oxidized and attach themselves to small aerosol particles. They partake in the bulk motion of the airmasses until they reach the lower troposphere. At cloud level they are efficiently removed by the scavenging action of water droplets, and reach the earth's surface as constituents of rainwater. Dry fallout, i.e., adhesion of windborne isotopes to ground surfaces, plays only a minor role so that, apart from decay in transit, wet precipitation represents for most isotopes the only effective removal mechanism.

This sink function, which characterizes the removal from the atmosphere, is at the same time the source function describing the input of isotopes into the oceans and into other geophysical reservoirs. It can be determined in any geographical region by measuring rainfall together with isotope concentrations in rainwater.

There are a few exceptions to this simple removal mechanism. The isotope C^{14}, f.i., is not removed by precipitation but by the exchange of radioactive and stable CO_2 at the ocean surface. A fairly complicated sink function applies to tritium, where molecular exchange as well as precipitation and re-evaporation play a role in introducing tritium into various terrestrial reservoirs.

The effect of geophysical mixing processes on the dispersion of isotopes is discussed in Sect. E. Some results of measurements which form the basis for the use of cosmic ray produced isotopes as tracers in geophysical investigations are discussed in Sect. F.

η) *Application for tracer studies.* Since one has now a fairly detailed knowledge as to when, where and at what rate cosmic ray produced isotopes are introduced into the biosphere, lithosphere, and hydrosphere one possesses an adequate basis for employing them as tracers. As seen from Table 1, the half lives of available isotopes range from days or weeks up to a few million years. Thus, suitable isotopes can often be found for investigating particular processes whose characteristic time scale falls within these limits. Rates of accumulation of polar ice, glacier motion, the size of subterranean water reservoirs and the rate of their replenishment from surface precipitation, the circulation pattern of water within and between the oceans, rate of sedimentation on the ocean floor and the well-known application of C^{14} for dating archaeological remains are examples of important problems which are being studied with the help of cosmic ray produced isotopes [$L0$], [$S1$], [$L1$], [$I1$], [$J1$], [$C3$].

As the techniques of measuring weak radioactivities improve, new possibilities may emerge.

Underground measurements of isotopic changes may make it possible to study time variations of high energy cosmic rays up to several 1000 GeV.

Solar and stellar neutrino flux in the present and in the past may become measurable.

It may become feasible to study the influx of cosmic dust in the geological past by studying particular isotopes in marine sediments.

A record of large sporadic particle emissions from the sun may still be preserved in the radioactivity of fossil ice on the polar caps, etc.

The present review article will concern itself mainly with the direct production of isotopes on earth and with their inventory in different parts of the geosphere. Since the inventories are sensitive to the time scale and nature of geophysical and geochemical processes, a brief discussion of some of these processes has been included. Some important references on experimental techniques used in geophysical tracer work are listed in the Appendix.

B. Corpuscular radiation.

I. Primary particles.

2. The atomic nuclei incident on the atmosphere may be divided into a galactic and a solar component. The two components differ considerably in chemical composition, energy distribution, intensity, and in the character of intensity variations with time.

Electrons and positrons, which also occur among the primary particles, do do not induce nuclear transmutations in the atmosphere and will therefore not be discussed further.

α) *Galactic particles.* Most prominent among the various components are the protons whose intensity as a function of energy outside the earth's magnetic field is shown [W7] in Fig. 1. Curve "a" refers to the intensity in 1954 when the emission of plasma clouds from the sun was near its minimum; the solar system was then comparatively free from travelling inhomogeneities of magnetic fields which obstruct the free passage of charged particles of low or moderate energies entering from the galactic space. Curve "b" refers to the intensity in 1958/59 during the last sunspot maximum and curve "c" refers to the lowest intensity recorded during that period. These curves show the effect of the "albedo", i.e. the reflection property of the solar system, when it is populated by outward travelling clouds of plasma with their internal currents and magnetic fields; the intensity reduction is large for galactic particles entering with low energies and tapers off in the region of 10—20 GeV. Above these energies all primary spectra are insensitive to the sunspot cycle and can be represented by an integral power law in momentum or energy with an exponent $\gamma = -\frac{5}{3}$.

Near the earth, spectra as shown in Fig. 1 are observable only in the polar stratosphere. At other latitudes the earth's field imposes certain cut-off restrictions which depend on geomagnetic co-ordinates, on angle of incidence, and on magnetic rigidity (i.e. the ratio of the particle's momentum and charge). The effect of the earth's field on the particle flux, arriving at various latitudes and longitudes and from different directions, is discussed by J. QUENBY in Chapter V of this book.

The complex nuclei of the galactic corpuscular radiation have the same rigidity spectrum as the protons; above a rigidity of 1.5 GV, ~14% of the primary particles are helium nuclei and ~1% consists of particles of greater atomic weight up to iron. Below 1.5 GV the different components may have different spectra; a decrease in the α/p ratio has been observed [B24] during a particular phase of the present sunspot cycle.

Details of the chemical composition are discussed by W. R. WEBBER in Chapter III of this book.

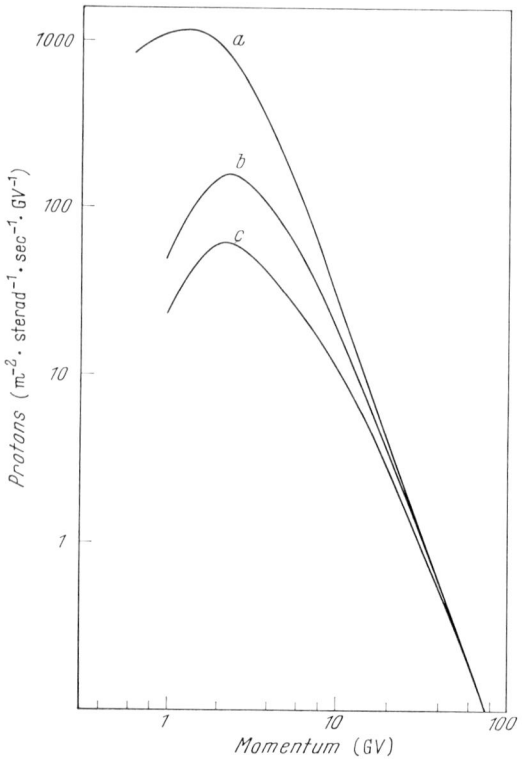

Fig. 1. Differential momentum spectrum of cosmic ray protons at three different periods during the last solar cycle (After WEBBER [W7]: curve a, Sunspot minimum; curve b, Sunspot maximum; curve c, Lowest observed flux (during a series of Forbush decreases in July 1959.)

At a given cut-off rigidity, protons represent 50% of the incident nucleons, the remainder consists of equal numbers of neutrons and protons arriving as constituents of complex nuclei. Time variations of the flux and energy spectra are discussed in Sect. C.II and more fully by J. QUENBY in Chapter V of this book.

β) *The solar component.* During certain solar flares, nuclei are accelerated near the sun's surface. If they are ejected from the sun, and if the earth intercepts their path of escape, a large burst of solar particles enters the atmosphere. Particles may continue to arrive up to a few days after the flare has subsided. Their direction of entrance is then no longer correlated with the sun's position but tends to become isotropic. This indicates that the late burst particles have been strongly scattered by magnetic inhomogeneities in the interplanetary medium both inside and outside the earth's orbit.

The intensity and energy spectra of burst particles can vary within wide limits from event to event as well as in the course of a single event. The total number of particles incident on the magnetosphere due to a single solar flare event may reach 10^{10} or more particles per cm² of energy above ~ 20 MeV [L6]. The rigidity spectra observed so far are well represented by exponential functions, i.e. $F = F_0 e^{-R/R_0}$ with R_0 ranging from 50 to 300 MV for different events and also varying in the course of a single event [F6].

On the other hand, the relative abundance among the complex nuclei does not seem to vary from one event to the next and probably represents the abundance ratios in the sun's photosphere; but the ratio of protons to complex nuclei above a given rigidity is subject to strong variations in time and from one burst to the next; p/α ratios ranging from 1 to ~ 30 have been observed [B14], [B15].

Existing observations on individual solar bursts have been discussed in detail in the literature [F6], [W8], [B14]. In view of the comparative rarity and variety of these events their average effect on the production of isotopes is difficult to evaluate. In this connection it is also important to note the further complication that, due to the magnetic disturbances which accompany solar bursts, the normal cut-off rigidities for various latitudes are often lowered, typically by a factor of order two [W8].

Because of the difficulty which one encounters when trying to obtain a reliable average value of solar particle flux at various latitudes by direct observations on the incoming particles, isotope measurements may be the most satisfactory method available at present of estimating long term averages of the solar contribution to the incident corpuscular radiation. This problem is discussed further in Sect. C.IV.

II. Propagation of cosmic radiation in the atmosphere.

3. α) *Interaction of nucleons.* The basic processes by which nucleons propagate and dissipate their energy in traversing the atmosphere can be studied in detail in the laboratory up to energies of 30 GeV with the help of existing particle accelerators and can be extrapolated to energies of several thousand GeV through the observation of interactions of cosmic ray particles in emulsions and cloud chambers. One finds that:

i) With good approximation, collisions of nucleons with light target nuclei (and also therefore collisions of protons with air nuclei) can be described in terms of collisions between free nucleons, provided that the energy lies sufficiently high above the nuclear binding energy.

ii) The large majority of collisions is peripheral, so that the incident nucleon retains a large fraction $\eta (\approx 0.55)$ of its kinetic energy and receives a small transverse momentum of order q ($\lesssim Mc/2$). This holds from a few hundred MeV to at least several thousand GeV [P5].

From α, i) it follows that, if the total energy of the incident nucleon is E_0, the recoil nucleon is ejected from the target with kinetic energy, W_r, given by

$$E_r = 1 + W_r = E_0 (1 + \eta W_0) \left[1 - \beta_0 \sqrt{1 - \frac{1+q^2}{(1+\eta W_0)^2}} \right], \quad (3.1)$$

where β_0 and W_0 are the velocity and kinetic energy of the incident nucleon. (All energies and momenta are expressed in proton mass units.) In Fig. 2, W_r is plotted as a function of W_0 for $q/Mc = \frac{1}{2}$ and three values of $\eta = (\frac{1}{3}, \frac{1}{2}$ and $\frac{2}{3})$; it illustrates that the energy of recoil nucleons is strictly limited and rarely exceeds 500 MeV, whatever the energy of the incident nucleon.

Thus, up to energies where nucleon-antinucleon production becomes important ($E_0 > 100$ GeV), incident nucleons on their way through the atmosphere produce a recoil nucleon of a few hundred MeV for every 70—80 g/cm² of atmosphere traversed, until they reach the ground or until their own energy is reduced to a value, ω, of order of 500 MeV.

[Primaries with $E_0 > 100$ GeV constitute about 1.5 promille of the total galactic corpuscular radiation and carry about 8% of its energy; the corresponding fraction for particles incident vertically at the equator is 4% and 25%, respectively. Because this part of the galactic radiation produces nucleon-antinucleon pairs, it is more efficient in giving rise to interactions in the atmosphere; the effect needs to be taken into account but will not be considered in detail, because its influence on isotope production is comparatively small.]

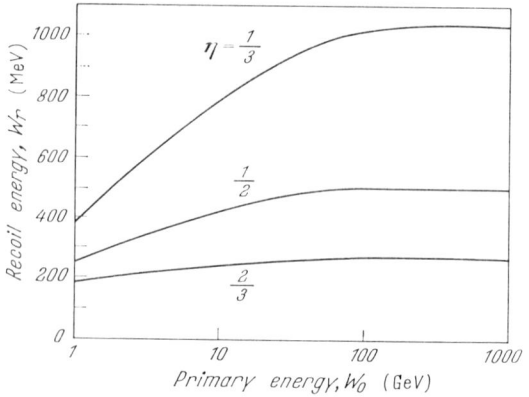

Fig. 2. Relation between the kinetic energy of the incident and the recoil particle in nucleon-nucleon collisions for different values of the elasticity η [Eq. (3.1)].

Before being reduced to an energy $\omega \approx 500$ MeV, a primary nucleon with kinetic energy W_0 makes therefore on the average

$$n' = (\log W_0/\omega)/(\log 1/\eta)$$

collisions, each leading to a recoil nucleon of a few hundred MeV. n' lies between zero and nine for primary energies lying between 0.5 and ~100 GeV.

A primary beam with an energy distribution following a power law with exponent γ, incident at latitude λ, from a direction defined by angles θ, φ and subject therefore to an energy cut-off, $E_c(\lambda, \theta, \varphi)$, produces per second and column of 1 cm² cross section perpendicular to the direction of incidence n_c high energy interactions (leading to $n_c + I_c$ nucleons of energy $W \lesssim \omega$). I_c is the total flux of particles above cut-off and n_c is given by

$$n_c \approx \frac{I_c}{\log 1/\eta} \left[\log \frac{E_c - 1}{\omega} + \frac{1}{\gamma} + \frac{1}{(\gamma+1) E_c} \right]. \tag{3.2}$$

The total number of interactions is obtained by integrating over all angles of incidence.

The resulting low energy nucleons ejected as recoils in high energy collisions are then slowed down further

by inelastic collisions (which may occur either within the same or in another target nucleus),

by elastic collisions, and

by ionization losses (in the case of protons).

On the average the neutrons produce 2—3 more spallation reactions before being captured in nitrogen, while most of the protons come to rest because of ionization losses.

Since the slowing down of nucleons from ~ 500 MeV to rest is essentially a local phenomenon, the energy spectra below ~ 500 MeV are independent of altitude (except near the top of the atmosphere where neutrons can escape before being slowed down fully).

This model of the nuclear cascade differs from that usually employed but incorporates the existing information on the behaviour of high energy nucleons; the simple calculation outlined above reproduces rather closely the upper curve in Fig. 16a, which is based exclusively on experimental data and represents the star production rates in the atmosphere as a function of latitude. As shown below, particles other than nucleons, i.e. the mesons and their decay products, do not contribute effectively to isotope production in the atmosphere.

β) *Creation of unstable particles.* Apart from spallation, the major feature of nuclear collisions above 500 MeV is the creation of pions. The production spectrum of pions in the atmosphere and the resulting flux of charged pions, muons and γ-rays have been calculated by many authors. (For a recent treatment and comparison with observations, see f.i. PAL and PETERS (1964) [P5]).

In the energy region covered by existing accelerators the majority of pions have an observed energy and angular distribution which can be reproduced accurately by assuming that they are created with low energy ($\varepsilon_\pi \approx 400$ MeV) in a reference frame which moves with a velocity which is close to either that of the incident or the recoil nucleon [D6]. The pion production process is therefore well represented by a model which assumes that in this energy range pion production takes place mainly through excitation of the colliding nucleons to low lying excited states and the subsequent decay of the resulting baryon isobars [L22], [D6], [P5]. Except for an extra contribution from primaries of very great energy, which will be discussed below, this isobar model can therefore be used for estimating the flux and energy distribution of created particles and of their decay products and also for estimating their effects on transmutation of nuclei in the atmosphere.

The energy of pions created in a nuclear collision lies then in the range

$$E_\pi = \frac{\varepsilon_\pi E_N}{M_N} (1 \pm \beta_N \beta_\pi), \qquad (3.3)$$

where E_N, β_N are the laboratory energy and velocity of the emerging nucleon and $\varepsilon_\pi, \beta_\pi$ the corresponding quantities for pions in the reference system in which the nucleon is approximately at rest. (Velocities are expressed in terms of the velocity of light.)

The probability that a pion interacts in the atmosphere before decay is of the order

$$\left(\frac{x}{\lambda_\pi}\right)\left(\frac{\tau_\pi E_\pi}{h_0 m_\pi c}\right) \approx \left(\frac{x}{\lambda_\pi}\right)\left(\frac{E_\pi}{130}\right), \qquad (3.4)$$

where x is the atmospheric pressure, λ_π the interaction mean free path of pions, both expressed in g/cm^2, $h_0 (\approx 7$ km) is the scale height of the atmosphere, and E_π is expressed in GeV.

From Eqs. (3.3) and (3.4), together with the experimental fact that $\varepsilon_\pi \approx 400$ MeV, it follows that only pions which are associated with a nucleon of energy $E_N > 100$ GeV (and which in addition are emitted approximately in the forward direction in the rest system of such a nucleon) have a reasonable probability of interacting before decaying into a muon and a neutrino. Since the flux of these high energy nucleons is very low, pions created in such processes do not participate effectively in nuclear transmutations.

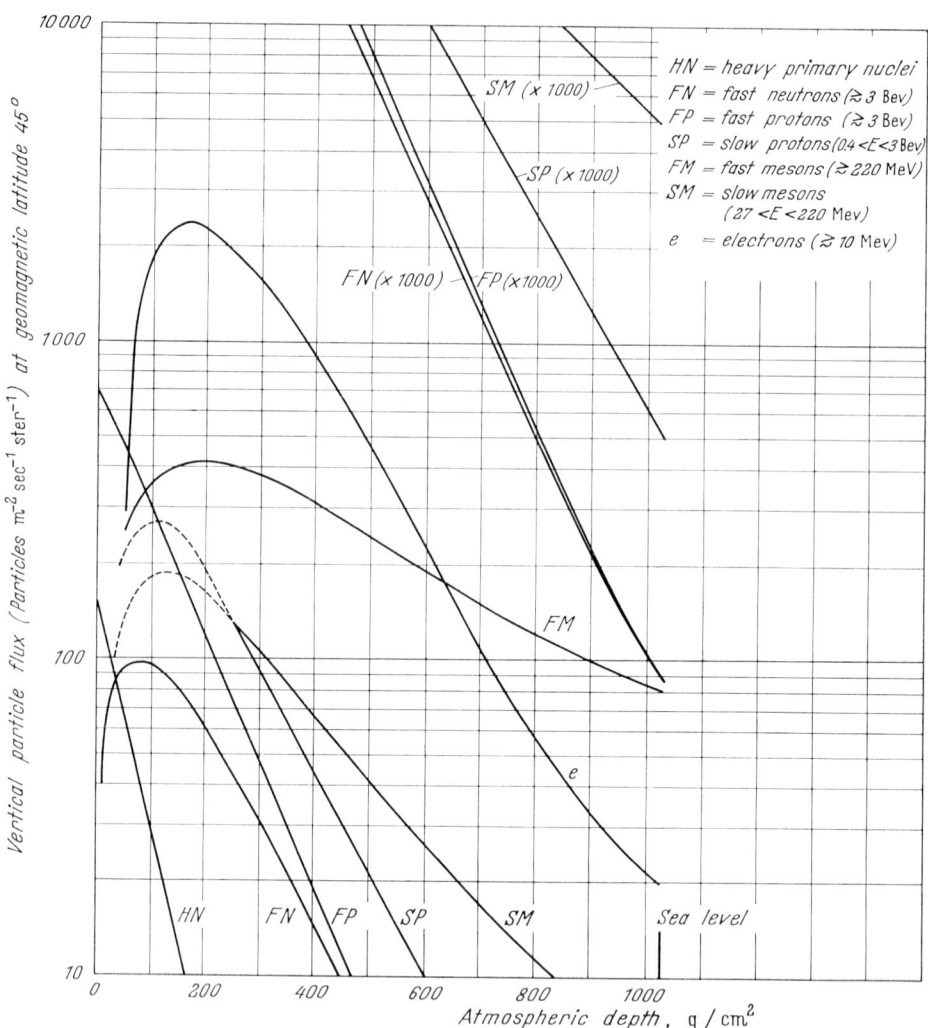

Fig. 3. Composition of cosmic radiation as a function of atmospheric depth. The flux values refer to geomagnetic latitude 45° and to longitude 80° West. (From PETERS [P3].)

In very energetic nuclear collisions (i.e. $E_N > 100$ GeV) another, additional process of pion creation plays an increasingly important role, the so-called fireball phenomenon, in which pions, again of energy $\varepsilon_\pi \approx 400$ MeV, are emitted in a reference frame which is not associated with either of the colliding nucleons but which is nearly at rest in the center-of-mass system of the colliding nucleons. These pions have laboratory energies in the range

$$E_\pi = \varepsilon_\pi \left[\sqrt{\frac{E_N + M_N}{2 M_N}} \pm \beta_\pi \sqrt{\frac{E_N - M_N}{2 M_N}} \right] \qquad (3.5)$$

and will decay in the atmosphere and not effectively participate in nuclear interactions unless the energy of the incident nucleon reaches a value of order $\sqrt{E_N M_N} \approx$ 100 GeV or $E_N \approx 10000$ GeV. Therefore, the pions created in the fireball process also do not contribute significantly to nuclear transmutations.

γ) *Meson decay products.* It remains to investigate whether any of the decay products of mesons (i.e. γ-rays and their resulting photon-electron showers in the case of neutral mesons, or muons in the case of charged mesons) can contribute to nuclear reactions.

Even a rough estimate shows that their contribution to isotope production in the atmosphere must be quite small.

Fig. 3 gives the altitude dependence of the vertical particle flux at 45° latitude [P3]. The γ-ray flux is not shown, but for the purpose of the estimate it can be taken to have the same altitude dependence and approximately the same intensity as the flux of electrons. (The photon: electron ratio is near unity at sea level and increases to a value of about two at 3500 m [P8].)

Multiplying the nuclear interaction cross-section for each kind of particle with its flux at the point where the flux reaches its maximum value, one can estimate the relative contribution to nuclear disintegrations:

$$\text{protons}: \gamma\text{-rays}: \text{muons} = 100:5:0.025.$$

It is shown in Sect. C.III that most of the isotope producing nuclear reactions in the atmosphere are due to slow neutrons; protons contribute only of an order of 10% to isotope production. It is then apparent that photo-nuclear reactions and the collision of muons with nuclei may be neglected when considering the atmosphere as a whole.

Locally, however, near the ground, where the absolute production rate of isotopes is very small, the relative contribution of γ-rays and muons is no longer negligible. This effect can be studied in the production of some of the very short-lived isotopes, such as Cl^{39} [W2], [L0].

C. Production of isotopes in the atmosphere.

4. A number of isotopes produced by cosmic rays in the atmosphere have been detected. In Table 1, we have listed in order of decreasing half-lives fourteen radioactive isotopes of half-lives longer than 1 day. The only stable isotope whose abundance on the earth can definitely be attributed to cosmic ray interactions is He^3; it has also been listed.

The table contains three radioisotopes which have not yet been detected, but whose half-lives are long enough to make them very useful from a geophysical point of view: Al^{26} (7.4×10^5 years), Ar^{39} (270 years) and Ar^{37} (35 days); slight improvements in low-level counting technique should make it possible to measure their concentration in the geosphere.

Not listed in the table are seven radioisotopes which have been detected [R9], [B6], [W2] but whose half-lives are shorter than one day: Mg^{28} (21 h), Na^{24} (15 h), S^{38} (2.9 h), Si^{31} (2.6 h), Cl^{39}(1 h), Cl^{38} (37 m), and Cl^{34m} (32 m). Because of their short half-lives, the application of these isotopes in geophysics is limited to the study of turbulent mixing and scavenging mechanisms in the lower troposphere; their production is not discussed in this article.

The atmospheric target nuclei primarily responsible for the production of the isotopes are shown in column 4 of Table 1. Since the abundance of argon in the atmosphere is low, ca. 1% by volume, isotopes of atomic weight less than 16 are produced mainly from interactions in N and O nuclei.

The atmospheric abundance of Ne, Kr and Xe is several orders of magnitude smaller than that of Ar; the detection of isotopes produced in argon spallations is possible though difficult, but existing technology is as yet adequate for measuring the concentration of isotopes produced from these less abundant rare gases.

However, an attempt was made recently to detect Kr^{81} (2.1×10^5 years) formed by neutron capture in Kr^{80}; the results are not yet conclusive [H3].

We shall now discuss the source functions of the various isotopes listed in Table 1.

Table 1. *Isotopes ($\tau_{\frac{1}{2}} > 1$ day) produced by cosmic rays in the atmosphere.*

Isotope	Half-life	Main radiation	Main target nuclide(s)	Reference
He^3	Stable	—	N, O	[F1]
Be^{10}	2.5×10^6 y	β^- — 550 KeV	N, O	[P1]
Al^{26}	7.4×10^5 y	β^+ — 1.17 MeV	Ar	Not detected
Cl^{36}	3.1×10^5 y	β^- — 714 KeV	Ar	[D2]
Kr^{81}	2.1×10^5 y	K — X ray	Kr	[H3]
C^{14}	5730 y	β^- — 156 KeV	N, O	[L16]
Si^{32}	500 y*	β^- — 100 KeV	Ar	[L14]
Ar^{39}	270 y	β^- — 565 KeV	Ar	Not detected
H^3	12.3 y	β^- — 18 KeV	N, O	[F1]
Na^{22}	2.6 y	β^+ — 540 KeV γ — 1.3 MeV	Ar	[M9]
S^{35}	87 d	β^- — 167 KeV	Ar	[G5]
Be^7	53 d	γ — 480 KeV	N, O	[A6]
Ar^{37}	35 d	K — X ray	Ar	Not detected
P^{33}	25 d	β^- — 250 KeV	Ar	[L12]
P^{32}	14.3 d	β^- — 1.7 MeV	Ar	[M8]

* Based on cross section measurements [H2].

I. Methods of evaluating the source functions.

5. The radioactive isotopes produced by cosmic rays are the only known atmospheric tracers whose introduction into the atmosphere occurs continuously at a rate which can be determined, both in space and time. Considerable efforts have therefore been made during the last few years towards determining these source functions in detail [L0]. Ideally one should have an accurate knowledge of the number and type of cosmic ray particles in the atmosphere, their energy spectrum, and the partial cross sections for the production of a particular nuclide in interactions which each type of atmospheric nuclide. The production rate, $C_j(x, \lambda)$, per gram of air of a nuclide, j, at an altitude corresponding to an atmospheric pressure of x gm cm^{-2} at latitude λ, is then given by

$$C_j(x, \lambda) = \sum_T \frac{N K_T}{A} \int_0^E \int_0^\theta \int_0^\varphi \sum_i J_i(x, E, \lambda, \theta, \varphi)\, \sigma_{i,j,T}(E)\, d\cos\theta\, d\varphi\, dE. \quad (5.1)$$

The function J_i represents the differential energy spectrum of the component, i, of cosmic radiation. N and A are the Avagadro's number and atomic weight of the target respectively. K_T is the fractional abundance by weight of a particular target nuclide, T, in the atmosphere, $\sigma_{i,j,T}(E)$ is the cross section for the production of nuclide, j, in the collision of a particle of component, i, and energy E with a target nuclide, T, (i.e. N^{14}, O^{16}, A^{40} etc.).

Strictly speaking, J_i is not only a function of geomagnetic latitude, λ, and the angles of incidence θ, φ, but also of longitude. However, the latter dependence is weak and has been neglected.

Furthermore, J_i is a function of time, being dependent on solar bursts and the general level of solar surface activity. The existing data on the distribution of the nucleonic, mesonic and the weakly interacting components in the atmosphere are adequate only for determining this function in the absence of major disturbances of the cosmic ray flux. The variation of J_i from its mean value can as yet be estimated only roughly, as will be done later in this section.

While the relevant cosmic ray data are adequate for evaluating nuclear reaction rates in the atmosphere, existing accelerator data on individual cross sections as a function of energy (i.e., cross sections for nuclear reactions leading to a particular end product) are, with few exceptions, too scanty to permit the integration of Eq. (5.1). Thus, more empirical methods for calculating the source functions had to be developed.

There exist several pieces of information on subsidiary, phenomenological properties of the cosmic radiation in the atmosphere, which can be used for constructing the source functions when nuclear data are lacking [L0]. The altitude and latitude variation of nuclear reaction rates can be obtained without reference to nuclear cross sections by using the measured altitude and latitude variation of bursts produced in ionization chambers and of stars in emulsions. Since these reactions occur in target materials different from those in the atmosphere, only relative but not absolute rates of reaction in air can be obtained in this manner and some procedure is required to normalize the data. Such a normalization can be achieved by using the rate of cosmic ray produced disintegrations observed in cloud chambers [B1], [B2] filled with nitrogen and argon at a particular geographic location on the earth's surface. Thus the rate of star production in each of the main constituents of the atmosphere can be evaluated quite accurately over the entire globe. Similarly, as will be shown, it is possible to evaluate the production rate of many isotopes without detailed information on spallation cross sections.

II. Altitude and latitude distribution of nuclear disintegrations in the atmosphere.

6. Throughout the lower atmosphere up to an altitude of about 12 km (a region comprising 80% of the air mass), there exists an equilibrium between star production and the number of neutrons and protons of energy $\lesssim 500$ MeV. Numerous experiments confirm this simple relationship between star production and the flux of low energy nucleons [S5], [S4], [S3], [S2]. As shown in Sect. B.II, nuclear reactions produced by high energy nucleons lead to recoil nucleons, i.e. low energy neutrons and protons, in equal numbers. The resulting steady state distribution of low energy nucleons is a purely local phenomenon since the final slowing down occurs in regions with dimensions of the order of a mean free path for nuclear encounters (50—100 gm·cm^{-2} of air). It is therefore to be expected that, throughout the equilibrium region in the atmosphere, the shape of neutron and of proton spectra below ca. 500 MeV remains the same at all latitudes and that the flux is proportional to the rate of star production. However, because the slowing-down mechanisms are different, the resulting steady state energy spectrum of protons is quite different from that of neutrons.

The total intensity of nucleons, the rate of star production, and also the shape of the nucleon spectra above ~ 500 MeV are of course latitude dependent because of the successively higher cut-off energies imposed on the incident primaries as one goes towards the equator. This is the reason why the flux of low energy nucleons and of nuclear disintegrations decrease less rapidly with altitude at lower than it does at high latitudes [S2]. The simple model of a nucleonic cascade discussed in Sect. B.II adequately explains these observed features.

In view of the simple relation which exists in the major part of the atmosphere between local disintegration rates and neutron flux, it is convenient to use the data on slow neutrons ($E \leq 30$ MeV) which are available for all latitudes and altitudes [S2], [S3], [L3] as an index for nuclear interaction rates.

The equilibrium between slow neutrons and the nuclear interactions in which they arise is disturbed as one approaches the uppermost region of the atmosphere[1], because some neutrons escape into space before being slowed down. Since this is essentially a geometric effect, the ratio between star production and slow neutrons increases with altitude but is not a function of latitude. To find the constant of proportionality at each height it is sufficient to compare the neutron flux at a given latitude and several altitudes with the star production rate measured in an arbitrary type of detector.

The star production rate in emulsions as a function of height at medium latitudes has been extensively investigated [L4], [R2]; with the help of these data it is possible to extend the relation between slow neutron flux and rate of total nuclear disintegrations into the upper regions of the atmosphere.

To complete the global map of nuclear disintegrations in the various atmospheric constituents it is still necessary to measure the absolute rate of disintegrations produced by cosmic rays in nitrogen, oxygen and argon separately at a particular locality. Such measurements have been carried out in nitrogen and argon filled cloud chambers [B1], [B2]. The data require corrections for experimental bias, but these corrections can be made fairly accurately [L3].

Experimental proof [L2] for the correctness of these star production rates in air used for normalization purposes has also been obtained by determining the rate of production of Be^7 in oxygen exposed at mountain altitude ($\lambda = 51°$ N, 685 gm·cm^{-2}). Since the inelastic cross section of oxygen and the excitation function of Be^7 in oxygen are known accurately [H4], [C4], the Be^7 production rate is a direct measure of the rate of nuclear disintegrations.

The number of nuclear disintegrations (stars) produced per unit time in the atmosphere as determined by the above procedure [L3] is shown in Fig. 4. The curves refer to disintegrations produced by particles of energy $E > 40$ MeV, because of the normalization data used. Furthermore, the altitude and latitude variation refers to a period of high solar activity, since the neutron curves are based on observations during 1948—1949.

The star production curves show a transition effect in the upper atmosphere between latitudes 0—50°. The position of the maxima is observed to shift to greater depths as one moves towards the equator. At the same time, the height of the transition maximum increases towards the equator. This behaviour must be expected and arises from the increasing hardness of the primary radiation as one proceeds to lower latitudes. At latitudes greater than 60°, the maxima vanish and the curves represent an initial exponential absorption of the star producing radiation with a mean absorption length of 70—80 gm·cm^{-2}, nearly equal to the interaction mean free path of fast nucleons in air. The near equality shows that most of the primary nucleons, at $\lambda > 60°$, may collide but produce no nucleons capable of further interactions. Since the majority of nucleons at these latitudes have energies less than 500 MeV, such a behaviour is to be expected.

---

[1] A disturbance in the simple proportionality relation occurs also near the surface of the earth; this is due mainly to neutron reflection, especially near large bodies of water. Since isotope production is very small at great atmospheric depth, this effect is not important for the evaluation of isotope production rates.

The general character of the curves of Fig. 4 can, therefore, be understood in terms of the known energy distribution of primary cosmic ray particles and present knowledge of high energy nuclear processes. Thereby they give support to the method used for obtaining the relative star production rates in the stratosphere.

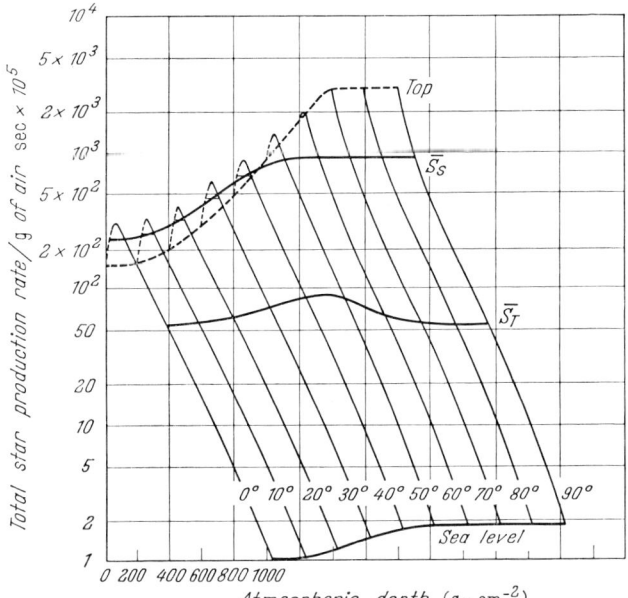

Fig. 4. The number of nuclear disintegrations (stars) per gram of air, per second (with an energy release above ∼40 MeV) plotted as a function of atmospheric pressure. In order to avoid overlapping, curves referring to different latitudes have been displaced with respect to each other along the abscissa. The production rate of H^3, He^3, Be^7 and Be^{10} is obtained by multiplying the ordinate scale with the yield factors 0.14, 0.12, 4.5×10^{-2} and 2.5×10^{-2}, respectively. (From LAL and PETERS [L0].)

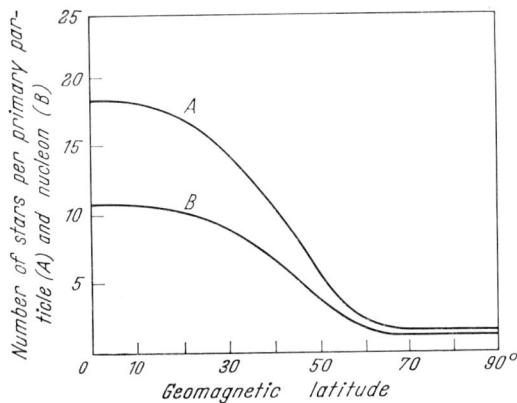

Fig. 5. The number of nuclear disintegrations per incident primary cosmic ray particle (A) and per incident primary nucleon (B) as a function of latitude. (From LAL [L13].)

Fig. 5 shows, as a function of geomagnetic latitude, the total number of stars produced in the atmosphere per primary particle and per primary nucleon. The approximate flatness of the curves between latitude 0° and 30° indicates that the increase in energy per primary particle above ∼3 BeV/nucleon is used up mainly in the production of mesons, most of which decay before interacting so that only a minor fraction of this additional energy appears in nuclear interactions. The drop in the number of stars per primary at medium latitudes corresponds to the fact that the large number of lower energy primaries admitted by the earth's field in these latitudes induce only a small number of nuclear interactions. The flattening of the curves for $\lambda > 60°$ corresponds to the fact that the additional primaries admitted at these high latitudes have so little energy that they are usually brought to rest by ionization, before making a nuclear collision.

III. Rates of production of isotopes in the atmosphere.

7. In this section we will discuss the production rates of all isotopes listed in Table 1. Most of these isotopes arise as a result of spallation, i.e. they are either emitted by the target nucleus or, more often, left behind as residue in nuclear disintegrations produced by nucleons. Some contributions come from nuclear reactions induced by photons and muons including capture of negative muons. For the isotopes listed in Table 1, however, these are negligible (Sect. B.II). The contribution from the capture of negative muons becomes important in the lower regions of the atmosphere, but only in those cases where the number of nucleons emitted from the target is very small, i.e. it affects mainly the short lived isotopes, Cl^{39}, S^{38}, whose production rates are not discussed in this article.

Apart from spallation reactions, isotope production in the atmosphere occurs by the capture of thermal neutrons. Two isotopes produced in this process are C^{14} and Kr^{81} arising from neutron capture in N^{14} and Kr^{80}, respectively.

The yield of various isotopes and their latitude-altitude distribution in the atmosphere will be discussed separately for the "spallation" and "neutron capture" produced isotopes.

α) *Yield of isotopes in spallation reactions.* The energy thresholds for the formation of the various nuclides listed in Table 1 differ considerably. The lowest energy is required in the case of C^{14} and Kr^{81}, i.e. the isotopes which arise chiefly not from ordinary spallation reactions but from the capture of slow neutrons through the exothermic reactions N^{14} (n, p) C^{14} and Kr^{80} (n, γ) Kr^{81}. All other reactions are endothermic with thresholds up to ~ 200 MeV. Only few isotopes, i.e. Na^{22}, Al^{26}, and also the short lived isotopes Na^{24}, Mg^{28} have thresholds close to the upper limit; for others it lies around 50 MeV.

In the lower part of the atmosphere the problem of determining the yield of various isotopes is particularly simple. As mentioned in the preceding section, the shape of the energy spectrum of neutrons and protons in the equilibrium region, i.e. at pressures greater than 200 g/cm^2, is independent of altitude and latitude. Therefore, throughout this part of the atmosphere the yield of all isotopes which are produced by spallation reactions (or by neutron capture) is strictly proportional to the star production rate, whatever the energy dependence of the excitation curves for individual reactions may be.

The general behaviour of yield functions, i.e. of production cross sections is fairly well known for isotopes with low production thresholds since a large number of cross sections in light and medium weight target nuclei have been measured in accelerator laboratories. The cross sections increase rapidly and reach peak values within a few tens of MeV above threshold, after which they decrease somewhat, usually by less than 30%, and then remain constant up to the highest energies. A typical curve is shown in Fig. 6 which represents Be^7 production in the bombardment of carbon with protons [H4], [C4]. Because of this behaviour of the yield functions and because of the comparatively low values of production thresholds, it is the interactions produced by neutrons which account for most of the isotope production.

Sect. 7. Rates of production of isotopes in the atmosphere. 567

This fact becomes immediately apparent when comparing the energy spectra of neutrons [C1], [H1], [R1] and protons shown in Fig. 7. In the high energy region the flux and energy spectrum of protons is the same as that of neutrons,

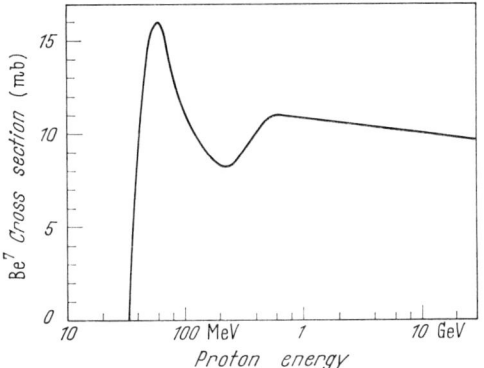

Fig. 6. Excitation function for the production of Be⁷ in carbon by fast protons.

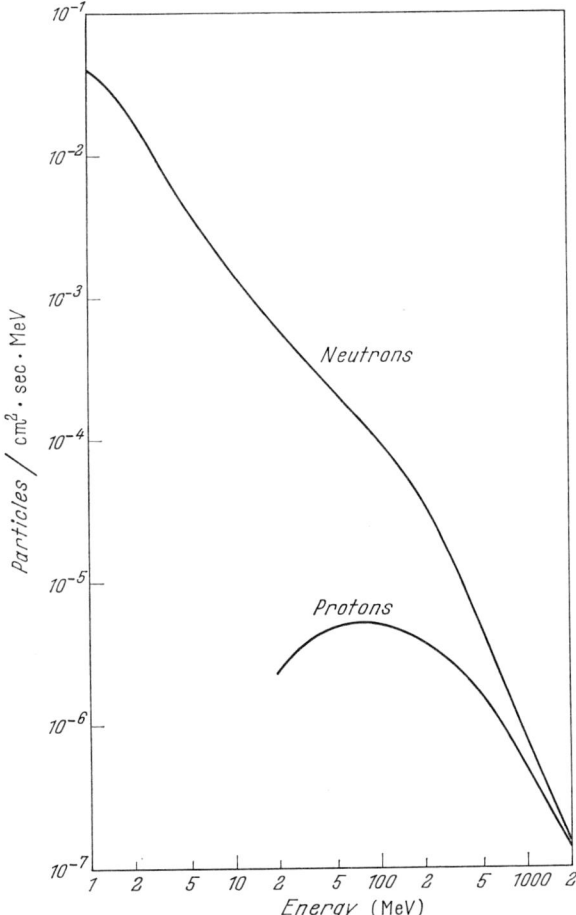

Fig. 7. The differential kinetic energy spectrum of nucleons in the atmosphere at $\lambda = 50°$, 680 gm cm⁻².

but below ~ 1 GeV, ionization loss causes the proton to neutron ratio to decrease. Already at 100 MeV, where the absolute proton flux in the equilibrium region reaches a maximum, the p/n ratio has fallen to about 3%.

It is clear from Fig. 7 and from the general behaviour of the cross sections that the major contribution to isotope production comes always from the part of the nucleon energy spectrum above, but near the threshold value for the particular reaction. This means that isotopes are mainly produced by the neutrons in the atmosphere; for isotopes with low production threshold the contribution of protons is relatively small ($\sim 10\%$), while for those with production thresholds near ~ 200 MeV the protons contribute approximately 25%.

In the upper part of the atmosphere, i.e. above the equilibrium region, the nucleon spectrum hardens and high energy interactions become relatively more frequent. This, as mentioned earlier, is due partly to the stars produced directly by particles of the primary cosmic radiation, partly to the escape of secondary neutrons into outer space before they had a chance to dissipate all their energy in collisions. Therefore, the formation of isotopes with high production thresholds becomes relatively more frequent as one approaches the upper boundary of the atmosphere. The magnitude of this effect can be estimated by studying the change in the prong size distribution of stars with altitude. Star prongs are caused by hydrogen and helium isotopes (fragments with atomic number higher than two occur rarely except as residue of the target nucleus).

As an example, one may consider the low threshold isotopes Cl^{36}, S^{35} and the high threshold isotope Na^{22}. All three arise from the spallation of A^{40}, the former in 1 or 1 and 2 prong stars, the latter in stars with 4, 5, 6 or 7 prongs. The variation in the prong size distribution with altitude permits therefore a first order correction to the yield per star measured at low altitude. This seems adequate since the corrections are always small. The altitude and latitude variation of isotope production (i.e. the product of yield and star production rate) can be established in this manner for all isotopes with atomic number $Z > 2$.

Normalization of the curves can be obtained by measuring the absolute rate of isotope production in the appropriate target material exposed to cosmic rays at a given altitude and latitude. Such data exist [L2] for the short lived isotopes Be^7, S^{35}, P^{33}, and P^{32}. Absolute production rates for isotopes of half-lives <1 day, Mg^{28}, Na^{24} etc. could be obtained in the same way, but this has not yet been done. Where applicable, this method for obtaining source functions is the most reliable one available at present. It is designated as method A.

It seems hardly feasible to extend this method to obtaining the yield function for long lived isotopes like Be^{10}, Cl^{36}, Al^{26} and Si^{32}, because of the weakness of the activity produced in an exposure of reasonable duration. Here the best method available seems to be to evaluate the total yield of a particular element using the prong number distribution of cosmic ray produced stars and then to distribute the yield among its various stable and radioactive isotopes by means of some semi-empirical formula describing spallation reactions [R3]. Such semi-empirical relations are not valid for spallation products close to the target nucleus [R3], [H2]. In such cases, f.i. Ar^{39}, Cl^{36}, it is better to base one's estimate on the general characteristics of excitation functions for the appropriate reactions in target nuclei of comparable mass. These methods of obtaining isotope yields are designated as B.

The procedure outlined above is not satisfactory for H^3 and He^3 because these isotopes occur not only as residual target material, but occur also among the star prongs so that their production depends sensitively on the nature of the star. Fortunately, one has in the case of tritium sufficient accelerator data [C1] on the

Sect. 7. Rates of production of isotopes in the atmosphere.

production cross section as a function of energy, so that the source function can be obtained with the help of Eq. (5.1) and the nucleon energy spectra. Since adequate cross section measurements, however, are only available for incident

Table 2. *Yield of isotopes in spallation reactions**.
(The yield refers to disintegrations in the target nuclei of interest, i.e. N+O, or Ar.)

Nuclide	Method	Yield/star	Latitude-altitude distribution curves	Nuclide	Method	Yield/star	Latitude-altitude distribution curves
He^3	C	1.2×10^{-1}	Fig. 4	Na^{22}	B	3.5×10^{-3}	Fig. 10
Be^{10}	B, C	2.5×10^{-2}	Fig. 4	S^{35}	A	9.5×10^{-2}	Fig. 8
Al^{26}	B	5.5×10^{-3}	Fig. 10	Be^7	A, C	4.5×10^{-2}	Fig. 4
Cl^{36}	B	7.7×10^{-2}	Fig. 8	P^{33}	A	4.2×10^{-2}	Fig. 9
Si^{32}	B	1×10^{-2}	Fig. 9	P^{32}	A	5.1×10^{-2}	Fig. 9
H^3	C	1.4×10^{-1}	Fig. 4				

* The values refer to $\lambda = 51°$, $x = 680$ gm·cm^{-2}.

protons, not for neutrons, and since the targets used do not correspond accurately to the nuclear composition of the atmosphere, a certain amount of extrapolation is necessary.

This method of deriving the source functions is designated as method C. It can also be applied in the case of Be^7 where sufficient accelerator data are available;

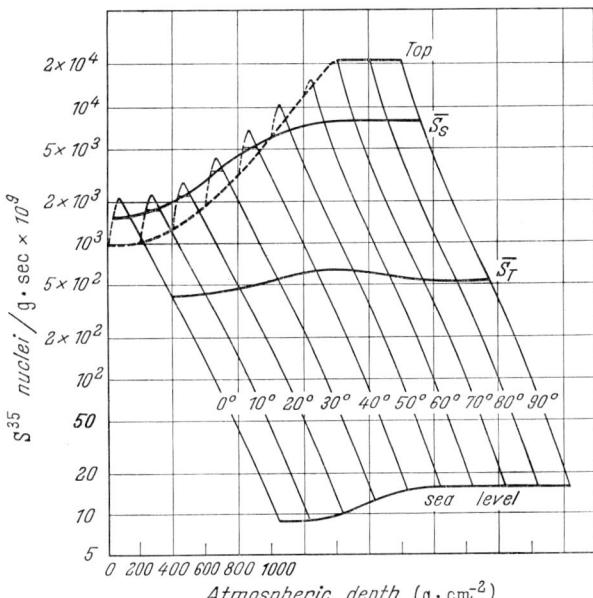

Fig. 8. The number of S^{35} atoms produced by cosmic rays per second per gram of air plotted as a function of atmospheric pressure. Curves belonging to different latitudes have been displaced with respect to each other along the abscissa. The production rate of Cl^{36} is obtained by multiplying the ordinate by 0.8. (From LAL and PETERS [L0].)

in the latter case the result has been used as a check on the source function derived by method A.

Applying one or the other of these procedures, the yields of isotopes in nuclear disintegrations have been estimated for different regions of the atmosphere [L0], [L7], [B6].

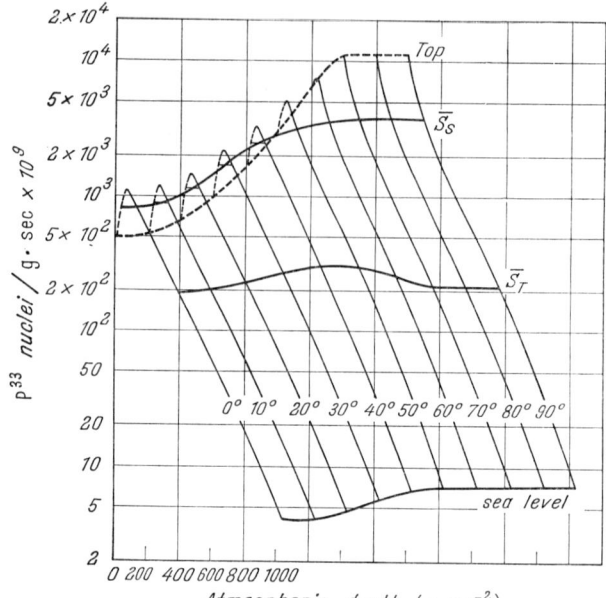

Fig. 9. The number of P^{33} atoms produced by cosmic rays per second per gram of air plotted as a function of atmospheric pressure. Curves belonging to different latitudes have been displaced with respect to each other along the abscissa. The production rates of Si^{32} and P^{32} are obtained by multiplying the ordinate with factors 0.24 and 1.2, respectively. (From LAL and PETERS [*L0*].)

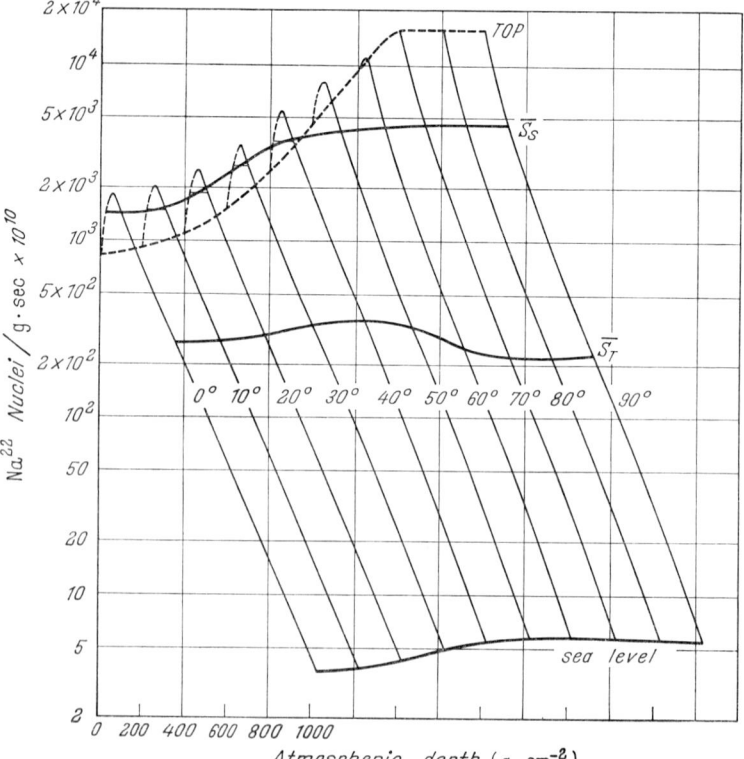

Fig. 10. The number of Na^{22} atoms produced by cosmic rays per second per gram of air as a function of atmospheric pressure. Curves belonging to different latitudes have been displaced with respect to each other along the abscissa. The production rate of Al^{26} is obtained by multiplying the ordinate by 1.6.

Sect. 7.   Rates of production of isotopes in the atmosphere.   571

The method used and the estimated yields per disintegration in the appropriate target gas, (i.e. N and O or Ar) in the equilibrium region of the atmosphere are summarized in Table 2.1II. The yield of He^3 in nuclear disintegrations is based on the results of cross section measurements available [B16], [F3], [G10], [K5] for carbon, oxygen, and medium weight nuclei; no measurements are yet available for nitrogen.

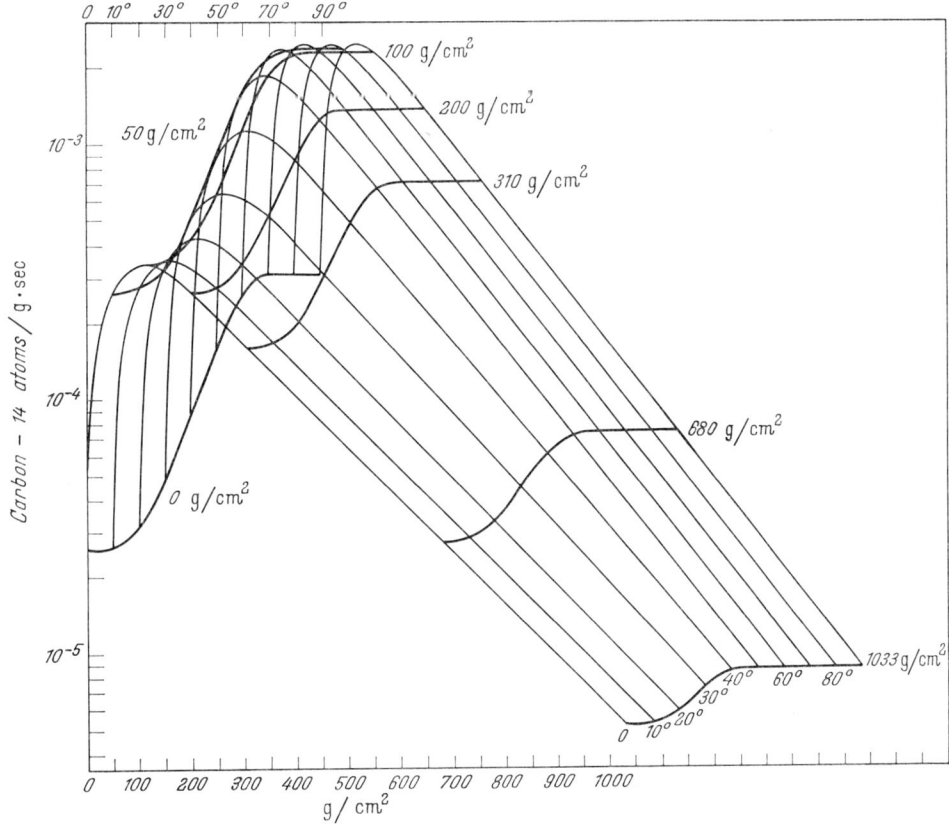

Fig. 11.

Figs. 11 and 12 show the highest and the lowest C^{14} production rates to be expected during a sun spot cycle. The number of C^{14} nuclei produced per gram of air per second is plotted as a function of atmospheric pressure for various latitudes. The curves have been normalized to a production rate of one neutron/sec cm² column at the magnetic poles during sun spot minimum. (From Lingenfelter [L10].)

Changes in the yield of various isotopes between the equilibrium region and the top of the atmosphere have been derived using both the changes in the star size distribution and the change in the energy spectra of the nucleonic component, based on direct observations in emulsions and in gas filled cloud chambers [B1], [B2], [R2], [L3]. If one groups together isotopes which have similar production thresholds so that their relative yields remain constant within 5% throughout the atmosphere, then four altitude-latitude curves suffice to describe the production rates of all isotopes listed in Table 1.

The nuclear disintegration rates given in Fig. 4 adequately describe the altitude and latitude variation of the isotopes H^3, He^3, Be^7, and Be^{10}. The production rates of S^{35}, Cl^{36}, and Si^{32}, $P^{33, 32}$ and Na^{22}, Al^{26} are given by Figs. 8, 9, and 10, respectively.

The global average rates of nuclear disintegrations in nitrogen plus oxygen, and in argon are estimated to be 1.78 and $1.6 \times 10^{-2}/\text{cm}^2 \cdot \text{column} \cdot \text{sec}$, respectively, from Fig. 4; stars in argon represent 0.9% of all disintegrations in the atmosphere.

β) *Yield of isotopes in reactions due to the capture of thermal neutrons.* Of all the isotopes listed in Table 1, the thermal capture process is of importance only for C^{14}, which is produced by the reaction N^{14} (n, p) C^{14}. The reaction is exothermic with a cross section of 1.8 barns for thermal neutrons.

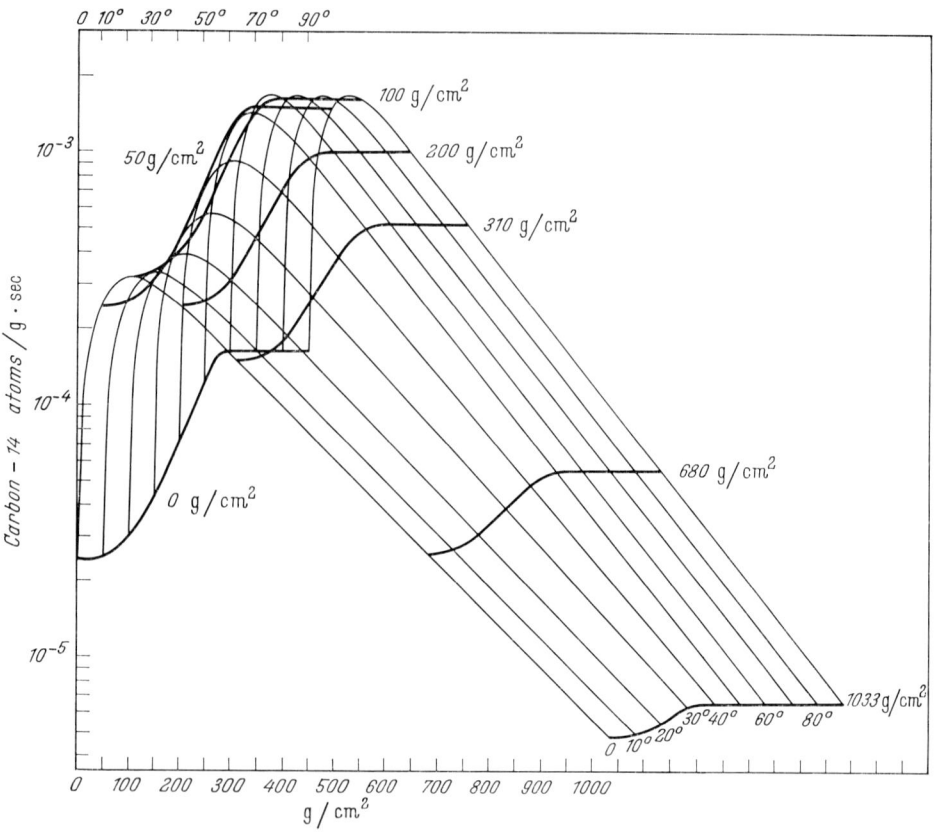

Fig. 12. Production rate of C^{14} in the atmosphere during sum-spot maximum; for other details see Caption to Fig. 11.

A detailed study of C^{14} formation at all altitudes and latitudes and over the entire energy region has been carried out [*L 10*]. The results are summarized in Figs. 11 and 12 which represent C^{14} production throughout the atmosphere for the two periods 1953/54 (solar minimum) and 1957/58 (solar maximum), respectively. The values include the comparatively small contribution from spallation reactions both in nitrogen and other constituents of the air.

Averaged over the last 10 solar cycles the global production rate is found to be 2.5 ± 0.5 atoms/cm² sec; the global rates during solar minimum and maximum of 1953/54 and 1957/58 are estimated to be 2.6 and 2.08, respectively [*L 10*].

The isotope Kr^{81}, expected to be produced by the capture of thermal neutrons in atmospheric Kr^{80}, has not yet been detected with certainty [*H 3*]. The latitude and altitude dependence of the production is expected to be identical with that of C^{14}. Its global production rate has been estimated [H3] to lie between 10^{-6} and 1.5×10^{-7} atoms/cm² · sec⁻¹, the uncertainty being mainly due to lack of knowledge of capture cross-section [M1], [R8] of thermal neutrons in Kr^{80}.

IV. Time variations in isotope production.

8. The production of isotopes in the earth's atmosphere may vary either because of changes in the flux and energy spectrum of particles in interplanetary space, or because of changes in the geomagnetic field which partially shields the earth from these particles.

It is convenient to discuss the changes under three different headings:

α) Long term variations, i.e. changes which may have occurred in earlier geological epochs or at least have preceded the period of direct cosmic ray observations.

β) Variations established by direct observations of the particle flux during the past ~ 30 years. The most prominent features here are the periodic changes which occur in anti-correlation with the 11 year cycle of sun spot numbers.

γ) Variations of a more local and temporary character caused by injection of low energy particles into planetary space during solar flare events.

While under β) and γ) we consider the effect of observed time variations in cosmic radiation on the production of isotopes, the problem to be discussed under α) is the inverse, namely, what can one learn from observations on isotope concentrations about the early history of the primary cosmic ray flux?

α) *Long term variations.* The intensity and energy spectrum of cosmic ray particles in the periods which preceded their discovery and more or less continuous registration can be deduced to some extent from the cumulative isotopic changes which have taken place in irradiated materials. In addition to terrestrial material one can examine meteorites and one may hope that samples of interplanetary dust and of the moon's surface will become available for similar studies in not too distant a future.

Information on prehistory which one may expect to derive from these different materials, though similar, is not identical.

The isotopic changes produced in meteorites at some depth below the skin represent presumably a fairly unadulterated record of the galactic component of the cosmic radiation. If the integrated effects of particles emitted in solar bursts are important they may be expected to appear prominently, mainly in the surface layer.

As yet there exists no clear evidence for accumulated effects of solar corpuscular radiation, which should manifest themselves by exceptionally high radioactivity near the surface. However, meteorites may not be the best detectors for past solar activity; those from the asteroidal belt are presumably irradiated at distances from the sun much larger than one astronomical unit, so that the effect of solar particles could well be less pronounced in meteoritic samples than f.i. on the surface of the moon. Thus, a comparison between the distribution of isotopic changes in meteorites and on the moon may eventually make it possible to separate effects of galactic and solar particles and thereby open the possibility of studying solar activity in the remote past.

The analysis of radiation effects in terrestrial materials involves additional complexities. The effects depend less sensitively on solar bursts than they do in lunar material, because of the shielding provided by the geomagnetic field. But for that very reason the isotopic changes produced on earth reflect a different set of variables: possible time variations in the cosmic radiation as well as major changes in the earth field in ancient times.

Thus, in order to explore the past history of cosmic radiation, it is clearly important to study cosmic ray induced isotopic changes in all types of samples which can be obtained.

The present state of knowledge on long term cosmic ray changes is summarized below, but it should be borne in mind that the interpretation of the available data is often based on uncertain though perhaps plausible assumptions and that the study of past history of cosmic radiation is still in its infancy.

(i) *Information derived from meteorites.* Recent investigations involve the simultaneous determination of the concentrations of a large number of cosmogenic isotopes in the same meteorite sample. The radioisotopes: Be^{10}, Mn^{53}, Al^{26} ... P^{32} form a fairly unbroken sequence with respect to half-lives and permit a comparison of cosmic ray intensity extending from the most recent past back to $\sim 10^7$ years. The observed concentrations suggest that, during this period, cosmic ray intensity has remained the same within an experimental uncertainty of a factor two [A 3], [A 4], [G 3]. Furthermore, from analyses of stable and radioactive isotopes in meteorites of different size, it can be concluded that the energy spectra of galactic cosmic radiation during the past hundreds of millions of years has also been essentially the same as at present [H 5].

The above conclusions do not exclude any short term changes, defined as changes which lasted for only a small fraction of the time which separates us from their occurrence. Thus, large bursts in the galactic radiation lasting for a few tens of thousand years may have occurred millions of years ago without having been detected by these means.

For studies going further back into the past, one has so far only one isotope which seems suitable. Spallation reactions in iron meteorites produce a measurable activity due to K^{40} ($\tau_{\frac{1}{2}}=1.2\times 10^9$ years). It seems possible to get information on the cosmic ray intensity averaged over billion year periods from the measured K^{40} activity, provided one can find samples which have been irradiated over such periods. While existing measurements in suitable meteorites seem to point towards a fairly constant level of cosmic ray flux even over billion year periods, such conclusions must be considered as preliminary and uncertain [V 2], [V 3], [H 5].

A detailed account of this subject is given by M. Honda and J. R. Arnold in Chapter VIII of this book.

(ii) *Information derived from terrestrial samples.* In meteorites the produced isotopes remain in place, but those produced in the earth's atmosphere are in motion and become dispersed among various geophysical reservoirs. The nature of the geophysical and geochemical processes responsible for this dispersion are often unknown, and it is necessary in that case to evaluate the entire global inventory of a particular isotope rather than to study its concentration in a sample of irradiated material.

This is a complication, but not always a very serious one. It can be done relatively easily in the case of Kr^{81}, because $\sim 98\%$ of its inventory is in the atmosphere [R 10], [B 4] where its concentration is expected to be uniform throughout. Similarly, the bulk of the C^{14} inventory resides in the oceans so that a fairly precise global inventory can be obtained from a limited number of sea water analyses [B 13]. Thus, a possible approach to a study of the time variations of cosmic ray intensity based on terrestrial materials consists in measuring the concentrations of isotopes in the atmosphere and in sea water.

The study of Kr^{81} seems to be particularly useful for an evaluation of the mean cosmic ray intensity, more precisely the global neutron source function in the atmosphere, during the last mean life of Kr^{81}, i.e. 0.3×10^6 yrs. The single measurement available so far refers to the specific activity of an atmospheric krypton sample collected prior to the nuclear weapon era. The measured value is found to be 0.007 d.p.m./litre with an estimated error of $\pm 100\%$. It corresponds to a mean production rate of 10^{-7} Kr^{81} atoms/cm² · sec. This value is not inconsistent with an assumption of constancy of the average cosmic ray flux and an estimated present day production rate of 1.5×10^{-7} atoms/cm² · sec based on a neutron capture cross-section of 12.5 barns [R8] (Sect. C.IIIβ).

Estimates of cosmic ray intensity based on the global inventory of C^{14} are more accurate but cannot be extended as far back into the past. Only 2% of the C^{14} inventory resides in the atmosphere, most of it is dissolved in the oceans [C5]. The average specific concentration in various oceans, i.e. the ratio C^{14}/C^{12}, lies between 80% and 90% of the atmospheric value. From this one obtains a global C^{14} inventory [C5], [F3] of 1.9 d.p.s./cm², which is accurate to about $\pm 10\%$ and is in good agreement with existing estimates based on the present cosmic ray flux (Sect. C.IIIβ). One concludes from these data that the primary flux averaged over the mean life of C^{14} (that is over ca. 8000 years) has the same value as today with an accuracy of $\sim 20\%$.

It is not always necessary to estimate the global store of the isotope in order to evaluate its cosmic ray intensity in the past. There exist terrestrial samples, known to have remained in place with the isotopic composition frozen in at some particular stage of history. An unbroken record of the atmospheric C^{14} concentration extending back to 3500 years exists in the rings of living trees [S8], [L18]. The record is extended further into the past (to ~ 5000 years) with the help of wood and of other organic material forming part of archaeological specimen of known age [L18].

The specific concentration of C^{14} in organic carbon is directly related to its value in the atmosphere at the time when the plant or animal was growing. Results based on measurements of the C^{14}/C^{12} ratio in tree rings and in archaeological specimen of known age show that C^{14} production has remained unchanged within $\sim 5\%$ during the past 5000 years [L18]. Within this limit, a real variation of the order of 2% seems to have occurred about 300 yrs. ago. Whether the observed variation should be attributed to a change in solar particle emission or to a change in the earth's magnetic field or to still other unsuspected causes is not known [L18], [D4].

Other undisturbed regions which contain a more or less unbroken record of isotope production may be found on the polar ice caps and in deep sea marine sediments. Although there are fewer suitable isotopes available than in meteorites, H^3, Si^{32}, Al^{26}, and Be^{10} represent a reasonably good sequence of half-lives to study cosmic ray variations within the last $\sim 10^7$ years. Compared to meteorites, the orderly arranged accumulations such as tree rings, fossil ice layers or undisturbed marine sediments have the important advantage of preserving also records of comparatively short lived disturbances in cosmic ray intensity.

Measurements of the specific activity of C^{14} in the biogenous fraction of certain marine sediments, dated by the Ionium-Protactinium method, indicate that the cosmic ray intensity has been the same during the last ca. 10^4 years [K6], [R11], a result in agreement with the "atmospheric" C^{14} record.

The amount of Al^{26} and Be^{10} in slowly accumulating sediments may reflect cosmic ray intensity variations in the more distant past [P7], [L7]. The only measurements [K1] of Be^{10} in a seemingly undisturbed sediment core estimated to represent deposition during the last 2×10^5 years appears to support the conclusion that the cosmic ray intensity has been essentially constant during this period. If further work should establish the reliability of this method, it could in principle lead to the reconstruction of an unbroken intensity record extending back for about 10^7 years.

Studies on the polar cap are particularly interesting from the point of view of possible solar contributions to isotope production, since there the earth's field is comparatively ineffective in shielding the atmosphere. The studies of H^3 in Greenland Ice have given an indication that the annual rate of its accumulation is inversely correlated with the sunspot cycle, which implies that, in general, the galactic component of primary radiation dominates the production of this isotope [B8], [R7]. However, in one of these experiments, there also seem to be a few examples of excess H^3 in thin sections of ice deposited during high solar activity, which may indicate a temporary but significant contribution from the sun [R7]. When such studies have been carried out at different locations and combined with an accurate account of structural discontinuities (due to melting during summer, which makes it possible to distinguish successive annual deposition layers), one should be able to put such conclusions on a firm footing. Because of the short half-life of H^3 these researches into the history of cosmic radiation are restricted to a period covering the last 3—4 sunspot cycles.

Recent experimental work indicates that in the near future it may be possible to extend the time scale [D5] to a few thousand years with the help of Si^{32}. Attempts to extend the time scale further by discovering Al^{26} or Be^{10} in fossil ice have as yet not been successful.

β) *Variations connected with recurrent changes in cosmic ray intensity.* The absolute intensity as well as the energy spectrum of particles registered by various instruments change in time far more than can be accounted for by statistical fluctuations. Some variations are due to changes in the number and energy of primaries which enter the atmosphere, others to changes of pressure and temperature [P3], [W7].

Many of these intensity fluctuations are very small or of so short duration compared to the half-lives or atmospheric residence time of isotopes that they do not affect concentration or inventory, and can be neglected. One may exclude from the discussion on these grounds all meteorological effects whether random, diurnal or seasonal because they are small ($\sim 1\%$); similarly one may omit variations which arise during the magnetic storms caused by the collision of solar plasma clouds with the earth's magnetosphere. Resulting changes in cut-off rigidity produce modulation in the primary flux at different latitudes, but such fluctuations are too small and too short-lived to be of practical importance for isotope production.

(i) *Forbush type decreases.* The same plasma clouds which are responsible for the geomagnetic storms frequently produce very substantial reductions in cosmic ray intensity at large distances from the earth. The clouds contain and carry with them magnetic fields and therfore are opaque to direct traversal by charged particles.

As they expand into the interplanetary space they sweep cosmic ray particles out of the volume which they occupy, but after some time particles diffuse back into the volume and restore the flux density. This phenomenon is believed to be the origin of the so-called Forbush decreases. The term designates a sudden drop of primary intensity observable simultaneously over the entire globe (and also at appreciable distance from the earth) followed by a gradual return to normal, usually within one or two days. Being connected with the emission of plasma clouds, the frequency of Forbush decreases is strongly correlated with the degree of solar surface activity. Furthermore, since the plasma streams leaving the sun are often associated with quasistationary active regions on the sun's surface, such disturbances of the primary flux have a tendency to recur with the 27 day period of solar rotation [D5], [S10].

The magnitude of Forbush decreases outside the magnetosphere is sometimes as large as 40% but it is less for energetic particles ($<5\%$ at equatorial latitudes); the restoration to normal intensity occurs more or less exponentially with a recovery time of a few days, but sometimes longer. The changes are strongly rigidity dependent and extend up to few tens of GV rigidity [W7]. The variability of Forbush decreases from event to event makes it difficult to estimate their effect on isotope production.

However, since the frequency of these short-lived decreases is correlated with solar activity, their effect on isotope production is included approximately when evaluating the changes of the intensity and energy spectra associated with the solar cycle.

(ii) *The 11 year cycle.* Even though the effect of individual Forbush decreases on isotope production is small, their cumulative effects are important. It seems that the recovery of cosmic ray flux in the interior of these plasma clouds is not quite complete, so that, as more and more plasma streams leave the sun during the ascending branch of the 11 year sunspot cycle, the regions of reduced transparency begin to fill an increasingly large volume of the solar system. Since the opacity of a cloud containing magnetic fields is largest for particles of low magnetic rigidity, the low energy part of the galactic primary spectrum is affected most severely. One would expect the intensity to fall in anticorrelation with the sunspot number, i.e. with an 11 year period and with some lag in phase. The higher the particle energy, the smaller should be the amplitude of the 11 year intensity variation and the greater the phase lag.

Such a causal relation between the observed 11 year variation of cosmic ray flux and sunspot number as has been sketched here is reasonably well established [S9], [S10]; the modulation phenomenon, as revealed by measurements extending by now over nearly two complete solar cycles, can be described as follows:

1. The amplitude of the intensity modulation is larger for low energy than for high energy particles, as can be seen from Fig. 1.

2. The modulation is rigidity dependent, which means it is identical for all particles which have the same ratio of momentum to charge (i.e. to atomic number). It was discussed earlier (Sect. B.I.α) that the rigidity spectra of singly and multiply charged galactic cosmic ray particles are identical within errors of measurements above 1.5 GV. The observations on modulation imply that the relative intensities of different cosmic ray components for $R>1.5$ GV remain the same at all times.

3. The modulation displays a phase lag with respect to the sunspot number, which depends on the rigidity interval and on whether the sunspot number is decreasing or increasing. Thus, during one complete solar cycle, the modulated cosmic ray intensity is a double valued function of sunspot number.

4. In the ascending branch of the sunspot number cycle the phase lag increases with increasing rigidity. As the sunspot number rises, modulation sets in with less than 3 months delay for particles of a rigidity of $1.5-2$ GeV, while for particles of rigidity >14 GeV, the time lag is about 1 year.

During sunspot minimum, the primary cosmic ray spectrum observed at the top of the atmosphere comes closest to the spectrum which exists in the interstellar space. Yet it is probable that even then, when solar activity is at its lowest, the flux of particles of rigidity less than ~1.5 GV observed within the solar system is lower than it is beyond the limits of the solar system.

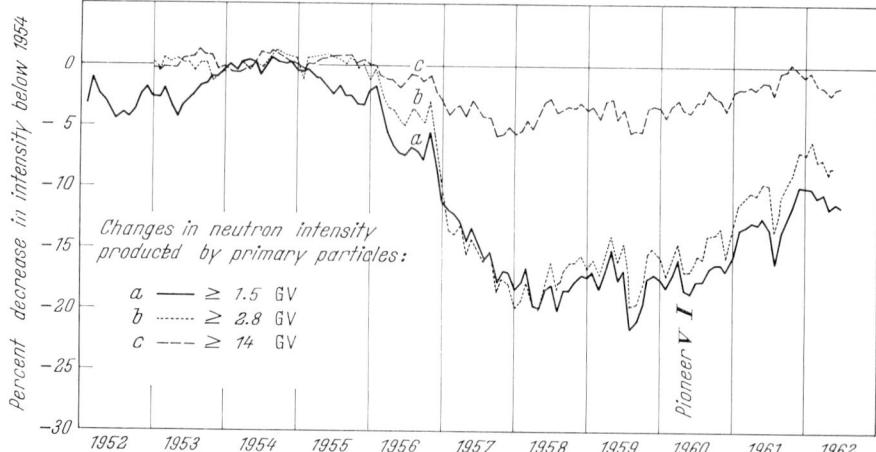

Fig. 13. Decrease of neutron intensity at sea level below the 1954 value (in percent), during the last solar cycle. (After SIMPSON [S9].)

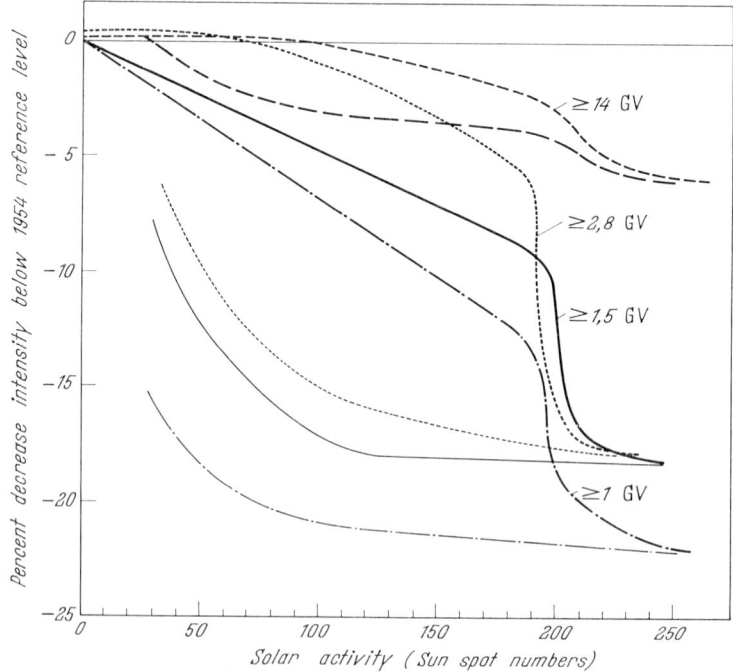

Fig. 14. Changes in the sea level neutron flux due to primaries in different rigidity intervals, as a function of sun spot number. The flux is plotted as a percentage decrease below the 1954 value. The graph exhibits the fact that the correlation between primary flux and sun spot number is different during the ascending and the descending branch of the sun spot cycle. (From SIMPSON [S9].)

The average monthly neutron intensities observed on the ground during the last solar cycle are shown in Fig. 13. In Fig. 14 these and similar observations at other latitudes have been used to illustrate the change of primary flux in different energy intervals with sun spot number. The curves exhibit clearly the hysterisis effect. The hysterisis loop becomes smaller as the rigidity increases.

The intensity of the nucleonic component during 1958 and 1954 is shown in Fig. 15 for different positions in the atmosphere and at different latitudes [$L0$]. Similar curves can be obtained from the observed spectral changes in the primary galactic radiation for other periods during the last decade. The changes between 1954 and 1958 are an illustration of extreme variation. The solar activity was at minimum in 1954; the year 1958 represents a period of unusually high solar activity, in fact the highest in the available record [$S11$] which extends over about 200 years.

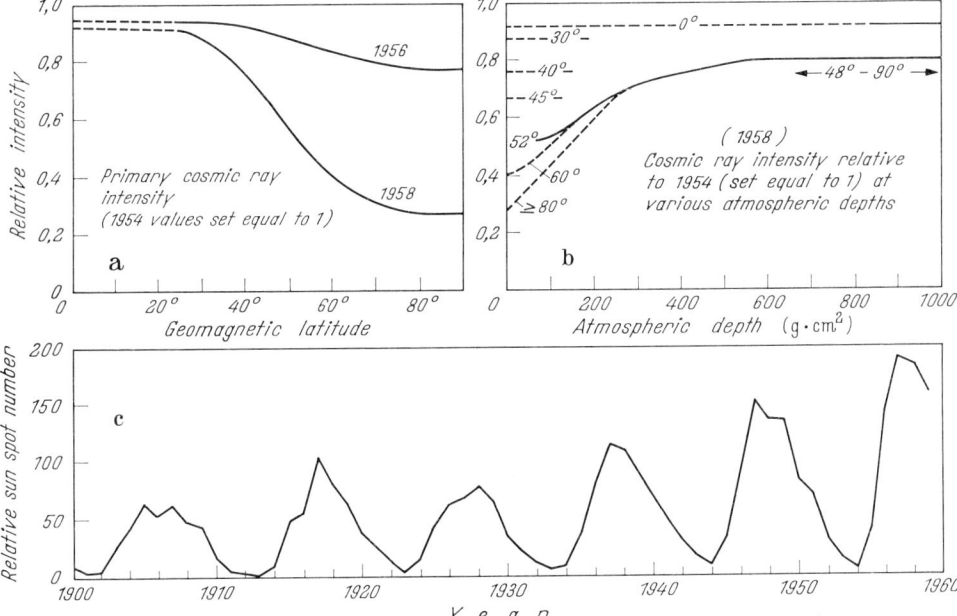

Fig. 15. a—c. a Primary cosmic ray flux in 1956 and 1958 relative to the flux in 1954 plotted as a function of geomagnetic latitude. b Cosmic ray intensity during 1958 at various latitudes plotted relative to the 1954 intensity as a function of atmospheric pressure. c Annual sun spot numbers since 1900. (From LAL and PETERS [$L0$].)

Fig. 15 illustrates that, because of the strong rigidity dependence of the modulation amplitude, the largest changes occur in the polar region and in the upper most 100 gm·cm^{-2} of atmosphere; during 1958 the nucleon flux dropped to ∼30% of the 1954 value. At depths greater than 300 gm·cm^{-2}, the decrease amounted to ∼25% at latitudes above 45°; below 45°, in the same depth interval, the effect diminished steadily, reaching a value of ∼7% at the equator.

Isotope production rates must be expected to follow closely the changes in nucleon intensities. Therefore, the observations which have been made during the last solar cycle make it possible to deduce the corresponding time variation in isotope production at all latitudes and altitudes with fair accuracy.

γ) *Sudden variations connected with solar particle bursts.* It is important to investigate whether particles emitted by the sun can contribute appreciably to the production of isotopes in the atmosphere and need to be considered in the estimate of total production rates.

As discussed in B. I.β, bursts of charged particles occasionally reach the earth. Since they consist predominantly of low energy protons (in the range of few tens of MeV), their arrival is confined essentially to the polar regions. These bursts, which are the result of acceleration and an emission of particles from the sun, last typically a few days, during which both the flux and energy spectra change. The number of bursts depends on the general level of solar activity; the average value over a solar cycle is of the order of ten per year. Integrated over a three day period, fluxes as high as 10^{10} particles/cm^2 ($E_{kin} > 20$ MeV) have been observed [W 8], [L 6]. This should be compared with the flux of galactic particles entering near the pole, which at its highest during solar minimum may reach a few times 10^5 particles/cm^2 in a comparable period. Therefore, the flux of low energy particles in the immediate vicinity of the geomagnetic poles is almost certainly dominated by the solar contribution but, due to the rapid decrease of solar particle flux with increasing magnetic rigidity, it is the galactic component which dominates in other regions.

The number of particles in solar bursts decreases so rapidly with energy ($I \sim E^{-4}$ to E^{-6}) that most of the isotopes with production threshold near ~ 40 MeV are created in a thin shell of the polar atmosphere of an effective thickness which corresponds to the range in air of protons with energies close to the threshold value, i.e. less than 2 gm·cm^{-2}. The corresponding thickness is 0.3 gm·cm^{-2} for protons above 10 MeV. Since the range-energy relation is considerably flatter than the number-energy spectrum of solar protons, isotopes which have the lowest threshold energy of production will be produced most copiously [L 0]. This means that the isotope whose inventory is most affected by solar bursts is C^{14}, since it is produced in the capture of slow neutrons by nitrogen, and the (p, n) process has the lowest threshold energy of all proton-induced reactions which can lead to transmutations. It has been shown [L 0] that, if the global production of C^{14} by solar particles were as high as three times that due to the galactic cosmic radiation, the corresponding solar particle contribution to H^3 (the isotope which has the next lowest production threshold) would amount only to about 25%.

Recent calculations [L 19] lead to an estimate of solar contribution to the global time averaged production rate of C^{14} of 0.03 to 0.06 C^{14} atoms/cm^2·sec during the period 1956/61, i.e. less than 3% of the galactic contribution (Sect. C.IIIβ). The calculation is based on an assumed average flux of solar protons during the last solar cycle, given by the number-rigidity relation (c.f. Sect. B.Iβ)

$$dN = 2e^{-R/R_0} dR \text{ protons/cm}^2 \cdot \text{sec} \cdot \text{MV} \qquad (8.1)$$

with

$$R_0 = 125 \text{ MV}.$$

This corresponds to a value of 4 protons/cm^2 sec for the integrated average flux above 100 MeV kinetic energy, an order of magnitude higher than that estimated during a typical solar cycle [W 8].

It is clear that an average solar flux with estimates based on a small number of events of widely varying characteristics must be treated with great caution. Nevertheless, the estimates support the information obtained from comparing the global C^{14} inventory with the value expected when neglecting the non-galactic contribution to the cosmic ray flux. One may conclude that, on the average, the solar contribution is at most a few per cent for C^{14} and much smaller for all other isotopes. One should remember, however, that large fluctuations do occur, during which, for short periods of time, the concentrations of isotopes in the polar stratosphere and the global inventory of the very short-lived isotopes will be upset appreciably.

V. Global averages of nuclear disintegrations and of isotope production in the atmosphere.

9. The integrated star and isotope production rates in the atmosphere are shown in Fig. 16. These are useful for an understanding of concentrations which one observes in various parts of the atmosphere and of the fall-out patterns, which will be discussed in Sects. E and F.

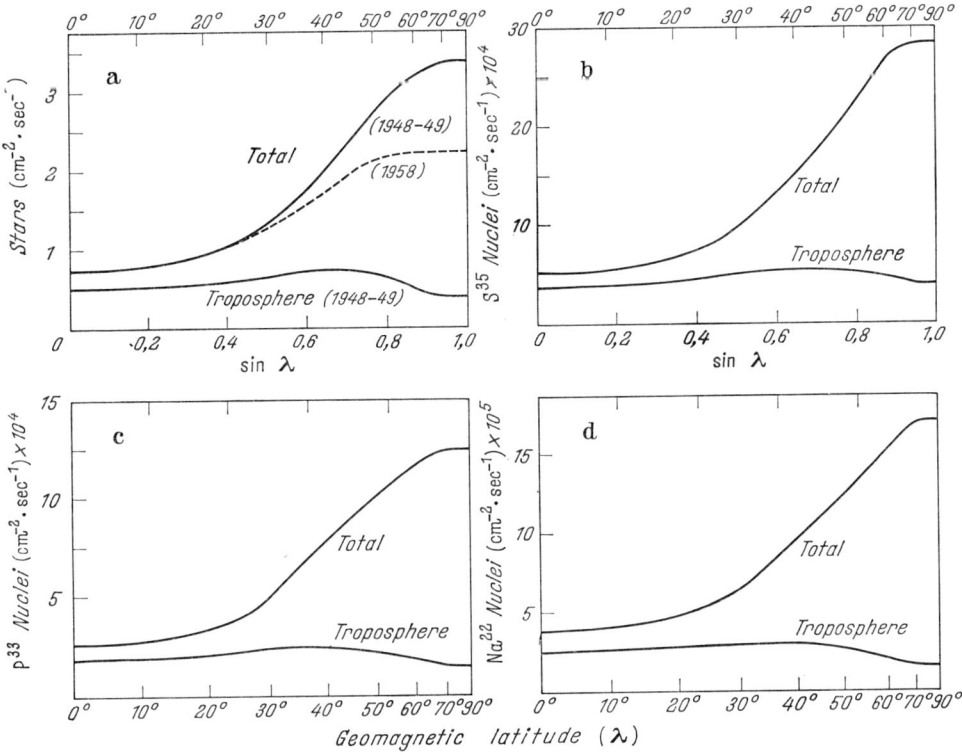

Fig. 16 a—d. The number of nuclear disintegrations (stars) and atoms S^{35}, P^{33} and Na^{22} produced per second in an air column with a cross section of 1 cm² plotted against geomagnetic latitude, λ. The lower curves show the fraction produced below the tropopause, the upper curves the total production. All curves (except the dotted curve in Fig. a) refer to the years 1948—49.

Uniform concentrations over the entire globe are expected only in the case of the radioisotopes C^{14}, Kr^{81} and Ar^{39}, because here half-lives as well as residence times in air are long compared to time scales involved in atmospheric exchange processes. A similar situation should exist for the stable isotope, He^3, which is produced directly in nuclear spallation reaction, as well as in the decay of H^3.

Other long-lived isotopes, e.g. Be^{10} and Si^{32}, are not expected to be uniformly dispersed in the atmosphere, because they are quickly removed as soon as they reach the cloud-bearing strata; their fall-out is therefore expected to be latitude dependent. However, most of these long-lived radioisotopes reach the oceans where large scale mixing eventually tends to wash out all latitude dependent features [L 7]. Thus global average production is a useful concept for long-lived isotopes.

For short-lived isotopes concentrations behave quite differently in the stratified fairly stable stratosphere and in the turbulent and well mixed troposphere.

In these cases, a knowledge of the global averages, separately for the troposphere and the stratosphere, is essential for discussing their applications to meteorological studies.

The integrated production rates/cm² column·sec for stars and the various isotopes for the troposphere as well as the total atmosphere are shown separately in Fig. 16. The tropopause heights used are the mean of the summer and winter values [B12]. All curves refer to the period 1948—49, when the solar activity was fairly high; since the neutron data used to obtain the isotope production, curves (Sect. C.II) refer to that period.

As stated earlier, the most intense period of solar activity on record occurred in 1958. We estimate that the global atmospheric star production in that year was decreased by 22% compared to 1948—49. The dashed curve in Fig. 16a shows the star production rate calculated for the atmosphere during 1958. During 1954, a period of solar minimum, the corresponding value is estimated to be about 9% higher than in 1948—49 [L0]. Based on the available data [S11] on sunspot number in the last 19 solar cycles, and assuming that the relationship between sunspot number and cosmic ray intensity has throughout been what was observed between 1954 and 1958, one estimates that the global star production during an average solar cycle should be about 4% higher than it was in 1948—49. With this correction, the calculated global averages of star and isotope production and an estimate of the global inventory of each isotope are given in Table 3.

Table 3. *Cosmic ray produced isotopes and stars.*

Isotope	Half-life	Production rate (atoms cm⁻² sec⁻¹)		Global inventory
		Troposphere	Total atmosphere	
He^3	Stable	6.7×10^{-2}	0.2	3.2×10^3 tons [R14]
Be^{10}	2.5×10^6 y	1.5×10^{-2}	4.5×10^{-2}	430 tons
Al^{26}	7.4×10^5 y	3.8×10^{-5}	1.4×10^{-4}	1.1 tons
Cl^{36}	3.1×10^5 y	4×10^{-4}	1.1×10^{-3}	15 tons*
C^{14}	5730 y	1.1	2.5	75 tons
Si^{32}	500 y	5.4×10^{-5}	1.6×10^{-4}	1.4 kg
H^3	12.5 y	8.4×10^{-2}	0.25	3.5 kg
Na^{22}	2.6 y	2.4×10^{-5}	8.6×10^{-5}	1.9 g
S^{35}	87 d	4.9×10^{-4}	1.4×10^{-3}	4.5 g
Be^7	53 d	2.7×10^{-2}	8.1×10^{-2}	3.2 g
P^{33}	25 d	2.2×10^{-4}	6.8×10^{-4}	0.6 g
P^{32}	14.3 d	2.7×10^{-4}	8.1×10^{-4}	0.4 g
Cosmic ray stars		0.6	1.8	—

* Inventory includes a rough estimate of Cl^{36} produced by the capture of neutrons at the earth's surface [D2].

It should be noted here that the effect of the solar cycle on isotope production in the stratosphere is strong but that the tropospheric production rates remain essentially unchanged during a typical solar cycle, even at high latitudes.

With the exception of C^{14}, ca. 70% of all stars and isotopes listed in Table 3 are produced in the stratosphere. This ratio is a global average; Fig. 16 shows that it varies strongly with latitude.

Table 3 does not include the production estimates for three geophysically useful radioisotopes: Kr^{81}, A^{39} and A^{37}, and a number of less useful ones which have half-lives shorter than 1 day. Reliable production estimates for the noble gas isotopes cannot be made at present because neither of the three methods discussed in Sect. II can be applied. Method A is not applicable because the activities resulting from cosmic ray exposures are not expected to be detectable. Method B cannot be used because all the three isotopes are produced in zero pronged stars. It is also impossible to use method C because the relevant cross-section data are not available. Estimates of their global inventories based on the general behaviour of the cross-sections in medium weight nuclei have been made for these isotopes [H3], [R12], but are uncertain within a factor of three. These estimates are listed in Table 4.

Table 4. *Approximate estimates for the global production rates of noble gas isotopes*, Kr^{81}, Ar^{39}, Ar^{37}.

Isotope	Half-life	Global atmospheric production rate atoms/cm². sec	Reference
Kr^{81}	2×10^5 y	1.5×10^{-7} — 10^{-6} a)	[H3]
A^{39}	270 y	5.6×10^{-3} b)	[R12]
A^{37}	35 d	8.3×10^{-4}	[R12]

a) The values quoted are based on two available cross-section measurements [M11], [R8] for the thermal neutron capture in Kr^{80} which, however, are not compatible with each other.
b) Includes contributions from the decay of cosmogenic Cl^{39} in the atmosphere.

VI. Possible additional sources of isotopes in the earth's atmosphere.

10. Certain aspects of the solar corpuscular radiation are not yet adequately explored so that it is not possible to evaluate their relevance for isotope production in the atmosphere.

Most important of these could be the role played by neutrons of solar origin.

Another uncertain feature of the solar corpuscular radiation is the isotopic composition of its hydrogen component. If there were a significant H^3 fraction in the solar particle flux, it would affect [L20] the tritium concentration as well as the global inventory of He^3.

Finally, one ought to consider the importance of He^3 and of radioactive isotopes brought into the atmosphere as a component of the primary galactic cosmic radiation.

α) *Solar neutrons.* Neutrons from the sun can and certainly do reach the atmosphere, but no reliable theoretical estimate exists as to how strong that flux ought to be, nor are there any direct observations in quiet periods or during solar bursts. A flux of low energy neutrons would be considerably more effective than a corresponding proton flux for producing isotopes because,

neutrons penetrate the earth's field so that their arrival is not restricted to the polar regions as is the case for protons;

ionization losses in the atmosphere and the existence of a Coulomb barrier reduce the effectiveness of protons for isotope production when compared to that of neutrons of comparable energy.

On the other hand, the effectiveness of solar neutrons for isotope production is reduced by decay in flight; at 20 MeV the probability for a neutron to traverse the sun-earth distance is ~10% and at 5 MeV it is ~10%.

Most of the neutrons leaving the sun must be produced near the surface in p, n reactions in light elements. It is possible that this process goes on at all times, i.e. even when the sun is comparatively quiet [S9], [L5]. In that case, one must expect also a quasi-stationary injection of protons into the interplanetary space, which results from the decay of those neutrons which have penetrated the boundary of the sun's magnetosphere [S9]. A quasi-stationary flux of solar protons of 80—350 MeV was in fact observed in 1960—61 [B3], [V4]. Evidence for a proton flux at still lower energies, up to 10 MeV, has been presented [S6]. If indeed these protons arise from decay of neutrons, the corresponding neutron flux at the top of the earth's atmosphere must be of order of 500/m²·sec [S9] and their contribution to isotope production would not be negligible.

A necessary consequence of the decay hypothesis is that the quasistationary particle flux from the sun contains protons but no helium, i.e. that it has a composition different from matter accelerated at the sun's surface. This question remains to be investigated.

Apart from the (as yet hypothetical) neutron emission from the quiet sun, there is also the possibility of occurrence of large neutron bursts in connection with solar flares. Since, unlike charged particles, they can escape easily, it is in fact possible that neutron bursts are associated even with those solar flares which do not lead to the emission of charged particles. Ground-based neutron monitors would detect such bursts only if the neutron energies exceeded ~ 1 GeV.

An upper limit on the long term average contribution of solar neutrons to terrestrial isotope production can be based at present only on estimates of the global inventory of C^{14}.

β) *Solar tritium.* The observed concentration of H^3 and He^3 in fragments of the satellite, Discoverer XVII (which was in orbit during a major solar cosmic ray event following the November 12, 1960 flare), have been interpreted as indicating the presence of these nuclides in the solar particle beam [T1], [S7]. However, it has also been suggested [L6] that the observation of H^3 in this satellite can be satisfactorily accounted for as due to interactions of solar alpha particles and protons in the body of the satellite. The observations of He^3, on the other hand, cannot be understood in terms of its production inside the vehicle. Thus, the question of direct injection of these isotopes in solar cosmic ray events must remain open at present. An upper limit for the time averaged direct influx of tritium, as distinguished from production within the atmosphere, can be obtained from the global inventory [C1], [L0].

γ) *Influx of isotopes from outside the solar system.* The contribution of isotopes arriving as part of the galactic primary radiation can be shown to be rather unimportant.

Short lived radioisotopes cannot be present among the primaries of galactic origin, unless produced shortly before arrival in a collision of a heavy primary nucleus in the interstellar gas. Since the density of interstellar matter along the primary trajectory is not likely to exceed 1 atom of hydrogen or 10^{-24} g/cm³, such collisions are extremely rare. The characteristic age of the collision products is the time necessary to traverse an amount of matter comparable to an interaction mean free path which even for very heavy primaries is of the order of at least 10^7 years. Thus, of the isotopes considered so far, only He^3 and Be^{10} could possibly have their terrestrial inventory enhanced by a direct influx of high energy particles from outside the solar system.

Helium nuclei of the galactic radiation with energies sufficiently low to come to rest in the upper atmosphere before being broken up by nuclear encounters are admitted only near the poles, where their flux is ca. 0.05 particles/cm² sec. Even if He^3 was the dominant isotope, this influx is about ten times smaller than the combined production of H^3 and He^3 isotopes by spallation reactions in the polar region and therefore constitutes only a few per cent of the global production rate.

Similar considerations apply to beryllium. Even if the entire galactic flux of beryllium nuclei consisted only of the radioisotope Be^{10}, the number of nuclei which can be brought to rest in the atmosphere is more than two orders of magnitude smaller than the number produced in spallation reactions.

D. Terrestrial isotopes of non-atmospheric origin.

11. So far we have discussed the influx and production of isotopes in the atmosphere. Additional sources of isotopic changes on the earth are:

1. Natural radioactivity: decay and fission of long-lived natural constituents of the earth, especially of elements of the uranium and thorium series and nuclear interactions produced by the resulting α-particles and secondary neutrons.

2. Nuclear interactions in the lithosphere by secondary cosmic ray particles and stellar neutrinos.

3. Influx of isotopes produced outside the earth's atmosphere, e.g. those carried in meteoritic material and cosmic dust accreted by the earth.

Influx of isotopes carried by the solar plasma cannot either be ruled out at present.

The natural radioactivity in the earth and the resulting nuclear reactions have been dealt with extensively in the literature [A 10], [R 14], [M 12] and will not be discussed here.

I. Production of isotopes in the lithosphere.

12. Since most of the cosmic ray energy is already dissipated in the atmosphere, isotopic changes induced in the crust, even in its outer layers, are small. This is the reason why investigations so far have been confined to the uppermost layers of the lithosphere. With improvements in techniques it should become possible to study deeper layers. Nuclear transmutations in the outer most layer of the crust are induced by fast nucleons, thermal neutrons and stopping negative muons. The component responsible for most of the disintegrations changes with depth: below a few meters, fast muons are the dominant source. At a depth exceeding $400 \text{ kg} \cdot \text{cm}^{-2}$, most of the interactions are due to neutrinos of the secondary cosmic radiation and to neutrinos of solar and stellar origin.

The rate of production of Cl^{36} as a result of the capture of thermal neutrons by chlorine present in rocks and sea water has been calculated. The estimated production seems sufficiently large to permit detection in rocks exposed at mountain altitude or sea-level; it is estimated that about 3×10^{-3} atoms/sec·kg Cl are produced at sea level in rocks containing as little as 0.35% chlorine [D 2].

Other production rates have been estimated by exposing various chemicals up to 30 meters below the surface. In favourable cases the production rates are of the order of 2×10^{-2} and 2×10^{-3} atoms/kg·sec of crustal materials due to neutron interactions and capture of negative muons, respectively [R 5].

Estimates of total nuclear disintegrations from sea level down to $\sim 500 \text{ kg} \cdot \text{cm}^{-2}$ have been made [L 8] and are shown in Fig. 17. Curve I shows the contributions of the nucleonic component; less than 10% of the reactions are due to protons. The total rate of thermal neutron capture due to neutrons in equilibrium with the fast nucleonic component is given by Curve II. The frequency of nuclear disintegrations due to interactions of fast muons and due to capture of negative muons is given by Curves III and IV, respectively.

It is seen from this figure that below ca. $0.7 \text{ kg} \cdot \text{cm}^{-2}$ muons are the chief source of nuclear disintegrations. The negative muon captures contribute about 10% to the total rate.

The expected rate of interactions due to solar neutrinos is $\sim 10^{-11}$ gm$^{-1}\cdot$sec^{-1}; these reactions should become discernible against the background of cosmic ray interactions at a depth greater than ca. 400 kg$\cdot$cm^{-2}.

Cosmogenic changes induced in the lithosphere could be useful for studying geophysical processes, such as erosion of surface rocks. They could also be useful for estimating, via the muon component, high energy primary cosmic ray flux in

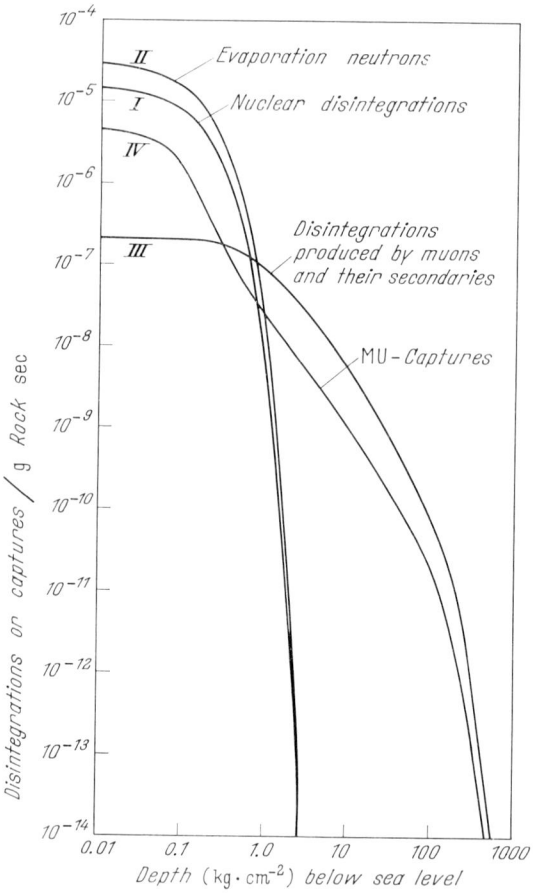

Fig. 17. The rate of nuclear disintegrations due to nucleons and muons in the upper layers of the earth's crust as a function of depth. (From LAL [L8].)

the past and for studying the intensity of neutrinos of stellar and cosmic origin. The experimental technique is not yet sufficiently advanced to permit an attack on the astrophysical problems connected with neutrino flux.

II. Accretion of isotopes contained in extraterrestrial matter.

13. The rate of influx of extraterrestrial dust is estimated to be $\sim 10^4$ tons/day [$A\,11$]. Certain cosmogenic isotopes brought in by this accretion process could be detectable with available techniques and serve as monitors for the dust accreted by the earth in earlier periods [$L\,21$]. The radioisotopes Mn^{53} and Ni^{59} which are not produced in the earth's atmosphere, and Al^{26} whose atmospheric production rate is probably smaller than its influx due to dust, may be useful for studying this problem [$K\,3$], [$W\,3$].

The isotopic changes in dust and in the outermost layers of meteorites are due mainly to the lowest energy cosmic particles, i.e. presumably those of solar origin. The outermost layers of meteorites usually melt and are blown off during atmospheric transit; cosmic dust seems to provide at present the best available sample for studying the integrated low energy solar proton flux. Such analyses require several grams of cosmic dust and have to be based on accumulations on the earth over long periods; they have to be taken from places where accumulation proceeds relatively free from terrestrial contamination. Ocean sediments and polar ice caps may satisfy this requirement.

The result of an experiment [$K3$] carried out to detect Al^{26} assumed to be contained in cosmic dust is discussed in Sect. F.IIIβ.

E. Circulation of isotopes in the geosphere.

14. Soon after creation many of the isotopes listed in Table 1 become oxidized; this applies to all except the noble gases He^3, $A^{37, 39}$ and Kr^{81}. Among the oxides, only CO_2 is a permanent gas and remains free, while the others become attached to aerosols within a very short time. The size distribution of aerosols present in the atmosphere and the attachment of radioactivity to these particles has been studied extensively [$J1$]. Attachment occurs mainly on particles of less than one micron diameter, for which gravitational settling is not important. Therefore the cosmic ray produced activity when attached to aerosol follows the motion of air molecules in the atmosphere just like that part of the activity which remains unattached (e.g. He^3, $Ar^{37, 39}$, Kr^{81} and $C^{14}O_2$ and water vapour molecules containing H^3).

Removal of the various cosmic ray produced isotopes from the atmosphere occurs when aerosol is collected on droplets during condensation of moisture in the lower parts of the troposphere. Wet precipitation removes these isotopes efficiently. The adherence of radioactive aerosol to ground surfaces without condensation of moisture seems to be rather unimportant; it can however play a role in the fall-out of bomb-produced radioactivity if it is imbedded in or becomes attached to comparatively large dust grains.

The radioactive permanent gases are removed from the atmosphere by a much slower process, namely molecular exchange at the surface of the ocean and other water bodies and, in the case of CO_2, also by being incorporated into the metabolism of vegetation. Thus, while the two groups have the same source functions and their dispersion in the air is governed by the same meteorological features, their sink functions are quite different.

After reaching the earth's surfaces, whether through precipitation or molecular exchange reactions, the isotopes are dispersed further through a variety of migratory and mixing processes.

The nature of the geophysical or geochemical processes which affect the distribution of the various isotopes can be discussed most conveniently if one divides the atmosphere and the outer layers of the earth into a number of zones or reservoirs within which there exists a certain degree of homogeneity with regard to composition, pressure, temperature and the general character of mass motion. A schematic description of isotope migration is shown in Fig. 18. This figure shows the four major modes of circulation and exchange which determine the dispersion of the isotopes listed in Table 1. The various boxes, marked A, M, D etc., are divisions or sub-divisions of the geosphere as shown in Fig. 19. They represent zones which can be considered reasonably homogeneous with respect to their composition and the transfer of material across their boundaries. These zones are separated by more or less well defined surfaces of discontinuity (f..i the air-sea interface), boundary regions across which distinct and drastic changes occur in the processes responsible for the mixing and dilution of isotope containing material.

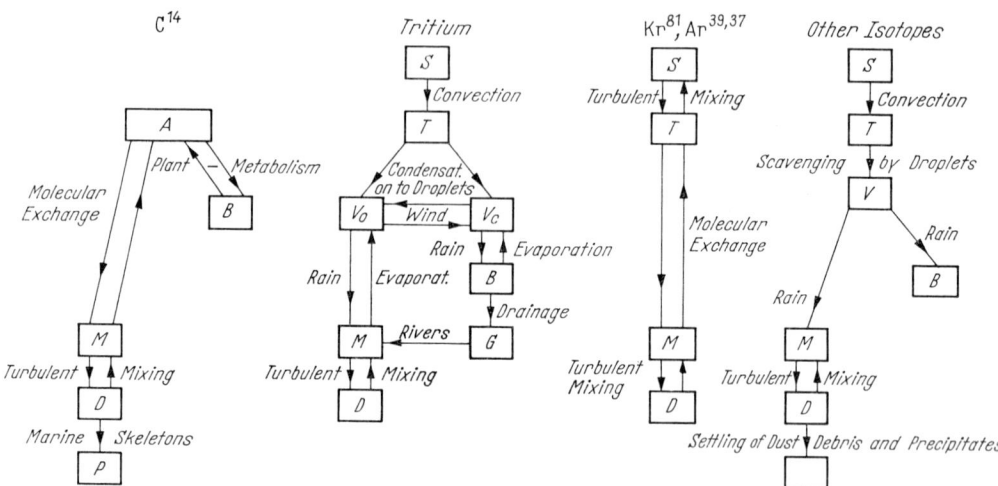

Fig. 18. Simple models giving the main avenues of flow for various cosmic ray produced isotopes in the geosphere; see Fig. 19 for explanation of symbols. (From LAL and PETERS [L0].)

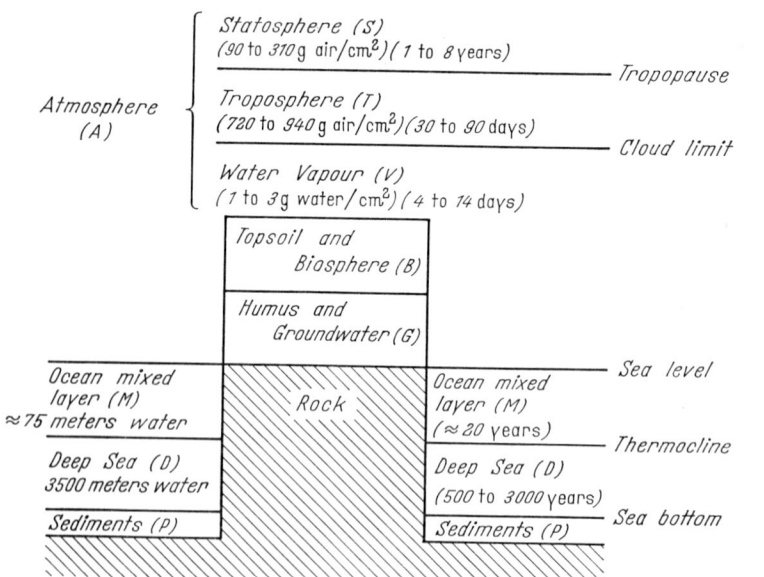

Fig. 19. The geosphere and its principal geophysically significant subdivisions. The numbers represent very rough estimates of the size of each reservoir and its turnover time. (From LAL and PETERS [L0].)

The principal aim of isotope work in geophysics is to obtain detailed information on the transfer processes involved, i.e. the mass of the reservoirs and the rates at which water, air or particular molecular compounds are transferred from one reservoir to another.

The rate of flow of isotopes across the interfaces depends on the chemical nature of the molecules which incorporate the isotope, and the coefficients of

I. Dispersion of isotopes in the atmosphere and fall-out.

15. α) *The distribution of isotopes in the stratosphere.* In any region where radioisotopes partake in the motion of the medium in which they are suspended, their concentrations can be expressed by the relation derivable from the conservation of bulk matter and tracer substance:

$$\frac{dc^*}{dt} = q - \lambda c^* - \boldsymbol{v} \cdot \boldsymbol{\nabla} c^* \tag{15.1}$$

where $\boldsymbol{v}$ is the velocity vector, λ the decay constant of the radioisotope and c^*, q are its concentration and its rate of production per unit mass. If turbulent motions are present, the concentration will be subject to short term fluctuations and its average value, $\overline{c^*}$, will obey the same equation as c^* with the addition of a term describing diffusion by eddy transport.

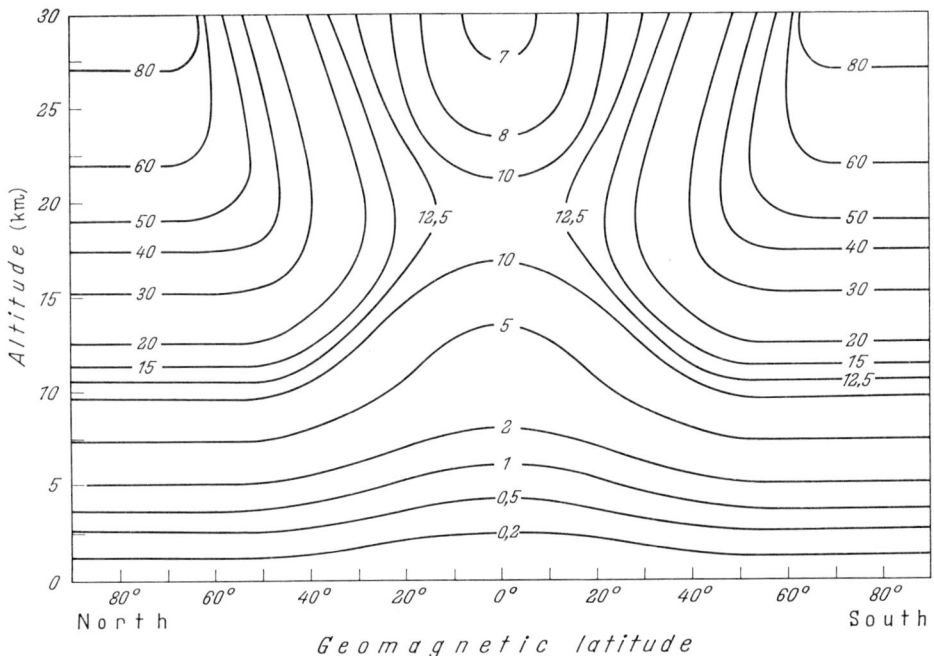

Fig. 20. A north-south section through the atmosphere showing surfaces of constant Be^7 production (nuclei/min, m³ air at S.T.P.).

However, in this section the discussion shall be confined to the stratosphere, the comparatively stable region of the atmosphere above the tropopause; for this region one may assume that transport by eddies is rare and unimportant compared to that by large scale orderly streaming motion. This certainly is a simplification with respect to the actual situation and is used here only for illustrative purposes.

The source function, q, in the stratosphere is simply the rate of isotope production discussed in Sect. C. For instance, Fig. 20 based on Fig 4 describes q for the radioisotope Be^7, as a function of latitude, θ, and height, h, above sea level. Thus, the production rate, q, is a known function for all parts of the atmosphere; it varies with time, as was discussed in Sect. C.IV, but apart from short-lived disturbances due to solar flares, it oscillates with a period long compared to the residence times of air in all but the very highest layers of the stratosphere. Therefore, in a particular phase of the solar cycle, q may be considered to be constant in time.

If the velocity field of stratospheric winds is stable for periods longer than the half-lives of isotopes of interest, a steady state is set up, and Eq. (15.1) reduces to

$$\boldsymbol{v} \cdot \boldsymbol{\nabla} c^* = q - \lambda c^* . \tag{15.2}$$

Since q is known everywhere, the isotope concentration c^* and its gradient determine the component of wind velocity along the gradients. Measurements on isotopes of different half-lives yield non-parallel components of the velocity vector.

The importance of the method lies in the fact that, even in the presence of the strong zonal wind velocities which prevail in the stratosphere, it remains sensitive to slow vertical drifts and small meridional components of the motion. For, let C^* be the equilibrium concentration of an isotope as a function of latitude and atmospheric pressure, i.e. the concentration which would be reached in the absence of all air motion, then C^* is also a known function

$$C^*(h, \theta) = \frac{q(h, \theta)}{\lambda} . \tag{15.3}$$

For small deviations from equilibrium Eq. (15.2) leads to

$$\frac{C^* - c^*}{C^*} = \frac{\Delta c^*}{C^*} = \frac{\boldsymbol{v} \cdot \boldsymbol{\nabla} \log q}{\lambda} . \tag{15.4}$$

One can use this approximation for any wind velocity provided one employs isotopes of sufficiently short mean life, i.e. large enough λ values. Since q is independent of longitude, φ, the term $\boldsymbol{v} \cdot \boldsymbol{\nabla} \log q$ and therefore also the deviation of concentrations from their equilibrium value are independent of the zonal wind component, v_φ.

To illustrate the sensitivity of stratospheric isotope concentrations to vertical air drift and meridional wind components, one may read off in Fig. 20 the logarithmic gradient of the source function at some particular point in the stratosphere. For instance, at a latitude $\theta = 30°$ and an altitude of 12 km corresponding to a pressure $x = 200$ g/cm², one finds

$$\frac{\partial}{\partial h} \log q = \frac{1}{7} \text{ (km)}^{-1}, \tag{15.5}$$

$$\frac{\partial}{R_{\text{earth}} \partial \theta} \log q = \frac{1}{3000} \text{ km}^{-1} . \tag{15.6}$$

A deviation of Be^7 concentration from its equilibrium value by an amount of 10—20% corresponds then to vertical airdrifts of $0.4 < v_h < 0.8$ meters per hour and meridional wind speeds of $4.5 < v_\theta < 9$ centimeters per second. The same deviation in the concentration of P^{32} corresponds to velocities larger by a factor 3.7.

Such quasi-stationary states with isotope concentrations which show a significant departure from their equilibrium values seem to exist in the stratosphere (see Sect. F.II).

β) *Motion of isotopes in the troposphere and their removal by rain or snow.* The rapidly varying wind conditions in the troposphere prevent all stationary or quasi-stationary distributions. Some information on the history of motion of individual air masses can be derived from the ratio between isotopes of different half-lives in the same sample. One may write the conservation law Eq. (15.1) in the co-ordinate system which moves with the air mass:

$$\frac{dc^*}{dt} = yq' - \lambda c^*. \tag{15.7}$$

The yield per star averaged over all constituents of the atmosphere, y, is a constant throughout the troposphere. The source function q' is now a function of time, and represents the changing rate of star production along various parts of the trajectory followed by the air mass.

Assuming that the air mass remains in the troposphere and that the circulation within the troposphere is rapid compared to the half-lives of two isotopes, $q'(t)$, may be replaced by some average value and one obtains a relation between concentrations of two different isotopes:

$$\frac{c_1^*(t) - c_1^*(0) e^{-\lambda_1 t}}{c_2^*(t) - c_2^*(0) e^{-\lambda_2 t}} = \frac{y_1}{y_2} \frac{\lambda_2}{\lambda_1} \frac{1 - e^{-\lambda_1 t}}{1 - e^{-\lambda_2 t}}, \tag{15.8}$$

This applies only to an air sample which is small enough and to a time which is short enough so that all parts of the sample may be considered to have the same radiation history.

If $t=0$ designates the time at which the air sample was last cleaned of all radioactivity by condensation of moisture in the cloud level, then $c^*(0)=0$. For an air mass which remains in the troposphere, the ratio $c_1^*(t)/c_2^*(t)$ changes with time from an initial value, y_1/y_2 to an asymptotic value, $y_1 \lambda_2/y_2 \lambda_1$, as shown in the lower curves (marked α) of Fig. 21. The concentration ratio of two isotopes in the same rain samples yield a distribution of radiation ages for different air masses. A third isotope can serve to test the consistency of relation (15.8) derived on the basis of this simple model.

If, on the other hand, an air mass entered the troposphere from above at $t=0$, after having being irradiated for a long time in the stratosphere, then

$$c^*(0) = \frac{y}{\lambda} \bar{q}_S', \tag{15.9}$$

where $\bar{q}_S'$ is the average star production rate in the lower region of the stratosphere,

In that case the ratio of isotopes changes in time according to the upper curves (marked β) of Fig. 21.

The difference between the curves are quite marked for isotope pairs with the appropriate half-lives. It is therefore possible to distinguish air masses which entered the troposphere from above and air masses which remained below the tropopause in the interval between successive removals of activity by condensation of moisture. Experimental results are discussed in Sect. F.II. The average circulation time, τ_F, between successive cleansing is of order of 30—40 days [*L0*]. Observed differences in isotope ratios in individual rains support the view that one is often justified, in spite of the large turbulence in the troposphere, to speak of air masses which had different radiation histories between successive cleansing by precipitation.

For those isotopes which are brought down with precipitation the pattern of fall-out can then be described approximately as follows:

a) Within each latitude belt the fall-out is roughly proportional to the amount of local precipitation. This is so because, within a period of order τ_F, zonal wind velocities distribute air masses over all longitudes.

b) Because of strong turbulence and since isotope production in the troposphere is approximately independent of latitude (see Fig. 16) the fall-out of the fraction of isotopes produced in the troposphere, F_T, is nearly uniform in latitude and can be written

$$F_T = \frac{Q_T}{1 + \lambda \tau_F}, \qquad (15.10)$$

where Q_T is the average tropospheric isotope production per unit area and the factor multiplying it represents the correction for decay.

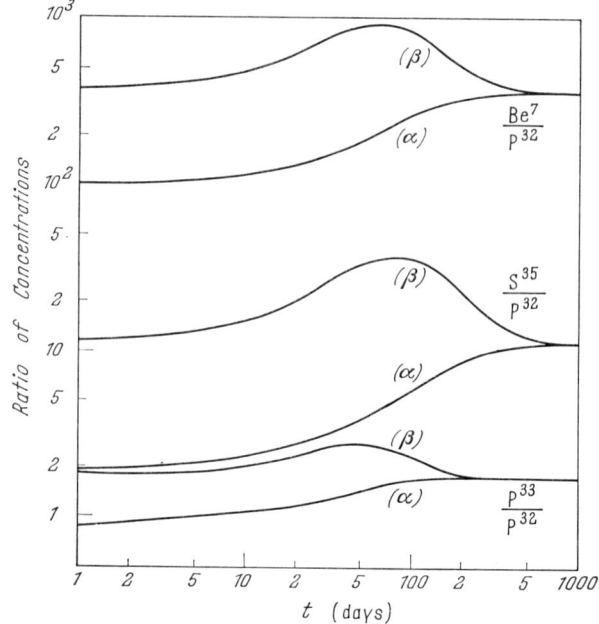

Fig. 21. Expected ratios between the concentrations of Be⁷, S³⁵, P³³ and P³² for the two hypothetical trajectories of air masses discussed in Sect. E.I.β (From Lal and Peters [L0].)

c) The average residence time of air in the stratosphere is so long that most of the isotopes whose mean life is short, ($\tau < 1$ year), decay before reaching the troposphere and the earth's surface. Therefore, there is little contribution from the stratosphere to the fall-out of short-lived isotopes.

The average production rate in the troposphere is 0.6 stars/cm² sec (Table 3). Thus

$$Q_T \sim 0.6 \, y \, (\text{cm}^2 \cdot \text{sec})^{-1}. \qquad (15.11)$$

The total fall-out for short-lived isotopes averaged over a particular latitude belt is then given by

$$F \sim \frac{0.6 \, y}{1 + 40 \, \lambda} \text{ atoms/cm}^2 \cdot \text{sec}. \qquad (15.12)$$

Here, y is the yield per star averaged over all constituents of the atmosphere, and the decay constant, λ, is expressed in (days)⁻¹. F is nearly independent of latitude; local fall-out is obtained by multiplying F with the ratio of local to total precipitation for the particular latitude belt.

The fall-out pattern for long-lived isotopes is quite different. The contribution from the stratosphere is important and is strongly dependent on latitude. This is not due to the original latitude dependence of production; data on fission products, especially data on Sr^{90}, have shown [M 10] that the fall-out pattern is practically independent of the latitude at which the activity was introduced into the stratosphere. The pattern is presumably determined primarily by the discontinuity in the tropopause at median latitudes, where most of the transfer of air from stratosphere to troposphere seems to take place [J 1], [I 1]. The fall-out from the stratosphere shows strong seasonal variations [J 1]. Isotopes produced by cosmic radiation in the stratosphere are expected to exhibit the same fall-out

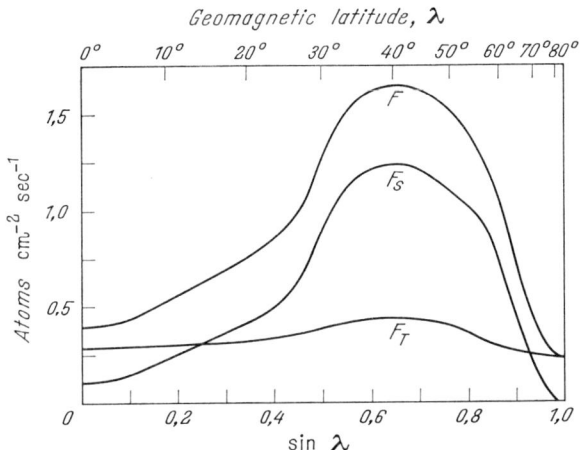

Fig. 22. The expected fall-out pattern for a long-lived cosmic ray produced isotope whose global production averages 1 atom/cm², sec is given by the curve marked "F". The curves marked F_T and F_S show the contribution to the fall-out of isotopes produced in the troposphere and stratosphere, respectively. (After Lal [L 7].)

pattern and seasonal variations as bomb debris; the pattern will of course differ from bomb fall-out in being symmetric in the two hemispheres (except for the fact that the seasonal variations are 180° out of phase).

The global average fall-out of radioisotopes produced in the stratosphere and attached to aerosol is given by

$$F_S = \frac{y Q'_S K_{ST}}{(\lambda + K_{ST})(1 + \lambda \tau_F)}, \qquad (15.13)$$

where Q'_S (1.2 cm^{-2}·sec^{-1}) is the star production in the stratosphere averaged over all latitudes; K_{ST}^{-1} is the partial residence time of air molecules or tracer substances in the stratosphere against removal to the troposphere and may be taken to be of the order of 1—5 years. Thus for isotopes with mean lives much longer than this, e.g. Si^{32},

$$F_S = 1.2 \, y \, \text{cm}^{-2} \cdot \text{sec}^{-1}. \qquad (15.14)$$

The latitude distribution of F_S must be expected to be the same as the distribution of Sr^{90} fall-out in the northern hemisphere measured in connection with stratospheric bomb tests [M 10].

In Fig. 22 is plotted the calculated [L 7] latitude distribution of total fall-out, $F = F_T + F_S$ as well as F_T alone (fall-out due to production of isotopes in the troposphere) for a hypothetical stable isotope with a global average production rate of 1 atom/cm². The expected local fall-out of individual isotopes can be calculated from this curve by multiplying with

the actual average global production rate,
the factor correcting for decay in the atmosphere,
the fraction of local to total precipitation at the appropriate latitude belt.

The fall-out values obtained in this manner represent the sink function for the atmosphere as a whole and at the same time the source function for the other geophysical reservoirs. As stated above, they apply only to those isotopes which are attached to aerosol and are carried to the ground in precipitation.

The other isotopes which can exist in the atmosphere unattached and in gaseous form have quite different fall-out pattern, except for tritium which forms HTO molecules and is removed by precipitation like the isotopes discussed above. But there are several complications [C1], [B17]:

Part of the tritium falling on land is carried to the ocean in river water.

The cloud layer contains not only the isotopes introduced from above but also HTO re-evaporated from the ground. Since precipitation stays close to the surface only on continents, re-evaporation of tritium from land is much more important than from the ocean.

Finally, there is molecular exchange between atmospheric HTO molecules and ordinary water molecules at the surface of large water bodies.

As a result of these three processes, there is a net transfer of tritium from the continents to the oceans, so that for a given amount of precipitation more tritium ends up in the ocean than on land. The processes which play a part in the fall-out of H^3 are indicated schematically in Fig. 18.

γ) *Removal of radioactivity by molecular exchange.* The only known mechanism for the removal of the cosmic ray produced helium, argon and krypton isotopes, listed in Table 1, is the molecular exchange at the air-sea interface leading to dissolution and thus establishing a gradient in their specific concentrations between the atmosphere and the sea. The exchange rates for argon and krypton are not known, but they are certainly slow so that at least all A^{37} decays in the atmosphere.

For $C^{14}O_2$, the absorption by vegetation is also of importance, but the residence time in the atmosphere is dominated by the exchange at the sea-air interface. These exchange processes are slow compared to the turnover time for air, so that the long-lived isotopes become uniformly distributed throughout the entire atmosphere.

Nevertheless, the geographical distribution of the input rate of C^{14} and noble gases into the ocean is presumably far from uniform since the molecular exchange rate depends strongly on temperature as well as on the amount of surface agitation [B5].

The exchange rate for C^{14} can be estimated in various ways. It can be based on the relative concentration of C^{12}, C^{13} and C^{14} in the atmosphere and the mixed upper layer of the ocean. This leads to a partial lifetime of C^{14} for absorption in sea water [L0].

$$\tau_{AM} = \frac{1}{K_{AM}} = 7.5 \text{ years}. \tag{15.15}$$

The range of values for τ_{AM}, based on this procedure and observations on the rate of change of propagation of man-made transient changes in the atmospheric C^{14}/C^{12} ratio, give residence times in the region 2 years $< \tau_{AM} <$ 10 years [B5], [A7], [F5], [R13], [B18]. This is rather short compared to the half-life of C^{14} so that decay in the atmosphere is unimportant, and the atmospheric sink function

Sect. 16.   Introduction of isotopes into the deep ocean.   595

is equal to the average source function. The same is expected to be true for the noble gases other than Ar^{37}.

The isotopes He^3, Kr^{81}, C^{14} and Ar^{39} are expected to be uniformly distributed throughout the atmosphere.

II. Introduction of isotopes into the deep ocean.

16. Since production by cosmic rays in the ocean is negligible, the isotopes enter from above. They first encounter a layer characterized by strong currents and rapid mixing (the so-called mixed layer), where the geographic non-uniformity of the input discussed in the previous section is obliterated to some extent. Thus, it seems reasonable to assume for the deep ocean a source function which is zero in the interior and has a uniform value at the surface. This is approximately correct for C^{14} which exists mostly in solution in sea water and follows the mass motion of the water.

There are, however, other isotopes such as phosphorus and silicon which do not follow the mass motion of the water; a large fraction is quickly taken up by the living organisms contained in the mixed layer. These isotopes enter the deep sea partly in solution, partly incorporated in debris such as scales, shells or bones. The debris often redissolve either after having reached the ocean bottom or in transit at various depths. The input depends on the amount and nature of biological material in the mixed layer which is geographically non-uniform. Therefore, the effective source function of these isotopes in the deep sea reservoir is a function of all three coordinates, longitude, latitude and depth.

The same conservation law applies for the ocean as for the atmosphere, but the source functions are different; q either is zero everywhere except on the surface (where it may be approximated by a uniform distribution), or else, for those isotopes which are concentrated in organic matter, it is complicated and largely unknown.

We consider first the isotopes which are not retained in the marine biosphere or scavenged in some other way but follow always the motion of the water. In oceanography one is often concerned with estimating the turn-over time of certain regions in the ocean. The problem is similar to that of determining the age of air masses on the basis of isotope ratios. One may use the conservation law in a frame of reference moving with the water:

$$\frac{dc^*}{dt} = q - \lambda c^*. \qquad (16.1)$$

The simplest model corresponds to the assumption that there exist throughout the ocean stationary current systems which carry water masses from the upper mixed layer to the deep sea and back again. Assuming no turbulence which can lead to an exchange between different current systems, the age of a water sample in the deep sea can be determined by measuring the isotope concentration, c^*, in the sample if one knows the concentration c_M^* in the mixed layer. Since the isotope concentration in the mixed layer is reasonably uniform, at least for major water bodies and far from the coastal regions, c_M^* can be measured. In that case the age, t, is given by

$$c^* = c_M^* e^{-\lambda t}. \qquad (16.2)$$

A somewhat more refined, though still greatly simplified model is to assume that each region of the ocean belongs to a stationary current system which at regular intervals, τ, and for a short period, σ, approaches the boundary of the mixed layer; here turbulence causes a certain amount of mixing, thereby adding some new radioactivity to the deep sea water, proportional to the time of contact, σ, and to the degree of undersaturation, $c_M^* - c^*$ (where c_M^* is the concentration in the mixed layer). The conservation law in the system moving with the current can then be written:

$$\frac{dc^*}{dt} = -\lambda c^* + q \begin{cases} = -\lambda c^*, & \text{when the water mass is far below the mixed layer} \quad (16.3\,\text{a}) \\ = -\lambda c^* + \alpha(c_M^* - c^*), & \text{when the water mass is in contact with the mixed layer,} \quad (16.3\,\text{b}) \end{cases}$$

In a quasi-stationary state the two solutions

$$c_1^* = a e^{-\lambda t}, \qquad (16.4\,\text{a})$$

$$c_2^* = b e^{-(\lambda+\alpha)t} + \frac{\alpha c_M^*}{\lambda+\alpha} \qquad (16.4\,\text{b})$$

must coincide when the water mass enters or leaves the region where it finds itself in contact with the mixed layer, i.e.

$$c_1^*(n\tau) = c_2^*(n\tau), \qquad (16.5\,\text{a})$$

$$c_1^*(n\tau - \sigma) = c_2^*(n\tau - \sigma) \qquad (16.5\,\text{b})$$

for any integer value of n, where n is the number of completed circulation beginning with arbitrary time, $t_0 = 0$.

Thus one obtains the concentration in the deep ocean

$$c^* = \frac{\alpha c_M^*}{(\lambda+\alpha)} \frac{e^{-\lambda(t-\tau)}(e^{\alpha\sigma} - e^{-\lambda\sigma})}{e^{\alpha\sigma} - 1}, \qquad (16.6)$$

where the time t is measured with respect to the most recent contact with the surface layer.

Since c_M^*, the concentration in the mixed layer, can be measured, and the decay constant, λ, is known, the circulation time τ, the age (i.e. the time of the last contact) t, the duration of contact, σ, and the coefficient of mixing, α, can be obtained by making simultaneous measurements on four isotopes in a deep sea sample.

If the isotopes have half-lives long compared to the time in which the current is in contact with the upper layer ($\lambda\sigma \gg 1$), Eq. (16.6) reduces to

$$c^* = \frac{\alpha c_M^*}{\lambda+\alpha} e^{-\lambda(t-\tau)} \qquad (16.7)$$

so that three isotopes suffice for determining circulation time, age, and mixing coefficient.

Next we consider isotopes which accumulate in the marine biosphere and are introduced into the deep sea not by turbulent mixing on the upper boundary of the reservoir but by dissolution of falling debris. The source function, q, has the same form for the radioactive isotope and for stable isotopes of the same element. It differs only by a constant factor R_0, the ratio of unstable to stable isotopes of the element in marine organisms. By measuring both c^*, the concentration of radioactive, and c, the concentration of the corresponding stable isotope, one can eliminate q from Eq. (15.1) and obtain

$$\boldsymbol{v} \cdot (R_0 \boldsymbol{\nabla} c - \boldsymbol{\nabla} c^*) = \lambda c^*. \qquad (16.8)$$

III. Conservation laws for a system of well mixed reservoirs.

17. Once the source and sink functions for the major reservoirs are known, it becomes possible to investigate particular exchange processes using cosmic ray produced isotopes as tracers. We shall therefore derive briefly the relations most frequently used in the application of these isotopes, following closely the presentation given in an earlier paper by LAL and PETERS [*L0*].

For any volume in the geosphere one can write equations expressing the conservation of stable and unstable isotopes or compounds. They are of the form

$$\int_V dV \frac{dn}{dt} = -\oint_S n\boldsymbol{v}\cdot d\boldsymbol{S}, \qquad (17.1)$$

and

$$\int_V dV \left(\frac{dn^*}{dt} - q + \lambda n^*\right) = -\oint_S n^* \boldsymbol{v}^* \cdot d\boldsymbol{S}, \qquad (17.2)$$

where n is the number of atoms or molecules per unit volume, and $\boldsymbol{v}$ their velocity near the surface. The left-hand integral is carried out over the entire volume of interest and the right-hand integral over all its boundary surfaces. An asterisk designates radioactive isotopes or their compounds and q and λ are the rate of production per unit volume and the decay constant, respectively.

In order to make use of these relations, it is necessary to begin by constructing a model, which means that one has to decide where to place the physically significant boundaries so that the volume is divided into well mixed reservoirs, and what are the principal physical processes to be considered for transferring molecules or atoms across these boundaries.

The boundaries are then considered to have been chosen such that the average concentration in the reservoir can be taken to be the same as the concentration at its boundaries (i.e. concentration within the reservoir is essentially uniform). Also the physical conditions along the boundaries between any two reservoirs are taken to be reasonably uniform with respect to the transport of material.

Further, it is assumed that a steady state has been reached so that the total number $N = nV$ of any given species of atoms or molecules in a reservoir remains constant. Such assumptions are fairly good approximations for the major reservoirs shown in Fig. 19. (In the reservoir, P, the precipitates and sediments on the floor of the ocean, a steady state condition does not of course apply to the stable isotopes which continue to accumulate, but even there it remains valid for radioactive isotopes.)

For such a steady state model, Eqs. (17.1) and (17.2) can be integrated. For a reservoir i, communicating with neighbouring reservoirs, j, one obtains:

$$\sum_j (N_i k_{ij} - N_j k_{ji}) = 0, \qquad (17.3)$$

$$Q_i = \lambda N_i^* + \sum_j (N_i^* k_{ij}^* - N_j^* k_{ji}^*), \qquad (17.4)$$

$$\sum_j (M_i K_{ij} - M_j K_{ji}) = 0. \qquad (17.5)$$

The first equation refers to a stable element, the second to its radioactive isotope, i.e. an atom with closely similar physical and chemical characteristics; the third equation refers to compounds with different properties, molecules of water or air or others which are representative of the bulk of the materials constituting the reservoir. N, N^* with their respective subscripts refer to the total number of stable and radioactive atoms of a given element present in the reservoir, i.e. concentration n times volume V. M_i is the corresponding number of molecules of the medium in which the isotopes are embedded. Q_i is the total rate of production of the cosmic ray produced radioisotopes in the reservoir, i.

The reciprocal of the transfer rate k_{ij}^*, defined by

$$\frac{1}{k_{ij}^*} = \frac{V_i}{\int_{S_{ij}} v^* \cdot d\mathbf{S}}, \qquad (17.6)$$

may be interpreted as a partial residence time (i.e. the time in which a number of nuclei in the reservoir, i, would be reduced by a factor $1/e$, if the nuclei were stable, if no new nuclei were permitted to enter the reservoir, and if none were permitted to leave except through the boundary surface S_{ij} which separates reservoir i from the neighbouring reservoir j).

$1/k_{ij}$ and $1/K_{ij}$ are the corresponding partial residence times for stable isotopes and molecules.

The inverse of the total residence time or mean age in reservoir i is given by

$$k_i = \sum_j k_{ij}, \qquad (17.7)$$

$$k_i^* = \lambda + \sum_j k_{ij}^*, \qquad (17.8)$$

$$K_i = \sum_j K_{ij}. \qquad (17.9)$$

The relation between the transfer rates k^* and k of different isotopes belonging to the same chemical element is

$$k_{ij}^* = \alpha_{ij} k_{ij}, \qquad (17.10)$$

where α_{ij} is the isotope fractionating factor applicable to crossing the boundary surface S_{ij} in the direction from i to j;

$\alpha_{ij} = 1$, if the isotopes are moving due to gravity or pressure gradients. (Examples are: falling drops or the gravitational settling of skeletal debris in the sea, and also isotopes carried by the mass motion of the medium such as isotopes attached to aerosol or isotopes dissolved in sea water.)

$\alpha_{ij} \neq 1$, if the isotopes move from one reservoir to another by evaporation, condensation, molecular exchange processes, fixation in plant or animal organisms, formation of authigenic minerals etc.

If the stable isotopes share the mass motion of the air or water constituting the reservoir, one has the additional relation

$$k_{ij} = K_{ij}. \qquad (17.11)$$

This applies, for instance, to stable hydrogen in water; here the isotope fractionating factor may be either unity, as in the case of transport by rivers and ocean currents, or different from unity, as in the case of transport by evaporation or condensation.

It is convenient to rewrite Eq. (17.4) in terms of the newly introduced symbols:

$$Q_i = R_i N_i \lambda + \sum_j (R_i N_i \alpha_{ij} k_{ij} - R_j N_j \alpha_{ji} k_{ji}), \qquad (17.12)$$

Sect. 18. The distribution of isotopes in the principal terrestrial reservoirs. 599

where the letter R with its subscript is used to designate specific activity, i.e. the relative number of radioactive and stable atoms of the same element in the same reservoir.

Eqs. (17.3), (17.4) and (17.5), together with (17.11), where applicable, form the basis of most of the geophysical applications of cosmic ray produced isotopes which have been made until now.

The specific activities R_i, R_j are measured directly. The ratio of isotope fractionating factors α_{ji}/α_{ij} can in principle be obtained from laboratory experiments. In practice, however, it is difficult to determine this quantity because it depends on the temperature at which the transfer takes place and on the relation between the time characteristic for mixing and homogenizing the reservoir and the residence time of the isotope. Furthermore, it is different for a continuous fractionating process and for a process which proceeds by discontinuous steps, each of which involves a large fraction of the contents of one or both reservoirs.

If possible, it is therefore preferable to estimate the enrichment of a radio-isotope in a fractionation process by determining the enrichment factor for a stable isotope of the same element in the same process. For instance, the enrichment of tritium can be estimated from the isotopic enrichment of deuterium with respect to protium, that of C^{14} from the enrichment of C^{13} with respect to the abundant isotope C^{12}.

The isotope production rate Q which appears in Eq. (17.12) is known from cosmic ray data (see Sect. C). Chemical analysis furnishes one additional quantity, namely the concentration $c_i = N_i/M_i$ of the stable isotope in the reservoir. Unknown and to be calculated from the conservation laws are usually the transfer rates K_{ij} and possibly the total mass M of the reservoir.

IV. The distribution of isotopes in the principal terrestrial reservoirs.

18. In this section an estimate will be made of the partial inventories of isotopes in the principal geophysical reservoirs from the point of view of investigating their usefulness as tracers for studying particular geophysical problems. We shall assume that a steady state has been reached so that these accumulations are the result of a balance between production, decay and global dispersion caused by a number of time-independent large scale exchange processes.

The distribution of an isotope between various terrestrial domains is determined by its half-life as well as by its physical and chemical properties. The usefulness of an isotope for investigating a particular geophysical process depends primarily on how sensitive this distribution is to the desired parameters characterizing the process; in addition it is of course necessary to decide whether the specific activity of the isotope in the relevant exchange reservoir is large enough to be measured with existing techniques and with a reasonable expenditure of time and money. The reservoir model calculations discussed in Sect. E.III, though not accurate and in some cases not even realistic, are nevertheless sufficiently descriptive of the actual geophysical situation to permit an evaluation as to whether a particular isotope is suitable for the study of a given geophysical problem.

The expected distributions of the various isotopes in the principal exchange reservoirs: stratosphere, troposphere, land surface (including the biosphere), mixed upper oceanic layer, deep ocean layer, and ocean sediments, are shown in Columns 1—6 of Table 5. The calculations are based on the following assumed values of the exchange constants.

Table 5. *Steady state fractional inventories of cosmic Radio-*

Exchange reservoir	Be^{10}	Al^{26}	Cl^{36}	Kr^{81}	C^{14}
Stratosphere (S)	3.7×10^{-7}	1.3×10^{-6}	10^{-6}	0.16	3×10^{-3}
Troposphere (T)	2.3×10^{-3}	7.7×10^{-8}	6×10^{-8}	0.82	1.6×10^{-2}
Land surface (B+G)	0.29*	0.29*	0.29*	0	4×10^{-2}
Mixed oceanic layer (M)	5.7×10^{-6}	1.4×10^{-5}	1.4×10^{-2}	4×10^{-4}	2.2×10^{-2}
Deep oceanic layer (D)	10^{-4}	7×10^{-5}	0.69	2×10^{-2}	0.92
Oceanic sediments (P)	0.71	0.71	0	0	4×10^{-3}
λ (yrs^{-1})	2.8×10^{-7}	9.4×10^{-7}	2.2×10^{-6}	3.3×10^{-6}	1.24×10^{-4}

* Part of this inventory may in fact be carried as silt or dust into the oceans before decay.

The average turnover time of stratosphere air $1/K_{ST}$ has been assumed to be 2 years [B23], [M2].

The exchange of $C^{14}O_2$ and of Kr^{81}, Ar^{39}, Ar^{37} across the air-sea interface has been based on the radiocarbon investigations [B5], [B13] and direct measurements of concentrations of the rare gases in sea water [R10], [B4].

The average turnover time of deep sea water, $1/K_{DM}$, has been assumed to be 1000 years [B5], [B13].

Table 6. *Specific activities of cosmic ray produced isotopes in the atmosphere and in the ocean.*

Radio-isotope	Average specific activity in the atmosphere (d.p.m./kg air)		Radio-isotope	Average specific activity in oceans	
	stratosphere	troposphere		d.p.m./ton water	d.p.m./g element
H^3	6	7×10^{-2}	Be^{10}	10^{-3}	1.6×10^3
Na^{22}	5×10^{-3}	6.7×10^{-5}	Al^{26}	1.2×10^{-5}	1.2×10^{-3}
S^{35}	0.28	7.8×10^{-3}	Cl^{36}	0.55	3×10^{-5}
Be^7	17	0.63	C^{14}	260	10
Ar^{37}	0.19	2.1×10^{-2}	Si^{32}	2.4×10^{-2}	8×10^{-3}
P^{33}	0.15	7.6×10^{-3}	Ar^{39}	2.9×10^{-3}	5×10^{-3}
P^{32}	0.17	1.4×10^{-2}	H^3	36	3.3×10^{-4}

Values for residence time of various elements in the ocean against removal to the bottom sediment, $1/K_{DP}$, have been taken from the literature [G1], [L14].

In view of the simplification implied in the use of reservoir models and the uncertainties of the relevant parameters, the fractional inventories given in Table 5 should be considered to be only approximate. In fact it is one of the main tasks of isotope work to reduce these underlaying uncertainties, particularly with reference to the size of reservoirs and the rate of exchange of materials between them. Nevertheless, the approximate values of the parameters which we have suffice for a discussion of the manner in which the various isotopes distribute themselves on the earth and for evaluating their usefulness for particular applications.

The calculated concentrations (d.m.p. per gm of bulk reservoir material) of short-lived isotopes in the atmosphere and of the long-lived isotopes in the oceans are given in Table 6. In the case of oceans, it is of considerable importance to know the specific activities, i.e. d.p.m. per gm of the purified chemical constituent extracted from the sea. These values, based on the known chemical composition of sea water [G2], are also shown in Table 6.

ray produced radioisotopes in the exchange reservoirs.

isotope								
Si^{32}	Ar^{39}	H^3	Na^{22}	S^{35}	Be^7	Ar^{37}	P^{33}	P^{32}
1.9×10^{-3}	0.16	6.8×10^{-2}	0.25	0.57	0.60	0.63	0.64	0.60
1.1×10^{-4}	0.83	4×10^{-3}	1.7×10^{-2}	8×10^{-2}	0.11	0.37	0.16	0.24
0.29*	0	0.27	0.21	0.10	0.08	0	5.6×10^{-2}	4.7×10^{-2}
3.5×10^{-3}	2×10^{-4}	0.35	0.44	0.24	0.2	0	0.13	0.11
0.68	3×10^{-3}	0.3	8×10^{-2}	4×10^{-3}	2×10^{-3}	0	7×10^{-4}	10^{-4}
2.8×10^{-2}	0	0	0	0	0	0	0	0
1.4×10^{-0}	2.3×10^{-0}	5.6×10^{-2}	0.27	2.9	4.8	7.2	10	17.7

It is apparent from Table 5 that the steady-state distribution among the five principal reservoirs differs enormously from one isotope to the other. Extreme cases are Be^{10}, Al^{26}, where more than 99% of the inventory is either in the ocean sediments or deposited in the lithosphere and $Ar^{37, 39}$, where more than 99% is present in the atmosphere.

V. The applicability of various isotopes to the study of particular geophysical processes.

19. Exchange processes may be unidirectional (f.i. the deposition of sediments on the ocean floor) or they may be cyclic, as f.i. the H_2O cycle illustrated schematically in Fig. 18; in either case the study of the rate of exchange of material requires at least one isotope whose half-life is comparable to the time scales involved.

There exist also processes in which isotopes are thrown out of circulation, i.e. when they reach the biosphere, ocean sediments, permanent ice or other semistable formations, they may become immobilized and isolated from the exchange cycles which transported them into these reservoirs. They can then diminish only by decay and become suitable for dating the time interval which has elapsed since they ceased to partake in the cycle. The best known example here is the role played by the isotope C^{14} in dating objects of archaeological interest.

Guided by such general considerations and by the concentrations given in Table 6, one is led to the following conclusions:

The isotopes Be^{10}, Al^{26}, C^{14} and Si^{32} are suitable for studying the chronology of bottom sediments of various oceans. The isotopes C^{14}, Si^{32} and Ar^{39} are well suited for the study of oceanic circulation, whereas the isotopes H^3 and Be^7 are useful for investigating short-term vertical mixing in the upper layers of oceans. H^3 is also finding application in the study of several problems connected with hydrology, e.g., the time scales involved in the mixing and the circulation of water between the atmosphere, the continents and the oceans, and the characteristics and sizes of the subterranean water reservoirs. H^3 and Si^{32} are useful for investigating the migration and ages of glaciers and polar ice deposits. The C^{14} activity in atmospheric CO_2 trapped during formation of a glacier has also been used in this field [S1].

The isotopes Na^{22}, S^{35}, Be^7, Ar^{37}, P^{33}, and P^{32} are useful, primarily, for meteorological investigations, the shorter lived ones for the study of troposphere circulation, cloud formation and scavenging processes, the longer lived ones for investigating the slower convection and mixing processes operative in the stratosphere. These isotopes with the exception of A^{37} can be measured easily in the atmosphere or in rain water.

Modern separation and low level counting techniques are adequate for measuring the concentration of the isotopes suitable for oceanographic work, except for Ar^{39}, Cl^{36} and for H^3 at great depths.

F. Some important observations on cosmic ray produced radionuclides in the geosphere.

20. In recent years many measurements have been made of the concentration of cosmic ray produced isotopes in the atmosphere, biosphere, hydrosphere and marine sediments. The data have been analysed with a view to check isotope production estimates and to obtain an insight into geophysical processes. It is beyond the scope of this article to discuss these results at length. Some examples have been given in earlier sections dealing with the production mechanism of certain isotopes and with time variations in the primary radiation. A few additional results will be discussed here, in order to bring out more clearly various ways in which these isotopes have been employed in the study of geophysical problems until now.

During the last decade the distribution of naturally produced isotopes has often been disturbed by the injection of artificially produced isotopes in nuclear test explosions. The largest part of present day inventories of C^{14}, H^3 and Na^{22} in the atmosphere are bomb produced [F4], [B19], [B20]. The concentration of some of the short lived isotopes, f.i. S^{35} and to a smaller extent Be^7 and P^{32}, were also modified by nuclear tests but have since returned to normal [L0], [D1], [B20]. Among the longer lived isotopes, the global inventory of C^{14} has been disturbed measurably and the inventories of Cl^{36} and Kr^{81} by an unknown amount; other isotopes still remain predominantly of cosmic ray origin. The discussion which follows is based on data which are believed to be free from errors due to contamination by artificial isotopes.

I. Global inventory studies.

21. The task of experimentally determining the global inventory of isotopes from a limited number of measurements in a few geophysical reservoirs presupposes the knowledge of the various processes responsible for their dispersion. This knowledge is usually not available a priori but has to be derived from studies involving the isotopes themselves, i.e. their relative concentration in various parts of the geosphere. If one obtains a good fit between the predicted and observed isotope distribution in the principal geophysical reservoirs, without violating any geophysical data derived from other sources, it is reasonable to assume that one has achieved a fair understanding of the relevant processes.

In the case of long lived isotopes the interpretation of the data depends also on the assumption that no substantial changes in the level of the cosmic ray intensity have taken place in the past. This assumption in turn can only be justified by observations on the distribution of these very isotopes themselves.

The case of tritium may serve as an illustration of the first requirement. Earlier results seemed to show that cosmic rays could produce only one-fifth to one-tenth of the tritium observed on earth, and this led to speculations regarding its possible accretion from the sun [C6], [B19]. However, a more detailed study of fall-out patterns of artificial and cosmic ray produced radioactivity showed that the discrepancy was due to incorrect assumptions about the residence time of air in the stratosphere and the latitude dependence of its intrusion into the troposphere [L0], [C1], [B21].

C^{14} measurements illustrate the second point. Studies of the concentration of C^{14} in wood samples of known ages have shown that the cosmic ray intensity has remained constant within narrow limits during the last few thousand years and thereby have established that a comparison between global inventory and rate of production based on present day cosmic ray intensities is meaningful [L18], [W9]. The inventories of Cl^{36}, Al^{26} and Be^{10}, when determined, could be used in an analogous manner.

II. Studies in the atmosphere.

22. Direct measurements of the concentration of radioactive nuclides in air have become possible recently due to the development of techniques for collecting sub-micron size aerosols from large amounts of air. Samples involving filtering of up to ca. 30 tons of air have been collected by high flying aircraft. The technique has been used extensively up to an altitude of ca. 20 km; at higher altitudes balloon borne samplers have been employed, but the number of measurements is still relatively small [D7].

α) *Stratospheric air.* The observed distribution of natural S^{35}, Be^7, P^{32}, P^{33} and Na^{22} in the lower region (<20 km) have revealed that

i) The concentrations of S^{35}, Be^7, P^{33} and P^{32} in stratospheric air are nearly in equilibrium with their production [B7], [R4], [F2] as calculated in Sect. III. Figs. 23a and 23b show the isoconcentration lines as observed [F2] during June-September 1961 and October 1959-June 1960, respectively. For comparison, the expected distribution of Be^7, in the absence of vertical and meridional winds, is also shown in Fig. 23c. Within the errors, some of which arise from uncertainties in the amount of air sampled, the experimental and the calculated distributions in the stratosphere are essentially similar.

ii) More precise information on the extent to which equilibrium has been established comes from the observed ratios of isotope concentrations; there are fewer data but they are free from sampling errors.

The observed disintegration ratios, P^{33}/P^{32} and Be^7/P^{32}, give mean values of 0.95 ± 0.1 and 75 ± 10, respectively. The few available ratios of Be^7/S^{35} centre around 50 ± 5. The predicted equilibrium ratios are 0.82, 100 and 55 with estimated errors of less than 30%. The measurements of ratios confirm the conclusion drawn from measurements of absolute concentrations, namely that the air motions are such as to allow these activities to build up close to a secular equilibrium with their production [R4], [R15].

Clearly, air motions between regions of appreciably different production rates, organized or turbulent, do not occur in the stratosphere on time scales short compared to the half-lives of these isotopes. There appears to be a tendency for a slight decrease in the ratio Be^7/P^{32} with altitude at low to medium latitudes; this indicates that the time scale involved in transferring air from the lower stratosphere to the troposphere is longer than the half life of Be^7, but not by a very large factor.

A clear indication of seasonal air motions is obtained from the Be^7/P^{32} ratios in the polar stratosphere during winter and spring. Recent results [B20] show that the P^{32} activity always remains near saturation in the polar stratosphere, but marked changes do occur in Be^7 concentration — a result which would be expected on the basis of present understanding of transport and mixing processes in the polar stratosphere during these months.

iii) The absolute concentrations of Na^{22} as well as the concentration ratio Na^{22}/Be^7 show considerable departure from secular equilibrium, being smaller by as much as an order of magnitude. The departure from equilibrium is found to be strongly altitude-dependent [B 7], as can be seen from Fig. 24.

The observed degree of under-saturation of Na^{22} activity in the stratosphere indicates that the apparent irradiation age of stratosphere air ranges from a few

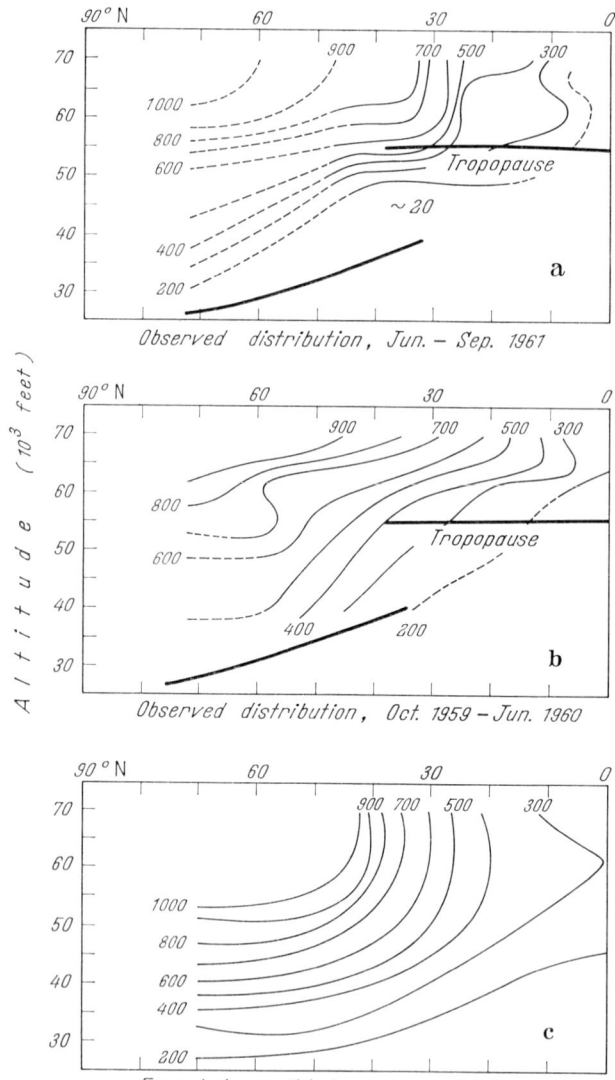

Fig. 23 a—c. Observed lines of equal Be^7 concentrations for two periods are shown in Figs. (a) and (b). The predicted distribution for a motionless atmosphere is given by Fig. (c). Units are d.p.m./10^3 cu. ft. S.T.P. (= d.p.m./36.4 kg air). The thick lines show the position of the "mean tropopause". (From FEELY et al. [F 2]).

months in the lowest layer just above the tropopause to a few years at altitudes of 18—20 km. These results have been discussed in terms of various atmospheric circulation models [B 7]. The Na^{22} data impose limits on admissible models for meridional transport of air in the stratosphere. On the assumption that vertical eddy diffusive processes alone are responsible for the observed altitude distribution of Na^{22}, one finds that the coefficient of vertical eddy diffusivity at 40° S is

3×10^4 cm² sec⁻¹ in regions close to the tropopause and 2×10^3 cm² sec⁻¹ at 18—20 km. The average value at the equator is 2×10^3 cm² sec⁻¹. The values [B7] are consistent with those derived from fission products and ozone data.

Such results which are based on Na²² data [B7], [B20] obtained prior to its "artificial" injection in the atmosphere show that this isotope is well suited for studying the nature of large scale circulation in the stratosphere, in particular when combined with measurements of one of the short lived isotopes (e.g. Be⁷, P³² or P³³) which are in near equilibrium throughout the stratosphere.

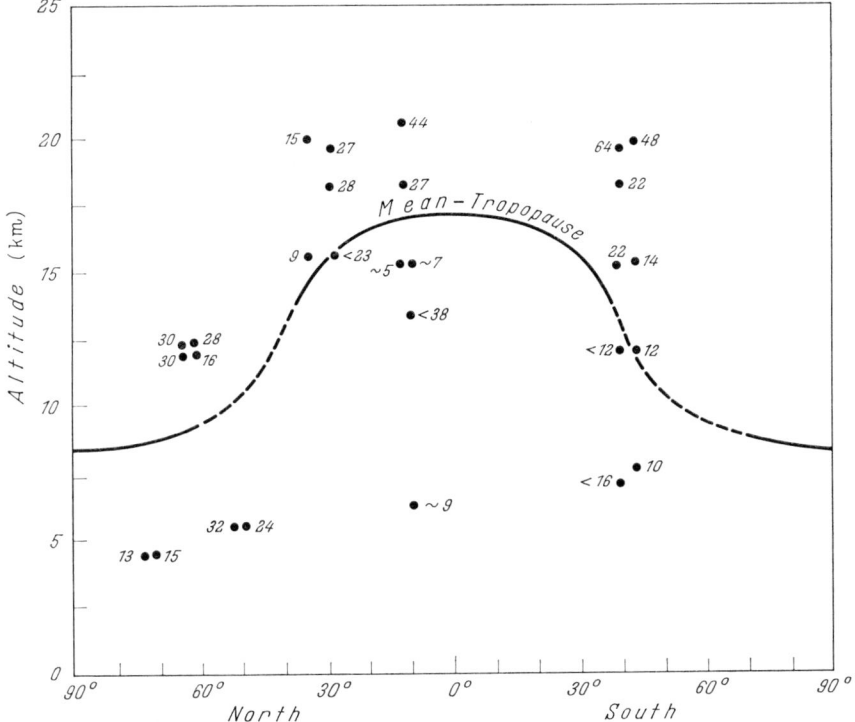

Fig. 24. Observed ratios of Na²² to Be⁷ activity in air, in a north-south section of the atmosphere. The numbers shown in the figure should be multiplied by 10⁻⁵. (After BHANDARI and RAMA [B7].)

β) *Tropospheric air.* In the troposphere at all latitudes, the concentration of Be⁷ is variable. The average concentration increases somewhat with altitude but the absolute value and the gradient are considerably smaller than those corresponding to secular equilibrium with local production [R15]. A still larger under-saturation [B7] is observed for the long-lived radionuclide Na²² (cf. Fig. 24). In contrast to the situation in the stratosphere the activity ratios Be⁷/P³² and P³³/P³² have a wide range of distribution between 17 and 54, and between 0.5 and 0.9, respectively [R15].

The above features can be understood in terms of a build-up of activities in the presence of (a) vertical mixing and (b) removal by condensation processes. The observed altitude distribution is indicative of the time scales involved in these two processes. Mean time scales of three to four weeks are indicated by the data.

γ) *Fall-out of isotopes.* The concentration of various radionuclides in rain water collected from widely different geographic locations lead to the following conclusions [L0], [R15]:

(i) Large variations in the concentration of an isotope occur between individual rains at a given location. Be^7 concentration ranging from less than 3 to more than 100 d.p.m./litre have been observed. This taken in conjunction with the tropospheric air data shows the great variability in the amount of water condensed from a given air mass, or more precisely, in the amount of air effectively cleansed by one litre of rain.

(ii) The average concentrations of the nuclides Be^7, P^{33} and P^{32} in rains are nearly independent of the geomagnetic latitude. The calculated fall-out values of these isotopes are found to be consistent with those expected from their production in the troposphere alone and removal by wash-out with mean intervals of 30—40 days (cf. Fig. 22). The fraction contributed by the stratosphere seems to be small (less than 20%) for the case of Be^7 at all latitudes and still smaller for isotopes with shorter half lives.

(iii) The activity ratio Be^7/P^{32} in rains is found to be similar to that observed in the tropospheric air. The distribution in this ratio indicates again that the mean wash-out period is of the order of 40 days, consistent with the value obtained from fall-out on the assumption of a well-mixed troposphere.

The fall-out of long lived isotopes is more difficult to measure; data [K1], [K7] exist only for Si^{32}. In contrast to the short-lived isotopes, where fall-out is found to be essentially latitude independent, the fall-out of Si^{32} is strongly dependent on latitude. This can be understood in terms of the model discussed in Sect. E.Iβ. The fall-out agrees well with that expected on the basis of its production by cosmic rays.

The production rate and the half life of tritium are such that this isotope also would be useful for studying global fall-out patterns. However, before sufficient measurements in natural waters could be carried out, large amounts of man-made tritium, produced in thermonuclear weapon tests, were released; they completely overshadowed its natural inventory. The best available estimate of the pre-bomb global inventory corresponds to an average production rate of 0.5 ± 0.3 atoms/cm^2 sec [C1] to be compared with a calculated production rate of 0.25 ± 0.08 atoms/cm^2 sec (Sect. C.V.).

III. Oceanographic studies.

23. Useful data have been obtained on the concentration of Be^{10}, C^{14} and Si^{32}. Most tritium measurements refer to the post-thermonuclear era. As was mentioned earlier, the other long-lived isotopes suitable for oceanographic studies, Cl^{36}, Al^{26} and Ar^{39}, have not yet been detected in the marine environment.

α) *Studies in sea water.* Extensive measurements on C^{14} concentrations have been made in the Atlantic [B13]; limited data exist for the Pacific ocean [B11]. The distribution of radiocarbon in the Atlantic subsurface and deep water is found to be in conformity with a general circulation pattern based on other oceanographic evidence [B13]. C^{14} data have provided the time scales in the circulation. C^{14}/C^{12} ratios have been measured at different latitudes (40° S—15° N) in the Pacific at a fixed depth of 3500 m where oceanographic data suggest a steady northward motion. These measurements yield [B10], [B13], for the northward component, a velocity of 0.06 ± 0.02 cm sec^{-1}. Reservoir model calculations based on the observed C^{14} distributions have been discussed in detail by BROECKER [B5], [B13]. Apparent residence times for different water masses and for the deep oceans as a whole have been obtained.

Based on these measurements, the global inventory [C5], [F3] of C^{14} is found to be in good agreement with the production rate of 2.5 ± 0.5 C^{14} atoms/cm$^2 \cdot$ sec, calculated on the basis of the galactic component of cosmic radiation alone (see

Sect. C.IIIβ). This places an upper limit of 20% on the contribution of solar corpuscular radiation to C^{14} production and much lower limits on the solar contribution to the production of other isotopes.

At present Si^{32} data are confined to concentrations in surface waters [K1], [K7]. Reservoir model calculations, using the estimated production rate of Si^{32} and the observed surface concentration in Pacific and Atlantic yield time scales for circulation which are consistent with results based on C^{14} data. A new technique [L9] for the extraction of silicon from sea water may soon make it possible to carry out both C^{14} and Si^{32} measurements in the same deep sea water samples. This should facilitate the study of circulation patterns.

β) *Studies in sediments.* The most extensive measurements are those of C^{14}. Its specific concentration in marine sediments containing appreciable biogenous calcite [A12], [B22] has been used to obtain sedimentation rates reaching back to $\sim$40000 years in the past.

A limited number of measurements is available of Si^{32} concentrations in rapidly accumulating sediments rich in biogenous silica [K7], [K1] and of Be^{10} in slowly accumulating red clay sediments [M4], [G4], [A5], [K1]. These measurements show that the concentrations are in the expected range, but they are not yet sufficient to obtain global inventories.

The activity of Al^{26} could not be detected* in a South Pacific core in which the observed concentrations of Be^{10} indicate a sedimentation rate of $\sim$5 metres/10^6yrs, in good agreement with the palaeontological estimates [K1], [K3]. The upper limit of Al^{26} was found to be 0.4 d.p.m./kg, which is not inconsistent with the expected value of $\sim$0.04 d.p.m./kg due to production of Al^{26} in the atmosphere. In sediments of lower accumulation rates, it should become possible to detect the activity of Al^{26} produced in the atmosphere. If the observed activity is in excess of the expected value, it should probably be attributed to an influx of Al^{26} contained in the cosmic dust accreted by the earth.

IV. Studies in the biosphere.

24. In the marine biosphere, C^{14} and Si^{32} have been studied. In land-based organic material only C^{14} has been detected so far. Rings of certain trees preserve a true record of the atmospheric C^{14}/C^{12} ratio. Supplementary data are available from archaeological samples of known ages up to ca. 5000. Deviations [L18] of less than $\pm 3\%$ seem to occur up to 3000 B.P.; slightly larger deviations, up to ca. 7%, are found for the period 4500—5000 B.P.

These small changes in the atmospheric concentration could represent very appreciable changes in C^{14} production if one considers the nature of its global circulation. The atmospheric changes are expected to be considerably less than those in production rate, f.i. it is calculated [W9] that a 22 year periodic oscillation with ca. $\pm 25\%$ amplitude would result in amplitudes in the atmospheric concentrations of only ca. 1.5%. However, as was discussed in Sect. C.IV, further study is required to establish the origin of the observed fluctuations.

G. Appendix.

Cosmic ray produced radioactivities in terrestrial samples are always weak and usually highly diluted; in many cases it is therefore necessary to employ special techniques for sample collection, for chemical purification and for low level

---

* Al^{26} activity has since been detected in Pacific Sediments [see, LAL and VENKATA-VARADAN, Science **151**, 1381 (1966), and WASSON, ALDER and OESCHGER, Science (in press)].

counting. It may be useful to assemble here references which can serve as a guide to these practical problems.

I. Sampling techniques.

a) *Air.* CHAGNON [C 7]; FERGUSSON [F 8]; FRIEND et al. [F 7]; HAGEMANN et al. [H 7]; WOOD et al. [W 10].

b) *Wet precipitation.* BHANDARI et al. [B 6]; GOEL et al. [G 6]; NILSSON et al. [N 1].

c) *Polar ice.* BEGEMANN [B 8]; LANGWAY et al. [L 1].

d) *Ocean water.* BIEN et al. [B 10]; BROECKER et al. [B 13]; GERARD and EWING [G 11]; SVEDRUP et al. [S 12]; SCHINK and ANDERSON [S 13].

e) *Sediments.* KULLENBURG [K 8]; SVEDRUP et al. [S 12]; ZENKEVICH [Z 2].

II. Chemical separation and purification procedures.

a) *Air and wet precipitation samples.* ARNOLD and AL-SALIH [A 6]; BHANDARI et al. [B 6]; GOEL et al. [G 6]; KHARKAR et al. [K 1]; MARQUEZ, COSTA and ALMEIDA [M 9]; RAMA and HONDA [R 4]; WINSBERG [W 2].

b) *Hydrosphere.* BIEN et al. [B 10]; BROECKER et al. [B 13]; KAUFMAN and LIBBY [K 9]; KHARKAR et al. [K 7]; LAL et al. [L 9].

c) *Sediments.* KHARKAR et al. [K 1]; KHARKAR and LAL [K 3]; MERRILL et al. [M 4].

III. Counting techniques.

ANAND and LAL [A 1]; DAVIS and SCHAEFFER [D 2]; LAL and SCHINK [L 11]; LIBBY [L 16]; OESCHGER [O 1]; VISTE and ANDERS [V 1]; WATT and RAMSDEN [W 4]; ZUTSHI [Z 1].

References.

[A 1] ANAND, J. S., and D. LAL: Nature, Lond. **201**, 775 (1964).
[A 2] ANDERS, E.: Geochim. et Cosmochim. Acta **19**, 53 (1960).
[A 3] ARNOLD, J. R.: Ann. Rev. Nuclear Sci. **11**, 349 (1961).
[A 4] ARNOLD, J. R., M. HONDA, and D. LAL: J. Geophys. Res. **66**, 3519 (1961).
[A 5] ARNOLD, J. R.: Science **124**, 584 (1956).
[A 6] ARNOLD, J. R., and H. A. AL-SALIH: Science **121**, 451 (1955).
[A 7] ARNOLD, J. R., and E. C. ANDERSON: Tellus **9**, 28 (1957).
[A 8] APPARAO, M. V. K., and S. RAMADURAI: On the Relative Characteristics of Proton and Helium Nuclei in Primary Cosmic Radiation. Preprint (1964).
[A 9] ANDERS, E.: Meteorite Ages. The Moon, Meteorites and Comets, ed. by MIDDLEHURST and KUIPER, p. 402. Chicago: Chicago University Press 1963.
[A 10] ADAMS, J. S.: Radioactivity of the Lithosphere. Nuclear Radiation in Geophysics, ed. ISRAEL and KREBS, pp. 1—17. Berlin-Göttingen-Heidelberg: Springer 1962.
[A 11] ALEXANDER, W. M., C. W. MCCRACKEN, L. SECRETAN, and O. E. BERG: Space Research, Vol. III, p. 891. Amsterdam: North Holland Publ. Co. 1963.
[A 12] ARRHENIUS, G., G. KJELLBERG, and W. F. LIBBY: Tellus **3**, 222 (1951).
[B 1] BROWN, W. W.: Phys. Rev. **93**, 528 (1954).
[B 2] BULLOCK, F. W.: Proc. Phys. Soc. Lond. A **70**, 134 (1957).
[B 3] BRYANT, D. A., T. L. CLINE, U. D. DESAI, and F. B. MCDONALD: J. Geophys. Res. **67**, 4983 (1962).
[B 4] BIERI, R., M. KOIDE, and E. D. GOLDBERG: Preprint (1964).
[B 5] BROECKER, W. S.: Mixing Phenomena Within the Atmosphere and Oceans as Determined from Radiocarbon Data. Lamont Geological Observatory Report (1961).
[B 6] BHANDARI, N., S. G. BHAT, D. P. KHARKAR, S. KRISHNA SWAMY, D. LAL, and A. S. TAMHANE: Tellus **18**, 504 (1966).
[B 7] BHANDARI, N., and RAMA: J. Geophys. Res. **68**, 1959 (1963).
[B 8] BEGEMANN, F.: Z. Naturforsch. **14a**, 334 (1959).
[B 9] BEGEMANN, F.: Chimia **16**, 1 (1961).
[B 10] BIEN, G. S., N. W. RAKESTRAW, and H. E. SUESS: Tellus **12**, 436 (1960).
[B 11] BIEN, G. S., N. W. RAKESTRAW, and H. E. SUESS: Radioactive Dating. (Proc. of Athens Symposium, 19—23, Nov. 1962.) Internat. Atomic Energy Agency, Vienna (1963), p. 159.
[B 12] BYERS, H. R.: The Earth as a Planet, ed. by G. P. KUIPER. Chicago: Chicago University Press 1954.
[B 13] BROECKER, W. S., R. D. GERARD, M. EWING, and B. C. HEEZEN: J. Geophys. Res. **65**, 2903 (1960).
[B 14] BISWAS, S., and C. E. FICHTEL: Astrophys. Jour. **139** (3), 941 (1964).

[B 15] BISWAS, S.: The Composition of Solar Particle Radiation. Proc. of the Internat. Conference on Cosmic Rays (Jaipur, Dec. 1963) **1**, 43 (1964).
[B 16] BAILEY, L. E.: Angle and Energy Distribution of Charged Particles from the High Energy Nuclear Bombardment of Various Elements. (Thesis) UCRL-3334 (1956).
[B 17] BOLIN, B.: Proc. Sec. Int. Conf. Peaceful Uses of Atomic Energy **18**, 336 (1958).
[B 18] BOLIN, B., and E. ERIKSSON: Rosby Memorial Volume, pp. 130—142. New York: Rockefeller Inst. Press 1959.
[B 19] BEGEMANN, F., and W. F. LIBBY: Geochim. et Cosmochim. Acta **12**, 277 (1957).
[B 20] BHANDARI, N., D. LAL and RAMA: Tellus **18**, 391 (1966).
[B 21] BEGEMANN, F.: The Tritium Content of Atmospheric Hydrogen and Methane. In: Earth Science and Meteorites, ed. J. GEISS and E. D. GOLDBERG, p. 169. Amsterdam: North-Holland Publ. Co. 1963.
[B 22] BROECKER, W. S., K. K. TUREKIAN, and B. C. HEEZEN: Amer. J. Sci. **256**, 503 (1958).
[B 23] BJORNERSTEDT, R., and K. EDVARSON: Ann. Rev. Nuclear Sci. **13**, 505 (1963).
[B 24] BALASUBRAMAHNIYAN, V. K., and F. B. MCDONALD: J. Geophys. Res. **69**, 3289 (1964).
[C 1] CRAIG, H., and D. LAL: Tellus **13** (1), 85 (1961).
[C 2] CURRIE, L. A.: Phys. Rev. **114**, 878 (1959).
[C 3] CRAIG, H.: The Natural Distribution of Radiocarbon. In: Earth Science and Meteorites, ed. by GEISS and GOLDBERG, p. 103. Amsterdam: North Holland Publ. Co. 1963.
[C 4] CUMMING, J. B.: Ann. Rev. Nuclear Sci. **13**, 261 (1963).
[C 5] CRAIG, H.: Tellus **9**, 1 (1957).
[C 6] CRAIG, H.: Phys. Rev. **105**, 1125 (1957).
[C 7] CHAGNON, C. W.: Editor: Joint Indo-United States Balloon Flight Program 1961. AFCRL-62-1135 (1962).
[D 1] DERVINSKY, P. J., J. T. WASSON, E. C. COUBLE, and N. A. DIAMOND: J. Geophys. Res. **69**, 1457 (1964).
[D 2] DAVIS, R., and O. A. SCHAEFFER: Ann. N.Y. Acad. Sci. **62**, 105 (1955).
[D 3] VRIES, HL. DE: Research in Geochemistry, ed. by P. H. ABELSON. New York: Wiley Inc. 1959.
[D 4] DAMON, P. E., and AUSTIN LONG: Carbon-14, Carbon Dioxide and Climate. Preprint (1963).
[D 5] DORMAN, L. I.: Cosmic Ray Variations. Moscow: State Publishing House for Technical and Theoretical Literature 1957.
[D 6] DAMGAARD, G., and K. HANSEN: Reported at the Sienna Conference on Elementary Particles 1963.
[D 7] DREVINSKY, P. J., E. A. MARTELL, and D. LAL: Joint Indo-United States Balloon Flight Program 1961. AFCRL-62-1135 (Dec. 1962).
[F 1] FALTINGS, V., and P. HARTECK: Z. Naturforsch. **5a**, 439 (1950).
[F 2] FEELY, H. W., B. DAVIDSON, J. P. FRIEND, R. J. LAGOMARSINO, and M. W. M. LOE: Project Star Dust. DASA 1309. New Jersey: Isotopes Inc. 1963.
[F 3] FULLER, M. O.: Disintegration of Oxygen by 300 MeV Neutrons. (Thesis) UCRL-2699 (1954).
[F 4] FERGUSSON, G. J.: Natural and Artificial Radiocarbon in the Atmosphere and Oceans. Preprint (1963).
[F 5] FERGUSSON, G. J.: Proc. Roy. Soc. Lond. A **243**, 561 (1958).
[F 6] FREIER, P. S., and W. R. WEBBER: J. Geophys. Res. **68**, 1605 (1963).
[F 7] FRIEND, J. P., H. W. FEELY, P. W. KREY, J. SPAR, and A. WALTON: High Altitude Sampling program. DASA 1300, Vol. 1 (1961).
[F 8] FERGUSSON, G. J.: J. Geophys. Res. **68**, 3933 (1963).
[G 1] GOLDBERG, E. D., and G. O. S. ARRHENIUS: Geochim. et Cosmochim. Acta **13**, 153 (1958).
[G 2] GOLDBERG, E. D.: Amer. Rev. Phys. Chem. **12**, 29 (1961).
[G 3] GEISS, J., H. OESCHGER, and U. SCHWARZ: Space Sci. Rev. **1**, 197 (1962).
[G 4] GOEL, P. S., D. P. KHARKAR, D. LAL, N. NARSAPPAYA, B. PETERS, and V. YATIRAJAM: Deep Sea Res. **4**, 202 (1957).
[G 5] GOEL, P. S.: Nature, Lond. **178**, 1458 (1956).
[G 6] GOEL, P. S., N. NARSAPPAYA, C. PRABHAKARA, T. RAMA, and P. K. ZUTSHI: Tellus **11**, 91 (1959).
[G 7] GOEL, P. S., and M. HONDA: Unpublished.
[G 8] GEORGE, E. P.: Observations of Cosmic Rays Underground and their Interpretation. Progress in Cosmic Ray Physics, Vol. 1, p. 395. Amsterdam: North-Holland Publ. Co. 1952.
[G 9] GOEL, P. S., and T. P. KOHMAN: Radioactive Dating. (Proceedings of the Athens Symposium, 19—23 Nov. 1962.) Internat. Atomic Energy Agency, Vienna 1963, p. 413.

[G 10] GOEBEL, K., H. SCHULTES, and J. ZÄHRINGER: Production Cross Sections of Tritium and Rare Gases in Various Target Elements. CERN report, 64—12 (1964).
[G 11] GERARD, R., and M. EWING: Deep Sea Res. **8**, 298 (1961).
[H 1] HESS, W. N., H. W. PATTERSON, R. WALLACE, and E. L. CHUPP: Phys. Rev. **116**, 445 (1959).
[H 2] HONDA, M., and D. LAL: Nuclear Phys. **51**, 363 (1964).
[H 3] HOUTERMANS, F. G., RAMA, H. OESCHGER, and S. AEGERTER: A Search for Cosmic Ray Produced Kr^{81} in the Atmosphere. Preprint (1964).
[H 4] HONDA, M., and D. LAL: Phys. Rev. **118**, 1618 (1960).
[H 5] HONDA, M., and J. R. ARNOLD: Science **143**, 203 (1964).
[H 6] HAWKINS, G. S.: Ann. Rev. Astronom. Astrophys. **2**, 149 (1964).
[H 7] HAGEMANN, F., J. G. RAY jr., L. MACHTA, and A. TURKEVICH: Science **130**, 542 (1959).
[I 1] ISRAEL, H., and KREBS, editors: Nuclear Radiation in Geophysics. Berlin-Göttingen-Heidelberg: Springer 1962.
[J 1] JUNGE, C. E.: Air Chemistry and Radioactivity. New York and London: Academic Press 1963.
[J 2] JUNGE, C. E., O. OLDENBERG, and J. T. WASSON: J. Geophys. Res. **67**, 1027 (1962).
[K 1] KHARKAR, D. P., D. LAL, and B. L. K. SOMAYAJULU: Radioactive Dating. (Proceedings of the Athens Symposium, 19—23, Nov. 1962.) Internat. Atomic Energy Agency, Vienna 1963, p. 175.
[K 2] KOUTS, H., and L. YUAN: Phys. Rev. **86**, 128 (1952).
[K 3] KHARKAR, D. P., B. S. AMIN, and D. LAL: Deep Sea Research **13** (in press) (1966).
[K 4] KOENIGSWALD, G. H. R. v.: Space Sci. Rev. **3** (3), 433 (1964).
[K 5] KELLOGG, D. A.: Phys. Rev. **90**, 224 (1953).
[K 6] KULP, G. L., and M. L. VOLCHOK: Phys. Rev. **90**, 713 (1953).
[K 7] KHARKAR, D. P., V. N. NIJAMPURKAR, and D. LAL: Geochim. Cosmochim. Acta **30**, 621 (1966).
[K 8] KULLENBURG, B.: Deep Sea Coring. Reports of the Swedish Deep Sea Expedition (1947/48), ed. by H. PETTERSON, Vol. IV, p. 37.
[K 9] KAUFMAN, S., and W. F. LIBBY: Phys. Res. **93**, 1337 (1954).
[L 0] LAL, D., and B. PETERS: Cosmic Ray Produced Isotopes and their Application to problems in Geophysics. Progress in Cosmic Ray Physics and Elementary Particle Physics, Vol. 6, p. 3. Amsterdam: North-Holland Publ. Co. 1962.
[L 1] LANGWAY, C. C. jr., H. OESCHGER, A. RENAUD, and B. ALDER: Sampling Polar Ice for Radiocarbon Dating. Preprint (Oct. 1964).
[L 2] LAL, D., J. R. ARNOLD, and M. HONDA: Phys. Rev. **118**, 1626 (1960).
[L 3] LAL, D., P. K. MALHOTRA, and B. PETERS: J. Atmosph. Terr. Phys. **12**, 306 (1958).
[L 4] LORD, J. J.: Phys. Rev. **81**, 901 (1951).
[L 5] LINGENFELTER, R. E., and E. J. FLAMM: Science **144**, 292 (1964).
[L 6] LAL, D., G. RAJAGOPALAN, and V. S. VENKATAVARDAN: Proceedings of the Internat. Conference on Cosmic Rays (Jaipur, Dec. 1963) **1**, 99 (1964).
[L 7] LAL, D.: On the Investigations of Geophysical Processes Using Cosmic Ray Produced Radioactivity. In: Earth Science and Meteoritics, compiled by J. GEISS and E. D. GOLDBERG, p. 115. Amsterdam: North Holland Publ. Co. 1963.
[L 8] LAL, D.: Proceedings of the Internat. Conference on Cosmic Rays (Jaipur, Dec. 1963) **6**, XX (1964).
[L 9] LAL, D., J. R. ARNOLD, and B. L. K. SOMAYAJULU: Geochim. et Cosmochim. Acta **28**, 1111 (1964).
[L 10] LINGENFELTER, R. E.: Rev. Geophys. **1**, 35 (1963).
[L 11] LAL, D., and D. R. SCHINK: Rev. Sci. Instrum. **31**, 395 (1960).
[L 12] LAL, D., N. NARSAPPAYA, and P. K. ZUTSHI: Nuclear Phys. **3**, 69 (1957).
[L 13] LAL, D.: Investigations of Nuclear Interactions Produced by Cosmic Rays. Thesis, Bombay University (1958).
[L 14] LAL, D., E. D. GOLDBERG, and M. KOIDE: Science **131**, 332 (1960).
[L 15] LAL, D., RAMA, and P. K. ZUTSHI: J. Geophys. Res. **65**, 669 (1960).
[L 16] LIBBY, W. F.: Radiocarbon Dating. Chicago: Chicago University Press 1952; 2nd. ed. 1959.
[L 17] LOHRMANN, E., and E. SCHOPPER: Nucleonen in der Atmosphäre. Handbuch der Physik, 1965.
[L 18] LIBBY, W. F.: Science **140**, 278 (1963); also Antiquity **37** (147), 213 (1963).
[L 19] LINGENFELTER, R. E., and E. J. FLAMM: J. Atmosph. Sci. **21** (2), 134 (1964).
[L 20] FLAMM, E., R. E. LINGENFELTER, G. J. F. MACDONALD, and W. F. LIBBY: Science **138**, 49 (1962).
[L 21] LAL, D.: J. Oceanogr. Soc. Japan, 20th Anniversary Volume, 600 (1962).

References.

[L 22] LINDENBAUM, S. J., and R. M. STERNHEIMER: Phys. Rev. **105**, 1874 (1957); **123**, 333 (1961).
[M 1] MEYER, P., and R. VOGT: Phys. Rev. **129**, 2275 (1963).
[M 2] MARTELL, E. A.: Science **129**, 1197 (1959).
[M 3] MCDONALD, F. B.: Phys. Rev. **116**, 462 (1959).
[M 4] MERILL, J. R., M. HONDA, and J. R. ARNOLD: Analyt. Chem. **32**, 1420 (1960).
[M 5] MCDONALD, F. B., and W. R. WEBBER: J. Phys. Soc. Japan **17**, (Suppl. A.II), 428 (1962).
[M 6] MIYAKE, S., K. HINOTANI, and K. NUNOGANI: J. Phys. Soc. Japan **12** (2), 113 (1957).
[M 7] MYLROI, M. G., and J. G. WILSON: Proc. Phys. Soc. Lond. A **64**, 404 (1951).
[M 8] MARQUEZ, L., and N. L. COSTA: Nuovo Cim. **2**, 1038 (1955).
[M 9] MARQUEZ, L., N. L. COSTA and I. G. ALMEIDA: Nuovo Cim. **6**, 1292 (1957).
[M 10] MACHTA, L., and R. J. LIST: Open Hearings of the Joint Committee on Atomic Energy on "Fall-Out from Nuclear Weapons Test", May 5—8 (1959). Washington; see also: Advances in Geophysics, Vol. 6, p. 273. New York: Academic Press 1959.
[M 11] MACNAMARA, J., and H. G. THODE: Phys. Rev. **80**, 296 (1950).
[M 12] MORRISON, P., and J. PINE: Ann. N. Y. Acad. Sci. **62**, 69 (1955).
[N 1] NILSSON, R., I. OLSSON, A. BERGGREN, and K. SIEGBAHN: Ark. Geophys. **3**, 111 (1959).
[O 1] OESCHGER, H.: Low Level Counting Methods. Proceedings of the Symposium on Radioactive Dating, held at Athens (19—23 Nov. 1962), publ. by I.A.E.A. Vienna, 13 (1963).
[P 1] PETERS, B.: Proc. Ind. Acad. Sci. **41**, 67 (1955).
[P 2] PETERS, B., J. Atmosph. Terr. Phys. **13**, 351 (1959).
[P 3] PETERS, B.: Cosmic Rays. Handbook of Physics, ed. by CONDON and ODISHAW, Chap. 12, pp. 201—244. McGraw-Hill Book Co. 1958.
[P 4] PETERS, B.: Some Problems Connected with the Chemical Composition of Galactic Cosmic Radiation. Semaine d'Étude sur le Problème du Rayonnement Cosmique dans l'Espace Interplanetaire (Pontificae Academiae Scientiarum Scripta Varia), p. 1 (1963).
[P 5] PAL, Y., and B. PETERS: Kgl. danske Vidensk. Selsk., Mat.-Fys. Medd. **33**, No. 15 (1964).
[P 6] PETERS, B.: Nature, Flux and Energy Distribution of Cosmic Ray Particles in Interplanetary Space. Semaine d'Étude sur le Problème du Rayonnement Cosmique dans l'Espace Interplanetaire (Pontifica Academia Scientiarum Scripta Varia), p. 532 (1963).
[P 7] PETERS, B.: Z. Physik **148**, 93 (1957).
[P 8] PUPPI, G.: The Energy Balance of Cosmic Radiation. Progress in Cosmic Ray Physics, Vol. 3, p. 341. Amsterdam: North Holland Publ. Co. 1956.
[Q 1] QUENBY, J.: Time Variations of Cosmic Rays. Handbuch der Physik (this volume).
[R 1] ROSSI, B.: High Energy Particles. New Jersey: Prentice-Hall, Inc., Englewad Cliffs 1956.
[R 2] RAY, E. C.: Experimental Results of Flights in the Stratosphere. Handbuch der Physik, Bd. XLVI/1, pp. 130—156. Berlin-Göttingen-Heidelberg: Springer 1961.
[R 3] RUDSTAM, G.: Phil. Mag. **46**, 344 (1955).
[R 4] RAMA, and M. HONDA: J. Geophys. Res. **66**, 3227 (1961).
[R 5] RAMA, and M. HONDA: J. Geophys. Res. **66**, 3533 (1961).
[R 6] RAMA: Bull. Nat. Geophys. Res. Inst. **1** (4), 241 (1963).
[R 7] RENAUD, A., E. SCHUMACHER, B. HUGHES, H. OESCHGER, and C. MUHLEMANN: J. Geophys. Res. **68**, 3783 (1963).
[R 8] REYNOLDS, J. H.: Phys. Rev. **79**, 886 (1950).
[R 9] RODEL, W.: Nature, Lond. **200**, 999 (1963).
[R 10] REVELLE, R., and H. E. SUESS: Gases. The Sea, ed. by M. N. HILL, p. 313. New York: Interscience Publ. 1962.
[R 11] ROSHOLT, N. N., C. EMIBIANI, J. GEISS, F. F. KOCZY, and P. J. WANGERSKY: J. Ged. **69**, 162 (1961).
[R 12] RAMA: Unpublished calculations.
[R 13] REVELLE, R., and H. E. SUESS: Tellus **9**, 18 (1957).
[R 14] RANKAMA, K.: Isotope Geology. London: Pergamon Press 1954.
[R 15] RAMA: J. Geophys. Res. **68**, 3861 (1963).
[S 1] SCHOLANDER, P. F., H. DE VRIES, W. DANSGAARD, L. K. COACHMAN, D. C. NUTT, and E. HEMMINGSEN: Meddelelser om Grønland **165**, (Nr. 1), 26 (1962).
[S 2] STAKER, W. P., M. PAVALOW, and S. A. KORFF: Phys. Rev. **81**, 889 (1951).
[S 3] SOBERMAN, R. K.: Phys. Rev. **102**, 1399 (1956).
[S 4] SIMPSON, J. A., H. W. BALDWIN, and R. B. URETZ: Phys. Rev. **84**, 332 (1951).
[S 5] SIMPSON, J. A.: Phys. Rev. **83**, 1175 (1951).

[S6] STONE, E. C.: J. Geophys. Res. **69**, 3939 (1964).
[S7] SCHAEFFER, O. A., and J. ZÄHRINGER: Phys. Rev. Letters **8**, 389 (1962).
[S8] SUESS, H. E.: Secular Changes in the Concentration of Atmospheric Radiocarbon. Problems related to interplanetary matter. NAS-NRC, Publ. 845 (1961).
[S9] SIMPSON, J. A.: Semaine d'Étude sur le Problème du rayonnement cosmique dans l'espace interplanetaire (Pontificale Academiale Scientiarum Scripta Varia), p. 323 (1963).
[S10] SIMPSON, J. A.: Astrophys. J., Suppl. **4**, 378 (1960).
[S11] SCHOVE, D. J.: J. Geophys. Res. **60**, 127 (1955).
[S12] SVEDRUP, H. U., M. W. JOHNSON, and R. M. FLEMING: The Oceans, Their Physics, Chemistry and General Biology. New York: Prentice Hall 1942.
[S13] SCHINK, D., and M. ANDERSON: J. Geophys. Res. **67**, 3596 (1962).
[T1] TILLES, D., J. DE FELICE, and E. L. FIREMAN: ICARUS, **2** (3), 258 (1963).
[V1] VISTE, E., and E. ANDERS: J. Geophys. Res. **67**, 2913 (1962).
[V2] VOSHAGE, H., and H. HINTENBERGER: Radioactive Dating (Proceedings of the Athens Symposium, 19—23 Nov. 1962). International Atomic Energy Agency, Vienna, p. 367 (1963).
[V3] VOSHAGE, H.: Z. Naturforsch. **17**a, 422 (1962).
[V4] VOGT, R.: Phys. Rev. **125**, 366 (1962). — MEYER, P., and R. VOGT: Phys. Rev. **129**, 2275 (1963).
[W1] WINSBERG, L.: Phys. Rev. **95**, 205 (1954).
[W2] WINSBERG, L.: Geochim. et Cosmochim. Acta **9**, 183 (1956).
[W3] WASSON, J. T.: ICARUS **2**, 54 (1963).
[W4] WATT, D. E., and D. RAMSDEN: High Sensitivity Counting Techniques (International Series of Monographs on Electronics and Instrumentation), Vol. 20. Oxford: Pergamon Press 1964.
[W5] WADDINGTON, C. J.: Progr. Nuclear Phys. **8**, 3 (1960).
[W6] WADDINGTON, C. J.: The primary Cosmic Radiation. Proc. Int. School of Phys. "Enrico Fermi", Course XIX (Como, 23 May—3 June 1961), p. 135. New York: Academic Press 1963.
[W7] WEBBER, W. R.: Time Variations of Low Rigidity Cosmic Rays during the Recent Sunspot Cycle. Progr. Cosmic. Ray Physics and Elem. Part. Phys. Amsterdam: North Holland Publ. Co. **6**, 77 (1962).
[W8] WEBBER, W. R.: An Evaluation of the Radiation Hazard due to Solar Particle Events. Report D2-90469, Boeing Company, Seattle, Dec. 1963; also P. FREIER and W. R. WEBBER, Science **142**, 1587 (1963).
[W9] WOOD, L., and W. F. LIBBY: Geophysical Implications of Radiocarbon Date Discrepancies. Isotopic and Cosmic Chemistry, p. 205. Amsterdam: North Holland Publ. Co. 1964.
[W10] WOOD, R., D. LUNDGREN, S. ROHRBOUGH, W. TORGESON, J. BORTH, J. UPTON, P. STROOM, S. JONES, and I. HALL: General Mills Report No. 2328 (1962).
[Z1] ZUTSHI, P. K.: Nucleonics **21**, 50 (1962).
[Z2] ZENKEVICH, L. A.: Deep Sea Res. **4** (1), 67 (1956).

Effects of Cosmic Rays on Meteorites.

By

M. Honda and J. R. Arnold.

With 6 Figures.

1. Cosmic rays and meteorites are two classes of material bodies that reach the earth from outer space. The study of their interactions has helped to increase our understanding of both.

A great many things are known about the cosmic radiation. In the region of space near the earth it consists mainly of high-energy protons, with an important component of alpha particles, and a smaller fraction of heavier nuclei. Typical energies are in the region of 10^9 electron volts (Gev) per nucleon, although particles of far higher energy occur. Between the discovery of the neutron in the early 1930's and the advent of very-high-energy machines a few years ago, most of the information available on fundamental particles came from cosmic-ray studies. Still, despite abundant data on composition, energy, and direction of motion, the place and mode of origin of cosmic rays is only partially understood.

Cosmic-ray bombardment affects terrestrial materials. The best-known example of such an effect is the production of carbon-14 by the $N^{14}(n,p)C^{14}$ reaction. The half-life of C^{14}, 5600 years, and the biological importance of carbon, make this nuclide most important for chronological studies. Carbon-14 is not the only radioactive nuclide produced by cosmic rays; other short- and long-lived nuclides are formed by interaction between the high-energy particles and the nitrogen, oxygen, and argon of the air. Cosmic-ray produced nuclides occur in very small amounts in the solid surface materials of the earth because of the absorption by the mass of air (1 kg/cm²) above us. However, they can be detected by modern methods of low-level counting. Extensive reviews covering recent advances in this field are available [3, 4].

The reactions which produce the observed radioactive species on earth occur mainly in the stratosphere and upper troposphere. What happens to each nuclide after formation depends on its chemical and physical properties. If it is long-lived or stable, it will be transported away from the point of origin, and usually it undergoes various geochemical reactions. In general, these changes are not quantitatively understood. As a result, it is difficult to decipher the fossil record of cosmic radiation which the data undoubtedly contain. The one exception is, again, C^{14}. The general success of the C^{14} dating method is very good evidence for the approximate constancy of the intensity of the cosmic rays striking the atmosphere over the last few half-lives of this nuclide. Apparent small variations in the C^{14} concentration may be due to past changes in intensity. However, other geochemical or geophysical phenomena (such as changes in the rate of CO_2 transport across the air-sea interface and magnetic field variations) may be involved.

Unlike the earth's atmosphere, the meteorites are physically and chemically stable targets. While their history presents problems of its own, they are much more suitable than terrestrial targets for an extended study of the fossil record of cosmic radiation. Until samples of the surface of the moon and of other bodies without atmosphere (including, for some purposes, artificial targets) become available, meteorites will be the most suitable bodies for such studies.

Course of radiation in a meteorite.

2. Let us follow the radiation into a meteorite. High-energy protons and other charged nuclei enter its surface from every direction. After passing throught a depth of matter equivalent (on the average) to about 100 grams per square centimeter, each particle undergoes interaction, giving rise to new active particles of various energies and types. These secondary particles undergo further interactions, and this cascade process continues until the energy is dissipated. Among the nuclear particles in which we are interested are high-energy protons of primary and secondary origin, neutrons of all energies, and, in the case of condensed bodies such as meteorites, important numbers of π-mesons also. The energy dependence of the primary flux is shown in Fig. 1 (1). Above 1 Gev it is well represented by a law of the form $(1+E)^{-2.5}$, where E is the energy in Gev. The mean kinetic energy per nucleon is in the neighborhood of 4 Gev. In space, the total flux over all angles amounts to about five nucleons per square centimeter per second. Inside the meteorite body the intensity of the high-energy particles decreases with depth in accordance with a mean absorption thickness (for a collimated beam) of about 150 g/cm². The low-energy secondary flux increases in the first few centimeters below the surface of the body. After that it passes through a maximum and then decreases, somewhat more slowly than the primary intensity. The total nuclear active particle flux increases rapidly at first, as the low-energy particles become much more numerous than the remaining high-energy ones. Below a depth of 150 g/cm² or so, the total flux decreases, the energy spectrum reaching an approximate equilibrium and changing (steepening) only slowly.

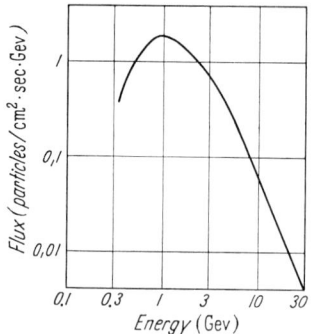

Fig. 1. Differential flux of primary nucleons near the earth [5]. From data of MacDonald and Webber.

A special effect arises in quite large bodies. Here an appreciable flux of low-energy, or thermal, neutrons makes an appearance. There are few experimental data showing the effects of thermal neutrons in meteorites (2). Products with which we are concerned here are mainly produced by spallation — used here as a generic term for nuclear reactions at energies above about 10 Mev. In these reactions the emission of one or more nucleons from the target nucleus is to be expected, and the products will range in mass from that of the target nucleus down to single neutrons or protons.

There are a number of possible sources of time variation in the intensity of the cosmic radiation reaching us on earth. Even in the last few decades, a cyclic change of the flux has been detected, with a period corresponding to the 11-year solar cycle (1). The lowest-energy portion of the primary cosmic radiation is much affected by the sun. There is a general decrease in this radiation at times of high solar activity, punctuated by occasional high-intensity bursts from the sun. The sources of primary cosmic radiation in the Gev region are generally believed to be in the galaxy, far from the solar system. Very energetic particles, carrying 10^{20} electron volts and even more, may be of extragalactic origin. However, the total flux appears to be mainly galactic.

It must be kept in mind that primary cosmic radiation in the energy region of 1 to 10 Gev contains most of the flux and is responsible for most of the transmutations observed. The data give information mainly about this energy region.

A number of sources for cosmic rays within the galaxy have been suggested (3). A supernova occurs in our galaxy every few centuries, and these catastrophes may well give rise to most of the particles observed. Other suggested sources — much more numerous, although less dramatic — are the contact binary stars and the flare stars among the red dwarfs. Whatever their origin, it is thought that the particles are accelerated and are trapped in the magnetic fields found in the spiral-arm structure of the galaxy. The intensity of the radiation in interplanetary space will be affected by (i) variations in source intensity; (ii) variations in the general magnetic field in the region of the galaxy through which the sun and meteorite are passing; and (iii) variations, like those observed during the sunspot cycle, in the intensity and the extension into space of the sun's magnetic field. It seems highly probable that over a span of many aeons (1 AE $=10^9$ years) the population of sources and the magnetic field of the galaxy in the region of the sun have changed considerably. Have there been important variations over shorter periods? The record of cosmic radiation in meteorites seems to offer the best hope of attacking this question. However, this requires that we know something about the history of the meteorites themselves.

History of meteorites.

3. It seems to be firmly established that meteorites originate inside the solar system. They are usually considered to be small bits of asteroid. The one meteorite whose orbit has been well determined, Pribram (Luhy) (4), indeed had an orbit passing out into the asteroid belt. However, major perturbations caused by passages close to the inner planets profoundly alter the original orbits of these bodies (5). The moon, the planets Mars and Mercury, and the comets have all been considered possible sources of meteoritic bodies. Because of the very large size and mass of Jupiter, bodies whose orbits extend out to its distance from the sun or beyond are likely to be captured by it or ejected from the solar system. On the basis of this and other arguments, it seems highly probable that meteorites have spent their careers in space closer than about 5 astronomical units to the sun.

The great majority of meteorites observed to fall are stones. In fact, about 80 percent of all observed "falls" are stones of one class, called chondrites. Chondrites contain several silicate minerals, chiefly pyroxenes and olivine; iron-nickel metal (of the order of 10 percent); and troilite, ferrous sulfide (about 5 percent by weight). The commonest elements (targets) are therefore oxygen, silicon, magnesium, and iron, with important amounts of sulfur, aluminum, calcium, sodium and nickel. Chondrites get their name from characteristic spherical inclusions called chondrules. A variety of different objects are lumped under the generic name "achondrite". Some are similar in chemical composition to the chondrites. Another group is much richer in calcium. About 10 percent of the observed falls are iron meteorites, some of which have been very large. Iron meteorites are composed of metallic iron, containing usually 5 to 10 percent of nickel. Inclusions of ferrous sulfide, iron-nickel phosphide, and graphite are usually present. Many of the common terrestrial elements are extremely rare in iron meteorites. These include not only obviously incompatible elements, such as potassium and calcium, but also some, such as manganese, silicon, titanium, and vanadium, which occur commonly in artificial iron alloys. An intermediate group, the stony irons, contain comparable quantities of metal and stone. A group of these, called pallasites, contain large and easily separated crystals of olivine embedded in metal phase.

Secular equilibrium.

4. Before discussing the radioactive products of cosmic-ray bombardment, let us review the notion of secular equilibrium. If a radioactive substance is produced at a steady rate for a time that is long compared to its half-life, the number of nuclei decaying will come to equal the number being formed, so that no further changes of concentration occur. The observation that the rate of decay of the radioactive species in a meteorite immediately after fall is equal, within the limits of experimental error, to the present rate of production will lead us to conclude, conversely, that the cosmic-ray intensity has been constant over periods of the order of the mean life of this species. This converse is not rigorous, since a coincidence of production and decay rates might be an accidental consequence of wide temporal variations in rate of production. Still, the presumption that the cosmic-ray intensity has been constant is strong, especially when the same result is observed for many radioactive species of different half-lives.

How can one calculate the absolute production rate for each species at each point in the target? This rate is dependent on the absolute intensity and the energy spectrum of the bombarding flux. This in turn is a function of the depth of the sample in a particular meteorite, and of the size and shape of the meteorite body in space. It is not easy to deduce these values quantitatively, but it is possible to make rough estimates. For instance, Na^{22} and Be^{10} are produced only by high-energy particles. The primary flux of five nucleons per square centimeter per second may be reduced by 75 percent at an effective depth of 25 centimeters in iron, or 200 g/cm^2. The cross sections for these species are about 3 to 5 millibarns, respectively (1 mbarn equals 10^{-27} cm^2), at 4 Gev, the mean energy of the primary beam. The cross section is rather independent of energy in the Gev region. Therefore, the production rate of Na^{22} in iron (atomic weight 56) for example is

$$5 \cdot (1/4) \cdot 3 \cdot 10^{-27} \cdot 6 \cdot 10^{23} (1000/56) \cdot 60 = 2.4 \text{ atom/kg min}.$$

This number is in rough agreement with the observed activity of these species in the freshly fallen meteorite Aroos (Table 3). On the other hand, Mn^{54} is a typical "low-energy product". Its rate of production is greatly increased by the large number of secondary neutrons produced in the iron block. Above about 50 Mev the cross section is in the vicinity of 30 millibarns, nearly 10 times that for Na^{22} production. But the observed activity is more than 100 times as high (Table 1). This indicates that more flux — about one order of magnitude more — is available for production of Mn^{54} at moderate depths, a result in agreement with what is known about the development of cascades in solid bodies. Between the two extremes there are many nuclides whose energy thresholds for spallation are between 0.1 and 1 Gev. Unless the production cross section is especially low, we expect the concentrations of these nuclides in iron meteorites to lie between those of Na^{22} and Mn^{54}. In a stone meteorite it is necessary to take account of the complex chemistry of the target. For example, Na^{22} in this target is produced mainly from magnesium and silicon. Here it is a typical "low-energy product", like Mn^{54}.

Table 1. *Measured and estimated production rates of nuclides in Aroos.*

The production cross section is given for the center of a 20-cm sphere (93 percent Fe, 7 percent Ni) from 3-Gev proton bombardment data. The estimated relative production rate is given for a depth of 100 g/cm, radius 200 g/cm, relative to $Cl^{36}=1$. Measurements of the Aroos meteorite are corrected to the time of fall; dpm, disintegrations per minute.

Nuclide	Half-life	Production estimated		Aroos [c]
		cross section [a] (mbarn)	relative production rate [b]	(dpm/kg)
H^3	12.3 yr		12	
Be^7	53 day	6.5	0.69	
Be^{10}	$2.5 \cdot 10^6$ yr		.31	4.1 ± 0.4
C^{14}	5600 yr		.13	1.8 ± 0.3 [d]
Na^{22}	2.6 yr	1.3	.11	2.1 ± 0.3
Al^{26}	$7.4 \cdot 10^5$ yr		.10	1.5 ± 0.4 [g]
P^{32}	14.3 day	5	.41	
Si^{32}	~300 yr		.05	0.8 ± 0.3
Cl^{36}	$3.1 \cdot 10^5$ yr		1	16 ± 1.6
Ar^{37}	34 day		0.65	20 [e]; 15 [f]
Ar^{39}	260 yr		.9	16 ± 1 [e]
Ca^{45}	165 day		.28	5 ± 1
Sc^{44m}	2.4 day	9		
Sc^{46}	84 day	18	1.5	30 ± 3
Sc^{47}	3.4 day	11		
Ti^{44}	48 yr		0.38	4.4 ± 0.4
V^{48}	16 day	46	7.3	90 ± 45
V^{49}	330 day	107	9.6	164 ± 16
Cr^{51}	28 day	120	19	260 ± 120
Mn^{52}	5.7 day		6.6	
Mn^{53}	$\sim 2 \cdot 10^6$ yr		33	515 ± 52
Mn^{54}	300 day	240	38	470 ± 47
Fe^{55}	2.6 yr	550	220	1600 ± 600
Fe^{60}	$\sim 10^5$ yr		0.1	
Co^{56}	77 day	20	4.5	120 ± 34
Co^{58}	71 day	77	17	
Co^{57}	270 day	55	5.5	89 ± 9
Co^{60}	5.3 yr		0.6	17 ± 2
Ni^{59}	$8 \cdot 10^4$ yr		1.4	60 ± 15

a [6].
b [5].
c M. Honda, and J. R. Arnold: Geochim. et Cosmochim. Acta **23**, 219 (1961).
d P. S. Goel, and T. P. Kohman: Science **136**, 857 (1962).
e E. L. Fireman, and J. de Felice: J. Geophys. Res. **65**, 3035 (1960).
f R. Davis, R. W. Stoenner, and O. A. Schaeffer in [9], p. 355.
g M. Lipschutz, P. Signer, and E. Anders: J. Geophys. Res. **70**, 1473 (1965).

Time variations.

5. Some conclusions about time variations of the cosmic radiation can be reached by making simple comparisons among pairs and groups of products, without a detailed laboratory or theoretical study of relative production rates. In each meteorite specimen we can compare, for example, (i) short-lived and long-lived nuclides produced in very similar nuclear reactions; (ii) radioactive and stable products; and (iii) products of widely different mass numbers. If the excitation function (cross section as a function of the energy of the bombarding particle) is closely similar in shape for two species, the relative production of the two is given by the ratio of the cross sections at any energy, independent of the detailed nature of the bombarding flux. Examples are given in Table 2. Nearly all those given have actually been measured in meteorites. The half-lives cover the range from

2 weeks to millions of years, and the differences in excitation functions (the energy dependence of cross section) in iron meteorite targets are small, since the mass numbers of the products are similar. As will be shown in other tables, the ratios, as actually observed in iron meteorites, seem to be essentially constant from one sample to another, and they are in agreement with the production rates deduced from observed (or sometimes estimated) cross-section data. These findings seem to indicate that the intensity of the cosmic radiation has, on the average, been preserved an within uncertainty of the order of a factor of 2.

If the cosmic-ray intensity has really been constant, stable products will have accumulated at a steady rate. If relative cross-section data are available, we can calculate a "bombardment age" by considering a suitable pair of nuclides, one radioactive and one stable. An ideal case is $Cl^{36}-Ar^{36}$. About 80 percent of all

Table 2. *Stable nuclides and short- and long-lived radioactive species produced in similar nulcear reactions in iron meteorites.*

Radioactive Nuclide (Half-life)			Stable Nuclide
short-lived	Intermediate	long-lived	
Mn^{52} (5.7 d)	Mn^{54} (300 d)	Mn^{53} (2·10^6 y)	Cr^{53}, Cr^{54}
V^{48} (16 d)	V^{49} (330 d)		V^{50}, Cr^{50}
Sc^{46} (84 d)	Ca^{45} (165 d)	Ti^{44} (48 y)	Sc^{45}, Ca^{43}, Ca^{46}
	Ca^{41} (8·10^4 y)	K^{40} (1.3·10^9 y)	K^{41}, Ca^{42}
Ar^{37} (35 d)	Ar^{39} (260 y)	Cl^{36} (3·10^5 y)	Ar^{38}, Ar^{36}, S^{36}
	Na^{22} (2.6 y)	Al^{26} (7.4·10^5 y)	Ne^{20}, Ne^{21}, Ne^{22}
Be^{7} (53 d)	C^{14} (5.6·10^3 y)	Be^{10} (2.5·10^6 y)	
	H^3 (12.3 y)		He^3, He^4

the Ar^{36} atoms formed in an iron meteorite result from the decay of radioactive Cl^{36} (6). By correcting for the independently produced Ar^{36}, we can arrive directly at a "current" production rate for Ar^{36}. The ratio of the number of atoms now present to the production rate gives us the age. In Tables 3 to 7[1] we present the observed data on several iron and stone meteorites, covering a wide range of radioactive and stable, volatile and non-volatile, nuclides.

Meteorite samples.

6. This field of study could not exist were it not for the availability of a wide variety of meteorite samples. From most of the great collections — particularly those of the Academy of Sciences of the U.S.S.R., of the Smithsonian Institution, of the American Museum of Natural History in New York, of the Harvard University Museum, and of the Nininger Collection at Arizona State University — samples have been generously made available to qualified investigators. Private collectors have also been most helpful. Freshly fallen meteorites are especially important, because of the possibility of measuring many short-lived radioactive species. News of new falls spreads rapidly across national boundaries. Samples of the Aroos, Bruderheim, and Ehole meteorites have been supplied from the countries in which they fell — the U.S.S.R., Canada, and South-West Africa. R. E. Folinsbee of the University of Alberta has collected and distributed samples of two freshly fallen meteorites, Bruderheim and (in the spring of 1963) Peace River. The utility of this international cooperation is obvious. The meteorite

---

[1] The data on the radioactive and stable rare gas isotopes are taken from a number of sources. Most of the data presented here on nonvolatile nuclides were obtained in our laboratory. [M. Honda, and J. R. Arnold: Science, **143**, 203 (1964).]

Harleton, recovered near Marshall, Texas, reached our laboratories in record time, arriving about 10 days after recovery. In this sample it was easily possible to measure P^{32}, which has a half-life of 14 days.

Measurement.

7. With activities in the range of several disintegrations per minute per kilogram, the use of massive samples and extremely sensitive counting techniques is a necessity. SHEDLOVSKY, HONDA and ARNOLD have processed as much as 3.5 kilograms of the Odessa iron meteorite and 1 kilogram of the Bruderheim stone. Even so, in many cases the counting rate for the sample is less than 0.1 count per minute. Low-level techniques require heavy iron shields, anti-coincidence and coincidence counting methods, and, where possible, pulse-height discrimination. In the case of some high-yield nuclides, a 50-gram sample may be sufficient. For the measurement of stable nuclides, a much smaller sample, 1 to 10 grams, is usually enough, because of the high sensitivity of the mass spectrometer. Neutron activation analyses have also been used in He^3, Sc^{45}, and Ca^{46} determinations, and also recently for the radioactive nuclide Mn^{53} (29).

It is often desirable to measure the production separately in different mineral phases. A rough separation of the magnetic and non-magnetic stone phases is quite easy. Still, the metal fraction separated from a chondrite by this method may contain a small percentage of silicates after separation. In the case of pallasites, a very clean separation is possible (8). A "wet" method is also useful. For example, the olivine phase in a chondrite can be dissolved in cold strong acid much more easily than the pyroxene phase can be (30). The metal phase can also be dissolved electrolytically or with heavy metal ions such as Cu^{++} and Hg^{++}, other phases being left behind. The real difficulties begin with the chemistry. Inconveniently large samples must be dissolved under very clean conditions, and chemical procedures must be designed for separating more than 14 elements from one sample (7, 8).

The chemical processing must be carried out quickly because of the decay of the short-lived species. In a number of cases the chemical procedures themselves have had to be devised in the few days between the first notice of the meteorite's fall and the arrival of the sample in the laboratory. After separation of the groups and of individual elements, each counting sample must be "recycled to constant activity". That is, the element must be repurified repeatedly until the ratio of radioactivity to mass becomes constant. This is the only sure means of determining the chemical identity of the radioactive source. It is best to carry out several different chemical procedures, in order to remove unknown impurities. For the nonvolatile elements, solvent extraction, ion exchange, the use of complexing agents, electrolysis, and co-precipitation are all useful.

In some cases it is possible to measure the radioactivity of particular nuclides in a nondestructive way (9, 10). The γ-ray-emitting species Mn^{54}, Al^{26}, Co^{60}, and Na^{22} can be measured in 1-kilogram samples by means of a large sodium iodide crystal and a multi-channel analyzer. It is possible to make many measurements of these species, at least in meteorities where their concentration exceeds about 10 disintegrations per minute per kilogram. Extraction, purification, and measurement of the volatile rare gases require still another set of techniques. Many nonvolatile stable nuclides have been measured in iron meteorities, because of the fortunate circumstance that they are nearly free of many common elements. On the other hand, no such nuclides have been measured in stones. Even in the iron

Table 3. *Long-lived radioactivities produced*

Meteorite	Location	Date of fall [b]	Weight at recovery (kg)	Sample No.	Radioactivity		
					Be^{10} [2.5×10^6]	Al^{26} [7.4×10^5]	Cl^{36} [3.1×10^5]
Aroos (syn., Jardymlynsky)	Azerbaijan, U.S.S.R.	1959	150		4.1 ± 0.4	3.6 ± 0.4	16 ± 1.6
Deep Springs (nickel-rich ataxite)	Rockingham County, N.C.	(Find)	11.5		—	—	6.0 ± 0.6
Mt. Ayliff	Cape Province, South Africa	(Find)	13.6		—	—	17 ± 2
Grant	New Mexico	(Find)	500		4.0 ± 0.3	3.6 ± 0.2	12.4 ± 1.2[h]
Williamstown	Grant County, Ky.	(Find)	31		3.5 ± 0.4	2.9 ± 0.4	3.8 ± 0.7
Clark County	Kentucky	(Find)	11		—	—	4 ± 1
Treysa	Hesse, Germany	1916	63		—	—	22 ± 2
Admire (metal phase, pallasite)	Lyons County, Kan.	(Find)	>50		1.7 ± 0.3	1.5 ± 0.5	7.4 ± 0.9
Odessa	Ector County, Tex.	(Find)	>1000	(I) (II)	2.1 ± 0.1 1.0 ± 0.7	1.1 ± 0.1 —	5.1 ± 0.5 6.5 ± 1.3
Sikhote-Alin (syne., Ussuri)	Eastern Siberia	1947	80 000		2.4 ± 0.3	—	8.8 ± 0.9
Carbo	Sonora, Mex.	(Find)	454		1.6 ± 0.3	—	5.5 ± 0.6
Canyon Diablo	Coconino County, Ariz.	(Find)	$>30 000$		0.8 ± 0.2	0.8 ± 0.3	1.1 ± 0.2
Brenham (metal phase, pallasiderite)	Kiowa County, Kan.	(Find)	4 320		0.2 ± 0.4	0.1 ± 0.5	0.1 ± 0.2

a Half-lives (in years) are given in brackets.
b "Find" indicates not an observed fall.
c P. Signer, and A. O. Nier: In: Researches on Meteorites (C. B. Moore, Ed.). New York: John Wiley and Sons 1961.
d E. L. Fireman, and J. de Felice: J. Geophys. Res. **65**, 3035 (1960).
e P. S. Goel, and T. P. Kohman: Science **136**, 875 (1962).
f R. Davis, R. W. Stoenner, and O. A. Schaeffer in [9], p. 355.
g H. Wänke, and E. Vilcsek: Z. Naturforsch. **14a**, 929 (1959).
h D. Heymann, and O. A. Schaeffer: J. Geophys. Res. **66**, 2535 (1961).
i E. L. Fireman, and J. de Felice: Geochim. et Cosmochim. Acta **18**, 183 (1960).
j E. Vilcsek, and H. Wänke in [9], p. 381; also Z. Naturforsch. **16a**, 379 (1961).
k M. A. van Dilla, E. C. Anderson, and J. R. Arnold: Geochim. et Cosmochim. Acta **20**, 115 (1960).
l At time of fall.
m E. L. Fireman, J. de Felice, and D. Tilles, in [9], p. 323.

phase of stones, measurement of such nuclides is very difficult because of the inevitable silicate contamination. The avoidance of laboratory contamination is particularly difficult with such elements as potassium and manganese, which are commoner in laboratory materials than in samples. Finally, as in other critical work, a rigid discipline of blanks and standards must be maintained, along with repeated interlaboratory comparisons. Each laboratory maintains a "library" of solutions and fractions from each meteorite processed . Frequently it is possible to return to these solutions when new ideas occur.

The data of Tables 3 through 7 present, to the experienced eye, a beautiful and orderly pattern. Apart from a few anomalies (some of them doubtless caused by cosmic rays in iron meteorites

(disintegrations/min kg [a])							Ar^{36} [c] $(10^{-8} cm^3/g)$
Ti^{44} [48]	Mn^{53} [~2×10^6]	Ni^{59} [8×10^4]	Co^{60} [5.26]	H^3 [12.3]	C^{14} [5600]	A^{39} [260]	
4.4 ± 0.4	515 ± 52	60 ± 15	17 ± 2	35 ± 7 [d]	1.8 ± 0.3 [e]	16 ± 2 [d] 16 ± 0.9 [f] 17.2 ± 0.5 [g]	28
—	570 ± 60	0 ± 34	—	—	—	—	58
—	—	40 ± 20	—	—	—	—	28
—	360 ± 40	56 ± 14	—	—	—	0.02 ± 0.09 [i]	18
—	360 ± 40	0 ± 14	—	—	—	—	17
—	320 ± 35	—	—	—	—	0.9 ± 1.3 [j]	46
3.0 ± 0.7	270 ± 30	60 ± 14	—	80 ± 12 [i,l] 100 ± 30 [j,l]	—	20 ± 0.4 [g,j] 13 ± 0.6 [i]	21
—	200 ± 20	300 ± 30	—	—	—	—	2.5
—	240 ± 24	70 ± 25	(<10) [k]	—	0.29 ± 0.11 [e]	0 ± 0.14 [j]	(I) 1.6
—	200 ± 25	—		—			(II) 3.5
1.8 ± 0.4	250 ± 25	45 ± 17	95 ± 10 [l]	0 ± 1 [i,m]	1.7 ± 0.4 [e]	7.1 ± 0.2 [i] 4.5 ± 0.3 [j]	6.4
0.3 ± 0.7	280 ± 50	74 ± 16	(0 ± 2)	—	0.71 ± 0.12 [e]	0 ± 0.14 [g]	12.6
—	120 ± 14	60 ± 15	(<10) [k]	—	—	0 ± 0.15 [i] 0 ± 0.7 [j]	
—	<15	<20	—	—	—	—	0.0

by experimental errors) and slight but significant variations in relative abundances, the data seem to be essentially consistent. The ratios of the neon isotopes Ne^{20}, Ne^{21}, and Ne^{22} are approximately 1:1:1, with systematic variations of a few percentage points. This is true for both iron and stone meteorites — a remarkable fact in view of the difference in the nuclear reactions involved. Similarly the ratio of Cr^{53} and Cr^{54} is found to be 1:1 in iron. In the case of the argon isotopes Ar^{36} and Ar^{38}, the ratio is always about 2:3. The He^3/He^4 ratio is about 1:3. The yield of helium nuclei in spallation is comparatively high; the helium content of meteorites is generally of the order of 10^{-6} cm^3/g. This, and the very high relative abundance of He^3, made possible the first discovery of cosmic-ray products in meteorites by PANETH, REASBECK, and MAYNE, in 1952 (11).

These examples, and others, show that the relative yields of the common spallation products, those falling in or close to the stability "valley" on the chart of nuclides, are rather uniform. The smooth distribution of products with charge and mass makes it much easier to predict the production rate of a new species. On the other hand, the natural abundances of these isotopes show a very irregular pattern, since they were produced in an entirely different way.

Some of the anomalies in observed concentrations have led to other discoveries. For example, in the Williamstown iron meteorite (Table 3), the Cl^{36} content is too low, by a factor of about 4, as compared to the Cl^{36} content of the other long-lived radioactive species. This discrepant result, first obtained by SPRENKEL (12), led to the conclusion that the Williamstown iron meteorite, which is a find rather than

Table 4. *Cosmic-ray-produced radioactive nuclides in stone meteorites.*

Nuclide	Half-life	Radioactivity (disintegrations/min kg at time of fall except for Finds)				
		Bruderheim [a] (Alberta, Canada)	Harleton [b] (Texas)	Ehole [c] (Angola, Africa)	Achilles [d] (Kansas)	Admire (stone)[e] (Kansas)
Be^{10}	2.5×10^6 yr	19 ± 2	21 ± 2	19 ± 2	19 ± 2	14 ± 2
Na^{22}	2.6 yr	90 ± 10	64 ± 7	84 ± 17		
Al^{26}	7.4×10^5 yr	60 ± 6	45 ± 5	70 ± 7	50 ± 5	43 ± 4
P^{32}	14.3 days		14 ± 2			
Cl^{36}	3.1×10^5 yr	7.5 ± 0.8	7.0 ± 0.7	7.8 ± 1.0	6.0 ± 0.6	
Sc^{46}	84 days	6.2 ± 0.6	5.4 ± 0.7			
Ti^{44}	48 yr	2.0 ± 0.2	1.4 ± 0.2			
V^{48}	16 days	34 ± 7	17 ± 2			
V^{49}	330 days	34 ± 2	20 ± 6			
Cr^{51}	28 days	110 ± 27	60 ± 20			
Mn^{53}	$\sim 2 \times 10^6$ yr	85 ± 17	44 ± 8	110 ± 20	60 ± 12	
Mn^{54}	300 days	100 ± 13	38 ± 5	90 ± 20		
Fe^{55}	2.6 yr	340 ± 80	≤ 180	500 ± 100		
$Co^{56}+Co^{58}$	~ 74 days	14 ± 4	4 ± 1			
Co^{57}	270 days	11 ± 1	6.5 ± 0.7	14 ± 6		
Co^{60}	5.3 yr	9 ± 1	1.5 ± 0.5	4.8 ± 1.2		
Ni^{59}	8×10^4 yr	12 ± 3	6 ± 6			
H^3	12.3 yr	260 ± 30 [f]	$\{310 \pm 20$ [g] $\\ 265 \pm 15$ [h] $\}$	$\{295 \pm 10$ [g] $\\ 302 \pm 10$ [h] $\}$		
C^{14}	5600 yr	$\{56 \pm 3.0$ [i] $\\ 63 \pm 5$ [j] $\}$	$\{38 \pm 2.4$ [j] $\\ 57 \pm 5$ [i] $\}$		56 ± 7.6 [j]	17 ± 1.5 [j]
Ar^{39}	260 yr	$\{10.5 \pm 1.0$ [f] $\\ 11.5 \pm 0.3$ [h] $\}$	$\{9.1 \pm 0.4$ [h] $\\ 7.5 \pm 0.5$ [g] $\}$	$\{7.8 \pm 0.3$ [g] $\\ 8.2 \pm 0.3$ [h] $\}$		
Ar^{37}/Ar^{39}		$\{1.33 \pm 0.13$ [h] $\\ 2.2 \pm 0.4$ [g] $\}$	1.02 ± 0.13 [h]	$\{1.65 \pm 0.29$ [h] $\\ 2.1 \pm 0.3$ [g] $\}$		

a Date of fall, 1960; weight at recovery, 300 kg; iron content, 22 percent.
b Date of fall, 1961; weight at recovery, 8 kg; iron content 22 percent.
c Date of fall, 1961; weight at recovery, 2.4 kg; iron content 30 percent.
d Find; weight at recovery, 16 kg; iron content, 22 percent.
e Find; weight at recovery >50 kg; iron content of stone phase, 8 percent.
f E. L. Fireman, and J. de Felice: J. Geophys. Res. 66, 3547 (1961).
g [9]. Fireman et al.
h [9]. Davis et al.
i H. E. Suess, and H. Wänke: Geochim. et Cosmochim. Acta 26, 475 (1962).
j P. S. Goel, and T. P. Kohman: Science 136, 875 (1962).

an observed fall, had been lying on the ground in Kentucky for about 6×10^5 years before its discovery. Many other meteorites of considerable terrestrial age have since been found. Two iron meteorites, Tamarugal and Ider, were found [9] to contain almost no Cl^{36} — a result which indicates an age greater than a million years. Some stone meteorites are observed to contains little or no C^{14}; this lack of C^{14} indicates that even stone meteorites can be preserved for tens of thousands of years on earth (31). For the measurement of terrestrial ages, Si^{32}, Ar^{39}, Ti^{44}, and Ni^{59} are also very useful.

Cosmic-ray age.

8. Let us return to the question of the cosmic-ray age. As already described, this number is based simply on a ratio of concentrations of stable and radioactive species. If a production rate can be estimated accurately, the stable-nuclide concentration is sufficient. If this age is to be interpreted as a real time period, we must assume not only that the cosmic-ray intensity in space and time has been effectively constant but that the bombardment conditions in the meteorite, in particular the depth of a particular specimen below the surface, have remained essentially unaltered since its formation. The data shed light on these questions. Typical bombardment ages for iron meteorites are in the range of 10^8-10^9 years, while those for chondrites are in the range of 10^7 years. Both these ages are quite short as compared to the ages determined by uranium-lead, potassium-argon, and rubidium-strontium methods for the minerals contained in the meteorites. All these methods agree well on ages close to 4.5×10^9 years. The simplest explanation of these findings is that meteorites were formed as small bodies by violent collisions between larger objects a relatively short time ago. If most objects were formed in a few large collisions, this implies the existence of clusters of ages at certain values. This clustering is seen for iron meteorites (coarse octahedrites) at 5.6×10^6 years, for H-group chondrites at 4.5×10^6 years, and for some rare types; in other cases it is suspected but not proven (13). Another possibility is that the reduction in size of meteoritic bodies has occurred rather gradually through many small collisions — for example, with cosmic dust (the erosion hypothesis) (14). A younger age for the stones might be related to the softness of these bodies relative to iron. A number of strong objections have been raised to this erosion hypothesis in its extreme form (15), but the effect must undoubtedly occur to some degree. The effect of gradual erosion, or of a succession of collisions that remove small parts of the meteorite, would be a gradual increase in the effective intensity of the bombardment received by an originally deep portion of the body. At present the suspicion is that both large and small collisions may be important. It seems safe to adopt a model in which a single big collision first exposes the meteorite to an appreciable cosmic-ray flux. After this, erosion (and multiple small collisions) continue to act on the body, while occasional big collisions may reduce its size again drastically. Finally, the meteorite is either captured by the earth or one of the other planets, ejected from the solar system, or broken into pieces so small that they cannot reach the earth.

In a large meteorite the high-energy products, such as Ne* and Al^{26} from iron, occur in highest concentration on the outside. Their concentration decreases rather rapidly with depth. Specimens that were once buried deep in a larger body show much higher relative concentrations of low-energy products than specimens from nearer the surface. For example, in Table 3, Treysa shows smaller amounts of low-energy products than Carbo. The former must be from near the surface and the latter from near the center of their respective original bodies. The radioactive isotopes that we find whose half-lives are short as compared to the bombardment age must have been produced while the body had more or less its present size. Hence, their distribution with depth need not be the same as that of the stable nuclides. One would expect the low-energy products to be more prominent among the stable species. The data presented in our tables do not show any such effect. An especially favorable case for such study is the iron meteorite Grant, which has been studied extensively by NIER and his co-workers (16). They have presented contour maps of the rare-gas concentrations in this nearly spherical body. GOEL and KOHMAN [9] have compared the depth dependence of Cl^{36} concentration with that for the stable Ar^{36}. The curve for depth dependence of Cl^{36} seems steeper, but at present the data do not fall into a consistent pattern. In other well-studied meteorites, such as Carbo and Casas Grandes, the contour lines for rare-gas concentration are not closed (17). Apparently these meteorites broke up in passing through the earth's atmosphere. Because of the great forces encountered by an irregular body in the atmosphere, this breaking-up process probably takes place in the great majority of cases, but often only one fragment is

Table 5. *Cosmic-ray produced stable nuclides and potassium-40 in some iron meteorites.*

Meteorite	10⁻⁸ ccNTP/g [a]			K⁴¹/K⁴⁰ [e]	10⁻⁹ g/g						k'_2
	He^3	Ne^{21}	Ar^{38}		K^{40} [b]	Ca^{43} [b]	Sc^{45} [c]	V^{50} [b]	Cr^{53} [d]	Cr^{54} [d]	
Aroos	655	8.15	43.5	1.928±0.023	0.49±0.03	1.7 ±0.15	—	4.79±0.40	25±2	25±1	2.45
Clark County	1095	15.2	69	2.144±0.027	0.82±0.03	2.9 ±0.5	4.4±0.3	7.2 ±0.8	—	—	2.36
Mt. Ayliff	868	8.9	—	—	0.60±0.02	2.0 ±0.2	3.4±0.3	5.17±0.34	—	—	2.39
Grant (Surface)	550	7.45	54.5	—	—	—	—	—	—	—	2.4
Williamstown	465	5.3	29	1.803±0.023	0.37±0.02	1.18±0.10	—	3.48±0.30	19±1	19±1	2.53
Treysa	580	8.0	33	1.743±0.026	0.38±0.02	1.15±0.06	1.8±0.2	2.47±0.20	13±1	13±1	2.13
Carbo	315	3.2	18.5	1.97 ±0.07	0.22±0.01	1.05±0.15	1.4±0.2	2.43±0.26	17±6	15±3	2.64
Sikhote-Alin	165	2.1	10.5	—	0.13±0.01	0.49±0.05	—	1.01±0.08	—	—	2.36
Bruderheim (Metal phase)	—	—	—	—	—	—	—	0.38±0.17	—	—	—

[a] P. SIGNER, and A. O. NIER: In: Researches on Meteorites (C. B. MOORE, Ed.). NewYork: John Wiley and Sons 1961.
[b] H. STAUFFER, and M. HONDA: J. Geophys. Res. **66**, 3584 (1961); **67**, 3503 (1962).
[c] H. WÄNKE: Z. Naturforsch. **13a**, 645 (1958); **15a**, 953 (1960).
[d] MASAKO SHIMA, and M. HONDA: Earth and Planetary Science Letters **1**, 65 (1966).
[e] H. VOSHAGE and H. HINTENBERGER: Z. Naturforsch. **16a**, 1042 (1961).

recovered. It would be very valuable to have more meteorites like Grant. The evidence for more than one major collision is particularly clear in the case of some very large iron meteorites, as was first pointed out by VILCSEK and WÄNKE (18). For example, in Table 3, in two specimens (I and II) of Odessa we find Cl^{36}/Ar^{36} ratios leading to two quite different ages, 100 and 300 million years. The same situation occurs in the large iron meteorite Sikhote-Alin. These second collisions exposed fresh or nearly fresh surfaces, and the high concentration of the radio-active species which built up thereafter led to an effective age intermediate between the times of the two collisions (if there were only two). In view of all these observations there is reason to believe that most iron meteoritic bodies in space have been substantially altered in size, on a time scale of 10^8 years or so.

Production rates.

9. A more detailed and quantitative treatment of relative and absolute production rates yields more results. These have been obtained in two ways. First, excitation functions in iron and other important elements have been measured for a good many species. By methods of nuclear systematics, a number of other excitation functions can be estimated with a good deal of confidence. Using the very extensive information available on primary cosmic rays and their interactions, one can estimate the flux of nuclear-active particles as a function of depth in spherical meteorites, and in the limiting case of a very large body. The production rates estimated in this way are reliable to within perhaps a factor of 2, and fortunately it has been possible to verify the predictions in a number of cases where these really did precede the measurements. Unfortunately, very few cross-section data are available for magnesium, silicon, and calcium targets. It must be assumed that the substitution of neutrons and pions for protons as bombarding particles at moderate or high energies does not affect the cross sections [5]. However, the smoothness of variation of cross-section with product charge and

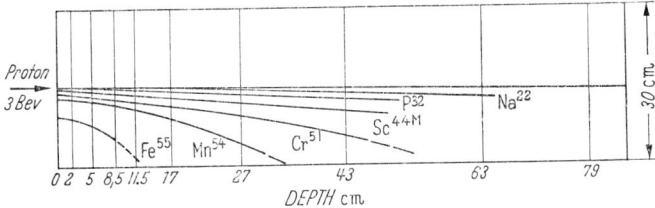

Fig. 2. Contours showing spread of the internal beam for different products [6]. Half of the production of each species at any depth takes place inside the circle generated by rotating the corresponding line. Lower-energy products show wider spread.

mass, as studied by RUDSTAM (19) and others, saves us from any really serious errors. One is on particularly safe ground in comparing isobars — that is, nuclei of the same mass number. The shape of excitation functions is generally very closely the same for all isobars, and a single cross-section ratio is sufficient, except for a few cases very near the target mass.

A more direct experimental method is to simulate the conditions in space and to bombard a thick target of iron or stony material with a beam of high-energy protons, available at one of our larger accelerators. In such bombardments, large blocks of iron [6] and also blocks of glass and metal (20) have been used to simulate the composition of chondrites. The products in the target were measured as a function of depth and lateral distance from the bombarding beam. The spread of the beam for different products may be seen in Fig. 2. Fig. 3 shows the radius dependence of production rates. Many short-lived radioactive species have been measured. The resulting data can be integrated over angle and depth to obtain the effect of bombardment by a uniform isotropic flux.

What can be learned from this more exhaustive analysis? First, the conditions of bombardment can be better understood. A useful empirical formula, already

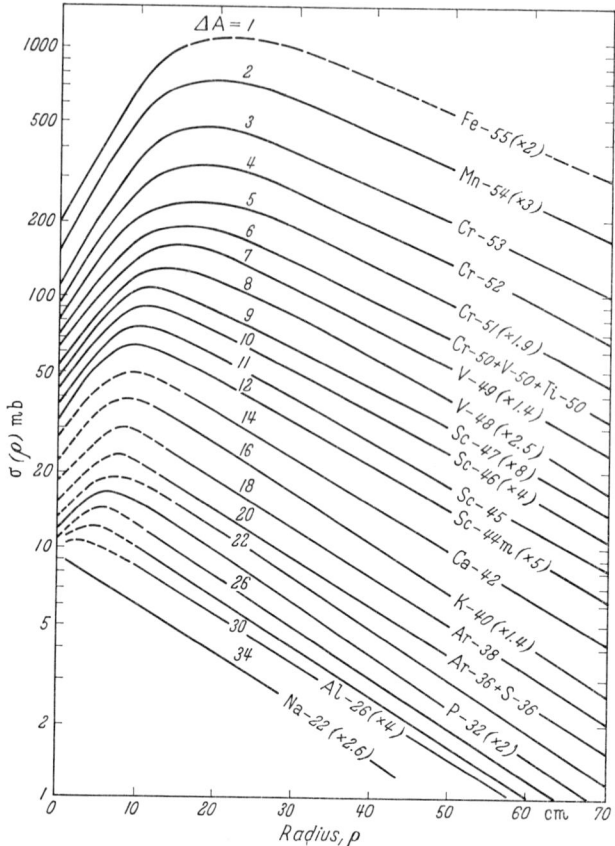

Fig. 3. Total isobaric yield from iron for each mass number as a function of radius. The isobars correspond to the production at the center of a spherical body. Light curves are interpolated between heavy curves derived directly from 3-Gev bombardment data [6].

inferred from the meteorite data themselves [7], is placed on a firm basis. This equation, for the total production rate of all species of mass number A, is

$$Q(A) = k(\Delta A)^{-k_2}.$$

The net number of nucleons emitted, $A_{\text{target}} - A_{\text{product}}$, is ΔA. The constant k is proportional to the flux intensity, and k_2 is found to be a constant for $\Delta A > 5$. It decreases for smaller values of ΔA. The total isobaric yield can be estimated rather well from any convenient product of given A. For bombardment by a soft or low-energy flux, k_2 is large, and the yield falls off rapidly with ΔA. This constant, then, tends to increase with depth and size. A semi-logarithmic plot of the total isobaric yields versus ΔA, for several nuclides, gives a comparatively accurate measure of k_2. This can be compared with the value of k_2 obtained by integrating the laboratory-bombardment data over all angles (Figs. 4, 5). Fig. 4

shows how k_2 at the center of an iron sphere varies with the radius of the sphere. Fig. 5 gives the variation in k_2 with depth for four different radii. The curve of Fig. 4 is the lower envelope of these curves.

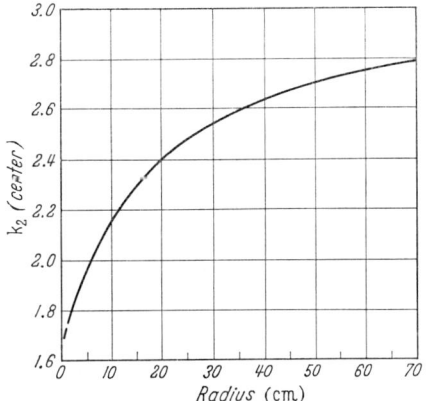

Fig. 4. The constant k_2 at the center of a spherical meteorite, as a function of radius.

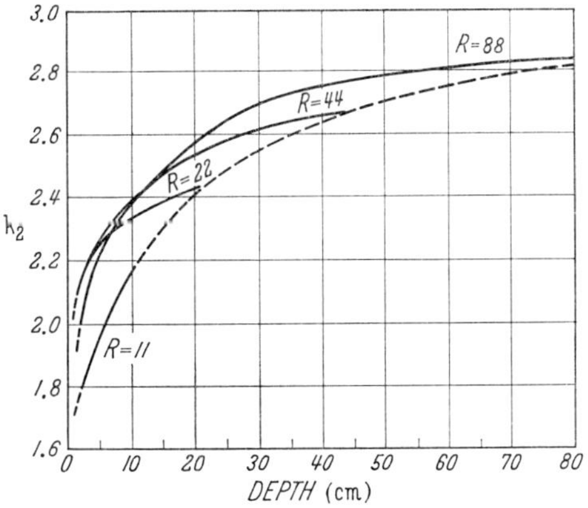

Fig. 5. The constant k_2 as a function of depth for meteorites of different radii. The dotted line represents the envelope (see Fig. 4).

Similarly, for the stable products, if the flux has been constant,

$$C(A) = k' \, T (\Delta A)^{-k'_2}$$

where T is the bombardment age and the constants are marked with primes to indicate that they refer to stable species. Now it is apparent that agreement between k_2 and k'_2 is a refined test of the constancy of bombardment conditions, while k and k' should agree if the flux has been constant. This is equivalent to the statement that the same T can be derived from each suitable pair of nuclides.

How much does k_2 vary? Its total observed range is from about 2.1, near the surface of small bodies (such as Treysa) to somewhat less than 2.7 for the interior of large ones (such as Carbo). In large bodies the variation with depth is rapid at first, as the low-energy secondaries build up; then it slows down. As for the total production, it rises at first, has a broad maximum, and then falls steadily. By using both parameters, we can hope to get some measure of both the depth and the size of the original body. All this is illustrated in Figs. 2—5. Fig. 6 shows a good example of the determination of k'_2. Is erosion important in iron meteorites? If so, the curves have been traversed to the left, and k_2 and k'_2 must differ, the former being lower. If erosion is dominant, k'_2 should be an average of values, varying continuously from ~ 2.8 down to k_2. This would mean that k'_2 would be somewhat greater than 2.4 for Treysa, as compared to k_2 at 2.1, a

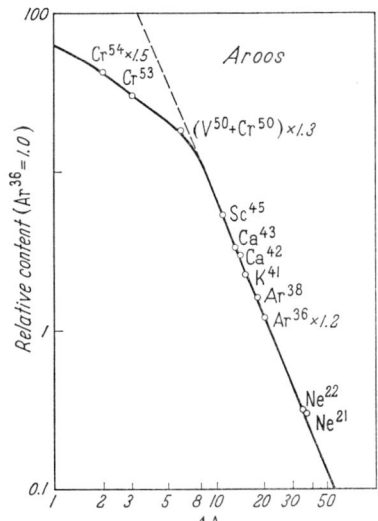

Fig. 6. Content of various stable nuclides in the iron meteorites Aroos illustrating the variation of production rate with ΔA in the determination of k'_2.

difference well outside the limit of experimental error. No such difference is observed, the two values being the same as far as we can see. For a big object like Carbo, the two values would differ by less than 0.1, because of the flatter slope in this mass region. A difference of this degree cannot be excluded by the

Table 6. *Estimated and measured cosmic-ray produced stable nuclides and potassium-40 in Aroos iron meteorite* (see Table 1 and Fig. 6).

Mass Number (A)	Estimated Production cross section of total Isobars [a] ($R=20$ cm) (mbarn)	Nuclide	Calculated relative production rate [b] ($Ar^{36}=1.2$)	Measured in Aroos (10^{-8} ccNTP/g)
3	—	He^3	27	655 [c]
4	—	He^4	135	2540 [c]
20	—	Ne^{20}	0.24	8.5 [c]
21	—	Ne^{21}	0.24	8.15 [c]
22	4	Ne^{22}	0.24	8.6 [c]
36	15	S^{36}	—	—
		Ar^{36}	1.2	28 [c]
38	20	Ar^{38}	1.6	43.5 [c]
39	23	K^{39}	1.8	—
40	27	Ar^{40}	0.38	—
		K^{40}	2.3 { 1.5	28 ± 1.7 [d]
		Ca^{40}	0.4	—
41	31	K^{41}	2.6	54 [d,f]
42	35	Ca^{42}	3.1	70 *,[g]
43	40	Ca^{43}	3.6	89 ± 8
44	47	Ca^{44}	4.2	100 *,[d]
45	56	Sc^{45}	5.0	140 *,[d]
46	70	Ca^{46}	0.05	1.9 ± 0.2 [d]
		Ti^{46}	—	—
47	87	Ti^{47}	—	—
48	110	Ti^{48}	—	—
49	140	Ti^{49}	—	—
50	180	Ti^{50}	—	—
		V^{50}	5.2	215 ± 17 [d]
		Cr^{50}	—	180 [e],*
51	240	V^{51}	—	<1000 *,[d]
52	330	Cr^{52}	—	—
53	480	Cr^{53}	—	1060 ± 80 [e]
54	740	Cr^{54}	—	1040 ± 80 [e]
		Fe^{54}	—	—

* Estimated values using data found in other meteorites.
a [6].
b [5].
c P. SIGNER and A. O. NIER: In: Researches on Meteorites (C. B. MOORE, Ed.). New York: John Wiley and Sons 1961.
d H. STAUFFER and M. HONDA: J. Geophys. Res. 66, 3584 (1961); 67, 3503 (1962).
e MASAKO SHIMA and M. HONDA: Earth and Planetary Science letters 1, 65 (1966).
f H. VOSHAGE and H. HINTENBERGER: Z. Naturforsch. 16a, 1042 (1961).
g H. WÄNKE: Z. Naturforsch. 15a, 953 (1960).

available data, and these can be used only to confirm the general correctness of the model. Fragmentary data on other small iron meteorites such as Charlotte and Bristol, appear to show a low value of k'_2, and they may be other good cases to study thoroughly.

Using the bombardment data (Table 1; Figs. 2 and 3), we can now demonstrate the constancy of the internal flux over millions of years to within 40 percent or so. Comparison of k_2 and k'_2 in Treysa shows that its size has been more or less constant over hundreds of millions of years. In this case, the internal flux can be directly related to the primary flux. Furthermore, the energy distribution of the primary flux cannot have changed enough, over the whole time of bombardment, to produce by itself a measurable change in k_2. For big meteorites, this says very little, but for small ones like Treysa, a large change in the "hardness" of the primary spectrum should show visibly.

The nuclide K^{40} is an entirely special case, with its half-life of 1.3×10^9 years. It is the only species available, even in principle, for checking the constancy of the cosmic radiation on a billion-year time scale. Unfortunately, there are difficulties. First, in the usual iron meteorites, whose cosmic-ray age is about 500 million

Table 7. *Cosmic-ray-produced stable nuclides in some stone meteorites* [8].

Meteorite	Weight at recovery (kg)	Amount of nuclide (10^{-8} cm^3/g [a])			Fe (%)
		He^3	Ne^{21}	Ar^{38}	
Chondrites					
Bruderheim	300	47	10.1	1.41	22.5
Ehole [b]	2.4	29	4.78		~30
Harleton [b]	8	70	12.6		22
Richardton . . .	90	32	9.5	0.96	30.6
Achondrites					
Norton County (Calcium-poor) . .	1050	220	63	2.6	1.6
Johnstown [b] (Calcium-poor) . .	40	40.7	6.74	0.7	13.1
Nuevo Laredo (Calcium-rich) . .	0.5	3.8	0.45	1.79	
Pallasite					
Admire (stone phase)	> 50	134	41.2	0.87	8

[a] Standard temperature and pressure.
[b] H. HINTENBERGER, H. KÖNIG u. H. WÄNKE: Z. Naturforsch. 17a, 1092 (1962). — H. HINTENBERGER et al.: Z. Naturforsch. 19a, 327 (1964).

years, only about 15 percent of the K^{40} will have decayed if the cosmic-ray intensity has been constant. This is not a very large change as compared with the differences already discussed. A very accurate estimate of the production rate relative to the neighboring stable species is required. Only one meteorite, Deep Springs, appears definitely to have a cosmic-ray age in excess of 1 billion years. This meteorite is a find, and apparently it has a considerable terrestrial age, so that one must make important corrections in comparing data for radioactive and stable species. VOSHAGE (21) has shown that the cosmic-ray intensity in these old iron meteorites was (at least in most cases) somewhat (perhaps as much as a factor of two) less in the past than it is at present. It is possible that this is due to the progressive reduction in size that we have discussed, or even that the primary flux was higher long ago. We can say only that the cosmic-ray intensity many hundreds of millions of years ago was not very different from that at present. More work, including detailed comparisons of k_2 and k'_2, is needed.

At the other end of our time scale, variations in the concentration of the very-short-lived species such as P^{32}, Ar^{37}, V^{48}, Cr^{51}, Mn^{52}, and others, can tell us something about the very recent history of meteorites. In the case of orbits extending deep into the asteroid belt, one might expect effects of the variation of cosmic-ray intensity in space, especially at the periods of maximum in the solar cycle. The shielding effect of the sun's magnetic field ought to be less marked at these great distances than at the earth. The evidence is ambiguous (22), (23). Our own data do not show this effect. It may well be possible to see the activation produced in meteorites by the occasionally very intense bursts of particles from the sun.

The most precise data yet available for time and space variations are those of SCHAEFFER and coworkers (28), who have compared the three species Ar^{37}, Ar^{39}, and Cl^{36} (Table 2), in six or seven freshly fallen chondrites and irons. The first two are an ideal pair for space variations, and can be compared accurately since they are counted in the same counter at the same efficiency. The longlived Cl^{36} is counted externally, but by careful work an accuracy of 5% is possible. SCHAEFFER'S results show a ratio of Ar^{39}/Cl^{36} consistent with a constant cosmic ray intensity in all cases, with an estimated accuracy of 15%. The ratio Ar^{37}/Ar^{39} is not constant. It indicates generally a higher flux for Ar^{39}, by as much as 20%. This may be ascribed to space variations, and to the 11-year solar cycle.

The experience gained in this field may be utilized in two other branches of cosmochemistry. One of these is concerned with the much older record of high-energy nuclear reactions, currently believed to have occurred at the end of the process of general synthesis of the elements. In the earliest stages of evolution of our solar system, there is reason to believe, the acceleration of particles to high energies may have occurred at a much greater rate than at present. There is evidence for this (for instance, the presence of technetium) in some stars. According to FOWLER et al. (24), the light elements deuterium, lithium, beryllium, and boron found in terrestrial materials and meteorites were produced at this stage. The total bombardment dose was perhaps 10^3 to 10^4 times higher than the "present" cosmic-ray bombardment in usual iron meteorites. Whatever the detailed mechanism may have been, these elements can only have been produced by some such nonthermal high-energy process. They must thereafter have remained at temperatures low on the million-degree scale of stellar interiors. The production process must have resembled, in important respects, present-day cosmic-ray production. In fact, the elements lithium, beryllium, and boron are very much enriched in the primary cosmic-ray beam itself, as compared to any general sample of galactic material available to us. This results from the bombardment of heavier cosmic-ray nuclei by the stationary interstellar hydrogen. If the conditions of high-energy bombardment differed to any considerable degree between the region of the earth and the region where meteorites originate, we might expect measurable differences, between the earth and meteorites, in the abundance pattern of elements and isotopes. Such differences have recently been observed (25).

Recovered artificial satellites.

10. The analysis of recovered artificial satellites is currently of great interest (26, 27). Such studies have yielded some real surprises. Radioactive and stable nuclides have been detected in unexpectedly high amounts in satellites of the Discoverer series. These include tritium, Ar^{37}, Co^{57}, Ag^{106}, Xe^{127}, and Bi^{205}. Even stable He^3 has been measured in the materials of Discoverer XVII, exposed for less than a day. In other flights very small activities were seen, as would be expected in view of the very short period of bombardment. These bursts of production present many puzzling features and must, presumably, be explained either as effects of solar flares or as the result of extensive passage through the VAN ALLEN belt. In the case of the Soviet satellite Sputnik IV, a piece of which was recovered in the United States after the satellite had been outside the earth's atmosphere for 843 days, the activation appears to have been produced mainly by cosmic rays. Since the satellite was close to the earth, the intensity was reduced by a factor of 4 to 5. There is some dispute about the extent of activation by solar particles (27).

The day when we will have samples of matter taken from the moon and even the planets is now near enough at least to be visible. At that point this aspect of cosmochemistry (or theochemistry as some call it) can hope to take another large step forward.

General references.

Meteorites.

[1] Krinov, E. L.: Principles of Meteoritics. New York: Pergamon Press 1960. A general introduction to the subject, with emphasis on history and descriptive aspects.
[2] Mason, B.: Meteorites. New York: John Wiley and Sons 1962. Another general work, with more stress on quantitative aspects, especially mineralogy and chemistry.

Cosmic-Ray Effects (Terrestrial).

[3] Libby, W. F.: Radiocarbon Dating, 2nd edit. Chicago: University Chicago Press 1955.
[4] Lal, D., and B. Peters: This volume. See also Progr., Elem. Particle and Cosmic Ray Physics 6, Chap. 1 (1962).

Cosmic-Ray Effects (Meteorites).

[5] Arnold, J. R., M. Honda, and D. Lal: Record of Cosmic-Ray Intensity in Meteorites. J. Geophys. Res. **66**, 3519 (1961).
[6] Honda, M.: Spallation Products Distributed in a Thick Iron Target Bombarded by 3-Bev protons. J. Geophys. Res. **67**, 4847 (1962).
[7] Geiss, J.: Experimental Evidence on the History of the Cosmic Radiation. Proc. Internat. Conference on Cosmic Rays, Jaipur, Dec. 1963, Vol. 3, p. 434.
[8] Kirsten, R., D. Krankowsky u. J. Zähringer: Edelgas- und Kaliumbestimmungen an einer größeren Zahl von Steinmeteoriten. Geochim. et Cosmochim. Acta **27**, 13 (1963).
[9] Radioactive Dating, Internat. Atomic Energy Agency, Vienna 1963. This symposium volume contains a number of papers in this field.
[10] Anders, E.: Meteorite Ages. Rev. Mod. Phys. **34**, 2 (1962).

References cited.

(1) McDonald, F. B., and W. R. Webber: Phys. Rev. **115**, 194 (1959).
(2) Eberhardt, P., J. Geiss, and H. Lutz: Helv. Phys. Acta **34**, 460 (1961).
(3) Shapiro, M. M.: Science **135**, 175 (1962). — Ginzburg, V. I.: Progr. Elem. Particle and Cosmic Ray Physics IV, Chap. 5 (1958). — Hayakawa, S.: Proc. Intern. Conf. Cosmic Rays Kyoto III, p. 181 (1962). [See also other papers in the same volume and in the Proceedings of the International Conference on Cosmic Rays, Jaipur (1963).]
(4) Ceplecha, Z., J. Rajchl, and L. Sehnal: Bull. Astr. Inst. Czech. **10**, Nr. 4, 137 (1959).
(5) Arnold, J. R.: Astrophys. J. **141**, 1536, 1548 (1966).
(6) Schaeffer, O. A., and J. Zähringer: Phys. Rev. **113**, 674 (1959).
(7) Honda, M., and J. R. Arnold: Geochim. et Cosmochim. Acta **23**, 219 (1961).
(8) Honda, M., S. Umemoto, and J. R. Arnold: J. Geophys. Res. **66**, 3541 (1961).
(9) Rowe, M. W.: Quantitative Measurement of Gamma-Ray Emitting Radionuclides in Meteorites. Los Alamos Rept. LA-2765 (1963).
(10) Dilla, M. A. van, E. C. Anderson, and J. R. Arnold: Geochim. et Cosmochim. Acta **20**, 115 (1960).
(11) Paneth, F. A., P. Reasbeck, and K. I. Mayne: Geochim. et Cosmochim. Acta **2**, 300 (1952).
(12) Sprenkel, E. L.: Thesis, University of Rochester 1959.
(13) Hintenberger, H., H. König, L. Schultz u. H. Wänke: Z. Naturforsch. **20a**, 984 (1965). — Anders, E.: Space Sci. Rev. **3**, 583 (1964).
(14) Whipple, F. L.: J. Geophys. Res. **68**, 4929 (1963). — Geiss, J., and H. Oeschger: Proc. 1st. Intern. Space Science Symposium, p. 1071. Amsterdam: North-Holland Publishing Company 1960.
(15) Urey, H. C.: J. Geophys. Res. **64**, 1721 (1959).

(16) HOFFMANN, J. H., and A. O. NIER: Phys. Rev. **112**, 2112 (1958). — SIGNER, P., and A. O. NIER: J. Geophys. Res. **65**, 2947 (1960).
(17) HOFFMANN, J. H. and, A. O. NIER: J. Geophys. Res. **65**, 1063 (1960).
(18) VILCSEK, E., u. H. WÄNKE: Z. Naturforsch. **16**a, 379 (1961).
(19) RUDSTAM, G.: Thesis, University of Uppsala, Sweden 1956. — See also M. HONDA, and D. LAL: Phys. Rev. **118**, 1618 (1960). — Nuclear Phys. **51**, 363 (1964).
(20) SHEDLOVSKY, J. P. et al.: J. Geophys. Res. **69**, 2231 (1964)
(21) VOSHAGE, H.: Z. Naturforsch. **17**a, 422 (1962).
(22) FIREMAN, E. L., and J. DE FELICE: J. Geophys. Res. **65**, 3035 (1960).
(23) STOENNER, R., R. DAVIS, and O. A. SCHAEFFER: J. Geophys. Res. **65**, 3025 (1960).
(24) FOWLER, W. A., J. GREENSTEIN, and F. HOYLE: Geophys. J. **6**, 148 (1962).
(25) REYNOLDS, J. H.: J. Geophys. Res. **68**, 2939 (1963).
(26) FIREMAN, E. L., J. DE FELICE, and D. TILLES: Phys. Rev. **123**, 1936 (1961). — KEITH, J. E., and A. L. TURKEVICH: J. Geophys. Res. **67**, 4525 (1962). — WASSON, J. T.: J. Geophys. Res. **66**, 2659 (1961); **67**, 3513 (1962). — SCHAEFFER, O. A., and J. ZÄHRINGER: Phys. Rev. Letters **8**, 389 (1962).
(27) FELICE, J. DE, E. L. FIREMAN, and D. TILLES: J. Geophys. Res. **68**, 5289 (1963). — SHEDLOVSKY, J. P., and J. H. KAYE: J. Geophys. Res. **68**, 5069 (1963). — ROWE, M. W., E. C. ANDERSON, and M. A. VAN DILLA: J. Geophys. Res. **69**, 831 (1964).
(28) SCHAEFFER, O. A., R. DAVIS, jr., R. W. STOENNER, and D. HEYMANN: Proceedings of the International Conference on Cosmic Rays, Jaipur, 1963, vol. 3, p. 480.
(29) MILLARD, H. T., jr.: Science **147**, 503 (1965).
(30) HINTENBERGER, H., E. VILCSEK u. H. WÄNKE: Z. Naturforsch. **20**a, 939 (1965).
(31) SVESS, H. E., and H. WÄNKE: Geoclin. et Cosmochim. Acta **26**, 475 (1962).

Ergänzung zum Beitrag
E. SCHOPPER, E. LOHRMANN und G. MAUCK.

Nukleonen in der Atmosphäre.

Ziff. 7. Statistische Auswertung der Sterne, S. 400, 2.—25. Zeile von oben.

Neuere Beiträge zur Monte-Carlo-Rechnung der intranuklearen Kaskade.

Mit den großen Beschleunigern hat man seit Jahren die Möglichkeit viele dieser Fragen unter eindeutigen und reproduzierbaren Bedingungen zu studieren. Ergebnisse früherer Untersuchungen mit Hilfe der Kosmischen Strahlung (K.S.) kann man den zusammenfassenden Darstellungen von TEUCHER [Te 53c], SYMANZIK [Sy 53] und GOTTSTEIN [Go 53] entnehmen. Zum quantitativen Verständnis der Sterne in Kernemulsion hat das bekannte Serbersche Modell für (hochenergetische) Kernreaktionen von Nukleonen (bzw. Mesonen) mit komplexen Kernen erheblich beigetragen. Der Reaktionsmechanismus läuft dabei in zwei Schritten ab. In einem ersten „schnellen" Schritt (von etwa 10^{-22} sec Dauer) löst das stoßende Teilchen eine intranukleare Kaskade aus, wobei angenommen wird, daß die Kaskadenwechselwirkungen im Kern zwischen individuellen (und wegen der vergleichsweise kleinen, meist vernachlässigbaren Bindungsenergie) freien Nukleonen stattfinden. Diese Wechselwirkungen werden durch die Wirkungsquerschnitte freier Teilchen im freien Raum bestimmt, jedoch modifiziert durch das Pauli-Prinzip im Kernvolumen. Man bezeichnet diesen Ansatz oft als „Stoßnäherung". In einem zweiten „langsamen" Reaktionsschritt (10^{-18} bis 10^{-16} sec) gibt der hochangeregte Zwischenkern nach WEISSKOPF durch Abdampfen von Nukleonen oder leichten Kernsplittern seine Anregungsenergie ab.

Die intranukleare Kaskade ergibt weitgehend Übereinstimmung mit experimentellen Befunden, und zwar einerseits bei direkter Berechnung — etwa nach SYMANZIK [Sy 53] oder andererseits bei den vielfach angewandten Näherungslösungen mit Hilfe der Monte-Carlo-Methode (MCM). Eine knappe Übersicht über Ergebnisse mit der MCM sei gegeben. BERNARDINI u. Mitarb. [Be 52a, b] haben für 340 MeV-Protonen mit der MCM für Sterne mit 0 bis 3 Spuren die prozentualen Häufigkeiten geladener, sekundärer Kaskadenteilchen mit verschiedenen Energien errechnet und finden gute Übereinstimmung mit ihren gemessenen Sterndaten. — METROPOLIS et al. haben in Los Alamos mit Hilfe der MCM intranukleare Kaskaden für leichte, mittlere und schwere Kerne gerechnet und die Ergebnisse mit K.S.-Daten (Emulsions- und Zählerdaten) bzw. Beschleunigerdaten verglichen: in der Arbeit ([Me 58a], S. 185; dort auch ausführliche Literatur über MCM und frühere Arbeiten) stimmen für Protonen- und Neutroneneinfallsenergien zwischen 85 bis 356 MeV die errechneten Größen wie inelastische Wirkungsquerschnitte, Transparenz, Anregungsenergien der Restkerne, Sternhäufigkeiten, Winkelverteilungen etc. gut mit den verschiedenen experimentellen Daten überein. Wichtige theoretische Ergebnisse sind weiter das Neutron/Proton-Verhältnis der emittierten Kaskadenteilchen im Verein mit den physikalisch möglichen Reaktionsendkernen. In [Me 58b, S. 204] sind für Protonen mit Energien von 450 MeV bis 1,8 GeV die mittleren Zahlen der emittierten Kaskaden-

Protonen und -Neutronen und Pionen bzw. deren Winkelverteilungen errechnet. Im Großen und Ganzen ergibt sich hinreichende Übereinstimmung mit experimentellen Daten, aber es werden Diskrepanzen bei den Spektren, den Winkelverteilungen und den Wirkungsquerschnitten der (p, pn)-Reaktionen (um einen Faktor 2—3 zu klein errechnet) ersichtlich. WHETSTONE (Los Alamos Report-LA-3206) gibt einen Überblick über den Stand weiterer Rechnungen nach der Methode von METROPOLIS et al. bis zum Jahr 1964; insbesondere werden Änderungen diskutiert, die an der ursprünglichen Kaskadenrechnung vorgenommen worden sind. Weiter sei eingefügt, daß für (p, xn)-Reaktionen A. CARETTO (Report-TID-22024) den Einfluß von Target-Dickeneffekten mit Hilfe der MCM bei Kernverdampfungen in hochenergetischen Kernreaktionen berechnet und diskutiert. H. W. BERTINI (Report-ORNL-3360, p. 199—203) hat intranukleare Kaskadenrechnungen mit MCM ebenfalls über die Mesonproduktionsschwelle hinaus fortgesetzt. Für die Stoßprozesse setzt er allein das $T=\frac{3}{2}$ Isobar an, so daß beim Nukleon-Nukleon-Stoß Ein- bzw. Zwei-Mesonerzeugung, beim Pion-Nukleon-Stoß nur Ein-Mesonerzeugung möglich ist. Ergebnisse sind dort angekündigt.

Ein Bericht (mit ausführlichen Referenzen) über Fluktuationen in intranuklearen Kaskaden mit MCM ist im Report ORNL-3714, Vol. II., p. 97—118 erschienen, und zwar für Energien von 25—350 MeV und von 250 MeV bis 2 GeV.

Die Neutronenproduktion bzw. -Multiplizität bei hochenergetischen Kernreaktionen in mittleren und schweren Elementen ist mit Hilfe der MCM von BERCOVITCH et al. [Be 60] berechnet worden; eine Ergänzung dieser Daten bietet Physics Division Progress Report-P-1964 der Atomic Energy of Canada Ltd, Chalk river.

Für Protonen mit Energien von 10 bis 100 GeV hat J. RANFT (CERN-Report-64-47) die Meson-Nukleon-Kaskade für verschiedene Abschirmmaterialien mit der MCM errechnet. — Für Meson-Nukleon-Stöße bis 25 GeV hat R. ANSARI [An 62] Rechenergebnisse der MCM mit solchen von Phasenraumintegralen verglichen.

Für Protonenenergien von $3 \cdot 10^{12}$ eV geben R. D. SETTLES und R. W. HUGGET [Se 64] mit Hilfe der MCM eine ausführliche Mesonschaueranalyse. Es werden „Monte Carlo Schauer" von jeweils $n_s = 16$ Pionen durchgerechnet, wobei sie auf das MCM-Modell von LOHRMANN, TEUCHER und SCHEIN [Lo 61], einer grundlegenden Arbeit, der weitere Daten und Literatur entnommen werden kann, zurückgreifen. Dagegen wurde der erweiterte Ansatz von VAN HOVE [Ho 63], der intermediäre Teilchenzustände in Form angeregter Bosonen bzw. isobare Zustände für die Kaskade mitberücksichtigt, nicht mituntersucht. Die Monte Carlo-Jets werden mit zwei verschiedenen Verteilungen des Transversalimpulses p_t (der „schiefen" Verteilung $\sim dp_t\, p_t \exp(-p_t/p_0)$ mit $p_0 = 0{,}38$ GeV/c im Sinne des COCCONI, KOESTER, PERKINS-Modell [Co 61], bzw. einer symmetrischen Gauß-Verteilung) und zwei verschiedenen Winkelverteilungen (Isotropie bzw. $1/\sin\Theta \cdot d\Omega$ im Schwerpunktsystem) berechnet. Die schiefe Verteilung zusammen mit der $1/\sin\Theta$ Winkelverteilung (vgl. wieder [Lo 61]) scheinen die experimentellen Daten am besten wiederzugeben. Wegen der ausführlichen Diskussion der Castagnoli-Formel bzw. -Energie E_c, der Energie E_{ch} der geladenen Teilchen und deren Zusammenhänge mit der Inelastizität im Spiegelsystem des Stoßes, sowie Bezug auf die Ergebnisse der Krakau-Warschau Zwei-Zentren-Jets sei auf die Originalarbeit verwiesen.

Es sei noch auf die experimentellen Daten von HANSEN und FRETTER [Ha 60] und KIM [Ki 64] in diesem Zusammenhang hingewiesen, ohne damit vollständig zu sein, da dieses Thema in anderen Beiträgen abgehandelt wird.

Sachverzeichnis.

(Deutsch-Englisch.)

Bei gleicher Schreibweise in beiden Sprachen sind die Stichwörter nur einmal aufgeführt.

Abkühlzeit, *cooling time* 293, 294.
Abschirmeffekt, *screening effect* 4—5, 13.
Abschirmquerschnitt, vollständiger, *screening cross section* 4, 13.
—, unvollständiger, *incomplete* 5.
Abschneidesteifigkeit, *rigidity, cutoff* 183.
Absorption durch Paarerzeugung, *absorption by pair production* 276.
— hochenergetischer Photonen im galaktischen und intergalaktischen Raum, *of high energy photons in galactic and intergalactic space* 276.
Absorptionslänge, *absorption mean free path (absorption length, see also attenuation mean free path)*, 78, 212, 386, 387, 394, 404, 430, 431, 484.
— für Sternerzeugung, *for star production* 394, 404.
— für Neutronen, *of neutrons* 437, 444, 446, 449, 450, 463.
— für Protonen, *of protons* 463.
Achondrite, *achondrites* 615.
Adiabatische Abbremsung, *adiabatic deceleration* 366.
— Invarianten, *invariants* 315, 355, 359, 360—364, 366, 367.
Ähnlichkeitsbeziehung, *similarity relation* 65
Äquator der kosmischen Strahlung, *cosmic ray equator* 314, 538.
Äquatorialer Ringstrom, *equatorial ring current* 320, 321, 322.
Aerosole, *aerosols* 587, 593, 603.
Albedo, *albedo* 195, 555.
Alfvén-Geschwindigkeit, *Alfvén velocity* 343.
Alphateilchen (s. auch Heliumkerne, Heliumkomponente), *alpha particles (see also Helium nuclei, Helium component)* 343.
Altersbestimmung, Kalium-Argon-Methode, *age determination, potassium-argon method* 623.
—, Rubidium-Strontium-Methode, *rubidium-strontium method* 623.
—, Uran-Blei-Methode, *uranium-lead method* 623.
Analytische Fortsetzung, Methode, *method of analytic continuation* 29, 90.
Anfall von bestrahltem extraterrestrischem Material, *accretion of irradiated extraterrestrial material* 552, 586, 587.
Anisotropien, *anisotropies* 340, 344, 346, 352, 357, 358.
Annahmekegel, *acceptance cone* 326.

Ansprechfunktionen, differentielle, *response functions, differential* 337, 339.
Anwachskurven der Primärkomponenten, *growth curves of primary components* 247.
Approximation A der Theorie der Elektronenschauer, *Approximation A in electron shower theory* 11, 15, 23—27.
Approximation B der Theorie der Elektronenschauer, *Approximation B in electron shower theory* 11, 15, 28—33.
Ariel I-Resultate, *Ariel I results* 337, 338, 363, 364, 365.
Aroos 616, 617, 618, 620, 624, 628.
Asymmetrie in hochenergetischen Wechselwirkungen, *asymmetry in high energy interactions* 130, 132.
Asymptotischer Annahmekegel, *asymptotic cone of acceptance* 326, 327.
Asymptotische Richtungen, *asymptotic directions* 325, 326, 327, 333, 341, 345.
— Breite, *latitude* 327.
— Länge, *longitude* 328, 333.
Atmosphäre, Gipfel, *top of the atmosphere* 373, 377, 483.
Atmosphärische Effekte, *atmospheric effects* 326, 327.
— Schwellwerte, *thresholds* 326.
Aurora-Zone (Polarlichtzone), *auroral zone* 320.
Ausbeute an Isotopen in Spallationsreaktionen, *yield of isotopes in spallation reactions* 566—573.
Ausbeutefunktionen, *yield functions* 187, 311, 328, 329, 331, 337.
Ausbreitung der kosmischen Strahlung, *propagation of cosmic rays* 239—251.
Ausfall, *fall-out* 589—595, 606.
Ausgedehnte Luftschauer, *extensive air showers* 75—84, 139, 144, 150, 224—227, 281.
— laterale Strukturfunktion, *lateral structure function* 77—82.
— Größenverteilung, *size spectrum* 224—225.
Austauschkoeffizient, *coefficient of exchange* 589.

Baryonen, angeregte (s. auch Isobarenmodell), *baryons, excited (see also isobar model)* 164—165, 559.
Bernsteins Lösung in der Elektronenschauertheorie, *Bernstein's solution in electron shower theory* 5, 34—35.

Beschleunigung, statistische (s. auch FERMIs Beschleunigungsprozeß), *acceleration, statistical (see also Fermi acceleration process)* 252, 255, 359.
Beschleunigungsprozesse, *acceleration processes* 239, 252, 254—256, 339.
Beschleunigungsprozesse, langsame, *slow acceleration processes* 254—256.
—, schnelle, *fast* 254—256.
Beschußalter, *bombardment age* 618, 627.
Betatron-Abbremsung, *betatron deceleration* 359—360, 361—362.
Betazerfall der Spaltprodukte, *beta decay of fission fragments* 316.
Bethe-Formel für Energieverluste, *Bethe formula for the stopping power* 270.
Bethe-Heitler-Formel, Korrektionen, *Bethe-Heitler formula corrections* 4—10.
Bilddipol, *image dipole* 324.
Biosphäre, *biosphere* 553, 588, 595, 596, 599, 601, 602, 607.
Bornsche Näherung, Korrekturen für Abweichungen, *Born approximation, corrections for deviations* 9—10.
Boson, intermediäres, *Boson, intermediate* 176.
Breiteneffekt, *latitude effect* 196, 199—200, 311, 313, 375, 427, 473, 562.
Breiteneffekt langsamer Neutronen, *latitude effect of slow neutrons* 436.
— schneller Neutronen, *of fast neutrons* 442, 451.
—, totaler, der Neutronen, *of neutrons, total* 444, 452, 477, 520.
—, der Neutronen bei Sonnenfleckenminimum und -maximum, *of neutrons during sunspot minimum and maximum* 438, 442, 451.
—, der primären kosmischen Strahlung, *of the primary cosmic radiation* 473.
— der Ionisation einer Restatmosphäre während eines Sonnenfleckenzyklus, *of the ionisation in the atmosphere during a solar cycle* 476.
Bremsstrahlung 266, 271, 274, 284, 290, 292, 294.
—, Elektronen-Elektronen, *electron-electron* 274.
—, innere, *inner* 266, 292.
—, im intergalaktischen Medium, *in the intergalactic medium* 284.
—, Verlustrate, *loss rate* 266, 271, 290.
—, Erzeugung durch nicht-thermische Elektronen, *production by non-thermal electrons* 292.
Bruderheim 618, 622, 624, 629.

C^{13}/C^{12}-Verhältnis, C^{13}/C^{12} *ratio* 231.
C^{14}/C^{12}-Verhältnis, C^{14}/C^{12} *ratio* 575.
C^{14}-Altersbestimmung, C^{14} *dating* 574—575, 607, 613.
C^{14}-Zerfallsrate, C^{14} *decay rate* 501, 503, 510.
C^{14}-Erzeugung, C^{14} *production*, 498—517, 554, 562, 566, 571—573, 613, 617.
—, absolute Normierung, *absolute normalisation* 498—503, 504—505.

C^{14}-Erzeugung durch solare Protonen, C^{14}-*production by solar protons* 510—517, 580.
— während des Sonnenfleckenzyklus, *during the solar cycle* 509, 571—573.
Castagnoli-Methode, *Castagnoli method* 130, 132.
Cauchy-Riemann-Gleichung, *Cauchy-Riemann equation* 88.
Chondrite, *chondrites* 615, 619, 623, 625, 630.
Chromosphärische Eruption, *solar flare* 315, 330, 341, 346, 353, 355, 366, 368, 590.
Clark County 620, 624.
Cocconi-Köster-Perkins-Modell (CKP-Modell), *Cocconi-Köster-Perkins model (CKP model)* 387, 389.
Compton-Prozeß (Compton-Streuung), *Compton process (Compton scattering)* 3, 16, 266, 272, 290.
Compton-Synchrotron-Strahlung, *Compton-synchrotron radiation* 291.
Cygnus A 302.

Deep Springs 620, 629.
Demographie der Neutronen, *demography of neutrons* 496, 507—508, 520.
Deuterium-Bildung, *deuterium formation* 266.
Deuteronen, primäre Häufigkeit, *deuterons, primary abundance* 204.
Differenz-Gleichungen in der Theorie der Elektronenschauer, *difference equations in electron shower theory* 96, 100—108.
Diffusion, regelmäßige, *diffusion, regular* 246—250.
Diffusionsextrapolation, *diffusion extrapolation* 209.
Diffusionsgleichungen für Elektronenschauer, *diffusion equations for electron showers* 11—18.
Diffusionskoeffizient, *diffusion coefficient* 244, 347—348.
Diffusionsmechanismus der Ausbreitung, *diffusion mechanism of propagation* 244—250.
Dipol-Annäherung für das Erdmagnetfeld, *dipole approximation of the geomagnetic field* 311—314.
Dipol-Modell, zentriert, *dipole model, centered* 311, 313, 314, 537.
—, exzentrisch, *eccentric* 311, 313, 538.
Dipolmoment, solarer, *solar dipole moment* 349.
Diskrete Quellen hochenergetischer Photonen, *discrete sources of high energy photons* 285—302.

e-Neutrinos, *e-neutrinos* 174—176, 307
Ehole 618, 622, 623, 629
Einfallszonen, *impact zones* 325, 344—346, 356, 368
Ein-Feuerball-Emission, *one-fireball emission* 158—161
Eingefangene Strahlung, *trapped radiation* 315, 316, 321
Einzelstreuung, *single scattering* 119
—, Beitrag zur Lateralausbreitung, *contribution to the lateral spread* 67, 92

Sachverzeichnis.

Eisenmeteorite, *iron meteorites* 615, 618, 619—621, 623, 625, 627, 629, 630.
Elektronen, atmosphärische sekundäre, *electrons, atmospheric secondaries* 74—75, 220.
—, galaktische sekundäre, *galactic secondary* 220—223, 274—275
—, —, primäre, *primary* 219—223, 267, 283
Elektronenerzeugung im intergalaktischen Medium, *electron production in the intergalactic medium* 273—275
Elektronen-Erzeugungsspektrum, *electron production spectrum* 270
Elektronenschauer, Theorie in Approximation A, *electron shower theory, approximation A* 11, 15, 23—27.
—, — in Approximation B, *Approximation B* 11, 15, 28, 33.
—, Diffusionsgleichungen, *diffusion equations* 11—18
—, elementare Lösungen, *elementary solutions* 19.
—, eindimensionale, *one-dimensional* 2, 14—15.
—, dreidimensionale, *three-dimensional* 2, 15, 38—49.
—, Energiemessungen, *energy measurements* 69—74, 112—113.
Elfjahre-Zyklus, *eleven-year cycle* 185, 332, 374, 577—579, 614, 630.
Emulsionskammer, *emulsion chamber* 70, 72—73, 113, 118, 125, 126, 138.
Emulsionsstapel, *emulsion stack* 119, 121, 132.
Energiedichte der Kosmischen Strahlung, *energy density of cosmic rays* 240.
Energiespektrum, Maximum im differentiellen, *peak in the differential energy spectrum* 238.
— der Primärteilchen, *of primaries* 143, 153, 195—227, 378—380.
— der Myonen, *of muons* 147, 149.
— der Neutrinos, *of neutrinos* 174—175.
— der Neutronen in der Atmosphäre, *of neutrons in the atmosphere* 417, 447.
— der Protonen in der Atmosphäre, *of protons in the atmosphere* 200—204, 382—391, 417.
— der Schauerelektronen und Photonen, *of shower electrons and photons* 22, 24, 33, 34, 142—143, 149.
—, Werte des Exponenten, *values of the exponent* 384.
Energieübertragung an geladene Sekundärteilchen, *energy transfer to charged secondaries* 130, 134.
— an γ-Strahlen, *to γ-rays* 130, 134.
Energieverluste durch Expansion, *energy loss due to expansion* 274.
Energieverlustrate der Elektronen in der Galaxie, *energy loss rate of electrons in the galaxy* 270, 273, 275.
— der Myonen, *of muons* 148.
Entweichen der Teilchen aus der Quellgegend, *escape of particles from the source region* 254.

Erdmagnetfeld, *earth magnetic field* 311—313, 325, 378, 537—538.
Erosionshypothese, *erosion hypothesis* 623, 627.
Erzeugungsspektrum von Neutronen und Protonen (Rossi-Modell), *production spectrum of neutrons and protons (Rossi model)* 416.
—, inhomogenes, *inhomogeneous* 389.
Explorer X-Resultate, *Explorer X results* 343.
Extragalaktische diskrete Quellen hochenergetischer Photonen, *extragalactic discrete sources of high energy photons* 301—302.

Familien von Elektronenschauern, *families of electron showers* 76, 126, 140, 142—147, 165.
Fermis Abbremsprozesse, *Fermi deceleration process* 361—362, 366.
Fermis Beschleunigungsprozeß (s. auch statistische Beschleunigung), *Fermi acceleration process (see also statistical acceleration)* 359.
Fermis thermodynamisches Modell der Vielfacherzeugung, *Fermi's thermodynamical model of multiple production* 155, 168.
Feuerbälle, *fireballs* 157—168, 169, 560.
Flügge-Yuansches Dichtemaximum langsamer Neutronen, *Flügge-Yuan density maximum of slow neutrons* 374, 426, 427, 438, 439, 483.
Fluß der breitenempfindlichen Primärteilchen, *flux of latitude sensitive primaries* 378.
—, integraler, der Primärteilchen, *integral flux of primaries* 380.
Fokker-Planck-Annäherung, *Fokker-Planck approximation* 12, 17, 91.
Forbush-Effekt, *Forbush effect (Forbush decrease)* 310, 320, 323, 330, 331, 335, 338—340, 341, 346, 347, 350—352, 354, 357, 362, 368, 576.
Fourier-Transformation, *Fourier transformation* 39, 85.
Fragmentation schwerer Kerne, *fragmentation of heavier nuclei* 209, 247—249.
Fragmentationsparameter, *fragmentation parameters* 210—212.
Fraktionierungsprozeß, *fractionation process* 599.
Funkenkammer, *spark chamber* 120.

Galaktische Gasdichte, *galactic gas density* 270.
Galaktische Magnetfelder, *galactic magnetic fields* 240.
Galaktisches Zentrum, *galactic center* 299
γ-Strahlen, s. Photonen, Photonenkomponente, *γ-rays, see photon, photon component*.
γ-Strahlspektrum, *γ-ray spectrum* 140, 142—143.

GAUNTs Annäherung für den Bremsstrahlungsquerschnitt, *Gaunt approximation to the bremsstrahlung cross section* 292.
Gauss-Koeffizienten, *Gauss coefficients* 313.
Geomagnetischer Äquator, *geomagnetic equator* 313.
Geomagnetische Effekte, *geomagnetic effects* 311, 30.
Geomagnetischer Hohlraum, *geomagnetic cavity* 320—321, 324, 343.
Geosphäre, *geosphere* 553, 587—607.
Globale Erzeugungsrate von Radioisotopen, *global production rate of radioisotopes* 572—573, 581—583.
Globales Inventar der Radioisotope, *global inventory of radioisotopes* 602—607.
Globale Karte von Kernumwandlungen, *global map of nuclear disintegrations* 564—566.
Grant 620, 623, 624, 625.
Greensche Funktion, Methode, *Green function method* 35, 97.
GREISENs Annäherung für die laterale Strukturfunktion, *Greisen approximation of the lateral structure function* 57—59, 150.

H/M-Verhältnis, *H/M ratio* 212, 218—219, 227, 232.
H-Quantum, *H-quantum* 165—168.
Häufigkeit, relative, der verschiedenen Ladungskomponenten, *relative abundance of various charge components* 228—237.
—, gerader und ungerader Kernladungen, *odd-even charge abundance* 230, 247.
—, von Be und B, *of Be and B* 231.
— von Kernen mit $Z \geq 30$, *of nuclei with $Z \geq 30$* 229.
Häufigkeitsüberschuß der L-Kerne, *over abundance of L-nuclei* 230.
Halbschatten, *penumbra* 312, 315, 318.
Hankel-Transformation, *Hankel transformation* 46, 85, 87.
Harleton 618, 619, 622, 629.
Harte Strahlung, durch Radioaktivität emittierte, *hard radiation emitted through radioactivity* 288.
HASEGAWAs Theorie der Viel-Feuerbälle-Emission, *Hasegawa theory of multiple-fireball emission* 165—168.
Hauptkegel, *main cone* 312.
Hauptphase, *main phase* 323, 324.
He/M-Verhältnis, *He/M ratio* 212.
He/S-Verhältnis, *He/S ratio* 232, 234—235.
Heliumkerne, primäre, *primary Helium nuclei* 204—208.
—, Energiespektrum, *energy spectrum* 205—207.
—, Isotopenverhältnis, *isotopic composition* 208.
—, Steifigkeitsspektrum, *rigidity spectrum* 207.
Hochenergieschauer, *jets* 118, 129.

Höhenabhängigkeit der harten Komponente in der Atmosphäre, *altitude dependence of the hard component in the atmosphere* 420.
— von Kernumwandlungen in der Atmosphäre, *of nuclear disintegrations in the atmosphere* 563—566.
— von Neutronen in der Atmosphäre, *of neutrons in the atmosphere* 427, 435, 440, 448.
— von Protonen in der Atmosphäre, *of protons in the atmosphere* 414, 418, 420.
— von Sternen in der Atmosphäre, *of stars in the atmosphere* 393—397.
,,hot universe'' kosmologisches Modell, *hot universe cosmological model* 284.
Hubble-Radius, *Hubble radius* 273, 277, 283.
Hydrosphäre, *hydrosphere* 553, 595—599, 602, 606—607.
Hyperon/Nukleon-Verhältnis, *hyperon/nucleon ratio* 391.
Hysteresis 185.

IGY 334, 338, 339.
Imp I-Resultate, *Imp I results* 341, 342.
Inelastizität in Nukleonenwechselwirkungen, *inelasticity in nucleon interactions* 132, 135, 151, 159, 162, 167, 385, 387, 388.
Inhomogenitäten, magnetische, *magnetic inhomogeneities* 342.
Integrale Darstellung, Methode, *method of integral representation* 29, 90.
Intensität, unidirektionale, *unidirectional intensity* 181.
Intergalaktisches Medium, Elektronenerzeugung und Energieverlust, *electron production and energy losses in the intergalactic medium* 273—275.
Interplanetare Felder, *interplanetary fields* 310, 318, 324, 325, 330, 341, 342, 347, 353, 357, 362, 368.
Interplanetares Medium als Röntgenstrahlenquelle, *interplanetary medium as X-ray source* 300.
Interplanetares Plasma, *interplanetary plasma* 310, 313, 325, 330.
Interstellarer Staub (s. auch kosmischer Staub), *interstellar dust (see also cosmic dust)* 552, 587.
Interstellarer Wasserstoff, Dichte, *density of interstellar hydrogen* 243, 270.
Ionisationskalorimeter, *ionisation calorimeter* 69, 384.
Ionisationskammer, *ionisation chamber* 188, 195, 318, 333, 340, 384.
Ionisationsverlust-Prozesse, *ionisation loss processes* 3, 13, 270.
Ionisations-Regressionskurve, *ionisation regression curve* 186—187.
Impulsspektrum der Protonen, *momentum spectrum of protons* 400, 463, 555, 556.
— der Deuteronen, *of deuterons* 400, 401.
— der Sekundärteilchen, *of secondary particles* 400, 401.
Isobar-Modell, *isobar model* 389.

Isotopenerzeugung in der Atmosphäre, *isotope production in the atmosphere* 553, 561—585.
—, zeitliche Veränderungen, *time variations* 573—580.
Isotopenhäufigkeiten in Meteoriten, *isotopic abundance in meteorites* 616—617, 621, 630.

K-Fluoreszenzausbeute, *K-fluorescence yield* 292.
K/π-Verhältnis, K/π *ratio* 121, 150, 157, 169.
K-Zahlen, *K-figures* 333, 343.
Kalorimetrische Methode (s. auch Ionisationskalorimeter), *calorimetric method (see also ionisation calorimeter)* 69, 130, 131, 384.
Kalos-Blatt-Methode, *method of Kalos and Blatt* 104—105.
Kaskadenschauer (s. auch Elektronenschauer, Nukleonenkaskade), *cascade shower (see also electron showers, nucleon cascade)* 125, 126, 139, 144, 198.
KEPLER's Supernova von 1604, *Kepler's 1604 supernova* 272.
Klein-Nishina-Formel, *Klein-Nishina formula* 272.
Kernwechselwirkungen, Verhältnis geladener zu ungeladenen Teilchen, *neutral-charged ratio in nuclear interactions* 121, 122, 457.
Kernumwandlungen, durch kosmische Strahlung ausgelöste, *cosmic ray induced nuclear transformations* 551—557, 571, 614—615, 625—630.
Korona, *corona* 300, 344, 350.
Kosmischer Staub, *cosmic dust* 552, 587, 623.
Kosmisches Strahlungsalter, *cosmic ray age* 622—625.
Kosmologische Theorien, Prüfungen, *tests of cosmological theories* 284—285.
Krebsnebel, *Crab nebula* 289, 290, 291, 294—299, 304—306.
—, Abdeckung durch den Mond, *lunar occultation of the Crab nebula* 295, 306.
Kritische Energie, *critical energy* 10—11.

L-Kerne, Unterschiede in den Spektren, *differences in the spectra of L-nuclei* 236.
—, differentielles Spektrum, *differential spectrum* 214.
—, integrale Intensitäten, *integral intensities* 213.
—, solare Modulation des Spektrums, *solar modulation of the spectrum* 215.
L/S-Verhältnis, L/S *ratio* 212, 213, 227, 232, 235, 248.
Ladungszusammensetzung der Primärstrahlung, *charge composition of the primary radiation* 195—210, 213—219, 223—224, 233—237.
Landau-Annäherung, Theorien in, *theories in the Landau approximation* 12, 14, 18, 44.
—, Theorien ohne, *theories without* 12, 15, 18, 44—49, 61, 64.
Landau-Effekt, *Landau effect* 6—8, 18, 44.

LANDAUs hydrodynamisches Modell der Vielfacherzeugung, *Landau's hydrodynamical model of multiple production* 156.
Laplace-Transformation, *Laplace transformation* 85, 86, 99, 101—104.
Larmor-Frequenz, *Larmor frequency* 272.
Larmor-Radius, *Larmor radius* 271, 352, 355, 356, 361.
Lateralausbreitung der Schauerteilchen, *lateral spread of shower particles* 17, 22.
Laterale Strukturfunktion der Elektronen und Photonen, *lateral structure function of electrons and photons* 43—44, 48—49, 62, 63, 73.
— — bei extrem hohen Energien, *at extremely high energies* 75—84.
— —, GREISENs Annäherung, *Greisen approximation* 57—59.
— — für den Energiefluß der Schauerteilchen, *of the energy flow of shower particles* 83—84.
Lebensdauer der kosmischen Strahlenteilchen, *life time of cosmic ray particles* 245.
Leitfähigkeit, elektrische, des interplanetaren Plasmas, *electrical conductivity of the interplanetary plasma* 344.
Leitisotopenuntersuchungen, *tracer studies* 554, 597, 599.
LIOUVILLEs Theorem, *Liouville's theorem* 184, 312, 354, 360.
Lithosphäre, *litosphere* 553, 585—587, 601.
log tang ϑ-Diagramme, *log tan* ϑ *plots* 130, 157, 158.
Luftschauer, s. ausgedehnte Luftschauer, *air showers, see extensive air showers*

McIlwain-Parameter, *McIlwain parameter* 315.
Magnetischer Knick, *magnetic kink* 352, 356, 362.
Magnet-Spektrometer, *magnetic spectrometer* 412.
Magnetisches Streuzentrum, *magnetic scattering centre* 350, 363.
Magnetische Stürme, *magnetic storms* 317, 318—320, 321, 323, 330, 364.
Magnetische Zunge, *magnetic tongue* 350, 351—352, 354, 356.
Magnetosphäre, *magnetosphere* 323, 324.
Mariner II-Resultate, *Mariner II results* 341—342, 344.
Mars I-Resultate, *Mars I results* 340.
Massenspektrometer, *mass spectrometer* 619.
Materiedicke, von Primärteilchen durchlaufene, *thickness of matter traversed by primary cosmic rays* 246, 584.
Mehrfachstreuung, *plural scattering* 119.
Mehrfachstreuungseffekte, *plural scattering effects* 92.
Mellin-Transformation, *Mellin transformation* 23, 85, 86.
Mesonenteleskop, *meson telescope* 325, 329, 330, 332.

Meteorite, *meteorites* 552, 573—574, 587, 613—631.
—, Strahlungsalter, *radiation ages* 553, 618, 622—625.
—, Erzeugungsrate von Radionukliden, *radionuclide production rate* 625—630.
Meteoroide, *meteoroids* 552.
Meteorologische Effekte, *meteorological effects* 330.
Minimum-Korndichte, *minimum grain density* 119, 372.
Mischungskoeffizient, *coefficient of mixing* 596.
Modellatmosphäre, *model atmosphere* 492.
Modellexperimente zu geomagnetischen Effekten, *model experiments on geomagnetic effects* 325.
Modulation, aperiodische, *aperiodic modulation* 431.
—, periodische, *periodic modulation* 431.
— der Nukleonen in der Atmosphäre, *of nucleons in the atmosphere* 374, 375.
— der Heliumkomponente, *of the Helium component* 191—192.
— der Neutronen in der Atmosphäre, *of neutrons in the atmosphere* 424—429, 471, 478—508, 567.
— der Protonenkomponente, *of the proton component* 189—190, 328.
— der L-Kerne, *of the L-nuclei* 215.
— der S-Kerne, *of the S-nuclei* 192.
—, solare, *solar* 186, 189—195, 374, 375.
Modulationsfunktion, *modulation function* 337—338, 362.
Modulationsmechanismus, *modulation mechanism* 330—368.
Molekularer Austausch, Entfernung von Radioaktivität, *removal of radioactivity by molecular exchange* 594, 595, 598.
Molière-Einheit, *Molière unit* 58.
MOLIÈREs Streutheorie, *Molière scattering theory* 44, 68, 93—96.
MOTTs Formel für die Streuwahrscheinlichkeit, *Mott formula of the scattering probability* 92.
Multiplizität der Sekundärteilchen-Erzeugung, *multiplicity of secondary production* 135, 151, 167, 169, 269, 385.
Myonen (μ-Mesonen), Energiespektrum, *muons (μ-mesons), energy spectrum* 147, 149.
— —, Energieverlustrate, *energy loss rate* 148.
— —, Intensität/Tiefe-Beziehung, *intensity-depth relation* 148.
— —, Ladungsüberschuß, *positive excess* 391.
— —, Reichweitenschwankungen, *range fluctuations* 108—112.
— —, Wechselwirkungen, *interactions* 135, 160, 585.
μ-Neutrinos, *μ-neutrinos* 119, 174—176.

Nebelkammer, *cloud chamber* 118, 120, 131, 136, 137.
Neon-Hodoskop, *neon hodoscope* 80, 151.

Neutrino-Astronomie, *neutrino astronomy* 266.
Neutrino-Erzeugungsspektrum, *neutrino production spectrum* 304.
Neutrino-Experimente mit kosmischen Strahlen, *cosmic ray neutrino experiments* 173—177.
Neutrino-Paarerzeugung, *neutrino pair production* 303.
Neutrinoquellen, *neutrino sources* 302—307.
Neutrinoteleskop, *neutrino telescope* 305.
Neutrinos, Energiespektrum, *energy spectrum of neutrinos* 174, 175.
—, solare, Radioisotopenerzeugung, *radioisotope production by solar neutrinos* 586.
Neutronenalbedo (s. auch Neutronenverlustfluß), *neutron albedo (see also neutron leakage flux)* 527.
Neutronendichte, Flügge-Yuansches Maximum, *Flügge-Yuan maximum of neutron density* 374, 426, 427, 438, 439, 483.
Neutroneneinfang, *neutron capture* 498—517.
Neutronenfluß in der Atmosphäre, *neutron flux in the atmosphere* 374, 434—453, 464—535.
— —, Gleichgewichtsspektrum, *equilibrium spectrum* 465.
— —, Simpsonsches Maximum, *Simpson maximum* 374, 450, 453.
— —, Berechnung mit der S_n-Methode, S_n-*method calculation* 497.
— —, unerklärte Schwankungen, *unexplained fluctuations* 443.
Neutronenverlustfluß, *neutron leakage flux* 427, 428, 452, 508, 518—535.
—, Winkelverteilung, *angular distribution* 521—527, 533.
—, aperiodisch emittierter, *aperiodically emitted* 532—535.
—, Breiteneffekt, *latitude effect* 524, 531, 533.
—, experimentelle Werte, *experimental values* 527—532.
—, Spektrum, *spectrum* 519.
— während des Sonnenfleckenzyklus, *during the solar cycle* 520.
Neutronenerzeugung durch solare Protonen, *neutron production by solar protons* 510—537.
— in der Atmosphäre, *in the atmosphere* 428, 472, 494.
Neutronenerzeugungsrate, *neutron production rate* 472, 474—478.
Neutronen-Monitor, *neutron monitor* 185, 313, 315, 325, 329, 332—336, 338—340, 345, 462, 463.
Neutronenreaktionen, Wirkungsquerschnitte in Luft, *neutron reaction cross sections in air* 499.
Neutronen, Alterungstheorie, *neutrons, age theory* 425, 482.
—, —, Demographie, *demography* 498, 507—508, 520.
—, —, Quellverteilung, *source distribution* 489.

Neutronen, Alterungstheorie, Winkelverteilung, *neutrons age theory, angular distribution* 463—464.
—, —, Zerfallsdichte im interplanetaren Raum, *decay density in interplanetary space* 526.
—, Diffusionstheorie, *diffusion theory* 425, 480—489, 493—495.
—, —, Energiespektrum, *energy spectrum* 434, 447, 448, 464—472, 567.
—, —, langsame, *slow* 434—439, 452, 466—468, 483, 524—526, 527—529.
—, —, schnelle, *fast* 432, 441—453, 463—464, 521—531.
Neutronen in der Atmosphäre, Absorptionslänge, *neutrons in the atmosphere, absorption mean free path* 444, 446, 449, 450.
— —, Absolutzahl, *absolute number* 502.
— —, Breitenabhängigkeit, *latitude dependence* 438, 451.
— —, Dichte, *density* 439, 440, 485.
— —, Modulation, *modulation* 424—429, 508.
— —, Quellspektrum, *source spectrum* 432.
— —, Ursprung, *origin* 424—429.
Neutronen, primäre solare, *primary solar neutrons* 426, 583—584.
—, — —, Abbremslänge und Abbremszeit, *slowing down length and slowing down time* 481.
—, — —, Geschwindigkeitsverteilung, *velocity distribution* 425.
—, — —, Übergangseffekt in Wasser, *transition effect in water* 441.
Neutronenquelle, Normierung, *neutron source, normalisation* 431, 496.
Neutronenquellspektrum, *neutron source spectrum* 428, 429, 432.
Neutronensterne, *neutron stars* 287, 288—290.
Neutronentransportgleichung, *neutron transport equation* 479—489, 495—498.
Nishimura-Kamata-Funktion, *Nishimura-Kamata function* 56—63, 150.
Nius Theorie der Zwei-Feuerbälle-Emission, *Niu's theory of two-fireball emission* 161—165.
Nicht-Dipolglieder des Erdmagnetfelds, *non-dipole terms of the geomagnetic field* 313, 325.
Nukleare Kaskade (Nukleonenkaskade), *nuclear cascade (nucleon cascade)* 151, 372, 383—391, 428, 559, 564.
Nukleo-aktive Komponente, *nuclear-active (nucleoactive) component* 136, 138, 379, 385.
Nukleonen, elastische Streuung, *nucleons, elastic scattering* 171—173.
—, Modulation in der Atmosphäre, *modulation in the atmosphere* 372—377, 385, 422, 431, 508, 557—559.
— in der Atmosphäre, geomagnetische Effekte, *in the atmosphere, geomagnetic effects* 374, 379, 422, 431, 508.
— —, Höhenabhängigkeit der inelastischen Streuung, *altitude dependence of inelastic scattering* 427.

Odessa 619, 620, 625.
Ohmsche Dissipationszeit des interplanetarischen Plasmas, *Ohmic dissipation time of the interplanetary plasma* 344.
Operatorenrechnung, *Operator calculus* 97.

Paarerzeugung, *pair production (pair creation)* 3, 6—9, 16, 276.
—, Effekt der Atomelektronen, *effect of atomic electrons* 8—9.
—, in Photon-Photonstößen, *in photon-photon collisions* 276.
—, Landau-Effekt, *Landau effect* 6—8.
—, Migdals Methode, *Migdal's treatment* 7.
— von Nukleonen, *of nucleons* 558.
Pallasite, *pallasites* 616, 619, 620.
Pfotzer-Maximum, *Pfotzer maximum* 117, 373.
Photonen, hochenergetische, in der Atmosphäre, *high energy photons in the atmosphere* 74—75.
—, primäre, *primary* 265.
Photonenastronomie, hochenergetische, *high energy photon astronomy* 267.
Photonenerzeugende Prozesse, *photon producing processes* 266, 267—285.
Photonenerzeugungsspektrum, *photon production spectrum* 277.
Photonenspektrum bei hohen Energien, *high energy photon spectrum* 279.
Photopionenerzeugung, *photopion production* 278.
Pinkaus Methode zur Bestimmung von Schauerenergien, *Pinkau's method for the determination of shower energies* 112.
Pion/Neutron-Verhältnis, *pion/neutron ratio* 463.
Pion/Proton-Verhältnis, *pion/proton ratio* 463.
Pionenerzeugung, *pion production* 121, 153—168, 268—270, 385, 557—561.
Pionenmultiplizität im Isobar-Modell, *pion multiplicity in the isobar model* 390, 559.
Pionenverteilung, *pion distribution* 388.
Pionen, neutrale, Erzeugungsspektrum, *production spectrum of neutral pions* 198, 269, 297.
Pionisation 385
Pioneer V, Resultate, *Pioneer V results* 340, 341, 342, 356.
Planeten als Röntgenstrahlenquellen, *planets as X-ray sources* 300.
Plasma, interplanetares, *interplanetary plasma* 310, 313, 349.
—, solares, *solar plasma* 313, 321, 323, 332, 341—342, 349.
Plasmafrequenzen, *plasma frequencies* 274.
Plasmaschwingungen, *plasma oscillations* 274.
Plateaukorndichte, *plateau grain density* 119, 392.
Plötzlicher Sturmeinsatz, *sudden commencement* 318, 332, 346, 353.
Polare Eiskappe, Radioisotope, *polar ice cap radioisotopes* 576, 587, 601.

Polarkappe, *polar cap* 320.
Polarlichtzone, *auroral zone* 320.
Positron/Elektron-Verhältnis, *positron/electron ratio* 222, 238, 282.
Positronenvernichtungs-Querschnitt, *positron annihilation cross section* 278.
Primärkerne, Gesamtfluß, *primary nuclei, total flux* 374.
Primärkomponente, *primary component* 311, 337, 377—383.
Primärspektrum (s. Energiespektrum, Steifigkeitsspektrum), *primary spectrum (see energy spectrum, rigidity spectrum)* 143, 153, 195—227, 337, 380—383, 387.
Projekt Argus, *project Argus* 316.
Proton/Helium-Verhältnis (P/He-Verhältnis), *proton/helium ratio* 197, 232, 234—235.
Protonen, differentielles Energiespektrum, *protons, differential energy spectrum* 200—204, 413, 414.
—, Gesamtfluß, *total flux* 374.
— in der Atmosphäre, Energiespektrum, *in the atmosphere, energy spectrum* 382—391, 410—422.
— —, experimentelle Methoden, *experimental methods* 200—204, 383, 411—413, 463.
—, Prozentsatz in der gesamten durchdringenden Komponente, *percentage in the total penetrating component* 419.
—, solare, *solar* 315, 317, 318, 320, 325, 365, 512, 556.
—, Vertikalintensität, *vertical intensity* 418—420.
—, Winkelverteilung, *angular distribution* 422—424.
Protonenspektrum beim Sonnenfleckenminimum, *sun spot minimum proton spectrum* 203.

Quellen und Senken hochenergetischer Elektronen, *sources and sinks of high energy electrons* 275.
Quellen- und Senkenfunktion für Isotopenerzeugung, *source and sink functions for isotope production* 553—555, 562—575, 587, 589, 591, 594—595, 597.
Quellgegenden, *source regions* 254, 257—260.
Quellspektrum der Verdampfungsneutronen, *source spectrum of evaporation neutrons* 427, 430.

Radioaktivität, durch kosmische Strahlung erzeugte, *cosmic ray induced radioactivity* 553, 561, 562, 602—607, 613, 614, 620—622, 630.
Radioemission, *radio emission* 241, 242, 282, 292.
Radiohelligkeitstemperatur, *radio brightness temperature* 282.
Radius des galaktischen Halo, *radius of the galactic halo* 277.
Reichweitenschwankungen hochenergetischer Myonen, *range fluctuations of high energy muons* 108—112.

Registrierstationen für kosmische Strahlung, *cosmic ray registration stations* 535—537.
Regressionskurve der atmosphärischen Ionisation, *regression curve of atmospheric ionization* 186, 187.
Ringstrom, äquatorialer, *equatorial ring current* 320—322, 324.
Riometer 317.
Röntgenstrahlen, charakteristische, *characteristic X-rays* 266, 292.
—, durch Elektron-Ionen-Strahlungsrekombination erzeugte, *X-ray emission by electron-ion radiative recombination* 293.
—, durch inelastische Stöße mit nachfolgender Strahlungsdeexzitation, *X-ray emission by inelastic collisions followed by radiative de-excitation* 293.
— von äußeren Galaxien, *X-rays from external galaxies* 281—282.
Röntgenstrahlenerzeugung in diskreten Quellen, *X-ray production in discrete sources* 290—294.
Röntgenstrahlfilme für Energiemessung an Elektronenschauern, *X-ray films for energy measurements on electron showers* 69, 139—140.
Röntgenstrahlenquellen, *X-ray sources* 286—302.
Ruhige kosmische Strahlung, *quiescent cosmic radiation* 377.

S-Kerne, *S-nuclei* 216—218.
Sattelpunktmethode, *saddle point method* 31, 88.
Säkulares Gleichgewicht, *secular equilibrium* 616.
Sco XR-1 285, 301, 306.
Schauer, s. ausgedehnte Luftschauer, Elektronenschauer, Kaskadenschauer, *shower, see cascade shower, electron shower, extensive air shower*.
Schauerachse, *shower axis* 60—69.
Schaueralter, *shower age* 21.
Schauerdiagramme, *shower diagrams* 398, 399.
Schauerfunktionen, *shower functions* 23—33.
Schauermaximum, *shower maximum* 19, 26.
Schauer-Sekundärteilchen, *shower secondaries* 121—122, 268, 385, 557—559.
—, Verhältnis neutraler zu geladenen, *neutral-charge ratio* 121—122.
Schauerteilchen, Gesamtzahl, *shower particles, total number* 18, 27, 66.
Schockwelle, *shock wave* 352—355, 357.
Schwellwertsteifigkeit, *threshold rigidity* 311, 313, 318, 321—325.
Sedimentation der Radioisotope, *sedimentation of radioisotopes* 575, 587, 601, 602, 607.
Sgr XR-1 286.
Sikhote-Alin 620, 624, 625.
Simpsons Maximum des Neutronenflusses, *Simpson maximum of the neutron flux* 374, 450, 483.

Simpson pile, s. Neutronenmonitor, *see neutron monitor*
Solare energiereiche Teilchen (s. auch Protonen), *solar energetic particles (see also protons)* 344—347.
Solare Modulation (s. auch 11-Jahre-Zyklus, 27-Tage-Zyklus), *solar modulation (see also 11-years cycle, 27-days cycle)* 374, 378.
Solare Neutrinos, *solar neutrinos* 586.
Solares Plasma, *solar plasma* 313, 321, 323, 341, 342, 349, 359, 368.
Solare Teilcheneinbrüche, Variationen der Erzeugung von Radioisotopen, *solar particle bursts, variations in radioisotope production* 579—580.
Sonne als Röntgenstrahlenquelle, *sun as X-ray source* 299—300.
Sonnenfleckenzyklus (s. auch 11-Jahre-Zyklus), *sunspot cycle (see also 11-years cycle)* 185, 329, 330, 333, 336—340.
Sonnensystem, Röntgenstrahlenquellen, *solar system X-ray sources* 299—300.
Sonnenwind, *solar wind* 320, 321, 323, 324, 342, 344, 349, 355, 359, 365.
Spallationsreaktionen, *spallation reactions* 551, 559, 566, 574, 614, 616, 621.
Spiegelpunkte, *mirror points* 315.
Spuren in Kernemulsionen, *tracks in nuclear emulsions* 392, 403.
Spurenlänge, differentielle, *differential track length* 18, 25, 38.
—, totale (integrale), *total (integral)* 16, 18—19, 25, 35—38.
Standardatmosphäre, *standard atmosphere* 376, 377, 491.
„steady state" kosmologische Theorie, *steady-state cosmological theory* 284.
Steifigkeit, magnetische, *magnetic rigidity* 182, 311, 312, 318, 555, 556.
Steifigkeitsspektrum, *rigidity spectrum* 183, 374, 556, 557.
Steinige Meteoriten, *stone meteorites* 615, 616, 620, 621.
Sterne in Kernemulsionen, *stars in nuclear emulsions* 391—410, 453—458, 565.
— —, Absoluthäufigkeit, *absolute frequency* 396—397.
— —, Höhenabhängigkeit, *altitude dependence* 393.
— —, Prozentsatz von Deuteron-, Triton- und He-Spuren, *percentage of deuteron, triton, and He tracks* 401.
Sternerzeugung in der Atmosphäre, *star production in the atmosphere* 394, 396.
— —, mittlere freie Weglänge, *mean free path* 454.
— —, Reaktionsraten, *reaction rates* 454, 564, 582.
Sterngrößenverteilung, *star size distribution* 397, 398, 402, 404, 568, 571.
Sternkoronae, harte Strahlung von, *hard radiation from stellar coronae* 300—301.
Störmer-Einheit, *Störmer unit* 311.

Störmers Theorie, *Störmer theory* 312, 313, 318, 321.
Stoßlänge, mittlere (s. auch Wechselwirkungslänge), *collision mean free path (see also interaction mean free path)* 151, 386, 387.
Strahlungslänge, *radiation length* 10—11.
Strahlungsprozeß, *radiation process* 3, 6—9.
—, Effekt der Atomkerne, *effect of atomic nuclei* 8—9.
—, Landau-Effekt, *Landau effect* 6—8.
—, Migdals Methode, *Migdal's treatment* 7.
Stratosphäre, *stratosphere* 396, 555, 589, 590. 599, 601, 603—605, 614, 616, 621.
Streuung, Methode der relativen, *method of relative scattering* 120, 175.
—, vorgetäuschte, *spurious scattering* 120.
Strömung, *convection* 357—363, 364, 365.
Strukturfunktionen in der Theorie der Elektronenschauer, *structure functions in electron shower theory* 2, 39, 43, 44, 47—49, 49—69.
—, achsenferne, *at large distance from the shower axis* 67—69.
—, achsennahe, *near the shower axis* 60—69.
—, normalisierte, *normalized* 56.
—, mittlere quadratische Abweichungen, *mean square deviations* 53—56.
Stufenerzeugung von Sekundärteilchen, *plural production of secondaries* 116.
Südafrikanische Anomalie, *South African anomaly* 337.
Supernovae 257, 258, 287, 288, 615.
Supernovae-Überreste, *supernovae remnants* 285.
Synchrotronstrahlung, *synchrotron radiation* 240, 266, 271, 290.

Tagesgang der Intensität, *daily intensity variation* 326, 327, 333, 350, 351, 354, 365.
Takagis Modell für Nukleonenwechselwirkung, *Takagi model of nucleon interactions* 161.
Tamm und Belenky, Lösungen für die Spurenlänge, *track length solutions of Tamm and Belenky* 37, 38.
Tau XR-1 285.
Thermische Neutronen, durch — induzierte Kernumwandlungen, *nuclear disintegrations induced by thermal neutrons* 566, 572, 573, 585, 614.
Thompson-Grenze, *Thompson limit* 272.
Trägerbaryonen, *persisting baryons* 390.
Transparenz, *transparency* 313.
Transversalimpuls, *transverse momentum* 76, 113, 125, 127—129, 145, 157, 166, 169, 170, 172, 386.
Treysa 620, 623, 624, 627, 628, 629.
Tritium, solares, *solar tritium* 584.
Troposphäre, *troposphere* 554, 591—595, 599, 601, 605, 606.

41*

Übergangseffekt von Neutronen in Wasser, *transition effect of neutrons in water* 441.
— von Sternen, *of stars* 405—410, 564.
Universales Spektrum der kosmischen Strahlung, *universal cosmic ray spectrum* 267.

Variation, kurzzeitliche, *short-term variation* 341.
—, langzeitliche, *long-term* 310, 332, 357, 363, 366.
—, Steifigkeitsabhängigkeit, *rigidity dependence* 332, 336—340, 362.
—, tägliche (Tagesgang), *daily* 326, 327, 333, 350, 351, 354, 365.
—, 27-Tage-, *27-day* 330.
Variationskoeffizient, *variational coefficient* 326.
Verdampfungsneutronen, Quellspektrum, *source spectrum of evaporation neutrons* 427, 430.
Verlustfluß aus dem galaktischen Halo, *leakage out of the galactic halo* 273.
Vernichtung (Zerstrahlung) von Materie und Antimaterie, *matter-antimatter annihilation* 266, 278, 284.
Verweilzeit, *residence time* 598.
Vielfacherzeugung von Sekundärteilchen, *multiple production of secondaries* 115—118, 153—169.
— —, FERMIs thermodynamisches Modell, *Fermi's thermodynamical model* 155, 268—270, 385.
Vielfacherzeugung von Sekundärteilchen, LANDAUs hydrodynamisches Modell, *multiple production of secondaries, Landau's hydrodynamical model* 156.
Vielfachstreuung, *multiple scattering* 91—96, 119.
Viel-Feuerbälle-Emission, *multiple-fireball emission* 165—168.
Viererimpulsübertrag, *four-momentum transfer* 158.

Wechselwirkungslänge (s. auch Stoßlänge), *interaction mean free path (see also collision mean free path)* 123, 211.
Wiederkehralbedo, *reentrant albedo* 197.
Williamstown 620, 621, 624.
Winkelausbreitung der Elektronenschauerteilchen (s. auch Winkel-Strukturfunktion), *angular spread of electron shower particles* 22.
Winkel-Strukturfunktion für Elektronen und Photonen, *angular structure function of electrons and photons* 43, 44, 47, 48, 100.
Wismut-Spaltungskammer, *Bismuth fission chamber* 461, 462.

X-Teilchen, *X-particles* 122, 124, 127.

Zerfall von π^0-Mesonen, *π^0-meson decay* 266.
Zerstrahlung von Materie und Antimaterie, *matter-antimatter annihilation* 266, 278, 284.
Zirkulationszeit, *circulation time* 596.
Zwei-Feuerbälle-Modell, *two-fireball model* 161—165.

Subject Index.

(English-German.)

Where English and German spelling of a word is identical the German version is omitted.

Absorption by pair production, *Absorption durch Paarerzeugung* 276.
— of high energy photons in galactic and intergalactic space, *hochenergetischer Photonen im galaktischen und intergalaktischen Raum* 276.
Absorption mean free path (absorption length, see also attenuation mean free path), *Absorptionslänge* 78, 212, 386, 387, 394, 404.
— — for star production, *für Sternerzeugung* 394, 404.
— — of neutrons, *für Neutronen* 437, 444, 446, 449, 450, 463.
— — of protons, *für Protonen* 463.
Abundance, relative, of various charge components, *Häufigkeit, relative, der verschiedenen Ladungskomponenten* 228—237.
—, —, odd-even charge, *gerader und ungerader Kernladungen* 230, 247.
—, —, of Be and B, *von Be und B* 231.
—, —, of nuclei with $Z \geq 30$, *von Kernen mit $Z \geq 30$* 229.
Abundances, isotopic, in meteorites, *Isotopenhäufigkeiten in Meteoriten* 616—617, 621, 630.
Acceleration processes, *Beschleunigungsprozesse* 239, 252, 254—256, 339.
— —, fast, *schnelle* 254—256.
— —, slow, *langsame* 254—256.
Acceleration, statistical (see also Fermi acceleration), *Beschleunigung, statistische (s. auch Fermis Beschleunigungsprozeß)* 252, 255, 359.
Acceptance cone, *Annahmekegel* 326.
Accretion of irradiated extraterrestrial material, *Anfall von bestrahltem extraterrestrischem Material* 552, 586, 587.
Achondrites, *Achondrite* 615.
Adiabatic deceleration, *adiabatische Abbremsung* 366.
— —, invariants, *Invarianten* 315, 355, 359, 360—364, 366, 367.
Aerosols, *Aerosole* 587, 593, 603.
Age determination, potassium-argon method, *Altersbestimmung, Kalium-Argon-Methode* 623.
— —, rubidium-strontium method, *Rubidium-Strontium-Methode* 623.
— —, uranium-lead method, *Uran-Blei-Methode* 623.

Air showers, see extensive air showers, *Luftschauer, s. ausgedehnte Luftschauer*.
Albedo 195, 555.
—, reentrant, *Wiederkehralbedo* 197.
Alfvén velocity, *Alfvén-Geschwindigkeit* 343.
Alpha particles (see also Helium nuclei, Helium component), *Alphateilchen (s. auch Heliumkerne, Heliumkomponente)* 343.
Altitude dependence of neutrons in the atmosphere, *Höhenabhängigkeit der Neutronen in der Atmosphäre* 427, 435, 440, 448.
— — of nuclear disintegrations in the atmosphere, *von Kernumwandlungen in der Atmosphäre* 563—566.
— — of protons in the atmosphere, *von Protonen in der Atmosphäre* 414, 418, 420.
— — of stars in the atmosphere, *von Sternen in der Atmosphäre* 393—397.
— — of the hard component in the atmosphere, *der harten Komponente in der Atmosphäre* 420.
Analytic continuation, method, *analytische Fortsetzung, Methode* 29, 90.
Angular spread of electron shower particles (see also angular structure functions), *Winkelausbreitung der Elektronenschauerteilchen (s. auch Winkel-Strukturfunktionen)* 22.
Angular structure functions of electrons and photons, *Winkel-Strukturfunktionen für Elektronen und Photonen* 43—44, 47—48, 100.
Anisotropies, *Anisotropien* 340, 344, 346, 352, 357, 358.
Annihilation, matter-antimatter, *Vernichtung (Zerstrahlung) von Materie und Antimaterie* 266, 278, 284.
Approximation A in electron shower theory, *Approximation A der Theorie der Elektronenschauer* 11, 15, 23—27.
Approximation B in electron shower theory, *Approximation B der Theorie der Elektronenschauer* 11, 15, 28—33.
Ariel I results, *Ariel I-Resultate* 337, 338, 363, 364, 365.
Aroos 616, 617, 618, 620, 624, 628.
Asymmetry in high energy interactions, *Asymmetrie in hochenergetischen Wechselwirkungen* 130, 132.

Asymptotic cone of acceptance, *asymptotischer Annahmekegel* 326, 327.
— directions, *Richtungen* 325, 326, 327, 333, 341, 345.
— latitude, *asymptotische Breite* 327.
— longitude, *asymptotische Länge* 328, 333.
Atmosphere, model, *Modellatmosphäre* 492.
—, standard, *Standardatmosphäre* 376, 377, 491.
—, top, *Atmosphäre, Gipfel* 373, 377, 483.
Atmospheric effects, *atmosphärische Effekte* 326, 327.
— — threshold, *Schwellwert* 326.
Attenuation length, *Absorptionslänge*, absorption length, attenuation length 78, 152, 212, 386, 387, 394, 404, 430, 431, 484.
Auroral zone, *Aurora-Zone (Polarlichtzone)* 320.

Baryons, excited (see also isobar model), *Baryonen, angeregte (s. auch Isobarenmodell)* 164—165, 559.
Baryons, persisting, *Trägerbaryonen* 390.
Bernstein solution in electron shower theory, *Bernsteins Lösung in der Elektronenschauertheorie* 5, 34—35.
Beta decay of fission fragments, *Betazerfall der Spaltprodukte* 316.
Betatron deceleration, *Betatron-Abbremsung* 359—360, 361—362.
Bethe formula for the stopping power, *Bethe-Formel für Energieverluste* 270.
Bethe-Heitler formula, corrections, *Bethe-Heitler-Formel, Korrektionen* 4—10.
Biosphere, *Biosphäre* 553, 588, 595, 596, 599, 601, 602, 607.
Bismuth fission chamber, *Wismut-Spaltungskammer* 461—462.
Bombardment age, *Beschußalter* 618, 627.
Born approximation, corrections for deviation, *Bornsche Näherung, Korrekturen für Abweichungen* 9—10.
Boson, intermediate, *Boson, intermediäres* 176.
Bremsstrahlung 266, 271, 274, 284, 290, 292, 294.
—, electron-electron, *Elektronen-Elektronen* 274.
—, inner, *innere* 266, 292.
— in the intergalactic medium, *im intergalaktischen Medium* 284.
— loss rate, *Verlustrate* 266, 271, 290.
— production by non-thermal electrons, *Erzeugung durch nicht-thermische Elektronen* 292.
Bruderheim 618, 622, 624, 629.

C^{13}/C^{12} ratio, C^{13}/C^{12}-*Verhältnis* 231.
C^{14}/C^{12} ratio, C^{14}/C^{12}-*Verhältnis* 575.
C^{14} dating, C^{14}-*Altersbestimmung* 574—575, 607, 613.
C^{14} decay rate, C^{14}-*Zerfallsrate* 501, 503, 510.
C^{14} production, C^{14}-*Erzeugung* 498—517, 554, 562, 566, 571—573, 613, 617.

C^{14}-production, absolute normalisation, C^{14}-*Erzeugung, absolute Normierung* 498—503, 504—505.
— — by solar protons, *durch solare Protonen* 510—517, 580.
— — during the solar cycle, *während des Sonnenfleckenzyklus* 509, 571—573.
Calorimetric method (see also ionisation calorimeter), *kalorimetrische Methode (s. auch Ionisationskalorimeter)* 69, 130, 131, 384.
Carbo 620, 623, 624, 627, 628.
Cascade showers (see also electron showers, nucleon cascade), *Kaskadenschauer (s. auch Elektronenschauer, Nukleonenkaskade)* 125, 126, 139, 144, 198.
Castagnoli method, *Castagnoli-Methode* 130, 132.
Cauchy-Riemann equation, *Cauchy-Riemann-Gleichung* 88.
Charge composition of the primary radiation, *Ladungszusammensetzung der Primärstrahlung* 195—210, 213—219, 223—224, 233—237.
Chondrites, *Chondrite* 615, 619, 623, 625, 630.
Circulation time, *Zirkulationszeit* 596.
Clark County 620, 624.
Cloud chamber, *Nebelkammer* 118, 120, 131, 136—137.
Cocconi-Köster-Perkins model (CKP model), *Cocconi-Köster-Perkins-Modell (CKP-Modell)* 387, 389.
Coefficient of exchange, *Austauschkoeffizient* 589.
Coefficient of mixing, *Mischungskoeffizient* 596.
Collision mean free path (see also interaction mean free path), *Stoßlänge, mittlere (s. auch Wechselwirkungslänge)* 151, 386, 387.
Compton process (Compton scattering), *Compton-Prozeß, (Compton-Streuung)* 3, 16, 266, 272, 290.
Compton-synchrotron radiation, *Compton-Synchrotron-Strahlung* 291.
Conductivity, electrical, of the interplanetary plasma, *Leitfähigkeit, elektrische, des interplanetaren Plasmas* 344.
Convection, *Strömung* 357—363, 364—365.
Cooling time, *Abkühlzeit* 293, 294.
Corona, *Korona* 300, 344, 350.
Cosmic dust, *kosmischer Staub* 552, 587, 623.
Cosmic ray age, *kosmisches Strahlungsalter* 622—625.
Cosmological theories, tests, *kosmologische Theorien, Prüfungen* 284—285.
Crab nebula, *Krebsnebel* 289, 290, 291, 294—299, 304—306.
— —, lunar occulation, *Abdeckung durch den Mond* 295, 306.
Critical energy, *kritische Energie* 10—11.
Cygnus A 302.

Daily intensity variation, *Tagesgang der Intensität* 326—327, 333, 350, 351, 354, 365.

Subject Index.

Decay of π^0-mesons, *Zerfall von π^0-Mesonen* 266.
Deep Springs 620, 629.
Demography of neutrons, *Demographie der Neutronen* 496, 507—508, 520.
Deuterium formation, *Deuterium-Bildung* 266.
Deuterons, primary abundance, *Deuteronen, primäre Häufigkeit* 204.
Difference equations in electron shower theory, *Differenzgleichungen in der Theorie der Elektronenschauer* 96, 100—108.
Diffusion, regular, *Diffusion, regelmäßige* 246—250.
Diffusion coefficient, *Diffusionskoeffizient* 244, 347—348.
Diffusion equations for electron showers, *Diffusionsgleichungen für Elektronenschauer* 11—18.
Diffusion extrapolation, *Diffusionsextrapolation* 209.
Diffusion mechanism of propagation, *Diffusionsmechanismus der Ausbreitung* 244—250.
Dipole approximation of the geomagnetic field, *Dipol-Annäherung für das Erdmagnetfeld* 311—314.
Dipole model, centered, *Dipol-Modell, zentriert* 311, 313, 314, 537.
— —, eccentric, *exzentrisch* 311, 313, 538.
Dipole moment, solar, *Dipolmoment, solares* 349.
Discrete sources of high energy photons, *diskrete Quellen hochenergetischer Photonen* 285—302.

e-neutrinos, *e-Neutrinos* 174—176, 307.
Earth magnetic field, *Erdmagnetfeld* 311—313, 325, 378, 537—538.
Ehole 618, 622, 623, 629.
Electron production in the intergalactic medium, *Elektronenerzeugung im intergalaktischen Medium* 273—275.
Electron production spectrum, *Elektronen-Erzeugungsspektrum* 270.
Electron shower theory, Approximation A, *Elektronenschauer, Theorie in Approximation A* 11, 15, 23—27.
— — —, Approximation B, *in Approximation B* 11, 15, 28—33.
— — —, diffusion equations, *Diffusionsgleichungen* 11—18.
— — —, elementary solutions, *elementare Lösungen* 19.
— — —, one-dimensional, *eindimensionale* 2, 14—15.
— — —, three-dimensional, *dreidimensionale* 2, 15, 38—49.
Electron showers, energy measurements, *Elektronenschauer, Energiemessungen* 69—74, 112—113.
Electrons, atmospheric secondaries, *Elektronen, atmosphärische sekundäre* 74—75, 220.

Electrons, galactic secondary, *Elektronen, galaktische sekundäre* 220—223, 274—275.
— —, primary, *primäre* 219—223, 267, 283.
Eleven-year cycle, *Elfjahre-Zyklus* 185, 332, 374, 577—579, 614, 630.
Emulsion chamber, *Emulsionskammer* 70, 72—73, 113, 118, 125, 126, 138.
Emulsion stack, *Emulsionsstapel* 119, 121, 132.
Energy density of cosmic rays, *Energiedichte der kosmischen Strahlung* 240.
Energy loss due to expansion, *Energieverluste durch Expansion* 274.
Energy loss rate of electrons in the galaxy, *Energieverlustrate der Elektronen in der Galaxie* 270, 273—275.
— — — of muons, *der Myonen* 148.
Energy spectrum, peak in the differential, *Energiespektrum, Maximum im differentiellen* 238.
— — of primaries, *der Primärteilchen* 143, 153, 195—227, 378—380.
— — of muons, *der Myonen* 147, 149.
— — of neutrinos, *der Neutrinos* 174—175.
— — of neutrons in the atmosphere, *der Neutronen in der Atmosphäre* 417, 447.
— — of protons in the atmosphere, *der Protonen in der Atmosphäre* 200—204, 382—391, 417.
— — of shower electrons and photons, *der Schauerelektronen und Photonen* 22, 24, 33, 34, 142—143, 149.
— —, values of the exponent, *Werte des Exponenten* 384.
Energy transfer to charged secondaries, *Energieübertragung an geladene Sekundärteilchen* 130, 134.
— — to γ-rays, *an γ-Strahlen* 130, 134.
Equator, cosmic ray, *Äquator der kosmischen Strahlung* 314, 538.
Equatorial ring current, *äquatorialer Ringstrom* 320, 321, 322.
Erosion hypothesis, *Erosionshypothese* 623, 627.
Escape of particles from the source region, *Entweichen der Teilchen aus der Quellgegend* 254.
Evaporation neutrons, source spectrum, *Verdampfungsneutronen, Quellspektrum* 427, 430.
Explorer X results, *Explorer X-Resultate* 343.
Extensive air showers, *ausgedehnte Luftschauer* 75—84, 139, 144, 150, 224—227, 281.
— —, lateral structure function, *laterale Strukturfunktion* 77—82.
— —, size spectrum, *Größenverteilung* 224—225.
Extragalactic discrete sources of high energy photons, *extragalaktische diskrete Quellen hochenergetischer Photonen* 301—302.

Fall-out, *Ausfall* 589—595, 606.
Families of electron showers, *Familien von Elektronenschauern* 76, 126, 140, 142—147, 165.

Fermi acceleration process (see also statistical acceleration), *Fermis Beschleunigungsprozeß (s. auch statistische Beschleunigung)* 359.
Fermi deceleration process, *Fermis Abbremsprozeß* 361—362, 366.
Fermi's thermodynamical model of multiple production, *Fermis thermodynamisches Modell der Vielfacherzeugung* 155, 168.
Fireballs, *Feuerbälle* 157—168, 169, 560.
Flare particles, *Teilchen aus chromosphärischen Eruptionen* 325.
Flügge-Yuan density maximum of slow neutrons, *Flügge-Yuansches Dichtemaximum langsamer Neutronen* 374, 426, 427, 438, 439, 483.
Flux of latitude sensitive primaries, *Fluß der breitenempfindlichen Primärteilchen* 378.
Flux of primaries, integral, *Fluß der Primärteilchen, integraler* 380.
Fokker-Planck approximation, *Fokker-Planck-Annäherung* 12, 17, 91.
Forbush effect (Forbush decrease) *Forbush-Effekt* 310 320, 323, 330, 331, 335, 338—340, 341, 346, 347, 350—352, 354, 357, 362, 368, 576.
Four-momentum transfer, *Viererimpulsübertrag* 158, 160, 162, 163, 169.
Fourier transformation, *Fourier-Transformation* 39, 85.
Fractionation process, *Fraktionierungsprozeß* 599.
Fragmentation of heavier nuclei, *Fragmentation schwererer Kerne* 209, 247—249.
Fragmentation parameters, *Fragmentationsparameter* 210—212.

Galactic center, *galaktisches Zentrum* 299.
Galactic gas density, *galaktische Gasdichte* 270.
Galactic magnetic fields, *galaktische Magnetfelder* 240.
γ-rays, see photons, photon component, *γ-Strahlen, s. Photonen, Photonenkomponente.*
γ-ray spectrum, *γ-Strahlspektren* 140, 142—143.
Gaunt approximation to the bremsstrahlung cross section, *Gaunts Annäherung für den Bremsstrahlungsquerschnitt* 292.
Gauss coefficient, *Gauss-Koeffizienten* 313.
Geomagnetic cavity, *geomagnetischer Hohlraum* 320—321, 324, 343.
Geomagnetic effects, *geomagnetische Effekte* 311, 330.
Geomagnetic equator, *geomagnetischer Äquator* 313.
Geosphere, *Geosphäre* 553, 587—607.
Global inventory of radioisotopes, *globales Inventar der Radioisotope* 602—607.
Global map of nuclear disintegrations, *globale Karte von Kernumwandlungen* 564—566.
Global production rate of radioisotopes, *globale Erzeugungsrate von Radioisotopen* 572—573, 581—583.

Grant 620, 623, 624, 625.
Green function method, *Greensche Funktion, Methode* 35, 97.
Greisen approximation of the lateral structure function, *Greisens Annäherung für die laterale Strukturfunktion* 57—59, 150.
Growth curves of primary components, *Anwachskurven der Primärkomponente* 247.

H/M ratio, *H/M-Verhältnis* 212, 218—219, 227, 232.
H-quantum, *H-Quantum* 165—168.
Hankel transformation, *Hankel-Transformation* 46, 85, 87.
Hard radiation emitted through radioactivity *harte Strahlung, durch Radioaktivität emittierte* 288.
Harleton 618, 619, 622, 629.
Hasegawa theory of multiple-fireball emission, *Hasegawas Theorie der Viel-Feuerbälle-Emission* 165—168.
He/M ratio, *He/M-Verhältnis* 212.
He/S ratio, *He/S-Verhältnis* 232, 234—235.
Helium nuclei, primary, *Heliumkerne, primäre* 204—208.
— —, energy spectrum, *Energiespektrum* 205—207.
— —, isotopic composition, *Isotopenverhältnis* 208.
— —, rigidity spectrum, *Steifigkeitsspektrum* 207.
hot universe cosmological model, „hot universe" *kosmologisches Modell* 284.
Hubble radius, *Hubble-Radius* 273, 277, 283.
Hydrosphere, *Hydrosphäre* 553, 595—599, 602, 606—607.
Hyperon/nucleon ratio, *Hyperon/Nukleon-Verhältnis* 391.
Hysteresis 185.

IGY 334, 338, 339.
Image dipole, *Bilddipol* 324.
Imp I results, *Imp I-Resultate* 341—342.
Impact zones, *Einfallszonen* 325, 344—346, 356, 368.
Inelasticity in nucleon interactions, *Inelastizität in Nukleonenwechselwirkungen* 132, 135, 151, 159, 162, 167, 385, 387, 388.
Inhomogeneities, magnetic, *Inhomogenitäten, magnetische* 342.
Integral representation method, *Integrale Darstellung, Methode* 29, 90.
Intensity, unidirectional, *Intensität, unidirektionale* 181.
Interaction mean free path (see also collision mean free path), *Wechselwirkungslänge (s. auch Stoßlänge)* 123, 211.
Intergalactic medium, electron production and energy losses, *Intergalaktisches Medium, Elektronenerzeugung und Energieverluste* 273—275.
Interplanetary fields, *Interplanetare Felder* 310, 318, 324, 325, 330, 341, 342, 345, 347, 353, 357, 362, 368.

Interplanetary medium as X-ray source, *Interplanetares Medium als Röntgenstrahlenquelle* 300.
Interplanetary plasma, *Interplanetares Plasma* 310, 313, 325, 330.
Interstellar dust (see also cosmic dust), *Interstellarer Staub (s. auch kosmischer Staub)* 552, 587.
Interstellar hydrogen, density, *Interstellarer Wasserstoff, Dichte* 243, 270.
Ionisation calorimeter, *Ionisationskalorimeter* 69, 384.
Ionisation chamber, *Ionisationskammer* 188, 195, 318, 333, 340, 384.
Ionisation loss processes, *Ionisationsverlust-Prozesse* 3, 13, 270.
Ionisation regression curve, *Ionisations-Regressionskurve* 186—187.
Iron meteorites, *Eisenmeteorite* 615, 618, 619—621, 623, 625, 627, 629, 630.
Isobar model, *Isobar-Modell* 389.
Isotope production in the atmosphere, *Isotopenerzeugung in der Atmosphäre* 553, 561—585.
Isotope production, time variations, *Isotopenerzeugung, zeitliche Veränderungen* 573—580.

Jets, *Hochenergieschauer* 118, 129.

K-figures, *K-Zahlen* 333, 343.
K-fluorescence yield, *K-Fluoreszenzausbeute* 292.
K/π ratio, *K/π-Verhältnis* 121, 150, 157, 169.
Kalos and Blatt method, *Kalos-Blatt-Methode* 104—105.
Kepler's 1604 supernova, *Keplers Supernova von 1604* 272.
Klein-Nishina formula, *Klein-Nishina-Formel* 272.

L-nuclei, differences in the spectra, *L-Kerne, Unterschiede in den Spektren* 236.
—, differential spectrum, *differentielles Spektrum* 214.
—, integral intensities, *integrale Intensitäten* 213.
—, solar modulation of the spectrum, *solare Modulation des Spektrums* 215.
L/S ratio, *L/S-Verhältnis* 212, 213, 227, 232, 235, 248.
Landau approximation, theories in, *Landau-Annäherung, Theorien in* 12, 14, 18, 44.
Landau approximation, theories without, *Landau-Annäherung, Theorien ohne* 12, 15, 18, 44—49, 61, 64.
Landau effect, *Landau-Effekt* 6—8, 18, 44.
Landau's hydrodynamical model of multiple production, *Landaus hydrodynamisches Modell der Vielfacherzeugung* 156.
Laplace transformation, *Laplace-Transformation* 85, 86, 99, 101—104.
Larmor frequency, *Larmor-Frequenz* 272.
Larmor radius, *Larmor-Radius* 271, 352, 355—356, 361.

Lateral spread of shower particles, *Lateralausbreitung der Schauerteilchen* 17, 22.
Lateral structure function of electrons and photons, *Laterale Strukturfunktion der Elektronen und Photonen* 43—44, 48—49, 62, 63, 73.
— — —, at extremely high energies, *bei extrem hohen Energien* 75—84.
— — —, Greisen approximation, *Greisens Annäherung* 57—59.
— — — of the energy flow of shower particles, *für den Energiefluß der Schauerteilchen* 83—84.
Latitude effect, *Breiteneffekt* 196, 199—200, 311, 313, 375, 427, 473, 562.
— — of the ionisation in the atmosphere during a solar cycle, *der Ionisation einer Restatmosphäre während eines Sonnenfleckenzyklus* 476.
— — of fast neutrons, *für schnelle Neutronen* 442, 451.
— — of slow neutrons, *für langsame Neutronen* 436.
— — of neutrons, total, *Breiteneffekt der Neutronen, totaler* 444, 452, 477, 520.
— — of neutrons during sunspot minimum and maximum, *der Neutronen bei Sonnenfleckenminimum und -maximum* 438, 442, 451.
— — of the primary cosmic radiation, *der primären kosmischen Strahlung* 473.
Leakage out of the galactic halo, *Verlustfluß aus dem galaktischen Halo* 273.
Life time of cosmic ray particles, *Lebensdauer der kosmischen Strahlenteilchen* 245.
Liouville's theorem, *Liouvilles Theorem* 184, 312, 354, 360.
Lithosphere, *Lithosphäre* 553, 585—587, 601.
log tan ϑ plots, *log tang ϑ-Diagramme* 130, 157, 158, 160, 164, 165.

Magnetic kink, *Magnetischer Knick* 352, 356, 362.
Magnetic scattering centres, *magnetische Streuzentren* 350, 363.
Magnetic spectrometer, *Magnet-Spektrometer* 412.
Magnetic storms, *magnetische Stürme* 317, 318—320, 321, 323, 330, 364.
Magnetic tongue, *magnetische Zunge* 350, 351—352, 354, 356.
Magnetosphere, *Magnetosphäre* 323, 324.
Main cone, *Hauptkegel* 312.
Main phase, *Hauptphase* 323, 324.
Mariner II results, *Mariner II, Resultate* 341—342, 344.
Mars I results, *Mars I, Resultate* 340.
Mass spectrometer, *Massenspektrometer* 619.
McIlwain parameter, *McIlwain-Parameter* 315.
Mellin transformation, *Mellin-Transformation* 23, 85, 86.
Meson telescopes, *Mesonenteleskope* 325, 329, 330, 332.

Meteorites, *Meteorite* 552, 573—574, 587, 613—631.
—, radiation ages, *Strahlungsalter* 553, 618, 622—625.
—, radionuclide production rate, *Erzeugungsrate von Radionukliden* 625—630.
Meteoroids, *Meteoroide* 552.
Meteorological effects, *meteorologische Effekte* 330.
Minimum grain density, *Minimum-Korndichte* 119, 372.
Mirror points, *Spiegelpunkte* 315.
Model experiments on geomagnetic effects, *Modellexperimente zu geomagnetischen Effekten* 325.
Modulation, aperiodic, *Modulation, aperiodische* 431.
— —, periodic, *periodische* 431.
Modulation of nucleons in the atmosphere, *Modulation der Nukleonen in der Atmosphäre* 374, 375
— of the proton component, *der Protonenkomponente* 189—190, 328.
— of the helium componente, *der Heliumkomponente* 191—192.
— of the L-nuclei, *der L-Kerne* 215.
— of the S-nuclei, *der S-Kerne* 192.
Modulation, solar, *Modulation, solare* 186, 189—195, 374, 375.
—, —, of neutrons in the atmosphere, *der Neutronen in der Atmosphäre* 424—429, 471, 478—508, 567.
Modulation function, *Modulationsfunktion* 337—338, 362.
Modulation mechanism, *Modulationsmechanismus* 330—368.
Molecular exchange, removal of radioactivity, *molekularer Austausch, Entfernung von Radioaktivität* 594—595, 598
Molière scattering theory, *Molières Streutheorie* 44, 68, 93—96.
Molière unit, *Molière-Einheit* 58.
Momentum spectrum of protons, *Impulsspektrum der Protonen* 400, 463, 555, 556.
— — of deuterons, *der Deuteronen* 400, 401.
— — of secondary particles, *der Sekundärteilchen* 400, 401.
Mott formula of the scattering probability, *Motts Formel für die Streuwahrscheinlichkeit* 92.
Multiple production of secondaries, *Vielfacherzeugung von Sekundärteilchen* 115—118, 153—169.
— —, FERMI's Thermodynamical model, *Fermis thermodynamisches Modell* 155, 268—270, 385.
— —, LANDAU's hydrodynamical model, *Landaus hydrodynamisches Modell* 156.
Multiple-fireball emission, *Viel-Feuerbälle-Emission* 165—168.
Multiple scattering, *Vielfachstreuung* 91—96, 119.
Multiplicity of secondary production, *Multiplizität der Sekundärteilchen-Erzeugung* 135, 151, 167, 169, 269, 385.

Muons (μ-mesons), energy loss rate, *Myonen (μ-Mesonen), Energieverlustrate* 148.
— —, energy spectrum, *Energiespektrum* 147, 149.
— —, intensity-depth relation, *Intensität/Tiefe-Beziehung* 148.
— —, interactions, *Wechselwirkungen* 135, 160, 585.
— —, positive excess, *Ladungsüberschuß* 391.
— —, range fluctuations, *Reichweitenschwankungen* 108—112.
μ-neutrinos, μ-*Neutrinos* 119, 174—176.

Neon hodoscope, *Neon-Hodoskop* 80, 151.
Neutrino astronomy, *Neutrino-Astronomie* 266.
Neutrino experiments, cosmic ray, *Neutrinoexperimente mit kosmischen Strahlen* 173—177.
Neutrino pair emission, *Neutrino-Paarerzeugung* 303.
Neutrino production spectrum, *Neutrino-Erzeugungsspektrum* 304.
Neutrino sources, *Neutrinoquellen* 302—307.
Neutrino telescope, *Neutrinoteleskop* 305.
Neutrinos, energy spectrum, *Neutrinos, Energiespektrum* 174, 175.
Neutrinos, solar, radioisotope production, *Neutrinos, solare, Radioisotopenerzeugung* 586.
Neutron albedo (see also neutron leakage flux), *Neutronenalbedo (s. auch Neutronenverlustfluß)* 527.
Neutron capture, *Neutroneneinfang* 498—517.
Neutron density, Flügge-Yuan maximum, *Neutronendichte, Flügge-Yuansches Maximum* 374, 426, 427, 438, 439, 483.
Neutron flux in the atmosphere, *Neutronenfluß in der Atmosphäre* 374, 434—453, 464—535.
— —, equilibrium spectrum, *Gleichgewichtsspektrum* 465.
— —, Simpson maximum, *Simpsonsches Maximum* 374, 450, 453.
— —, S_n-method calculation, *Berechnung mit der S_n-Methode* 497.
— —, unexplained fluctuations, *unerklärte Schwankungen* 443.
Neutron leakage flux, *Neutronenverlustfluß* 427, 428, 452, 453, 508, 518—535.
— —, angular distribution, *Winkelverteilung* 521—527, 533.
— —, aperiodically emitted, *aperiodisch emittierter* 532—535.
— —, during the solar cycle, *während des Sonnenfleckenzyklus* 520.
Neutron leakage flux, experimental values, *Neutronenverlustfluß, experimentelle Werte* 527—532.
— —, latitude effect, *Breiteneffekt* 524, 531 533.
— —, spectrum, *Spektrum* 519.

Neutron monitor, *Neutronen-Monitor* 185, 313, 315, 325, 329, 332—336, 338—340, 345, 462—463.
Neutron production by solar protons, *Neutronenerzeugung durch solare Protonen* 510—517.
Neutron production in the atmosphere, *Neutronenerzeugung in der Atmosphäre* 428, 472, 494.
Neutron production rate, *Neutronenerzeugungsrate* 472, 474—478.
Neutron reaction cross section in air, *Neutronenreaktionen, Wirkungsquerschnitt in Luft* 499.
Neutron source distribution, *Neutronen, Quellverteilung* 489.
Neutron source, normalisation, *Neutronenquelle, Normierung* 431, 496.
Neutron source spectrum, *Neutronenquellspektrum* 428, 429, 472.
Neutrons, age theory, *Neutronen, Alterungstheorie* 425, 482.
—, angular distribution, *Winkelverteilung* 463—464.
—, decay density in the interplanetary space, *Zerfallsdichte im interplanetaren Raum* 526.
—, demography, *Demographie* 498, 507—508, 520.
Neutron stars, *Neutronensterne* 287, 288—290.
Neutron transport equation, *Neutronentransportgleichung* 479—489, 495—498.
Neutrons, diffusion theory, *Neutronen, Diffusionstheorie* 425, 480—489, 493—495.
Neutrons, energy spectrum, *Energiespektrum* 434, 447, 448, 464—472, 567.
Neutrons, fast, *Neutronen, schnelle* 432, 441—453, 463—464, 521—531.
Neutrons in the atmosphere, absolute number, *Neutronen in der Atmosphäre, Absolutzahl* 502.
— —, absorption mean free path, *Absorptionslänge* 444, 446, 449, 450.
— —, density, *Dichte* 439, 440, 485.
— —, latitude dependence, *Breitenabhängigkeit* 438, 451.
— —, origin, *Ursprung* 424—429.
— —, modulation, *Modulation* 424—429, 508.
— —, source spectrum, *Quellspektrum* 432.
Neutrons, primary solar, *Neutronen, primäre solare* 426, 583—584.
Neutrons, slow, *Neutronen, langsame* 434—439, 452, 466—468, 483, 524—526, 527—529.
—, slowing down length and slowing down time, *Neutronen, Abbremslänge und Abbremszeit* 481.
—, transition effect in water, *Übergangseffekt in Wasser* 441.
—, velocity distribution, *Geschwindigkeitsverteilung* 425.

Nishimura-Kamata function, *Nishimura-Kamata-Funktion* 56—63, 150.
Niu's theory of two-fireball emission, *Nius Theorie der Zwei-Feuerbälle-Emission* 161—165.
Non-dipole terms of the geomagnetic field, *Nicht-Dipolglieder des Erdmagnetfelds* 313, 325.
Nuclear cascade (nucleon cascade), *nukleare Kaskade (Nukleonenkaskade)* 151, 372, 383—391, 428, 559, 564.
Nuclear-active (nucleoactive) component, *nukleoaktive Komponente* 136, 138, 379, 385.
Nuclear interactions, neutral-charged ratio, *Kernwechselwirkungen, Verhältnis neutraler zu geladenen Teilchen* 121, 122, 457.
Nuclear transformations, cosmic ray induced, *durch kosmische Strahlung ausgelöste Kernumwandlungen* 551—557, 571, 614—615, 625—630.
Nucleons, elastic scattering, *Nukleonen, elastische Streuung* 171—173.
Nucleons, modulation in the atmosphere, *Nukleonen, Modulation in der Atmosphäre* 372—377, 385, 422, 431, 508, 557—559.
Nucleons in the atmosphere, geomagnetic effects, *Nukleonen in der Atmosphäre, geomagnetische Effekte* 374, 379, 422, 431, 508.
— —, inelastic scattering, altitude dependence, *Höhenabhängigkeit der inelastischen Streuung* 427.

Odd-even abundance ratio, *Häufigkeitsverhältnis gerader und ungerader Ladungen* 230, 247.
Odessa 619, 620, 625.
Ohmic dissipation time of interplanetary plasma, *Ohmsche Dissipationszeit des interplanetarischen Plasmas* 344.
One-fireball emission, *Ein-Feuerball-Emission* 158—161.
Operator calculus, *Operatorenrechnung* 97.
Overabundance of L-nuclei, *Häufigkeitsüberschuß der L-Kerne* 230.

Pair creation (pair production), *Paarerzeugung* 3, 6—9, 16, 276.
— —, effect of atomic electrons, *Effekt der Atomelektronen* 8—9.
— —, in photon-photon collisions, *in Photon-Photonstößen* 276.
— —, Landau effect, *Landau-Effekt* 6—8.
— —, Migdal's treatment, *Migdals Methode* 7.
— —, of nucleons, *von Nukleonen* 558.
Pallasites, *Pallasite* 616, 619, 620.
Penumbra, *Halbschatten* 312, 315, 318.
Persisting baryons, *Trägerbaryonen* 290.
Pfotzer maximum, *Pfotzer-Maximum* 117, 373.
Photon astronomy, high energy, *Photonenastronomie, hochenergetische* 267.

Photon producing processes, *Photonen-erzeugende Prozesse* 266, 267—285.
Photon production spectrum, *Photonenerzeugungsspektrum* 277.
Photon spectrum, high energy, *Photonenspektrum bei hohen Energien* 279
Photons, high energy, in the atmosphere, *Photonen, hochenergetische, in der Atmosphäre* 74—75.
Photons, high energy primary, *Photonen, hochenergetische primäre* 265.
Photopion production, *Photopionenerzeugung* 278.
Pinkau's method for determination of shower energies, *Pinkaus Methode zur Bestimmung von Schauerenergien* 112.
Pion/neutron ratio, *Pion/Neutron-Verhältnis* 463.
Pion/proton ratio, *Pion/Proton-Verhältnis* 463.
Pion distribution, *Pionenverteilung* 388
Pion multiplicity in the isobar model, *Pionenmultiplizität im Isobar-Modell* 390, 559.
Pionisation 385.
Pion production, *Pionenerzeugung* 121, 153—168, 268—270, 385, 557—561.
Pions, neutral, production spectrum, *Pionen, neutrale, Erzeugungsspektrum* 198, 269, 297.
Pioneer V results, *Pioneer V, Resultate* 340, 341—342, 356.
Planets as X-ray sources, *Planeten als Röntgenstrahlenquellen* 300.
Plasma frequencies, *Plasmafrequenzen* 274.
Plasma, interplanetary, *Plasma, interplanetares* 310, 313, 349.
Plasma oscillations, *Plasmaschwingungen* 274.
Plasma, solar, *Plasma, solares* 313, 321, 323, 332, 341—342, 349.
Plateau grain density, *Plateaukorndichte* 119, 392.
Plural production of secondaries, *Stufenerzeugung von Sekundärteilchen* 116.
Plural scattering effects, *Mehrfachstreuungseffekte* 92.
Polar cap, *Polarkappe* 320.
Polar ice cap, radioisotopes, *polare Eiskappe, Radioisotope* 576, 587, 601.
Positron annihilation cross section, *Positronenvernichtungs-Querschnitt* 278.
Positron/electron ratio (P/e ratio), *Positron/Elektron-Verhältnis* 222, 238, 242.
Precipitation, *Ausfall* 587, 591—594, 606.
Primary component, *Primärkomponente* 311, 337, 377—383.
Primary nuclei, total flux, *Primärkerne, Gesamtfluß* 374.
Primary spectrum (see energy spectrum, rigidity spectrum), *Primärspektrum (s. Energiespektrum, Steifigkeitsspektrum)* 143, 153, 195—227, 337, 380—383, 387.
Production spectrum, inhomogeneous, *Erzeugungsspektrum, inhomogenes* 389.

Production spectrum of neutrons and protons (Rossi modell), *Erzeugungsspektrum von Neutronen und Protonen (Rossi-Modell)* 416.
Project Argus, *Projekt Argus* 316.
Propagation of cosmic rays, *Ausbreitung der kosmischen Strahlung* 239—251.
Proton/helium ratio (P/He ratio), *Proton/Helium-Verhältnis* 197, 232, 234—235.
Proton spectrum, sunspot minimum, *Protonenspektrum beim Sonnenfleckenminimum* 203.
Protons, angular distribution, *Protonen, Winkelverteilung* 422—424.
Protons, differential energy spectrum, *Protonen, differentielles Energiespektrum* 200—204, 413, 414.
Protons in the atmosphere, energy spectrum, *Protonen in der Atmosphäre, Energiespektrum* 382—391, 410—422.
Protons in the atmosphere, experimental methods, *Protonen in der Atmosphäre, experimentelle Methoden* 200—204, 383, 411—413, 463.
Protons, percentage in the total penetrating component, *Protonen, Prozentsatz in der gesamten durchdringenden Strahlung* 419.
Protons, solar, *Protonen, solare* 315, 317, 318, 320, 325, 365, 512, 556.
Protons, total flux, *Protonen, Gesamtfluß* 374.
Protons, vertical intensity, *Protonen, Vertikalintensität* 418—420.

Quiescent cosmic radiation, *ruhige kosmische Strahlung* 377.

Radiation length, *Strahlenlänge* 10—11.
Radiation process, *Strahlungsprozeß* 3, 6—9.
— —, Landau effect, *Landau-Effekt* 6—8.
— —, Migdal's treatment, *Migdals Methode* 7.
— —, effect of atomic electrons, *Effekt der Atomelektronen* 8—9.
Radioactivity, cosmic ray induced, *Radioaktivität, durch kosmische Strahlung erzeugte* 553, 561—562, 602—607, 613—614, 620—622, 630.
Radio brightness temperature, *Radiohelligkeitstemperatur* 282.
Radio emission, *Radioemission* 241—242, 282, 292.
Radius of the galactic halo, *Radius des galaktischen Halo* 277.
Range fluctuations of high energy muons, *Reichweitenschwankungen hochenergetischer Myonen* 108—112.
Reentrant albedo, *Wiederkehralbedo* 197.
Registration stations, cosmic ray, *Registrierstationen für kosmische Strahlung* 535—537
Regression curve of atmospheric ionization, *Regressionskurve der atmosphärischen Ionisation* 186—187.
Residence time, *Verweilzeit* 598.

Subject Index.

Response functions, differential, *Ansprechfunktionen, differentielle* 337, 339.
Rigidity, magnetic, *Steifigkeit, magnetische* 182, 311, 312, 318, 555—556.
Rigidity, cut off, *Abschneidesteifigkeit* 183.
Rigidity spectrum, *Steifigkeitsspektrum* 183, 374, 556—557.
Rigidity, threshold, *Schwellwertsteifigkeit* 311, 313, 318, 321—325.
Ring current, equatorial, *Ringstrom, äquatorialer* 320—322, 324.
Riometer 317.

S-nuclei, *S-Kerne* 216—218.
Saddle point method, *Sattelpunktmethode* 31, 88.
Scattering, single, *Einzelstreuung* 119.
—, multiple, *Vielfachstreuung* 119.
—, plural, *Mehrfachstreuung* 119.
—, spurious, *Streuung, vorgetäuschte* 120.
—, method of relative, *Streuung, Methode der relativen* 120, 125.
Sco XR-1 285, 301, 306.
Screening effect, *Abschirmeffekt* 4—5, 13.
Screening cross section, complete, *Abschirmquerschnitt, vollständiger* 4, 13.
— — —, incomplete, *Abschirmquerschnitt, unvollständiger* 5.
Secular equilibrium, *säkulares Gleichgewicht* 616.
Sedimentation of radioisotopes, *Sedimentation der Radioisotope* 575, 587, 601, 602, 607.
Sgr XR-1 286.
Shock wave, *Schockwelle* 352—355, 357.
Shower, see cascade shower, electron shower, extensive air shower, *Schauer, s. ausgedehnte Luftschauer, Elektronenschauer, Kaskadenschauer*.
Shower age, *Schaueralter* 21.
Shower axis, *Schauerachse* 60—69.
Shower diagrams, *Schauerdiagramme* 398, 399.
Shower functions, *Schauerfunktionen* 23—33.
Shower maximum, depth of the, *Schauermaximum, Tiefe des* 19, 26.
Shower particles, total number, *Schauerteilchen, Gesamtzahl* 18, 27, 66.
Shower secondaries, *Schauer-Sekundärteilchen* 121—122, 268, 385, 557—559.
— —, neutral-charge ratio, *Verhältnis neutraler zu geladenen* 121—122.
Sikhote-Alin 620, 624, 625.
Similarity relation, *Ähnlichkeitsbeziehung* 65.
Simpson maximum of neutron flux, *Simpsons Maximum des Neutronenflusses* 374, 450, 483.
Simpson pile, see Neutron monitor, *s. Neutronenmonitor*.
Single scattering, contribution to the lateral spread, *Einzelstreuung, Beitrag zur Lateralausbreitung* 67, 92.
Solar energetic particles (see also protons) *solare energiereiche Teilchen (s. Protonen)* 344—347.

Solar flare, *chromosphärische Eruption* 315, 330, 341, 346, 353, 355, 366, 368, 590.
Solar modulation (see also 11-years cycle, 27-days variation), *solare Modulation (s. auch 11-Jahre-Zyklus, 27-Tage-Zyklus)* 374, 378.
Solar neutrinos, *solare Neutrinos* 586.
Solar particle bursts, variation in radioisotope production, *solare Teilcheneinbrüche, Variationen der Erzeugung von Radioisotopen* 579—580.
Solar plasma, *solares Plasma* 313, 321, 323, 341—342, 349, 359, 368.
Solar system X-ray sources, *Sonnensystem, Röntgenstrahlquellen* 299—300.
Solar wind, *Sonnenwind* 320—321, 323—324, 342, 344, 349, 355, 359, 365.
Source regions, *Quellgegenden* 245, 257—260.
Source and sink functions for isotope production, *Quellen- und Senkenfunktionen für Isotopenerzeugung* 553—555, 562—575, 587, 589, 591, 594—595, 597.
Sources and sinks of high energy electrons, *Quellen und Senken hochenergetischer Elektronen* 275.
Source spectrum of evaporation neutrons, *Quellspektrum der Verdampfungsneutronen* 427, 430.
South African anomaly, *Südafrikanische Anomalie* 337.
Spallation reactions, *Spallationsreaktionen* 557, 559, 566, 574, 614, 616, 621.
Spark chamber, *Funkenkammer* 120.
Standard atmosphere, *Standardatmosphäre* 376—377, 491.
Stars in nuclear emulsions, *Sterne in Kernemulsionen* 391—410, 453—458, 565.
— —, absolute frequency, *Absoluthäufigkeit* 396—397.
— —, altitude dependence, *Höhenabhängigkeit* 393.
— —, percentage of deuteron, triton, and He tracks, *Prozentsatz von Deuteron- Triton- und He-Spuren* 401.
Star production in the atmosphere, *Sternerzeugung in der Atmosphäre* 394, 396.
Star production mean free path, *Sternerzeugung, mittlere freie Weglänge* 454.
Star production reaction rates, *Sternerzeugung, Reaktionsraten* 454, 564, 582.
Star size distribution, *Sterngrößenverteilung* 397, 398, 402, 404, 568, 571.
Steady state cosmological theory, 'steady state' *kosmologische Theorie* 284.
Stellar coronae, hard radiation from, *Sternkoronae, harte Strahlung von* 300—301.
Störmer theory, *Störmers Theorie* 312, 313, 318, 321.
Störmer unit, *Störmer-Einheit* 311.
Stone meteorite, *steiniger Meteorit* 615, 616, 620—621.
Stratosphere, *Stratosphäre* 396, 555, 589—590, 599, 601, 603—605, 614, 616, 621.

Structure functions in electron shower theory, *Strukturfunktionen in der Theorie der Elektronenschauer* 2, 39, 43—44, 47—49, 49—69.
— —, mean square deviations, *mittlere qudratische Abweichungen* 53—56.
— —, at large distance from the shower axis, *achsenferne* 67—69.
— —, near the shower axis, *achsennahe* 60—69.
Structure functions, normalized, *normalisierte* 56.
Sudden commencement, *plötzlicher Sturmeinsatz* 318, 332, 346, 353.
Sun as X-ray source, *Sonne als Röntgenstrahlenquelle* 299—300.
Sunspot cycle (see also 11-years cycle), *Sonnenfleckenzyklus (s. auch 11-Jahrezyklus)* 185, 329, 330, 333, 336—340.
Supernovae 257—258, 287—288, 615.
Supernovae remnants, *Supernovae-Überreste* 285.
Synchrotron radiation, *Synchrotronstrahlung* 240, 266, 271, 290.

Takagi model of nucleon interactions, *Takagis Modell für Nukleonenwechselwirkung* 161.
Tamm and Belenky, track length solutions of, *Lösungen für die Spurenlänge* 37—38.
Tau XR-1 285.
Thermal neutrons, nuclear disintegrations induced by, *durch thermische Neutronen induzierte Kernumwandlungen* 566, 572—573, 585, 614.
Thickness of matter traversed by primary cosmic rays, *Materiedicke, von Primärteilchen durchlaufene* 246, 584.
Threshold rigidity (see also cut off rigidity), *Schwellwert der Steifigkeit (s. Abschneidesteifigkeit)* 538.
Thompson limit, *Thompson-Grenze* 272.
Tracer studies, *Leitisotopenuntersuchungen* 554, 597, 599.
Track length, differential, *Spurenlänge, differentielle* 18, 25, 38.
— —, total (integral), *totale (integrale)* 16, 18—19, 25, 35—38.
Tracks in nuclear emulsions, *Spuren in Kernemulsionen* 392, 403.
Trapped radiation, *eingefangene Strahlung* 315, 316, 321.
Transition effects of stars, *Übergangseffekte der Sterne* 405—410, 564.
— — of neutrons in water, *von Neutronen in Wasser* 441.

Transparency, *Transparenz* 313.
Transverse momentum, *Transversalimpuls* 76, 113, 125, 127—129, 145, 157, 166, 169—170, 172, 386.
Treysa 620, 623, 624, 627, 628, 629.
Tritium, solar, *Tritium, solares* 584.
Troposphere, *Troposphäre* 554, 591—595, 599, 601, 605—606.
Two-fireball model, *Zwei-Feuerbälle-Modell* 161—165.

Universal cosmic ray spectrum, *universales Spektrum der kosmischen Strahlung* 267.

Variation, daily, *Variation, tägliche (Tagesgang)* 326—327, 333, 350, 351, 354, 365.
Variation, 27-day, *Variation, 27-Tage-* 330.
Variation, long-term, *Variation, langzeitliche* 310, 332, 357, 363, 366.
—, rigidity dependence, *Steifigkeitsabhängigkeit* 332, 336—340, 362.
Variation, short-term, *Variation, kurzzeitliche* 341.
Variational coefficient, *Variationskoeffizient* 326.

Williamstown 620, 621, 624.

X-particles, *X-Teilchen* 122, 124, 127.
X-rays, characteristic, *Röntgenstrahlen, charakteristische* 266, 292.
X-ray emission by electron-ion radiative recombination, *Röntgenstrahlen, durch Elektron-Ionen-Strahlungsrekombination erzeugte* 293.
X-ray emission by inelastic collisions followed by radiative de-excitation, *Röntgenstrahlen, durch inelastische Stöße mit nachfolgender Strahlungsdeexzitation* 293.
X-ray films for energy measurements on electron showers, *Röntgenfilme für Energiemessungen an Elektronenschauern* 69, 139—140.
X-rays from external galaxies, *Röntgenstrahlen von äußeren Galaxien* 281—282.
X-ray production in discrete sources, *Röntgenstrahlenerzeugung in diskreten Quellen* 290—294.
X-ray sources, *Röntgenstrahlenquellen* 286—302.

Yield functions, *Ausbeutefunktionen* 187, 311, 328, 329, 331, 337.
Yield of isotopes in spallation reactions, *Ausbeute an Isotopen in Spallationsreaktionen* 566—573.

Druck der Universitätsdruckerei H. Stürtz AG., Würzburg